McGRAW-HILL YEARBOOK OF SCIENCE & TECHNOLOGY

1988

McGRAW-HILL YEARBOOK OF SCIENCE & TECHNOLOGY

1988

COMPREHENSIVE COVERAGE OF RECENT EVENTS
AND RESEARCH AS COMPILED BY THE STAFF OF THE
McGRAW-HILL ENCYCLOPEDIA OF SCIENCE & TECHNOLOGY

McGRAW-HILL BOOK COMPANY

New York   St. Louis   San Francisco

Auckland   Bogotá   Caracas   Colorado Springs
Hamburg   Lisbon   London   Madrid   Mexico   Milan
Montreal   New Delhi   Oklahoma City   Panama   Paris
San Juan   São Paulo   Singapore   Sydney   Tokyo   Toronto

McGraw-HILL YEARBOOK OF SCIENCE & TECHNOLOGY
Copyright © 1987 by McGraw-Hill, Inc.

1234567890 DOW/DOW 8943210987

Library of Congress Cataloging in Publication data

McGraw-Hill yearbook of science and technology.
1962–   . New York, McGraw-Hill Book Co.

    v. illus. 26 cm.
    Vols. for 1962–   compiled by the staff of the
McGraw-Hill encyclopedia of science and
technology.
    1. Science—Yearbooks. 2. Technology—
Yearbooks. 1. McGraw-Hill encyclopedia of
science and technology.
Q1.M13   505.8   62-12028

ISBN 0-07-046183-X
ISSN 0076-2016

INTERNATIONAL EDITORIAL ADVISORY BOARD

EDITORIAL STAFF

Sybil P. Parker, Editor in Chief

Arthur Biderman, Senior Editor
Jonathan Weil, Editor
Betty Richman, Editor

Edward J. Fox, Art Director
Patrick J. Aievoli, Art Production Supervisor
Pablito M. Darden, Designer

Joe Faulk, Editing Manager
Frank Kotowski, Jr., Editing Supervisor
Stephen M. Smith, Editing Supervisor
Patricia W. Albers, Editorial Administrator
Barbara Begg, Editing Assistant
Hazel P. Carter, Editing Assistant

Thomas G. Kowalczyk, Production Supervisor

Art suppliers: Eric G. Hieber, EH Technical Services, New York, New York; Melvin Grainger, New Jersey.

This book was set in Times Roman. It was composed by the McGraw-Hill Book Company Professional & Reference Division composition unit.

This book was printed and bound by R. R. Donnelley & Sons Company, The Lakeside Press at Willard, Ohio.

CONSULTING EDITORS

CONSULTING EDITORS (continued)

Dr. Carl N. McDaniel. *Associate Professor, Department of Biology, Rensselaer Polytechnic Institute.* DEVELOPMENTAL BIOLOGY.

Dr. N. Karle Mottet. *Professor of Pathology, University of Washington School of Medicine.* MEDICINE AND PATHOLOGY.

Prof. Jay M. Pasachoff. *Director, Hopkins Observatory, Williams College, Williamstown, Massachusetts.* ASTRONOMY.

Dr. William C. Peters. *Professor Emeritus, Mining and Geological Engineering, University of Arizona.* MINING ENGINEERING.

Prof. Don S. Rice. *Department of Anthropology, University of Chicago.* ANTHROPOLOGY AND ARCHEOLOGY.

Prof. D. A. Roberts. *Plant Pathology Department, Institute of Food and Agricultural Sciences, University of Florida.* PLANT PATHOLOGY.

Prof. W. D. Russell-Hunter. *Professor of Zoology, Department of Biology, Syracuse University.* INVERTEBRATE ZOOLOGY.

Dr. Anthony M. Trozzolo. *Charles L. Huisking Professor of Chemistry, University of Notre Dame.* ORGANIC CHEMISTRY.

Dr. David Turnbull. *Gordon McKay Professor of Applied Physics, Harvard University.* SOLID-STATE PHYSICS.

Prof. Joan S. Valentine. *Department of Chemistry and Biochemistry, University of California, Los Angeles.* INORGANIC CHEMISTRY.

Dr. Richard G. Wiegert. *Department of Zoology, University of Georgia.* ECOLOGY AND CONSERVATION.

Dr. W. A. Williams. *Department of Agronomy and Range Science, University of California, Davis.* AGRICULTURE.

CONTRIBUTORS

A list of contributors, their affiliations, and the titles of the articles they wrote is given on pages 501–506.

PREFACE

The 1988 *McGraw-Hill Yearbook of Science and Technology*, continuing in the tradition of its predecessors, presents the outstanding recent achievements in science and technology. Thus it serves as an annual review and also as a supplement to the *McGraw-Hill Encyclopedia of Science and Technology*, updating the basic information in the sixth edition (1987) of the Encyclopedia.

The Yearbook contains articles reporting on those topics that were judged by the consulting editors and the editorial staff as being among the most significant recent developments. Each article is written by one or more authorities who are actively pursuing research or are specialists on the subject being discussed.

The *McGraw-Hill Yearbook of Science and Technology* provides librarians, students, teachers, the scientific community, and the general public with information needed to keep pace with scientific and technological progress throughout the world. The Yearbook has long served this need through the ideas and efforts of the consulting editors and the contributions of eminent international specialists.

SYBIL P. PARKER
EDITOR IN CHIEF

A-Z

1988

Accretion tectonics

Tectonics is the subdiscipline in geology that is concerned with fundamental earth forces and the effects these have on the distribution and configuration of rocks and strata. Movement along faults, uplifting of mountains, and folding of strata are common phenomena that traditionally have been the subject of tectonic studies. With the advent of plate tectonics in the mid-1960s, geologists began to grasp the full extent of the mobility of the Earth's outer layer of rock. This outer layer is composed of plates, each 60 to 90 mi (100 to 150 km) thick. The relative motion between adjoining plates produces most landforms and crustal configurations. On a global scale, oceanic crust is relatively young, averaging only 55 million years (m.y.) in age, with no ocean floor older than 200 m.y. The age range of continental rocks, however, is quite different, extending from the present age to an age of 3.8 billion years (b.y.). Exactly how all these rocks of diverse ages and compositions gathered to form continents is the concern of accretion tectonics.

Ocean crust forms as plates move away from each other. A 34,000-mi-long (55,000-km) ridge system, which defines the axis along which spreading and crustal generation are taking place, wanders across the modern ocean floor. Because the Earth maintains a constant volume, the spreading motion along oceanic ridges must be compensated by regions of crustal convergence where crust is subducted. In these regions, called subduction zones, one plate passes beneath another plate. A complex series of chemical reactions that results in explosive volcanism is assumed to be associated with subduction zones. The "ring of fire" surrounding the Pacific Ocean is the best-known segment of the 22,000-mi-long (35,000-km) volcanic ring that lies above the world's subduction zones. A plate does not always move directly away from or toward an adjoining plate. Motion ranging from oblique to perfectly tangential is also common. Here, plates slide past one another, defining earthquake faults such as the San Andreas Fault of California or the Alpine Fault of New Zealand.

Terranes. This mobility of the Earth's outer layer is manifested also by the juxtapositioning of rock bodies that formed apart from one another but have subsequently been amalgamated as a consequence of collision forces between two or more plates. The study of these phenomena is called accretion tectonics. With this concept, geologists assert that much of the cordillera of North and South America, the Himalayas of Asia, or the Alpine system of Europe is best represented as an agglomeration of discrete rock packages known as tectonostratigraphic terranes (commonly abbreviated to terranes; see **Fig. 1**). Continents therefore represent long-term effects of numerous accretion episodes. Some continents, for example North America, possess a crude radial symmetry, with the oldest rocks in the interior and successively younger belts of rocks that ring the older core. Australia, however, displays a west-to-east succession, with 3- to 4-b.y.-old rocks cropping out along the west coast. The rocks become progressively younger toward the east. This trend continues all the way to New Zealand, interrupted only by the superposed crustal spreading in the Tasman Sea that has had the effect of partially disrupting the accretion symmetry (see Fig. 1). The symmetric pair to this succession may exist on another continent (or parts of it may appear on several continents) which has been rifted from the west coast of Australia. In a similar fashion, the breakup of Gondwana 100 million years ago (m.y.a.) fractured the accretion symmetry that surrounded the previously amalgamated continents of Australia, Antarctica, India, Africa, and South America.

The symmetry in Asia is particularly difficult to discern because the region from India to Malaysia

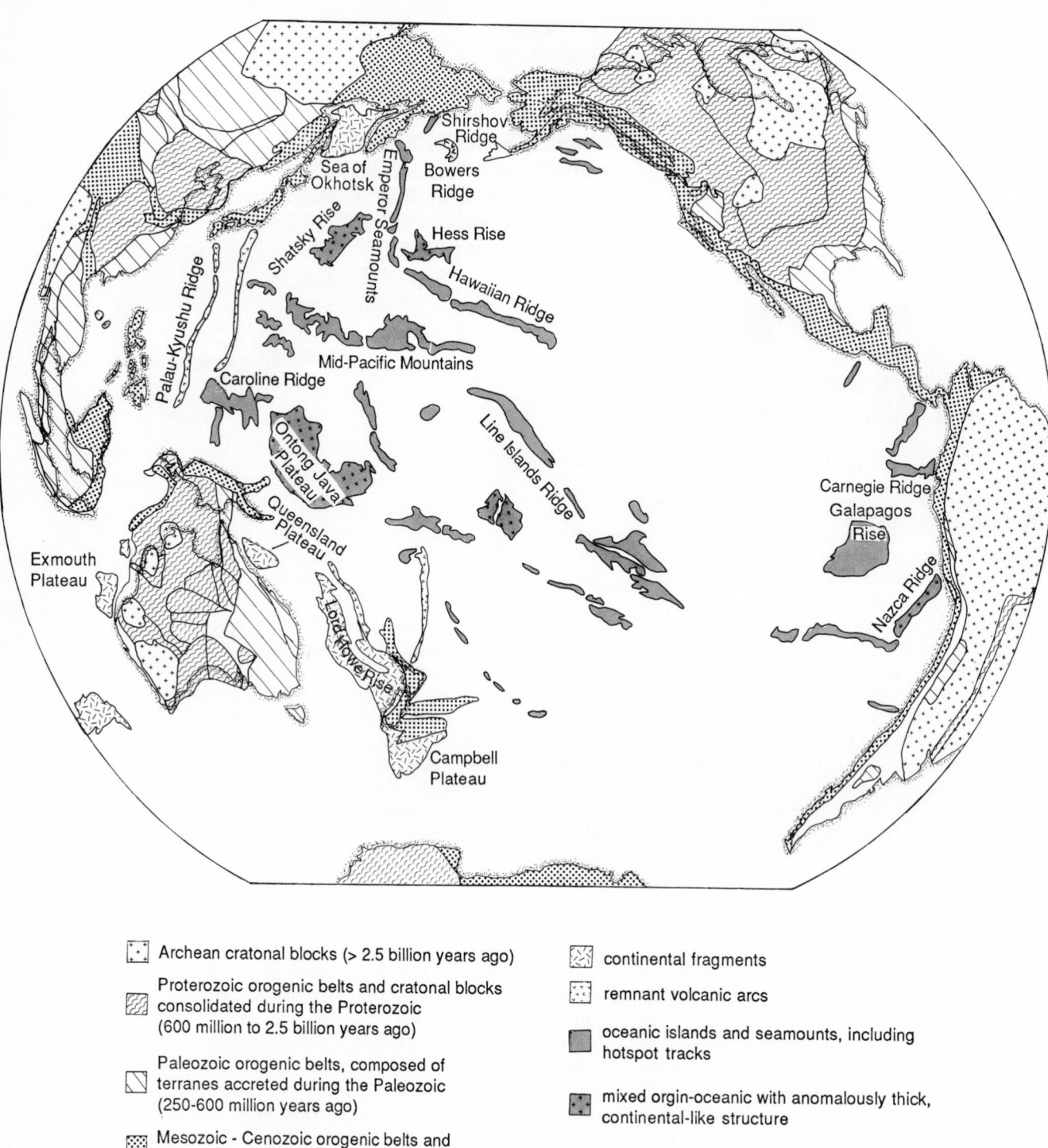

Shirshov
Ridge
Bowers
Ridge
Sea of
Okhotsk
Emperor Seamounts
Hess Rise
Shatsky Rise
Hawaiian Ridge
Palau-Kyushu Ridge
Mid-Pacific Mountains
Caroline Ridge
Line Islands Ridge
Oniong Java
Plateau
Carnegie Ridge
Galapagos
Rise
Queensland
Plateau
Exmouth
Plateau
Nazca Ridge
Lord Howe Rise
Campbell
Plateau

☐ Archean cratonal blocks (> 2.5 billion years ago)

☐ Proterozoic orogenic belts and cratonal blocks consolidated during the Proterozoic (600 million to 2.5 billion years ago)

☐ Paleozoic orogenic belts, composed of terranes accreted during the Paleozoic (250–600 million years ago)

☐ Mesozoic - Cenozoic orogenic belts and accreted terranes (0–250 million years ago)

☐ continental fragments

☐ remnant volcanic arcs

☐ oceanic islands and seamounts, including hotspot tracks

☐ mixed orgin-oceanic with anomalously thick, continental-like structure

Fig. 1. Generalized map of circum-Pacific tectonostratigraphic terranes.

and northward across China to the Siberian Platform composes as many as five discrete older continental fragments, Tarim, Sino-Korea, Yangtze, Indochina, and India, each with some vestige of an accretionary rim formed somewhere other than in their current setting (see **Fig. 2**). The Siberian Platform has acted as a geologic backstop against which large crustal fragments have swept in from the south. In this way, the belts of accretion have become progressively younger toward the south, with India being the last to arrive, and Australia and the smaller crustal fragments of Indonesia soon to follow. Furthermore, within each accretionary zone the rocks often have a complex history inherited from previous episodes of crustal accretion and dispersion.

The geometry of individual terranes as well as assemblages of terranes in a tectonic collage is the product of the history of plate movements and associated tectonic interactions. Terranes that form on an oceanic plate (for example, basaltic seamounts or volcanic island arcs) generally retain their shape until they collide and accrete. The process of tectonic accretion and the events that follow commonly result in crustal movements that modify the shapes of the terranes. In Asia the direction of accretion has been mostly northward head-on with the Siberian Platform. The shapes of the microcontinental terranes retain the configuration inherited from an earlier continental rifting episode and do not reflect contortions as a result of the accretion tectonics; however, 42 m.y.a. India began to collide with Asia. The resulting north-south compression is causing the previously accreted terranes to the north to extend east and west, absorbing the crustal strain. The terranes of the cordillera of North America reflect an elongation oriented toward the northwest, the principal motion vector between the North American Plate and oceanic plates to the west. These oceanic plates have carried the terranes to the cordillera and also exerted stresses that act to reshape previously accreted terranes or extant portions of the North American craton.

Crustal movements. The movement history or kinematics of individual terranes is not always known. The study of fossils to establish ancient paleogeographic settings may provide evidence for the displacement. Also, for some rocks the preserved paleomagnetic signal can specify the latitude at which a rock was deposited. (The Earth's magnetic field can be modeled as a dipole or bar magnet coincident with the Earth's spin axis. The force lines of this field steepen with respect to the Earth's surface as given reference points approach the North and South poles. Thus, for rocks deposited at the Equator, the magnetic inclinations are zero, tangential to the surface, while rocks in the Arctic possess magnetic orientations of nearly 90°, perpendicular to the Earth's surface.)

Despite the uncertainty involving the specific kinematic history of an individual terrane or an assemblage of terranes, plate tectonics provides a means and a necessity for both small and large amounts of crustal movements. The average modern plate speed is approximately 1.6 in. (4 cm) per year, plate speeds up to 6.2 in. (16 cm) per year are known from the South Pacific, and some paleomagnetic data suggest that speeds up to 12 in. (30 cm) per year may have occurred in the past. At the rate of 1.6 in. (4 cm) per year, a 4-b.y.-old rock would have moved 99,000 mi (160,000 km) or a distance equivalent to four circumglobal transits. Transoceanic travel should not be a surprise, because in 200 m.y., which is the longevity of most ocean basins, a terrane

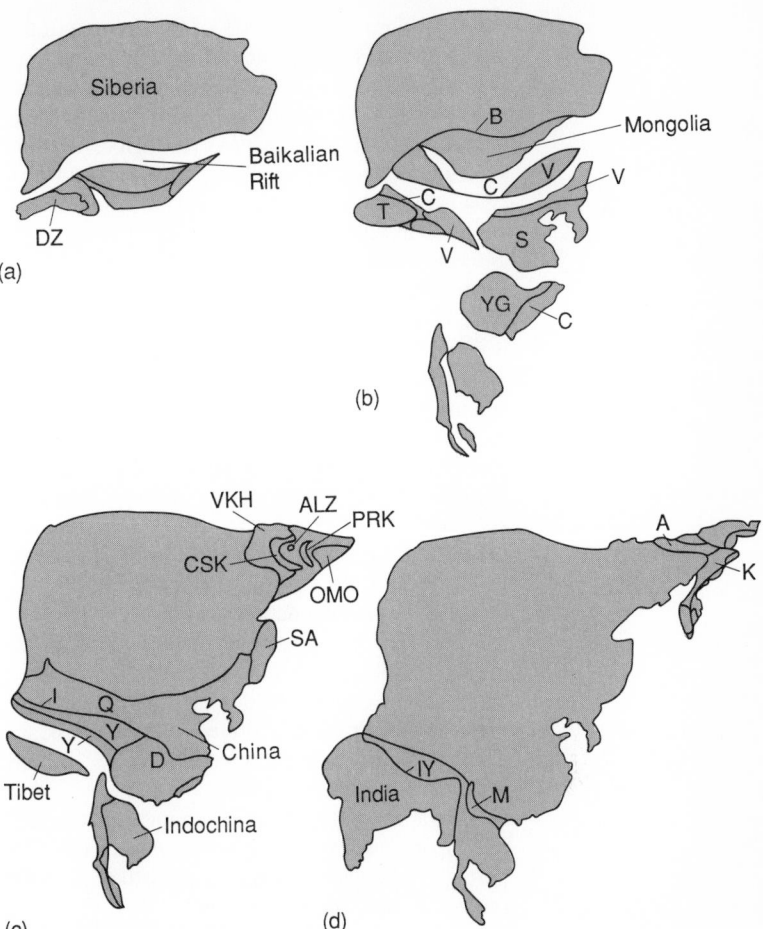

Fig. 2. Schematic representation of accretionary evolution of Siberia and Asia. (*a*) About 550 million years ago. (*b*) About 300 m.y.a. (*c*) About 200 m.y.a. (*d*) About 35 m.y.a. DZ = Dzungaria; C = Caledonian fold belts; B = Baikalian suture; V = Variscan fold belts; T = Tarim; S = Sino-Korea; YG = Yangtze; CSK = Cherskiy terrane; ALZ = Alazeya; PRK = Prikolymsk; VKH = Verkhoyansk fold belt; OMO = Omolon terrane; SA = Sikhote Alin; I = Indosinides sutures; Q-D = Qilian-Dabie Shan suture; Y = Yenshanides fold belts; A = South Anyui; K = Koryak Highlands; M = Mekong fold belt; IY = Indus-Yaluzangbu suture. (*After E. R. Schermer, D. G. Howell, and D. L. Jones, The origin of allochthonous terranes: Perspectives on the shaping of continents, Annu. Rev. Earth Planet. Sci., 12:107–131, 1984*)

traveling at 1.6 in. (4 cm) per year could move 5000 mi (8000 km).

Terrane classification. The composition and stratigraphy of a terrane are the key factors for identifying its spatial limits as well as for inferring possible kinematic histories for a given terrane. A host of terrane classifications is possible. A simple scheme which clearly displays the accretionary tectonic aspects on a global scale classifies terranes into the five categories described below:

Oceanic rocks such as ocean crust or seamounts. Even though ocean crust represents nearly 70% of the rock covering the surface of the globe and is constantly being created along the ocean-ridge spreading centers, ocean crust is rarely preserved as a terrane. These rocks seem to be too dense to become accreted; alternatively, in the process of subduction the ocean crust is the conveyor belt that continues to flow into the deeper parts of the mantle. As many as 25 global ocean floors have been created and destroyed (subducted) since plate tectonic processes began, but all of this material represents only 4% of the volume of the mantle that lies above and around the liquid core. Seamounts, on the other hand, are more buoyant. On a global basis they are growing at approximately 0.05 mi³ (0.2 km³) per year. Because they protrude above the sea floor, the accretion process is facilitated once a seamount enters into a subduction zone.

Oceanic rocks mixed with continentally derived sediments. The world's river systems carry large volumes of detritus to the ocean. Much of this material is deposited in the deltas and on the shelves of the submerged parts of the continents. The volume of sediment that lies wholly on the oceanic crust is approximately 34,000,000 mi³ (142,000,000 km³). This volume includes an amount resulting from the porosity of the sediment, but if this amount is subtracted the equivalent is 23,000,000 mi³ (94,000,000 km³) of sedimentary rock on the sea floor. Because the oceanic crust averages 55 m.y. in age, a continental denudation of 0.396 mi³ (1.65 km³) per year is represented by this material. Accretion tectonics, in effect, transfers this material back into the continental domain, though the sediment may be drastically altered in the process; for example, some is melted and becomes igneous rock that intrudes the continental framework, some is offscraped at the toe of a subduction zone and piles up in great masses, and other portions may be partially subducted and plated onto the underside of the leading edge of a continent. Subsequent vertical Earth motions reveal these rocks in an altered or metamorphic state as schists and gneisses. Exactly how much sediment is subducted into the mantle along with the ocean crust is not known; this is an important question regarding the budget for continental growth.

Oceanic volcanic arcs. These explosive volcanic edifices grow above subduction zones (for example, the arcs of Sumatra, New Hebrides, Marianas, Aleutian, and Lesser Antilles). The global production of these volcanic arcs is approximately 0.26 mi³ (1.1 km³) per year. Because they stand high and are composed of relatively light material, they are readily accreted once they have entered into a subduction zone or have obliquely converged against a continental margin.

Volcanic arcs with continental basement. These types of terranes represent areas where a subduction zone lies along the edge of a continent (for example, the Andes of South America or the Cascade volcanoes of the western United States). The volume of new crustal material isincluded in the global production figure of 0.26 mi³ (1.1 km³). This new continental material is incorporated initially as an igneous event (liquid magma intruding into the host rock), but because of the spatial proximity to the dynamic motions of an active continental margin these kinds of volcanic arcs are unstable. They commonly are translated along the margin and ultimately become incorporated into the terrane collage as a result of accretion tectonics.

Continental fragments. The pieces of old continental blocks in Asia admixed with the other types of terranes is an example. Many such terranes, however, are much smaller flakes that have been ripped off a continental margin elsewhere. The repositioning of continental material dramatically changes the shapes of continents, but it is essentially a zero-sum proposition with regard to total continental growth.

Summary. Thus accretion tectonics is a corollary to the plate tectonic paradigm. The outer layer of the Earth is composed of mobile plates. The continuous movement of these plates shifts the continents and results in the creation of new ocean crust along the axes of the globe-girdling system of spreading ridges. In zones where a component of collision exists between two or three plates, crustal fragments may amalgamate to form a collage of accreted terranes. In this way the debris that is constantly being eroded from the continents and deposited in the deep ocean is returned to the continental framework. Newly created volcanic rock in the form of seamounts and volcanic ridges or arcs may also accrete to the continents. This has the effect of increasing the overall mass of continents. Preliminary estimates of the rates of these phenomena suggest denudation of the continents at about 0.396 mi³ (1.65 km³) per year, whereas newly created volcanic rock is produced at about 0.31 mi³ (1.3 km³) per year. Depending upon the efficacy of accretion tectonics, continents will grow or shrink.

For background information SEE CONTINENTAL DRIFT; MARINE GEOLOGY; PLATE TECTONICS in the McGraw-Hill Encyclopedia of Science and Technology.

David G. Howell

Bibliography. P. J. Coney, D. L. Jones, and J. W. H. Monger, Cordilleran suspect terranes, *Nature*, 288(5789):329–333, 1980; D. G. Howell (ed.), *Tectonostratigraphic Terranes of the Circum-Pacific Region*, 1985; D. G. Howell, Terranes, *Sci. Amer.*, vol. 253, no. 5, 1985; D. L. Jones et al., The growth of western North America, *Sci. Amer.*, 247(5):70–84, 1982; A. Reymer and G. Schubert, Phanerozoic addition rates to the continental crust and crustal growth, *Tectonics*, 3(1):63–77, 1984; N. J. Silbering and J. Hillhouse, Wrangellia: A displaced terrane in northwestern North America, *Can. J. Earth Sci.*, 14(11):2565–2577, 1977; H. Williams and R. D. Hatcher, Jr., Suspect terranes and accretionary history of the Appalachian orogeny, *Geology*, 10(10):530–536, 1982.

Acoustic noise

Antinoise or antisound refers to noise deliberately created to mimic an existing noise field in antiphase so that the two fields cancel each other, resulting in silence. Antinoise, heard on its own, would be the same as noise. The idea of mutual destruction of interfering wave fields originated over 100 years ago; however, practical realization has been made possible only recently by high-speed electronic technology. Research is in progress to develop the ability to silence the low-frequency rumble from jet engines as well as other unwanted noise sources. In a few years, many such applications will be commonplace.

Principles of antinoise. The idea of making silence by adding a secondary, antiphase wave field to cancel an existing one was first patented in 1934. In fact, the phenomenon of destructive interference has long been appreciated for its effect on acoustic fields. It also plays a central role in James Lighthill's theory of aerodynamic sound production, formulated in 1952. According to this theory, aerodynamic sources are naturally arranged in a closely packed quadrupole array in which only a small fraction of the acoustic potential available from any single element of the array escapes destructive interference. But the term antinoise is used only for those deliberately created waves that are produced by a controlled source that is superposed on an existing noise field for the purpose of artificially creating a destructive interference.

Source ambiguity. The source of sound waves is largely a question of a point of view. For example, just as the light field reconstituted in a region through a hologram is indistinguishable from that of the real event, so does the ambiguity carry over to sound; perfectly silent source fields are possible. For this reason the definition of a sound source is rather arbitrary. However, the wave field written in terms of ϕ, the velocity potential, produced by a particular source field q is unambiguously calculated from the equation below,

$$\frac{\delta^2 \phi}{\delta t^2} - c^2 \nabla^2 \phi = q$$

where ∇^2 is the laplacian operator and c is the speed of sound, together with the radiation condition stating that a source cannot anticipate the field it produces. There are many physical processes, described by this equation, which have a nonzero q in some region. This is the source region, and q describes the source distribution.

The significance of this observation about source ambiguity in the context of active noise control is that, if two different source distributions can generate the same wave field, and one of the source distributions is under human control, then a simple change in sign makes the primary noise field subject to extinction by the presence of the secondary field. Furthermore, even though the source of antinoise can be of a completely different construction from that of the primary field, the silence is, in principle, achievable everywhere outside the source distribution.

Requirements for canceling wave fields. In order to achieve complete silence, the two canceling wave fields must match exactly, not only continuously for all time, but also everywhere in space. Any degree of mismatch will result in some residual sound. But the closer the match is made, the more sound attenuation can be achieved.

It follows from the fundamentals of acoustics that the reproduction of a sound beam would require sources spaced closer than half a wavelength apart and distributed over the entire cross section of the beam. This is essentially the spatial form of the Nyquist frequency criterion. In fact, the closer the sources are spaced, the better the matching. Sounds in the audible frequency range, from 20 to 20,000 Hz, have wavelengths from 60 ft (20 m) to 0.8 in. (2 cm). Current transducer technology limits active sound control to the low-frequency (long-wavelength) regime.

The control system that generates the signals for the secondary sources which create the antisound must respond to the temporal variation of the source or wave field being controlled. Thus the higher the frequency of the sound, the faster the processing required. It is here that recent advances in high-speed electronic processing have allowed the technology to progress rapidly. Current systems of simple geometry now operate up to frequencies of a few kilohertz.

Operation of antinoise system. In order to ascertain what sound the control system must produce, it is necessary to have some knowledge of the source process, that is, of how the source fluctuates in time. This information is found from a transducer located near the source which responds to the source activity. Most commonly used are microphones that respond to the sound's air-pressure fluctuations, or accelerometers that sense the vibrations induced by the action of the source. The signal from the transducer or, for more complex spatial source distributions, the signals from an array of transducers are sampled by the digital control system to produce a continuous high-speed stream of numbers. Complex real-time processing algorithms are used to simulate the effects of the above equation for the wave field within the complex geometrical constraints of the particular spatial environment. This simulation requires extremely fast numerical calculations, and most systems use computational units in parallel to speed up the process. Typical multiplications, the basis of calculations, are carried out in less than 100 nanoseconds. New advances in emitter coupled logic (ECL) technology will allow calculations to proceed at 10 times this rate. However, the possible frequency bandwidth of the active sound control system will increase by only a factor of 3, since the number of calculations that must be accomplished increases in proportion to the frequency bandwidth, and the rate at which they must be undertaken is proportional to the highest frequency.

The results of the calculations, again a set of streams of numbers, are converted into analog electrical signals, which are used to drive the

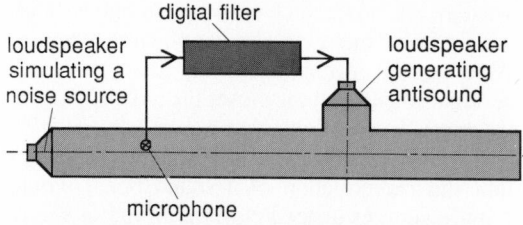

Fig. 1. System for testing sound control in ducts through antinoise. One loudspeaker simulates a source of noise in the duct, and the microphone senses noise in the duct. The other loudspeaker generates antisound to cancel the noise from the source.

secondary sources. Up to now, these sources have usually been specially adapted forms of conventional loudspeakers, but as the sources under control have become larger, with a consequent increase in the sound levels, research has been increasingly directed toward developing alternative sources of sound. Some of the most interesting alternatives come from deliberately causing the flow of a jet to fluctuate by introducing a modulating mechanism. The modulation produces a fluctuating volume of sound, the essence of the simplest source form. Devices of this kind can produce intense sound fields, and have been used to communicate over many miles. A variant of this source mechanism causes the rate of burning of a flame to fluctuate by modulating the fuel supply rate. Experiments in 1984 showed that this device can produce sound levels much more intense than those of the gas modulation process itself.

Applications of antinoise. The antinoise technique has been used to control acoustic noise generated in the cockpits of fighter aircraft, in long narrow ducts, in the exhaust of a gas turbine, and in various forms of rotating machinery.

Cockpit noise. One of the simplest geometries,

where the technique is highly advanced, is that of so-called active ear defenders, which the crew members of some modern fighter aircraft now wear. The cockpit noise levels can be so high as to cause fatigue and interfere with communication among the crew members. Good communication can be achieved only if the signals are substantially above the noise; thus reduction of the noise is essential if crew members are not to be harmed by the high level of these signals. Lightweight ear defenders employ active sound control to maintain the small cavity of the ear at constant pressure by using the earphones themselves. The sound that penetrates the headset from outside creates pressure fluctuations in the cavity which are matched, in antiphase, by sound from the earphones. The communication signal can then be fed at a low level into this relatively quiet environment.

Duct noise. Another relatively simple situation where sound can be controlled actively is in long narrow ducts, such as air-conditioning ducts or pipework (**Fig. 1**). The sound wave, traveling down the duct from a distant noise source, is monitored by a microphone which measures its pressure fluctuations. The digital filter processes this signal to produce the loudspeaker drive signal. The sound from the loudspeaker and the propagating wave combine beyond the loudspeaker to produce silence.

Exhaust control. The first major antisound suppressor was installed in 1981 in the exhaust of a gas turbine which drives a compressor to pump gas through underground gas pipelines (**Fig. 2**). The gas turbine engine is housed in a sound-deadening enclosure, and the hot, spent gases are exhausted through the stack. The low-frequency rumbling noise that escapes from the stack may be heard in the neighborhood around the gas compressor site. The antisound system is very similar to that of the duct controller. It has microphones near the bottom of the stack where the sound is generated, but it uses 72 large loudspeakers driven by a microprocessor-based controller to reduce the noise exhaust rumble by 15 dB.

Periodic noise. Special control systems can be used to reduce the noise generated by rotating machinery of all forms, such as diesel and gasoline engines, gearboxes, fans, compressors, pumps, and propellers. The signals used to drive the loudspeakers for periodic noise are generated in a very simple way. Since the same waveform is required for each cycle of the machine, it can be stored in electronic memory and a microprocessor used to recall and send it to the loudspeaker in synchronism with the machine. The synchronization is maintained by an electrical signal taken from a tachometer. The system can adapt to changes and, typically, noise reductions in excess of 20 dB can be maintained. The next generation of propeller-driven aircraft will have the internal cabin noise field controlled in this manner.

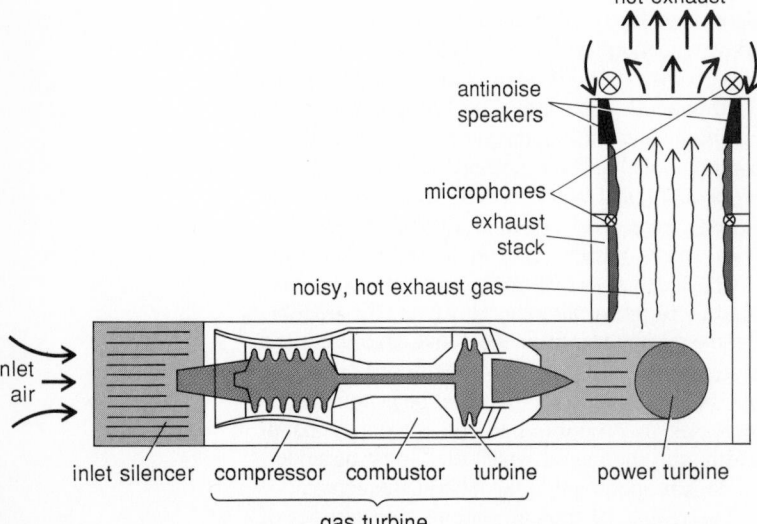

Fig. 2. Antinoise system at gas compressor station. The exhaust stack has a conventional, sound-absorbent surface. Antinoise speakers prevent the escape of rumbling noises.

For background information SEE ACOUSTIC NOISE; INTERFERENCE OF WAVES; SOUND in the McGraw-Hill Encyclopedia of Science and Technology.

Colin F. Ross

Bibliography. J. E. Ffowcs Williams, Anti-sound: A review lecture, *Proc. Roy. Soc. Lond.*, A395:63–88, 1984; J. E. Ffowcs Williams and C. F. Ross, *Anti-sound*, 1987; C. F. Ross, I. Roebuck, and J. E. Ffowcs Williams, Anti-phase noise reduction, *Phys. Technol.*, 16:19–31, 1985.

Adhesive bonding

Adhesive bonding may be defined as the holding of materials together by surface attachment. This technique of attachment is very ancient. The original glues were based on naturally occurring materials such as coal and pine tars, animal protein, and blood albumin. From the bonding of papyrus in ancient Egypt for use in writing to the bonding of elastomers to metal for tanks in World War II, these same materials were of primary importance to the adhesive technologist.

Not until the mid-1930s did the chemistry and technology of polymer and resins based primarily on principles of organic chemistry begin to yield new raw materials for new generations of adhesives. The adhesive industry continued to develop new theories of adhesion and appropriate raw materials, based on phenolic resins, urethanes, polyvinyl esters, and, in the 1950s, the epoxy resins.

Formation of adhesive bond. Among the parameters that must be considered for an adhesive product to have practical applications are contact and transition.

Contact. The adhesive must be brought into intimate, continuous contact with the substrate or surface that is to be joined. To accommodate this requirement, the adhesive at some point must be liquid during the formation of the bond in order to wet, spread, and form a continuous film. This requirement is generally fulfilled by application of a solution of solid polymer in a volatile solvent, by subjection of a solid polymer to sufficient heat or pressure, or both, during contact with the substrate to cause a liquid flow, and by application of liquid reactive components that can be made to react chemically in place to form a solid adhesive polymer after wetting and spreading.

Transition. Once the adhesive has been brought into intimate, continuous contact with the substrate and has at some point been in a liquid state during contact with the two adherends, the adhesive must pass through a transition to a tough, nonflowing, nonliquid, load-bearing interlayer in the bonded assembly.

In the case of permeable or porous substrates the assembly can be made while the adhesive is still wet with solvent. The transition from liquid to solid may then be realized via simple evaporation of solvent through the porous substrate. This process is used in many adhesive areas such as wood bonding, fabric bonding, and carton sealing. In some cases these techniques can also be used to advantage with essentially nonporous substrates.

In the case of nonporous or nonpermeable substrates the transition of the adhesive interlayer from liquid to solid must be accomplished via techniques other than solvent evaporation. There are two methods most commonly used. The method involving heat to flow–cool to set is best represented by thermoplastic adhesives or hot-melt adhesives. The solid adhesives are heated to their flow temperature, the substrates are joined and held in contact with the molten system, and the assembly is then allowed to cool to form the bond. The other technique utilizes an in-place chemical reaction. Normally liquid co-reactive moieties are placed on the substrate, the substrates are joined, and the adhesive passes from a liquid to a solid via a chemical reaction in place. Representative adhesive systems are epoxy-amines, isocyanate-polyols, anaerobic dimethacrylates, acrylic-peroxides, and alpha-cyanoacrylates.

A technique that may also be considered in this respect is the slow crystallization of certain polymers that progress from liquids to solids via a crystallization phenomenon; this is surely at work in the case of neoprene-phenolic contact cements.

Other attributes. The adhesive products themselves must necessarily have additional attributes to be commercially practical. Some of these additional attributes are measurable shelf or can stability, reasonable cure times and temperatures, relatively nontoxic and nonenvironmentally polluting constituents, relative versatility in substrate affinity, and in many cases ability to gap-fill (to tolerate the bonding of imperfect mating surfaces).

Theoretical considerations. There are many different theories of adhesion. The original theories suggested that a true chemical bond between the adhesive and adherend was necessary and responsible for the formation of the bond. At about the same time, it was theorized that the bond was a result of intermolecular attraction between molecules in close proximity or between the adherend and adhesive. These forces are known as van der Waals or London forces of intermolecular attraction. Most adhesive technologists conjectured that probably combinations of both true chemical bonds (ionic, covalent, or coordinate) and van der Waals forces come into play during the bonding operation.

These theories gave rise to considerable controversy and additional research since the actual bond strengths accomplished in practice did not approach the theoretically calculated strengths possible through van der Waals forces alone—let alone a true chemical bond.

One view held that the adhesive polymers were not properly selected to wet and spread on the substrate to allow the intimate contact for van der Waals forces to come into play. Via measurement of contact angles and bond strengths, W. A. Zisman developed his theory of critical surface polarity. In simplified form, it states that if the molecular polarity (or surface tension) of the liquid adhesive is less than the surface polarity of the substrate, wetting and spreading will occur and a superior bond will be formed; and conversely, if the liquid adhesive polarity is higher than the polarity of the substrate surface, wetting and spreading will not occur and therefore a weak bond is formed.

Metals are considered to have a high surface polarity, and also strongly bonded molecular layers of water on their surfaces. A molecular model of a metal surface is shown in **Fig. 1**.

The bonding of metals has been approached with some success by using reactive ingredients to react with chemisorbed water layers as depicted in Fig. 1. Metals are considered to be high-energy surfaces, while at the other end of the spectrum polyethylene is not wet by water and seems to abide by Zisman's theory as applied to adhesive bonding. The surface of polyethylene has an extremely low surface polarity.

Low-energy surfaces have been effectively bonded when the surface is converted by oxidation techniques to offer greater polarity to the substrate in order to allow the adhesives to wet and spread effectively.

A similar technique of modifying polymeric substrates, such as cured elastomers via subject-ing the cured elastomer surface to chlorination prior to bonding with adhesive systems, has proved to be of practical utility in improving the bond strength by allowing more effective wetting and spreading of the adhesive constituents.

It is reasonable to conclude that the practical application of this theory has been essentially substantiated. However, research has offered evidence that some cases of adhesive bonding tend to refute the teachings of critical surface polarity under closely controlled laboratory conditions that indicate surface morphology of a polyethylene surface may be a greater factor than surface polarity. Still, the actual bond strength, while abiding by Zisman's theory, although improved, did not approach the theoretical levels calculated via van der Waals forces.

Another theory by J. J. Bikerman suggests that a weak boundary layer is the reason that bond strengths are below the theoretical level. Again, in simplified terms, this theory states that the probability that an adhesive-adherend failure will propagate across a heterogeneous interface is practically zero; essentially, there is no such thing as adhesive failure; all failures are cohesive within a single phase; and this single phase is labeled as a weak boundary layer. Research to substantiate or refute the practicality of this hypothesis has generally proven the validity of the theory.

The CASING (crosslinking at surfaces in inert gases) technique developed from studies on polyethylene lends credence to the presence of a weak low-molecular-weight layer on the surface as perhaps more detrimental than the actual surface polarity in obtaining acceptable bond strengths. The formation or presence of weak boundary layers on substrate surfaces has been recognized in adhesive bonding, as well as the less obvious formation of weak boundary layers on the adhesive side of the adhesive–adherend interface via migration of weak layers from constituents in the adhesive to the interface.

Other recent research on theories of adhesion considers the subject of fracture mechanics. It is suggested that during the formation of an adhesive bond, points of stress are formed and concentrated in proximity to the interface because of the essential heterogeneous nature of the adhesive and the adherend in most cases. When an exterior force is applied to the bonded assembly, a failure initiates at a point or several points of high stress concentration. A fracture or fractures can therefore initiate (or are already present) at these built-in, concentrated points of stress far below the force value expected by theoretical calculation. The fracture then propagates via stress relief along a path of points of concentrated stress in close proximity to the adhesive–adherend interface.

In the field of vulcanization bonding of elastomers and elastoplastics, an important theory has

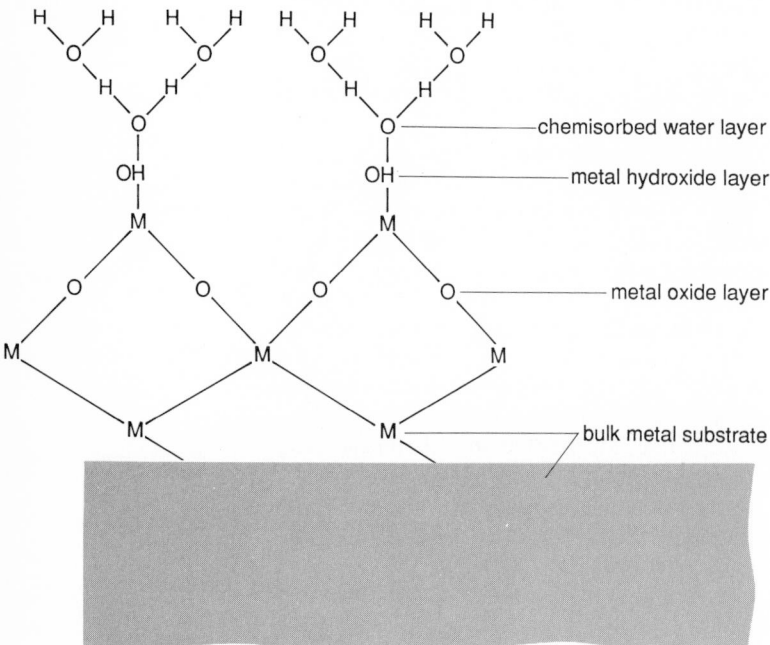

Fig. 1. Molecular model of a metal surface. M = metal.

chemisorbed water layer

metal hydroxide layer

metal oxide layer

bulk metal substrate

been advanced that the adhesive bond is dependent upon the diffusion and mutual solubility and compatibility of the adhesive polymer with the substrate polymer. Therefore the bond is, to a great extent, due to the molecular entanglement of the adhesive and adherend polymer chains or segments.

Contemporary trends. Worldwide the automotive industry is accelerating the process of replacing metal components with engineering plastics. Many of the plastic components must be adhesively bonded, while the metal components that are replaced could be mechanically fabricated via welds, rivets, or crimped seams. **Figure 2** shows an automobile that has exterior body sheet entirely of plastic. The rear deck, hood, and roof sections are all-bonded assemblies of flat exterior sheet to plastic reinforcing hat-shaped section beams. Approximately 5% of all automotive exterior panels are plastic today. It has been forecast that by 1995 nearly 50% of all exterior automotive body sheet will be plastic. In addition, bonded composite leaf springs, thermoplastic headlamps, and plastic lift gates are now in production. All of these adhesively bonded assemblies have gone into production since 1982.

Removal of spot welds from metal exterior automotive body sheet is also being accomplished by newer metal-bonding structural adhesives as represented by the all-bonded metal doors of the automobile shown in **Fig. 3.**

The worldwide food packaging industry is also accelerating the process of replacing metal and glass rigid food containers with flexible multilayer laminated plastic containers. Representative examples are the flexible adhesive-bonded fruit juice packages, the flexible "brick-pack" vacuum-sealed plastic containers for fresh-ground coffee, and multiple applications of laminated lid stock for rigid plastic containers. The adhesives are used to laminate the multi-plies of the flexible package. Combinations of metal foils to polyethylene terephthalate to linear low-density polypropylene, as well as many other combinations of plastic and metal foils, are bonded with high-quality laminating adhesives.

Fig. 3. Automobile with all-bonded hem flange door assembly. (*Volkswagen Corp.*)

This market is expected to grow at a 10% annual rate for the foreseeable future.

For background information SEE ADHESIVE; POLYMER in the McGraw-Hill Encyclopedia of Science and Technology.

James A. Graham

Bibliography. J. J. Bikerman, *Surface Chemistry, Theory and Applications*, 1958; R. J. Good, *Adv. Chem. Ser.*, Theory for the estimation of surface and interfacial energies, 43:74, 1964; B. Miller (ed.), Composite structures: Next wave in Detroit, *Plast. World*, November 1986; G. L. Schneberger (ed.), *Adhesives in Manufacturing*, 1983; L. H. Sharpe and H. Schonhorn, Surface energetics, adhesion, and adhesive joints, *Adv. Chem. Ser.*, 43:189, 1964; W. A. Zisman, in P. Weiss (ed.), *Adhesion and Cohesion*, 1962.

Aeronomy

Over the last few hundred years, human activities have perturbed the chemical composition of the atmosphere on a global scale. This relatively rapid change in the abundance of many of the trace gases has sparked considerable interest in assessing the environmental consequences of these perturbations. The most notable of these debates has focused on the trace gas and climate problem, or the greenhouse effect.

Carbon dioxide. The familiar part of the greenhouse effect is the role played by carbon dioxide (CO_2) in potential global warmings. Together with water vapor and clouds, CO_2 reflects to the Earth part of the infrared surface radiation that would otherwise escape into space, thereby helping to establish the current surface temperatures. However, it is now clear that the global average concentration of CO_2 has increased 25% during the industrial era, largely due to the burning of carbon-containing fossil fuels. Furthermore, it is likely that CO_2 concentrations will double by the latter part of the twenty-first century. Since these enhanced CO_2 concentrations imply that more radiation would be reflected to Earth, higher surface temperatures are predicted as a result.

Fig. 2. All-plastic-body automobile. (*General Motors Corp.*)

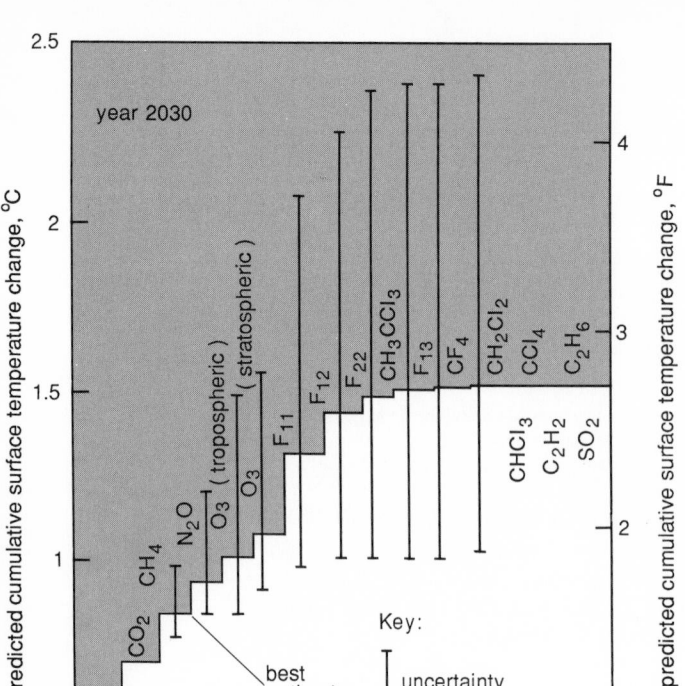

Fig. 1. Cumulative surface temperature warming predicted in 2030 due to projected increases in the radiatively important trace gases. (*After V. Ramanathan et al., Trace gas trends and their potential role in climate change, J. Geophys. Res., 90:5547–5566, 1985*)

The past two decades have seen a substantial research effort aimed at quantifying the prediction of global warming due to the CO_2 growth rate. Increasingly accurate representations of the interactions of radiation, clouds, oceans, and CO_2 abundance have been incorporated into climate models. Current predictions are that a doubling of CO_2 may result in an average warming of 5.4°F (3°C). Such an alteration in mean surface temperature would probably induce considerable changes in precipitation distributions, agricul-

tural patterns, and sea-level habitation, all of which would be accompanied by disruptive societal and economic consequences. Understandably, sharpening the CO_2-climate picture is a major goal of future research.

Other radiatively important trace gases. Carbon dioxide constitutes only part of the trace gas and climate phenomenon. There is evidence that the abundances of other radiatively important trace species—primarily methane (CH_4), nitrous oxide (N_2O), tropospheric ozone (O_3), and two chlorofluorocarbons known as F_{11} and F_{12}—are also increasing in the atmosphere. The combined greenhouse effect of these other infrared-active trace species may cause a global warming comparable to that expected from CO_2. **Figure 1** gives the cumulative surface warming expected in the year 2030 from projected increases in several trace gases. The stepped line shows that the temperature increase, 2.7°F (1.5°C), expected from all of the trace gases is about twice that from CO_2 alone. The vertical bars reflect the uncertainty in extrapolating the abundances to 2030.

The non-CO_2 trace gas and climate problem is distinctly different from and more complex than that of CO_2 alone for several reasons. While the increases in CO_2 are due largely to fossil fuel and biomass burning, the other radiatively important trace species arise from a wide variety of both anthropogenic and natural sources. Furthermore, the abundances of many of these species, unlike CO_2 which is relatively inert chemically, are altered in the atmosphere by photochemistry. Lastly, the global distributions of some of the shorter-lived radiatively important trace species are strongly influenced by dynamical processes, which are of less importance for CO_2. While the radiation physics of the trace gases are similar, the approaches to understanding their respective roles in climate change differ substantially.

Methane. In the last several years, one of the most dramatic discoveries regarding atmospheric trace gases has been the realization that the abundance of CH_4 is increasing about 1% per year. Furthermore, from analyses of air trapped in dated ice cores, it appears that the concentrations of CH_4 may have doubled in the past 350 years. The reasons for such an increase are not clear, since the possible sources are diverse and poorly quantified: ruminant animals, organic-rich sediments, rice paddies, termites, and natural gas leakage. Global distributions and trends are becoming available, as shown in **Fig. 2**. Two features provide clues that dictate future research directions: most of the CH_4 is in the Northern Hemisphere, which indicates a link to land-related processes; and there is a clear seasonal cycle, which suggests a biological influence. The dominant loss of CH_4 is by chemical reaction with the hydroxyl radical, OH, which itself is attacked by numerous pollutants. An intriguing line of inquiry is that the CH_4 abundance may be

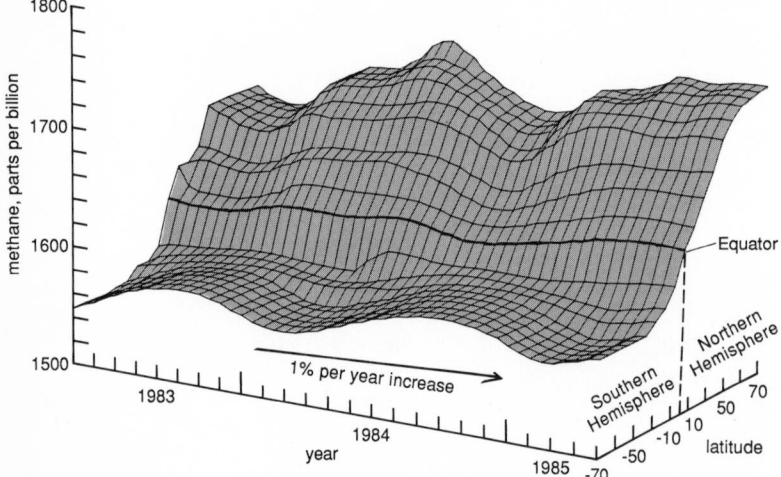

Fig. 2. Global distribution and trends of methane (CH_4). (*After L. P. Steele et al., The global distribution of methane in the troposphere, J. Atmos. Chem., 5:125–171, 1987*)

increasing, not just due to increasing CH_4 emissions, but because elevated pollution levels may be suppressing the hydroxyl radical concentrations in the atmosphere.

Nitrous oxide. Nitrous oxide is increasing about 0.25% per year. Its residence time in the atmosphere is very long (more than 100 years), since the only loss is by photodestruction in the stratosphere (more than 7 mi or 12 km above the Earth). The sources of N_2O are thought to be microbial production in soils and the ocean's combustion processes. Both are very poorly quantified at present, however, which is reflected by the rather large uncertainty range in the 2030 extrapolation in Fig. 1.

Tropospheric ozone. About 10% of the ozone in the atmosphere resides in the troposphere. This fraction plays a direct greenhouse role, since the higher pressures at these lower altitudes broaden the absorption spectrum of ozone so that it partially overlaps the wavelength window through which most of the outgoing infrared surface radiation escapes to space. Hence, it was of some concern to climatologists to learn that tropospheric ozone may be increasing.

The classical view of this fraction of ozone in the troposphere is that it is a naturally occurring residue resulting from downward transport from the large abundance of ozone in the stratosphere. However, there is growing evidence that the direct photochemical production of ozone in the global troposphere may rival the stratospheric source. The chemical production of ozone by nitrogen oxides and hydrocarbons in urban areas from smog chemistry is well known. Such production can be occurring significantly even in rural areas, as **Fig. 3** shows. The relations between ozone and its precursor, NO_x (the sum of nitric oxide and nitrogen dioxide, $NO + NO_2$), have been determined by winter and summer measurements at a rural mountain site. In the winter, when low temperatures suppress the chemistry, ozone shows little relation to the abundance of its chemical precursor. However, in the summer there is a clear increase of ozone corresponding to the higher NO_x concentrations. Since it is possible that the average NO_x levels may be elevated in the Northern Hemisphere because of global-scale pollution, it is suspected that ozone may be increasing in the global troposphere as a result.

Indeed, limited data suggest that such an increase is occurring, perhaps as much as 15–30% in the Northern Hemisphere over the past few decades. The data are limited, however, and are of a nonuniform and uncertain quality. In contrast to the longer-lived radiatively important trace species, no network of regular measurements of tropospheric ozone currently exists, primarily because of a dearth of unequivocal measurement techniques.

Chlorofluorocarbons. Figure 1 shows that a num-

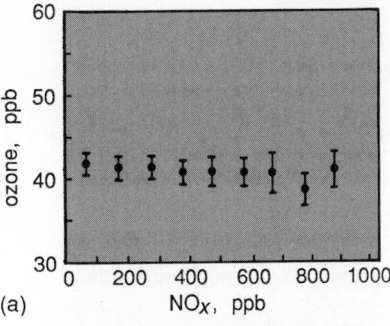

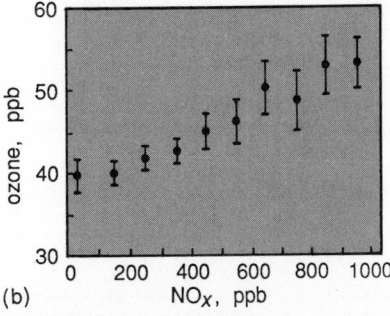

Fig 3. Relation between ozone (O_3) and its precursors, nitric oxide plus nitrogen dioxide ($NO + NO_2$), at a rural site in the Colorado Mountains in (*a*) winter and (*b*) summer. $NO_x = NO + NO_2$. (*After D.D. Parrish et al., Background ozone and anthropogenic ozone enhancement at Niwot Ridge, Colorado, J. Atmos. Chem., 4:63 – 80, 1986*)

ber of halogenated hydrocarbons can contribute to the greenhouse effect, the primary two being F_{11} ($CFCl_3$) and F_{12} (CF_2Cl_2). Because these two compounds are also believed to be involved in the destruction of stratospheric ozone, their sources, trends, and distributions have been scrutinized keenly over the past decade. The sources of these species are purely anthropogenic, largely spray propellants and refrigerants. F_{11} and F_{12} have very long residence times in the atmosphere. During the first several years after their initial use, the atmospheric concentrations of these compounds were increasing at a rate of several percent per year. However, a partial ban on their nonessential uses (because of the potential stratospheric ozone loss) and a general economic downturn in several industrialized nations have lessened this increase in recent years. The added awareness of this new climate role has renewed calls for a more complete ban, and hence the uncertainties in the projections in Fig. 1.

Future research. An adequate understanding of the complete trace gas and climate phenomenon will require a better definition of the characteristics of the non-CO_2 greenhouse gases. In particular, the long-term trends in methane and tropospheric ozone must be determined in more detail, particularly the latter. The emission sources of N_2O and CH_4 are major unknowns. The chemical processes whereby ozone is formed and CH_4 is lost must be characterized for the major global chemical regimes. The natural source of ozone from the stratosphere needs to be better quantified, as must its losses to surfaces. Lastly, models of the radiative, transport, and chemical properties of the trace gases must be developed to test the conceptual understanding of these gases and to predict possible future climatic effects arising from their changing abundances.

For background information SEE AERONOMY; ATMOSPHERIC CHEMISTRY; ATMOSPHERIC OZONE; GREENHOUSE EFFECT; PHOTOCHEMISTRY in the McGraw-Hill Encyclopedia of Science and Technology.

Daniel L. Albritton

Bibliography. D. D. Parrish et al., Background ozone and anthropogenic ozone enhancement at Niwot Ridge, Colorado, *J. Atmos. Chem.*, 4:63–80, 1986; V. Ramanathan et al., Trace gas trends and their potential role in climate change, *J. Geophys. Res.*, 90:5547–5566, 1985; L. P. Steele et al., The global distribution of methane in the troposphere, *J. Atmos. Chem.*, 5:125–171, 1987.

Agricultural soil and crop practices

In agricultural production the energy of sunlight is converted through photosynthesis into plant products that are utilized by humans either directly or after conversion into animal products. In addition to the direct input of solar energy, agricultural crop production requires substantial energy inputs from numerous combinations of human labor, animal power, and mechanical or chemical energy. Human input can vary greatly in how it is applied. In primitive agricultural systems, most human input is in the form of labor required for seeding, cultivating, and harvesting. In technologically advanced agricultural production systems such as in the United States, the amount of human labor expended per hectare is much less, but this input is greatly multiplied by use of stored energy (especially fossil energy) in the forms of machinery, fertilizers, irrigation water, and pesticides. Thus, inputs into an agricultural production system can be compared by converting them to energy equivalents (such as megajoules per hectare or diesel fuel per hectare).

Inputs. The degree to which inputs are used to supplement human energy in agricultural production systems varies widely. In the progression from the more primitive to the more advanced systems, inputs generally increase. In primitive systems, energy is provided basically by human labor. In more advanced systems, other forms of energy are substituted, enabling one individual to cultivate a larger area. Use of animal power increases severalfold the area that can be cultivated by one individual; use of small tractors increases the cultivated area even more; and use of large tractors again increases severalfold the cultivated area per individual. Thus, in technologically developed nations such as the United States, a small percent of the population is producing food and fiber for the entire

nation; by contrast, in some developing nations over 80% of the population is employed in agricultural production.

In order for the investment in the large equipment required for highly developed agriculture to be economically feasible, crop yields must be maintained at a relatively high level. Thus additional inputs into the crop production system become necessary. These include energy for pumping and distributing irrigation water; energy for drying, storing, and transporting products; and energy for fertilizers and chemicals. Nitrogen, the primary fertilizer input, is manufactured largely from natural gas and atmospheric nitrogen. Consequently, while crop yields and production per hour of human labor in technologically advanced agricultural systems may be extremely high, efficiency of energy utilization is often less than that for primitive farming systems. In the development and intensification of farming practices, one form of energy has been substituted for another.

Technology transfer. A recent example of the evolution of agricultural inputs and production is a series of developments in southeast Asia during the last few decades. Prior to the 1950s, agriculture in most southeast Asian nations was relatively primitive, with the exception of some large commercial operations specializing in production of specific crops. Inputs to the rest of the agriculture in that region were confined largely to those available to a family operation in a local community; these consisted of human labor and limited animal power with little, if any, mechanical or electrical power. Fertilizers and other agricultural chemicals were not available, and very few improved cultivars existed. Most of the population in many nations was involved in agriculture, with famine and starvation occurring in years of low production.

In the 1950s the Ford and Rockefeller foundations and other organizations became involved in the advancement of agriculture in less developed nations. These groups provided the training needed to improve the systems of agricultural production so that these nations could become self-sufficient in food production. As a result of these efforts, numerous high-yielding wheat and rice cultivars were developed; production packages including new varieties and the required fertilizer, water management, and pest control techniques were passed on to farmers. This technology is now used on over 120 million acres (50 million hectares); in the last few decades grain production in many southeast Asian nations has increased severalfold, and some nations are now exporting grain rather than importing it.

The widespread use of advanced techniques of agriculture is not without problems, however, especially in terms of economic returns and environmental quality. As energy prices increase, the cost of added inputs also increases, reducing

Average energy input into corn production in the United States, MJ/ha*		
Category	1945	1980
Labor	300	100
Machinery	4,900	11,400
Fuel	14,800	21,700
Fertilizers	1,900	27,300
Seed	800	1,600
Irrigation	1,100	2,100
Pesticides	0	600
Drying	100	3,300
Electricity	900	8,500
Transportation	500	1,900
Total	25,300	78,500
Corn yield (output)	83,000	197,800
MJ output/MJ input	3.28	2.52

* 1000 MJ/ha = 0.2 gal diesel fuel equivalent per acre.

These inputs are about evenly divided between energy for mechanical practices (tillage, seeding, cultivating, harvesting, drying) and energy in agricultural chemicals (fertilizers and pesticides). Additional energy is needed for irrigated farming. The amount required may be near zero for some gravity systems, or may exceed the combined energy of all other inputs when the water is pumped from deep wells.

Although output/input ratios, when calculated on an energy-equivalent basis, are very useful in comparing agricultural production systems, efficiency of individual inputs is also of concern. For example, it is of interest to study the use efficiency of irrigation water and nitrogen fertilizer inputs for irrigated winter wheat production in Nebraska, as shown in the **illustration**. With no irrigation water the input of nitrogen/fertilizer had little or no effect on wheat yields (illus. *a*). The addition of irrigation water without added nitrogen resulted in a very modest yield response. When both inputs were added, the effects were cumulative, giving highest yields from the highest

the profitability of the higher-input approach. Also, the use of more chemicals in production practices increases the potential for these chemicals or their degradation products to affect the food chain and the environment, either directly through biological uptake or indirectly by escaping into the atmosphere or water. Increasing inputs increase the potential for nitrates and pesticides to enter drinking water sources or for nitrous oxides to escape into the atmosphere to react with ozone. The possibility of environmental degradation may become increasingly important in deciding the amount and types of inputs to be used in future agricultural production systems.

For the more primitive production systems, such as the slash-and-burn cultivation utilized in the Amazon basin and other tropical areas, it has been estimated that inputs of about 36,000 MJ/ha (7.4 gal diesel fuel/acre) are required. With an output of about 504,000 MJ (103 gal diesel fuel), the output/input ratio is about 13.8. Research has demonstrated that energy inputs increase with technological development. For corn production in the United States, inputs, when calculated as energy equivalents, averaged 25,000 and 78,000 MJ/ha (5.1 and 15.9 gal diesel fuel/acre) in 1945 and 1970, respectively. Corresponding energy outputs were 83,000 and 198,000 MJ (17.0 and 40.5 gal diesel fuel), for output/input ratios of 3.28 and 2.52, respectively (see **table**). These comparisons show that, as production increases with the addition of more inputs, the efficiency of energy utilization is reduced. If those inputs used in the United States for storing, processing, and preparing food are added to the inputs required for production, it is found that over 10 MJ are expended for each megajoule of energy in the food consumed. This value has increased from about 1 MJ/MJ in 1910.

Efficiency. Inputs into agricultural production originate from many sources. Those in present-day American agriculture often total approximately 200,000 MJ/ha (40.8 gal diesel fuel/acre).

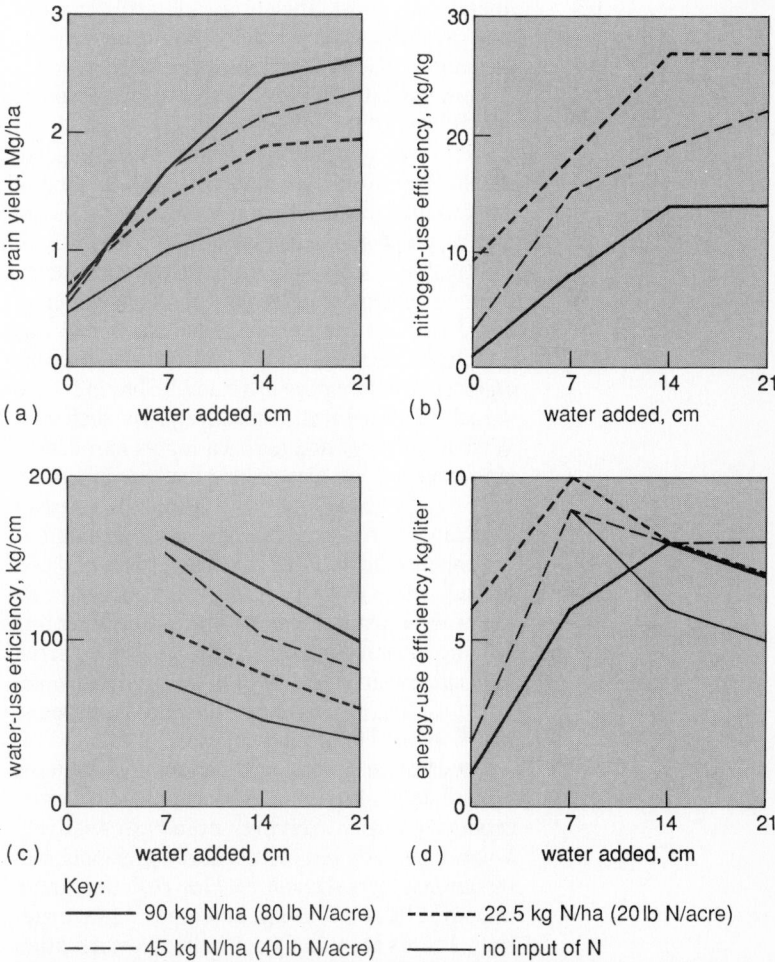

Use efficiency of inputs for producing winter wheat in Nebraska. Graphs show (*a*) grain yields, and increase in wheat yield per unit input of (*b*) fertilizer nitrogen, (*c*) irrigation water, and (*d*) their energy equivalents. 1 Mg/ha = 893 lb/acre. 1 kg/cm = 2.2 lb/in. 1 kg/liter = 7.64 lb/gal. 1 cm = 0.4 in.

Key:
— 90 kg N/ha (80 lb N/acre)   ---- 22.5 kg N/ha (20 lb N/acre)
— — 45 kg N/ha (40 lb N/acre)   — no input of N

input rates of both water and nitrogen.

When the wheat yield data in the illustration are considered in terms of efficiencies of each variable input, a somewhat different picture emerges. Water-use efficiency, calculated as the increase in yield per unit input of irrigation water, decreased as the quantity of the water input increased (illus. *c*); however, water-use efficiency improved by increasing nitrogen inputs. Most efficient use of the added irrigation water occurred when the highest fertilizer nitrogen rate was applied with the least amount (7 cm or 2.8 in.) of irrigation water. If the objective of this production enterprise were to maximize efficiency of use of a very limited water supply, these results suggest using relatively high fertilizer nitrogen inputs with limited irrigation. Relative prices would determine the most economical nitrogen rate because maximum economic return and maximum water-use efficiency may not be synonymous.

Nitrogen-use efficiency is also presented in terms of increased yield per unit of fertilizer nitrogen applied (illus. *b*). Like water-use efficiency, nitrogen-use efficiency was greatest for the lowest input of fertilizer nitrogen, and was increased by increased inputs of water. Consequently, the most efficient use of fertilizer nitrogen occurred when a small amount of fertilizer was applied along with adequate water.

Both the fertilizer nitrogen and the water inputs can be converted to an energy-equivalent input basis. Conversion factors vary somewhat, depending on form of nitrogen fertilizer used, method of application, type of irrigation used, depth of pumping (if pumping was involved), and other factors. For purposes of comparison in the illustration, however, energy required to apply 1 cm (0.4 in) of irrigation water and energy for manufacture of 1 kg (2.2 lb) of nitrogen fertilizer was calculated as the equivalent of the energy in 8.5 and 1.5 liters (2.3 and 0.4 gal) of diesel fuel, respectively. By such conversions of water and nitrogen inputs into energy equivalents, it was determined that energy-use efficiency decreased only slightly as the rates of application of nitrogen and water were increased. A maximum value of 10 kg of grain per liter of diesel fuel equivalent (8.4 lb per gallon) was obtained with the minimum inputs of 7 cm (2.8 in.) of water and 22.5 kg of nitrogen per hectare (20 lb per acre). However, with highest rates of nitrogen and water inputs, the increase in wheat production dropped to only 7 kg per liter (55 lb per gallon) of fuel.

Low-input systems. Such agricultural systems require few off-farm purchases. Seed is often produced locally. Fertilizer needs are relatively low because soil fertility needs are generally met by utilizing animal manures, crop residues, green manures, legumes, ashes, and other waste materials. Weeds are usually controlled by cultivation and crop rotation, and insects by rotations, biological agents, and hand picking or similar techniques. Gravity irrigation is often used, and crops

are dried in the field to the extent possible before harvesting. Local marketing of products is often promoted to reduce the need for transportation.

Reducing inputs to such an extent is a drastic departure from conventional agricultural production methods commonly used in the United States. The low-input system is often advocated by proponents of organic farming, which employs agricultural production systems that severely restrict or eliminate the use of agricultural chemicals. Organic farming methods can be successfully employed on a limited scale; however, there are insufficient organic wastes and other required resources, or markets for the expanded livestock enterprises that would follow, for a broad expansion of organic farming.

The energy crisis of the 1970s resulted in renewed interest in techniques that can be used to reduce cost of agricultural production. As a result, several practices that help reduce inputs and enhance profitability have emerged or are emerging. For example, it is possible to convert from clean tillage to no-till, reducing energy inputs by the equivalent of 30 to 40 liters of diesel fuel per hectare (3.6 to 4.8 gal per acre). For irrigated agriculture, this reduction may be even greater because of less evaporation and reduced irrigation input. By using a winter cover crop with continuous corn, especially if the cover crop is a legume, fertilizer nitrogen inputs frequently can be reduced by 50–100 kg/ha (45–90 lb/acre). If grain crops are produced in rotation or double-cropped with legume crops, need for nitrogen fertilizer is greatly reduced or entirely eliminated (especially if forage legumes are used). Likewise, need for pesticide inputs is often greatly reduced by using a crop rotation. Animal manures and other organic wastes, when available, may further reduce fertilizer inputs. One recent study demonstrated that by using legumes and rotations, energy input for corn production was less than half that required with a conventional corn-soybean rotation utilizing fertilizers and pesticides (96,000 vs. 228,000 MJ/ha or 19 vs. 45 gal fuel/acre).

Consideration of alternatives to relatively high-energy-input crop production systems as presently used in many parts of the world must include study of the effect of changes in total quantity of production. For commercial agriculture the first priority is that the production system provide sufficient profit to enable the producer to remain in business. In most nations price and marketing of most agricultural commodities is controlled as much by government policies as by the relative values of inputs. These policies dictate to a large degree the profit and loss in an enterprise. Conversion from a continuous grain to a legume-based rotation, while maintaining yields of the grain crop, would reduce total production of grain because of reduced land area used for production. Likewise, production of animal products would increase if a livestock enterprise were associated with the legume rotation. If such a conversion were practiced extensively on a national scale, volume of grain products on the market would decrease and volume of livestock products would increase. With no interference from government policy, based on supply and demand, grain prices should increase and livestock prices should decrease. In practice, however, as a result of government policies, changing demands, and other factors, it is questionable if this simplistic model would apply for large areas for an appreciable time period.

Advances in agricultural production in North America since the end of World War II have resulted from improvements in four production factors: yield potential (breeding), disease and stress resistance, cultural and management practices, and mechanization of agriculture. Advances in each of these production technologies probably contributed 20 to 30% to the yield increases in the United States. These changes in agricultural production have drastically changed many facets of the United States economy. Only about 2% of the United States population is presently employed in agricultural production, compared to 20% or more before World War II. No other society in history has had such an ample supply of food at a cost that is such a low percentage of average annual income.

Future advances. It is anticipated that these advances will include development of improved cultivars, "smart" field machinery, computer-assisted agricultural production, and other novel technology.

Improved cultivars. Research indicates that continuing increases of inputs will yield corresponding increases in production. Developments in the fields of biotechnology and genetic engineering suggest that the desired quality can be engineered into food products and that cultivars of crops that are highly tolerant of adverse climatic stresses, salinity, soil acidity, diseases, insects, and other production-limiting factors can be developed. These cultivars may also have greatly enhanced production potentials. For example, plant scientists are finding that many corn cultivars that retain green leaves during the period for grain development have a much greater yield potential than those that begin to senesce the upper leaves during this period. Such cultivars, when grown in a favorable soil environment with the adequate quantity and type of fertilization, may yield harvests double those of present cultivars. However, it would be necessary to greatly increase fertilizer and other inputs to achieve that potential. *See Plants, saline environments of*.

Field machinery. Advances in soil science and agricultural engineering are expected to yield "smart" field machinery—machinery that automatically adjusts to changing soil environments within the field by feeding information from sensors for soil water, compaction, nutrient status, and other parameters into onboard computers that adjust machine settings accordingly. By this technique, planting rate and depth, fertilizer rate and material, pesticides, and other inputs can vary across the field as soil conditions vary. This will ensure that each plant is growing in a soil environment that approaches the optimum as nearly as possible.

Computer assistance. Such an accomplishment requires more research information, but could result in optimizing production inputs to maximize economic return. By incorporating the cost of each production practice and the commodity price trends into these computer programs, machines might be adjusted so that net return would be maximum even though the actual yields might be less than maximum. Likewise, programs needed to ensure environmental quality control could be added to interact with the production and economic programs, thereby assuring that the inputs needed for maximum net return would also protect groundwater quality, control soil erosion, control salinity, and so forth.

Novel technology. More research is needed before all of the above scenarios are possible, but the required technology is being developed. When fully operational, this approach to agricultural production may require no more, or even fewer, inputs than are used today. For example, if the cost of fossil fuels increases rapidly again, computer programs may indicate that fertilizer and tillage inputs should be greatly decreased or eliminated for a given field. Thus, level of input will be dictated not only by production potentials but also by economic and environmental considerations.

Development of this technology has been initiated in recent years because the availability of low-cost computers has made this approach economically feasible. A few retailers of fertilizer have already equipped their field distribution equipment with computers into which information from soil maps and soil test results is entered. As the machine travels across the field, kind and rate of fertilizer material applied are automatically varied in accord with this stored information on soil properties and soil nutrient availability. As research continues to clarify crop growth processes and requirements, and as technology develops improved on-board sensors, the current simplified computer programs used to regulate fertilizer application will be expanded to include control of most, if not all, inputs into the crop production systems. Then it will become possible to optimize inputs to achieve whatever goals are selected, such as various combinations of maximum yield, maximum economic return, environmental protection, and minimum input. Given the goals of the enterprise, the producer can then prescribe the best combination of treatments (inputs) to achieve the goals.

For background information SEE AGRICULTURAL SOIL AND CROP PRACTICES; AGRICULTURE; AGRONOMY in the McGraw-Hill Encyclopedia of Science and Technology.

J. F. Power

Bibliography. P. H. Abelson (ed.), *Energy: Use, Conservation, and Supply*, American Association for the Advancement of Science, 1974; D. F. Bezdicek (ed.), *Organic Farming: Current Technology and Its Role in a Sustainable Agriculture*, American Society of Agronomy, 1984; T. C. Edens, C. Fridgen, and S. L. Battenfield (eds.), *Sustainable Agriculture and Integrated Farming Systems*, 1985.

Allelopathy

Higher plants, immobilized as they are, have developed a chemistry to protect themselves against environmental stresses as well as encroachments by other plants and organisms. Presently, allelopathy denotes the deleterious effects that higher plants of one species (donors) have on the germination, growth, or development of another species. It is also recognized that microbes associated with the donors may be involved in the generation and release of the chemicals responsible for allelopathy (allelochemicals).

Three areas of interest in recent studies on allelochemicals are classification (source, target species), characterization, and mechanisms (plant succession, agricultural implications, ecological studies). Though the primary emphasis of studies of allelopathy is on higher terrestrial plants, some attention is being given to the role of this phenomenon in managing noxious aquatic plants, fresh-water as well as marine. It seems likely that an understanding of the red tides will involve allelopathy.

Classification. Some chemicals synthesized by plants serve a self-protection function. Chemicals that provide the donor with an interspecies advantage are known as allomones. For example, phytoalexins are natural fungitoxic substances that are synthesized by plants as a response to microbial infection.

Allelochemicals can be classified on the basis of the source and modes of release: they have been found in nearly all plant tissues (leaves, stems, roots, rhizomes, seeds, flowers, fruits) and modes of release include volatilization, root exudation and leaching, and decomposition of plant residue.

Volatile allelopathic compounds typically are found in more arid regions, and are produced by such genera as *Artemisia*, *Eucalyptus*, and *Salvia*. Substances produced are monoterpenes and sesquiterpenes. Vapors of the toxic allelopathic compounds (allelotoxins) may be absorbed directly by surrounding plants, or the toxin may reach the plants indirectly through its absorption by dew that is in contact with the plants, or the toxin-laden dew may travel through the soil and be taken up by plant roots.

Many allelopathic compounds are released by plant roots through active exudation, leakage, or sloughing of dead cells. A special adsorption column has been used to trap selectively organic compounds from root exudates while it transmitted nutrient ions and hydrophilic compounds. Over 15 compounds are produced by roots of *Hemartha altissima* (*Bigalta limpograss*); these have been identified as derivatives of benzoic, cinnamic, and phenolic acids.

Above-ground portions of the plants may be leached by rainwater or by dew or fog condensation, and a variety of substances have been identified, including amino acids, alkaloids, gibberellic acids, organic acids, pectic substances, phenolic compounds, sugars, and terpenes. Recent research indicates that soybeans (*Glycine max*) may be inhibited by leachates from velvetleaf (*Abutilon theophrasti*).

When the plant dies, chemicals are released directly, through leaching, or by the reaction of microbes that effect enzymatic degradation of material present in the plant tissue. It has been found that management of selected crop residues can reduce the germination and growth of weeds. An example of the degradation of plant residues is degradation of cyanogenic glycosides of *Prunus* species or Johnson grass (*Sorghum halepense*) to produce hydrogen cyanide and benzaldehydes. Increasingly, weed management will involve appropriate utilization of crop residues. So-called sick soil is probably a reflection of poor management of crop residues.

Characterization. It has been estimated that over 10,000 low-molecular-weight compounds have been isolated from higher plants and fungi; the total number may be over 400,000. Some are secondary products of plant metabolism, and some are degradation products from the microbial enzymatic action. About a dozen types of compounds are known, and examples of some of the types are given below.

Organic acid. Abscisic acid (I), an endogenous

I

growth hormone inhibitor, is synthesized in relatively large amounts during environmental stresses, for example, drought. Because this inhibitor restricts those growth hormones responsible for extension growth and cell division, growth almost ceases, and the plant conserves energy during critical periods. In tomato plants, water stress of 3 days leads to a marked increase in abscisic acid, which causes the leaves to close, thus diminishing water loss. The abscisic acid is an allelolchemical in the sense that it defends the donor against itself.

Lactones. Patulin (II) is synthesized by *Penicil-*

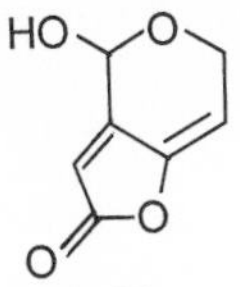

II

lium sp. growing on wheat. It inhibits the seed germination and seedling growth of several cultivated plant species, including corn. Other lactones include psilotin and psilitonin (III) from

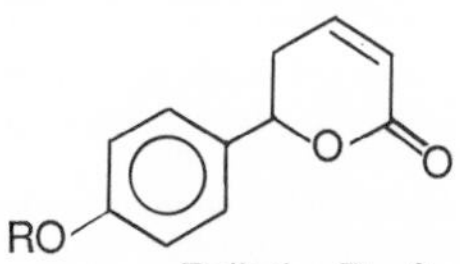

Psilotin: R = β-D-glucose
Psilotinin: R = H
III

Psilotum nudum; both inhibit germination and seedling growth.

Russian knap weed (*Centurea repens*), a perennial herb, has become a major threat in some parts of the United States. It becomes well established in western rangeland, leaving the land nonproductive or even counterproductive, owing to phytotoxins that are sesquiterpene lactones; some 14 of these belong to the guaianolide series. The structural formulas of an epoxy derivative (IV), a chlorohydrin (V) that would react to form the epoxy derivative (IV) upon elimination of hydrogen chloride, and a triolefin derivative (VI) are shown.

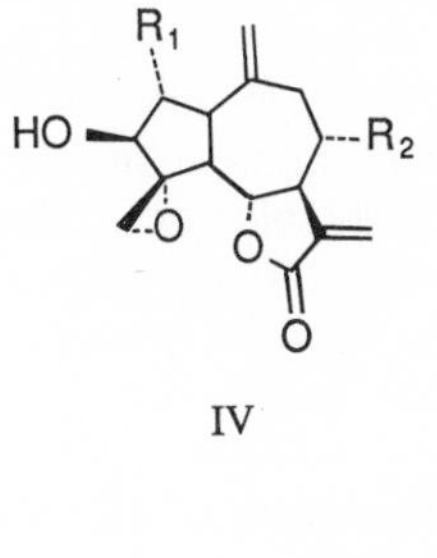

IV

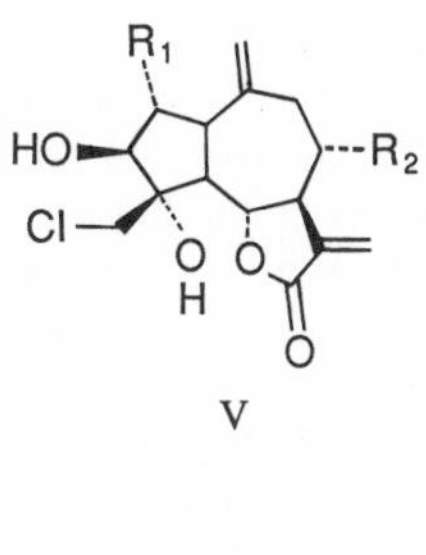

V

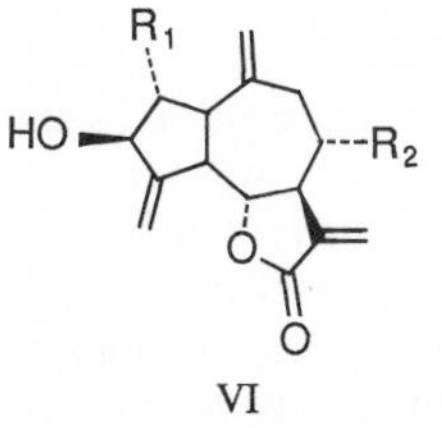

VI

Coumarins are lactones of *o*-hydroxycinnamic acids. Coumarins occur as glucosides in plants and are leached into the environment. Examples of such allelopathic materials are aesculetin, scopoletin, and aesculin (VII).

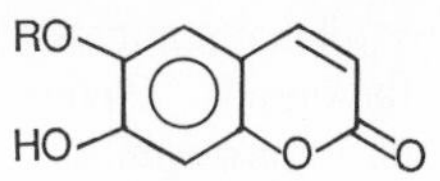

Aesuletin: R = H
Scopoletin: R = CH$_3$
Aesculin: R = β-D-glucose
VII

Terpenes. Volatile monoterpenes such as α-pinene, β-pinene, camphor, and 1/8-cinecole are allelopathic agents found in arid zones.

Phenols. Flavonoids comprise a large group of phenolic compounds, a number of which are allelopathic, including sayanediene (4′-hydroxy-7,3′-dimethoxyflavone).

Other phenol compounds include juglone (5-hydroxy-1,4-naphthoquinone), which occurs in the genus *Juglans* and is reportedly responsible for allelopathy by walnut trees.

Ferulic acid (4-hydroxy-3-methoxycinnamic acid), as well as other phenolic acids, is found in a variety of crop residues, and it is produced from intermediates of respiratory metabolism via the shikimic acid pathway. Ferulic acid inhibits wheat and rye root growth, and affects soybean growth adversely. Studies have indicated that prickly sida seed carpel enhances the inhibitory activity of ferulic acid, apparently because a bacterium present effects decarboxylation and produces 4-hydroxy-4-methoxystyrene.

Other compound types. These include alkaloids, tannins, terpenoids, and steroids. The category of miscellaneous and unknown compounds is expanding as more examples of allelopathy are recognized. Oxygenated fatty acids, for example, have been recognized as being responsible for allelochemical effects of aquatic macrophytes on algae; these materials are C$_{18}$-hydroxycyclopentenone and C$_{20}$-trihydroxycyclopentyl fatty acids, extracted from the submerged aquatic plant *Eleocharis microcarpa* (Torr).

Mechanisms. Allelochemicals have indirect and direct effects on plants. Indirect effects on vascular plants influence nitrogen fixation, nitrification, disease resistance, and susceptibility of plant roots to invasion by mycorrhizal fungi. Water-soluble inhibitors have direct effects on plant-water relationships. Phenolic compounds can cause membrane depolarization, which would have a number of consequences. Ferulic acid lowers water potential in grain sorghum by reducing turgor and osmotic pressure, and causes partial stomatal closing in young seedlings. Effects on water balance should impede other physiological processes. Other, more specific mechanisms are being investigated, including membrane disruption in the marine red tide organism *Ptychodiscus brevis* (formerly *Gymnodinium breve*) by substances released by the green alga *Nannochloris* sp. Here also, plant-water relationships seem to be involved.

Other direct effects involve inhibition of mineral absorption by allelochemicals, chiefly phenolic acids and flavonoids. Phenolic acids increase membrane permeability by being incorporated into the membrane and affecting the structure, by altering the anionic field (as anions of phenolic acids), or by some other property of the phenolic acid such as the chelating tendency.

For background information SEE ABSCISIC ACID; ALLELOPATHY; PLANT GROWTH; PLANT PHYSIOLOGY; PLANT-WATER RELATIONS in the McGraw-Hill Encyclopedia of Science and Technology

Dean F. Martin

Bibliography. M. B. Green and P. A. Hedin (eds.), *Natural Resistance of Plants to Pests*, ACS Symp. Ser. 296, 1986; E. L. Rice, *Allelopathy*, 2d ed., 1984; A. C. Thompson (ed.), *The Chemistry of Allelopathy*, ACS Symp. Ser. 268, 1984.

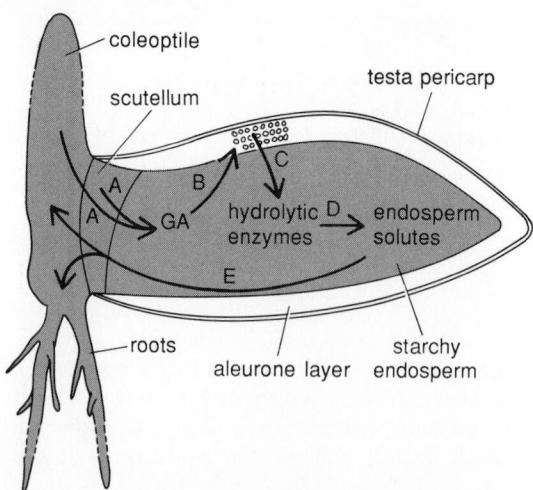

Fig. 1. Germinating barley seed. (Step A) Gibberellic acid, or GA, produced by the coleoptile and scutellum (B) migrates into the aleurone layer where (C) synthesis and release of hydrolytic enzyme is induced. These enzymes serve (D) to hydrolyze the reserves in the endosperm which function (E) to nourish the growing embryo. (After R. L. Jones and J. E. Armstrong, Evidence for osmotic regulation of hydrolytic enzyme production in germinating barley seeds, Plant Physiol., 48:137–142, 1971)

Amylase

The control of α-amylase synthesis and secretion in cereal grains has been intensively studied as a model system for investigating the mechanism of action of the plant-growth regulator gibberellic acid.

Molecular mechanisms. Recent work has focused on the molecular mechanisms of this control. The α-amylase gene was cloned to facilitate studies of its expression in isolated aleurone layers. It was found that gibberellic acid regulates α-amylase gene transcription. However, additional control at a subsequent step of α-amylase synthesis has not been ruled out. For example, gibberellic acid may also regulate the stability of α-amylase messenger ribonucleic acid (mRNA) or the rate of its translation. The plant-growth regulator abscisic acid, which counteracts the effect of gibberellic acid in aleurone cells, has recently been shown to prevent gibberellic acid–induced α-amylase gene transcription.

The details of secretion of α-amylase have also been investigated. Secretion of the protein seems to occur by a mechanism similar to that observed for many animal proteins. The ion Ca^{2+} is required to obtain high levels of α-amylase production by aleurone layers; it controls the synthesis and secretion of one of the two groups of α-amylase proteins. This control of protein synthesis is exerted at a step after mRNA synthesis and processing.

In barley and other cereal seeds, α-amylase and several other hydrolytic enzymes are secreted from a specialized layer of cells, the aleurone layer, into the starchy endosperm (**Fig. 1**). The secreted enzymes hydrolyze the food stores of the endosperm into sugars and amino acids. These sugars and amino acids can then be absorbed by the embryo via the scutellum and be used for the young plant's growth. It has been known since 1960 that gibberellic acid can induce the production of α-amylase in barley endo-

sperm; it is believed that gibberellic acid in the embryo is responsible for control of hydrolytic enzyme synthesis in the aleurone layer.

For experimental work, aleurone layers are manually stripped from the starchy endosperm, and since they comprise a uniform population of cells, biochemical studies are greatly simplified. Synthesis of α-amylase in isolated aleurone layers begins 4 to 8 h after application of gibberellic acid; enzyme secretion lags, beginning 8 to 12 h after the application. α-Amylase constitutes up to 60% of the newly synthesized protein in aleurone layers that have been incubated in gibberellic acid for 24 h. Translation of mRNA in cell-free extracts showed that α-amylase mRNA increased after gibberellic acid treatment, accounting for about 20% of the translatable mRNA after 24 h. While it is possible with cell-free translation techniques to estimate the amount of functional, processed mRNA present in a cell, the importance of mRNA processing or stability cannot be analyzed. Therefore, the α-amylase mRNA was cloned and used as a probe to test for levels of α-amylase mRNA sequences present in the cells.

Control of α-amylase mRNA synthesis. The abundance of α-amylase mRNA in gibberellic acid–treated aleurone layers made cloning the gene relatively easy. To clone α-amylase mRNA, a complementary deoxyribonucleic acid (cDNA) library was made from mRNA derived from gibberellic acid–induced aleurone layers. Clones of cDNA which hybridized strongly to mRNA from gibberellic acid–induced aleurone and weakly or not at all to mRNA from noninduced aleurone were selected. Of these, clones hybridizing to mRNA which produced α-amylase when translated in cell-free extracts were identified as α-amylase clones.

Cloning of the α-amylase gene allowed analysis of α-amylase mRNA accumulation. This was accomplished by studies of hybridization of the radiolabeled clone to mRNA derived from aleurone layers that had been subjected to various treatments. By this technique it has been shown that only low levels of α-amylase mRNA sequences are present in aleurone layers before gibberellic acid application; increased levels of α-amylase mRNA can be detected as early as 1 h after addition of gibberellic acid to aleurone layers. The α-amylase mRNA increases up to 50-fold after 16 h of incubation in gibberellic acid.

Experiments in which nuclei were isolated from aleurone cells, and the mRNAs in the process of being synthesized were identified, have shown that gibberellic acid controls amylase mRNA accumulation in aleurone cells, at least in part, by inducing the transcription of the amylase gene. The role of mRNA turnover in the regulation of α-amylase mRNA accumulation has not been addressed.

Abscisic acid, a second class of plant growth regulators, inhibits the production of α-amylase from aleurone layers, overriding the effect of gibberellic acid. Hybridization of α-amylase cDNA clones to aleurone mRNA revealed that abscisic acid prevents the accumulation of α-amylase mRNA; analysis of transcripts produced by isolated nuclei showed that abscisic acid inhibits gibberellic acid–induced α-amylase gene transcription.

The α-amylase produced by barley aleurone cells is actually a mixture of several polypeptides which can be separated by nondenaturing gel electrophoresis. The barley cultivar Himalaya used in most isolated aleurone work contains four α-amylase components as detected by agar gel electrophoresis. These proteins fall into two groups based on isoelectric point (pI) and other biochemical properties. The concentration of gibberellic acid required to induce the two groups of α-amylases is also different. Production of α-amylase with low pI is stimulated by 10^{-8} M gibberellic acid, while the high-pI group is not induced until gibberellic acid is at a concentration of at least 10^{-7} M. By using chromosome addition lines of wheat it has been shown that the

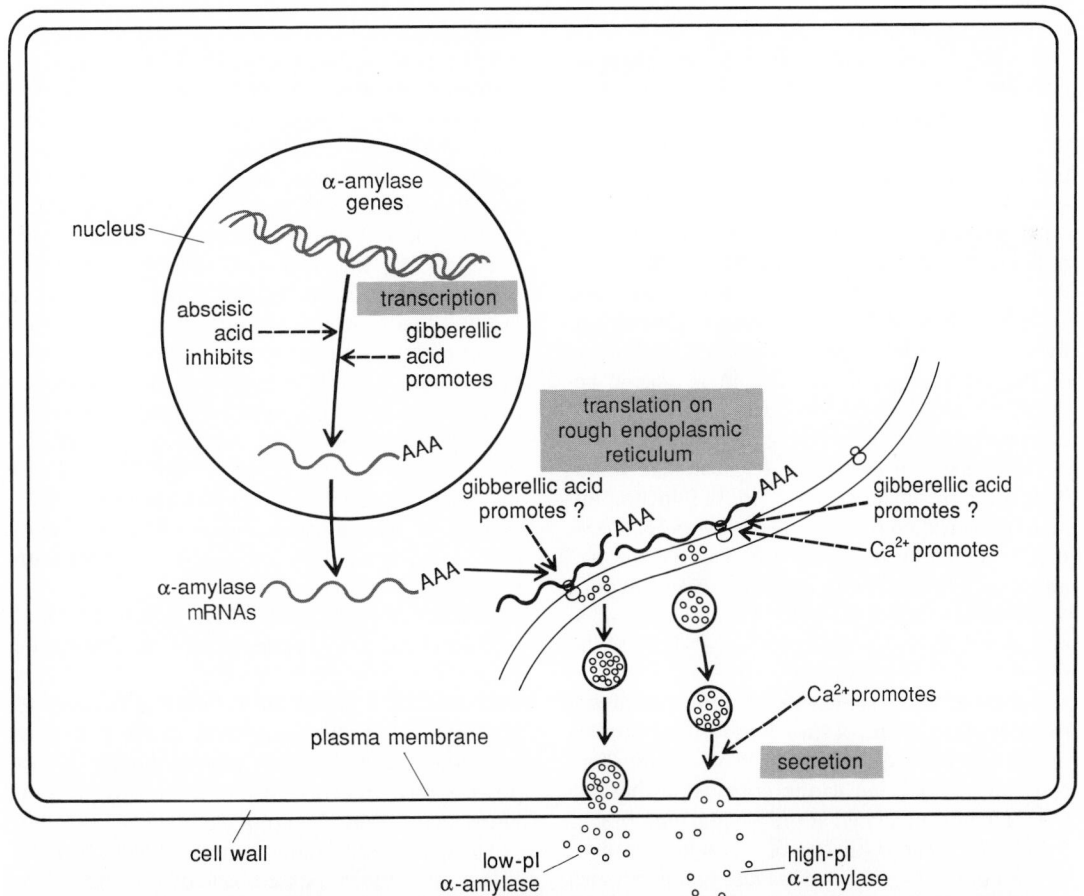

Fig. 2. Diagram of control of α-amylase synthesis and secretion in an aleurone cell. The transcription of α-amylase genes is induced by gibberellic acid and inhibited by abscisic acid. Translation of α-amylase mRNA occurs on the rough endoplasmic reticulum. This process may be promoted by gibberellic acid. Translation of high-pI α-amylase mRNA requires Ca^{2+}. Both isoenzymes are secreted into the endosperm. Secretion of high-pI α-amylases requires Ca^{2+}. The question marks signify that promotion of these steps by gibberellic acid has not been established. AAA = poly-A tail of an mRNA molecule.

high-pI α-amylases are coded for on chromosome number 6, and the low-pI α-amylases are coded for on chromosome number 1 of barley. Complementary DNA cloning has resulted in two types of α-amylase clones corresponding to each group of amylase proteins. Hybridization of radiolabeled α-amylase clones to barley nuclear DNA digested with restriction enzymes, separated by electrophoresis and blotted onto nitrocellulose paper, suggests the presence of several α-amylase genes or pseudogenes, perhaps as many as eight of them.

The temporal patterns of high- and low-pI α-amylase synthesis and secretion are different. In isolated aleurone layers, the low-pI group has a higher basal level and it responds more rapidly to gibberellic acid than does the other group. However, after 24 h of gibberellic acid treatment the relative levels of the two groups shift. This behavior of the α-amylase isoenzymes is paralleled by the accumulation of their mRNAs, which was determined by hybridization of specific cDNA clones to aleurone mRNA.

Presecretory processing. Barley aleurone α-amylase is synthesized on and sequestered inside the endoplasmic reticulum. It is assumed that the enzyme is secreted by means of membrane-bound vesicles. However, such vesicles are not observed in abundance in barley aleurone cells with electron microscopy. Barley and wheat α-amylase are synthesized in cell-free extracts as precursors that are 2000 and 1500 daltons larger than the mature enzymes. The barley precursor has been processed to the mature form in cell-free extracts which contain enzymes responsible for modifying secretory enzymes. This behavior of the translatable α-amylase mRNA conforms to the current model for protein secretion in animal cells. In this model, secretory proteins are synthesized with a leader sequence at the N-terminus of the polypeptide which acts in a "signal sequence," directing the ribosome-nascent protein complex to the endoplasmic reticulum.

Ca^{2+} control of α-amylase synthesis and secretion. The presence of Ca^{2+} in the incubation medium is required by aleurone layers for maximum production of α-amylase. Levels of Ca^{2+} affect only the secretion and synthesis of high-pI α-amylases. Other α-amylases are synthesized and secreted at the same rate regardless of the external Ca^{2+} concentration. The Ca^{2+} effect on secretion is very rapid, with secretion declining to 25% within 15 min of Ca^{2+} withdrawal. Studies of the secretion of α-amylase from single aleurone layers have led to the conclusion that the Ca^{2+} effect is mediated at the plasma membrane.

Summary. The regulation of α-amylase production is complex, involving at least two hormones and Ca^{2+}. Controls exist at several steps of the protein biosynthetic pathway. **Figure 2** presents an overview of the biosynthetic process. Gibberellic acid induces transcription of the α-amylase genes and may also promote α-amylase mRNA translation. Abscisic acid blocks transcription of α-amylase genes. The ion Ca^{2+} is important in controlling secretion of one group of α-amylases, and synthesis of this group of α-amylases is tightly coupled to secretion.

For background information SEE ABSCISIC ACID; AMYLASE; GIBBERELLIN; PLANT METABOLISM in the McGraw-Hill Encyclopedia of Science and Technology.

Jill Deikman

Bibliography. J. Deikman and R. L. Jones, Control of α-amylase mRNA accumulation by gibberellic acid and calcium in barley aleurone layers, *Plant Physiol.*, 78:192–198, 1985; J. V. Jacobsen and L. R. Beach, Control of transcription of α-amylase and rRNA genes in barley aleurone protoplasts by gibberellin and abscisic acid, *Nature*, 316:275–277, 1985; R. L. Jones, Protein synthesis and secretion by the barley aleurone: A perspective, *Israel J. Bot.* 34:377–395, 1985; J. C. Rogers, Two barley α-amylase gene families are regulated differently in aleurone cells, *J. Biol. Chem.*, 260:3731–3738, 1985.

Animal virus

Many different kinds of viruses that invade animal cells have been described. Each kind is distinguished from others in one or more ways, such as by the type of nucleic acid that it contains, the strategy by which the nucleic acid is replicated, the size or shape of the virus, or its ability to react with specific antibodies. Each kind of virus possesses a distinct structure defined by the interactions between its specific proteins and nucleic acids. Some viruses also possess lipids, which are not viral products but are compounds taken from the host cell at the time of exit or budding. However, the problem of initiating infection is similar for all types of viruses: the viral nucleic acid, which constitutes the virus's genome, must traverse both the viral and the cellular membrane barriers to enter the cell's cytoplasm. The subsequent fate of the viral genome is determined by the particular replication strategy of the invading virus.

Structure of enveloped viruses. Every virus that contains lipid has it arranged in a membranelike bilayer surrounding the genome. A viral bilayer must fuse with a cellular membrane in order to accomplish the infectious entry of the viral genome into the cell. Some of the details of this process have recently been learned for several viruses containing lipid envelopes.

The enveloped animal viruses that have been extensively studied possess several common properties: (1) a small size, 70–200 nanometers in diameter; (2) a ribonucleic acid (RNA) genome of limited coding capacity, specifying only 4 to 12 different viral proteins; (3) a membrane consisting of a con-

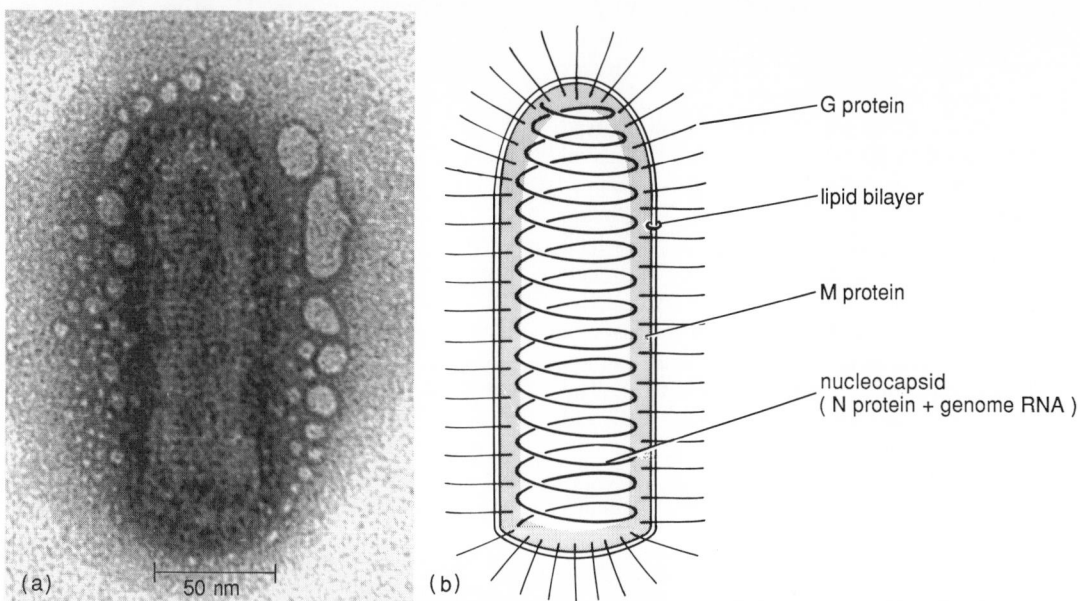

Fig. 1. Vesicular stomatitis virus. (*a*) Photograph of virus negatively stained with 1% phosphotungstic acid. (*b*) Diagram of virus, indicating RNA, lipid bilayer, and three major structural viral proteins. (*From J. Lenard and D. K. Miller, Entry of enveloped viruses into cells, in P. Cuatrecasas and T. Roth, eds., Receptor-Mediated Endocytosis, Receptors and Recognition, Series B, vol. 15, pp. 119–138, Chapman and Hall, 1983*)

tinuous lipid bilayer surrounding the genome; and (4) one to three different transmembrane glycoproteins, all coded by the viral genome.

The structure of one such virus, the vesicular stomatitis virus, is shown in **Fig. 1**. Three major structural proteins are indicated: the G protein, which is the virus's sole glycoprotein, embedded in the lipid bilayer; the N protein, which combines with the RNA genome to form the nucleocapsid; and the M protein, which is associated with the bilayer and the nucleocapsid, and perhaps with the G protein.

Binding of virus to cell surface. The first function of the viral glycoproteins in initiating infection is to bind the virus to the outer surface, or plasma membrane, of the target cell. The specificity of this binding is different for different viruses. Some viruses can only bind to very specific proteins on the cell surface. Such specificity limits the kinds of cells that the virus can infect. Myxoviruses, such as influenza, and paramyxoviruses, such as Sendai, bind to sialic acids, 9-carbon sugars attached to either proteins or lipids of the host-cell plasma membrane. For other viruses the site of attachment is less specific, and may include a variety of different membrane proteins or lipids.

Fusion of viral and cell membranes. After attachment of the virus to the target cell, the viral glycoproteins must promote fusion between the lipid bilayers of the viral envelope and one or another of the cell's membranes. Some enveloped viruses fuse directly with the cell's plasma membrane (**Fig. 2***a*). The paramyxoviruses, notably Sendai virus, enter the cell in this way. Other viruses, however, take a more complicated

route. They bind at various sites on the cell's plasma membrane, but are unable to fuse with it. Instead, the sites to which the viruses bind mediate the latter's internalization into endosomes inside the cell. This is accomplished through coated pits by the process of endocytosis (Fig. 2*b*). Internalization of the virus in this manner is not sufficient to initiate infection, however, since the viral genome is still separated from the cell cytoplasm by both the viral and the endosomal membranes. Fusion between these

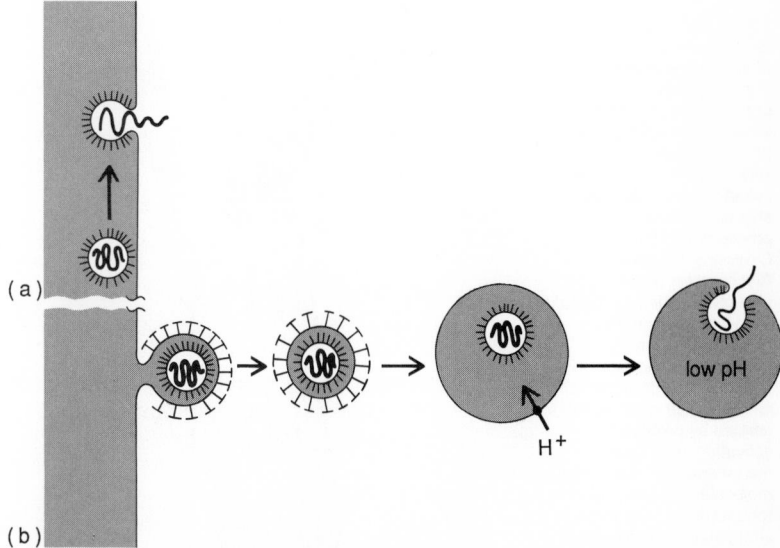

Fig. 2. Pathways of enveloped virus entry. (*a*) Direct fusion of viral membrane with cell's plasma membrane, followed by release of nucleocapsid into cytoplasm. (*b*) Endocytosis of viruses through coated pits into endosomes inside the cell, followed by fusion of viral and endosomal membranes. (*After J. White, M. Kielian, and A. Helenius, Membrane fusion proteins of enveloped animal viruses, Quart. Rev. Biophys., 16:151–195, 1983*)

two membranes completes the process. Vesicular stomatitis virus, influenza virus, and Semliki Forest virus are examples of well-studied viruses that infect cells by this pathway.

Which of the two entry pathways will be used is determined by specific properties of the fusion glycoproteins of the different viruses. Viruses that fuse with plasma membranes possess fusion glycoproteins that are active at neutral pH, which is the pH of tissue-culture medium or of extracellular fluid. Viruses that enter the cell by endocytosis and subsequently fuse with endosomal membranes possess fusion glycoproteins that are active only at low pH, generally below 5.5. The pH inside endosomes has been measured and found to be around 5.0–5.2, low enough to activate the viral fusion glycoprotein.

Endosomes are not the only acidic organelles within mammalian cells. Lysosomes also possess highly acidic interiors but differ from endosomes in being filled with a wide variety of potent digestive enzymes. Both lysosomes and endosomes are acidified by proton (hydrogen ion) pumps, which transfer hydrogen ions from the cytoplasm into the organelle, at the expense of metabolic energy. These acidic structures are parts of an elaborate and important system that links the cell interior to the world outside. Extracellular material such as hormones and nutrients are transferred through this system from the cell surface, sequentially through coated pits, endosomes, and lysosomes. Some of the enveloped viruses have acquired the ability to use this essential cellular pathway to gain entrance into the cell. Virus particles that enter this pathway but do not fuse with the endosomal membrane are shortly transferred to lysosomes, where they are rapidly degraded.

Hemagglutinin structure and function. Membrane fusion induced by a glycoprotein of the influenza virus has been most intensively studied. The influenza fusion glycoprotein, hemagglutinin, exists on the surface of the virus as a trimer of subunits having a molecular weight of about 75 kilodaltons each. Each subunit crosses the viral bilayer once, and is arranged with its amino terminus on the outside and its carboxyl terminus on the inside. Over 90% of hemagglutinin is external to the viral bilayer, and this region mediates the fusion process. The external domain of hemagglutinin has been crystallized after separating it from the transmembrane and cytoplasmic domains by proteolysis. Structural analysis by x-ray diffraction has revealed the molecular structure of the protein at a resolution of 0.3 nm, which is sufficient to trace the path of the polypeptide backbone through the overall structure (**Fig. 3**).

Each 75-kilodalton subunit of hemagglutinin must be activated by proteolytic cleavage at a specific site before it is competent to catalyze fusion. Two proteins are thus created from hemagglutinin, HA_1 and HA_2. At the new amino terminus of HA_2, there exists a hydrophobic peptide of about 20 amino acid residues which possesses some homology to sequences found in other viral fusion glycoproteins. This peptide is thought to play an important role in mediating the virus fusion reaction, although its precise function is not yet clear.

When the environment of the influenza hemagglutinin molecule is changed from pH 7, where it is inactive, to pH 5, where it can mediate fusion, its conformation is greatly altered. The trimeric form in which it exists at pH 7 undergoes a loosening and a rearrangement on the viral surface. The hydrophobic amino terminus of HA_2 becomes more highly exposed and may interact directly with the target (nonvirus) membrane. Sensitivity to proteolytic digestion changes, as does antigenicity when tested with particular monoclonal antibodies. The structure of the low-

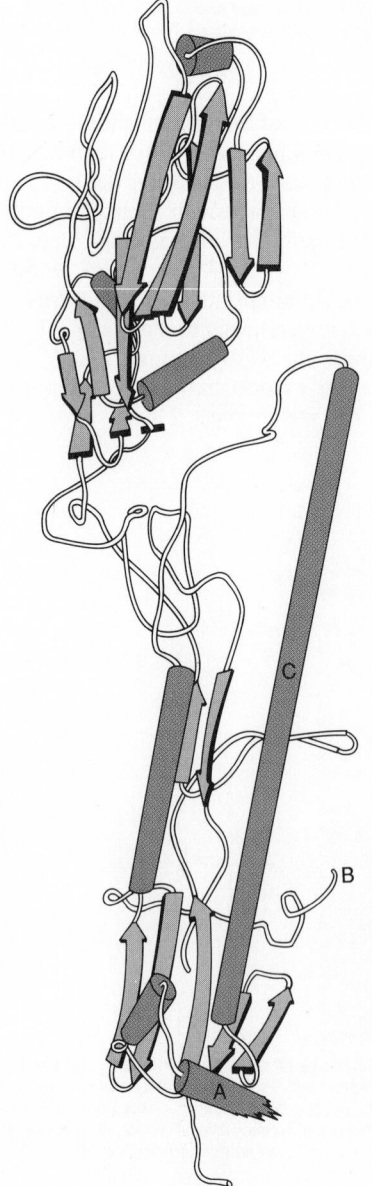

Fig. 3. Schematic three-dimensional structure of fusing or external portion of hemagglutinin glycoprotein of influenza virus, as deduced by x-ray crystallography. α-Helical segments are shown as cylinders, and β-pleated sheets as arrows, proceeding in N- to C-terminal direction. Glycoprotein penetrates viral membrane near broken cylinder at bottom (A). Fusion with a target membrane is thought to be mediated by interaction of an N-terminal segment (B), created by proteolytic activation, with target membrane. Three of these molecules are associated to form a trimer on surface of viral particles. Trimerization occurs by association between membrane-spanning portions of molecule (not shown), and between stem portions, centered on long helical segment (C). (*After D. Wiley*)

pH, fusion-active form of hemagglutinin has not been determined by x-ray crystallography, since the protein has not been crystallized in this form. Still less is known about the mechanism of action of other viral fusion glycoproteins.

Viruses without lipid. Viruses without lipid in their structures face problems similar to those of enveloped viruses in initiating infection. Their genomes are usually surrounded by a protective shell of viral protein. To initiate infection, the viral nucleic acid must pass through this shell and must also, as with the enveloped viruses, cross a cell membrane into the cell's cytoplasm. Specific viral proteins may be modified, by proteolysis or altered pH, so as to interact with cell membranes and thereby aid the passage of the viral genome. Membrane-active proteins have been identified in reovirus and poliovirus as well as other viruses. The precise mechanism by which these proteins work, however, is not yet known.

For background information SEE ANIMAL VIRUS; CELL MEMBRANES; ENDOCYTOSIS; VIRUS in the McGraw-Hill Encyclopedia of Science and Technology.

John Lenard

Bibliography. B. N. Fields et al. (eds.), *Virology*, 1985; J. White, M. Kielian, and A. Helenius, Membrane fusion proteins of enveloped animal viruses, *Quart. Rev. Biophys.*, 16:151–195, 1983.

Anorthosite

Anorthosites are intrusive igneous rocks containing greater than 90% plagioclase, which is a solid solution between albite ($NaAlSi_3O_8$) and anorthite ($CaAl_2Si_2O_8$). They grade into gabbroic anorthosites containing 78–90% plagioclase. The most widely distributed anorthosites and gabbroic anorthosites are encountered in layered basic intrusions resulting from the fractional crystallization of parental gabbros. These anorthosites, which have occurred in all geologic periods, consist of layers of calcic plagioclase (60–85% anorthite) together with minor pyroxene. The rocks themselves exhibit variations in whole-rock and mineral chemistry that are well explained in terms of fractionation of gabbroic liquids. On a much larger scale, these same processes appear to have given rise to the very calcic anorthositic crust of the Moon, in which plagioclase contains about 95% anorthite.

Massif. In contrast to the calcic anorthosites associated with layered basic intrusions are the distinctive anorthosite massifs that are mainly of Proterozoic age. In general, these large bodies consist of intermediate plagioclase containing 45–60% anorthite, show very little chemical variation, and are not obviously related to the typical mafic differentiates of layered basic igneous intrusions. Instead, they appear to be intimately associated with monzonitic, syenitic, and granitic rocks. The entire suite tends to be relatively rich

in alkalies and iron, reflecting the evolved nature of the magmas.

The origin of these massif anorthosites and their associated rocks has remained a source of debate for decades. Petrologically they are puzzling because their approximately monomineralic nature would require extremely high temperatures if they were to have existed as anorthositic liquids. Also, their relatively sodic composition has been difficult to explain in magmas containing little, if any, water. Geophysically they are enigmatic because of their almost exclusive restriction to several mid-Proterozoic belts, especially the belt traversing North America, Greenland, and the Baltic Shield. Within these belts, ages of anorthosite emplacement range from approximately 1700 to 950 million years ago (m.y.a.) with close to 70% of the ages falling in the 1400–1500-m.y.a. range. Although Archean and Phanerozoic massif-type anorthosites are known, they tend to be more calcic than the Proterozoic massifs and are far less voluminous. Clearly the mid-Proterozoic was a time in which conditions were optimal for anorthosite genesis.

Laramie model. During 1985 and 1986, several new studies offered additional insights into the evolution of Proterozoic anorthosite massifs. One of these efforts focused on the approximately 1400-m.y.a. Laramie Anorthosite Complex, which consists of andesine anorthosite as well as orthopyroxene-bearing monzonites and syenites. Detailed chemical studies of the solid-solution minerals in these rocks demonstrate that continuous plagioclase, olivine, and low-calcium pyroxene solid solutions exist from the anorthosites through the monzonitic and syenitic rocks. The feldspars pass from plagioclase through ternary feldspars to alkali feldspar, and the olivines vary from 53 to 98% of the iron end-member fayalite.

The chemical continuity of the mineral species was considered to reflect a comagmatic history for rocks in the complex. It was postulated that a mantle-derived high-potassium gabbro, or even trachyandesite, is the parental magma. Such melts would have densities greater than the density of the continental crust and would tend to form a pool (pond) at the crust–mantle interface, where crystallizing olivine and pyroxene would sink and plagioclase would float. Eventually plagioclase diapirs would form and rise into the heated, weakened crust to form anorthosite massifs. Continued, deep fractionation of the residual potassium- and iron-rich magma would yield monzonitic and syenitic cumulates, which would form additional diapirs that would rise up and intrude the anorthosites. Because the monzonites and syenites are cumulates, they would not necessarily show europium-depleted rare-earth patterns expected of liquids from which plagioclase has been extracted. It is believed that the anorthositic, monzonitic, and syenitic rocks were emplaced at relatively shallow depths (approximately 3–10 km or 1.8–6 mi) and evolved under anhydrous, low-oxygen fugacity conditions.

Bimodal model. In 1986 it was suggested that the anorthosite massifs of the Adirondacks represent the mafic roots of bimodal caldera complexes of mid-Proterozoic age. These massifs are partially or wholly surrounded by orthopyroxene-bearing monzonites, syenites, and granitic rocks. Also associated with the suite are magnetite-bearing alaskites, some of which are believed to represent metamorphosed rhyolitic ash-flow tuffs. Associated with these are conformable iron oxide ores often interlayered with calc-silicates. The entire complex has been highly deformed and metamorphosed so as to obscure the original configurations, and radioactive clocks have been reset. However, the originally anhydrous nature of the suite has resulted in preservation of original whole-rock compositions, thus facilitating chemical interpretations. Although rubidium/strontium, uranium/lead, and neodymium/samarium isotopic systems appear to have been disturbed by the 1100–1000-m.y.a. granulite facies metamorphism that affected the area, a neodymium/samarium isochron age of 1288 ± 36 m.y.a. has been obtained for rocks from the southern Marcy anorthosite massif. Younger ages have been found as well, but it is argued that the 1288-m.y.a. isochron, which is based on the least metamorphosed samples, gives the best minimum estimate for the emplacement of Adirondack anorthosite. This age would place the anorthosites relatively close to the 1450-m.y.a. average for the North American belt. However, recent uranium/lead zircon ages strongly suggest an emplacement of about 1130 m.y.a. for the anorthosite-charnockite suite of the Adirondacks. This is similar to ages in southern Greenland and the Baltic Shield.

The model that was developed emphasizes an earlier observation that Proterozoic anorthosite massifs develop in anorogenic tectonic regimes such as those associated with abortive or incipient rifting. It is only within such settings that parental magmas of broadly gabbroic composition can undergo protracted ponding and high-pressure fractionation at the crust–mantle interface, thus evolving toward anorthositic differentiates. If rifting is rapid or pronounced, the magmas will rise to the surface as typical gabbros. If the tectonic environment is compressional, any mafic magmas present may become mixed and hydrated. Only a regime of incipient rifting is consistent with slowly evolving mantle diapirs giving rise to mafic magma ponded at the crust–mantle interface for long periods of time.

It has been pointed out that the ponded mafic magmas will precipitate olivine and pyroxenes, which will sink. The removal of aluminous high-calcium pyroxene will deplete the residual liquid in its potential anorthite component. When plagioclase begins to precipitate, it will be of an intermediate composition and, supposedly, will float or remain in suspension.

Ultimately the heat of intrusion and crystallization from the ponded magmas causes a melting of lower crustal rocks at, or near, the crust–mantle interface. The net effect is an overall weakening of the crust, which eventually is pierced by low-density diapirs of crustal melts cored by magmas of massif anorthosite. The magmas are believed to be highly charged with crystals of andesine plagioclase which continue to grow from the interstitial melt. The envelope of crustally derived melts is of syenitic to granitic composition and anhydrous, although a halogen-rich fluid phase may be present.

As the diapirs rise, plagioclase continues to crystallize, and the noritic magma drives toward silica-poor, iron- and titanium-rich compositions, now represented as ferrodiorite sheets and dikes or immiscible iron-titanium oxide bodies. Meanwhile, the acidic envelope drives toward highly evolved iron-enriched, high-silica rhyolites whose liquidus temperatures and viscosities are lowered by halogens. It has been proposed that because of the anhydrous nature of the melts, the complexes are able to penetrate to shallow crustal levels. At their site of emplacement, they evolve into caldera complexes underlain by zoned magma chambers. Surface volcanic activity gives rise to voluminous rhyolitic and dacitic ash-flow tuffs now represented by many of the alaskitic rocks of the Adirondacks.

Comparison of models. The principal difference between the bimodal model and the Laramie model lies in the relationship between the anorthositic-noritic rocks on the one hand and the monzonitic-syenitic-granitic rocks on the other. The bimodal model emphasizes the striking absence of dioritic rocks from these suites, and has drawn attention to the similarity between them and the bimodal associations of Cenozoic caldera complexes of the southwestern United States. According to this view, the continuity in solid-solution compositions within the Laramie Complex may be fortuitous, and the absence of europium anomalies in the monzonitic and syenitic rocks is not due to their cumulate nature but results from the fact that they did not evolve from the same source as the anorthosites.

The principal similarities between the models lie in the importance attributed to ponding at the crust–mantle interface within anorogenic regimes as well as the high level of final emplacement and its possible culmination in caldera-style volcanism. It is notable that anorthositic rocks of Paleozoic age have been recognized in west-central Africa at the northern end of the Younger Granite belt of Niger and Nigeria. These occur within calderas and ring complexes associated with a broad zone of abortive rifting.

It is possible that mid-Proterozoic time was

characterized by widespread crustal stability in which subcontinental diapirs underwent ponding and differentiation as discussed above. One locus of this activity is now represented by the 1000–1700-m.y.a. (peak approximately 1450 m.y.a.) belt of anorogenic extrusive and intrusive magmatism that traverses North America and the Baltic Shield. Although exposed in only a few instances, anorthosites are demonstrably related to many of these complexes, and anorthositic xenoliths are common. It is proposed that the anorthosites of the Adirondacks, as well as those in the rest of the Grenville Province, are part of this belt but have been overprinted by the Grenville Orogeny. As such, these anorthosites and their associated rocks link the 1400–1500-m.y.a. high-level anorogenic complexes of Missouri and Wisconsin with those of Labrador, Greenland, and the Baltic Shield. Intrusion of the anorthosites may have dehydrated the surrounding crust, thus providing an already anhydrous setting for the 1000–1100-m.y.a. granulite facies event of the Grenville orogeny.

For background information SEE *ANORTHOSITE;* *MAGMA; MASSIF; PETROGRAPHIC PROVINCE; PROTEROZOIC* in the McGraw-Hill Encyclopedia of Science and Technology.

James M. McLelland

Bibliography. R. Emslie, Proterozoic anorthosite massifs, *NATO ASI Series*, 158:39–60, 1985; D. Lindsley et al., *Geol. Soc. Amer. Abstr. Programs*, 17:664, 1986; J. McLelland, Pre-Grenvillian history of the Adirondacks as an anorogenic, bimodal caldera complex of mid-Proterozoic age, *Geology*, 14:229–233, 1986; J. Valley, Polymetamorphism in the Adirondacks, *NATO ASI Series*, 158:217–236, 1985.

Anthropology

Human skeletal remains from archeological sites provide a unique opportunity to study human biological and cultural adaptation from a long-term perspective. Commonly included in this research are topics such as health status, genetic relationship, diet, and disease evolution. While the importance of these topics has long been recognized in anthropological investigations of the past, recent reintegration of human biological concepts with ecological archeology and the application of chemical, medical, and genetic methodologies have added new approaches and new perspectives to the study of humankind's recent prehistory. The midcontinental United States is an area remarkably rich in the archeological resources necessary to conduct such integrated bioarcheological studies.

Changing subsistence patterns. Within recent years, bioarcheological scholarship in the eastern United States has focused upon evaluating the impact of the shift from a broad-based hunting

Chronology for West-Central Illinois

Cultural period	Approximate age, radiocarbon years before present
Mississippian	900
Late Woodland	1400
Middle Woodland	2000
Early Woodland	2550
Late Archaic	3000
Middle Archaic	8000
Early Archaic	9000

and gathering adaptation to maize (corn) agriculture. The former is characteristic of Archaic populations dating from approximately 9000 years before present (BP) to 3000 years BP (see **table**). The development of maize agriculture culminated in the Mississippian Period about 900 years BP. The Mississippian is associated with the development of ancient cities such as Cahokia, near present-day St. Louis, Missouri. During the intermediate Woodland times (3000–900 years BP), local groups intensified their encouragement and then cultivation of local weedy seed crops, including such plants as maygrass (*Phalaris*), goosefoot (*Chenopodium*), and marshelder (*Iva*) as well as adopted maize. Since local populations, even Archaic hunter-gatherers, became sedentary rather early within this sequence, these groups and their environment make an ideal natural laboratory in which to investigate the impact of changing subsistence patterns upon the human condition. In addition, these studies of hunter-gatherers in temperate climates provide an instructive contrast to those of contemporary groups which tended to occupy marginal desertic and arctic environments.

Chemical techniques. While the broad outlines of dietary and demographic changes can be read from patterns of village site distribution and from the refuse left by paleopopulations, it is the skeletal remains themselves that afford the most precise estimates of dietary differences. Two recently developed chemical techniques for estimating diet from ancient tissues, carbon isotope and trace-element analyses, allow relatively precise measures of dietary differences within and between earlier human groups.

Carbon isotope analysis. This technique has virtually revolutionized the perception of the importance of maize as an economic resource within the ancient Midwest. Archeological debate has long centered upon the degree to which corn cultivation was a necessary precursor to the development of the relatively large sedentary villages that characterize the Middle Woodland Period of prehistory (about 2000–1400 years BP). This was the time of deposition of attractive artifacts (Hopewell items), finely crafted artifacts commonly discovered by archeologists as grave goods in Middle Woodland burial mounds in the eastern United States. They were deposited with the dead in the elaborate log crypts which are found in the burial mounds characteristic of the period. Further, it has been suggested that maize cultivation had figured prominently in the early development of Mississippian, as a necessary precursor to the appearance of complex societies and cities such as Cahokia, where as many as 35,000 people may have lived during the 11th millennium of the Christian Era.

Many tropical plants, such as maize, fix carbon through a photosynthetic pathway that is distinct from that of temperate climate plants. As a result, the ratio of stable carbon isotopes $^{13}C/^{12}C$ in the tissues of maize differs from that of most other native North American plants. This patterned difference is passed along the food chain and thus makes it possible to distinguish between prehistoric midcontinental peoples who consumed significant amounts of maize and those who did not.

Recent studies have demonstrated that maize was not a significant food for Middle Woodland peoples. Instead, maize gradually assumed importance during the Late Woodland and Emergent Mississippian periods. In the Cahokia area, by 1000 years BP, as much as 25–50% of dietary carbon may have been derived from maize. The most extreme use of maize, as determined from carbon isotope analysis, occurred among Late Mississippian peoples from the Ohio and Cumberland River regions of Ohio, Kentucky, and Tennessee. Estimates of dietary carbon range from 65 to 80%, clearly defining a subsistence strategy heavily dependent upon maize cultivation during the period from 900 to 500 years BP. Maize obviously became an important component in the diet of prehistoric North American peoples, though the period of greatest maize-dependence appears to postdate the development of major centers, such as Cahokia.

Trace-element analysis. This second chemical technique focuses upon the analysis of trace elements within the mineral fraction of skeletal and dental remains. For example, stable strontium is known to concentrate in plants and then in the bony tissues of animals that consume plants. Animals that do not consume plants but rather flesh of other animals receive relatively little strontium. Thus, within a given ecosystem, the bones of herbivores and carnivores can be distinguished; within an omnivorous species such as humans, relative herbivory and carnivory can be estimated.

Of the numerous potentially informative trace elements, stable strontium has received the greatest attention, due both to its clear signal of plant consumption and to its stability within the bone matrix after remains have entered the archeological record. Zinc, a complementary signature for meat, is also frequently analyzed. Studies of these elements have documented dietary differences between status groups during the Middle Woodland Period, as well as a gender-based difference in access to meat during the Late Woodland Period. In at least some midwestern groups, males apparently had privileged access to meat during the period of transition to maize agriculture.

Population dynamics. Studies of health status and demographic change have developed in parallel to documentation of dietary change in the Midwest. Sedentary villages of groups characterized as hunter-gatherers appear relatively early in the archeological sequence. Habitation and mortuary site distribution studies suggest that population density grew rather slowly throughout the Archaic and Early and Middle Woodland periods, with a notable increase occurring later, in tandem with the intensification of seed cultivation and maize agriculture. Recent statistical analyses of demographic data indicate that an important factor in this population increase during the Woodland Period was increased fertility due to decreased birth intervals. The use of a weaning food derived from the cultivated local starchy seeds has been implicated in this process.

Fundamental to inferences about population dynamics and long-term human adaptation is knowledge of the genetic background of the paleopopulations under investigation. Archeologists have tended to equate major prehistoric adaptive or cultural changes with population movement or migrations, on the assumption that population replacement is a major force in culture change.

Due to the richness of the archeological record for prehistoric mortuary sites in the Midwest, there exist diachronic sequences of skeletal samples of large size which can be used to investigate the degree to which population replacement should be viewed as a major explanatory variable in regional prehistories. Size and shape of skeletal remains, as well as other inherited characteristics of both bones and teeth, have been studied and compared across Woodland and Mississippian populations. These studies have failed to identify convincing evidence of significant population movement that correlates with the periods of transition between Middle and Late Woodland or Late Woodland and Mississippian.

Local groups, at least in the central Midwest, adopted new adaptive strategies and styles of material culture without significant population replacement. This does not mean, however, that genetic change is not observable in the archeological record. For instance, one cemetery in the lower Illinois River region, about 60 mi (100 km) north of Cahokia, spans the Late Woodland to Mississippian periods and thus constitutes an ideal circumstance for investigating genetic change in temporally sequential paleopopulations. Although there is no significant difference between the Late Woodland mound sample and the earliest Mississippian cemetery, the most recent part of the Mississippian cemetery provided evidence for a different genetic configuration, obvious only in the males. This gender-based difference is interpreted to mean that the process of "Mississippianization"—indeed characterized by distinctive shifts in adaptive strategies and material culture—led to changes in residence and likely marriage patterns. Genetic change is thus inferred to have resulted from culture change, rather than being a cause.

The rich archeological sequence that documents the shift from hunting and gathering to maize agriculture also permits unique insight into the biological costs and benefits of changing subsistence patterns. These sequelae can be measured by a variety of skeletal attributes, both demographic and health-related.

Recent years have seen increased emphasis upon the use of so-called nonspecific indicators of populations stress in the evaluation of health status for ancient human groups. Rather than being attributable to specific diseases or nutritional inadequacies, these features represent episodes of ill health of indeterminate cause. Thus, quality of life in prehistoric groups is measured by counting the number of periods of episodic nutritional and disease upset, by comparing the rates of growth, and by investigating age-specific mortality. None of these by itself identifies the disease agent or dietary inadequacy, although sometimes these can be inferred through other avenues of archeological investigation.

Lines of growth arrest. Examples of one such nonspecific indicator of ill health are lines of growth arrest (Harris lines) that are visible on radiographs of long bones. These lines are in fact the transverse projections of radiopaque calcified plates of bone that are the result of the initiation of long-bone growth after a period when growth was halted. In that these lines can resorb, and their correspondence to a stress episode is not assured, they are an incomplete chart of an individual's health status. Yet, their patterning can be informative. For instance, among prehistoric hunting and gathering populations in the Midwest and in California, and historic Caribou Eskimo remains, these lines commonly assume a repetitive pattern (**Fig. 1**). This configuration is interpreted to represent a predictable annual seasonal stress. A late-winter hunger time is documented for the Caribou Eskimo, and this seasonal stress would be expected for hunter-gatherers in more southern climes as well.

Dentition. Analogous to the developmental lines of arrest in long bones are enamel defects (hypoplasias) observable as linear depressed bands in the dentition (**Fig. 2**). These have the advantage of permanency and age specificity, though their presence is apparently dependent upon the stage of dental development at the time of an appropriate disease or nutritional stress. Hypoplastic line frequency has been observed to increase with the advent of maize agriculture among Mississippian populations from Dickson Mounds site, located in the Central Illinois River valley approximately

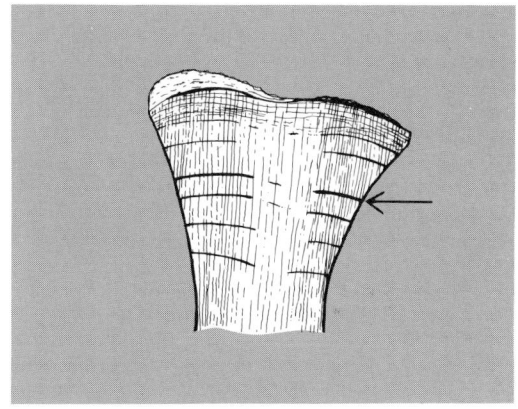

Fig. 1. Example of Harris-line repetitive pattern characteristic of hunter-gatherer populations. The arrow indicates the location of a Harris line (or line of growth arrest). The regularly spaced sequence of lines is interpreted as an indication of periodic, seasonal stress commonly found in Archaic North American peoples.

110 mi (180 km) north of Cahokia.

Microscopic analysis of sections taken through dental enamel have also proved instructive in interpreting biocultural differences during Middle Woodland times. High-status males who are buried with elaborate Hopewell items in special log crypts have been observed to be taller than other males in the same burial mounds. Although explanations that emphasize status specially achieved by tall individuals during their lifetimes have been advanced, dental microscopic study supports an alternative, dietary explanation. Since the tall males show fewer microscopically visible indicators of stress, it is concluded that nutrition was better for these privileged individuals during their youth. Thus, an argument for familial control of resources and inherited status is supported.

Other indicators. In addition to elevated rates of dental hypoplasias, the Dickson Mounds Mississippian skeletal sample also shows increased numbers of other nonspecific stress indicators when compared with earlier Late Woodland remains. Mississippian youth appear to be growing

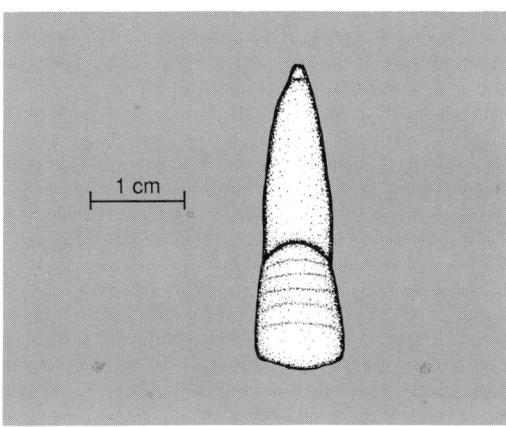

Fig. 2. Dental hypoplasia from a lower-left canine (cuspid) of a prehistoric Late Woodland skeleton.

less rapidly, as indicated by decreased age-specific rates of long-bone increase. Rates of infectious pathology and of age-specific mortality also increase, as does skeletal evidence of anemia.

The Dickson pattern appears to be the extreme when compared with other central-midwestern Mississippians. Studies completed in an area south of Dickson Mounds in the lower Illinois River region provide evidence for decreased growth and increased age-specific mortality among Late Woodland populations during the time of transition to maize agriculture, with an apparent rebound among late Mississippian groups. Juveniles, who also show elevated rates of hypoplasias, appear to be especially disadvantaged during this transitional period. Nearer to Cahokia, relative good health for juveniles from the farmstead communities associated with Cahokia has been reported. Even further south, among maize agriculturalists in the Central and Lower Mississippi valley, evidence seems to suggest poor health associated primarily with increased population size and nucleation rather than with the development of maize agriculture. Rates of infectious pathology and anemia soar, however, for the heavily maize-dependent Upper Mississippian populations from Ohio, Kentucky, and Tennessee.

Interpretation. Although it is difficult to generalize from these data at this time, it seems apparent that the health-related sequelae of maize agriculture in the eastern United States vary considerably. In frontier contexts, such as Dickson Mounds, where living space and maize horticulture may be constrained by a number of social and environmental variables, evidence of ill health is clear. Closer to the major centers, such as Cahokia and Moundville in Alabama, negative biological sequelae of agriculture are less apparent, though the period of transition appears to have been difficult. Population size, an important variable in infectious disease transmission, seems to be of key importance in predicting health status. Among the most recent Upper Mississippian groups, where both maize dependence and population size probably exceeded that for Middle Mississippian, ill health—especially evidence of anemia and infectious pathology—is quite evident.

Increased population size and interaction among Mississippian peoples is clearly associated, however, with the appearance of a specific pathology which resembles the expected pattern for skeletal tuberculosis. This disease, characterized by erosive lesions in the spinal column, often leading to vertebral collapse and death among young adults, appears relatively late in the prehistoric sequence, though it is widespread throughout the eastern United States after 900 years BP. Though the origins and exact nature of the tuberculosislike pathology remain the subject of discussion, it is clear that this is a health stress newly maintained in agricultural populations of large size that characterized the recent prehistoric period in eastern North America.

Data from these same Woodland and Mississippian series also bears upon a frequently disputed issue in the history of disease: the evidence for a treponemal infection among prehistoric North American populations that could have served as the basis for the outbreak of epidemic venereal disease reported in Europe soon after Columbus's initial return from the New World. The most recent studies of long-bone and cranial lesions that resemble those expected of venereal syphilis and its close treponemal relatives, yaws and bejel, confirm earlier reports that identify the observed New World pattern with the yaws-bejel-syphilis constellation. What is missing at this time is conclusive evidence for venereal transmission among North American natives. The age-progressive nature of the lesions among Woodland populations that has been reported resembled more closely the pattern expected for bejel or yaws. Though individual Mississippian cases do approximate those noted in clinical studies of individuals suffering from a venereal treponemal infection, at present there is no convincing means to differentiate these from the nonvenereal (yaws-bejel) forms.

For background information SEE ANALYTICAL CHEMISTRY; ANTHROPOLOGY; ARCHEOLOGICAL CHEMISTRY; ARCHEOLOGY in the McGraw-Hill Encyclopedia of Science and Technology.

Jane E. Buikstra

Bibliography. J. E. Buikstra (ed.), *Prehistoric Tuberculosis in the Americas*, Northwest. Univ. Archeol. Prog. Sci. Pap. 5, 1981; J. E. Buikstra and D. C. Cook, Palaeopathology: An American account, *Annu. Rev. Anthropol.*, 9:433–470, 1980; M. N. Cohen and G. J. Armelagos (eds.), *Paleopathology at the Origins of Agriculture*, 1984; R. I. Gilbert, Jr., and J. H. Mielke (eds.), *The Analysis of Prehistoric Diets*, 1985.

Archeology

Pottery, a common artifact at archeological sites in North, Central, and South America after approximately 2500 B.C., is studied not only through its shape and decorative styles but also in terms of its chemical and mineralogical composition. One objective of such compositional analysis is to determine the provenience (also referred to as provenance) of the pottery, that is, the geological source of the raw material, or the actual location of manufacture, or both. Provenience information is useful for reconstructing the economic processes of production and trade in antiquity.

Aboriginal pottery of the New World has a number of characteristics that distinguish it from Old World ceramics, and some of these traits are important for provenience studies. New World pottery was built by manual shaping rather than by being thrown on a wheel. In addition, the pottery was not given a glaze, an impermeable, glassy coating. Finally, it was fired without kilns, and the relatively brief, low-temperature firings did not vitrify the ceramic body; instead, these earthenwares are crystalline, porous, and often coarse-textured. Wheels, glazes, and kilns did not come into widespread use in the New World until they were introduced by Europeans in the midsixteenth century or later.

Provenience studies. These are based on a concept known as the provenience postulate, which states that geochemical differences between individual sources of the raw material for pottery making can be detected analytically and that compositional variations between sources will be greater than those within a given source. The analysis generally begins with the descriptive physicochemical characterization of multiple specimens of the pottery of interest, using mineralogical methods, chemical methods, or preferably some combination of both. Multivariate statistical transformations of the resultant data, such as discriminant function or cluster analysis, create groupings of pottery specimens that exhibit similarities in the kinds and quantities of constituents present. These groupings may permit limited provenience interpretations, such as suggesting the composition of the most typical pottery at a particular site, or they may facilitate distinguishing local products from nonlocal or imitation products. More commonly, it is desirable to make a statistical comparison of the characterization data on the unknown pottery and similar data on material of known origin, either raw resources (for example, clay and added inclusions, known as temper) or pottery from a known locale (for example, a known workshop, kiln, or waster dump—a refuse deposit of discarded vessels). Because provenience studies are increasingly common, many rigorous physical and chemical analyses of pottery and raw materials have been carried out all over the world by archeometric laboratories. The major facilities of this type have data banks of tens of thousands of such measurements to use in statistical comparisons.

A variety of mineralogical and chemical characterization techniques have been used in provenience studies, and each has its advantages and disadvantages for different kinds of pottery.

Petrographic analysis. The mineral constituents of pottery are usually identified by means of petrographic techniques, which have come into archeology from geology. Their application to ceramic materials is justified through an analogy between ceramic materials and metamorphosed sedimentary rock (pottery may be considered a type of artificial stone); identifications are based on the optical (light-transmitting) properties of minerals as observed in a special polarizing microscope. Pottery is studied petrographically by using either thin sections—cross-sectional slices of a ceramic that are cut and ground to a thickness of 0.03 mm (0.001 in.)—or individual grains separated from the ceramic body. These techniques are particularly well suited to the low-fired, coarse-textured, unvitrified New World pottery. In these materials, individual mineral grains are typically of sufficient size, variety, or abundance to allow identification of the products of particular raw materials or the various formulations used by potters of different communities in preparing their clays. For example, an early application of petrographic analysis in New Mexico focused on glaze-painted pottery. In this geologically diverse region, potters in different communities used different tempers in their wares, so that correlation of the temper types with the geographical spread of the paint style through time led to an understanding of the history of production of this ware. First the ware was traded into the upper Rio Grande basin from outside; then one site in the basin began to control production; after a time the technique of producing the paint was widely learned, and several new centers of production developed.

X-ray diffraction. Another mineral identification technique is x-ray diffraction, which is based on the regular array of planes or layers of atoms unique to each mineral's crystalline structure. Monochromatic x-rays are aimed onto a powdered pottery specimen, diffracted by the planes of the different mineral crystals, and observed with an x-ray detector. A unique series of peaks (the diffraction pattern) produced by a particular mineral constitutes the basis for the mineral's identification.

X-ray diffraction has some disadvantages for ceramic characterization and provenience objectives. The diffraction pattern of a low-fired ceramic sample can yield a complex series of overlapping peaks that is difficult to interpret because the sample is usually a complex mixture of numerous minerals; these minerals include clays; larger grains, such as quartz, feldspar, calcite, and mica; and a variety of accessory minerals. The structure of the clay minerals themselves is usually destroyed in firing, so they can rarely be identified by x-ray diffraction. Furthermore, it is impossible to determine from the diffraction pattern alone whether the coarser constituents were naturally present in the clay or were added by the potter as temper.

X-ray diffraction is therefore often used as a supplemental technique, and it is particularly useful if a low-fired specimen is experimentally refired to high temperatures above 1000°C (1800°F) in the laboratory. The new minerals formed after such refiring are easily analyzed and can permit identification of both clays and inclusions. X-ray diffraction analysis has been used (without high-temperature refiring) in efforts to determine whether the inclusions in some lowland Maya pottery were imported volcanic ash or a mineral substance local to the limestone region.

Chemical analyses. These are usually differentiated in terms of the relative abundances of elements that are to be analyzed in the sample, for example, those techniques determining major constituents (elements present in pottery in amounts greater than 2%, for example, aluminum, silica, oxygen, and sometimes iron, calcium, and potassium); minor constituents (between 0.1 and 2%, for example, calcium, iron, potassium, titanium, magnesium, manganese, and sodium); and trace elements (less than 0.1% down to parts per million or parts per billion). The kinds and amounts of trace elements are uniquely characteristic of individual clays and related minerals, as well as the pottery made from them; thus the trace elements, together with minor elements, form the basis for establishing the so-called chemical fingerprints of pottery and for carrying out most provenience studies of artifacts.

Chemical methods of analysis are less well suited to characterization of coarse, sandy pottery pastes than are mineralogical techniques, because chemical methods are highly sensitive to minute compositional variations, and coarse pottery is typically extremely inhomogeneous. Chemical characterization is better suited to fine-textured, apparently untempered pottery, which is easier to sample reliably. This point can be illustrated by comparing analysis of pottery with analysis of obsidian, a volcanic glass used for making stone tools. Obsidian is noncrystalline and compositionally homogeneous; the manufacture of tools requires only the removal of flakes of this raw material until the desired shape is achieved. Pottery, by contrast, is crystalline and compositionally heterogeneous, consisting of a variety of clay and nonclay minerals, each with a distinct chemical composition. In manufacturing pottery, clays may be sieved, washed, or settled; they may be mixed with water to develop plasticity or with a variety of mineral or nonmineral materials to modify the characteristics of forming, drying, and firing. Finally, they are heated to high temperature. Any of these procedures, which are commonly employed by potters, has the potential for adding elements to or subtracting them from the original chemical fingerprint of the raw clay. They can thereby distort the presumed relationship between a fired pot and the raw resources, and they have the potential to subvert the principles underlying the provenience analyses. The manufacturing procedures followed by present-day Shipibo-Conibo potters of lowland Peru are an extreme example: these potters employ three kinds of clay and three kinds of temper, varying the combinations not only to make different kinds of pots but also to form different parts of one pot. Analytically, then, the body, neck, and base of a single vessel could register as three distinct compositional (mineral and chemical) entities.

Physical methods. Although the microchemical procedures of classical wet chemistry were used through the 1930s and 1940s to identify the constituents of pottery, particularly decorative pigments on ceramic vessels, the most commonly used techniques today are the physical methods of analysis. These base the identification of chemical elements on detectable changes in the elements when energy is introduced into them, for example, by heating or bombarding with x-rays, electrons, or neutrons. Spectroscopic methods, which depend on analysis of the electromagnetic spectrum of radiation emitted or absorbed by the elements, are typically employed in provenience analyses, and these include optical emission spectroscopy (OES), x-ray fluorescence (XRF), proton-induced x-ray emission (PIXE), neutron activation analysis (NAA), microprobe analysis, and atomic absorption spectrometry (AAS).

Although optical emission spectroscopy was one of the earliest instrumental techniques applied to provenience studies of pottery, neutron activation analysis has become more widely and successfully used since the 1970s. In this procedure a tiny sample of pottery is removed, ground into a fine powder, and subjected to a beam of neutrons in a nuclear reactor. The elements in the specimen become radioactive and emit detectable quantities of gamma radiation; detection and analysis of this radiation permits identification of the kinds and amounts of elements present in the sample. Neutron activation analysis is highly sensitive to minute amounts of individual constituents in the sample and can detect as many as 25 or 30 elements at a time, making it the preferred technique for provenience studies in most circumstances. With respect to New World pottery, neutron activation analysis has been applied to a variety of research problems in North America and Central America; the most extensive of these has been the study of the composition and distribution of Fine Orange pottery in the Maya area, a very fine-textured ware found in the lowlands after about A.D. 800. Neutron activation analysis, supplemented with petrographic studies, suggested that this widespread ware was made in a number of localized production centers along a major river drainage in the western periphery of the lowlands. Neutron activation analysis has also been used to study samples of New World Colonial tin-glazed majolicas, a type of richly colored and ornamented pottery. It was possible to distinguish the earthenware imported from Spain from similar products manufactured after a comparable potting industry had been established in Mexico by the Spaniards.

In spite of all its attributes, neutron activation analysis has two disadvantages: it requires operators trained in working with nuclear reactors, and it is extremely costly. As a consequence, other techniques of instrumental analysis are being used and improved for pottery provenience applications. One of these is atomic absorption spectroscopy. This method has several disadvantages (as compared with neutron activation analysis, for example) in that it can be used to determine only one element at a time and does not have equal sensitivity to elements present in extremely small quantities. It is, however, low in cost and relatively simple to perform, and advances in the instrumentation have overcome many of the earlier operational problems. Atomic absorption spectroscopy is being used increasingly for provenience analysis of pottery, one example being an investigation of the provenience of the much-studied Maya Fine Orange pottery from Mexico.

For background information SEE ACTIVATION ANALYSIS; ARCHEOLOGY; ATOMIC SPECTROSCOPY; PETROGRAPHY; X-RAY DIFFRACTION in the McGraw-Hill Encyclopedia of Science and Technology.

Prudence M. Rice

Bibliography. M. Maggetti, H. Westley, and J. S. Olin, Provenance and technical studies of Mexican majolica using elemental and phase analysis, in J. B. Lambert (ed.), *Archaeological Chemistry—III*, Advances in Chemistry Series 205, pp. 151–191, 1984; J. A. Sabloff et al., Analyses of fine paste ceramics, in *Excavations at Seibal, Department of Peten, Guatemala*, Peabody Mus. Archaeol. Ethnol. Mem. 15(2), 1982; A. O. Shepard, Rio Grande glaze-paint pottery: A test of petrographic analysis, in F. R. Matson (ed.), *Ceramics and Man*, pp. 62–87, 1965; L. M. Torres, A. W. Arie, and B. Sandoval, Provenance determination of Fine Orange Maya ceramic figurines by flame atomic absorption spectrometry, in J. B. Lambert (ed.), *Archaeological Chemistry—III*, Advances in Chemistry Series 205, pp. 193–213, 1984.

Asymmetric synthesis

An interesting aspect of organic chemistry is that some compounds have configurations of the constituent atoms that are lacking in symmetry elements. As a result, these compounds have two, mirror-image forms and are termed chiral. Each mirror-image form is called an enantiomer, while an equal molar mixture of the two enantiomers is known as a racemate or racemic mixture. The most common source of this asymmetry is a carbon atom bearing four different substituents.

The chemical and physical properties of enantiomers are identical, except that they rotate the plane of plane-polarized light in opposite directions. However, since many biological receptors are, in themselves, chiral molecules, there may exist nonequivalent interactions with enantiomeric molecules. The preparation of a chiral molecule in one enantiomeric form is then often required; for example, it may be necessary to separate the desired biological properties present in that compound from undesirable toxicological properties possessed by its mirror image. (The tragedy involving the drug thalidomide, a racemate, would have been avoided had the precaution of separation of the enantiomers been taken. The desired sedative activity resides in one enantiomer, while the other enantiomer, devoid of sedative properties, is highly teratogenic.)

Asymmetric synthesis, a method for preparing a high yield of one enantiomer with minimal production of its mirror image, has often provided technically viable industrial processes. Most of the examples reported involve preparation of small-volume, high-value products such as pharmaceuticals. However, in at least one case (*l*-menthol) asymmetric synthesis is the method of choice for preparing a chemical sold in large volume.

Factors in process choice. An industrial process is chosen after evaluation of many factors, including chemical yield, raw material cost, technical feasibility, capital investment, and safety and environmental issues. The sum of these factors is used to decide whether to employ an asymmetric synthesis rather than a more classical process.

Typical of the older approaches to enantiomeric chiral molecules that are now being supplanted by asymmetric synthesis is resolution. A racemic chiral molecule containing, for example, a carboxylic acid function is reacted with one enantiomer of a chiral amine (the resolving agent). The product of this reaction consists of an equal mixture of two salts with the same gross chemical structure. The salts are not mirror images (and thus not enantiomers) and are called diastereomers. Diastereomers have differing physical properties and can be separated on that basis, usually by selective crystallization. The desired salt is then treated with a strong mineral acid to liberate the enantiomeric carboxylic acid and recover the resolving agent (as its mineral acid salt) for reuse. Unless a method for use of the wrong enantiomer, such as racemization, is available, the maximum possible yield in a resolution is 50%. An asymmetric synthesis, on the other hand, can provide a 100% yield of one enantiomer.

To date, the instances in which the process decision has been in favor of an asymmetric synthesis have some common features: The asymmetry-inducing reaction provides a good chemical yield of product that consists mainly of the desired enantiomer (high enantiomeric excess). Since the enantiomeric excess is never 100%, a facile enantiomeric enrichment, usually through crystallization, is available. The reactions are metal-mediated. The metal presumably orders the substrate and reagent in the reaction transition state, creating the steric environment which leads to the selective preparation of one enantiomer. Finally, the reagents used to carry out the asymmetric synthesis often involve expensive metals or chiral ligands of these metals. For this reason, recovery of the valuable reagents must be achieved, or else they must be used as catalysts. The more successful examples have employed chirality-inducing reagents in a catalytic mode.

Industrial applications. At least five asymmetric syntheses are now practiced, providing enantiomerically pure compounds in commercial quantities ranging from kilogram to ton lots.

Amino acids. Amino acids occur in nature almost exclusively in one enantiomeric form, designated (S), as opposed to the unnatural (R) forms. The (R) and (S) nomenclature system, devised by R. S. Cahn, C. K. Ingold, and V. Prelog, unambiguously defines the absolute configuration of a molecule based upon a ranking of the substituents, for example, on a carbon atom. When a molecule is viewed from the side opposite the lowest-ranking substituent, a clockwise orientation of the remaining substituents is (R), while a counterclockwise orientation is (S). For amino acids, RCH(NH$_2$)COOH, the priorities are: 1, NH$_2$; 2, COOH; 3, R; and 4, H.

Aromatic amino acids, specifically phenylalanine and its derivatives, are obtained in the natural form by hydrogenation of an unsaturated precursor over a soluble rhodium(I) catalyst to which is ligated a chiral phosphine. Two examples involve (S)-phenylalanine methyl ester (structure I), by reaction (1), and (S)-3,4-dihydroxyphenylalanine (III; L-dopa), by reaction (2). The former compound is an intermediate for the nonnutritive sweetener aspartame (II), while the latter is a pharmaceutical employed in the treatment of Parkinson's disease.

To be economically viable, these hydrogenations must make effective use of the catalyst systems; for example, each molecule of catalyst must lead to the production of many molecules of product. Using relatively low hydrogen pressures, which is a desirable feature from a process viewpoint, the goal is achieved with over 10,000 molecules of protected amino acids being obtained for each molecule of catalyst.

The structural requirements for the substrates and catalysts in these hydrogenations are rather specific, with variations usually resulting in products of much lower enantiomeric excess or in unusably slow rates of hydrogenation. For example, the indicated olefin configuration and protection of the amine as an amide are necessary for both reaction rate and high product enantiomeric excess. This is presumably because the olefin and the amide carbonyl groups are ligated to the rhodium atom during the reaction. On the other hand, substituents on the aromatic ring or protection of the carboxylic acid as an ester has little effect on the course of the reaction.

$$(1)$$

$$(2)$$

The best chiral, asymmetry-inducing ligands have two phosphorus atoms connected by a two- or four-carbon bridge. Chirality at phosphorus, as in DIPAMP (IV), or in the connecting bridge, as in PPNP (V), leads to high enantiomeric excess products. In reactions (1) and (2) the hydrogenations proceed in 83% (phenylalanine) and 94% (L-dopa) enantiomeric excess.

As noted above, this hydrogenation reaction is relatively intolerant of changes in substrate structure. As a result, extension of the method to other types of olefins usually leads to products of significantly lower enantiomeric excess.

Cyclopropanes. Decomposition of a diazoacetic acid ester in the presence of an olefin leads, via carbene insertion into the double bond, to a cyclopropane. When the reaction is catalyzed by a chiral copper complex, one enantiomer of the product predominates. Ethyl (*S*)-2,2-dimethyl-cyclopropane carboxylate is thus obtained in 92% enantiomeric excess when ethyl diazoacetate is added to isobutylene in the presence of the dimeric copper catalyst R-7644 (VI), as in reac-

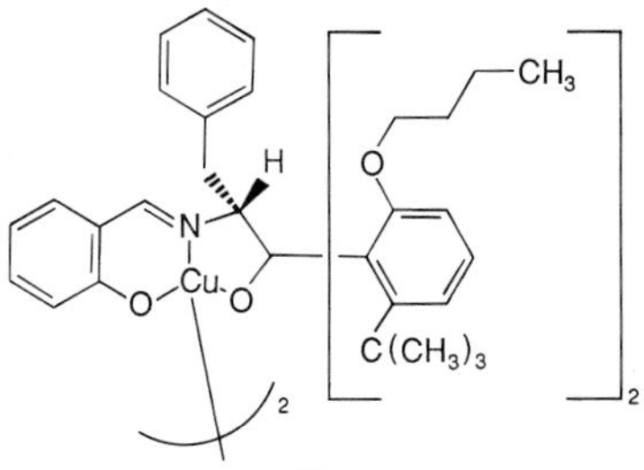

(VI)

tion (3). The reaction product is used in synthesis

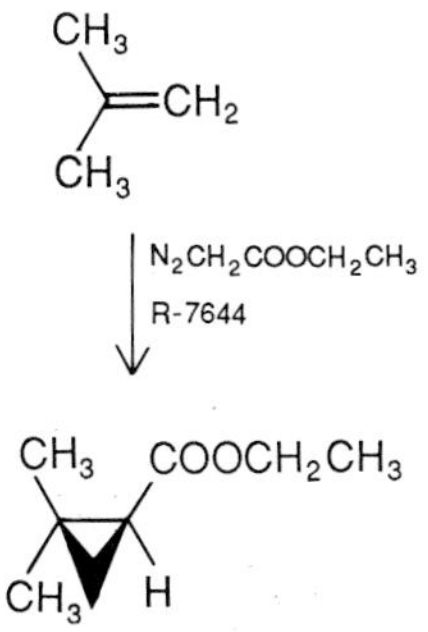

(3)

of an enzyme inhibitor which enhances the clinical usefulness of a β-lactam antibiotic. Other cyclopropanes are also available in high enantiomeric excess by this method; the industrial preparation of the pyrethroid insecticides in the near future by this approach seems likely.

l-Menthol. Soluble rhodium catalysts of the type described above for amino acid synthesis will also cause, in certain cases, olefin migrations. One example is the conversion of an allylic, β,γ-unsaturated amine to the corresponding vinylic, α,β-unsaturated amine. When the process is carried out with chiral catalysts such as (*R*)- or (*S*)-BINAP-ligated rhodium(I), very high enantiomeric excesses can be obtained. Structure (VII) shows (*S*)-BINAP.

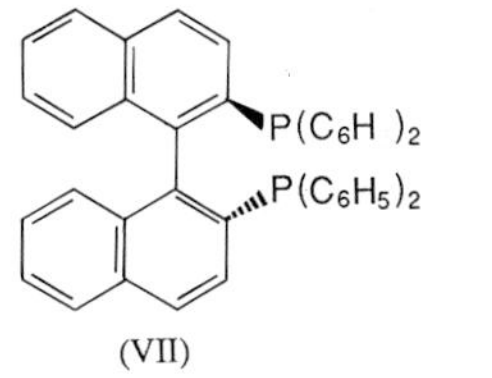

(VII)

Metric-ton lots of *l*-menthol (XI) are prepared in a synthesis using this method [reaction (4)]. Diethylgeranylamine (X) is prepared from myrcene (IX), which is itself obtained by cracking the naturally occurring but enantiomerically impure β-pinene (VIII). When the geranylamine is

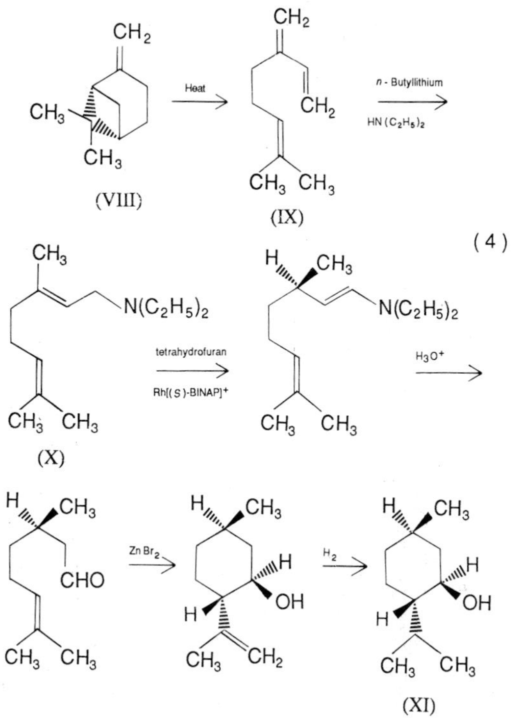

(4)

heated in the presence of the rhodium-BINAP catalyst system, complete isomerization occurs, yielding product of 98% enantiomeric excess. Catalyst usage, as in amino acid preparation, is very effective, and substrate-to-catalyst ratios of up to 10,000:1 can be employed. Hydrolysis of the crude isomerization product gives (*R*)-citronellal. This material is subjected to acid treatment, which causes cyclization to isopulegol. The isopulegol is crystallized to chemical and enantiomeric homogeneity and then hydrogenated to *l*-menthol. Over 1000 metric tons/year of *l*-menthol are made by using this technology.

It appears that other substrates, for example, allylic alcohols, may be employed in this isomerization reaction. Other industrial applications thus seem likely.

(7R,8S)-Disparlure. This compound is the pheromone of the gypsy moth. To be an effective lure in traps for the male gypsy moth, the material must be of very high enantiomeric excess. A synthesis achieving this goal involves asymmetric epoxidation of *Z*-2-tridecen-1-ol [reaction (5)]. An epoxidation reagent consisting of an equimolar mixture of titanium isopropoxide and (−)-diethyl tartrate with *tert*-butylhydroperoxide as primary oxidant gives crystalline epoxy alcohol of 91% enantiomeric excess. This compound is

converted in three straightforward steps to the pheromone, (7R,8S)-disparlure (XII). The asym-

(5)

(XII)

metric epoxidation reaction is quite general, working well with most types of allylic alcohols. In the case of secondary allylic alcohols, the reagent system effects highly selective kinetic resolutions. Careful reaction optimization or the use of molecular sieves in the reaction mixture often allows the use of as little as 5 mol % of the titanium reagent. This advance paves the way for catalytic asymmetric epoxidation to be used in large-scale industrial processes.

Prostaglandins. The hydroboration reaction can be employed to convert olefins to alcohols of high enantiomeric excess. One application [reaction (6)] has resulted in the preparation of a bicyclic lactone (XIII) which can be elaborated into most

(6)

(XIII)

of the primary prostaglandins. Diisopinocampheylborane, obtained by the reaction of β-pinene with diborane, reacts with (carbomethoxymethyl)cyclopentadiene to give, after oxidation, a hydroxy ester of at least 95% enantiomeric excess. Conversion of the alcohol to its methanesulfonate ester is followed

by saponification, which is accompanied by ring closure with inversion, to give the desired lactone. Crystallization gives enantiomerically homogeneous product.

The hydroboration reaction requires 1 mole of organoborane for each mole of olefin. This stoichiometric relationship does not allow a catalytic use of the chirality-inducing reagent. As a result, it is unlikely that asymmetric hydroboration will be employed in the preparation of chemicals other than high-value fine chemicals needed only in small volume.

For background information SEE ASYMMETRIC SYNTHESIS; HETEROGENEOUS CATALYSIS; STEREOCHEMISTRY; STEREOSPECIFIC CATALYST in the McGraw-Hill Encyclopedia of Science and Technology.

John W. Scott

Bibliography. J. D. Morrison, *Asymmetric Synthesis*, vols. 1–5, 1983–1985.

Autoimmunity

One of the key issues in biology has been the question of how autoimmunity arises and how this process leads to disease. This has several implications in many chronic diseases, including rheumatoid arthritis, systemic lupus erythematosus, cardiovascular disease (rheumatic fever), autoimmune kidney disease, and multiple sclerosis. In many of these diseases, neither the initiating agent nor the mechanisms that are involved in the breaking of tolerance to self (the organism's own tissues) are known. Viruses or other microorganisms have always been suspect as causative or initiating factors.

Viral initiation of autoimmune disease. Viruses may initiate autoimmune events in several ways. First, viruses can alter the immune response by infecting lymphoid cells that generate cell-mediated or humoral responses. For example, the measles virus can infect T cells and B cells, as well as macrophages. This infection of cells can cause altered immune responses to other antigens. Therefore, autoimmunity could occur by virus infection of a subset of T cells that regulate the immune response. These regulatory cells (suppressor T cells) dampen or limit the degree of the immune response. Without regulated control of the immune response, a chronic nonlimiting response to self could result. This response to self would not be limited by the normal functioning of the checks and balances of the immune system.

Second, viruses can cause polyclonal responses in which B cells are stimulated to produce antibodies. An example is the Epstein-Barr virus infection. Here, infected B cells are driven to differentiate into plasma cells, the cells that secrete antibodies of a preprogrammed specificity. The specificity of the antibody may be to a viral protein or cellular component. Antibodies with a variety of specificities, including self, are found in infected individuals. The self antibodies may be sufficiently numerous to cause or promote autoimmune disease because they are directed against critical cellular sites or because immune complexes are formed and deposited.

Third, viruses may initiate autoimmunity through the concept of molecular mimicry. Molecular mimicry is the sharing of an antigenic determinant between a viral product and a host protein or cellular component. An antiviral immune response against the cross-reacting determinant would lead to reactivity with the shared self site. This cross-reacting immune response could then initiate disease production. These types of cross-reacting immune responses leading to disease are the focus of this article.

Evidence for common antigenic sites. The first evidence that viruses share common antigenic sites or determinants with self components came from the development of monoclonal antibody technology. This technology allowed the production of large amounts of an antibody with one specificity. During virus infection, antibodies to self components were often detected in the circulation of infected individuals. However, there was no easy way to determine if these antibodies arose through molecular mimicry. The advent of monoclonal antibodies solved this problem. By using monoclonal antibodies, common sites on viral proteins and host cell proteins have been identified. For example, one of the measles virus proteins, phosphoprotein, has a common site with one of the cytokeratin proteins found in normal cells. This was demonstrated by two techniques. *See Monoclonal antibodies.*

First, such dual reactivity was observed by immunofluorescent staining of cells that were both infected and uninfected with the measles virus. This technique allows direct visualization of the structures to which the monoclonal antibody binds the infected and uninfected cells. Here, the antibody is allowed to react with the target cell, and the binding of that antibody is monitored by the binding of a second labeled antibody that is directed against the first. The binding of the labeled second antibody is visualized by inspection of the reacted cells with an ultraviolet microscope. By screening various measles virus monoclonal antibodies, an antibody was discovered that reacted with both viral inclusion bodies and intermediate filament proteins.

Second, western blotting was used to identify the viral and normal cellular protein for which the cross-reactive monoclonal antibody was specific. In this technique, viral or cellular proteins are separated by molecular weight by using sodium dodecyl sulfate polyacrylamide gel electrophoresis. The separated proteins are then transferred electrophoretically to nitrocellulose paper. The protein-bound strips of nitrocellulose paper are then reacted with monoclonal antibody. The binding of the monoclonal antibody can be monitored by using a radioactive or enzyme-linked second antibody directed against the first. It was found that the monoclonal antibody reacted with measles virus phosphoprotein (with a molecular weight of 70,000) and one of the cytokeratin proteins (with a molecular weight of 54,000) from normal cells. Thus, this indicates there is a common antigenic site between a viral protein and a normal host-cell protein. Similar reactivities using monoclonal antibodies have been identified in many viral systems. It appears that roughly 3–4% of all antiviral monoclonal antibodies react with components of host cells.

Identification of common regions. Monoclonal antibodies are excellent reagents to describe common antigenic sites between virus and self. However, this method does not readily allow the identification of the specific region on the common proteins or shared determinants. To directly approach this problem, peptide stretches of known disease-producing areas from self proteins were determined. These amino acid sequences cause autoimmune disease when injected with adjuvant into a suitable animal. The disease-producing sequences were analyzed by a computer, and the computer identified viral proteins sharing regions of homology with the disease-producing amino acid sequences.

Effects of cross-reacting viral peptide. One of the homologies to be described was a region from myelin basic protein (amino acids 66–75) and the hepatitis B virus polymerase (amino acids 589–598). The amino acid stretch from this region of myelin basic protein is encephalitogenic for rabbits; that is, when injected into a rabbit with adjuvant these amino acids will induce an autoimmune disease known as experimental allergic encephalitis (EAE). From the computer analysis, hepatitis B virus polymerase was found to share six amino acids in tandem with the encephalitogenic region for rabbits. The viral polymerase region sharing the common amino acids with the rabbit encephalitogenic site was synthesized. This viral peptide was then injected with Freund's complete adjuvant into rabbits, and the animals were monitored for autoantibody production (antibody to myelin basic protein), cellular reactivity to myelin basic protein, and disease production.

When the sera from seven rabbits immunized with one injection of viral peptide were tested for antibody to myelin basic protein, five had significant levels of antibody. The binding of this antibody to myelin basic protein could be inhibited by use of the viral peptide, thereby demonstrating the specificity of the antibody. Therefore, sensitization of a rabbit using a viral peptide that cross-reacts with a self protein can lead to autoantibody reproduction.

To test for cellular reactivity, peripheral blood lymphocytes from eight rabbits sensitized with the viral peptide were tested for their ability to respond to myelin basic protein or viral peptide. Peripheral blood lymphocytes from all the sensitized rabbits proliferated when cultured in the presence of viral peptide, indicating a positive reaction. Peripheral blood lymphocytes from four of the eight rabbits responded positively to myelin basic protein. Thus, half the animals tested reacted positively to the self protein, myelin basic protein.

The brains and spinal cords of eleven rabbits sensitized with viral peptide were examined for histologic lesions characteristic of experimental allergic encephalitis. In four of these animals, lesions were observed in the central nervous system that were consistent with histologic changes observed in this autoimmune disease. Therefore, in some of the animals sensitized with viral peptide, cellular infiltrates were found in brains and spinal cords indicating disease production.

Molecular mimicry and disease process. These types of experiments provide clues to the nature of molecular mimicry and autoimmune disease. Infection by a virus can lead to an antiviral immune response. Should this response be against a determinant on the virus that is similar or identical to a site on a host protein, tissue or cell-specific injury could result. The responsible mechanism for injury would be the generation of cytotoxic cells specific for the cross-reacting sites or the production of an autoantibody. This antibody would bind to the common site, and a complement cascade—a series of 20 proteins which, when activated by an antibody, lead to the production of pores in the membranes of cells—would then ensue and result in cellular destruction. In addition, immune complex deposition would initiate disease in the kidneys, arteries, or choroid plexuses.

Once the cross-reacting immune response is set in motion, the initiating agent need not be present. The virus may be cleared or eliminated from the body, yet the humoral or cellular elements continue to attack self components, resulting in injury. The injured tissue releases more self antigen, and the cycle of destruction continues. Thus the probability of virus being recovered from the actual sites of damage is unlikely.

Autoimmune disease probably occurs only when the actual shared site between the virus and the host is a disease-inducing site. In the earlier example of homology between the encephalitogenic (disease-inducing) site and viral polymerase, actual disease production would not occur if the shared site, that is, the encephalitogenic site, did not take part in the disease production. Should the cross-reaction take place at a site not involved in disease production, autoantibody might be formed, but no tissue damage would be observed. An example of autoimmune central nervous system disease may be what is sometimes observed in patients with measles virus

infection. This virus can cause a postinfectious encephalopathy on rare occasions where peripheral blood mononuclear cells from such patients proliferate when cultured with myelin basic protein. Measles virus proteins may share common elements with myelin basic protein or other central nervous system proteins. This possibility remains to be confirmed. Similar sets of events could play a role in Guillain-Barré syndrome, myasthenia gravis, thyroiditis, arthritis, diabetes, and multiple sclerosis.

Lastly, molecular mimicry may be advantageous for the virus. If the virus is regarded as self, an effective immune response that would normally clear the virus is not mounted, and so the virus can persist. Since viruses are intracellular parasites, their possession of a site or region on their proteins in common with host proteins may enable them to be governed by cellular signals. These intracellular signals would direct viral proteins to assemble or congregate in appropriate cellular compartments. Once there, virion particles would assemble and viral maturation could take place.

For background information SEE AUTOIMMUNITY; ELECTROPHORESIS; IMMUNE COMPLEX DISEASE; IMMUNO-FLUORESCENCE; MONOCLONAL ANTIBODIES in the McGraw-Hill Encyclopedia of Science and Technology.

Robert S. Fujinami

Bibliography. R. S. Fujinami et al., Molecular mimicry in virus infection: Crossreaction of measles virus phosphoprotein or of herpes simplex virus protein with human intermediate filaments, *Proc. Nat. Acad. Sci. USA*, 80:2346–2350, 1983; R. S. Fujinami and M. B. A. Oldstone, Amino acid homology between the encephalitogenic site of myelin basic protein and virus: Mechanism for autoimmunity, *Science*, 230:1043–1045, 1985; R. T. Johnson, *Viral Infections of the Nervous System*, 1982; J. Srinivasappa et al., Molecular mimicry: Frequency of reactivity of monoclonal antiviral antibodies with normal tissues, *J. Virol.*, 57:397–401, 1986.

Battery

Batteries are electrochemical devices that allow electric energy to be stored and transported. Rechargeable, or storage, batteries perform this storage by converting electrical energy into chemical energy upon charging and by converting chemical energy into electrical energy upon discharging. In recent years, research has been conducted on more than a dozen new battery technologies: chromium/iron redox; zinc/chlorine; zinc/bromine; aluminum/air; zinc/air; iron/air; sodium/sulfur (ceramic electrolyte); sodium/sulfur (glass electrolyte); lithium/iron sulfide; lithium/iron disulfide; lithium/molybdenum disulfide; zinc/ferricyanide; and hydrogen/

nickel oxide. Many of these advanced storage batteries hold the promise of better performance or a lower cost than conventional technologies, such as lead-acid and nickel/cadmium batteries. Two systems in particular, sodium/sulfur (ceramic electrolyte) and zinc/bromine, are currently under intensive research and development both in the United States and in other countries. These two will be discussed in detail.

Advanced storage battery applications fall into three general categories: electric vehicle, load leveling, and storage in solar energy systems (photovoltaic and wind). These applications can play an important role in improving energy productivity and facilitating the substitution of coal, nuclear, and alternative energy sources for petroleum and natural gas. While considerable research and development is still required, advanced battery systems hold great promise for satisfying future energy storage needs.

Application requirements. Each advanced storage battery application has unique requirements for technical performance and economy. Typical requirements for five major mobile and stationary applications are shown in the **table**.

Mobile applications of current interest are commuter and fleet-operated passenger electric vehicles, urban-delivery electric vehicles, full-performance electric passenger vehicles, and hybrid passenger vehicles. Fleet-operated electric vehicles are those driven on a relatively fixed mission every working day and then returned to a home base each evening for charging. Full-performance electric vehicles perform with characteristics equivalent to those of internal-combustion-engine vehicles. Hybrid vehicles contain both batteries and a small gasoline or alcohol engine. Batteries for mobile applications require the most stringent technical characteristics. Vehicle performance and packaging requirements dictate that a battery have high specific energy (measured in Wh/kg or Wh/lb) for range, high specific power (measured in W/kg or W/lb) for acceleration, and high energy density (measured in Wh/L or kWh/ft^3) for packaging. These technical characteristics must be achieved in a battery that is also low in cost, long-lived, capable of random depth-of-discharge cycling, and relatively tolerant to vehicle environments. Fleet-operated electric vehicles are already in operation and will probably be the first commercial application of advanced storage betteries.

Stationary applications of current interest are utility load leveling, customer-side-of-the-meter load leveling, and storage of energy generated by solar-powered and wind-powered electrical energy systems. Batteries for stationary applications have less stringent technical characteristics than do electric vehicle batteries. The most important technical requirements are a small footprint (measured in kWh/m^2 or kWh/ft^2), which correlates to a good battery energy density (measured in Wh/L or kWh/ft^3), and long life with deep-discharge cycling. These characteristics must be achieved at low cost and high reliability.

Load leveling by storage batteries at the utility site has two major advantages. The first involves reducing the utility load variation. Storage batteries can be charged during nighttime hours by using base-load generation and can be discharged during the daytime and early evening heavy-load periods. The second advantage concerns peak-load shaving to reduce the required generating capacity and displace petroleum and natural gas

Typical advanced storage battery requirements

	Electric vehicle			Solar	
Characteristic	Fleet	Advanced	Utility load leveling	Stand-alone	Utility-connected
Capacity, kWh	20–40	20–50	100,000	100	25
Energy efficiency, %	>70	>70	>65	>70	>70
Direct-current voltage, V	100	100–200	2000–3000	110	220
Specific energy, Wh/kg (Wh/lb)	50 (23)	110 (50)	*	*	*
Specific power, W/kg (W/lb)	100 (45)	130 (59)	*	*	*
Energy density, Wh/L (kWh/ft^3)	80 (2.3)	140 (4.0)	30 (0.85)	*	*
Cycle life	800	800	2000	2000	2000
Life, years	3	3	8	8	8
Discharge	Specific energy and power delivered under actual driving conditions; acceleration from 0 to 30 mi/h (48 km/h) in 14 s	Specific energy and power delivered under actual driving conditions; acceleration from 0 to 55 mi/h (88 km/h) in 20 s	10 MW for 10 h (load leveling); 0.5–5 h at max power (peak shaving)	Daily cycle to 10% of rated capacity at 20-h rate	Daily cycle to 80% depth of discharge at >5-h rate
Charge	<8 h	<8 h	7 h at constant power	<20 h	<8 h

* Not applicable.

used in generating units that are employed in periods of peak demand. Utilities could possibly also use batteries to reduce spinning reserve requirements: the capacity required for instantaneous power in the event of a sudden power plant outage, and also for system regulation.

Storage battery load leveling could also occur at a residential, commercial, or industrial site. The technical performance and economic requirements of this customer-side storage are similar to those for utility load leveling. A utility rate structure that includes high peak-demand charges or time-of-day energy charges (energy charges that vary significantly during the course of the day) favors customer-side-of-the-meter battery storage. A major advantage of using batteries for both utility and customer-side-of-the-meter load-leveling operations is the building-block nature of batteries, which allows the storage to be matched to the load.

As a final category, batteries for solar applications may operate in either stand-alone or grid-connected systems. Conventional, flooded, lead-acid batteries have been used in stand-alone applications for many years. More recently, however, starved-electrolyte, maintenance-free, sealed lead-acid batteries have been used in remote stand-alone solar applications. Battery storage in solar systems connected to a utility grid has value under economic conditions characterized by a low sell-back ratio (the ratio of the amount paid to cogenerators for energy sold back to utilities to the amount charged by utilities for energy) or high time-of-day rates. However, this economic scenario causes the battery operation to be very similar to the customer-side-of-the-meter load-leveling operation and results in an operation that could be separate from the solar electric generation system.

Sodium/sulfur battery. The sodium/sulfur battery (**Fig. 1**) is a high-temperature system that operates at 300–350°C (570–660°F). The discharge equation is given by Eq. (1), and the

$$2Na + 5S \longrightarrow Na_2S_5 \qquad (1)$$

open-circuit voltage is 2.08 V. The theoretical specific energy, based solely on the weights of the anode and cathode materials, is 758 Wh/kg (344 Wh/lb). The actual specific energy of a sodium/sulfur battery is expected to be three to four times that of present lead-acid batteries. In addition to this enhanced performance, it has the advantage of using inexpensive and abundant raw materials.

Specific cell designs are strongly influenced by the power and energy requirements of the intended application. A high power-to-energy ratio is required for electric vehicle cells, whereas load-leveling cells are designed more for energy storage than for power capability. Higher energy is achieved by adding more reactants to the cell while keeping the electrolyte area constant. Higher power capability is achieved by increasing the area of the electrolyte surface relative to the volume of the reactants. Currently, sodium/sulfur electric vehicle cells (whose nominal capacity is 10 ampere-hours) are capable of achieving a specific energy of 165 Wh/kg (75 Wh/lb) and a specific power of 200 W/kg (91 W/lb). Cycle lifetimes in excess of 500 cycles are routine, and lifetimes over 1000 cycles have been reported frequently. Cell energy efficiencies are typically 80%. Performance goals for a full-scale electric vehicle battery (having a nominal voltage of 200 V and energy of 40 kWh) are a specific energy of 150 Wh/kg (68 Wh/lb) and a specific power of 150 W/kg (68 W/lb). These goals include scale-up losses and efficiency losses due to auxiliary components.

At operating temperature the anode material is liquid sodium and the cathode material is liquid sulfur. These chemically active electrodes are separated by a solid electrolyte that is a conductor of sodium ions. Typically, this electrolyte material is a tube of beta″-alumina ceramic whose nominal composition is $Na_2O \cdot 11Al_2O_3$. The ionic conductivity of beta″-alumina at temperatures above 260°C (500°F) is almost as high as that of the sulfuric acid electrolyte used in common lead-acid batteries at 25°C (77°F). The design of the sodium/sulfur system is unique in that it reverses the features of conventional batteries by using liquid reactive materials and a solid separating electrolyte. Most often (Fig. 1) the cell design involves a central tube of beta″-alumina containing sodium on the inside and surrounded by sulfur on the outside. The sulfur is impregnated in a layer of carbon fiber or felt to facilitate the flow of electrons through the cathode material. The cell assembly is encased in a metal housing that also acts as the positive current collector. During discharge, sodium metal is oxidized at the anode/beta″-alumina interface. The

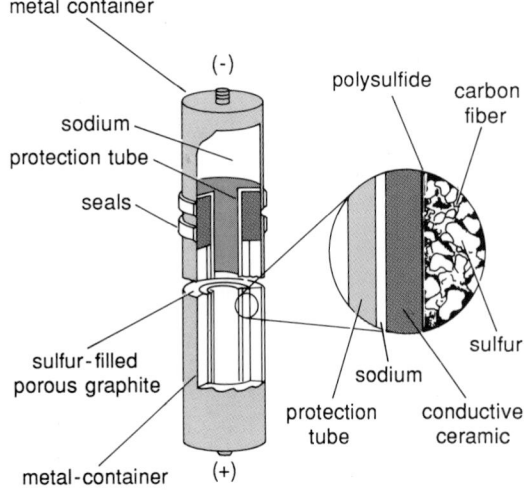

Fig. 1. Schematic of a sodium/sulfur cell. (*Ford Aerospace and Communcations Corp.*)

metal container

(−)

sodium

protection tube

seals

sulfur-filled porous graphite

metal-container backing electrode

(+)

polysulfide

carbon fiber

sodium

protection tube

conductive ceramic

sulfur

sodium ions move through the electrolyte to the sulfur cathode. Free electrons generated by the oxidation of sodium move from the anode through the external electrical load to the wall of the metal container and the sulfur cathode. Sulfur is reduced at the carbon-felt surfaces and combines with the sodium ions to form sodium polysulfide compounds.

The chemical and physical processes that occur in the sulfur electrode are complex. Many sodium polysulfide compounds are produced during discharge, and this results in several phase changes in the electrode. The initial discharge reaction product is liquid sodium pentasulfide (Na_2S_5). The sodium pentasulfide is not soluble in molten sulfur. Consequently, throughout almost 60% of the discharge the sulfur electrode contains two immiscible liquid phases. Further discharge causes the sulfur to be consumed entirely. Then the sodium pentasulfide is progressively converted, in a single liquid phase, to a composition nearing sodium trisulfide (Na_2S_3). As discharge continues, a two-phase mixture again forms as solid sodium disulfide (Na_2S_2) precipitates out of the liquid. Discharge of cells into this region, however, should be avoided because of low cell voltage and the difficulty of recharging from this solid phase. Molten sodium polysulfides are extremely corrosive. Therefore, care must be taken in the selection of a material for the metal container (that is, current collector) associated with the sulfur electrode.

Recently a 36-kWh experimental electric vehicle sodium/sulfur battery was designed, fabricated, and evaluated. The key areas identified as requiring improvement were cell reliability, recharging strategy, and thermal management. Important remaining technical issues at the cell level of development are electrolyte durability and corrosion-resistant sulfur containers. Beta″-alumina durability may be limited because of current-induced degradation and fracturing. Resolution of these problems through a better understanding of the beta″-alumina microstructure and the use of zirconia additives to increase its strength may hold promise. The sulfur container must be (1) a good electrical conductor to reduce current losses; (2) mechanically strong, as it is the primary external cell structure; (3) lightweight, so as not to penalize specific energy performance; and (4) relatively inert to resist attack by the sodium polysulfide compounds at 350°C (660°F). Coatings and platings on inexpensive materials (for example, low-alloy steel, aluminum, and possibly stainless steel) appear to be the most promising solutions to this chemical attack.

Zinc/bromine battery. The zinc/bromine battery (**Fig. 2**) is an ambient-temperature circulating-electrolyte advanced battery system. A full battery consists of multiple bipolar cells held together in stacks by end plates and a clamping mechanism. The battery's discharge equation is

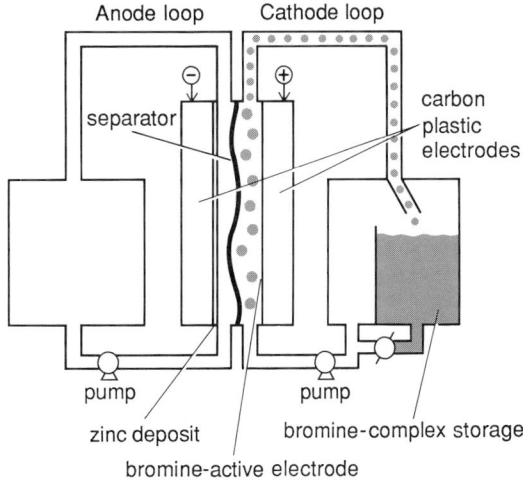

Fig. 2. Schematic of a zinc/bromine cell. (*Exxon Research and Engineering Company*)

given by Eq. (2), the open-circuit voltage is 185 V,

$$Zn + Br_2 \rightarrow ZnBr_2 \qquad (2)$$

and the theoretical specific energy is 428 Wh/kg (194 Wh/lb). Most of the active materials in the zinc/bromine battery are stored external to the cells and are pumped into the stack as needed.

On charge, zinc is electrodeposited on the anode and bromine is evolved at the cathode. In one advanced design the bromine which is generated reacts with an organic complexing agent and forms an immiscible second phase with the electrolyte. As the electrolyte is circulated, the second phase is swept from the cells into the catholyte storage tank, where it separates gravimetrically. On discharge, complexed bromine is carbureted into the catholyte flow stream, where it is returned to the cell. In the cell stack the bromine complex is reduced with the accompanying oxidation of zinc and the release of electrical energy. Separators are used in the cells to prevent the direct reaction of the bromine-rich flow stream with the deposited zinc.

Most cell designs include a carbon-plastic bipolar electrode that has an enhanced surface-area coating on its cathode side and an injection-molded flow frame. The flow frame contains channels for the distribution of both the anolyte and catholyte electrolyte streams, and a microporous separator. The end electrodes in the cell stack serve as current collectors and have embedded metal screens to increase their lateral conductivity. All external plumbing and storage tanks are plastic. Inexpensive construction materials, coupled with low fabrication and reactant costs, result in a very favorable projected cost for this technology.

The zinc/bromine development effort currently is aimed at both the electric vehicle application and the load-leveling application. Tests on a 200-V 15-kWh module designed for electric vehicle applications indicate that the battery is capable of producing a specific energy of 75 Wh/kg (34 Wh/lb) and a peak power of 40 kW, measured for 30 s, at up to 75% depth of discharge. This power translates into approximately 95 W/kg (43 W/lb). Recently a 200-V 30-kWh battery successfully powered an advanced experimental test vehicle using a drivetrain with an alternating current motor. This was the first advanced secondary battery to successfully provide such power. Performance goals for a full-scale electric vehicle battery, after scale-up losses, are a specific energy of 60 Wh/kg (27 Wh/lb) and a specific power of 150 W/kg (68 W/lb). Several stationary batteries in the 10–20-kWh range have also been built. The longest-lived of the 20-kWh batteries achieved 178 cycles, although smaller modules have lasted as long as 650 cycles. A laboratory prototype 0.5-kWh battery has achieved more than 2000 cycles. Typical energy efficiencies for a multicell stack are 60%.

The zinc/bromine system, as with all flow batteries, has the advantages of ease of thermal and electrolyte management and virtually no capacity loss with increasing cycle life. Cell voltage equalization is not a problem since the battery requires periodic full discharge to prevent formation of zinc dendrites. Flow batteries are, however, mechanically complex and generally less energy-efficient than other advanced battery technologies. This complexity has the potential for reducing the system's reliability, making the selection of the specific application important. Pump operations and the shunting of conductive paths through the electrolyte contribute to lower system efficiencies.

For background information *SEE BATTERY; ELECTRIC POWER GENERATION; ENERGY STORAGE; SOLAR CELL; STORAGE BATTERY* in the McGraw-Hill Encyclopedia of Science and Technology.

Kevin D. Murphy; Robert P. Clark

Bibliography. N. J. Magnani et al., *Exploratory Battery Technology Development and Testing Report for 1985*, SAND86-1266, Sandia National Laboratories, 1986.

Biomedical engineering

Replacement of damaged or diseased body parts with prosthetic devices is a common practice in medicine and dentistry. The devices vary significantly in degree of effective function and of biocompatibility of the material. Historically the design of prosthetic devices was dictated by the need to match the physical properties of the natural body part with a material which elicited a minimal toxic response from the host.

A key property when considering a material as a candidate for a prosthesis has been the inertness. When an inert material is placed in the body, the natural response is to initiate the growth of a thin fibrous capsule between it and the normal tissue. The thinner the capsule, the more inert and acceptable is the biomaterial.

More recently it has been established that no material is totally inert within the body environment; all materials elicit one of four possible responses when placed in living tissue: a toxic material produces tissue death; a material that is nontoxic and resorbable dissolves and is replaced by normal tissue; a material that is nontoxic and nonresorbable (that is, inert) is surrounded by a fibrous tissue capsule of variable thickness; and a material that is nontoxic and biologically active forms an interfacial bond with living tissue.

Surface-active materials. Nontoxic and biologically active materials, designated as surface-active, were developed to address a problem often encountered with nontoxic, nonresorbable materials. With inert materials there is no adhesion between the capsule and the implant material, and any stress may produce movement of the implant within the capsule. The body's natural response to movement of the implant is to increase the thickness of the capsule in order to isolate the source of irritation. Problems associated with a thickened fibrous capsule include impingement upon the blood supply to adjacent normal tissue, creation of a possible site for tumor formation, accumulation of corrosion products, calcification and hardening of the capsule, and a continuous buildup of local stress concentration leading to implant or tissue fracture.

In an effort to address the interfacial problems associated with inert biomaterials, a significant amount of research has been directed toward employing surface-active materials. The composition of these materials is such that the surface interacts with adjacent normal tissues, resulting in the formation of a chemical bond between the tissue and the implant surface. Currently there are four major classes of surface-active biomaterials either undergoing clinical trials or in actual clinical use: dense bioactive glasses, bioactive glass-ceramics, surface-active composites, and hydroxylapatite ceramics.

As a class, surface-active materials represent a unique advancement in the design of prosthetic devices. Their composition is selected to produce biologically active surfaces which influence the behavior of cells so that normal tissue is synthesized and incorporated into the surface of the material. This tissue bond serves to prevent motion at the implant–tissue interface, providing an effective mechanism of stabilization.

Bioactive glasses. The earliest evidence for the development of a chemically bonded implant–tissue interface was obtained at the University of Florida by using a series of specially designed bioactive glasses which are termed Bioglass (a trademark). Bioglasses are silicate-based and contain calcium oxide (CaO), sodium oxide (Na_2O), and phosphorus pentoxide (P_2O_5) along with silicon dioxide (SiO_2). Implanted Bioglasses undergo a series of reactions producing a compositional graded interface consisting of bulk glass, a layer rich in SiO_2, a layer rich in calcium and phosphorus, and bone. The Ca,P-rich layer forms as soon as the Bioglass is exposed to a physiologic environment. It is initially amorphous or noncrystalline, but it crystallizes into a mixed hydroxyl–carbonate apatite structure with time. Collagen fibers which are present become incorporated within the crystallizing apatite agglomerates.

This type of reaction is similar to that found in normal bone formation. A SiO_2-rich layer forms beneath the Ca,P-rich layer when hydrogen ions (H^+) in the tissue fluids exchange with sodium ions (Na^+) from the glass. An effect of this ion exchange is to shift the local pH to an alkaline level, which is more compatible with bone growth. The thickness of the bonding zone can reach 200–300 micrometers. An interfacial bonding zone of this magnitude provides for a gradual transition in elastic modulus across the interface, allowing for the effective transfer of stress through the implant device into the surrounding tissues.

There is a specific compositional range of glasses which are capable of forming a bond with bone. These are designated as area A in the **illustration**. Variations in the Na_2O/CaO ratio and the $SiO_2/(Na_2O + CaO)$ ratio which fall

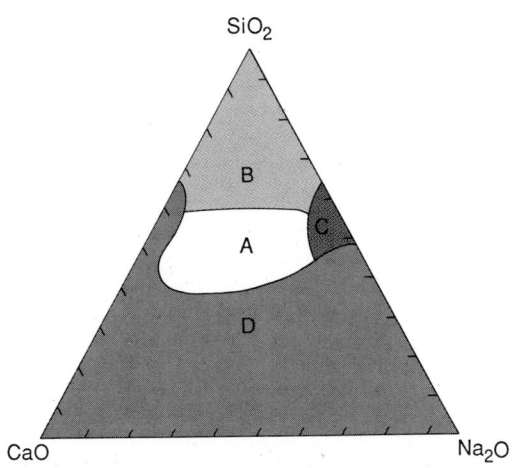

SiO₂

Bioglass-bone bonding boundary; behavior of bioactive glass of different compositions. All compositions have a constant 6 wt % P_2O_5. Area A: bonding at 30 days or less; area B: nonbonding, reactivity too low; area C: nonbonding, reactivity too high; and area D: nonbonding, non-glass-forming.

outside area A produce compositions which do not bond (area B), which will dissolve (area C), or which do not form glass (area D).

Clinical application of the bioactive glasses can be divided into two major categories based on the inherent weakness of the glass. In applications where the implant will not be mechanically stressed, solid or particulate forms of the surface-active glass can be utilized.

Clinical human studies have demonstrated the successful use of solid surface-active glass implants for reconstruction of the ossicular chain and for maintaining mandibular ridge height. Applications in which an implant will be subjected to significant

mechanical loading require a combination of the bone-bonding surface activity of the bioactive glass with a substrate material to provide the necessary strength. Surface-active glasses have been coated on substrates such as stainless steel, titanium alloys, and aluminum oxide ceramics. An alternative method to overcome the inherent mechanical limitations of these glasses involves the construction of a composite of a surface-active glass and stainless steel fibers. Animal studies with coated devices and the composite have demonstrated that load-bearing orthopedic devices and dental implants can have excellent mechanical behavior. An important aspect which has to be carefully monitored is the consequence of compositional changes which may occur as the surface-active glass is coated onto the substrate. For example, the addition of aluminum to the glass can prevent bonding. Also, the presence of oxides of zirconium, titanium, and tantalum can impair bone development at the interface.

Applications of bioactive glasses either in use or being tested include buried conical implants to preserve alveolar bone, load-bearing dental implants (coatings on metals or Al_2O_3), particulate forms to treat periodontal defects, coatings on orthopedic devices for noncement fixation, solid implants for restoration of the ossicular chain in the middle ear, and coatings on orthodontic appliances for tooth movement.

Bioactive glass-ceramics. Another class of surface-active materials, glass-ceramics, is produced through the nucleation and growth of crystals in the glass. By subjecting the glass to an appropriate heating cycle, the clear glass becomes an opaque glass-ceramic consisting of crystals in a glassy matrix which have improved mechanical strength. A surface-active glass-ceramic has been used clinically for restoring the ossicular chain. Animal studies have demonstrated successful application of a glass ceramic for bone augmentation procedures. A glass ceramic has also been utilized as a coating on a substrate for cementless fixation of orthopedic prostheses in animals. A potential drawback of surface-active glass-ceramics is dissolution of the material along the crystal boundaries.

Applications of bioactive glass-ceramics either in use or being tested include coatings on orthopedic devices for cementless fixation, solid implants for jaw augmentation, and solid middle-ear prostheses to restore the ossicular chain.

Surface-active composites. A third class of surface-active materials is made up of designed composites. A composite is a material composed of two materials bonded together. One material serves as a matrix surrounding the particles or fibers of the second material. Bone is an example of a natural composite with collagen fibers cemented together via the inorganic mineral component (hydroxylapatite). A primary failure mode of many biomedical composites occurs at the bond between the matrix and the filler under the influence of tissue fluids and cellular enzymes. In spite of these potential problems, the attractiveness of combining materials with the desired combination of bioactivity and specific mechanical properties has led to successful surface-active composites. Applications of surface-active composites either in use or being tested include particulate hydroxylapatite and finely ground autologous bone for maxillofacial reconstruction and for alveolar ridge augmentation, bioactive glass and stainless steel composite for single-tooth implant, and polymethylmethacrylate and bioactive glass-ceramic as a bone cement.

Hydroxylapatite ceramics. The final class of surface-active materials comprises the hydroxylapatite ceramics. These materials are utilized as implants primarily in the dental area. Their chemical composition is identical to the mineral component of bone: $Ca_{10}(PO_4)_6OH_2$. The fact that they not only are biocompatible but also can form a chemical bond with bone has been demonstrated in numerous studies. While there are similarities between the bonding mechanisms of Bioglass and hydroxylapatite ceramics, the bonding zone of the hydroxylapatite ceramic implants is on the order of 1 μm as compared to 200 μm for the Bioglasses. A consequence of this thin bonding zone is a very high gradient in the elastic modulus.

Applications of hydroxylapatite ceramics either in use or being tested include buried conical implants to preserve alveolar bone; load-bearing transmucosal implants (solid hydroxylapatite and hydroxylapatite-coated titanium); solid implants to restore the bony conduction system and the canal wall of the middle ear; and particulate forms to augment the alveolar ridge, to perform maxillofacial reconstruction or to treat periodontal defects.

For background information SEE BIOMECHANICS; BIOMEDICAL ENGINEERING; BIORHEOLOGY; COMPOSITE MATERIAL; PROSTHESIS in the McGraw-Hill Encyclopedia of Science and Technology.

Arthur E. Clark; Larry L. Hench

Bibliography. L. L. Hench and A. E. Clark, in D. F. Williams (ed.), *Biocompatibility of Orthopaedic Implants*, vol. 2, pp. 129–170, 1982; L. L. Hench and J. Wilson, Surface active biomaterials, *Science*, 226:630–636, 1984.

Blood

Research on the system for dissolving blood clots has accelerated recently because of the outstanding progress that has been made in gene cloning and in expression and engineering of the enzymes that catalyze activation of clot-dissolving mechanisms. It is anticipated that these biotechnological developments will allow commercial production of extremely effective agents for dissolving intravascular clots. In order to design proper agents for this purpose, an understanding of the interactions that are important in the naturally occurring fibrinolytic system is necessary; the most relevant of the interactions are summarized in the **illustration**.

Plasminogen activators. Plasmin, the enzyme that is capable of catalyzing dissolution of the fibrin clot, exists in solution in an inactive form, plasminogen. The conversion of plasminogen to plasmin is catalyzed by agents called plasminogen activators, which are of several different classes.

Tissue plasminogen activator. One type, normally extrinsic to blood, is termed tissue plasminogen activator (t-PA), several representatives of which have been isolated from normal and neoplastic cells. Tissue plasminogen activator is synthesized and secreted from cells as a single-chain protein, containing 527 amino acid residues and a leader sequence of 35 amino acid residues, the first 25 of which are a signal polypeptide. Certain variants of this molecule have been discovered in various cell types after processing: a form that is three residues longer (L form) on the amino terminus than the remainder of the molecules (S form); a disulfide bond-stabilized, two-chain molecule which arises from cleavage of the peptide bond linking arginine-278 with isoleucine-279; a molecule that contains carbohydrates attached to three of the four potential asparagine-based glycosylation sites (type I); and a molecule that is glycosylated only on two of these four sites (that on asparagine-187 is never present). The enzymatic differences between these variants have not been fully investigated, but it is known that the enzymatic activity of the one-chain form is equivalent to that of the two-chain form in the presence of the positive effector molecule, fibrin. More basic research is needed in defining the relationships between the structure and function of t-PA in order to produce the most efficacious drug for blood clot dissolution. An excellent approach toward such enlightenment is the characterization of strategically generated deletion and insertion variants of t-PA by recombinant deoxyribonucleic acid techniques.

An important property of the t-PA class of enzymes is its ability to interact with the blood clot and, concomitantly, to stimulate plasminogen activation on the clot surface. The plasmin that is formed in this process catalyzes clot digestion, after which plasmin is released into blood, where it is inactivated as a result of binding to its primary plasma inhibitor, α_2-antiplasmin. Another physiological inhibitor of plasmin is present in plasma, α_2-macroglobulin. While not the major plasma inhibitor of plasmin, it nonetheless may function to some extent in this capacity. The t-PA that is also released from the clot surface, after clot lysis, is most likely neutralized in plasma by another, less understood protein inactivator. None of these inhibitors are believed to inactivate plasmin or tissue plasminogen activator when these two enzymes are bound to the clot surface.

Urokinase. A separate class of extrinsic plasminogen activators is represented by the widely studied enzyme urokinase (u-PA). While this protein is synthesized as a single-chain molecule (scu-PA), it has been isolated in various forms. High-molecular-weight, two-polypeptide-chain u-PA arises from scu-PA by cleavage of the peptide bond between lysine-158 and lysine-159, the two resulting peptide chains stabilized by disulfide bonds. Another variant, low-molecular-weight, two-chain u-PA, is formed as a result of cleavage of the peptide bond in the high-molecular-weight, two-chain form that links lysine-135 and lysine-136. Here, a 135-residue portion of the original protein is excised from the molecule.

Like t-PA, the u-PA classes are present in a variety of cells and tissues, but the two-chain forms possess at least one important difference from the t-PA class, that is, their relative inability to bind to the fibrin clot and to have their enzymatic activities stimulated by fibrin. The result of such a property is that u-PA catalyzes activation of plasminogen in the fluid phase of plasma, a process that provides plasmin for its other functions, for example, those involved in extracellular proteolysis. In such cases, in order to function, both plasmin and u-PA must be available at concentrations that overcome the levels of naturally occurring inhibitors that are present in the physiological fluid, cell, or organ wherein plasmin exerts its relevant activity. Recently, an extremely valuable discovery was made that the scu-PA form of urokinase is stimulated in plasma by fibrin. It is not certain at this stage of development of this specific area whether this is due to direct interaction of scu-PA with fibrin or to another, less understood mechanism. However, the fibrin-induced stimulation of the clot-dissolving ability of scu-PA in plasma provides a great deal of encouragement that fibrin-stimulated drugs that are alternatives to t-PA may be discovered.

Intrinsic activators. There are activators of plasminogen that are naturally present in blood plasma. Some, such as factor XII (Hageman factor)–dependent activators, are generated as a result of the surface interactions that are necessary to form blood clots, via the clotting system intrinsic to plasma, and others exist that are factor XII–independent. The former enzymes possess poor plasminogen activation properties, and their importance is questionable. The latter activators may well be small amounts of t-PA and u-PA that are released into plasma from cells or tissues. In fact, a recent study indicates that one factor XII–independent activator found in plasma was related to scu-PA.

Streptokinase. A final class of plasminogen activators is best represented by the bacterial protein streptokinase. This protein is a catabolic by-product of certain strains of β-hemolytic streptococci and is present in conditioned media of bacterial cultures. Streptokinase catalyzes activation of

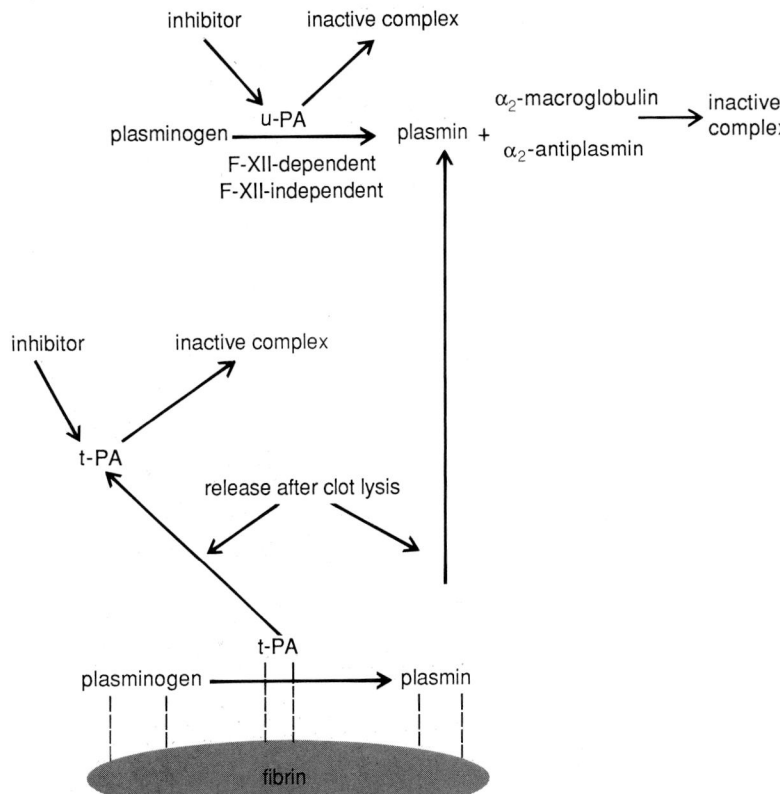

Important interactions in the naturally occurring fibrinolytic system. u-PA = urokinase. t-PA = tissue plasminogen activator.

plasminogen by an indirect mechanism and functions with plasminogen both bound to fibrin and present in the fluid phase. This agent has been useful in the past in thrombolytic therapy.

Thrombolytic therapy. For a number of reasons the most desirable clot-dissolving agent would be a substance that occurs naturally in humans, since immune reactions would be avoided. In addition, a suitable activator would allow specific activation of plasminogen bound to the clot, permitting maintenance of plasma fibrinogen at near-normal levels and preventing the plasma fibrinogen from being digested by high levels of circulating plasmin that would result from systemic activation of plasminogen. Other desirable properties of a suitable activator for thrombolytic therapy include a long circulatory lifetime, minimizing the cost of therapy; prevention of reocclusion of the vessel after clot dissolution; and rapid action, since fresh clots are most responsive to such therapy.

All of the plasminogen activators that have been described thus far possess some, but not all, of these properties. Two-chain urokinase is a naturally occurring and effective plasminogen activator, but as a result of its infusion into patients, a lytic state occurs wherein circulating fibrinogen is depleted and a tendency to bleed is established. Streptokinase has been employed effectively as a thrombolytic agent, but in addition to its deficiencies similar to those of two-chain urokinase, repeated infusion produces immune reactions. Single-chain urokinase has great potential, since it is stimulated by the fibrin clot and has very low activity toward plasminogen activation in solution. However, work is just beginning on this recently discovered agent, and its ultimate effectiveness in clot dissolution awaits clinical trials. To date, the most promising thrombolytic agent is tissue plasminogen activator, since it possesses the important property of clot dissolution with minimal systemic fibrinogen depletion. The agent that is currently undergoing clinical trials has a disadvantage; when it is infused in patients, some degree of fibrinogen depletion occurs, although less than with other activators.

Second-generation forms of tissue plasminogen activator are presently at various stages of development. Possible improvements include production of a protein that has been cloned from a human nontransformed cell line, is not as susceptible to circulating inhibitors, possesses higher clot-binding tendencies, and has enhanced plasminogen activation capability. Many of these features may be genetically engineered into the protein, and much work is under way based on this approach. Other desirable ways to obtain effective thrombolytic agents include development of drugs that stimulate, in the patient, endothelial cells to produce higher quantities of tissue plasminogen activator or lower inhibitor concentrations; discovery of molecules that are synergistic with tissue plasminogen activator and allow it to function at lower concentrations; and development of monoclonal antibodies that direct plasminogen activators to the clot or away from the plasma inhibitors. Given the large array of methods available and the great attention this field is currently receiving, it is likely that an effective thrombolytic agent will be available in the near future.

Chemical basis of plasminogen activation. Human plasminogen (HPg) is a large protein containing 790 amino aids in known sequence and one or two carbohydrate chains. This molecule is inactive and is converted to the clot-dissolving enzyme plasmin (HPm) as a result of activator-catalyzed cleavage of the peptide bond linking arginine-560 and valine-561 in plasminogen. As a result of this reaction, conformational alterations occur in the protein that provide the enzyme active site. All activators described above directly possess the enzymatic activity needed for this cleavage reaction, except streptokinase (SK), which does so in an indirect manner after forming a complex with plasminogen (activator formation). Small levels of this latter complex undergo a series of reactions, all occurring within the complex, yielding an enzyme that converts plasminogen to plasmin (plasmin formation). This activation scheme is shown. Here, after formation of the initial SK-HPg complex, a conformational alteration occurs in the HPg moi-

$$SK + HPg \leftrightarrow SK\text{-}HPg \leftrightarrow SK\text{-}HPg^* \leftrightarrow$$
$$SK\text{-}HPg' \to SK\text{-}HPm \leftarrow SK + HPm$$

ety, yielding one form of a plasminogen activator, SK-HPg*. A further conformational rearrangement occurs, providing another activator, SK-HPg'. Both activators are relatively unstable, and a final form of the activator complex, containing equimolar levels of SK and HPm, is produced, which can also arise from combination of SK and HPm. Catalytic levels of any of these activator species are capable of converting plasminogen to plasmin in a second stage of the reaction.

For background information *see* Blood; Enzyme; Fibrinogen; Plasmin; Thrombosis in the McGraw-Hill Encyclopedia of Science and Technology.

Francis J. Castellino

Bibliography. F. J. Castellino, An example of a unique enzyme-protein-substrate modifier reaction: Plasmin-streptokinase interaction, *Trends Biochem. Res.*, 4:1–5, 1979; B. A. K. Chibber et al., Rapid formation of an anion-sensitive active site in stoichiometric complexes of streptokinase and human plasminogen, *Proc. Nat. Acad. Sci. USA*, 83:1237–1241, 1986; D. Collen et al., Activation of plasminogen by pro-urokinase, II. Kinetics, *J. Biol. Chem.*, 261:1259–1266, 1986; D. P. Malinowski et al., Characterization of a complementary deoxyribonucleic acid coding for human and bovine plasminogen, *Biochemistry*, 23:4243–4250, 1984; D. Pennica et al., Cloning and expression of human tissue-type plasminogen activator, *Nature*, 310:214–221, 1983.

Branhamella

The genus *Branhamella*, in the family Neisseriaceae, has only one species, *B. catarrhalis*. The species was previously called *Neisseria catarrhalis*, but because of its low degree of deoxyribonucleic acid (DNA) homology with other *Neisseria* species, it was transferred to a new genus, *Branhamella*.

Branhamella catarrhalis is a common colonizer of the upper respiratory tract in humans. Quite recently the suggestion was made that it has pathogenic properties as well, and it has been implicated as a causative agent of a variety of respiratory infections both in children and in adults.

It is an aerobic gram-negative diplococcus whose morphological characteristics resemble those of the Neisseriae. It does not oxidize any carbohydrates, but possesses a deoxyribonuclease and a nitratase. *Branhamella catarrhalis* strains are sensitive to several antibiotics, including cephalosporins, erythromycin, chloramphenicol, and tetracycline. Conventionally *B. catarrhalis* strains have also been considered to be sensitive to penicillin and ampicillin. However, at present most disease isolates in different parts of the world produce β-lactamase and are therefore resistant to these β-lactame antibiotics.

Human infections. *Branhamella catarrhalis* is found to be a relatively common culture in middle-ear aspirates obtained from children with acute otitis media. Several studies have suggested that it is the causative agent of from 6 to 25% of all cases of otitis media. This would make it the third most common bacterium in this infection, following the pneumococcus and *Haemophilus influenzae*. *Branhamella catarrhalis* has also been frequently isolated from sinus aspirates of patients with acute maxillary sinusitis; in this infection, as well, *B. catarrhalis* follows the pneumococcus and *H. influenzae* in importance.

The significance of *B. catarrhalis* in lower respiratory infections has been more difficult to evaluate. Sputum cultures for etiological diagnosis have been of questionable value, and transtracheal aspiration is seldom performed. However, several recent reports have demonstrated that *B. catarrhalis* is a causative agent of lower respiratory infections. *Branhamella catarrhalis* has been found mostly in immunocompromised patients or in patients with some underlying disease such as chronic pulmonary disease, although there have been reports of pneumonia in otherwise healthy adults. Some recent studies suggest that *B. catarrhalis* is a significant pathogen of the lower respiratory tract in young children and elderly patients, but is found infrequently in young adults, for example, military recruits. Occasionally *B. catarrhalis* has been associated with a variety of other infections such as urethritis, meningitis, septicemia, endocarditis, or conjunctivitis.

The pathogenic mechanisms of *B. catarrhalis* infections are still poorly understood. Among the bacteriological features, deoxyribonuclease production is considered important for initiating inflammation on the mucous membranes.

When *B. catarrhalis* has been isolated from middle-ear aspirates, sputa, or sinus aspirates, these samples were typically purulent, containing a profusion of polymorphonuclear leukocytes. In addition, the bacteria were frequently seen intracellularly in the leukocytes. The most convincing evidence of the pathogenic role of *B. catarrhalis* is the demonstration of an antibody response to *B. catarrhalis* antigens during an ongoing infection. In addition to an antibody response in serum, the presence of *B. catarrhalis*–specific IgG and IgA antibodies in the middle ear during acute otitis media has also been determined.

A lower respiratory infection can be attributed to *B. catarrhalis* on the basis of a positive culture from the sputum or transtracheal aspirate or a demonstrated serum antibody response. The clinical and laboratory findings for such patients very clearly point to a bacterial infection, and include respiratory symptoms; moderate or high fever; pathological chest x-ray findings; and elevated levels of the white blood cell count (WBC), erythrocyte sedimentation rate (ESR), and serum C-reactive protein (CRP).

Bacteriological and biochemical characteristics. The specimens believed to contain *B. catarrhalis* should be cultured on blood or enriched chocolate agar plates and incubated for 1 to 2 days at 95°F (35°C) in an atmosphere containing 5% carbon dioxide (CO_2). After having incubated for 1 day, the *B. catarrhalis* colonies are grayish, nonpigmented, and 1 to 2 mm in diameter. *Branhamella catarrhalis* is oxidase-positive and catalase-positive and does not produce acid from glucose, maltose, sucrose, fructose, or lactose. The best identification tests for differentiating *B. catarrhalis* from *Neisseria flavescens* are nitrate reduction and production of deoxyribonuclease; *B. catarrhalis* reduces nitrate and produces deoxyribonuclease, whereas *N. flavescens* does not.

For background information SEE ANTIBODY; MEDICAL BACTERIOLOGY in the McGraw-Hill Encyclopedia of Science and Technology.

M. Leinonen; P. H. Mäkelä

Bibliography. B. W. Catlin, Transfer of the organism named *Neisseria catarrhalis* to *Branhamella* gen. nov., *Int. J. Syst. Bacteriol.*, 20(2):155–159, April 1970; J. J. Christensen, O. Gadeberg, and B. Bruun, *Branhamella catarrhalis*: Significance in pulmonary infections and bacteriological features, *Acta Pathol. Microbiol. Immunol. Scand.*, Sect. B, 94:89–95, 1986; G. V. Doern, *Branhamella catarrhalis*: An emerging human pathogen, *Clin. Microbiol. Newsl.*, 7(11):75–78, June 1985; M. Leinonen et al., Preliminary serologic evidence for a pathogenic role of *Branhamella catarrhalis*, *J. Infect. Dis.*, 144(6):570–574, December 1981.

Celestial mechanics

The solar system, the "birthplace" of classical mechanics, is generally perceived as evolving with clockwork regularity. Yet Newton's equations of motion possess irregular solutions as well as regular solutions. There are situations in the solar system where irregular solutions play an important role. Saturn's satellite Hyperion has a highly irregular rotation, the distribution of asteroids has been significantly modified in regions where trajectories are irregular, and meteorites are transported from the asteroid belt to Earth along irregular trajectories.

Chaotic behavior. Irregular trajectories of deterministic equations of motion are termed chaotic. A distinguishing feature of chaotic behavior is that chaotic trajectories show a sensitive dependence on the initial conditions. The distance between chaotic trajectories which are initially near each other grows, on the average, exponentially with time. This is to be contrasted with an average linear growth in the separation of nearby regular trajectories.

For conservative dynamical systems (those without friction or other sources of energy dissipation), chaotic trajectories are also distinguished from regular trajectories by the number of independent constants of the motion. Chaotic trajectories possess fewer constants of the motion. In conservative systems, some initial conditions lead to regular behavior while others lead to irregular behavior. Almost all conservative dynamical systems behave irregularly for some initial conditions.

While the orbits and spins of the natural satellites are modified over very long times by tidal friction, on shorter time scales the motions of the planets and their satellites are well approximated by conservative equations of motion.

Chaotic tumbling of Hyperion. The chaotic tumbling of Hyperion, one of Saturn's more distant satellites with an orbital period of 21 days, offers one of the most dramatic physical examples of chaotic behavior. The rotation rate and spin-axis orientation undergo significant changes in only a few orbital periods. This chaotic tumbling is primarily a consequence of the highly aspherical shape of Hyperion, which was determined from images transmitted by *Voyager 2* to measure 255 mi × 160 mi × 135 mi (±10 mi) or 410 km × 260 km × 220 km (±15 km), and to a lesser extent a consequence of the large eccentricity of Hyperion's orbit (approximately 0.1). Weak tidal friction acting over the age of the solar system is responsible for bringing Hyperion to this chaotic state.

The Earth's Moon always points the same face toward the Earth; the equality of the rotation period and orbital period of the Moon is a natural consequence of the action of tidal friction. Tidal friction tends to bring the spin axis into coincidence with the axis of largest moment of inertia, and over longer times brings the spin axis perpendicular to the orbital plane as the rotation rate is slowed until the rotation period equals the orbital period. Except for Hyperion, all satellites in the solar system which are sufficiently close to their host planet for the tidal torques to have been strong enough to significantly affect the rotation rate over the age of the solar system are observed to be in this state where the spin period is locked to the orbital period. Estimates of the time scale for tidal despinning indicate that Hyperion should have a rotation rate comparable to that required to always point one face toward Saturn; Hyperion's rotation rate has been significantly affected by tidal friction.

Gravitational forces decrease with the inverse square of the distance between the masses. Thus the part of the satellite nearer the planet is attracted more strongly than the part on the far side. If the satellite is spherically symmetrical, no net torque can be placed on the satellite. However, planets and satellites are never perfectly spherical, and the deviation from spherical symmetry together with the nonuniform gravitational field of the planet gives rise to a torque on the satellite. If the rotation axis is fixed perpendicular to the orbital plane (the usual outcome of tidal

evolution), the equation of motion for the orientation of a satellite, given below, is obtained by

$$C \frac{d^2\theta}{dt^2} = -n^2(B - A) \frac{3}{2} \left(\frac{a}{r}\right)^3 \sin 2(\theta - f)$$

setting the applied torque equal to the product of the moment of inertia and the acceleration of the orientation. In this equation, θ is the angle between the principal axis of least moment of inertia (usually the longest axis) and the major axis of the orbit, a is the length of the semimajor axis of the orbit, r is the distance from planet to satellite, f is the angular position of the satellite in its orbit measured from the point of closest approach to the planet, n is the mean angular motion of the satellite in its orbit, and $A \leqq B \leqq C$ are the principal moments of inertia of the satellite (**Fig. 1**). That the torque depends on the

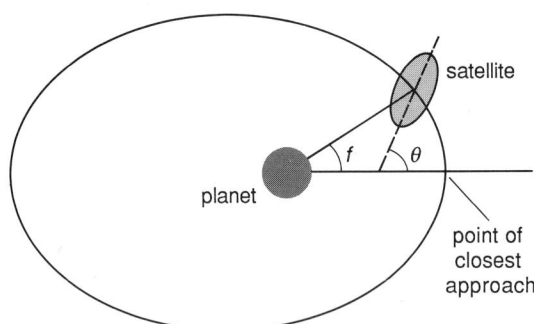

Fig. 1. Specification of the orientation and position of a satellite whose spin axis is fixed perpendicular to the orbital plane. The orientation of the satellite is specified by the angle θ, and the position in the orbit is given by the angle *f*.

inverse cube of the distance reflects the fact that the torque arises from the differential strength of the gravitational force across the satellite. The difference of principal moments of inertia, $B - A$, is larger for more out-of-round bodies. Hyperion, which is nearly twice as long as it is wide, is subject to especially large torques.

The character of a particular trajectory and the structure of the space of initial conditions is most easily determined by computing a Poincaré surface of section. Rather than examining the full time development of a trajectory, the trajectory is sampled periodically. For the spin-orbit coupling problem the surface of section is generated by plotting the rate of change of the orientation versus the orientation at each passage of the satellite through the point in its orbit closest to the planet. The sequence of points which emerges reveals the character of the trajectory. If the trajectory is regular, the constants of the motion will constrain the points to fall on a curve; if the trajectory is chaotic, the points will fill an area in an apparently random manner. The surface of section for Hyperion (**Fig. 2**) shows an intermingling of regular and irregular trajectories. All of the scattered points in the large chaotic zone in the center of the figure belong to the same trajectory.

Stability analysis shows that the orientation of the spin axis perpendicular to the orbital plane is unstable in the large chaotic zone. The state in which all other tidally evolved satellites in the solar system are found, the synchronous state, lies within this zone and is also attitude-unstable for Hyperion. Over the age of the solar system, tidal friction brought the spin axis of Hyperion perpendicular to the orbital plane, and slowed the spin to a value nearly synchronous with the orbital mean motion. Once the large chaotic zone was reached, the orientation of the spin axis perpendicular to the orbital plane became unstable and Hyperion began to tumble. Examination of the rate of separation of nearby trajectories shows the tumbling motion to be fully chaotic.

Observations of Hyperion are not yet adequate to fully confirm the chaotic tumbling of the satellite, though they are all consistent with it. The most convincing evidence for chaotic tumbling comes from the *Voyager* pictures, which show that the long axis of Hyperion is out of the orbit plane and the spin axis is near the plane. This is consistent with chaotic tumbling but inconsistent with other known regular rotation states.

Distribution of asteroids. The existence of gaps in the distribution of the asteroids near resonances with Jupiter has puzzled astronomers for over 100 years. A resonance occurs when the orbital period of the asteroid is commensurate with the orbital period of Jupiter. For example, there is a gap in the distribution of asteroids where the asteroid period would be close to one-third the orbital period of Jupiter. Since the gaps are associated with resonances, it has always been clear that resonance dynamics must play a role in their formation, but the precise mechanism eluded discovery. The basic difficulty was the lack of analytic understanding of the dynamics of complex resonances which could not be alleviated through numerical simulations because of the great amount of computer time required. The development of an approximate numerical method for following the evolution of an asteroid trajectory which is on the order of 1000 times faster than conventional methods enabled essential new features in the dynamics to be discovered. A large portion of the trajectories near the 3/1 resonance are chaotic, and when viewed over very long times the eccentricities of the chaotic trajectories have rather remarkable behavior (**Fig. 3**). The eccentricity may remain low for several million years and then suddenly increase to large values (greater than 0.3). After a burst of high-eccentricity behavior the asteroid may again spend a period of time at low eccentricity. The behavior of the eccentricity is important since asteroids on such orbits cross the orbit of Mars for eccentricities above 0.3. The true character of asteroid trajectories becomes apparent only when they

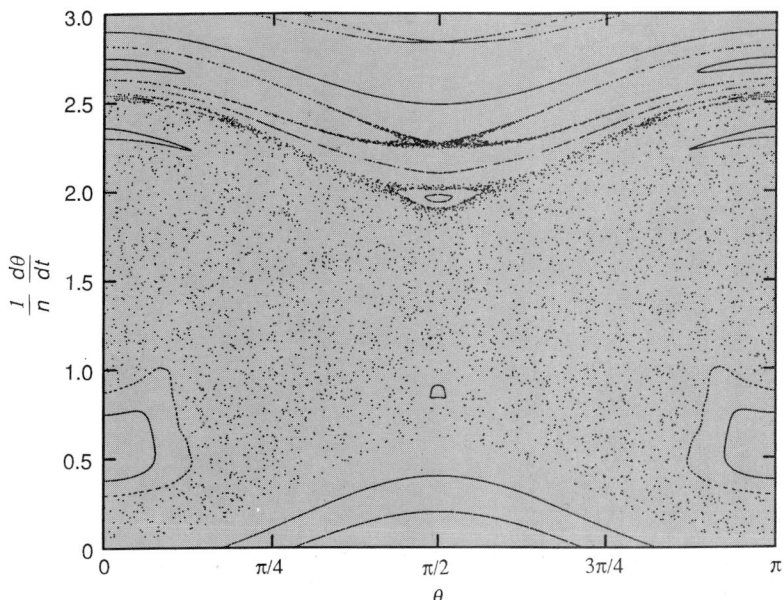

Fig. 2. Surface of section for Hyperion. The rate of change of the orientation is plotted versus the orientation every time the satellite passes through the point of closest approach to the planet.

are studied over hundreds of thousands or millions of years. This gap in the distribution of asteroids can be explained by the removal of the Mars- and Earth-crossing asteroids through collisions or close encounters. A detailed examination of the distribution of asteroids near the 3/1 resonance shows excellent agreement with the boundary of the chaotic zone (**Fig. 4**). It is not yet

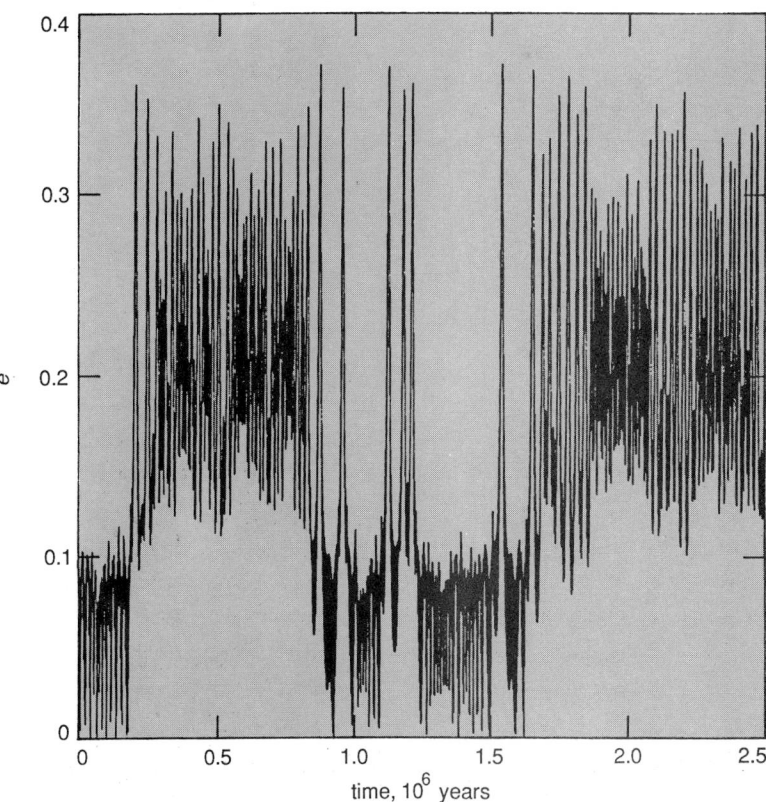

Fig. 3. Irregular behavior of the eccentricity e of an asteroid orbit near the 3/1 resonance with Jupiter. For this computation the motion is restricted to the plane of Jupiter's orbit.

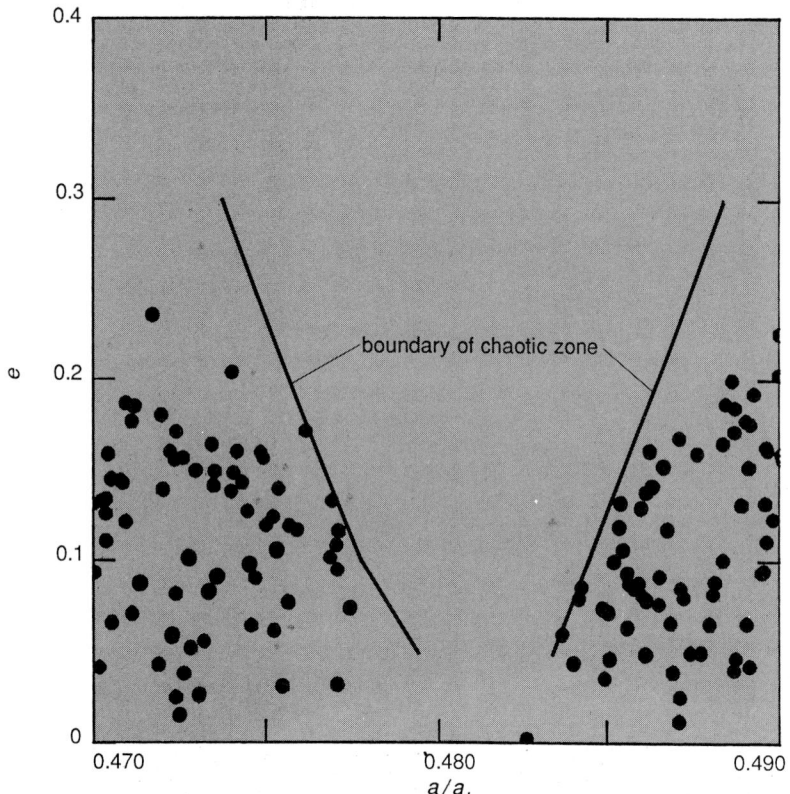

Fig. 4. Comparison of the actual distribution of asteroids to the outer boundary of the chaotic zone, showing excellent agreement. The eccentricity *e* of the asteroids is plotted versus the ratio of the semimajor axis *a* to the semimajor axis of Jupiter a_J. Regular and irregular trajectories within this boundary are removed through collisions or close encounters with Mars or Earth.

For background information SEE ASTEROID; CELESTIAL MECHANICS; DETERMINISM; METEORITE; MOON; SATURN in the McGraw-Hill Encyclopedia of Science and Technology.

Jack Wisdom

Bibliography. J. Wisdom, Chaotic behavior and the origin of the 3/1 Kirkwood gap, *Icarus*, 56:51–74, 1984; J. Wisdom, Meteorites may follow a chaotic route to Earth, *Nature*, 315:731–733, 1985; J. Wisdom, S. J. Peale, and F. Mignard, The chaotic rotation of Hyperion, *Icarus*, 58:137–152, 1984.

known whether a similar mechanism can account for all of the gaps in the distribution of asteroids, though preliminary evidence indicates that this should be possible.

Transport of meteorites. It is rather surprising that the origin of the meteorites, those stones which contain so many clues concerning the formation of the solar system, is still not definitively known. It is widely believed that meteorites originate in the asteroid belt; yet until recently a dynamical mechanism for transporting the meteorites from the asteroid belt to the Earth which is consistent with the meteorite data eluded discovery. In computing the trajectory shown in Fig. 3, only the effects of Jupiter were taken into account and the motion was artificially restricted to the plane of Jupiter's orbit. If full three-dimensional motion is allowed and the effects of the major planets are included, the eccentricities of chaotic trajectories near the 3/1 resonance reach even larger values (greater than 0.6), large enough that the trajectories directly cross the orbit of the Earth. These chaotic trajectories provide the necessary dynamical mechanism for bringing debris from asteroidal collisions near the 3/1 resonance to the Earth. This mechanism has been shown to be consistent with the known properties of the meteorites. The larger fragments partially account for the observed population of Earth-crossing asteroids.

Cellular immunology

The immune system has evolved to protect the organism from foreign pathogens such as bacteria or viruses or from cells that have become transformed. Lymphocytes recognize foreign antigens by their difference from self. B lymphocytes, constituting the humoral immune response, proliferate in response to specific antigens and differentiate to secrete antibodies called immunoglobulins. The cellular immune response is mediated by different populations of T (thymus-derived) lymphocytes. In response to the recognition of specific antigens, helper and suppressor T cells regulate the response of B lymphocytes and other T cells. Cytolytic T lymphocytes recognize foreign antigens on the surface of a cell and are able to destroy the cell. Thus cytolytic T lymphocytes play a critical role in the host immune response to virally infected cells, to cancer cells, and in allograft rejection.

T cells express an antigen-specific receptor on their surface. Cytolytic T lymphocytes recognize either foreign (allogeneic) proteins encoded in the major histocompatibility complex or foreign antigens in the context of self (syngeneic) major histocompatibility complex proteins. On the cell surface the T-cell receptor is physically associated with a nonpolymorphic glycoprotein termed CD3 (T3, Leu-4) which is thought to transduce a transmembrane signal from the cell surface to the cytoplasm of the cell. Triggering of the cytolytic T lymphocyte involves not only T-cell receptor binding to its specific ligand but interactions of a number of cell surface molecules. The functional importance of these surface molecules has been inferred from the ability of monoclonal antibodies directed against them to inhibit cytolysis. These molecules include lymphocyte function–associated antigen-1 (LFA-1), the sheep red blood cell receptor (CD2), LFA-3, CD4, and CD8, and will be discussed in detail below. Each may contribute to the avidity (strength) with which a cytolytic T lymphocyte binds to its target cell, although the mechanism by which each molecule functions may differ.

The mechanism by which cytolytic T lymphocytes destroy or lyse target cells may be divided into several independent stages (**Fig. 1**). Initially, cytolytic T lymphocytes bind to the target cell, an event termed adhesion. Antigen recognition depends on the cytolytic T-lymphocyte T-cell receptor interacting with a specific ligand on the target cell. Multiple receptor–ligand interactions regulate adhesion and antigen recognition and may contribute to T-cell triggering. Concurrent with cytolytic T-lymphocyte activation, unidirectional T-lymphocyte cytoplasmic rearrangement may be observed. This results in the release of lytic granules—a calcium-dependent process, referred to as effector granule exocytosis. Once the granules have been released, the cytolytic T lymphocyte is no longer required for target cell lysis. The target cell is lethally injured; an increase in membrane permeability to small molecules results in cell swelling with eventual rupture and cell death. The cytolytic T lymphocyte may detach and sequentially lyse other target cells.

Adhesion. Effective cytolysis mediated by cytolytic T lymphocytes requires intimate cell–cell contact. Conjugate formation between cytolytic T lymphocytes and target cells occurs independently of recognized specific antigen. This adhesion is regulated by a number of cell surface interactions at the interface between the T cell and the target cell. Conjugate formation may be inhibited by monoclonal antibodies directed against LFA-1, CD2, and LFA-3.

LFA-1 is a broadly distributed leukocyte antigen composed of a 180,000-dalton alpha chain and a 95,000-dalton beta chain. Antibody to LFA-1, in the absence of complement, inhibits a variety of T-cell functions, including alloantigen- and mitogen-stimulated T-cell proliferation, and cytolytic T lymphocyte–mediated and natural killer–mediated cytotoxicity. Anti-LFA-1 monoclonal antibody inhibits conjugate formation between cytolytic T lymphocytes and target cells, as well as natural killer cells and target cells; and may be shown to inhibit homotypic adhesions between an Epstein-Barr virus–transformed B-cell line, suggesting that the molecule may function to strengthen cell adhesion. A ligand for LFA-1 has not definitively been identified.

CD2 is a 50,000-dalton molecule expressed on all cells of T-cell lineage and is the earliest thymic differentiation antigen identified. The functional consequences of anti-CD2 monoclonal antibody are critically dependent on the specific antigenic determinant (epitope) recognized by the monoclonal antibody. Certain anti-CD2 monoclonal antibodies (anti-LFA-2, Leu-5, 9.6) may inhibit T-cell conjugate formation and T-lymphocyte–mediated cytolysis, while other monoclonal antibodies (9.6) inhibit natural killer–mediated cytolysis. A number of monoclonal antibodies inhibit T-cell proliferation induced by mitogens or alloantigens.

LFA-3 is a broadly distributed glycoprotein of approximately 60,000 daltons initially identified by the ability of anti-LFA-3 monoclonal antibody

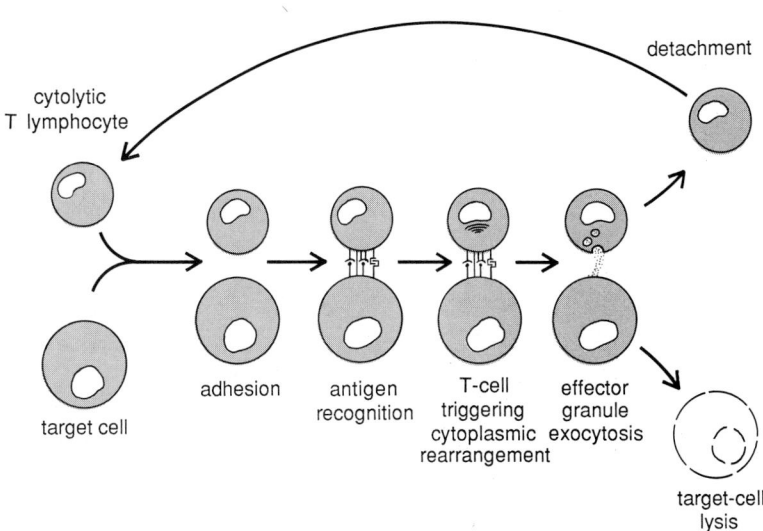

Fig. 1. Stages in the destruction (lysis) of target cells.

to inhibit T-lymphocyte–mediated cytotoxicity. Anti-LFA-3 monoclonal antibody inhibits conjugate formation, T-lymphocyte–mediated cytotoxicity, and T-cell proliferation by binding to the target (or stimulator) cell, not by binding to the effector cell, suggesting it may be a ligand for a T-cell surface molecule. Murine homologs for CD2 and LFA-3 have not yet been identified.

Recent evidence suggests that CD2 on the T cell may be the receptor for LFA-3 on the target cell. Antigen-nonspecific conjugate formed in the absence of divalent cations (principally magnesium) and at 4°C may be blocked by either anti-CD2 or anti-LFA-3 monoclonal antibody. Inhibition by anti-LFA-1 monoclonal antibody may be seen by conjugates formed in the presence of divalent cations and at 37°C. This suggests that two adhesion-dependent pathways for conjugate formation exist: CD2 and LFA-3 molecules are involved in one interaction and are distinct from LFA-1–dependent interactions.

Antigen recognition. Cytolytic T lymphocyte–target cell adhesion either precedes or occurs concurrently with antigen recognition (**Fig. 2**). Antigen

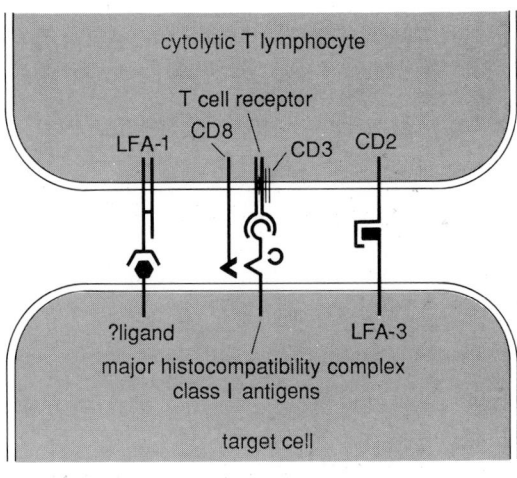

Fig. 2 Schematic diagram of cytolytic T lymphocyte–target cell adhesion and recognition.

binding to the T-cell receptor/CD3 complex is required for cytolytic T lymphocyte triggering. Monoclonal antibody directed against the T-cell receptor or CD3 may be shown to inhibit recognition and lysis. Monoclonal antibodies directed against appropriate target cell antigens (major histocompatibility complex class I or class II) also inhibit recognition and subsequent lysis. Monoclonal antibody directed against CD4 (T4, Leu-3 in the human; L3T4 in the mouse) and against CD8 (T8, Leu-2 in the human; Lyt 2 in the mouse) may also, under certain conditions, inhibit antigen recognition. CD4 and CD8 are expressed on mutually exclusive populations of mature T cells. The phenotypic expression of CD4 or CD8 on the T cell appears not to correlate with T-cell function (cytolytic T lymphocyte versus helper cell) but with the major histocompatibility class that the T-cell receptor recognizes. In other words, $CD8^+$ T cells generally recognize foreign antigens in the context of class I proteins and may be inhibited by anti-CD8 monoclonal antibody. $CD4^+$ T cells are restricted by class II proteins and may be blocked by anti-CD4 monoclonal antibody. This has led to the hypothesis that the CD8 and CD4 molecules are receptors that bind to nonpolymorphic determinants of class I and class II antigens, respectively. The T-cell receptor interacts with polymorphic regions of the major histocompatibility complex molecule.

T-cell triggering. Cytolytic T lymphocyte activation results from appropriate multivalent interactions at the cell surface, leading to a transmembrane signal. Experimentally, antigen recognition may be bypassed by incubating the cell with anti-CD3 monoclonal antibody, directly triggering the cell. It is also possible to activate the cell via the CD2 molecule. Upon activation, the microtubule-organizing center orients between the cytolytic T lymphocyte nucleus and the target cell. Other cytoplasmic proteins, notably talin, have been shown to orient nonrandomly in the direction of lysis. T-cell triggering and cytoplasmic rearrangement is dependent on the presence of calcium and sensitive to inhibitors of energy metabolism.

Following reorientation of the microtubule-organizing center, granules are released into the space between the cytolytic T lymphocyte and the target cell. These granules contain a pore-forming complex, termed perforins or cytolysins, and several proteases, including serine esterase. Granule exocytosis is sensitive to chloroquine and monensin and to protease inhibitors.

Target cell lysis. Perforins released from the granules into the cleft are thought to intercalate, in the presence of calcium, in the membrane of the target cell. These molecules may form ring-like structures which alter the membrane permeability of the target cell, allowing the membrane to become permeable to small ions. Whether insertion of perforins or cytolysins is sufficient or whether the actual formation of the ringlike structures is necessary for target cell lysis is presently under investigation. Colloid osmotic lysis then ensues with irreversible swelling of the target cell. Lysis is unidirectional: the cytolytic T lymphocyte may detach or de-adhere and participate in repeated lytic interactions.

For background information SEE ANTIBODY; ANTIGEN; CELLULAR IMMUNOLOGY; IMMUNOLOGY; MONOCLONAL ANTIBODIES in the McGraw-Hill Encyclopedia of Science and Technology.

Steven Burakoff; Barbara E. Bierer

Bibliography. S. J. Burakoff et al., A molecular analysis of the cytolytic T lymphocyte response, *Adv. Immunol.*, 36:45, 1984; P. A. Henkart, Mechanism of lymphocyte-mediated cytotoxicity, *Annu. Rev. Immunol.*, 3:31, 1985; E. Martz, Mechanisms of specific tumor cell lysis by alloimmune T lymphocyte: Resolution and characterization of discrete steps in the cellular interaction, *Contemp. Top. Immunobiol.* 7:301, 1977; W. E. Paul, *Fundamental Immunology*, 1984.

Cephalopoda

Cuttlebone is the vernacular name for the internal, calcified, chambered shell of species of *Sepia* and *Sepiella*, genera of the decapodan family Sepiidae of the exclusively marine molluscan class Cephalopoda. Like the swimbladder of fish, the cuttlebone contains gas and provides buoyancy to counterbalance the tissue weight of the animal in water. In contrast to the swimbladder, however, the cuttlebone is rigid and can be crushed by high external pressures. This limits the depth distribution of cuttlefishes in the sea; according to the specific strength of their bone, they can live in greater or lesser depths. Some species like *Sepia officinalis*, the common cuttlefish of the European coasts, must stay within the upper 200 m (640 ft), whereas others can live at depths of several hundred meters.

Phyletic precursors. In order to understand the particular design of the cuttlebone, it is useful to glance at other shell constructions found in both recent and fossil cephalopods. Indeed, fossil shells seem to indicate that the evolution of cephalopods started in the Upper Cambrian with a morphological remodeling (based on a thoroughly altered developmental program) of the shell complex: the inner, apical part of the roughly conical, limpetlike shell became subdivided into distinct chambers by the periodical formation of septa produced by a specialized epithelium of the visceral mass. The apical attachment of this epithelium to the inner surface of the embryonic shell was maintained throughout shell growth, so that all subsequent chambers were built around a tissue extension of the visceral mass, known as siphuncular tissue. The tubular extensions of the septa, known as septal necks, together form the siphuncular tube of the chambered part (phragmocone) of the shell.

Among the extant cephalopods, only *Nautilus* has an external (not covered by the integument of the mantle) chambered shell similar to the coiled shells of the fossil ammonites. All other living cephalopods form a closed shell sac of varying size and morphology; however, in only two families, the Spirulidae and the Sepiidae (both of the decapodan order Sepiodea), does the shell-forming epithelium secrete calcium compounds along with the organic shell matrix known as conchiolin, which makes up all of the internal horny feather-shaped shell (pen or gladius) of living squids. The calcium carbonate crystals (in the form of aragonite) are built into the organic substrate to constitute lamellar, prismatic, and spherulitic structures.

In *Spirula* the calcified shell is truly chambered (starting with a roughly globular initial chamber) and coiled. More rapid growth on the dorsal side results in so-called endogastric coiling, which contrasts with the exogastric coiling of the *Nautilus* shell. In the cuttlebone of the Sepiidae, there are no actual chambers as in the shell of *Spirula*. There are only narrow spaces, less than 1 mm (0.04 in.) in height; the dividing septa are held apart by meandering walls or "pillars" (**Fig. 1**). Within these spaces, organic lamellae lying parallel to the calcified septa form intracameral floors. This arrangement looks very different from the *Spirula* shell with its spacious chambers. However, when compared with that part of a *Spirula* shell chamber where the tubular septal neck reaches into the opening of the next septal neck (which belongs to the preceding chamber), the architectural features appear strikingly similar. In expanding the morphological comparison to *Nautilus* shells and to the chambered shells of fossil caphalopods, there is a characteristic combination of closely spaced septal extensions held apart by supporting structures and accompanied by thin lamellae all along the siphuncular tube, that is, in that area where the siphuncular tissue is in contact with the septal necks. Thus, apparent homologies indicate that the lamellar part of the cuttlebone of *Sepia* and *Sepiella* is made entirely of septal necks flattened out against the so-called dorsal shield [the remainder of the ancestral conotheca (thin wall) surrounding the phragmocone]. The slight longitudinal curvature of the cuttlebone is the only vestige of the endogastric coiling of the (more proximate) ancestral shell type known from the fossil *Spirulirostra*.

Buoyancy control. The functioning of the siphuncular complex in living cephalopods has been analyzed, and it has been demonstrated that fossil chambered shells can be interpreted as buoyancy regulators. Although the cuttlebone of the Sepiidae lacks the spacious chambers typical of calcified cephalopod shells, it presents the active component of the weight-adjusting apparatus, the siphuncular complex, in a form that is particularly interesting for experimental work. The siphuncular tissue forms a wide membrane covering the posterior, so-called striated zone of the lower bone surface

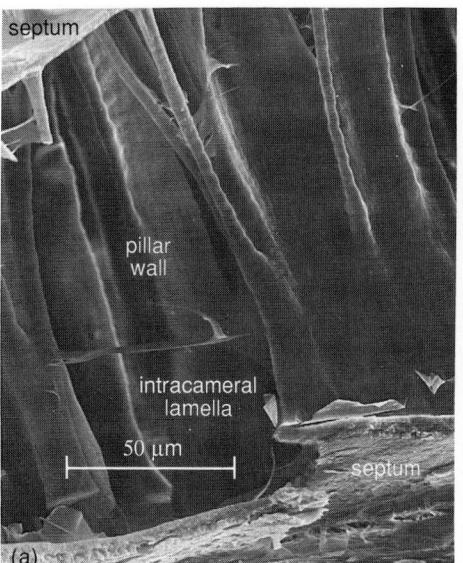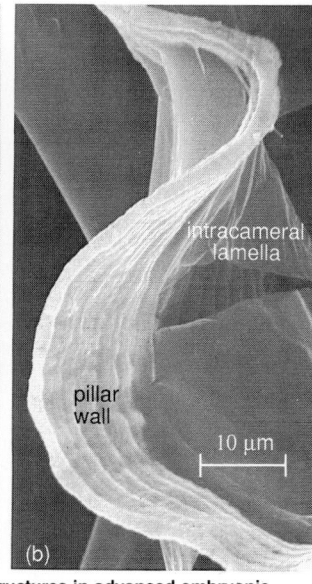

Fig. 1. Scanning electron micrographs of cuttlebone structures in advanced embryonic stage of *Sepia officinalis*. (*a*) Lateral view of a chamber opened up to expose the pillar wall and septum, both calcified; the intracameral lamella is purely organic. In adult cuttlebones, there are about a half dozen of these lamellae at regular intervals between the septa. (*b*) Surface view, at higher magnification, of a meandering pillar wall after removal of the overlying septum. Note the partly torn intracameral lamella.

(**Fig. 2**). The striae are the free posterior edges of the septa (or septal necks); they are entirely organic and allow fluids to move across them, in contrast to the calcified major part of the septum which forms a tight barrier between the chambers. When a new septum is formed by the secretory action of the epithelium adjacent to the siphuncular epithelium anteriorly, the liquid contained in the newly closed chamber will be extracted through the organic rim forming the posterior limit of the compartment. The newest complete chambers, which lie centrally along the length of the animal, are thus completely

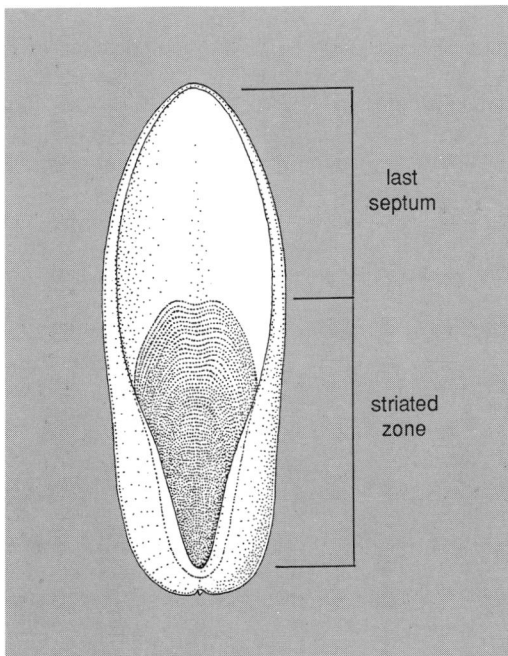

Fig. 2. Semischematic presentation of the lower (ventral) surface of a cuttlebone. In the living animal the bone is entirely enclosed in the so-called shell sac. The striated zone is covered by the siphuncular epithelium, which is continuous with the epithelium covering the last septum produced. Note the rim of the dorsal shield visible all around the chambered part of the cuttlebone.

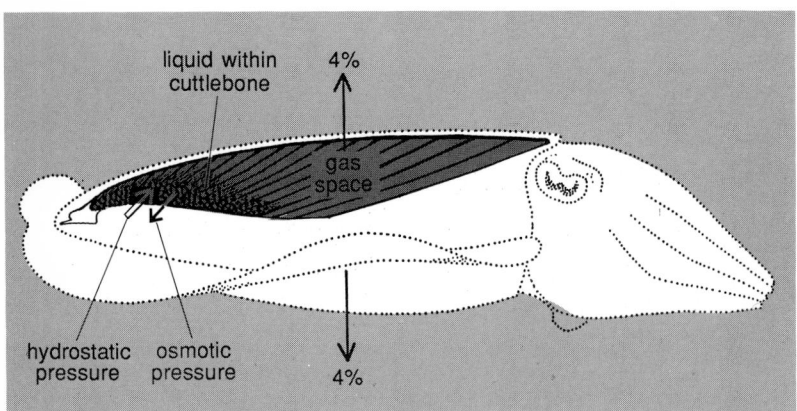

Fig. 3. Diagram showing a cuttlebone. The oldest and most posterior chambers are almost full of liquid. (*After E. J. Denton and J. B. Gilpin-Brown, Floatation mechanisms in modern and fossil cephalopods, Adv. Mar. Biol., 11:197–268, 1973*)

filled with gas. The hydrostatic pressure of the sea water is balanced by the osmotic pressure between cuttlebone liquid and the blood. In sea water the cuttlebone gives a net lift of 4% of the animal's weight in the air, thus balancing the excess weight of the rest of the animal (**Fig. 3**).

Experiments have shown that liquid is extracted from a chamber by so-called osmotic pumping achieved by the different concentrations of salts at the shell-tissue interface. In animals caught at 70 m (230 ft) depth and placed quickly under a hydrostatic pressure corresponding to that depth, it was found that the liquid within the chambers was everywhere markedly hyposmotic (that is, had a lower salt concentration in comparison) to the animals' blood. The difference in concentration between the liquid within the chambers and the blood was judged to be that which would give an osmotic pressure of approximately 7 atm (700 kilopascals) if placed across a suitable semipermeable membrane. This is the osmotic pressure that would be required to match the pressure of the sea at 70 m (230 ft) depth. Liquid thus removed is replaced by gas (97% nitrogen) which passively diffuses into the chamber and always remains at pressures lower than atmospheric, generally about 0.8 atm (80 kPa). The organic intracameral lamellae remain wettable in all chambers and allow readmission of liquid. Indeed, the earliest chambers are generally refilled in adult cuttlebones, and it has been suggested that this is important for the maintenance of the horizontal position of the body when the animal hovers above the bottom. This liquid filling may also serve to protect the early chambers against implosion when the adult animal descends to depths below which the same chambers would have imploded in juveniles. This problem is distinct from that of bringing the whole animal close to neutral buoyancy, however.

Design-related limits of habitation depth. The chambered cephalopod shell can be compared to modern submarines, which also have limited mechanical strength to withstand compression. The Bathyscaphe has been developed for extremely deep diving. In this vehicle, only the chamber that houses the navigators and observers is air-filled and required to resist high pressures; the necessary buoyancy is provided by thin-walled tanks containing a liquid lighter than sea water. This method of achieving neutral buoyancy is used by many modern-era squids, which store metabolic waste products lighter than sea water. Compared with the "old-fashioned" chambered shell design, fluid-derived buoyancy and the corresponding freedom that certain squids have to move into great depths may appear superior. However, the price paid for this advantage is the limitation imposed upon muscular development entailing limited swimming capacity. Indeed, some chambered-shell cephalopods are clearly competitive, where representatives of the alternative mode are absent. The Sepiidae are especially successful competitors of sepiolid and teuthoid squids and bottom-living octopuses in inshore waters as well as on deeper bottoms of the continental slopes. Nevertheless, there is a limit set by the shell design, and two questions arise: How can the mechanical strength of the cuttlebone be optimized? Do the animals actively avoid descending to implosion depth?

Mechanical strength. The first question can be answered partly on the basis of comparative observations involving implosion depth and implosion morphologies in different species of *Sepia*. It appears that mechanical strength of the cuttlebone increases with septal curvature and closer spacing of septa. Thus the relatively narrow cuttlebones of *S. orbignyana* and *S. elegans* withstand higher hydrostatic pressures (about 55 and 40 atm or 5500 and 4000 kPa, respectively) than the rather broad, flat bone of *S. officinalis*. In the last-mentioned species, adults withstand pressures of 15 to 20 atm (1500 to 2000 kPa), whereas early juvenile bones implode at pressures between 5 and 10 atm (500 and 1000 kPa). The importance of close spacing of septa can also be appreciated in a single species when implosion morphologies are compared between individuals that have grown under different alimentary regimes. Indeed, underfeeding in *S. officinalis* results in septal crowding that enhances mechanical strength. With increasing resistance against compression, the efficiency of the cuttlebone in buoyancy regulation will probably decrease. In evolutionary terms the ultimate mechanically desirable product, a very slender bone with nearly circular cross section, will provide little buoyancy and will still be more fragile than the highly elastic organic gladius of squids.

Active avoidance of implosion. The second question cannot yet be answered. However, coherent hypotheses may be attempted. There may be proprioceptors in the shell sac epithelium that record pressure-induced deformations of the bone, triggering a behavioral response of the animal to move upward; but such a mechanism would probably be imprecise and possibly keep the animal very far above the implosion depth. A conceivable alternative would be a bioacoustic alarm system that would process vibrations occurring just before compression failure. This would allow the animal to stay rather close to implosion depth, but would also carry a high risk related to navigational errors in rapid swimming. Elucidation of the problem of depth recording by cuttlefishes and other chambered-shell cephalopods is certainly one of the most challenging tasks in cephalopod biology today.

For background information *see* Cephalopoda; Decapoda (Mollusca); Nautilus; Sepioidea in the McGraw-Hill Encyclopedia of Science and Technology.

S. von Boletzky

Bibliography. K. Bandel and S. v. Boletzky, A comparative study of the structure, development and morphological relationships of chambered cephalopod shells, *Veliger*, 21:313–354, 1979; E. J. Denton and J. B. Gilpin-Brown, Floatation mechanisms in modern and fossil cephalopods, *Adv. Mar. Biol.*, 11:197–268, 1973; A. Naef, *Die fossilen Tintenfische*, Gustav Fischer Verlag Jena, 1922; P. Ward and S. v. Boletzky, Shell implosion depth and implosion morphologies in three species of *Sepia* (Cephalopoda) from the Mediterranean Sea, *J. Mar. Biol. Ass. U. K.*, 64:955–966, 1984.

Ceramics

Ceramics and ceramic composites are a rapidly evolving category of materials capable of being tailored to have unique combinations of electrical, optical, mechanical, and chemical properties. Such advanced or high-performance ceramics are inorganic, nonmetallic materials having combinations of various characteristics: fine-scale microstructures are generally present; the materials can have complex compositions and crystal structures; the composition is generally of high purity; and accurately controlled additives are frequently used. Advances in this field have resulted from bringing together three special fields of science and engineering: design of ceramic microstructures, design with ceramics as components of systems, and sophisticated processing to make the required materials.

The utility of advanced ceramics in a variety of electronic, optical, and structural applications is based on their strong chemical bonds and the mixed ionic-covalent character of these bonds. The high values of strength, stiffness, and melting point typical of ceramics derive from this bond character. This cluster of properties leads to the capability of structural ceramics for mechanical use at high temperatures. Another result of this fundamental bond character is the typical empty electronic conduction band and wide band gap with its accompanying insulating and semiconducting properties, as well as the characteristic transparency and ability of ceramics to interact with light. This second cluster of properties leads to the very diverse capabilities of electromagnetic ceramics, which include electronic, optical, and magnetic ceramics. A third cluster of properties resulting from the bond character is chemical in nature. Many uses of ceramics, such as corrosion-resisting applications, biological applications, and catalyst applications, derive from these properties.

Structural ceramics. High-performance structural ceramics uniquely combine strength, strength retention at high temperatures, hardness, dimensional stability, good corrosion and erosion behavior, high elastic modulus, and low mass density. Structural ceramics are used as monolithic parts, as composites, and as protective coatings.

Monolithic structural ceramics are currently based primarily on silicon nitride, silicon carbide, partially stabilized zirconium oxide, or alumina systems. Ceramic matrices combined with particulates, whiskers, or fibers of a different ceramic compound or metal for enhanced performance have yielded composites with five times the resistance to fracture (toughness) of monolithic ceramics.

Structural ceramics are used in many protective modes, especially as coatings. Wear and bearing applications involve differing combinations of adhesion, wear, and friction requirements. An important special application of ceramic coatings is to metal-cutting tools, where the resulting increase of tool life can be as much as a factor of three.

The largest potential market for structural ceramics is in automobiles and trucks. Ceramics are already being used by some automobile manufacturers to improve performance and efficiency of engines. In 1981 one manufacturer introduced silicon nitride glow plugs for diesel engines and later introduced ceramic swirl chambers within the cylinder and ceramic parts into the intake heater of heavy-duty trucks. Such incremental introduction of ceramic parts into engines is under study by many automobile firms. Ceramic parts under development include rocker arm fulcrums, valve seats, and turbocharger rotors. A silicon nitride turbocharger rotor was introduced in one engine model in 1985. A more ambitious approach seeks to develop insulated, low-heat-rejection engines (especially diesels) that do not need water cooling. In such engines, ceramics would be used extensively to insulate all parts of the cylinders, valves, and exhaust system, as well as to make a moderate-temperature turbine to extract energy from the hot exhaust gas. Another ambitious approach is to build high-temperature gas turbines with all hot parts made of ceramic. Two programs under partial sponsorship of the United States government, one to develop a two-shaft gas turbine and the other to develop a single-shaft turbine, have made considerable progress. Still another promising opportunity for ceramics is in aircraft turbines, both for the most demanding service in the stators and rotors and for less demanding service in intermediate-temperature parts where weight savings are important.

In general, silicon nitride and silicon carbide appear most promising for use at temperatures above 1800°F (1000°C). Fiber-reinforced ceramic-ceramic composites presently work well up to 1800°F (1000°C), and new matrices are under development to raise the upper temperature limit. Transformation-toughened zirconia and transformation-toughened alumina have excellent strength and toughness, but progressively lose the advantage of transformation toughening as the temperature is raised. In the present forms of these materials, this advantage is largely lost above 1500°F (800°C).

Ceramics with low thermal expansion, such as silica, are used for applications where thermal shock resistance combined with radar transparency is needed, such as nose cones for missiles. Carbon-carbon composites are used for rocket nozzles. These are typically made by weaving carbon fibers into the shape of the desired part, infiltrating with pitch, and heating to convert the pitch into carbon.

Electromagnetic ceramics. Electromagnetic ceramics are used in the electrical, electronics, and electrooptics industries. Major applications include low-loss optical fibers, which are causing a revolution in broadband information transfer; multilayer ceramic-to-metal interconnecting and mounting packages for critical silicon semiconductor integrated circuits (**Fig. 1**); ceramic multilayer chip capacitors of exceptional volumetric efficiency, required to decouple integrated circuits; piezoelectric ceramic transducers for sonar and medical ultrasonic tomography equipment; and chemical, mechanical, and thermal sensors

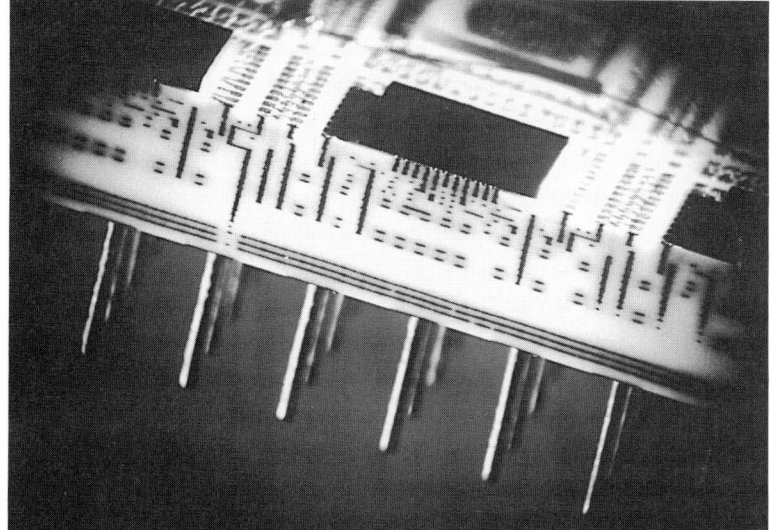

Fig. 1. Multilayer ceramic substrate showing a three-dimensional network of connections. The assembly serves as a mechanical support and complex wiring harness for electronic chips which are soldered onto the upper surface. (*IBM Corp.*)

for automobiles and automated manufacture and control. *See* Transducer.

Electromagnetic composites constitute a growing field, and, in particular, composites are becoming increasingly important to electronic ceramics. Layer structures with alternating levels of ceramics and metals, for use in both electronic substrates and multilayer capacitors, are made by cofiring. Challenges include developing materials with low permittivity in combination with acceptable values of dielectric loss, thermal expansion, thermal conductivity, and strength. Piezoelectrics with very high figures of merit can be fabricated by proper composite design.

Interfaces are an increasingly critical aspect of ceramics. Grain boundary phenomena are the basis of the behavior of varistors, including those used for overload protection through controlled breakdown. Many sensors, especially chemical sensors, are based on surface phenomena.

Optical ceramics (using the term ceramics in its broad sense) includes the rapidly growing field of optical communication fibers. Several generations of this technology already exist, but new developments are still being made and processing science is still being challenged. Planar optical technologies are in an earlier stage of development but offer great promise for optical signal processing, including optical computing. *See* Optical information systems.

Chemical and thermal ceramics. Many structural and electronic applications of ceramics depend on the chemical and thermal characteristics as important secondary properties. In addition, these properties constitute the primary basis of many uses of ceramics. Space shuttle tile, a fibrous mat of ceramics, provides very high thermal insulation resistance combined with high thermal shock resistance. Ceramic catalyst carriers are used in chemical processing and air-pollution control. Ceramics adapted to be compatible with the chemistry of the human body are used as dental implants and are entering use as bone replacements.

Designing ceramics. Ceramics are often made as composites of grains, layers, or fibers with different compositions and crystal structures in order to achieve superior properties. For many applications a large number of properties must have values within specified ranges. For example, ceramic substrates used in packaging electronic chips (Fig. 1) must have specified values of thermal expansion, thermal conductivity, permittivity, electrical resistivity, strength, and surface flatness, and chemical properties allowing coating with metals. These properties depend in different ways on composition and crystal structures. Computer-assisted mathematical modeling of properties which can be achieved through design of composites is increasingly essential to guide experimental optimization of a complete set of properties.

Ceramics present special requirements for successful application under high stress to avoid brittle and catastrophic fracture. A new field of designing for brittle materials, including ceramics, includes computer-assisted analysis of the stresses and temperatures throughout parts in great detail. This analysis is combined with knowledge of the mechanical behavior so that probabilities of failure can be computed for different loading conditions and different designs. The design is varied and stress calculations are repeated to assist in finding the design that will minimize the failure probability.

Processing of ceramics. In contrast to metals and alloys, which are usually made by casting from a melt, ceramics are usually made from powder mixtures by heating to a temperature which is below the melting point but sufficiently high to permit sintering. This process involves reaction, bonding, consolidation by diffusion, and sometimes other processes which transport matter. Ceramic processing typically involves a series of steps: production of fine powder, suspension of this powder in a liquid or wax, forming of an unsintered part of the same shape as the final part but oversized to allow for shrinkage, heating to sinter, and final machining for critical dimensions. Traditional ceramics, dating from prehistory, are made in a similar manner. The distinction of modern, advanced ceramics is in the special compositions and structures used and in the degree of control of composition, structure, and microstructure which is required. Unless carefully controlled, any stage of the processing sequence can introduce flaws which may seriously degrade properties, especially strength.

Advanced ceramics are usually made from synthetic powders produced by chemical synthesis rather than from natural raw materials used for traditional ceramics. Such powders are usually very fine and are required to have a precise distribution of particle sizes. The suspending media used in making unsintered ceramic shapes are typically complex mixtures of inorganic and organic chemicals to perform the functions of dispersion, viscosity control, drying control, and strength retention in the unsintered part. The unsintered part is shaped by dry pressing, slip casting, or injection molding. The ability to make ceramic parts very close to final shape is an important feature in making complex parts, such as turbocharger rotors (**Fig. 2**), at reasonable cost with a minimum of final machining. Control of shrinkage during the sintering stage is thus essential. Final machining, when needed, must be done by grinding and abrasive sawing, often with diamond as the medium because of the great hardness of ceramics, and is therefore expensive.

Modern trends in ceramic processing include novel chemical processes, involving both colloidal chemistry and gas-phase reaction routes, to synthesize powders with particle diameters below 1 micrometer. Sol-gel techniques permit the combination of the synthesis and forming stages

Fig. 2. Silicon carbide turbocharger rotor mounted on a steel shaft. The complex blade shapes are produced by injection molding and require no machining. (*Sohio Engineered Materials Co.*)

and can produce very fine and uniform microstructures. Another approach is to process through a glass stage and then go through a controlled crystallization process, yielding the so-called glass-ceramics. Still another approach involves chemical vapor deposition to form fibers or layers, or to bond particles or fibers into a composite.

Sintering aids are necessary to fully densify silicon nitride. These aids are oxides which form a glassy phase at the sintering temperature. Unfortunately, they promote creep and ultimately failure under long-time loads at characteristic service temperatures, despite the fact that these temperatures are hundreds of degrees lower than the sintering temperature. Much effort is being directed into tailoring the oxide additives to form a glass at the sintering temperature and yet crystallize at the service temperature to extend service life.

For background information *SEE ATMOSPHERIC ENTRY; CAPACITOR; CERAMICS; COMPOSITE MATERIAL; GRAIN BOUNDARIES; HIGH-TEMPERATURE MATERIALS; INTEGRATED OPTICS; OPTICAL COMMUNICATIONS; OPTICAL INFORMATION SYSTEMS; PIEZOELECTRICITY; PROSTHETICS; SILICON; TRANSDUCER; VARISTOR* in the McGraw-Hill Encyclopedia of Science and Technology.

John B. Wachtman, Jr.

Bibliography. W. D. Kingery, H. K. Bowen, and D. R. Uhlman, *Introduction to Ceramics*, 1976; H. Krockel, M. Merz, and O. Van der Biest (eds.), *Ceramics in Advanced Energy Technologies*, 1984; National Academy of Sciences, *Report of the Research Briefing Panel on Ceramics and Ceramic Composites*, 1985; D. W. Richerson, *Modern Ceramic Engineering*, 1982.

Chromosome

The deoxyribonucleic acid (DNA) within the cell nucleus is intricately complexed with the histones, a small group of very basic proteins, and the nonhistone chromosomal proteins, a much more diverse group of proteins. These three components, in a 1:1:1 mass ratio, together constitute the chromatin. Recent studies employing visual and biochemical techniques have made it possible to map the spatial distribution of the nonhistone chromosomal proteins and to better understand their function.

Function of histones. A fairly good understanding has been obtained of how the five different types of histones work together to fold and organize the DNA into an array of repeating subunits, called nucleosomes. Each nucleosome consists of 146 base pairs of DNA wound in 1¾ turns around the outside of a histone octamer containing two each of the four core histones (H2A, H2B, H3, and H4). Each nucleosome is connected to the next by the continuing DNA strand; this linker DNA is from 20 to 100 base pairs in length. The 10-nanometer chromatin fiber so formed is further supercoiled to generate a 30-nm fiber; the fifth histone, H1, plays a critical role in stabilizing this structure. While the DNA of one chromosome is believed to be a single continuous molecule, the chromatin fiber appears to be subdivided into distinct topological domains with an average size of several tens of thousands of base pairs. As the cell enters metaphase, a further compaction of the chromatin is required. At the return to the interphase state a selective decondensation occurs, leaving inactive regions of the genome in the form of dense heterochromatin, while active and inducible genes are released into a more dispersed form called euchromatin.

Function of nonhistone chromosomal proteins. There are many roles in the organization and differential compaction of the genetic material and in the regulation of gene expression that cannot be carried out by the histones. Recent investigations have identified some of the nonhistone chromosomal proteins that are involved in these functions. There are estimated to be approximately 500 different types of nonhistone chromosomal proteins in the average mammalian cell nucleus, counting those present at a few hundred copies or more per cell. This fraction includes proteins that play structural roles in chromatin organization, enzymes of chromosomal metabolism [for example, DNA polymerase, histone acetyltransferase, and ribonucleic acid (RNA) polymerase], and regulatory proteins required for the controlled expression of subsets of genes. These categories overlap; for example, topoisomerase II, an enzyme capable of changing the degree of supercoiling of the DNA fiber, is a major structural component of the metaphase chromosome.

Visual studies of nonhistone chromosomal proteins. Techniques that visualize the spatial distribution of nonhistone chromosomal proteins in nuclei and chromosomes have been found particularly useful in classifying these proteins.

Types of chromosomes used. Polytene chromosomes and lampbrush chromosomes have been particularly useful for such studies because they are well organized and large enough to allow the visualization of individual domains (or chromomeres) using a light microscope. Polytene chromosomes are generated in certain large, nondividing cells by successive rounds of replication without separation of the chromatin replicas, which remain aligned side by side in perfect register. A pattern of differential condensation, called banding, allows reproducible identification of specific positions along the chromosome arms. In *Drosophila* (a fruit fly) the centromeres of all four chromosomes remain associated in the chromocenter with the chromosome arms [one from the X chromosome, right and left arms from the second chromosome (2R and 2L), and right and left arms from the third chromosome (3R and 3L)] radiating from this centric heterochromatin mass. (The fourth chromosome is quite small and is usually not visualized.) Each of the euchromatic arms contains approximately 1000 DNA fibers; the heterochromatic material, including the entire Y chromosome, is severely underreplicated relative to the euchromatin. Regions along the chromosome where the genes are being transcribed at a rapid rate become enlarged and diffuse in appearance, and are referred to as puffs (**Fig. 1***a*).

Results of immunofluorescent staining. Staining of the polytene chromosomes by using antibodies specific for a given nonhistone chromosomal protein can be used to assess the pattern of association within the chromosomes for that protein. The antibodies are visualized by tagging (directly or indirectly) with a fluorescent dye. While the histones, as expected, are observed to be located throughout the polytene chromosomes (Fig. 1*b*), certain nonhistone chromosomal proteins are preferentially associated with only those loci that are known to puff at some time in this cell type, the salivary gland cell of *D. melanogaster* (Fig. 1*c*). Thus this subset of nonhistone chromosomal proteins appears to be preferentially associated with loci that are active or inducible. In contrast, other nonhistone chromosomal proteins have been identified which are preferentially associated with the inactive heterochromatin (Fig. 1*d*). With an antibody specific for one of these proteins the heterochromatic chromocenter is prominently stained, while only a few sites in the euchromatic arms are stained. The sites along the arms are loci previously identified as inactive, containing highly repetitious "satellite" DNA, and therefore cataloged as a type of heterochromatin. The nonhistone chromosomal protein with this distribution probably plays a role in generating an inactive, compact heterochromatin structure.

Immunofluorescent staining of the polytene chromosomes has also been used to assess the distribution of certain enzymes. RNA polymerase, for example, is found concentrated at the puff sites and in the nucleolus, the sites of high levels of transcriptional activity (Fig. 1e). A similar distribution is observed for topoisomerase I (Fig. 1f). This "nicking-closing" enzyme temporarily cuts one strand of the DNA, allowing the double helix to unwind, and then precisely reseals the cut so that the DNA molecule is not permanently damaged. The topoisomerase I may assist in "unpackaging" the chromatin fiber and presumably acts to relieve torsional stress in the DNA generated by transcription by RNA polymerase.

Transient association of enzymes. RNA polymerase and topoisomerase I are only transiently associated with specific loci. A small set of genes, including in *Drosophila* those identified as loci 63BC, 64EF, and 67B, are activated by a heat shock, caused when the temperature of the organism is raised several degrees. Little or no RNA polymerase or topoisomerase I is associated with these loci when they are inactive, but a substantial amount of each enzyme is present following a heat shock stimulus (**Fig. 2**). These enzymes must be present in the nucleus in excess, with a pool in the nucleoplasm, to allow their recruitment as needed for gene expression.

Studies of metaphase chromosomes. Important information about the structure of metaphase chromosomes has been obtained by an analogous approach. In particular, distribution studies using antibodies specific for topoisomerase II show the enzyme to be an important component of the proteinaceous core structure, or scaffold (**Fig. 3**). Indeed, biochemical studies have found that the bulk of the topoisomerase II is so engaged at metaphase. Genetic studies in yeast have found that topoisomerase II activity is absolutely essential to untangle the DNA strands after replication. In the absence of this enzyme, the two daughter molecules remain wrapped around each other, and cannot separate to segregate properly into the two daughter cells. Thus topoisomerase II is believed to play a dual role, both as an essential enzyme and a structural protein involved in maintaining appropriate chromosome organization.

Biochemical studies. The low-resolution results obtained by these visual techniques can be confirmed and extended by biochemical techniques. If cells are exposed to ultraviolet irradiation, a small percentage of the nonhistone chromosomal proteins will become covalently crosslinked to the DNA with which they are associated. After the chromatin has been extracted and the DNA has been cut into fragments a few kilobases in length, a protein-DNA adduct can be selectively purified by using an antibody specific for that protein. The associated DNA sequences can be identified by using specific radioactive recombinant DNA probes. Experi-

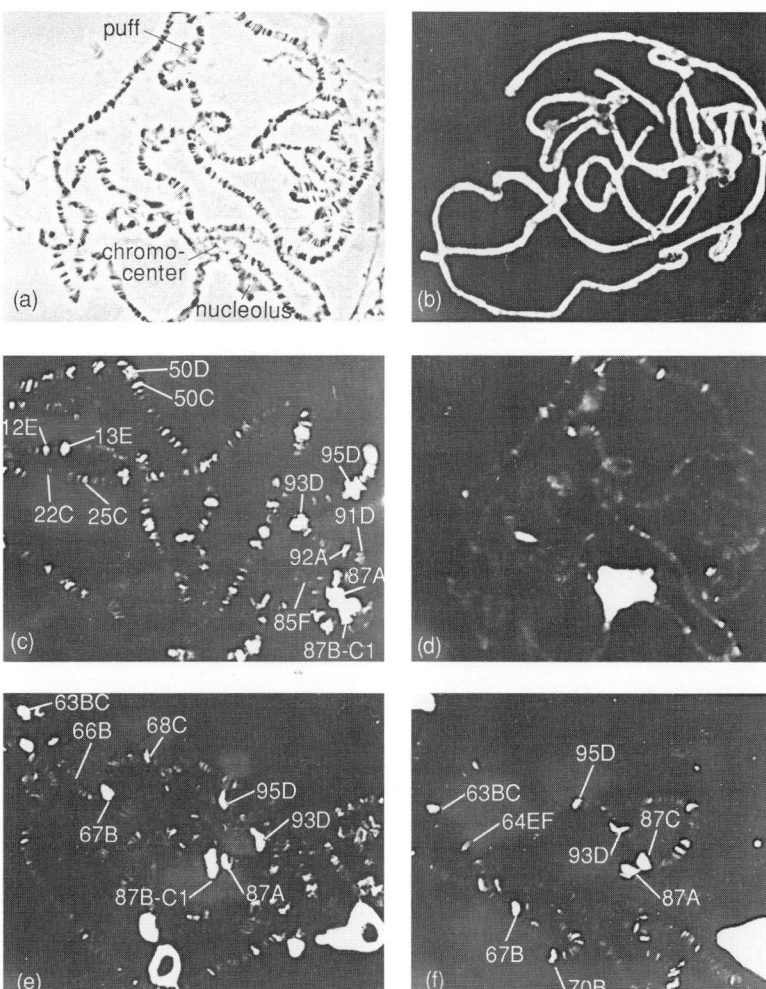

Fig. 1. Distribution patterns of various chromosomal proteins in the polytene chromosomes of *Drosophila melanogaster* salivary gland cells. (*a*) Phase contrast picture, with the chromocenter, nucleolus, and typical puff indicated. (*b*) Immunofluorescent staining pattern obtained by using antibodies to histone H2A; (*c*) a specific nonhistone chromosomal protein of 60,000 molecular weight; (*d*) a specific nonhistone chromosomal protein of 18,000 molecular weight (with same chromosomes as in part *a*); (*e*) RNA polymerase; and (*f*) topoisomerase I. Specific bands, referred to as chromomeres or loci, are designated by a number-letter system. The chromosomes used in parts *c*, *e*, and *f* were from animals that had been subjected to heat shock, leading to activation of loci 63BC, 64EF, 67B, 70B, 87A, 87C, 93D, and 95D.

ments of this type have also shown that the heat-shock genes are associated with topoisomerase I after, but not before, activation by the heat-shock stimulus. Other biochemical techniques can be applied in specific cases. For example, the drug camptothecin inhibits topoisomerase I so that, when the DNA is purified, a nick (a cut in one strand) is found wherever the enzyme was bound to the DNA in the nucleus. The position of these nicks can be mapped at high resolution by using recombinant DNA probes. Such an analysis shows that the topoisomerase I interacts with both strands of the DNA throughout the region of transcription. A dynamic picture of the active gene is thus obtained, as the topoisomerase I must be constantly moving on and off to carry out its role (presumably unpackaging and unwinding the DNA) without impeding the passage of the RNA polymerase.

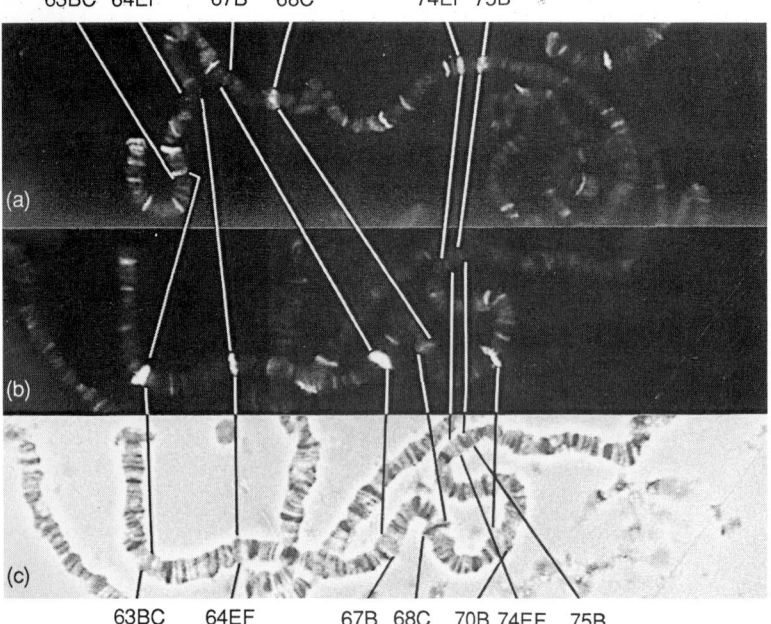

63BC 64EF 67B 68C 74EF 75B

63BC 64EF 67B 68C 70B 74EF 75B

Fig. 2. Change in distribution of topoisomerase I in response to heat shock. (*a*) Immunofluorescent staining of chromosome arm 3L before heat shock. (*b*) Results after heat shock. (*c*) Phase contrast picture of the chromosomes shown in part *b*. Upon heat shock the topoisomerase I becomes associated with the newly activated loci (63BC, 64EF, 67B, 70B), and is less abundant at other loci (68C, 74EF, 75B), which are relatively repressed under these conditions. (*From G. Fleischmann et al., Drosophila DNA topoisomerase I is associated with transcriptionally active regions of the genome, Proc. Nat. Acad. Sci. USA , 81:6958–6962, 1984*)

Prospects. Techniques now being developed are allowing biologists to look for the specific binding of proteins to defined DNA sequences in the test tube. Studies of this type have identified nuclear protein fractions that bind specifically to DNA sequences which are located at specific

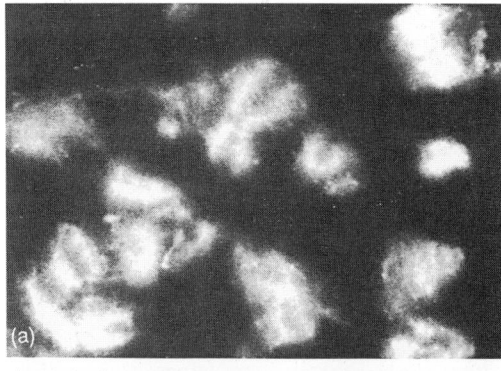

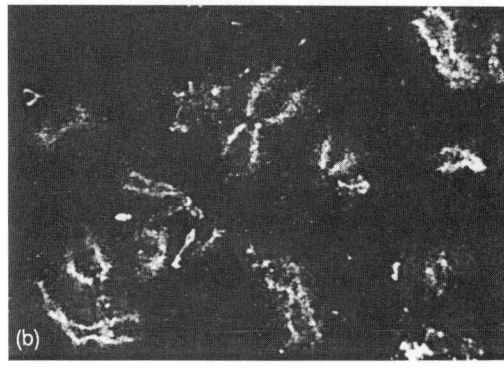

Fig. 3. Distribution of topoisomerase II in the metaphase chromosomes of a chick cell. (*a*) Chromosomes swollen and stained with DAPI, which shows the pattern of the DNA. (*b*) Chromosomes stained with Texas Red-labeled antibodies, which show the distribution of the topoisomerase II. (*Courtesy of Magarete M. Hech and William C. Earnshaw*)

chromosome structures, such as the chromocenter or the telomere (chromosome end), or which play a role in regulating gene transcription. The use of these binding assays should make possible the purification of the proteins for further study. Additional experiments, with techniques such as those described above, will be needed to confirm the pattern of interactions of these proteins in the living cell. As seen above, the best indicators of the presence of a protein are antibodies specific for that protein. New techniques for generating specific monoclonal antibodies are very helpful. The study of human biology has also revealed that some people, suffering from autoimmune diseases, generate antibodies against particular components of the nucleus. Antibodies against both the B and Z forms of DNA, the histones, the topoisomerases, and many structural nonhistone chromosomal proteins have been identified in the serum of different patients of this type. These antibodies are being used to isolate and characterize the proteins and to obtain recombinant DNA clones for the genes which encode them. While much work remains to be done, the necessary tools are now available to identify specific nonhistone chromosomal proteins, and to determine how these components help to shape chromosome structure and function.

For background information SEE AUTOIMMUNITY; CHROMOSOME; IMMUNOFLUORESCENCE; MONOCLONAL ANTIBODIES; NUCLEOPROTEIN in the McGraw-Hill Encyclopedia of Science and Technology.

Sarah C. R. Elgin

Bibliography. W. C. Earnshaw and M. M. S. Heck, Localization of topoisomerase II in mitotic chromosomes, *J. Cell Biol.*, 100:1716–1725, 1985; G. Fleischmann et al., *Drosophila* DNA topoisomerase I is associated with transcriptionally active regions of the genome, *Proc. Nat. Acad. Sci. USA*, 81:6958–6962, 1984; D. S. Gilmour et al., Topoisomerase I interacts with transcribed regions in *Drosophila* cells, *Cell*, 44:401–407, 1986.

Color vision

Recent advances in color vision include new insights into the role of the photopigments of the eye and the processing within the central nervous system.

ROLE OF PHOTOPIGMENTS

Because most neuroscientists interested in color vision are currently studying the processing of color in the brain, little recent progress has been made concerning chromatic function in the primate retina proximal to the receptors. However, remarkable advances have recently taken place in the understanding of receptoral photopigments and their role in normal and defective color vision.

Visual receptors. The action spectrum of a visual receptor is defined as the reciprocal of the radiance required to produce a criterion magnitude of response, as measured continuously throughout the visible electromagnetic spectrum. Each class of cone is color-blind because any two spectral distributions of light, if suitably adjusted in radiance, can produce equivalent responses from any one of the classes. Within each small region of the retina of the eye, the initial basis for color perception is provided by the ratio of excitation of three classes of photoreceptors. A knowledge of the cone action spectra is crucial for understanding color vision, because signals from three kinds of cones provide the sole input to the higher stages of visual processing which are concerned not only with color but also with spatial perception.

The response of a visual receptor depends nonlinearly upon the action spectra of its constituent photopigment molecules, the radiance of light flux reaching the cornea of the eye that is imaged in the appropriate retinal area, and pre-receptoral absorption of light in the eye media. The action spectra depicted in **Fig. 1** are based on indirect evidence of the 1970s obtained from psychophysical experiments using intact human observers with normal and abnormal color vision. The observers were asked to make various kinds of discrimination and color-matching judgments under well-controlled physical conditions. These functions are consistent with the facts of protanopic and deuteranopic (red- and green-blind) color deficiency and the laws of color matching.

Microspectrophotometry. Direct evidence concerning the cone action spectra has been obtained in recent years by microspectrophotometry, in which are made measurements of light absorption in the outer segments of cones, where the photoreactive molecules are contained. This technique, first developed in the 1960s, was used effectively to study the receptors of fishes. However, it met with only limited success when applied to the much smaller primate photoreceptors, the rods and cones, in which photopigment densities are low. Microspectrophotometry of primate receptors was subsequently abandoned for about 15 years. The method has been revived recently; with improvements in technique it has been used to measure the absorbance of receptors from Old World primates, whose vision is very similar to that of humans, as well as the absorbance of one human eye.

Uncertainties exist concerning preretinal absorption, which varies among observers and selectively affects the results of psychophysical experiments. Also, small variations appear to exist in the action spectra of human cones of a given class. For these reasons, no set of three functions can hope to describe anything more than the average data of normal humans. Within this limitation the microspectrophotometric results are consistent with the action spectra of Fig. 1.

Electrophysiological measurement. Primate receptors have proved to be too small to impale with microelectrodes, thereby frustrating efforts to determine the action spectra of cones electrophysiologically. Recently, by developing and using an ingenious suction electrode into which the cone outer segment is drawn, action spectra of a few primate cones have been obtained which, as far as they go, are also consistent with those of Fig. 1.

Cone distribution. Knowledge of the distribution of the three kinds of cones in the retinal mosaic is also important for an understanding of color vision. The three kinds of cones are found in unequal proportions, which vary as a function of retinal location. Inspection of cones under the microscope prior to microspectrophotometric measurement reveals no apparent differences between them. Yet within each sample, B cones are always found in much smaller numbers than the long-wave–sensitive cones (R and G).

For unknown reasons certain dyes, when injected into the vitreous body, are taken up selectively by B cones, whose arrangement in the mosaic can then be examined histologically. **Figure 2** shows a section of macaque monkey retina stained with Procion black. Although anatomical methods by themselves cannot reveal cellular function, identification of the darkly stained elements in Fig. 2 as B cones is strongly supported by psychophysical evidence from humans showing punctate regions highly sensitive to short wavelengths which agree with the spacing of B cones of Fig. 2. Both findings are consistent with other evidence that B cones, when functioning in isolation, are incapable of providing a high level of spatial resolution. Figure 2 shows that each B cone forms the centerpiece of a hexagonal array in which it is surrounded by six R or G cones. This arrangement prohibits contiguity between B

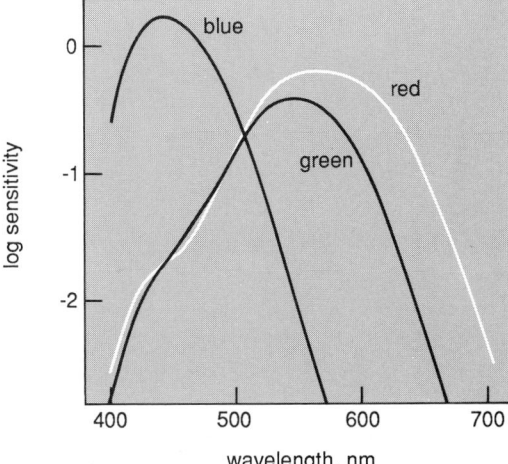

Fig. 1. Action spectra of human cones, based on psychophysical observations with normal and color-deficient human observers, plotted from equations. (*After V. C. Smith and J. Pokorny, Spectral sensitivity of the foveal cone photopigments between 400 and 500 nm, Vision Res., 15:161–171, 1975*)

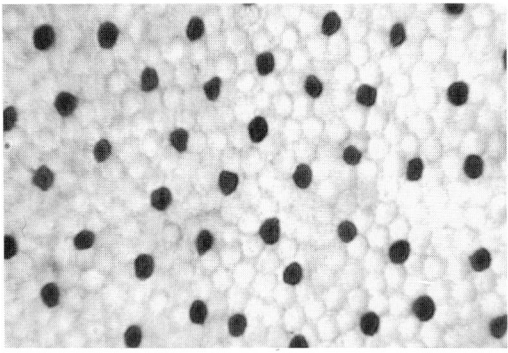

Fig. 2. View of macaque retina stained with Procion black, which is taken up selectively by B cones. R. and G. cones which make up the rest of the array, do not stain and cannot be discriminated from one another. (*F. M. de Monasterio, National Eye Institute, NIH; published in E. P. McCrane, et al., Non-fluorescent dye staining of primate blue cones, Investig. Ophthalmol. Visual Sci. 24:1449–1455, 1983*)

cones and eliminates any possibility of interaction between them mediated by electrical signals across tight junctions; such signals are believed to occur between neighboring long-wave–sensitive cones.

To date, no staining method has been found which can discriminate between R and G cones. The recent data confirm that B cones are absent in the central fovea (the region of highest visual acuity to which the images of fixated objects are delivered) and show that their maximum density occurs at an eccentricity of about 1° of visual angle. In contrast, R and G cones have their highest density in the central fovea. Even in the region where they are most densely represented, B cones constitute a very small percentage of the receptor population; and unlike the long-wave–sensitive cones, which do double duty, they appear to have no important role to play in spatial vision.

Genetics. Application of the methods of molecular biology has recently revealed spectacular details concerning the genetic basis of the visual photopigments. Their deoxyribonucleic acid (DNA) sequences have been determined; and it is found that, whereas R- and G-cone pigments are 99% homologous, the percentage of sequence homology between either of these and the B-cone pigment is less than 50%. **Figure 3** illustrates these

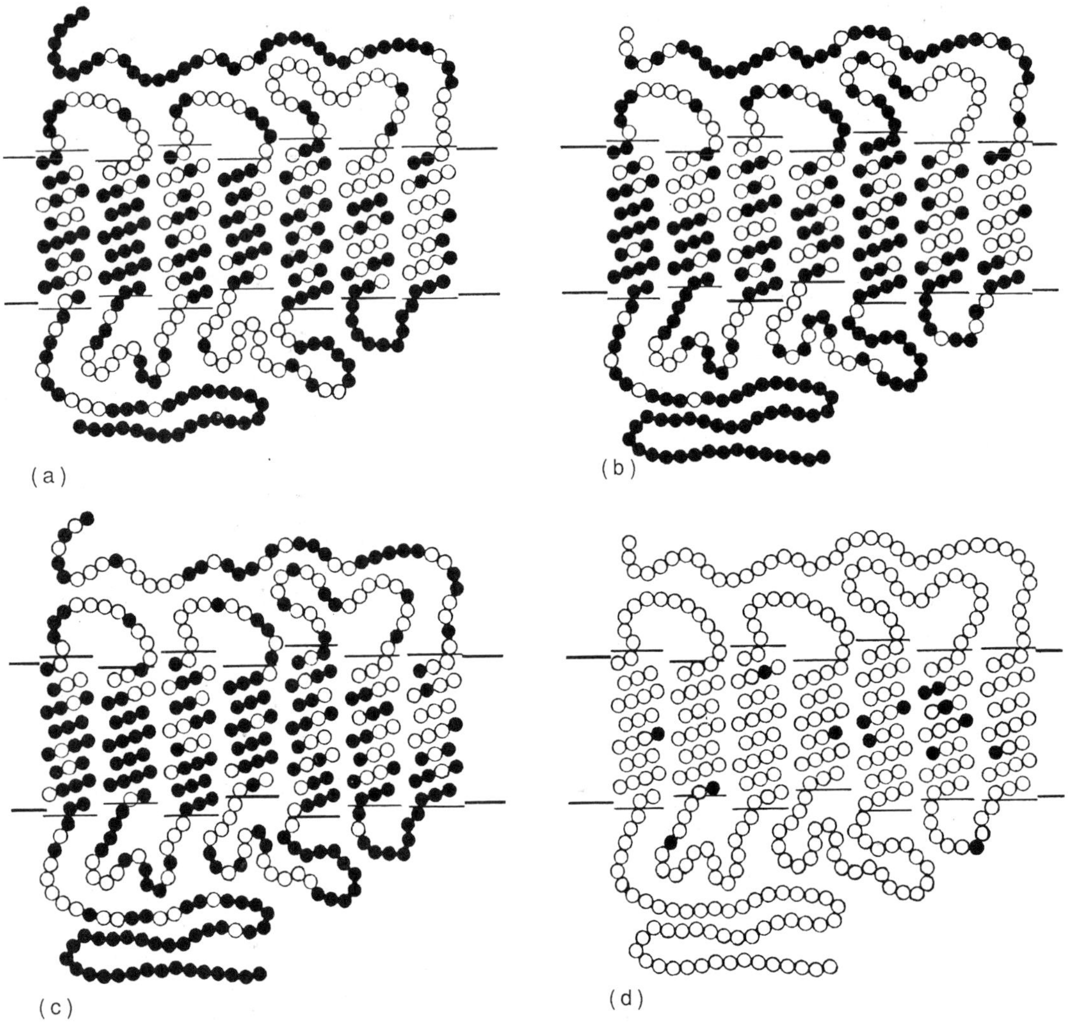

(a)

(b)

(c)

(d)

Fig. 3. Comparisons of human visual pigments showing identities (white) and differences (black) in their amino acids when their sequences are arranged for maximum concordance. (*a*) Blue versus rhodopsin. (*b*) Green versus rhodopsin. (*c*) Green versus blue. (*d*) Red versus green. (*After J. Nathans, D. Thomas, and D. S. Hogness, Molecular genetics of human color encoding blue, green, and red pigments, Science, 232:193–202, 1986*)

differences by depicting pairwise comparisons of human visual pigments in which amino acid identities are shown in white and differences in black.

Confirming what has long been deduced from the study of human pedigrees, which show a much higher incidence of color deficiency in males than in females, the genes for R- and G-cone pigments are found on the X chromosome, and one of these types is missing or altered in cases of red-green color deficiency. Not anticipated was the finding that the X chromosomes of males (and presumably females) with normal color vision may have more than one G-cone pigment gene, the significance of which is unclear. It has also been demonstrated that the gene for human B-cone pigment derives from chromosome 7, consistent with the autosomal mode of inheritance of visual deficiencies ascribed to B-cone abnormality. (The gene for rhodopsin, the very abundant rod pigment not directly concerned with color vision, derives from chromosome 3.)

Because the genes for red and green pigments are so similar in their DNA sequence, it is to be expected that recombinations between them will occur. Depending upon the exact site of occurrence, recombination may cause the loss or alteration of either the R-cone or the G-cone pigment gene. Analysis of genomic DNA from 25 males with a variety of red-green color deficiencies has identified several hybrid genes among both the R- and G-cone pigment genes, and it has been concluded that these genotypes are the result of unequal intragenic recombination. Unequal intergenic recombination or gene conversion may account for the loss of G-cone pigment genes in some color-deficient males, as well as for the multiple G-cone pigment genes observed in color-normal males.

Other studies. Additional evidence relating cone type to color vision has recently been reported by an international team of investigators. New World monkeys, unlike their Old World counterparts, exhibit large individual differences in their normal color vision. Female squirrel monkeys are either trichromatic or dichromatic, but so far as is known, all male squirrel monkeys are red-green dichromats. In a double-blind study, behavioral measurements of thresholds, wavelength discrimination, and color matching were obtained for 10 monkeys, following which their eyes were subjected to microspectrophotometric analysis by a different team of investigators. The absorbance spectra of the monkey cones were found to predict the behavioral data, confirming that the basis of the behavioral variation lies in the visual photopigments.

Robert M. Boynton

NEURAL PROCESSING OF COLOR INFORMATION

The stimulus for vision consists of the intensity of light (number of photons) and the wavelength of the light (energy level of the photons) at each point in space. (Some animals, excluding hu-

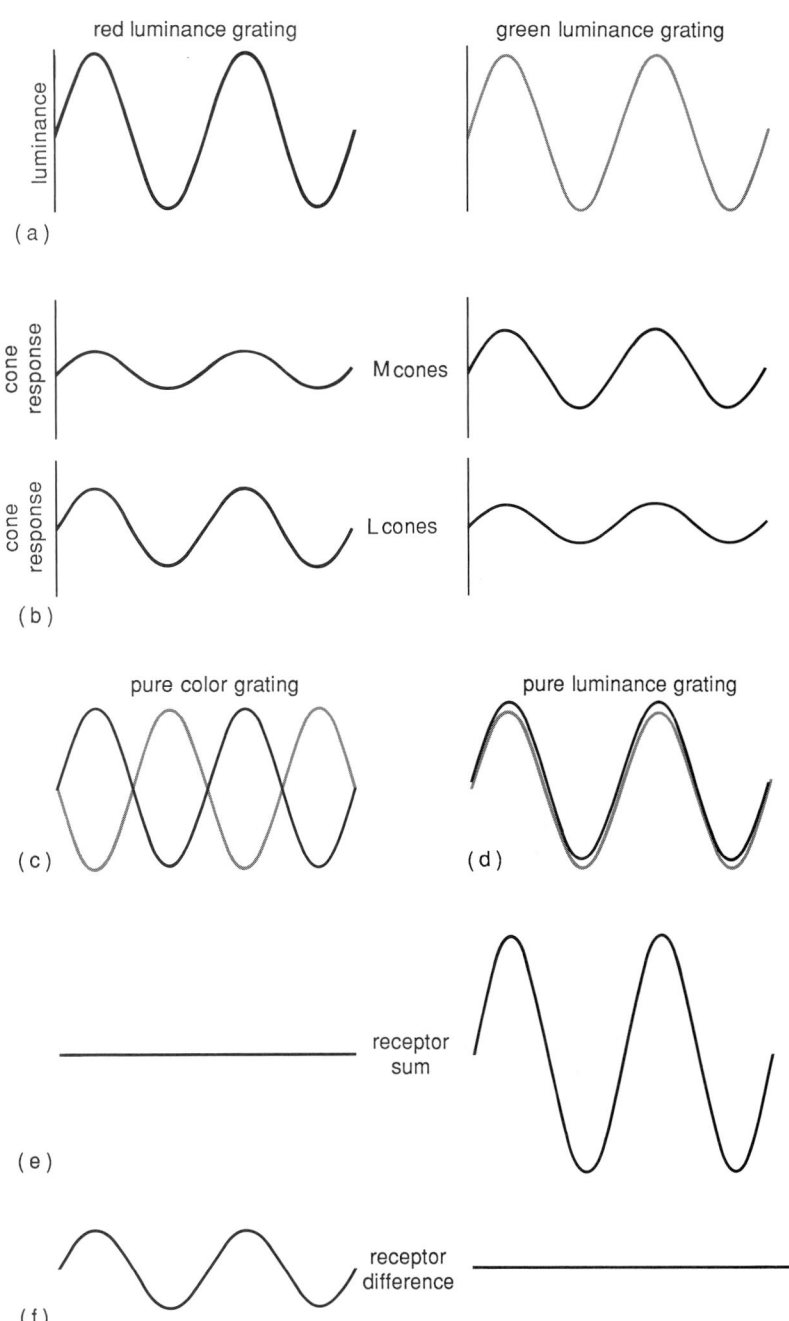

Fig. 4. Luminance and color patterns. (*a*) Two luminance-grating cross sections, one red and one green. In each case the luminance varies sinusoidally across space. (*b*) The relative responses of arrays of M and L cones to each of the individual gratings. Note that both cone types would respond to both gratings, but with different amplitudes. (*c*) The red and green luminance gratings have been added out of phase to produce a pattern which varies in chromaticity but is constant in luminance across its extent. (*d*) The two components have been added in phase to produce a luminance grating. (*e*) Reflection of the sums of the cone responses to each of the component patterns. The receptor sums vary with luminance contrast. (*f*) The receptor differences vary with color contrast.

mans, can also detect variations in the polarization of the light.) Although intensity and wavelength differences covary to some extent (objects often differ from their backgrounds along both dimensions), they do not always do so. Separate information about intensity and wavelength at each point in space is thus very useful in defining objects. All animals with vision respond to variations in both intensity and wavelength, but only some can differentiate so as to respond independently to each. Those that can, including most birds, fishes, and primates, have color vision. Thus an animal with color vision has available two separate maps of the world: one related to variations in the intensity of the light (seen as brightness), and the other to variations in the wavelength of the light (color) across a visual scene.

Retinal receptors. At the initial visual stage the retinal receptors, intensity, and wavelength are confounded. An increase in photon catch by a receptor, regardless of the wavelength of the photons, leads to a hyperpolarization of the receptor and a decrease in its synaptic output to the next neural stage; a decrease in photon catch leads to depolarization and an increase in synaptic output. A receptor, then, can signal only the

number of photons it has captured. It cannot separate the number and the wavelength of the arriving photons. Thus a 100-photon catch might be caused by a light of 550 nanometers at a particular intensity or a 650-nm light of 10 times that intensity.

The receptor pigments, peaking at about 420, 530, and 560 nm, respectively, in the three types of cones, are broadly tuned in wavelength. The two long-wavelength receptors, in what can be termed the long-wavelength (L) and middle-wavelength (M) cones, absorb across the whole spectrum and do not differ greatly in their spectral sensitivity (Fig. 1). These two cone types, in fact, contribute to all visual functions and form the basis for human black-and-white vision, in daylight conditions, as well as color vision. The short-wavelength (S) cones, on the contrary, seem to contribute only to color vision.

Differentiation. How, then, does the nervous system differentiate color and luminance information? **Figure 4** shows that intensity variations drive the L and M cones in the same direction, whereas wavelength variations, when equated for equal total photon catch, drive receptors in opposite directions. A system which summed the output of the different receptors, independent of receptor type, would detect the intensity variations. A system which responded only to receptor differences, on the other hand, would detect only wavelength variations. This is the tactic taken by the visual nervous system in separating these two types of information.

Evidence that some cells in the visual nervous system difference the outputs of different receptor types (spectrally opponent cells) while others sum receptor outputs (spectrally nonopponent cells) was first found more than 25 years ago in recordings of the responses of cells in both fish retina and primate lateral geniculate nucleus to chromatic flashes of light. The experiments on the primate visual system provided evidence for some six different functional cell types in the visual path to the brain (**Fig. 5**). The two types of spectrally nonopponent cells fire to increases or to decreases of light in a region with respect to its surround. These increases and decreases presumably signal luminance changes. Two of the spectrally opponent cell types difference the L and the M cones, in opposite pairs, and two difference the S cones from the L + M cones. These opponent cells presumably signal chromatic changes.

Recent research. This problem of how color and luminance are analyzed and encoded in the macaque monkey (and presumably human) geniculate nucleus has been reexamined in experiments in which responses to variations in light in different directions through color space were recorded. Perceptually, color space is three-dimensional (among trichromats), with the three color axes being black-white, red-green, and blue-yellow. All the thousands of colors seen by

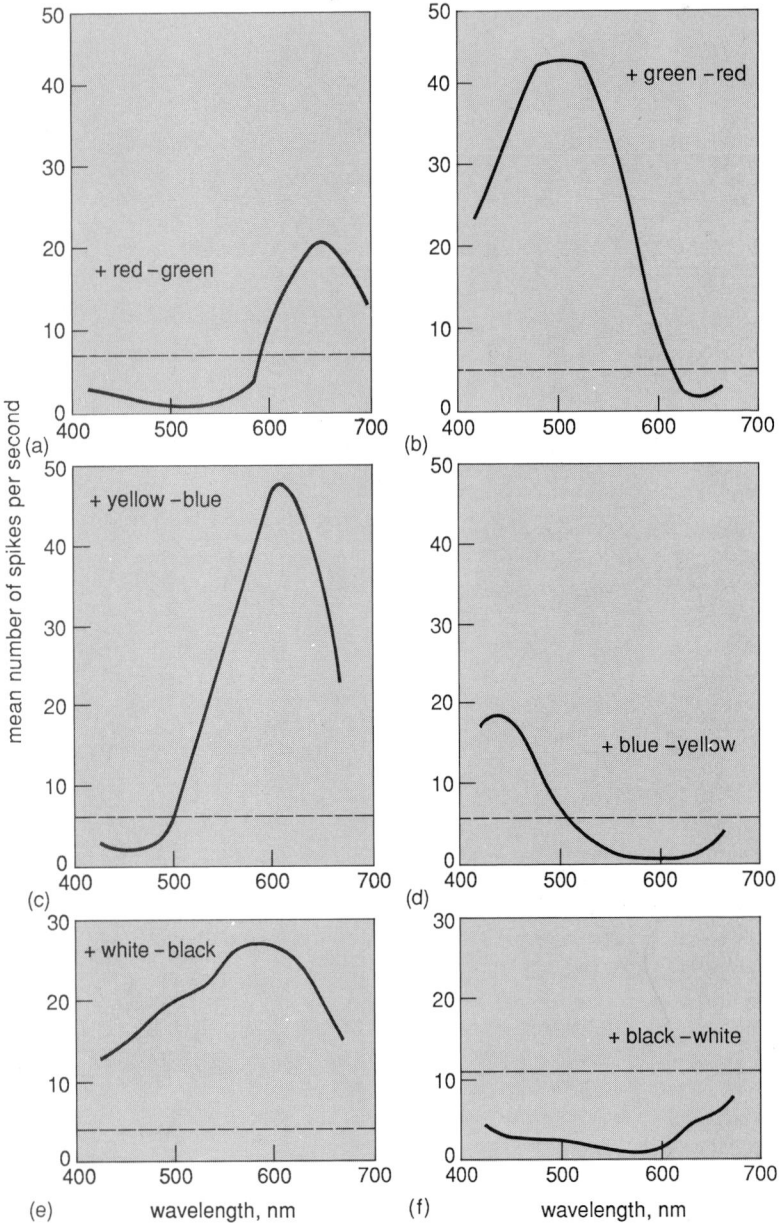

Fig. 5. Responses of different macaque lateral geniculate cells to flashes of light of different wavelengths. (*a–d*) Functions show averages of the four basic color-opponent cell types. (*e–f*) Responses of cells that are sensitive primarily to luminance contrast rather than to color contrast. (*After R. L. De Valois, I. Abramov, and G. H. Jacobs, Analysis of response patterns of LGN cells, J. Opt. Soc. Amer., 56:966–977, 1966*)

humans can be described in terms of their values along these three axes. The recording data shown in Fig. 5 suggest that the perceptual color axes may correspond to the different opponent and nonopponent cell types in the primate geniculate nucleus. However, this is not precisely so; much of the luminance information is multiplexed with the L − M opponent cell color information, and the color axes formed by the L − M and S − (L + M) cells do not correspond exactly to the red-green and blue-yellow perceptual axes. There must be a further transformation of color information at cortical levels; such evidence is being obtained from recordings in the macaque striate cortex.

Recordings of the color-selective responses of cells in striate cortex (the first cortical visual area) of the macaque monkey show two new characteristics of cells at this level. One characteristic is the presence of new varieties of opponent cells, including some whose responses correspond to the perceptual color axes of red-green and yellow-blue. The other is that most cells at this level are responsive only to patterns containing components within a particular orientation and spatial frequency range. This is true whether the patterns are luminance-varying or color-varying. Thus most striate cells utilize both luminance and color variations to gain information about the shape of objects.

Information processing. The perceived color of objects is determined largely by the wavelength of the light reflected from them. However, to some extent it is also a function of the history of stimulation and of the illumination in surrounding regions. This is even truer for the brightness of objects. The main function of such modifications of brightness and color information is to preserve object constancy in the presence of changes in the level and wavelength composition of the illuminant: light from the Sun is millions of times dimmer and somewhat more reddish at dawn and dusk than in the middle of the day. These variations need to be compensated for if objects are to maintain a constant appearance. Much of this compensation takes place in adaptational changes at the level of the receptors, the receptor gain being adjusted as light intensity varies, and differentially in different cone types as the wavelength of the illuminant varies. However, more global interactions are also involved in color and brightness constancy.

The early processing of visual information is quite local, however. Cells up through the level of the first (striate) cortical processing level deal only with the information within very local retinal regions. This is true for all types of visual processing: only information about local movement, depth, color, brightness, shape, and so forth is computed. Clearly, one of the primary functions of later visual processing regions (which are at present poorly understood) must be to modify the local information on the basis of what is happening elsewhere in the visual field. Evidence for a more global type of processing in the case of color information has been reported to take place in a later visual area, V4.

For background information SEE COLOR; COLOR VISION; PHOTORECEPTION; VISION in the McGraw-Hill Encyclopedia of Science and Technology.

Russell L. De Valois

Bibliography. A. M. Derrington, J. Krauskopf, and P. Lennie, Chromatic mechanisms in lateral geniculate nucleus of macaque, *J. Physiol. (Lond.)*, 357:241–265, 1984; F. M. de Monasterio et al., Density profile of blue-sensitive cones along the horizontal meridian of macaque retina, *Investig. Ophthalmol. Visual Sci.*, 26:289–302, 1985; R. L. De Valois, I. Abramov, and G. H. Jacobs, Analysis of response patterns of LGN cells, *J. Opt. Soc. Amer.*, 56:966–977, 1966; J. D. Mollon and L. T. Sharpe (eds.), *Colour Vision: Physiology and Psychophysics*, 1983; J. Nathans et al., Molecular genetics of inherited variation in human color vision, *Science*, 232:203–210, 1986; D. Ottoson and S. Zeki (eds.), *Central and Peripheral Mechanisms of Colour Vision*, 1985; L. G. Thorell, R. L. De Valois, and D. G. Albrecht, Spatial mapping of monkey V1 cells with pure color and luminance stimuli, *Vision Res.*, 24:751–769, 1984; S. M. Zeki, Colour coding in the cerebral cortex: The reaction of cells in monkey visual cortex to wavelengths and colours, *Neuroscience*, 9:741–765, 1985.

Communications satellite

Recent advances in communications satellite service include (1) work on the Advanced Communications Technology Satellite (ACTS) system that will incorporate new technology to reduce costs and make more efficient use of space resources; (2) the launching of the Galaxy satellites, which combine the trends toward domestic use of satellites and the sale or lease of portions of a satellite; and (3) the construction of second-generation INMARSAT satellites to serve ships and offshore structures.

Advanced Communications Technology Satellite. The ACTS system, sponsored by the National Aeronautics and Space Administration and scheduled for operation in 1990, embodies new technology necessary to reduce satellite communications costs and improve the efficiency with which space resources—geosynchronous orbit and the radio-frequency spectrum—are used.

The satellite will operate in geosynchronous orbit, that orbit on the equatorial plane at an altitude of 22,300 mi (35,900 km) which has the same period of rotation as the Earth. Satellites in this orbit therefore appear, from Earth, to be stationary, and the direction in which antenna beams from ground stations are pointed to these satellites need not be changed with time. A number of satellites, spread out over the geosynchronous orbit, can provide complete coverage of the Earth except for regions in the extreme north and south. By assigning a different part of the frequency spectrum to each signal, many signals may be transmitted simultaneously. Placing satellites far enough apart in orbit to preclude interference with one another allows the frequency spectrum to be reused a number of times.

Frequency reuse techniques. The ACTS system uses several other techniques for frequency reuse. The polarization properties of radio waves allow signals transmitted with orthogonal polarization to be electronically separated at the receiving antenna even though they occupy the same frequency spectrum. The frequency spectrum can be further reused if narrow, multiple, independent beams covering different geographic areas can be formed by the satellite's antenna. A key advantage is increased strength of signals as the beams are narrowed, allowing smaller antennas to be used at Earth stations. *See Satellite communications*.

Fixed and scanning beams. The narrow beams do, however, illuminate only a small portion of the Earth's surface. The antenna beams utilized by ACTS provide usable signals over an area of about 8000 mi^2 (20,000 km^2). Many beams would be required to contiguously cover the entire continental United States. ACTS employs two approaches to reducing the number of beams required. A practical number of beams can be achieved by providing spot beams only on the major communications centers. These major centers connect to local, Earth-based telephone switching centers and trunks to extend service to the users.

Alternatively, a few beams capable of being rapidly scanned, in a stepped fashion, can service the entire continental United States. During the time the beam dwells on an area, information is transmitted and received in short signal bursts. Each user in that dwell area transmits in a different portion of the dwell time and uses only the received signal from a portion of the dwell time. This approach, while allowing contiguous coverage with a practical number of beams, requires that information be in a form which can be stored during the intervals between antenna beam dwells. A digital signal format is used in the ACTS system.

Equipment on the satellite determines not only in which area a user is located but also in which time period the user's information must be transmitted, a mode of operation referred to as time-division multiple access.

Adaptive techniques. Signals in the 30-GHz uplink and 20-GHz downlink (Ka-band) frequency bands being used by ACTS are degraded more by rain and other atmospheric phenomena than the more commonly used C-band (6-GHz uplink and 4-GHz downlink) and Ku-band (14-GHz uplink and 12-GHz downlink) allocations. ACTS employs techniques that allow the system to adapt to the degradations. Transmitters capable of operating efficiently in both a low- and a high-power mode allow signal power to be adjusted to overcome atmospheric losses. Additional adaptability is achieved by applying forward error correction coding to signals and by reducing transmission rates when signals are degraded.

Antenna coverage. The use of Ka-band frequencies makes the very narrow beams of ACTS possible with antennas small enough to allow launch from the space shuttle. The transmit antenna (20 GHz) is approximately 11 ft (3 m) in diameter, while the receive antenna (30 GHz) is approximately 7 ft (2 m) in diameter. **Figure 1** shows projections on the Earth's surface of the beams generated by the ACTS antennas. Three fixed beams are centered on Cleveland, Ohio; Atlanta, Georgia; and Tampa, Florida; and two scanning beams provide contiguous coverage of the northeastern United States. Coverage of

Fig. 1 ACTS multibeam antenna coverage.

Key : ⊚ fixed beam   ◯ isolated scanning beam   ▨ ▤ scanned regions

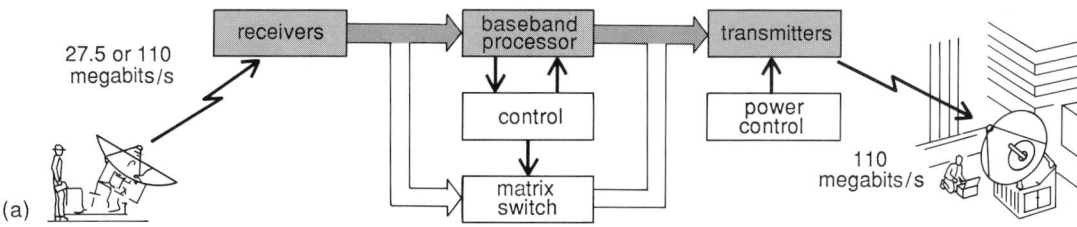

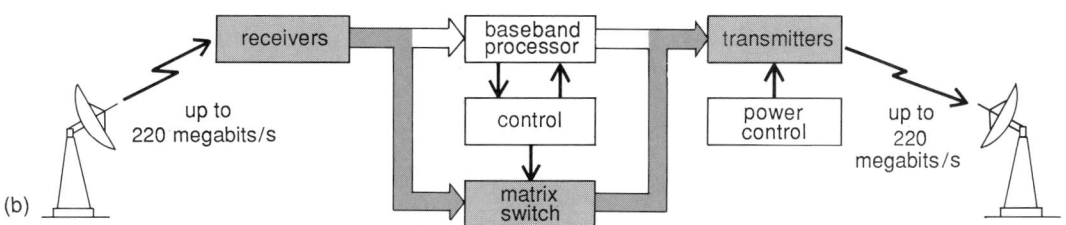

Fig. 2. ACTS data flow. (*a*) Baseband processor mode, using two scanning beams and time-division multiple access. (*b*) Matrix switch mode, using three active fixed beams and either time-division multiple access or frequency-division multiple access.

other major communications centers is provided by an isolated dwell of one of the scanning beams. A 3-ft (1-m) steerable antenna provides coverage of Alaska and Hawaii and other areas not accessed by the primary antennas. Future operational satellites will have additional scanning beams to provide contiguous coverage of the continental United States.

Data flow. The ACTS scanning beam configuration with direct access to small terminals at user locations will provide an effective switchboard in the sky for its users. The flow of signals through the satellite can take place in either of two alternative modes. In the baseband processor mode (**Fig. 2***a*), the baseband processor derives user digital information such as voice or images in digital form or computer data from the received signals. The baseband processor then stores the information, reads a portion of the information regarding destination, and, when the antenna beam is pointed to that destination, transmits the information to the receiving user.

The matrix switch mode (Fig. 2*b*) does not store information in the spacecraft. The path between three fixed beams can be altered by means of the matrix switch. This switch can be used to establish a long-term connection between two ground terminals, duplicating current frequency-division multiple-access practices, or it can, at a very high rate, alter sequentially the interconnections so that each terminal can communicate with each of the other ones in successive time periods by using time-division multiple-access techniques.

Satellite linkage. Linking of several satellites in geosynchronous orbit, using highly directional antennas, will allow worldwide communications without the need to relay through terrestrial facilities in overlapping coverage areas, further preserving the frequency spectrum. ACTS will use a laser communication payload integrated with the Ka-band equipment to allow experi-

ments with such links. These experiments will employ both space-to-ground and space-to-orbiting vehicle links.

Control. Control of the communications functions of the ACTS will be exercised from a master control station located at NASA's Lewis Research Center in Cleveland, Ohio. This station will provide instructions to the baseband processor or matrix switch and periodically update these instructions as required to meet service demands of users.

Spacecraft configurations. **Figure 3** shows the ACTS with solar panels and antennas deployed

Fig. 3. Operational configuration of ACTS flight system. (*RCA Corp.*)

in orbit. At launch the solar panels are folded against the spacecraft faces and the antenna reflectors are folded into the subreflector structure above the spacecraft so that the craft can be mounted in the space shuttle cargo bay. The structural body of the spacecraft measures 84 in. × 80 in. × 70 in. (approximately 2.5 m × 2.5 m × 2 m). The craft measures 48 ft. (14.5 m) across the solar panels and 30 ft (9 m) across the antenna reflectors. Beginning-of-life in-orbit weight will be 2800 lb (1270 kg).

Robert S. Lawton

Galaxy satellite. Satellites have relayed television, telephone, and data signals between continents since the 1960s. Declining cost and increased demand for such services led to the development of domestic communications satellites in the 1970s and early 1980s. Since about 1983, sales or leases covering portions of a satellite have also become prevalent. Combining these two trends, Hughes Communications' Galaxy satellite system provides point-to-point and point-to-multipoint C-band satellite services to a variety of domestic communications users in the continental United States, Alaska, Hawaii, and the Caribbean Basin. Users include broadcasters, cable television program distributors, long-haul carriers, resellers, value-added carriers, and private corporations.

First launched in 1983, the Galaxy system currently consists of three geosynchronous satellites, located at 74°, 93.5°, and 134° West longitude. *Galaxy 1* is the primary United States domestic commercial video relay, while *Galaxy 2* and *Galaxy 3* carry mostly data, private video, and telephone traffic. A fourth C-band satellite at 140°W and Ku-band satellites at 130° and 71°W have been authorized by the Federal Communications Commission.

Spacecraft design. The Galaxy satellites are members of the HS 376 family of spacecraft developed by the Hughes Aircraft Company (**Fig. 4**). All were launched by Delta 3920 vehicles, which placed the spacecraft in an elliptical transfer orbit with its apogee, or high point, at geosynchronous altitude. In its stowed transfer-orbit configuration, each spacecraft spun passively and required no active control. A slit-shaped Sun sensor and two Earth-horizon sensors provided the necessary attitude references. Firing of the internal solid motor at apogee placed the spacecraft in a nearly geosynchronous drift orbit, which was trimmed once the satellite reached its assigned longitude. Deployment of the antenna reflector and telescoping aft solar drum resulted in a configuration 21 ft 8 in. (6.5 m) in height and 7 ft (2 m) in diameter. At the end of its 10-yr life the spacecraft mass will be 1091 lb (495 kg).

The spacecraft consists of the two main sections—spinning and despun—held together at launch by pyrotechnic locks. Once on station, the locks fire and allow the motor connecting the sections to counter the satellite's spin rate and point the despun section at the Earth. The despun section includes the antenna support structure, which is made of composites to minimize thermal distortions, and the main flat equipment shelf holding most of the communications hardware. An annular vertical shelf encircles the equipment shelf and carries the high-power traveling-wave-tube amplifiers. A dual load-path structure carries the despun platform launch loads through a central tube and outboard struts of the lower section to the upper-stage interface ring. The flat spun-section equipment shelf holds the attitude control, power control, and spinning telemetry and command electronics, as well as the batteries.

Spinning the lower portion of the satellite at 50 revolutions per minute provides gyroscopic attitude stability and a scanning motion for the horizon and Sun attitude sensors and the radial thrusters. A counterrotating bearing assembly

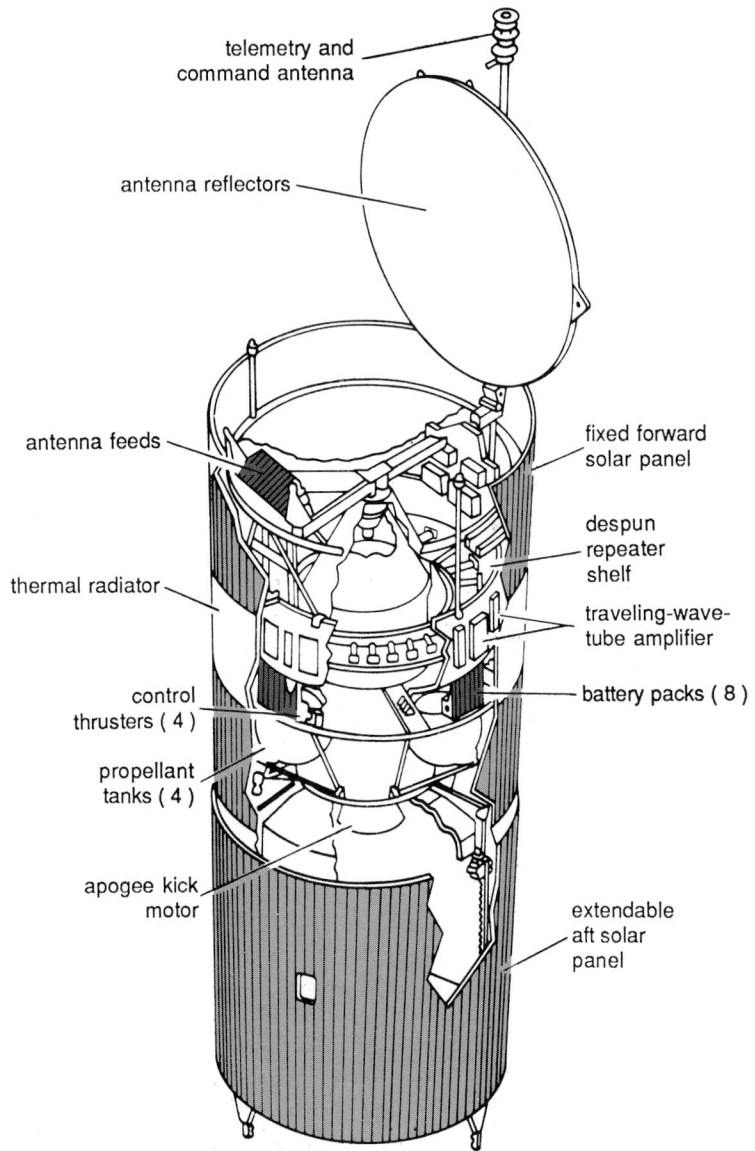

telemetry and command antenna

antenna reflectors

antenna feeds

thermal radiator

control thrusters (4)

propellant tanks (4)

apogee kick motor

fixed forward solar panel

despun repeater shelf

traveling-wave-tube amplifier

battery packs (8)

extendable aft solar panel

Fig. 4. Configuration of Galaxy satellite.

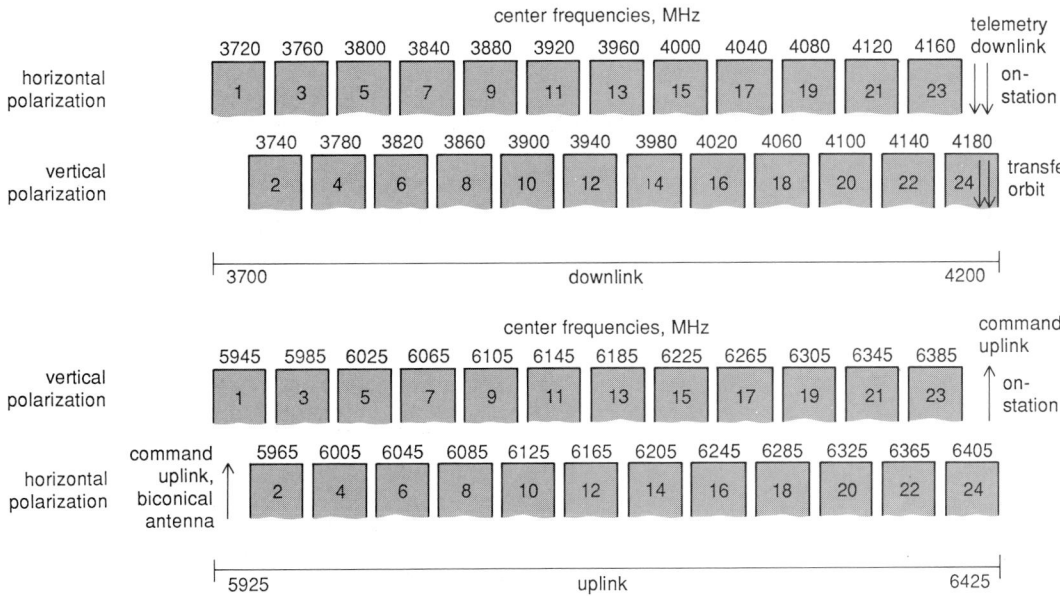

Fig. 5. Frequency plan of Galaxy satellite.

despins the upper portion containing the payload so that the antennas view the Earth; closed-loop tracking of a ground beacon maintains the antenna pointing to within 0.05°. In this gyrostat configuration, nutation, or wobbling motion, tends to accumulate as liquid propellant sloshing in the tanks dissipates energy. To continuously correct this undesired motion, small changes in bearing speed, commanded by the attitude control electronics, act as a torque on the product of inertia on the despun platform. This product of inertia couples the torque so as to continuously damp nutation without using propellant. A thruster-actuated nutation controller is also available for coarse corrections.

The telescoping solar drum supplies 743 W of direct-current power at the end of the satellite's life. Two 32-cell, 26.4-ampere-hour nickel-cadmium batteries, controlled by a voltage regulator during discharge, allow full operation of the equipment connected to the dual power buses when the Sun is eclipsed by the Earth.

Command units on the spun and despun sides of the spacecraft demodulate the frequency-shift-keyed commands sent from the ground control stations and distribute them to the other units and the payload. Companion telemetry encoders multiplex spacecraft data for transmission to the monitoring ground stations. The telemetry and command subsystem also relays the ground beacon to allow ranging to accurately determine the spacecraft's orbit, and provides two-axis communications antenna-pointing information to the attitude control subsystem.

Hydrazine propellant flows from four 19-in.-diameter (0.5-m) conispherical tanks inside the spinning rotor to the two axial and two radial 5-lbf (22-newton) thrusters under helium pressure. Ground operators command the firing of these thrusters to maintain the on-orbit locations to within 0.1° in both east-west and north-south directions. Dual half-systems, each with two thrusters, ensure redundancy.

Passive thermal control using paint and multilayer blankets maintains the spacecraft's temperature balance. A mirror band radiator midway up the upper solar panel rejects waste heat from the communications payload equipment on the internal despun shelf. Heaters augment the passive control on the batteries, apogee motor, and propellant lines.

Communications payload. Each satellite carries a complement of 24 transponders, each with a bandwidth of 36 MHz. They are arranged in a frequency plan (**Fig. 5**) in which the 12 odd-numbered channels have vertical polarization on the uplink from the Earth to the satellite and horizontal polarization on the downlink. Even-numbered channels have this polarization scheme reversed. This approach results in twofold reuse of the 500-MHz spectrum allocated to the satellite at a given orbit location without interference. All channels cover the entire United States and Caribbean service area.

Video, telephony, and data signals are transmitted at 6 GHz from the user's terminal to the satellite, where they are collected by the receive antenna (**Fig. 6**). The single antenna reflector has two surfaces with embedded parallel conductors in them. Each surface reflects only one of the two polarizations and passes the other, allowing separation of the two signals sharing each frequency. Leaving the antenna, the signals pass to one of the four redundant broadband receivers, where they are converted downward in frequency to about 4 GHz. The odd and even channels are separated by filters in the input muliplexers and fed to individual 9-W traveling-wave-tube amplifiers. Following amplification in the traveling-wave-tube amplifiers, output multiplexers combine the odd- and even-numbered channels and pass them to the dual-mode transmit antenna, which radiates them earthward.

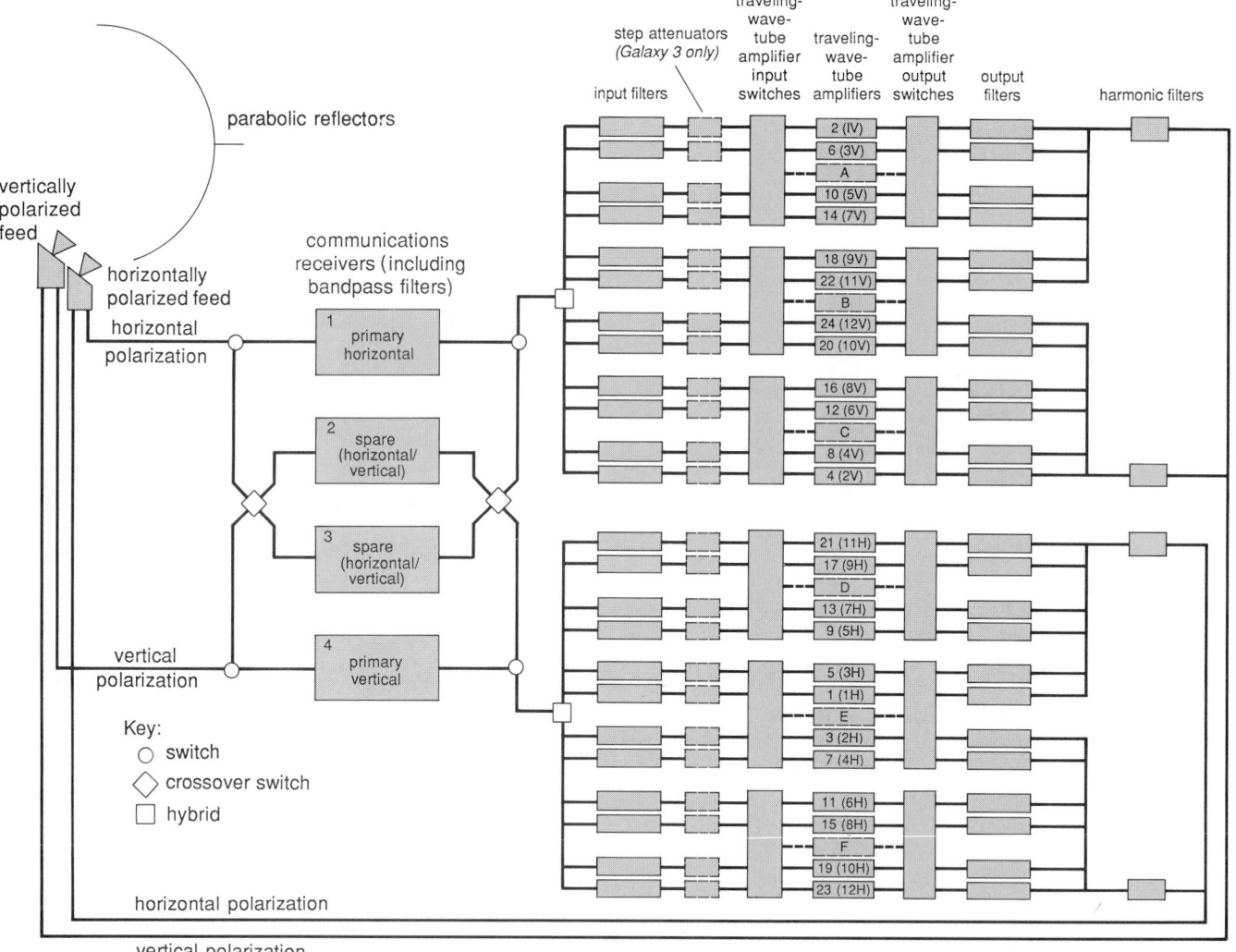

Fig. 6. Design of communications payload of Galaxy satellite. Numbering of traveling-wave-tube amplifiers corresponds to that of channels in Fig. 5. (The alternative scheme, in which channels that have horizontal polarization on the downlink are labeled 1H through 12H and those with vertical polarization are labeled 1V through 12V, is also given.) Spare amplifiers are labeled A through F.

The 24 traveling-wave-tube amplifiers are separated into six groups. Each group of four has a spare traveling-wave-tube amplifier, which can replace any of the four amplifiers. Spare receivers also can switch into the incoming signal path to back up the primary ones. This approach prevents users from being affected by the failure of any single spacecraft component.

David A. Steffy

INMARSAT 2 satellite. Since the introduction of satellite communications in the 1960s, a dramatic increase in the capacity of communications channels between continents and remote locations has taken place, and a large proportion of national and international telephone, video, and data channels are now supported by satellites. The benefits of satellite communications are finally being applied to link ships and offshore rigs with land-based facilities and homes, enabling people at these locations to communicate by telephone and telex with the ease of those served by terrestrial communications links.

Maritime satellite communications were pioneered in 1976 by COMSAT General, with the deployment of the three geosynchronous MARISAT satellites providing 12 communications channels with global coverage. These satellites were later leased from COMSAT to the International Maritime Satellite Organization (INMARSAT), which officially came into being in February 1979 and is now supported by 45 nations.

As maritime traffic rapidly expanded, two dedicated satellites were also leased from the European Space Agency in 1980, each with a capacity of 60 voice channels. Additional maritime communications packages were subsequently implemented on the *INTELSAT V* satellites, augmenting the overall service capacity needs.

Second-generation satellites are now being procured by the INMARSAT organization from the British Aerospace Corporation as the main contractor. The spacecraft platform is being constructed by British Aerospace in cooperation with several European countries, while the com munications module is being designed by the Hughes Aircraft Company.

Each spacecraft will be able to support the equivalent of at least 200 simultaneous voice channels in the shore-to-ship and ship-to-shore directions. In addition, the satellites will provide voice and data communication to commercial, domestic, and transoceanic airlines, a new service authorization recently acquired by the INMARSAT organization.

Three spacecraft are now under construction, with the launch date for the first satellite planned for late 1988 and the subsequent two to be launched at 4-month intervals. The second-generation space segment will thus comprise three satellites in geosynchronous orbit, located over the Atlantic Ocean (at approximately 26° West longitude), the Indian Ocean (at approximately 65°E), and the Pacific Ocean (at approximately 177°E). The physical characteristics of the satellites are designed to be compatible with both the Ariane launcher and the Space Transportation System using the Payload Assist Module (PAM-D).

Satellite design. The operational design life of each satellite, including the fuel capacity to maintain the orbital position, is 10 years, with the initial on-station mass to be 2800 lb (1270 kg). The spacecraft configuration (**Fig. 7**) is of the three-axis stabilized type with a main body that consists of a boxlike structure, and solar power panels that extend from the north and south sides of the body. The solar panels will rotate once per day relative to the spacecraft body in order to face the Sun. The so-called antenna farm is located on the side of the spacecraft facing the Earth.

The tracking, telemetry and command subsystem, using C-band frequencies, provides command reception, telemetry transmission, and ranging functions. During the transfer orbit, a deployed C-band antenna is used, in conjunction with fill-in telemetry horns, to provide pseudo-omnidirectional coverage, while Earth-facing, higher-gain, wide-beam horn antennas are utilized during on-station operation.

The attitude determination and control subsystem uses a momentum-bias, three-axis stabilization, employing independently powered momentum wheels that provide gyroscopic rigidity and torque control. Redundant infrared Earth sensors provide the required reference data.

The primary structure of the satellite comprises a central cone and cylinder with the external panels that form a rectanguloid configuration. The liquid apogee engine is mounted within the center structure supporting the propellant tanks.

Primary solar power for the spacecraft is provided by two fold-out array wings which are attached to the opposite faces of the spacecraft body by a yoke and a solar array drive mechanism. The solar array is designed to provide 1200 W of direct-current power at the end of the 10-year mission. Energy during eclipse periods is

Fig. 7. Deployed configuration of *INMARSAT 2* satellite. (*INMARSAT*)

derived from two nickel-cadmium batteries. The array and battery outputs are conditioned to provide a single regulated bus voltage.

Thermal control design is based upon passive techniques augmented by electric heaters. Primary heat rejection to space is accomplished by radiating surfaces located on the north-south external spacecraft panels.

The propulsion subsystem uses a bipropellant consisting of hydrazine and nitrogen to provide orbital transfer, attitude, and orbital control maneuvers. During the on-station normal operating mode, solar sailing, a novel feature, will be employed to minimize the east-west stationkeeping maneuver requirements, thus increasing the satellite operational lifetime.

Communications subsystems. The communications subsystem consists of two functionally independent, shore-to-ship (C-band to L-band) and ship-to-shore (L-band to C-band) transponders. The communications link between the shore stations and the spacecraft is accomplished by using C-band frequencies, while the link between the spacecraft and ships uses the L-band (1.6-GHz uplink and 1.5-GHz downlink).

The shore-to-ship (C-to-L) transponder uses a C-band, receiving-array, global-coverage antenna to receive the multiple uplink signals from the coastal Earth stations at a frequency of 6 GHz, followed by a low-noise receiver with frequency down-conversion to the L band. This in turn is followed by a high-power amplifier assembly, utilizing four parallel and linearized L-band traveling-wave-tube amplifiers that deliver 180-W power at a frequency of 1.5 GHz.

The combined output is fed through the output filters to the L-band transmitting planar array. The antenna, utilizing 61 radiating elements, provides for a shaped-beam, global-coverage pattern allowing an increased gain at the edge of the area of coverage, to compensate for the low elevation angle of the satellite when viewed from this region.

The ship-to-shore (L-to-C) transponder uses an L-band receiving array antenna to receive multiple low-level uplink signals at a frequency of 1.6 GHz from ships located anywhere within the hemispherical coverage of the satellite. The antenna is followed by a low-noise receiver that performs frequency conversion to an intermediate frequency for filtering. At the intermediate frequency, selective filters are used to constrain the transmission to the allocated bandwidth. The bandwidth-defined signals are frequency-up-converted to the C-band (3.6 GHz) and applied to a linearized 8-W, C-band traveling-wave-tube amplifier. The output, after further filtering, is applied to a C-band, global-coverage, planar array for retransmission to the Earth stations. All the communications antennas utilize right-hand-circular polarization with the exception of the satellite-to-shore transmitter, which uses left-hand-circular polarization.

For background information SEE COMMUNICATIONS SATELLITE in the McGraw-Hill Encyclopedia of Science and Technology.

Edmund Jurkiewicz

Bibliography. J. F. Farrell, Hughes Communications' Galaxy satellite system, *AIAA Conf.*, March 1986; Inmarsat satellite order, *Aviation Week Space Tech.*, 122(16):30, April 22, 1985; M. Kachmar, Switchboards in the sky, *Microwaves and RF*, pp. 35–38, June 1986; L. Moy, Hard ACTS to follow, *Space*, 2(1):4–12, March–May 1986; *Ocean View*, monthly publication of INMARSAT; J. Poyser, Ka-band is coming, *Satellite Dealer*, 3(11):60–62, July 1986; R. Stoddard, Maintaining as no. 1, *Satellite Commun.*, 10(10):20–22, September 1986.

Complex numbers and complex variables

A major mathematical event occurred in 1984 when L. de Branges proved the Bieberbach conjecture, a problem in complex analysis which had stood unresolved since 1916. Although the result has few consequences, the problem was so easily understood and so elegant that it attracted the efforts of many mathematicians. In attempting to find a solution, they devised powerful methods which have led to results in geometric function theory more significant than the Bieberbach conjecture itself. This article describes the problem

and its history, and indicates some of the ideas involved in the proof.

Conformal mappings. A complex-valued function f of a complex variable $z = x + iy$ is said to be analytic at a point z_0 if it has a derivative there. If f is analytic near z_0, then it has derivatives of all orders and a (unique) power series expansion at z_0 given by Eq. (1). A function f is said to be univalent in a region D of the complex plane if $f(z) \neq f(\zeta)$ for z and ζ in D unless $z = \zeta$. A func-

$$f(z) = f(z_0) + f'(z_0)(z - z_0) +$$
$$\frac{1}{2!}f''(z_0)(z - z_0)^2 + \cdots \quad (1)$$

tion which is analytic and univalent in D is said to provide a conformal mapping of D onto another region, called the range of f or the image of D under f. A conformal mapping $w = f(z)$ from the z plane to the w plane preserves the angles between curves (**Fig. 1**). The derivative $f'(z_0)$ is a nonzero complex number which indicates the behavior of f near z_0. Its modulus $|f'(z_0)|$ is the local magnification factor of arc length, while $|f'(z_0)|^2$ is the local magnification factor of area, usually called the jacobian. Its argument arg $f'(z_0)$ gives the angle of local rotation of curves passing through z_0.

Conformal mappings are useful in two-dimensional problems of fluid flow or electrostatics, because they allow a problem given in a complicated region to be transformed to a simple region such as an infinite strip or a circular disk, where the problem is easily solved. Thus, conformal mappings provide appropriate changes of coordinates for many mathematical and physical problems. For this reason and for their intrinsic interest, they have been studied actively for many years.

Background of the Bieberbach conjecture. A function f is said to belong to the class S if it is analytic and univalent in the unit disk (the set of all z with modulus $|z| = \sqrt{x^2 + y^2} < 1$), and if it is normalized so that $f(0) = 0$ and $f'(0) = 1$. This amounts to requiring that f have a power series expansion of the form given by Eq. (2) for certain complex coefficients given by Eq. (3).

$$f(z) = z + a_2 z^2 + a_3 z^3 + \cdots \quad (2)$$
$$|z| < 1$$

$$a_n = \frac{f^{(n)}(0)}{n!} \quad (3)$$

Some simple examples of functions of class S are the identity mapping $f(z) = z$; the function $f(z) = z(1 - z)^{-1}$, which maps the disk onto the half-plane Re $\{z\} > -\frac{1}{2}$; and the Koebe function, given by Eq. (4), which maps the disk onto the

$$k(z) = z(1 - z)^{-2} = z + 2z^2 + 3z^3 + \cdots \quad (4)$$

entire complex plane minus the part of the nega-

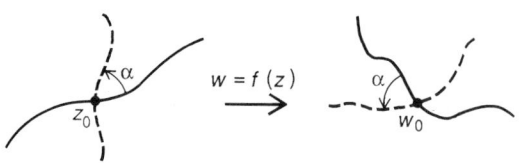

Fig. 1. Angle-preserving property of conformal mappings.

tive real axis from $-\frac{1}{4}$ to $-\infty$ (**Fig. 2**). The existence of much more general examples is assured by the classical Riemann mapping theorem, which asserts that every simply connected domain in the complex plane (except for the whole plane) can be mapped conformally onto the unit disk. Indeed, if g maps the unit disk conformally onto a given domain D, then $f = [g - g(0)]/g'(0)$ belongs to S and has a range obtained from D by simple translation, magnification, and rotation. Because these operations are reversible, any conformal mapping of the disk is similarly obtained from a function of class S.

The study of univalent functions as a special branch of complex analysis began around 1910 with the ideas of P. Koebe. His work led to the discovery that suitable rotations of the Koebe function, having the form given by Eq. (5) for

$$e^{-i\theta}k(e^{i\theta}z) = z(1 - e^{i\theta}z)^{-2} \qquad (5)$$

some value of θ, maximize or minimize the local distortion $|f'(z_0)|$ of a function f in S at a given point z_0 of the disk. Another basic result, the Koebe one-quarter theorem, says that the range of each function in S must contain the disk $|w| < \frac{1}{4}$. The Koebe function, which omits the value $-\frac{1}{4}$, again plays an extremal role. This evidence led L. Bieberbach to make his conjecture that the Koebe function and its rotations (uniquely) maximize the modulus of each coefficient; that is, $|a_n| \leq n$ for all n ($n = 2, 3, \ldots$) for every function f of class S. Bieberbach proved that $|a_2| \leq 2$.

Attempts at proof. In 1923 K. Löwner (later known as C. Loewner) introduced a basic representation of slit mappings in terms of a partial differential equation of special form. (Here a slit mapping means a function in S which maps the disk onto the entire complex plane minus an unbranched arc extending to infinity. The Koebe function is the simplest example. The slit mappings form a dense subfamily of S in a certain natural sense.) Loewner used his method to prove that $|a_3| \leq 3$, but was unable to proceed to higher coefficients. Subsequently Loewner's method became a standard tool in the theory of univalent functions, with diverse and impressive applications. For example, the method was used in 1936 to obtain the sharp bounds on the local rotation $\arg f'(z_0)$ under a conformal mapping of class S. But subsequent progress on the Bieberbach conjecture followed other approaches.

The conjecture was easily verified for functions with special properties, such as real coefficients or starlike range. (A region is said to be starlike if the line segment joining any of its points to the origin lies entirely in the region.) In 1925 the correctness of the full conjecture in order of magnitude was established when it was shown that $|a_n| \leq en$, where $e = 2.718 \ldots$ is the base of natural logarithms. The value of the constant in this inequality was gradually whittled down from e to a number slightly less than 1.07 by 1978.

Beginning in the 1930s, M. M. Schiffer devel-

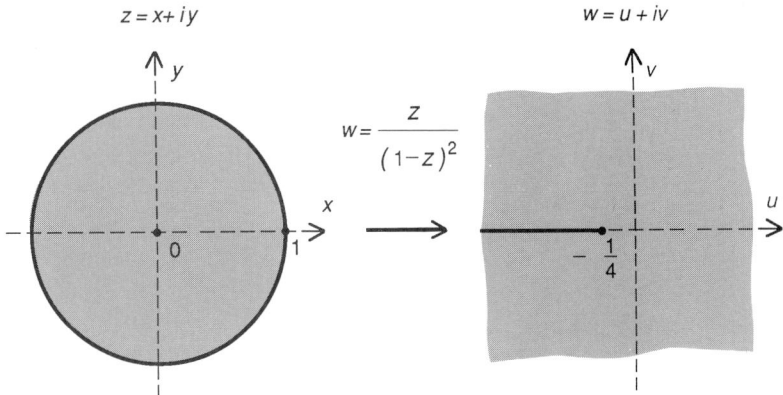

Fig. 2. Mapping property of the Koebe function.

oped a calculus of variations for univalent functions. His method has enjoyed great success in geometric function theory and in mathematical physics. When applied to extremal problems over the class S, Schiffer's method gives geometric information about the extremal functions. Typically it makes it possible to infer that each extremal function satisfies a certain differential equation (analogous to the Euler-Lagrange equation in the classical calculus of variations) and maps the disk onto the complement of a system of smooth arcs with special properties. For Bieberbach's problem it had shown that each function maximizing $|a_n|$ must map the disk onto the complement of a single arc extending to infinity with increasing modulus and having other nice properties. The resolution of Bieberbach's conjecture was equivalent to proving that this arc is a radial half-line. In 1955 the variational method, in combination with Loewner's method, led to a proof of the Bieberbach conjecture for $n = 4$. The cases $n = 6$ and $n = 5$ were settled in 1968 and 1972 by techniques which seemed to offer little hope of a general solution.

During the 1960s, I. M. Milin focused attention on the logarithmic coefficients of a function in S, defined by Eq. (6). The logarithmic coefficients

$$\log \frac{f(z)}{z} = 2(\gamma_1 z + \gamma_2 z^2 + \cdots) \qquad (6)$$

of the Koebe function are $\gamma_n = 1/n$. Milin's work led him to conjecture that inequality (7) holds for

$$\sum_{k=1}^{n} (n - k + 1)\left(k|\gamma_k|^2 - \frac{1}{k}\right) \leq 0 \qquad (7)$$

$$n = 1, 2, \ldots$$

all functions f in S. He showed that this inequality, if true, would imply the Bieberbach conjecture and several intermediate conjectures which had been proposed over the years.

De Branges's proof. In 1984 de Branges announced a proof of the Milin conjecture (and hence of the Bieberbach conjecture). His proof used operator theory and special functions and was embedded in a long manuscript with much extraneous material, making confirmation difficult. By coincidence, he was then preparing for an official scientific visit to the Steklov Mathematical Institute in Leningrad. There, univalent function theorists, notably Milin, E. G. Emelianov, and A. Z. Grinshpan, helped to extract the essential elements of the proof and to simplify the presentation so that it could be readily confirmed by specialists in the field. C. H. FitzGerald and Ch. Pommerenke then found a further simplification. Ironically, the final form of the proof does not introduce essentially new ideas but relies only on Loewner's method and an inequality for Jacobi polynomials discovered previously in a totally different context by R. Askey and G. Gasper. The basic approach is simply to show that the sum to be proved negative is the initial value $\varphi(0)$ of a real-valued function $\varphi(t)$ which increases to zero as t increases from zero to infinity. The derivative $\varphi'(t)$ is calculated with the help of Loewner's differential equation, and the Askey-Gasper inequality is applied to show that $\varphi'(t) \geq 0$.

For background information SEE COMPLEX NUMBERS AND COMPLEX VARIABLES; CONFORMAL MAPPING in the McGraw-Hill Encyclopedia of Science and technology.

Peter L. Duren

Bibliography. A. Baernstein et al. (eds.), *The Bieberbach Conjecture: Proceedings of the Symposium on the Occasion of the Proof*, American Mathematical Society, 1986; L. de Branges, A proof of the Bieberbach conjecture, *Acta Math.*, 154:137–152, 1985; P. L. Duren, *Univalent Functions*, 1983; C. H. FitzGerald and Ch. Pommerenke, The de Branges theorem on univalent functions, *Trans. Amer. Math. Soc.*, 290:683–690, 1985; Ch. Pommerenke, The Bieberbach conjecture, *Math. Intelligencer*, 7(2):23–25, 32, 1985.

Computer-integrated manufacturing

The technology of flexible manufacturing systems has been developing since the late 1960s. Broadly defined, flexible manufacturing systems are programmable and flexible manufacturing processes that utilize unattended processing stations. Such a system converts manufacturing processes which are elemental and batch-oriented to random continuous processes. The general goal is to achieve the production control ideal of having a workpiece at the right time at the right place. The technology converts discrete, batch-oriented workpiece manufacturing to the equivalent of a continuous process by controlling programmable machine tools and work stations, with workpieces arriving randomly.

Metal-cutting applications. Flexible manufacturing systems were first applied to the metal-cutting

process, primarily to prismatic or box-shaped workpieces, such as transmissions housings, pumps, and engine blocks. The systems were initially designed to solve the problems of midvolume producers, manufacturers of products such as agricultural equipment or machinery for construction.

Figure 1 shows a matrix created in the early 1970s to convey the concept of flexible manufacturing systems as a solution to the needs of the midvolume manufacturers. The midvolume manufacturers' needs were not being addressed from the standpoint of providing maximum automation, that is, making a part in one place, as were the high-volume producers with transfer lines (used in highly dedicated processes for high-volume production) or the low-volume producers with numerically controlled machining centers on a stand-alone basis. In flexible manufacturing technology, programmable machine tools are integrated with automated materials handling, and a control system allows random delivery of parts for a family of workpieces (**Fig. 2**).

Basically, new developments have accompanied the ever-expanding role of flexible manufacturing systems. The midvolume producer has applied the system concepts to broader uses, while the concepts have been expanded to both the high-volume and low-volume ends of the production spectrum (Fig. 1).

The scope of flexible manufacturing systems has expanded. From the receipt of raw material to the delivery of finished workpieces, the system now includes a greater extent of the manufacturing process, involving not only the storage and acquisition of raw materials but also quality assurance and automated assembly.

The initial applications of flexible manufacturing systems to large, expensive prismatic workpieces are now being extended to other metal-cutting processes, such as those used to manufacture round parts, for example the shafts and gears that go inside the large prismatic workpieces. Also under development are sheet-metal applications. The technology for handling workpieces that require elemental processing on a random basis in metal cutting is now being extended to processes other than metalworking, such as manufacturing of rubber products.

As flexible manufacturing systems are developed and expanded, less direct human attention is required in the manufacturing process. This intensifies the development of the technology for many types of sensors for monitoring of equipment used in manufacturing, including application of heuristic techniques and artificial intelligence.

The midvolume producers were the first to apply flexible manufacturing system concepts, and they continue to use the primary applications. The trend in this area is to even greater flexibility in the application of the processing work stations. For instance, in metal cutting, the more dedicated process of machining by means

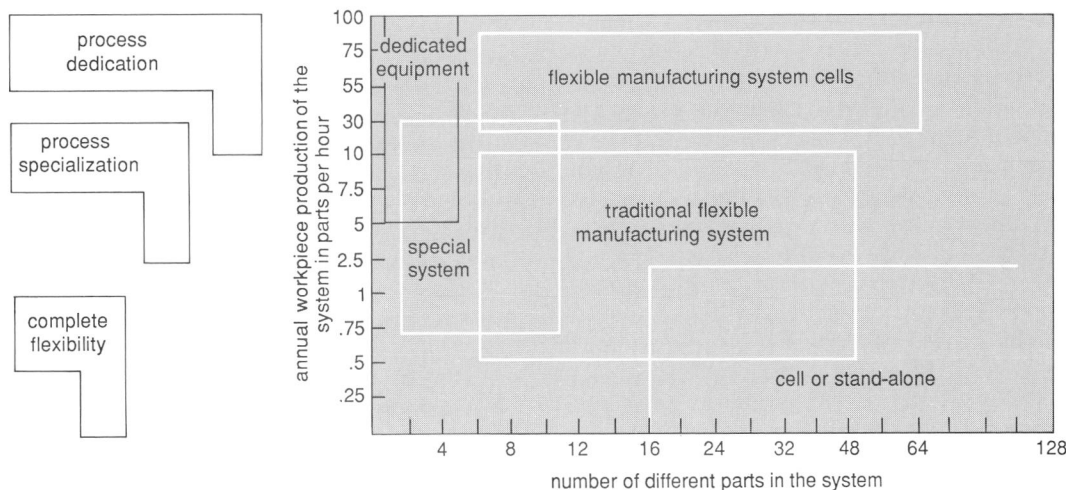

Fig. 1. Matrix that shows the relationship between product volume and product variety. As the annual production of a workpiece increases, the process becomes more dedicated.

of multiple-spindle heads has given way to machining using single-spindle numerically controlled machining centers.

Extended production-volume applications. The successes in flexible manufacturing systems enjoyed by the midvolume producers have led to development of applications in both the high-

volume and low-volume ends of the production spectrum. Manufacturers in the high-volume end, for example, manufacturers of automotive and other consumer-oriented goods, see flexible manufacturing systems as a means of reacting more efficiently and more rapidly to the demands of the marketplace in terms of both production

Fig. 2. Flexible manufacturing system for the midvolume producer. (*Kearny and Trecker Corp.*)

volume and production variety. Rather than dedicated processes (such as transfer lines), the flexible systems use universal machine tools (for example, numerically controlled lathes and machining centers) in conjunction with robots to form cells, which are further linked with automatic guided-vehicle systems and automated storage and retrieval systems. This configuration provides very high-volume production.

At the low-volume end of the production spectrum, flexible manufacturing systems are being designed for the manufacture of a lot size of a single part. This system is capable of bringing the workpiece, fixtures, tooling, and associated programs to the machine tool; manufacturing a single part; and then bringing in a totally different part and all of its associated tooling, fixtures, and programs. Thus, the flexible manufacturing system technologies provide high levels of tool management control and tool migration, aided by system software management tools which facilitate the simulation and production management of the system. This capability is added to the basic flexible manufacturing system control software of workpiece routing, scheduling, tool management, fixture management, human system interfacing, reporting, materials handling, maintenance, and internal inventory management.

Extension of scope. The scope of early flexible manufacturing systems extends from the receipt of raw materials to the completion of finished workpieces. As users analyzed their total factory requirements more thoroughly, the systems were extended to include raw-materials acquisition, storage, delivery to machining systems, quality monitoring, and assembly operations. Quality assurance has become an integral part of the flexible manufacturing system, through either postinspection of the workpiece with computer-controlled coordinate measuring machines or in-process quality monitoring by operation. The systems are now extended to materials handling and storage. Automatic guided vehicles and automated storage and retrieval systems are used in these applications as well as extended control communications among the stations for materials handling, storage, processing, measurement, and assembly.

Application to other processes. Flexible manufacturing systems were initially applied to prismatic workpieces, because they were the largest, required the most materials-handling effort, and were the most expensive components. As the applications of these systems were extended to reduce the inventory level of the prismatic workpieces, by moving from a batch inventory management system to a random real-time processing, it seemed logical to extend this inventory philosophy to shafts, gears, and other round parts. Systems of numerically controlled lathes,

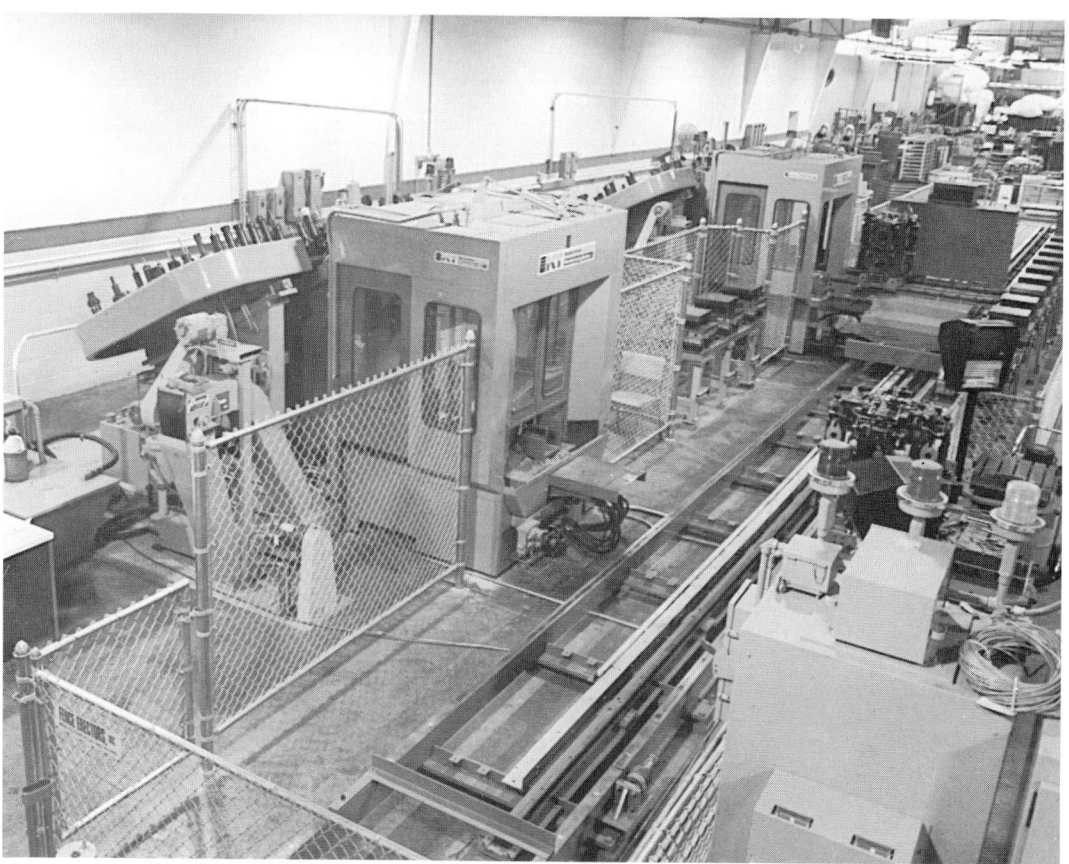

Fig. 3. Flexible manufacturing cellular system for metal cutting. (*Kearny and Trecker Corp.*)

or total gear manufacturing systems, are being developed. These systems integrate the processes of heat treating and grinding with turning in order to provide the capability of finishing the workpiece within the system. This same strategy of developing systems to manufacture a complete workpiece on a random basis is also being applied to sheet-metal workpieces that require metal forming.

Because metal cutting is a relatively expensive process, the metal-cutting industry has been the first to apply the technology of flexible manufacturing systems to its products. Obviously, the control systems developed can control any process that involves a discrete-element batch orientation. Thus applications of flexible manufacturing system controls and software are beginning to be made to processes other than metal cutting. For instance, a manufacturer of V-belts for power transmission is applying this system to its manufacturing needs. The technology can also be extended to woodworking, plastic processing, and others.

Cellular system. As flexible manufacturing technology has matured, generalized solutions have begun to appear. An example is the flexible manufacturing system cell, or cellular system for metal cutting (**Fig. 3**). This type of cellular system normally has one to six processing stations linked with a simplified materials handling system, that is, a shuttle car and a simplified level of control. The user can then develop a large-scale flexible manufacturing system in increments.

Technological advances. Flexible manufacturing systems require a minimum of human attention to the various processes. Sensors are used to duplicate human controls. Probe or touch technology, for quality control, tool monitoring, and workpiece identification or other identification requirements, has been widely applied in flexible manufacturing systems. Machine vision for these applications is also being developed.

Research is also concerned with the use of voice commands to the machine tool, to the system, to the engineering computers for processing, or for feedback from the machines.

Computer integrated manufacturing (CIM) connections are appearing in the transfer of engineering data to flexible manufacturing systems directly, and also in connections to corporate management information systems such as the material resource planning (MRP) system. Communications are being developed to control processes other than metal cutting and metal forming; they are being extended into total factory communication between all processes through protocol applications like manufacturing automation protocol (MAP).

The use of computers and artificial intelligence in these systems leads to the development of heuristic models that allow the processing station, such as a machine tool, to learn from its environment, to store the learned information within itself, and to make the appropriate calibrations for more precise performance of the machines. This capability, in turn, will lead to greater predictability in processing. Techniques for automatic loading of of the workpieces into the system and unloading out of the system are being pursued and some applications exist.

For background information SEE AUTOFACTUR-ING; COMPUTER-INTEGRATED MANUFACTURING; PLANT FACILITIES (INDUSTRY); TOOLING in the McGraw-Hill Encyclopedia of Science and Technology.

Paul R. Haas

Bibliography. E. Teicholz and J. N. Orr, *Computer Integrated Manufacturing Handbook*, 1987.

Coordination chemistry

Recent research involving the chemistry of metal-thiolate coordination compounds represents an area of inorganic chemistry which has created much interest and activity. Inspiration for this work has come from the discovery of the significance of metal-sulfur centers in metalloproteins as well as the occurrence of metal-sulfur centers in heterogeneous catalysts, solid-state materials, and minerals. In general, organic thiols [RSH; **Fig. 1**] react with metal ions [M] to form metal-thiolates [$M(SR)_2$], as shown in reaction (1), where R represents an organic functional

$$2RSH + MCl_2 \rightarrow M(SR)_2 + 2HCl \qquad (1)$$

group. The thiolate ligand (RS^-) can coordinate to the metal as either a terminal ligand or a bridging ligand. In comparison, the inorganic equivalent of a thiolate (HS^-) usually reacts with metals to form metal sulfides (MS) because of the enhanced acidity of the proton [reaction (2)]. The

$$H_2S + MCl_2 \rightarrow MS + 2HCl \qquad (2)$$

sulfide anion (S^{2-}) frequently coordinates as a

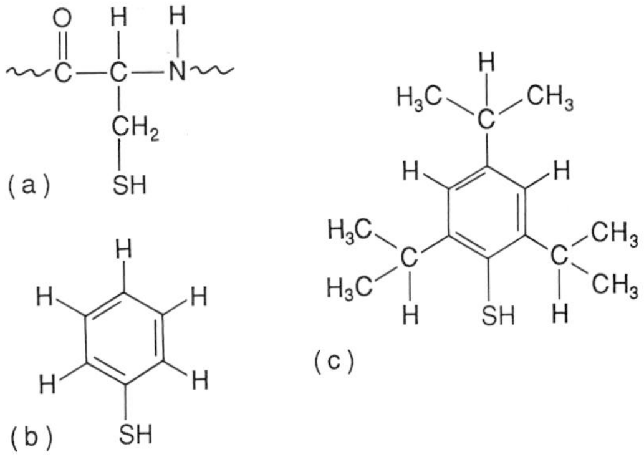

Fig. 1. Organic thiols. (*a*) Cysteine. (*b*) Benzenethiol. (*c*) 2,4,6-Triisopropylbenzenethiol.

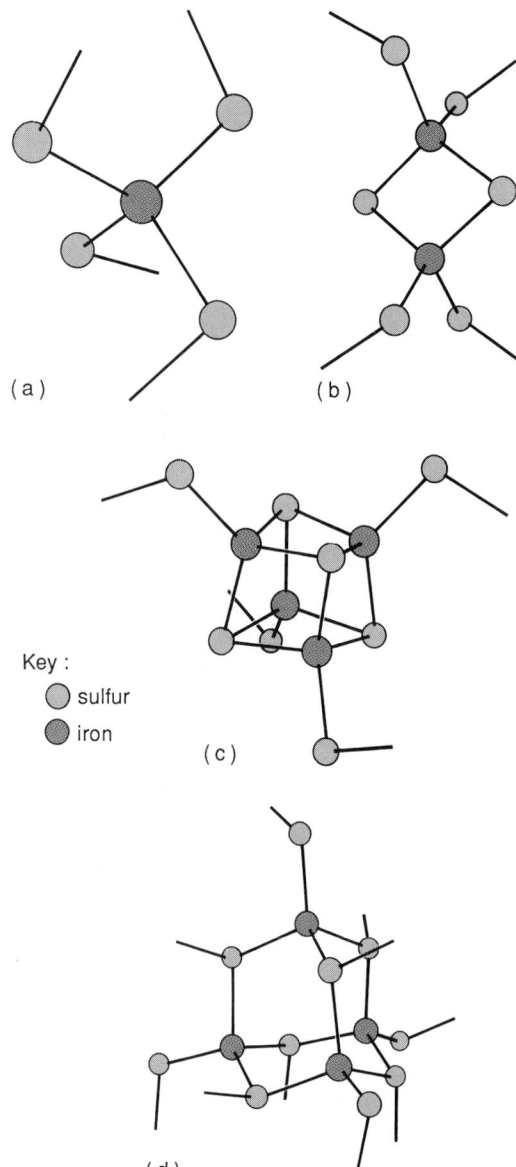

Key :

⚪ sulfur

⚫ iron

(a)

(b)

(c)

(d)

Fig. 2. X-ray structural diagrams of various Fe-S centers. (a) $[Fe(SR)_4]^-$. (b) $[Fe_2S_2(SR)_4]^{2-}$. (c) $[Fe_4S_4(SR)_4]^{2-}$. (d) $[Fe_4(SR)_{10}]^{2-}$.

terization of these proteins revealed that their Fe-S centers are coordinated to thiolate (cysteinyl sulfurs of the protein) and to sulfide ligands in a number of combinations. These Fe-S centers contain one, two, three, or four iron atoms. Moreover, an Fe-S protein can have more than one center or one type of center.

Classification. The Fe-S proteins can be classified according to the number of iron atoms contained at the active site. The structure of the three major types of Fe-S centers are shown in **Fig. 2**a, b, and c. The rubredoxins (Rd) contain one iron atom coordinated by four cysteines in a distorted tetrahedral array. The ferredoxins (Fd) contain either two or four iron atoms in the Fe-S center. In the 2-Fe ferredoxins, the iron atoms are connected by two bridging sulfide ligands, and each iron atom is also coordinated to two cysteinyl sulfides with a tetrahedral coordination geometry about each iron. The 4-Fe ferredoxins contain four iron atoms and four sulfide ligands joined in an approximate cubic arrangement, with each iron atom coordinated to three sulfides; the fourth ligand is the sulfur of a cysteine residue. **Figure 3** shows a schematic representation of an Fe-S protein which contains two 4-Fe clusters.

Synthesis. The evolution of the discovery and characterization of the unique Fe-S centers in proteins was complemented by fundamental research by bioinorganic chemists trying to synthesize and study simple coordination compounds as models for the Fe-S sites in proteins. Extremely successful results have come from the quest to accurately reproduce Fe-S active sites of the proteins in simple inorganic analogs for the purpose of elucidating the structural, electronic, and reactivity properties of the Fe-S centers in proteins. By using simple organic thiolates (such as benzene thiolate, PhS$^-$, where Ph represents the phenyl group, C_6H_5) to mimic the cysteine sul-

bridging ligand to two or more metal centers. Compounds of the type $M(SR)_2$ and MS are often insoluble, polymeric materials.

Current efforts in this area have centered upon the synthesis and characterization of new types of soluble metal-thiolate and metal-sulfide-thiolate compounds. During the course of this research, significant new transition-metal compounds and new chemistry have been discovered. To illustrate the importance and the extent to which the chemistry of metal-thiolate compounds has been developed recently, this article focuses on the chemistry of just one of the transition metals, iron.

Metalloproteins containing iron-sulfur (Fe-S) centers constitute a large class of proteins which are found in all known living organisms. The discovery of this very important group of electron-transfer proteins did not occur until the 1960s. Subsequent x-ray crystallographic charac-

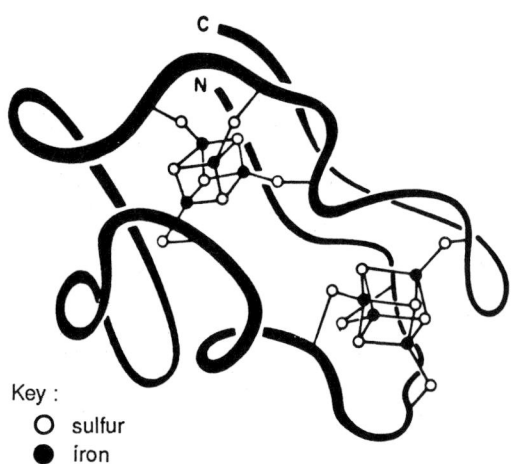

Key :

○ sulfur

● iron

Fig. 3. Schematic structure of *Peptococcus aerogenes* ferredoxin (oxidized). (*After L. H. Jensen et al., Structures of two non-heme iron-sulfur proteins: Rubredoxin at 0.15 nm (1.5 Å) resolution and ferredoxin at 0.20 nm (2.0 Å) resolution, Biochem. Soc. Trans., 1:27–29, 1973*)

furs from the protein backbone and by carefully controlling reaction conditions, the three basic Fe-S centers in proteins have been reproduced by a number of synthetic schemes.

The simplest routes to the synthesis of the inorganic analogs of the Fe-S proteins are as follows: The reaction of greater than four equivalents of lithium thiolate [(RS)Li] with $FeCl_2$ leads to the formation of the analog for rubredoxin, $[Fe(SR)_4]^{2-}$ [reaction (3)]. If less than four

$$FeCl_2 + 4RS^- \rightarrow [Fe(SR)_4]^{2-} + 2Cl^- \quad (3)$$

equivalents of (RS)Li are used, then the anion isolated, $[Fe_4(SR)_{10}]^{2-}$ (Fig. 2d), is a cluster containing four irons which are held intact by six bridging thiolate ligands [reaction (4)]. The reac-

$$4FeCl_2 + 10RS^- \rightarrow [Fe_4(SR)_{10}]^{2-} + 8Cl^- \quad (4)$$

tion of $[Fe(SR)_4]^{2-}$ with sulfur in acetonitrile generates $[Fe_2S_2(SR)_4]^{2-}$, which is an analog for the 2-Fe ferredoxins [reaction (5)]. The analogs

$$2[Fe(SR)_4]^{2-} + 2S \rightarrow$$
$$[Fe_2S_2(SR)_4]^{2-} + RSSR + 2RS^- \quad (5)$$

for the 4-Fe ferredoxins, $[Fe_4S_4(SR)_4]^{2-}$ (Fig. 2c), are synthesized by the reaction of $[Fe_4(SR)_{10}]^{2-}$ with sulfur in methanol [reaction (6)] or by the

$$[Fe_4(SR)_{10}]^{2-} + 4S \rightarrow [Fe_4S_4(SR)_4]^{2-} + 6RSSR \quad (6)$$

reduction of $[Fe_2S_2(SR)_4]^{2-}$ [reaction (7)]. All of

$$2[Fe_2S_2(SR)_4]^{2-} \rightarrow$$
$$[Fe_4S_4(SR)_4]^{2-} + RSSR + 2RS^- \quad (7)$$

the anionic Fe-S compounds are precipitated as air-sensitive crystals with organic cations such as R_4N^+.

Synthesis of models for the three major types of Fe-S centers in proteins has allowed chemists to obtain precise crystallographic, spectroscopic, and magnetic data on pure, well-characterized Fe-S compounds. Comparisons of data acquired on the proteins and synthetic analogs have provided a means of gaining insight into the behavior of the different types of Fe-S centers.

Redox series. Since the Fe-S proteins function primarily as electron-transfer centers, the identity and properties of the discrete redox levels of each of the Fe-S centers are significant. Electrochemical studies have demonstrated the extent of the electron-transfer series for each Fe-S center [reactions (8)–(10), where the subscripts ox and

$$[Fe(SR)_4]^{1-} \rightleftharpoons [Fe(SR)_4]^{2-}$$
$$Rd_{ox} \rightleftharpoons Rd_{red} \quad (8)$$
$$[Fe(III)] \qquad [Fe(II)]$$

$$[Fe_2S_2(SR)_4]^{2-} \rightleftharpoons [Fe_2S_2(SR)_4]^{3-}$$
$$Fd_{ox} \rightleftharpoons Fd_{red} \quad (9)$$
$$[2Fe(III)] \qquad [Fe(II) + Fe(III)]$$

$$[Fe_4S_4(SR)_4]^{1-} \rightleftharpoons [Fe_4S_4(SR)_4]^{2-} \rightleftharpoons [Fe_4S_4(SR)_4]^{3-}$$
$$HP_{ox} \rightleftharpoons HP_{red}; Fd_{ox} \rightleftharpoons Fd_{red} \quad (10)$$
$$[3Fe(III)+Fe(II)] \quad [2Fe(III)+2Fe(II)] \quad [Fe(III)+3Fe(II)]$$

red represent oxidized and reduced, respectively, HP_{ox} represents the oxidized level of high-potential iron-sulfur proteins, and HP_{red} represents the reduced level of high-potential iron-sulfur proteins].

The members of each redox series are related by a one-electron-transfer reaction, with the formal electron count tallied in the number of Fe(II) and Fe(III) ions. Correlation of physical properties has established the correspondence between the redox-active levels of the Fe-S center in the proteins and the synthetic analogs. The physiological redox potential of the $Rd_{ox} \rightleftharpoons Rd_{red}$ couple occurs at about -0.05 V versus the standard hydrogen electrode. The redox potential for the $Fd_{ox} \rightleftharpoons Fd_{red}$ couple of the 2-Fe ferredoxins occurs between -0.25 and -0.45 V. Interestingly, the 4-Fe ferredoxins use two different redox couples, depending upon the particular protein. The redox couple of usual ferredoxins, $Fd_{ox} \rightleftharpoons Fd_{red}$, occurs around -0.40 V, corresponding to the $[Fe_4S_4(SR)_4]^{2-} \rightleftharpoons [Fe_4S_4(SR)_4]^{3-}$ transformation, whereas the redox couple for high-potential (HP) ferredoxins ($HP_{ox} \rightleftharpoons HP_{red}$) occurs around $+0.35$ V, corresponding to the $[Fe_4S_4(SR)_4]^- \rightleftharpoons [Fe_4S_4(SR)_4]^{2-}$ transformation.

The redox potentials of the analogs can be modified substantially to more negative values by increasing the electron-donating properties of the R substituent of the thiolate (RS^-). For instance, the $[Fe_4S_4(SR)_4]^{3-} \rightleftharpoons [Fe_4S_4(SR)_4]^{2-}$ couple measured in the same solvent (DMF) is -0.45 V when R = p-nitrobenzene and is -1.17 V when R = t-butyl, an absolute difference of 0.72 V. Initially, comparison of the redox potentials of the Fe-S proteins (in H_2O) and the synthetic analogs (in DMF) containing electronically similar thiolate ligands revealed bothersome discrepancies of 0.5–0.9 V for identical redox couples. However, studies have shown that these differences in potential diminish (to 0.1–0.2 V) for the 1-Fe and 4-Fe cases when the measurements are taken in the same aqueous solutions; the residual differences are attributed to the influence of the protein backbone.

Isolation. The outstanding advantage of studying the simple iron-thiolate analogs over the proteins lies in the ready accessibility and characterization of the analogs as well as the higher degree of precision to which structural and physical parameters can be measured. Therefore, one synthetic objective is the isolation of pure, crystalline analogs of each of the biologically relevant oxidation levels of protein Fe-S centers. The analogs for the most stable biological oxidation levels proved to be the most difficult to obtain.

Rubredoxin is isolated and crystallographically characterized as the stable Fe(III) protein. Attempts to oxidize the analog compound, $[Fe(SR)_4]^{2-}$, to the Fe(III) species or to prepare $[Fe(SR)_4]^-$ directly from $FeCl_3$ were not initially successful (in compounds containing monodentate thiolates) since Fe(III)-thiolate species readily decompose to Fe(II) and disulfide (RSSR). This problem was overcome by using bulky organic thiolates. The rationale behind this approach came from the thought that, in a certain sense, proteins function as sterically hindered ligands. By using 2,4,6-triisopropylbenzenthiolate, stable $[Fe(SR)_4]^-$ could be isolated from protic solvents. Apparently, the protein backbone as well as the steric ligands protect the Fe-SR center from protic solvents which induce the autoredox reaction. These results encouraged the execution of other studies which showed that $[Fe(SR)_4]^-$ compounds containing simple thiolates could be isolated by rigorously excluding protic solvents. The x-ray diffraction structures of both the Fe(III) and Fe(II) analogs contain approximately tetrahedral $[FeS_4]$ units, with each compound having four equivalent Fe-S bonds. As expected, the Fe-S bond distance increases upon reduction of the iron center; the Fe-S bond length is 0.230 nanometer in $[Fe(III)(SPh)_4]^-$ and 0.236 nm in $[Fe(II)(SPh)_4]^{2-}$.

Characterization. Of the two synthetic analogs for the biologically relevant oxidation levels of the 2-Fe center, only $[Fe_2S_2(SR)_4]^{2-}$ has been characterized by x-ray crystallography. It has not yet been possible to isolate the reduced analog, $[Fe_2S_2(SR)_4]^{3-}$, although species of this type have been generated in solution. Spectroscopic studies indicate that reduced 2-Fe centers are trapped-valence species with one Fe(II) and one Fe(III).

Two of the three analogs for the biologically relevant oxidation levels for the 4-Fe center, $[Fe_4S_4(SR)_4]^{3-}$ and $[Fe_4S_4(SR)_4]^{2-}$, were readily synthesized and characterized. Although the high-potential Fe-S proteins are isolated in their oxidized form, synthesis and characterization of stable models for the oxidized form of high-potential iron-sulfur proteins, $[Fe_4S_4(SR)_4]^-$, was achieved only recently by using the sterically encumbered thiolate and nonpolar solvents (CH_2Cl_2). The properties associated with the stability of the analog, $[Fe_4S_4(SR)_4]^-$, are relevant to the situation in the biological systems where the unique stability of the oxidized high-potential proteins has been attributed to the hydrophobic environment of the Fe-S center.

Detailed spectroscopic measurement of the analogs for the three oxidation levels for the 4-Fe compounds revealed the unusual and interesting situation in which the oxidation state of the individual iron atoms is neither +2 nor +3, but is an intermediate, average value which depends upon the oxidation level of the $[Fe_4S_4]$ core. For instance, in the $[Fe_4S_4(SR)_4]^{2-}$ case which formally contains two Fe(II) and two Fe(III) ions, the irons behave spectroscopically as if there are four identical iron atoms, each with an oxidation state of +2.5. X-ray crystallographic studies of the $[Fe_4S_4(SR)_4]^n$ series ($n = -1, -2, -3$) have established that the Fe-S and Fe-SR distances increase by 0.002 and 0.004 nm, respectively, as the Fe-S core is reduced ($n = -1$ to -2 to -3). For $[Fe_4S_4(SR)_4]^-$, the average Fe-S bond is 0.226 nm and the average Fe-SR bond is 0.221 nm.

To date, no synthetic analog has been prepared for the recently discovered 3-Fe centers in proteins. Conversely, a number of Fe-S coordination compounds (for which there are as yet no examples in biology) have been characterized. Some of those Fe-S compounds include $[Fe_6S_6(SR)_6]^{3-/2-}$, $[Fe_8S_6I_8]^{3-}$, and $[Fe_6S_9(SR)_2]^{4-}$.

For background information SEE BIOINORGANIC CHEMISTRY; COORDINATION CHEMISTRY; ORGANOSULFUR COMPOUND; OXIDATION-REDUCTION; PROTEIN; STEREOCHEMISTRY; X-RAY CRYSTALLOGRAPHY in the McGraw-Hill Encyclopedia of Science and Technology.

Michelle Millar

Bibliography. I. G. Dance, The structural chemistry of metal thiolate complexes, *Polyhedron*, 5:1037–1104, 1986; W. Lovenberg (ed.), *Iron-Sulfur Proteins*, vols. 1–3, 1977; T. G. Spiro (ed.), *Iron-Sulfur Proteins*, 1982; S. A. Koch, L. E. Maelia, and M. Millar, Syntheses and structures of the $[Fe(SPh)_4]^-$ and $[Fe(SEt)_4]^-$ anions: A general route to iron(III) tetrathiolate complexes, *J. Amer. Chem. Soc.*, 105:5944, 1983; T. O'Sullivan and M. Millar, Synthesis and study of an analogue for the $[Fe_4S_4]^{3+}$ center of oxidized high-potential iron-sulfur proteins, *J. Amer. Chem. Soc.*, 107:4096–4097, 1985.

Cosmology

On a large scale, matter in the universe is distributed not smoothly but in lumps, the galaxies and clusters of galaxies. An outstanding puzzle in cosmology is what started the process of clumping that led to galaxy formation. The answer may possibly involve one of the more exotic structures hypothesized by particle theorists, the cosmic string. This theory, though still quite speculative, does have considerable indirect observational support and can explain some otherwise mysterious features of the galaxy distribution.

The universe still bears the imprint of its violent birth, the big bang that set it expanding in all directions. As its evolution is traced backward from the present toward that singular event, progressively higher energies are involved. In the earliest stages, the first fraction of a second after its birth, the universe was immensely hotter and denser than even the densest stars. The relevant physics in such extreme conditions is unknown but is closely tied to fundamental theories of particle interactions. Among the various strange objects that may have been formed at that time are cosmic strings. These long, thin structures are predicted by many theories of subnuclear particle physics. Once formed, they could survive long after the initial hot, dense phase, long enough in fact to play a crucial role in the process of creating the large-scale structures now observed: galaxies and clusters of galaxies.

Strings and vortices. Strings are closely akin to vortices. Many physical systems at very low temperatures exhibit so-called linear defects of this kind. Examples are vortex lines in superfluid helium and tubes of magnetic flux in so-called type II superconductors.

The typical example occurs in a theory where, in the minimum-energy state, there is an arbitrary angular variable θ. For example, this can occur if the energy associated with some pair of variables ϕ_1 and ϕ_2 has a so-called Mexican hat shape (**Fig. 1**), with rotational symmetry in the ϕ_1-ϕ_2 plane, so that its minimum occurs on a circle of given radius. In the lowest-energy state the radius is fixed, but the angular position θ around the circle may vary.

As the system cools to the temperature where the energy of random thermal fluctuations becomes less than the height of the central peak, the values of ϕ_1 and ϕ_2 will tend to settle down into the circular valley, eventually reaching the circle of minima. However, the angular position θ may vary from one part of the system to another.

It is possible for θ to vary in such a way that the value of θ changes smoothly from 0 to 360° as the point at which θ is evaluated circles around a large imaginary loop in space (**Fig. 2**). It is then impossible to keep ϕ_1 and ϕ_2 in the valley everywhere inside the loop. Somewhere it is necessary to go over the crown of the Mexican hat; there must be a point where ϕ_1 and ϕ_2 are both zero. In fact, this must happen all along a line. This line is the string. It represents a concentration of excess energy, localized in a very thin tube.

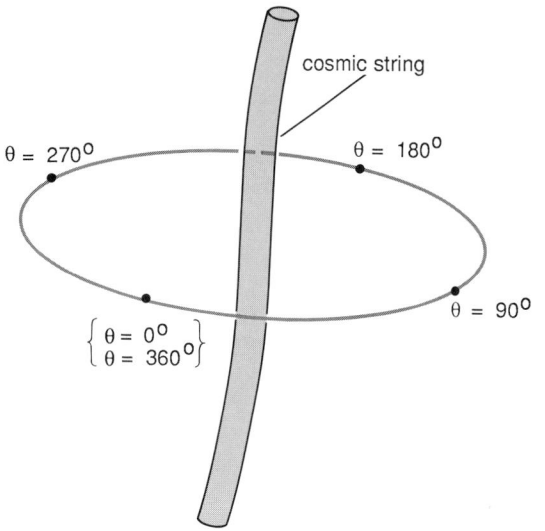

Fig. 2. **Structure of cosmic string.**

Cosmic strings. For liquid helium the transition temperature at which vortices form is only 2.2 K (−456°F). Many theories of fundamental particle interactions show similar behavior, though at an immensely greater temperature, accessible only in the very early universe, perhaps 10^{30} K (or 10^{30} °F).

If one of these theories is correct, then initially the angular variable θ can be expected to vary randomly from place to place. There would then be occasional trapped defects around which θ changes by 360°, that is, strings. These strings cannot have free ends; they either form closed loops or are infinitely long. The strings would tend to shorten under tension but would be stretched by the universal expansion. The very convoluted initial configuration of strings would become simpler as it evolved, but some strings would survive.

A particularly important role is played in this scenario by isolated loops of string. Occasionally, as the strings move, one may intersect itself; then a closed loop of string may break off. Once formed, a loop can disappear only by shrinking to a point. This would be a very slow process; the loops would oscillate many times, gradually losing energy by gravitational radiation, before finally vanishing.

Observational effects of loops. Even now, some loops would still be present. Their effects could be seen in a variety of ways. One of the most intriguing is a string's imaging property. Cosmic strings have very unusual gravitational effects. A particle at rest near a straight string experiences no gravitational acceleration, but there is a gravitational effect nonetheless. A light ray traveling past the string would be bent inward by a small angle, so that double images of distant objects could be observed, one on either side of the string, separated by a few seconds of arc (**Fig. 3**). Some of the so-called double quasars that are observed, pairs with very similar characteristics

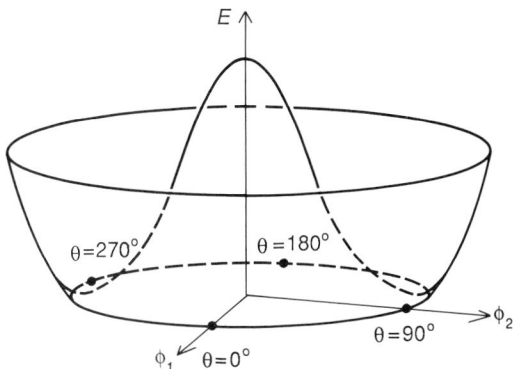

Fig. 1. **So-called Mexican hat energy function E associated with a pair of variables, ϕ_1 and ϕ_2.**

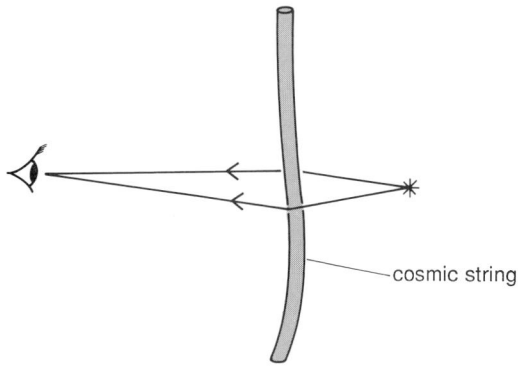

Fig. 3. Lensing by a cosmic string.

cosmic string

a few seconds apart, may be examples of this effect. Other, more prosaic explanations are possible, but gravitational lenses of other types would usually be expected to yield an odd number of images, one or three but not two.

A more definitive test of the cosmic string idea may come from another prediction: there should be a sharp discontinuity in the measured temperature of the microwave background radiation where a cosmic string intervenes. The predicted magnitude of this effect is not far below present observational limits, a discontinuity of perhaps 50 microkelvins.

Galaxy correlations. Perhaps the main reason why cosmic strings have excited many cosmologists is that they can explain some puzzling features of the observed distribution of galaxies. The original suggestion that strings might provide the initial perturbations from which galaxies grow is due to Ya. B. Zel'dovich, but the critical discovery of the importance of loops was made by A. Vilenkin.

The critical epoch, so far as the origin of galaxies is concerned, is the time when the electrons and photons in the universe combine to form atomic hydrogen, when the universe is a few tens of thousands of years old. Prior to that time, the pressure of radiation would prevent the formation of gravitational condensations. Loops present then would act as seeds on which the surrounding matter would start to condense. Consequently, the observed correlations of positions of galaxies today can be expected to reflect the correlations of loops in the early universe. Numerical simulations have been done to estimate the correlation function of loops. The results agree remarkably well with observation and make it possible to estimate the one adjustable parameter in the theory, the energy scale associated with strings.

One particularly encouraging result concerns the correlations of rich clusters of galaxies, the so-called Abell clusters. Rich clusters may be expected to form around particularly large, massive loops. These rare large loops are also highly correlated, and the same will therefore be true of the rich clusters. This explains rather well the otherwise puzzling fact that rich clusters are distributed in an even less random way than are galaxies themselves.

Properties of strings. The parameters of these strings are truly astronomical. The most likely value of the energy scale appears to be around 10^{16} GeV, roughly the rest-mass energy of a bacterium. The corresponding mass per unit length of the strings is 10^{21} kg m^{-1} (10^{21} lbm ft^{-1}), or 1000 tons in a nuclear diameter. This means that a string passing through the Earth from pole to pole would be 1000 times as massive as the Earth itself. Fortunately, strings are so rare that the Earth is unlikely to encounter one. The string tension is enormous: 10^{38} newtons (2×10^{37} lbf), strong enough to lift the weight of several million Suns.

Strings come in several varieties, some with properties even stranger than those already described. For example, some theories predict that the strings should be superconducting, able to carry large electric currents without loss. If so, there would be interesting effects when they move through large-scale magnetic fields.

Cosmic strings are certainly not an established fact, but they have introduced a very interesting new possibility into cosmological debates and there does seem to be substantial circumstantial evidence for their existence. Within a few years it should be known with certainty whether or not these structures exist.

For background information SEE BIG BANG THEORY; COSMIC BACKGROUND RADIATION; COSMOLOGY; GRAVITATIONAL LENS; QUANTIZED VORTICES; UNIVERSE in the McGraw-Hill Encyclopedia of Science and Technology.

T. W. B. Kibble

Bibliography. N. Turok, Cosmic strings and the correlation of Abell clusters, *Phys. Rev. Lett.*, 55:1801–1804, 1985; A. Vilenkin, Cosmic strings and domain walls, *Phys. Rep.*, 121:263–315, 1985; A. Vilenkin, Cosmological density fluctuations produced by vacuum strings, *Phys. Rev. Lett.*, 46:1169–1172, 1981; Ya. B. Zel'dovich, Cosmological fluctuations produced near a singularity, *Mon. Not. Roy. Astron. Soc.*, 192:663–667, 1980.

Crossing-over (genetics)

The human X and Y chromosomes contain a region of very similar deoxyribonucleic acid (DNA) sequence at the tips of their short arms in which an obligate crossover results in an exchange of DNA between these chromosomes. At least one gene which escapes X inactivation is found in this region, and somatic effects in XO individuals suggest that other genes may be present here. The site of crossover varies from meiosis to meiosis and in rare cases can transfer the male-determining locus from the Y to the X chromosome.

Sex chromosomes. The human sex chromosomes are thought to have evolved from a pair of identical ancestral chromosomes to their present state in which the X chromosome contains five times as much DNA as the Y chromosome. Many X-linked genes are known, but the Y chromosome has only one known gene, *MIC2Y*, which codes for a cell surface protein, 12E7. Additionally, the Y chromosome must carry genes which determine maleness. In spite of these differences, it was suggested in the 1930s that these chromosomes might still be sufficiently homologous to be capable of recombination with each other. Whether or not this recombination is possible has important consequences for the understanding of the mechanisms involved in ensuring correct segregation of chromosomes during meiosis, and could possibly explain cases in which individuals are male in the absence of a Y chromosome.

Crossing-over is the process by which combinations of different alleles of genes on homologous chromosomes are altered by exchange of DNA from one chromosome to its homologue during meiosis. The frequency with which this exchange is detected between pairs of genes defines the genetic distance between these genes (or other genetic markers) and allows a map to be constructed. If two markers are on different chromosomes, they will be transmitted together into the progeny cells after meiosis one-half of the time. This recombination frequency of 50% is diagnostic for "unlinked" markers; recombination frequencies of less than 50% represent linkage.

Early observations. The first suggestion that a recombination between the X and Y chromosomes occurred in many, if not all, meioses arose in the 1930s from light microscope observations in which partial pairing of the X and Y chromosomes was first observed. With the advent of the electron microscope the X and Y chromosomes could be seen to pair and form a structure comparable to that which was found between homologous autosomes during meiosis. It was clear, however, from the extent of this pairing that DNA sequences on the X and Y chromosomes which were known to be different were present in this paired region. In as much as pairing was thought to require a region of identical DNA sequence, the significance of the observation was disputed.

Genetic evidence in support of recombination between the X and Y chromosomes in the mouse came from the discovery that the *Sxr* (sex reversed) mutation was located on the sex chromosomes. This mutation is transmitted via males and gives rise to XX individuals which are phenotypically male. In a mating between a normal female and a male carrying the *Sxr* mutation, one-quarter of the offspring are normal males, one-quarter carrier males, one-quarter normal females, and one-quarter XX males. This pattern of inheritance is typical of a dominant mutation present on an autosome, but no linkage to a mouse autosome could be demonstrated. The

Sxr mutation was shown to represent a duplication of part of the Y chromosome which was transferred to the X chromosome by crossing-over between the two DNA molecules during male meiosis. Again, the relevance of this observation was disputed because the Y chromosome involved was clearly abnormal.

Obligatory crossing-over hypothesis. The hypothesis that a recombination between the mammalian X and Y chromosomes was obligatory and a region of these chromosomes therefore showed a pattern of inheritance which appeared autosomal, that is, was pseudoautosomal, was put forward to tie together several observations. The model could account for pairing, the pattern of inheritance of *Sxr*, somatic effects in individuals with a normal autosomal constitution but only a single X chromosome and no Y chromosomes (XO individuals), and the lack of X inactivation in some genes found on the X chromosome.

Genes on the mammalian X chromosome are present in females in twice the dose found in males. This difference is compensated in mammals by the random inactivation of one X chromosome in female somatic tissues. The level of gene products from most X-linked genes is thus the same in both males and females. Some genes in humans have been shown to escape this inactivation. These genes—for steroid sulfatase and those coding for the antigens *Xg* and 12E7—are all located in the terminal part of the short arm of the human X chromosome which is close to, and perhaps within, the pseudoautosomal region. In humans the gene for 12E7 has been shown to exist on both the X and Y chromosomes and to

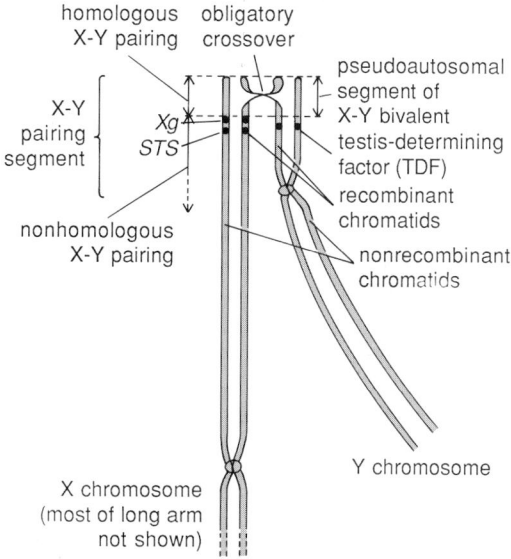

Fig. 1. Pseudoautosomal model. Crossing-over between one pair of DNA molecules results in a region of the X and Y chromosomes being inherited as if autosomal. Probable positions of genetic markers in this region of the human chromosomes are indicated. (*After P. S. Burgoyne, Sex chromosomes: Mammalian X and Y crossover, Nature, 319:258–259, 1986*)

exchange between them infrequently. In mice the steroid sulfatase gene is present in both sex chromosomes. Other genes present on this region of the sex chromosomes would normally be present in two copies in both sexes and would not need to be dosage-compensated. In XO individuals there would be only one copy of such genes, which could explain the somatic abnormalities observed, such as short stature and webbing of the skin of the neck.

Direct tests of crossing-over. A direct test of the occurrence of a recombination event between the X and Y chromosomes at each meiosis can be provided in this region by using DNA sequences which are polymorphic (different forms of the sequence being present at the same locations in DNA of different individuals). The inheritance of such a sequence could be followed in families and would be expected to show a lack of sex linkage. The first such sequence found had been localized to the extreme tip of the X and Y chromosomes, within 20,000 base pairs of the end of the DNA molecule in these chromosomes. In family studies this sequence has been shown to be unlinked in genetic terms to the region of the Y chromosome which determines maleness. A copy of this sequence from the father's Y chromosome is inherited with equal probability by a son or daughter because of a crossover event as shown in the model in **Fig. 1**. In **Fig. 2** an example of this analysis in one family is presented.

Such a pattern of inheritance could be caused if a recombination event always occurred at a single site. This would result in genetic markers on one side of this point showing complete sex linkage while those on the other side would show a total lack of sex linkage. Alternatively there could be many points at which crossover could occur in such a way that one or many always occurred between the marker determining sex and the end of the Y chromosome involved. One way of distinguishing between these possibilities is to look at recombination between multiple markers on this region of the X and Y chromosomes. This examination shows that the second possibility is correct. A study of three linked markers in the pseudoautosomal region of the human Y chromosome showed that they all have different recombination frequencies with the sex-determining gene and with each other rather than the all-or-none linkage predicted by the first possible scheme. Genetic maps of this region constructed from these data are shown in **Fig. 3**. The distances measured between pairs of markers in this figure are additive. This suggests that multiple recombination events are not taking place, a conclusion supported by the observation that double recombination events have not been seen. One simple explanation for this phenomenon would be that the physical size of this region precludes the occurrence of more than one recombination event.

In recombination between the X and Y chromosomes the genetic map distances are ten times larger than in the same region when the distances are measured in recombination between two X chromosomes. Clearly the DNA present in this region is the same in amount and base sequence in both situations. A possible explanation for this difference depends on two assumptions: first, that a recombination event is necessary because it provides a topological linking against which the spindle microtubules can act to segregate the homologous chromosomes into daughter nuclei; and second, that recombination can take place only between identical DNA sequences. If these assumptions are correct, then recombination

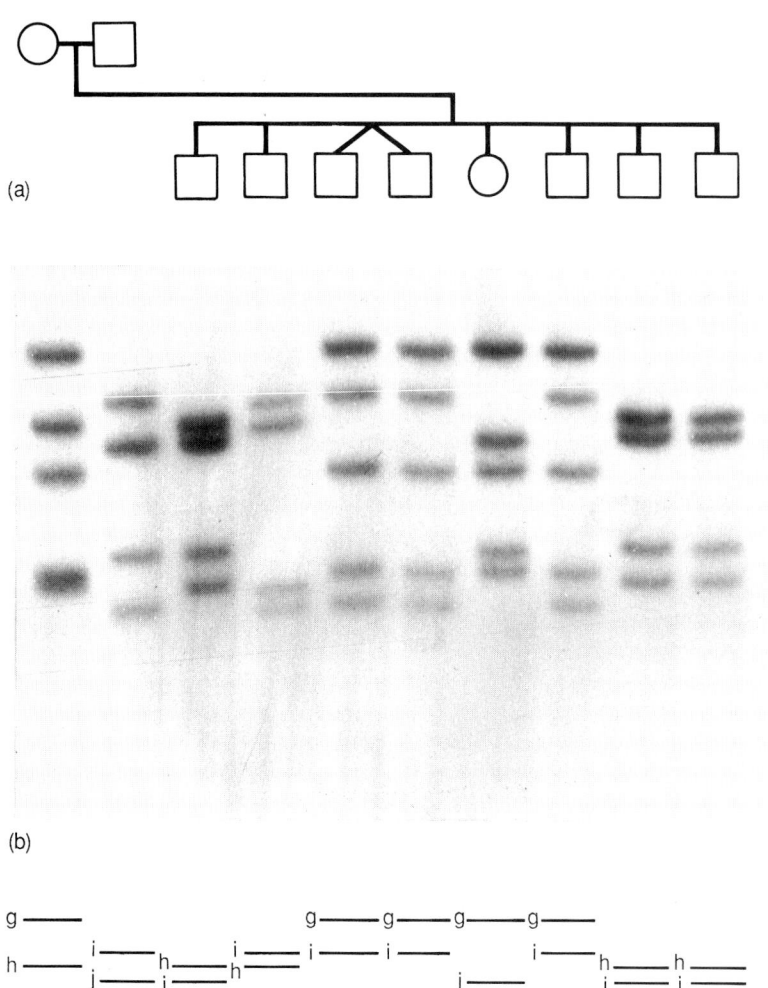

(a)

(b)

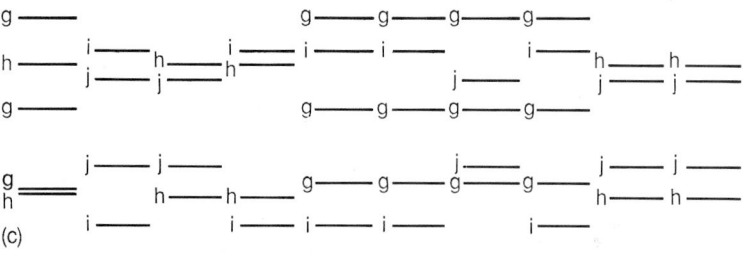

(c)

Fig. 2. Analysis of inheritance of a DNA marker from the pseudoautosomal region. This polymorphic marker detects DNA fragments of different sizes in different individuals. In each individual two alleles are present: allele g consists of three fragments; the other alleles, h, i, and j, each have two fragments. (a) Pedigree of the family used. Circles represent females and squares males. (b) Display of fragments by gel electrophoresis and DNA hybridization. (c) Interpretation of the data. The father's alleles, i and j, are inherited by equal numbers of sons and therefore show no sex linkage.

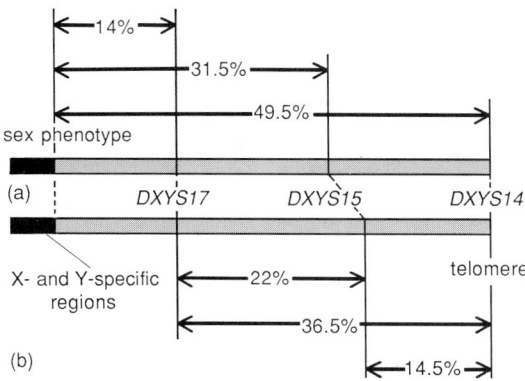

Fig. 3. Genetic map of the human pseudoautosomal region. This map is constructed from a large number of analyses of the type shown in Fig. 2. Markers are at loci labeled *DXYS14*, *DXYS15*, and *DXYS17*. Recombination frequencies are given as percentages. (*a*) Map derived from measurement of recombination frequencies between markers and the sex-determining gene. (*b*) Map derived from measurement of locus distances between pairs of markers. (*After F. Rouyer et al., A gradient of sex linkage in the pseudoautosomal region of the human sex chromosomes, Nature, 319:291–295, 1986*)

must take place in the XY pair only in the pseudoautosomal region (which probably amounts to about 4×10^6 base pairs), whereas in the XX pair it can take place throughout the whole length of the chromosome so that the observed frequency in the small pseudoautosomal region is low.

Abnormal sex determination. The ability of recombination to occur at a wide range of sites within the pseudoautosomal region provides an explanation for some abnormalities of sex determination. In humans, XX males occur with a low frequency. It seems most likely that a fraction of these individuals represent recombination events between the X and Y chromosomes which have occurred close to the pseudoautosomal region but between this region and the centromere. This positioning results in transfer of a male-determining gene or gene cluster to the X chromosome. In support of this model, studies with DNA probes suggest that many XX males contain Y-chromosomal DNA in varying amounts and that in those cases analyzed this DNA is present on the tip of the paternal X chromosome.

For background information SEE CROSSING-OVER (GENETICS); HUMAN GENETICS; LINKAGE (GENETICS); RECOMBINATION (GENETICS); SEX-LINKED INHERITANCE in the McGraw-Hill Encyclopedia of Science and Technology.

Howard J. Cooke

Bibliography. P. S. Burgoyne, Genetic homology and crossing over in the X and Y chromosomes of mammals, *Hum. Genet.*, 61:85–90, 1982; H. J. Cooke et al., Hypervariable telomeric sequences from the human sex chromosomes are pseudoautosomal, *Nature*, 317:687–692, 1985; F. Rouyer et al., A gradient of sex linkage in the pseudoautosomal region of the human sex chromosomes, *Nature*, 319:291–295, 1986.

Cyclosporin

Progress over the past 30 years in the surgical techniques of organ transplantation has been remarkable and has required the development of new immunosuppressive drugs that prevent the rejection of foreign tissue grafts. Cyclosporin A is the forerunner of a new generation of immunosuppressive agents which selectively and reversibly inhibit graft rejection by acting on immunocompetent cells. Cyclosporin A has also proven to be a useful tool for the experimental immunologist in the dissection of the complex process of T-cell activation, and holds promise as a major therapeutic agent for the treatment of autoimmune disease.

Cyclosporin A was discovered in 1970. The pioneering studies of J. Borel and coworkers established the basis for cyclosporin A as a selective immunosuppressive agent that appeared to act only on certain immunocompetent cells. In contrast to other immunosuppressive drugs (for example, Azathioprine), it is free of myelotoxicity and does not impair the proliferative capacity of hemopoietic stem cells. Cyclosporin A was successfully synthesized in 1980; it is a neutral hydrophobic, cyclic peptide composed of 11 amino acid residues.

Effects on lymphocyte function. Cyclosporin A inhibits the ability of T lymphocytes to proliferate in tissue culture in response to a number of different stimuli, including plant lectins such as concanavalin A or phytohemagglutinin, specific foreign protein antigens to which the host had been sensitized, and foreign histocompatibility antigens found on the surface of allogeneic cells (the mixed leukocyte reaction).

In general, the process of T-cell activation by any of these ligands can be divided into two steps. The first step involves the induction of a receptor for a growth factor which is called interleukin 2 (IL-2), while the second step involves the production of the growth factor itself. Cyclosporin A has a selective inhibitory effect on the accumulation of IL-2 messenger RNA but has only a partial inhibitory effect on the accumulation of mRNA for the IL-2 receptor. In the absence of IL-2, no T-cell proliferation is observed even though large numbers of growth factor receptors are present. Cyclosporin A has no effect on the proliferation of T cells to preformed IL-2, provided that the responding cells already have intact growth factor receptors. In addition to inhibiting IL-2 production, cyclosporin A has an inhibitory effect on the accumulation of mRNA for several other lymphokines, including IL-3 (a growth factor for bone marrow stem cells), IL-4 (a growth and differentiation factor for both T and B lymphocytes), and interferon γ. However, cyclosporin A does not inhibit the production of interferon α or β by virus-infected human or mouse lymphocytes.

Although cyclosporin A has no effect on the proliferative response of B lymphocytes to lipopolysaccharide, recent studies have shown that the proliferative response of B lymphocytes to anti-immunoglobulin (anti-Ig) antibodies was highly susceptible to inhibition by cyclosporin A. Cyclosporin A appears to block an early event resulting from crosslinking of surface Ig receptors, but not the later stages of B-cell growth and differentiation which are mediated by T-cell–derived lymphokines. Although the crucial step in the lymphocyte activation cascade on which cyclosporin A exerts its inhibitory effect has not been defined, it is likely that the mode of action of cyclosporin A on T and B lymphocytes will prove to be identical. The nature of the specific intracellular receptor for cyclosporin A remains controversial; the immunosuppressive effects of cyclosporin A may be secondary to its ability to bind to calmodulin or to a distinct cytosolic binding protein, cyclophilin.

Effects on alloreactivity. The role of cyclosporin A as a potent suppressor of organ allograft rejection has been demonstrated for many different organ allografts in a range of species including mice, rats, guinea pigs, rabbits, dogs, pigs, and nonhuman primates. In all cases, cyclosporin A has been more potent and has had fewer side effects than other chemical immunosuppressive agents in clinical use. The grafts protected from rejection by cyclosporin A therapy include kidneys, hearts, lungs, liver, pancreas, islets of Langerhans, skin, bone marrow, nerve, muscle, small bowel, veins, cornea, and joints. Transplantation of many of these organs was performed in situations that previously proved impossible with standard immunosuppressive therapy. In humans, the aim of clinical organ allografting is to achieve complete allograft acceptance without significant depression of immunological defenses against infection. Cyclosporin A has proven to be extremely useful in the transplantation of human kidneys, hearts, liver, and bone marrow.

One of the most intriguing observations made with cyclosporin A is that for many transplanted organs a limited course of cyclosporin A results in a very prolonged allograft survival. There is a considerable body of data in experimental animals that the critical mechanism of action of cyclosporin A involves a relative sparing of suppressor T lymphocytes which in turn render the host tolerant to allografted tissue. For example, the adoptive transfer of T lymphocytes from heart graft recipients treated with cyclosporin caused significant prolongation of test cardiac allografts placed into untreated, immunologically virgin rats. Transfer of cells from normal rats, cyclosporin A–treated but ungrafted animals, or grafted but cyclosporin A–untreated recipients all failed to prolong test graft survival.

In the rat, cyclosporin A–mediated transplantation tolerance can be subdivided into three stages. Stage 1 coincides with the cessation of cyclosporin A therapy at day 7 following transplantation of an allogeneic heart. A nonspecific and unstable state of graft tolerance is present. The survival of a second graft is prolonged regardless of its antigen specificity. However, rejection of the second graft is invariably followed by loss of the original transplant. In stage 2, 14 days after the original transplant, second grafts from the original donor strain have a more prolonged survival, whereas all third-party grafts are rapidly rejected. In stage 3, 50 days after the original transplant, a state of stable specific unresponsiveness is achieved, with second grafts from the original donor strain surviving indefinitely, but third party grafts being rejected rapidly without affecting the indefinite survival of the original transplant. It is likely that a specific and stable state of cyclosporin A–induced unresponsiveness in the organism develops with time, possibly owing to a gradual proliferation of T suppressor cells.

The successful use of cyclosporin A to prolong allograft rejection in both experimental animals and humans has prompted a large number of laboratory studies to investigate the mechanisms involved in the tolerant state. Several studies have demonstrated that suppressor cells with different properties can be generated in tissue culture during the course of sensitization to foreign histocompatibility antigens in the presence of cyclosporin A. Although it is difficult to translate the results of these laboratory studies to the situation in the intact animal, one must assume that certain of the suppressor T-cell activities seen in cell cultures are also responsible for the long-term immunosuppression induced by cyclosporin A in the animal. However, it must be emphasized that suppressor cells may be only partially responsible for the ability of cyclosporin A to inhibit allograft rejection, since cyclosporin A need be given only as a short course of therapy with certain allografts in some species (for example, kidney grafts in the rat) but must be administered for prolonged periods in other species (for example, kidney grafts in humans).

Effects on autoimmune disease. The ability of cyclosporin A to inhibit T-cell activation in the laboratory strongly suggests that this drug has the potential to function as a potent immunotherapeutic agent in autoimmune disease. Cyclosporin A has been shown to be highly effective in the prevention of a number of experimental autoimmune diseases induced by the administration of an antigen in adjuvant, including experimental allergic encephalomyelitis and experimental autoimmune uveitis. When cyclosporin A is administered during the sensitization phase, the onset of the disease is suppressed as long as the treatment lasts. Once the drug is withdrawn, the disease frequently recurs. In certain disease states, delayed or therapeutic administration of cyclosporin A also results in impressive improvement. It is likely that cyclosporin A functions in these experimental models by inhibiting the expansion of antigen reactive cells and the production of lymphokines and other inflammatory mediators. When drug treatment is terminated and antigen persists in the host, there is a prompt resumption of effector T-cell function.

Cyclosporin A has also proven to be quite useful in the treatment of spontaneous autoimmune disease in the NZB/W mouse (an antibody and immune-complex–mediated disease) and in the treatment of diabetes in the BB rat, a strain in which insulin-dependent, nonobese, ketotic diabetes mellitus develops spontaneously.

Preliminary studies on the use of cyclosporin A to treat autoimmune disease in humans are quite encouraging. Certain patients with posterior uveitis respond rapidly to cyclosporin A, with improvement in visual acuity and in ocular inflammatory activity. Cyclosporin A has also been used to treat certain children with type-1 diabetes mellitus. The initial results are promising, but it is still unknown whether complete relapse will occur in most patients when cyclosporin A administration is discontinued. The major drawback to the widespread use of cyclosporin A for the treatment of other autoimmune diseases such as rheumatoid arthritis or systemic lupus erythematosus, which are usually not life-threatening, is the high incidence of nephrotoxicity seen after prolonged administration of the drug.

For background information *see* A*UTOIMMUNITY;* I*MMUNOSUPPRESSION;* T*RANSPLANTATION BIOLOGY* in the McGraw-Hill Encyclopedia of Science and Technology.

Ethan M. Shevach

Bibliography. J. F. Borel (ed.), *Ciclosporin*, 1986; E. M. Shevach, The effects of cyclosporin A on the immune system, *Annu. Rev. Immunol.*, 3:397–423, 1985; D. J. G. White (ed.), *Cyclosporin A*, 1982.

Cytokinins

Cytokinins were originally defined by F. Skoog as a class of plant hormones or plant growth regulators that can initiate cell division in tobacco pith culture in the presence of optimal auxin. Today many more regulatory roles have been assigned to cytokinins as their interactions with other plant hormones are studied. The roles include delay of senescence, chloroplast biosynthesis, pigment synthesis, pathogenic interactions between microorganisms and higher plants, organ development, and tissue differentiation. However, a primary role of cytokinins still appears to be the regulation and stimulation of cell division. Recent research has led to the hypothesis that cytokinins trigger cell division by opening calcium channels on the plasma membrane of plant cells and raising the intracellular calcium ion concentration, which in turn stimulates the cell to divide.

Biosynthesis. Cytokinins are structurally related compounds that contain an adenine nucleus with various substitutions at N^6 (the nitrogen attached to the sixth position of a purine molecule). The most commonly occurring free cytokinins are zeatin and its derivatives and $N^6\text{-}(\Delta^2\text{-}$isopentenyl)-adenine and its derivatives. Very active synthetic cytokinins include 6-benzylaminopurine and its derivatives. Cytokinins occur as free compounds in both higher plants and microorganisms and as nucleosides in the transfer ribonucleic acid (tRNA) of plants, animals, and microorganisms.

Twenty-eight naturally occurring cytokinins have been identified, but the level of activity that these compounds exhibit can vary widely. Several technical problems have impeded elucidation of the mechanism of biosynthesis of free cytokinins. For example, unambiguous detection of the very low levels of free cytokinins in plant tissue is extremely difficult with currently available analytical techniques. A further complication is the presence of cytokinins in tRNA, which does not appear to be part of the biosynthetic pathway for the biologically active free cytokinins. In addition, the most probable precursor molecules (adenine and its nucleoside or nucleotide) are found in much higher levels than cytokinins. They also play a central role in cellular metabolism, which presents problems in labeled precursor studies. Newly developed quantitative analysis techniques using monoclonal antibodies, which provide identical antibody molecules to cytokinins, hold new promise for highly specific detection. While the mechanism of biosynthesis is still to be worked out by plant physiologists and biochemists, the details of how cytokinins trigger cells to divide are becoming clearer.

Mode of action. There is increasing evidence that induction and regulation of cell division in both plants and animals are mediated by an increase in the intracellular calcium ion (Ca^{2+}) concentration. Normally the cytoplasmic calcium concentration is kept very low (about 10^{-8} M) by a variety of Ca^{2+} pumps on both the plasma membrane and internal membranes such as the endoplasmic reticulum and the mitochondria. When Ca^{2+} channels are opened on the plasma membrane or Ca^{2+} is released from internal stores, cytoplasmic free calcium rises and can activate calcium-binding proteins such as calmodulin or calcium-regulated proteins such as those found in microtubules (the major component of the mitotic spindle). It is thought that this triggers some cells to divide. Recently scientists have proposed that cytokinin may trigger plant cells to divide by opening calcium channels on the plasma membrane, causing an increase in intracellular calcium ion (Ca^{2+}) concentration.

Experimental evidence. An ideal model system that has been used to investigate the complex interaction between hormonal signaling, activation and aggregation of ion channels, and initiation of cell division is the cytokinin-induced mitosis and bud formation in the protonema of the moss *Funaria hygrometrica*. Research on this system has led to the proposal that the action of cytokinin as a mitotic regulator may be mediated by a localized increase in intracellular calcium ion concentration.

The *Funaria* protonema grows as a filamentous mat of cells with two cell types: chloronema cells with large chloroplasts and transverse cross walls; and caulonema cells with small, spindle-shaped chloroplasts and oblique cross walls. The latter are the target cells for bud induction by exogenous cytokinins. The elongate target cells respond to cytokinins by a localized outgrowth at the distal end of the cells followed by nuclear migration to that region after 20–22 h of treatment. After 24 h of cytokinin treatment a small initial cell has been cut off from the original target cell. This initial cell divides in three planes, producing a mass of cells termed a bud.

Research using a fluorescent calcium-membrane indicator dye, chlorotetracycline, has shown that cytokinin-induced division in target cells is preceded by localized changes in membrane-associated Ca^{2+} that predict the site of the asymmetric division within the cell (**Fig. 1**). In addition, the Ca^{2+} ionophore A23187 (which carries Ca^{2+} into cells across the plasma membranes down its concentration gradient), in the presence of exogenous Ca^{2+}, can induce the initial division in the absence of cytokinins. Calcium-free medium, the extracellular Ca^{2+} antagonist La^{3+} (lanthanum), the Ca^{2+} channel inhibitors D 600 and verapamil, and the intracellular Ca^{2+} antagonist TMB-8 all block bud formation. The essential source of Ca^{2+} appears to be extracellular, because blocking Ca^{2+} uptake with Ca^{2+}-transport inhibitors stops both nuclear migration and subsequent division. It seems plausible that cytokinins may exert their effect on

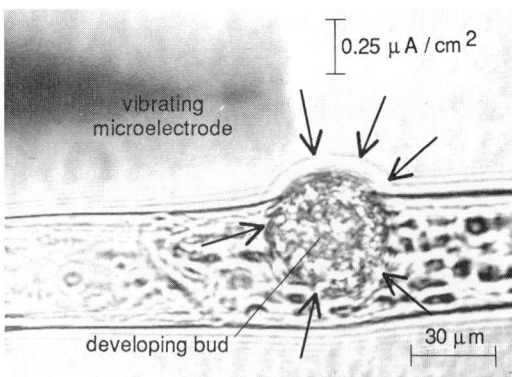

Fig. 2. Representative image of current flowing into developing bud on *Funaria* measured by a vibrating microelectrode. Arrow lengths are proportional to amount of current measured. (*From M. J. Saunders, Cytokinin activation and redistribution of plasma-membrane ion channels in Funaria: A vibrating microelectrode and cytoskeleton-inhibitor study, Planta, 167:402–409, 1986*)

asymmetrical cell division in *Funaria* by activating or concentrating Ca^{2+} channels at the presumptive bud site.

Additional support for this hypothesis is derived from studies on ion currents (in particular, Ca^{2+} currents) in a variety of polarized developing systems. Localized ion currents across plasma membranes preceding and predicting future growth zones have been reported in germinating zygotes of fucoid brown algae, germinating lily pollen grains, branching hyphae of the fungus *Achlya*, and branching filaments of the alga *Vaucheria*.

Vibrating microelectrode studies. Changes in the distribution and activation of plasma-membrane ion channels after cytokinin treatment in caulonema cells of *Funaria* have been examined by using a nonintrusive vibrating microelectrode. This instrument can map the electric current (flow of positively charged ions) around cells. A metal-filled glass micropipette with an electroplated platinum ball (10 micrometers in diameter) on its tip is attached to a piezoelectric element, which generates electricity in dielectric crystals subjected to mechanical pressure, and vibrated at a constant frequency of 200 Hz with a tip vibration of 20 micrometers. The potential differences measured by the electrode tip at the extremes of its vibration are then converted to current density. The electrode is moved with a micromanipulator to various places around the cell to map the current pattern. By changing the ions in the media around the cell or by using compounds that block the uptake of specific ions, the flow of specific ions into and out of the cell can be traced.

The vibrating microelectrode studies have yielded several findings. They have demonstrated that untreated caulonema cells have maximum inward current at the nuclear region; addition of cytokinins induces an increase in inward current along the length of target cells and that increased current subsequently shifts to their distal ends (presumptive bud site); and after

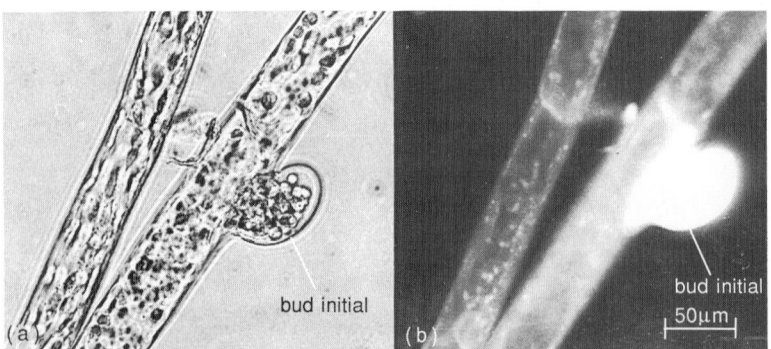

Fig. 1. *Funaria* 30 h after cytokinin treatment, stained with chlorotetracycline, shown in (*a*) phase contrast and (*b*) fluorescence micrographs. The bud initial is bright, showing a zone of membrane-associated calcium. (*From M. J. Saunders and P. K. Hepler, Localization of membrane associated calcium following cytokinin treatment in Funaria using chlorotetracycline, Planta, 152:272–281, 1981*)

establishment of the growth zone at the presumptive bud site, inward current falls and is similar to current detected at growing tips (**Fig. 2**). This current has a Ca^{2+} component since it falls to zero after the treatment with a Ca^{2+}-uptake inhibitor. It appears as if the calcium channels move in the plasma membrane to the end of the cell, where mitosis will take place, and collect there. If microfilament inhibitors are used, the channels do not aggregate and the bud forms over the nucleus instead of over the distal end of the cell. These observations lend support to the hypothesis that an initial mode of action of cytokinins is activation of plasma-membrane ion channels.

For background information SEE *CYTOKININS; PLANT GROWTH; PLANT HORMONES* in the McGraw-Hill Encyclopedia of Science and Technology.

M. J. Saunders

Bibliography. A. Dorn and M. H. Weisenseel, Advances in vibrating probe techniques, *Protoplasma*, 113:89–96, 1982; P. J. Hofman, C. Forsyth, and J. Van Staden, A radioimmunoassay for dihydrozeatin and dihydrozeatin riboside, and its application to a study of the in vitro metabolism of dihydrozeatin by soybean callus, *J. Plant Physiol.*; 121:1–12, 1985; D. S. Letham and L. M. S. Palmi, The biosynthesis and metabolism of cytokinins, *Annu. Rev. Plant Physiol.*, 34:163–197, 1983; M. J. Saunders, Cytokinin activation and redistribution of plasma-membrane ion channels in *Funaria*: A vibrating microelectrode and cytoskeleton-inhibitor study, *Planta*, 167:402–409, 1986.

Data compression

Information in a computer is represented as a sequence of bits. Usually, this representation is derived in a straightforward way, such as by using a 7- or 8-bit code (for example, ASCII or EBCDIC) to represent each character in the alphabet. Data compression is the technique of finding a shorter bit representation for the same information. Compression is desirable in computer and communications systems for two principal reasons: first, current databases contain vast amounts of data, and compression can reduce by two-thirds the amount of computer storage occupied. Second, large quantities of data are often transmitted from one location to another, and reducing the number of bits for a given amount of information increases the effective rate of transmission.

All data compression schemes take advantage of the fact that some bit sequences are more common than others. The common sequences are given shorter representations at the expense of the uncommon ones which must be given longer representations. Overall, the average length of the data is less, although some sequences may be longer. Compression is of little use without expansion, the ability to reproduce the original information from its compressed form.

The field of mathematics known as information theory determines that the minimum number of bits required to represent a sequence (its entropy or information content) is given by expression (1), where p is probability of the sequence in a

$$-\log_2 p \qquad (1)$$

given context. For example, in English, following the letter q it may be predicted that the letter u appears with 93.75% probability and that a period appears with 6.25% probability. In this context the information content of u is 0.093 bits and the information content of a period is 4.0 bits. In an abstract sense, the problem of data compression is twofold: estimating the probability of each sequence, and determining a coding scheme whose length approximates the information content. In practical data compression methods, it is often difficult to separate the two.

Huffman's method. In a popular compression method due to D. Huffman, a sample of data is examined to count the number of times each character occurs. These counts are assumed to be proportional to the probability of each character, and a bit representation for each character is determined that is as close as possible to the predicted information content. Expansion is done in sequence; that is, the compressed bits are examined one at a time until the representation for a character is recognized. This process con-

Table 1. Huffman codes for "the cat saw the rat"				
Character	Count	Probability, %	Information content	Huffman code
space	4	21	2.25	000
a	3	16	2.66	001
c	1	5	4.25	0100
e	2	10	3.25	100
h	2	10	3.25	101
r	1	5	4.25	0101
s	1	5	4.25	0110
t	4	21	2.25	11
w	1	5	4.25	0111

tinues to recognize the next compressed character, and so on.

From the sample of data (2), the characters are

$$\text{``the cat saw the rat''} \qquad (2)$$

counted and the Huffman codes in **Table 1** are determined. The compressed representation of (2) is therefore sequence (3). Huffman's method

$$\begin{array}{c} \textbf{111011000000100001110000110} \\ \textbf{01011100011101100000010100111} \end{array} \qquad (3)$$

as described here can be expected to compress English text (including uppercase and lowercase, punctuation, and spacing) to between 4 and 5 bits per letter (as opposed to 7 or 8 bits for ASCII or EBCDIC).

The Huffman codes are calculated backward. The two least probable characters (in this example, **s** and **w**) will eventually be given codes that differ only in the last bit (the code for **s** ends in **0** and that for **w** ends in **1**). The code for the common prefix (**011**) is unknown at this stage, but its probability is equal to the sum of the two least probabilities (10.5%). In the remaining steps, this prefix is treated as an imaginary character (denoted **sw**) for which a Huffman code is to be found; the two individual characters are not considered further. With **sw** substituted for **s** and **w**, the above process is repeated until only two characters remain. The complete sequence of steps is given in **Table 2**. (The *space* character is denoted by β.) After step 7, there are only two remaining prefixes: β**acrsw** and **eht**, which are given codes **0** and **1**, respectively. Next, β**a**, **crsw**, **eh**, and **t** are assigned **00**, **01**, **10**, and **11**. Once these codes are known, they can be constructed for β, **a**, **cr**, **sw**, **e**, and **h**: **000**, **001**, **010**, **011**, **100**, and **101**. Finally, the codes for **c**, **r**, **s**, and **w** are determined to be **0100**, **0101**, **0110**, and **0111**.

Dictionary encoding. In most text, such as English, longer sequences such as words can easily be recognized. Assuming these words are stored in a dictionary, each word in the data to be compressed may be replaced by a number which is the position of that word in the dictionary. Typically, these numbers have a shorter representation than the words themselves. For example, with dictionary (4) and data (5), the compressed representation would be sequence (6). In

this example, **the** is replaced by **7** because it is

$$\textbf{cat, frog, garden, in, rat, saw, the} \qquad (4)$$

$$\text{``the cat saw the rat in the garden''} \qquad (5)$$

$$\text{7 1 6 7 5 4 7 3} \qquad (6)$$

the seventh word in the dictionary; **cat** is replaced by **1** because it is the first word in the dictionary; and so on. Since there are 7 words in the dictionary, each number can be represented as 3 bits, for a total size of 24 bits. In a large amount of text, such as a book, no more than a few thousand different words are expected; if there were 65,536 words, each word would be represented by 16 bits. For typical English text, the dictionary representation uses 3 to 4 bits per character. It is possible to apply Huffman's method to the codes produced by the dictionary method to achieve further compression.

A number of methods build a dictionary dynamically while compressing a sequence of data, rather than relying on an existing one. One approach is to keep a dictionary of only the most recent n words that have appeared in the data, where n is some constant, usually between 100 and 10,000. All commonly used words will remain in the dictionary, and exceptional words that are not in the dictionary are represented in some other manner (such as ASCII or EBCDIC). A very simple variant on this method is to store the entire sequence of data that has already been compressed. When a new sequence is to be compressed, the stored sequence is searched to see if it contains the new one. If it does, the new sequence is represented as its position in the stored one. For example, if the data processed so far were given by (7), the next data **the** could be

$$\text{``the cat saw the rat in ''} \qquad (7)$$

represented either as **1** or as **13**.

Ziv-Lempel compression. It is time-consuming to search the entire stored sequence, and a fast method has been developed by J. Ziv and A. Lempel that stores only some of the previous sequences but achieves results slightly better than the dictionary schemes. This method, unlike the dictionary methods, requires no prior knowledge about what constitutes a word in the data. For many practical applications, Ziv and Lempel's is the method of choice.

Ziv-Lempel compression begins with a dictionary containing every character in the alphabet. During compression, the dictionary is searched to find the longest string that is a prefix of the text yet to be compressed. The compressed representation of the string is its position in the dictionary. Then a new string, formed from the string just compressed and the upcoming character in the text, is added to the dictionary. Compression continues as described above, using the enlarged

Table 2. Steps in calculation of Huffman code

Step	Least probable codes		Common prefix
1	**s** (5.25%)	**w** (5.25%)	**sw** (10.5%)
2	**c** (5.25%)	**r** (5.25%)	**cr** (10.25%)
3	**e** (10.5%)	**h** (10.5%)	**eh** (21%)
4	**cr** (10.5%)	**sw** (10.5%)	**crsw** (21%)
5	β (21%)	**a** (16%)	β**a** (37%)
6	**eh** (21%)	**t** (21%)	**eht** (42%)
7	β**a** (37%)	**crsw** (21%)	β**acrsw** (58%)

dictionary and the remainder of the text. For example, with the initial dictionary given by (8), the string (9) is encoded by the Ziv-Lempel method as sequence (10). The first letter **t** is en-

$$\beta \text{ a c e h r s t w} \qquad (8)$$

$$\text{"the cat saw the rat"} \qquad (9)$$

$$\mathbf{8\ 5\ 4\ 1\ 3\ 2\ 8\ 1\ 7\ 2\ 9\ 1\ 10\ 12\ 6\ 15} \qquad (10)$$

coded as **8** and the string **th** is added to the dictionary. The second letter **h** is encoded as **5** and **he** is added to the dictionary, and so on. Eventually, the string **th** is encountered again (after compressing "the cat saw ") and is represented as **10** (its position in the dictionary). Then the string **the** is added to the dictionary. Following **th**, the string **eβ** is encoded as **12** and the new string **eβr** is added to the dictionary. Finally, **r** and **at** are encoded as **6** and **15**. At the end of the text the dictionary contains the entries given by (11).

$$\beta \text{ a c e h r s t w th he eβ βc ca}$$
$$\text{at tβ βs sa aw wβ βt the eβr ra} \qquad (11)$$

Arithmetic coding. Compression methods have been developed that separate the two aspects of data compression, that is, estimating the probability of each character and producing an optimal code based on the estimate. These methods hinge on arithmetic coding which, for any probability, can produce a representation that is arbitrarily close to the corresponding information content. Arithmetic coding involves representing a long string of characters as a single number.

Arithmetic coding is most simply illustrated by using a two-character alphabet (**a** and **b**) with fixed probabilities (80 and 20%, respectively). The text to be compressed is (12). Proceeding

$$\text{abaabaaabaaaab} \ldots \qquad (12)$$

one character at a time, as many characters as possible are represented as an 8-bit number between 0 and 255. The range of possible values is partitioned in proportion to the probabilities of the characters: any number in the lower 80% of the range (0–204) represents **a** and any number in the upper 20% (205–255) represents **b**. Since the first character is **a**, the lower part of the range must be chosen, and the upper part is discarded. Again, this reduced range (0–204) is partitioned: the values 0–163 represent **a** and the values 164–204 represent **b**. The next character in the text is **b**, so the lower half of the range is discarded, and the encoded value must be in the range 164–204.

This process of partitioning the range of values and encoding characters continues until the range contains only one number; that single number represents the string of characters.

This example results in the sequence of ranges 0–255, 0–204, 164–204, 164–196, 164–189, 185–189, 185–188, 185–187, 187. The final value 187 represents string (13). From the character

$$\text{abaabaaab} \qquad (13)$$

probabilities, the information content of this string is calculated to be 7.9 bits. Decoding takes place by reproducing the partitioning steps. At each step the number 187 is compared to see which partition it fits in: if it is in the lower range, the next decoded character is **a**, otherwise **b**. The range not containing the decoded character is discarded, and the process continues.

If each character is given a static probability, the amount of compression is nearly identical to that achieved by Huffman coding. However, it is not necessary to use unvarying estimates for the character probabilities; any knowledge whatsoever about the text may be used to predict upcoming characters. If a more accurate prediction is made, compression is improved. In example (12), it appears quite likely that the upcoming characters are given by (14). The problem of

$$\text{aaaaabaaaaaabaaaaaaab} \ldots \qquad (14)$$

making accurate probability estimates is a subject of current research. Current methods involve building a model to represent the data processed so far, and using that model to predict upcoming bits. Models that predict the next character based on the previous few characters can now compress English text to between 2 and 3 bits per character. C. Shannon conjectured that the actual information content of English is about 1.4 bits per character. Recent data compression techniques are challenging that bound.

Special-purpose techniques. In addition to the general techniques described above, a number of data compression methods developed with a particular purpose are in common use. These techniques generally rely on the data having a special form but still take advantage of the principles of information theory. Digram encoding relies on the fact that there are unused characters in the alphabet, and uses these characters to represent common pairs of characters. For example, English text could be reduced in size by replacing the digrams **th**, **an**, and **ie** by single characters. Run-length encoding encodes strings of the same character as a single number (a repetition factor). In English, run-length encoding is of little use, but in computer data, runs of special characters such as *space* or *underscore* might be expected. Difference encoding takes advantage of a sequence of data that differs little from one value to the next (for example, hourly temperature readings at a fixed location). Each value is encoded as the difference from the previous value. The advantage is that, because the differences are smaller than the actual values, they may be represented in fewer bits. Leading character elimination is used in dictionaries stored in alphabetical order. If the dictionary is large, adjacent entries tend to have several initial characters in common. The coding for each subsequent word has two parts: the number of characters in common with the previous word, and the unique suffix. These special-purpose techniques, while they do not outperform general methods, are used because they are simple and efficient to implement and because they aptly model some common forms of data.

For background information SEE INFORMATION THEORY in the McGraw-Hill Encyclopedia of Science and Technology.

Gordon V. Cormack

Bibliography. D. Severance, A practitioner's guide to data base compression, *Inform. Syst.*, 8(1):51–62, 1983; T. A. Welch, A technique for high-performance data compression, *IEEE Comput.*, 17(6):8–19, June 1984; I. H. Witten and J. G. Cleary, Foretelling the future by adaptive modelling, *Abacus*, 3(3):16–73, 1986.

Decapoda (Crustacea)

In polygamous species a male's reproductive strategy is to inseminate as many females as possible during his lifetime. When his success is affected by competition and aggression from other males, as is frequently the case in polygamous species, an individual which is at a competitive disadvantage sometimes adopts an alternative mating strategy. For example, small males of many species take advantage of their size and inconspicuous behavior to intercept females that are attracted to a larger, courting male. Such a phenomenon has been observed and documented in many organisms: insects (field crickets, flies), fishes (Gila topminnow, Atlantic salmon), and amphibia (bullfrogs). In any species in which wide physical variations exist, alternate male mating strategies can be expected.

In most cases a male practices only one reproductive behavior throughout his lifetime. However, in some species mating patterns are changed in the course of the animal's lifetime, which is the case in the fresh-water prawn.

Prawn morphotypes and mating strategies. Mature males of the fresh-water prawn *Macrobrachium rosenbergii* can be divided into three distinct morphological types. Two of the morphotypes demonstrate distinct mating strategies: the largest, dominant males actively court and protect the females prior to mating, and the smallest males practice a form of sneak-mating consistent with their size and high mobility. Males of the third morphotype are in an intermediate phase with a reduced rate of reproductive activity.

A male population of *M. rosenbergii* shows a wide size distribution. Such a distribution is the outcome of the coexistence of the three distinct morphotypes, each distinguishable by its size, claw color, and length of the claw in relation to the length of the body (defined as relative claw length). When the male population is divided according to claw color, three distinct categories of relative claw length become apparent. Small males, which are most numerous, are characterized by clear or light pink claws and a relative claw length of 0.4 to 0.9. Orange claw males are larger, characterized by the color of their claws and a relative claw length of 0.8 to 1.4. Blue claw males, the largest of the males, are characterized by thick, spiny, dark-blue claws and a relative claw length of 1.4 to 2.0. The proportion of small males to orange claw males to blue claw males is roughly 5:4:1.

Blue claw males. The blue claw male is dominant and sexually active, with a relatively large reproductive system. A determined territory, which provides for shelter and access to food and females, is agonistically defended by the blue claw male from the invasion of other dominant males. Females which are ready to mate evidently seek males that can protect them during their subsequent molt and vulnerable period. The long claw and the distinctive blue color of the male attract the female and aid in defending the territory. The blue claw male surrounds the female with his claws, simultaneously probing and cleaning her abdomen and thoracic sterna with his other legs (**Fig. 1**). At most, eight to ten females may be associated with a single blue claw male. The blue claw male is the culmination of the male developmental pathway. At this stage, growth has almost completely ceased.

Orange claw males. Orange claw males are subdominant, nonterritorial, and almost sexually inactive. Their reproductive system is relatively small. Aquaria observations reveal that, in the absence of blue claw males, orange claw males may approach and even injure a newly molted and receptive female. Orange claw males have

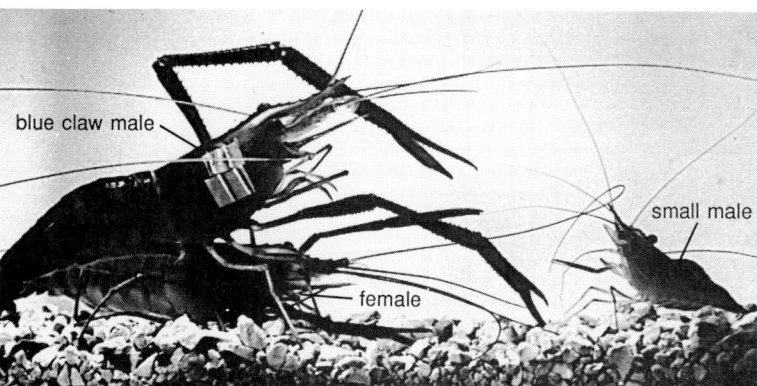

Fig. 1. Small male attempting to sneak in between a courting dominant blue claw male and a receptive female.

blue claw male

female

small male

never been observed protecting females, although mating apparently does occur occasionally, as verified by the fact that, as a result, the females carry fertile eggs. This suggests that orange claw males may be physiologically capable of mating and fertilizing but may lack some behavioral patterns associated with sexual maturation. It is likely that under natural conditions orange claw males never have an opportunity to mate. They continue to grow, however, and, if they survive, they eventually may undergo a transformation into blue claw males. Such a transformation would occur after an orange claw male succeeds in occupying a territory containing females, either by displacing a smaller or weakened blue claw male or by replacing a deceased blue claw male.

Small males. Small males are nonterritorial. They avoid competition by retreating into the body of water (as opposed to the floor of the pond or aquarium) where they swim or hover, and by hanging onto rocks and vegetation. They are sexually active and have a relatively large reproductive system. Small males are observed to be strongly attracted to premating molted females and occasionally succeed in copulating by sneaking in between the female and her blue claw mate (Fig. 1). Females have been successfully fertilized when only small males have been present, indicating that small males are sexually virile.

Significance. Blue claw males and small males demonstrate alternative mating strategies. Orange claw males are considered an intermediate developmental stage between the two.

Blue claw males seem to possess most of the attributes which contribute to reproductive success: they are highly attractive to females, their larger size is an advantage in agonistic encounters with other males, and they are able to protect fertilized females. These benefits are evident in the high fertilization rate (85–100%) observed in aquaria whenever a normally functioning blue claw male is present. The main costs associated with the blue claw mating strategy involve energy expended in defending a territory and reduced flexibility to adapt to changing environmental conditions.

Small males are less attractive to females. However, the small male morphotype has two options: (1) to remain small and occasionally succeed at sneak fertilization, or (2) to continue on the developmental pathway, becoming an orange claw male and eventually a blue claw male (**Fig. 2**). The fitness of the first option decreases as the frequency of small males increases, while the fitness of the second option decreases as the frequency of orange claw males increases.

The concept of an evolutionarily stable strategy is based on the fact that certain behavior patterns (such as feeding, courting, and mating) are inherited. It implies that in cases where individuals of the same species display different mating strategies, an equilibrium is maintained in

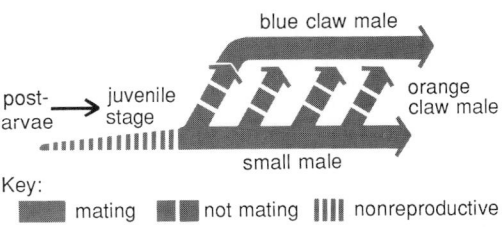

Fig. 2. Reproductive life history pathways for *Macrobrachium rosenbergii* males.

which the relative frequency of each strategy is fixed. Under these conditions the fitness of both strategies is identical. Thus, in an evolutionarily stable strategy the fitness of each small male option may be expected to be equal, and the ratio of small males to orange claw males may be expected to remain constant.

In the case of *M. rosenbergii*, neither genetic differences nor parental manipulation can account directly for male polymorphism and the two alternative mating strategies. The former possibility is unlikely since a small male is capable of undergoing a transformation into an orange claw male and eventually into a blue claw male. However, it is possible that genes which direct the individual growth rate may determine the preliminary size hierarchy observed even in early juvenile stages. From the juvenile stage onward, the size of the animal may dictate its chances of continuing to grow through the developmental pathway to the dominant blue claw stage. A direct relationship between relative body size and mating behavior has also been described in the dung fly species *Sepsis cynipsea* and *Scatophaga stercoraria*. In both, most of the size variation is attributable to environmental causes.

A highly competitive environment, in which there is a high frequency of large males, may favor small, sneaking males, since territorial males primarily are occupied with agonistic behavior that consumes a significant amount of their time and energy. In contrast, if the frequency of large males were low, the opposite would be true. Energy normally expended in agonism would be released for sexual activity, and the blue claw reproductive advantage over the small males would increase. In an evolutionarily stable strategy, the proportion of each male type in the population should remain constant. This hypothesis is consistent with field observations of *M. rosenbergii* populations in which the frequency of the three male morphotypes was found to be constant despite wide variation in ecological conditions.

For background information SEE DECAPODA (CRUSTACEA); POPULATION GENETICS; REPRODUCTIVE BEHAVIOR in the McGraw-Hill Encyclopedia of Science and Technology.

Amir Sagi

Bibliography. D. G. Constantz, Behavioral ecology of mating in the male gila topminnow, *Poeciliopsis occidentalis* (Cyprinodontiformes: Poecilidae), *Ecology*, 56:966–973, 1975; M. R. Gross, Sunfish, salmon, and the evolution of alternative reproductive strategies and tactics in fishes, in G. W. Potts and R. J. Wootton (eds.), *Fish Reproduction*, pp. 55–75, 1984; Z. Ra'anan and D. Cohen, Ontogeny of social structure and population dynamics in the giant freshwater prawn *Macrobrachium rosenbergii* (de Man), in A. Wenner (ed.), *Crustacean Issues 3: Crustacean Growth: Factors in Adult Growth*, pp. 277–311, 1985; Z. Ra'anan and A. Sagi, Alternative mating strategies in male morphotypes of the freshwater prawn *Macrobrachium rosenbergii* (de Man), *Biol. Bull.*, 169:592–601, 1985.

Design engineering

Silicon micromechanics and micromachining is a relatively new technology that uses integrated-circuit manufacturing techniques and materials, not to build electronic devices, but to fabricate miniature mechanical devices and components.

Mechanical properties of silicon. The complex electronic integrated-circuit devices that continue to have great influence have been made possible as a result of the growing ability to exploit the unique chemical and electronic properties of single-crystal silicon. While the electronic virtues of silicon have been reviewed and

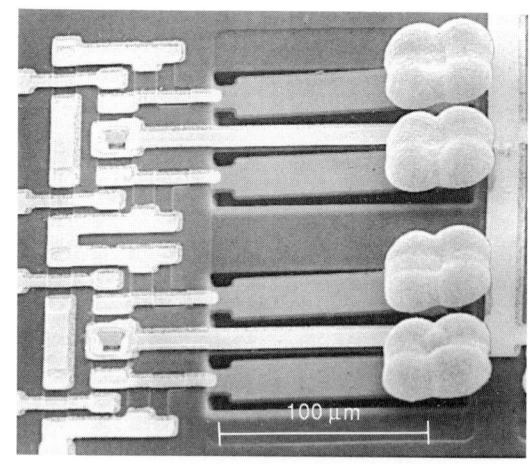

Fig. 2. Cantilever-beam accelerometer integrated on silicon chip alongside metal-oxide-semiconductor detection and amplifying circuitry. Large gold masses on small ends of four beams make them more sensitive to forces. This accelerometer is smallest such device fabricated.

analyzed extensively, only recently has it become accepted that silicon can also be employed as a superior mechanical material. Single-crystal silicon has a strength-to-weight ratio that is at least three times as high as that of stainless steel. Silicon technology regularly employs thin films such as silicon dioxide and silicon nitride, which are well known as hard, corrosion-resistant, abrasion-resistant, protective mechanical materials. Silicon is, by far, the most abundant element in the Earth's crust, and high-purity, single-crystal silicon wafers are readily available and inexpensive. While the electronics industry has invested billions of dollars to perfect fabrication materials and processes to improve the electronic quality and performance of silicon-integrated circuits, the mechanical qualities of silicon have been simultaneously optimized. Thin films of silicon dioxide grown by integrated-circuit manufacturing methods, for example, have been shown to be 10 to 100 times stronger than bulk materials.

Silicon micromachining. One of the most important characteristics of any mechanical material is its machinability. Research on silicon over the years has resulted in the development of many chemical and plasma techniques for the so-called etching of unique, miniature, precise shapes, patterns, and mechanical structures in silicon wafers. The process of photolithographically transferring two-dimensional patterns onto silicon wafers, universally practiced by integrated-circuit manufacturers, can be combined with these deep chemical or plasma etching processes to produce a great variety of three-dimensional shapes and structures in single-crystal silicon. Many of these shapes cannot be produced in any other way. The etch rates of several of these chemical etchants depend not only on the crystallographic orientation of the silicon wafer but

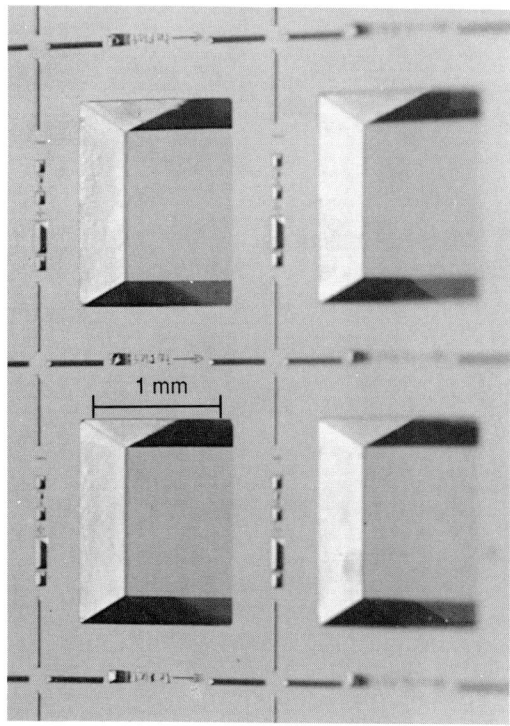

Fig. 1. Array of cavities etched on the back side of a silicon wafer to form the major features of a micromachined silicon pressure sensor. Piezoresistors on front side detect strains in thin silicon diaphragms as they are flexed when pressure is applied. Smaller etched patterns are alignment marks to assure proper registration between front and back sides.

also on the level of minute layers of impurities in the crystal (called dopants by integrated-circuit manufacturers). This property provides an additional dimension of control for creating micromachined devices. As an example, thousands of precisely controlled holes can be micromachined completely through a silicon wafer; these holes can be as small as 25±1 micrometers on a side. No other technology can produce similar results. Micromachining methods have been employed to fabricate microminiature cantilever beams, thin diaphragms, thermally isolated bridges, high-precision grooves and holes, valves, tubes, vibrating elements, needle-sharp probes, miniature vacuum chambers, and other unique structures.

Mallory bonding. One of the powerful aspects of micromachining technology is the ability to build extremely small mechanical devices. This ability arises naturally from the use of the same equipment and general processes employed by integrated-circuit manufacturers to build extremely small electronic devices and circuits. In order to handle and package such small mechanical components, a novel bonding technique is used to mount micromechanical chips onto mechanically isolating glass substrates. The Mallory company discovered in the late 1960s that polished silicon chips can be hermetically sealed to polished glass plates by placing the two pieces together, heating them to about 662°F (350°C), and applying approximately 800 V across the assembly. No intermediate glues, epoxies, or metal layers are required. Since complementary patterns can be micromachined into both the silicon wafer and the glass plate, then aligned and bonded together, Mallory bonding is a versatile and indispensable tool for the fabrication of silicon micromechanical devices. This tool is now used as a mechanical isolation method for pressure sensors, and as a hermetic sealing technique for miniature plumbing systems such as gas chromatographs and fluid dispensers.

Silicon pressure sensors. Since the late 1960s the science of micromachining has been used primarily in the production of mechanical sensing devices such as pressure sensors. Silicon-based pressure sensors are made by etching depressions in the back of a wafer that are aligned with diffused resistors on the front side (**Fig. 1**). If a resistor is located on a thinned silicon region, then the strain caused by a pressure differential across the thin diaphragm will induce changes in resistance that are proportional to pressure. The change in resistance with pressure is known as piezoresistance. For higher-performance applications the silicon chip is Mallory-bonded to a glass substrate, which isolates packaging stresses from the thin silicon diaphragm. Many millions of silicon pressure sensors are manufactured each year for applications as diverse as intake-manifold pressure sensors in automobiles (virtually every automobile now manufactured in the

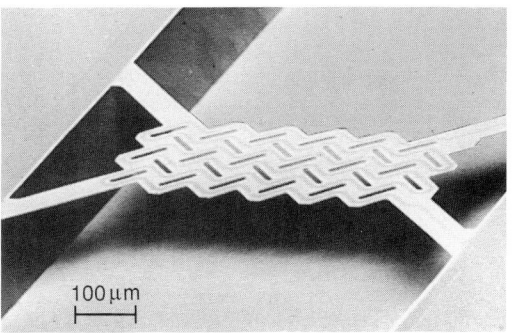

Fig. 3. Micromechanical bridge, 0.5 mm long, that forms sensing element in a mass-flow sensor chip. Thin-film resistor on surface of bridge senses flow rates as gases flow down channel micromachined under bridge.

United States contains a silicon pressure sensor), aerospace altitude sensors, disposable pressure sensors for blood-pressure monitoring during medical operations and intensive care, underwater depth gages, pressure and flow sensors for all types of industrial fluids and gases in refineries and chemical plants, and high-reliability aerospace fuel and hydraulic pressure indicators. These products span the ranges of disposability, reliability, and expense from inexpensive commercial sensors to high-performance, expensive transducers.

Other sensors and components. Although they are the first to become successfully commercialized, pressure sensors are only one example of the types of sensors and precision mechanical components which can be fabricated by silicon micromachining. Other mechanical sensors include accelerometers (**Fig. 2**), vibration sensors, strain sensors, mass-flow sensors (**Fig. 3**), tactile sensors, chemical sensors, acoustic sensors, high-sensitivity infrared detectors, and vacuum sensors. With the same micromachining principles, the following mechanical components have

Fig. 4. Ink-jet nozzles micromachined in silicon.

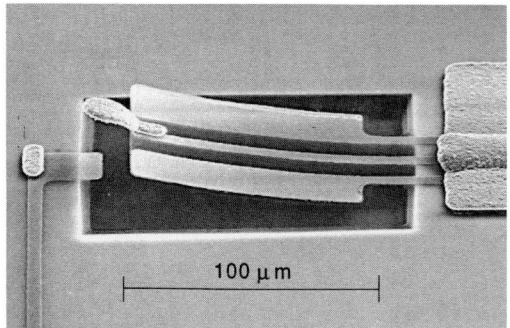

Fig. 5. World's smallest mechanical electrical switch. Cantilever beam, 100 micrometers long, was fabricated by silicon micromechanical techniques. Voltage applied to beam causes it to deflect downward so that metal tab on end of beam makes electrical contact with metal line on substrate.

been demonstrated and some have been commercialized: precision nozzle arrays for ink-jet printing (**Fig. 4**), print heads for thermally sensitive paper, miniature tubes and valves for fluid flow and plumbing, high-frequency vibrating mirrors for laser scanning, light-modulating elements, micromechanical electrical switches (**Fig. 5**), neuron-sized brain-probe arrays, and high-efficiency cooling plates for integrated-circuit chips. Most of these structures represent the smallest mechanical devices of their type ever demonstrated.

Prospects. Silicon micromechanics and micromachining are very active areas of research and development in Europe, Japan, and the United States. The growing use of computers and microprocessors in current electronic equipment, appliances, and toys has expanded the need for sensors to provide input and feedback to the computer system. Since it uses the same technology that has made complex integrated circuits generally available, silicon micromachining has the potential for decreasing the costs and increasing the performance of all types of physical sensors and miniature mechanical components. The broad applicability of silicon micromachining technology has not yet been fully exploited, and numerous new devices and uses await future development.

For background information SEE ACCELEROMETER; INTEGRATED CIRCUITS; SILICON in the McGraw-Hill Encyclopedia of Science and Technology.

Kurt E. Peterson; Phillip Barth

Developmental genetics

Developmental genetics is the study of the genetic elements which influence or control the development of an organism. These elements, namely the genes and their regulatory accessories (for example, promoters, enhancers, and receptors), receive and process the environmental and cellular signals and give rise to the regulated expression of genes. Gene transfer techniques which allow the introduction of foreign genetic materials into both plant and animal cells have provided the necessary tools for the study of developmentally regulated gene expression. Integration of the introduced genetic material into the recipient plant or animal chromosomes results in genetically altered (transformed) organisms which can pass on the new traits to their offspring. Gene transformation techniques have been utilized to investigate the regions of deoxyribonucleic acid (DNA) which flank the gene itself and are thought to contain the DNA sequences which control when and in what cells the gene is expressed. The methods of gene transfer into higher plant and animal cells will be discussed, as will the molecular approaches of investigating developmental gene regulation using hybrid genes to distinguish tissue-specific expression in higher plants and animals.

Transformation. The introduction of foreign genes into either plant or animal cells and the transformation of these cells depend on the successful delivery of DNA into the cell nucleus. In plant cells, techniques have mainly focused on the use of a soil bacterium, *Agrobacterium tumefaciens*, which harbors a large tumor-inducing plasmid (Ti plasmid). Bacterial cells attach to plant cell walls, usually as a result of a breach in the epidermis of the plant cell caused by physical wounding. The Ti plasmid DNA is mobilized into the plant cell nucleus, and a portion of the plasmid DNA (tDNA) is transferred to the plant chromosome (**Fig. 1**). This transfer is manifested in the formation of a tumor due to the overproduction of auxins and cytokinins (plant hormones) by enzymes encoded in the tDNA. (The vir region of the Ti plasmid encodes the genes responsible for the transfer and integration of the tDNA.) This natural gene-transferring system has been modified to deliver other foreign DNA, thereby streamlining the bacterial system for plant cell transformation by using genes of interest. SEE CYTOKININS.

In animal cells, techniques have been developed that utilize another means of introduction: the microinjection of foreign DNA into the nucleus of a fertilized egg (usually of a mouse) using microscopic needles (**Fig. 2**). The injected eggs are implanted in the uterus of pregnant female mice and allowed to develop normally. The resulting litters are assayed for the presence of the injected genes (for example, by southern DNA blot analysis). A relatively high percentage of the mouse eggs (approximately 25%) will integrate the injected foreign DNA into the chromosome, consequently producing stably transformed mice.

The insertion or integration of the foreign DNA sequences into the recipient host plant or animal

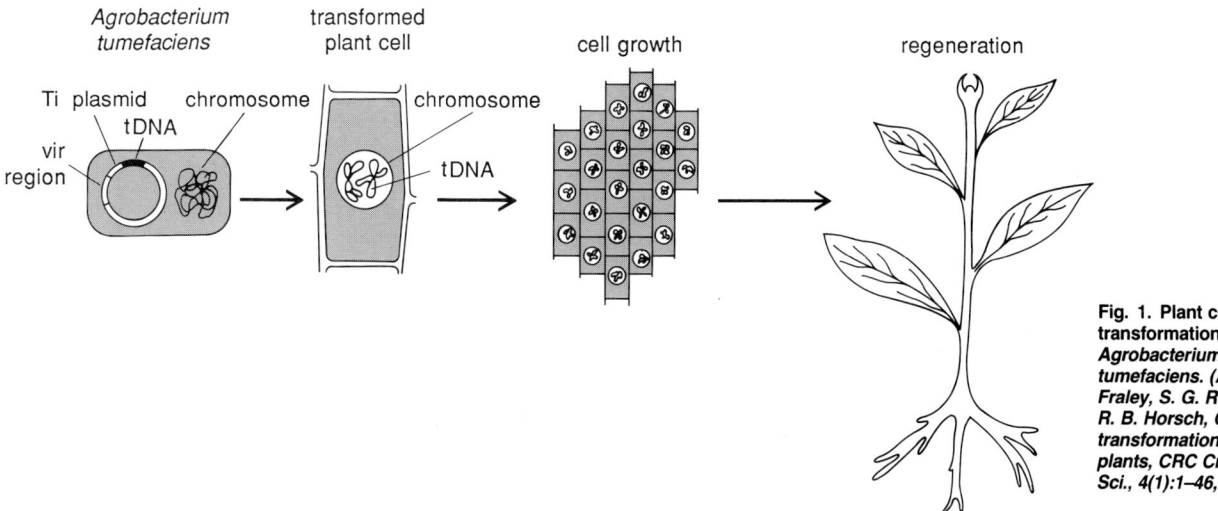

Fig. 1. Plant cell transformation by *Agrobacterium tumefaciens*. (After R. T. Fraley, S. G. Rogers, and R. B. Horsch, Genetic transformation of higher plants, CRC Crit. Rev. Plant Sci., 4(1):1–46, 1986)

cell chromosomes produces a modified or transformed organism referred to as transgenic, that is, possessing the genetic traits of the transferred gene in the organism's chromosomal composition. These transferred traits are stable and can be passed on to subsequent offspring by means of normal fertilization. The insertion into the recipient cell chromosomes occurs randomly both by the natural vehicle of *A. tumefaciens* and by mechanical means using microinjection of DNA into fertilized mouse eggs. At present, insertion into specific sites or chromosomes is not possible by either system.

Specificity of gene expression. Genes that are expressed selectively or preferentially in a certain cell or specific organs are referred to as cell- or tissue-specific. In plants a classic example of tissue-specific genes is the photosynthetic genes found preferentially expressed in the leaves in response to light. Studies of this light-regulated response using characteristic wavelengths (that is, red light to establish the response and far-red light to abolish the response) have shown that the light-regulated response of these genes in plant leaves is linked to the photoreceptor protein phytochrome. Other tissue-specific genes which are responsible for the utilization and transfer of energy harvested by the photosynthetic genes are also found in the leaves. In animal cells, tissue-specific gene expression can be demonstrated by the enzyme elastase I, a digestive enzyme which is synthesized, stored, and secreted by mammalian pancreatic tissues and can be found only in the pancreas.

Most gene products are not confined to any particular cell or tissue. Their presence must be induced either by a cellular signal or hormone or in response to an environmental stress situation. For example, metals like cadmium or zinc, which are toxic to cells in elevated levels, will cause the protein metallothionein to be synthesized.

The difficulty in understanding the above-mentioned systems and the complex interrelationship between organismal development, the cellular environment, and the chromosomal components has been greatly alleviated by the use of

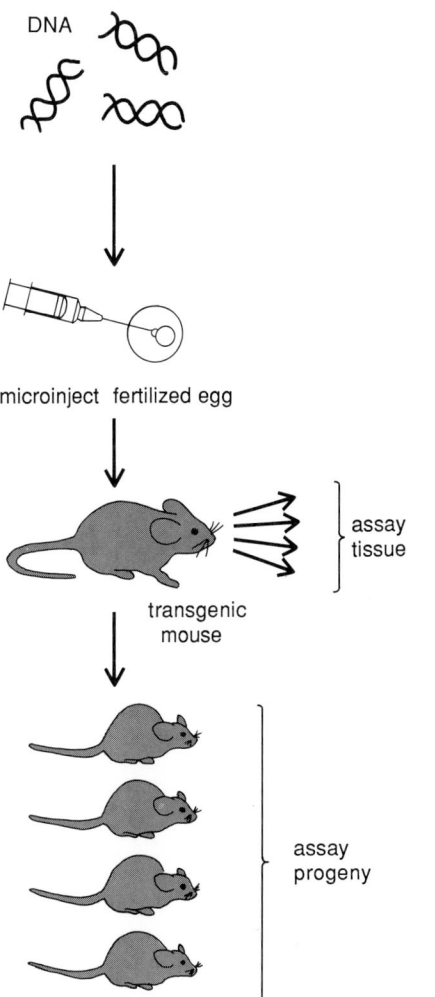

DNA

microinject fertilized egg

assay tissue

transgenic mouse

assay progeny

Fig. 2. Animal cell transformation by microinjection of fertilized mouse eggs. (After J. H. Kelly and G. J. Darlington, Hybrid genes: Molecular approaches to tissue-specific gene regulation, Annu. Rev. Genet., 19:273–296, 1985)

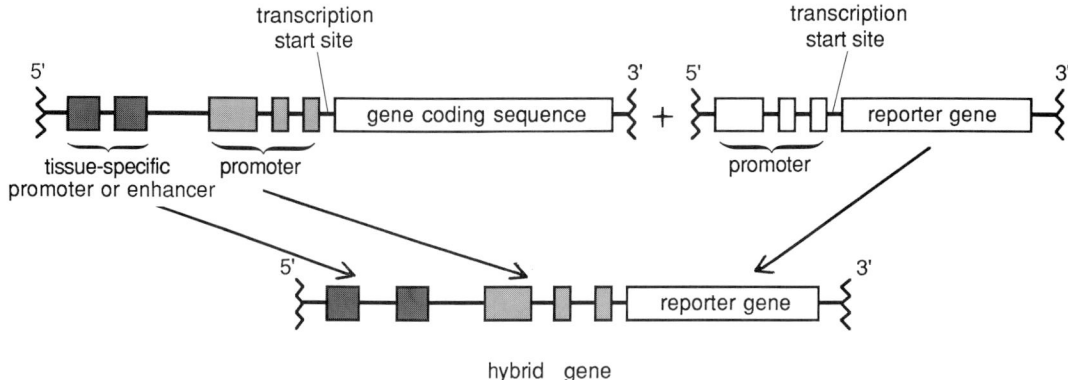

Fig. 3. Diagrammatic representation of a hypothetical gene, showing salient regulatory features and coding sequences for the structural gene. The construction of a hybrid gene consisting of a reporter gene sequence and regulatory elements of an experimental gene is shown, with so-called cassettes of different regulatory regions having been maneuvered adjacent to the reporter gene sequence.

molecular techniques such as molecular cloning and gene manipulation which permit the construction of hybrid genes.

Hybrid genes. The universal structure of a gene consists of a structural coding region containing the unique DNA sequences which specify a protein or regulatory receptor (binding site), flanked by regions of DNA which function to modulate the expression of this gene sequence (**Fig. 3**). These regions are the regulatory elements referred to as promoters, enhancers, and tissue-specific promoters and enhancers.

DNA is synthesized by enzymes which add nucleic acid subunits from one end (the 5′ end) toward the other end (the 3′ end). This 5′ to 3′ directionality of the DNA can be carried over to describe the relative physical positions of the gene sequence and its regulatory genetic elements. That is, the genetic elements positioned 5′ to the gene sequence are closer to the 5′ end than is the structural gene sequence and are referred to as being upstream from the gene. Likewise those elements positioned 3′ to the gene are referred to as lying downstream from the gene, and are closer than the gene sequences they regulate to the 3′ end. These regulatory elements have been studied in both plant and animal genes and can be manipulated by using molecular cloning techniques to investigate specific regions or sequences and their role in developmental gene expression. The basic logic employed in the study of these regulatory elements is to treat these entities as being analogous to cassettes that can be manipulated or maneuvered in order to examine the characteristic properties that affect the expression of the gene in the developing organism. Another modification which is routinely utilized is the replacement of the normal gene coding sequence with a more easily monitored gene sequence called a reporter gene (Fig. 3). These reporter genes code for an enzymatic marker or a protein which confers resistance to an antibiotic or cell toxin and thereby provides a distinguishing result of hybrid gene expression in the recipient cell.

Hybrid genes are constructed by fusing regulatory sequences with reporter sequences and are used as the foreign DNA transferred into plant or animal cells. Initially, experiments were performed to determine whether: (1) specific regions can be described which have properties to direct gene expression (promoters); (2) regions can be found which modulate cell- or tissue-specific expression (cell- or tissue-specific promoters); and (3) regions responsible for enhanced or elevated expression (enhancers) can be distinguished. The elevated expression of a particular gene sequence is manifested in larger amounts of either gene product or enzymatic activity. This result of enhancer activity is often referred to as high-level expression.

Subsequent experiments were directed toward the investigation of the boundaries and domains of these regulatory elements by using chemical modification (for example, synthetically made DNA); or addition, deletion, or substitution of specific bases (subunits) within the regions of interest to further define the role of these specific sequences in the expression of genes throughout development.

Results. In the investigation of photosynthetically active genes in plants, two genes were chosen for study, the small subunit of the carbon dioxide–fixing enzyme ribulose bisphosphate carboxylase (rbcS) from pea, and the chlorophyll a/b–binding protein (Cab) isolated from wheat. Upstream regions of these genes were used in the study. The hybrid gene was constructed by using these upstream sequences fused to a reporter gene, the coding sequence from the bacterial transposon conferring resistance to the antibiotic chloramphenicol. Subsequent expression was monitored in transgenic tobacco or petunia plants. The level of expression of the hybrid genes (rbcS- or Cab-reporter) was monitored relative to an internal control or reference gene.

In order that the study focus on the activity confined to a finite region, successive deletions or enzymatically shortened regions were tested until activity could no longer be detected. Experiments using these two genes revealed two regulatory regions found upstream of the reporter gene which direct expression of the reporter gene. One region is responsible for light-regulated induction and for tissue-specific (leaf) expression. The other, a region with enhancer-like activity, could be localized within the photoregulatory–tissue-specific controlling region. Preferential expression of the rbcS-reporter hybrid gene could be shown in the leaves of the transgenic petunia and tobacco plants, a significantly lower level of expression could be detected in the stem, and little to no expression could be detected in root tissues. Similar experiments using Cab-reporter genes in transgenic tobacco or petunia plants demonstrated identical results. In both cases, this region exhibited both tissue-specific and light-inducible properties. These experiments show that regulatory elements (promoter, tissue-specific promoter, and enhancer) can be distinguished by their activities, but they cannot as yet be separated from one another. This inseparability of regulatory elements may be a property associated with these particular plant genes.

Similarly, the use of hybrid genes in transgenic mice has helped to define regions in the upstream flanking sequences of the mouse gene coding for metallothionein, in response to cellular substances which stimulate expression. Upstream regulatory regions from the mouse metallothionein gene were fused to a reporter sequence consisting of the human growth hormone gene. In the recipient animals the experiment scored for larger transgenic mice on the basis of growth stimulated by the presence of increased levels of growth hormone. The activity of the hybrid gene, metallothionein–growth hormone, was examined biochemically in different tissues in the transgenic mouse. Increased levels of activity of the hybrid gene were found in three tissues, the liver, heart, and testes; this illustrates a select or preferential response in expression by the hybrid gene to the environmental stimulus found in these tissues.

In transgenic experiments utilizing the elastase I gene, found in the pancreas, the rat elastase I gene was detected biochemically in different tissues of transgenic mice. The expression of the rat elastase I gene in transgenic mouse tissue mimics for the most part the tissue-specific response observed in rats. But although the regulatory regions which flank the rat elastase I gene are sufficient to direct transgenic expression in mice, they are not as stringent or tightly controlled as they are in rat tissue where they restrict expression of the gene to the pancreas. Minor levels of expression could be detected in other tissues

(liver, kidney, and intestine) of the transgenic mice, albeit several orders of magnitude less than the predominant expression observed in the mouse pancreatic tissues.

For background information SEE DEVELOPMENTAL GENETICS; GENE; GENE ACTION; GENETIC ENGINEERING in the McGraw-Hill Encyclopedia of Science and Technology.

Nathan M. Chu

Bibliography. R. Fluhr et al., Organ-specific and light induced expression of plant genes, *Science*, 232:1106–1112, 1986; R. T. Fraley, S. G. Rogers, and R. B. Horsch, Genetic transformation in higher plants, *CRC Crit. Rev. Plant Sci.*, 4(1):1–46, 1986; J. H. Kelly and G. J. Darlington, Hybrid genes: Molecular approaches to tissue-specific gene regulation, *Annu. Rev. Genet.*, 19:273–296, 1985; R. Palmiter et al., Metallothionein–human GH fusion genes stimulate growth of mice, *Science*, 222:809–814, 1983; R. Palmiter and R. L. Brinster, Transgenic mice, *Cell*, 41:343–345, 1985; G. Swift et al., Tissue-specific expression of the rat pancreatic elastase I gene in transgenic mice, *Cell*, 38:639–643, 1984.

Dielectric materials

Dielectric materials are of vital importance in telecommunications equipment, electronic circuit components, and quasi-optical elements. The latter two categories include insulators, supports, beam dividers, lenses, resonators, substrates, windows, dielectric waveguides, radomes, and radiation-absorbing materials. For all these applications, high-precision measurements of dielectric and magnetic parameters are required. In some technologies, such as the construction of microwave ovens, diathermy, and industrial heating, the dielectric loss is a critical parameter. The study of the dielectric properties of a material is essentially the quantitative characterization of its interaction with electromagnetic fields. The interaction of externally applied electromagnetic fields with the molecular, polycrystalline, or crystalline system provides a sensitive experimental probe to investigate the system's behavior.

Until recently there was almost no reliable dielectric data available in the millimeter and submillimeter wavelength range (frequencies of 60–1000 GHz). Measurements of the dielectric properties of materials at these wavelengths, which lie between the microwave and optical regions, are extremely difficult. These wavelengths are too short for the practical use of a microwave single-mode resonant cavity, while a familiar blackbody (optical) source such as a mercury-vapor lamp normally provides too little energy for millimeter-wave measurements with a Fourier spectrometer. Nevertheless, the Fourier technique has now been immensely improved to provide routine, high-precision, continuous data from 60 GHz to many thousands of gigahertz. The Fabry-Perot open resonator technique has also been improved to provide reliable comparative single-frequency points on continuous broadband Fourier data.

Theory of dielectrics. Dielectric physics can be approached from two directions: the macroscopic theory, which formulates general relations between the properties of dielectric materials, and the microscopic theory, which shows how the motion of the constituent entities on an atomic scale leads to the observed dielectric properties. Although the macroscopic theory is simpler, it is somewhat limited in the information it provides. On the other hand, the microscopic theory is more interesting and exact, but its full elaboration involves computational difficulties.

Macroscopic theory. The construction of any device that is to be operated at high frequencies requires knowledge of the quality of the materials to be employed. The quality of a material is defined by its properties such as relative electrical permittivity, $\hat{\varepsilon} = \varepsilon/\varepsilon_0$, and relative magnetic permeability, $\hat{\mu} = \mu/\mu_0$, where ε and μ are the permittivity and magnetic permeability of the material, and ε_0 and μ_0 are the permittivity and magnetic permeability of free space. Since lossy components are present, both $\hat{\varepsilon}$ and $\hat{\mu}$ are complex, so that $\hat{\varepsilon} = \varepsilon' - j\varepsilon''$ and $\hat{\mu} = \mu' - j\mu''$ (where $j = \sqrt{-1}$). When an alternating field of frequency ν and circular or angular frequency $\omega = 2\pi\nu$ is applied, the displacement current is determined by ε' and μ', and an in-phase or lossy current is determined by ε'' and μ''. Another quantity, the complex conductivity σ, is directly related to the complex dielectric permittivity through Eq. (1). The complex refractive index,

$$\hat{\sigma} = \sigma' + j\sigma'' = \omega\varepsilon'' + j\omega\varepsilon' \qquad (1)$$

given by Eq. (2), determines the phase change

$$\hat{n} = n - jk = n - \frac{i\alpha c}{4\pi\nu} \qquad (2)$$

and attenuation of an electromagnetic wave passing through the medium. Here $c = (\mu_0\varepsilon_0)^{-1/2}$ is the speed of light and α is the absorption coefficient in nepers/m. In the general case, where the medium may be magnetic and conducting, these quantities are related by Maxwell's relation, Eq. (3). At high frequencies, for most common materials, the relative magnetic permeability $\hat{\mu}$ is close to unity and, for a dielectric medium, there are no free charges, so that the conductivity at zero frequency is zero. Maxwell's relation then simplifies to Eq. (4).

$$\hat{n}^2 = \hat{\varepsilon} \qquad (4)$$

Conductors and dielectrics must often be combined in electrical applications; there are thus wide variations in the electrical properties of the materials used. The nature of a material can be distinguished by the relative values of $\sigma' = \omega\varepsilon''$ and $\omega\varepsilon'$. For a conductor, the σ' values are dominant, whereas for a dielectric, $\omega\varepsilon'$ values become dominant. In a semiconductor, σ' and $\omega\varepsilon'$ are comparable in magnitude. Together, the quantities mentioned above suffice to describe the overall electrical behavior of a particular material medium at high frequency, but the parameters are also functions of field strengths and of frequency.

At frequencies above 30 GHz, it is convenient to employ optical or quasi-optical measurement techniques. Optical parameters such as the refractive index $\hat{n}$ and the absorption coefficient α are therefore measured directly. The real (ε') and imaginary (ε'') parts of the relative permittivity $\hat{\varepsilon}$ are then calculated by Eqs. (5) and (6).

$$\varepsilon' = n^2 - (\alpha c/4\pi\nu)^2 \qquad (5)$$

$$\varepsilon'' = 2n(\alpha c/4\pi\nu) \qquad (6)$$

Microscopic theory. P. Debye made the first attempt to explain the radio-frequency absorption of a dipolar molecule in a liquid or solid. In his theory, the rate of dipole reorientation in the liquid state is based on a molecular model in which the rigid molecule rotates with a viscous drag leading to an angular velocity ω. On removal of a static field at time $t = 0$, the dipole polarization P decays with a unique rate coefficient k according to Eq. (7). According to Debye, dipo-

$$P(t) = P(0)\, e^{-kt} = P(0)\, e^{-t/\tau} \qquad (7)$$

lar activity should be limited to compounds for which the angular frequency ω is less than the reciprocal of the molecular relaxation time τ, which is typically 10 GHz. J. Poley subsequently predicted the existence of a new dispersion process (absorption) for polar molecules in the millimeter- and submillimeter-wave frequency regions.

The behavior of dielectric media can be modeled by simple resistance-capacitance (RC) circuits connected in series so that Eq. (8) is satis-

$$V(t) - [dV_c(t)/dt]RC = V_c(t) \qquad (8)$$

fied, where $V(t)$ is the applied voltage and $V_c(t)$ is the output voltage across the capacitor. In the simple RC circuit where $V(t)$ is constant, the output voltage takes the elementary form given by Eq. (9), where $\tau = RC$. The function $r(t)$

$$V_c(t) = r(t) = \exp(-t/\tau) \qquad (9)$$

measures the response of the circuit to changes in $V(t)$. This leads to Eq. (10), where the complex

$$\frac{\hat{\varepsilon}(\omega) - \varepsilon_\infty}{\varepsilon_s - \varepsilon_\infty} = \int_0^\infty r(t)[\exp(-j\omega t)\,.dt] \qquad (10)$$

Fourier transform of the time derivative of the response function $r(t)$ gives the scaled complex relative permittivity, $[\hat{\varepsilon}(\omega) - \varepsilon_\infty]/(\varepsilon_s - \varepsilon_\infty)$. Here ε_∞ represents the relative permittivity at the highest frequency of the measurement range, and ε_s represents the relative static permittivity or the relative permittivity at very low frequencies. Substitution of the correlation function $f(t) = \exp(-t/\tau)$ for the response function $r(t)$ in Eq. (10) yields the Debye equation (11). There is a single

$$\frac{\hat{\varepsilon}(\omega) - \varepsilon_\infty}{\varepsilon_s - \varepsilon_\infty} = \int_0^\infty f(t)[\exp(j\omega t)\,dt] = \frac{1}{1 + j\omega\tau} \qquad (11)$$

relaxation time τ, and the correlation function behaves like an exponential for large values of time t, representing relaxation processes at the low-frequency end of the spectrum and resonance processes at high frequencies.

Various oscillator models have been considered in which a molecule oscillates with simple harmonic motion at small amplitude in a deep potential well but then suddenly undergoes a large-amplitude jump into a different potential well.

Measurement techniques. A broadband Fourier-transform spectroscopic technique is ideally suited to millimeter- and submillimeter-wave dielectric measurements. Dispersive or asymmetric Fourier-transform spectroscopy gives phase information in addition to amplitude information which, after analysis, directly provides the absorption coefficient α and the refractive index n as continuous functions of frequency. The real part (ε') and the imaginary part (ε'') of the complex relative permittivity ($\hat{\varepsilon}$) and the loss tangent ($\tan\delta = \varepsilon''/\varepsilon'$) are calculated by Eqs. (5) and (6). **Figure 1** is an example of these data, comparing the real parts of the permittivities of polyethylene and polypropylene at millimeter wavelengths. This high-precision data reveal detailed properties of materials and make it possible to distinguish nominally identical materials.

Various other measurement techniques, such as time-domain spectroscopy, the frequency-domain transmission-line (waveguide bridge), microwave closed-cavity methods, untuned or stochastic cavity methods, microwave free-space methods, and calorimetric methods, can be employed at microwave and radio-frequency ranges. When high power is required, in particular for a lossy material, a laser and a Mach-Zehnder interferometer can be utilized.

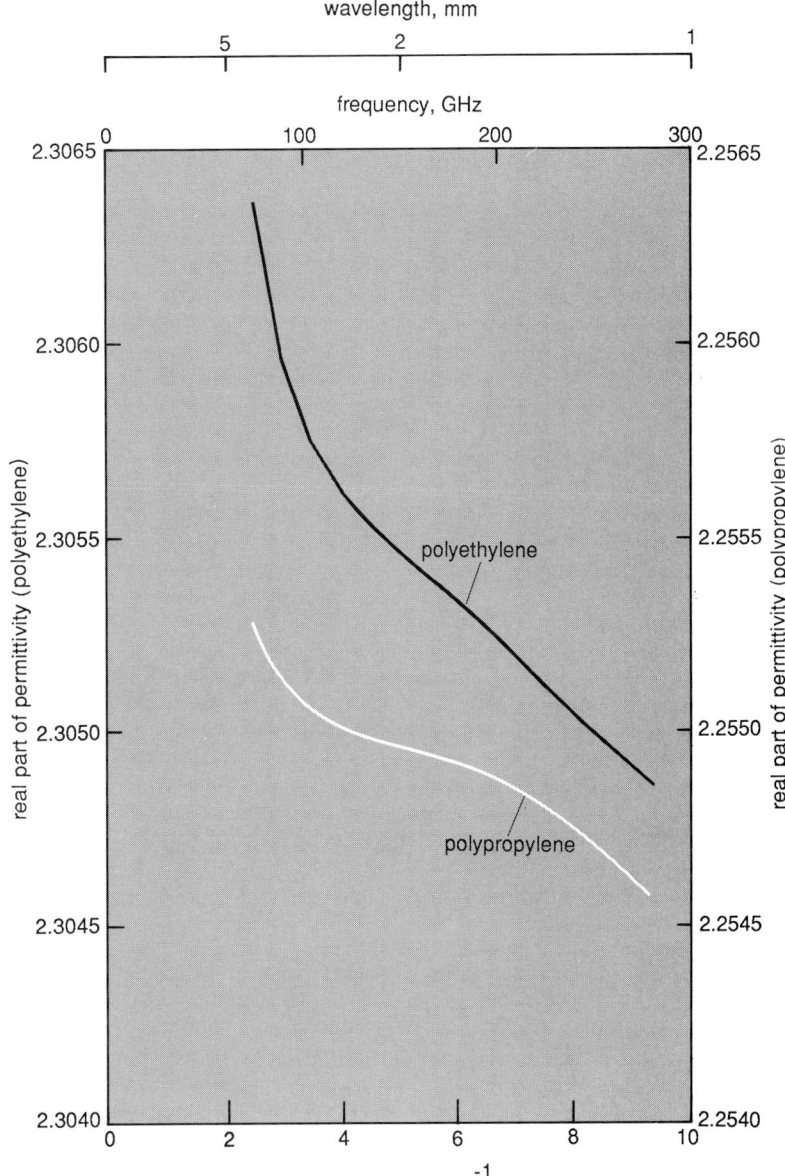

Fig. 1. Comparison of real part of permittivity of polyethylene and polypropylene at millimeter wavelengths, at 26°C (79°F).

Properties of materials. Common dielectric materials can be grouped into various classes, for example, polymers, ceramics, glassy or fused-silica–based insulating and amorphous materials, semiconductors, magnetic and ferrite materials, and polar and nonpolar liquids.

Glasses. These have always been attractive as optical materials. Unfortunately, their good transparency is mostly restricted to the visible and near-visible frequency regions. However, in the millimeter-wave region below 300 GHz, many glasses regain moderate transparency. It is found that the millimeter-wave absorption in glasses is dominated by the water content in the material.

Polymers. Polymeric solids are in extensive use in microwave devices and in millimeter- and

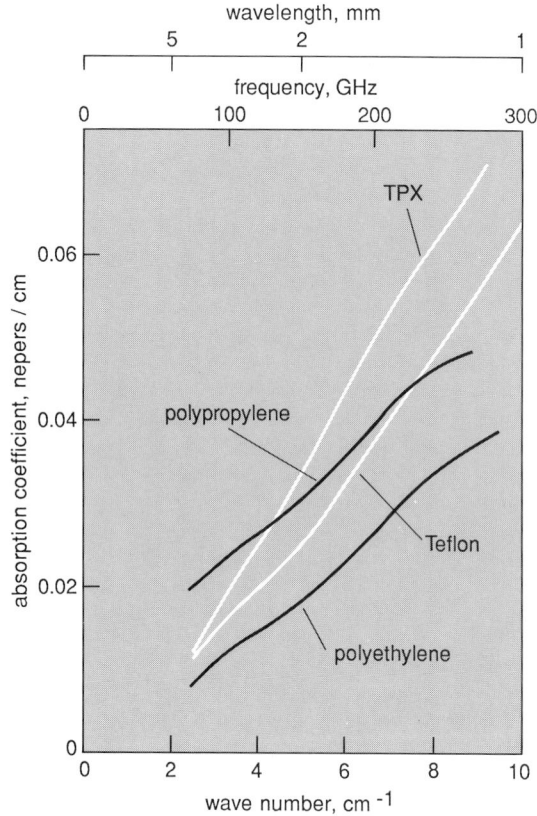

Fig. 2. Millimeter-wave absorption spectra of the nonpolar polymers polyethylene, polypropylene, polytetrafluoroethylene (Teflon), and poly-4 methylpentene-1 (TPX), at 26°C (79°F).

tivity values to be achieved. In the millimeter and microwave frequency regions, absorption in semiconductors is dominated by free-carrier absorption. A deep-trap impurity such as Cr^- for gallium arsenide (GaAs) reduces the free-carrier absorption significantly (**Fig. 3**). Silicon is a good submillimeter-wave material, because the lattice vibration bands in silicon appear only at 18,000 GHz. Single-crystal materials are always attractive because they exhibit minimal absorption loss, but they are also birefringent. The birefringent effect is an advantage in some applications such as the polarimeter and phase rotator.

Magnetic and ferrite materials. These are attractive for use as radar or radiation-absorbing materials, and as isolators and modulators in microwave, millimeter-wave, and submillimeter-wave frequency ranges. The measurement of all of the four quantities, ε', ε'', μ', and μ'', presents difficulties; two of the quantities must be suppressed in order to measure the other two. Millimeter- and submillimeter-wave data for these materials are not yet available. Such data would make

submillimeter-wave instrumentation as substrates, dielectric waveguides, windows, and lenses. Nonpolar polymers such as polyethylene, polypropylene, and polytetrafluoroethylene (Teflon) exhibit very low attenuation through the entire radiofrequency, microwave, millimeter-wave, and submillimeter-wave spectral regions (**Fig. 2**). Common polar polymers such as Plexiglas, acrylic, and nylon exhibit at least 20–30% higher absorption loss than nonpolar polymers. Plexiglas and acrylic are used as substitute materials for glass because of their transparency in the visible frequency region. The absorption of all polymers increases monotonically up to 2000 GHz, and then isolated lattice vibration modes appear. The low-frequency monotonic increase in absorption is due to the amorphous continuum.

Ceramics. These materials are very valuable for high-temperature work. They are almost opaque in the submillimeter-wave region but, like glasses, again become transparent in the microwave and millimeter-wave regions. Ceramics such as alumina have permittivities with large real parts, which enable them to be used in dielectric waveguides and as windows where high-temperature application is required. SEE CERAMICS.

Semiconductors. These are irreplaceable millimeter- and submillimeter-wave materials, but very high-resistance specimens are needed in those applications which require transparency. Controlled doping of impurities in pure-compound semiconductors enable these high-resis-

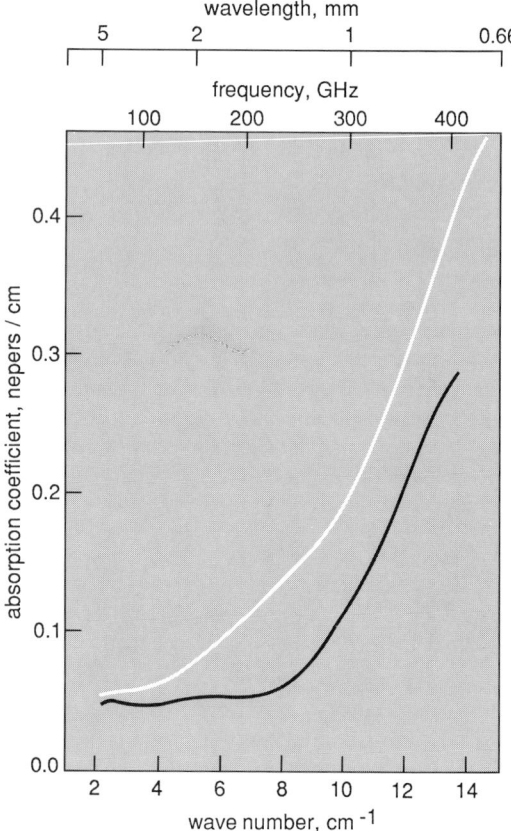

Fig. 3. Comparison of absorption coefficient spectra of two gallium arsenide specimens in the millimeter-wavelength range, at 27°C (81°F). Rise of absorption of both specimens with frequency is due to multiphonon absorption peak centered near 600 GHz. Lower absorption of one of the specimens is due to its higher chromium content and hence deep-trap concentration that suppresses free carriers. (*After M.N. Afsar, Dielectric measurements of millimeter wave materials, IEEE Trans. Microwave Theory Tech., MTT-32:1598–1609, 1984*)

possible the fabrication of specific ferrite materials for specific applications.

Liquids. Most polar (dipolar) liquids exhibit Debye-type (relaxation) broad absorption in the radio-frequency range. An additional broad absorption band centered around 1000 GHz and isolated sharp resonance-type absorption bands beyond 3000 GHz are observed in all polar liquids. Nonpolar (nondipolar) liquids also show a broad weak absorption band centered around 2100 GHz and isolated weak but sharp absorption bands at higher frequencies. Attempts have been made to correlate this broad absorption in nonpolar (nondipolar) liquids with their multipolar (for example, quadrupole or octupole) nature. Water and hydrogen-bonded molecules have always been of great interest to physicists and physical chemists. These liquids are nearly opaque for beam transmission applications.

Poley absorption was first discovered in a polar liquid, chlorobenzene. Since then, a wide range of polar liquids has been investigated and a Poley band has been observed in every case.

Liquids can be obtained in a very pure form and are ideally suited for calibration of microwave, millimeter-wave, and submillimeter-wave measuring instruments. As standard dielectric reference materials, liquids surpass solids because of the control that is possible over manufacturing processes and the availability of microqualitative chemical analysis. Nonpolar liquids, such as cyclohexane and decalins, are extremely low-loss materials and are ideal as solvent materials in dilution studies at microwave, millimeter-wave, and submillimeter-wave frequencies.

For background information SEE DIELECTRIC MATERIALS; DIELECTRIC MEASUREMENTS; INFRARED SPECTROSCOPY; INTERFEROMETRY; PERMITTIVITY in the McGraw-Hill Encyclopedia of Science and Technology.

Mohammed Nurul Afsar

Bibliography. M. N. Afsar, Dielectric measurements of millimeter wave materials, *IEEE Trans. Microwave Theory Tech.*, MTT-32:1598–1609, 1984; M. N. Afsar et al., The measurement of the properties of material, *Proc. IEEE*, 74:183–199, 1986; M. N. Afsar and K. J. Button, Millimeter wave dielectric measurements of materials, *Proc. IEEE*, 73:131–153, 1985; R. H. Cole, Evaluation of dielectric behavior by time domain spectroscopy, *J. Phys. Chem.*, 79:1459–1474, 1975.

Ecological interactions

Among the various interactions between organisms in an ecosystem, perhaps the most fascinating are symbioses. The term symbiosis refers to close and permanent associations between organisms of different species. Symbioses generally demonstrate a very high degree of specificity and are excellent examples of coevolution, in which intimate associations have formed between species that have evolved together for millions of years. Various types and examples of symbioses exist in nature.

While a significant amount of information has been obtained on symbiotic associations, only recently has it become possible to study these associations at the molecular and cellular levels. The first section of the article discusses the *Rhizobium* system, a nitrogen-fixing plant-microbe symbiosis that has been studied and characterized most extensively at these levels, in part because of its considerable agricultural importance. In this symbiosis, both species benefit from the relationship. The combined cellular and molecular changes in this system are crucial for the formation of the symbiosis. It appears that these types of cellular and molecular changes are also important in many other types of ecological interactions.

The second section of the article discusses parasitism, a form of symbiosis in which one member of the association is harmed.

Types of symbiotic relationships. Depending on how each participating species is affected by the association, symbiotic relationships can be classified into three categories: parasitism, commensalism, and mutualism.

In parasitism, one member of the association benefits while the other is harmed. One organism, called the parasite, feeds off the other, called the host. The host is not helped in any way; it is only harmed. Examples of parasitic relationships range from ticks on vertebrates to mistletoe on trees. Parasitism is discussed in greater detail below.

In commensalism, one member benefits and the other is neither harmed nor helped by the association. The so-called commensal organism gains food, protection, shelter, or any combination thereof from the host organism. Examples of commensalism include small crustaceans that live in the mouths of fishes, where they pick up bits of the fishes' leftover food; and orchids that live on tree limbs in the tropics, and grow better because they are raised off the ground.

Both members benefit by the association in mutualism. This type of symbiosis is quite common among plants and animals. The best-understood symbiotic association in this category occurs with the lichen. Here a very close association between an alga and a fungus allows them to grow together under conditions in which neither could survive alone. The alga (through photosynthesis) provides food and the fungus provides support and protection.

Symbioses can also be characterized by the degree of interdependence between the two participating species, regardless of how each is actually affected by the association. Interdependence between different species is a basic law of nature, and the various degrees of interdependence are well illustrated in the plant-microbe world. There is no such thing as an axenic plant (a plant completely isolated from other living organisms) in nature. So, while all plants are associated with other living organisms, the degree of intimacy between these organisms varies.

For the most part, interactions between organisms are still poorly understood at the molecular and physiological levels. Nevertheless, progress has been made with the *Rhizobium* system, where a bit of the molecular picture is emerging. With the rapidly increasing sophistication of recombinant deoxyribonucleic acid (DNA) technology, along with the awareness that many associations are beneficial or crucial for the host plant, a surge of interest in understanding and manipulating these processes is now occurring.

Rhizobium system. Most microbes interact with plants in order to obtain nutrition. Relatively few microbial species cause harm to plants. This is probably due to the fact that plants have evolved numerous defense mechanisms for dealing with harmful microorganisms. One of the most important plant-microbe symbioses from an ecological point of view exists between certain types of plants (primarily legumes) and bacteria of the genus *Rhizobium* that infect their roots. This symbiotic association is crucial to the important process of nitrogen fixation. *See* NITROGEN FIXATION.

Rhizobium is a group of gram-negative soil bacteria that can infect and nodulate legume roots, thereby forming a nitrogen-fixing symbiosis. The infection of legume root hairs by *Rhizobium* is a highly specific process that involves many steps of cellular recognition, eventually culminating in the formation of a root nodule that reduces atmospheric nitrogen into ammonia for the host plant. This symbiosis contributes significantly to the nitrogen economy of agricultural crops because nitrogen is the nutrient that most often limits crop productivity.

Research techniques. Research has been facilitated by use of techniques that allow the formation of this symbiotic association to be studied under laboratory conditions. The Fahraeus slide technique (**Fig. 1**), in which the bacteria and host plant are grown between a cover slip and slide under controlled conditions, has frequently been used to study the *Rhizobium* infection process.

By using this technique the steps of the infection process have been elucidated. (1) The bacteria attach to the root hairs of the plant, (2) the root hair tip curls, (3) bacteria entrapped within the curl penetrate the root hair cell wall and form a so-called infection thread, (4) the infection thread grows into the root hair base and penetrates the root cortex, and (5) cortical cells in the root grow in front of the infecting bacteria and eventually form a root nodule. Within root nodules the bacteria are contained within the plant cells.

The advancement of mutant analysis techniques has also greatly facilitated the study of

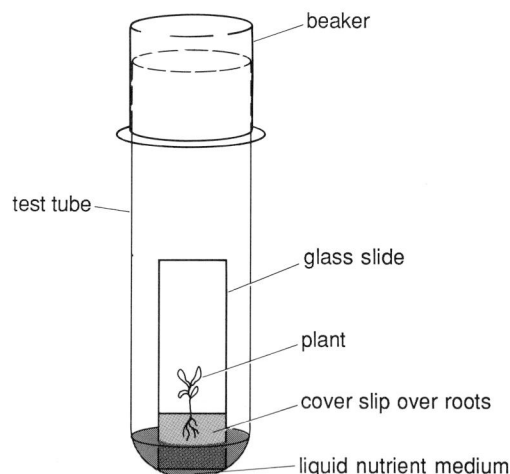

Fig. 1. Fahraeus slide technique used to study the *Rhizobium*-legume infection process. Inverted beaker is placed over test tube to prevent contamination by airborne microorganisms. (*After F. B. Dazzo and A. E. Gardiol, Host specificity in Rhizobium-legume interactions, in D. P. S. Verma and T. A. Hohn, eds., Plant Gene Research, pp. 3–31, Springer-Verlag Wien, 1984*)

symbiosis formation, as discussed below.

Cellular recognition. Cellular recognition between microorganisms and higher plants is a crucial factor in the formation of a symbiotic association. This phenomenon is also important because it affects plant morphogenesis, nutrition, and protection against infectious disease. In general, cellular recognition is believed to arise from a specific union, reversible or irreversible, between chemical receptors on the surfaces of the interacting cells. This idea implies that communication occurs when cells that recognize one another come into contact. Thus, most of the recent work has focused on studying the complementary components of cell surfaces at the biochemical level. The infection of legume roots by a bacterial symbiont has been examined most extensively in the *Rhizobium* system. Factors that influence cellular recognition are difficult to study unless the complete process can be continuously monitored under controlled growth conditions (that is, laboratory conditions). It is now possible to use information obtained about cellular recognition in the *Rhizobium* system as a model against which new information from other systems can be compared.

There is a high degree of host specificity in *Rhizobium* symbioses; that is, specific *Rhizobium* species preferentially form associations with specific legume species. The infection process is really an elegant sequence of cellular recognition steps. The first step is highly specific, involving the lectin recognition hypothesis. This hypothesis states that recognition between the rhizobial cell and the legume host root involves a binding of the plant lectins to unique carbohydrate molecule receptors found exclusively on

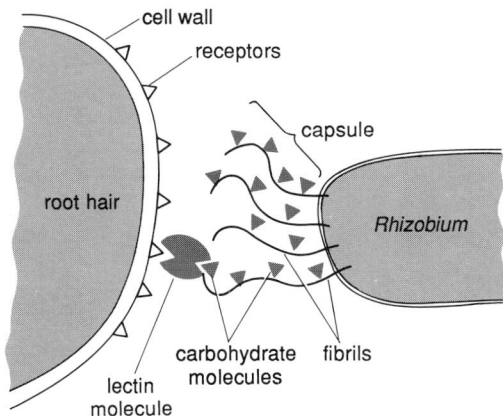

cell wall

receptors

capsule

root hair

Rhizobium

carbohydrate
molecules

fibrils

lectin
molecule

Fig. 2. Proposed lectin cross-bridging in *Rhizobium*-legume model for cellular recognition. (*After F. B. Dazzo and A. E. Gardiol, Host specificity in Rhizobium-legume interactions, in D. P. S. Verma and T. A. Hohn, eds., Plant Gene Research, pp. 3–31, Springer-Verlag Wien, 1984*)

the bacterial symbiont (**Fig. 2**). The plant lectin serves as a cross-bridging molecule. In the *Rhizobium*-legume symbiosis, firm bacterial attachment to host roots is considered the first or an early indication of cellular recognition. It is clear, however, that lectin-mediated bacterial attachment is only one of a number of events necessary for successful infection because root hair infection can be blocked at other steps after attachment.

Interaction at molecular level. The rhizobial nodulation process involves a series of complex, highly evolved interactions between a prokaryote and its eukaryotic partner. The complex nature of these interactions has made it difficult to analyze the individual steps. Recently, however, it has become possible to learn more about the entire process by studying bacterial and plant mutants. The mutants used in these studies have altered phenotypic (or somatic) traits which directly affect their ability to successfully complete the nodulation process; that is, they are defective at a specific step of the process. Elucidation of exactly why the mutant is defective at a given step is one way to learn more about the entire process.

The expression of specific bacterial and plant genes is temporally and spatially coordinated during the development and maintenance of a nodule in the *Rhizobium* system. Plant genes that encode nodule-specific proteins are usually not expressed in uninfected plants. This is because the expression of these genes is repressed. During the nodulation process, however, expression of plant genes encoding for nodule-specific proteins is no longer repressed (that is, expression is derepressed). Derepression of various bacterial genes also occurs. The *Rhizobium* bacteroids exhibit derepression of nitrogenase genes, hydrogenase genes, specific cytochrome genes, and

genes responsible for changes in the outer membrane.

Throughout the nodulation process the bacteria undergo many biochemical and morphological changes. Once they are inside the host cell, *Rhizobium* cell division and differentiation are influenced by the host. Following infection and formation of the root nodule, the bacteria stop dividing, become swollen, and take on pleomorphic shapes. At this point in development, the bacteria are referred to as bacteroids. The bacteroids are capable of fixing atmospheric nitrogen and are able to utilize some compounds provided by the plant as energy sources. They perform unique functions in close coordination with the host. Thus, the bacteroids in some ways behave like cellular organelles. The *Rhizobium* bacteroids are actually physiologically dependent on the host for several processes.

Sandra A. Nierzwicki-Bauer

Parasitism. Parasitism is an interspecific association in which the reproductive success of one individual (the parasite) is increased at the expense of the reproductive success of the other (the host). Parasitism differs from other types of symbiosis by virtue of the parasite's potential pathogenicity, and from predation by virtue of its intimate and extended nature: parasites rarely kill their hosts outright. From an ecological point of view, parasites generally exhibit much higher rates of reproduction than their hosts, whereas predators reproduce slowly in comparison with their prey.

Parasites may be divided into two broad categories. The microparasites (including viruses, bacteria, and protozoa) undergo many generations inside the same host individual. With exceptions, they tend to induce relatively efficient protective immunity and may therefore be associated with short-term or acute disease. In contrast, the reproduction of macroparasites (including helminths and arthropods) usually involves transmission stages and obligatory transfer from one host individual to another, either of the same species (direct life cycle) or of a different species (indirect life cycle). Although macroparasites do not escape immunological responses, immunity is in general not completely protective; macroparasites thus tend to be associated with continual reinfection and with chronic disease.

Because parasites affect humans as well as every other living species, parasite ecology encompasses medical, veterinary, agricultural, botanical, and zoological fields of research. Recently, theory, experiment, and field study have all contributed in attempts to understand temporal and spatial patterns of infection, that is, to determine how many hosts are infected, when, for how long, and why. In addition to the impact of parasites on the evolution and ecology of animal and plant communities, research in parasite ecology is inevitably focused on the control of parasites of medical and agricultural significance. The optimization of control is dependent upon an understanding of the undisturbed dynamics of the interaction between parasite and host populations. In the following, examples are taken from medical helminthology to illustrate concepts of current interest in parasite ecology as a whole.

In the past few years, two general features of parasite-host systems, immunology and host heterogeneity, have become prominent in explanations of ecological patterns. Immunological responses imply that infection dynamics may be influenced by history of exposure. Heterogeneity among hosts implies that mean parameter values may be insufficient to explain observed trends, since individual hosts may differ, both in susceptibility and in their immunological responses to infection.

Reinfection following treatment. The significance of acquired immunity and heterogeneity may be illustrated by reference to the results shown in **Fig. 3**. Rapid reinfection following the distribution of anthelminthic drugs within a community has been shown to be a universal feature of helminth ecology. In this example the prevalence of *Ascaris lumbricoides* is reduced from 98% to nearly zero after treatment, but returns exactly to its pretreatment level of 98% within a period of 12 months. The observed pattern of reinfection is a direct consequence of the feedback mechanisms controlling parasite population abundance. It has been found that prolonged control of infection requires repeated treatment at intervals determined in general by the life expectancy of the parasite, its rate of reproduction, and its distribution within the host community. The quantitative details of the results, however, depend upon whether or not acquired immunity is important in the regulation of parasite population size, and whether or not there exists heterogeneity between hosts in their susceptibility to infection.

Role of acquired immunity. As illustrated by the above example, recent developments in medical

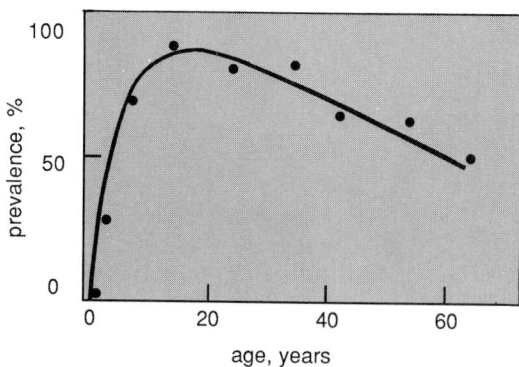

Fig. 4. Prevalence of *Schistosoma mansoni* infection in a human community. Children are uninfected at birth. Prevalence rises to a peak of 90% at 20 years of age and then declines in the older age-classes of the population. Only 50% of those alive at 70 years still harbor the infection. (*After R. M. Anderson and R. M. May, Helminth infections of humans: Mathematical models, population dynamics and control, Adv. Parasitol., 24:1–101, 1985*)

parasitology include the ecological interpretation of field data and the assessment of the significance of acquired immunity in the generation of observed patterns. Of particular interest is the relationship between host age and parasite prevalence or intensity, the curve being convex in shape for many common helminth infections, including the blood fluke, *Schistosoma mansoni* (**Fig. 4**). Both the proportion of the host population suffering infection and the average number of parasites per host increase during infancy, peak during childhood, and then decline in the adult age-classes of the population. Although this phenomenon is well established, its cause is still unclear. It could be generated by acquired immunity to infection, but age-related changes in exposure (caused, for example, by behavioral differences between children and adults in hygiene or habitat use) could be equally responsible. Field surveys alone provide little information on potential causes, so that additional methods of investigation are necessary, especially since the underlying mechanisms may be of critical significance in relation to the prognosis for control. Successful vaccination, for example, is difficult to achieve if acquired immunity is of no significance in natural infection.

This particular problem has been addressed with a combination of theoretical and experimental techniques. Repeated infection with *S. mansoni* in laboratory mice was used to generate experimental age-prevalence and age-intensity curves. In contrast to most immunological studies, where the degree of host resistance is determined by a large primary infection and subsequent challenge, these trickle-infection experiments represent an ecological approach to immunology. The results (**Fig. 5**) illustrate, first, the generation of a plateau in the average number of parasites per host as a result of the balance between the rates of parasite immigration and parasite mortality. Second, they show that convexity may result at high exposure levels in the

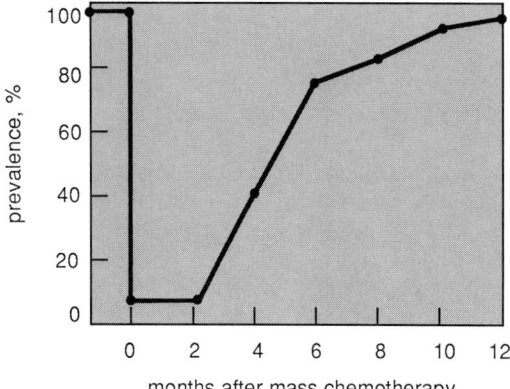

Fig. 3. Reinfection in a human community after treatment for *Ascaris lumbricoides* infection. At time 0 on the graph, all subjects were given anthelminthic treatment to remove all intestinal parasites. (*After R. M. Anderson and G. T. Medley, Community control of helminth infections of man by mass and selective chemotherapy, Parasitology, 90:629–660, 1985*)

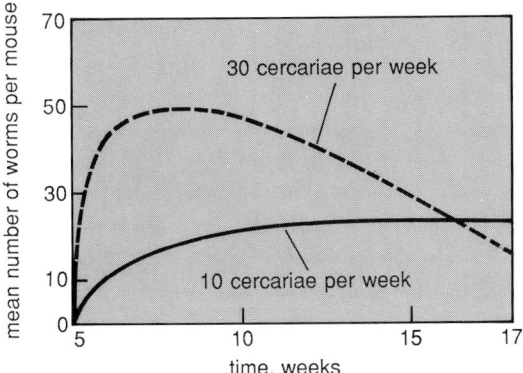

Fig. 5. Intensity of *Schistosoma mansoni* during experimental trickle infection in laboratory mice. The 10 mice are uninfected at the beginning and are then infected with either 10 or 30 cercariae per week for the duration of the experiment. At the lower rate of infection, intensity rises to a plateau. At the higher rate of infection, the curve is convex, reaches an early peak, and then declines. (*After J. A. Crombie and R. M. Anderson, Population dynamics of Schistosoma mansoni in mice repeatedly exposed to infection, Nature, 315:491–493, 1985*)

absence of age-related changes in the rate of infection. The dynamics of the experiment may thus be the result of immunological responses mounted by the host, even though no sterile or protective immunity is seen. Exposure to the parasite simply induces responses in the host which have an adverse effect on either the establishment or survival of future infections. The clarification of the exact nature of these responses and their precise effects on infection dynamics will require the integration of both immunological and ecological experi-ments. Results of this kind should be of interest to those making policy decisions for schistosomiasis control.

Effects of host heterogeneity. Host heterogeneity may exert considerable influence on the ecology of host-parasite interactions. Although the existence of host variation has been recorded for some time, the exploration of the phenomenon is relatively recent. Work on helminth parasites indicates that a substantial part of this heterogeneity has its origins in differential immune responsiveness to infection and is under genetic control. Surprisingly, it has been shown in many instances that differences in resistance to infection can be inherited in simple fashion, suggesting that relatively few genes are involved. The aim of a number of recent field studies has been to uncover the degree to which genetic heterogeneity can be related to patterns of parasitic disease observed in the real world. Parasites are aggregated in almost all host communities, with a few individuals harboring relatively large numbers of parasites (**Fig. 6**). The question is whether this aggregation is the result of predisposition to infection, with some organisms being particularly susceptible as a result of their genetic makeup. Present evidence suggests genetic predisposition, although its extent and exact nature are un-

known. Clearly, the discovery of genetic markers could result in substantial benefits for the diagnosis and control of parasitic disease of both medical and economic significance.

Evolutionary considerations. In evolutionary terms the maintenance of an efficient defense against disease is very costly. Animals are by necessity designed within certain physical and energetic constraints, and so investment in the development and maintenance of immunological systems must take place at the expense of investment in other processes such as growth or reproduction. The observation that animals from all phyla do, in fact, maintain very complex defenses must thus be indicative of the selective force of parasitic infection. Both immunity and host heterogeneity are therefore of interest from evolutionary as well as ecological points of view.

Much interest has been generated by the suggestion that parasitism may have been responsible for the evolution of sex, and for the maintenance of sexually selected characters that appear at first glance to be detrimental to the survival of the animal. Sexual reproduction is inefficient (in terms of the maximum number of offspring to be produced) in comparison with asexuality, and the reason for its evolution is thus not immediately apparent. The hypothesis recently put forward is that sex may have evolved in response to parasitic infection as a means of increasing genetic diversity, thus allowing hosts with relatively long generation times to keep evolutionary pace with their more rapidly evolving parasites. In addition, sexually selected characters such as the peacock tail may have evolved as a means by which females chose mates on the basis of their resistance to disease. (Genetically resistant males would be "brighter" than their infected competitors.) Deceptively simple and elegant, the theory underlying this hypothesis is complex and controversial. The degree of interest it has initiated, however, is indicative of the extent to which parasites are fundamental to many areas of biological research.

Even the evolution of host-parasite systems is

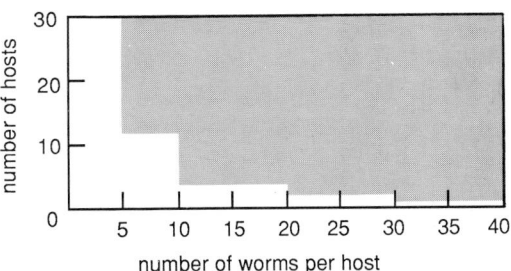

Fig. 6. Distribution of *Ascaris lumbricoides* infection in a human community. The majority of people are infected with fewer than 10 parasites. A few people (about 5% of the population) have relatively high levels of infection and harbor more than 30 parasites each. (*After J. Martin et al., The prevalence and intensity of Ascaris lumbricoides infections in N. Bangladesh, Trans. Roy. Soc. Trop. Med. Hyg., 77:702–706, 1983*)

not without potential application. This is perhaps best exemplified by the level of concern and uncertainty surrounding the design of programs for the control of parasitic disease in the face of the distressingly rapid evolution of parasite resistance to standard agents. Parasites appear all too often to have the upper hand. *See* Mammalia.

For background information *see* Ascariasis; Ecological interactions; Medical parasitology; Nitrogen cycle; Parasitology; Schistosomiasis in the McGraw-Hill Encyclopedia of Science and Technology.

Anne E. Keymer

Bibliography. R. M. Anderson, J. A. Crombie, and R. M. May, Predisposition to helminth infection in man, *Nature*, 320:195–196, 1986; R. M. Anderson and R. M. May, Vaccination and herd immunity to infectious diseases, *Nature*, 318:323–329, 1985; W. D. Bauer, Infection of legumes by rhizobia, *Annu. Rev. Plant Physiol.*, 32:407–449, 1981; F. B. Dazzo and G. L. Truchet, Interactions of lectins and their saccharide receptors in the *Rhizobium*-legume symbiosis, *J. Membrane Biol.*, 73:1–16, 1983; D. Rollison and R. M. Anderson, *Ecology and Genetics of Host-Parasite Interactions*, 1985; W. D. Sutton, C. E. Pankhurst, and A. S. Craig, The *Rhizobium* bacteroid state, in K. L. Giles and A. G. Atherly (eds.), *Int. Rev. Cytol.*, Suppl. 13, pp. 149–177, 1981; D. Wakelin, Genetic control of immunity to helminth infection, *Parasitol. Today*, 1:17–23, 1985.

Electric filter

In an age when semiconductor devices are being made smaller and less expensive, engineers continue to search for filtering technologies that are compatible in size and price. A method of filtering electronic signals that meets the objectives of small size and low cost uses mechanical resonators in place of electrical inductors and capacitors. These resonators are mechanically coupled by wires of small diameter to form a so-called mechanical filter.

Mechanical filters are used in many areas of communications, from inexpensive amplitude-modulation (AM) radios to telephone systems. They are also used in applications as diverse as wristwatches and train-control systems. The mechanical filter provides each system with very stable narrow-bandwidth slectivity. In the case of radio and telephone filters, the desired audio information passes through the mechanical filter while adjacent stations or phone conversations are rejected by it. In other words, a mechanical filter passes a wanted band of frequencies, or a single frequency, and rejects all others.

Frequency response characteristics. A typical response of a four-resonator mechanical filter is shown in **Fig. 1**. The graph shows the signal

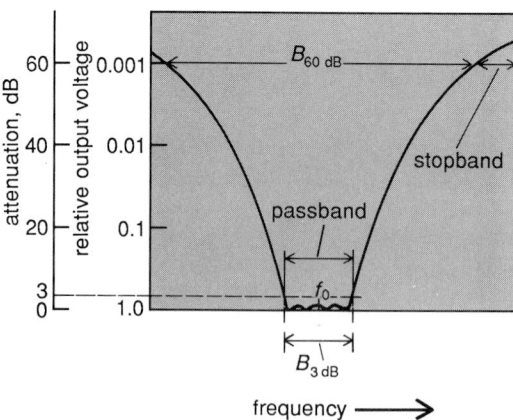

Fig. 1. Typical frequency response of a four-resonator mechanical filter.

amplitude at the output of a filter referenced to the maximum amplitude in the passband; this signal amplitude is plotted versus frequency. In the passband the relative output voltage stays close to a value of 1; this means zero attenuation. As the frequency of the signal applied to the filter moves away from the passband, the relative output voltage drops; that is, the attenuation increases. The stopband is often defined as the inverse of the region where the attenuation is more thant 60 dB greater than in the passband. The selectivity of the filter is usually defined as the ratio of the 60-dB bandwidth (B_{60dB}) to the 3-dB bandwidth (B_{3dB}); this ratio is called the shape factor. The frequency midway between the limits of the passband is called the center frequency F_0. The wideness of the filter is referred to as the fractional bandwidth; the fractional bandwidth is the ratio of the 3-dB bandwidth to the center frequency.

Basic filter elements. Figure 2 shows a schematic block diagram of a mecanical filter. A voltage V_1 and a current I_1 are applied to the input transducer. The transducer converts the electrical energy to mechanical energy in the form of forces f_1 and f_2 and velocities v_1 and v_2, and stresses and strains. The transducers may be of the magnetostrictive or the piezoelectric types shown in the figure. Electric current through the coil of the magnetostrictive transducer produces a magnetic field which passes through a magnetostrictive ferrite rod, causing it to expand and contract at the frequency of the applied signal current. A voltage applied to the piezoelectric transducer produces an electric field across the piezoelectric ceramic disk which causes it to expand and contract, again at the frequency of the applied signal. The magnetostrictive and piezoelectric transducers are usually bonded to metal to form a composite resonator. The vibrating composite resonators drive interior mechanical resonators by means of wires of small diameter called coupling wires.

The mechanical resonators are of various

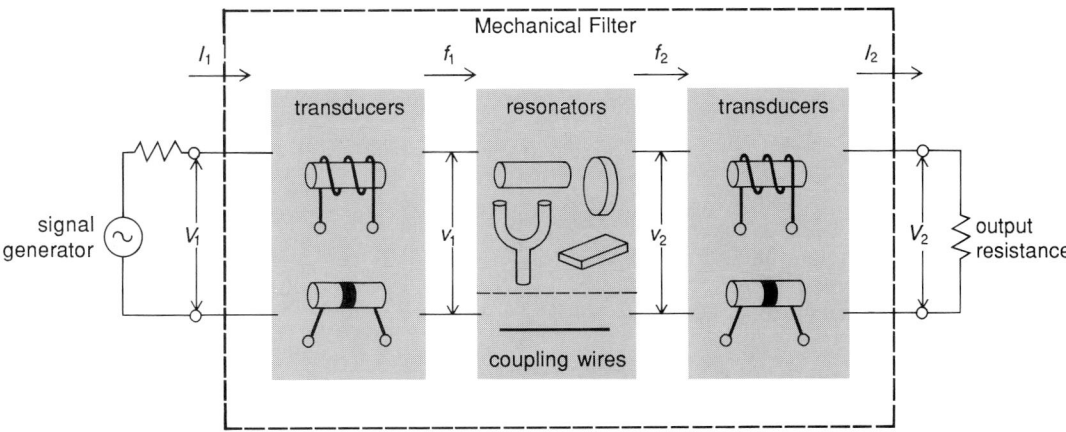

Fig. 2. Elements of a mechanical bandpass filter.

types depending on the particular filtering characteristics needed. At low frequencies, in the audio range, the resonators may be tuning forks or simple flat bars that vibrate in a flexural mode. At frequencies above the audio range, up to frequency limits of 500–600 kHz, the resonators can be in the form of rods or disks. Mechanical energy is also coupled between the interior resonators by means of the coupling wires.

At the output end of the filter the mechanical vibrations are converted back to electrical signals by means of magnetostrictive or piezoelectric transducers. Stresses and strains in the materials produce magnetic or electric fields which, in turn, result in a current I_2 or a voltage V_2 being applied to the electrical resistance (resistor) across the output of the filter.

Example. **Figure 3** shows a torsional mechanical filter used in telephone signaling circuits such as dialing circuits. The filter is composed of input and output composite transducer-resonators and two interior resonators. The resonators are coupled by a single coupling wire.

The signal generator voltage V_0 is applied across the source resistance R_S and the piezoelectric ceramic transducer. The transducer is composed of two oppositely polarized half-disks bonded together. An electric field between the plated end surface of the ceramic and the metal causes the ceramic to twist. In turn, the entire ceramic plus the metal resonator twists, causing a torsional motion about a model line around the center of the resonator. The arrows show how a rod moves in opposite directions on opposing sides of the nodal line. At the nodal line there is no motion, and that is therefore the point where a wire support is welded to the resonator. The wire support is bonded to the base of the filter enclosure or to a bracket or circuit board which is mounted on the base.

The composite transducer-resonator is coupled to the first interior resonator by the coupling wire between the resonators. In other words, the mechanical energy in the form of motion and force in the first resonator is transmitted to the second resonator by the wire, which acts as a spring. The signal is coupled from the second resonator to the third to the last by means of the coupling wire. Torsional motion in the output composite transducer-resonator causes an electric field to be generated between the flat surfaces of the piezoelectric ceramic, resulting in an output voltage across the load resistor R_l.

Filtering process. The filtering process that allows a band of frequencies to be passed and the remaining frequency spectrum to be rejected can be described intuitively. A resonator driven at its natural frequency will achieve a maximum amplitude of vibration. This holds true when someone is pushed on a swing, when a soprano breaks a glass with a note (frequency) equal to the resonance frequency of the glass, or when a mechanical bar, disk, or fork is driven at its resonance. If the driving signal is not at the

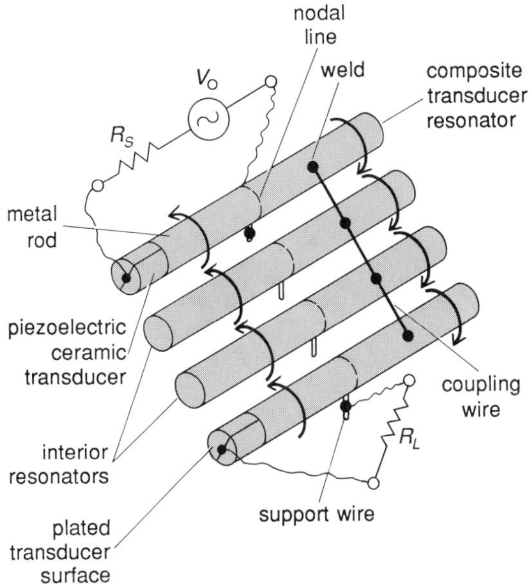

Fig. 3. Torsional, four-resonator mechanical filter used on telephone signaling circuits.

frequency of resonance of the resonator, the amplitude is less. The farther the frequency of the driving signal is from that of the resonator, the lower the amplitude of the vibration.

Coupled resonators. To pass a band of frequencies, two coupled resonators are required—for example, the two interior resonators shown in Fig. 3 with no wire coupling them to the end resonators. The two resonators are tuned to the same frequency. If one of the resonators is driven, by some means, at this frequency, they will both vibrate in the same phase relationship (for instance, in the direction of the arrows), and the coupling wire between them will have no effect because the entire wire is moving with the resonators. The amplitude of the vibration will be at a maximum value. If the signal driving one of the resonators is now increased in frequency, a second frequency is found at which the amplitude of the vibration again reaches a maximum value, but the two resonators turn in opposite directions.

The higher-frequency resonance occurs because the coupling wire is now being elongated and compressed, and therefore acts as a spring. The stiffening effect of the spring raises the natural frequency of vibration of the coupled pair of resonators. It can thus be seen that a wire-coupled pair of identically tuned resonators will have two natural resonances. The number of natural resonances will increase by the number of resonators added, but the width of the passband shown in Fig. 1 is essentially determined by the stiffness of the coupling wire.

Terminating resistance. The second concept of vital importance to the filtering process is the need for some kind of electrical or mechanical resistance or damping to flatten the amplitude peaks corresponding to the natural resonances of the system of coupled resonators. This is accomplished, in the case of the mechanical filter of Fig. 3, by the electrical resistance of resistors R_S (source) and R_L (load) being transferred through the composite transducers and causing damping of the four response peaks. The result is the frequency response of Fig. 1.

Another example of the concept of an electrical circuit affecting a mechanical circuit is the much greater torque which is needed to turn the shaft of a generator when the electrical circuit draws more current. The smoothing of the amplitude peaks is similar to the damping of the resonances in an automobile suspension by the shock absorbers.

Selectivity. As discussed above, the term selectivity refers to the steepness of the sides of the response curve of Fig. 1. Assuming that the width of the passband B_{3dB} remains constant, then the width of the response at greater attenuation levels decreases as more resonators are used. For example, B_{60dB} will decrease.

In summary, the factors that determine the

Factors that determine the shape of the frequency response curve of a mechanical filter	
Determining factor	Characteristic determined
Frequency of resonators	Low edge of filter passband
Stiffness of coupling wire	Width of passband (bandwidth)
Number of resonators	Steepness of response curve
Source and load resistance	Passband variations (ripples)

frequency response curve of a mechanical filter are given in the **table**.

Mechanical filter characteristics. Center frequencies as low as 200 Hz can be achieved by using tuning-fork resonators. Even lower frequencies can be achieved, but the vibrating parts of the tuning fork become either very long or impractically thin. The highest practical frequencies are about 600 kHz. At the high end of the spectrum, disk or rod resonators are used. The high-frequency limitation is primarily due to the small size of a resonator at high frequencies. The frequency of a torsional or extensional resonator is inversely proportional to its length. The frequency of a disk vibrating in a flexural mode is inversely proportional to the diameter squared and directly proportional to the thickness. As the resonators become very small, it is more difficult to weld the wires accurately and assemble the composite transducer-resonators.

Bandwidth is usually described as a percentage of center frequency and is called the fractional bandwidth in percent ($B_{3dB} \times 100/f_0$). Typical values range from 0.1 to 10%. Very narrow mechanical filters suffer from the fact that the shift of the center frequency with changing temperature and age may be a value that is a high percentage of the filter bandwidth, and the desired signal may be lost. Very wide filters often have unwanted resonances called spurious resonances in or near the filter passband.

The selectivity depends on the number of resonators, as discussed above. Filters such as watch crystals use a single resonator, whereas highly selective filters for voice communication radios and telephone systems use as many as 12 resonators. In the latter case the ratio B_{60dB}/B_{3dB} may be as low as 1.3 to 1. The passband ripples for these high-performance, multiresonator filters can be made as low as 0.2 dB.

For background information SEE ELECTRIC FILTER; MAGNETOSTRICTION; MECHANICAL VIBRATION; PIEZOELECTRICITY; RESONANCE (ACOUSTICS AND MECHANICS); VIBRATION in the McGraw-Hill Encyclopedia of Science and Technology.

Robert A. Johnson

Bibliography. R. A. Johnson, *Mechanical Filters in Electronics*, 1983; D. F. Sheahan and R. A. Johnson, *Modern Crystal and Mechanical Filters*, 1977; G. C. Temes and S. Mitra, *Modern Filter Theory and Design*, 1973.

Electric power systems

Since 1974, economic conditions combined with legislative and regulatory actions have encouraged the use of electric power lines and other power-transmitting facilities by parties other than the owners of the lines. Such use is called transmission access. The terms wheeling and common carriage are synonyms. Transmission access can produce a number of problems in the planning and operation of the bulk power systems. These include overall heavier system loading; degraded voltage conditions, poor control of system frequency and power interchanges; high line loading due to parallel or loop flows; uncertainty as to the size, location, and timing of new lines and generation; and an attendant reduction in overall system reliability. Power system engineers and researchers are addressing these problems with new computer programs to quickly analyze the large number of planning alternatives created by cogeneration, improved methods to analyze the steady state and dynamic performance of the system for both planning and real-time operation, new measures of system reliability to ensure a proper balance between reliability and cost, and improved methods of on-line system analysis and control.

Transmission access incentives. Economies of scale and geographic imbalances of electric generating resources such as hydroelectric power, coal fields, and oil wells have always created economic incentives to transmit power from energy-rich areas to areas with high energy costs and heavy electrical demands. Hydroelectric power transfers from Canada to the eastern seaboard of the United States and from the northwestern United States to southern California are typical of such interchanges. Thus, wheeling, the transfer of one utility's energy over another utility's lines for delivery to a third utility, has been a widespread but little publicized practice for many years. Furthermore, since each utility had a franchised responsibility to provide reliable service to its customers, utilities are careful to monitor the effects of wheeling on their systems and keep flows within safe limits. Utility planning and operating activities are coordinated to ensure that power transfers can be safely carried for the system conditions anticipated.

Advent of cogeneration. The 1974 oil embargo and subsequent tenfold increase in the price of oil brought about passage of the Public Utility Regulatory Policy Act (PURPA) of 1978. That act encourages development of alternative energy sources (wind, solar, and so forth) and more efficient use of existing electrical facilities. The response to this act took the form of cogeneration, that is, generation by nonutility entities, and transfer of electrical energy by nonowners over utility-owned transmission lines. In one sense, PURPA introduced a form of competition in that a utility no longer enjoyed a monopoly in supplying power to an area.

By the early 1980s the impact of PURPA was clearly visible. States had established economically attractive cogeneration rates and the number of applications from nonutility entities mounted into the thousands. The total new capacity provided through these applications soon exceeded the most optimistic forecasts of growth, in some cases by a factor of 2.

Utility concerns grew as to (1) how to plan and construct the necessary transmission facilities to accommodate hundreds of separate entities with diverse sizes and locations, (2) how to schedule the day-to-day and hour-by-hour operation of such large numbers of units with diverse operating characteristics, and (3) how to maintain the reliability of the bulk power system and local area supply under these conditions when the cogenerators had no franchised requirement to provide a given level of reliability.

Loop flows. The increased use of transmission systems for economic interchange of power and wheeling highlights the relevance of certain laws of physics that constrain the ways such a system can be used. Individual circuits are operated as part of a large network, bringing into play Kirchhoff's laws of parallel or loop flow. Under these principles, electric current does not take a single path, but divides along multiple parallel paths having different carrying capacities in a way that is proportional to the admittance of all the available paths.

The electrical phenomenon of loop flow means, for example, that if Ontario Hydro is transmitting 1000 MW to the New York Power Pool, the amount of power that actually flows over lines directly linking the two systems may be only 500 MW; the other 500 MW may flow in two other electrically parallel paths through utility systems in Michigan, Ohio, Kentucky, West Virginia, Virginia, Maryland, Pennsylvania and New Jersey (see **illus.**).

Thus, actual paths of power flow can be quite different from the contractual paths. What may appear on paper to be a simple transaction between neighboring utilities can, in fact, require significant coordination with numerous other nearby systems. Utilities routinely accommodate the complexities of loop flows, but the prospect of substantially increasing the loop-flow numbers and schedules with more wheeling and third-party generation causes many utilities serious concern.

Analysis tools and procedures. Engineers use several tools to study transmission system performance. The two most basic are power flow or load flow and transient stability. These are large-scale computer programs that require some of the most powerful computers available.

The power flow provides a "snapshot" of a particular hour in time, usually a summer or

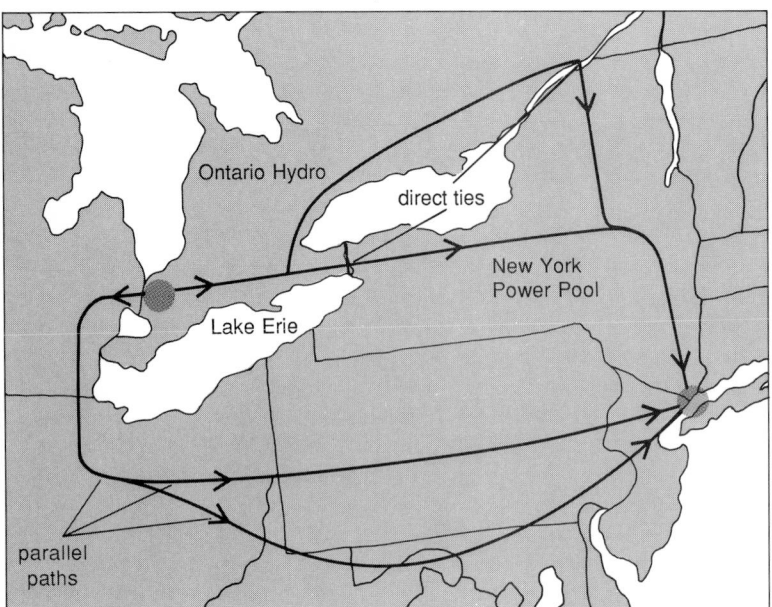

Example of loop flows in transmission networks.

winter peak. It calculates voltages and power flows for the particular year and season that the engineer wishes to analyze. By running a series of these tests that analyze both normal and outage conditions on several competing alternatives, the planning engineer is able to determine which reinforcement is most effective.

The second basic tool, transient stability, is used to determine the response of the power system to specific severe disturbances. This information is used to reduce the level of system transfer, reschedule maintenance conditions, or take other actions to reduce the impact of such disturbances on the system and maintain stability.

Operations personnel in power control centers utilize sets of tools similar to those employed in planning. In addition, the power control center utilizes a set of on-line programs with powerful computers to perform the similar functions of economic dispatch, contingency (what-if) analysis, and system monitoring in real time as system conditions actually change.

Technical corrective measures. The first technique that engineers use to balance flows in parallel lines is the planning of compatible routes and voltage levels. In addition to good design practices, there are three primary tools by which the engineer can balance line flows: phase-shifting transformers, series capacitors, and high-voltage direct current. Each has its advantages and disadvantages. All add cost and complexity to the transmission system.

Phase-shifting transformers. Phase-shifting transformers, or phase shifters, can control the flow of power. These devices can effectively increase or decrease the impedance of a transmission line in which they are installed, thus serving to reduce or increase flow in that line. This action then causes a corresponding increase or reduction in the flow on any parallel facilities. Although phase shifters are used extensively at 115 and 138 kV, they have had a poor record of performance at 230 and 345 kV, and none have been built at 500 and 765 kV. Also their high cost has made utilities hesitant to install them. Moreover, they may become the thermally limiting devices in the system. (Thermal limits result from the heating effects of the power carried by the devices.) Phase shifters can be purchased either with a fixed setting to permanently control flow in a given direction and amount or with adjustable settings so as to maintain a desired flow throughout a range of operating conditions.

Series capacitors. A series capacitor has an electrical characteristic which effectively reduces the impedance of the transmission line in which it is installed. Series capacitors are usually installed in circuits of 345 kV and above. They are usually installed as a fixed percentage of circuit impedance. They have two primary disadvantages: they can be thermally limiting, and in certain installations they can contribute to unwanted and sometimes destructive oscillations of large generators.

High-voltage direct current. The vast majority of transmission facilities are alternating current. For direct current, the energy is first converted from alternating current to direct current in a converter station, transmitted as direct current, then reconverted from direct current to alternating current in an inverter station, and pumped back into the alternating-current grid. The controls of the converter and inverter can be set to carry a prescribed amount of power on the direct-current line in either direction independent of changes on the alternating-current system. Thus the direct-current line is not constrained by the rules of alternating-current power flow as determined by impedance. Furthermore, the controls can be set to react very quickly so that the direct-current line can help to minimize or even counterbalance power oscillations in the alternating-current system. Lastly, the cost of the actual direct-current line is slightly less than that of a comparable alternating-current line.

One disadvantage of high-voltage direct current is the high cost of the converter and inverter terminals. A second disadvantage is that care is required in setting the controllers for fast action so that they will respond correctly. The combination of load levels, generation conditions, and facilities out of service, coupled with all possible contingencies, can produce a set of conflicting requirements which are sometimes difficult, if not impossible, to meet.

Current research. Research which will assist in solving the problems of transmission access has been conducted in software tools used by utilities to both plan and operate the power systems, as well as in development of hardware to control the flow of power and improve system performance.

On the equipment side, key technical achievements have included modern direct-current technology, including circuit breakers, control systems, and less expensive converter and inverter terminals. A recently developed subsynchronous damping device allows series capacitors to be applied to long high-voltage alternating-current lines in order to improve their power-carrying ability. Areas of possible future research include semiconductor-based phase-shifting transformers and series capacitors.

Better computer-based tools that can meet the growing analytic requirements of system planners and operators continue to emerge. A first-of-its-kind transmission-system reliability program is moving from prototype to production grade. Codes for analyzing the integration of multiterminal high-voltage direct-current transmission into large alternating-current systems are now available for utility use. A recently completed program is used to analyze transmission-access conditions. Applying regression analysis, it allows the user to consider large numbers of alternative plans to screen the most feasible alternatives.

The concurrent processing approach (now emerging in the next generation of computer systems) has been shown to be feasible and is thought to be cost-effective for reducing the computation burden and boosting the speed of power system simulations. Actual tests of such multiprocessor architecture are being pursued, including the hypercube design, in realistic power system environments.

Nearing completion are several computer programs which will enable the power system operator to deal with complex and fast-moving situations. Software modules are now available to incorporate effects of cogeneration into existing real-time control software used in energy management systems. A sophisticated contingency analysis package to study what-if conditions, based on real-time data, is being implemented in several control centers.

As the number of connections to a power system increases, it becomes increasingly difficult for an operator to diagnose the results of a particular set of switch operations resulting from a system problem and perform the necessary restorative actions to minimize customer interruptions and prevent further damage to equipment. Investigatory research shows this to be a good application for an expert system. In this area of artificial intelligence, the knowledge and skill of experienced operators is captured and incorporated in step-by-step logic statements and used to provide, in effect, an operator's assistant to aid system operators during complex, fast-moving events. Recent research has shown that such diagnosis can be completed and the best strategy for restoration can be displayed to an operator in about 1 s.

For background information SEE CONCURRENT PROCESSING; DIRECT-CURRENT TRANSMISSION; ELECTRIC POWER SYSTEMS; ELECTRIC POWER SYSTEMS ENGINEERING; EXPERT SYSTEMS; KIRCHHOFF'S LAWS OF ELECTRIC CIRCUITS; REAL-TIME SYSTEMS; TRANSFORMER in the McGraw-Hill Encyclopedia of Science and Technology.

Robert Iveson

Bibliography. J. A. Casazza, Understanding the transmission access and wheeling problem, *Public Util. Fortnightly*, 116(9):35–42, 1985; T. Moore, Network access and the future of power transmission, *EPRI J.*, 11(3):4–13, 1986; J. L. Pfeffer, Policies governing transmission access and pricing: The wheeling debate revisited, *Public Util. Fortnightly*, 116(9):26–33, 1985; Washington and the Utilities, PURPA: Still hazy after all these years, *Public Util. Fortnightly*, 118(1):33–35, 1986.

Electrical communications

The increasing demand for computer communications and advances in integrated circuit technology have led to the rapid development of digital transmission equipment for voice-bandwidth telephony, line-of-sight terrestrial radio, satellites, and subscriber loop circuits. Bandwidth-efficient data transmission over these circuits is made possible by the use of adaptive equalization to compensate for time dispersion of received signals resulting from imperfect transmission characteristics. Adaptive equalization and the more general related fields of reception of digital signals, adaptive filtering, and system identification have been areas of active research and development since the 1960s.

Principles of adaptive equalization. A synchronous digital transmitter sends symbols or pulses at a regular signaling rate, where each symbol represents a number of bits of information. Time dispersion or multipath propagation in the channel causes overlap of received symbols, known as intersymbol interference. An equalizer is a signal-processing device designed to combat this phenomenon.

Figure 1 shows the major components of a typical data transmission system with a modulator-demodulator (modem). A baseband equivalent model, which represents the transmission characteristics of the passband system translated by the carrier frequency used by the modulator and demodulator, can be derived for any linear modulation system, for example, phase-shift keying or quadrature-amplitude modulation. A simplified model is shown in **Fig. 2** along with a tapped delay line or linear transversal equalizer. In this model the box labeled "channel" includes the effects of the transmitter filter, the modulator, the transmission medium, and the demodulator. A pulse with a discrete amplitude level is transmitted through the

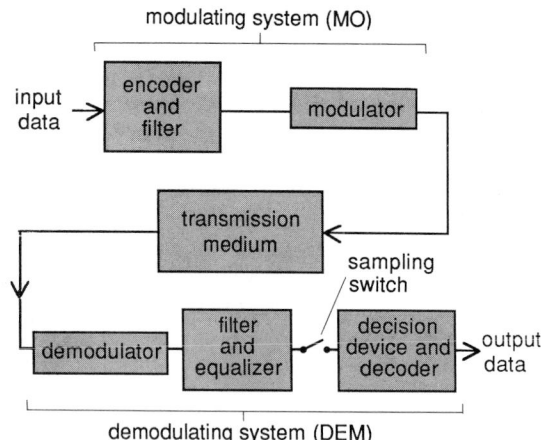

Fig. 1. Major components of a typical data transmission system.

channel every T seconds, where $1/T$ is the signaling rate. The received signal is the superposition of the impulse response of the channel to each transmitted symbol and additive noise. A sample of the received signal at instant $kT + t_0$, where t_0 accounts for the channel delay and sampler phase, carries information about the pulse amplitude at time kT. The received sample, however, is corrupted by additive noise as well as interference from neighboring symbols. The total intersymbol interference is zero if and only if the channel impulse response has zero crossings at T-spaced intervals. This condition is difficult to satisfy even approximately when the signaling rate is increased to be a large fraction of the available channel bandwidth. The purpose of an equalizer, placed in the path of the received signal, is to reduce the intersymbol inter-

ference as much as possible in order to maximize the probability of correct decisions.

A functional block diagram of a commonly used digital transversal equalizer is shown in Fig. 2. Samples of the received signal acquired at the signaling rate are stored in a digital delay line (shift register or memory), and the products of these samples and the respective equalizer coefficients are digitally computed and summed to produce an equalizer output. During each signaling interval a new received sample is shifted into the delay line, and an output is computed. The difference between the equalizer output and the expected transmitted symbol is the error. The coefficients of a least-mean-square equalizer are chosen to minimize the mean-square error, that is, the sum of squares of all the intersymbol interference terms plus the noise power, at the output of the equalizer.

In most practical situations the channel characteristics are not known beforehand. Therefore, prior to regular data transmission, a known signal may be transmitted during a training period to acquire information about the channel characteristics. If a synchronized version of the known training signal is provided, a sequence of error signals can be computed at the equalizer output and used to adjust the equalizer coefficients to reduce the sum of the squared errors. During each signaling interval, each coefficient is adjusted in a direction opposite to an estimate of the gradient of the mean-square error with respect to that coefficient. After training, the coefficients of an adaptive equalizer are continually updated in a decision-directed mode. In normal operation, receiver decisions and equalizer error estimates based on these decisions are correct with high probability. This allows the adaptive equalizer to maintain precise equalization and track relatively slow variations in the channel characteristics. In general, the need for precise adaptive equalization increases with the desired bandwidth efficiency of the transmission system measured in (bits/s)/Hz.

Alternative equalizer structures. An improved transversal equalizer is obtained by increasing the sample rate at the equalizer input to a value greater than twice the bandwidth of the received signal. The delay line taps of such an equalizer are spaced at an interval which is a fraction of the signaling interval T. For example, for a $T/2$ equalizer the received signal is sampled and shifted into the equalizer delay line at a rate of $2/T$, and one output is produced during each symbol interval for every two input samples. An error signal is computed corresponding to the output, and each coefficient may be updated once for that symbol interval. The sensitivity of the minimum mean-square error achieved with a fractionally spaced equalizer, with respect to the sampler phase (the position within a symbol interval at which samples of the analog waveform are obtained at the equalizer input), is typically far less

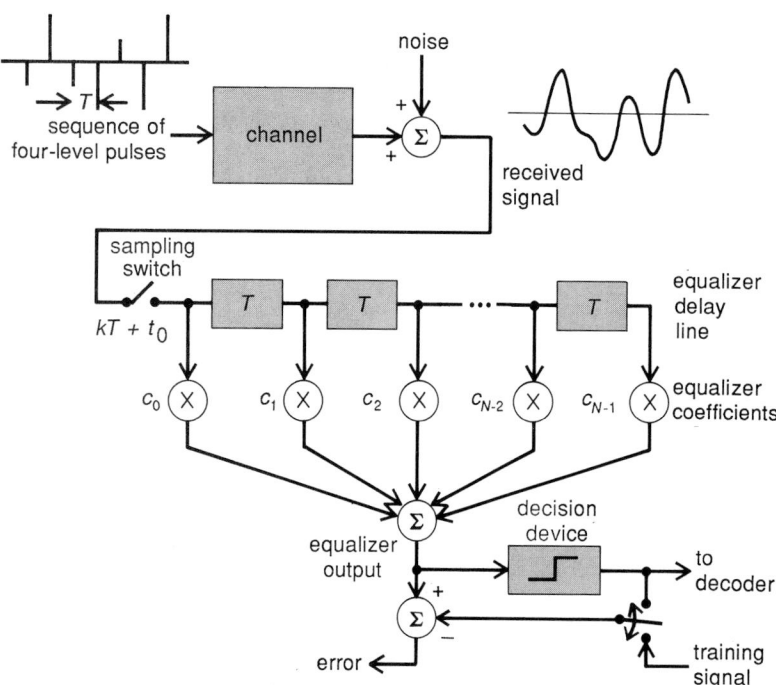

Fig. 2. Baseband channel model of digital transmission system, and digital transversal equalizer.

than that with a T-spaced equalizer can. A fractionally spaced equalizer can also effectively compensate for more severe channel delay distortion and can deal with amplitude distortion with less noise enhancement than a T-spaced equalizer can.

A simple nonlinear equalizer, which is particularly useful for channels with severe amplitude distortion, uses decision feedback to cancel the interference from symbols which have already been detected. The forward part of a decision feedback equalizer (**Fig. 3**) resembles the linear transversal equalizer discussed above. In normal operation the symbols already detected are correct with high probability; therefore, the postcursor intersymbol interference contributed by these symbols can be canceled by subtracting from the equalizer output the past decisions appropriately weighted by the feedback coefficients. The forward and feedback coefficients may be adjusted simultaneously to minimize the mean-square error. Since the output of the feedback section of the decision feedback equalizer is a weighted sum of noisefree past decisions, the feedback coefficients do not contribute to the noise power at the equalizer output.

Applications. Advanced equipment incorporating adaptive equalization or related techniques includes 9.6-kbit/s full-duplex modems for two-wire, dial-up telephone connections; 19.2-kbit/s modems for four-wire, conditioned, leased telephone lines; 140-Mbit/s microwave digital radios; 144-kbit/s full-duplex transceivers for two-wire subscriber loops; and echo cancelers for telephone calls over satellite connections.

Voice-band modems. An adaptive equalizer is an essential component of all modern voice-band telephone line modems at speeds from 2.4 to 19.2 kbits/s. These modems use phase-shift keying for 2.4- and 4.8-kbit/s transmission, and combined phase and amplitude modulation or quadrature-amplitude modulation for higher speeds. In quadrature-amplitude modulation, two double-sideband, suppressed-carrier, amplitude-modulated (in phase and quadrature) signals are superimposed on each other at the transmitter and separated at the receiver by using quadrature or orthogonal carriers (90° out of phase) for modulation and demodulation. So-called complex

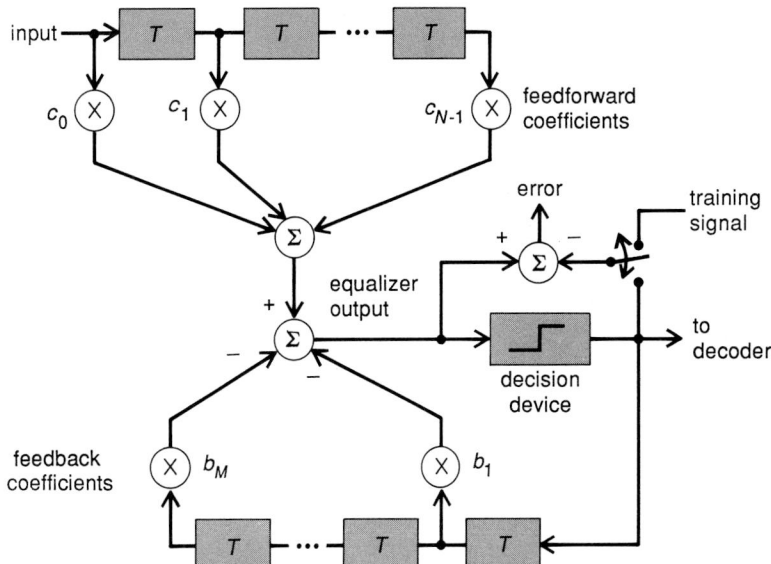

Fig. 3. Decision feedback equalizer.

adaptive equalizers for these modems comprise a set of four transversal filters (with cross coupling) for two inputs and two outputs, namely the in-phase and quadrature signals mentioned above. Such equalizers combat not only intersymbol interference in each signal but also the cross interference between the signals caused by asymmetry in the channel transmission characteristics around the carrier frequency.

Voice-band telephone modems may be classified into one of three categories based on intended application: two-wire public switched telephone network connections; four-wire, point-to-point, leased lines; and four-wire, multipoint, leased lines. The **table** lists recent developments in each category. (The trellis-coded modulation scheme adds redundant code bits to the data and combines coding and modulation in a sophisticated manner to realize more reliable data transmission at high rates over bandwidth-limited channels.)

Microwave digital radio. Digital line-of-sight radio transmission has been an important step in the move to establish integrated digital networks. Spurred by the need to increase the capacity of radio channels in the 4-, 6-, and 11-GHz bands, recent quadrature-amplitude-modulation radio

Recent developments in voice-band telephone modems			
Application	Highest data rate, kbits/s	Special characteristics	Year introduced
Two-wire, public switched telephone network (PSTN)	2.4	Full-duplex using frequency division	1979, 1983
	9.6	Full-duplex using echo cancellation	1986
Four-wire, leased, point-to-point	14.4	Quadrature-amplitude modulation (QAM)	1981
	16.8	Trellis-coded QAM	1984
	19.2	Trellis-coded QAM	1985
Four-wire, leased, multipoint	9.6 (inbound)	Startup time 10–20 ms	1982–1984
	14.4 (outbound)	Trellis-coded QAM	1985

systems have achieved bandwidth efficiencies as high as 4.5 (bits/s)/Hz. This is in part due to the application of sophisticated equalization techniques to combat multipath fading. Adaptive transversal or decision feedback equalizers are used in combination with amplitude slope equalizers (which compensate for amplitude versus frequency distortion such that the equalizer amplitude varies approximately linearly on a decibel scale over a certain range of frequencies), and space diversity receivers to increase transmission reliability during frequency-selective fades. Systems using 64-quadrature-amplitude modulation have been deployed, and prototype systems using 256-quadrature-amplitude modulation have been reported. (A 64-quadrature-amplitude modulation system uses 8 distinct amplitude levels to modulate each carrier resulting in an $8 \times 8 = 64$-point two-dimensional signal constellation, capable of carrying 6 bits of information per signal interval. Similarly, a 256-quadrature-amplitude modulation system uses 16 levels to modulate each carrier.)

Digital subscriber loops. The basic interface to the evolving Integrated Services Digital Network (ISDN) will allow a common subscriber (who uses ordinary telephone service and is directly connected to the local central office) to access two 64-kbit/s channels and one 16-kbit/s channel. This has created the need for 144-kbit/s full-duplex transmission over many of the existing two-wire subscriber loops which now carry voice signals from telephone sets to local central offices. The twisted-pair local loop cables introduce frequency-dependent loss which is a function of the gage mix (mixture of gages among the sections of twisted pairs that comprise the cable), the length of the cable, and the presence of open-circuited cable pairs (or bridged taps) of unknown length bridged onto the main cable pair. Some form of adaptive equalization is necessary to compensate for the time dispersion introduced by these characteristics. In digitally implemented integrated-circuit transceivers, a decision feedback equalizer with a few forward taps can be used with enough feedback coefficients to cancel the effect of the postcursor intersymbol interference due to bridged taps. Experimental integrated transceivers are already available, and final designs and volume production await agreement throughout the industry on a standard method for transmission over two-wire digital subscriber loops for the Integrated Services Digital Network.

Echo cancelers. Adaptive echo cancellation is a related application of adaptive filtering technology. Impedance mismatches at two- to four-wire converters in the telephone network create reflections or echoes. Even a low-level echo becomes objectionable to the talker as the echo delay increases. Therefore, adaptive echo cancelers have become essential components of satellite links which add an approximately 0.6-s round-trip delay to a telephone connection. The input to the adaptive echo canceler is the sequence of voice samples received via satellite for transmission to the near-end subscriber. Some of this signal is reflected to the remote subscriber. An adaptive transversal filter with sample-interval tap spacing models the echo-path impulse response. The filter output is a replica of the echo which can be subtracted from the near-end signal containing the echo before sending it over the satellite link to the remote subscriber. The echo canceler coefficients are adjusted to reduce the mean-square value of the residual echo. Integrated-circuit adaptive-echo cancelers with 128 taps (16-ms time span) have been developed and deployed in satellite and long terrestrial links.

Adaptive echo cancellation is also required for full-duplex data transmission over long two-wire circuits, for example, the 9.6-kbit/s public switched telephone network modem and the digital subscriber loop transceiver mentioned above.

For background information *SEE BANDWIDTH REQUIREMENTS (COMMUNICATIONS); DATA COMMUNICATIONS; ELECTRICAL COMMUNICATIONS; MODULATION* in the McGraw-Hill Encyclopedia of Science and Technology.

Shahid Qureshi

Bibliography. D. Chakraborty and C. J. Woljsza, A survey of modem design and performance in digital satellite communication, *IEEE J. Selected Areas Commun.*, SAC-1:5–20, 1983; J. K. Chamberlain et al., Receiver techniques for microwave digital radio, *IEEE Commun. Mag.*, 24(11):43–54, 1986; D. G. Messerschmitt, Design issues in the ISDN U-interface transceiver, *IEEE J. Selected Areas Commun.*, SAC-4:1281–1293, 1986; D. G. Messerschmitt, Echo cancellation in speech and data transmission, *IEEE Selected Areas Commun.*, SAC-2:283–296, 1984; S. U. H. Qureshi, Adaptive equalization, *Proc. IEEE*, 73:1349–1387, 1985.

Electrical utility industry

The year 1986 was noteworthy for the electrical utility industry for two reasons. The first was the occurrence of a nuclear reactor accident in Chernobyl, in the Soviet Union, that released a radioactive cloud that was transported throughout western Europe and Scandinavia, and severely contaminated a wide area surrounding the site. The second was a merger between two major, investor-owned utilities in the United States, the first such merger in a decade.

Although the Chernobyl incident was seen by some as an indictment of nuclear power in general, experts were quick to point out that the accident was due to the unique design of the Soviet reactor and could not have occurred in a reactor of the types used elsewhere throughout the world. *SEE NUCLEAR POWER*.

The merger of the Toledo Edison Co. and the Cleveland Electric Illuminating Co. in Ohio was noteworthy both because of the rarity of mergers among investor-owned utilities and because industry analysts have broadly declared that in the future many such mergers will be inevitable and in the best interests of the industry. The two utilities were joint owners of a troubled nuclear plant which represented a major portion of the total assets of the two companies, operated in the same state, and had worked closely together in system planning and operations of their interconnected systems, all of which made such a merger logical.

Ownership. Despite a relatively high and sustained rate of growth in recent years among customer-owned cooperatives, the United States industry is dominated in essentially all measures by investor-owned corporations. Such privately held utilities serve 76.5% of the 103.5 million electric customers in the United States. Municipal-, district-, and state-owned entities serve 13.3% of total electric customers, and rural electric cooperatives serve 10.2%. Federal utilities are basically wholesalers to other utilities and serve an inconsequential number of retail customers directly.

Investor-owned utilities also own and operate 76.6% of the installed generating capacity, in the United States while publicly owned utilities own 10.3%, federal agencies 9.4%, and cooperatives 3.7%.

The seeming discrepancy between the high percentage of customers served by rural electric cooperatives and the low percentage of ownership of generating facilities reflects the fact that most such utilities are primarily distribution companies. These buy their power wholesale from large generating and transmission cooperatives, federal agencies, or privately owned utilities.

Cooperatives are the fastest-growing sector of the industry, gaining new customers at the rate of almost 3% per year, compared to about 1.5–1.6% for the dominant investor-owned sector. This disparity exists because the cooperatives tend to serve exurban and rural areas where much of the current industrial, commercial, and residential growth is concentrated. Also, because cooperatives can borrow capital for expansion at preferential rates through government-backed agencies and, in addition, are not subject to regulation to the same degree as investor-owned utilities, they are less reluctant to commit to expansion. Expansion among distribution cooperatives commonly takes the form of buying into joint ownership of large-scale, central-station generating capacity planned and built by investor-owned utilities, to the extent that rarely is a large new unit now built that is not jointly owned.

Capacity additions. United States utilities at the end of 1985 had a total installed generating capacity of 688,400 MW. They added a net total of 21,900 MW during 1986, raising total industry capacity to 710,300 MW. Capacity represents the actual aggregated design capacity of all generating units connected to utility systems. In actuality, these units may not be capable of achieving their design rating for various reasons. Involved are such factors as lower-than-normal levels at hydroelectric reservoirs, boiler tube leaks, derating due to required maintenance, and actual outages, either scheduled or forced. The North American Electric Reliability Council therefore prefers to assess the ability of a system to meet demand in terms of available capability—the actual ability to perform at periods of maximum demand—rather than capacity. Capability of the industry in 1986 stood at 638,700 MW.

The composition of the capacity additions during 1986 was 380 MW of conventional hydroelectric; 1074 MW of pumped storage hydroelectric; 5005 MW of fossil-fueled (coal) steam; 13,653 MW of nuclear steam; 571 MW of geothermal; and 1524 MW of internal combustion, solar, wind, and waste-fired capacity. There was a net decrease of 271 MW in combustion turbine capacity. The composition of total generating plant by type at the end of 1986 is given in the **table** (see also **illus.**).

Fossil-fueled capacity. The Fuels Use Act of 1974

United States electric power industry statistics for 1986*		
Parameter	Amount	Change compared with 1985, %
Generating capacity, MW		
Conventional hydro	66,467 (9.4%)	0.57
Pumped storage hydro	17,947 (2.5%)	6.37
Fossil-fueled steam	472,241 (66.5%)	1.07
Nuclear steam	94,050 (13.2%)	16.98
Combustion turbine	51,002 (7.2%)	(0.53)
Internal combustion and other renewables	6,536 (0.9%)	30.4
Geothermal	2,081 (0.3%)	37.8
TOTAL	710,324	3.19
Noncoincident demand, MW‡	488,226	3.9
Energy production, TWh†	2,588.1	2.2
Energy sales, TWh†		
Residential	821.8	3.5
Commercial	634.7	3.5
Industrial	830.2	1.0
Miscellaneous	92.1	3.1
TOTAL	2,378.8	2.6
Revenues, total, × 10⁶ dollars	160,000	5.0
Capital expenditures, total, × 10⁶ dollars	31,398	(7.8)
Customers, 10³		
Residential	91,752	1.7
TOTAL	103,500	1.7
Residential usage, kWh/ customer	9,144	1.8
Residential bill, cents/kWh (average)	7.58	(0.13)

* After 37th annual electric utility industry forecast, *Elec. World*, 200(9):45–52, September 1986; 1986 annual statistical report, *Elec. World*, 200(4):49–64, April 1986; and extrapolations from monthly data of the Edison Electric Institute–Association of Electric Companies.
† TWh (terawatthour) = 10^{12} Wh.
‡ Noncoincident demand is the sum of the peak demands in the individual utilities, regardless of the time at which they occurred.

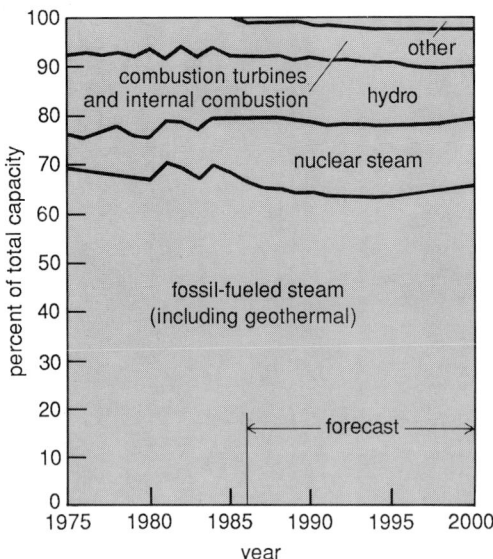

Probable mix of net generarting capacity. (After 37th annual electric utility industry forecast, Elec. World, 200(9):45–52, September 1986)

is still in effect, proscribing the use of oil and gas as the primary fuel for new fossil-fired capacity, although legislation to lift the ban on gas was introduced in 1986. Therefore, all new units now entering service have been designed for the firing of coal as a primary fuel.

Historically, utilities have retired fossil-fired units at an age of roughly 30 years, their normal economic life. Some utilities are now actively planning to extend that life by comprehensive maintenance and refurbishment of older units. Not only can the economic life of a unit be doubled in many cases, but by retrofitting new technology such as advanced-design blading in the turbine, capacity can be increased 2 to 5%. Utilities forecast that such upratings will provide as much as 5500 MW of additional capacity by 1995 at an incremental cost of about $300/kW, compared to $850 to $1000/kW for entirely new units.

Utilities plan to install an additional 25,450 MW of fossil-fueled capacity between 1987 and 2000. Utilities spent $8.75 billion on construction of fossil-fired units in 1986. Investor-owned utilities spent $5.1 billion, publicly owned utilities spent $2.8 billion, and cooperatives spent $850 million.

Nuclear capacity. No new order for a nuclear unit has been placed since 1978, and there are no plans by utilities to order such a unit for at least another decade. However, plants on which construction began as long as 15 years ago are now reaching completion and are entering service in large numbers. During 1986, 13 units either entered commercial service or were granted licenses by the Nuclear Regulatory Commission to load fuel or conduct low-power tests prior to entering service. When these enter full service, there will be 98 units in 65 different locations.

Eight of the new units were pressurized-water reactors (PWR) and the remainder were boiling-water reactors (BWR). The total number of each type now in service or in preliminary testing is 65

pressurized-water reactors and 32 boiling-water reactors. There is one high-temperature, gas-cooled reactor on utility lines, at Fort St. Vrain, Colorado.

Combustion turbines. Combustion turbines are used primarily to meet peak demands. Their relatively low capital cost of about $350/kW, versus $850 to $1000/kW for a modern coal-fired unit, offsets their higher fuel consumption. These machines burn gas or oil and have heat rates of about 12,000 Btu/kWh (3.5 joules of heat per joule of electrical energy), compared to about 9000 Btu/kWh (2.6 joules of heat per joule of electrical energy) for a modern fossil-fired unit. They can be started from a remote location and can go from a cold shutdown to full load in only 2 to 5 min, which makes them especially suitable for supplying power in emergencies. Combustion turbines are used also to supply startup power for other types of units after a complete system shutdown, since they can be fully self-contained.

In 1986, 200 MW of capacity of this type entered service, but retirements resulted in a net loss. However, some utilities forecast a shortage of capacity during the next decade because of the lack of construction of other types of generation, and are planning a total of 6800 MW of new combustion turbine capacity between 1987 and 1995.

Combustion turbines can be used in tandem with steam turbines in combined-cycle operation. The 900–1000°F (482–538°C) exhaust gases produce steam in a heat-recovery boiler to supply a steam-turbine–generator set. The heat rate in this mode may be as low as 8500 Btu/kWh (2.5 joules of heat per joule of heat energy). Utilities now operate 5780 MW of combined-cycle units.

Utilities spent $200 million on combustion turbines in 1986, of which $69 million was by investor-owned utilities, $96 million by publicly owned utilities, and $35 million by cooperatives.

Hydroelectric. At the end of 1986, installed capacity of conventional hydroelectric power totaled 66,467 MW at adverse reservoir conditions. Median conditions would increase this capacity. Forecasts call for an additional 4000 MW of these units to be constructed over the next 10 years, although appropriate sites are becoming increasingly difficult both to locate and to license because of environmental constraints.

Small hydro installations with capacities of 0.51–50 MW are increasingly popular. Under the incentives provided by the Public Utilities Regulatory Policy Act, hundreds of such units have been and are being installed in abandoned mill dams, irrigation canal locks, and even in area water supply lines where small turbines can be used to reduce the transmission pressure for use in distribution lines, and to generate electricity at the same time. Several utilities have formed subsidiaries to design and install such units throughout the country.

During 1986, utilities spent $638 million for

conventional hydro. Investor-owned companies spent $175 million, public bodies spent $140 million, and cooperatives spent $323 million.

Pumped storage. Pumped storage is one of the only practical ways to store large quantities of electrical energy. During periods of low demand, large, efficient base-load generators drive pumps that move water from a lower to an upper reservoir. During subsequent periods of high demand, the water from the upper reservoir is released through the pumps, now reversed to act as hydro-turbines, and these drive the pumping motors, now in a generating mode. About 65% of the energy used to pump the water to the upper reservoir is recovered in the generating mode. The scheme is economically justified by making more efficient use of the large, expensive base-load units, especially the highly capital-intensive nuclear units, that might otherwise be idle or lightly loaded during off-peak periods.

Utilities spent $80 million on pumped storage installations during 1986, with investor-owned utilities contributing $60 million and public utilities the remainder.

Rate of growth in demand. After an average growth of 2.5% in peak demand in 1985, growth from the summer of 1985 to the summer of 1986 jumped 3.9%. To a large degree, this strong surge can be attributed to extraordinarily hot weather throughout the Midwest, Southeast, and Mid-Atlantic regions, coupled with continuing strength in the national economy. The entire Southwest and West Coast regions, however, experienced much-cooler-than-normal summer weather, and the Northwest, in particular, endured a weak economy.

Long-term growth in summer peak demand has reached a consensus level of from 2.0 to 3.0% per year, compounded annually. This rate of growth is expected to decline gradually over the next 20 years. The declining trend arises from demographic factors such as the decreasing rate of population growth, and the aging of the population past the prime years for family formation and for purchasing major goods. Price-induced conservation will continue to be a suppressant, as will the growing body of legislative and regulatory efficiency standards.

The pattern of load growth must be examined in terms of regions, rather than on a national level. In 1986, for example, large utilities in the Midwest and Southeast experienced double-digit growth, while the western utilities, particularly those in the Northwest, saw their system loads actually decrease from 1985.

The current low level of yearly increases in demand has sharply increased the reserve margin, that is, the excess of installed capacity over actual demand. This is because generating units on which construction was begun as long as 10 years ago continue to come into commercial service. A rule of thumb is that national reserve margin should be about 25%, and it is now about 38%. Although utilities are, in response, canceling or delaying construction of additional capacity, construction of units started years ago and now nearing completion will support this high national reserve until about 1995. However, despite the high national margin, individual utilities, especially those in areas of high growth, will fall below margins required for reliable operation by the late 1980s.

The North American Electric Reliability Council prefers to use a different measure of adequacy of capacity, a factor called capacity margin. This is calculated by dividing the difference between actual peak demand and available capacity at time of peak by the available capacity. This capacity factor reflects a more realistic picture of spare capacity by accounting for capacity that is out of service at time of peak for whatever reasons. Currently, capacity margin is about 27%, compared to 38% for reserve margin.

Usage. Sales of electricity rose at a lower rate than did peak demand in 1986. Overall sales to all classes of customers gained 2.6% over 1985 compared to a gain of only 1.4% from 1984 to 1985, sustained by the continued strong economy and the recovery of housing starts. Commercial sales held up well in 1986, rising 3.5%; this represents a slight slowdown after growth of 6.9% and 5.1% in the previous 2 years. Industrial sales rose only 1% over 1985, despite strong economic indicators; nonetheless, it was a substantial gain over the 1984–1985 period when sales dropped 1.7%. Residential sales shared in the general increase, rising at a rate of 3.5% over 1985. (Sales in each category are given in the table.)

Despite the continual increases in the cost of electricity, electric heating for residences, driven by the growing popularity of the heat pump, continues to gain. Over one-half of all new residences built in the United States for more than a decade have been electrically heated, and this continues to be true in 1986. Heating energy sales rose to 187.2×10^{12} Wh in 1986.

Residential use per customer on a national basis continued recovery from the decline experienced in the 1984–1985 period, rising from 8978 kWh per customer in 1985 to 9144 kWh in 1986. This produced revenues of 7.6¢/kWh and an average annual residential bill of $693.42. In contrast to this average, each cooperative residential customer used 10,824 kWh and paid only 6.4¢/kWh. Rates across the country varied widely from national averages. For example, a resident of New York City paid 15.60¢/kWh, while a resident of Seattle paid only 2.33¢/kWh.

Total revenues for the investor-owned sector of the industry for 1986 were $148.9 billion, and $11.1 billion for the cooperatives.

Fuels. Utilities have shifted to coal as fuel from oil and gas ever since the energy crisis of the early 1970s, and in fact have been forced to do so by the provisions of the Fuels Use Act of 1974. The consumption of coal in 1986 rose 4.4% to 693.5×10^6 tons (627.3×10^6 metric tons). Oil use again declined, by 15.3%, to 173.3×10^6 barrels (27.5×10^3 m^3). Although gas was in good supply, usage decreased 2.3%, to 3030×10^{12} ft^3 (85.7×10^{12} m^3).

The same pattern repeated in total net energy generated. Coal generated 1475.1×10^{12} Wh, accounting for 56.9% of the total. Oil was used for 105.8×10^{12} Wh, or 4.1%. Gas was the fuel for 287.6×10^{12} Wh, or 11.1% of the total. Nuclear generation continued to gain as new stations came on line for full base-load operation, and generation from this source rose to 16.6% of total energy produced, or 428.9×10^{12} Wh. About 280×10^{12} Wh, or 10.8%, was contributed by hydroelectric, and 10.8×10^{12} Wh, or 0.4%, by sources such as wind, waste, and solar.

Distribution. Distribution capital expenditures for 1986 amounted to $6.9 billion, with an additional $2.2 billion expended in maintenance. During the year 15,400 mi (24,807 km) of three-phase equivalent overhead lines and 9970 mi (16,060 km) of three-phase equivalent underground lines came into service at voltages ranging from 4.16 to 35 kV. The majority of this construction was 15 kV, which accounted for 10,258 mi (16,511 km) of three-phase equivalent overhead and 6560 mi (10,568 km) of three-phase equivalent underground circuitry. The percentages of total overhead construction held by the other voltages were 13.5%, 16.7%, and 3.1% for 35 kV, 25 kV, and 4 kV, respectively. For underground construction the percentages were 14.3%, 18.5%, and 1.4%, respectively.

During 1986, utilities energized 15,241 megavolt-amperes of distribution substation capacity and expended $870 million in capital for distribution substation construction.

Transmission. Utilities spent $2.4 billion in capital accounts for transmission lines in 1986. They spent $559 million for overhead lines operating at 345 kV and above, and $615 million on lines operating at voltages of 220 kV and below. For underground construction, which can cost on average eight times more than equivalent overhead construction, capital expenditures were $9.3 million at voltages of 220 kV and above, and $25.3 million for circuits of 161 kV and below. Utilities installed only 881 mi (1419 km) of overhead lines at 345 kV and above, but 5076 mi (8177 km) at 220 kV and below. Looking at the power-transmitting capability of these lines rather than just length gives a different perspective. New overhead lines at 345 kV and above have an estimated capability of 875 gigawatt-miles (1410 GW-km), compared to only 690 GW-mi (1111 GW-km) for those operating at lower voltages. The proportions of lines at various voltages are different for underground cable because of the technology involved. In 1986, only 9 mi (14.5 km)

of underground circuitry operating at or above 230 kV was built. There were 14 mi (22.6 km) at or below 161 kV.

Utilities brought into service 39.6 GVA of transmission substation capacity in 1986 and expended $1.1 billion for substation construction. Maintaining existing transmission plant cost $630 million.

Capital expenditures. Total capital expenditures in 1986 were $33.8 billion. Of this total, $20.9 billion paid for generating facilities, $3.8 billion for transmission, $7.8 billion for distribution, and $1.3 billion for miscellaneous facilities.

Total assets held by investor-owned utilities at the end of 1986 were $397.8 billion. Electrical cooperatives had assets of $11.3 billion.

For background information SEE ELECTRIC POWER GENERATION; ELECTRIC POWER SYSTEMS; ENERGY SOURCES; TRANSMISSION LINES in the McGraw-Hill Encyclopedia of Science and Technology.

William C. Hayes

Bibliography. Edison Electric Institute, *Statistical Yearbook of the Electric Utility Industry*, 1985; 1986 annual statistical report, *Elec. World*, 200(4):49–64, April 1986; North American Electric Reliability Council, *1986 Annual Data Summary Report*, 1986; 37th annual electric utility industry forecast, *Elec. World*, 200(9):45–52, September 1986.

Electrochemistry

Electrochemical systems usually use unmodified metal, carbon, or semiconductor electrodes; for example, the familiar flashlight battery (the Le-Clanché primary cell) contains zinc and carbon electrodes. Recently there has been a growing interest in modifying the surface of the substrate by attaching specific molecules to it or by coating it with layers of polymers or other materials. Such modified electrodes are being investigated for application as displays based on electrochromism (color changes induced by electrical potential changes) or electrogenerated chemiluminescence (light emission from electrically generated species), as analytical sensors, as new types of microelectronic devices, and as electrocatalytic surfaces. They also provide a useful structure for obtaining fundamental information about the nature of charge and mass transport in surface films and at interfaces. The goal of this research is the purposeful design of surface structures on electrodes to carry out a particular electrochemical reaction efficiently and with high selectivity. In these surface structures the passage of electric current across the substrate/solution interface can be used both to grow novel structures on the surface and, with electrochemical techniques, to characterize them once they are produced. Some of these surface structures are examples of so-called integrated chemical systems. The latter are heterogeneous systems containing several components (such as semiconductor, polymer, metal, and catalyst) designed to carry out a specific process, and often involve synergistic effects among the components.

Types of modified surfaces. Surfaces can be modified by a number of different methods (**Fig. 1**). Some molecules adsorb spontaneously and strongly to a surface (Fig. 1a). They are held more effectively through chemical bonding to substrate surface structures (covalent attachment; Fig. 1b). Both adsorption and covalent bonding produce a monolayer of surface molecules, consisting of about 6×10^{13} molecules per square centimeter of surface. Thicker layers (up to a few micrometers thick) can be produced by coating with a film of polymer. These polymers may contain electroactive groups that can be oxidized or reduced by passing current through the electrode substrate; typical molecules of this type are poly(vinylferrocene) and various forms of polymerized methyl viologen (Fig. 1c). Other polymer films are polyelectrolytes that take up solution species by an ion-exchange process. For example, layers of the cation-exchange polymer Nafion (a trade name) can absorb positively charged ions from solution (Fig. 1d). In layers in which the electroactive groups are held on the polymer backbone or are held by electrostatic forces in a polyelectrolyte (Fig. 1c–d), charge propagation through the polymer layer occurs by electron (e^-) hopping from site to site. Charge neutrality is maintained by movement of counterions into or out of the solution phase. It is also possible to form layers of polymers that are true electronic conductors; such polymers have a band structure like metals or semiconductors, and electron transport through the film does not involve net movement of counterions (Fig. 1e). These films can be produced directly on the electrode surface by oxidation or reduction of the monomer. Polymers such as polypyrrole and polyaniline can undergo oxidation and reduction processes and thus be converted from an insulating to a conductive state.

More complicated structures can be formed by growing small metallic, semiconductive or conductive zones within a different polymer. **Figure 2** shows an example of small needles of a conductive material, tetrathiafulvalenium bromide, grown electrochemically in a thin film of Nafion on an electrode surface. Other types of surface films on electrodes have included such different materials as clay particles and enzymes or other proteins.

Electrochromic surfaces. An example of a modified electrode that involves an electrochromic material is shown in **Fig. 3a**. A film of the polymer bearing the viologen group (N,N'-dialkyl-4,4'-bipyridinium) is formed electrochemically on the electrode surface. In its oxidized

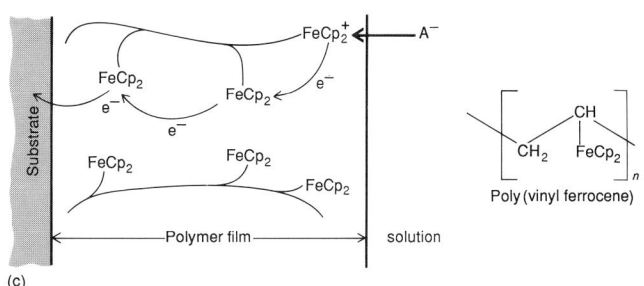

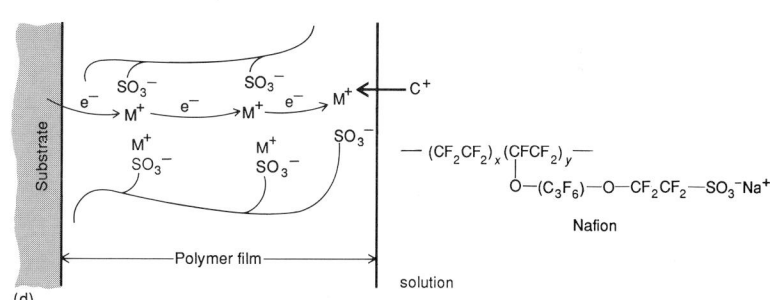

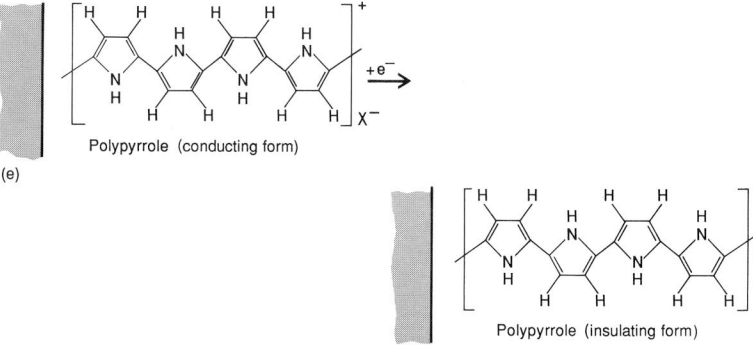

Fig. 1. Types of modified electrodes. (a) Irreversible adsorption of monolayer; in this example, Q = 9,10-phenanthraquinone. (b) Covalent attachment of ferrocene ($FeCp_2$) via silane linkage. (c) Film of electroactive polymer—here, poly(vinylferrocene)—showing charge transport through the film by hopping between centers. A^- is an anion in solution. (d) Film of polyelectrolyte—here, Nafion—showing charge hopping between incorporated cations, M^+. C^+ is a cation in solution. (e) Film of electronically conductive polymer—here, polypyrrole—in oxidized (conducting) and reduced (insulating) forms.

form ($\sim PQ^{2+}$, where the symbol $\sim$ represents the polymer chain) it is colorless, but when a negative potential is applied to the electrode and the film is reduced to the $\sim PQ^+$ form, the film becomes a deep blue-violet color. It can be

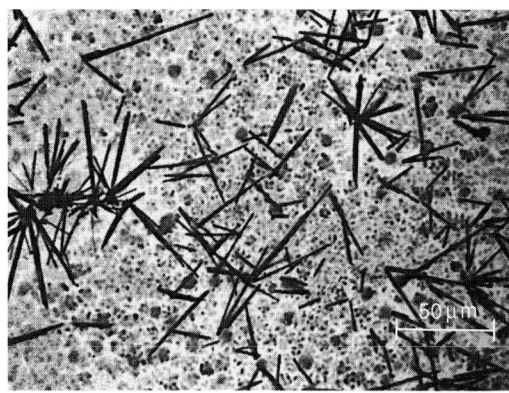

Fig. 2. Needles of conducting tetrathiafulvalenium (TFF$^+$) bromide grown electrochemically in a Nafion film on a tin oxide electrode surface. The needles shown are less than a micrometer wide and 20–100 micrometers long. The darker zones between the needles also represent TTF$^+$ that can be reversibly reduced to colorless TTF. (*From T. P. Henning, H. S. White, and A. J. Bard, Polymer films on electrodes, 10. Electrochemical behavior of solution species at Nafion-tetrathiafulvalenium bromide polymers, J. Amer. Chem. Soc., 104:5862, 1982*)

reoxidized to the colorless form by application of a more positive potential. Thus the electrode surface can be repeatedly switched between two states of different color. Such electrochromic materials can be the basis of display devices (for example, in calculators or computers) or for windows whose transparency can be controlled

electrically. This same feature involving electrical switching of a surface film between two different states can be applied as a digital memory. By forming films of different polymer layers on the electrode surface (a bilayer arrangement), bistable switching between the states can be accomplished. In bistable switching, different applied potentials are needed to charge and discharge the element.

Electrocatalysis. Polymerized viologen can also be used as a component of an electrocatalytic layer (Fig. 3*b*). Carbon (graphite) electrodes provide inefficient surfaces for evolving hydrogen (H$_2$) in the electrolysis of water. Such unmodified electrodes require application of a relatively large voltage to split water into hydrogen and oxygen (O$_2$). Platinum (Pt) is a catalyst for hydrogen evolution, and a carbon electrode covered with a layer of the polymer ~PQ$^+$ and containing finely divided platinum dispersed throughout the layer will evolve hydrogen at a lower applied voltage. In the reduction process, electrons are passed from the carbon to the ~PQ^{2+} (to form ~PQ$^+$) and then, via the Pt, to H$_2$O. This same type of layer can be coated on a semiconductor surface (for example, silicon) to cause evolution of hydrogen under irradiation in a photoelectrochemical reaction; this may find application in conversion of solar energy to a fuel.

Carbon is also a poor material for oxygen evolution in the electrolysis of water. Its performance can be improved significantly by coating it with ruthenium dioxide (RuO$_2$), which is an electrocatalyst for oxygen evolution. Such RuO$_2$ layers, mixed with other oxides and coated on titanium substrates, are widely used as dimensionally stable anodes for chlorine evolution in commercial chloralkali cells.

Sensors. Modified electrodes are also being investigated for use in analytical devices. Here one electrode serves as an indicator or sensor for a component of interest, while a second electrode, called the reference electrode, is employed to complete the circuit. While operation is in the amperometric mode, a potential is applied between the indicator and reference electrode, generating a current that is proportional to the concentration of the species of interest. Alternatively, in the potentiometric mode, the potential that arises spontaneously between the indicator and reference electrode can be used to detect and measure the desired component. *See Transducer.*

Modified electrodes can serve as sensors and reference electrodes. For example, the enzyme glucose oxidase can be immobilized in a membrane on an electrode surface and used to detect glucose. In this system the underlying electrode is an oxygen sensor that measures the amount of O$_2$ reaching the surface by reducing the O$_2$ to H$_2$O. In the presence of glucose, oxygen is consumed by the reaction below, which is cata-

Fig. 3. Applications of modified electrodes. (*a*) Electrode coated with a film of polymerized viologen (PQ^{2+}); schematic diagram showing an electrochromic polymer layer that can be switched electrically between colored and bleached states. (*b*) Schematic diagram of water electrolysis at carbon electrodes modified with catalytic layers [polymerized viologen (~PQ^{2+}) and RuO$_2$] to increase the rate of hydrogen (H$_2$) and oxygen (O$_2$) evolution.

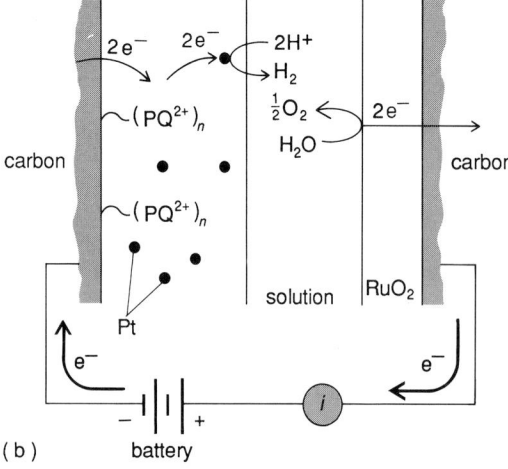

$$Glucose + \tfrac{1}{2} O_2 \xrightarrow{\text{glucose oxidase}} gluconic\ acid$$

lyzed by the enzyme; the flux of oxygen to the electrode surface, and hence the current, is decreased. Other sensors have been developed to measure dissolved gases, ions, and biomolecules.

Other applications. Modified surfaces can be used to control the flow of ions, and in some ways they are analogous to biological membranes. For example, a layer of polypyrrole (Fig. 1e) can be deposited on a gold mesh and used to separate two ionic solutions. In its reduced (uncharged) form the layer is impervious to ions, while in its oxidized (cation) form small anions (such as Cl^-) will flow, under a superimposed field, from one solution to the other through the separator. Thus the film acts as an ion gate that can be opened and closed electrically. A related concept is involved in using such surfaces for the controlled release of ions into a solution. For example, a polypyrrole film can be charged with glutamate anions by oxidizing it in a solution containing sodium glutamate. This glutamate-loaded electrode can then be transferred to a different medium into which the release of glutamate is controlled by the flow of current to the electrode. These types of electrodes are being explored for the controlled release of physiologically active agents, such as drugs and neurotransmitters.

Modified electrodes are also being investigated as components of so-called molecular electronic devices. One form of these involves microelectrode arrays that contain individual electrode elements with sizes and spacings of micrometer dimensions covered with polymer layers whose electronic conductivity can be changed by oxidation or reduction. These electrodes can produce devices equivalent in function to solid-state diodes and transistors. For example, when a pair of gold electrodes deposited on an insulating substrate is bridged by a nonconductive polymer layer, no current will flow between them under an applied voltage. When the polymer layer is electrochemically converted to a conductive state, current will flow. Hence, the current flows between elements (which can be relatively large) can be controlled by small currents that oxidize or reduce the films; transistorlike behavior results. Other devices use pairs of different polymers in a bilayer or sandwich arrangement to form structures in which switching is possible.

Modified electrodes have also been studied as means of improving the selectivity in synthesis of materials at electrodes and in devices for concentration of small quantities of ions from solution, and as components in batteries. Moreover, quantitative studies with such electrodes have produced a better understanding of how ions and electrons move through polymers and across interfaces. The study of modified electrodes remains a field of high activity. While there have been few actual practical applications, many promising possibilities lie on the horizon.

For background information *SEE ELECTROCHEMISTRY; ELECTROCHROMIC DISPLAYS; ELECTRODE; ELEC-* *TRODE POTENTIAL* in the McGraw-Hill Encyclopedia of Science and Technology.

Allen J. Bard

Bibliography. A. J. Bard, Chemical modification of electrodes, *J. Chem. Educ.*, 60:302–304, 1983; L. R. Faulkner, Chemical microstructures on electrodes, *Chem. Eng. News*, pp. 28–45, February 27, 1984; R. W. Murray, Chemically modified electrodes, in A. J. Bard (ed.), *Electroanalytical Chemistry*, vol. 13, pp. 191–368, 1984; M. S. Wrighton, Surface functionalization of electrodes with molecular reagents, *Science*, 231:32–37, 1986.

Electronics

Molecular electronics involves the systematic use of molecular materials to produce new or improved electronic devices. Conventional devices mostly rely on inorganic semiconductors, metals, and conductors, but molecular materials can fulfill many similar functions. Molecular materials also have specific advantages. These include controllability of molecular size, shape, and properties obtainable by chemical modification; reproducibility of material dimensions and properties implicit in the molecular structure; and selectivity of physical and chemical behavior governed by the molecular constitution and packing. A growing number of devices therefore employ molecular materials to produce novel or significantly enhanced performance.

Polymers. Many devices use polymers, which are generally strong and easily processed by low-energy methods. Conventional technology for very large-scale integration uses polymer resists to mask selected areas of a silicon wafer while other areas are processed. Progress toward smaller, faster, low-power devices depends on reducing feature sizes, which requires improved resists with controlled response to the light or ionizing radiation which delineates the mask. Resists are essential for the processing step but are absent from the final device. However, conventional devices often incorporate polyimide dielectrics (I). This was the first, and probably

Ar, Ar' = aryl groups

(I)

remains the most widespread, use of a molecular material as a constituent element of a device; polymers also find extensive use as encapsulating layers around devices.

Electroactive polymers are of considerable importance. Aromatic and heteroaromatic materials are often effective photoconductors, conducting electric current when exposed to light. They can be adapted for use in photocopiers if attached to a polymer backbone, allowing large-area thin films to be prepared; a commercial example is poly(*N*-vinylcarbazole) [PVK; II]. Piezoelectric

(II)

and pyroelectric polymers are also found, the main example being poly(vinylidene fluoride) [PVDF; III]. These two classes of polymers

$$-\left[CH_2CF_2 \right]_n$$

(III)

respond electrically to pressure and temperature changes, and conversely they develop a mechanical distortion or a temperature change when subjected to an electric field. They are therefore used as transducers in compact lightweight microphones, infrared detectors, and sensors.

Much attention is being devoted to conducting polymers. Polymer structures allowing electron delocalization along the backbone can often be doped (chemically oxidized or reduced) to achieve conductivities up to metallic levels of 1000 siemens/cm in the best cases, for example, poly(acetylene) [$(CH)_n$; IV]. Polymers of aro-

$$-\left[CH{=}CH \right]_n$$

(IV)

matic species such as poly(phenylene) [V], poly-(phenylene sulfide) [VI], poly(pyrrole) [VII], and poly(thiophene) [VIII] have advantages in ease of

(V)   (VI)

(VII)   (VIII)

preparation, tractability, or stability. Conducting polymers are expected to find applications in lightweight electromagnetic screening, novel battery systems, and the connection of active molecular elements to conventional inorganic elements in hybrid devices. In one demonstration device, poly(pyrrole) immersed in an electrolyte exhibits behavior analogous to that of a transistor: the current between source and drain microelectrodes increases greatly once the potential of a central gate electrode exceeds that required to oxidize the poly(pyrrole), so that the gate voltage is amplified (see **illus.**).

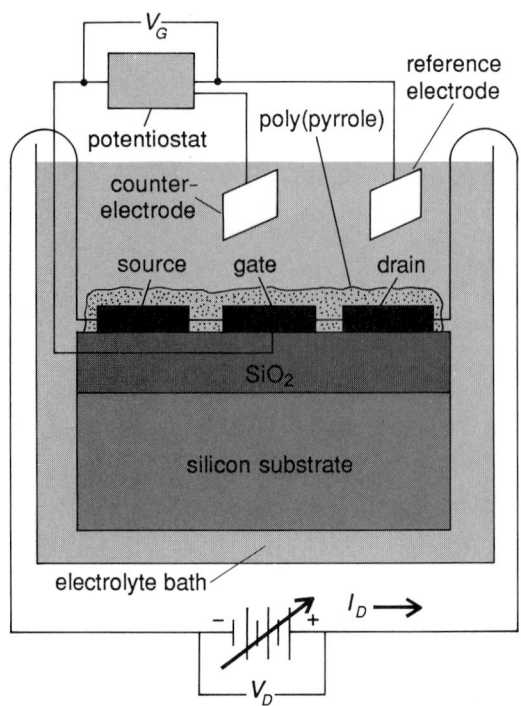

Molecule-based transistor. The silicon dioxide (SiO_2) layer is about 0.45 micrometer thick on a 0.3-mm-thick silicon substrate. The gold source, drain, and gate electrodes are about 3 μm wide, 140 μm long, and 0.12 μm thick, and are coated with about 10^{-7} mol/cm^2 poly(pyrrole). The whole assembly is immersed in electrolyte solution. I_D = current between source and drain; V_D = voltage between source and drain; V_G = gate potential. (*After H. S. White, G. P. Kittelsen, and M. S. Wrighton, Chemical derivatization of an array of three gold microelectrodes with polypyrrole: Fabrication of a molecule-based transistor, J. Amer. Chem. Soc., 106:5375–5377, 1984*)

Langmuir-Blodgett films. Amphiphilic molecules containing both hydrophilic and hydrophobic groups may adsorb at a water-air interface. The surface layer can then be compressed into a floating monolayer and transferred to a substrate by dipping. The resulting film is called a Langmuir-Blodgett (LB) film. These films are highly ordered and free of voids or pinholes, but noncrystalline, with many features useful for electronics applications. The thickness of one layer is determined reproducibly by the molecular dimensions, and multilayers (possibly of different species) can be prepared by repeated dipping. Specific properties can be incorporated by using molecules with suitable electric, magnetic, optical, or chemical behavior.

Polymerizable Langmuir-Blodgett films have been developed. Uniform polymerization improves stability through covalent bonding in the film while retaining the film thickness precisely; for this purpose, diacetylene Langmuir-Blodgett films are attractive because they polymerize with minimal disruption. Selective polymerization defines microstructures for integrated circuits or integrated optics. For instance, the terminally unsaturated long-chain molecule ω-tricosenoic acid (IX) makes a good negative resist under

$$H_2C = CH(CH_2)_{20}CO_2H$$

(IX)

electron-beam polymerization.

Langmuir-Blodgett films find uses in processing of enhanced-performance devices as insulating layers of precisely determined thickness; their fabrication requires much lower temperatures than does the oxidation of silicon, and they are available for semiconductors that lack oxides suitable for use in electronic devices. These films can improve the performance of metal-insulator-semiconductor (MIS) devices. Examples are a solar cell based on cadmium telluride and an electroluminescent diode based on gallium phosphide, each markedly more efficient with two or three Langmuir-Blodgett monolayers between the metal and the semiconductor than with none. A metal-insulator-semiconductor-semiconductor (MISS) structure, such as metal–Langmuir-Blodgett film–n-type gallium arsenide (GaAs)–p^+-type gallium arsenide, acts as a switch, and will serve as a sensor if the Langmuir-Blodgett film contains an organic species sensitive to some external influence which affects the switching characteristics.

Other sensors are based on measurable changes that occur in a surface Langmuir-Blodgett film when it adsorbs a target species, even at the parts-per-million level. These changes are detected by surface plasmon resonances on a metal, surface acoustic waves on a quartz oscillator, or surface electronic states on an inorganic semiconductor. Such sensors have been demonstrated for the detection of several molecular species, including nitrogen oxides and acetic acid. *SEE SURFACE PHYSICS; TRANSDUCER.*

Pyroelectric Langmuir-Blodgett films serve as infrared detectors. Such devices require a noncentrosymmetric sample, achievable as a single layer or as alternating layers of different species. As the figure of merit for such detectors depends inversely on the film thickness, excellent response is obtained as compared with thicker polymer or ceramic detectors. Noncentrosymmetric Langmuir-Blodgett films are also employed for their nonlinear optical properties such as second-harmonic generation or parametric oscillation, important in optical communications. The response of such films increases quadratically with the number of layers, and two different layers may enhance one another's response.

Optical materials. Molecular materials possess optical properties valuable in electronics applications such as information storage and display, optoelectronics, and communications. The nonlinear response already mentioned for Langmuir-Blodgett films can be achieved also in doped polymers and in molecular crystals. The molecules can be tailored to optimize the nonlinear response, optical absorption, and crystal packing. A classic example occurs in the nitroanilines,

where a single isolated molecule of the para compound (X) has high nonlinear response but

(X)

the crystal structure is inactive for second-harmonic generation, while the meta compound (XI) has a lower molecular response but an active

(XI)

crystal structure; adding a methyl group at the meta position in *para*-nitroaniline to form methylnitroaniline (XII) retains a high molecular

(XII)

response and an active crystal structure. The resulting material has an excellent figure of merit with respect to both second-harmonic generation and the Pockels linear electrooptic effect; moreover, like most molecular materials, it is highly resistant to optical damage.

Liquid-crystal devices fall within the above definition of molecular electronics, although they are often excluded from this field and regarded as a separate topic. Their well-known use in low-power displays has required effective systematic molecular design to meet the technological objectives of these devices; and their development has continued. Recently, liquid-crystal side-chain polymers have been applied to information storage. An unaligned scattering state can be selectively aligned electrically above the polymer glass-transition temperature T_g to give negative contrasting clear regions which persist on cooling toward T_g. Alternatively, an electrically aligned material can have positive contrasting regions written in by a laser beam which provides temporary local heating; this gives good contrast and spatial resolution with simple erase and rewrite capabilities.

Information can be stored also as colored spots produced by a laser beam in a photochromic material. Here individual molecules change color on illumination, so that grain size no longer limits resolution as in conventional storage media, and developing and fixing are unnecessary. Practical photochromic storage media must be thermally stable in both colorless and colored forms, while being optically readable and preferably optically erasable. Molecules tailored to meet these conditions include the fulgides (XIII), derivatives of

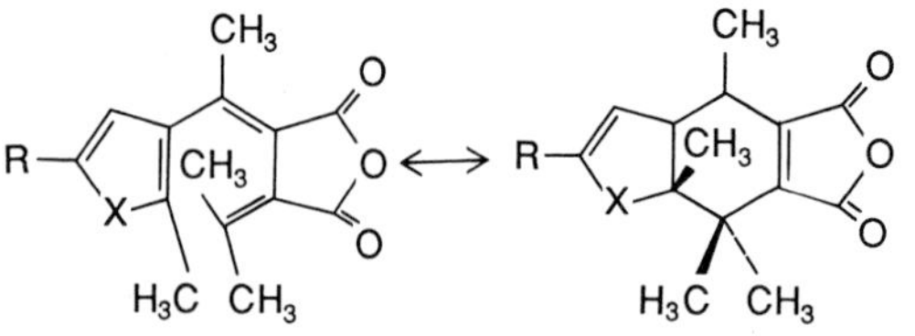

X = O, S, NCH$_3$ R = H, CH$_3$, C$_6$H$_5$

(XIII)

bismethylene succinic anhydride, which function by exploiting the Woodward-Hoffman rules governing electrocyclic reactions: optical and thermal reactions proceed by different stereochemical pathways, and the thermal one is blocked by bulky side groups. Optical techniques are limited in resolution by the wavelength of the laser light, but this restriction can be circumvented by storing information at various wavelengths in the same physical region of a sample consisting of photoactive molecules that are dispersed in an amorphous medium and have a distribution of absorption maxima. This technique, persistent spectral hole burning, has been demonstrated in systems including the phthalocyanines (XIV).

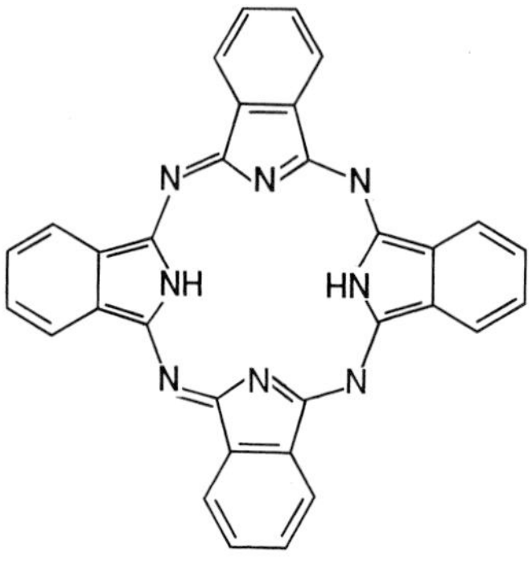

(XIV)

See Optical information systems.

Molecular electrochromic media change color on oxidation or reduction at a suitable potential and can be used for displays. One type uses viologens (XV), whose color changes depend on

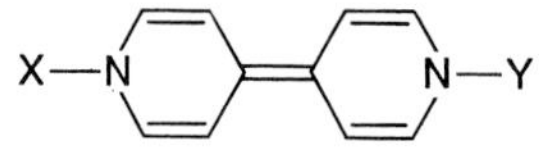

(XV)

the substituents X and Y (typically alkyl or aryl groups) and counterions, which also modify the display response times and durability. Another possibility is use of poly(pyrrole) [VII], which is dark in the conducting form prepared electrochemically but transparent green in the insulating form, while poly(isothianaphthene) [XVI] is transparent in its conducting form.

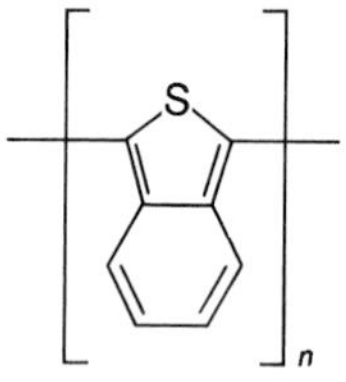

(XVI)

Prospects. Molecular electronics is already successful in the areas outlined above, in most of which devices are at the prototype stage or beyond, and active development continues. The components in microelectronic circuits are now so small that they will soon be reduced to molecular dimensions. Molecular electronics will then have to treat whole systems of molecules rather than individual molecular elements. Such systems will have to largely assemble themselves specifically and spontaneously under the influence of their intermolecular forces, as biological systems do. After assembly, the constituent molecules will interact nonlinearly (and not necessarily electronically), leading ultimately to supermolecular arrays for information processing.

For background information *see* Electrochromic displays; Integrated circuits; Liquid crystals; Monomolecular film; Nonlinear optical devices; Polymer; Woodward-Hoffmann rule in the McGraw-Hill Encyclopedia of Science and Technology.

R. W. Munn

Bibliography. F. L. Carter (ed.), *Molecular Electronic Devices*, 1982; *J. Mol. Electr.*, 1985– ; G. G. Roberts, An applied science perspective of Langmuir-Blodgett films, *Adv. Phys.*, 34(4):475–512, 1985.

Elementary particle

The picture of the fundamental constituents of matter and the interactions among them that has emerged in recent years is one of great beauty and simplicity. All matter appears to be composed of quarks and leptons, which are pointlike, structureless particles with a spin of ½. If gravitation, which is a negligible perturbation at the energy scales usually considered, is set aside, the interactions among these particles are of three types: electromagnetism, the weak interaction responsible for radioactivity, and the strong interaction that binds quarks into protons, neutrons, and other subnuclear particles. All three interactions are described by gauge theories based on symmetry principles, and are mediated by force-carrying particles with a spin of 1 called gauge bosons. Thus, quantum chromodynamics describes the strong interactions, and the electroweak theory provides the so-called unified description of the weak and electromagnetic interactions. The quarks experience all three interactions; the leptons participate only in the weak and electromagnetic interactions. The quarks and leptons are arranged in families, and there is a hint of extended families containing both quarks and leptons. The standard model (see **illus**.) thus has an appealing simplicity and an impressive generality. The picture has a pleasing degree of coherence and holds the promise of deeper understanding—in the form of further unification of the interactions—still to come.

Success of standard model. Remarkable progress has been made in applying quantum chromodynamics, the theory of strong interactions, to experimental situations and in extracting detailed and precise information about the high-energy interactions of quarks and gluons (the carriers of the strong force) from observations of the products of high-energy collisions. The most telling information has come from the study of the jets of pions and other strongly interacting particles produced in violent proton-antiproton collisions in the Super Proton Synchrotron Collider at the European Laboratory for Particle Physics (CERN) in Geneva. Still more incisive experimental studies are expected soon from the 2-TeV tevatron collider at Fermilab in Illinois.

Intermediate bosons. The CERN experiments have also refined knowledge of the properties of the intermediate bosons, W^+, W^-, and Z^0, the mediators of the weak interactions. Present experimental information about the masses, disintegrations, and production rates of these force carriers is in excellent agreement with the predictions of the electroweak theory and entirely consistent with the understanding of a wealth of low-energy weak-interaction phenomena. The masses of the intermediate bosons are now known to be $M(W) = 81.8$ GeV/c^2 and $M(Z) = 92.6$ GeV/c^2, each with an uncertainty of about 1.5 GeV/c^2, where c is the speed of light.

Charmed particles. Another area of important progress is the study of particles containing charmed quarks. The first charmed particles were found in 1976, but detailed studies of their properties have been impeded by the small numbers of charmed particles observed. At issue is the interplay between the strong and weak interactions in the decay of charmed particles which requires, for detailed study, the observation of thousands of examples of charmed particles. Until recently, most information about charmed particle decays has come from the study of electron-positron annihilations into strongly interacting particles. There, relatively small numbers of charmed particles are produced in a rather benign experimental environment. In contrast, it has been known for several years that charmed particles are produced copiously in collisions involving hadrons. The complication is that they are produced amid a sea of background particles uninteresting for this purpose but difficult to sift through.

An experiment undertaken at Fermilab in 1986 has succeeded in extracting very large samples of

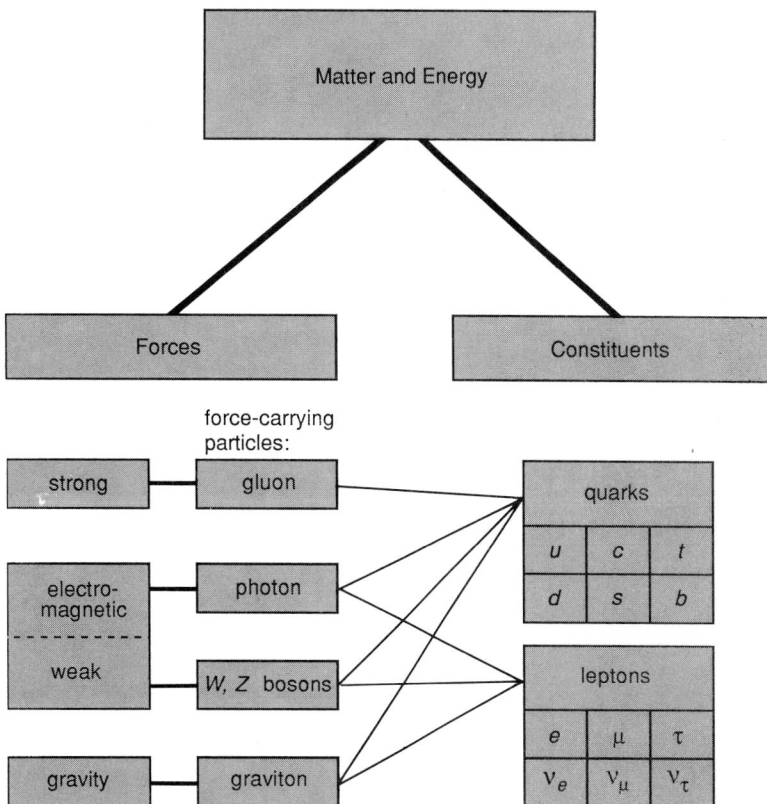

Standard model of elementary particle physics.

charmed particles from collisions of high-energy photons with protons. In addition to instrumental advances made possible by a decade of effort in this field, the Fermilab experiment has exploited a novel parallel-computing architecture based on so-called microprocessor farms to analyze 10^8 collision events for charmed particles. The emerging data will bring maturity to this important field. The development of inexpensive, large-scale parallel-computing resources will change the way that experiments at high-energy accelerators are conceived and analyzed.

Beyond the standard model. The continuing success of the standard model gives added motivation to examine its self-consistency and to create a more complete so-called theory of the world, encompassing the insights it has given. To approach this challenge, the standard model itself may be used for guidance. The shortcomings of the standard model may be expressed as follows. The most sharply posed issue has to do with the incomplete understanding of why electromagnetism acts over very long (astronomical) distances while the effects of the weak interaction are restricted to less than 10^{-17} m. A second area of concern has to do with the meaning of quark-lepton generations. The idea that quarks and leptons should be grouped together in generations seems required for the internal consistency of the electroweak theory, but there is no fundamental understanding of why this should be so.

Third, it is even possible to inquire into the origin of the gauge symmetries themselves. Such questions—and this is but a partial list—are stimulated by the standard model itself and by the desire to find ever simpler descriptions of nature of ever more general applicability.

Supersymmetry. One possible solution to the Higgs problem involves introducing a completely new set of elementary particles with spins differing by one-half unit from those of the known quarks, leptons, and gauge bosons. These postulated new particles are the consequences of a conjectured so-called supersymmetry which relates particles of integer and half-integer spin. Evidently, supersymmetry cannot be an exact symmetry of the world, for that would imply a spinless counterpart of the electron with the same mass as the electron, a particle that would long since have been observed. Supersymmetry would stabilize the mass of the Higgs boson at a mass below 1 TeV/c^2, and the supersymmetric partners of the known particles are likely themselves to have masses less than about 1 TeV/c^2. Part of the theoretical appeal of supersymmetry is that the local form of supersymmetry leads to quantum theories which have as their classical limit Einstein's gravitation. This raises the possibility of obtaining a quantum theory of gravity and of incorporating gravity into a unified description embracing all the fundamental interactions.

Extensive but still exploratory searches for superpartners have been carried out in a variety of accelerator experiments. The negative results obtained so far provide constraints on specific models of supersymmetry but do not rule on the general idea that supersymmetry may provide the resolution to the problem of electroweak symmetry breaking. Negative results become decisive only after a thorough exploration of the 1-TeV scale, one of the principal goals of the Superconducting Super Collider, a proposed 40-TeV high-luminosity proton-proton collider.

The machines that are coming into operation include the Fermilab proton-antiproton collider, the Stanford Linear (electron-positron) Collider, and, somewhat later, the Large Electron-Positron Collider at CERN, and HERA, the electron-proton collider under construction at the Deutsches Electronen Synchrotron (DESY) near Hamburg, West Germany. These machines will extend the search into the regime of a few tenths of a TeV/c^2, where the first indications of superpartners might well be found.

Technicolor. A second possible solution to the Higgs problem is based on the idea that the Higgs boson is not an elementary particle at all but is in reality a composite object made out of elementary constituents analogous to the quarks and leptons. Although they would resemble the usual quarks and leptons in certain respects, these new constituents would be subject to a new type of ultrastrong interactions (often referred to as technicolor) that would confine them within about 10^{-19} m. The new phenomena would include a rich spectrum of bound states, akin to the spectrum of known hadrons.

Again, there is no evidence yet for any of these new particles. The technicolor scenario is very appealing because it goes beyond the Higgs mechanism of the standard model in much the same way as the Bardeen-Cooper-Schrieffer theory of superconductivity gives a microscopic interpretation of earlier phenomenological descriptions of the superconducting state.

Superstring theory. Other theoretical inventions respond to questions about the meaning of generations and the origin of gauge symmetries. These too imply new kinds of matter and force particles to be looked for in the coming generation of experimentation.

A particularly audacious line of argument is represented by the search for a so-called superstring theory which would solve all problems at once, while unifying gravitation and the other forces of nature. According to superstring theories, the known laws of physics are approximate versions of a far richer and more comprehensive theory that takes account of structure on a fantastically small distance scale. Elementary particles are to be regarded not as structureless points but as excitations of minute strings about 10^{-35} m in length. The familiar fundamental particles are to be understood as different modes of vibration of a single string. Superstring theory combines earlier notions of strings as the fundamental objects with the mathematical structure of supersymmetry. The observation of supersymmetric partners on or below the 1-TeV scale would therefore provide powerful encouragement for this program.

Although theoretical speculation of this sort is helpful and often indispensable, the decisive role played by experimentation, without which theory is sterile, should never be forgotten. While it is important to follow up the hints given by theory, it is also important to explore completely new areas. This is a primary motivation for the Superconducting Super Collider.

For background information *see* Elementary particle; Fundamental interactions; Intermediate vector boson; Lepton; Particle accelerator; Quantum chromodynamics; Quarks; Supersymmetry in the McGraw-Hill Encyclopedia of Science and Technology.

Chris Quigg

Bibliography. E. Eichten et al., Supercollider physics, *Rev. Mod. Phys.*, 56:579–707, 1984; M. B. Green, Superstrings, *Sci. Amer.*, 255(3): 48–60, September 1986; J. D. Jackson, M. Tigner, and S. Wojcicki, The Superconducting Supercollider, *Sci. Amer.*, 254(3):66–77, March 1986; C. Quigg, Elementary particles and forces, *Sci. Amer.*, 252(4):84–95, April 1985.

Fate maps (embryology)

Much of the study of plant development has centered on the activity of the root and shoot apical meristems. One of the fundamental questions concerns the means by which the apical meristem gives rise to the rest of the plant. Within this broad question are secondary questions relating to the fate of cells present in the apical meristem at any particular stage in development: whether differential function is partitioned in the meristem (for example, whether there is a separate subpopulation of cells in the shoot meristem that will give rise exclusively to the reproductive parts of the plant); and how the cells in a radially symmetrical leaf primordium formed by the meristem contribute to the formation of a bilaterally expanded leaf blade. Other important questions relate to the number of cells in the embryonic shoot meristem and the clonal relationship between those cells and the cells that give rise to other parts of the embryo. Recent studies of cell fate in plants at critical stages of development are leading to a reconsideration of some of the traditional views of shoot and leaf development.

Method of mapping. In order to answer questions relating to the fate of cells, it is necessary to mark cells at one stage of development so that their progeny can be identified at subsequent stages. The most effective means in plants involves cell-autonomous genetic markers that confer a visually distinct phenotype to the marked cell and all of its derivatives. Individuals are exposed to ionizing irradiation, or sometimes chemical mutagens, to produce chromosomal mutations that result in distinct phenotypic alterations, usually pigment (chlorophyll or anthocyanin) deficiencies. In a plant heterozygous for a visible, cell-autonomous mutation, such as albino, loss of the chromosome part carrying the dominant wild-type allele from a cell results in the expression of the albino mutation in the affected cell and all of its progeny, producing a sector (clone) of white tissue in the developing green plant (**Fig. 1**). Individuals composed of tissues of two (or more) genetically different kinds are called chimeras or mosaics. Chimeras in which the mutant portion of the plant is derived from a single mutant cell generated at a particular known stage of development make it possible to map the fate of individual cells and their derivatives in the subsequent stages of development. Thus the cell lineage of a single mutant cell can be traced unequivocally in the mature plant.

The fate of the cells present in the meristem (or organ primordium) at the time of mutagenesis can be deduced from the pattern of sectors obtained in the mature plant (or organ). The size of a sector shows the extent of the cell lineage in the plant or organ. It is a measure of the amount of proliferation of the mutant cell and its derivatives. A sector generated from a permanent initial

Fig. 1. Sunflower shoot with a chlorophyll-deficient sector extending through several vegetative nodes. The sector represents the lineage of a single mutant cell present in the embryo meristem at the time of mutagenesis. Mutagenesis was accomplished by exposure of dry seeds to cobalt-60.

cell would persist throughout the lifetime of the shoot; it would extend the entire length of the shoot. The width of a sector relative to the circumference of the organ which it occupies provides an estimate of the number of cells present in the meristem or organ primordium at the time of mutagenesis. For example, if the widest epidermal sectors commonly occupy one-third of the circumference of some portion of a shoot derived from an irradiated meristem, it would indicate that approximately three cells in the surface layer of the meristem are responsible for producing the epidermis of that portion of the shoot. Such analysis of sectors representing cell lineages (clones) of genetically marked cells in chimeric individuals is called clonal analysis.

Plant studies. Plant species for which this method has been used to construct fate maps relating to the entire plant or to an organ include maize, sunflower, tobacco, and cotton.

Maize. A fate map of cells in the outermost meristem layer of the dormant maize (corn, *Zea mays*) embryo has been derived from the clonal analysis of plants grown from x-irradiated seed (**Fig. 2**). Sector patterns strongly suggest that cells which will ultimately form the male inflorescence (tassel) are set apart in the embryo meristem from cells which will form the vegetative portion of the plant. Sectors that are visible in the leaves never extend into the tassel; sectors that affect the tassel are limited to the tassel. In

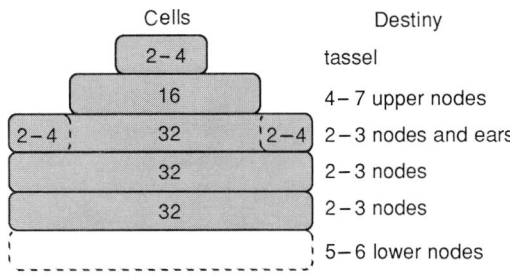

Fig. 2. Simple fate map of the outermost layer of the dormant maize embryo meristem. The map is based on sector patterns generated in the epidermis of plants grown from x-irradiated seed. Anthocyanin production was used as the genetic marker. (*After E. H. Coe, Jr., and M. G. Neuffer, Embryo cells and their destinies in the corn plant, in S. Subteiny and I. M. Sussex, eds., The Clonal Basis of Development, pp. 113–129, Academic Press, 1978*)

addition, since sectors that affect the tassel divide the tassel vertically into halves, thirds, or fourths, it can be inferred that 2-4 cells at the summit of the apex give rise exclusively to the tassel. Sector patterns also suggest that there are several subpopulations of cells that give rise to successive portions of the vegetative part of the plant. Above the 5-6 lowest nodes, which are already present in the dormant seed embryo, there appear to be three successive vertical classes of sectors, each extending 2-3 nodes up the shoot, and a fourth class extending 4-7 nodes. From these sector patterns and from the mean width of sectors in each class, it is apparent that there must be four tiers of cells in the embryo meristem with approximately 32 cells in circumference in each of the first three tiers, and 16 cells in the fourth. Ear branches (female reproductive parts) arise from subsets of 2-4 cells located at opposite sides of the third 32-cell tier. These results indicate that the cell lineage patterns in the maize shoot are much more regular than had been previously thought.

Sunflower. A similar clonal analysis of the sunflower (*Helianthus annuus*) embryo meristem shows somewhat less regularity in classes of shoot sectors. Because there is a wider variation in the extent of sectors, as well as a greater overlap of sectors, subpopulations of cells within the embryo meristem must be less clearly defined in sunflower than in maize. There appear to be approximately three or four imprecisely defined tiers of cells in each of the three layers of the embryo meristem, each of which gives rise to a highly variable number of nodes, the uppermost giving rise to some vegetative nodes, as well as to the terminal inflorescence. This investigation concerns the fate of cells in the embryonic apical meristem. Meristems are being investigated at various ages after germination to establish whether the fate of cells in the sunflower meristem becomes more precisely defined in later stages of development, particularly whether the fate of cells at the summit of the meristem will become limited to eventual formation

of the inflorescence at some later time during vegetative development.

Tobacco. The fate of cells in the developing leaf primordium of tobacco (*Nicotiana tabacum*) also has been studied in detail by using clonal analysis. Results of the study make it necessary to reevaluate the traditional view of leaf development, which holds that the entire radially symmetrical leaf primordium is derived from a cell or small group of cells located at the apex of the developing leaf axis, with the lamina (blade) of the leaf then developing through the activity of marginal meristems along the two lateral edges of the leaf axis. Analysis of sectors generated at successive stages of leaf development are not consistent with this view. **Figure 3***a* shows a model of the type of sector (cell lineage) predicted by this traditional view of early leaf development. Sectors radiate from the apex of the leaf axis and occupy a constant proportion of the circumference of the axis. Figure 3*b* is a model of the pattern of sectors actually generated in the leaf axis when the meristem is irradiated just before the leaf is initiated. Only a small number of sectors extend to the tip of the leaf, sectors are more or less parallel to the leaf axis, and sectors do not change in size in proportion to the circumference of the axis. Therefore, the leaf primordium must arise from a "line source" of cells, each cell of which contributes to a different longitudinal section of the axis, rather than from a small group of apical cells that give rise to the entire axis.

The data from clonal analysis also establish that the expanding lamina are not produced by the activity of a marginal meristem. If the cells along the margins of the axis were fated to give rise to the entire lamina, then sectors generated by irradiating the leaf primordium just prior to formation of the lamina would extend from the margin to the axis (midrib) of the blade. Instead, sectors either abut the leaf margin or abut the midrib; or they occur isolated in the region between the margin and the midrib. Thus the actual sector pattern shows that at least three files of

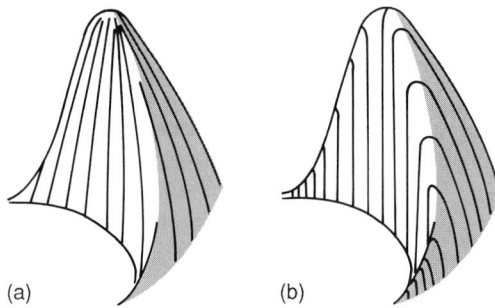

(a)   (b)

Fig. 3. Two models for the fate of cells in the formation of the axis of a tobacco leaf. Lines represent boundaries of cell lineages. (*a*) Pattern of cell lineages predicted by the traditional view that the leaf axis is derived from a group of apical initial cells. (*b*) Pattern of cell lineages that actually occurs as shown by clonal analysis. (*After R. S. Poethig and I. M. Sussex, The cellular parameters of leaf development in tobacco: A clonal analysis, Planta, 165:170–184, 1985*)

cells on each side of the axis must give rise to the lamina. Clearly the traditional view of cell fate in the developing leaf primordium must be reconsidered.

Cotton. A different type of pigmentation chimera has recently been used to map the organizing shoot apex of the cotton (*Gossypium*) embryo. In certain interspecific crosses, chimeric seedlings are produced that can have separate haploid clonal lines of either parental genotype or the expected diploid genotype, or both. When a chimera is generated so early in development, the clones of differing genotype are very large. If a structure is derived from two or more cells or cell clones, the structure can be made up of both genotypes, with a distinct phenotype boundary visually separating the cell types. But if a structure derives from a single cell or from a group of cells belonging to a single clone, then the structure will be entirely of a single phenotype (for example, all green or all yellow, never both).

The number of cells or cell clones from which the structure derives can be deduced from the sector patterns. The chimeric cotton seedlings produce plants in which sectors occupy two-thirds or one-third of the circumference of the shoot—evidence that at the time of its organization the meristem must arise from the lineages of three cells. Sector patterns in the cotyledons reveal eight, roughly fan-shaped, clones of cells—evidence for a more or less linear arrangement of eight cells at each side of the embryo that will give rise to the cell lineages from which the cotyledons develop. Further evidence from sector patterns suggests that the first and second leaves ultimately arise from the lineages of two cells and one cell, respectively, not clonally related to the three cells that will give rise to the meristem. This means that the developmental origin of the first two leaves differs fundamentally from that of all subsequent leaves and aerial structures, which are formed by the apical meristem. Previously it had been thought that all true leaves are formed by the apical meristem.

Utility and limitation. Mapping the fate of cells in the meristem, leaf primordium, or developing embryo indicates the part of the developing plant or organ to which a cell or group of cells will contribute in the normal course of development. This mapping reveals only what a cell or group of cells will become by virute of its position at a particular stage in development. It does not indicate a developmental commitment of cells to a particular fate, nor does it indicate a functional significance of cell lineages in the developing organ or organism. Nevertheless, understanding the cell lineage patterns in the developing organism clearly does provide information that necessitates the reevaluation of some long-accepted ideas about plant development. Moreover, fate maps provide a basis for asking important questions about the developmental commitment (de-

termination) of cells present at critical stages of development.

For background information SEE APICAL DOMINANCE; APICAL MERISTEM; FATE MAPS (EMBRYOLOGY); PLANT GROWTH in the McGraw-Hill Encyclopedia of Science and Technology.

Dorothy E. Jegla

Bibliograpy. M. L. Christianson, Fate map of the organizing shoot apex in *Gossypium*, *Amer. J. Bot.*, 73:947–958, 1986; E. H. Coe, Jr., and M. G. Neuffer, Embryo cells and their destinies in the corn plant, in S. Subtelny and I. M. Sussex (eds.), *The Clonal Basis of Development*, pp. 113–129, 1978; D. E. Jegla, Clonal analysis of meristem development, in J. G. Atherton (ed.), *The Manipulation of Flowering*, 45th Nottingham (England) Easter School in Agricultural Science, pp. 101–108, 1987; R. S. Poethig and I. M. Sussex, The cellular parameters of leaf development in tobacco: A clonal analysis, *Planta*, 165:170–184, 1985.

Feeding mechanisms (invertebrate)

Many animals feed on aquatic organisms much smaller than themselves that are present in low concentration in the water. Such animals may capture their food by a process of filtration. This process involves three components: the movement of large volumes of particle-laden water, the interception of particles from the water, and the transfer of these particles to the mouth for digestion. Different mechanisms are available in different animals for each of these components, but only in a few cases have adequate quantitative descriptions been produced for the development of physical models of the mechanisms. The bivalve mollusks are large and important filter feeders; data on the structures and physical parameters of filter feeding in the sea mussel *Mytilus* are now adequate for assessment of proposed mechanisms of filtration and water propulsion, and for the development of improved models.

In general, particles may be removed from the water by mechanical sieving, in which case only particles larger than the pores of the filter are retained. Alternatively, components of the filter may be adhesive (usually with sticky mucus), in which case a proportion of particles smaller than the filter mesh may be intercepted by sticky strands and may also be retained. Particles caught on an adhesive mucous net filter are eaten when the whole net is ingested. In many cases, particles concentrated from the water by mechanical sieving are also brought into contact with strands of mucus which are ingested complete with adherent food particles; in other cases, particles may be accumulated and swallowed directly.

Bivalve mollusks feed on the abundant microscopic algae and bacteria that exist in fresh and salt water. The size of food particles to be filtered

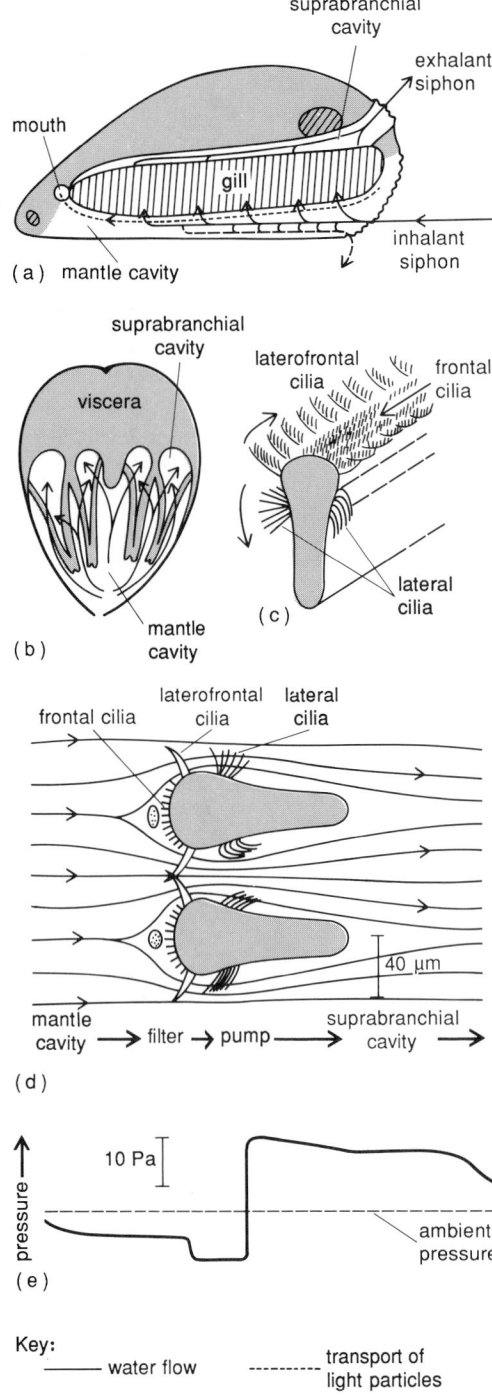

Fig. 1. Gill structure in *Mytilus*. (*a*) Position of the gill and pathways of water flow and transport of heavy and light particles, seen after removal of the left shell. (*b*) Cross section through the body showing the direction of water flow between gill filaments. (*c*) Arrangement of cilia on a gill filament; the frontal surface always faces the mantle cavity; the arrows indicate the direction of effective strokes of cilia. (*d*) Flow of water between adjacent filaments in relation to the position of filter and pump components. (*e*) Changes in pressure associated with components shown in *d*.

may be as small as a few micrometers for many planktonic algae or less than half a micrometer for marine bacteria. In bivalves the filter that sieves out particles of these dimensions is formed from arrays of cilia, but setal bristles or mucous strands are used by other animals. The pore size of the sieve must be less than the diameter of the food particles. Because of the viscosity of water, considerable pressures are required to move water through such small pores at a reasonable rate. In hydrodynamic terms, the system operates at a very low Reynolds number. The need to pass large volumes of water also requires that the filter bars should be thin enough to provide a large pore area, but the filter bars must be strong enough to resist the water pressures required for filtration.

Organization of feeding structures. There is a large space between the two shells of a mussel called the mantle cavity (**Fig. 1***a* and *b*). Sea water is circulated through the cavity by cilia on the gills (ctenidia) when the shells are open. The two gills comprise four longitudinal sheets which are made of slender gill filaments, each folded at the center of its length; a cross section of the body thus shows two W-shaped curtains separating the larger, lower mantle cavity from a smaller, upper suprabranchial space (Fig. 1*b*). Lateral cilia on the gill filaments propel water between adjacent filaments from the inhalant to the exhalant part of the mantle cavity (Fig. 1*c* and *d*). These cilia form a pump which draws water into the mantle cavity through a wide inhalant siphon and propels it out of the body in a jet from the narrow exhalant siphon (Fig. 1*a*). This current of water carries particles into the mantle cavity, only some of which are ingested. Large, dense particles carried into the mantle cavity in the fast inhalant current may settle out by gravitation as the water speed falls within the cavity; such particles are removed by a mucociliary cleansing system (Fig. 1*a*). Smaller, lighter particles, which are carried with the water flow in all but very slow streams, are intercepted by the laterofrontal cilia as the water flow enters the space between adjacent filaments of the gill. The pumping action of the lateral cilia is required not only to generate the pressure to draw water through the laterofrontal ciliary filter but also to overcome viscous drag in the inhalant siphon, between the gill filaments, and in the exhalant siphon, as well as to drive the exhalant jet.

Data indicate that a mussel 35 mm (1.5 in.) long might have a total gill area of about 800 mm^2 (1 in.^{2}). About 200–1000 mm^3 (0.01–0.06 in.^{3}) of water might be pumped through such a mussel every second. Since about half the gill area is interfilament space, the rate of flow between the filaments must average 0.5–2.5 mm s^{-1} (0.02–0.1 in. s^{-1}). Water pressures associated with this flow have been measured and found to be about 5 pascals (7 × 10^{-4} lbf/in.^{2}) below ambient in the mantle cavity and about 12 pascals (1.7 × 10^{-3} lbf/in.^{2}) above ambient in the suprabranchial cavity. The former pressure draws water into the mantle cavity, and the latter drives water out of the suprabranchial cavity. As the pressure profile in Fig. 1*e* shows, there must be a pressure drop across the laterofrontal filter, estimated as about 4 Pa (6×10^{-4} lbf/in.^{2}) at a flow rate of 0.8 mm s^{-1} (0.03 in. s^{-1}). The pressure generated by the lateral ciliary pump must therefore be of the order of 20 Pa (3×10^{-3} lbf/in.^{2}). It has been found that the ciliary pump continues to operate if substantial back pressures are applied, and it is estimated that pump pressures of up to 40 Pa (6×10^{-3} lbf/in.^{2}) can be produced.

Gill pump. The lateral cilia are about 15 micrometers long and occur closely packed in a band about 10 micrometers deep along either side of each filament. They beat regularly at frequencies in the range 10–20 Hz, and are coordinated to form very regular waves, each about 10 μm long (**Fig. 2**a–c), called metachronal waves because the cilia of which they are composed show a small progressive phase difference along the length of each wave. Examination of the waveform in frontal (Fig. 2a and b) and lateral (Fig. 2c) views allows the beat pattern and construction of the wave to be elucidated. Cilia are extended during their effective strokes, and during the recovery strokes each curved cilium bends toward the left in the views seen in Fig. 2c and d. About 600 cilia make up each metachronal wave, with, at any instant, about 150 cilia at different stages of the effective stroke, and the remainder in the recovery stroke. The metachronal waves move to the left when the gill filament is seen in side view with effective strokes downward (Fig. 2c), so that the waves at opposite sides of each interfilament space move in opposite directions (Fig. 2a).

At the level of the lateral cilia the space between adjacent filaments is about 40 μm wide. Clusters of cilia performing their effective stroke project from either side of this space at about 10 μm intervals and to a distance of about 15 μm. Mounds of cilia performing recovery strokes project to a distance of about 10 μm between these clusters, and these mounds restrict the width of the space through which water can be propelled by effective strokes to about 20 μm, thereby doubling the average water flow to speeds of 1–5 mm s^{-1} (0.04–0.2 in. s^{-1}). The tip speed of cilia during their effective stroke is about 4 mm s^{-1} (0.2 in. s^{-1}) at a frequency of 20 Hz. In order to generate the high flow rates and sustain the high pressure differences that have been reported, this ciliary pump must be operating with high efficiency, beyond that predicted by present models.

Filter. Each laterofrontal cirrus is a compound cilium composed of 40–50 cilia (**Fig. 3**a) which are joined at their bases but diverge at their tips to form a feather-shaped structure (Fig. 3b). Such cirri occur at intervals of about 3 μm in single rows along either side of each filament. They are about 25 μm long and 7 μm wide, and adjacent cirri beat alternately at about 7 Hz. At any instant, half of the cirri form a net across the interfilament space (Fig. 3a) and the other half

are bent toward the frontal face of the filament. The ciliary branches of the cirri form a mesh with pores about 1 μm wide which extends across most of the interfilament space for most of the time. Measurements of particle retention by mussel gills suggest that particles down to about 1 μm can be retained, and more than half of 2-μm particles are caught (Fig. 3c). The small pore size explains not only the retention of small particles but also the high pressures required to pass the observed water flows. Most workers agree that these cirri act as filters and that during their effective strokes they carry the retained particles to the frontal face of the gill. An alternative

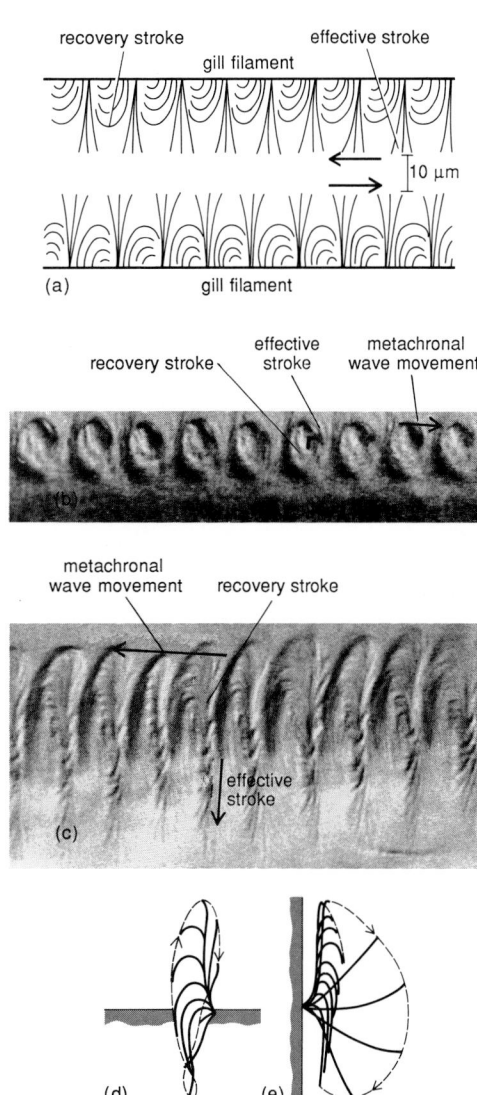

Fig. 2. Gill pump consisting of lateral cilia. All parts of the figure are to same scale, indicated in a. (a) Form and movement (arrows) of metachronal waves at either side of an interfilament space, seen from the frontal surface, with lateral cilia propelling water into the page. (b) Light micrograph of a frontal view of a band of active lateral cilia. (c) Light micrograph of a lateral view of a band of active lateral cilia, with effective stroke, and thus water flow, downward. (d) Sequence of ciliary profiles during a beat cycle seen from above the cell surface, as in c. (e) Sequence of ciliary profiles during beat cycle in d, as seen by an observer looking from the left in e. Effective strokes and water flow are again downward.

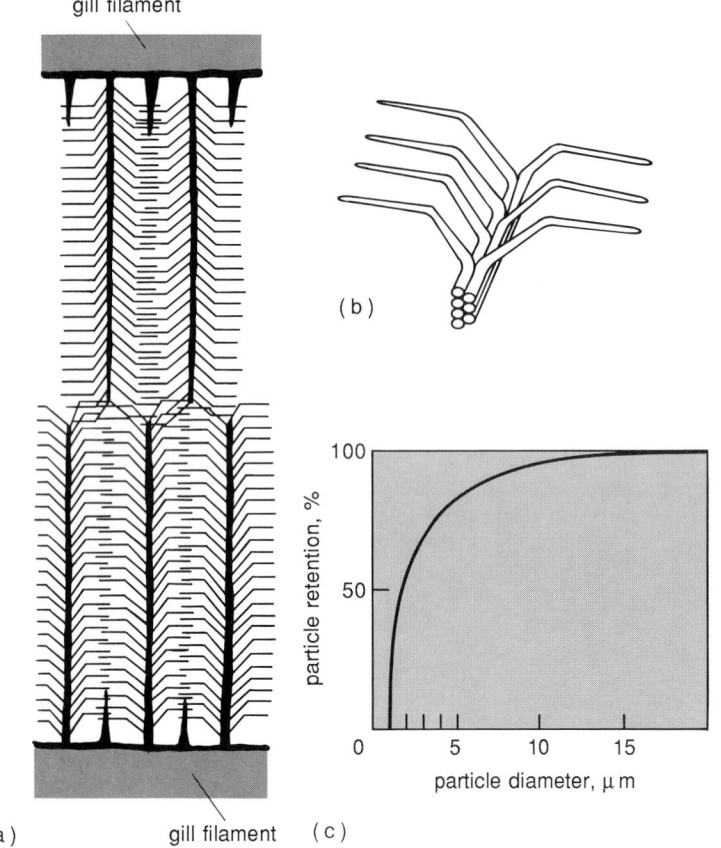

gill filament

(b)

100

particle retention, %

50

0   5   10   15
particle diameter, μm

(a)   gill filament   (c)

Fig. 3. Laterofrontal ciliary filter. (*a*) Alternate laterofrontal cirri seen in frontal view; intervening cirri are bent out of focus by their beating. (*b*) Structure of part of a laterofrontal cirrus. (*c*) Percentage retention of particles of different sizes by *Mytilus*.

hydrodynamic explanation for separation of particles from the water flow has been proposed, but this remains to be fully evaluated.

Trapping and transport of particles. Mucous secretion occurs along the frontal surface of gill filaments, and most workers believe that the particles passed to the frontal surface by the laterofrontal cirri are trapped there by adhesion to a strand of mucus. This is propelled by short frontal cilia to the gill margin, where other cilia in a food groove transport the mucous strands toward the mouth (Fig. 1*a*). The extent of participation of mucus in the trapping and transport of particles has been questioned, and it has been suggested that experimental disturbance may stimulate mucous secretion that is not required for normal functioning. This process again requires further study.

For background information *SEE* B*IVALVIA*; F*EEDING MECHANISMS* (*INVERTEBRATE*); L*AMELLIBRAN-CHIA*; M*OLLUSCA* in the McGraw-Hill Encyclopedia of Science and Technology.

M. A. Sleigh

Bibliography. H. D. Jones and J. R. Allen, Inhalant and exhalant pressures in *Mytilus edulis* (L.) and *Cerastoderma edule* (L.), *J. Exp. Mar. Biol. Ecol.*, 98:231–240, 1986; C. B. Jorgensen, A hydromechanical principle for particle retention in *Mytilus edulis* and other ciliary suspension feeders, *Mar. Biol.*, 61:272–282, 1981; N. R. Silvester and M. A. Sleigh, Hydrodynamic aspects of particle capture by *Mytilus*, *J. Mar. Biol. Ass. U.K.*, 64:859–879, 1984.

Fertilizer

Recent advances in fertilizer use include utilization of organic materials in the humid tropics and research involving the application and management of inorganic fertilizers in low-input farming.

ORGANIC MATERIALS

Advances in the use of animal wastes and plant residues as fertilizers in the humid tropics vary widely among regions. In tropical and subtropical Asia, major emphasis is placed on improving the efficiency of traditional systems. Biogas digesters have achieved notable success, although not all the problems associated with their use have been resolved. In Africa and Central and South America, most efforts involve less complex technologies. Recent research shows that traditional techniques are often difficult to use because animals are not penned, labor demands are excessive, soil conditions differ from those where use of organics has been more fully studied, and cultural acceptance of these practices is low. Most research today is aimed at overcoming these problems.

Animal wastes. The most advanced, and probably some of the longest-standing, examples of the use of organic materials as fertilizers are found in Southeast Asia, including subtropical China. One notable aspect of the systems used there is the high level of integration between plant and animal production. Animals consume crops, crop residues, and crop by-products, and in turn they play a critical role in crop production by supplying organic fertilizers.

Such systems are often complex. For example, swine and other small animal species may be housed over fishponds, their manure forming a feed supply for the fish. When the ponds are drained for fish harvest, the accumulated muck is collected and applied directly to crops or used in making compost.

In tropical Africa, organic materials are also used in some areas, but generally much less intensively than in Asia. In Malawi (southeast Africa), for example, farmers with animals collect and apply dung. In other parts of Africa, domestic animals are housed at night in corrals. The corrals are placed where intensive crop production is planned for the following cropping season in order to take advantage of nightly manure production.

In Central and South America, even these rudimentary systems are rarely seen. Two factors probably account for this. First, large domestic animals, with the exception of the American cameloids in the upper Andes, were nonexistent in the New World tropics prior to European colonization. Achieving intensive use of organic fertilizers without animal wastes is very difficult. Second, population pressure throughout this region has been low historically. With access to large areas for cultivation, including the ability to leave areas fallow for 10 or 20 years or even longer, agriculturists were not highly concerned with maintaining soil fertility.

The major issues which need to be addressed by researchers working with the use of organic materials as fertilizers in the humid tropics flow directly from the status of their use in different regions of the world. In Asia, recent research has emphasized improving the efficiency of traditional systems. In Central and South America, and to a lesser degree in Africa, the problem remains of finding ways to use organic materials that will meet the cultural, economic, and labor constraints of societies that have not traditionally used such practices. For example, the norm in much of Asia is the consumption of fish fed on human excreta. In Central America, the fish would be considered unfit for human consumption by most residents.

Digesters. Notable success has been achieved in Asia, particularly in India, in the introduction of biogas digesters. Both animal and plant wastes can be used in the digesters, although a very high C/N (carbon to nitrogen) ratio must be maintained. The organic materials are decomposed by anaerobic microorganisms, which produce methane gas as a by-product of respiration.

The great advantage of methane digesters is that they permit use of wastes both as fuel and as fertilizer. Carbon and nitrogen are removed and used as fuel during the process. However, nitrogen, phosphorus, potassium, and other essential plant nutrients remain in the decomposed sludge. In fact, their relative concentration increases because of the removal of other elements.

Problems in using the sludge have largely stemmed from two sources. First, sludge is wet and difficult to transport. Second, as it dries or is stored, it is susceptible to loss of nitrogen as a gas. Improved methods of storing and preparing the material for field application would represent significant advances in the use of this technology.

Plant residues. In South America, some recent effort has been devoted to using plant materials for composting. A purely plant-based system would, of course, eliminate the need for manure. Yields comparable to those obtained with chemical fertilizers can be achieved, but only at extremely high rates of application of the decomposed plant material. Also, labor demands are very high.

In Central America, efforts to reduce labor have been made by attempting to eliminate the necessity to turn compost heaps. Adequate aer-

Fig. 1. A compost heap interlaced with bamboo chimneys greatly increases aeration and can make turning the heap unnecessary.

ation, sufficient to achieve this goal, was obtained by interlacing an aboveground compost heap with bamboo from which the middle was removed. Holes were made in the bamboo stalks every 18 to 24 in. (46 to 61 cm) [**Fig. 1**]. Stable C/N ratios of 25/1 to 30/1, considered an adequate indicator of stabilization of the decomposition process, were achieved in 60 days.

Practical problems. In Africa and Central and South America, most emphasis today is placed on introducing well-known methods of using animal wastes and crop residues to local populations. The practical problems that must be overcome stem from several factors. First, confinement of animals is often not the norm. Collecting manure, by far the most concentrated organic source of plant nutrients, becomes very labor-intensive and often impossible where animals roam free.

Second, virtually all methods of using organic fertilizers are labor-intensive. While overpopulation may be widely viewed as a major problem in many countries, family, kin, or social group labor is often not extensively available in farm communities.

Third, the cultural heritage which encourages the use of such practices is often virtually nonexistent. In Central America, for example, using compost or other organic fertilizers is often regarded as primitive, if not demeaning.

Technical problems. Three major technical problems have remained unresolved in the use of organics in humid tropical areas. First, nutrient losses by leaching, particularly of nitrogen and potassium, are often very high when materials are stored or during composting. For example, after a few weeks of storage manure may contain less than 1% nitrogen under high rainfall conditions. Pit systems for storage or for composting, often highly recommended in the literature, are not effective. They encourage nutrient loss. Overground systems should be used. An enclosed structure is preferable but adds considerably to the cost of the technology.

Second, plant deficiency of phosphorus is the greatest single factor limiting yields on many tropical soils, both on highly weathered soils and on some volcanic soils. Virtually all organic fertilizers, with the exception of bone meal or fishmeal, are low in phosphorus. This aspect of organic fertilizers is rarely addressed in literature describing their properties and uses in midlatitude areas, where phosphorus deficiency is less common.

Third, many of the long-term beneficial effects of applying organic materials, so frequently described in the literature, appear to be inoperative in the humid tropics. Data from Costa Rica show that soil organic matter levels reverted to preapplication levels in 90 days at three sites. This is explainable by the high level of biological activity in the tropical environment. Very few studies have examined in detail the effects of organic fertilization on physical and chemical soil properties in the tropics. These data would indicate that long-term soil improvement may be difficult to achieve in many areas.

Finally, the practice of capping the compost heap with soil or of incorporating soil into organic fertilizers should be avoided in many tropical areas. Soils that are high in iron and aluminum are unsuitable for this purpose. These elements interact with phosphorus in the organic material to make that element relatively unavailable to plants once the organic material is applied as a fertilizer. Many soils of the humid tropics are high in iron and aluminum.

Future research. Three areas hold special promise for future research in the use of organic fertilizers in the humid tropics. The first is further examination of the potential of applying mixes of chemical and organic fertilizers. The use of chemically derived phosphorus should receive attention. Adding phosphorus to compost during decomposition did not prove successful in Costa Rica, but addition at the time of field application was successful. Research in this area could help overcome a significant drawback to the use of organics in much of the humid tropics.

Second, all systems which reduce labor and permit the user to derive more than one benefit from the use of organics should receive high priority. For example, on steep hillsides it may be possible to use leguminous plants as retainers for terraces. Also, plant trimmings can spread on the ground to supply nutrients, aid in moisture retention, and reduce weed competition.

Finally, increased attention to the use of human wastes is needed both to take advantage of their potential as a fertilizer and to reduce health risks. Human waste treatment is rarely practiced in rural Africa or Latin America. Overcoming cultural re-

Fig. 2. A self-composting latrine requires a lined tank, a vent, and easy access to the tank.

straints is a major problem. The use of self-composting latrines may offer the highest potential because this technology is relatively low in labor demand, results in an easily handled product, and overcomes problems of disease and parasite transmission. However, the cost of construction can be relatively high. The reception tank must be lined, be vented, and provide easy access (**Fig. 2**). At least two latrines are required to provide facilities on a continual basis since each must be closed for use during the decomposition process.

Only recently has modern science turned its attention to the use of organic fertilizers in the humid tropics. After World War II, with the widespread use of chemical fertilizer, scientists paid relatively little attention to improving use of organics in the humid tropics or elsewhere. With rising costs of all petroleum-based chemicals, renewed interest in organics has developed. To date, science has not made great strides, but the problems that have come to light do indicate fruitful areas for future advances.   *M. E. Swisher*

INORGANIC MATERIALS

Fertilizer constitutes one of the inputs critical to the achievement of crop yield levels sufficient to satisfy the food needs of a growing world population. It has been estimated, for example, that approximately 50% of the grain yield increase which has been achieved in recent years is attributable to the increase in the use of inorganic fertilizer. Still, many farmers in developing countries do not have access to the resources necessary to optimize crop production from either an agronomic or economic point of view. Considerable emphasis has been placed on research to identify the best ways to provide crop requirements for nutrients consistent with the constraints faced by low-income farmers. The use of indigenous inorganic resources represents a method through which developing countries can provide, in many cases, fertilizers to farmers who would otherwise be unable to obtain them. Locally available phosphate rock is now a well-recognized means through which the cost of inorganic fertilizer can be reduced.

Phosphate rock approach. Phosphorus is only one of the essential nutrients required for crop growth, but it is one of the most expensive for farmers in developing countries to obtain. Phosphate rock is used as the basic raw material for most commercial phosphorus-containing fertilizers. While the majority of the phosphate rock is mined in the United States, the Soviet Union, and North Africa, numerous deposits are also located throughout the developing world. Most of these deposits have not been used commercially because many are considered to be problem ores in that their chemical properties are not suited to conventional process technology or the quantity of easily extractable ore is not sufficient to warrant the investment for large-scale production.

Instead of using these resources for processed fertilizer, many developing countries are now considering the option of simply grinding the rock to a small particle size for direct application to the soil. This approach represents the quickest and cheapest method of providing phosphorus to the agricultural sector while minimizing the loss of foreign exchange and providing local employment opportunities.

Phosphate rock effectiveness. While the simplicity of using finely ground phosphate rock is inviting to many developing countries in which both low-input farming systems and indigenous deposits of phosphate rock are encountered, the degree to which this resource can replace or supplement the use of processed fertilizer depends upon a number of agronomic and economic factors. From an agronomic point of view, it must be recognized that the chemical processes to which phosphate rock is subjected in conventional fertilizer production convert the rock from sparingly soluble apatite to highly soluble forms, such as monocalcium phosphate; these are more readily available for plant utilization. The actual agronomic effectiveness of the finely ground phosphate rock is, therefore, generally less than that of processed fertilizer and is also more difficult to predict. However, the lower cost can compensate in many cases for some reduction in agronomic effectiveness.

The factors which strongly influence the performance of phosphate rock relative to that of processed fertilizer include the mineralogy of the rock, the soil properties, and fertilizer management.

Mineralogy. This determines the chemical reactivity of the apatite mineral in the rock. The reactivity of the phosphate rock varies from deposit to deposit and even at different locations within the same deposit. The variability is related to the degree of carbonate substitution for phosphate in the mineral structure. Apatite minerals with such substitution are classified as francolites; as the degree of substitution increases, the chemical stability of the mineral decreases and the potential rate of dissolution increases. There is an extreme difference in the capability of various Latin American phosphate rocks to correct phosphorus deficiency when the rocks are applied to the same soil. This difference is the result of the extreme differences in chemical reactivity among these sources; the fact that one source of phosphate rock is agronomically ineffective does not mean that other sources will be so.

Soil properties. This factor determines the degree to which the potential rate of phosphate rock dissolution will be attained. Dissolution of the phosphate rock in soil solution, using fluorapatite as an example, may be simply expressed by the reaction below. From the mass action law, it can

$$Ca_{10}(PO_4)_6F_2 + 12H \rightarrow 10Ca^{2+} + 6H_2PO_4^- + 2F^-$$

be seen clearly that the dissolution of phosphate rock in soil solution would be favored under conditions of low soil pH, low exchangeable calcium, and low concentration of solution phosphorus. These are conditions that are often found in developing countries, but the selection of soils to which phosphate rock is applied becomes critical, since the phosphate rock would not be expected to dissolve rapidly after application unless these conditions are met.

An additional soil property that has been found to influence strongly the relative agronomic effectiveness of phosphate rock is the phosphorus-retention capacity of the soil. As phosphorus enters the soil solution, it reacts with iron and aluminum oxides in acid soil, which retains the phosphorus in plant-unavailable forms. This process results in counteracting phosphate rock effectiveness. On the positive side, the removal of phosphorus from the soil solution is a chemical driving force which increases the rate at which the phosphate rock dissolves. Despite this, phosphate rock is incapable of supplying high concentrations of phosphorus for even a short period of time in the soil solution. This can be observed with soluble fertilizer, and the net result is a relatively lower effectiveness of phosphate rock in those soils exhibiting the highest phosphorus-retention capacity. In these soils the indigenous phosphate rock can still be used, but it may be necessary to apply the rock in a mixture containing a more soluble phosphorus fertilizer or with a partial acid treatment. Fortunately, many acid soils in the tropics are not as high in phosphorus-retention capacity as had been originally thought.

This is especially true in the savannahs of sub-Saharan Africa and Latin America.

Management. The constraints of mineralogy and soil properties require some modification in fertilizer management for utilization of natural phosphate rock to its maximum efficiency. Four of the most important management considerations are selection of appropriate conditions, rate of application, method of application, and time of application. Phosphate rock should be the phosphorus fertilizer of choice only when the soil conditions described above are present. Also, even with these conditions, achievement of maximum crop yields requires use of the phosphate rock in combination with additional fertilizers or practices necessary to eliminate other impediments to plant growth. Rate of application should be selected on the basis of soil test (if available) and will generally be higher than the rates for soluble fertilizer. This requirement is still consistent with low-input agriculture, however, because of the low cost of the material.

The chemical reactions leading to the dissolution of phosphate rock in the soil suggest that maximum availability of phosphorus from phosphate rock can be obtained through broadcast application and incorporation into the soil. This maximization of contact between fertilizer and soil increases the probability of both dissolution of the phosphate rock and interception by plant roots. This management has been confirmed by many single-crop experiments. It must be kept in mind, however, that release of plant-available phosphorus from both the phosphate rock and the reaction products in the soil continues for a long time. It is now suspected that phosphate rock can be used effectively with point placement in a long-term program if dictated by labor constraints or erosion hazards. While soil response to phosphate rock in the first year of the program may be low, this can be compensated by the addition of a small quantity of soluble phosphorus. The availability of phosphorus from subsequent phosphate rock applications should increase year by year, however, because of the redistribution of the phosphate rock during tillage operations and the high residual effectiveness of the phosphate rock due to its slow but continual dissolution.

Finely ground, natural phosphate rock is a low-cost, inorganic fertilizer adaptable to many low-input farming systems in developing countries. However, its effectiveness is not universal, and it should be recommended only where the rock properties and soil properties are suitable. Under those conditions and with proper management, low-income farmers have the opportunity to increase production, improve their standard of living, and with time, move into a position where the utilization of more costly but more efficient fertilizer sources will become feasible.

For background information SEE AGRICULTURAL SOIL AND CROP PRACTICES; FERTILIZER; FERTILIZING in the McGraw-Hill Encyclopedia of Science and Technology.

Lawrence L. Hammond

Bibliography. Food and Agricultural Organization, *Organic Materials as Fertilizers*, Soils Bull. 27, 1975; L. L. Hammond and L. A. Leon, Agronomic effectiveness of natural and altered phosphate rocks from Latin America, *Proceedings of the 3d International Congress on P Compounds*, Institut Mandial du Phosphate, Brussels, pp. 503–518, 1984; L. A. Leon, W. E. Fenster, and L. L. Hammond, Agronomic potential of eleven phosphate rocks from Brazil, Colombia, Peru, and Venezuela, *Soil Sci. Soc. Amer. J.*, 50:798–802, 1986; A. U. Mokwunye and P. L. G. Vlek (eds.), *Management of Nitrogen and Phosphorus Fertilizer in Sub-Saharan Africa*, p. 229, 1986; National Research Council, *Food, Fuel and Fertilizer from Organic Wastes*, BOSTID Rep. 31, National Academy of Sciences, 1981; M. E. Swisher, An investigation of the potential for the use of animal wastes on small, mixed farms in Costa Rica, unpublished Ph.D. dissertation, Department of Geography, University of Florida, 1982.

Flower

Flowering is a complex, multiple-phase process about which relatively little is known. The process can be divided into at least five sequential phases: flower induction, differentiation of flower parts, flower opening, pollination and fertilization, and flower senescence.

Flower induction is widely described in textbooks since much of the work on this phase was carried out early in the twentieth century. Many species use photoperiodic mechanisms to initiate flowering. A better term might be noctoperiodic, since night length is the critical environmental factor. Some species use temperature-sensitive induction mechanisms, and other species seem to flower without regard to environmental cues. A universal flower initiation hormone, florigen, was proposed many years ago; while applied growth regulators were found to initiate flowering in some species, no truly universal florigen has been isolated.

The differentiation of flower parts has also been linked to plant growth regulators over several years. In species with rather variable sex expression (hemp and squash, for example), the percentage of male flowers in a population may be increased by enhanced gibberellin access. Enhancing the relative balance of auxins, ethylene, and cytokinins stimulates differentiation of female flowers. These manipulations may be made by either surgical excision of growth regulator sources or applications of plant growth regulators.

Flower-opening processes have been examined cursorily in the past, but recent advances have been more comprehensive. The process may be divided into growth steps (corolla expansion and unfolding, filament elongation, style elongation, and so forth) and nongrowth steps (nectar, fragrance, and pigment production). The mechanisms for growth steps are now being rapidly elucidated. Models of flower opening have been developed recently for *Gaillardia* (a member of the aster family having both ray and disc flowers in each inflorescence) and *Ipomoea* (a member of the morning glory family having solitary bisexual flowers).

Gaillardia. Applied gibberellins induce rapid *Gaillardia* corolla expansion in laboratory preparations, and the concentration of gibberellins in the corolla increases 20-fold at the time of corolla expansion. The exact gibberellins involved have not been identified. The idea that increased gibberellin concentration stimulates corolla expansion is rather old and has been indicated in several species. Other growth regulators have been examined in *Gaillardia*, but they fail to alter growth when applied in laboratory preparations and do not change in concentration in the corolla.

In *Gaillardia* (**Fig. 1**), as in many of the Asteraceae, flat bladelike flowers (rays) are arranged in a circle around small, tubular flowers (discs). Filament elongation in *Gaillardia* disc flowers occurs later than corolla expansion in ray flowers. It is therefore not surprising that a different hormonal signal might elicit filament elongation. Of the several growth regulators, only applied auxin greatly stimulates filament growth in the laboratory preparations, and only auxin increases in concentration with enough magnitude to explain filament growth. The source of the auxin appears to be the anther, and polar (occurring mainly in one direction) basipetal transport of the indoleacetic acid (IAA) from anther to filament (faster than acropetal transport) is apparently the basis for the timing of filament growth.

Style elongation in the *Gaillardia* disc flower occurs a few hours later than filament elongation; yet styles use the same signal (auxin) to initiate elongation. The delay is explained by the spatial separation of the auxin source (pollen) from the style tissues (Fig. 1). The auxin transported down the filament apparently cannot be transported acropetally from the base of the carpel to the style. Instead, style elongation does not occur until after filament elongation and anther dehiscence, when the pollen is deposited directly upon the folded stigma (a distinctive event in Asteraceae). The auxin apparently then diffuses from the pollen into the stigma and is transported basipetally down the style, inducing style growth. Isolated styles elongate in response to applied auxin, and the auxin concentration increases in a correlative manner at the time of style elongation. These correlations cannot be demonstrated

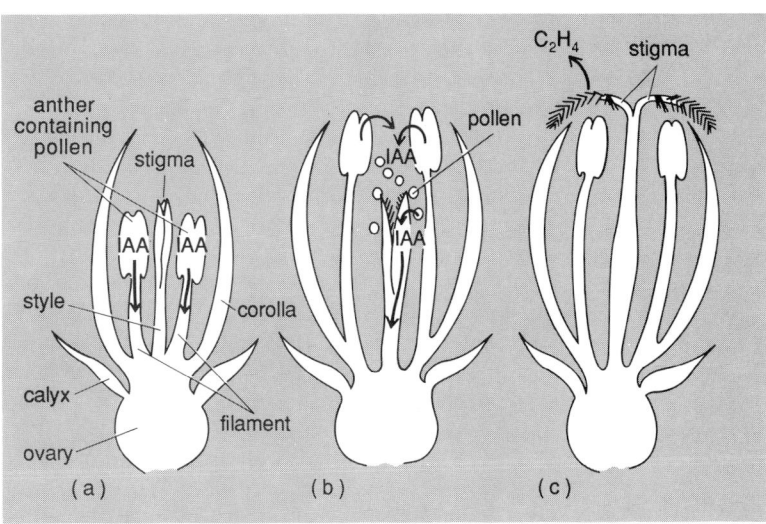

Fig. 1. *Gaillardia* disc flower development. (*a*) Corolla opens at the top. (*b*) Filaments have elongated. (*c*) Style has elongated and pollen is removed by pollinators. IAA = indoleacetic acid. C_2H_4 = ethylene.

for other growth substances.

The stigma of *Gaillardia* does not unfold until the pollen is removed; the lower concentrations of auxin then stimulate unfolding. Unfolding is enhanced by ethylene production, which may be a function of the high auxin concentrations that have passed through these tissues. The timing of increases in ethylene production correlate with stigma unfolding, and isolated stigmas unfold rapidly in response to low concentrations of applied auxin or moderate concentrations of ethylene gas. The receptive surface is thereby exposed only after self-pollen removal; nevertheless, *Gaillardia* is biochemically self-incompatible (that is, its own pollen cannot complete fertilization in that individual). A physical (or structural) self-incompatibility mechanism would fail in a species with compact inflorescences and acropetal maturation within the inflorescence. The physical exclusion mechanism may have been more important to a primitive solitary-flowered ancestor.

The evolution of flower-opening mechanisms cannot be determined from the few existing analyses of largely unrelated species. While this determination is a desirable goal, the tremendous amount of work required for the analyses will prevent any systematic understanding of flower opening for some time.

Ipomoea. Even more recently, flower opening in *Ipomoea nil* (morning glory; **Fig. 2**) has been studied. The responses are quite different from those of *Gaillardia*. Corolla expansion also involves gibberellins; they elicit growth in laboratory preparations and increase in concentration during corolla expansion. Again, the analyses of exact gibberellin forms have yet to be completed. Corolla expansion in morning glory is also inhibited in early stages by ethylene production, but production declines as the gibberellin concentrations increase, permitting corolla expansion. Thus it appears that corolla expansion involves a shifting balance of two growth regulators, one a promoter and the other an inhibitor.

The filaments of morning glory elongate simultaneously with corolla expansion, so it is not surprising that similar signals are used for both steps. Indeed, native gibberellin concentrations increase as the filaments begin to elongate. The production of inhibitive ethylene decreases suddenly at the time of filament growth initiation. Filament elongation in the laboratory is enhanced by application of gibberellin or inhibited ethylene biosynthesis (achieved by application of aminoethoxy vinyl glycine or cobalt ions).

The actual unfolding of the corolla is a very rapid process. It has been determined that the flowers can measure the length of the night, and if the night is longer than 5 h, the corolla unfolds over a 1-h period. If the night is shorter, or if a longer night is interrupted with red light, the corolla fails to unfold. The red light effect is photoreversible with far-red light. Applied abscisic acid elicits ethylene production and corolla unfolding in the laboratory, as does amino cyclopropane carboxylic acid (the ethylene biosynthetic precursor). The abscisic acid effect can be negated by blocking ethylene biosynthesis. The unfolding of the flower is naturally accompanied by large increases in ethylene production. This information indicates the involvement of a biological clock which uses a phytochrome-switched elapsed-time counter to measure the length of the night. Under successful conditions, abscisic acid is produced (synthesized, transported, or released from conjugates), and this step elicits ethylene production. Ethylene production, in turn, brings

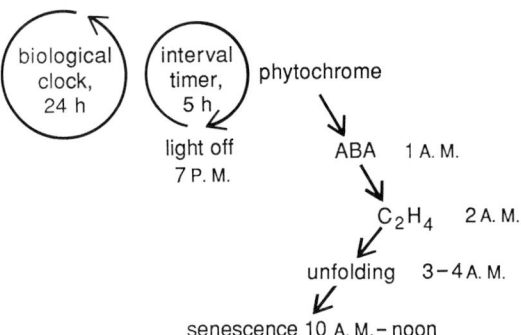

Fig. 3. Mechanism of corolla unfolding in *Ipomoea nil*. ABA = abscisic acid. C_2H_4 = ethylene.

about corolla unfolding at dawn (**Fig. 3**).

The earlier literature indicates that ethylene production in open morning glory flowers brings about corolla senescence by midday. In fact, the unfolding and senescence processes share the ethylene burst as the initiating signal. Unfolding begins with the first 10-fold increase in ethylene production, but senescence processes are not noticed until further 10–100-fold increases are observed. The delay between the events seems to be caused by either the different sensitivity of the two processes or a long lag time for the senescence symptoms (Fig. 3).

The pollination–fertilization events have been structurally studied for many decades. Self-incompatibility is described elsewhere. Recent work has elucidated some of the development of fragrances to attract pollinators, but the controls on timing of nectar, pigment, and fragrance production remain undescribed.

The senescence processes mentioned above have been extensively studied. In morning glory, for example, there are thorough descriptions of the ethylene initiation signal, the responses in respiration-rate increase, hydrolytic enzyme production, corolla wilting and inrolling (inward rolling of the corolla tissue), and processes associated with senescence. Other flowers that have been extensively studied include roses and carnations. In each, ethylene seems to be a senescence initiator.

For background information SEE ABSCISIC ACID; AUXIN; FLOWER; GIBBERELLIN; PLANT GROWTH; PLANT HORMONE in the McGraw-Hill Encyclopedia of Science and Technology.

Ross E. Koning

Bibliography. R. E. Koning, The role of ethylene in corolla unfolding in *Ipomoea nil* (Convolvulaceae), *Amer. J. Bot.*, 73:152–155, 1986; R. E. Koning, The roles of plant hormones in the growth of the corolla of *Gaillardia grandiflora* (Asteraceae) ray flowers, *Amer. J. Bot.*, 71:1–8, 1984; R. E. Koning and M. M. Raab, Parameters of filament elongation in *Ipomoea nil* (Convolvulaceae), *Amer. J. Bot.*, 74:510–516, 1987; Y. Murakami, The role of gibberellins in the growth of floral organs of *Pharbitis nil*, *Plant Cell*

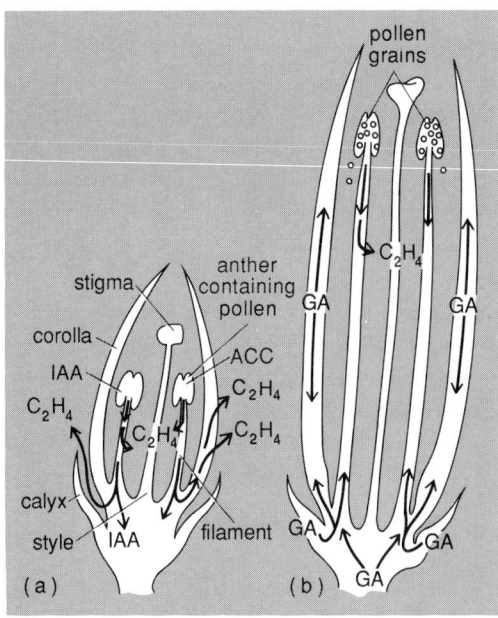

Fig. 2. Growth of corolla and filament of *Ipomoea*. (*a*) Flower bud at about one-fourth final size. (*b*) Flower slightly older. ACC = aminocyclopropane carboxylic acid. GA = gibberellic acid. IAA = indoleacetic acid. C_2H_4 = ethylene.

Physiol., 14:91–102, 1973; M. M. Raab and R. E. Koning, Interacting roles of ethylene and gibberellins in corolla expansion of *Ipomoea nil*, *Amer. J. Bot.*, 74:6, 1987.

Food engineering

This article surveys some recent advances in food engineering that involve novel aspects of drying, solvent extraction, and extrusion cooking.

ACOUSTICALLY AUGMENTED DRYING

The term acoustically augmented drying covers the general area of drying by air movement and heat, with the addition of sonic or ultrasonic sound to the drying environment. Included are processes known as sonic drying, ultrasonic drying, jet drying, pulse drying, and pulse combustion drying. The sound source may be a pulse combustor, a pneumatic horn, or an electronic horn. The pulse combustor, the most common, is a very efficient type of burner; along with heat it produces a very intense sound in the audible range.

This is a new area of technology, and it is still under development. The first reference in the literature to the possibility of using sonic energy for drying appeared in 1936, but little was done with the concept until the late 1970s. Several dryers have been built by using this concept, but the scientific study of the technology is still in the early stages. The technology appears to have wide potential, however, and can be applied to various configurations of dryers, which can handle solutions, slurries, and solid particles. Dryer design is not considerably advanced over that of the original dryers.

Theory. Several basic definitions will be helpful in discussing the concepts of acoustically augmented drying. First, the term drying implies the removal of a liquid from a solid-liquid system. For food use, the liquid is usually water, but it may also be a solvent such as alcohol. While the rest of this discussion will use the term water, it is understood that other solvents may be substituted.

Removal of water from a solid usually involves two different classifications of water. Free water is water that is easily removed from the solid, primarily water on the surface of the solid. If the drying conditions are kept constant, the rate of removal of free water remains constant as water is removed. This is called the constant-rate period.

Bound water, the second classification, is more difficult to remove from a solid than free water. It includes water in the pores of the solid, water adsorbed onto the solid, and water chemically bound to the solid. Even with drying conditions kept constant, the rate of removal of bound water decreases as additional water is removed. This is termed the falling-rate period.

The theoretical basis for acoustically aug-mented drying is not fully understood. Some basic relationships have been shown to hold, however. These involve Dalton's law of additive pressures of gases in a mixture. The simplest mode of drying is evaporation of the water at the interface between free water and the surrounding air. For evaporation during the constant-rate period, Dalton's law approximates the evaporation rate, as shown in the equation below, where dm/dt is the mass rate of evaporation, k is the

$$dm/dt = kA\,(P - p)/H$$

rate constant, A is the area of water surface, P is the saturation vapor pressure of water at the surface temperature, p is the vapor pressure of the water in the surrounding air, and H is the total pressure of the surrounding air. The important factors in acoustically augmented drying which affect the evaporation rate are rate coefficient (rate is proportional), total air pressure (rate is inversely proportional), and surface area (rate is proportional).

The rate constant k is strongly dependent on the air velocity at the air-water interface. Strong sound waves at the interface can produce high local air velocity. For example, at 1000 cycles and 160 dB the velocity amplitude of the air due to the sound is 250 in./s (630 cm/s). This significantly increases the value of k.

Reducing the total air pressure is an effective means of increasing drying rate. This is the basis for vacuum drying, where the drying is done under a partial vacuum. In the standard vacuum dryer the partial vacuum is constant throughout the drying period. In an acoustically augmented dryer, the pressure at any local point in the dryer cycles above and below atmospheric pressure because of the compression and rarefaction of the air due to the sound waves. This results in a sinusoidal variation in pressure. During each lower-pressure half-cycle, it is theorized that a fraction of the water flashes to vapor. During this half-cycle, the dryer behaves like a vacuum dryer. This allows drying at a lower temperature than normal atmospheric drying, even though the overall average pressure is still 1 atm (10^5 pascals). During the higher-pressure half-cycle, the drying rate decreases, but the water is heated. This facilitates drying during the next half-cycle.

Once the surface moisture is removed from the material, drying decreases during the falling-rate period, because the rate becomes limited by how rapidly the water inside the material can migrate to the surface. One proposed mechanism by which sound increases this migration involves water in the pores of the material. During the lower-pressure half-cycle, some water in the pores of the material flashes to vapor. This vapor forces part of the liquid water to the surface, where it evaporates rapidly.

Limited studies have been made on the effects

of intensity and frequency of sound on drying. In general, increasing the intensity increases the drying rate. The effect of frequency is not well defined, and preliminary studies have given conflicting results. It appears that the optimum frequency may be a function of the drying chamber design, the material being dried, and the material particle size. At least one university is continuing scientific studies on the mechanism and on the effect of intensity and frequency. As this technology continues to develop, such information will become increasingly important.

Advantages. The use of acoustically augmented drying methods has several advantages, based on the increase in moisture removal rate. If the drying temperature is kept the same, the time for drying is reduced. Faster drying means greater throughput for a given dryer. Alternatively, a smaller dryer may be sufficient, at less capital cost.

If the drying time is kept the same, the drying temperature is reduced. Any combination of reducing time and temperature will generally result in reducing the change in the product being dried. Other than cooking, any change in food is usually undesirable. The ideal result would be a dried food which reconstitutes exactly to the original food. Acoustically augmented drying may better approach this ideal than other drying methods.

Applications. Acoustically augmented drying has some specific applications. Several products have been made with this technology which cannot be made with any other conventional technology. Examples are dry high-fructose corn syrup and dry citrus juices produced without added drying aids.

Another category of products especially suited to this technology is the large group of materials which are temperature-sensitive. Many foods and pharmaceuticals fall into this category, along with numerous industrial materials. These products processed by using acoustically augmented drying can demonstrate enhanced quality such as better flavor, aroma, and color, and improved nutrient level.

Once the technology is mature, acoustically enhanced drying has the potential to compete with standard drying methods in terms of capital cost and operating cost. All of the potential advantages should lead to further development and application in the near future.   *Jay S. Marks*

SUPERCRITICAL FLUID EXTRACTION

The food industry is always seeking innovative separation processes that permit careful control of product characteristics such as chemical and nutrient content, flavor, aroma, color, and texture. Such a new technology is supercritical fluid extraction. This novel separation process exploits the unusual solvent power of a supercritical fluid, that is, a fluid at a temperature and pressure above its critical point. The solvent characteristics are easily adjustable, allowing selective extraction and fractionation of desirable compounds from a mixture in a single step. The use of supercritical fluids gains impetus as the number of safe solvents continues to decrease, tighter environmental constraints are imposed, higher energy costs restrict conventional energy-intensive techniques, and the concern about the thermal stability of many ingredients grows.

A generic process scheme will incorporate a certain sequence of events. The state of the extracting fluid is altered to a supercritical state; then the fluid is contacted with a feed to obtain selective extraction of one or more compounds; the extract phase is then withdrawn and its state altered, changing the solubility parameters, to separate solute and solvent. This separation can also be very selective if the state of the system is altered in a controlled manner.

Solvent properties. The efficiency of an extraction process depends on transport properties, interfacial phenomena, contact time and patterns, and solubilities; thus phase behavior is a paramount concern. The characteristics of supercritical fluids and the ease with which they are adjustable makes supercritical fluid extraction an attractive possibility. The density, hence solvent strength, of a supercritical fluid is extremely sensitive to both temperature and pressure in the near-critical region. While the density of a supercritical fluid approaches that of a liquid, for higher solvent capacity, the transport properties are typically one or two orders of magnitude larger, allowing greater penetration and thus higher mass-transfer rates. The fact that a supercritical fluid exhibits properties intermediate between a liquid and a gas can make it an ideal solvent, particularly since supercritical fluid extraction can be carried out at relatively moderate temperatures (104–$140°F$ or 40–$60°C$), minimizing problems with heat-labile materials. Also, the ease of separation of the solute and solvent further enhances the viability of supercritical fluid extraction.

The use of carbon dioxide (CO_2) as a supercritical solvent is of special interest to the food industry. It is physiologically harmless, germicidal, nonflammable, noncorrosive, readily available, and inexpensive, and has a low critical point ($88°F$, 73 atm, or $31°C$, 7.4 megapascals). Other supercritical solvents of interest to food processing are ethane, ethylene, and fluorinated hydrocarbons, since they have low critical temperatures and high solvent power; and the fluorinated hydrocarbons are chemically inert.

The solubility and selectivity of a supercritical fluid can be altered drastically by the addition of small amounts of an entrainer, a compound of intermediate volatility. Acetone, methanol, and ethanol are commonly used. Mixed solvents have produced many interesting results and further illustrate the flexibility of supercritical solvents. For example, supercritical CO_2 may be doped with several mole percent of certain polar cosolvents to increase the solubility of sterols by as much as two orders of magnitude, particularly if hydrogen bonds are formed. Alcohol cosolvents (second solvents added to the original solvents, generally in small concentrations) can enhance the selectivity of a particular sterol from a mixture of cholesterol, stigmasterol, and ergosterol by the formation of a specific complex in the solid phase.

Applications. At present there are no commercial-scale separation facilities for food applications in the United States. All commercial-scale plants are in Europe, and basically they all utilize the technology for two applications: decaffeination of coffee and tea, and extraction of alpha acids from hops. The step in the process at which supercritical fluid extraction is applied can have a major effect. The early work on decaffeination focused on treatment of the roasted beans. Flavor, color, and aroma, all of which are developed during the roasting process, were lost during decaffeination. When the sequence was reversed, with decaffeination of the green coffee beans prior to roasting, the desirable coffee properties were largely retained. For the extraction of alpha acids from hops, the stable extracts (bittering aroma chemicals important to the flavor of beer) expedite the brewing process since raw hops are no longer needed. Energy savings also contributed to the success of these applications.

Low-alcohol wines are produced in Switzerland and Germany by using supercritical CO_2 first to remove flavor compounds and then to remove the alcohol; finally the flavor components are returned to the wine product. Near-commercial applications are in flavor and aroma usage, involving removal of key aroma and flavor chemicals from one source to be used subsequently in another unrelated product. Examples are coffee and tea aroma oils, citrus oils, and fruits. Other promising areas include fermentation reactions, where continuous removal of the product from the aqueous phase into the supercritical phase may alleviate the problem of low yield associated with product inhibition; extraction and fractionation of lipids from various sources, for example, cholesterol from oils, neutral lipids from phospholipids, and saturated fatty acids from unsaturated ones; extraction of natural antioxidants from foodstuffs (such as α-tocopherol); the defatting of grains; and the deodorization of fats, with the removal of organoleptically objectionable oxidized components.

Advantages. Supercritical fluid extraction processes provide a number of advantages: (1) gentle treatment of natural substances sensitive to temperature, (2) fast extraction rates (due mainly to high diffusivity), (3) products free of solvent, (4) facile adjustment of the dissolving characteristics of the supercritical fluid, (5) selective separation of nonvolatile substances, (6) separation of natural substances without aroma loss, (7) low heats of adsorption, (8) energy requirements lower than with distillation, (9) lower cost of supercritical fluid solvents compared to conventional solvents, (10) simple solvent recovery, and (11) reduction, and in some cases elimination, of environmental problems.

Disadvantages. There are a number of disadvantages associated with supercritical fluid extraction: (1) high initial investment costs due mainly to high-pressure requirements, (2) process problems associated with materials handling in high-pressure equipment, (3) difficult scale-up since the design equations are not well known, and (4) incomplete thermodynamic understanding of the process. Research and development efforts are being extended to overcome these disadvantages, particularly in the area of multicomponent phase equilibria and the fundamental aspects of phase behavior; this information is needed for prediction and design. Recent efforts to model phase equilibria of inherently complex biomaterials has been moderately successful.

Robert J. Fisher

TWIN-SCREW EXTRUSION COOKING

Currently, the food industry is evaluating twin-screw extrusion as a key technology for providing process flexibility critical to strategic viability of the food factory of the future. Twin-screw extrusion cooking offers several advantages over single-screw extrusion, including increased process performance and flexibility and increased scope of product applications. A strategically important aspect of twin-screw extrusion is its potential for creating innovative products with improved nutritional performance and with unique appearance, flavor, texture, shape, and so forth.

Basic technology. The primary function of extrusion cooking is to texturize or restructure native starch or protein flours into particulates of various shapes and sizes with a host of unique properties, including chewy or crunchy texture, fibrous or porous appearance, significant water-holding capacity, and flavor. Extrusion cooking combines physical shearing stresses with thermal heat to convert the native molecular structure of food particles or macromolecules, such as starch granules and protein bodies, into a hot continuous molten material which produces fibrous structures during extrusion, puffing (optional), and drying. Usually, these physicochemical reactions require addition of water during a preconditioning step or via injection within the extruder. The excess water is then removed to the atmosphere by vapor flashing during extrusion and subsequent drying. Engineering expertise for food extruders has evolved from the polymer extrusion processing field. A screw extender consists of three basic parts: one or two long, threaded (flighted) screws, a hollow cylinder or barrel, and a die plate with one or more holes

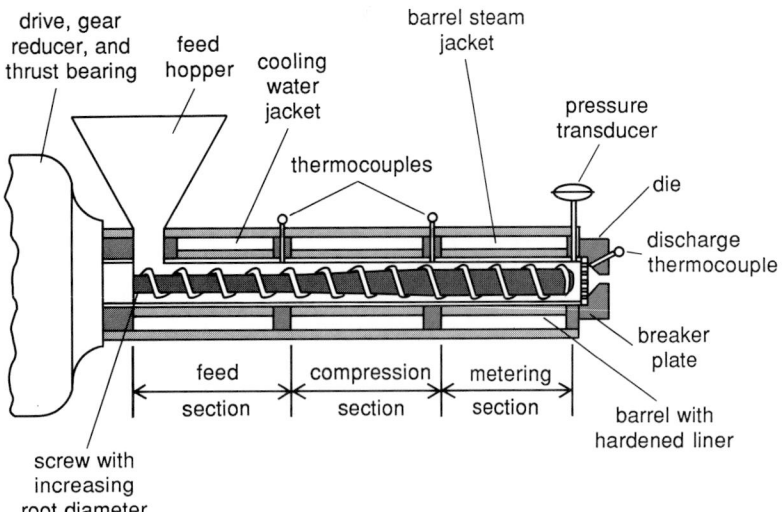

Fig. 1. Cross section of a typical food extruder. (*After J. M. Harper, Extrusion of Foods, vol. 1, 1981*)

(**Fig. 1**). The flighted screw shaft is fitted within the barrel with a close tolerance between the flight edges and barrel surface. The die plate is bolted to the output end of the barrel. The other end of the barrel has an opening through which dry or moist ingredients are fed. The rotating screw serves as a mixer, conveyor, viscous heat generator, and pressure pump, while the barrel serves as a conduit and a heating or cooling surface. The die assembly serves as a forming device in which the food product usually obtains its final shape, size, texture, and so forth. The uniqueness of an extruder derives from its ability to efficiently generate so-called local heat via dissipation of mechanical work during shearing between the screw and barrel. Simultaneously with this heat generation, material is mixed and conveyed through the device to the die, where it is shaped and sized.

Early applications. By definition, extrude means to shape a substance by forcing it through a specifically defined opening after the material is properly conditioned. Food extrusion dates back to the mid-1930s, during which single-screw extruders were used to shape or form pasta products, where there was little need for heat. In the late 1930s, cereals were made by extruding precooked ingredients. In the 1940s, cooking extruders were used for producing puffed snack products, and precooking animal feeds in order to improve nutritional value. Dry expanded extruded pet foods have developed since the 1940s. During the 1960s, ready-to-eat breakfast cereals were developed with medium- to high-shear cooking single-screw extruders.

Over the past 40 years, single-screw extruders have been applied to a wide range of food products. In most cases, these machines and processes are tailored to produce a single or narrow range of product types. With the increased dy-namics of consumer demands, the food industry is attempting to commercialize more flexible process technologies. Thus, twin-screw extrusion (**Fig. 2**) has recently received much attention as a key technology for flexible and efficient continuous processing. Many leading food companies anticipate that twin-screw extruders not only will provide a flexible means of commercializing new products but also will serve as a revolutionary means of replacing certain traditional batch-type processes used in manufacturing.

Twin-screw technology. Twin-screw extruders are broadly classified by the relative rotational direction of the screws: counterrotating or corotating (**Fig. 3**). A secondary classification is often assigned with respect to the extent of intermeshing of the screws: fully, partially, or not intermeshed. Counterrotating twin-screw extrusion produces a rolling-squeezing effect which tends to spread apart the screw shafts, increasing wear on the barrel lining. Thus, operation is limited to low speeds, resulting in reduced throughputs and less mechanical energy dissipation per unit mass. For these reasons, most twin-screw food extrusion applications involve intermeshing (or partial intermeshing) corotating screws. These systems usually operate at 100 to 500 revolutions per minute and range from 10 to 40 in ratio of length to screw diameter. Screw diameters for food applications range from 1.2 in. (30 mm) for lab units, to 2.0–2.8 in. (50–70 mm) for pilot or market development units, to 3.2–8 in. (80–200 mm) for production-scale operations. Typical throughputs are 11–110 lbm/h (5–50 kg/h) for lab units, 111–880 lbm/h (50–400 kg/h) for pilot units, and 1110–11,100 lbm/h (500–5000 kg/h) for most commercial-scale units. Raw materials usually enter at ambient temperature, and typical operating conditions are exit temperature 185–338°F (85–170°C), die pressure 100–1000 psig (700–7000 kilopascals), and moisture content 18–50% (wet basis).

Corotating intermeshed twin-screw food extruders offer key advantages over single-screw counterparts: self-wiping ability, improved residence times, increased flexibility in screw (geometric) configurations, increased resistance to process upset, more uniform shear rates and mixing, improved heat transfer, and a wider range of raw materials. Close intermeshing of the screws provides a self-cleaning aspect, which is critical to maintaining high-quality products for longer periods of continuous operation. Corotating twin screw units also exhibit more narrow residence time distributions, thus enhancing product quality.

The most flexible twin-screw extruder designs consist of splined (or keyed) screw shafts and various types of screw elements which slide on to the shafts in paired configurations. These screw elements, usually 0.25–4.0 diameters in length, are used to configure a particular screw design appropriate for a specific product. The screw

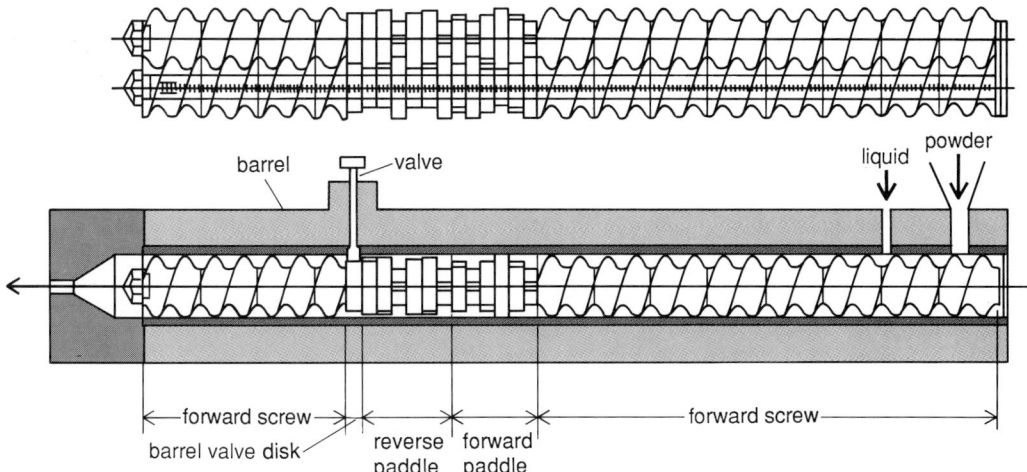

Fig. 2. Example of a screw profile in a twin-screw cooking extruder. (*After W. A. Yacu, Extrusion cooking analysis: Processing aspects of a twin-screw co-rotating extruder, Baker Perkins BCS Ltd. Paper, 1983*)

elements are usually classified into one of three categories: feeding screws, multiple-flighted screws used for mixing, and paddles or lobes used for kneading. The ability to generate an almost infinite number of geometric configurations from the same basic machine and array of screw elements is the key to the flexibility.

The complex cross-flow pattern of mixing-shearing created by corotating intermeshed screws significantly enhances process stability and thus lends itself to computer-aided process control. In most food extrusion cooking processes, 70–100% of thermal energy is derived from viscous heat generation via mechanical energy. Often, heating or cooling at various barrel locations is required to achieve a desired thermal history. The improved mixing of corotating intermeshed twin screws improves the transfer of heat across the material-barrel interface.

Process uniqueness. Twin-screw extrusion offers tremendous potential as a means of obtaining desired reductions in microbial populations while creating desirable product texture through physicochemical changes such as starch gelatinization and protein denaturation. Twin-screw extruded products may offer nutritional and shelf-life stability advantages over conventionally processed foods. Potential advantages stem primarily from the fact that during extrusion the food macromolecules are simultaneously subjected to local shear stresses and thermal energy, which accel-

erate certain physicochemical reactions. In conventional processses such as canning or batch cooking, primary physicochemical changes are almost exclusively derived from thermal energy. Current studies indicate that subjecting native starch granules and protein bodies to specified shear stress histories during thermal heating significantly reduces time required to obtain desired physicochemical attributes. This reduction in total heating time may significantly enhance nutritional aspects as well as improve flavor and extend shelf life. Commercialization of twin-screw extrusion cooking is technically constrained by three major barriers: natural variability in physical properties of raw food materials, lack of adequate knowledge of process effects on the complex and irreversible physicochemical changes, and lack of engineering analysis methods due to the complex extruder geometry and physical properties of foods.

Innovative designs. Twin-screw extrusion provides process flexibility which offers vast potential for the creation of innovative food products. Some extruder manufacturers offer barrel designs to accommodate venting and addition of liquids or solids at various locations along the screw. Thus, flavors and vitamins sensitive to heat or shear can be added just prior to the die assembly. Stability of the relationships of twin-screw extruders to heat and pressure offers potential for postdie process innovation. In fact, postdie processing is projected as an area where technology breakthroughs which could revolutionize segments of the food industry will next occur. Examples include online controlled atmospheric packaging of extrudate and coextrusion of multiple food materials.

For background information *SEE* D_RYING_; F_OOD_ _ENGINEERING_; F_OOD MANUFACTURING_; S_OLVENT EXTRACTION_ in the McGraw-Hill Encyclopedia of Science and Technology.

Ronnie G. Morgan

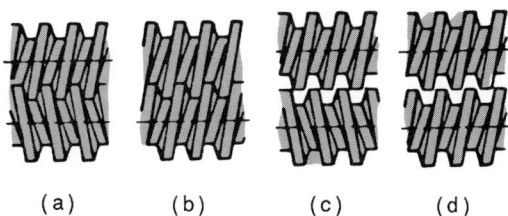

Fig. 3. Types of twin-screw extruders. (*a*) Counterrotating, intermeshing. (*b*) Corotating, intermeshing. (*c*) Counterrotating, nonintermeshing. (*d*) Corotating, nonintermeshing. (*After J. M. Harper, Extrusion of Foods, vol. 1, p. 155, 1981*)

Bibliography. J. M. Harper, Processing characteristics of food extruders, *International Confer-*

ence on Engineering and Foods, Edmonton, Alberta, Canada, 1985; M. A. McHugh and V. J. Krukonis, *Supercritical Fluid Extraction: Principles and Practice*, 1986; S. S. H. Rizvi et al., Supercritical fluid extractions: Fundamental principles and modeling methods, *Food Tech.*, 40(6):55–65, 1986; E. W. Schuler, Twin-screw extrusion cooking systems for food processing, *Cereal Foods World*, 31(6):413–416, 1986; J. M. Wong and K. P. Johnston, Solubilization of biomolecules in carbon dioxide based supercritical fluids, *Biotech. Prog.*, 2(1):29–39, 1986.

Food manufacturing

Recent advances in preservation of food include development of processes for food irradiation and the use of ultrahigh pressures during processing.

FOOD IRRADIATION

Food irradiation is the collective common or popular name for a diverse group of applications of ionizing energy to foods, feeds, and their raw materials and ingredients to accomplish a broad range of specific objectives as summarized in the **table**. Industrial food irradiation, a steadily growing worldwide processing activity, was given major impetus in July 1983 with the formal adoption of an international general irradiated foods standard and code of practices by the United Nations World Foods Standards Program. Since then, member countries have been incorporating elements of this standard and code into their national food laws and regulations, thereby opening the way for the expansion of commercial food irradiation and international trade in products so treated. Three forms of ionizing energy or radiation have been approved for such applications: gamma rays, accelerated electrons, and x-rays. The precise energy level of each is measured in megaelectronvolts.

Gamma rays. These are generated by radioactive decay of the manufactured radioisotope cobalt-60, each atom of which emits two gamma rays (1.173 and 1.332 MeV) in decaying to stable nickel-60. The same manufactured cobalt-60 sources used industrially for gamma processing are also employed in cancer radiotherapy units in hospitals worldwide for the treatment of certain cancers. The only other potentially feasible radioisotope source of gamma rays for processing and treatment applications is cesium-137, each atom of which emits one 0.662-MeV gamma ray

Food applications of ionizing energy	
Dose	Application
Representative food applications up to 100 krad (1 kGy)	
5–15 krad	Inhibition of postharvest sprouting of root crops (potato, onion, garlic, and so forth)
10–15 krad minimum	Plant protection quarantine treatment for postharvest fruit fly eradication from fruit and produce (for example, citrus, papaya, mango)
Up to 100 krad	General insect disinfestation of agricultural commodities, including grains and grain products
30 krad minimum	Control of *Trichinella spiralis* in raw pork, and other food-borne parasites as well
50–100 krad	Fresh-market-life extension through delay of normal fruit ripening, maturation, and senescence (for example, papaya, mango, banana)
	Significant reduction in levels of common spoilage microorganisms in and on perishables (suboptimal range in most cases)
Representative food applications from a 100-krad (1-kGy) average or minimum dose to a 1000-krad (10-kGy) dose	
100–300 krad	Fresh-market life extension and delayed spoilage through reduction in levels of common spoilage microorganisms (as in meats, poultry, fishery products, berries, mushrooms, and so on); significant reduction in levels of pathogenic food-borne infectious and toxin-producing microorganisms, and control of any parasites that may be present
100–1000 krad	General elimination of non-spore-pathogenic microorganisms in fresh, frozen, and dehydrated foods; feeds; and ingredients (for example, *Salmonella* and *Campylobacter* contamination of poultry and feed); elimination of common spoilage microorganisms and parasites as well
	Improvement of physical and functional properties of certain foods and ingredients (for example, increased juice press yields and bread loaf volume)
	Synergistic combination of irradiation with other processes and treatments (for example, thermal processing)
	Sanitization of food packaging materials prior to filling and sealing
	Interruption of undesirable physiological processes of certain fresh produce (for example, mushroom cap opening, asparagus stalk elongation and fiberation)
Representative food applications above 1000 krad (10 kGy or 1 Mrad)	
500–3000 krad	Sanitization-to-sterilization of dry-food raw materials and ingredients (for example, spices and vegetable seasonings), and processing aids (for example, food-grade enzymes)
1000–5000 krad minimum	Sterilization of precooked or processed low-acid meat, poultry, and fish products in hermetically sealed packaging for long-term nonrefrigerated storage
	Tenfold reduction in rehydration time of dehydrated vegetables (as in dry soup mixes), together with sanitization and sterilization
	Sterilization of food packaging materials prior to filling and sealing (for example, dairy product packaging, aseptic packaging materials)
1000–10,000 krad	Reduction-to-elimination of virus contamination (most practical for waste streams)

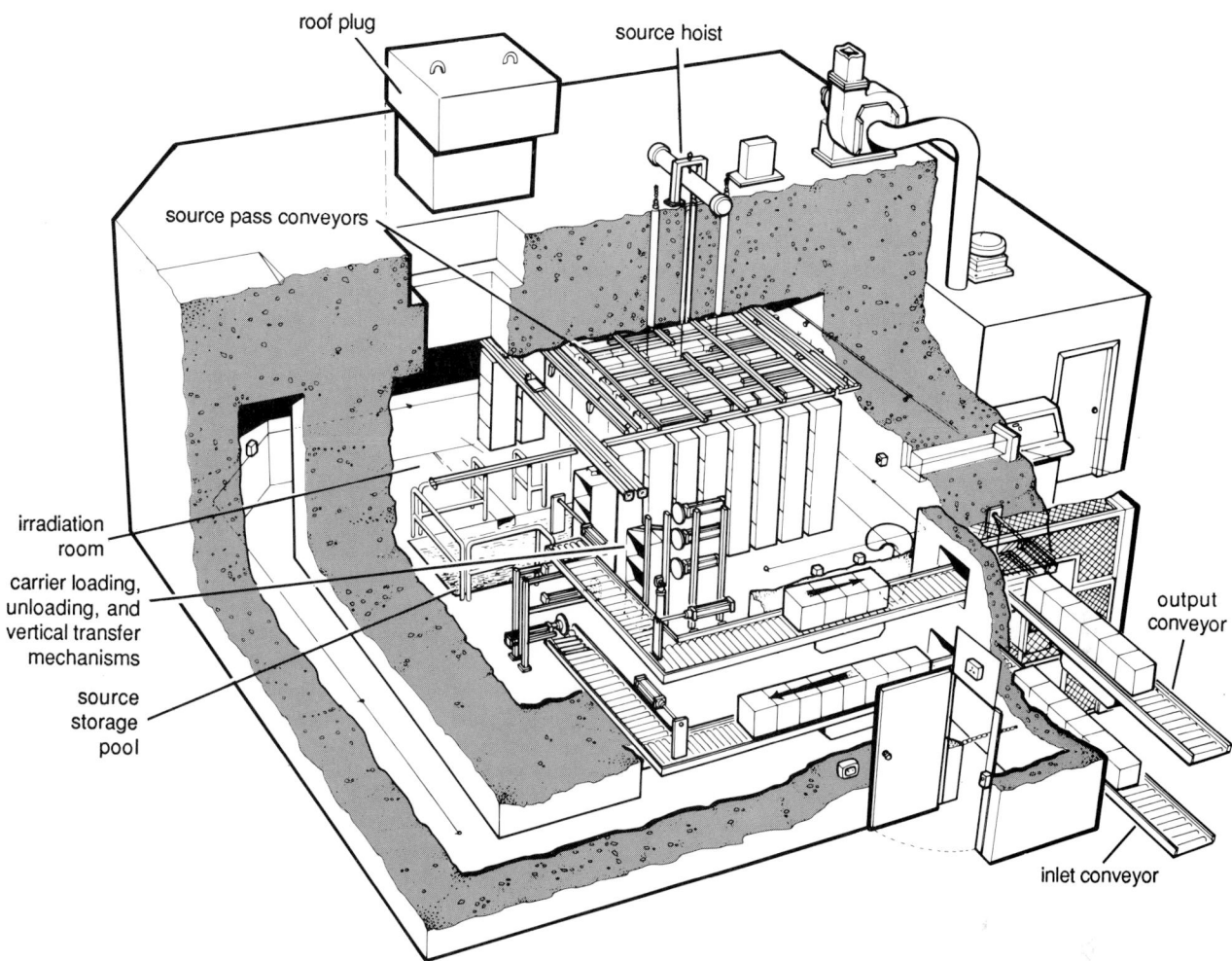

Fig. 1. Industrial cobalt-60 gamma processing facility. (*Atomic Energy of Canada Limited, Radiochemical Company*)

as it decays to stable barium-137. A typical industrial cobalt-60 gamma processing facility is shown in **Fig. 1**.

Accelerated electrons. Other common sources of ionizing energy for industrial processing which have been approved for food are electrically powered machines known as electron accelerators, which accelerate electric current (electrons) to very high energy levels, in the hundreds of thousands to millions of electronvolts. There are several types of electron accelerators employed in industrial processing. Applications include manufacturing of shrink wrap (plastic packaging film), strengthening of wire and cable insulation, and curing of polymeric coatings. Linear accelerators are probably the most suitable for food processing (**Fig. 2**). Electron accelerators designed for food processing and treatment may not be operated above 10 MeV to assure that noradioactivity is induced in the products.

Accelerated electrons have a much greater linear energy transfer to the absorbing materials than do gamma rays. In practical processing terms, this translates to much higher penetrabil-

ity potential for gamma rays (up to several inches to feet into materials depending upon their makeup and density) as compared to accelerated electrons of equivalent energy (10-MeV electrons can penetrate into water or other near-unit-density materials only up to about 2 in.). Thus gamma radiation can be applied to cartons and even full pallet loads of materials, whereas 10-MeV electron irradiation is limited to thin targets of no more than a few inches' depth or thickness for fairly dense materials. Electron accelerators, on the other hand, can deliver a prescribed amount of ionizing energy (a dose) to a product many times faster than a feasibly sized radioisotopic gamma emitter, and they can be turned off when not in use. Thus, where the limited depth of penetration is not a constraint (in treatment of surfaces or thin, shallow-depth targets), electron accelerators can be the source of choice for the ionizing radiation, although their operation and maintenance are more complex and sophisticated than that of gamma irradiators.

X-rays. This type of ionizing energy is permitted for food irradiation of up to 5 MeV. The

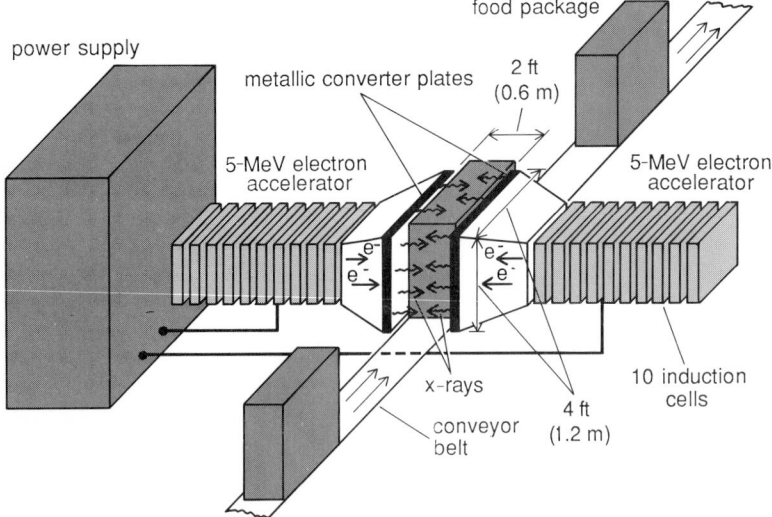

power supply

food package

metallic converter plates

2 ft
(0.6 m)

5-MeV electron
accelerator

5-MeV electron
accelerator

e⁻

e⁻

e⁻

e⁻

x-rays

conveyor
belt

4 ft
(1.2 m)

10 induction
cells

Fig. 2. Schematic diagram of a two-sided linear induction electron accelerator.

5-MeV maximum-energy restriction precludes any possibility of radioactivity induction in the product being processed (cobalt and cesium gamma rays are, of course, well below the 5-MeV energy limit).

Being pure electromagnetic energy, x-rays and gamma rays of equal energy levels are identical in all respects, except that gamma rays are emitted by certain decaying radioisotopes such as cobalt-60 and cesium-137 whereas x-rays are produced by modified electron accelerators. Unlike cobalt-60 and cesium-137, which emit gamma rays at specific wavelengths, x-ray machines deliver a spectrum of x-ray energies up to the operating voltage (for example, 5 MeV). Also, while radioisotopes such as cobalt-60 emit their decay energy isotropically (in all directions), much of it being unavoidably wasted, machine-generated electrons and x-rays are much more directed, and therefore can be utilized more efficiently once emitted. A difficulty in developing industrial-scale x-ray emitters lies in achieving reasonably high conversion of electron energy into x-ray energies in the first place.

Both gamma rays and x-rays, being electromagnetic energies of distinct energy levels, belong to the so-called electromagnetic spectrum. Of the regions of the electromagnetic spectrum used for food treatment (microwave for thawing and heating, infrared for heating, ultraviolet for surface disinfection, gamma rays and x-rays), gamma rays and x-rays are of sufficiently high energy levels to qualify as ionizing energy or radiation.

X-ray machines for processing are essentially industrial-scale versions of the small x-ray units commonly used in medicine and dentistry. Reasonably efficient, cost-effective industrial models are under development worldwide, but commercial use is at least a few years away. Such machines should offer the advantage of operation either in the electron-beam mode or in the x-ray mode when, for example, greater penetration is needed.

Equipment. Although the equipment can be complex (for example, linear accelerators), radiation processing in general, including food irradiation, is basically a rather straightforward process wherein the product is carried past the energy source by a standard conveying system. As the product passes through the radiation field, it plus any accompanying packaging continually absorbs the energy until the prescribed dose has been delivered; the product is then removed from the field. The dose is measured by an empirical (typically spectrophotometric or colorimetric) method wherein liquid or solid-state dosimeters are measured for color change (absorption change at a specific wavelength of visible light), which is directly proportional to the quantity of ionizing energy absorbed (in kilorads or kilograys). Such dosimetry is analogous to thermocouple measurement of temperature changes inside a retort during thermal processing of canned foods, the temperature change being directly relatable to the quantity of thermal energy absorbed by the product. Just as with thermal processing during which the entire product must be heated in order to destroy the contaminating microorganisms that are the real target of the process, during radiation processing the entire product absorbs the ionizing energy, although the real targets may be contaminating insects, parasites, bacteria, or fungi.

Applications. Because the entire product absorbs the energy no matter what the objective, a great deal of research had to be done to assure public health authorities worldwide that radiation-processed foods are safe and wholesome. This has been established. In fact, one of the more attractive aspects of the process is that it can rid products of parasites and microorganisms that cause illnesses without changing the nature and state of the product (for example, meat, poultry, and fish remain raw and indistinguishable from the nonirradiated). It is a cold process that does not inactivate naturally occurring enzymes.

If long-term nonrefrigerated shelf stability is the objective (radiation sterilization, analogous to thermal canning), then products must be first heated sufficiently to inactivate endogenous enzymes, which would otherwise continue to catalyze chemical reactions in the product, causing gradual deterioration of texture, flavor, and so forth during long-term storage. If, on the other hand, radiation is applied to achieve only short-term fresh-market-life extension of several days to a few weeks (for example, through delay of ripening or reduction of spoilage microorganisms), the product can remain in the raw state without any need for thermal enzyme inactivation, which in effect results in a cooked product.

The exceptions are root crops, notably potatoes and onions, which are irradiated in the raw state for prevention of sprouting, following which they may be stored raw for months.

Another very attractive food-related application of ionizing energy is microbial reduction in dry raw materials and ingredients that typically are rather heavily contaminated, such as animal feeds, and especially spices and seasonings and other dry ingredients. Microbial disinfection of spices and seasonings is a growing industrial radiation application in North America and elsewhere, partly because the process is so efficient and effective with dry materials which, because of the absence of free moisture, are among the least radiation-sensitive of all products. Nevertheless, under appropriate conditions many other products, including high-moisture ones, tolerate irradiation very well at doses which can provide important benefits. Such applications are being achieved increasingly on an industrial scale in a growing number of countries. Food irradiation is taking its place among the more traditional food preservation and treatment methods as it continues to gain regulatory approvals and to find niches in which it is the method of choice to accomplish practical, beneficial objectives.

Currently, the only food-related irradiation applications being done on a routine industrial basis in North America were the sterilization of spices and seasonings such as black pepper, paprika, and onion powder; the presterilization of certain food packaging materials in bulk prior to shipment to food plants where they are filled with product. It is anticipated that by the end of 1988 commercial quantities of certain raw fruit and produce items grown in and imported into North America will be treated with gamma radiation to eliminate infesting insects such as the various fruit flies and to extend their fresh-market life through a delay of normal spoilage of a few days to a few weeks, depending upon the item. Subsequently, it is expected that some other perishables (such as certain meat, poultry, and fish products) and processed foods and ingredients will be treated by irradiation for one or another purpose, including reductions in harmful food-poisoning microorganisms and parasites.

On a worldwide basis, much more food irradiation is being done in parts of the world other than North America, notably South Africa, Israel, certain countries of eastern and western Europe (such as Hungary and the Netherlands), the Soviet Union, and certain Asian countries such as Japan, People's Republic of China, and India. The Food and Agriculture Organization of the United Nations estimates that as of the end of 1986 approximately 1 billion pounds of food products of all kinds were being irradiated annually. This total is increasing each year as more approvals and facilities are being put in place worldwide. Nevertheless, food irradiation will not become a technological tidal wave engulfing the world food supply. Rather, its uses in food-related fields will continue to be a gradual exercise in finding niches in which either it is the only practical method of achieving an essential or desirable objective, or it is the all-around best alternative method.   *George G. Giddings*

ULTRAHIGH-PRESSURE FOOD PRESERVATION

Advances have been made recently in preservation of biological materials and foods using ultrahigh pressures.

Treatments. Protein denaturation involves loss of solubility and biological activity; an example is the change that takes place when an egg is cooked. Proteins can be denatured irreversibly at room temperature by hydrostatic pressures above 1000 atm (100 megapascals). Heat-shocked (158°F or 70°C for 30 min) spores of *Bacillus coagulans*, *B. cereus*, *B. subtilis*, and *Clostridium sporogenes* could be inactivated at increasing rates at temperatures between 77 and 167°F (25 and 75°C) and pressures between 500 and 8000 atm (50 and 800 MPa). Typically, 10 million cells of *B. coagulans* could be inactivated in 30 min at 4000 atm (400 MPa) in 0.1-*M* sodium phosphate at 167°F (75°C). Heating at 158°F (70°C) for 30 min after pressure treatment resulted in further inactivation.

Vegetative bacterial forms are easier to inactivate by pressure. Typically, exposure to 3000 atm (300 MPa) for as little as 10 min at 77°F (25°C) can result in the inactivation of 1 million cells.

Hydrostatic pressure in the range of 1500 atm (150 MPa) can be used to prepare active vaccines from viable cells. Treated cells, while not capable of reproducing, retain their antigenic effect. Yeasts appear to be slightly more pressure-resistant to inactivating than bacteria at room temperature.

Milk treated at 5000 atm (500 MPa) for 30 min at 95°F (35°C) showed 10,000 less bacteria per gram. Viruses can be inactivated at significant rates at 2000 atm (200 MPa) at room temperature; however, higher pressures will increase the rate. Bacteriophages appear to be more susceptible to pressure inactivation than bacteria.

In general, at pressures above about 1000 atm (100 MPa), increasing the exposure temperature will increase the inactivation or denaturation rate of proteins, microbes, and toxins; however, if the inactive form has a greater volume than the active form, pressure may reduce the rate of inactivation at temperatures between room temperature and 140°F (60°C).

Pressure will lower the pH of a water solution by about 0.5 per 1000 atm (100 MPa). The ionization constants of weakly ionized acids, for example, can be increased severalfold at 1000 atm (100 MPa) over those observed at 1 atm (10^5 Pa). The effect of water activity on the survival of microbes in food as a function of pressure, temperature, pH, and media composition must be studied on an individual basis.

Microbes held at pressures above 1000 atm (100 MPa) will not grow or reproduce at room temperature. At pressures between about 500 and 1000 atm (50 and 100 MPa), some microbes may grow in size or reproduce slowly, depending on temperature, gas atmosphere, and media. Oxygen pressures greater than 1 atm (10^5 Pa) have been shown to be toxic to selected microbes. The effects of exposing foods to carbon dioxide, oxygen, or mixtures of these gases under pressure to promote more rapid inactivation of microbes must be studied on an individual basis, since adverse effects on food quality could negate the advantages of microbial inactivation. When foods surrounded by or containing air or other gases are hydrostatically compressed, they tend to absorb the gas. If the absorbed gas is oxygen or carbon dioxide, it can react with food pigments, lipids, vitamins, and flavors. Reaction rates can be increased by pressure if reaction products occupy a volume smaller than that of the reacting agents. For this reason, foods to be treated by hydrostatic pressure should be degassed (vacuum-packed), unless tests show no effect of occluded gases. Normal precautions must be taken to prevent the outgrowth of *C. botulinum* in foods that have been vacuum-packed or flushed with inert gas, because under anaerobic conditions spores are highly resistant to hydrostatic pressures at temperatures up to 212°F (100°C).

Effects. Pressure effects on enzymes of interest to food processors appear to be minimal for hydrostatic pressure up to several thousand atmospheres. Because protein denaturation is accompanied by an increase in volume, pressures between 500 and 1000 atm (50 and 100 MPa) may increase the rate of enzyme activity at or slightly above room temperature over comparable conditions of temperature, substrate concentration, and so forth at 1 atm (10^5 Pa). Thus pressures of 500–1000 atm (50–100 MPa) can partially protect enzymes from thermal inactivation at temperatures in the range of 120–160°F (50–70°C).

Raw foods treated by hydrostatic pressure to reduce microbial or parasite contamination would retain their enzyme activity. Raw foods susceptible to enzymatic spoilage could be affected adversely by hydrostatic pressure.

Studies were recently reported concerning the effect of hydrostatic pressure treatments to 1000 atm (100 MPa) as a method for tenderizing beef prior to rigor mortis and after. Beef prior to rigor mortis can be tenderized significantly at 1000 atm (100 MPa) in 60 min at 95°F (35°C). However, postrigor treatment at 1000 atm (100 MPa) required a temperature of 140°F (60°C) to achieve significant tenderization. Pressure could be responsible for increasing the proteolytic enzyme activity in the beef, resulting in improved tender-

ness through partial breakdown of tough connective tissue.

Above 2000 atm (200 MPa), hydrostatic pressure will irreversibly denature proteins at room temperature at rates proportional to pressure. Thus, protein foods such as milk, egg, beef, poultry, and seafood will appear to be cooked to a progressively greater extent as pressures and treatment times are increased even at room temperature. Thus, above 2000 atm (200 MPa) the effects of hydrostatic pressure appear to parallel mild heat treatments.

The gelation temperature of starch is elevated and its gelation rate slowed with increasing pressures above 1000 atm (100 MPa). This follows from the increase in volume during starch gelation. The effect of hydrostatic pressure on other food components requires further study.

Applications. Early workers investigating the effects of hydrostatic pressure on milk, egg white, peaches, tomatoes, and low-acid vegetables observed that pressures up to 7000 atm (700 MPa) would not preserve low-acid products in a predictable manner. In addition, protein foods showed signs of coagulation at pressures above several thousand atmospheres. However, a combination of 140°F (60°C) and 1000–2000 atm (100–200 MPa) allowed the preservation of peaches in commercial-size cans.

Hydrostatic pressures to several thousand atmospheres have little effect on food structures. Fruits, meats, shellfish, and vegetables subjected to pressures up to several thousand atmospheres appear unchanged. Foods containing gases change from opaque to translucent, reflecting the absorption of the gas by the liquid portion of the food. Dissolved gas will be released on standing. A major advantage of pressure treatment is that foods packed in large pieces such as whole hams or cuts of beef can be treated uniformly, since pressure effects are transmitted instantly and equally to all parts of the food.

Current knowledge suggests pressure treatment can best be used in conjunction with refrigeration or frozen storage to reduce the numbers of vegetative microbes in food.

Packaging. Flexible packaging provides an ideal material for treatment of foods by hydrostatic pressure. Water is compressed several percent per thousand atmospheres, and gases in or around the food are compressed and dissolved in the liquid portion of the food. Depending on shelf life and handling needs, foods can be vacuum-packed by using conventional equipment and can be treated after packaging.

To date, little work has been done on the effects of hydrostatic pressure on flexible packaging materials. The effect of food materials on the structure of flexible packaging systems held at several thousand atmospheres remains to be studied.

Equipment. Equipment for hydrostatic treatment of solids up to 6 in. (15 cm) in diameter and

30 in. (75 cm) in length at pressures to 4000 atm (400 MPa) is available commercially. Foods to be treated are packaged to exclude as much gas as possible, brought to the desired treatment temperature, and placed in the treatment pressure vessel containing water or oil at the desired treatment temperature. The vessel is sealed and is brought to the desired pressure by means of a multiplier pump, usually at a pressurization rate of several thousand atmospheres per minute. Under these conditions, little heating due to mechanical work takes place, and the rate of pressurization appears to have little effect on the lethality of hydrostatic pressure on microbes. Depressurization can take place at several thousand atmospheres per minute. Again, the rate appears to have little lethal affect. The batch treatment process described can be automated by using multiple chambers to obtain production rates in the range of hundreds of pounds of packaged food per hour.

Costs for treating packaged food by hydrostatic pressure will depend on the treatment pressure and treatment time. For treatments where cycle times can be reduced by higher treatment pressures, increased costs associated with higher-pressure equipment can be weighed against potential lower cost per pound from greater throughput.

In principle, pumpable foods such as juices and purees can be treated continuously. The pressurized product would be held for the needed time and temperature in a holding tube and decompressed so as to recover as much mechanical work as possible. Treated juice, for example, could be filled aseptically into sterile packages.

Laboratory equipment that is capable of treating gram (0.4-oz) quantities of food at pressures to 6000 atm (600 MPa) is available off the shelf. Batch systems operating to 4000 atm (400 MPa) which are capable of treating up to a kilogram (2.2 lb) of sample are available for under $20,000.

Food and process safety. Hydrostatic pressure is a mechanical process and should not affect the nutritional value or the safety of foods and ingredients.

Microbial changes during posttreatment storage and handling must be considered as with any food. Thus, pressure-treated foods must carry handling and storage instructions which reflect their degree of sterility. For example, foods pressure-treated to eliminate *Salmonella* sp. and to reduce total microbial counts would require refrigerated or frozen storage, depending on the desired shelf life.

Commercial pressure-treatment equipment is designed to operate safely over a large number of pressure-treatment cycles. The cost of maintenance, repair, and safety checks must be considered in the application of pressure treatment. Because the system is liquid-filled, appropriate shielding for the treatment area is recommended by equipment manufacturers.

For background information *SEE* ELECTROMAGNETIC RADIATION; FOOD MANUFACTURING; GAMMA RAYS; PARTICLE ACCELERATOR; X-RAYS in the McGraw-Hill Encyclopedia of Science and Technology.

Daniel F. Farkas

Bibliography. P. J. Elias and A. J. Cohen (eds.), *Recent Advances in Food Irradiation*, 1983; G. G. Giddings, Food irradiation: The reluctant food additive for all agendas, *Food Rev. Int.*, 2(1):109, 1986; E. S. Josephson and M. S. Peterson (eds.), *Preservation of Food by Ionizing Radiation*, vols. 1–3, 1983; J. J. Macfarlane, High pressure technology and meat quality, in R. Lawrie (ed.), *Developments in Meat Science*, vol. 3, ch. 6, 1985; A. J. H. Sale, G. W. Gould, and W. A. Hamilton, Inactivation of bacterial spores by hydrostatic pressure, *J. Gen. Microbiol.*, 60:323–334, 1970; Society for Experimental Biology, *The Effects of Pressure on Organisms*, 1972; W. M. Urbain, *Food Irradiation*, 1986; A. M. Zimmerman (ed.), *High Pressure Effects on Cellular Processes*, 1970.

Food web

A food web is a network of pathways that connect food reservoirs in a manner that facilitates the flow of energy and the cycling of nutrients throughout an ecosystem. The flow in energy dissipates at each stage by respiration, with just a small fraction of the energy reaching the final consumer. This flow of energy establishes the pattern of trophic levels. Superimposed on this scheme of energy flow is the nutrient cycle. Generally, nutrient elements pass from reservoir to reservoir in a cyclic manner that returns most of the nutrient to the primary reservoir, with only a small loss from the system.

Heavy metals appearing in food webs can have significant effects on ecosystems. This article discusses recent advances in characterizing the impact of lead from industrial sources and of cadmium transferred from sludge.

Lead. A new approach has been developed for measuring the full impact of industrial lead on the food webs of natural ecosystems. In this method the change in the ratio of calcium to lead in the nutrient pools of several successive trophic levels is measured. The resulting data indicate that the contribution of anthropogenic lead to both terrestrial and marine food webs may be 10- to 400-fold greater than the contribution of natural lead.

Within the nutrient cycle, there is evidence that a process called nutrient biopurification maintains pools of nutrient substances at optimum concentrations, to the exclusion of nonnutrient substances. Nutrient biopurification may be illustrated by the example of calcium and lead. If the calcium in soil moisture is not sufficient to support the nutrient requirements of a plant, the plant must form a nutrient pool within its tissues that optimizes the calcium concentration for its requirements. Although often measured with respect to water, the concentration of the nutrient can be expressed with respect to any substance. By measuring the ratio of calcium to lead at several trophic levels, it can be shown that the pool of natural calcium becomes progressively freer of nonnutrient elements at each successive stage.

Natural sources. The use of nutrient biopurification analysis has aided scientists in the investigation of the lead cycle in terrestrial ecosystems. Lead is a relatively common element that is found in the mineral constituents of rocks. During the weathering process of soil formation, natural lead is slowly released from parent rock material to become a constituent of other soil components. The rate of release depends on the chemical environment of the soil and the resistance of the individual minerals to the weathering process. This release begins the natural cycle of lead through the food web. In the simplest of these pathways, lead passes directly to secondary minerals, normally clays, that are formed as the soil matures. These clays eventually become a major constituent of the sedimentary rock that completes the cycle, and so this pathway is not a direct part of the food web. Lead may also become attached to the surfaces of organic particles by a chemical process called chelation, or adsorbed on ferromanganese films that surround soil particles. In aqueous solution, lead may be bound to humic and fulvic acids or simply dissolved as an inorganic salt, most commonly a sulfate or a carbonate.

Lead that is dissolved in soil moisture may be taken up by plants. However, plants do not normally take up much lead because of its low solubility and because of the ability of roots to exclude lead during calcium uptake. Most of the lead that is taken up by the plants is retained in the roots. Stems, leaves, and fruits of plants contain very little natural lead and pass only a small amount on to grazing food chains.

Because the salts of lead are very insoluble and the chelation bonds with organic molecules are strong, very little lead passes through the soil system into the groundwater. Consequently, most of the pathways of lead in soil ultimately converge on the formation of secondary minerals. The cycle is complete when these minerals are compressed by geologic processes to form sedimentary rocks.

Anthropogenic sources. Not all lead in soils derives from the natural lead of parent rocks. Anthropogenic lead, which enters the biosphere as a result of human activities, may be mixed with the natural lead in soil. Most atmospheric lead is anthropogenic, from automobile exhausts, smelters, and coal-fired power plants. In the atmosphere, lead is found on particles that are very small, usually less than 1 micrometer in diameter. These particles enter terrestrial ecosystems carried by rain or snow or directly deposited on plant or soil surfaces. Consequently, the leaves and fruits that are normally almost free of natural lead may have a relatively large amount of anthropogenic lead. Some of this lead is eaten by herbivores, and the rest passes to the soil as plant litter.

In the grazing food chain, anthropogenic lead that is deposited on the surfaces of plants and on animal fur is consumed at each trophic level, overriding the natural process of nutrient biopurification by reducing the ratio of calcium to lead. Although this process is of some concern, there is no evidence that it has resulted in any toxic effects in natural ecosystems that are not near smelters or roadways.

Through plant litter and animal refuse, atmospheric lead ultimately reaches the soil surface to become incorporated into the pathways followed by natural lead in the soil. Because this atmospheric lead is strongly bound to the organic matter, it typically is more concentrated in the organically rich top layers of the soil. In the soil system the decomposer food chain is particularly vulnerable to atmospheric lead. In a complex sequence of stages, groups of soil microorganisms, insects, and earthworms progressively break down or decompose the energy-rich organic litter. Because the lead does not decompose, it becomes more concentrated as the organic matter is removed. There is some evidence that, in areas where the atmospheric input of lead is high, the lead in the decomposer food chain reaches concentrations that are toxic for the decomposers. The net effect is that the decomposer food chain is broken, and further decomposition is inhibited. If this were to happen on a wide scale, terrestrial ecosystems would become unbalanced by the buildup of litter. This unbalance is usually rectified in a natural manner, however, by the substitution of a decomposer population that is more tolerant to lead, thus restoring the chain.

Until recently, most of the lead in the atmosphere could be attributed to automobile exhaust. But the use of lead in gasoline is being phased out in the United States; as of 1987, only a small amount of lead (0.1 g/gal or 0.1g/3.785 liters) was allowed even in leaded gasoline. As a result of this decision, lead in air will be drastically reduced. But there will be little change in the amount of lead in soils. Some scientists estimate that it takes from 700 to 6000 years for half of the lead in soils to be removed by natural processes. Consequently, the anthropogenic lead that has already been added to soils will be around for a long time.

Differences. It is possible to distinguish atmospheric lead from natural lead. Industrial lead is mined from an ore body; natural lead comes from crustal rock. An unusual feature of lead is that it occurs as several stable isotopes. Some of these isotopes are daughter products of the radioactive decay of heavier elements. These radioactive elements have half-lives ranging from several thousand years to a few million years. Thus the isotopic composition of lead may exhibit little change over a time scale of thousands of years, while over a period of tens of millions of years a difference can be detected between lead from the crust and lead from ores. The radioactive decay process was interrupted when lead ore bodies were formed, and the ore retained the ratio of lead isotopes that was present at the time. However, the process continued in the crustal rock that is the precursor of the soil, and the ratio of lead isotopes continues to change up to the present. Thus, several million years after the formation of a lead ore body, the ore lead can be distinguished from the soil lead by the difference in the ratio of the isotopes, even though both leads are chemically identical. By determining the sources of the lead, scientists can trace the cycle of atmospheric lead through the food web and distinguish this cycle from the natural lead that originated from crustal rock.

Robert Elias

Cadmium. Cadmium, a heavy metal, is a potential hazard to animals and humans. With the increased use of sludge application to land, there is an increased cadmium burden placed on the soil-plant-animal system. Cadmium levels in sludge are increasing and will continue to increase, depending on the source; when sludge is applied to land, the cadmium accumulates in plants grown on these sites. Animals consuming plant material grown on sludge-amended soils accumulate cadmium in kidney and liver tissues. This heavy metal, then, is found in minimal levels in milk and meat. It has been found that plant type and soil characteristics affect the level of cadmium and can be managed so as to reduce the hazard.

Sewage sludge, the precipitated solid matter produced by water and sewage treatment processes, has been spread on land for many years. Throughout the world the use of land for the disposal of sludge has been practical for reasons of sanitation and fertilization. Because of the cost of incineration, the potential increase in air and water pollution, and the available nutrients for plant utilization, there is renewed interest in the application of sewage sludge on land.

The composition of sewage sludge varies widely depending on the extent of treatment, type of community, and time period. The composition of sludge at a given facility varies from hour to hour, day to day, month to month, depending on seasonal changes and habits of population. One of the greatest potential hazards in applying sewage sludge on land is the accumulation of heavy metals in the soil. Buildup of heavy metals such as zinc, copper, nickel, and cadmium may reduce plant growth, concentrate in plant tissue, and thus enter the feed and food chain. There is concern for cadmium specifically because of this potential entry into the food chain. Cadmium is toxic to virtually every system in the animal body; it tends to accumulate in animals and humans from low-level exposure, so that the greater hazard from increased concentration of cadmium in plants is to humans, rather than to plants directly.

Cadmium concentration in sewage sludge ranges from 3 to 3000 parts per million (ppm), with a mean value of 106 ppm, compared to soil with concentrations of 0.01 to 7 ppm, with 0.06 ppm of cadmium being typical. Thus, the application of sewage sludge to soil would result in increased concentration of total cadmium in the soil. Application of superphosphate fertilizer containing cadmium as a contaminant can also add the metal to soil in a quantity equivalent to that added in application of sludge. However, because of the differences in rates of application of fertilizer and sludge, cadmium added to a particular plot is greater from sludge addition.

Soil properties. The behavior of cadmium in soil is not well understood. However, soil pH, soil organic matter, clay content and type, hydrous oxide content, and cation-exchange capacity appear to influence cadmium activity. Soil pH is the major soil characteristic that affects the uptake of cadmium by growing plants. Heavy metals are more soluble under acid conditions than under neutral to alkaline conditions; thus it is important that soil pH be maintained at 6.5 or above after sludge application. Increasing the pH of soils typically reduces the cadmium content of crop tissues and grains. However, pH adjustment on sludge-amended soils will not reduce plant cadmium content to that found in plants grown on untreated soils. Undoubtedly, cadmium uptake by plants varies with different soil types and pH values.

Cadmium uptake is related to the ratio between cadmium added to the soil and the cadmium-absorbing capacity of the soil. Further, an increase in soil temperature leads to an increased uptake of cadmium. When cadmium is applied to the soil it is more available (readily absorbed) by the plant initially, and over that time the cadmium is apparently converted to less available forms. This is suggested by the fact that application of animal sludge to corn in a given year increased cadmium content in leaves more than did overall accumulative amounts of cadmium applied. When cadmium addition to land is discontinued, there is a measurable decrease in cadmium uptake by corn the first year, with smaller reductions the following years. Soil properties and their influence on cadmium uptake from the soil-water system need further investigation.

Plant factors. Uptake of sludge-borne cadmium depends on plant type as well as cadmium application rate, soil pH, soil temperature, nitrogen and phosphorus fertilization, and the addition of elements such as zinc and copper. Research indicates that different species, varieties, and tissues of plants contain different cadmium concentrations for similar rates of application. Species differ in uptake and distribution of cadmium within the plant. Vegetative tissue tends to accumulate cadmium, leaving storage tissue relatively free of contamination. The addition of 200 tons of sewage sludge per acre (450 metric tons per hectare) increased the cadmium concentration of corn grain from 0.02 to 0.05 ppm; by contrast, the cadmium content of leaf tissue was increased from 0.26 to 1.32 ppm.

Cadmium concentrations in corn grain are typically only 3 to 15% as high as in leaves, while the grain of soybeans, wheat, oats, and sorghum reaches 30 to 100% of the vegetative tissue concentration. Crops may contain undesirable concentrations of cadmium without visible symptoms of toxicity to the plant. Leafy vegetables such as lettuce, chard, spinach, and turnip greens may contain in excess of 100 ppm of cadmium with no toxicity symptoms in the plant.

Sludge processing influences availability of cadmium to the plant. Application of high-lime sludge cake may cause soils to become alkaline, thus reducing cadmium availability. Anaerobic digestion of sludge may be necessary to reduce the uptake of cadmium by plants grown on soils with repeated sludge application; however, results are not consistent.

Cadmium and other metals interact, and this affects cadmium uptake by plants. Copper additions increase cadmium content of corn and rye shoots; additions of zinc alone, zinc plus copper, selenium, or molybdenum reduce cadmium content of soybean grain, even though vegetative cadmium content increases. Ammonium fertilization may increase plant cadmium uptake, while phosphorus fertilization appears to decrease cadmium content of corn leaves.

It is difficult to separate the effect or interaction of soil and plants concerning cadmium uptake. Obviously, both factors act together, resulting in food or feed that may be enriched with cadmium. However, cadmium concentration in plant tissues can be reduced by maintaining soil pH at or above 6.5; growing crops that accumulate low concentrations of cadmium or, for crops harvested for grain, selecting types with low cadmium grain levels; applying low levels of sludge annually to soils used for growing feed or food crops; and using sludges low in cadmium on cropland. In addition, cultivation of nonedible crops, for example, corn grain for animal feed, in sludge-treated soils will help limit cadmium accumulation in animal feeds and human food.

Animal metabolism. At birth, animal tissues are devoid of cadmium and accumulate cadmium with time and exposure. There is apparently no effective homeostatic control mechanism to limit cadmium absorption and retention below a nontoxic threshold. Once cadmium is absorbed, the body tends to retain it in tissues. In humans the half-life is estimated at 16 to 33 years. Tissue cadmium is turned over slowly regardless of dietary cadmium challenges, and continued absorption occurs irrespective of tissue cadmium burden. Cadmium excretion that does occur is primarily via feces.

Air, water, and soil are all pathways for cadmium to enter animals or humans either directly or via the food chain. Ingested cadmium is poorly absorbed from most diets, and absorption is influenced by the levels of calcium, zinc, manganese, and copper, possibly because of similar absorptive mechanisms.

A specific metal-binding protein, metallothionein, is responsible for the accumulation of cadmium in animals. The concentration of this protein increases following continued exposure to cadmium. As cadmium level increases, the amount of the metallothionein available to deactivate the cadmium seems to increase also. Estimates are that kidney and liver account for half of the absorbed cadmium. Numerous investigators have shown that cadmium is rapidly concentrated in the liver and kidney and that cadmium is shifted from other tissues to the kidney. Cadmium concentrations in the kidney are approximately 10 times those observed in the liver. See METALLOPROTEIN.

In mammals, the placenta and mammary glands effectively limit cadmium transfer into fetus and milk. Further, cadmium taken up by plants and ingested by animals does not appear to accumulate in heart or muscle tissue. Thus, foods (milk and meat) consumed directly by humans accumulate minimal levels of cadmium.

For background information SEE CADMIUM DISEASE; ECOLOGY; FOOD WEB; LEAD ISOTOPES (GEOCHEMISTRY); LEAD TOXICOLOGY in the McGraw-Hill Encyclopedia of Science and Technology.

Douglas Pamp

Bibliography. B. J. Bray et al., Trace metal accumulation in tissues of goats fed silage produced on sewage sludge-amended soil, *J. Environ. Qual.*, 14:114–118, 1985; Council for Agricultural Science and Technology, *Application of Sewage Sludge to Cropland: Appraisal of Potential Hazards of Heavy Metals to Plants and Animal*, EPA Publ. MCD-33, 1976; R. H. Dowdy et al., *Effects of Sewage Sludge on Corn Silage and Animal Products*, USDA Rep. 600/S2-84-075, Environmental Protection Agency, 1984; H. Jenny, *The Soil Resource*, 1980; A. Kabata-Pendias and H. Pendias, *Trace Elements in Soils and Plants*, 1984; National Research Council, Subcommittee on Mineral Toxicity in Animals, *Mineral Tolerance of Domestic Animals*, 1980; J. O. Nriagu and C. I. Davidson, *Toxic Metals in the Atmosphere*, 1986; T. Y. Toribara, M. W. Miller, and P. E. Morrow, *Polluted Rain*, 1980.

Forensic chemistry

Following the discovery of hypervariable minisatellite deoxyribonucleic acid (DNA) and the demonstration of its use in paternity testing by A. J. Jeffreys, the United Kingdom Forensic Science Service carried out a research program to evaluate the potential of genetic fingerprinting in forensic science. Genetic fingerprinting enables virtually 100% discrimination between individuals from small samples of blood or semen. No two people have the same genetic fingerprint, unless they are identical twins.

Hypervariable minisatellites consist of tandem repetitive regions of DNA and are dispersed throughout the human genome. They share a common core sequence of 10–15 base pairs. The polymorphism arises from variation in the number of minisatellite repeats found at a particular locus. A hydridization probe containing a tandem repetition of the core sequence can detect many highly variable loci simultaneously. The restriction enzyme used to detect minisatellites cuts the DNA on either side of the gene. If it cuts within the minisatellite sequence itself, the DNA will be cut into very small pieces and will not be visualized. A high mutation rate accounts for the high polymorphism. It is thought that the core sequence may function as a meiotic recombination signal.

A number of different minisatellite probes have been described. Two of these are identified by the codes 33.15 and 33.6. Different minisatellite probes have differences in their sequences. For example, 33.15 has a sequence of 16 bases which is repeated 29 times, and 33.6 consists of a trimeric core sequence of 37 base pairs which is repeated 25 times. These two probes will detect different sets of minisatellites. Generally, the greater the differences between the arrangements of core sequences in probes, the greater will be the differences of resultant fingerprints.

Since hypervariable minsatellites are completely specific to an individual, they provide a powerful tool for paternity testing. Genetic fingerprinting has potential in criminal investigations, since DNA of high molecular weight can be isolated from bloodstains as old as 4 years made on cotton cloth and digested with the restriction enzyme *Hin*fI. Fingerprints obtained from different tissues or body fluids of a particular individual are identical. Furthermore, sperm nuclei can be separated from vaginal cellular debris obtained from semen-contaminated vaginal swabs, so that the suspected male donor can be positively identified. The technique of DNA fingerprinting will certainly revolutionize forensic biology, particularly with regard to the identification of rape suspects.

DNA extraction. Biological samples for forensic analysis consist mainly of dried blood or semen stains on cloth or other surfaces, and vaginal swabs taken after an alleged rape. The blood and semen specimens may be several days or even weeks old. Hair roots may also be examined for the presence of polymorphic markers. DNA fingerprints can be obtained only from undegraded DNA of high molecular weight. Previous attempts to isolate DNA from dead or aged material have been reported for museum specimens (muscle from a 140-year-old guagga, an extinct relative of the zebra, and 2400-year-old mummies). In both cases, only DNA of low molecular weight could be obtained from blood-stained material provided it was kept dry. Such samples were stable for as long as 4 years, but if samples are wet or in a humid atmosphere, degradation of the DNA rapidly ensues, and fingerprints cannot be obtained after 4 or 5 days. Nevertheless, DNA can be isolated from samples which may be typically encountered in forensic laboratories.

DNA can be extracted from whole blood, whole semen, vaginal fluid, hair roots, buccal cells, bloodstains, and semen stains by overnight incubation in a mixture of detergent, proteinase K, and dithiothreitol. The detergent lyses cell nuclei; proteinase K digests protein which would otherwise interfere with the recovery of DNA at a later stage; and dithiothreitol reduces thiol or sulfur groups which are found in many proteins.

After purifications in mixtures of phenol and chloroform and in ethanol, samples are digested with the restriction enzyme *Hin*fI, and electrophoresis is carried out through a 20-cm-long 0.6% agarose gel. Afterward the DNA is denatured in place by incubating the gel in 0.25 M hydrochloric acid (HCl) followed by 0.5 M sodium hydroxide (NaOH). After neutralization in *tris*-HCl [a buffer solution formed from *tris* (hydroxymethyl) aminomethane and hydrochloric acid, pH = 7.5], the DNA is transferred onto a nitrocellulose membrane by Southern blotting. That is, the gel is placed onto a filter-paper base in a tray containing a salt solution. A wet nitrocellulose filter is placed on top of the gel, and on top of this are placed dry absorbent filter paper and paper towels. DNA is transferred onto the nitrocellulose membrane by capillarity. When transfer has been completed, the nitrocellulose filter is baked in an oven at 176°F (80°C). The DNA is now permanently bound to the membrane, which can be subsequently hybridized in solution to radioactively labeled minisatellite probes. Hybridization takes place at 142°F (61°C) overnight. Excess probe is washed away and the nitrocellulose filter is dried. Autoradiography is then carried out by placing the filter in an x-ray cassette in contact with a film sensitive to beta radiation. After a suitable period of time (2–7 days) the film is developed and the fingerprints can be visualized (**Fig. 1**).

Because preliminary experiments showed that semen-contaminated vaginal swabs contained large amounts of DNA from the female, which tend to obscure many of the bands from the

control
(reference
sample)

control
(reference
sample)

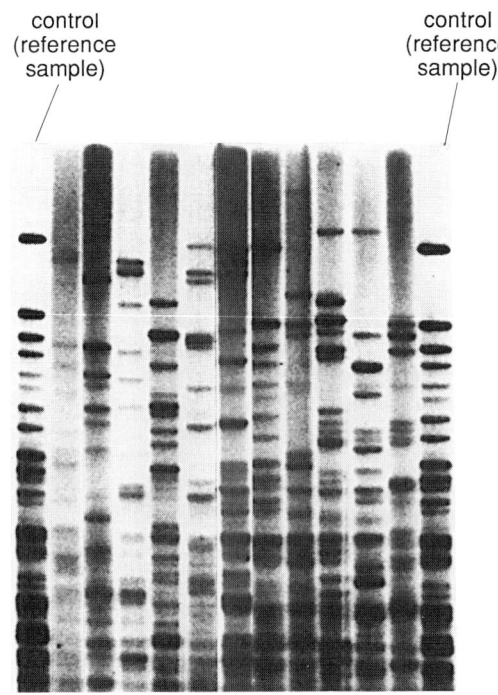

Fig. 1. DNA fingerprints of blood stains from different individuals.

sperm, female cell nuclei are preferentially lysed by preliminary incubation in a mixture of detergent and proteinase K. Sperm nuclei are impervious to this treatment because they are surrounded by a membrane of cross-linked thiol-rich proteins which maintains their integrity, and they can therefore be separated from the female component by centrifugation. Sperm nuclei are sub-

sequently lysed by treatment with a mixture of detergent, proteinase K, and dithiothreitol. Since the resulting fingerprints are predominantly free of female contaminants, an unambiguous detection of sperm DNA can be made (**Fig. 2**).

Electrophoresis of samples on 0.5% agarose gels showed that DNA of high molecular weight could be isolated from blood and semen stains that were at least 2 years old, from fresh hair roots, and from buccal cells lining the mouth. However, no DNA could be extracted from hair shafts or from dead skin cells. The ability to extract sufficient DNA from buccal cells may be useful if there is an objection to taking blood samples from a suspect.

DNA fingerprints obtained from bloodstains, semen stains, and hair roots have been shown to be specific to individuals when compared with whole blood and semen samples. Spurious bands have not been observed, even from old samples or from samples which have been partially digested by the restriction enzyme. Fingerprints can be obtained from bloodstains up to 4 years old, and no differences have been observed between patterns from fresh hair roots or bloodstains and semen stains up to 4 years old and those from fresh blood or semen from the same individual.

Approximately 5 microliters of semen or equivalent semen stain and 60 microliters of blood or equivalent bloodstain are required to produce a DNA fingerprint.

Advantages of DNA. Typically, a battery of polymorphic protein and blood group marker systems are currently used in forensic identification. The Metropolitan Police Laboratory in London uses 12 different enzyme and blood-grouping systems. Detection of some systems such as phosphoglucomutase 1 and rhesus is difficult after 26 weeks. Esterase D cannot be reliably detected after 1 month.

Only five systems are currently in common use for identifying genetic markers in semen (ABO, phosphoglucomutase 1, glyoxylase, peptidase A, and the Lewis factor), and all must be used with considerable caution because the semen samples analyzed are often contaminated with vaginal fluid which itself has enzyme and blood group activities. Semen contains proteolytic enzymes which further reduce the amounts of proteins recovered. Furthermore, bacterial activity may be responsible for producing erroneous results, particularly in the identification of ABO blood groups. Because DNA hydridization probes recognize a core sequence of 10–15 base pairs, so far found only in chordates, bacterial or yeast contamination does not affect the results of a DNA analysis.

Probabilities. DNA fingerprints are destined to completely change the emphasis of blood grouping in forensic science. By using eight polymorphic protein systems together, the probability that two randomly chosen individuals share the

vaginal swab

donor male

donor female

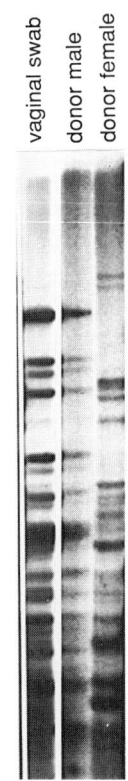

Fig. 2. DNA fingerprint from semen-contaminated vaginal swab taken 2 h after intercourse, compared with blood from the donor male and blood from the donor female. The DNA extraction methods employed remove virtually all of the female DNA from the swab.

same phenotype has been calculated as 0.014. In practice, the degree of characterization is often much lower, particularly when semen is grouped. In contrast, the condition of nonassociation has been absolute with traditional blood-grouping tests; that is, if the phenotypic expression does not match, a common origin is not possible. The probability that a particular band in a DNA fingerprint from one individual is also present in another individual is dependent upon molecular weight, and for the DNA minisatellite probe 33.15 this has been calculated as 0.07 for 10–20 kilobase fragments, 0.2 for 6–10 kilobase fragments, and 0.26 for 4–6 kilobase fragments. Approximately 11 bands between 4 and 20 kilobase are scored in any genetic fingerprint, giving a probability of chance association of 5.8×10^{-8}. By using two DNA minisatellite probes, 33.15 and 33.6, the degree of association can approach certainty; a probability of 5×10^{-19} of chance association has been estimated.

For background information SEE DEOXYRIBONU-CLEIC ACID (DNA); ELECTROPHORESIS; FORENSIC CHEMISTRY in the McGraw-Hill Encyclopedia of Science and Technology.

Peter Gill

Bibliography. P. Gill, A. J. Jeffreys, and D. J. Werrett, Forensic application of DNA "fingerprints," *Nature*, 318:577–579, 1985; A. J. Jeffreys, J. F. Y. Brookfield, and R. Semeonoff, Positive identification of an immigration test-case using human DNA "fingerprints," *Nature*, 317:818–819, 1985; A. J. Jeffreys, V. Wilson, and S. L. Thein, Hypervariable "minisatellite" regions in human DNA, *Nature*, 316:76–79, 1985.

Forestry

Forest biotechnology, commonly called genetic engineering, is a relatively new forest science. In less than 10 years it has been demonstrated that the basic tools of biotechnology can be successfully applied to forest trees. The next decade promises to be very exciting in forestry, with major changes in the way trees are grown and wood and wood chemicals are produced.

Biotechnology has provided forestry with a unique set of tools for overcoming the conventional biological constraints, that is, delayed sexual processes and the long life and large size of trees. The techniques of biotechnology provide a set of working tools for identifying and isolating genes, the basic inheritance units of life, and transferring them from one organism to a closely related or even a nonrelated organism. There has become available in forestry a means for directing useful genetic changes at the molecular level by careful selective manipulation without disrupting useful features of trees. If these efforts are successful, timetables for tree improvement can be established that are similar to those of agriculturists who produce new varieties in 3 to 5 years. Where formerly it required decades to produce trees of superior growth, wood quality, or pest resistance, it now will be possible to modify selected trees within a much shorter time. More effective biological control agents will be developed by the direct modification of natural pests of forest insects, and cell factories will be created for the purpose of producing useful chemicals from wood and waste-wood residues.

Such a program requires manipulation of a broad genetic base. As many useful genes as possible must be available, since at present the specific requirements of future societies are unknown. Modern forestry must therefore minimize genetic losses by a vigorous and active genetic conservation effort. Unlike the agriculturist, the forester still has the opportunity to maintain this natural forest genetic base in the United States and has access to appropriate methods.

Tools of biotechnology. Biotechnology uses aspects of many different disciplines focused on a set of common problems. It has borrowed and modified techniques from biochemistry, plant physiology, and genetics, and has adapted them to forestry. The basic tools currently employed in forest biotechnology include gene splicing (recombinant DNA), cell fusion (parasexual hybridization), and somaclonal selection (cell screening).

Gene splicing. After desirable genes have been identified, they are isolated and transferred to a host tree. Isolation of the genes is accomplished by means of enzymes (restriction enzymes); they are then multiplied by using cloning techniques. Within the last several years, systems have been developed for transferring known fragments of genetic information from one organism to a forest tree. The most common carrier is the Ti plasmid from the common gall bacterium (*Agrobacterium tumefaciens*). In a few cases the genetic information is transferred directly by microinjection, a new technique developed to accelerate gene transfers in cells with large nuclei, in which the genetic material is inserted directly into the nucleus of an isolated cell by means of a very fine needle controlled by a special instrument, a micromanipulator.

Cell fusion. This involves forest tree cells. Genetic information is transferred by combining the cell protoplasts, plant cells in which the cell walls have been carefully removed by the use of enzymes. By selected chemical or electrical means the protoplasts are induced to fuse with each other. This type of pairing is most commonly attempted with related tree species that will not naturally sexually mate to produce seeds, such as the various species of elms. While this method currently is of very limited value because of the difficulties encountered in obtaining whole plants from fused cells, it does offer a useful technique for both nuclear and cytoplasmic inheritance.

Somaclonal techniques. These methods expose individual cells, groups of cells, or tissues to various environments under sterile laboratory conditions. Somaclonal techniques provide, in some cases, more efficient screening for stress, herbicide tolerance, and disease resistance than conventional field techniques. Most frequently, single cells are used, although new research suggests that other types of plant material should be tested also, and an array of tissues might well be used.

Current research. The insertion of the Aro-A-gene into selected tissues of poplars is a recent example of the success of forest biotechnology. The Aro-A-gene, a patented gene isolated from bacteria, has been demonstrated to confer tolerance to the herbicide glyphosate in higher plants. This experiment indicates that genes from organisms other than trees can be used by trees and that it is possible to transfer individual genes. While the concept will require some additional testing, there appears to be no need for long-term testing. The herbicide tolerance can be tested in young trees under greenhouse control conditions. *See PLANT PHYSIOLOGY*.

An example of the successful application of somaclonal technology is the acceleration of the identification of genetic resistance in poplars to certain leaf diseases. Results of recent investigations indicate that poplars resistant to *Septoria musiva* can be identified in a few months in the laboratory as compared with years by the traditional field testing practiced by foresters.

There is active research involving use of restriction enzymes to identify the genetic composition of the nuclear polyhedrosis virus (NPV). This virus is a commonly used biological control agent for the harmful gypsy moth. Over time, forest pests frequently develop natural resistance to both chemical pesticides and biological control agents. There is a constant need for new or improved control systems. Once the basic virus genetic structure is known, biotechnological techniques will permit its rapid manipulation in order to respond quickly to changing host resistance.

In the last few years, powerful enzymes called ligninases have been identified. This group of enzymes has the ability to break down lignin, an expensive waste product of the pulp and paper-making process. The genes for producing ligninase have been identified, and genetic engineering research is seeking to create an organism that produces ligninase without consuming the lignin biodegradation products. If this effort is successful, it would lead to the establishment of a new major forest chemical products industry.

Forest resource conservation. The basic element of forest biotechnology, the gene, is stored in the living tree. It should not be assumed that nature alone will protect this vital resource. Many current forest practices are actually narrowing the forest genetic resource—for example, high grading, the common practice of harvesting only the very best trees. As forest management becomes more intensive, fewer natural stands remain untouched; more natural forests will be replaced by nonlocal planting stock. With the expansion of tree improvement programs, more genetically improved seedlings will replace natural populations. Yet the primary goal of forest management remains the production of a broad range of forest products and services on a sustained basis under very different environmental conditions. This requires a reliable and highly varied genetic reservoir for current and future use, as well as for unforeseen contingencies. Even with the techniques of modern forestry, environmental events that shape the long-lived forest stands cannot be controlled.

Genetic conservation strategies. A number of feasible methods for forest genetic conservation are being applied in the United States. Natural regeneration systems are encouraged in many areas by modifying current harvesting systems to provide a sufficient number, distribution, and type of parent or seed trees, thus ensuring a broad genetic base which is representative of the local population.

When artificial regeneration systems are employed, local seed sources representative of local conditions are used, with seeds frequently collected from local stands prior to harvesting.

When a given area has certain unique characteristics such as unusual species composition, growth rate or form, or the ability of forests to grow under unusual conditions, such as high or low soil moisture content and low nutritional levels, the forest stand can be identified and protected as a special-use area, such as a Research Natural Area or a Wilderness Area. Research Natural Areas are special administrative areas that can be established by a land managing organization. Wilderness Areas include large forested areas, unique in their biological compositions or scenic and outdoor values, and they are established by congressional legislation. Research Natural Areas can be established quickly to protect threatened or endangered species or even small populations. However, because of their relatively small size they cannot be relied upon to conserve representative populations of large forested areas.

A typical genetic conservation unit is the Lake Twenty-Two Research Natural Area, located in Mount Baker Snoqualmie National Forest, Washington. This is an old-growth western red cedar–western hemlock forest on the west side of the Washington Cascades. Such areas can be successfully employed to protect and maintain unique genetic resources.

For large forested areas in which intensive management is being or will be practiced, special Genetic Resource Management Units are being identified and established. A Genetic Resource

Management Unit is a representative area in which only local seed sources are used and disgenic selection practices are avoided; the removal of the best trees or the removal of certain size classes is avoided. Normal forest uses and management practices can still be conducted. The basic requirement is maintenance of a high degree of natural genetic diversity.

These methods are designed to maintain small or large populations and whole forest communities. Individual trees of many species are being maintained by storing their seeds at very low temperatures until needed. For some species, for example, pines, seeds have been successfully stored for as long as 40 years. However, such storage may result in some genetic losses over time and should not be relied upon as the only means of genetic conservation.

For background information SEE FOREST AND FORESTRY; FOREST PLANTING AND SEEDING; GENETIC ENGINEERING in the McGraw-Hill Encyclopedia of Science and Technology.

Stanley L. Krugman

Bibliography. J. C. Barber and S. L. Krugman, Preserving forest tree germplasm, *Amer. For.*, 80(1):8–11, 1974; S. L. Krugman, The ethical question, *J. For.*, 84(1):40–41, 1986; S. L. Krugman, Policies, strategies and means for genetic conservation in forestry, in C. W. Yeatman et al. (eds.), *Plant Genetic Resources: A Conservation Imperative*, pp. 71–78, 1984; S. L. Krugman, Traditional forest genetics vs biotechnological and physiological approaches, in F. Caron, A. G. Corriveau, and T. J. B. Boyule (eds.), *Proceedings of the 20th Meeting of the Canadian Tree Improvement Association*, pp. 62–67, 1986.

Fractals

The concept of fractal structure has great utility because structures which appear random can be described, in fact, within a precise geometric mathematical framework. Fractal concepts may describe not only static geometrical properties but also dynamical properties. Recent observations of the dynamical properties of materials have yielded results that are in agreement with the predictions for the dynamical behavior of fractal structures. These observations suggest that fractal structures may be a rather widespread phenomenon, and that this concept may apply to a large variety of amorphous solids (for example, composites, glasses, and ceramics). Specifically, recent measurements of neutron scattering in epoxy resins, and light scattering of dehydrated gels both show the properties expected for fractal networks.

Statics. The most easily understood fractal concept is density. For fractal structures, density means that there is no constant relation between mass and volume as the length scale is changed. The amount of mass $M(r)$ inside a sphere of radius r for a crystalline material scales as in Eq. (1), where A is a numerical constant and d is the

$$M(r) = Ar^d \qquad (1)$$

so-called euclidean dimension (d is 3 in the usual world). However, for fractal materials the amount of mass inside a sphere of radius r scales as in Eq. (2), where B is a constant which

$$M(r) = Br^D \qquad (2)$$

depends upon the openness of the fractal structure, and D is the so-called fractal dimension. In general, D is less than or equal to d because of the open character of fractal structures. Such structures tend to exhibit inhomogeneous arrangements of particles, with large amounts of open spaces (voids) and irregular atomic arrangements. The mass density ρ of a sphere of radius r is the ratio of the amount of mass $M(r)$ inside the sphere to the volume of the sphere $V(r)$, given by Eq. (3), where C is a constant. Because of the

$$V(r) = Cr^d \qquad (3)$$

peculiar mass scaling for fractal structures, this mass density is given by Eq. (4). This equation

$$\rho(r) = \frac{M(r)}{V(r)} = \frac{Br^D}{Cr^d} \propto r^{D-d} \qquad (4)$$

immediately implies that fractal materials can be expected to be very light, since D is less than d in general.

Dynamics. The dynamical properties of fractal networks can be most easily visualized by the process of diffusion on a fractal. An example of such a process is an imaginary walker who can step only on the atoms which make up the fractal but who can walk in any direction on the fractal at a constant rate. The walker's average displacement is, of course, zero, but the mean-square displacement is not. It is usually specified by a diffusion constant $\mathcal{D}$. For stystalline materials, $\mathcal{D}$ is a constant and the walker's mean-square displacement $r^2(t)$ in time t is given by Eq. (5),

$$\langle r^2(t) \rangle = \mathcal{D}t \qquad (5)$$

where $\langle \ldots \rangle$ means that an average is taken over many different starting points. The quantity $\mathcal{D}$ is called the diffusion constant. Because a fractal has so many open spaces and the atoms are connected to one another in such a tortured manner, the so-called diffusion constant on a fractal is not really a constant. Instead, it has been shown that the diffusion constant varies with radius r according to Eq. (6), where θ is a

$$\mathcal{D}(r) \propto r^{-\theta} \qquad (6)$$

constant (usually between 1 and 2 for $d = 3$). If Eq. (6) is substituted into Eq. (5), it can be seen that a much longer time is required for a walker to travel the same mean-square distance on a fractal structure than on a crystalline structure. These two equations can be combined to yield Eq. (7), so that Eqs. (8) and (9) follow, where the new

$$\langle r^2(t) \rangle = 2t \propto r^{-0}t \qquad (7)$$

$$\langle r^{2+0}(t) \rangle \propto t \qquad (8)$$

$$\langle r^2(t) \rangle \propto t^{2/(2+0)} \equiv t^{\bar{d}/D} \qquad (9)$$

concept of fracton dimension $\bar{\bar{d}}$ is defined by Eq. (10). The fracton dimension plays a central role in

$$\bar{\bar{d}} = \frac{2D}{2 + \theta} \qquad (10)$$

the dynamics of fractal networks.

An example is provided by a model in which atoms are removed at random from a perfect crystalline array. The resulting structure is known as a percolation network. If the percentage of atoms remaining exceeds a critical value p_c, it can be shown that there remains a connected path through the material from one side to another. Such a structure is termed an infinite network. It can be imagined that this network is allowed to vibrate, and an important dynamic property of the network is the number of vibrational modes with frequency f between f and df, where df is an infinitesimally small range of frequencies. This so-called density of states was shown to be proportional to $f^{\bar{d}-1}$. The analogous result for the crystalline array before any atoms were removed would be f^{d-1}.

The reason that $\bar{d}$ is referred to as a dimension can now be understood. The density of vibrational states for a fractal can be obtained from a crystalline array by replacing the euclidean d by the fracton dimension $\bar{d}$. It has become common practice to label the vibrations of a crystalline array as phonons, reflecting their quantum-mechanical character. Similarly, the quantized vibrations of a fractal network are referred to as fractons. The fracton dimensionality determines the relationship between the vibrational frequency and the spatial extent of a quantized vibration of a fractal network. This is known as the fracton dispersion law. Thus, the fracton dimension is sufficient to determine the density of states of a vibrating fractal network, and the dynamical dispersion law for excitations on that network. It is the power of fractal geometry that a single quantity $\bar{d}$ is sufficient for both these properties.

The last concept which is necessary to characterize fractal structures is the crossover length ξ. In fractal dynamics, length scales are short compared with ξ, while in crystalline dynamics, length scales are long compared to ξ. For a percolation network the crossover length ξ depends upon the concentration of occupied sites p, in terms of the fundamental interatomic distance a, according to Eq. (11), where v is termed a

$$\xi = a(p - p_c)^{-v} \qquad (11)$$

critical exponent and equals $\frac{4}{3}$ for percolation networks in $d = 2$.

Amorphous structures. It is possible that amorphous structures may exhibit fractal dynamics at short length scales (or high vibrational frequencies). These structures have a mass which is nearly equal to that of crystalline materials, so that Eq. (4) for the mass density does not appear to apply. However, their vibrational properties may depend upon those atoms which are connected in an infinite network. The material can be thought of as being composed of an infinite network of atoms with the voids filled in with atoms which do not participate in the long-length-scale vibrational properties. In this way the material does not exhibit fractal mass properties [as given by Eq. (4)] but does exhibit fractal dynamics (where the density of vibrational states is proportional to $f^{\bar{d}-1}$). The validity of this description can be examined by measuring either the density of states or the character of the vibrational states. At low frequencies f, the states may be described by sound waves, with a wavelength that is greater than ξ. However, as the frequency f is increased, the wavelength diminishes. Eventually, for sufficiently large frequency, the wavelength will equal ξ. This occurrence marks a crossover into the regime of fractal dynamics, and the vibrational excitations are then referred to as fractons. The fractons are known to be localized in space, and so are distinctly different from sound waves, which may propagate freely through even random media. Localized vibrations (here, fractons) would appear to be overdamped if measured in a scattering experiment; that is, their mean-free path would be comparable with their wavelength, or,

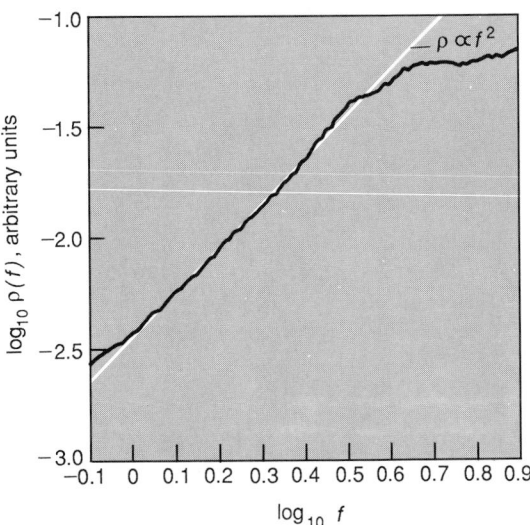

Fig. 1. Vibrational density of states $\rho(f)$ for crystalline diglycidyl ether of bisphenol-A (DGEBA) plotted on a log-log scale against the frequency f of vibrational excitation. (Frequency f is expressed in terms of the corresponding energy $E = hf$, in millielectronvolts, where h is Planck's constant. A frequency of 2.42×10^{11} Hz corresponds to 1 meV.) The straight line is drawn with a slope of 2, showing the behavior expected for a crystalline material. The deviation from this line at high frequencies is caused by molecularlike excitations.

equivalently, their energy-width would be comparable with their frequency f.

Experimental observations. Recently, both of these predictions have been observed experimentally. Diffuse inelastic neutron scattering experiments were performed on crystalline and amorphous diglycidyl ether of bisphenol-A (DGEBA). **Figure 1** exhibits the measured vibrational density of states for the crystalline material. This density increases smoothly as f^2, as predicted for crystalline structures with $d - 1 = 2$. The deviation at the high-frequency end of the scale is caused by molecularlike excitations. When DGEBA is melted and allowed to set, an amorphous material is obtained. **Figure 2** exhibits the measured vibrational density of states for the amorphous material. This density initially increases as f^2, but then rolls over into a much lower slope, suggesting a fracton dimensionality $\bar{d} \leq d$. The deviation from the f^2 behavior can be interpreted as the crossover between normal sound waves and fractons as the vibrational excitation energy increases.

Recently, an even more direct example of this crossover has been observed in light-scattering experiments on silica aerogels. The gels were prepared in solution and then dried by using hypercritical solvent evacuation. **Figure 3** exhibits the measurement of the half-linewidth Γ, plotted as a function of the frequency of the vibrational excitation. For purposes of comparison the scales of both quantities have been divided by the crossover frequency f_{co}, the frequency at which the sound wavelength equals ξ. It can be seen that the linewidth becomes comparable with the frequency at values of frequency slightly below f_{co}, providing a strong indication of crossover from sound waves to

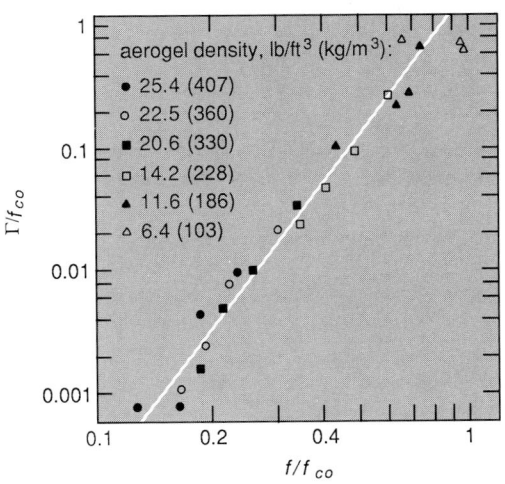

Fig. 3. Measured half-linewidths of vibrational excitations for six silica aerogels with differing density, plotted on a log-log scale against their frequency. Both quantities are normalized to the crossover frequency f_{co}, at which the sound wavelength equals the crossover length ξ.

fracton vibrational excitations.

These examples suggest that fractal dynamics may apply to a wide variety of amorphous structures, perhaps even to the most common of such structures, the glasses. If this is true, the acoustic and heat transport properties of these materials could be accounted for with the relatively simple geometrical concept of fractal dynamics. Experiments are in progress to test this idea, and will probably be decisive in answering the question of whether fractal dynamics describes the vibrational properties of amorphous materials.

For background information SEE DIFFUSION IN GASES AND LIQUIDS; FRACTALS; LATTICE VIBRATIONS in the McGraw-Hill Encyclopedia of Science and Technology.

Raymond Orbach

Bibliography. S. Alexander and R. Orbach, Density of states on fractals: "Fractons," *J. Phys. (Paris) Lett.*, 43:L-625–L-631, 1982; Y. Gefen, A. Aharony, and S. Alexander, Anomalous diffusion on percolating clusters, *Phys. Rev. Lett.*, 50:77–80, 1983; B. B. Mandelbrot, An explicit fractal model of percolation clusters, *Ann. Israel Phys. Soc.*, 5:59–80, 1983; B. B. Mandelbrot, *The Fractal Geometry of Nature*, 1983; B. B. Mandelbrot, Fractals in physics: Squig clusters, diffusions, fractal measures, and the unicity of fractal dimensionality, *J. Stat. Phys.*, 34:895–930, 1984; R. Orbach, Dynamics of fractal networks, *Science*, 231:814–819, 1986.

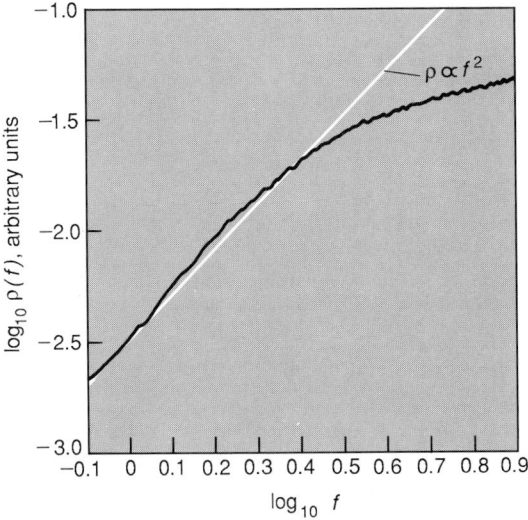

Fig. 2. Vibrational density of states $\rho(f)$ for amorphous diglycidyl ether of bisphenol-A (DGEBA) plotted on a log-log scale against the frequency f of vibrational excitation. The straight line is drawn with a slope of 2. The measured density of states deviates from this line at frequencies considerably below those in Fig. 1, which correspond to molecular motions.

Genetic engineering

The ability to identify, isolate, clone, manipulate, and express the genes of any organism, prokaryotic or eukaryotic, is the basis of modern genetic engineering. By using recombinant deoxyribonucleic acid (DNA) techniques, a gene from a foreign source, such as a plant or mammal, can be inserted into a stably replicating plasmid vector which is then introduced into living cells, and the gene is expressed therein to produce high levels of the encoded peptide or protein. This permits the production of high-value or medically useful substances in microorganisms, such as bacteria, yeast, and fungi, that normally do not produce them.

However, isolation of a genomic DNA sequence or messenger ribonucleic acid (mRNA) sequence necessary for generating complementary DNA (cDNA) clones is sometimes difficult. Moreover, it is often difficult to achieve useful expression levels from cDNA clones, and precise manipulation of cDNA and genomic constructs can be quite laborious. As a result, total gene synthesis has become increasingly useful to the biotechnologist.

Rationale. The primary rationale for wanting to make a synthetic gene is the ability to tailor the DNA sequence to suit both the desired expression system, that is, the plasmid vector and the target microorganism, and the goals of the work. For example, the goals might be high-level expression and secretion of a peptide hormone, understanding the regulatory mechanism of gene expression, or studying the structure–function relationships of an enzyme via protein engineering. This gives the modern gene designer considerable flexibility and several advantages over the original designer.

For example, new restriction sites can be added at the ends or in the middle of the gene's coding sequence to facilitate construction of the expression vector and subsequent modification of the gene by mutagenesis outside the organism. Alternatively, existing restriction sites that are incompatible with the cloning strategy can be removed. Any or all noncoding intron sequences present in the natural gene may be deleted in the synthetic gene. Control-element sequences that regulate expression of the gene, determine the start and stop points of protein translation, or target the protein product for extracellular secretion can be built into the synthetic gene. The codon sequences that specify particular amino acid residues can be selected for optimized expression according to the preferences of the target organism. Finally, DNA sequences from different genes can be fused to generate hybrid or multifunctional proteins with altered or novel biological activities. Thus, gene synthesis enables complete customization or improvement of the original, naturally occurring gene.

Requirements. This points up an obvious but significant limitation of gene synthesis. The complete DNA sequence of the target gene must be known a priori in order for it to be synthesized. Thus, if it is not already available, the natural gene has to be isolated and cloned for determina-

tion of its sequence. Although this requires additional time and labor, rapid methods for cloning and sequencing DNA have been developed. In some instances, even though the DNA sequence of the gene is not known, the amino acid sequence of the encoded peptide or protein is available. Back translation of the protein sequence via the genetic code, that is, amino acid to trinucleotide codon, provides a partially ambiguous DNA sequence. By using the codon preferences of the source organism, a putative gene sequence can be determined without isolating the natural gene.

Total gene synthesis is dependent on two key technologies: the chemical synthesis and purification of defined-sequence oligonucleotides, and the enzymatic assembly of the short oligonucleotides into long, double-stranded DNA fragments. The pioneering developments in both of these areas were done by H. G. Khorana and coworkers in the 1960s, and culminated in the total synthesis of the 77-base-pair (bp) yeast alanine transfer RNA (tRNA) gene in 1970. This effort involved the chemical synthesis, purification, and enzymatic ligation of 14 different oligonucleotides 5–16 residues in length, and required 3 years of work by 13 researchers. Since then and primarily since 1982, several advancements and refinements in DNA synthesis chemistry, automated instrument design, and DNA fragment assembly strategy have greatly reduced the time and labor required for whole-gene construction. The total synthesis of the 1057-bp bovine rhodopsin gene was completed in 1986 in Khorana's laboratory, using only 72 oligonucleotides (15–40mers), four researchers, and one year's time. It is now feasible to synthesize very long, complex genes in a short time (see **table**).

Building blocks. Synthesis of DNA involves oligonucleotide synthesis, work-up and purification, and characterization.

Oligonucleotide synthesis. The chemical synthesis of DNA requires the stepwise, directed formation of phosphate bonds between the 3′-hydroxyl group (3′-OH) of one nucleoside and the 5′-OH of an adjacent nucleoside. The two synthetic methods used most often are the phosphotriester and the phosphoramidite methods. Both employ the same nucleoside-protecting group methodology, that is, the acid-labile dimethoxytrityl group (DMTr) at the 5′-OH's, and base-labile acyl groups (Ac) at the heterocyclic amino functions. In addition, both use the solid-phase approach almost exclusively. In most cases the first 5′-DMTr-nucleoside is covalently attached through its 3′-OH to a chemically derivatized, insoluble polymeric support, for example, polystyrene, cellulose, silica gel, or controlled-pore glass.

Synthesis proceeds stepwise by successive removal of the 5′-DMTr group from the terminal nucleoside, followed by coupling of a 5′-DMTr-3′-phosphorylnucleoside to the free 5′-OH, and so on. The older, more established phospho-

Major gene syntheses

Year	Gene	Length, bp	Number and size of oligonucleotides used	Strategy
1970	Yeast alanine tRNA	77	14/5–16mers	Block ligation
1976	*Escherichia coli* tyrosine suppressor tRNA	206	40/4–13mers	Block ligation
1977	Somatostatin	54	8/11–16mers	Block ligation
1981	Human IFN-α	514	66/14–21mers	Block ligation
1982	IFN-α C-terminus	144	4/39–43mers	Enzymatic fill-in
1983	Human IFN-γ	454	62/11–16mers	Block ligation
1984	Eglin C protease inhibitor	232	6/34–61mers	Enzymatic fill-in
1984	Human growth hormone	584	78/7–26mers	Block ligation
1984	Human proinsulin	286	42/11–17mers	Block ligation
1985	SV40 enhancer element	125	15/14–18mers	Shotgun ligation
1985	Somatomedin C	226	18/16–31mers	One-step ligation
1985	Calmodulin	455	61/9–20mers	Block ligation
1986	Connective tissue–activating peptide	288	20/14–32mers	One-step ligation
1986	Cerebratalus toxin BIV	180	6/55–65mers	Block ligation
1986	Human lysozyme	418	56/11–19mers	Block ligation
1986	Human c-Ha-ras protein	576	38/17–42mers	Block ligation
1986	Bovine rhodopsin	1057	72/15–40mers	Block ligation
1986	Calf prochymosin	1142	28/59–102mers	Shotgun and block ligation

triester method employs 5′-DMTr-nucleoside-3′-arylphosphate compounds in the presence of a condensing agent to generate aryl-protected phosphate-triester linkages (**Fig. 1a**). In the newer phosphoramidite method, 5′-DMTr-nucleoside-3′-dialkylaminophosphites in the presence of tetrazole activator are the coupling agents, producing alkyl-protected phosphite-triester linkages. The nascent phosphite linkages are quite labile and are subsequently oxidized to the more stable phosphate prior to the next synthetic cycle (Fig. 1b).

The entire process is now automated, and the labor-intensive manual methods of oligonucleotide synthesis have been supplanted by computer-driven, fully programmable DNA syn-thesizers or "gene machines." Although the phosphotriester method is fast and efficient (10–20 min per nucleotide addition cycle, 97–98% stepwise coupling efficiency), it can be used to synthesize only up to 30–40mers. The solid-phase phosphoramidite chemistry is currently the method of choice for oligonucleotide synthesis. Its faster cycle time (6–10 min/cycle) and higher coupling efficiency (98–99.5% stepwise yield) make the synthesis of 50–100mers fairly routine.

Work-up and purification. Upon completion of synthesis, the fully protected oligonucleotides are cleaved from the support, and the protected phosphate and amino groups are deblocked to give the free oligonucleotides. The next step is to purify the major oligonucleotide product away

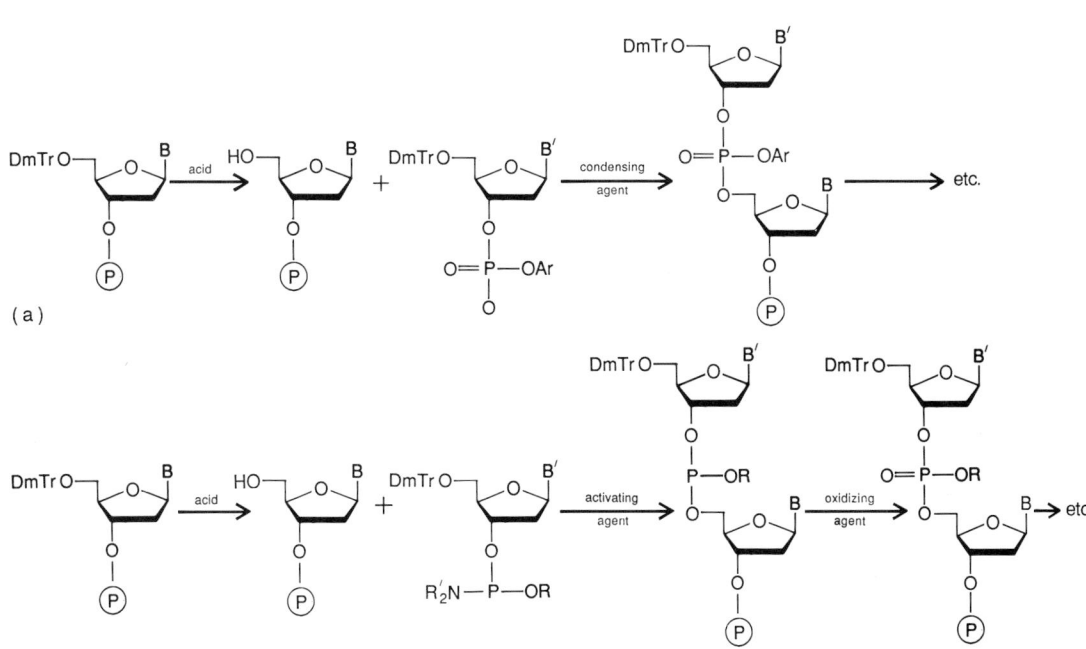

Fig. 1. Reaction schemes for solid-phase oligonucleotide synthesis by the (*a*) phosphotriester method and (*b*) phosphoramidite method. B, B′=protected bases (A, C, G, T), DMTr = dimethoxytrityl, P = polymer support (for example, controlled-pore glass), Ar = ClC$_6$H$_4$—, R = CH$_3$— or NCCH$_2$CH$_2$—, R′ = CH$_3$— or (CH$_3$)$_2$CH—.

from the shorter, failure sequences generated during synthesis. The two methods used most often are high-performance liquid chromatography (HPLC) and polyacrylamide gel electrophoresis (PAGE).

The advantages of HPLC purification are its rapidity and large capacity. By using anion-exchange HPLC, several OD_{260} units (100–200 micrograms) of oligonucleotide can be purified in 20–30 min. However, resolution of the product from truncated sequences of similar size, for example, 50mer from 49mer, 48mer, and so on, becomes more difficult as the chain length increases. Additional fractionations by anion exchange or reversed-phase HPLC are sometimes necessary in order to obtain products of very high purity. Purification by HPLC is used primarily for oligonucleotides in the 16–40mer size range.

Alternatively, preparative PAGE in the presence of 7 M urea readily resolves oligonucleotides that differ by only a single nucleotide. Adjustment of the polyacrylamide concentration from 20 to 6% permits the purification of 16–100mers to single-band homogeneity in a single electrophoresis. However, the DNA must be recovered from the polyacrylamide slices cut out of the gel. For <30mers the oligonucleotide is simply soaked out of the gel fragment, giving 1–2 OD_{260} units of product (3–6 nanomoles). Longer sequences must be electroeluted from the gel slice, and yield 0.1–1 OD_{260} units (100–1000 picomoles of 100mer). At this point, whether HPLC or PAGE was used, a final desalting step is carried out to give pure, ready-to-assemble oligonucleotide.

Characterization. Prior to the use of the purified oligonucleotides in fragment assembly, it is advisable to characterize them. Comparison of the oligomer's size versus known size markers is the fastest method. This is carried out by 5'-radiolabeling samples of the sequences in question, then running analytical PAGE and autoradiography. In addition to sizing the oligonucleotide, this method shows the level of its homogeneity.

For compounds prepared on a reliable DNA synthesizer, the sequence fidelity is quite high, and sizing of the product is sufficient. If there is any question about the reliability of the oligonucleotide, the complete sequence may be determined. For <20mers this is done by using the mobility-shift or wandering-spot method. Longer oligonucleotides (20–100mers) are sequenced by using a modified Maxam-Gilbert chemical cleavage method. Alternatively, the oligonucleotides can be assembled into a double-stranded fragment, the fragment cloned into a suitable sequencing vector for bacterial amplification (for example, M13), and the entire sequence determined by the primed Sanger dideoxy method.

Assembly strategies. The two basic methods for gene synthesis are block ligation and enzymatic fill-in.

Block ligation. This is the most frequently used method for constructing synthetic genes. This approach involves the stepwise building of successively longer double-stranded fragments via enzymatic ligation. An elegant example of the block ligation strategy is the synthesis of the 514-bp human leukocyte interferon (IFN-α) gene (**Fig. 2***a*). Sixty-six oligonucleotides (14–21mers) were synthesized, purified, and 5'-phosphorylated by using the enzyme T4 polynucleotide kinase and adenosinetriphosphate (ATP). These were then annealed in groups of 5–8 and joined together by using T4 DNA ligase. Each resulting double-stranded fragment had a single-strand 3'-tail uniquely complementary to the 3'-tail of the adjacent fragment. These fragments were then annealed 2–3 at a time and ligated again to form four large blocks. In a final annealing–ligation step the four blocks were joined together into the complete gene containing unique restriction sites at each end to facilitate its subsequent cloning.

While this approach is conceptually simple, it has some drawbacks. For a gene of average size (0.5–1.0 kilo-base pairs, kbp) numerous oligonucleotides are required. These must be designed with unique overlap sequences, then synthesized and purified. A second problem is that the efficiency of oligonucleotide kinasing can vary considerably, for example, only 50–80% of the 5'-ends phosphorylated. This causes significant nonconstructive annealing and leads to incomplete ligation. As a result, each fully ligated fragment must be gel-purified away from the nonligated impurities to prevent improper ligation during subsequent constructions. This purification is carried out by polyacrylamide gel or agarose gel electrophoresis, a process that is both time-consuming and yield-reducing as recovery of the fragments from the gel matrices is often low.

In the last 5 years, several refinements and variations of block ligation strategy have been employed which greatly enhance its efficiency and utility for gene synthesis. The simplest change has been the use of longer oligonucleotides for fragment construction. This decreases both the number of oligonucleotides and the number of subfragment ligations needed for a complete gene assembly. The recent synthesis of the 576-bp human c-Ha-ras protein gene required only 38 oligonucleotides (17–42mers) and six annealing–ligation reactions. The longer oligonucleotides also form more stable complementary pairs which anneal more rapidly and ligate more efficiently. As a result, assembled subfragments can be ligated without gel isolation, and short genes can be completely assembled in one step; for example, 20 oligonucleotides (14–32mers) were annealed and ligated in a single reaction to generate the complete 288-bp gene for connective tissue–activating peptide.

Recently, rapid shotgun cloning methods have

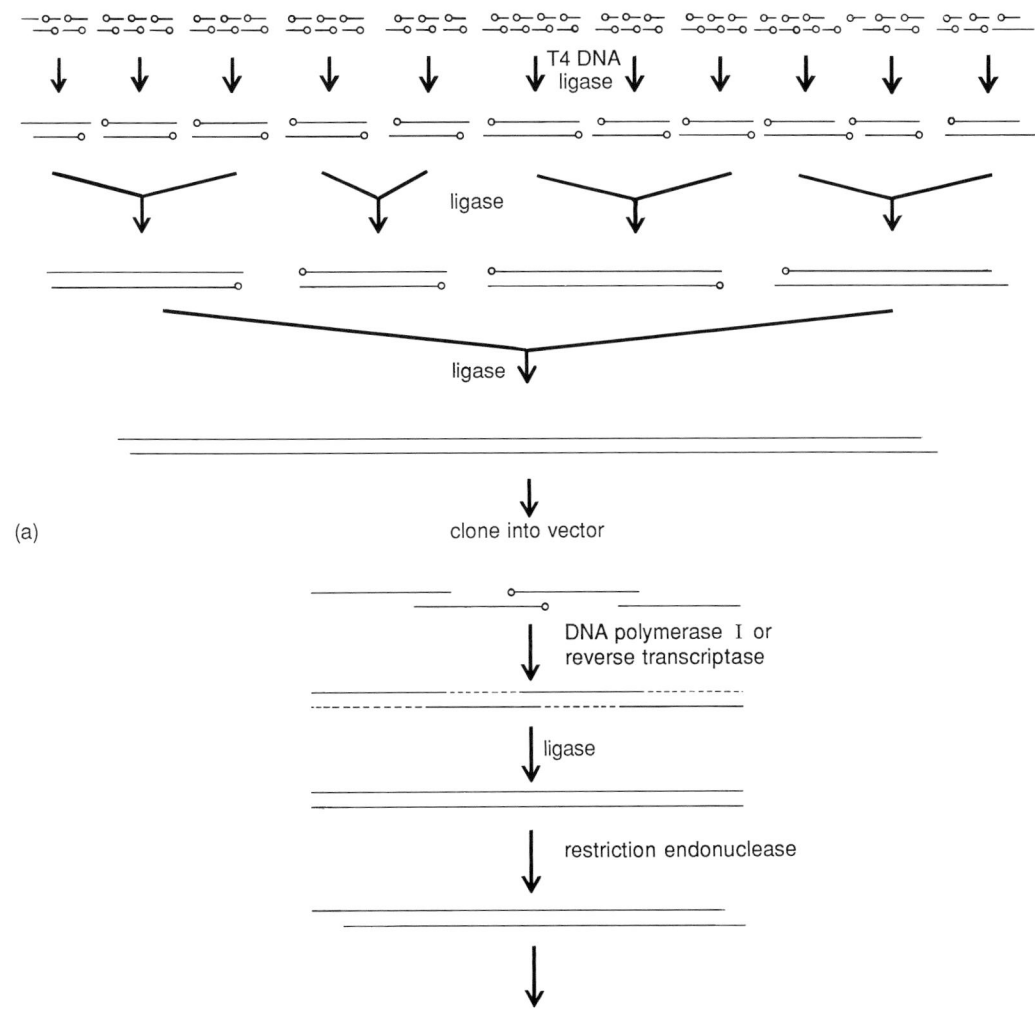

Fig. 2. Strategies for gene assembly using the (a) block ligation approach and (b) enzymatic fill-in approach. The small open circle denotes a 5′-phosphate.

been applied also to gene synthesis. In this approach, 15 oligonucleotides coding for the SV40 enhancer element (125 bp total) were mixed together with the appropriately digested vector plus DNA ligase. The crude mixture was used to transform bacteria which were then plated out and screened for the presence of the gene.

The recent introduction of high-efficiency phosphoramidite reagents for the chemical 5′-phosphorylation of support-bound oligonucleotides has also found application in gene synthesis, eliminating the troublesome kinase step from the assembly scheme. The chemically phosphorylated oligonucleotides ligate very efficiently as evidenced by the one-step construction of a 476-bp prochymosin gene fragment from 10 oligonucleotides in the 71–102mer size range.

Enzymatic fill-in. The availability of longer oligonucleotides, such as 40–60mers, and enzymes having efficient primer-extension activity led to the development of an alternate gene synthesis strategy. In the enzymatic fill-in approach, oligonucleotide pairs are designed to anneal at their 3′-ends, forming a short 10–15-bp double-stranded segment with extended single-strand tails. This primed-template structure is then enzymatically "filled in" from both 3′-ends by using avian myeloblastosis virus (AMV) reverse transcriptase or DNA polymerase I (Klenow fragment), plus deoxynucleoside-3′-triphosphate substrates (dNTPs). The resulting flush-ended, double-stranded fragment (60–100 bp) is gel-purified, then cut with restriction enzymes to produce sticky ends for directional cloning into a vector or ligation to an adjacent fragment generated by the same procedure.

The fill-in approach has the obvious advantage of requiring only half the number of synthetic oligonucleotides that the block ligation strategy requires. Thus, a 132-bp segment of the human leukocyte interferon (IFN-α) gene was synthesized in two fragments by using only four oligonucleotides (~40mers). This greatly reduces the time-consuming oligonucleotide synthesis–purification phase of gene assembly. However, the efficiency of the enzymatic extension can vary considerably, depending on the enzyme and the secondary structure of the DNA. In addition, the overlaps must be carefully screened for uniqueness over the entire fragment sequence in order to eliminate potential secondary priming sites that could lead to mutations or variant constructions.

Recently, the enzymatic fill-in strategy has been extended to four overlapping oligonucleotides (46–50mers), yielding a 164-bp fragment of bovine growth hormone gene (bGH) in a single fill-in–ligation step (Fig. 2*b*). The application of even longer oligonucleotides, for example, 100mers, to this strategy would permit the rapid construction of longer DNA fragments (200–350 bp), thereby facilitating the assembly of very long gene sequences (1–3 kbp).

For background information *see* Gene; Genetic engineering in the McGraw-Hill Encyclopedia of Science and Technology.

Richard W. Barnett

Bibliography. S. P. Adams and G. R. Galluppi, DNA synthesis and applications to molecular biology, *Med. Res. Rev.*, 6:135–170, 1986; H. G. Gassen and A. Lang (eds.), *Chemical and Enzymatic Synthesis of Gene Fragments: A Laboratory Manual*, Verlag Chemie, Weinheim, 1982; K. Itakura, J. J. Rossi, and R. B. Wallace, Synthesis and use of synthetic oligonucleotides, *Annu. Rev. Biochem.*, 53:323–356, 1984; B. E. Kaplan, The automated synthesis of oligodeoxyribonucleotides, *Trends Biotech.*, 3:253–256, 1985.

Geochemistry

Many chemical, mineralogical, and physical changes take place as aqueous pore fluids interact with rocks in the Earth's crust. Not only are aqueous solutions of critical importance in the dissolution, transportation, and deposition of chemical species, but they substantially affect the mechanical behavior of the rocks that contain them. In addition, many subsurface processes take place in open, dynamic systems and include complex fluid-rock interactions under changing conditions of temperature, pore-fluid pressure, lithostatic pressure, deviatoric stress, and chemical environment. New experimental systems have been designed to control or monitor as many of these parameters as possible in order to simulate closely subsurface conditions. This article discusses some recent experimental studies of water-rock interactions.

Laboratory investigations represent one valuable method of increasing understanding of water-rock interactions. Although laboratory data cannot completely replace field-derived data, experimental studies can control or monitor many of the variables. The value of the experiment, however, is directly related to how closely the natural environmental conditions are simulated. Because subsurface processes can take place under a wide variety of conditions, experimental systems and approaches to the study of water-rock interactions have been quite diverse.

Although some early, leaching-type experiments were conducted in open systems, most early experimental work on the chemical and mineralogical aspects of water-rock interactions was done in closed systems. The rock and the fluid were usually sealed inside relatively inert gold or platinum capsules which were placed in pressure vessels and heated. At the end of the experiment the pressure vessel was quenched, the capsule was removed, and the reacted fluids and solids were analyzed. Although systems of this type provided much useful information, only the final fluid composition could be determined in such an experiment. In some cases the analyses were complicated by back reactions with the rock during cooling.

Extraction-type experimental system. In order to monitor the chemistry of the fluid during reactions, an extraction-type vessel can be used to study high-temperature water-rock interactions (**Fig. 1**). The reaction chamber consists of a gold bag with a titanium closure attached to a titanium-lined sampling valve via a gold-lined exit tube. With this apparatus, solution can be drawn off for analysis during the course of an experiment. As solution is removed through the sampling valve, the gold bag collapses and the temperature-pressure conditions remain unchanged. Also, because the solution is not in contact with the solid during the quench, problems with back reactions are eliminated.

This extraction-type reaction chamber has been used extensively to study the chemistry of heated sea water and high-temperature sea water–basalt interactions as they relate to submarine hydrothermal processes. These studies have demonstrated the importance of the water-rock mass ratio and the critical role played by dissolved magnesium (Mg) in controlling pH and metal abundances in solution. Magnesium is removed rapidly from sea water, resulting in acidic solutions enriched in metals as long as the water-rock ratio or the temperature is high. Hydrolysis reactions overwhelm H^+-generating reactions when the water-rock ratio is low.

This extraction-type system has been used to study diagenesis, ore deposition, geothermal energy, and the isolation of nuclear waste. Although this type of system is very useful in tracking chemical evolutionary trends, it is limited to conditions with relatively large water-rock ratios and with pore pressure equal to confining pressure.

Flow-through hydrothermal system. Other recent experimental studies have been conducted in open, flow-through systems. One such system is schematically shown in **Fig. 3**. This system is similar to the extraction-type system in that it is constructed of relatively inert materials, and the composition of the fluid can be monitored during the reaction. However, reactions can be studied at very low instantaneous water-rock mass ratios, and the pore-fluid pressure (P_p) and confining-fluid pressure (P_c) are controlled independently. The ability to control the effective stress ($P_c - P_p$) in chemically oriented studies is very valuable, because it strongly influences the mechanical behavior of rocks and controls temperatures and pressures of volatilization-devolatilization reactions that are so important in the upper crust. Consequently, fluid-rock interactions can be studied in environments representing normal burial conditions ($P_c > P_p$) or abnormally high pore-pressure conditions (geopressured zones). In addition, because the average flow rate can be controlled by the volume of sample taken and the time interval between samples, the apparatus is particularly well suited to studies of both static and dynamic systems.

This flow-through system has proven to be very useful in the study of diagenetic processes. For example, it has been used to investigate the interactions between sea water and feldspathic sand at 200°C (390°F), 1000 bars (100 megapascals), and an instantaneous water-rock ratio of 0.3. In this study, dramatic and rapid changes in fluid chemistry occurred (**Fig. 2**), even at this relatively low temperature. Rapid removal of magnesium from solution by smectite formation was accompanied by increases in concentrations of calcium (Ca), potassium (K), and silicon dioxide (SiO_2). Iron (Fe) and manganese (Mn) concentrations increased rapidly early in the experiment as magnesium was being removed, and continued to increase throughout the experiment as the effective water-rock ratio increased because of fluid flow. Kaolinite and anhydrite also formed as alteration products.

The changes in pore-fluid chemistry observed early in this experiment may reflect those that occur in the subsurface under rock-dominated conditions where reaction kinetics are fast, fluid flow rates are slow, or residence times are long. Trends observed later may simulate water-dominated conditions where reaction kinetics are slow, fluid flow rates are fast, or residence times are short.

Rock-deformation experiments. The physical effects of pore fluids on the deformation of rocks have been studied extensively; however, the chemical effects have received much less attention. It is well known that the presence of water, even in trace amounts, lowers the strength of rocks. In some cases, various chloride solutions cause further reductions in strength. Some early experiments showed that specimens of Coconino Sandstone that were saturated with aluminum chloride and ferric chloride solutions had compressive strengths 8–15% lower than similar specimens saturated with distilled water. More recent experimental studies have shown that the mechanical behavior of a quartzose sandstone depends to a large extent on the ionic strength of

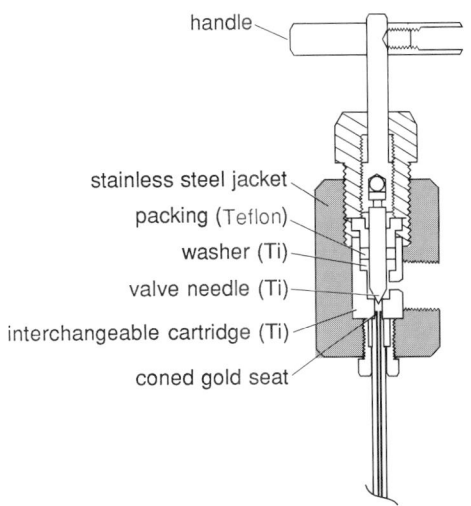

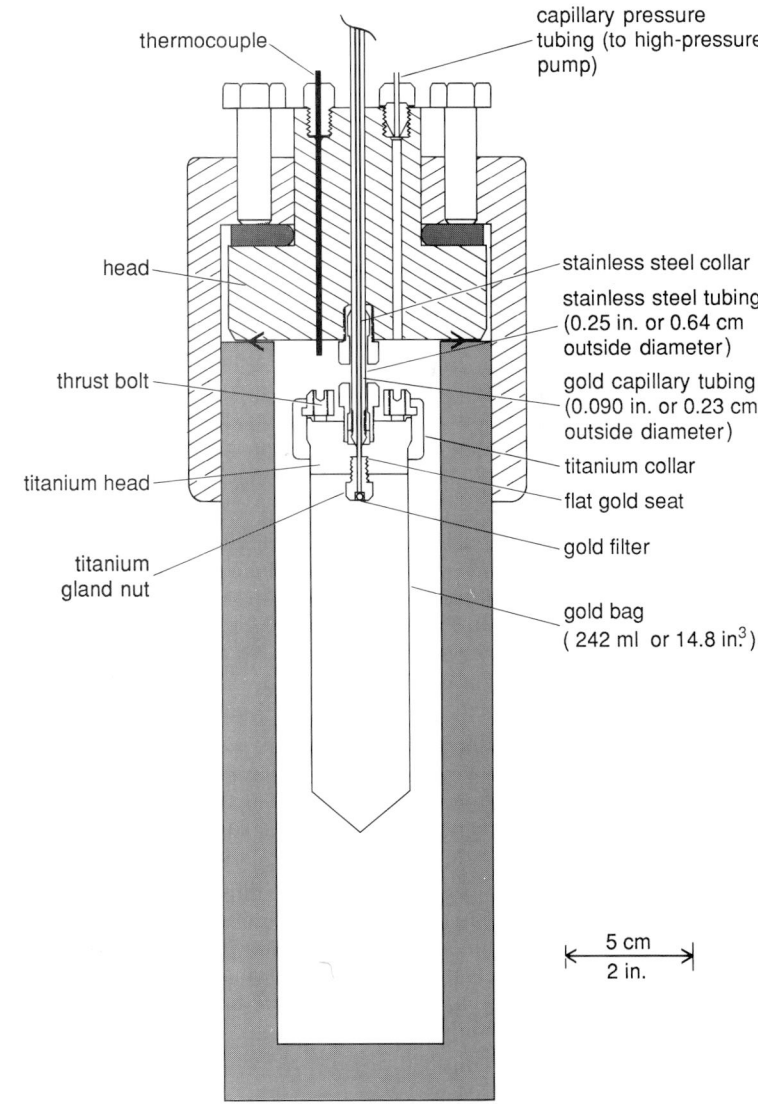

Fig. 1. Schematic illustration of extraction-type hydrothermal solution equipment and gold-titanium reaction cell. (*After W. E. Seyfried, Jr., P. C. Gordon, and F. W. Dickson, A new reaction cell for hydrothermal solution equipment, Amer. Mineralog., 64:646–649, 1979*)

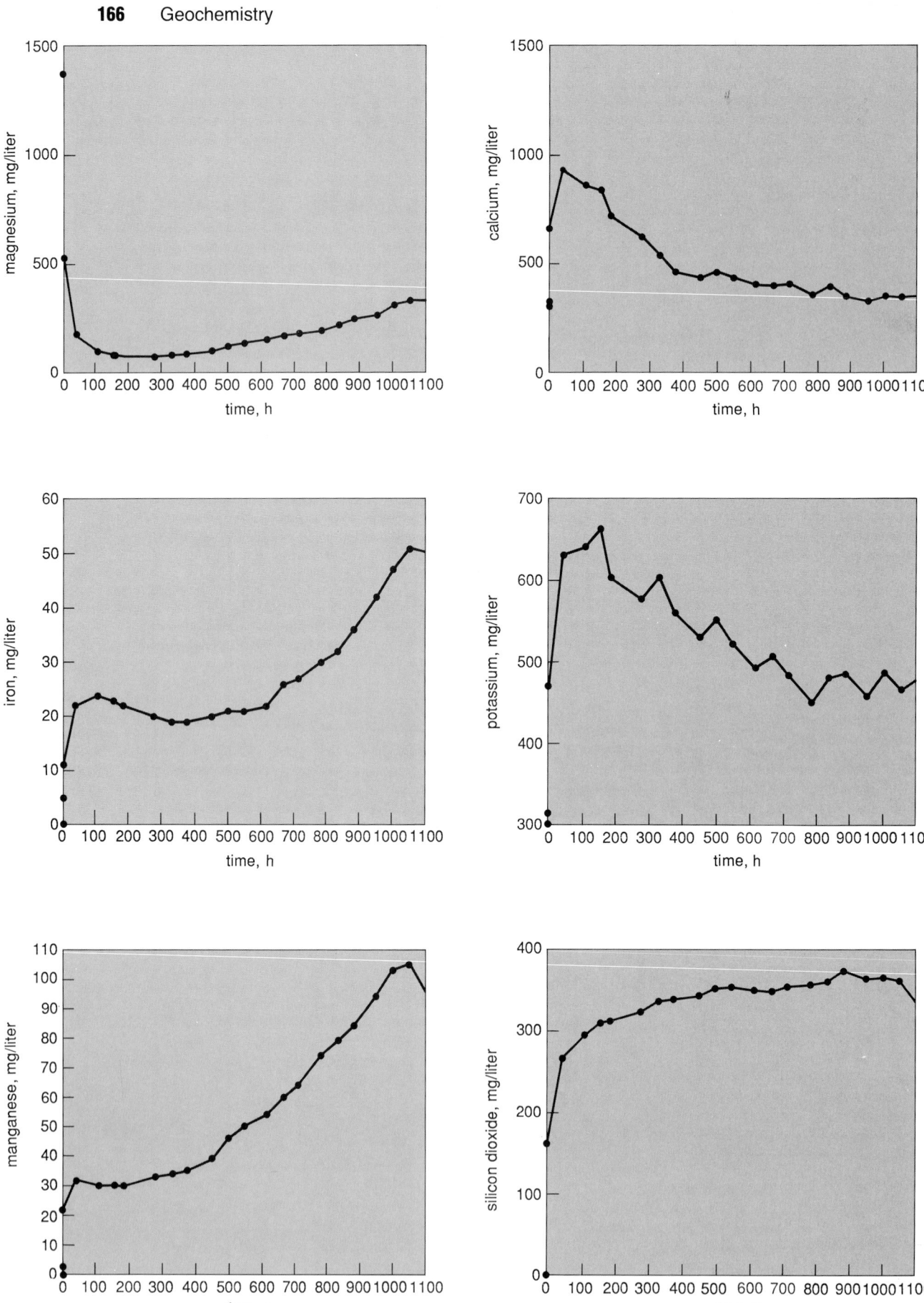

Fig. 2. Fluid compositions as a function of time as artificial sea water reacts with feldspathic sand in a flow-through reaction at 200°C (390°F), 1000 atm, (10² MPa). 1 mg/liter = 5 × 10⁻⁷ oz./in.³ (*After M. A. Bloom, Open-system studies of water-feldspathic sand interactions at 200°C and 1 kilobar: An experimental investigation, unpublished thesis, Texas A&M University, 1985*)

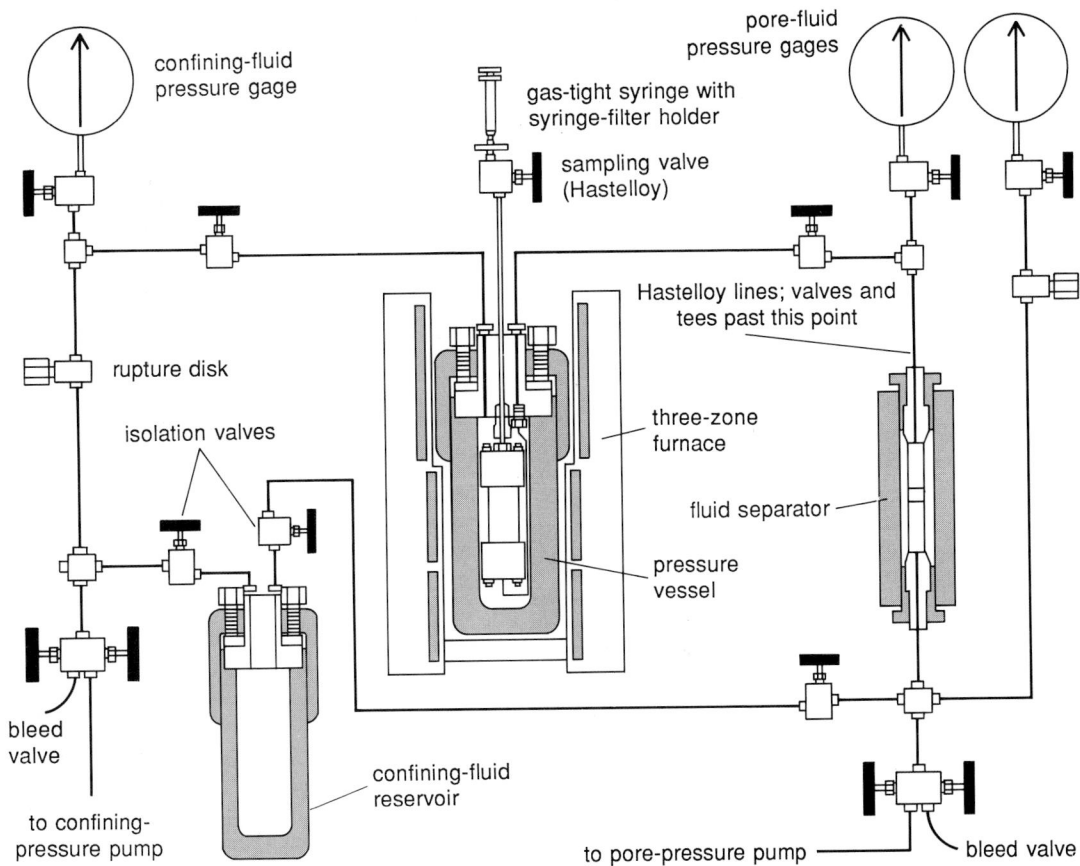

Fig. 3. Schematic illustration of a flow-through hydrothermal reaction system. (*After M. A. Bloom, Open-system studies of water-feldspathic sand interactions at 200°C and 1 kilobar: An experimental investigation, unpublished thesis, Texas A&M University, 1985*)

the pore fluid. Distilled water alone lowered the ultimate strength by 33%; low-ionic-strength (0.2-M) solutions of sodium chloride (NaCl) produced similar reductions in strength. Samples saturated with high-ionic-strength (5.0-M) solutions showed the greatest weakening; however, the samples saturated with intermediate-ionic-strength (1.0-M) solutions were stronger, by up to 20%, than those samples saturated with distilled water.

Pore-fluid chemistry also appears to affect the frictional characteristics of geologic materials. The presence of distilled water has been found to lower the coefficient of friction for a number of sandstones. In addition, iron chloride and calcium chloride solutions further reduce the coefficient of friction for relatively nonporous sandstone but have no additional effect on porous sandstone. It appears that for relatively friable rocks, such as the Berea and Coconino sandstones, chemically active solutions weaken the rock, resulting in rapid production of very fine-grained gouge; this increases the real area of contact and consequently increases the coefficient of friction. However, highly indurated, well-cemented rocks such as the Tennessee sandstone, which do not readily produce gouge, show no increase in the real area of contact and hence no increase in the coefficient of friction.

Summary. Numerous types of experimental water-rock systems exist, and others are being developed. Some are single-pass types such as the ones described here, and others are multiple-pass, recirculation-type systems designed primarily for geothermal studies. However, almost all chemically oriented systems are now constructed of nonreactive materials, such as gold, platinum, titanium, Hastelloy, and Inconel, and have the ability to monitor the composition of the fluid as reactions take place. In addition, some incorporate gas-backed pressure reservoirs in the confining- and pore-fluid systems so that pressure is essentially constant during flow; this allows estimates of the permeability of the sample to be made at elevated temperature and pressure. Also, chemically inert, rock deformation equipment is being constructed that will concentrate on the effects of fluid composition on rock behavior and will permit the compositional evolution of a fluid to be monitored during deformation.

The physics, chemistry, and hydrology of subsurface processes are closely interrelated. It is clear that future experimental studies of water-rock interactions will integrate mechanical effects, pore-fluid flow, and geochemistry.

For background information *see* GEOCHEMISTRY; PETROLOGY in the McGraw-Hill Encyclopedia of Science and Technology.

Andrew Hajash

Bibliography. M. A. Bloom, Open-system studies of water-feldspathic sand interactions at

200°C and 1 kilobar: An experimental investigation, unpublished thesis, Texas A&M University, 1985; J. D. Dunning and M. E. Miller, Effects of pore fluid chemistry on stable sliding of Berea sandstone, *Pure Appl. Geophys.*, 122:447–462, 1985; J. Logan and L. J. Feucht, The effects of chemical environment on the frictional properties of a quartzose sandstone, *Trans. Amer. Geophys. Union*, 66:1101, 1985; W. E. Seyfried, Jr., P. C. Gordon, and F. W. Dickson, A new reaction cell for hydrothermal solution equipment, *Amer. Mineralog.*, 64:646–649, 1979.

Geophysical exploration

The determination of the speed of propagation of elastic waves within the Earth has been a primary geophysical objective ever since the advent of seismic prospecting. In recent years, seismic tomography has developed into a viable technique for solving this problem. Seismic tomography produces an image of the elastic (or acoustic) wave speed from the recorded data.

Geophysical tomography. Tomography describes the reconstruction or imaging of object slices by using the measurements of waves passing through the object. Mathematically, these measurements represent line integrals or weighted sums of a medium's physical properties. The study of tomography is widely used outside the realms of geophysics as an imaging method. In x-ray tomography, for example, the densities of body tissues are reconstructed from measurements of x-ray absorption.

Although the term tomography has only recently come into wide use in geophysics, it can be applied to conventional imaging processes such as surface seismic reflection profiling, as well as to more recent processes such as borehole seismology. Geophysical tomography includes the imaging of seismic and electromagnetic waves. Recently, major seismic research efforts have been directed toward the acoustical imaging problem. Seismic tomography involves the estimation of wave velocity models by using the arrival times of either transmitted or reflected seismic waves. In exploration seismology the two types of transmission tomography problems are the surface-to-borehole and the borehole-to-borehole experiments. The borehole-to-borehole seismic tomographic survey (**Fig. 1a**), which is relatively new in exploration seismology, can be used to aid in the geologic problems of modeling rock formations between boreholes. The surface-to-borehole problem (Fig. 1b) is essentially vertical seismic profiling, in which a set of seismic sources are initiated at the Earth's surface and the resulting waves are recorded by seismometers in the borehole.

The reflection tomography experiment (Fig. 1c) describes the transmission of seismic waves to and from a reflecting boundary. In order to describe the ray paths for such waves, it is necessary to define the reflecting boundary or acoustical impedance contrast which creates the reflection. In this sense the reflection tomography problem represents a more difficult modeling problem than that of transmission tomography. The tomographic model generally represents the unknown acoustic wave velocity by velocity cells or elements which cover the region of interest (Fig. 1c).

Frequency domain methods. Medical tomography involves taking observations or views of an object around its entire surface. This viewing aperture allows the projection slice theorem to be applied. For two-dimensional slices of a material, this theorem states that the two-dimensional Fourier transform or frequency representation of an object can be constructed by taking one-dimensional Fourier transforms of the projected wave fields at viewing angles around the object.

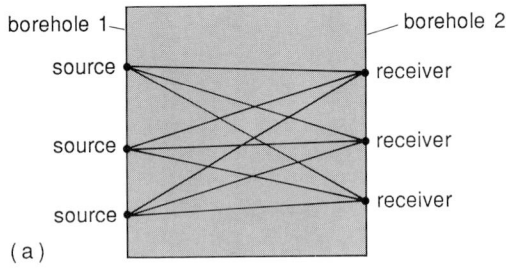

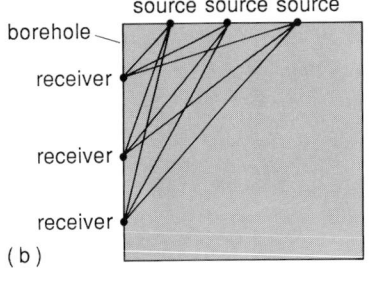

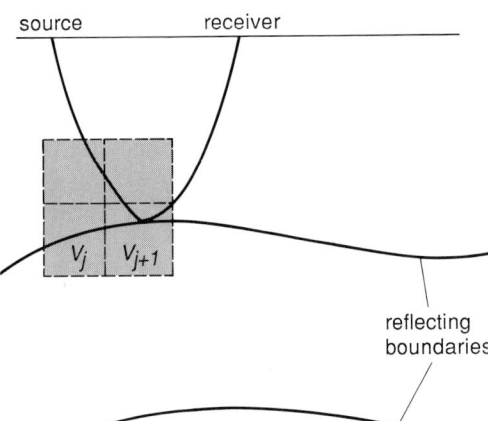

Fig. 1. Various source-receiver configurations for seismic exploration tomography. (*a*) Borehole-to-borehole experiment. (*b*) Surface-to-borehole experiment. (*c*) Reflection tomography (v_j = velocity of jth cell).

This allows the object to be constructed directly in the frequency domain. Then a two-dimensional inverse Fourier transform is used to reconstruct variations of the body's material properties. These methods work best with a complete angular view around an object. Unfortunately, such techniques are inappropriate in geophysics since geophysical data have an incomplete set of viewing angles formed by observations which are recorded only at the Earth's surface or in a few boreholes. This requires an approach to the problem different from that in the more easily controlled experiments of medical tomography. The seismic tomography problem is usually one of estimating wave velocities from a knowledge of acoustic travel times from source to receiver.

Seismic travel-time inversion. As in other tomographic problems, seismic tomography involves the evaluation of line integrals in which surface measurements represent weighted sums of a body's material property. In seismic travel-time tomography the observed travel times for a two-dimensional Earth model are written in line integral form as Eq. (1). In this representation, dl is

$$t = \int s(x,y)\, dl \qquad (1)$$

a differential length of the ray path and $s(x,y)$ is the slowness or the reciprocal of the velocity of the seismic wave. Because of refraction and reflection, the ray path depends on the unknown quantity $s(x,y)$. The model definition for travel-time tomography is given in **Fig. 2**, which displays the travel-time representation for a surface-to-borehole problem. In this example the travel time of the ray for an incoming wave is given by Eq. (2). That is, the travel time for the illustrated ray, t_1, is simply the sum of products of the

$$t_1 = d_{12}s_2 + d_{11}s_1 + d_{14}s_4 \qquad (2)$$

wave's slowness in a cell with the distance traveled in each cell. The model is conceptually very simple since the ray angle is constant as defined by the source-receiver geometry in this case. The travel-time model is computed by a use of a bookkeeping scheme which accumulates the products of ray segments and slowness values. Although the straight-ray approximation is valid for slowly varying velocity regimes, it is necessary to apply Snell's law to the seismic-ray representation to evaluate the true path of seismic waves. That is, the ray paths are not straight but are bent according to Snell's law, Eq. (3), for

$$\sin \theta_i / V_i = \text{constant} \qquad (3)$$

each velocity contrast, where V_i is the velocity (in cell i) and θ_i is the ray's angle of incidence. Since ray paths are dependent on velocities, tomographic solutions generally require an iterative approach. The general travel-time equations may be written in matrix notation as Eq. (4). D is the matrix of ray distances whose elements d_{ij}

represent the distance traveled by the ith ray in the jth cell, s is the slowness vector, and t is the vector containing observed travel times. Equation (4) is the statement of Eq. (1) in discrete form

$$Ds = t \qquad (4)$$

for an ensemble of rays.

The solution to the travel-time equations is given by a matrix inversion of the D matrix. In general, D is a large sparse matrix whose size is determined by the number of observed travel times and the number of cells in the model. These are often solved by algorithms that exploit the sparseness of the D matrix (that is, the fact that most of its entries are zeros). Since present tomography problems can involve the inversion of a sparse, ill-conditioned D matrix (that is, one whose inverse is overly sensitive to small changes in its numerical values) with 10^6 rows and 10^5 columns, numerical analysis plays an extremely important role in the solution process.

An iterative solution to the travel-time equations is needed for the case of refracted rays because the distance matrix depends on the slowness model. In this case, $D = D(s)$. The iterative or bootstrapping approach would involve solving for the change in the slowness vector, δs, relative to some initial guess s_0. This requires a solution to Eq. (5), where D is the distance matrix for the

$$D\, \delta s = \delta t \qquad (5)$$

previous slowness model and δt is the difference between observed and modeled travel times. An updated slowness model is computed by setting $s = s_0 + \delta s$. The rays for this new model are then traced according to Snell's law to produce a new D matrix, and a new solution to Eq. (5) is computed. The process is repeated iteratively until a suitable model is found. That is, a model is found so that the difference between the observed and modeled travel times is considered to be sufficiently small.

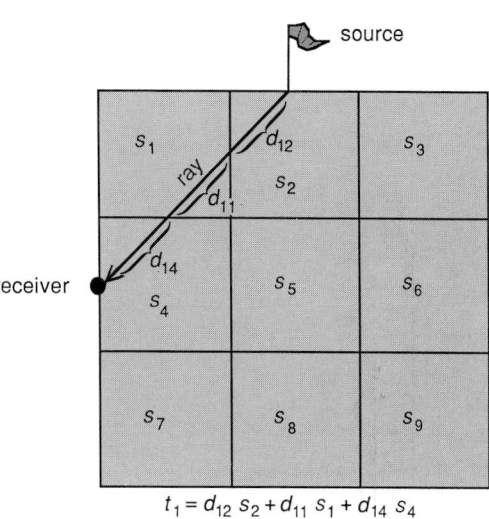

$$t_1 = d_{12}\, s_2 + d_{11}\, s_1 + d_{14}\, s_4$$

Fig. 2 Tomographic notation for a ray with travel time t_1.

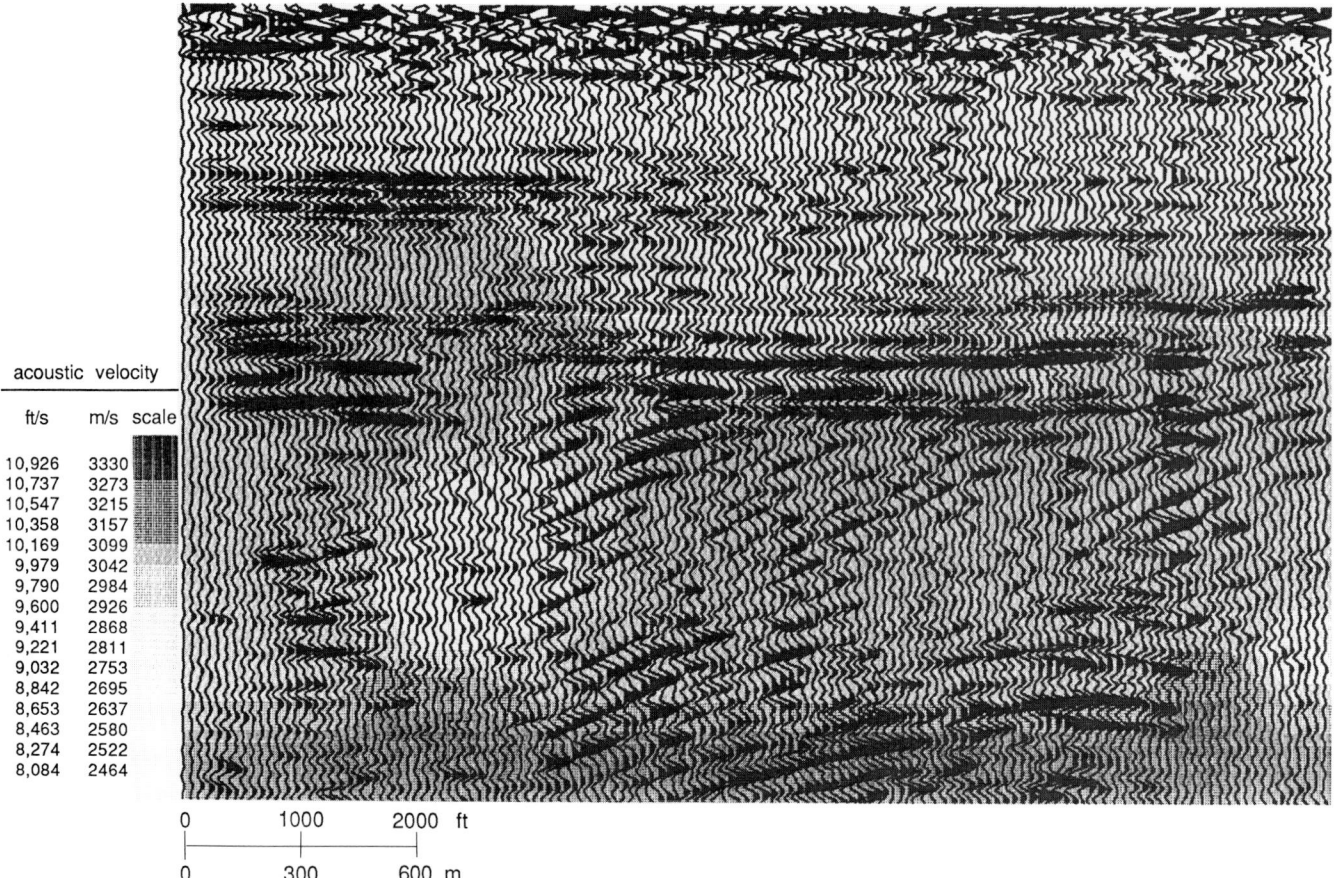

acoustic velocity

ft/s	m/s	scale
10,926	3330	
10,737	3273	
10,547	3215	
10,358	3157	
10,169	3099	
9,979	3042	
9,790	2984	
9,600	2926	
9,411	2868	
9,221	2811	
9,032	2753	
8,842	2695	
8,653	2637	
8,463	2580	
8,274	2522	
8,084	2464	

```
0        1000      2000  ft
|---------|---------|
0         300       600  m
```

Fig. 3. Overlay plat of a seismic tomogram (the gray scale) on a reflection seismogram (the series of curves) from the Wyoming overthrust belt. The total vertical dimension is 6562 ft (2000 m) and the horizontal dimension is 9843 ft (3000 m). (*From S. Treitel et al., Iterative tomographic migration: A few words and a picture, Leading Edge (Geophys.), 5(12):9, December 1986*)

Applications. The seismic tomography process has been applied to a number of data cases in order to estimate velocities. An example in **Fig. 3** shows a comparison of velocities derived from seismic tomography with a reflection seismogram. The reflection seismogram displays the reflection of seismic waves due to variations in acoustic impedance, the product of rock density and acoustical velocity. If seismic reflections were due to changes in velocity alone, then the velocity variations could be expected to coincide with the reflections. The seismogram is processed by using the velocity estimates and the measured travel times of the reflected waves to form a depth image. The seismogram is displayed as a series of curves which represent the amplitude of seismic wave echos caused by changes in acoustic impedance, while the background gray-scale shading displays the acoustic velocities computed from seismic tomography. Figure 3 illustrates how the reflection image or seismogram of the subsurface generally correlates with the velocity field of rocks in the media. This particular example was used in the interpretation of geology in the overthrust belt of Wyoming.

Travel-time tomography has also been successfully applied to borehole seismology, including surface-to-borehole and borehole-to-borehole recordings. As described earlier, these applications are generally transmission tomography problems. Unlike the reflection problem, these applications involve a relatively easy identification of travel times. In the transmission problem the end points of the transmitted seismic rays are known, whereas ray paths in the reflection problem are not well determined because of the unknown reflector depth. Therefore, the ray modeling is less ambiguous for transmission tomography than for reflection tomography applications. With the advanced development in borehole seismic sources and receivers, transmission tomography promises to be a worthwhile application for borehole seismic imaging.

For background information SEE ACOUSTIC TOMOGRAPHY; GEOPHYSICAL EXPLORATION; SEISMIC EXPLORATION FOR OIL AND GAS; SEISMOLOGY in the McGraw-Hill Yearbook of Science and Technology.

Larry R. Lines

Bibliography. T. Bishop et al., Tomographic determination of velocity and depth in laterally varying media, *Geophysics,* 50:903–923, 1985; R. P. Bording et al., Applications of seismic travel time tomography, *Geophys. J. Roy. Astron. Soc.,* in press; G. A. McMechan, Seismic tomography in boreholes, *Geophys. J. Roy. Astron. Soc.*, 74:601–612, 1983; M. H. Worthington, An introduction to geophysical tomography, *First Break (Geophys. Prosp.)*, 2:20–26, 1984.

Gravitational radiation

While radio observations of a binary pulsar have given compelling, albeit indirect evidence of the existence of gravitational radiation, direct detection of gravitational waves awaits experiments of improved sensitivity planned for the next few years. If present theoretical ideas about the numbers and strengths of astrophysical sources are correct, gravitational-wave antennas will be able to detect the passage of such waves.

Nature of gravitational waves. The existence of gravitational radiation was one of the early predictions of Albert Einstein's revolutionary theory of gravity, the general theory of relativity (1916). Gravitational waves are in many ways analogous to the electromagnetic waves (light) predicted by J. C. Maxwell's theory of electricity and magnetism. Just as electromagnetic waves carry information about the changes generated in the electric and magnetic fields of an electrically charged particle which experiences an acceleration, gravitational waves arise from the (accelerated) motion of the mass-energy, which is the source of a gravitational field. These wavelike variations of the (electromagnetic or gravitational) fields carry energy outward from the source. The waves travel at the speed of light, with their amplitudes

falling as their inverse distance from the source. Because the force of gravity is relatively weak, only gravitational waves generated by astronomical objects with masses not much smaller than the mass of the Sun and densities at least that of nuclear matter (about 10^{14} g/cm^3 or 10^{14} times the density of water) will be detectable. Such compact objects will necessarily have strong static gravitational fields, which often result in the rapid, nonuniform motion of large amounts of matter moving with nearly the speed of light. Examples of such compact objects include neutron stars and black holes.

Sources. Sources of gravitational radiation can be divided into three categories: impulsive, periodic, and stochastic.

Impulsive sources. An impulsive burst of gravitational radiation will result from astrophysical events such as the formation of a neutron star or of a black hole. These objects are formed in events in which massive stars are said to die. When massive stars have exhausted their nuclear fuel and can no longer support themselves against the force of gravity, they undergo core collapse and a subsequent supernova explosion. In this process the outer portions of the star are blown off while the central core implodes, forming a high-density neutron star or a black hole. It is during the implosion and the subsequent oscillations of the newly formed compact remnant that most of the gravitational radiation is generated. Supernova explosions are observed to occur about once every 10 to 30 years in the galaxy to which the Sun belongs, the Milky Way. The

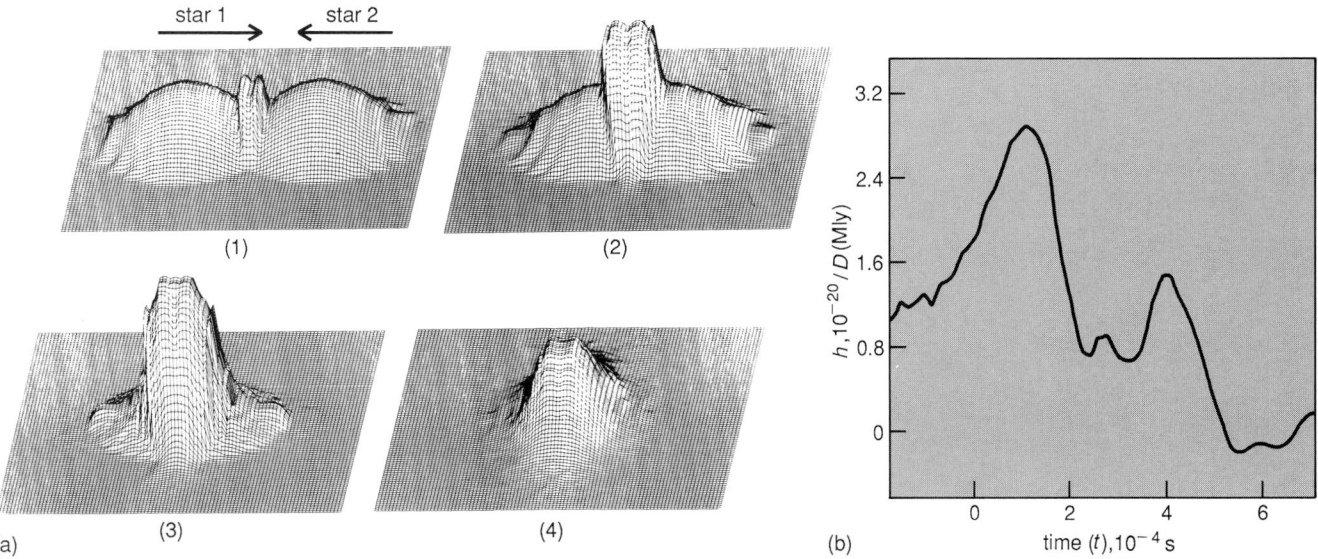

Fig. 1. Approximate simulation, using newtonian theory, of head-on collision of two identical neutron stars, each of 1 solar mass, released from rest at infinity. Following initial contact, recoil shocks propagate through stars, decelerating the matter still approaching collision. After several oscillations, stars eventually settle down and coalesce into a single object of 2 solar masses. (a) Surface densities plotted at sequential moments 1–4 during collision and coalescence. (b) Wave amplitude h of the resulting burst of gravitational radiation perpendicular to the collision axis, plotted against time, with time t=0 being the first instant of contact. Quantity h, explained in text, is measured in units of $10^{-20}/D$ (Mly), where D is the distance to the source in units of 10^6 light-years (6×10^{18} mi or 10^{19} km). (From D. L. Gilden and S. L. Shapiro, Gravitational radiation from colliding compact stars: Hydrodynamical calculations in two dimensions, Astrophys. J., 287:728–744, 1984)

best chance of detecting gravitational waves would occur if a gravitational wave detector were in operation during one of these relatively nearby galactic explosions, typically at a distance of 10^4 light-years (6×10^{16} mi or 10^{17} km). A more sensitive detector (discussed below) would be able to detect the gravitational waves produced by supernovae in distant galaxies out to 10^7 light-years (6×10^{19} mi or 10^{20} km). Although the amplitudes of these waves will be 1000 times smaller, the rate of distant supernovae explosions is much greater since there are many more stars undergoing collapse at large distances. Rough estimates suggest that about one such supernova occurs per month.

Somewhat more energetic but less frequent are collisions between neutron stars (**Fig. 1**). This kind of event is likely to occur in the dense stellar cores of active galactic nuclei and in binary systems containing two neutron stars.

Impulsive bursts of gravitational radiation are generated also when supermassive black holes (with masses 10^8 times that of the Sun) are formed or when they collide with each other. These black holes are believed to power quasars (the most prolific sources of energy in the universe) and active galactic nuclei.

Periodic and stochastic sources. Periodic sources continuously emit radiation. Possible sources include rotating (nonaxisymmetric) neutron stars or binary systems containing compact objects. Stochastic sources result in a so-called background noise of radiation. Such noise could arise from large numbers of uncorrelated impulsive sources or from the irregular motions of matter which may have occurred in the early universe. This gravitational radiation is somewhat analogous to the cosmic microwave background—the electromagnetic remnant of the big bang.

Binary pulsar. The binary pulsar mentioned above was discovered in 1974. It consists of a rapidly rotating neutron star (the pulsar) which is in orbit about another, unseen compact object, probably another neutron star. The presence of the rotating neutron star's strong magnetic field gives rise to a beam of electromagnetic radiation which sweeps past the Earth during each rotation. This beaming produces the pulses that are received (hence the term pulsar). Because the rotation rate is very steady, the pulses serve as a very precise clock which can be used to monitor the motions of the neutron star. In particular, the orbital motion of the pulsar modulates the pulses via the Doppler effect: the pulse period decreases when the pulsar is approaching, and increases when it is receding. In this way the orbital period of the pulsar can be measured. The period is about 8 h. The orbital motion of this compact binary system generates gravitational radiation, and the associated energy loss results in a slow decrease in the orbital period as the stars spiral toward each other. The gravitational waves are

too weak to detect directly, but the change in the orbital period is easily observed. The change predicted by general relativity theory due to the emission of gravitational radiation is very close to the observed change in the orbital period. It is worth noting that the binary pulsar has also provided a laboratory for observing several other predictions of Einstein's theory, including the change in the position of the perihelion (the point of closest approach), an effect first observed in Mercury's orbit about the Sun. In addition, it has provided the first accurate measurement of the mass of a neutron star: 1.4 solar masses.

Detection. Just as an understanding of how electromagnetic waves can be generated led to H. Hertz's direct detection of the waves in 1887, the present (somewhat incomplete) understanding of possible sources of gravitational waves provides convincing evidence that such waves are likely to be detected directly during the next decade.

Detectability of a gravitational wave signal depends on several factors. The wave amplitude, which depends both on the intrinsic strength of the source and the source's distance from the detector, must be sufficiently large. Considerable theoretical work has been done to estimate the expected wave amplitudes of plausible astrophysical sources of gravitational radiation. Another key parameter is the so-called event rate, the number of events per year. For example, as described above, supernovae in the Milky Way Galaxy occur at a rate of about one per 10 years, with a wave amplitude h of approximately 10^{-19} for a supernova occurring near the center of the Galaxy, 3×10^4 light-years (2×10^{17} mi or 3×10^{17} km) away. For supernovae as distant as the Virgo cluster of galaxies, the event rate increases to one per month, while the amplitude h drops to approximately 10^{-22}. Detectors sensitive to these latter sources will be needed to detect signals in a reasonable amount of observing time.

A gravitational wave will distort any object through which it passes; the wave's gravitational field alters the relative positions of particles. If the initial separation between two particles is x, a wave of amplitude h induces a time-varying change Δx approximately equal to hx. This is similar to the effect that electromagnetic waves have on electrically charged particles. For a wave with an amplitude h of approximately 10^{-22} and a detector a few meters in size, it is necessary to be able to detect deformations 10^7 times smaller than the size of an atomic nucleus.

Bar detectors. Large bar detectors were first used by J. Weber in the 1960s. A suspended cylindrical bar weighing several tons acts as a harmonic oscillator. The relative motion of the ends of the bar induced by the passage of a gravitational wave is detected with the use of transducers or superconducting quantum interference devices (SQUIDs). Because local sources of excitation in the bar would cause motions that would easily overwhelm those induced by a gravitational wave, the detector must be acoustically isolated from its surroundings. Thermally induced oscillations present a more formidable problem. Detectors are cooled to near absolute zero to reduce these oscillations and are used in widely separated pairs. The thermal motions of the two detectors will be uncorrelated, while the motions induced by a gravitational wave will be separated in time only by the light travel time between the two sites. Use of two sites also allows some directional determination.

Laser interferometers. A new kind of detector has also been under development. The laser interferometer uses two freely suspended mirrors at opposite ends of an L-shaped apparatus (**Fig. 2**). A laser beam is split between the two arms and then recombined after reflection to give an interference pattern. Any change in the relative path lengths (caused by oscillations induced by a passing wave) will result in a change in the interference pattern. Present designs should allow detection of waves with an amplitude h of approximately 10^{-22} once technological problems (such as amplifier noise) are solved. A larger interferometer now being considered, the Laser Interferometer Gravitational-wave Observatory, would consist of a vacuum system 2.5 mi (4 km) in length along each arm supporting several lasers simultaneously. Two sites separated by more than 600 mi (1000 km) would work in tandem. Eventual sensitivities should allow detection of wave amplitudes h of approximately 10^{-23}. Similar experiments are under consideration in several European countries. Another advantage of using interferometers as detectors is that they are broadband detectors; that is, they respond to a wide range of wave frequencies (10^2–10^4 Hz). Bar detectors are limited to a narrow band near the bar's resonant frequency, about 10^3 Hz. This broadband frequency range is the expected spectral region in which radiation from stellar collapse

and supernovae, as well as from neutron star collisions, is likely to peak.

Prospects. The direct detection of gravitational radiation will open a new window on the universe. Because of the relatively weak interaction between gravitational waves and matter, direct and detailed information will be received for the first time from the inner regions of some astrophysical events from which electromagnetic radiation (which interacts strongly with matter) is unable to emerge. Hence, the very property that makes the gravity waves so difficult to detect— their weak coupling to matter—enable them to emerge unimpeded to transmit this information.

For background information SEE GRAVITATION; GRAVITATIONAL COLLAPSE; INTERFEROMETRY; NEUTRON STAR; PULSAR; RELATIVITY; SQUID; SUPERNOVA in the McGraw-Hill Encyclopedia of Science and Technology.

Richard A. Saenz; Stuart L. Shapiro

Bibliography. N. Derulle and T. Piran (eds.), *Gravitational Radiation, Les Houches 1982*, 1983; S. L. Shapiro, R. F. Stark, and S. A. Teukolsky, The search for gravitational waves, *Amer. Sci.*, 73:248–258, 1985.

Halley's Comet

Important new information was collected when six interplanetary space probes visited Comet Halley (or Halley's Comet in informal usage) in March 1986. They obtained thousands of images, spectra, and measurements, including the first photographs and temperature measurements of the nucleus, or "dirty iceball," of a comet and numerous determinations of the chemical compositions of cometary dust grains, which allowed investigators to classify the grains into several basic categories.

Comet Halley travels in an elongated elliptical orbit with a perihelion roughly halfway between the Sun and the Earth's orbit and an aphelion between the orbits of Neptune and Pluto. The comet's average period is 76 years, and its favorably placed orbit, together with its size and activity, makes it visible to the naked eye at each apparition (approach to the Sun). Unfortunately, the 1985–1986 apparition of Comet Halley was the most unfavorable for Earth-based observers in millennia because the comet was on the far side of the Sun when perihelion occurred on February 9, 1986. The comet was thus not the spectacular object for public viewing that it had been during the 1910 apparition, and the interest in this apparition lay chiefly in the massive cooperative effort conducted by scientists to observe the comet.

Until 1985, information about comets was based on observations made from the Earth or from rockets and satellites in the Earth's vicinity. Strong indirect evidence pointed to the existence

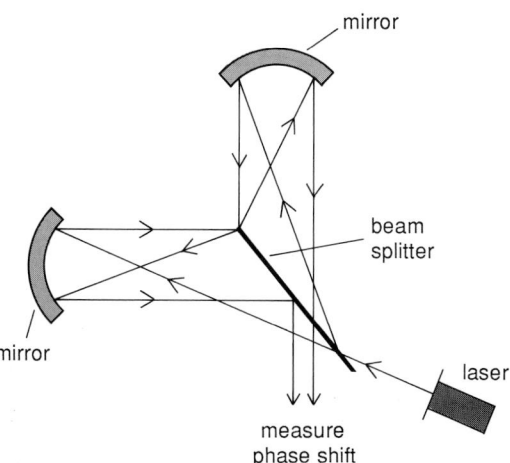

Fig. 2. Laser interferometer. A passing gravitational wave will cause relative displacement in the two arms, producing a shift in the interference pattern.

Space probe encounters with Comet Halley			
Spacecraft name	Closest approach to nucleus		Encounter date (1986)
	mi	km	
Giotto	375	605	March 14
Vega 2	4,990	8,030	March 9
Vega 1	5,520	8,890	March 6
Suisei	94,000	151,000	March 8
Sakigake	4,340,000	6,990,000	March 11
ICE	17,500,000	28,100,000	March 25

Spacecraft observations. Details of the encounters of spacecraft with Comet Halley are summarized in the **table**. All of the probes passed on the sunward side of the nucleus. The relative positions of the spacecraft *Giotto* and Comet Halley during their encounter, together with the directions of the Sun and Earth, are shown in **Fig. 1**. One of the craft, NASA's *International Cometary Explorer* (*ICE*), had already passed through the tailward (anti-Sun) side of the coma of Comet Giacobini-Zinner in September 1985.

Three probes, the European Space Agency's *Giotto* and the Soviet Union's *Vega 1* and *2*, passed through the visible coma of Comet Halley, each suffering significant damage from the impact of microscopic particles of comet dust at velocities of up to 47 mi/s (78 km/s). The principal purpose of each of these craft was to investigate the nature of the nucleus.

Two probes, Japan's *Suisei* and *Sakigake*, passed only through the hydrogen cloud. They investigated the interaction of the cometary plasmas and magnetic field structures with those of the solar wind. In addition, *Suisei* carried an ultraviolet camera, with which a long time series of images of the evolution of the hydrogen cloud was obtained. Both Japanese probes detected waves and energetic particles derived from the comet, and *Suisei*, quite improbably, was struck by two large cometary dust particles with masses estimated in the milligram range, although it was located very far from the dust coma and tail. *ICE* passed outside the hydrogen cloud of Comet Halley and was not intended to directly investigate the comet. However, even at a distance of about 2.2×10^7 mi (3.5×10^7 km) from the nucleus, *ICE* detected plasma waves and energetic ions that probably were associated with the comet.

Other spacecraft observed Comet Halley from positions in orbit around the Earth and the planet Venus. Observations were also made from aircraft and sounding rockets and from a large number of optical and radio telescopes on the ground.

Nucleus. The image sequence that was obtained with the *Suisei* ultraviolet camera revealed periodic intensity variations in the inner hydrogen cloud of Comet Halley and confirmed estimates from photographs made in 1910 that the spin period of the nucleus is about 53 h. The size and shape of the nucleus were found from images taken by *Giotto* (**Fig. 2**) and the *Vegas* from many directions. The nucleus's albedo was determined from the images, and its surface temperature was measured by infrared spectrophotometry.

The nucleus is an ellipsoidal or perhaps potato-shaped object, measuring about 9 mi × 4.5 mi × 4.5 mi (15 km × 7.5 km × 7.5 km). Surface relief of the order of 1000 ft (hundreds of meters) in elevation, undulating over length scales of about 0.6 mi (1 km), was apparent. The nucleus is exceptionally dark, with a geometric albedo of 0.04, comparable with the blackest known sur-

of a central nuclear body or nucleus from which all cometary material, both gas and dust, would originate, although such a nucleus had never been photographed or directly observed. As a comet approaches the Sun it develops a coma, an essentially spherical cloud of gas and dust that can extend hundreds of thousands of miles from the nucleus. Extending even farther outward is the hydrogen cloud, observable only in ultraviolet light (in the Lyman-alpha line at wavelength 121.6 nanometers). The dust and gas flowing from the nucleus also form dust tails and gas tails (also called plasma tails) that can extend millions or tens of millions of miles.

In the early 1950s, F. L. Whipple predicted that the nucleus is an agglomerate of frozen water and other frozen materials interspersed with nonvolatile particles or "dust." When the comet approaches the Sun, the frozen materials in the nucleus sublime under solar heating, and gas and dust are released to form the coma and the gas and dust tails.

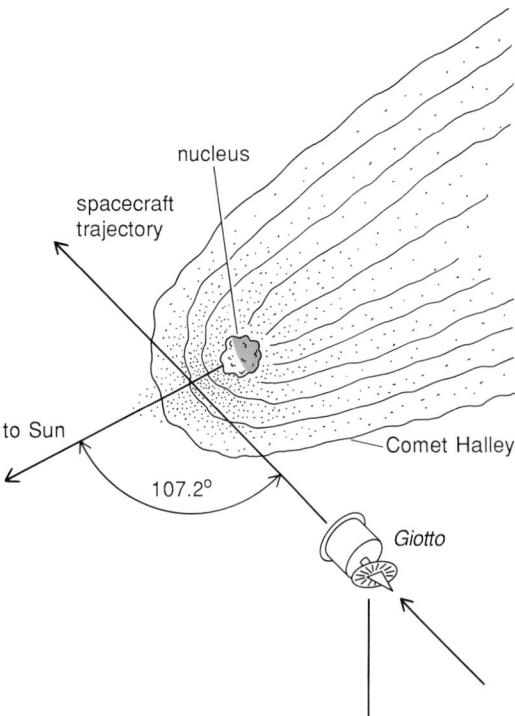

Fig. 1. Relative positions of spacecraft *Giotto* and Comet Halley during their encounter. Directions of Sun and Earth are also shown.(*After European Space Agency*)

Fig. 2. Composite of four images of nucleus of Comet Halley taken by Halley Multicolor Camera on the *Giotto* spacecraft at 12,394, 6119, 3074, and 1547 mi (19,947, 9847, 4947, and 2490 km) from the nucleus. As spacecraft approached nucleus, resolution increased and size of frames, indicated by rectangles, decreased due to camera's fixed field of views. Innermost frame is 2.5 mi (4 km) across. Image shows dark night side of nucleus to right, bright dust jets on day side to left, faint background light of dust coma, and several craterlike features of daylight side of nucleus in two innermost frames. (*Copyright by Max-Planck-Institut für Aeronomie, Lindau/Harz, FRG, 1986*)

faces on natural objects in the solar system. It backscatters light in a manner reminiscent of the lunar surface; that is, there is enhanced reflectivity at low phase angles, close to the vertical. This dark surface cannot consist of ice, as its temperature was 170°F (350 K) at the time of the spacecraft encounters, which compares with an estimated −120°F (190 K) for an admixture of ice and nonvolatile particulates at the same distance from the Sun. The hot surface is interpreted as a porous crust or regolith of nonvolatile material, with low thermal conductivity, perhaps just 0.4 in. (1 cm) thick. An alternate view is that the apparent surface is actually a thin, dense layer of gas-suspended solid particles.

Jets—collimated streams of outflowing matter—illuminated by scattered sunlight, were observed to emerge from several locations on the subsolar hemisphere of the nucleus. Jet lengths were typically 3 to 9 mi (5 to 15 km); the diameter of the active spot on the nuclear surface at the base of a jet on the northern tip of the nucleus was about 2 mi (3 km). An individual jet appears to consist of several or many smaller streams, which emerge from adjacent spots with sizes of the order of 500 ft (150 m). Dust-counter observations of multiple streams or structures with characteristic widths of about 500 ft (150 m), made from the probes, may represent a related phenomenon.

Coma. The coma gases expand at about 0.6 mi/s (0.9 km/s). The principal parent molecule

released from the nucleus is water, about 80% by volume. Carbon dioxide is the next most abundant parent molecule. Parent molecules, by definition, are molecules that are present as solids in the nucleus; after subliming they are soon broken down into smaller molecules, radicals, ions, and atoms by the action of solar ultraviolet radiation and by other processes. Ions detected in the comet by mass spectrometry from the probes were H_3O^+, H_2O^+, OH^+, $^{12}C^+$, $^{12}CH^+$, O^+, Na^+, $^{12}C_2^+$, $^{32}S^+$, $^{34}S^+$, and $^{56}Fe^+$. Infrared spectroscopy from the spacecraft detected, besides spectral features often observed in comets from the ground, the 1.38-micrometer and 2.7-μm bands of water, a 3.4-μm broad emission band attributed to C—H bonds, the 4.25-μm carbon dioxide band, and a 7.5-μm emission band that is ascribed to C—C bonds.

Ionized carbon was present over a large range of distance from the nucleus in greater quantity than expected from the dissociation of identified molecules; another source of atomic carbon may be required to explain this observation. Perhaps carbon is released from dust grains in the coma.

Total gas production from the nucleus was 7×10^{29} molecules/s or about 40 tons/s at the time of encounter. The dust-to-gas-production ratio, by mass, was in the range 0.1 to 0.25.

The dust had a flatter mass function (number of particles per unit mass interval versus mass) than anticipated. Grains with very small masses (less than 10^{-14} g) were much more abundant than expected from the interpretation of ground-based photometry of previous comets, and dust grains were detected in large numbers down to particle masses as small as 10^{-16} g.

The chemical compositions of many dust grains were measured by mass spectrometry of plasma generated by high-speed grain impacts on silver and platinum targets in the instruments. Characteristic dust types, by composition, were (1) particles that chemically resemble type I carbonaceous chondrites (the most primitive class of stone meteorites); (2) particles that resemble those of 1 but are richer in carbon, nitrogen, and sulfur; (3) particles with high abundances of hydrogen, carbon, and oxygen, perhaps derived from water ice and water–carbon dioxide clathrates; and (4) particles with pure silicate composition. The elemental abundances from the dust composition measurements cannot be reliably established because the impacts of measured dust particles occurred at much higher velocities than can be attained in laboratory calibration experiments.

Plasmas, fields, and accelerated particles. The comet was accompanied by a bow shock, located about 6×10^5 mi (1×10^6 km) from the nucleus. The bow shock, formed because of the supersonic motion of the comet with respect to the solar wind, was marked by a local intensification of the interplanetary magnetic field and by the emission of intense plasma waves. Heavy ions thought to be derived from the comet as neutrals, then ionized by solar ultraviolet light and swept up by the solar wind (so that they are termed pickup ions), reaching velocities up to twice the solar wind velocity, were observed outside the bow shock at distances of at least 6×10^6 mi (1×10^7 km) from the nucleus. They stimulate plasma waves of several types, which were observed over a wide range of frequencies. In addition, there appear to be charged particles that are accelerated to energies around 540 keV by wave interactions in the hydromagnetically disturbed environment of the comet.

Interplanetary magnetic field lines pile up in the ionosphere of the coma, reaching a field strength of about 75 nanoteslas, several times higher than the undisturbed interplanetary magnetic field. However, well inside the coma, the pressure of the ionized gas prevents the field from penetrating farther, and there is a small zero-field cavity, bounded by the contact surface at about 2850 mi (4600 km) from the nucleus. Outside the contact surface, ion temperatures were of the order of 10,000 K (18,000°F), while inside they were of the order of a few hundred kelvins to 1000 K (1300°F).

Fields in the tail were not observed since no probe flew through the tail of Comet Halley, but in Comet Giacobini-Zinner the *ICE* spacecraft found that the field in the tail takes a hairpin configuration, with oppositely directed, parallel lines of force separated by a thin, neutral sheet. This phenomenon, field draping, was predicted by H. Alfvén and accounts for the typical narrow shape of cometary plasma tails. *ICE* also found that the tail plasma is cool and contains much hydronium (H_3O^+), a substance not previously detected in space.

For background information *SEE* COMET; HALLEY'S COMET; SOLAR WIND in the McGraw-Hill Encyclopedia of Science and Technology.

Stephen P. Maran

Bibliography. J. C. Brandt, Space observations of comet Halley, *Nature*, 321:391–393, 1986; K. Hirao and T. Itoh, The Planet-A Halley encounters, *Nature*, 321:294–297, 1986; R. Reinhard, The Giotto encounter with comet Halley, *Nature*, 321:313–318, 1986; R. Z. Sagdeev et al., Vega spacecraft encounters with comet Halley, *Nature*, 321:259–262, 1986.

Hemerythrin

Hemerythrin is an iron-containing respiratory protein found in four marine phyla. Most of the studies of the protein have been carried out on protein derived from the sipunculids *Phoscolopsis gouldii* and *Themiste zostericola*. Hemerythrin is responsible for transporting and storing oxygen for use by the organism, and in this respect it behaves like the only other known respiratory proteins, hemoglobin and myoglobin, and like hemocyanin.

Structural aspects. In spite of its name, hemerythrin contains no heme group. Instead, its active site consists of a pair of iron atoms, which in the iron(III)-iron(III) state are bridged by an oxy group and two amino acid residues. This is so-called methemerythrin. As isolated, and at low pH (~6), one of the irons in methemerythrin is five-coordinate, something quite unusual in iron(III) complexes (**Fig. 1***a*). The vacant coordination site is believed to be occupied by OH^- at high pH (~9) and by anions such as azide, thiocyanate, and fluoride, with which the oxidized (met) protein combines readily. The structure of the active site in azidomethemerythrin is shown in Fig. 1*b*. Methemerythrin does not react with oxygen. Deoxyhemerythrin, in which both irons are in the +2 oxidation state, reacts with one mole of oxygen per two irons to form oxyhemerythrin, the structure of which is shown in Fig. 1*c*. Not much is known about the structure of deoxyhemerythrin because it is nearly invisible to many of the spectroscopic techniques (ultraviolet and visible, resonance Raman, electron paramagnetic resonance) which have been used so effectively in diagnosing the structure of the other forms. In addition, the crystals required for x-ray crystallography are very sensitive to traces of oxygen. Deoxyhemerythrin is now believed to have an —OH bridging group and to contain one five-coordinate iron(II), which oxygen attacks.

The two iron atoms in a subunit are strongly coordinated to amino acid residues from at least three of four helices which are part of the polypeptide chain (see Fig. 1). The folding of the polypeptide chain, whether in a subunit of the octamer from the coelomic fluid of *T. dyscritum* or *P. gouldii*, or from the trimer in *Siphonisoma funafuti*, or in the monomer from the retractor muscle of *T. zostericola*, is the same. There are four nearly parallel sections of α-helix in the so-called hemerythrin fold which has been observed in other proteins with entirely different sequences and functions. The molecular weight of the octamer is 108,000 daltons, with each subunit being 13,500 daltons, about the same as in the monomer form (myohemerythrin).

Chemical aspects. A good deal of the chemistry of hemerythrin has been explored. The reactions of oxygen with deoxyhemerythrin and deoxymyohemerythrin are very rapid (formation rate constants $\sim 10^7 \ M^{-1} \cdot s^{-1}$) and require specialized techniques (stopped-flow, temperature-jump relaxation, and laser photolysis) for their measurement. In the stopped-flow method reactants are rapidly mixed, then abruptly stopped after a very short time. The course of the reaction is followed with a detector placed quite near the mixer. This high reactivity is paralleled by hemo-

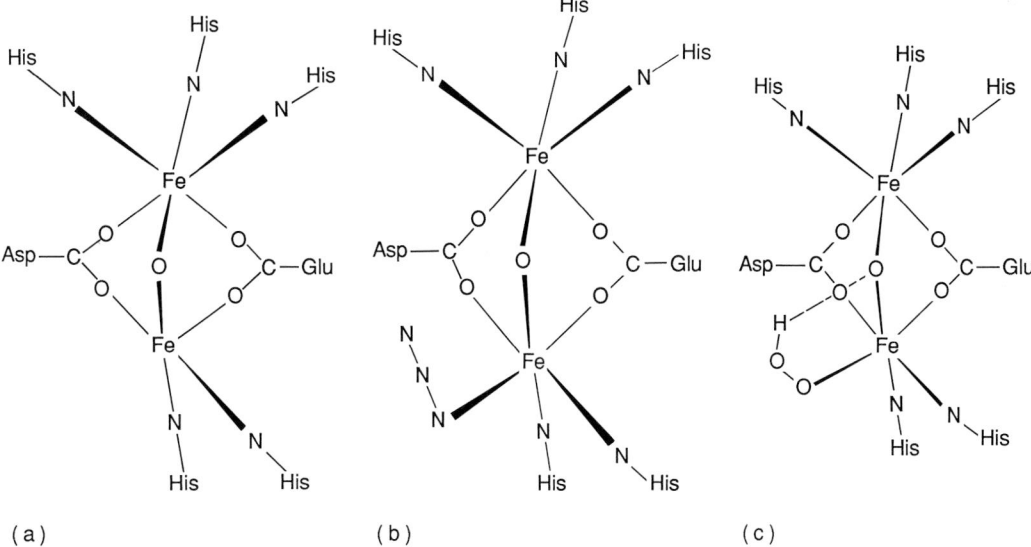

Fig. 1. Structures of the binuclear iron sites in (a) methemerythrin, (b) azidomethemerythrin, and (c) oxyhemerythrin. His = histidine; Asp = aspartic acid; Glu = glutamic acid. (*After R. E. Stenkamp, L. C. Sieker, and L. H. Jensen, Binuclear iron complexes in methemerythrin and azidomethemerythrin at 2.0-A resolution, J. Amer. Chem. Soc., 106:4951–4956, 1984*)

globin and hemocyanin and is an integral feature of respiratory proteins. It is found that a number of other small molecules—NO, HN_3, HNCO, and HF—also bind to deoxyhemerythrin with strength comparable to that of O_2. There is no evidence of cooperativity among the protein subunits, or of a Bohr effect when oxygen reacts with deoxyhemerythrin except for protein derived from the brachiopodal *Lingula unguis* and *L. reevii*. The reaction with oxygen leads only slowly to irreversible oxidation to methemerythrin (requiring days at pH ~7 and room temperature), a fact critical to the organism since the oxidized (met) form no longer binds oxygen. The components of a methemerythrin reductase system have been found in the hemerythrocytes of *P. gouldii* and five Australian sipunculids. Cytochrome b_5 isolated from *P. gouldii* reduces methemerythrin to deoxyhemerythrin.

Semi-methemerythrin. An exciting recent finding in hemerythrin chemistry has been the char-acterization of two semi-methemerythrins, one obtained from one-electron reduction of methemerythin, (semi-methemerythin)$_R$, and the other from one-electron oxidation of deoxyhemerythrin, (semi-methemerythrin)$_O$. The relationships between these and the other forms of hemerythrin are depicted in **Fig. 2**. The semi-met forms have characteristic spectra and kinetic properties and, most significantly, electron paramagnetic resonance spectra at liquid helium temperatures (3–10 K). The *g* values (measures of the size of the magnetic moments of the paramagnetic species) are 1.96, 1.88, and 1.66 (R form) and 1.95, 1.72, and 1.69 (O form). They arise from the single unpaired electron which must exist in the mixed oxidation state, namely the Fe(III)-Fe(II) unit present in the semi-methemerythrin forms. These spectra are useful fingerprints for the particular type of binuclear iron site found in hemerythrin. They have also been noted in the enzymatically active pink reduced forms of acid

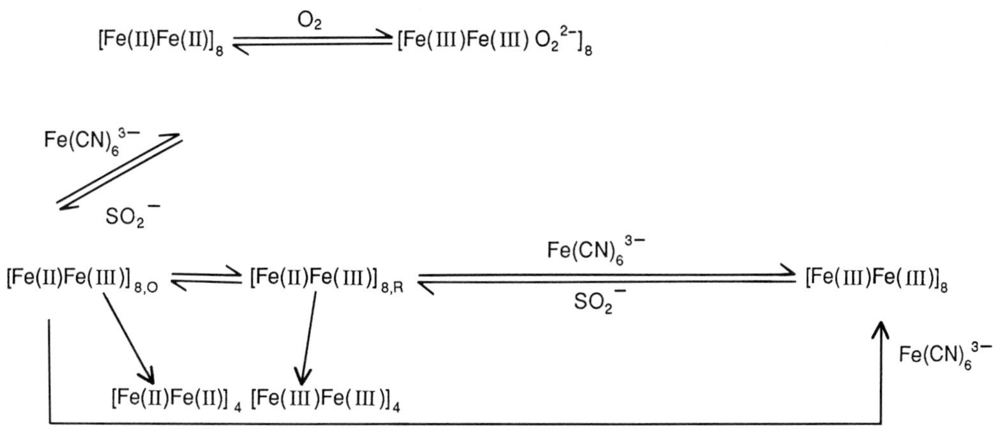

Fig. 2. Reaction scheme showing the relationships of the various forms of hemerythrin.

phosphatases from porcine uterus and from beef spleen, as well as in the reduced component A of methane monoxygenase from *Methylococcus capsulatus*.

Conformational change. Several examples exist of reactions of hemerythrin which are controlled by a slow conformational change (rate constant $\sim 10^{-3}\ s^{-1}$). These include the transformation of the acid form of methemerythrin to the basic one, dithionite reduction of methemerythrin at pH 9.0, addition of thiol reagents to methemerythrin, and denaturing of the protein by the addition of sodium dodecylsulfate. Also, the conversion of (semi-metmyohemerythrin)$_O$ to (semi-metmyohemerythrin)$_R$, which occurs spontaneously and can be monitored by the electron paramagnetic resonance changes, has a rate constant quite close to that for the conformational change. The elucidation of the detailed structure of the oxy-bridged binuclear iron center in hemerythrin and its occurrence in other proteins have encouraged the synthesis of model iron complexes for both met and deoxy forms. The spectral and magnetic properties of the model compounds are close to those of their hemerythrin analogs, but none of the chemical reactions, including reversible oxygenation, has as yet been duplicated.

For background information *SEE COMPLEX COMPOUNDS; COORDINATION CHEMISTRY; CYTOCHROME; RESPIRATORY PIGMENTS (INVERTEBRATE)* in the McGraw-Hill Encyclopedia of Science and Technology.

Patricia C. Wilkins; Ralph G. Wilkins

Bibliography. I. M. Klotz and D. M. Kurtz, Binuclear oxygen carriers: Hemerythrin, *Acc. Chem. Res.*, 17:16–22, 1984; S. J. Lippard, The bioinorganic chemistry of rust, *Chem. Britain*, 22:222–229, 1986; A. K. Shiemke, T. M. Loehr, and J. Sanders-Loehr, Resonance Raman study of oxyhemerythrin and hydroxomethemerythrin: Evidence for hydrogen bonding of ligands to the Fe-O-Fe center, *J. Amer. Chem. Soc.*, 108:2437–2443, 1986; R. G. Wilkins and P. C. Harrington, The chemistry of hemerythrin, *Adv. Inorg. Biochem.*, 5:51–85, 1983.

Herpes

During the past decade, herpes simplex infections of the genital tract have attracted increasing attention because of wider recognition of the symptoms, increasing prevalence in society, the alarm associated with a persistent and chronically recurring disease, the risk of transmission to sexual partners, and the concern of the pregnant woman about transmission of the virus to her offspring. Enhanced public knowledge will help allay fears associated with genital herpes simplex infection and provide a realistic understanding of the risks and problems associated with it. This article focuses on herpes simplex virus infections of pregnancy and the newborn.

Maternal infection. The epidemiology and management of genital herpes simplex virus infections during pregnancy have not been thoroughly studied However, two patterns of maternal infection are recognized as having fetal consequences: disseminated infection and genital maternal infection, either primary or recurrent, which remains localized.

Disseminated infection. In pregnancy, herpes simplex virus infection may be disseminated, that is, may extend well beyond the usual sites of disease, namely the oropharynx and genital tract. A disseminated infection can lead to involvement of other organs of the body or widespread cutaneous lesions, somewhat resembling chickenpox. In a limited number of cases, dissemination after primary pharyngeal or genital infection has led to such severe manifestations as hepatitis, thrombocytopenia, leukopenia, coagulopathy, and encephalitis. The clinical manifestations can include a tendency toward bleeding disorders and seizures with altered levels of consciousness when the central nervous system is involved. Although only a small number of patients have been reported, mortality among pregnant women with disseminated infection is greater than 50%. Fetal death has also been reported in greater than 50% of cases, although fetal deaths were not necessarily associated with the mortality of the mother. Surviving fetuses have generally been delivered by cesarean section either during the acute illness or at term; none, so far, have had evidence of disease. Accumulative experience, then, suggests that factors associated with pregnancy place the mother and fetus at increased risk for severe infection. However, such disseminated infection is an extremely uncommon manifestation of the disease and cannot be considered a major risk factor for maternal or fetal death.

Localized genital infection. The mother is the usual source of virus for transmission to the fetus or newborn, although a few cases of neonatal disease are the consequence of postnatal acquisition. The route of transmission of the virus favors contact between the fetus and infected maternal genital secretions during birth. Ascending infection can also occur, particularly in the presence of prolonged rupture or leakage of membranes, and transplacental infection can occur as a consequence of primary maternal infection. Transplacental infection leading to intrauterine infection of the fetus is poorly understood at the present time, although an increasing body of evidence indicates that this occurs more commonly than was once thought. Infection of the fetus in the uterus, associated with primary maternal infection, can lead to spontaneous abortion, stillbirth, and congenital malformations, particularly hydranencephaly, skin scarring, and chorioretinitis. Infections that occur after 20 weeks of gestation have not been generally associated with unusual problems in the newborn; these babies are generally born free of infection. It should be remembered that infection in the uterus must be extremely uncommon since the actual number of babies with herpes simplex infection at birth is extremely rare; only 4% of all deliveries involve babies with proven neonatal herpes simplex virus infection.

Recurrent maternal infection is the most common form of infection during gestation. Transmission of infection to the fetus appears to be directly related to the actual shedding of virus at the time of delivery. While prospectively acquired data regarding the actual number of women who shed virus at delivery are limited, it is an infrequent event, occurring in 0.4 to 1% of all deliveries. For women with a known history of genital herpes simplex infection, reccurent infection during pregnancy was found in 84% of these individuals. Moreover, asymptomatic viral shedding developed in only approximately 0.5% of infections. These data are similar to those obtained in other populations, particularly those from nonpregnant women. One incidence of cervical shedding without symptoms of infection in women has been reported. However, the observed rates of shedding among asymptomatic pregnant women vary from 0.2 to 7% from those of nonpregnant women. Thus, the frequency of cervical shedding is low, rendering the risk of transmission of virus to the infant exceedingly low when disease is recurrent, probably no greater than 1 in 30 to 1 in 40 babies. The frequency of viral shedding does not appear to vary during gestation. There is no increased incidence of premature onset of labor.

An important problem in the mothers of babies who develop neonatal herpes simplex infection is the asymptomatic nature of their disease at the time they go into labor and bear the babies. Among women without symptoms or a history of genital herpes, a subgroup can be identified whose sexual contacts have vascular lesions indicative of genital herpes. Yet even among these women who are especially at risk, more than half are asymptomatic and have no history of infection. Furthermore, approximately one-third of infected infants born to these asymptomatic women have no antibodies, strongly suggesting primary maternal infection. Thus, measures to identify women at risk for transmitting virus to their offspring must include late-trimester screening, especially at the time of delivery.

Neonatal infection. The estimated number of new cases of neonatal herpes simplex infection is thought to be between 1 in 2500 and 1 in 10,000 deliveries per year, although the number may be increasing. Overall in the United States, it is anticipated that approximately 700 to 1000 cases of neonatal herpes develop each year. Clearly, the occurrence of disease is far less common than would be anticipated from suggested prevalence data for genital herpes in adults of childbearing age in general, indicating the existence of some protective mechanisms for the fetus or the failure of virus to transmit from mother to baby. Clinical evidence of infection is usually present at between 8 and 17 days of life, depending upon the manifestation of the disease. Although skin vesicles are the hallmark of infection, between 25 and 30% of children never have evidence of skin lesions, especially when the disease is localized in the central nervous system.

The infection can be limited to the skin, eyes, or mouth, or it can involve the central nervous system or other visceral organs (lungs, liver, and adrenals). This latter form can be manifested clinically as hepatitis, pneumonitis, intravascular coagulopathy, encephalitis, or skin disease. Death is unusual when the disease is limited to the skin, but it occurs in 15% of babies whose disease remains localized to the central nervous system, and increases to 50% in babies with disseminated infection, even with an antiviral drug. Despite antiviral therapy, there is evidence of impairment in approximately 10% of children and debility in more than 50% with central nervous system infection and visceral disease.

Practical considerations. Women most likely to transmit virus to their offspring—those with a history of genital herpes, preferably laboratory-confirmed, or having a sexual partner with genital lesions—should undergo a series of virologic cultures and clinical assessment. These gestations should be monitored 34–36 weeks to termination of pregnancy by weekly cultures so that recommendations on cesarean section can be supported. If cultures are negative within 1 week of delivery and no lesions are present, vaginal delivery is considered appropriate. If viral cultures are not available, Papanicolaou cytologic studies are an alternative, although these are lower in sensitivity and specificity. Cesarean section is the procedure of choice for women with active primary or recurrent infection at delivery. Should membranes be ruptured for longer than 12 h, the value of cesarean section for prevention of fetal exposure is less clear because of the potential for ascending infection from the cervix. Decisions regarding cesarean section for either primary or recurrent infection should be based on fetal viability.

For background information SEE HERPES; VIRUS; VIRUS CHEMOPROPHYLAXIS in the McGraw-Hill Encyclopedia of Science and Technology.

Richard J. Whitley

Bibliography. C. Hutto et al., Congenital herpes simplex virus infections, *J. Pediatr.*, in press; R. J. Whitley, The natural history of herpes simplex virus infection of mother and newborn, *Pediatrics*, 66:489–494, 1980; R. J. Whitley et al., Neonatal herpes simplex virus infection: Follow-up evaluation of vidarabine therapy, *Pediatrics*, 72:778–785, 1983.

Holography

The use of holography has continued to increase, with display holography probably having the greatest use. This expansion, which began in the late 1970s, is expected to continue.

Holography, an imaging process invented by D. Gabor in 1947, is a radical departure from conventional photography. Here an image is not recorded at all; instead, a nonfocused pattern of light waves is recorded. This pattern is formed by interference between light reflected from the object and a second beam of light that arrives at the recording plate without illuminating the object. Later, when a beam of light illuminates the developed photographic plate, called a hologram, the light emerging from the other side has been refashioned so that it is essentially identical to the light waves that had originally been reflected from the object. Observers whose eyes receive these waves will be unable to distinguish them from the waves that their eyes would have received if they had been actually looking at the object itself. Consequently, observers are deceived into thinking that they are actually looking at the object whereas, in reality, they are looking only at an image. However, this image will have all the optical characteristics of the actual object; it will not appear to lie on the plate, but will appear to hang suspended in space in the vicinity of the plate—in front of it, behind it, or perhaps even straddling it. Furthermore, the image will be in three dimensions, with all the usual optical characteristics of actual objects, characteristics such as parallax, as the observers move their heads. In short, the image is, to a startling degree, a recreation of reality.

Display holography. This three-dimensional realism was first observed in the laser-viewed display holograms introduced in 1963. But the evolution of those early holograms into the highly perfected holograms that are now available entailed a long process of development. For display holography to be commercially successful, both the holograms and the sources used for viewing them had to be inexpensive. Ideally, holograms should be viewable with ordinary white light instead of the expensive lasers and mercury-arc sources that were originally needed. The white-light capability was first achieved with the Denisyuk hologram, which uses the depth dimension of the emulsion; later, the Benton hologram was developed, which achieved white-light viewing capability by giving up the parallax in the vertical plane.

In the Denisyuk technique, the reference beam passes through the recording plate onto the object, and light that is reflected from the object and travels back through the plate serves as the object beam. The beams. traveling in opposite directions, form fringes that run nearly parallel to the emulsion surface. The fringe spacing is typically about 0.2 μm, whereas the emulsion thickness is typically about 8 μm; hence, about 40 fringes are formed and recorded across the depth of the emulsion.

When the hologram is illuminated with a beam of white light, each of the fringe surfaces reflects a small portion (perhaps 0.1%) of the incident light. The reflections from successive surfaces add, giving a relatively large total reflection but only for a narrow bandwidth of light, for which the reflections add in phase. Hence, the Denisyuk hologram receives white light, but it selects only a narrow band of wavelengths for forming the holographic image.

The Benton hologram receives a beam of white light and in the vertical plane behaves like a diffraction grating, dispersing the incident light into its spectral components. Like any hologram, the Benton hologram forms a sharp image in light of any one wavelength. The Benton hologram, in dispersing the light, causes only a narrow bandwidth of light to reach the viewer's eye; consequently, in this narrow band a sharp image is formed. A viewer who changes position in a vertical direction will still see a sharp image, but in a different color. This white light viewing capability requires each point on the image to send only one wavelength of light in a given direction, and in order to bring this condition about, it is necessary to place a narrow horizontal slit in the system when the hologram is made; however, this narrow slit eliminates the vertical parallax.

A hologram could now be viewed with a very inexpensive, white light source. However, the hologram itself remained expensive, which inhibited the widespread use of display holography. This final barrier was overcome by the hot-stamping process that has now become virtually perfected. By this process, Benton-type holograms have been mass-produced in the millions. Holograms now appear on magazine covers, credit cards, and advertising brochures.

A third form of display hologram is the integral hologram, which is synthesized from a large collection of ordinary photographs, each taken from a slightly different position. The process of synthesis of a hologram from many photographs is automated; otherwise it would be laborious. The flexibility of this kind of hologram is considerable; a hologram can be made from any object that can be photographed from multiple directions. Since the original pictures are taken from positions that differ only in the horizontal direction, the holograms, like the Benton type, will be lacking in vertical parallax. Consequently, like the Benton hologram, they can be viewed in white light. Some very large holograms, for example, 3 × 3 ft (1 × 1 m), have been produced in this way, and the integral hologram is now a major element of display holography.

Of less commercial importance but more dramatic is the one-of-a-kind hologram made on a conventional photographic plate as an art object to be viewed in galleries or in private collections. Also, holographic portraiture has become a highly refined technique.

Research in display holography continues. The union of holography and computer graphics has

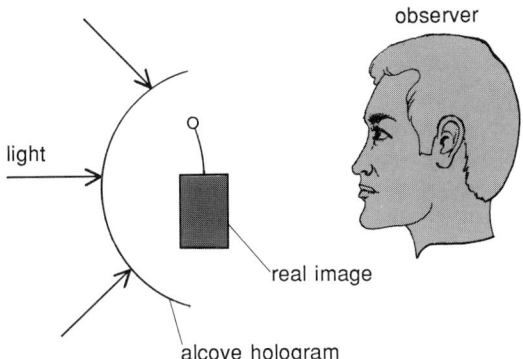

Fig. 1. Alcove hologram. The real image formed in front of the hologram is viewable over an angle of 180°.

led to the creation of holograms that are technically superb, and to the production of a new kind of hologram, the alcove hologram, whose surface is bent into an arc that is concave as seen by the viewer (**Fig. 1**). With this geometry real images have been formed that are viewable over a 180° angle. This was accomplished by processing the pictorial data by computer to compensate for the severe image distortions that occur when the hologram surface is bent into such an arc.

Holographic optical element. Essentially, this is an optical element that operates on the principles of diffraction rather than refraction, as do most optical elements such as lenses. With holographic technique, it is possible to make arbitrary and unusual optical elements. Holographic optical elements were first developed many years ago, but only recently have they been produced in significant numbers. There are various reasons for this increased activity: First, the technology of making high-quality holographic elements having high diffraction efficiency has only recently become refined. Second, a number of excellent applications have been found. The holographic optical element is essentially a specialty item that is best suited to applications where the traditional optical element has shortcomings; the holographic element does not compete very well head-on with established methods. Among the holographic optical elements that are coming into widespread use are the supermarket scanner, the holographic diffraction grating, and the beam combiner for head-up displays.

The operation of a hologram scanner, used, for example, in supermarket checkout, is shown in **Fig. 2**. The disk contains a holographically produced diffraction grating, constructed so as to be highly efficient; that is, most of the light incident on the grating is diffracted into one order. The laser beam, upon passing through the grating, is thus redirected in some chosen direction. The grating spacing (the spacing of the recorded fringes) varies from one portion of the grating to another; thus the direction of the diffracted beam depends on which part of the grating the beam strikes. As the disk rotates, the diffracted beam

direction changes and scans across the entirety of the bar pattern. Light reflected from the bar pattern is sensed by a photocell, which changes the reflectivity pattern of light and dark into an electrical signal. This signal is fed to a digital computer that identifies the item and gives the price. The computer information can also be used for inventory purposes.

The holographic greeting, which has now become a standard product with nearly all grating manufacturers, is nearly as diffraction-efficient as the conventional grating ruled with a diamond cutting stylus. In fact, the holographic rulings can be formed with greater accuracy than is possible with ruling engines.

The head-up display, a device used in aircraft, enables the pilot to view the instrument panel while still looking out the cockpit window. The devise projects an image of the panel in the direction of the window and forms the image at infinity; thus pilots need neither redirect nor refocus their eyes, and can continue to view the airspace in front of them. The peculiar optical requirements of this system can make the holographic solution attractive. The device must transmit most of the incoming light from outside, yet give an acceptably bright image of the instrument panel, and it must also operate over a fairly broad viewing angle.

In addition, holographic optical elements are finding a great number of rather specialized applications, such as light-coupling devices for use with fiber optics.

Hologram interferometry. During the mid-1960s, this technique emerged as the first major application area of holography. In hologram interferometry a hologram of a moving object is made. The object may be vibrating, or the hologram may be formed in two separate exposures, with the object stressed, bent, heated, or otherwise disturbed during the time between the two exposures. In either case, the result is that those portions of the object that move the most reconstruct the weakest image. Thus, the holographic image will depict how the object reacted to the vibration, force, and so forth. For example, the

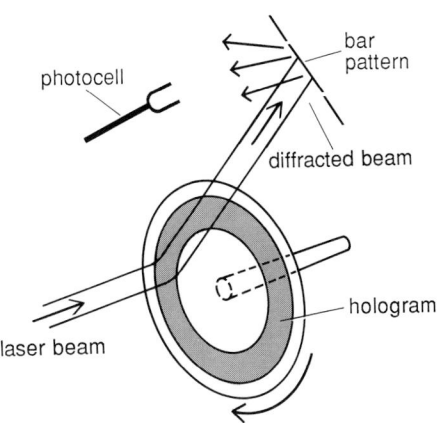

Fig. 2. Hologram scanner, used in supermarket checkout.

entire vibration pattern of the vibrating object is superimposed onto the holographic image.

Hologram interferometry is thus a valuable engineering and scientific tool. The principal problems in expanding the use of this technique are twofold: making it easy to use by technical personnel who are not holographers, and interpreting the complicated brightness patterns that often appear on the image. Over the years, progress has been made in both areas. Various companies now produce so-called holocameras that perform hologram interferometry with little more difficulty than the taking of an ordinary photograph.

Other areas. Other areas of research in holography include methods for making (and not just viewing) holograms in white light; the formation of holograms in real time (that is, instantaneous holography), where the conventional photographic recording materials are replaced by materials such as liquid crystals; and holography in the construction of logic devices that can be used for the construction of all-optical digital computers.

For background information SEE COMPUTER GRAPHICS; HOLOGRAPHY; OPTICAL INFORMATION SYSTEMS; SPECKLE in the McGraw-Hill Encyclopedia of Science and Technology.

Emmett N. Leith

Bibliography. P. Hariharan, *Optical Holography*, 1984; *Proceedings of the International Conference on Display Holography*, Lake Forest College, June 1985; Society of Photo-Optical Instrumentation Engineers, *Proceedings of the SPIE Conference on Holography*, January 24–25, 1985.

Hormone

Increases in urine volume (diuresis) and in urinary sodium excretion (natriuresis) are observed following rapid expansion of blood volume or actual distension of the heart atria. Both hormonal and nervous mechanisms have been implicated in this phenomenon. Investigations of hormonal mechanisms have centered on two substances referred to as natriuretic hormone and atrial natriuretic factor.

Natriuretic hormone has been partially purified from the plasma and urine of both humans and experimental animals, and it is thought to induce natriuresis by inhibiting the enzyme Na,K-ATPase. Inhibition of this enzyme in vascular smooth muscle is presumed to increase intracellular calcium concentration, leading to vasoconstriction and hypertension. Natriuretic hormone has a molecular weight of about 500 daltons and is resistant to boiling in acid and to proteolytic enzymes. Despite many years of intensive laboratory investigations, the nature and source of this hormone remain elusive. In fact, the existence of this hormone as a defined chemical entity is controversial.

Atrial natriuretic factor. This is a potent diuretic and natriuretic agent with hypotensive actions and inhibitory effects on the secretion of renin and aldosterone. These actions, as well as central nervous system actions which include effects on water and salt appetite, suggest that atrial natriuretic factor plays a major role in both the long- and short-term regulation of salt balance and blood pressure. Viewed within the context of general integratory mechanisms (those that interrelate to maintain bodily functions), these properties of atrial natriuretic factor appear to be directed to counteract the actions of the renin-angiotensin-aldosterone system. That is, in very general terms, atrial natriuretic factor is a hormone that opposes the mechanisms known to save water and salt and to elevate blood pressure.

Atrial natriuretic factor is produced by the bulk of muscle cells that form the atrial chambers of the mammalian heart. These cells possess all of the characteristics of heart muscle cells in terms of general cytological characteristics, including contractile elements (myofibrils) and elements common to secretory cells. The latter include granules referred to as specific atrial granules (**Fig. 1**), which store atrial natriuretic factor.

Purification and chemical analysis of atrial natriuretic factor from several mammalian species has shown that this hormone is a peptide. Although peptides of different sizes have been isolated from atrial muscle, it is now generally acknowledged that the main chemical form in tissue is a 126–amino acid peptide (ANF 1-126) found within the specific granules. The main circulation form is a 28–amino acid peptide (ANF 99-126; **Fig. 2**). Essential to the biological activity of atrial natriuretic factor is the presence of the disulfide loop found between the amino acid residues 105 and 121 of this peptide. The peptide shown in Fig. 2 has been variously referred to as cardionatrin I, natriuretic peptide, atriopeptin 28, and so forth. This peptide has been chemically

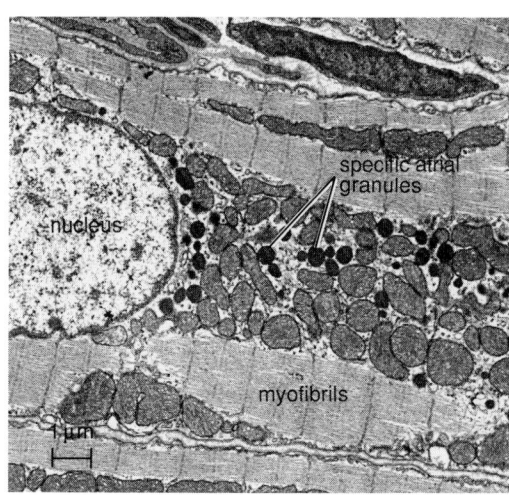

Fig. 1. Electron micrograph of a muscle cell from the heart atria.

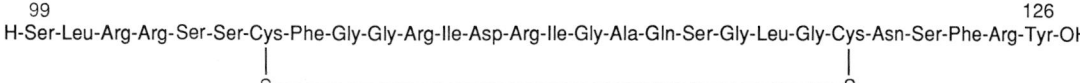

```
99                                                                              126
H-Ser-Leu-Arg-Arg-Ser-Ser-Cys-Phe-Gly-Gly-Arg-Ile-Asp-Arg-Ile-Gly-Ala-Gln-Ser-Gly-Leu-Gly-Cys-Asn-Ser-Phe-Arg-Tyr-OH
                  |                                                         |
                  S——————————————————————————————————————————S
```

Fig. 2. Amino acid sequence of ANF 99–126.

synthesized and is available commercially.

Physiological action. Intravenous injection of a few micrograms of atrial natriuretic factor peptide into an anesthetized, otherwise normal rat produces a dramatic diuresis and natriuresis that commences seconds after injection and subsides after about 10 min. Blood pressure characteristically drops 10–15 mmHg, and the hematocrit significantly increases. The blood pressure effect is likely due to a combination of three effects: direct vasodilation, volume depletion through diuresis, and shift of fluid from the intravascular to the extravascular space. The vasodilatory effect takes place through a direct smooth muscle relaxant property of atrial natriuretic factor which may be demonstrated in the laboratory through the use of standard pharmacological preparations such as aortic strips precontracted with noradrenaline, angiotensin, and so forth. Evidence of fluid shifts comes from the observation that the hematocrit rises even in animals from which the kidneys have been removed.

The effects of atrial natriuretic factor occur together with increases in the levels of the cyclic form of the nucleotide guanosine monophosphate (cGMP). However, the relationship, if any, between atrial natriuretic factor, cGMP, and observed effects is not understood.

Specific receptors for atrial natriuretic factor peptides have been found in all of its known target organs, including the zona glomerulosa of the adrenal cortex and various locations within the kidney, and in vascular smooth muscle. Current studies of the chemical identity of these receptors are expected to elucidate the mechanism of interaction between atrial natriuretic factor and its target sites.

Several lines of evidence suggest that sites of production for atrial natriuretic factor exist in the central nervous system. Although the atrial natriuretic factor levels found in these sites are quantitatively much smaller that those found in the cardiac atria, these sites of production are deemed to be of importance because they are in locations known to be involved in cardiovascular control.

Release. The main stimuli for release of atrial natriuretic factor from the atria and into the bloodstream are increases in volume load and increases in intra-atrial pressures. Direct stretch of the atria in laboratory preparations leads to increased release of atrial natriuretic factor. For example, an isolated rat atria bathed in a balanced salt solution will release atrial natriuretic factor at a rate of about 300 picograms/(ml)(min) when beating spontaneously and subjected to a load of 200 mg. If this load is increased to 5 g, the rate of atrial natriuretic factor release abruptly

changes to about 700 pg/(ml)(min). Returning the load to the original level of 200 mg brings about a decrease in the rate of release of atrial natriuretic factor to the initial rate. This suggests that the muscle cells in the atria possess an intrinsic stretch–secretion coupling mechanism.

The control of atrial natriuretic factor release from the intact heart into the bloodstream in human disease states is poorly understood. However, several clinical entities have been shown to be associated with atrial natriuretic factor blood levels departing significantly from normal values; this value is estimated to be between 10 and 50 pg per milliliter of plasma. These minute levels require sophisticated methods of measurement, which include radioimmunoassay or radioreceptor assay. Particularly dramatic increases in atrial natriuretic factor plasma levels have been observed in chronic congestive heart failure, with the highest levels being associated with the more advanced stages of failure. These high circulating levels of atrial natriuretic factor are obviously ineffective to prevent fluid retention, hyperaldosteronism, and other humoral and hemodynamic derangements which should be counteracted by atrial natriuretic factor. This apparent paradox is the subject of investigation in many laboratories.

Therapy. As a therapeutic agent, atrial natriuretic factor holds much promise because of its multiple desirable actions, including its diuretic and antihypertensive properties with few or no side effects. In practice, atrial natriuretic factor has the disadvantage of being a peptide and thus oral administration is precluded because it would be readily degraded in the digestive system. Preliminary reports indicate that intravenously injected atrial natriuretic factor is beneficial in heart failure and in other fluid-retaining states. The antihypertensive effect of atrial natriuretic factor has been demonstrated in experimental animals by injection of small doses. Current experimentation with new drug delivery systems and with the production of atrial natriuretic factor derivatives that are digestion-resistant may lead to orally active forms. Such forms of atrial natriuretic factor would likely have a wide application as diuretic and antihypertensive drugs.

For background information SEE HORMONE; HYPERTENSION in the McGraw-Hill Encyclopedia of Science and Technology.

Adolfo J. de Bold

Bibliography. A. J. de Bold, Atrial natriuretic factor: A hormone produced by the heart, *Science*, 230:767–770, 1985; H. E. De Wardener and E. M. Clarkson, The natriuretic hormone: Recent developments, *Clin. Sci.*, 63:415–420, 1982; J. H. Laragh, Atrial natriuretic hormone, the renin-aldosterone axis, and blood pressure–electrolyte homeostasis. *N. Engl. J. Med.*, 313: 1330–1340, 1985.

Immunology

Vaccination for the prevention of infectious diseases represents one of medicine's great triumphs in the twentieth century. Although most of the progress has been made in the area of viral vaccines, much has also been accomplished in the fight against bacterial pathogens. The incidence of diphtheria, tetanus, and pertussis (whooping cough) has been reduced to virtually negligible levels as a result of immunization programs throughout the developed countries. Tuberculosis has ceased to be a problem in many areas, in some cases because of improved living standards, but in others it may have been due to vaccination with the attenuated Bacille Calmette-Guérin (BCG). Killed vaccines for typhoid fever and cholera have also been widely used for the prevention of those diseases, but their safety (measured as reactogenicity or side effects) and efficacy are less than optimal. After World War II the use of antibiotics delayed further bacterial vaccine development until the problem of antibiotic resistance in common bacterial pathogens emerged. Prevention, rather than cure, is once more the goal of researchers in infectious diseases. Classical approaches to vaccine development continue, but the technology that has arisen with recombinant deoxyribonucleic acid (DNA) techniques is also being applied vigorously to the search for new, more effective, and safer vaccines.

Current vaccines. Commercial preparations currently available for immunization against bacterial diseases include the long-standing BCG, DPT (diphtheria, pertussis, and tetanus), cholera, and typhoid fever vaccines, the relatively new pneumococcal vaccines (Pneumovax and Pnu-Imune), and the most recent product, *b*-CAPSA I, developed for use against *Haemophilus influenzae*. These vaccines, their compositions, and the diseases against which they are deployed are listed in **Table 1**.

Within the last 10 years, vaccines for the prevention of bacterial meningitis and pneumonia have been approved by the Food and Drug Administration and released onto the commercial market. The four vaccines (*b*-CAPSA I, against *H. influenzae* type b; Menomune, against *Neisseria meningitidis* Groups A, C, Y, and W-135; and Pneumovax 23 and Pnu-Imune, against 23 of the most prevalent and invasive types of *Streptococcus pneumoniae*) are composed of purified polysaccharides derived from the bacterial capsules. These preparations are safe and immunogenic in both children over 2 years old and adults. Their efficacies have been demonstrated in extensive field trials. The major drawback is their lack of immunogenicity in children less than 2 years old, the group most at risk for contracting these diseases; research is now concentrating on enhancing their immunogenicity by coupling the polysaccharides to proteins such as diphtheria and tetanus toxoids. The most recent developments in bacterial vaccine research involve both recombinant DNA technology and classical microbial genetic approaches to the attenuation of live organisms. One such vaccine for typhoid fever is already on the market in Europe.

Research. Vaccine research in the last 20 years has addressed the childhood pathogens *H. influenzae*, *N. meningitidis*, *Streptococcus pneumoniae*, *Escherichia coli*, *N. gonorrhoeae*, *Salmonella typhi*, *Vibrio cholerae*, and *Pseudomonas aeruginosa*. Most studies in the past have

Table 1. Currently available bacterial vaccines

Vaccine	Description	Disease	Schedule
BCG	Live, attenuated *Mycobacterium tuberculosis* var. *bovis*	Tuberculosis	As required
DP<u>T</u>*	Toxoid from *Corynebacterium diphtheriae*	Diphtheria	2, 4, and 6 months, booster at 5 years
D<u>P</u>T*	Toxoid from *C. tetani*	Tetanus	Same as for DPT
DP<u>T</u>*	Killed cells of *Bordetella pertussis*	Whooping cough	Same as for DPT, unless contraindicated†
dT (diphtheria-tetanus vaccine for adults)	*C. diphtheriae* and *C. tetani* toxoids	Diphtheria, tetanus	Adults, every 10 years‡
Cholera	Killed cells of *Vibrio cholerae*	Cholera	As required
Typhoid	Killed cells of *Salmonella typhi*	Typhoid fever	As required
Pneumovax, Pnu-Imune	Purified polysaccharide from 23 capsular types of *Streptococcus pneumoniae*	Pneumonia	As required
Menomune	Purified polysaccharide from A, C, Y, and W-135 capsules of *Neisseria meningitidis*	Meningitis	Special use, for example, military
b-CAPSA I	Purified polysaccharide from *Haemophilus influenzae* type b capsule	Meningitis, epiglottitis	Children 2 years or older

* DPT is a trivalent vaccine containing toxoids from *C. diphtheriae* and *C. tetani*, and killed whole cells of *B. pertussis*. For ease of description they have been listed separately, with the relevant component underlined.
† Severe reactions, including convulsions and encephalopathies, have been attributed to the pertussis component of DPT. The frequency is approximately 1 in every 5 million doses. Any reaction to the first dose would mandate not including pertussis in the second and third DT immunizations.
‡ The dose of diphtheria toxoid is lower than that present in the childhood vaccine DPT.

Table 2. Experimental component vaccines

Organism	Disease(s)	Component(s)
Pseudomonas aeruginosa	Acute and chronic pneumonia	Pili, toxoids, polysaccharide surface proteins
Escherichia coli	Diarrhea, neonatal meningitis, cystitis	Colonization factors, pili, toxoids
Salmonella typhi	Typhoid fever	Polysaccharide, oligosaccharides linked to proteins*
Haemophilus influenzae	Meningitis, epiglottitis, pneumonia	Polysaccharide linked to proteins*
Neisseria meningitidis, Group B	Meningitis	Polysaccharide modified or linked to proteins*
N. gonorrhoeae	Gonorrhea	Pili
Bordetella pertussis	Whooping cough	Toxoid, filamentous hemagglutinin

* Proteins used as carriers for polysaccharides and oligosaccharides include diphtheria and tetanus toxoids, and cross-reacting material from mutant diphtheria toxins.

been directed toward major virulence factors such as capsules (which help the invading organism evade the phagocytic cells of the host's defense system), pili and adhesins (which help the organism adhere to mucosal surfaces of the body), and toxins (which are responsible for the symptoms of disease). Most approaches have included purification of capsular polysaccharides and pili, and detoxification of purified toxins by formalin inactivation. Examples of these types of vaccines are listed in Table 1, and other experimental vaccines falling in this class are shown in **Table 2**.

Other approaches have utilized the techniques of molecular biology to produce so-called mutant toxins—molecules which possess the immunogenicity and antigenicity of the native toxin but lack the enzymatic activity responsible for toxicity. Examples include the cross-reacting mutant toxins of *P. aeruginosa* and *Corynebacterium diphtheriae*.

The rationale behind the development of purified component vaccines is that when a major virulence factor (such as capsule or toxin) is known, antibodies formed against that particular

molecule should protect against disease. The major disadvantages with this approach are the expense involved in purifying the components and the relatively low immunogenicity of, for example, purified polysaccharides. Even the very successful and highly immunogenic tetanus and diphtheria toxoids require multiple administrations. Immunization with component vaccines rarely induces the strong protection seen after recovery from natural infection.

The first bacterial vaccines, developed by Louis Pasteur in the 1880s, were live, attenuated organisms derived from virulent isolates of *Pasteurella multocida* and *Bacillus anthracis*, but because of problems of quality control and reproducible efficacy they fell into disuse. In the first part of the twentieth century, the attenuated BCG was developed and continues today as the only vaccine for tuberculosis. The clear superiority of live vaccines over killed or component vaccines with respect to immunogenicity and duration of immunity has refocused attention on research into their development. Avirulent mutants of a number of pathogens have been isolated by using both classical mutagenesis and sophisticated molecular techniques. Examples of such mutant vaccine strains are listed in **Table 3**.

The galactose-sensitive strain of *Salmonella typhi* (the *galE* Ty21A mutant) developed in Switzerland has been tested extensively in field trials in Egypt and Chile; although the efficacy was excellent in the Egyptian trial, a change in the formulation has apparently led to disappointing results in Chile. The vaccine is available commercially in Europe and is tolerated well by recipients. In contrast to the formalin- and acetone-inactivated whole cell *S. typhi* vaccines which are administered by intramuscular inoculation, the live Ty21A vaccine is given orally in order to induce immunity at the mucosal surface of the intestine. An *aroA* mutant (a strain which requires aromatic amino acids which are not present in mammalian tissues) and temperature-sensitive mutants of *S. typhi* under development would also be given orally, to induce antibodies that would prevent attachment of the pathogen to

Table 3. Genetically attenuated, live bacterial vaccines

Mutant	Form of attenuation	Route	Comments
Salmonella typhi	Streptomycin dependence	Oral	Potency lost after lyophilization
S. typhi	Galactose-sensitive	Oral	Commercially available in Europe
S. typhi	Aromatic amino acid dependence	Oral	In development
S. typhi	Temperature-sensitive	Oral	In development
S. flexneri	Streptomycin dependence	Oral	Significant reversion to virulence
Haemophilus influenzae	Temperature-sensitive	Naso-pharyngeal	In development
Vibrio cholerae	Toxin-negative	Oral	Still causes diarrhea*
V. cholerae	Hybrid strains; *V. cholerae* antigens on avirulent *Escherichia coli*	Oral	In development
Pseudomonas aeruginosa	Temperature-sensitive	Naso-pharyngeal or oral	In development

* It was subsequently discovered that *V. cholerae* secretes a second, "Shiga-like" toxin which also causes diarrhea.

the mucosa and subsequent invasion of the lymphatic and vascular systems.

A *V. cholerae* mutant (Texas Star SR) unable to synthesize complete cholera toxin was isolated and tested for immunogenicity and reactogenicity in human volunteers. Although capable of inducing protection, the mutant apparently produces a shiga-like toxin which resulted in unacceptable levels of diarrhea. Current approaches to cholera vaccine development are utilizing recombinant DNA techniques to create hybrid avirulent organisms which express *V. cholerae* antigens on their surface.

Temperature sensitivity as a form of attenuation for the development of live vaccines has been studied since 1970, and the problem of reversion to virulence has recently been overcome by the combination of multiple mutations of identical phenotype in one strain, thus reducing reversion frequencies to negligible levels. Safe, live vaccines for *H. influenzae*, *P. aeruginosa*, and *S. typhi* are currently being developed by using this approach.

One of the major problems confronting vaccine researchers is the very low immunogenicity of some capsular polysaccharides, particularly that of the Group B *N. meningitidis*, the organism responsible for up to 50% of meningococcal meningitis in the United States. Many different approaches have been attempted to enhance the immunogenicity of the Group B meningococcal polysaccharide, including alteration of the chemical structure and covalent linkage to proteins, but so far success has been elusive.

For background information SEE ANTIBODY; BACTERIAL CELL CHEMISTRY; IMMUNOLOGY; INOCULATION; POLYSACCHARIDE; VACCINATION; VIRULENCE in the McGraw-Hill Encyclopedia of Science and Technology.

Anne Morris Hooke; Joseph A. Bellanti

Bibliography. R. Germanier (ed.), *Bacterial Vaccines*, 1984; J. B. Robbins, J. C. Hill, and J. C. Sadoff (eds.), *Bacterial Vaccines*, Seminars in Infectious Disease, vol. 4, 1982.

Ink

The flexographic printing process is characterized by the use of a flexible, elastomeric printing plate and a very simple ink-feeding mechanism. It involves an ink pick-up roller, an engraved ink-metering roller, a plate cylinder, and an impression cylinder. It contrasts sharply with the lithographic and letterpress processes which use multiple rollers to distribute ink in a thin, regulated film. The short ink train of the flexographic press has permitted the use of highly volatile printing ink, traditionally formulated with solvents such as alcohol. However, water was substituted as a solvent in some applications of flexographic printing as long as 20 years ago, particularly in the printing of certain paper goods.

Recent events have accelerated the conversion of solvent-based flexographic inks to water-based inks, not only in packaging printing but also in some areas of publication printing.

Packaging printing. The increasing pressure on the flexographic industry for compliance with recent air-pollution regulations has greatly expanded the applications of water-based inks. For example, 85% of the corrugated boxes, 75% of the paper cups and plates, and 55% of the milk cartons manufactured in the United States are now printed with water-based flexographic inks.

Labels, wrappings, envelopes, bags, and so forth are predominantly printed with water-based flexographic inks. Conversely, when these products are prepared by rotogravure, a large portion of them are still printed with solvent-based inks.

During the past several years the application of water-based inks on plastic substrates has increased in an attempt to reduce the problem of solvent emissions. A technique known as reverse printing has been developed in which the printing is done on the inside surface of a clear plastic film (the overwrap). The plate is reversed to provide a mirror image of that used for surface printing. The reverse printing of plastic overwraps with water-based inks for products such as paper tissues or towels is now standard. In these applications, gloss, abrasion resistance, and product resistance are not factors because the overwraps are reverse printed.

Inks may contain as much as 25% by weight of the total volatile components of an organic solvent and still comply with the Environmental Protection Agency (EPA) regulations for water-based inks. In fact, most of these inks do contain substantial amounts of lower-molecular-weight alcohols such as *n*-propanol or isopropanol in order to enhance wetting and increase adhesion and drying rate. The vehicles of the inks used on impervious substrates are mainly acrylic, acrylic copolymer, or styrene copolymer dispersions in water. In some cases, rosin-based resins that are saponifiable by alkaline materials are included among the components in the binder, and then the pH of the final ink may be in the 8.0–9.0 range.

The major application for surface printing by flexography is on polyolefin film for food packaging. Acceptance of water-base inks has been limited, mainly because of the slower drying of these inks compared with the solvent-based inks. The latent heat of vaporization of water is much higher than the solvents it replaces, and thus more energy input per unit time is required to accomplish evaporation and drying of the ink film. This fact limits the speed of the press and thus the production rates, unless the dryer temperature is increased. Good functional properties of the water-based ink films have been achieved, although these inks are still not quite equal in every respect to the current solvent inks.

The difficulties that still exist involve achieving

excellent gloss, adhesion, film resistance properties, and printability comparable with those provided by the polyamide-based solvent inks that are currently used in high-volume polyolefin printing for products such as bags for bread and other foods. The approaches include fortifying the binder of the water-based inks with other high-strength polymers. The amount of polymers is limited by their compatibility and solubility in the high-aqueous-content solvents permitted by the EPA. Thus most of the inks still have somewhat deficient film properties and slower drying rates, even though they contain the maximum amount of volatile organic solvents permitted by law. Research is continuing in this area to achieve better binder formulations and inks with higher solids content.

Publication printing. A major application for water-based flexographic inks developed during the past several years is the printing of newspapers and comics. The simplicity of the flexographic press compared with the complex web offset operation and the other advantages are a driving force for this development.

Offset lithography has been steadily displacing letterpress in the United States newspaper industry during the past 20 years, and it is estimated that about 20% of newspaper printing, representing about 40% of newspaper circulation, is still done by letterpress using photopolymer plates. These letterpress operations may be replaced by the flexographic process if it can be designed to provide quality and production speeds equal to those of the offset process at a competitive or lower cost.

Nine installations of flexographic presses for printing newspapers or comics are in use in various parts of the United States, and there are also several installations in England and Europe. These installations differ from one another in equipment, number of units, and usage, but all are actively printing with water-based flexographic inks.

Advantages of flexography. The major advantages of the flexographic process in printing newspapers and such are reduction of paper waste; use of water-based inks, which do not rub off; keyless inking, giving greater density uniformity; less expensive printing press units compared with offset lithography; and better printability, better contrast, and less show-through.

Paper waste is dramatically reduced compared with offset printing since there is no need to achieve an ink-water balance; nor is there any way to adjust ink feed keys due to the fixed anilox inking roll. The third or fourth print from a flexographic press is generally saleable copy, while with a web offset press the first one hundred to several hundred copies are usually wasted.

Water-based inks dry completely by evaporation of the water and cannot rub off as the non-drying-oil-based offset inks do. This may also be an advantage in terms of health as well as esthetics, since contamination of the reader's hands is eliminated.

The offset and letterpress inking systems have ink feed and dampening feed adjustments which must be set constantly to maintain a uniform ink density. In the flexographic process the keyless anilox inking roll assures a uniform and constant ink feed to the plate.

It is estimated that the cost of a flexographic press unit will be only 75% of that of a web offset unit, because of its simpler inking system, lack of dampening system, and lighter drive motor. Because of the lack of penetration of the water-based ink into the paper, the absence of an oily vehicle, and the fact that the ink dries completely, the printed sheet exhibits good contrast, less staining of the white areas by the ink vehicle, and less show-through from the reverse side of the paper. In addition, the print sharpness is superior to oil ink printing.

Problems. Several years ago when flexography for publication printing was introduced, most of the available inks were derived from the water-based inks developed for packaging printing. Since these inks were being used successfully in packaging printing, particularly on paper substrates, it was assumed that they would work satisfactorily in newspaper printing. Newspaper printing, however, takes place in an entirely different environment which generates several problems.

The initial problem was second-impression set-off. This occurs when the web which has been printed on the first side must be turned and printed on the reverse side. Most packaging is printed on only one side of a web. In printing the reverse of the web, as is required in a newspaper, the ink from the first side has a tendency to transfer to the impression cylinder of the second printing unit. If this transference does not reach equilibrium quickly, the ink will tend to build up and become "caked," providing a potential for web breaks. In addition, transference affects the quality of the printing. This transfer is caused by the ink on the first side becoming too tacky as it dries, causing it to strongly adhere to the second-impression roll surface. Many of the polymers used in the packaging inks tended to go through a very tacky stage while drying and, therefore, proved unsuitable for this application.

The second problem encountered was poor wetting of the plastic plate surface and lack of transference of the ink to the plate. This was caused in some cases by the surface tension of the inks being high relative to the surface energy of the printing plate. A third problem was rewetting of the ink on the plate after a stoppage occurred. It is not practical to clean newspaper plates after such stops.

Many of these problems have been solved by a combination of ink reformulation, changes in plate surface chemistry, and mechanical redesign of the press equipment. Inks containing some of the newer acrylic resins and styrene acrylic copolymers, which are water-dispersible, have provided better pigment dispersibility, printability, and drying. A recently developed formulation uses natural products such as modified starches and other water-dispersible binders in water-based newspaper flexographic inks. While these natural materials have been used as the sole binder in some inks, they have been most useful as modifiers for the acrylic-type materials. Other new formulations for newspaper flexography use dye-based inks which depend on the dye's affinity for the cellulose in the paper to prevent bleeding by the ink in water.

Proper selection of the anilox roller pattern, depth, and fineness is also critical in optimizing print quality, as is the correct ink viscosity and its control during the press run. The training and experience of the press-operating personnel in the pecularities of flexographic equipment can also contribute to improved printing quality. This is particularly important because very few newspaper press operators have ever run a flexographic press, having worked mainly on letterpress or lithographic presses.

There are still problems to be solved. The major one is the filling in of the halftone areas in plastic photopolymer plates, especially where the dots are close together and partially dried ink tends to pick up lint. Process colors have also posed some problems due to the poorer rheology of the water-based colored inks. These problems are being addressed through proper pigment and vehicle selection.

The use of flexography in the newspaper market is expected to expand. The process is also being actively investigated for use in other types of publication printing on uncoated stock. These applications should provide additional impetus for the continued development of water-based flexographic inks.

For background information SEE INK; PRINTING in the McGraw-Hill Encyclopedia of Science and Technology.

R. W. Bassemir

Bibliography. M. Capristo, Can we clear the air with water inks?, *Flexo Mag.*, vol. 11, no. 2, February 1986; R. Gamblin, Dye-based inks enhance color in flexo newspaper printing, *Flexo Mag.*, vol. 11, no. 2, February 1986; D. Graham, Flexo has the potential to become dominant in the newspaper industry, *Flexo Mag.*, vol. 9, no. 7, July 1984; G. MacLain, What is working with water inks?, *Flexo Mag.*, vol. 11, no. 2, February 1986.

Ion implantation

During implantation a stream of high-velocity ions is directed ballistically toward a substrate where the ions stop a short distance below the surface because of collisions with the target atoms. Because ion beams can be produced from virtually any element, any element can be injected into the near-surface region of a solid at concentrations that can be well above the classical solubility limit. The effect is to allow the controlled modification of the physical, chemical, mechanical, or electrical properties of the substrate.

Semiconductor applications. Historically, ion implantation has been used extensively by the semiconductor industry to change the electrical properties of silicon or gallium arsenide by the introduction of dopants, such as boron, silicon, or arsenic. The usefulness of implantation stems mainly from the accuracy with which doping levels can be controlled to provide precise impurity concentrations between 10^{14} and 10^{21} atoms/cm^3. Also, the high speed of the incident dopant atoms allows them to pass unimpeded through any surface impurity layer or thin screen coating that must be laid down on the substrate to isolate it from impurities. While implantation is a more expensive and complicated process than diffusion technology, its technical advantages are so great that the process is gradually being substituted for diffusion doping throughout the semiconductor industry.

Subsurface insulating layers. Recently, much attention has been given to the creation of subsurface insulating layers by the direct implantation of oxygen atoms to create buried silicon dioxide (SiO_2). During such production, oxygen ions are implanted at energies between 100 and 350 keV to form a more or less gaussian depth concentration of oxygen atoms within the silicon matrix which, although somewhat defective because of radiation damage, still retains its single-crystal structure to a useful extent. A subsequent treatment for several hours at temperatures close to 2450°F (1350°C) anneals this damage and causes the oxygen atoms to congregate in a well-defined layer consisting of stochiometric silicon dioxide within a (100) single-crystal silicon matrix. The thickness of these layers scales linearly with dose, and at doses of 0.8×10^{18} O^+ ions/cm^2 a layer of buried silicon dioxide is formed which has a measured thickness of 0.165 micrometer, very close to the theoretical value of 0.173 μm expected for silicon dioxide of normal density. Because these dose levels are about three orders of magnitude greater than the concentrations needed for conventional semiconductor fabrication, the construction of commercial devices incorporating these layers has necessitated the development of a new generation of implanters that produce O^+ ion currents above 150 mA.

The practical usefulness of such silicon-on-insulator integrated circuits derives from the excellent electrical isolation and the low parasitic capacitance that becomes possible between adjacent transistors. Low capacitance allows circuits to operate at higher speeds with reduced power dissipation. Electrical isolation causes silicon-on-insulator devices to be significantly more resistant to pulses of nuclear radiation than conventional integrated circuits, and is important in many military applications.

MeV implantation. Interest is rapidly developing in the use of ion implantation at energies above 1 MeV for semiconductor fabrication. At these energies, dopants penetrate several micrometers into the substrate and the range is always larger than the straggle length, so that the crystalline damage in the surface region is greatly reduced. **Figure 1** compares the ranges and induced damage for 200-keV and 1-MeV boron ions. For the low-energy case (boron at 200 keV), the damage distribution extends throughout the depth of the implant and the damage peaks at a depth less than the ion range distribution. In contrast, the damage distribution produced by 1-MeV boron ions is much more coincident with the range, and the near-surface region is relatively undamaged. This tendency of MeV implants to concentrate damage near the end of the ion range leads to a substantial region of nearly defect-free material near the surface which can be used for planar circuit fabrication.

It is expected that MeV implantation will ultimately displace the current diffusion technology for the production of most high-density complementary-metal-oxide-semiconductor (CMOS) wells when the feature size approaches 1 μm and below. One reason for this shift in processing procedures is that diffusion technology requires that the wafer be exposed for extended periods to very high temperatures which can induce wafer warpage and shrinkage of device geometries. These dimensional changes are particularly troublesome for the larger wafer sizes used today (diameters of 6–8 in. or 150–200 mm) and can lead to focusing and registration difficulties during subsequent lithographic steps. Another use of MeV implantation is the formation of low-resistivity dopant layers well below the active circuits, causing parasitic currents to flow well away from the sensitive surface circuitry and thereby reducing the possibility of forward-biasing stray bipolar junctions. Avoidance of this phenomenon, known technically as latch-up, is a major concern for device designers.

A novel application of MeV implantation is the fabrication of a conducting grid a few micrometers beneath the memory cells of random-access memory chips (**Fig. 2**). The purpose of such a grid is to prevent any mobile charges, which are formed by interactions of ionizing cosmic nuclear radiation or natural alpha particles with the substrate, from diffusing into individual memory cells or their associated read-write transistors.

Focused ion implantation. A recent innovation in implantation has been the development of ion beams which can be focused to submicrometer

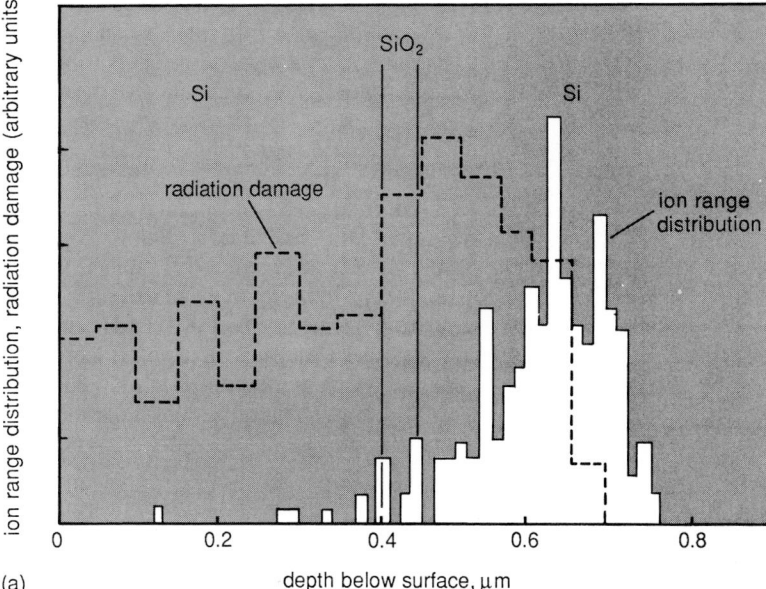

(a)

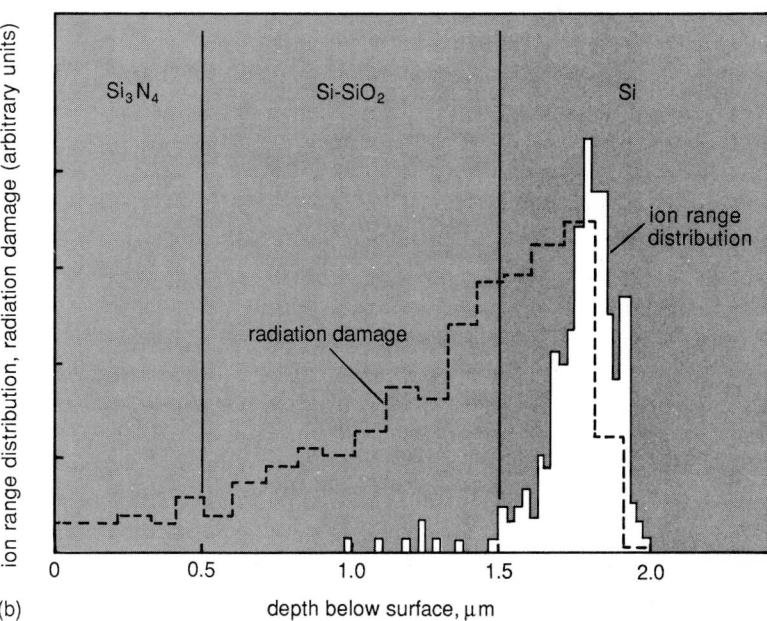

(b)

Fig. 1. Calculations of ion range distribution and radiation damage, showing comparative effects of range straggling and damage. (*a*) 200-keV B$^+$ ions incident on 0.4 μm of silicon (Si) and 0.05 μm of silicon dioxide (SiO$_2$) coating over silicon. (*b*) 1-MeV B$^+$ ions incident on 0.5 μm of silicon nitride (Si$_3$N$_4$) and 1.0 μm of silicon and silicon dioxide (Si-SiO$_2$) over silicon. (*After D. Pramanik and M. Current, Applications of MeV implantation, Solid State Technol., 27:211–219, 1984*)

diameters at current densities of the order of 1 A/cm^2. Such small-diameter beams make possible the direct writing of implanted circuits without the need for implantation masks. Unfortunately, while these current densities are high, the writing speed limits throughput, and the usefulness of this technique of direct implantation for most industrial fabrication is questionable. At present the applications being emphasized are the direct exposure of resist on wafers, where the photoresist is very sensitive and speed is often not critical; the fabrication of high-resolution

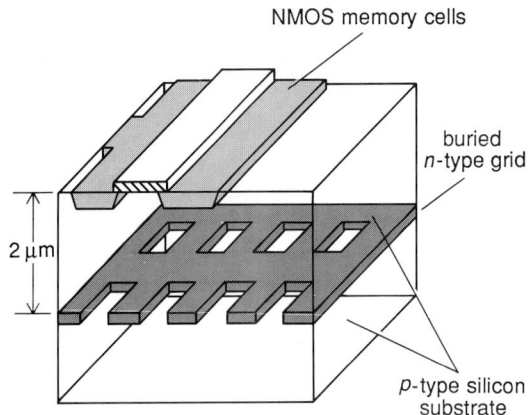

NMOS memory cells

buried
n-type grid

2 μm

p-type silicon
substrate

Fig. 2. Buried grid
structure below
negative-metal-oxide-
semiconductor (NMOS)
memory cells.

masks; and those applications where small specialized areas of a larger chip must be customized by implantation. An example of focused ion implantation is the production of high-resolution optical sensors which are directly incorporated within a conventional digital or analog circuit.

Metals implantation. Considerable effort has been made to produce new surface alloys and films by ion implantation into metals. It has now been demonstrated that a variety of superhard surfaces, low-friction bearings, wear-resistant regions, catalytic coatings, and corrosion-resistant alloys can be produced. An important feature of such surface modification is that the effects occur solely in the near-surface region, so that there is no change in the structural properties or the strength of the underlying bulk material. A protective alloy or chemical coating can be applied to finished components such as precision ball bearings with no measurable change in dimensions or strength.

For example, if palladium is implanted into titanium metal, the corrosion rate of the implanted titanium can be reduced by a factor of 1000 below the corrosion rate for untreated titanium. One application is the production of chemical passivation in titanium hip and knee balls that are used to replace natural joints that have been damaged by arthritis or accident. In the past the effective lifetime of such artificial joints had been 5 to 10 years; with palladium implantation it is expected that the useful lifetimes will be increased by as much as an order of magnitude.

Superconductivity. Since the mid-1970s the technique of ion implantation has been used to modify the superconducting properties of materials by introducing radiation damage sites or by forming new alloys that can change the transition temperature or the critical current. A variety of new and metastable superconducting alloys can be produced by implantation.

Light guides. Another nonsemiconductor application of ion implantation is the production of light guides. A channel of material can act as a light guide when the refractive index decreases as a function of distance from the central region of the guide. If active chemical species such as nitrogen ions are implanted into quartz (SiO_2), large changes in the refractive index are observed; with nitrogen ion doses in the range 10^{15} ions/cm^2 to 10^{16} ions/cm^2, the refractive index is found to increase steadily from 1.46 to 1.96 due to oxynitride formation.

Gallium arsenide (GaAs) appears to be particularly useful as a substrate for optical waveguides because it is possible to also fabricate on the same substrate a variety of light-producing lasers, light guides, and detectors, and the analog and digital circuitry needed to generate and interpret the data. Such a composite structure can have very high reliability.

For background information SEE INTEGRATED CIRCUITS; INTEGRATED OPTICS; ION IMPLANTATION; SUPERCONDUCTIVITY in the McGraw-Hill Encyclopedia of Science and Technology.

Kenneth H. Purser

Bibliography. M. I. Current and D. K. Sadana (eds.), *Advanced Applications of Ion Implantation*, Proceedings of SPIE—The International Society of Optical Engineering, 1985; J. K. Hirvonen (ed.), *Ion Implantation*, 1980; D. Pramanik and M. Current, Applications of MeV implantation, *Solid State Technol.*, 27:211–219, 1984; H. Ryssel and H. Glawischnig (eds.), *Ion Implantation: Equipment and Techniques*, 1983; S. M. Sze (ed.), *VLSI Technology*, 1983.

Kawasaki syndrome

Originally called mucocutaneous lymph-node syndrome, Kawasaki syndrome is a childhood illness first described approximately 20 years ago. The diagnosis is based on the fulfillment of five of six criteria: (1) fever lasting 5 or more days and unresponsive to antibiotics; (2) blood vessels of the lining (conjunctiva) of both eyeballs congested and more prominent; (3) lips dry with redness and cracking, tongue red with prominent papillae, lining of the mouth and pharynx diffusely red; (4) hands and feet (in the early stage) red with swollen palms and soles, while in the later stage peeling of the skin, often beginning under the tips of fingernails and toenails; (5) flat or raised red spots of varying sizes on the skin, mainly on the trunk; and (6) moderate enlargement of lymph nodes in the neck without pus formation.

Although not part of the syndrome's diagnostic criteria, damage to the heart's arterial circulation (coronary arteries) is the most significant aspect of this illness. Recognized manifestations of heart involvement vary from mild and reversible changes to severe disease associated with sudden death or chronic debilitating dysfunction.

Epidemiology. T. Kawasaki established the basic clinical criteria for the syndrome in Japan in 1967; M. Melish was the first to recognize the syndrome in the United States. Over 60,000 cases were reported in Japan prior to 1985, more than 2000 cases were confirmed in the United States before 1986, and cases have been recognized worldwide. Kawasaki syndrome is almost exclusively a disease of children, with only rare cases occurring after 8 years of age. Fifty percent of the patients are younger than 3 years of age. Males are affected slightly more frequently than females. Persons of Eastern ancestry, especially Japanese, have the highest risk of acquiring the disease, but any ethnic group may be affected. Blacks have an intermediate risk and whites have the least risk. Ordinarily the cases appear sporadically, but epidemics, with incidence rates of up to 150 cases per 100,000 children, have occurred in widespread areas of the United States and elsewhere. Person-to-person spread or common-source exposure has not been documented. Appearance of cases is not restricted by seasonal, geographic, socioeconomic, or environmental barriers. Recurrent disease in individuals and second cases in the same family have been reported but are uncommon. The case fatality rate is between 1 and 2%.

Etiology. The etiology of Kawasaki syndrome has eluded intensive investigation. Attention has been focused on toxic, allergic, immunologic, and, especially, infectious causes. Toxic agents, such as mercury, have been considered and then dismissed for lack of evidence. Substantive support for an allergic or immunologic cause has not been found. Infectious etiologies, including streptococcal, staphylococcal, rickettsial, leptospiral, and viral, have been scrutinized and can be associated with only a small minority of cases.

Clinical findings. Kawasaki syndrome is a three-phase illness that evolves from an initial phase of approximately 10 days to a subacute phase of approximately 25 days and subsides with a convalescent phase that ends between the sixth and tenth weeks of illness.

Initial phase. The initial or acute phase is characterized by fever, congestion and prominence of the blood vessels of the conjunctiva of both eyeballs, abnormalities of the lips and oral cavity, reddening and swelling of the palms and soles, skin rash, and lymph-node enlargement.

Fever, the initial symptom, is present in all cases, and lasts throughout the initial phase and often into the subacute phase. Temperatures range between 101 and 104°F (38 and 40°C) and may be associated with toxicity. The conjunctival blood vessels become prominent a few days into the illness in almost all patients, but no discharge is seen. The eyes generally clear as the fever subsides. Virtually every patient develops dry, red, and cracked lips; prominent tongue surface protuberances; or redness and dryness of the lining of the mouth and pharynx. The changes of the lips and mouth appear 1 to 3 days after the onset of fever, and most disappear as the fever subsides. The cracked lips heal 1 to 2 weeks later. Red palms and soles appear on the third to fifth day in 9 of 10 patients and remain for 1 to 2 weeks. Seventy-five percent also have swelling of the hands and feet.

Almost every patient develops a rash 1 to 5 days after the fever, and the rash remains for less than a week. The character of the rash varies in shape, feel, and intensity of its redness in different patients but never includes blister formation. It usually involves the trunk and occasionally the face, arms, and legs.

Lymph-node enlargement can appear early in the course and can persist for weeks; usually a single node in the neck is involved. Only one-half to three-fourths of patients have such a finding, and tenderness, redness, warmth, or firmness of the node is minimal.

Subacute phase. The subacute phase includes peeling of the skin, but more crucially is characterized by joint symptoms and signs, heart dysfunction, and elevation of blood platelet counts. The peeling occurs in over 90% of the children and usually begins under the tips of the fingernails, often extending over the palmar surface within a week. The feet are similarly involved, but usually follow the peeling of the fingers by a few days.

Joint involvement, most frequently of the knees, hips, and elbows, is experienced by about half of the patients. The pain, swelling, redness, or increased warmth may take weeks to resolve, but permanent damage does not occur.

The cardiac manifestations constitute the most significant aspect of Kawasaki syndrome and account for the mortality rate of 1 to 2%. The changes begin with coronary artery inflammation, progress to focal artery ballooning (aneurysm) and to blood clotting, and may terminate as scarring and severe narrowing of the involved arteries. Clinically, at least 20% of patients will develop evidence of heart involvement, including coronary artery enlargement or aneurysm formation, rhythm disturbances, heart muscle dysfunction or destruction, heart valve incompetence, or sudden death. Patients at greatest risk for serious heart dysfunction are those with coronary artery aneurysms; therefore, clinical evaluation focuses on identifying patients with coronary artery changes. Even though certain clinical findings, such as high fever or prolonged fever, are indicative of high risk for aneurysm formation, serious cardiac involvement is possible without specific signs or symptoms.

All patients should have cardiac evaluation, including chest x-rays, electrocardiograms, and echocardiograms. Noninvasive two-dimensional echocardiograms are considered almost as sensitive in detecting small vessel changes as angiograms which necessitate insertion of a catheter through a blood vessel and into the heart, where dye is injected prior to obtaining multiple x-rays. Serial echocardiograms can be performed over time to follow the progression or resolution of the aneurysms or changes in overall cardiac performance.

The aneurysms of approximately 50% of involved patients will disappear within 1 to 2 years. On the other hand, some patients, even without evidence of coronary artery involvement, will exhibit abnormalities of heart size and function long after their acute illness. Also of concern regarding long-term outlook is the possibility of early-onset coronary artery atherosclerosis in adulthood.

Platelet counts rise during the second week and can remain abnormal for weeks. It is believed that the elevated counts can lead to thrombus formation in the diseased coronary arteries and contribute to increased heart damage.

Convalescent phase. The convalescent phase begins when the clinical symptoms subside, and ends quietly when all laboratory evidence of inflammation disappears; however, cardiac dysfunction may continue to manifest itself. Transverse lines in the fingernails or toenails may appear initially during this phase.

Additional clinical features that may appear in the course of the disease include aseptic meningitis; diarrhea, abdominal pain, and swelling of the gallbladder; liver dysfunction; urethral inflammation and ulcers of the opening of the urethra; ear infection; inflammation of parts of the inner eyeball; elevated blood pressure; pneumonia; behavioral changes; nerve palsies; acute renal failure; aneurysms of major and medium-size noncoronary arteries; muscle inflammation; death of tissue in abdominal organs, testes, or digits of the hands; and loss of hair.

Laboratory studies. Laboratory abnormalities are nonspecific and may include mild anemia, increased white cell counts, marked increase in platelet counts, and increased levels of tests measuring inflammation; pus cells, protein, and red cells in urine; white cell elevations in cerebrospinal fluid; and elevated liver enzymes and bilirubin.

Therapy. Therapy is directed at the inflammatory activity of the disease in order to decrease the toxicity seen during the acute phase and to prevent aneurysm formation later in the disease. The therapy is also expected to decrease platelet activity in order to prevent thrombus formation resulting from the increased number of platelets. The lack of an identifiable etiologic agent removes the possibility of specific therapy, and antibiotics do not modify the course. The antiinflammatory effects of steroids have no demonstrable benefit and may increase the likelihood of aneurysm formation. Anticoagulants have been used, but they do not appear to be beneficial. Until recently, aspirin alone was recommended as standard therapy. Early utilization of high doses decreases the duration of fever in the majority of patients and somewhat lessens the likelihood of cardiac involvement and aneurysm formation. When the platelet count rises, the

dose is reduced to low levels that are believed to better inhibit the platelet clumping which could lead to thrombus formation in the abnormal arteries of the heart.

Recently, in studies from Japan and the United States, high doses of intravenous gammaglobulin were tested in conjunction with aspirin and this therapy was more effective than aspirin alone in preventing cardiac vessel damage, as measured by echocardiograms.

Surgical therapy of the cardiac artery aneurysms has been performed successfully on a small number of patients after their convalescent course. However, surgery during the subacute phase has been found to be hazardous and to have no demonstrable benefit.

For background information SEE MEDICINE; PUBLIC HEALTH in the McGraw-Hill Encyclopedia of Science and Technology.

Jacob A. Lohr

Bibliography. T. Kawasaki et al., A new infantile acute febrile mucocutaneous lymph node syndrome (MLNS) prevailing in Japan, *Pediatrics*, 54:271–276, 1974; G. Koren et al., Probable efficacy of high-dose salicylates in reducing coronary involvement in Kawasaki disease, *JAMA*, 254:767–769, 1985; M. Melish, Kawasaki syndrome (the mucocutaneous lymph node syndrome), *Pediatr. Annu.*, 11:255–268, 1982; J. Newburger et al., The treatment of Kawasaki syndrome with intravenous gamma globulin, *N. Engl. J. Med.*, 315:341–347, 1986.

Kondo effect

The Kondo model of a single magnetic impurity in a metal has remained a problem of great importance in many-body physics since it was formulated in 1964. Directly relevant to alloys with dilute concentrations of magnetic impurities, the model has stimulated development of powerful new theoretical tools which provided exact calculations of thermodynamic properties such as the magnetic susceptibility $\Delta\chi(T)$ (the induced magnetization per unit applied magnetic field per impurity, a function of the thermodynamic temperature T). However, only in the past few years has it been possible to reliably calculate static and dynamic properties, such as the extra electrical resistivity per impurity, $\Delta\rho(T)$, within a unified theoretical framework. This article describes the new approach and results obtained.

Basic model. The basic model contains a localized quantum-mechanical spin (with colinear magnetic moment) which has N allowed discrete directions (or degeneracy). This spin is embedded in the metal and couples with so-called exchange strength J to conduction electrons, each having the same spin magnitude as the impurity. For negative J, it is energetically favorable for the local moment and conduction moments to align antiparallel, while positive J favors parallel alignment. When such an exchange coupling is present between two localized moments with $N = 2$, the absolute value of J gives the splitting in energy between parallel and antiparallel states. For the interesting anomalies discussed below, J must be negative.

The model allows spin-flip scattering processes wherein conduction electrons exchange one unit of angular momentum $\hbar$ (Planck's constant divided by 2π) with the impurity while maintaining the direction of the total spin of the system. The conduction electrons obey the Pauli exclusion principle, so only one electron may occupy a state of given momentum, energy, and spin direction. This leads (at $T = 0$) to a distribution of occupied conduction states cut off sharply at an energy E_F (the Fermi energy) above the lowest energy state. While E_F is large (in temperature units, of order 10^4 K or more), properties from room temperature down are dominated by the spectrum of electronic states near E_F.

Dilute alloys with cerium (Ce) are well described by the model. Free Ce^{3+} ions possess a degeneracy $N = 6$ from a coupling of the single $4f$ electron's pure spin to its large orbital angular momentum. This may be reduced when the ion is embedded in a crystal, depending upon how strongly the $4f$ electron couples to the effective crystalline electric field, which reduces the symmetry from isotropic to that of the crystal host. Apparently $N = 6$ is realized for cerium in thorium (Th). For cerium in lanthanum hexaboride (LaB_6), $N = 4$; and for cerium in lanthanum aluminide ($LaAl_2$), $N = 2$. In magnitude, $|J|/E_F$ appears to be about 0.05–0.1 for these alloys.

While they often share many qualitative features with dilute cerium alloys, alloys that contain, say, manganese, iron, uranium, or thulium impurities require generalization of the above, simple model. Full solutions of model forms appropriate to these impurities have not yet been obtained.

Exact static results. Exact static results were derived in the 1970s from solutions for many-body quantum-mechanical states, by means of two independently developed approaches. In one approach, renormalization group ideas were applied to the $N = 2$ Kondo model, yielding effective Schrödinger equations related by renormalization-group transformations at a discrete sequence of logarithmically spaced temperatures. Solution at each temperature yielded the energy eigenvalues and hence thermodynamics at that temperature.

In the other approach, the $N = 2$ Kondo model was mapped to a related one-dimensional problem. The so-called Bethe-ansatz technique for one-dimensional many-body problems, first applied by H. Bethe in 1931, was then used to yield the states and the thermodynamics.

Subsequently, many workers contributed to both the renormalization group and Bethe-ansatz approaches. Among the rich physics obtained are two key results, asymptotic freedom and universality.

Asymptotic freedom. At high temperatures, the impurity acts like a free ion: $\Delta\chi(T)$ is proportional to $1/T$. Antiparallel coupling to the conduction electron moments causes a slow crossover near a temperature T_K to a so-called compensated many-body ground state in which $\Delta\chi(0)$ is proportional to $1/T_K$ with no net moment (one net unit of conduction spin cancels out the impurity spin). In contrast, the free ion is fully polarized by an arbitrarily small applied magnetic field at $T = 0$. This property of weak coupling at high energy or temperature and strong coupling at low energy or temperature is known as asymptotic freedom, and is shared with quarks, the fundamental constituents of nuclear matter.

Universality. Physical properties for a given value of N are universal functions of T/T_K, when J/E_F is small. This is illustrated in **Fig. 1**, where calculated $\Delta\chi(T)/\Delta\chi(0)$ curves for two widely different values of T_K coincide when plotted against T/T_0, where T_0 equals T_K to within a factor of two, depending on the degeneracy.

The Kondo temperature is given by the equa-

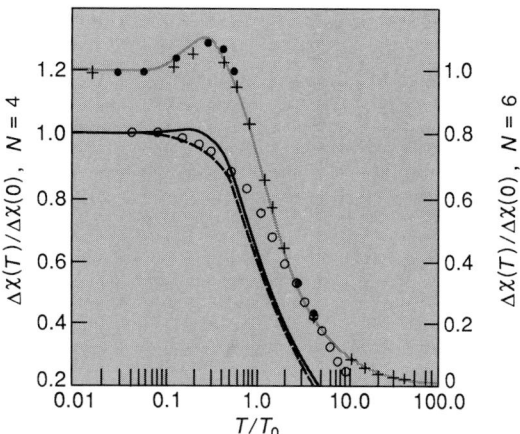

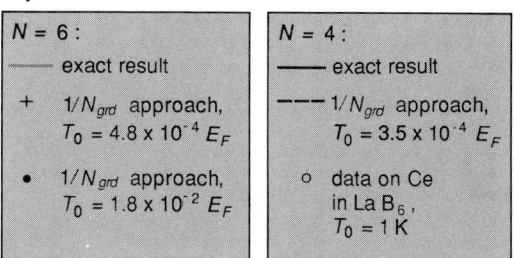

Key :

$N = 6$:	
——	exact result
$+$	$1/N_{grd}$ approach, $T_0 = 4.8 \times 10^{-4} E_F$
$\bullet$	$1/N_{grd}$ approach, $T_0 = 1.8 \times 10^{-2} E_F$

$N = 4$:	
——	exact result
- - -	$1/N_{grd}$ approach, $T_0 = 3.5 \times 10^{-4} E_F$
$\circ$	data on Ce in $La B_6$, $T_0 = 1$ K

Fig. 1. Universal magnetic susceptibility curves $\chi(T)$ of a Kondo impurity. Solid curves are exact Bethe-ansatz results. Open circles are from data on cerium (Ce) in lanthanum hexaboride (LaB_6) with $T_0 = 1$ K. Other points and curves are from the new $1/N$ approach described in the text, with values of T_0 given in the key. Points and curves with $N = 4$ are measured on the scale shown to the left of the graph, while those with $N = 6$ are measured on the scale to the right.

tion below, where k_B is Boltzmann's constant.

$$k_B T_K \approx E_F \left(\frac{|J|}{2E_F} \right)^{1/N} \exp \left(\frac{2E_F}{N|J|} \right)$$

Through the exponential dependence, small variations of N or J can sweep the value of T_K through orders of magnitude. The equation indicates that large values of N enhance T_K. Experimentally, Kondo temperatures range from the order of a millikelvin [as in the case of manganese (Mn) in copper (Cu)] to the order of 1000 K (as in the case of cerium in thorium).

Perturbative approaches. The exact approaches discussed above are not technically amenable to calculation of dynamical properties such as $\Delta\rho(T)$. More useful in this regard are perturbative approaches where elementary scattering processes are added together in a systematic fashion. Within such procedures there are well-known prescriptions for computing $\Delta\chi(T)$ and $\Delta\rho(T)$.

The simplest single spin-flip processes [giving terms proportional to $(J/E_F)^3$ in $\Delta\rho(T)$] cause $\Delta\rho(T)$ to increase logarithmically as T is lowered. This effect competes with a well-known high-temperature increase due to scattering off lattice vibrations and produces a resistivity minimum. This explanation of the minimum by J. Kondo (using the $N = 2$ model) in 1964 resolved an experimental puzzle that had existed since the 1930s. Subsequently, the body of anomalies associated with this problem became known as the Kondo effect. Kondo's work also showed that parallel exchange coupling or simple charge scattering leads to no such anomalies.

Approximate extensions of the work that encompassed an infinite number of multiple scattering processes (including all powers of J/E_F, with the most divergent logarithmic term kept at each order) gave reliable results above T_K (another manifestation of asymptotic freedom) and the correct (nonzero) value for $\Delta\rho(0)$, but otherwise gave unreliable results below T_K (as indicated by the broken curve of **Fig. 2**). Indeed, the failure of these early perturbative approaches stimulated the exact approaches described above. Importantly, the perturbative approaches suggested that the Kondo effect arose from the formation of a long-lived scattering resonance (the Abrikosov-Suhl or Kondo resonance) for $T < T_K$ and for electronic states within energies $k_B T_k$ of E_F. At a heuristic level, the assumption of such a resonance in the electronic spectrum near E_F is sufficient to provide the qualitative behavior seen in $\Delta\chi(T)$ and $\Delta\rho(T)$.

The 1/N expansion. A useful way of organizing the scattering processes is in a series in $1/N$, the degeneracy (for cerium impurities, at least) being rather large. This new approach, which has been applied to the Kondo model in the last few years, is well known in theories of ordered spin systems and from the theory of quarks (where a series in

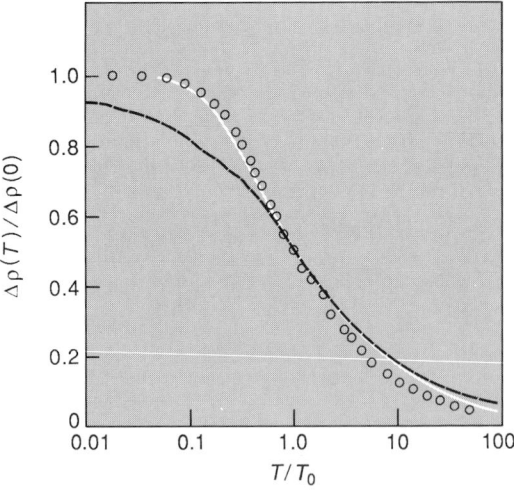

Fig. 2. Universal resistivity curve $\rho(T)$ of a Kondo impurity. The solid curve is the new 1/N calculation for N = 4. The broken curve is the best previous calculation, based on perturbative approaches. Open circles are experimental points for cerium (Ce) in lanthanum hexaboride (LaB$_6$) with T_0 = 1 K. Not shown is the lattice vibration scattering, which competes with the Kondo resistivity and produces a minimum.

$1/N_c$ is used, where N_c is the number of colors).

Implementation. The $1/N$ expansion is implemented formally by taking N large while holding the redefined exchange constant $J = NJ$ fixed. At zeroth order, an infinite summation like that of the perturbative approach described above is included. At this order, the Kondo resonance appears incorrectly as a stable quantum-mechanical state with energy $E_F + k_B T_0$. At order $1/N$, relaxation processes enter in which conduction electrons induce decay of a particular orientation of the impurity moment, and the Kondo resonance acquires a small energy uncertainty, approximately $\pi k_B T_0/N$, with a corresponding long lifetime. This effect was not included in the earlier perturbative treatments.

Results. In Fig. 1, calculations of $\Delta\chi(T)$ for $N = 4$ and $N = 6$ are shown. The $N = 6$ curves compare favorably with the exact Bethe-ansatz results. The $N = 4$ curve misses the weak maximum of the exact results, but otherwise agrees well with the exact curve and moderately well with experimental data for cerium in lanthanum hexaboride, with T_0 set to 1 K. The discrepancies visible in $\Delta\chi(T)$ between theory and experiment at higher temperatures are not understood. On the basis of overall favorable agreement, between theory and experiment for other properties discussed below, new measurements are called for.

In Fig. 2, the resistivity calculations for $N = 4$ are shown. The data for cerium in lanthanum hexaboride agree with theory for $N = 4$ over a temperature range whose maximum exceeds its minimum by a factor of over 10^3, again for $T_0 = 1$ K. Good agreement with cerium in lanthanum hexaboride has also been obtained for the specific heat (the heat gain per unit temperature rise,

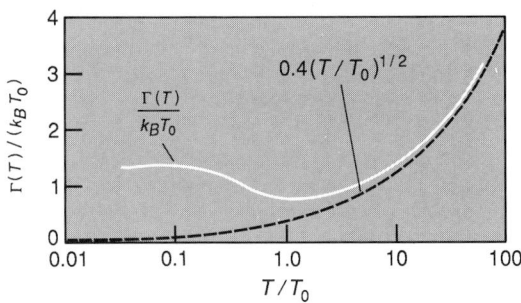

Fig. 3. Universal spin relaxation rate curve $\Gamma(T)$ of a Kondo impurity. The relaxation rate is interpreted as the width of the peak in the calculated neutron-scattering absorption spectrum.

a static property) and thermopower (the voltage change per unit induced temperature difference along the sample, a dynamic property) with this choice of T_0. Hence, the $1/N$ expansion technique provides a comprehensive one-parameter theory for all properties of dilute magnetic impurity systems, valid far below T_K.

Dynamical spectra can also be calculated with this method. For example, the spectrum of spin fluctuations, probed experimentally by neutron scattering or by nuclear magnetic resonance, can be obtained by using this approach. The width of the peak in the magnetic absorption spectrum is a measure of the relaxation rate (inverse lifetime) of the impurity spin. The universal curve for the effective relaxation rate of the impurity spin for $N = 6$ is shown in **Fig. 3**. In contrast to a non-Kondo magnetic impurity, the low-temperature limit is of the order of $2\pi k_B T_0/N$, rather than 0; there is a minimum near T_0; and at high temperatures the relaxation rate is roughly proportional to $T^{1/2}$ rather than T. All these features have been verified experimentally.

Prospects. The $1/N$ expansion has provided qualitatively useful ideas for understanding the so-called heavy-electron cerium compounds, which in some respects behave as though there is a Kondo impurity in every unit cell of the crystal lattice. Specific heat and susceptibility data for these compounds reveal apparent Fermi energies E_F of the order of 10 K; since E_F is proportional to the inverse of the electronic effective mass and normally E_F is about 10 K, the term heavy electron has been deemed appropriate. It has been shown that intersite interactions (which favor magnetic ordering and thus a net moment on each site at $T = 0$) are suppressed relative to the on-site Kondo effects by $1/N^2$ which is consistent with the predominant lack of magnetic order observed in these materials. Detailed quantitative work is still required to extend these preliminary ideas.

For background information SEE EXCHANGE INTERACTION; KONDO EFFECT; MAGNETIC SUSCEPTIBILITY; NEUTRON DIFFRACTION; PERTURBATION (QUANTUM MECHANICS); QUARKS; RENORMALIZATION in the McGraw-Hill Encyclopedia of Science and Technology.

Daniel L. Cox

Bibliography. N. Andrei, K. Furuya, and J. H. Lowenstein, Solution of the Kondo problem, *Rev. Mod. Phys.*, 55:331–402, 1983; D. L. Cox, N. E. Bickers, and J. W. Wilkins, Calculated properties of valence fluctuators, *J. Magn. Mag. Mat.*, 54–57:333–337, 1986; G. Grüner and A. Zawadowski, Properties of Kondo alloys, in D. F. Brewer (ed.), *Progress in Low Temperature Physics,* vol. 12, pp. 591–647, 1978; P. A. Lee et al., Theories of heavy electron systems, *Comments Condensed Matter Phys.*, 12:99–161, 1986.

Life, origin of

Although the origin of life has been a subject of conjecture for hundreds of years, the experimental investigation of this problem had its beginnings only in 1953 with the report by S. L. Miller that amino acids were produced when an electric discharge was passed through a mixture of methane, ammonia, hydrogen, and water vapor. Since then there have been hundreds of experiments which demonstrate that amino acids (the building blocks of proteins) and purines, pyrimidines, and sugars (constituents of nucleotides) can be prepared under conditions which may have existed on the primitive Earth. There has been much less success in condensing amino acids to proteins and purines, pyrimidines, sugars, and phosphate to nucleic acids under plausible primitive Earth conditions. In most of the attempted condensation reactions, the experimenter tried to effect polymerization in aqueous solution or by heating a mixture of the reactants in the dry state. It has become apparent that these polymers will not form spontaneously if the polymer building blocks are merely subjected to heat or light. Catalysis must have had an important role if biological polymers with the complexity required for originating life were formed on the primitive Earth. This article focuses on recent results concerning the possible role of clay mineral catalysis in the origin of life.

Role of clay. In 1947, 6 years before the results of Miller's amino acid synthesis were published, J. D. Bernal proposed that clay minerals may have had a central role in the prebiological formation of the first biological polymers. In addition, he recognized that the primary difficulty of imagining processes going to polymers would be the extreme dilution of the system. He suggested that a much more favorable condition for concentration, and one that must certainly have taken place on a very large scale, would be adsorption in fine clay deposits. Small molecules attached to the clays would not be fixed at random but would be held in specific positions favoring interaction and formation of more complex molecules. In addition to adsorption and orientation, the surface acidity of clays is an important property which contributes to their catalytic activity. Although the prebiological importance of clays was emphasized in his discussion, Bernal noted that other minerals, such as quartz, may also have been important in the formation of polymers of biological molecules on the primitive Earth.

Bernal's concept of adsorption and catalysis by clay minerals has been discussed and extended since it was first suggested. In the most radical extension, it has been proposed that the first life was based completely on clay and that in the process of evolution the "clay life" was taken over by the organic molecules associated with it. There are no well-defined experimental studies which provide direct support for this clay-life postulate, but some recent research has demonstrated a catalytic role for clays in the formation of biological polymers.

There are many other inorganic substances besides clays that may have participated in the formation of polymers on the primitive Earth. Iron- and manganese-containing minerals which have the potential to bind organic molecules of specific size are currently under investigation. However, experimental studies using these minerals have begun only recently.

Thermal condensation of amino acids. Clays enhance the rate of formation of the amide bond that links amino acids into proteins. Variations in temperature and water content have resulted in more efficient formation of this protein link when the amino acids were put through wet-dry cycles in the presence of clays than in their absence. The amino acids were dissolved in water, con-

centrated to dryness, and heated in the dry state at 140°F (60°C) for 1 or 2 days; then the amino acids were again dissolved and the cycle was repeated. The dissolution of the reaction products from the surface of the clay after heating and subsequent condensation allows other amino acids to occupy the reactive sites on the clay so that the catalytic formation of additional amide bonds can take place. The combination of the absence of water and the acidity of the clay surface is believed to be responsible for the formation of these bonds. When dimers of the amino acid histidine or certain nucleic acid derivatives are added to the mixture, the extent of amide bond formation is enhanced. Even higher yields and larger chains of condensed amino acids can be obtained if the metal ion associated with the clay is copper (Cu^{2+}) rather than sodium (Na^+).

Synthesis of RNA oligomers. It has been discovered recently that ribonucleic acid (RNA) catalyzes reactions in much the same way as do protein catalysts (enzymes). This suggests that RNA-like polymers may have had a central role in the earliest living systems. Before this was learned, it had already been established that RNA can be a repository for genetic information; thus the more recent finding suggests that information storage and catalysis, the fundamental properties of life, may be observed in a system consisting principally of RNA-like molecules. Consequently, the prebiotic synthesis of RNA and its analogs is of great interest, since RNA may have been the essence of the first life.

Initial studies on the polymerization of nucleotides, the building blocks of RNA, in the presence of montmorillonite clays were not promising. Dry heating of activated nucleotides in the presence of montmorillonite did not result in a significant enhancement of the degree of polymerization over that observed in the absence of the clay. It has been observed that the formation of longer-chain oligomers of RNA is enhanced when the template-directed condensation of the activated monomer is performed in the presence of the calcium phosphate mineral hydroxylapatite. The template is a chain of RNA which directs the synthesis of the new chain. The RNA template used to direct the synthesis of its complementary strand of RNA is bound to the hydroxylapatite, and the longer chains that are formed in the condensation reaction also bind to the mineral. The short chains stay in solution and continue to undergo polymerization until their chain length is sufficiently long that they remain bound to the hydroxylapatite most of the time. Binding to the hydroxylapatite results in the formation of higher-molecular-weight RNAs in preference to short chains. The role of the hydroxylapatite is only adsorption of oligmers; no catalysis of the formation of the phosphate ester bond was observed. This condensation procedure can be repeated by solubilizing the newly

Montmorillonite-catalyzed conversion of 3'-AMP to 2',3'-cAMP, a model system for the prebiotic formation of the phosphate ester bond. Nucleotides such as 3'-AMP bind reversibly to the montmorillonite clay [step B]. While bound to the clay, they react with a condensing agent (DISN) that is formed at the clay surface from the tetramer of HCN (DAMN) [step A]. The formation of 2',3'-cAMP is catalyzed by montmorillonite, and the 2',3'-cAMP is released from the clay [step C]. Steps A–C are similar to processes catalyzed by contemporary enzymes. The A in the structures represents adenine, a purine base present in both DNA and RNA.

formed RNA chains from the hydroxylapatite, activating them so that they can react further, and then adjusting the pH of the solution phase so that they again bind to the template and condense to form even longer chains. These experiments demonstrate that the minerals associated with a pool of water on the primitive Earth could have been the site for the template-directed synthesis, elongation, and ligation of oligonucleotides. The degree of adsorption and desorption of the templates and the synthesized RNA chains from minerals would have varied with both the pH and the concentration of salts in a pond or tidal pool. Thus, the process of binding, release, and activation of a growing chain is one that is likely to have taken place on the primitive Earth.

One systematic study of the possible role of montmorillonite clays in the formation of nucleotide polymers involved an investigation of the factors which affect the binding of nucleotides to the clays. Metal ions that are electrostatically bound to the clay surface play an important role: The divalent ions of zinc and copper are the most effective for the binding of nucleotides. Above pH 4 the binding to the clay is believed to be due to complexation of the nucleotide to the metal ion, which then serves as a link between the clay and the nucleotide. In an extension of these studies, it was observed that other organic molecules which complex to the metal ion along with the nucleotide can markedly enhance or inhibit the binding of the nucleotide to the clay.

It has since been discovered by studying the reactions of these bound nucleotides that clays could exhibit complex catalytic behavior, comparable to that observed with contemporary enzymes. Previously it had been observed that the ferric ion (Fe^{3+}) located in the lattice of montmorillonite is able to convert a tetramer of hydrogen cyanide (HCN) to a compound which activates the phosphate group of a nucleotide so that the phosphate group will undergo condensation reactions. When the hydrogen cyanide tetramer is added to a suspension of montmorillonite clay, to which a nucleotide is bound, the tetramer is activated on the surface of the clay in proximity to the bound nucleotides. Montmorillonite clay catalyzes the condensation of the activated tetramer with the bound nucleotides in a reaction that results in the formation of adenosine 2′,3′-cyclic monophosphate (2′,3′-cAMP; see **illus.**). The clay performs the following functions in this system: (1) it binds the nucleotide, (2) activates the condensing agent, (3) catalyzes the reaction between the activated tetramer and the nucleotide, and (4) releases the product so that another substrate molecule can bind to the clay. The complexity of the catalysis in this example approaches that of contemporary enzymatic reactions.

The use of montmorillonite clay for the formation of 2′,3′-cAMP has been successfully extended to the synthesis of dimers and trimers of AMP. This provides strong experimental support for Bernal's postulate that clay minerals may have had a central role in the formation of biopolymers on the primitive Earth. In addition, it suggests that the formation of RNA-like polymers took place spontaneously under prebiotic conditions.

For background information SEE CATALYSIS; ENZYME; LIFE, ORIGIN OF; RIBONUCLEIC ACID (RNA); RIBOZYME in the McGraw-Hill Encyclopedia of Science and Technology.

James P. Ferris

Bibliography. O. L. Acevedo and L. E. Orgel, Template-directed oligonucleotide ligation on hydroxylapatite, *Nature*, 321:790–792, 1986; T. R. Cech, RNA as an enzyme, *Sci. Amer.*, 235(5): 64–75, 1986; J. P. Ferris and W. J. Hagan, Jr., The adsorption and reaction of adenine nucleotides on montmorillonite, *Origins of Life*, 17:69–84, 1986; C. Ponnamperuma, A. Shimoyama, and E. Friebele, Clay and the origins of life, *Origins of Life*, 12:9–40, 1982.

Linear programming

In 1984 N. K. Karmarkar announced a new method for solving linear programming problems and showed that theoretically it had a polynomial time bound. Such a polynomial algorithm is guaranteed to find a solution in a time bounded by a polynomial function of the length of the input data required to describe it. The standard algorithm for such problems is the simplex method devised by G. B. Dantzig. While exhibiting exponential behavior on artificially constructed examples, the simplex method performs very well in practice and has been used with great success for almost four decades. Claims that the new algorithm solved large problems in a small fraction of the time required for the simplex method caused considerable excitement and some skepticism. While further testing needs to be done, reports by Karmarkar and his colleagues and computational results for variants of the new method on medium-sized problems indicate that for many large problems Karmarkar's algorithm may indeed be faster than the simplex method but very careful implementation is required to achieve these gains.

Motivation. The excitement caused by the new algorithm is due to the great significance of linear programming, which is concerned with the minimization (or maximization) of a linear function of several variables subject to linear equations and inequalities. Such problems arise throughout industry and government, in resource allocation, production scheduling, inventory control, and distribution. For a firm with a number of plants producing a variety of products over a time interval of several months, finding an optimal schedule can easily lead to such a problem with thousands of constraints and tens of thousands of variables. Problems of this size can be solved routinely on mainframe computers in a matter of minutes. However, the desire to model more detailed and complex systems can lead to vast formulations which require faster computers or new techniques. Karmarkar's algorithm was developed in the hope of solving these huge problems. This article will describe the ideas behind the algorithm and give a flavor of its theoretical analysis.

Karmarkar's was not the first polynomial algorithm for linear programming; that was developed in 1979 by L. G. Khachian, based on the work of D. B. Yudin and A. S. Nemirovskii. While the theoretical bounds are very similar and there are strong resemblances in parts of the algorithms, it appears that Khachian's ellipsoid method has little or no practical significance, although its theoretical implications in computational complexity are very broad.

Karmarkar's algorithm. Problems in general form can be solved by the new algorithm, but it is simpler to describe the method for problems in Karmarkar's canonical form. In this form the linear programming problem is to find a vector $x = (x_1, x_2, \ldots, x_n)$ to minimize expression (1)

$$c \cdot x = c_1 x_1 + c_2 x_2 + \cdots + c_n x_n \quad (1)$$

over all vectors x satisfying Eqs. (2) and also

$$a_{11} x_1 + a_{12} x_2 + \cdots + a_{1n} x_n = 0$$
$$\cdots\cdots\cdots\cdots\cdots\cdots\cdots\cdots\cdots\cdots\cdots\cdots \quad (2)$$
$$a_{m1} x_1 + a_{m2} x_2 + \cdots + a_{mn} x_n = 0$$

Eq. (3) and inequalities (4).

$$x_1 + x_2 + \cdots + x_n = n \quad (3)$$

$$x_j \geq 0 \qquad j = 1, 2, \cdots, n \quad (4)$$

The constraints of Eqs. (2) define a subspace Ω, while those in Eq. (3) and inequalities (4) define an $(n-1)$-dimensional simplex S (for $n = 3$, a triangle). This minimization problem is called (P). The data of the problem are the vector $c = (c_1, c_2, \ldots, c_n)$ of the objective function coefficients and the matrix $A = (a_{ij})$ of constraint coefficients. The n-dimensional vector e, given by Eq. (5), with each component equal to one, is

$$e = (1, 1, \ldots, 1) \quad (5)$$

key in the analysis; intuitively, it is the center of the simplex S. It is assumed that e is feasible in the problem (P), so that $a_{i1} + a_{i2} + \cdots + a_{in} = 0$ for all $i = 1, 2, \ldots, m$. Finally, it is assumed that the minimal value of $c \cdot x$ in (P) is zero.

Sequence of feasible solutions. Dantzig's simplex method (simplex here refers to the same geometrical figure as S above, although in a different context) would generate a sequence of extreme-point feasible solutions to (P) [similar to the vertices of a polygon or polyhedron], each one a vector with at most $m + 1$ of its n components positive. By contrast, Karmarkar's algorithm generates a se-

quence $\varepsilon x^k = (x_1^k, x_2^k, \ldots, x_n^k)\nabla$ of feasible solutions to (P), each of which has every component positive. Thus, geometrically these solutions correspond to interior points of the feasible region. This is analogous to various methods for nonlinear programming. While the simplex algorithm is finite, Karmarkar's algorithm gives an infinite sequence $\{x^k\}$ with objective function values converging to the optimal value zero. Indeed, these values satisfy inequality (6).

$$c \cdot x^l \leq \exp(-l/5n)c \cdot e \quad (6)$$

This key inequality guarantees that in a number of iterations of the order of nq a feasible solution that satisfies inequality (7) can be found.

$$c \cdot x \leq 2^{-q} c \cdot e \quad (7)$$

If the data of (P) are rational and the total length of the data is L (in bits), then by setting q to be a small multiple of L, an exact optimal solution to (P) can be determined from such an x. This generates a polynomial-time–bounded algorithm.

Potential function. Inequality (6) is trivial for $l = 0$ if x^0 is chosen to equal e. The general result would follow if it were possible to obtain a sequence $\{x^k\}$ that satisfied inequality (8) for

$$c \cdot x^{k+1} \leq \exp(-1/5n)c \cdot x^k \quad (8)$$

each k, but this appears to be difficult to achieve. Indeed, the algorithm described below does not assure a monotonic decrease in $c \cdot x^k$. Instead, it relies on Karmarkar's potential function, defined by Eq. (9). It can be shown that guaranteeing

$$f(x;c) := \sum_{j=1}^{n} \ln (c \cdot x/x_j)$$
$$= n \ln c \cdot x + \sum_{j=1}^{n} (-\ln x_j) \quad (9)$$

inequality (10) for all k implies that inequality (6) holds.

$$f(x^{k+1}; c) \leq f(x^k; c) - \tfrac{1}{5} \quad (10)$$

Nonlinear transformation. Making substantial progress from a feasible solution x^k can be hindered by the proximity of the inequality constraints $x_j \geq 0$. This problem is addressed by making a transformation of the space that maps x^k into the point e, which is as far from the nonnegativity constraints as possible, given that x lies in S. The transformation can be carried out only from a feasible solution with $x^k > 0$ (that is, $x_j^k > 0$ for all j).

The transformation used is the projective mapping T defined by $T(x) = \hat{x}$, where the components of $\hat{x}$ are given by Eq. (11); with inverse

$$\hat{x}_j = \frac{n(x_j/x_j^k)}{\sum_{i=1}^{n} (x_i/x_i^k)} \quad (11)$$

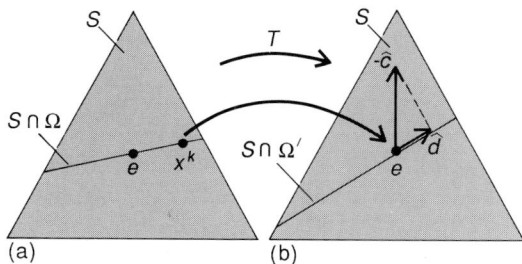

Transformation T in the case $n = 3$, $m = 1$. (a) Simplex **S** before the transformation. (b) Simplex **S** mapped into itself by T. Projected negative-gradient direction **d** is shown.

$T^{-1}(\hat{x}) = x$, where the components of x are given by Eq. (12). The transformation T has the follow-

$$x_j = \frac{nx_j^k \hat{x}_j}{\displaystyle\sum_{i=1}^{n} x_i^k \hat{x}_i} \qquad (12)$$

ing properties: (1) T maps the simplex S into itself, and each of its faces into itself. (2) T maps the current iterate x^k into the center e of the simplex. (3) T maps the subspace Ω into Ω', which consists of all vectors x that satisfy equations similar to Eqs. (2), but with a_{ij} replaced by $\hat{a}_{ij}$ defined by Eq. (13) [see **illus.**].

$$\hat{a}_{ij} = a_{ij} x_j^k \qquad (13)$$

In the transformed space, (P) becomes the fractional linear programming problem ($\hat{P}$) of minimizing expression (14) over all vectors $\hat{x}$

$$\frac{n\,\hat{c} \cdot \hat{x}}{\displaystyle\sum_{j=1}^{n} x_j^k \hat{x}_j} \qquad (14)$$

that lie in both $\hat{\Omega}$ and S, where the components of $\hat{c}$ are given by Eq. (15). Since the optimal

$$\hat{c}_j = x_j^k c_j \qquad (15)$$

value of ($\hat{P}$), and hence of (P), is assumed to be zero, optimal solutions of ($\hat{P}$) also minimize the linear function $\hat{c}\, d\, \hat{x}$.

Calculation of next iterate. The next iterate in the transformed space is obtained from $\hat{x} = e$ by ignoring the constraints $\hat{x} \cdot 0$ and moving in the direction of the negative gradient of $\hat{c} \bullet \hat{x}$, $-\hat{c}$, projected into the null space of the matrix of constraints of ($\hat{P}$). If this projection operation is denoted by Π, the direction of search is defined by Eq. (16). Thus, in the transformed space the

$$\hat{d} := -\Pi\hat{c} \qquad (16)$$

vector $\hat{x}$ is moved to $\hat{x} = e + \gamma\hat{d}$, and the next feasible solution is given by Eq. (17) [see illus. b].

$$x^{k+1} = T^{-1}(e + \gamma\hat{d}) \qquad (17)$$

There remains the problem of choosing the step size γ. Clearly, γ must be chosen so that $e + \gamma\hat{d} \geq 0$, and $x^{k+1} \geq 0$. Since $\hat{d}$ is a descent direction for $\hat{c} \bullet \hat{x}$, it might seem advantageous to

choose γ as large as possible with this constraint. The result would be an iterate x^{k+1} with a zero component, and this would prevent a transformation from being carried out at the next step. In addition, the direction $\hat{d}$ might not be a descent direction for the objective function (14) of ($\hat{P}$). A more careful analysis uses the potential function f. One of its remarkable properties is that it is invariant (up to an additive constant) under the projective transformation T: if x and $\hat{x}$ correspond, the transformation of f is given by Eq. (18). This is in sharp contrast to the way that

$$f(\hat{x};\hat{c}) = f(x;c) + \sum_{j=1}^{n} \ln x_j^k \qquad (18)$$

the linear objective function of (P) transforms into the fractional objective function (14) in ($\hat{P}$).

The implication of Eq. (18) is that it is sufficient to find a vector $\hat{x}$ that satisfies inequality (19);

$$f(\hat{x};\hat{c}) \leq f(e;\hat{c}) - \tfrac{1}{5} \qquad (19)$$

then $x^{k+1} = T^{-1}(\hat{x})$ will satisfy inequality (10). An ingenious geometrical argument of Karmarkar, using the fact that the radii of the inscribed and circumscribing spheres centered at e for the simplex S are in ratio $1:(n-1)$ shows that inequality (19) is guaranteed by choosing the vector $\hat{x}$ given by Eq. (20), where $\alpha = \tfrac{1}{3}$ and $\|\hat{d}\|$ is the

$$\hat{x} = e + \alpha\hat{d}/\|\hat{d}\| \qquad (20)$$

euclidean length of the vector d given by Eq. (21). Since the bound in inequality (19) is based

$$\|d\| = \left(\sum_{j=1}^{n} d_j^2 \right)^{1/2} \qquad (21)$$

on worst-case estimates, it is preferable in practice to make a line search for the best step size α.

In summary, the algorithm consists of the following steps at each iteration: (1) Given x^k, define the transformation T. (2) Compute $\hat{d}$, the projection of $-\hat{c}$ in the transformed space. (3) Set $\hat{x} = e + \alpha d/\|\hat{d}\|$. (4) Set $x^{k+1} = T^{-1}(\hat{x})$ from Eq. (12).

Efficiency of algorithm. By far the greatest computational work at each iteration is involved in the calculation of $\hat{d}$; this is equivalent to the solution of a positive-definite system of linear equations which form the normal equations of a weighted least-squares problem, such as arise in regression problems in statistics. The least-squares solution is closely related to a solution to the dual linear programming problem to (P). The efficiency of a computer implementation of Karmarkar's algorithm depends crucially on how this vector is calculated, and therefore on the sparse matrix techniques used for the positive-definite equation system or the least-squares problem above.

While the theoretical analysis yields a bound of

the order of n on the number of steps for a prescribed accuracy, the number required in computational tests seems to grow very slowly, if at all, with the size of the problem: between 20 and 40 iterations appear to give 5–10-digit accuracy even for problems with thousands of variables. Thus the efficient calculation of ds the crux of the whole algorithm.

Karmarkar's method is likely to remain a basic tool in optimization for a number of years. Much work remains to be done in providing sensitivity analysis for perturbed problems and extending the method for integer and combinatorial optimization problems.

For background information SEE LINEAR PROGRAMMING; NONLINEAR PROGRAMMING; STATISTICS in the McGraw-Hill Encyclopedia of Science and Technology.

Michael J. Todd

Bibliography. G. B. Dantzig, *Linear Programming and Extensions*, 1983; N. K. Karmarkar, A new polynomial time algorithm for linear programming, *Combinatorica*, 4:373–395, 1984; L. G. Khachian, Polynomial algorithms in linear programming, *USSR Computational Math. Math. Phys.*, 20:53–72, 1980; M. J. Todd, Polynomial algorithms for linear programming, H. Eiselt and G. Pederzoli (eds.), *Proceedings of Optimization Days 1986*, 1987.

Liquid helium

Since its discovery in 1972, superfluidity in liquid helium-3 (^{3}He) has provided one of the most fascinating fields of study in the area of condensed-matter physics. Only recently, however, have advances in experimental techniques brought about the possibility of studying two-dimensional superfluidity in ^{3}He films. Preliminary reports of superflow in such films are encouraging. The complex tensorial structure of the order parameter in superfluid ^{3}He indicates that a richness and variety of two-dimensional phenomena are possible.

Superfluid state in thin films of ^{3}He. The superfluidity of ^{3}He arises through the pairing of the ^{3}He atoms into "molecules" of a special kind, called Cooper pairs, which collectively form a coherent, macroscopic quantum state. As opposed to the case of ordinary superconductors, where formation of Cooper pairs of electrons is induced through exchange of phonons, the superfluid state in ^{3}He results from a direct atomic interaction. In the wave function of these Cooper pairs, both the spin S and the orbital angular momentum L differ from zero: $L = 1, S = 1$. Superfluid ^{3}He differs from superconductors, where $L = 0$ and $S = 0$, in this respect as well. The nonzero internal angular momentum of Cooper pairs is responsible for the fascinating anisotropic properties of the superfluid state.

Superfluid order parameter. An important concept is that of the superfluid order parameter, which is closely related to the Cooper-pair wave function

and describes macroscopic properties of the superfluid state. In ^{3}He the order parameter consists of nine complex functions versus only one in superconductors or superfluid helium-4 (^{4}He). These functions describe projections of the $L = 1, S = 1$ Cooper-pair wave function onto some arbitrarily chosen z axis. According to the laws of quantum mechanics, these projections can assume the values $L_z = -1, 0, +1$ and $S_z = -1, 0, +1$ and correspond to distinct quantum states of the superfluid. These states are called phases. In nature, in zero magnetic field, two such phases occur. The high-pressure phase, or the A-phase, which contains $L_z = -1, +1$ and $S_z = -1, +1$ Cooper pairs has anisotropic properties, with the orbital angular momentum $\vec{l}$ of the Cooper pairs being the anisotropy axis. In many ways this high-pressure phase is similar to ordinary liquid crystals. The low-pressure phase, or B-phase, contains a symmetric combination of different L_z and S_z states and has total angular momentum $J = L + S$ equal to zero. Consequently, the properties of the B-phase appear isotropic under all conditions which do not involve separate rotations in the real and spin space.

The nonvanishing orbital angular momentum of the order parameter implies that the superfluid will be very sensitive to the geometry of a container. Near an impenetrable, smooth wall parallel to the x-y plane, the $L_z = 0$ component of the order parameter will be suppressed to zero. The reason can be pictured as follows: ^{3}He atoms forming the $L_z = 0$ Cooper pairs rotate in a plane perpendicular to the wall. Since the size of a Cooper pair is a coherence length ξ (several tens of nanometers at zero absolute temperature), those Cooper pairs that are within distance ξ from the wall will tend to bump into it, losing the quantum coherence necessary for superfluidity. However, once the distance to the wall is larger than ξ, the $L_z = 0$ component will recover. In a thin film, with thickness d less than ξ, such recovery is not possible, and the $L_z = 0$ component of the order parameter vanishes in this case. Two other components, $L_z = \pm 1$, describe Cooper pairs rotating in the x-y plane, thereby avoiding scattering off the wall. As a result, as long as the surface is smooth, these two components will be basically unaffected. This means that in thin films $\vec{l}$ is always perpendicular to the surface (**Fig. 1**). The above discussion suggests that in thin films the superfluid should be expected to be always in the A-phase.

Spatial fluctuations. The phase transition and superfluid state in thin films differ from those of a bulk system in one additional aspect. The spatial fluctuations in the value of the complex order parameter, which are irrelevant in a three-dimensional system with all dimensions of a container much longer than ξ, become important in a thin film. Strictly speaking the long-range, macroscopic order that characterizes the superfluid state cannot exist at all in a two-dimensional sample. Instead, a new type of order develops, arising from the binding of oppo-

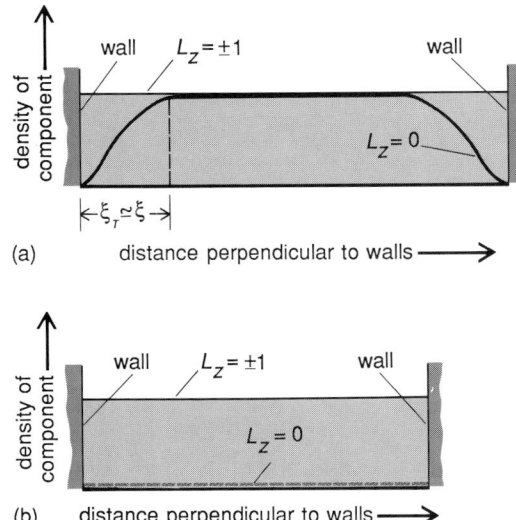

(a)   distance perpendicular to walls ⟶

(b)   distance perpendicular to walls ⟶

Fig.1 Superfluid state in a parallel-plate geometry. The magnitudes of the components of the order parameter are plotted as functions of the distance perpendicular to walls. The $L_z = \pm 1$ components are unaffected by the walls, while the $L_z = 0$ component vanishes at the boundaries. (a) Superfluid slab with thickness d much greater than ξ. The $L_z = 0$ component is restored to its bulk value over a distance $\xi_T \approx \xi$. (b) Thin film with thickness d less than ξ. Only the $L_z = \pm 1$ components are nonzero.

sitely charged vortices in the phase of the order parameter. (These vortices are analogous to those in superfluid ^{4}He. The charge of the vortex refers to the direction of rotation about its axis.) The phase transition from the superfluid to the normal state proceeds then through unbinding of these vortices and is called a Kosterlitz-Thouless transition. In superfluid ^{4}He, where the order parameter is a single complex function and has a simple U(1) symmetry, the phase fluctuations cause the phase transition in thin films to be of the Kosterlitz-Thouless type. In ^{3}He films the order parameter has more components and, in addition to the U(1) symmetry, it has a discrete Z_2 symmetry corresponding to two different directions of orbital angular momentum $\vec{l}$, $+\mathbf{z}$ and $-\mathbf{z}$, where $\mathbf{z}$ is the unit vector along the z axis. Therefore, fluctuations in the direction of $\vec{l}$ will be present alongside the U(1) phase fluctuations. The fluctuations in $\vec{l}$ arise through formation of regions with $\vec{l}$ parallel or antiparallel to the z axis, which are separated by domain walls. An example of how such a domain

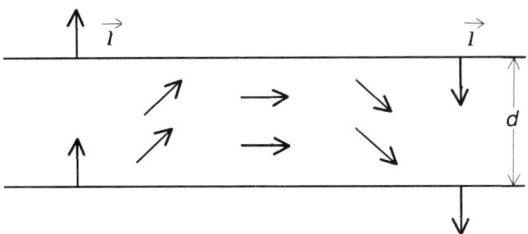

Fig. 2. Cross section of a ^{3}He film, showing a domain wall. Arrows indicate the vector l, which rotates about an axis parallel to the domain wall.

wall might appear is given in **Fig. 2**. The situation is quite similar to that found in two-dimensional ferromagnets, where the magnetization vector can point up or down. It is not clear which of the two types of fluctuations will ultimately dominate the superfluid phase transition. If phase fluctuations dominate, the transition will be of the Kosterlitz-Thouless type, while in the other case the transition will be very similar to a ferromagnetic transition found in spin systems. Because these two transitions belong to different universality classes with different behavior of thermodynamic quantities near the transition temperature, the question is of great interest. Theoretical calculations indicate that the competition is very close, and a definite answer cannot be given because of the numerical uncertainties involved. It appears that the determination of the type of phase transition in superfluid ^{3}He films must await experimental resolution.

Quantum size effects. As discussed below, incoherent boundary scattering plays an important role in determining the superfluid properties of thin films. But even if the surface of the film is atomically smooth, and scattering completely coherent, the superfluid properties of the film are influenced solely by the fact that ^{3}He atoms are enclosed in a finite volume. In this situation the quantum-mechanical uncertainty relations require ^{3}He atoms to carry some minimal momentum, of the order of $\hbar/d$, where $\hbar$ is Planck's constant divided by 2π, and d is the thickness of the film, in the z direction. These quantum size effects can become observable in some superfluid properties if the thickness of the film is smaller than some d_{QS} given by Eq. (1), where E_F

$$d_{QS} \simeq \frac{2E_F}{\omega_0} k_F^{-1} \qquad (1)$$

is the kinetic energy of ^{3}He atoms (the Fermi energy), ω_0 is the average strength of interactions between atoms forming a Cooper pair, and k_F^{-1} (the inverse Fermi wave vector) is of the order of the interatomic separation. The distance d_{QS} roughly corresponds to 40 layers of ^{3}He atoms. The quantum size effects tend to decrease the transition temperature T_c in superfluid ^{3}He and to produce an oscillatory thickness dependence in some other quantities, notably the nuclear magnetic resonance response. The reduction in transition temperature is rather small and will become significant only in very thin films, but the oscillatory character of the nuclear magnetic resonance response could be measured and several experiments are in progress to search for this sign of quantum size effects in ^{3}He films.

Influence of the substrate. The above discussion seems to indicate that, while the $L_z = 0$ component of the order parameter vanishes in thin films, the other two components remain largely unaffected, as long as d is much greater than k_F^{-1}. This is true only if the surfaces of the film are perfectly smooth.

Any type of disorder, arising from the presence of crystalline defects in the walls or magnetic and nonmagnetic impurities, will act to destroy even the $L_z = \pm 1$ Cooper pairs. This is another respect in which ^{3}He differs from superconductors. There, the disorder destroys Cooper pairs only if it involves magnetic impurities. Since in any realistic situation the substrate used to support ^{3}He film will contain some defects and impurities, it is necessary to investigate the effects of this disorder on the superfluid state.

The effect of a substrate disorder can be described by assuming that the ^{3}He atoms move between two parallel, impenetrable walls separated by a distance which varies in the x-y plane, and is thus given by Eq. (2), where d is the

$$d(x,y) = d + w(x,y) \qquad (2)$$

nominal average thickness, and $w(x,y)$ is some function with magnitude much smaller than d and zero average. The effect of variation in thickness can be treated as a small perturbation on a film of uniform thickness d. The existence of this surface roughness causes incoherent scattering of Cooper pairs from the walls. This will tend to reduce the superfluid transition temperature. This reduction can be evaluated by comparing the gain in energy through formation of the superfluid state to the finite lifetime of a Cooper pair in the presence of a rough surface. The result is shown in **Fig. 3**, where the transition temperature has been plotted as a function of d for several values of the roughness parameter $y = k_F^2 w^2$, which measures the root-mean-square fluctuations in thickness in units of the interatomic separation. [Here, w^2 is the average of $w^2(x,y)$ over all possible surface profiles.] In the case of a smooth surface ($y = 0$) the downward trend and the wiggles in the curve for the transi-

tion temperature are due to quantum size effects. An increase in the surface roughness causes a rapid reduction in the transition temperature, and the transition temperature drops to zero at nonzero values of the film thickness. The critical thickness d_c at which the transition temperature is reduced to zero for a specified degree of surface roughness is found to be given by Eq. (3), where $\eta^{-1} = 14.1$ in the A-phase and ξ_0 is the

$$d_c(y) \simeq 0.75 y \eta \xi_0 \qquad (3)$$

coherence length ξ at zero absolute temperature. For $y = 1$ it is found that $d_c \approx 35 k_F^{-1}$. It follows that superfluidity in ^{3}He films of moderate thickness (say 20 to several hundred layers) will not be destroyed by surface roughness except when y is much greater than 1, which would represent a very rough substrate.

Many low-temperature properties of the superfluid are governed by the character of the excitation spectrum of ^{3}He atoms. In both superconductors and superfluid ^{3}He the formation of the Cooper pairs results in a gap in the excitation spectrum; finite energy of order kT_c, where k is Boltzmann's constant, is needed to break up a Cooper pair. Although the gap is uniform in a superconductor, in the A-phase it is anisotropic and vanishes in the direction of the orbital angular momentum $\vec{l}$. The number of Cooper pairs $N(E)$ that can be broken apart at any given energy E is therefore always different from zero and is proportional to E^2 for small E. This is reflected in the specific heat of the A-phase, which at low temperatures decreases as T^3, rather than displaying the exponential decay characteristic of a superconductor.

In the presence of a rough substrate the number of Cooper pairs is reduced and their energy distribution is altered. As long as the disorder is small, the thermodynamic properties will be only slightly affected. For increasing roughness, however, there will be a drastic change in the excitation spectrum, and it will become possible to break up Cooper pairs at no cost in energy. Then $N(E)$ will approach a constant value at low E. This is the so-called gapless state, which also occurs in superconductors that contain magnetic impurities. The specific heat will now be linear in T, just as in ordinary liquid ^{3}He, but with a different numerical coefficient.

The appearance of the gapless state does not mean the destruction of superfluidity. Even though the Cooper pairs can be broken apart with essentially zero energy, they still exist, and the gapless superfluid will exhibit many phenomena that cannot be observed in the normal liquid. Thus, films of superfluid ^{3}He offer the unique opportunity of studying a gapless anisotropic superfluid, a form of matter that has not been realized in any other system. In order to observe this state, the surface roughness must be large enough to create the gapless superfluid, but not

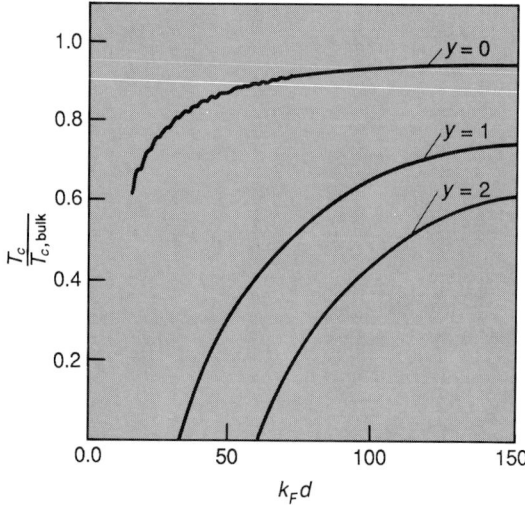

Fig. 3. Transition temperature of the superfluid ^{3}He film as a function of thickness. Ratio of film transition temperature T_c to transition temperature of bulk liquid $T_{c,\text{bulk}}$ is plotted against products of thickness d and Fermi wave vector k_F. Surface roughness parameter y is defined in the text.

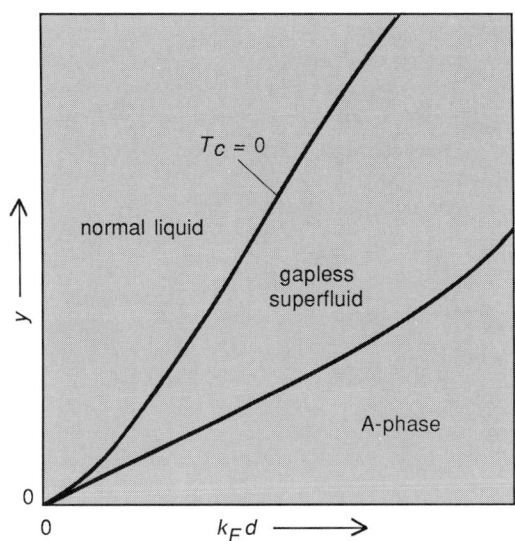

Fig. 4. Phase diagram of superfluid ^{3}He film at absolute temperature $T = 0$. The detailed boundaries between different regions will depend on a particular model for surface roughness, but the qualitative features are expected to persist.

so large as to destroy superfluidity. The situation is shown in the schematic phase diagram of **Fig. 4**, which indicates the value of film thickness and surface roughness, at absolute temperature $T = 0$, at which the A-phase, the gapless superfluid, and the normal liquid are found. By controlling the thickness, or by carefully choosing the substrate material, it should be possible to experimentally open the window with a view on properties of a "dirty" anisotropic superfluid.

For background information *SEE LIQUID HELIUM; QUANTIZED VORTICES* in the McGraw-Hill Encyclopedia of Science and Technology.

Zlatko B. Tešanović

Bibliography. T. C. Halsey, Topological defects in the fully frustrated XY-model and in ^{3}He-A films, *J. Phys. C.*, 18:2437–2454, 1985; A. J. Leggett, A theoretical description of the new phases of liquid ^{3}He, *Rev. Mod. Phys.*, 47:331–414, 1975; D. Stein and M. C. Cross, Phase transitions in two-dimensional superfluid ^{3}He, *Phys. Rev. Lett.*, 42:504–507, 1979; Z. Tešanović and O. T. Valls, Gapless superfluidity in ^{3}He films, *Phys. Rev.*, B34:7610–7616, 1986.

Lithosphere

Understanding of the nature of the solid layers of the Earth has grown as new information has become available. Recent studies have yielded insights into the geochemistry of the subcontinental mantle and the chemical character of the Archean crust.

GEOCHEMISTRY OF SUBCONTINENTAL MANTLE

Major chemical distinctions between the subcontinental and suboceanic lithosphere originate through two controlling factors. First, unless the subcontinental mantle is made less dense through removal of its pyroxene and garnet, that is, its basaltic component, it eventually will cool and become relatively so dense that its destruction by delamination and mixing with the remaining mantle is assured. Second, once stabilized by its compositionally induced lower density, over time the subcontinental mantle accumulates incompatible trace elements contributed by the passage of melts and fluids from below. The greater chemical heterogeneity of the subcontinental mantle compared with the suboceanic mantle thus appears to be both cause and effect of preservation of the former mantle type beneath continents.

Recent evidence from geochemical and isotopic studies of volcanic rocks and from rare exposures of upper-mantle material now at the Earth's surface shows that the continents are underlain by regions distinct from the mantle beneath the ocean basins. In certain cases these subcontinental mantle keels have remained attached to the overlying crust for billions of years. The presence of isolated mantle blocks beneath continents is important both to the preservation of the continents and to the general chemical evolution of the Earth.

The mantle generally is considered to be composed largely of peridotite, a rock type dominated by the mineral olivine with lesser amounts of pyroxene and garnet (see **table**). When peridotite melts, its pyroxene and garnet complement is extracted preferentially to produce a basaltic composition melt, leaving behind a residue richer in olivine.

Characteristics. An important consequence of this variation in the ratio of olivine to pyroxene and garnet is that the extraction of melt leaves behind a residue less dense than the starting material. As density differences control the slow convective flow in the mantle, the difference in density between mantle materials with varying amounts of pyroxene and garnet serves to separate and isolate chemically distinct regions in the mantle.

Based largely on physical characteristics, particularly the transmittal of seismic energy, the outer part of the Earth has been divided into three layers: the lithosphere, asthenosphere, and mesosphere (**Fig. 1**). The lithosphere is the outermost layer and contains both the crust and uppermost mantle. It is characterized by a number of rigid plates floating above the more plastic asthenosphere. The defining feature of the asthenosphere is the low speed at which seismic waves travel through it. The low velocity signature indicates that the material in the asthenosphere is at or perhaps slightly above its melting point. Beneath the asthenosphere, seismic velocities increase again going into the mesosphere.

Lithosphere-asthenosphere relationships. These can be relatively straightforward or complex, depending on location.

Chemical and mineralogical abundances in the mantle and in melts derived from it*

	Undepleted mantle	Average continental mantle	Ocean ridge basalt	Continental flood basalt
SiO$_2$	46.1	45.5	50.2	54.5
Al$_2$O$_3$	4.3	1.4	14.9	13.6
FeO	8.2	7.6	11.3	12.1
MgO	37.6	43.6	7.1	3.5
CaO	3.1	1.0	11.4	6.9
K$_2$O	0.03	0.11	0.16	1.7
Rb	0.9	2.3	1.0	47.
Ba	7.6	25.	12.	675.
(La/Sm)$_n$	1.0	1.5	0.5	2.5
% Olivine	53	67	0	0
% Pyroxene	31	27	43	38
% Garnet	15	6	0	0
% Plagioclase	0	0	49	56

* Concentrations of major-element oxides are given in weight percent, trace elements in parts per million by weight, and mineralogic compositions in volume percent. (La/Sm)$_n$ is the lanthanum-to-samarium ratio of the particular composition divided by the lanthanum-to-samarium ratio of the un-depleted mantle. Garnet is the stable silicate phase containing aluminum at depths below 36 mi (60 km), and plagioclase the stable aluminous phase above 18 mi (30 km) depth.

Under ocean basins. In the ocean basins the relationship between lithosphere and asthenosphere is relatively straightforward. The asthenosphere rises to very shallow depths under the ocean ridges, where the melt is extracted to form the thin oceanic crust. Moving away from the ridge, the ocean floor subsides, and the lithosphere thickens at the expense of the asthenosphere as a function of the square root of the age of the ocean floor. This relation is determined by cooling of the lithosphere by simple heat conduction to the ocean. Through heat conduction, the oceanic lithosphere grows downward by cooling the materials of the upper asthenosphere well below their melting temperature. Except for the uppermost 6 mi (10 km) or so of the oceanic mantle affected by melt removal, the lithospheric mantle under the oceans is similar to the underlying asthenosphere in chemical composition but is cooler than it. Continued cooling of the lithosphere causes its thickness to increase to the order of 60 mi (100 km) and its average density to rise substantially. After sufficient time (less than 200 million years) the lithosphere becomes denser than the underlying asthenosphere; it begins to sink back into the mantle, initiating the process of plate subduction.

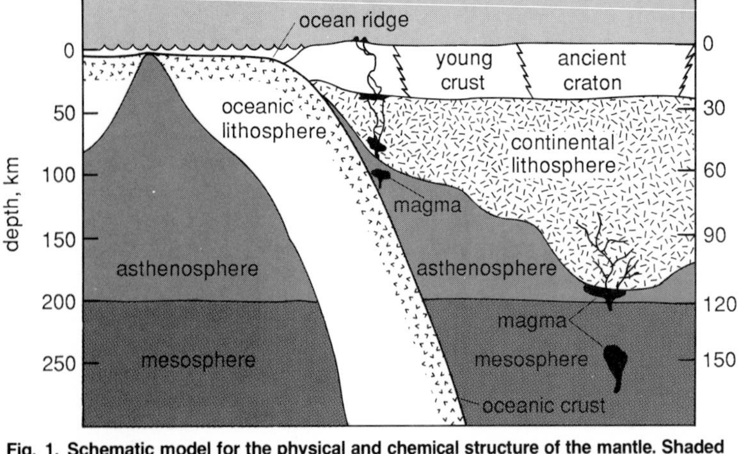

Under continents. In this case the relation between lithosphere and asthenosphere is not so straightforward. The oldest parts of the continents, the so-called Archean Cratons, are 2.5 to 4 billion years old. If the mantle beneath the cratons had the same composition as that in the asthenosphere, conductive cooling for billions of years would have resulted in a very thick, relatively cold root. Because the overlying continental crust is about six times thicker and, on average, about 5% less dense than oceanic crust, a thick, high-density continental root would not cause the overlying crust to subduct. Rather, the cold mantle would delaminate from the overlying crust and sink back into the deeper mantle, leaving only a thin lithosphere.

Seismic studies of the lithosphere and asthenosphere beneath continents reveal a much more complex picture than that beneath the oceans. In active continental areas such as volcanic arcs and rift valleys, well-developed seismic low-velocity zones begin at depths less than 60 mi (100 km). Under stable regions such as the old continental cratons, low-velocity zones are poorly defined, with the depth to the top of such layers estimated to be from 120 to 240 mi (200 to 400 km) beneath some cratons. Since the removal of melt from the mantle lowers the density of the residual peridotite, it has been suggested that the subcontinental mantle is composed of an olivine-rich residual mantle left over from the extraction of the mate-

Fig. 1. Schematic model for the physical and chemical structure of the mantle. Shaded areas indicate regions of mantle that have experienced extraction of partial melts. Beneath continents, the thick keel of lithosphere varies in thickness in relation to the age and geologic history of the overlying crust. Small, rising regions of magma are also shown. Depths are measured from sea levels.

rials now residing in the continental crust (Fig. 1). Because of the compositionally induced lower density of this residual mantle, cooling would not lead to a significant density difference with the underlying, less depleted mantle. Consequently, preservation of a thick lithosphere beneath continents not only would be dynamically stable but also could serve to "buoy up" the old sections of continent and make them less susceptible to destruction by plate subduction.

Element enrichment. While the concept of a depleted peridotite keel below the continents is appealing geophysically, volcanic rocks erupted through the continents, in general, tend to be more enriched in certain trace elements compared with their oceanic counterparts. This is especially true when comparing the most abundant rock type erupted in the oceans, the mid-ocean ridge basalt (MORB), with large-volume basalt eruptions on the continents, such as continental flood basalts (see table). Basaltic rocks are good tracers of mantle characteristics because they are derived by partial melting in the mantle, are abundant throughout Earth history, and are erupted in a wide variety of tectonic settings. With proper understanding of the compositional changes that take place during the melting process and the subsequent fractionation of the melts, the chemical characteristics of basalts can be used to determine the composition of their source materials in the mantle.

Compared with mid-ocean ridge basalt, continental basalts tend to be enriched in incompatible elements like potassium (K), rubidium (Rb), barium (Ba), and the light rare-earth elements. The term incompatible derives from the selective partitioning of these elements out of residual solid phases during melting, thereby concentrating them in the melt. Some of the relative enrichment of continental basalts, compared with oceanic basalts, is related to different degrees of concentration of the incompatible elements during melting and fractionation processes. Another component of the enrichment possibly derives from mixing between mantle-derived melts and the highly incompatible-element-enriched materials of the continental crust. In many cases, however, when the effect of these processes can be understood and subtracted from the observed compositions, the parent magmas of many continental basalts still appear to be derived from mantle sources that are relatively enriched in incompatible elements as compared with the sources of mid-ocean ridge basalt.

Multistage chemical processes. The relative enrichment in incompatible elements of continental basalts seems at odds with the suggestion that the subcontinental mantle is more depleted than the suboceanic mantle. Examination of the geochemical characteristics of the rare samples (xenoliths) of upper mantle sometimes ripped off their conduit walls and carried to the surface by explo-

sive volcanism offers a solution to this apparent contradiction. Xenoliths provide small but very important samples of mantle from as deep as about 120 mi (200 km). Recent data for xenoliths, in particular those from southern Africa, seem to support the idea of an upper layer of olivine-rich residual mantle overlying less depleted mantle. In southern Africa, the boundary between depleted and undepleted compositions varies from 100 to 120 mi (170 to 190 km) beneath the 3.5-billion-year-old craton, to about 80 mi (140 km) beneath the much younger crustal sections surrounding the craton.

One particularly important discovery provided by the xenolith data is that trace- and major-element compositions in mantle materials are not necessarily directly coupled. Since trace-element variations have little effect on the bulk density of the mantle, the determination of the degree to which major- and trace-element variations are decoupled is critical to the interpretation of trace-element data for volcanic rocks and its implications for buoyancy-driven motion in the mantle. Some xenoliths showing depletion in basaltic component also are depleted in incompatible trace elements (**Fig. 2**), as would be expected if the depletion occurred by melt removal. Other xenoliths, showing equal or more extreme depletion in basaltic component, are enriched in incompatible trace elements; this case clearly reflects the importance of multistage chemical processes in the mantle. For example, one explanation for samples that have a low calcium oxide

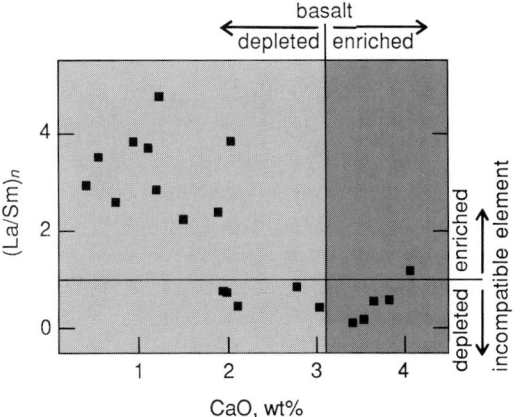

Fig. 2. Diagram of xenolith $(La/Sm)_n$ ratio and CaO content, broken into quadrants by the estimated CaO and $(La/Sm)_n$ of undepleted mantle. Xenoliths from the subcontinental mantle (shown by solid squares) generally show depletion in basalt component as indicated by low CaO content, with CaO contained primarily in pyroxene in the mantle. $(La/Sm)_n$ traces the abundance of incompatible elements compared with undepleted mantle where $(La/Sm)_n$ is defined as 1. As lanthanum is a more incompatible rare-earth element than samarium, a sample with a $(La/Sm)_n$ above 1 must be relatively enriched in incompatible elements compared with undepleted mantle. A significant portion of xenoliths show basalt component depletion but incompatible-element enrichment, indicative of a decoupling of major and incompatible trace-element abundances in the subcontinental mantle.

(CaO) content but a very high lanthanum/samarium (La/Sm) ratio is that the depletion in basaltic component (that is, CaO content) reflects the extraction of a partial melt that also removed lanthanum. Following this, at some unspecified later time, the addition of a small amount of a melt or fluid that was very enriched in incompatible elements greatly raised the lanthanum abundance of this type of sample but increased only slightly its CaO content. Multistage processes clearly are important in creating compositional variability in the subcontinental mantle, as materials depleted of the basalt component and enriched in incompatible elements are very common in the xenolith record.

Multistage chemical processes are not unique to the subcontinental mantle, as examples also can be cited from oceanic localities. Perhaps the more important distinction between oceanic and continental mantle is the time period over which enriched and depleted regions can be preserved. Oceanic lithosphere is recycled continually through plate subduction and appears to be completely homogenized into the mantle on time scales of 1 to 2 billion years. Subcontinental mantle apparently can survive for as long as the overlying continent. This is most clearly shown by isotopic model ages in excess of 3.2 billion years for inclusions in diamonds contained in volcanic rocks erupted through the 3.5-billion-year-old craton of southern Africa. These model ages indicate that diamond formation accompanied stabilization of the continental lithosphere in this area roughly 3 to 3.5 billion years ago. Similarly old model ages are found for many continental volcanic rocks and xenoliths.

Richard W. Carlson

ARCHEAN CONTINENTAL CRUST

The Earth's continental crust has a bulk composition broadly approaching that of volcanic rocks with intermediate chemical composition, and contains a significant portion of the terrestrial inventory of the radioactive heat-producing elements uranium (U), thorium (Th), and potassium. As much as 70 to 80% of the present continental crust may have been generated in the Archean between 3.8 and 2.5 billion years ago. No crust older than 3.8 billion years has yet been identified, and either crustal generation did not commence until this time or crust produced in the 800-million-year period after Earth accretion (4.55 billion years ago) was efficiently recycled into the mantle. Subsequent to its generation, the continental crust undergoes profound internal chemical differentiation which results in the enrichment of radioactive heat-producing elements in the upper crust. Removal of these elements from the deeper crust thermally stabilizes newly formed continental crust.

Crustal components. The Archean continental crust is characterized by marked chemical heterogeneity, both laterally and vertically through the crust. On both a local and regional scale the Archean crust is compositionally bimodal, the dominant component having a composition broadly corresponding to that of tonalitic igneous rocks, with variable but lesser proportions of a basic component. Rock types with intermediate chemical compositions are poorly represented. In general, the proportion of the basic end member increases with depth in the Archean continental crust, so that the lower crust has a considerably more basic composition than the upper crust. Small but highly variable proportions of sedimentary rocks, ultramafic igneous rocks, and granitic rocks occur at all levels in the Archean continental crust.

Low-grade terranes. In the low-grade terranes representative of Archean upper-crust levels, tonalitic batholiths predominate. These batholiths typically occur in association with later-stage potassium-rich granites and intrude or form the basement to the ultrabasic-basic volcanic sequences of greenstone belts. The unusually magnesium-rich lava flows (komatiites) typical of Archean greenstone belts are relatively rare in younger volanic associations.

High-grade terranes. Archean high-grade terranes are representative of the middle to lower levels of the continental crust and are dominated by tonalitic gneisses. These gneisses are typically intimately mixed with variable amounts of metasediments and metamorphosed layered ultramafic-gabbro-anorthosite complexes and their associated volcanics. Lower-crust gneisses metamorphosed under granulite-facies metamorphic conditions (8–10 kilobars or 800–1000 megapascals and 700–900°C or 1290–1650°F) are typically anhydrous and, relative to otherwise compositionally similar lower-grade amphibolite-facies rocks, depleted in Th, U, K, and Rb. Consequently, the Archean lower crust is refractory, with a very low radiogenic heat production. While depletion in Th, U, K, and Rb is often severe (for example, the Scourian granulites of northwestern Scotland), and is invariably associated with granulite-facies metamorphism, not all granulite-facies gneisses are so depleted (for example, the charnockites of southern India). Undepleted Archean granulites generally show evidence of extensive partial melting in place. In southern India, incipient granulite-facies metamorphism (charnockitization) of amphibolite-facies gneisses is common, with metasomatism and migmatization of amphibolite-facies gneisses preceding granulite-facies metamorphism. This feature appears to be characteristic of lower-pressure (4–6 kilobars or 400–600 megapascals) granulites from intermediate crustal levels.

Generation of continental crust. A distinctive feature of the geochemistry of Archean tonalites (upper- and lower-crust) as compared with younger tonalites is severe depletion in abundances of the heavy rare-earth elements [for example, ytterbium

(Yb) and lutetium (Lu)] relative to those of the light rare-earth elements [for example, lathanum and cerium (Ce)]. Depletion in elements such as niobium (Nb) and tantalum (Ta) in Archean tonalites is comparable with that observed in more modern tonalites generated in subduction zone environments. Both the major- and trace-element compositions of Archean tonalitic batholiths and gneisses are consistent with an origin by partial melting of amphibole- and garnet-bearing basic sources under hydrous conditions at elevated pressure. Suitable sources could be basaltic material underplated to the base of the continental crust or, if some form of plate tectonics operated in the Archean, subducted basaltic oceanic crust. Most ultramafic and mafic components of the Archean crust have a geochemistry indicative of an origin by partial melting of a mantle source at relatively low-pressure conditions.

Crustal processes. The mechanism by which radioactive heat-producing elements are redistributed within the continental crust can be constrained by evidence from Archean terranes. The samarium-neodymium (Sm-Nd), rubidium-strontium (Rb-Sr), and uranium-thorium-lead (U-Th-Pb) isotope systematics of Archean lower-crust rocks require that granulite-facies metamorphism and extraction of Rb, Th, U, and K occur within 100–200 million years of crustal generation. Two main mechanisms have been advocated for producing the depleted nature of the Archean lower crust. The first is that under the high pressure and temperature conditions prevailing during granulite-facies metamorphism, most rocks would undergo extensive partial melting. Such melting would result in the production of liquids enriched in Rb, Th, U, and K. Removal of these melts to higher crustal levels would leave a refractory, anhydrous lower crust depleted in Rb, Th, U, and K, with complementary enrichment of these elements in upper-crust granitoids. The alternative mechanism considers the role of the fluid phase during granulite-facies metamorphism. Fluid inclusions in granulite-facies gneisses are rich in carbon dioxide (CO_2), in contrast to water-rich fluid inclusions in amphibolite-facies gneisses. The presence of a fluid with a high CO_2/H_2O ratio during granulite-facies metamorphism has important consequences for melting relations in the lower crust; dehydration reactions are promoted, while melting temperatures are raised relative to hydrous systems, allowing granulite-facies metamorphism to proceed without significant melting necessarily occurring. This mechanism, often termed carbonic metamorphism, requires flushing of the lower crust by large volumes of CO_2-rich fluids. In general, field relationships and the chemistry of granulite-facies gneisses suggest that extensive partial melting has not occurred in the Archean lower crust and favor the dominance of carbonic metamorphism in determining the chemical characteristics of granulites.

Carbonic metamorphism. An ascending front of CO_2-rich fluid passing through the lower crust will remove Rb, Th, U, and K during granulite-facies metamorphism. The leading edge of the fluid front will develop the composition of a water-rich metasomatic fluid enriched in the Rb, Th, U, and K purged from deeper crustal levels. Influx of this fluid into intermediate-crust level amphibolite-facies gneisses causes metasomatism and migmatization immediately prior to the influx of fluids rich in CO_2 and granulite-facies metamorphism. Extensive crustal melting and granitoid production occurs in the middle crust as a precursor to granulite-facies metamorphism, and not within the deep crust during granulite-facies metamorphism. The source of the large volumes of carbonic fluids required to pass through the Archean lower crust is still enigmatic. The possible sources of the CO_2 are decarbonation reactions in the upper-mantle, crystallizing basalts which have underplated the continental crust, marine carbonates deeply subducted under plate interiors, and carbonate strata emplaced into the deep crust by overthrusting.

For background information SEE ASTHENOSPHERE; EARTH; FACIES (GEOLOGY); LITHOSPHERE; PERIDOTITE; ROCK; STRATIGRAPHIC NOMENCLATURE in the McGraw-Hill Encyclopedia of Science and Technology.

Barry L. Weaver

Bibliography. F. R. Boyd and J. J. Gurney, Diamonds and the African lithosphere, *Science,* 232:472–477, 1986; F. A. Frey and D. H. Green, The mineralogy, geochemistry, and origin of lherzolite inclusions in Victorian basanites, *Geochim. Cosmochim. Acta,* 38:1023–1059, 1974; T. H. Jordan, The deep structure of the continents, *Sci. Amer.,* 240:92–107, 1979; S. H. Richardson et al., Origin of diamonds in old enriched mantle, *Nature,* 310:198–202, 1984; S. R. Taylor and S. M. McLennan, *The Continental Crust: Its Composition and Evolution,* 1985; B. L. Weaver and J. Tarney, Empirical approach to estimating the composition of the continental crust, *Nature,* 310:575–577, 1984.

Magnetic bearing

A magnetic bearing is a device incorporating magnetics to levitate a shaft and float it in a magnetic field without any contact between the rotating and stationary elements. Magnetic bearings may be categorized into two general configurations; passive bearings which make use of permanent magnets to generate the field required, and active bearings which make use of electromagnets to generate the magnetic field. These two technologies have also been used simultaneously to provide a so-called hybrid magnetic bearing system.

History. The concept of suspending a rotor magnetically is not new. It was not until the midnineteenth century, however, that much serious investigation into this area took place. Experiments were undertaken with both passive and active systems, each with its own advantages and disadvantages. United States patents dating back to the mid-1800s describe the use of "magnetic bearings" to support rotating shafts, reduce mechanical friction, and improve the efficiency of electrical meters. While the basic theories involved have not changed significantly over the years, advancements in magnetic materials and power electronic components have provided many enhancements of magnetic bearing technology.

Passive configurations. Passive magnetic bearings make use of permanent magnets to generate the magnetic field required for levitation. This type of bearing system may be configured to operate in either attraction or repulsion. Both schemes typically require permanent magnets to be used in both the stator and the rotor. In the repulsion mode, similar poles must be present between the rotor and the stator; a magnetically north pole in the rotor must be across from a magnetically north pole in the stator, and likewise for the south poles. In attraction, a north pole in the rotor is across from the south pole in the stator, with the rotor's south pole across from the stator's north pole. However, if a ferromagnetic rotor component is used, the system operating by attraction may use a permanent magnetic as the stator only.

It is recognized that the attractive system must be extremely well balanced to provide even forces around the perimeter of the rotor. The magnetic forces attracting the rotor must be equal to all points to keep the rotor centrally located. Since the magnetic force varies with the air-gap distance between the rotor and stator, once the rotor has moved off center it will be more strongly attracted by one particular point of the stator and will continue to be pulled off center. Because of this inherent instability, the repulsion configuration is more widely utilized.

With the rotor and stator repelling each other, if the rotor should move off center respective to the stator it will be pushed back. As the rotor comes closer to the stator, the repulsion force of the magnets increases, tending to drive the rotor back toward its central location. While this is one of the advantages of the repulsion system over that of attraction, it is also one of the main disadvantages of the overall system. When a disturbance force occurs, the rotor tends to oscillate as it is pushed from one side of the stator to the other.

Passive magnetic bearings may be designed to accommodate both radial and axial loads as well as linear motion found in frictionless slides. Most applications to date have been on small-diameter shafts with relatively low speeds and low unbalance forces, and have been related to aerospace or light production. Passive configurations are more attractive economically than active magnetic bearing systems, since no supporting electronics are required. Since there are no contacting parts between the rotor and the stator, the only limiting factors for speed are the centrifugal forces generated by rotation. With the introduction of higher-flux-density magnetic materials such as samarium-cobalt and neodymium-iron-boron, the relative loading factor of passive magnetic bearing systems has substantially increased over the past 10 years. Unfortunately, the constant and controllable bearing stiffness cannot be maintained, because the stiffness value varies with the air-gap distance of rotor to stator. There is no means for energy absorption; consequently the damping of such a system is minimal. The inability of the passive magnetic bearings to handle moderate-to-high dynamic loads has significantly limited the practical applications of strictly passive bearing systems.

Active configurations. Active magnetic bearings operate by attraction, and they use electromagnetic stators with an associated electronic control source. The electromagnetic stators are fabricated by using slotted laminations, thin electrically insulated sheets of ferrous steel, which are stacked and bonded together. Coils of magnet wire are wound around the teeth of the lamination stack, giving the bearing an appearance similar to that of an electric motor.

While there are some systems that make use of alternating current to create a magnetic field, most active magnetic bearing systems operating in attraction incorporate direct current to generate the magnetic field in the stator. Pulse-width modulation is often used in the amplifier scheme to increase the efficiency of the system relative to linear amplifiers. The laminated stator is wound so as to provide individual magnetic quadrants, each capable of being independently supplied with current. By varying the current separately in each of the electromagnetic stator quadrants, the rotor position relative to the stator can be either varied or maintained.

Like the stator, the rotor is also constructed with a stack of laminations. Since the rotor does not require the use of wound coils, its laminations are not slotted, and they look like doughnut-shaped rings. This provides a good ferrous material to be attracted by the stator and also keeps eddy-current limits (open primary magnetic losses) to a minimum. Similar to that in permanent magnets, the loading factor of the bearing is based on the magnetic flux saturation limit of the material. Advanced materials such as vanadium Permadur have allowed substantially increased bearing loading levels. Speed limits are also related to materials and are based on the centrifugal stress levels generated in the rotor components.

As with attractive passive systems, once the rotor moves off center it is more strongly attracted by one part of the stator than the others; hence it becomes unstable. For this reason the active system makes use of position sensors to detect the rotor location, and servo feedback control to keep the rotor on center.

The position sensors generate a voltage signal proportional to rotor location. This voltage (where the shaft is) is compared to a position reference signal (where the shaft should be). Any differences between these two signals generates a so-called error signal. The error signal is conditioned, processed, and used to command the appropriate power amplifier either to maintain or to correct rotor location. This process of rotor location detection and correction takes place continuously and is known as a position-controlled closed-loop servo feedback system.

The active control system associated with the attractive electromagnetic bearing allows the stiffness and damping characteristics to be tailored to the exact rotor configuration. Stiffness is synonymous with the primary gain of the electronics, and damping is provided by the use of phase lead networks. The bearing system can respond to high dynamic loading and is capable of supplying sufficient damping to pass through the critical speeds of the rotor. The control electronics provide an immediate response to any rotor disturbances, which makes the active magnetic bearing system the most stable system for dynamic-load compensation.

Future developments. In general, the era of industrial magnetic bearings is just beginning; the use of these bearings grow as advances in both materials and electronics are achieved. Applications for the more exotic types of bearings systems will increase as designers push present technology to its limits.

For background information SEE ELECTROMAGNET; MAGNETIC MATERIALS in the McGraw-Hill Encyclopedia of Science and Technology.

David A. Weise

Bibliography. Papers presented at 27th IEEE Machine Tool Conference, October 1985.

Magnetic suspension and levitation

Since the mid-1960s, passenger-carrying vehicles using air cushions at first, and then electromagnetic suspension and levitation techniques, have been widely publicized. Equally impressive are advances in various industrial applications of magnetic suspension and levitation. Although each of the different techniques is virtually a technology in its own right, they all became collectively labeled maglev. The two most developed techniques are superconducting systems and suspension systems using controlled direct-current (dc) electromagnets. Several other methods have been researched extensively and have been proposed for application, but they have not

been found entirely suitable.

Superconducting magnet systems. Superconductivity is a property of electrically conducting materials which cease to offer any resistance to the flow of electric current when cooled below certain temperatures approaching those of liquid helium (4 K or −452°F). In magnets whose winding coils consist of superconducting materials, the high current densities that can be employed without any power dissipation make it possible to generate magnetic fields of intensities that otherwise could not be achieved. Superconducting magnets are, in effect, coils of alloy materials such as vanadium and gallium (Va_3G), or niobium and tin (Nb_3Sn), kept in well-insulated containers slightly below their critical temperatures (18 K or −427°F in the case of niobium-tin).

Lift generation. Sideways relative motion between a permanent magnet and a conducting sheet, for example, one made of aluminum, results in induced currents which repel the magnet. If conventional permanent magnets are used, the force of repulsion is no more than barely sufficient to support the weight of the magnet itself, even at relatively high speeds. The use of superconducting magnets, which have intense magnetic fields, enables enough lift force to be generated at speeds of the order of 100 mi/h (160 km/h) and above to permit levitation of high-speed passenger-carrying vehicles. The superconducting magnets are an integral part of the vehicle and the track is of aluminum sheets. Any heat entering the magnets has to be removed by on-board refrigeration.

Propulsion. In a vehicle employing contactless levitation, it is clearly advantageous to achieve propulsion in the same manner as levitation. The conducting sheet on the track is therefore replaced by conducting loops. The lift is obtained by the mechanism of electrodynamic levitation discussed above, but the loops are fed (**Fig. 1**) with currents from trackside installations in such a manner that they produce a pull on the vehicle-mounted magnets. This type of propulsion system is known as a linear synchronous motor. It is capable of producing the requisite thrust by electromagnetic interaction through air gaps of 4 in. (10 cm) or more.

Electromagnetic drag. One drawback in this scheme is that the currents induced in the track loops as the superconducting magnets approach them persist after the magnets have passed over the loops. In effect, this leaves an eddy-current wake and leads to electromagnetic drag. Curiously, though, the drag is high at low speeds and diminishes at higher speeds—the only known instance of a drag force diminishing with speed. Thus, electromagnetic drag is not a serious problem if the linear motor can overcome the high initial drag. The lift magnets can also be utilized to provide side guidance forces by interaction

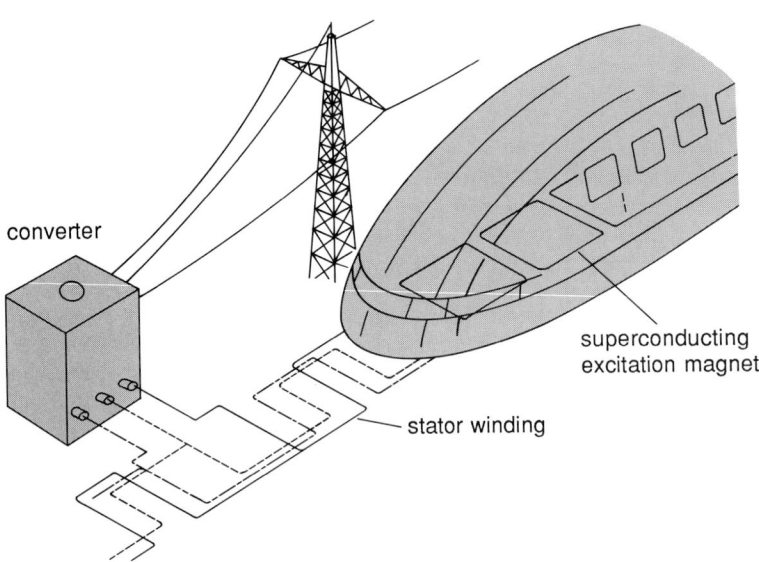

Fig. 1. Passenger-carrying vehicle with superconducting magnets for levitation and linear synchronous motor for propulsion. The converter converts three-phase current at the power line frequency (50 or 60 Hz) to three-phase current at a frequency that matches the motion of the vehicle over the stator windings.

with vertical conducting structures mounted on the sides of the vehicle track.

Applications. In the 1960s and 1970s, several centers in the United States, Canada, Germany, Japan, and Great Britain conducted research on superconducting vehicle schemes. Only the Japanese have persisted with research and experimental work and have demonstrated several vehicles weighing 5.5 short tons (5 metric tons) and above traveling at speeds of 310 mi/h (500 km/h). Since 1980 they have operated 22-short-ton (20-metric-ton) two-coach vehicle on a 8-mi (12.5-km) track at speeds of 150 mi/h (250 km/h).

Controlled dc electromagnet systems. By itself, the force of attraction between a magnetized body and one or more permanent magnets or electromagnets of constant strength is insufficient to provide stable suspension of the body. For example, if an attempt is made to suspend a steel ball against gravity by using a magnet, even an extremely slight disturbance of the ball from the position where the electromagnetic force on it is equal to its weight will cause the ball either to be attracted upward to the magnet or to fall to the floor. However, the ball can be stabilized by the system shown in **Fig. 2a**. The position of the ball is determined by means of an optical sensor composed of the light and the photocell. If the ball is attracted upward and moves toward the magnet, the signal from the optical sensor into the amplifier is such that the magnet current (and therefore the magnetic force of attraction) is reduced; and the converse occurs if the ball moves away from the magnet. In any such closed-loop (feedback) system of position control, there are time delays, and the box labeled

"compensation" therefore attempts not only to compensate for these delays but also to introduce some phase advance or anticipatory action. The force-distance characteristic of a system with closed-loop control is compared with the characteristics of some open-loop control systems (no feedback) in Fig. 2b.

The basic principles of such systems have been known for many years. A configuration capable of suspending a 460-lb (210-kg) weight was demonstrated in 1936. However, its electronic amplifiers were bulky and unreliable. The advances in power semiconductor devices and other devices since about 1960 have allowed the technology and applications of controlled direct-current electromagnet suspension to progress rapidly.

The principal elements of a suspension system employing controlled dc electromagnets are magnets, amplifiers, transducers or sensors, and control systems.

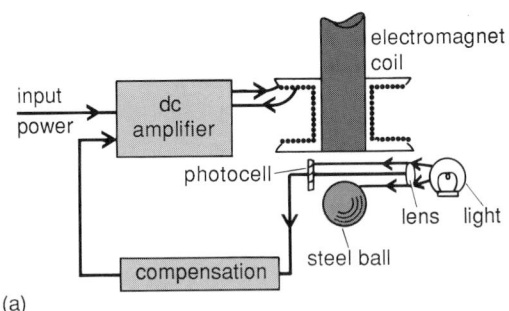

(a)

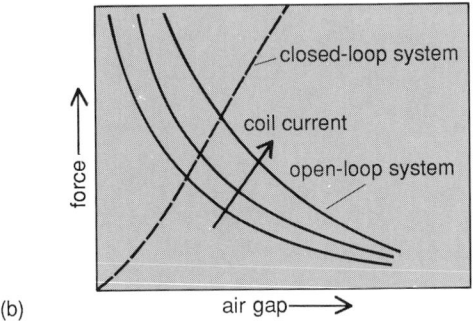

(b)

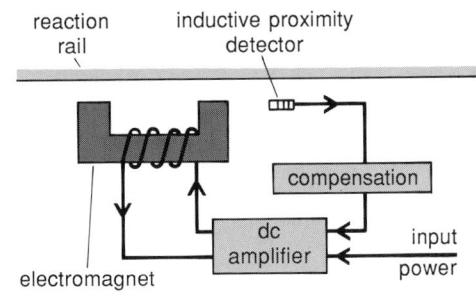

(c)

Fig. 2. Principle of magnetic suspension using controlled direct-current electromagnets. (a) Magnetic suspension with optical transducer. (b) Force-distance characteristics of open- and closed-loop systems. (c) Suspension of magnet from under a steel rail, using an inductive proximity detector.

Magnets. The magnets usually are made of mild steel or punched steel laminations. Depending on the available space and the voltage insulation required, the magnet windings are made of copper or anodized aluminum foil. The design of the magnets requires careful study since the width of the air gaps between the stationary and suspended members can be more significant here than in other electrical machines such as motors.

Amplifiers. The electronic amplifiers are usually transistorized and capable of supplying substantial and variable direct current. The principle on which they operate is one of switching the supply on for a given period (mark) and then off for the remainder of a controlling cycle (space). By varying the mark-space ratio, it is possible to control the effective current flowing in the load. The load in this case however, is, highly inductive, and great care must be taken to provide protection for the transistors from the surges of the stored energy in the magnets. Another property of the system that arises from the inductance of the magnets is that, although the steady voltage drop across the magnet coils may be as low as half a volt, the current in the system must change very rapidly, which requires application of a voltage several times greater than the quiescent voltage. The ratio of available voltage to the quiescent voltage, called the forcing ratio, may be 50 to 200, depending on the required response or the closed-loop bandwidth of the control system. The direct-current amplifiers are therefore quite substantial.

Transducers or sensors. The sensors or transducers must be noncontact. In a vehicle or bearing type of application, the optical sensor shown in Fig. 2a is unlikely to be practicable. As in the case of the amplifiers, considerable research has gone into the development of suitable transducers. The main ones used are inductive, capacitive, or magnetic. They are proximity sensors (Fig. 2c), and care is needed to make them rugged, immune from extraneous signals, and free from drift, since it is sometimes necessary to control the width of the air gaps to a fraction of a millimeter.

Control. The control of position of the suspended object is effected by direct measurement of the position. Also used are the rate of change of position of the object and the integral over time of the error between the desired and actual positions. In multiple magnet systems, the control system strategies become complicated because each system is highly nonlinear, so that interactions between systems can lead to conflict and instability. One type of system measures the magnetic flux density B in the air gap and provides another loop to control B. The force of attraction is proportional to B^2. The unstable force-distance characteristic shown in Fig. 2b is attributable to this nonlinearity. Introduction of a flux-control loop makes the system conditionally stable and considerably eases the problems of overall stability.

Applications to vehicles. The advantages of controlled direct-current electromagnetic systems over superconducting ones are: greatly reduced complexity, compactness of equipment, and low power requirements (sometimes less than 1 kW per metric ton). A great deal of effort has gone into developing passenger-carrying vehicles capable of traveling at 280 mi/h (450 km/h) or more. Since 1971 German vehicles have been produced that weigh 5.5 short tons (5 metric tons) and more. At present, a 132-short-ton (120-metric-ton) vehicle is being tested on a 20-mi (31-km) track at speeds of 280 mi/h (450 km/h). In Japan a 5.5-short-ton (5-metric-ton) magnetically suspended vehicle was demonstrated in 1976, and a 55-short-ton (50-metric-ton) vehicle in 1985. Both the Japanese and the German vehicles use linear synchronous motors for propulsion. While research has concentrated on the high-speed systems, the controlled direct-current electromagnet technique, unlike the superconducting one, is also applicable to low-speed (15–50 mi/h or 25–80 km/h) vehicles suitable for urban transport. At the high-speed end, the relatively small air gaps (0.3–0.4 in. or 7–10 mm) might present problems of unrealistically accurate track alignment.

Industrial applications. Industrial applications for frictionless and contactless bearings, conveyor systems, the synthetic fiber industry, and vibrators have also made enormous strides. A typical application involving magnetic bearings is a machine tool spindle capable of cutting metal at 30,000 revolutions per minute (**Fig. 3**). It has two radial bearings and one thrust bearing, and the drive (induction) motor is located between the two radial bearings. Similar magnetic bearings are now in use for centrifuges, pumps, and compressors wherever there are problems of high

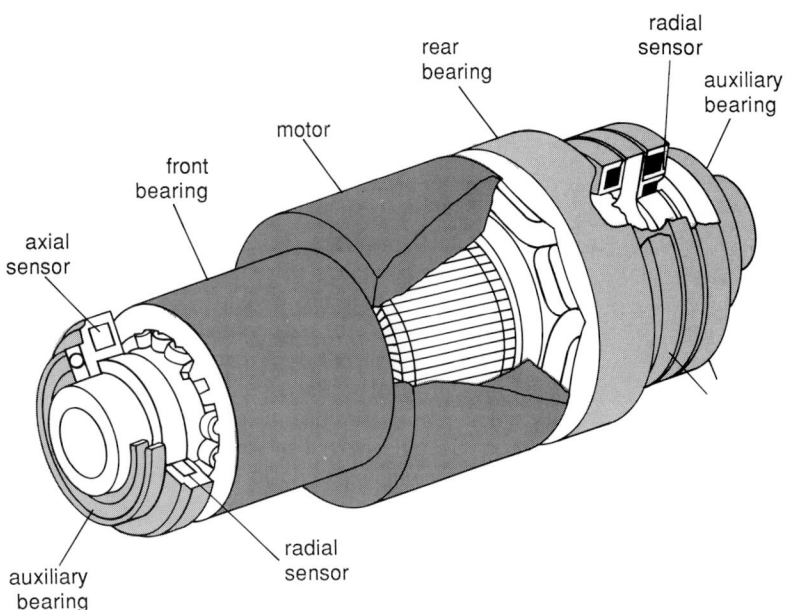

Fig. 3. High-speed (30,000 revolutions per minute) machine tool spindle with magnetic bearings.

speeds, hostile environment, access for servicing, or reliability. A bearing for a centrifuge-type application has been built in which the suspended rotor weighs 9 short tons (8 metric tons) and is driven by an arch-shaped, 40-kW linear induction motor. The power required to support the drum, however, is only 1400 W, less than that required for a 2-kW domestic electric heater.

Another innovation in the field of magnetic bearings, applicable to machines in which the suspended rotor weights are less than 10 lb (5 kg) or so, is to use only permanent magnets for the radial bearings and controlled direct-current electromagnets for the axial or thrust bearing. In models now being tested, the drive motor is an induction motor, but it is intended to replace this motor with a permanent magnet motor in the near future. Rotors spinning at 100,000 revolutions per minute or more are now a practical possibility.

For background information SEE ELECTROMAGNET; INDUCTION MOTOR; SUPERCONDUCTING DEVICES in the McGraw-Hill Encyclopedia of Science and Technology.

B. V. Jayawant

Bibliography. R. H. Borcherts et al., Baseline specification for a magnetically suspended high speed vehicle, *Proc. IEEE*, 61:569–578, 1973; B. V. Jayawant, *Electromagnetic Levitation and Suspension Techniques*, 1981; F. London, *Superfluids*, 1961.

Mammalia

Coevolution is the process of evolutionary change in two or more species resulting from reciprocal selective forces exerted upon one another. The prefix co- implies that the organisms are evolving together. Coevolutionary theory seeks to explain patterns of coadaptation among species living in close ecological or evolutionary association. Coevolution can lead to unique levels of ecological and evolutionary interactions, such as the compromise between parasite virulence and host defense that is apparent in many host-parasite systems.

Parasitism as a life-style is widespread in the biological world. It is thought that there are more species of parasites than of free-living organisms. Parasites can be defined as those organisms which spend a part or all of their life in or on a host from which food and other biological necessities are derived. Ectoparasitic arthropods on mammals spend much of their adult life in close association with the habitat created by the skin and hair of the host, or in its nest. These arthropods usually possess adaptations to this habitat and derive their food exclusively from their host. SEE ECOLOGICAL INTERACTIONS.

Arthropods have evolved to occupy a wide array of niches on mammals, with a wide diversity of relationships being seen between parasitic arthropods and their hosts. Facultative parasites (that is, fleas, ticks, and some mites) usually feed upon the host but spend most of their life cycle off the host, often in the host's nest. Obligate parasites (that is, chewing and sucking lice) spend their entire life cycle, including feeding, reproduction, egg laying, and all developmental stages, on the host. Usually they cannot survive long if off the host.

Cost of parasites to hosts. There are two levels of assessing parasites' cost to hosts. The proximate effect is the tissue damage and the pathological effect on the individual host. The ultimate effect is the energy loss, the lowered survival rate, and the reduced reproductive potential in the host population.

Parasites do not normally occur in high numbers that do severe damage. The selective forces on parasites dictate that they do not destroy the host or its offspring, their future supply of hosts. To do so would create an evolutionary dead end for the parasite itself. Thus, parasites tend to evolve to be less virulent. Although ectoparasites generally occur in small populations and there is little obvious outward damage to hosts, the evolutionary effect of parasites could be severe. Healthy, adult hosts are often resilient to the minor effects of low numbers of parasites. However, young or weak mammals may suffer severe damage from high populations of parasites. Also, an animal showing high parasite levels may not be as attractive to potential mates. The evolutionary cost of parasitism is measured in terms of hosts' producing a smaller number of offspring than nonparasitized animals, or offspring that are less fit to survive and reproduce.

Parasite and host adaptations. The process of coevolution may affect the morphology, behavior, and physiology of both parasites and hosts. Examples of changes in morphology are the extreme dorsal-ventrally flattened bodies of lice and the laterally flattened bodies of fleas that facilitate their movement through the fur and allow them to avoid being removed during host grooming. The distance between setae on fleas corresponds to the width of the host's hair, forming an effective lock mechanism for the flea and preventing the host from dislodging it. Additional adaptations which allow easy movement of the flea through the host's pelage and make the flea difficult to dislodge include posteriorly pointed setae and strong tarsal claws. The mouthparts are modified for piercing and sucking, an adaptation for feeding on blood. In many species of fleas the eyes are greatly reduced and in some species are even absent. In the highly specialized bat flea, *Porribius caminae* (see **illus.**), the dorsal setae are further modified to form so-called combs that act to prevent dislodgement of the flea by the host. Structural features such as reduced eyes and wings, legs that are adapted for grasping hair, and flattened

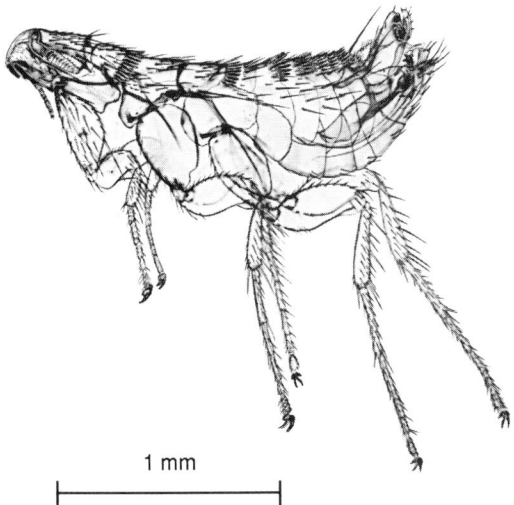

Highly specialized bat flea, *Porribius caminae*. Dorsal combs prevent dislodgement by host, and eyes are absent. (*From K. C. Kim, ed., Coevolution of Parasitic Arthropods and Mammals, John Wiley, 1985*)

bodies are found in most parasitic arthropods.

Behavioral examples include selection of egg-laying sites by lice on areas of the host's body that are the most difficult for it to groom. In hosts, behavioral responses are perhaps best expressed by the specialized grooming behavior directed toward the parasites, which is the primary factor controlling their numbers.

Physiological adaptations include the response of parasites to changes in blood chemistry of the host. An example is seen in the European rabbit flea's use of the reproductive steroids it ingests from the rabbit's blood to trigger its own reproductive hormones, thus synchronizing its life cycle to reproduction in its host.

Relation of parasite and host phylogeny. It has long been postulated that parasites, because of their conservative rates of evolution with respect to their hosts, may aid in studies of host taxonomic relationships. The hypothesis that the natural classification of parasites parallels that of their hosts was first proposed by H. Fahrenholz based on his studies of feather mites. According to Fahrenholz's Rule, in groups of permanent parasites the classification of the parasites usually corresponds directly to the natural relationships of the hosts. The basis of the theory is that at some point in the evolutionary history of the host and parasite the ancestral parasite entered a close association with the ancestral host, after which both evolved and speciated together. Thus, speciation and degree of divergence in the parasite taxa are parallel to those of the hosts. By understanding the systematic relationships between the parasites, taxonomists can better understand the relationships of the hosts. Such parallel evolution has been demonstrated for mesostigmatid bat mites, sucking lice on seals, streblid flies on bats, chewing lice on rodents, and harpyrhynchid bird mites. Those species that

conform to Fahrenholz's Rule usually are blood feeders.

While it is generally believed that there is a high degree of host specificity among arthropod ectoparasites and that parasite phylogeny reflects host phylogeny, studies of quill mites of birds led to the conclusion that noncongruent host-parasite relationships, indicative of resource tracking, are much more common than previously realized. The Resource Tracking theory is also supported by the chewing lice on alcids (aquatic diving birds, found on northern oceans). In Resource Tracking, it is believed that parasites specialize on a particular resource on the host, such as certain types of skin, hair, or feathers, or thicknesses of quill walls. Thus, the parasites are not tied closely to the exact host species, but to a resource that may be similar on several host species. It has been demonstrated that the Fahrenholz's Rule and Resource Tracking hypotheses are not conflicting, but represent the ends of a continuum based on dispersal opportunities and niches available to the parasites. Because of the morphological and behavioral differences between mammals and birds, Fahrenholz's Rule most often occurs with mammalian ectoparasites, and Resource Tracking most often occurs with avian ectoparasites.

Magnitude of parasite and host effects. Mammals are free-living, provide virtually all the parasite's requirements, and represent a large part of the total habitat of the parasite. However, parasites are only one of many extrinsic factors impinging upon the host. Other factors include predator avoidance, food finding, and habitat requirements. Thus, hosts are expected to have a greater evolutionary effect upon parasites than parasites have upon hosts.

Often the host exerts selective pressure on the parasite, but the parasite does not have a corresponding effect on the host. Such host-parasite systems cannot be considered to be true coevolution. In the pocket gophers (*Geomys*) and their chewing lice (*Geomydoecus*), the lice have evolved numerous behavioral and morphological specializations for living on the host. However, the parasites, which feed on dead skin, seem to have little effect upon the pocket gopher.

Rodent-botfly coevolution. One of the best-documented cases of mammalian-parasitic arthropod coevolution is seen in the botflies of the genus *Cuterebra*, parasitic on North American rodents. The larval botflies, as large parasites, have the potential of inflicting serious damage on their hosts. However, they have evolved mechanisms to scatter the eggs so that individual hosts are not overly parasitized. The larvae secrete an antibiotic agent that protects the host from bacterial infection. The rodent hosts have evolved a physiological tolerance to these parasites. Thus, the presence of an average parasite load of botflies is of little cost to hosts.

New research areas. Some of the directions for study in host-parasite coevolution involve determining: the selective forces exerted by hosts on parasites, and conversely, the selective forces exerted by parasites on hosts; the cost of parasitism to hosts; the manner in which the reproductive cycles of parasites are cued to reproductive cycles of hosts; the role of parasites in the epidemiology of diseases among hosts; whether parasites can affect the behavior of hosts; whether there is a genetic basis for resistance to parasite infections; whether parasites can play a role in regulating host populations; and whether parasites and the diseases transmitted by them can be responsible for delineating geographic ranges of species. Much remains to be learned about host-parasite coevolution, and in the next several years there should be major advances.

For background information SEE ACARINA; ANOPLURA; IXODIDES; MALLOPHAGA; MAMMALIA; PARASITOLOGY; SIPHONAPTERA in the McGraw-Hill Encyclopedia of Science and Technology.

Robert M. Timm; Barbara L. Clauson

Bibliography. K. C. Kim (ed.), *Coevolution of Parasitic Arthropods and Mammals*, 1985; A. G. Marshall, *Ecology of Ectoparasitic Insects*, 1981; M. Rothschild and R. Ford, Hormones of the vertebrate host controlling ovarian regression and copulation of the rabbit flea, *Nature*, 211:261–266, 1966; J. N. Thompson, *Interaction and Coevolution*, 1982; R. M. Timm, Fahrenholz's Rule and Resource Tracking: A study in host-parasite coevolution, in M. H. Nitecki (ed.), *Coevolution*, pp. 225–265, 1983.

Marine sediments

Sedimentary marine environments have been an important area of study for a number of years. Recent work has defined the variation of the sea-water strontium isotope ratio ($^{87}Sr/^{86}Sr$) for the Phanerozoic and has yielded information concerning the various forms of calcium carbonate in the sedimentary environment.

VARIATION OF STRONTIUM ISOTOPE RATIO

The variation of the $^{87}Sr/^{86}Sr$ ratio in sea water has recently been defined for the Phanerozoic (the last 590 million years). This was accomplished by using a high-resolution mass spectrometer to analyze strontium-bearing marine rocks and minerals of known age. The variation with time is caused by the contribution of strontium from two different sources: continental and oceanic rocks. The variation is therefore a record of oceanic-continental plate interaction that leads to mountain building and the formation of new crust. During times of rapid change of this ratio in sea water, strontium isotope measurements have the capability of solving a variety of stratigraphic problems relating to correlation or age assignments of rock units.

Strontium-87 increase. The natural radioactive decay of rubidium-87 (^{87}Rb) has led to the steady increase of ^{87}Sr through the 4.5 billion years of Earth history. There has not been a large increase, in part because of the exceptionally long half-life of ^{87}Rb (about 48.5 billion years). The increase in ^{87}Sr can be measured by comparisons with other isotopes of strontium that do not have long-lived parent isotopes of other elements (**Fig. 1**). The strontium isotope ratio is usually given as $^{87}Sr/^{86}Sr$. Variation in the ratio is totally due to the radiogenic contribution of ^{87}Rb decay, any fractionation in the isotopes being corrected by defining a constant $^{86}Sr/^{88}Sr$ ratio.

Strontium forms no common minerals but substitutes readily for calcium in most naturally formed rocks and minerals. The amount of strontium that substitutes for calcium in marine minerals varies enormously. Modern sea water itself contains only about 8 parts per million of strontium. Calcite ($CaCO_3$) in ancient limestone generally contains between 300 and 800 ppm strontium, whereas calcite's metastable polymorph, aragonite, commonly contains 8000–10,000 ppm in modern settings. Dolomite ($CaMg[CO_3]_2$) is much less tolerant of strontium in the crystal lattice and generally contains less than 200 ppm in ancient rocks, with some samples containing much less (about 30 ppm). Anhydrite ($CaSO_4$) and gypsum ($CaSO_4 \cdot 2H_2O$), two minerals precipitated from hypersaline waters, commonly contain 500–1500 ppm strontium.

Analyses. Rubidium is widely distributed as a proxy for potassium. Potassium, and therefore rubidium, concentrates in rocks and minerals found in continental crust.

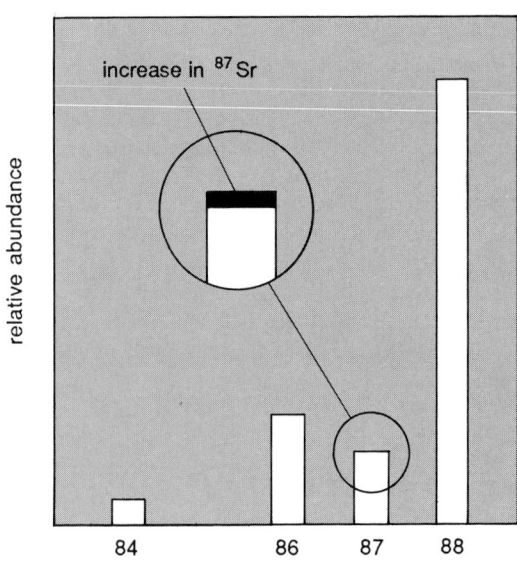

Fig. 1. Relative abundances of the various strontium isotopes, shown as a mass spectrometer scan.

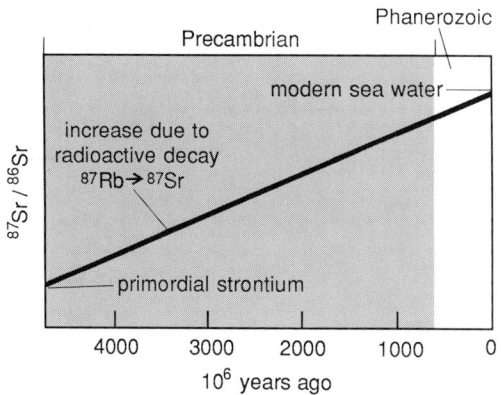

Fig. 2. Model for the evolution of $^{87}Sr/^{86}Sr$ ratio in sea water through geologic time. (*After F. E. Wickman, Isotope ratios: A clue to the age of certain marine sediments, J. Geol., 56:61–66, 1948*)

In 1948 F. E. Wickman suggested that ^{87}Rb-^{87}Sr dating could be used to determine the age of calcium-bearing minerals precipitated from sea water. He reasoned that the change in ^{87}Sr would be reflected in the $^{87}Sr/^{86}Sr$ ratio of sea water as a monotonic increase (**Fig. 2**). The ratio of ^{87}Sr to ^{86}Sr in sea water at any time would be reflected by calcium-bearing minerals precipitated from sea water and would remain unchanged through geologic time because of the exclusion of potassium and, therefore, rubidium from common minerals formed in sea water. Thus, it would be possible to collect a sample of Devonian limestone (about 370 million years old), separate the strontium, determine the $^{87}Sr/^{86}Sr$ by mass spectrometry, and define the $^{87}Sr/^{86}Sr$ ratio of Devonian sea water.

Early work in the 1950s showed that strontium isotope changes were much less than Wickman had calculated, on the basis of poorly understood crustal abundances of rubidium and strontium. The mass spectrometers of the time were incapable of measuring meaningful differences.

A study in 1970 reported the results of $^{87}Sr/^{86}Sr$ analysis of unaltered shell material from 25 intervals of known geologic age for the last 450 million years. These results showed very small variations in the isotope ratio and, unlike the monotonic increase predicted by Wickman, periods of increase and decrease in the $^{87}Sr/^{86}Sr$ ratio were found. By this time it was generally understood, on the basis of meteorite work, that the Earth's initial $^{87}Sr/^{86}Sr$ value (before a contribution from ^{87}Rb) was very close to 0.699. Measurements on modern sea water showed a value of about 0.709. There was, indeed, very little difference between the primordial ratio and the modern sea-water ratio of ^{87}Sr to ^{86}Sr.

In 1982 a research group used the results of 786 analyses of strontium from marine rocks and minerals of known age to define the $^{87}Sr/^{86}Sr$ ratio in sea water through Phanerozoic time. They used a multiple-collection high-resolution mass spectrometer that was designed and built for strontium analysis and yielded results with greater analytical precision. A best line and band of uncertainty were drawn through these data to give the best estimate of the sea-water $^{87}Sr/^{86}Sr$ ratio at any given time during the last 590 million years (**Fig. 3**). The research group's results showed about the same isotopic range as found by previous workers (0.7068–0.7091), but the curve documented many more episodes of increasing and decreasing values than had been apparent from earlier work. This group also analyzed 42 modern marine samples to show that the seas are perfectly mixed within the error of measurement ($^{87}Sr/^{86}Sr = 0.70907 \pm 4$).

Interpretation. The reasons for the increases and decreases were understood even before accurate documentation. The continents have potassium-rich rocks and the oceans are rich in calcium. Rubidium substitutes readily for potassium and has contributed to the reservoir of ^{87}Sr on the continents, so that the average $^{87}Sr/^{86}Sr$ ratio carried in solution to the oceans by rivers is radiogenic, with a value of about 0.715–0.720. The oceanic rocks, being potassium-poor, have remained closer to primordial strontium (0.699) and now have $^{87}Sr/^{86}Sr$ values between about 0.703 and 0.704. The strontium from this source is introduced into the oceans largely by subsea alteration of rocks, hydrothermal activity, and volcanic activity. Both continental and oceanic sources contribute strontium of different isotopic ratios all the time. It is when one source is dominant that the sea-water strontium curve rises or falls.

The increased contribution from oceanic sources is almost certainly the result of sea-floor spreading, which in turn leads to the movement of both continental and oceanic plates. The variation in strontium isotopes in sea water, therefore, records the sum of the results of plate tectonics and mountain building, although the separate results are partially hidden because of the overlapping influences. Some events are clear: the steep fall in the strontium isotope ratio during the Ordovician records the oceanic movement associated with the Taconic orogeny following a long period of stability, and the very rapid rise of the curve during the last 35 million years is due to widespread continental emergence exposing high-ratio strontium sources.

Age assignment. There are some periods when rapid changes in the $^{87}Sr/^{86}Sr$ ratio make the indicated technology a potentially powerful method to solve stratigraphic problems. One of the most promising periods is the last 35 million years, during which there was a dramatic increase in the ratio. Strontium isotope measurements can be used to make age assignments in this interval with an error of ± 0.5 million years. Very high-quality paleontologic age assignment can do this as well or better, but it is comparatively rare that the faunal suites necessary for this precision are present in the rock record. Unfossiliferous limestones, shell frag-

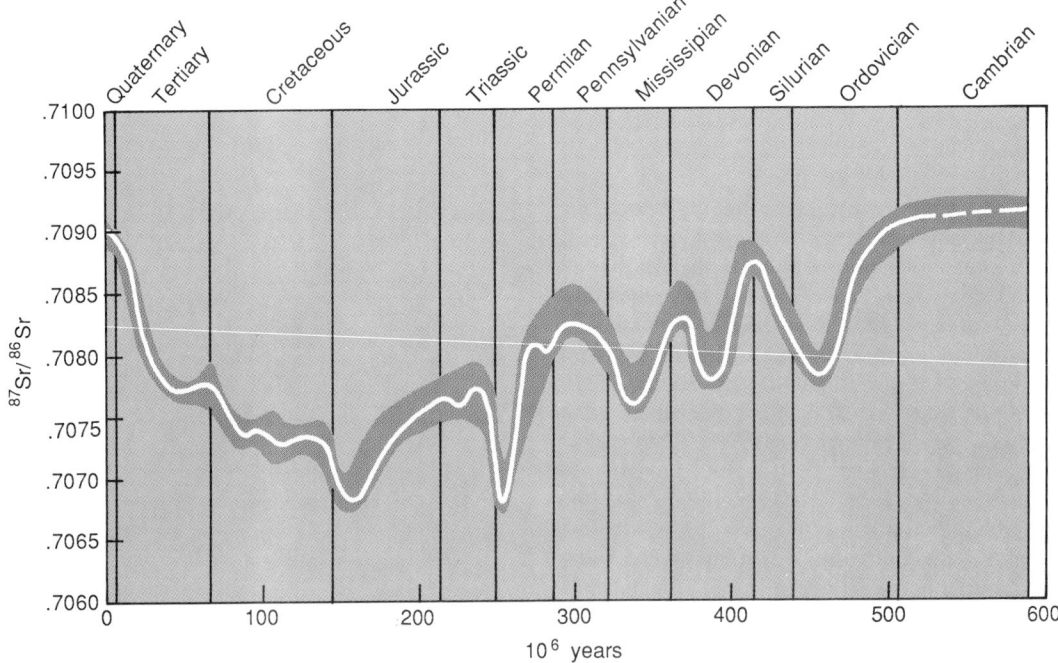

Fig. 3. Variation in ^{87}Sr/^{86}Sr ratio through Phanerozoic time. The curve represents the best estimate of the true composition at any one time with the shaded area representing the potential error. (*After R. B. Koepnick et al., Construction of the seawater* ^{87}*Sr/*^{86}*Sr curve for the Cenozoic and Cretaceous: Supporting data, Chem. Geol. (Isotope Science section), 58:58–81, 1985*)

ments, or shell material from long-ranging fossils can be analyzed to make high-resolution age assignment, as long as the material was formed in contact with open sea water. Later recrystallization of calcium-bearing marine minerals can significantly alter the original isotope ratio. This is particularly a problem in rocks that have undergone a complex geologic history of burial, heating, and deformation. If the calcium minerals remain a closed system during recrystallization, the original ratio is maintained. However, sedimentary rock sequences have numerous sources of foreign strontium, and this can be mobilized to displace the original seawater strontium in the calcium minerals during recrystallization at elevated temperature and pressure. This is less common but does occur during recrystallization at lower temperature, involving flow of substantial volumes of nonmarine waters through porous rocks.

Problems of open systems are by no means restricted to strontium isotopes. Virtually any geochemical measurement on ancient rocks and minerals must take into account the possibility of elemental and isotopic exchange. The consistency of strontium results from worldwide localities indicates that problems of exchange can be overcome with properly chosen samples.

There are limitations in the use of strontium isotopes to determine the age of marine calcium-bearing minerals; there is no unique ^{87}Sr/^{86}Sr value (except initially), and there are periods when little or no change occurs in the isotope ratio. During times of isotopic ratio change, when the general age of a rock sequence is known, the method is flexible as to the amount and quality of

material required to make a well-defined age assignment. The method can be used for rocks deposited in water of any temperature having a wide range of salinities, as long as the body of water is in contact with the open ocean. Strontium isotopes have also been used to solve or constrain the origin of minerals formed after deposition during diagenetic changes in sandstones, limestones, and dolomites.

The key to the use of strontium isotopes in the solution of geological and geochemical problems is the definition of the variation in sea water with time. This has been accomplished for the last 560 million years.

The definition for the approximately 4 billion years in Precambrian time will be exceptionally difficult. The Phanerozoic documentation was possible because of a very detailed time definition based on fossil age assignment. The diversity of fossilized life-forms is lacking in the Precambrian, and the assignment of the age of deposition has been, and remains, a significant problem during this long time period. By carefully choosing samples, there is every reason to believe that the ^{87}Sr/^{86}Sr ratio can be defined for the Precambrian seas. This definition will undoubtedly increase understanding of the very old global processes that precede the well-understood tectonic development of more recent geologic history.

Rodger E. Denison

CALCIUM CARBONATE MINERALS

Calcium carbonate minerals in marine sediments are receiving increased attention due to their ability to buffer the impact on the environment of fossil fuel–derived carbon dioxide. Because of these efforts, established theories are being challenged. A renewed recognition of the importance of the different chemical properties of these highly reactive minerals is resulting in a better understanding of the behavior of these minerals and the factors that control their accumulation in sediments.

Shallow water. There is generally a greater diversity of calcium carbonate minerals in shallow-water marine sediments than in deep-sea sediments. Aragonite (orthorhombic $CaCO_3$) is usually the most abundant carbonate mineral in shallow-water sediments. Calcite (rhombohedral $CaCO_3$) is the next most abundant shallow-water carbonate mineral. Calcite is subdivided into two types based on magnesium content. The dividing line between high- and low-magnesian calcites is at 4 mol % $MgCO_3$. However, the composition of these two types of calcite in shallow-water sediments is generally distinctly different, with low-magnesian calcites often having less than 1 mol % $MgCO_3$ while high-magnesian calcites generally have 10 mol % or more $MgCO_3$. The two types of calcite are often referred to simply as calcite and magnesian calcite. Magnesian calcites can contain in excess of 20 mol % $MgCO_3$. A third polymorph of calcium carbonate, vaterite (hexagonal $CaCO_3$), is rare in sediments. Dolomite [$MgCa(CO_3)_2$] is found in a number of environments as a minor component of the sediment. It is the thermodynamically most stable carbonate phase in sea water and constitutes a major portion of ancient sedimentary carbonate rocks. Most of the nondetrital dolomite, which is formed from marine waters, is protodolomite. Protodolomite contains an excess of calcium and is not ideally ordered.

Aragonite. In shallow-water sediments, aragonite is derived from both organisms and nonbiogenic processes. Many macroscopic aragonitic components of sediments such as coral, shells, and algal plates are clearly derived from organisms. Small (about 0.04 in. or 1 mm) spherical aragonite grains, called oöids, and aragonitic cements are generally believed to be of nonbiogenic origin. However, fine (about 2 micrometers) aragonite needles are of less certain origin. The mode of their formation has been of considerable controversy, and a great deal of research has been directed at solving this problem. From the mid-1960s to the early 1980s, fine-grained aragonite needles were generally accepted as being of algal origin. More recent studies, in which associated organic matter was analyzed, have indicated that a major portion of the aragonite needles may be of inorganic origin.

An associated problem is so-called whitings. These are patches of water containing substantial amounts of calcium carbonate which appear white relative to the surrounding water. The suspended calcium carbonate is predominantly aragonite needles. Whitings have long been held to be the result of fishes stirring up carbonate sediments from the sea floor in shallow waters. More recent studies have brought this theory into question. It now appears more likely that the calcium carbonate is suspended by other processes or forms from direct precipitation of aragonite from the water.

Magnesian calcite. This type of calcite is often the second most abundant carbonate mineral in shallow-water sediments. Magnesian calcites can also be formed by organisms (for example, sea urchins or benthic forminifera) and by inorganic processes. The inorganically precipitated magnesian calcites occur as intergranular cements and oöids. The magnesian calcite oöids are different from the aragonitic oöids in that they are radially fibrous, while the aragonitic oöids are usually composed of tangentially oriented needles.

A major problem in understanding the mechanism of magnesian calcite formation in marine sediments has been that the natural magnesian calcites contain about two times too much magnesium, on the basis of careful laboratory studies of the precipitation reactions of these magnesian calcites. This has led to speculation that they may not be of nonbiogenic origin or that adsorbed organic material may seriously alter their chemistry in natural systems.

Another major problem with understanding the behavior of magnesian calcites has been the determination of the influence of magnesian concentration on solubility. It has long been known that calcite solubility increases with increasing magnesium concentration, and that magnesian calcites with high magnesium concentrations are more soluble than aragonite. However, recent studies have indicated that the influence of magnesium on solubility is considerably less than previously believed. New data indicate that aragonite and magnesian calcite containing about 12 mol % $MgCO_3$ have the same solubility.

Low-magnesian calcites found in marine sediments (for example, mollusk shells or barnacles) are usually of biogenic origin. In most sediments they are only a very minor component. However, low-magnesian calcite is the major product of the early diagenesis of marine carbonates, and the most abundant form of calcium carbonate in sedimentary rocks.

Dolomite. This mineral forms by nonbiogenic processes in shallow-water marine sediments. Even when present, dolomite is usually only a very minor component of the sediment. It precipitates at very slow rates and is often associated with special environments such as algal mats and hypersaline lagoons. Since sea water is about 200 times supersaturated with respect to dolomite, its lack of common occurrence in recent sediments has been an area of active study. One recent investigation involving sediment samples from both the Deep Sea Drilling Project and laboratory experiments indicated that sulfate, which is a major component of sea salt, can severely interfere with dolomite formation.

Pore waters. A major question about the behavior of carbonate minerals in sediments is what controls the chemistry of the pore waters with which they are in contact. Until recently it was assumed that the pore waters were in equilibrium with low-magnesian calcite, which is more stable than high-magnesian calcite or aragonite. A reexamination of older data and new studies in the Bahamas have shown that this is not generally true and that a wide variety of pore-water compositions are possible. Most pore waters appear to be in a dynamic equilibrium with high-magnesian calcites. The probable reason for this is that adsorbed organic matter, such as humic acids, strongly inhibits precipitation of the more stable phases. If it were not for this inhibition, most calcium carbonate–rich sediments would rapidly become cemented into a limestone rock. The inhibition is also believed to be the reason that calcium carbonate–poor shallow water can reach very high supersaturations with respect to carbonate minerals.

Deep sea. The sources, mode of deposition, and factors controlling the accumulation of calcium carbonate minerals in deep-sea sediments are distinctly different than those for shallow-water sediments. For most deep-sea sediments, the calcium carbonate is monomineralic, being composed of low-magnesian calcite (< 1 mol %). This calcite is predominantly derived from pelagic animals (foraminifera) and plants (coccoliths). Aragonite, primarily in the form of pteropod and heteropod shells, is found in sediments of intermediate depth.

High-magnesian calcites have been observed in sediments flanking seamounts (for example, Bermuda) and carbonate banks (for example, the Bahama Banks). This material is washed off the shallow banks during storms such as hurricanes. The hydrated calcium carbonate mineral ikaite ($CaCO_3 \cdot 6H_2O$) has been observed in Antarctic sediments for the Bransfield Straight. The formation of this mineral is favored by subzero temperatures

and active microbiologic decomposition of organic matter in sediments. Recent studies of the composition of pore waters from deep-sea sediments indicate that upon burial reduced manganese may interact with the calcite to produce a mixed calcium and manganese carbonate with lower solubility than pure calcite; in addition, in some sediments rhodochrosite ($MnCO_3$) may dominate the pore-water chemistry of the sediment.

The occurrence of calcium carbonate minerals in deep-sea sediments is strongly influenced by the chemistry of the overlying water. The saturation state of deep-sea waters with respect to calcium carbonate minerals is primarily influenced by the addition of carbon dioxide from the decomposition of organic matter and pressure. Temperature has a secondary influence. The general pattern of calcium carbonate mineral occurrence in deep-sea sediments is presented in **Fig. 4**.

A major controversy existed for many years over whether calcite could accumulate in sediments below the depth at which the overlying water went undersaturated with respect to calcite. However, there is now a general consensus that this occurs and that calcite accumulation in deep-sea sediments is governed by complex reaction kinetics. The case for aragonite is less certain since recent work indicates that in many areas the saturation depth for aragonite and the depth at which it is no longer observed in sediments are close to coincident. A recent complicating hypothesis is that the decay of organic matter near the sediment–water interface can strongly influence these processes.

For background information SEE CARBONATE MINERALS; MARINE SEDIMENTS; ROCK AGE DETERMINATION; SEA WATER in the McGraw-Hill Encyclopedia of Science and Technology.

John W. Morse

Bibliography. R. Andersen and A. Malahoff (eds.), *The Fate of Fossil Fuel CO₂ in the Oceans*, 1977; R. G. C. Bathurst, *Carbonate Sediments and Their Diagenesis*, 1975; W. H. Burke et al., Variation of seawater ⁸⁷Sr/⁸⁶Sr throughout Phanerozoic time, *Geology*, 10:516–519, 1982; H. E. Cook and P. Enos (eds.), *Deep-Water Carbonate Environments*, based on a Symposium sponsored by the Society of Economic Paleontologists and Mineralogists, Spec. Publ. 25, 1977; R. B. Koepnick et al., Construction of seawater ⁸⁷Sr/⁸⁶Sr curve for the Cenozoic and Cretaceous: Supporting data, *Chem. Geol.* (Isotope Geoscience section), 58:58–81, 1985; J. D. Milliman, *Marine Carbonates*, pt. 1: *Recent Sedimentary Carbonates*, 1974; Z. E. Peterman, C. E. Hedge, and H. A. Tourtelot, Isotopic composition of strontium in seawater throughout Phanerozoic time, *Geochim. Cosmochim. Acta*, 34:105–120, 1970; F. E. Wickman, Isotope ratios: A clue to the age of certain marine sediments, *J. Geol.*, 56:61–66, 1948.

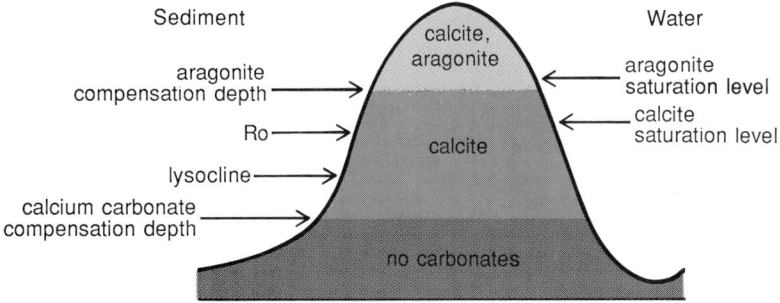

Fig. 4. Distribution of calcium carbonate minerals in deep-sea sediments and their relation to water chemistry. At the aragonite compensation depth, aragonite no longer accumulates in the sediment. Ro indicates the depth at which definite signs of calcite dissolution first appear. The lysocline is the depth at which major dissolution of calcite takes place.

Medical imaging

Applied potential tomography is a method which enables images of the electrical impedance of tissues to be produced. It involves the application of potentials to the body via skin electrodes. The currents which flow give rise to measurable potentials elsewhere on the body, and from these the impedance of the organs and tissues can be determined. The technique is new and there is no consensus as to the best name; alternative terms are electrical impedance tomography, conductivity imaging, impedance computed tomography, and electrical impedance tomography. Although tissue impedance has both resistive and capacitive components, the currents which can flow as a result of the capacitive components are only significant at high frequencies. Present applied potential tomography systems use relatively low frequencies, typically 50 kHz, and neglect capacitive components. Images are therefore of tissue resistivity.

A major attraction of applied potential tomography arises from the fact that the electrical resistivity of body tissue varies widely. This fact has been known for many years, but the technical and computational problems involved in mapping tissue resistivity were considered too difficult for this variation to be of any practical value. The body is an ionic conductor, and hence the resistivity of particular organs depends upon electrolyte concentrations and varies by more than an order of magnitude for soft tissues. Typical values for tissue resistivity are: blood, 1.6 ohmmeter; muscle, 5.0 Ωm; fat, 20 Ωm; bone, 160 Ωm. This wide range of tissue parameters is in marked contrast to the very small range of x-ray linear attenuation coefficients which are used for x-ray imaging. Because of this wide resistivity range, applied potential tomography gives good contrast separation of soft tissues, and it has the additional advantages of being much cheaper than x-ray computerized tomography and foregoing hazardous ionizing radiation.

On the other hand, x-ray computerized tomography gives excellent resolution and, because x-rays travel in straight lines, the computation of tomographic images is relatively easy. An electric current spreads throughout the body, however, and so provides a major problem in producing good-impedance images. It is only since 1983 that images of living organisms have been obtained, and it seems unlikely that applied potential tomography will ever offer the high-resolution images that x-ray computerized tomography and magnetic resonance imaging can produce.

Principles. Nerves and muscles within the body carry electrical impulses and can therefore be disturbed by an externally applied electric current. Alternating currents can be used to stimulate neural tissue, and it is found that the optimal frequency is around 50 Hz. However, because nerves take some time to initiate an impulse, the body is relatively insensitive to high-frequency currents. Alternating currents at frequencies above about 40 kHz can be used to probe the body without any risk of interfering with the heart, nerves, and muscles.

If a potential is applied between two electrodes placed on the skin, a current will flow throughout the body. This current will give rise to a potential distribution over the whole surface of the body. By using Poisson's equation below, where V is

$$\nabla \cdot (C \nabla V) = 0$$

the potential distribution and C is the distribution of conductivity, this potential distribution can be calculated, but the problem is not an easy one. It is necessary to know the position and resistivity of all the tissues within the body. This problem is often referred to as the forward problem. If the drive potential is applied between two electrodes elsewhere on the body, the potential distribution on the body surface will also change.

Electrical impedance imaging involves the inverse problem, that is, measuring the potential distribution over the surface of the body, and calculating the tissue resistivity distribution within the body from these data. In practice the calculation is difficult and requires more information on the potential distribution over the body surface than it is practical to record.

In **Fig. 1** the problem is shown in just two dimensions for a circular object. Sixteen electrodes are placed around the edge of the object, and a current I is passed between the first two electrodes. This current will give rise to a potential between all the remaining electrodes, and to illustrate this the isopotentials have been drawn. If the resistivity distribution within the object changes, the peripheral potentials will change.

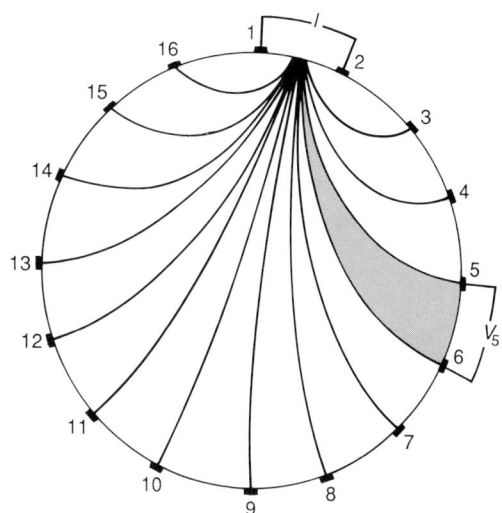

Fig. 1. Method of applied potential tomography in two dimensions for a circular object. Potential measurement V_5 is used to modify the assumed resistivity of the shaded area between curved isopotentials. By making modifications for many potential measurements, an image is built up by superposition.

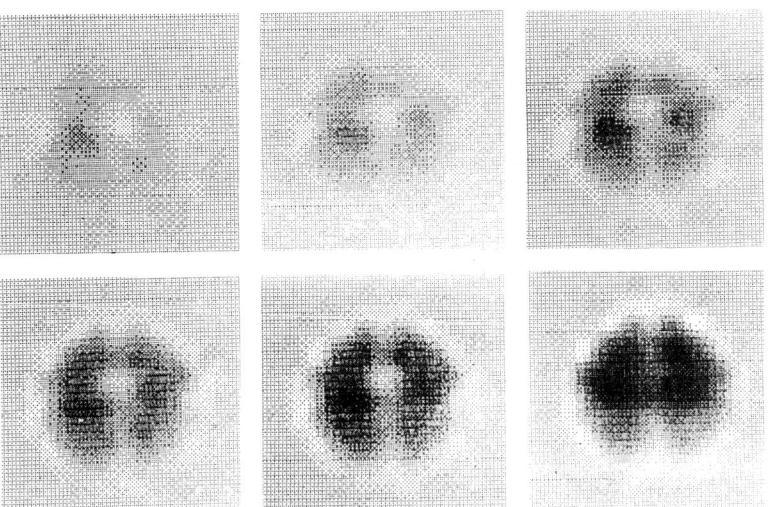

Fig. 2. Six lung images (at six levels of inspiration) produced from data collected from 16 electrodes placed around the thorax. In each case the front of the body is at the bottom and the right side is shown on the left.

This simplified description indicates how an image is produced; in practice various weighting factors and filters are used to improve the quality. In **Fig. 2** a series of lung images is shown. By comparing the resistivity distribution before and after taking a breath, an image of the lungs can be obtained. Sixteen electrodes placed around the chest provide a series of images for different volumes of inspired air which show the two lungs as high-resistivity areas. The images are not very sharp (as they are limited by the use of only 16 electrodes), but the changes in resistivity as air enters the lungs can certainly be seen. In principle the spatial resolution could be improved with more electrodes, but the practical and computational problems are considerable.

Advantages and disadvantages. Applied potential tomography has a number of advantages over other imaging modalities. These include good tissue contrast, high sensitivity to changes in resistivity, rapid data collection, relatively low cost, and the absence of any known hazard. However, the method has a number of disadvantages, which include poor spatial resolution and the need for surface electrodes.

The significance of any particular advantage or disadvantage depends upon the context. In order to localize a small tumor, high spatial resolution is required, but if it is desired to investigate the perfusion of the lungs with blood when the heart contracts, then high sensitivity and rapid data collection are much more important than spatial resolution.

Now the potential V_5 will depend upon the whole of this resistivity distribution, but the greatest sensitivity will be to changes in the area shown shaded between the isopotentials which end on electrodes 5 and 6. This is the basis of the method of applied potential tomography proposed by D. C. Barber and B. H. Brown. The initial assumption is that the resistivity distribution is uniform. The potential measurement V_5 is then used to modify the assumed resistivity of the shaded area between the curved isopotentials. If V_5 is higher than predicted for a uniform distribution, the assumed resistivity is increased, and, if it is lower, the assumed resistivity is reduced. This process is repeated for potentials recorded between all other adjacent electrode pairs and for current applied between all adjacent electrode pairs. This method is often referred to as one of back projection because it involves measuring a peripheral potential and back-projecting this measurement along curved lines into the image. With 16 electrodes, 208 (13 × 16) measurements can be made and used to build up an image.

The only published images of living organisms using applied potential tomography are from a 16-electrode system which collects 10 images each second. This system can show changes in resistivity of approximately 1% and offers a spatial resolution of 5–10% of the diameter of the body. Considerable effort is being devoted to improving the system, but a slow improvement is probable, rather than any sudden leap forward. More rapid data collection is certainly possible by making potential measurements from many pairs of electrodes at the same time. Sensitivity is more difficult to improve unless a larger drive current is used. Present systems use currents of about 1 mA. There are no technical problems attached to increasing this, but for currents above 10 mA, tissue heating and the possibility of biological interactions may become significant.

It has been suggested that by collecting data at two or more frequencies information on specific organs may be collected. For example, the lungs show a significant change in impedance between 50 and 100 kHz. By collecting data at the two frequencies and back-projecting one against the other, it has been suggested that the lungs could be imaged. This would result in a static image of the lungs as opposed to the images of Fig. 2, which show only the changes in resistivity during respiration.

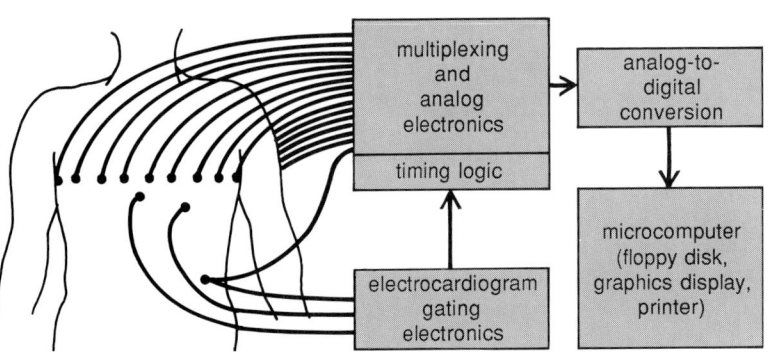

Fig. 3. Block diagram of 16-electrode electrocardiogram-gated applied potential tomography imaging system which allows cardiac gated images to be produced.

A radical change to avoid the problems of electrodes would be to induce currents magnetically and to record the perturbations caused by tissue currents. This possibility has been investigated but has not yet been shown to be feasible.

Clinical applications. At this early stage in development when commercial equipment is only just becoming available for research, clinical experience is very limited. All the applications developed so far exploit the ability of applied potential tomography to show changes in resistivity. By placing electrodes around the abdomen, the changes in resistivity of stomach contents following a meal or following administration of a pharmaceutical which stimulates acid secretion can be investigated. This method of investigating dysfunctions of gastric emptying and acid secretion has considerable advantages over existing methods which rely upon the use of radioactive labels and tubes passed into the stomach.

Lung imaging is relatively easy because lung resistivity rises when air is inhaled. This increase is due to the electric current having to flow around the thousands of air-filled alveoli. These images should indicate the presence of fluid within the lungs because the resistivity should fall rapidly with increasing fluid content. Fluid on the lungs (pulmonary edema) is a common consequence of both cardiovascular and respiratory diseases but is difficult to detect by any other means.

By gating the applied potential tomography data collection with an electrocardiogram (**Fig. 3**), it has been possible to image the changes in resistivity which occur when the heart contracts. Blood has a relatively low resistivity; hence when the heart ejects blood which then perfuses the lungs, the resistivity of the heart increases and that of the lungs decreases. Also when major arteries dilate on cardiac contraction, their resistivity falls. These cardiac-gated images can be displayed at normal heart rate and show the perfusion of the lungs with blood. Pulmonary embolism might well be investigated by this means. It is also possible, although not yet proven, that the cardiac output of blood can be calculated from these images.

Yet another clinical application under investigation is that of applied potential tomography to measure temperature changes during hyperthermia. In hyperthermia, localized heating, often induced by microwaves, is used to destroy a tumor. However, in order to destroy the tumor a temperature greater than 109°F (43°C) must be produced without undue heating of the surrounding normal tissue. Measurement of the temperature distribution is difficult, and applied potential tomography is being investigated as a means of accomplishing this. The resistivity of tissue changes by approximately 1% per °F (2% per °C), and these changes can be detected by using a ring of applied potential tomography electrodes placed around the tumor.

Applied potential tomography is unlikely to compete with high-quality medical imaging using x-ray computerized tomography or magnetic resonance imaging. However, its advantages of high speed and good sensitivity at low cost, combined with an ability to give information on physiological function, should find it a place in medical diagnosis.

For background information SEE COMPUTERIZED TOMOGRAPHY; MEDICAL IMAGING; POTENTIALS in the McGraw-Hill Encyclopedia of Science and Technology.

Brian H. Brown

Bibliography. D. C. Barber and B. H. Brown, Applied potential tomography, *J. Phys. E: Sci. Instrum.*, 17:723–733, 1984; D. C. Barber and B. H. Brown, Recent developments in applied potential tomography—APT, *Information Processing in Medical Imaging: Proceedings of the 9th Conference*, Washington, D.C., pp. 106–121, 1985; B. H. Brown, D. C. Barber, and A. D. Seagar, Applied potential tomography: Possible clinical applications, *Clin. Phys. Physiol. Meas.*, 6(2):109–121, 1985; B. H. Brown, D. C. Barber, and L. Tarrasenko (eds.), Electrical Impedance Tomography—Applied Potential Tomography: Proceedings of the European Workshop, *Clin. Phys. Physiol. Meas.*, vol. 8, suppl. A, 1987.

Medical parasitology

Recent advances in the field of medical parasitology have involved development of specific deoxyribonucleic acid (DNA) probes for use in diagnosing certain parasitic infections, and major improvements in chemotherapy for parasitic diseases.

DNA PROBES

Despite considerable efforts to control the major parasitic infections of humans (for example, malaria), the number of afflicted individuals has increased. New strategies are urgently needed to diagnose these diseases. These strategies will facilitate application of appropriate treatment and provide a means for measuring the efficacy of disease control programs. Achievement of these objectives will require development of sensitive, simple, and inexpensive means of identifying parasites.

A DNA probe is a labeled DNA molecule used in DNA hybridization experiments to locate target DNA or RNA whose sequence is complementary to that of the probe DNA. Target DNA must be single-stranded in order for the probe DNA to hybridize. The target DNA or RNA can either be immobilized on a support (such as nitrocellulose) or can be in solution. The probe DNA can either be radiolabeled or be linked to an enzyme. Recent advances in the development of specific DNA probes for detecting parasite DNA have allowed introduction of new methods for diagnosing certain parasitic infections.

Rationale. Use of DNA probes to detect parasitic infection is based upon a few general biological properties of DNA molecules. All living organisms contain DNA, which encodes the information necessary for cell development and replication. This information is determined by the sequential order of the four nucleotide bases which form the backbone of the DNA molecule. Biological species are differentiated by the variation in the sequence of these nucleotides. Each organism contains sequences of DNA which are unique to that species, sequences which are used in development of specific DNA probes.

An important feature of the DNA molecule concerns the pairing of nucleotides. Two strands of DNA are joined through base pairing to form a double-stranded molecule, which resembles a ladder. The ''rungs'' or base pairs of DNA are formed by chemical bonds between nucleotide bases on opposite strands of DNA. These bonds can form only between complementary pairs of nucleotide bases; thus, adenosine (A) always pairs with thymidine (T), while cytidine (C) always pairs with guanosine (G). The bonds are relatively weak, and the two strands of DNA can be separated in a process called denaturation. Under appropriate conditions the denatured single strands can join again to form a double-stranded molecule, provided the correct base-pair matches are formed between the strands.

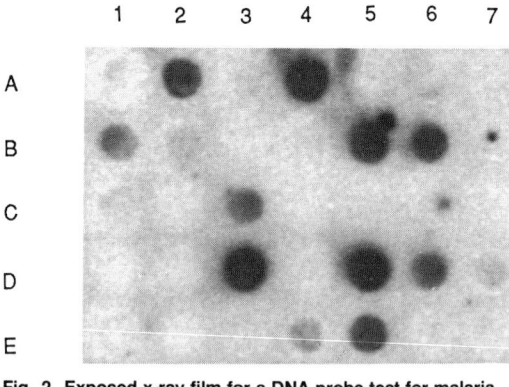

Fig. 2. Exposed x-ray film for a DNA probe test for malaria infection. Letters and numbers indicate grid divisions.

The process by which separate strands of DNA come together to form a double strand is called DNA hybridization. If the order of bases is incorrect (that is, if the pairing yields mismatches), hybridization cannot occur.

DNA hybridization and sequence specificity form the basis for use of DNA probes in diagnosis. **Figure 1** illustrates the general method. A sample of material which may contain parasites is spotted onto filter paper (Fig. 1*a*). In the case of malaria, which infects erythrocytes (red blood cells), the sample consists of a few drops of blood. If the sample could be magnified sufficiently, the double-stranded DNA molecules could be seen lying on the filter (Fig. 1*b*). The DNA is then denatured into single-component strands (Fig. 1*c*). Next, single-stranded probe DNA which has been labeled by incorporation of radioactive atoms is added to the structure of the probe DNA. The probe is then allowed to hybridize with the DNA on the filter. If the sample on the filter contains parasite DNA, it will bind with the probe DNA to become double-stranded (Fig. 1*d*). Then the filter is washed to remove unbound probe DNA and is exposed to x-ray film. Wherever the DNA probe has hybridized to parasite DNA, the x-ray film will be exposed (Fig. 1*e*).

Results from a typical test for malaria infection are shown in **Fig. 2**. Here the DNA probe was specific for *Plasmodium falciparum*, a major malaria-causing species in humans, but not for *P. vivax*, which also causes human malaria. Wherever a sample contained *P. falciparum* DNA, the probe hybridized, and this was detected by an exposed spot on the x-ray film. Samples which did not contain *P. falciparum* parasites (for example, at position A7) did not hybridize with the *P. falciparum*–specific DNA probe. In order to detect other malaria species, such as *P. vivax*, it would be necessary to retest the samples by using a DNA probe specific for those species.

This kind of test also shows how the technology can be used to quantitate the level of infection. The more parasites present in a sample, the more DNA they will have, which in turn will bind

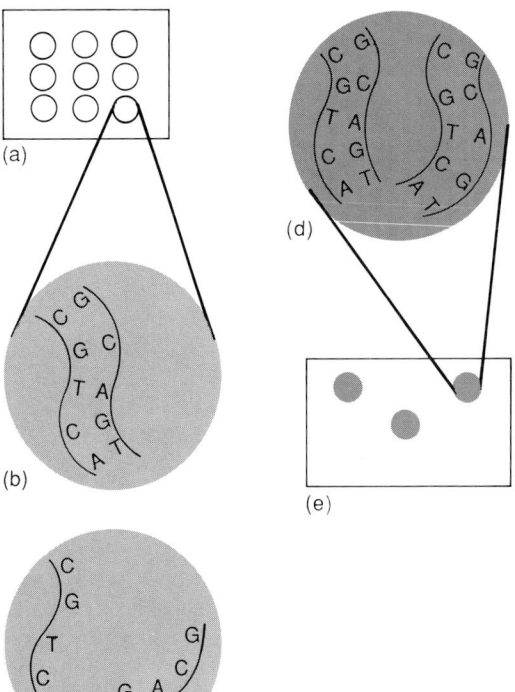

Fig. 1. Diagram of general method for using a DNA probe in diagnosis. (*a*) Samples on filter. (*b*) Expanded view of a single sample containing double-stranded DNA molecules. (*c*) DNA separated into single strands following denaturation. (*d*) Samples hybridized with radiolabeled probe. (*e*) Dark spots on exposed x-ray film indicating DNA homologous with probe DNA. A = adenosine. T = thymidine. C = cytidine, G = guanosine.

more probe, yielding darker spots on the film (Fig. 2, positions A4 and D3). Microscopic examination of blood indicated that these samples contained about 3000 parasites per microliter of blood. By contrast, sample D7 (Fig. 2) contained only about 40 parasites per microliter. Thus DNA hybridization can be used both to identify the species of infecting parasite and to quantitate the intensity of infection.

Preparation. Where does the probe come from? Although the details vary for development of specific DNA probes for different species, the general strategy is shown in **Fig. 3**. Total DNA is extracted from the species of interest (in this case, *P. falciparum*). With recombinant DNA technology, the malaria DNA (Fig. 3*a*) is cut with restriction enzymes into short pieces of random length (Fig. 3*b*) and introduced into bacterial plasmids (Fig. 3*c*). Plasmids are extrachromosomal pieces of bacterial DNA which can be readily isolated from, or reintroduced into, host bacteria. As a host bacterial cell divides, it replicates plasmid DNA in addition to host chromosomal DNA. Host bacteria can also be induced to replicate plasmids independently, thereby accumulating many copies of the plasmid DNA. Thus the bacteria are used as factories for producing large amounts of plasmid DNA. If the plasmid contains an extra piece of DNA (as in malaria DNA), the bacteria will produce large amounts of that sequence of DNA.

In order to select a bacterial colony producing the appropriate DNA for a species-specific DNA probe, duplicate bacterial colonies are hybridized with radiolabeled total DNA from either *P. falciparum* (Fig. 3*d*) or *P. vivax* (Fig. 3*e*). In some cases, bacterial plasmids are hybridized with DNA from both species (for example, positions A1, A2, B1, B2). This is not surprising, since the two species are closely related and share many conserved nucleotide sequences. However, in

two cases (positions B3, C2), DNA from *P. vivax* did not hybridize (Fig. 3*e*), whereas DNA from *P. falciparum* did (Fig. 3*d*). Therefore, these bacterial colonies contain inserted sequences of *P. falciparum* DNA which are specific for *P. falciparum* parasites. In practice, it is necessary to analyze further these candidate DNA probe sequences to be certain that they do not hybridize with DNA from other potentially contaminating species (in this case, human DNA and DNA from other infectious agents potentially found in human blood). In addition, it is necessary to test the sensitivity of the probe DNA to discover the minimum number of parasites which are detectable. This sensitivity will vary, depending upon the number of copies of the encoded sequence contained in the target (malarial) genome. Some sequences are repeated many times and are more useful as probes; for example, a DNA probe containing a sequence encoded 10 times within the malarial genome can be used to detect 10 times fewer parasites than a probe containing a sequence which is encoded only once.

Thus far the discussion has focused on the DNA probe hybridization assay and the general strategy for developing specific DNA probes by using recombinant DNA technology. While the details differ for various species, the general approach is similar throughout; it has yielded probes which are highly sensitive, specific, and simple to use.

Applications. Many questions can be answered through specific DNA probes. Clearly a major use lies in the clinical identification of infectious agents. At present, most clinical diagnosis of malaria infection is done by microscopic examination of blood. This procedure requires a highly trained technician to identify correctly the malarial species present. This method is time-consuming, especially when small numbers of parasites are present. A good microscopist can probably examine no more than 70 slides per day, whereas a technician can process 2000 samples per day by DNA hybridization, a clear advantage in processing a large number of samples. Accurate species diagnosis is important because different malarial species can be treated specifically with different drugs. *Plasmodium falciparum* has evolved many strains which are drug-resistant and therefore can be treated successfully with only a limited range of drugs. By limiting drug use to people who are actually infected with a sensitive species, the rate at which drug resistance develops can be substantially reduced.

A further use of DNA probes lies in their value for epidemiological work. Changing agricultural practices are causing deforestation in many parts of the world, with concomitant change in the ecology. As a result, the distribution of vector mosquitoes which can transmit malaria has been altered. One strategy for malaria control involves use of pesticides to limit the spread of mosquitoes. Large-scale screening of the population at

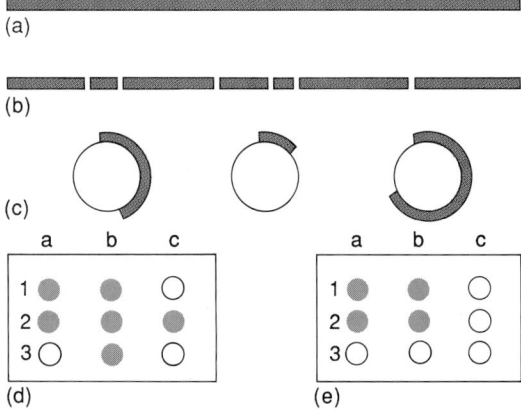

Fig. 3. Diagram of preparation of specific DNA probes. (*a*) Malaria DNA. (*b*) DNA after cutting with restriction enzymes. (*c*) Pieces of DNA inserted into bacterial plasmids. (*d*) Exposed x-ray film with dark spots indicating *Plasmodium falciparum* and (*e*) *P. vivax*; letters and numbers indicate grid divisions.

risk for malarial infection following use of insectides can be useful in measuring whether mosquito control effectively reduces infection in a given area.

Large-scale screening also aids in the development of an antimalarial vaccine. Although a vaccine is currently used in limited clinical trials, it is anticipated that before general use the vaccine will be tested in areas of high or moderate malaria risk. This will require constant monitoring of the study population to establish base-line levels of infection at the outset, to establish the rate at which vaccinated and nonvaccinated people become infected, and finally to assess the results at the end of the experiment.

Therefore, DNA probes assist clinicians in treatment decisions and help them measure the effectiveness of various control programs. The discussion presented here has focused on use of DNA probes for diagnosing *P. falciparum* malaria infection, but the same strategies have been employed to develop the appropriate assays for diagnosing other parasitic infections as well.

R. H. Barker, Jr.

ANTIPARASITIC DRUGS

Since about 1980, there have been a number of significant developments in the field of antiparasitic drugs. These include nematocidal drugs, trematocidal drugs, cestocidal drugs, ectoparasiticides, antimalarial drugs, trypanosomicidal drugs, and anticoccidial drugs.

Nematocidal drugs. During the late 1970s, eight benzimidazole carbamate drugs became available as veterinary anthelmintics: albendazole, cyclobendazole, fenbendazole, flubendazole, mebendazole, oxibendazole, oxfendazole, and parbendazole. They all have general structure (I), with a substi-

(I)

tuted functional group at position 5 (indicated by R). The phenylguanidine febantel and netobimin are transformed metabolically to benzimidazole carbamates. All benzimidazoles are highly effective against intestinal nematodes. They differ in their efficacy against lungworms (*Dictyocaulus*) and nematode larvae. In addition to mebendazole, other benzimidazoles are finding medical application. Recently, flubendazole and albendazole have been tested clinically and are available as human anthelmintics. Albendazole may become an important single-dose broad-spectrum oral nematocidal drug, and flubendazole has properties very similar to those of mebendazole and requires repeated administration.

Progress in nematocidal drugs has been in the direction of improving the ease and techniques of application. In the "pour-on" application, 10 to 30 ml of levamisole solution is poured onto the backs of cattle. The drug penetrates the skin and then acts systemically against lungworms.

Controlled release of low levels of anthelmintics, another new development, provides an opportunity to increase the convenience and certainty of effective treatment. A bolus introduced into the rumen releases morantel over a period of 2 months and provides an extended period of protection against reinfection for cattle turned out to pasture. A pulsed-release bolus which releases five successive 750-mg doses of oxfendazole to cattle at 3-week intervals through galvanic erosion of a steel-magnesium core was introduced in Europe in 1986.

Since the mid-1970s the main achievement has been the successful development of a potent nematocidal agent, ivermectin (II), a macrocyclic

(II)

lactone produced by an actinomycete. Ivermectin affects a variety of nematodes and arthropod parasites, apparently by its action on the mediation of neurotransmission by gamma aminobutyric acid (GABA). In invertebrates, ivermectin stimulates inhibitory GABA-ergic neurons (which employ GABA as a chemical transmitter) or neuromuscular synapses, lowers muscle membrane resistance, and causes an increase in chloride ion permeability. These effects are lethal to nematodes and arthropods but are not toxic to mammals because GABA-ergic neurons are restricted to the central nervous system, which is protected by the blood-brain barrier.

Ivermectin is effective against virtually all intestinal nematodes and their larvae at oral or subcutaneous doses 10 to 100 times less than those required for the benzimidazoles. It has been used since 1983 as a veterinary anthelmintic in many countries.

Ivermectin is also effective against the microfilariae stage of tissue-dwelling nematodes, the filariae, but does not appear to affect the adults. Only preliminary studies have been made using this drug in onchocerciasis patients. Whether the drug will also be useful in human medicine is not yet known.

Trematocidal drugs. Since the mid-1970s, substantial progress has been made in the chemotherapy of trematode infections. In the late 1970s, new injectable formulations of niclofolan and rafoxanide, two drugs that formerly had to be administered orally, were developed to treat cattle for liver fluke.

Closantel has been available since 1980. It is an excellent fasciolicide and is also effective against hematophagous nematodes, some parasitic maggots, cattle ticks, and mange mites. Closantel is a potent and selective uncoupler of mitochondrial phosphorylation in *Fasciola*.

A recent development is the benzenedisulfonamide clorsulon, registered in the United States in 1985 for use against liver fluke in cattle. It must be given orally and is almost equally effective against 8-week-old and adult liver flukes. Clorsulon inhibits two enzymes of the glycolytic pathway (3-phosphoglycerate mutase and phosphoglucomutase) of liver fluke, thus interfering selectively with energy formation from glucose. Another new fasciolicide, first marketed in Ireland in 1983, is the benzimidazole derivative triclabendazole. It is virtually as effective against immature flukes as against mature ones and possesses a safety index (maximal tolerated dose to effective dose) of 20 in sheep.

Beginning in 1980, chemotherapy of human trematode infections has been improved considerably by praziquantel. This pyrazinoisoquinoline is effective against all trematode parasites of humans and animals. In particular, praziquantel is effective against all species of blood fluke (*Schistosoma*) affecting humans, all intestinal flukes, human liver flukes (*Clonorchis*, *Opisthorchis* and also *Fasciola*), and lung flukes. Praziquantel is well tolerated, is given orally, and is highly effective against virtually all trematodes at a single dose or up to a 2-day course of treatment. Praziquantel obviates the use of older and less tolerated schistosomicidal drugs and provides the first effective treatment of human liver fluke infections. Praziquantel acts selectively on the surface membranes of trematodes by increasing their permeability to calcium ions (Ca^{2+}). The resulting influx of Ca^{2+} causes rapid structural damage to the surface of the parasite, and musculature contraction and paralysis occurs within seconds after contact with the drug. Subsequently, and more slowly, regulatory processes within the parasites are deranged, culminating in death of the parasites.

Cestocidal drugs. The discovery of praziquantel has also changed the chemotherapy of cestode infections. It became available as a cestocide for dogs and cats in 1977, and its medical use for humans began in 1980. A single dose is fully effective against all intestinal infections caused by adult tapeworms, including the dwarf tapeworm *Hymenolepis nana*, and in veterinary medicine *Dipylidium*, *Mesocestoides*, and, most important, *Echinococcus* infections of dogs. The mode of action of praziquantel against cestodes is identical to that against trematodes.

Infections with larval tapeworms were incurable and often fatal until praziquantel provided the first reliable chemotherapeutic tool for the cure of cysticercosis. For a second type of larval tapeworm infection, hydatidosis, which is caused by the larva of the dog tapeworm *Echinococcus*, high doses of benzimidazoles (mebendazole, flubendazole, albendazole) are being used in clinical trials. Even with treatment over many months, it has not been possible to effect a complete cure, although improvement can be achieved. Praziquantel is ineffective against hydatidosis but successfully eliminates *Echinococcus* from dogs, the source of human infection in sheep-rearing countries.

Ectoparasiticides. Until the 1970s, ectoparasite control had to rely on the application of organophosphates, carbamates, and (for tick control) amidines. During the 1970s, light-stable derivatives of the naturally occurring ingredients of pyrethrum extract have been synthesized by substitution of the naturally occurring pentyl alcohol by 3-phenoxybenzyl alcohol.

Permethrin, the first of these new synthetic pyrethroids, was a potent insecticide. Incorporation of an α-cyano group into the 3-phenylbenzyl alcohol further improved the usefulness of the next generation of synthetic pyrethroids (cypermethrin, deltamethrin, and fenvalerate), with which it became possible to control ticks in addition to flies. The latest synthetic pyrethroids have a fluorosubstitution (cyhalothrin, flucythrinate, cyfluthrin). The latest pyrethroid is flumethrin, a potent tickicide when used conventionally in a spray or dip. Beyond that, flumethrin is effective as a dermal application to cattle, thus obviating the need of moving cattle to the dip or spray, which is costly and laborious.

The synthetic pyrethroids break the resistance mechanisms which insects and ticks have acquired against organophosphates, carbamates, and amidines; however, resistance to pyrethroids is already a problem, especially in fly control. Synthetic pyrethroids possess a greater safety margin than the conventional insecticides. They became available commercially about 1980. Pyrethroids act very rapidly, affecting the nervous system of insects and acarines. There is no general agreement yet on what happens in detail after target organisms have been treated with pyrethroids.

Ivermectin (II) is not only an excellent veterinary anthelmintic but also a very good ectoparasiticide. At nematocidal doses it is very effective against parasites which have a close and long-lasting contact with the treated animal, such as parasitic maggots, mange mites, and sucking lice. It is also effective against the larvae of manure-breeding flies but less effective against ticks. It is effective against ectoparasites that have become resistant to organophosphates, carbamates, amidines, DDT, and pyrethroids.

Antimalarial drugs. For about three decades, chemotherapy and chemoprophylaxis of malaria could be achieved with quinoline derivatives: mainly chloroquine and, to some extent, hydroxychloroquine and amodiaquine. The use of proguanil and pyrimethamine has always been limited because of resistance to these drugs in some areas. In the late 1960s, chloroquine resistance (directed also against amodiaquine) of the most important malaria parasite, *P. flaciparum*, became important in some foci in Latin America and Southeast Asia. A combination of sulfonamide (sulfadoxine) with pyrimethamine has solved the problem temporarily. Resistance to this second-line drug combination is already occurring in the so-called hard-core areas of Southeast Asia and South America. Sulfadoxine, an inhibitor of the dihydropteroate synthetase, blocks the synthesis of folic acid by the parasite. Pyrimethamine acts later in the same pathway by inhibiting the enzyme dihydrofolate reductase.

Mefloquine, marketed in Switzerland in 1984, is effective against all four malaria parasites of humans. It interacts with the parasite's DNA in a way not yet understood. In order to delay the emergence of resistance to mefloquine, a combination of mefloquine with sulfadoxine and pyrimethamine has been developed.

There are two candidate antimalarial drugs under investigation. The phenanthrene derivative halofantrine has been studied intensively in the United States since 1946. Qinghaosu, the active principle of the herb *Artemisia annua*, was isolated in China in 1972. Qinghaosu and two derivatives are under clinical investigation in China and the United States. These trioxane derivatives are active against chloroquine-resistant malaria. They affect the plasma membrane of the parasite and inhibit protein synthesis, but the mode of action is not well understood.

Trypanosomicidal drugs. Treatment of the late stage of sleeping sickness is still unsatisfactory because of the toxicity of the employed arsenic compound Melarsoprol and because of drug resistance. Recently, a structural analog of ornithine, D,L-difluoromethylornithine, has been used to treat more than 100 cases of *Trypanosoma gambiense* infections. This drug inactivates the enzyme ornithine decarboxylase and blocks polyamine synthesis. Side effects seen in the clinical trial, for example, diarrhea and anemia, were reversible.

Nifurtimox, which allowed treatment of Chagas' disease for the first time, was followed by the 2-nitroimidazole benznidazole early in the 1980s. Both drugs inhibit the enzymes superoxide dismutase and peroxidase of the parasites, which then are killed by the accumulation of highly toxic hydrogen peroxide.

Anticoccidial drugs. Monensin, lasalocid, and the recently marketed drugs salinomycin, narasin, and maduramycin are fermentation products of fungi of the genera *Streptomyces* or *Actinomadura* and are used prophylactically in poultry.

The drugs, given in the feed at doses ranging from 110 parts per million (monensin) to 5 ppm (maduramycin), induce ion movement across the cell membranes of the parasite. This interferes with maintenance of the physiological proton gradient required for energy generation by the parasite.

Arprinocide, an adenosine analog, interacts with the purine salvage pathways by inhibition of hypoxanthine transport. The activity is based partly on the metabolic transformation within the host of arprinocide to arprinocide-1-*N*-oxide.

Halofuginone, derived from an extract of the plant *Dichroa febrifuga*, is effective at a level of only 3 ppm in the feed. It is also effective against bovine infections caused by parasitic protozoa of the genus *Theileria*. The biochemical mode of action has not been clarified.

Very recently, it has become possible to treat East Coast fever (theileriosis), the often fatal cattle disease, with parvaquone. The supposed mode of action is an inhibition of electron transport.

For background information SEE DEOXYRIBONUCLEIC ACID *(DNA)*; MALARIA; MEDICAL PARASITOLOGY; PARASITOLOGY in the McGraw-Hill Encyclopedia of Science and Technology.

P. Andrews; A. Haberkorn

Bibliography. R. H. Barker, Jr., et al. Specific DNA probe for the diagnosis of *Plasmodium falciparum* malaria, *Science*, 231:1434–1436, 1986; J. P. Leahey (ed.), *The Pyrethroid Insecticides*, 1985; *The Merck Index*, 10th ed., 1983; W. Peters and W. H. G. Richards (eds.), *Antimalarial Drugs: Handbook of Experimental Pharmacology*, vol. 68 I and II, 1984; United Nations Development Program, World Bank, WHO, *Tropical Disease Research (TDR)*, 7th Programme Report, 1985; H. Vanden Bossche, D. Thienpont, and P. G. Janssens (eds.), *Chemotherapy of Gastrointestinal Helminthis: Handbook of Experimental Pharmacology*, vol. 77, 1985; D. F. Wirth et al., Leishmaniasis and malaria: New tools for epidemiologic analysis, *Science*, 234:975–979, 1986.

Membrane mimetic chemistry

The term membrane mimetic chemistry was coined in 1982 to describe new types of processes and reactions whose developments had been inspired by the biological membrane. Faithful modeling of the biomembrane is not an objective of membrane mimetic chemistry. Rather, only the essential components of natural systems are recreated from relatively simple, synthesized molecules. Mimetic systems are designed for well-defined purposes. The insight gained into the functioning of real membranes is considered a bonus. (The term membrane mimetic is more restrictive than the term biomimetic. Biomimetic chemistry is directed at the mechanistic elucidation of biochemical reactions and at the development of new compounds modeled on, and expected to mimic, specific biological systems. The focal point of most biomimetic chemistry is enzyme modeling.)

Surfactant aggregates. These systems—micelles, monolayers, organized multilayers (Langmuir-Blodgett films), bilayer lipid membranes, and vesicles—have been used, for the most part, in membrane mimetics. Mimicking natural photosynthesis is a good example of the philosophy and approach of membrane mimetic chemistry. Therefore, the use of mimetic systems as hosts (compartments) for semiconductors will be highlighted.

Surfactants, or detergents, contain distinct hydrophobic (apolar) and hydrophilic (polar) regions. Depending on the chemical structure of their hydrophilic polar head groups, surfactants can be neutral, positively charged, or negatively charged. The hydrophobic part can vary in length (typically 8 to 20 carbon atoms), contain multiple bonds, or consist of two or more hydrocarbon chains. Head group-to-tail ratio determines, to a first approximation, the type of aggregates formed from a given surfactant. Various types of aggregates formed from surfactants are shown in **Fig. 1**.

Aqueous micelles. These spherical aggregates, 4–8 nanometers in diameter, are formed dynamically from surfactants in water above a characteristic concentration, the critical micelle concentration. Micelles break up and reform rapidly by

two known processes. The first process occurs on a microsecond time scale and is due to the release of a single surfactant and its subsequent reincorporation in the micelle. The second process occurs on a millisecond time scale and is ascribed to the dissolution of the micelle and to the subsequent reassociation of the monomers. Substrate interaction with the micelle is also dynamic.

Monolayers. These monomolecular layers are formed by spreading naturally occurring lipids or synthetic surfactants, dissolved in volatile solvents, over water in a trough. The polar head groups of the surfactants are in contact with water, the subphase, while their hydrocarbon tails protrude above it. Monolayers are characterized by surface area–surface pressure curves, surface potentials, and surface viscosities.

Techniques have been developed for transferring the monolayer onto a solid support and for building up organized multilayer assemblies (Langmuir-Blodgett films) in controlled topological arrangements. SEE ELECTRONICS.

Bilayer lipid membranes. These are formed by brushing an organic solution of a surfactant (or lipid) across a pinhole (2–4 mm or 0.08–0.16 in. in diameter) separating two aqueous phases. Alternatively, bilayer lipid membranes can be formed from monolayers by the Montal-Mueller method. In this method the surfactant, dissolved in an apolar solvent, is spread on the water surface to

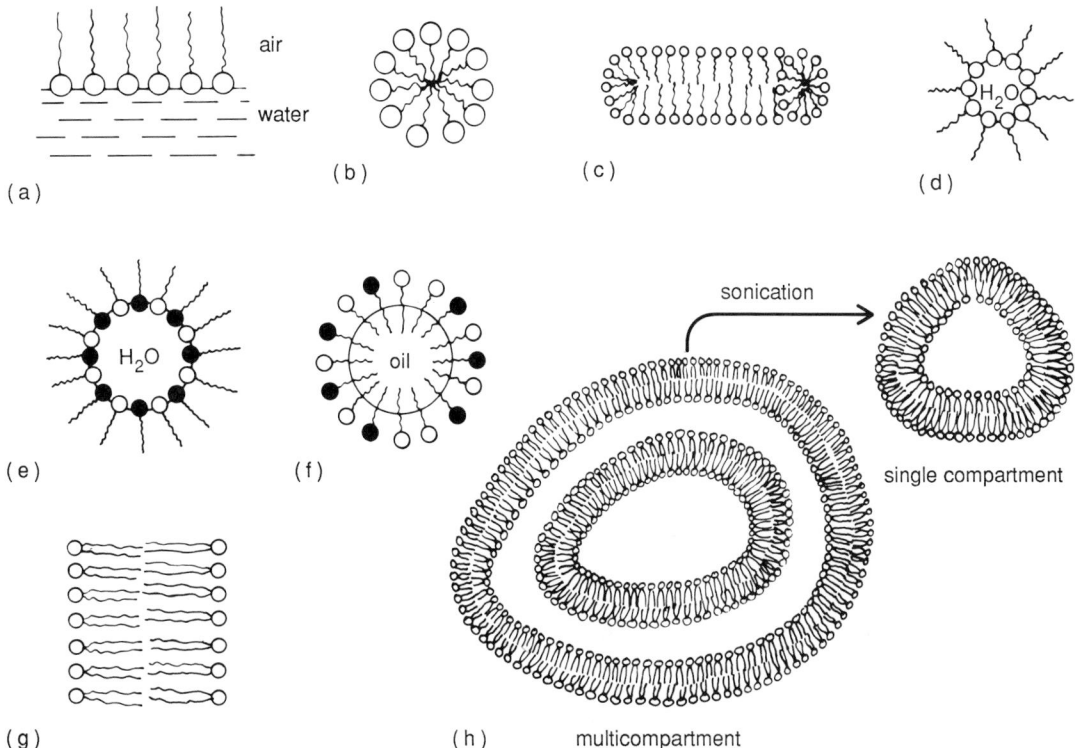

Fig. 1. Simplified representation of organized aggregates formed from surfactants. (*a*) Monolayer. (*b*) Spherical micelle. (*c*) Rodlike micelle. (*d*) Reversed micelle. (*e*) Water-in-oil microemulsion. (*f*) Oil-in-water microemulsion. (*g*) Bilayer lipid membrane. (*h*) Vesicles.

form a monolayer below the Teflon partitioning which contains the pinhole (0.1–0.5 mm or 0.004–0.02 in. in diameter). Careful injection of an appropriate electrolyte solution below the surface raises the water level above the pinhole and brings the monolayer into apposition to form the bilayer lipid membrane.

Vesicles. These are prepared by sonication from naturally occurring phospholipids or from such simple surfactants as dioctadecyldimethylammonium bromide (DODAB) or dihexadecylphosphate (DHP). They are single bilayer spherical aggregates with diameters of 50–100 nm and bilayer thickness of around 5 nm. Unlike micelles, vesicles do not break down on dilution. Nevertheless, they are dynamic structures. They undergo phase transition, fuse, and are osmotically active. Molecular motions of the individual surfactants in the vesicles involve rotations, kink formation, lateral diffusion on the vesicle plane, and transfer from one interface of the bilayer to the other (flip-flop). Vesicles are capable of organizing a large number of molecules in their compartments. Hydrophobic molecules can be distributed among the hydrocarbon bilayers of vesicles. Polar molecules may move about relatively freely in vesicle-entrapped water pools, particularly if they are electrostatically repelled from the inner surface. Small charged ions can be electrostatically attached to the oppositely charged vesicle surfaces. Species having charges identical with those of the vesicles can be anchored onto the vesicle surface by a long hydrocarbon tail.

The need for increased stabilities, controllable sizes, and permeabilities led to the development of polymerized surfactant vesicles. Vesicle-forming surfactants have been functionalized by vinyl, methacrylate, diacetylene, isocyano, and styrene groups in their hydrocarbon chains or at their head groups. Accordingly, surfactant vesicles could be polymerized in their bilayers or across their head groups. In the latter case, either the outer or the inner vesicle surfaces could be linked separately. All polymerized vesicles show enhanced stabilities compared with their unpolymerized counterparts. They have extensive shelf lives and remain unaffected by the addition of up to 30% alcohol.

Membrane mimetic systems as hosts. Substrate organization in membrane mimetic systems leads to altered solvation, ionization, and reduction potentials. Compartmentalization provides a method for the in situ generation and stabilization of monodispersed, ultrasmall, colloidal particles. The size of the particles is determined deductively by the compartment size and by the amount of the precursors placed in it. In practice, many factors influence nucleation, particle growth, and stability.

Controlled-size, monodispersed, colloidal catalysts (platinum, palladium, rhodium, and so forth) have been prepared in situ in water-in-oil microemulsions and vesicles. The rationale for this method is to prepare more efficient and selective catalysts. Since catalytic efficiency is related to particle size and uniformity (the smaller the particles and the larger the surface area the better the catalytic efficiency), this method fulfills some of these expectations.

Dispersed semiconductors. These are used increasingly in photosensitized water reduction. Photoexcitation of the semiconductor (SC) at a wavelength corresponding to the band gap promotes an electron (e^-) from the valence band to the conduction band, and hence separates the charges, as shown in Eq. (1), where h^+ represents a

$$SC \rightarrow e^- + h^+ \tag{1}$$

hole in the valence band. The charge carriers move rapidly in the semiconductor, frequently hitting each other and the surface. Electron and hole recombinations, in the absence of quenching, result in radiationless transition Δ (emission of heat) and, in some cases, in fluorescence $h\nu$ (emission of light), as shown in Eq. (2).

$$e^- + h^+ \begin{array}{c} \nearrow \Delta \\ \searrow h\nu \end{array} \tag{2}$$

To be useful, charge carriers must escape to the surface of the semiconductors where, at the semiconductor interface, they can transfer an electron to an acceptor (A) [Eq. (3)] or accept one from a donor (D) [Eq. (4)]. Thus, these inter-

$$e^- + A \rightarrow A^- \tag{3}$$

$$h^+ + D \rightarrow D^+ \tag{4}$$

facial electron transfers can mediate catalytic, sacrificial photosensitized water reduction (**Fig. 2**). Colloidally dispersed TiO_2, CdS, CdSe,

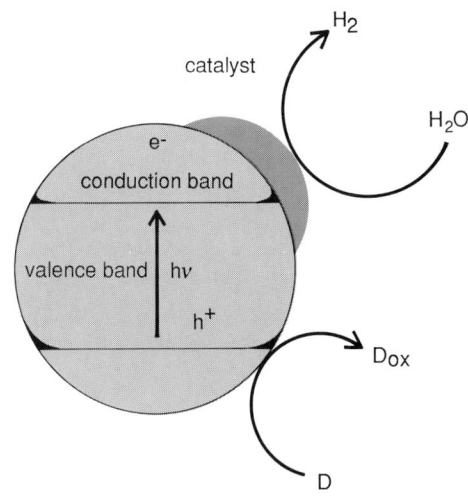

Fig. 2. Sacrificial water reduction systems involving catalyst-coated, colloidal semiconductor, valence band, and conduction band subsequent to excitation by light ($h\nu$). Upon irradiation in the present sacrificial electron donor (D), water is reduced to hydrogen (H_2). D_{ox} = oxidized donor.

Fe_2O_3, $SrTiO_3$, and their mixtures have been used as semiconductors either in the absence or in the presence of sensitizers. Sacrificial electron donors have included ethylenediaminetetraacetic acid (EDTA), thiols and alcohols; platinum (Pt), palladium (Pd), and rhodium (Rh) have been used as catalysts.

The use of colloidal semiconductors in artificial photosynthesis offers a number of advantages: the semiconductors are relatively inexpensive, they have broad absorption spectra and high extinction coefficients at appropriate bandgap energies, they can be made optically transparent enough to allow direct flash photolytic investigations of electron transfers, and they can be modified by derivatization or sensitizer absorption. In addition, electrons produced by band-gap excitation can be used directly without relays for catalytic water reduction (Fig. 2).

Unfortunately, colloidal semiconductors also suffer from a number of disadvantages. They are notoriously difficult to reproduce as small (<20 nm in diameter), monodispersed particles, which are needed to prevent nonproductive, electron-hole recombinations. The smaller the semiconductor particles, the greater the probability that the charge carriers can escape to the semiconductor surface, where electron transfer can occur. Semiconductors are equally difficult to maintain in solution without stabilizers, whose effects on the photoelectrical behavior are unpredictable.

Vesicle semiconductors. Methods have been developed for forming in situ, controllable-size (2–5 nm in diameter), uniform, rhodium-coated, CdS semiconductors in both polymerized and unpolymerized surfactant vesicles. In the presence of benzylalcohol as the sacrificial electron donor,

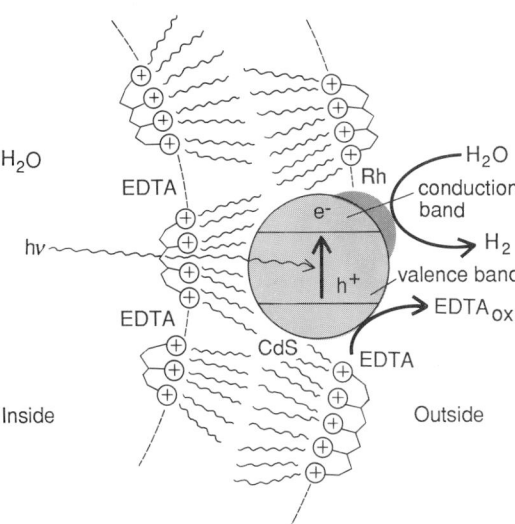

Fig. 3. Idealized model for CdS-sensitized photoreduction of water by ethylenediaminetetraacetic acid (EDTA) in polymerized vesicles prepared from surfactants containing styrene on their quaternary nitrogen head groups. The location and morphology of the semiconductor and the catalyst have not been confirmed. ox = oxidized.

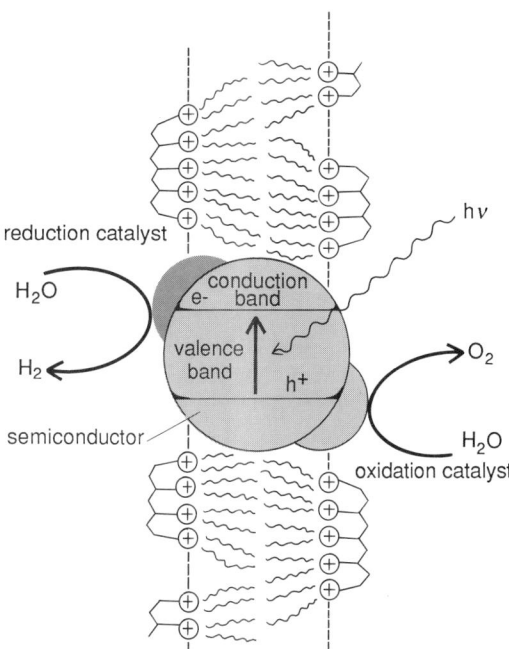

Fig. 4. Hypothetical model for a cyclic, water-splitting system, based on a semiconductor particle immobilized in a polymerized membrane with access to aqueous solutions on each side of the membrane. Specific and selective coating by catalysts leads to simultaneous and separate hydrogen and oxygen generation on each side of the polymerized membrane.

irradiation with visible light (**Fig. 3**) resulted in hydrogen production with a quantum efficiency of over 10%.

Other advantages of membrane mimetic systems involve the generation and stabilization of semiconductors. Membrane mimetic systems provide sites for semiconductor formation and shift the equilibrium between nucleation and particle growth in favor of the former. The charged microenvironments are responsible for altered band gaps and oxidation and redox potentials, as well as for the facilitation of charge separation.

The ultimate goal in using vesicle-stabilized semiconductors is to split water upon irradiation (**Fig. 4**). A comparison of Fig. 4 with natural photosynthesis illustrates the ideas of membrane mimetics. A polymerized surfactant aggregate is used instead of the thylakoid membrane. Energy is harvested by a semiconductor rather than by photosystem I and II. Electron transfer is rather simple. In spite of these differences, the basic principles are similar. Components of both natural and artificial photosyntheses are compartmentalized. The sequences of events are identical in the two systems: energy harvesting, vectorial charge separation, water oxidation and reduction.

Small magnetic particles of Fe_3O_4 have also been stabilized in surfactant vesicles. Each particle acts as a single-domain magnetite. These particles exert a magnetic effect which corresponds to the application of an external magnet with a strength of 3000 gauss.

Other applications. Within a remarkably short time, membrane mimetic chemistry has become a versatile chemical tool. Additional applications of compartmentalization of reactants in membrane mimetic systems may involve altered reaction rates, products, stereochemistries, and isotope distributions. Monolayers and organized multilayers can be employed profitably as molecular electronic devices. Opportunities also exist for using different surfactant aggregates with polymeric membranes for the control and regulation of reverse osmosis and ultrafiltration. The combination of judiciously selected polymeric membranes and functionalized surfactants holds the key for efficient membrane reactors.

Well-constructed surfactant vesicles and polymerized surfactant vesicles can be used as compartments for drugs, offering the possibility of target directing. The exclusive delivery of a given drug in a controllable manner has long been a goal of pharmacists. Much of the current activity in this area is based on liposomes (vesicles prepared from phospholipids of biological origin) together with immunological methodologies. Chemists are investigating the use of polymerized vesicles as hosts for drugs. On-command depolymerization (by an enzyme, for example) could trigger the release of drugs at the desired targets. It is anticipated that the realization of these and other, not yet invented applications of membrane mimetic chemistry will occur within a short time.

For background information SEE CELL MEMBRANES; DETERGENT; MICELLE; MONOMOLECULAR FILM; PHOTOCHEMISTRY; PHOTOSYNTHESIS in the McGraw-Hill Encyclopedia of Science and Technology.

Janos H. Fendler

Bibliography. J. H. Fendler, *Membrane Mimetic Chemistry*, 1982; J. H. Fendler and P. Tundo, Polymerized surfactant aggregates: characteristics and utilization, *Acc. Chem. Res.*, 17:3–8, 1984; M. Gratzel, *Energy Resources through Photochemistry and Catalysis*, 1983.

Memory

Memory allows a person's past experience to influence current thoughts, perceptions, and behavior. There has long been an assumed relationship between aging and memory; memory is believed to decline in the elderly. In recent decades, psychologists have begun a systematic, empirical attack on the question of how memory is related to age, especially memory for verbal materials. Most often, memory differences for young and old adults reflect different processing strategies during study. A somewhat simplified generalization is that young adults generally engage in a greater amount of active processing of the to-be-remembered event, and so lay down a memory trace that is more retrievable later.

This discussion of memory and aging focuses primarily on memory for verbal and pictorial materials in "normal" aged individuals, that is, healthy adults typically in their 60s and 70s. The notable memory losses characteristic of the roughly 5% of the elderly who suffer from senile dementia will not be considered here.

Models. Currently popular models of memory distinguish between short-term memory and long-term memory. Short-term memory contains the information of which the person is consciously aware at present; for example, it holds a phone number long enough for the person to use it. Long-term memory is traditionally viewed as "memory"; it holds things of past experience, which are not at the present moment part of conscious awareness. When a to-be-remembered item is in short-term memory, it may be rehearsed in ways that increase the probability that the information gets transferred to long-term memory. The sort of "deep" processing during rehearsal that can create a strong trace in long-term memory requires that the person engage in "elaboration" of the attended semantic information; this involves embedding the to-be-remembered item into a larger memory network by, for example, thinking of the item in varied ways which reflect various features, or by forming associations between the item and well-known things which can later help to gain access to that item during recall. This article deals only with age effects in long-term memory.

Age as a variable. Age is the type of·variable that cannot be directly manipulated. Since age cannot be randomly assigned and can be manipulated only by selection of individuals who possess the desired levels, people of different ages may differ in other ways as well, for example, in education level or verbal intelligence. Thus, when persons of different ages perform differently on a memory task, the question arises whether the performance difference is related to just age or to some other factor which happens to co-vary with age in the samples. Researchers often try to equate the young and old groups on variables such as level of education and verbal aptitude. Recent studies looking at aging and memory for discourse, for example, have demonstrated that a substantial part of the observed "age effect" may indeed be due to the differences in verbal and reading skills. If other variables are not measured, and their effects removed, the apparent age effects on memory are spuriously inflated.

Memory phenomena and aging. If memory processes are unchanged with aging, then variables that have substantial impact on memory for young adults should have similar effects for older adults. Often this similarity has been found, but those cases where the effects of a variable are different for young and old adults may be instructive about potential memory changes with aging.

If a list of words to be remembered contains

only nouns that could be readily depicted by simple pictures or drawings, the number of items named in a later recall is consistently greater for young adults if the items are presented in picture form rather than word form. Recent evidence suggests that this "picture superiority" effect is not observed in older adults, particularly if the individuals do not verbalize the name of each picture aloud at time of presentation.

An important finding for young adults is a category recall relationship. Individuals sort a stack of cards, each containing a single word, into a fixed number of piles, each comprising cards that "somehow belong together"; then the stack is shuffled and the person tries to sort again into the same piles. This procedure continues until the piles are consistent across two consecutive sortings. The number of words that a person can then recall is well predicted from the number of categories into which the person sorted the deck. A recent study shows that this relationship did not occur for an aged sample. With two to eight categories, the correlation between the number of words recalled (r) and number of categories was very strong for young adults ($r = .82$); there was no significant relationship for an aged group ($r = .10$). This finding was especially striking because there was no significant difference in the mean levels of overall recall for the young and aged.

Another interesting memory phenomenon is called the generation effect. If an individual is given a set of cues from which a to-be-remembered word can be generated, the word is remembered better than if the word is presented along with the cues. In both cases the person is remembering the same words with explicit knowledge that these words will be used for memory testing. The cues for generation of the item have ranged from use of synonymic or antonymic relations, to category membership, to sentence completions. Nearly all subjects generate the intended items. One recent study found that old normals benefited somewhat less than young adults from the generation of the to-be-remembered items. Another study found no generation effect for older adults on the initial trial, but the effect emerged across three study trials; for young adults, however, it was there from the first trial.

In a similar study the nature of the generation was manipulated so as to make it a more or less effective elaboration. The generation task took the form of finishing a sentence. "Precise" elaborations bore a clear and meaningful relation to the stem of the sentence, whereas "imprecise" elaborations simply gave more information which was not clearly tied to the semantic content of the sentence stem. In a second condition, precise or imprecise elaborations were provided by the experimenter; the sentence stem was the same in both conditions. For experimenter-provided elaborations, more was remembered of these sentence stems with precise elaborations than with imprecise; while the young remembered more than the old, the benefits of precise over imprecise elaboration were the same for both groups. Thus both demonstrated the usual effect that greater elaboration has positive memory consequences. For the self-generated conditions, however, the old were much less likely to generate precise elaborations. Whereas the young did particularly well under conditions designed to induce the generation of precise elaborations, the old did no better than they did for experimenter-provided elaborations. Thus, in this study older adults failed to demonstrate the generation effect under any conditions. They benefited from relevant elaboration as did the young, but they did not spontaneously engage in the same degree of such elaboration of the to-be-remembered material during study as did the young.

When memory is tested by recognition, the individual is often asked to say yes or no as to whether a test item was studied. Recognition performance for young adults has been shown to be better if the item is presented for test in the same context as during study, for example, following the same items. It has been reported that when the prior word context is semantically related to the target word, these context effects are potent for both young and old. For example, if the word "iron" is studied in the context of "copper," iron is best recognized on a later test if it again follows copper; it is less recognized if it follows "bronze," and least recognized if it follows a word which induces a different semantic content, in this case "clothes." When unrelated words provide the context for the target word, the young still show a significant benefit of similar context. The aged, on the other hand, show no context effect with such unrelated words.

Amount of study time is a strong determinant of later recall. One argument as to why the aged show poorer memory is a postulated general cognitive slowing. In that case, increasing the study time per item from 1.5 to 3–6 s should be particularly beneficial to the aged. It has been reported that the age-related recall deficit actually increases significantly with such increases in presentation times. However, when semantic orienting questions are used to guide the sort of elaboration of words during encoding, the benefit of additional study time is equal for young and old.

The common thread in these studies seems to be that the aged fail to engage in as much elaboration of the trace at time of encoding an event, despite evidence that they can benefit as much as young adults from such elaboration. The absence of the usual category-recall relationship in performance of the elderly suggests that they engage in less elaboration despite the task, or that they do something different by way of the strategies that they employ. The greater amount of information and detail in pictures permits them to be processed in a more elaborated fashion than words, yet older persons appear not to do so, since picture superiority effects are less reliable in the aged. The self-generation effect is also less stable for the elderly; they are less likely to generate the precise completions that facilitate memory. The elderly do benefit from such elaborations, however, when they are provided. A similar lack of elaboration during study was seen in research on context effects in recognition memory. The young spontaneously engage in extensive elaboration of the target word, including elaborating features relating to the prior item; the aged, on the other hand, do not engage in such extensive elaboration unless the context is so semantically related as to induce the obvious elaboration.

Memory in everyday life. While it is more difficult to study memory events outside the laboratory, some research of this type has been undertaken. One tack is to study "autobiographical" memory, which people have for events they have experienced in life. An approach to this was suggested by Francis Galton in 1879. The person is presented single words and asked to think of a particular event in her or his life which that word elicits and to describe it. A question of interest concerns the relative accessibility of events in memory as a function of how long it has been since they were experienced. For young adults, there is a preponderance of events from relatively recent memory and a reduction in the frequency of memories the farther back the events occurred in time. This relationship has been found to be very similar across ages. The similarity of forgetting functions is of interest since it contradicts another common stereotype about memory and aging, namely that while aged individuals may remember events from long ago very clearly they cannot remember recent events as well. While that stereotype is true of many individuals with senile dementia, there is no evidence to support it for normal, healthy aged individuals.

Another type of real-world study is concerned with memory for spatial information as it relates to the formation of "cognitive maps" of one's environment. Even for long-time residents of an urban area, the aged have less accurate verbal recall of urban landmarks; furthermore, they are less accurate in the location of such landmarks from memory. Older adults are less accurate than young adults in way-finding tasks where they are attempting to follow a route along which they have been previously led through an unfamiliar building. Also, sketch maps of the neighborhood produced by older adults are judged to be relatively disorganized, simple, inaccurate, and smaller in area than those produced by younger adults.

A common anecdotal report is that older persons have greater word-finding difficulty during conversations. In a study in which the task was to express a word after being given its definition—a reversed vocabulary test—significant age effects of the expected type were found, that is, older persons were less able to give the words. This result was not due to limitations of the vocabulary of the aged individuals; on a forward vocabulary test they performed as well as the younger adults.

There has been some success in attempts to develop training programs through which the elderly can improve their memory skills for daily activities, and these will perhaps be of benefit to individuals with memory complaints. Such programs will not be of much interest to the typical older person, however; a general observation of many researchers in memory and daily functioning of the elderly is that, despite a widespread belief among the elderly that their memory has deteriorated somewhat with age, very few of them feel that forgetfulness has produced any notable handicap in terms of their everyday functioning. The ability to sustain the same level of daily functioning may be due to an increased strategic use of certain external memory aids, such as writing notes to oneself or leaving something in a conspicuous location ;in order to prompt memory.

For background information SEE ALZHEIMER'S DISEASE; MEMORY in the McGraw-Hill Encyclopedia of Science and Technology.

Eugene A. Lovelace

Bibliography. J. E. Birren and K. W. Schaie (eds.), *Handbook of the Psychology of Aging*, 1985; N. L. Bowles and L. W. Poon, Aging and retrieval of words in semantic memory, *J. Gerontol.*, 40:71–77, 1985; J. C. Cavanaugh, J. G. Grady, and M. Perlmutter, Forgetting and use of memory aids in 20 to 70 year olds' everyday life, *Int. J. Aging Human Develop.*, 17:113–122, 1983; J. D. Sinnott, Prospective/intentional and incidental everyday memory: Effects of age and passage of time, *Psychol. Aging*, 1:110–116, 1986.

Metal forming

The forming of metals is associated with processes which change the shape of the metal by the application of force, as opposed to the casting of metals into molds. Certain properties are necessary to permit forming of metal sheets by pressworking. Further, for the process to be economical the metal must be available in sufficient quantity at a reasonable price. Iron, aluminum, copper, tin, nickel, zinc, and other metallic elements have commercial importance due to distinct properties such as appearance, density, conductivity, strength, and ductility. These elements are used individually or in combination in the metal-forming process.

Alloys. Steel is an alloy of iron and carbon and is widely used for its great range of properties, availability, and relatively low cost. Steel rusts, forming a loosely adhering reddish residue. Pure iron has more resistance to corrosion than steel. Aluminum also oxidizes, especially when it is alloyed with other elements. However, with pure aluminum, a dense, tightly adhering coating develops which protects the base metal from attack. A recently developed product for high-temperature applications, such as automobile exhaust systems, is a coating of aluminum on a stainless steel basis metal.

Metallic coatings. For low-cost protection of steel sheet products over a period of time, both zinc and aluminum metallic coatings have been developed. The protection afforded by zinc is limited to the time required to oxidize away the coating; then the steel starts rusting.

The goal of providing low-cost corrosion protection has been a main influence on process development in the steel industry in recent years. The problem has been approached from two directions: (1) improved corrosion resistance to road ice–control materials and (2) improved formability of metals with corrosion barrier coatings. The common metals for corrosion protection of steel sheet are zinc and aluminum, either singly or in combination. These can be applied by a hot-dip process or, in the case of zinc, also by electroplating. When the iron-zinc alloy layer of hot-dipped zinc coatings is minimized, the ductility of the product is improved and the coating has better adherence. There is no alloy layer in the electroplated zinc coating. Both processes are currently being used to benefit the automobile manufacturing industry. However, there is a preference for the electrogalvanized finish, and several steel producers have recently installed such lines.

Metal stamping. Presses for forming automotive parts are mechanical to accommodate the high production rates. They are grouped into press lines to perform a series of related operations. The number of steps necessary to achieve a desired shape depends not only on the complexity of the shape and the required final properties but also on the grade and manufacturing method used in making the metal sheet. Emphasis in stamping plants has been in the direction of automation; computers are used to program repetitive operations of metal forming, which are performed by robots in place of people. It has been found that placing strain gages on the stamping press columns allows better measurement of loads on the dies and permits close control of the stamping process. Changes in the press performance due to temperature, lubrication, the metal sheet being formed, the condition of the tooling, and associated operations all affect performance.

Roll forming. This is a high-speed forming process increasingly used in automobile manufacturing and other high-volume operations. Roll forming uses a flat strip shaped into a coil; the strip is formed simultaneously as it is moved longitudinally by pairs of rotating tools, usually in the form of contoured rolls. Production rates of up to 200 ft (5000 mm) per minute are common. Most materials that can be formed by other methods can be roll-formed.

High-strength low-alloy steels are an example. It is necessary to design the process for the specific material being used, because the springback characteristics of the strip must be accommodated. High-strength materials require larger bend radii of from two to five times the metal thickness. Most roll formers are flexible in the tooling they can accommodate and are programmable. A major advantage, in addition to the high speed of forming, is that practically no scrap is produced, and little trimming is necessary. Recent developments include the capability for interrupted forming such as stopping grooves a certain distance from the end of a roll-formed piece, variable radii curving, ring forming, dogleg shapes, and short pieces.

Metal spinning. For symmetrical bowl-shaped parts, metal spinning has advantages over deep drawing. The process has been improved to permit holding close tolerances of 0.002 in. (0.05 mm), which are required for some aircraft components. It is possible to form parts by spinning when the center of a cylindrical shape has a greater diameter than the ends. The most common metal for spinning is standard low-carbon steel; however, the interstitial-free (extra-low carbon content) steels have much greater capacity to form complex shapes. A cup shape with final dimensions of 4 in. (100 mm) diameter and 12 in. (300 mm) depth has been spun. Metals in sheet form such as aluminum, brass and other copper alloys, nickel, inconel, and titanium can be formed by spinning. New developments include the capability of spinning secondary forms and parts that have been preformed by welding. Improvements in lubricants and their application have increased tool life and the range of shapes possible by spinning. Automation has been done where the volume of parts and the repetitive nature of the form makes it worthwhile. It is possible to program the operation during a preliminary run and have the tooling follow the same process for succeeding parts.

Improved steels. New interstitial-free steels do not have the severe aging problems or the earing problems of the rimmed steels. Today's zinc-coated steel sheets have formability that is superior to uncoated steel.

The demand for light weight and more fuel-efficient automobiles during the 1970s encouraged the steel companies to develop high-strength low-alloy products. The yield strength of high-strength low-alloy steels ranges from 45 ksi (ksi = 1000 pounds per square inch) or 300 megapascals to over 200 ksi or 1400 MPa. Dual-phase (high-strength low-alloy) steels are of special interest because they are in a soft condition when received, which makes them easy to form; and they can be hardened after forming by a low-temperature bake.

Continuous annealing. Continuous annealing lines produce specific properties in low-carbon steel, such as drawing quality in rimmed and killed steels, in much shorter times than are necessary when box annealing is used.

Technological improvements. For fabricators using large tonnages of steel, the coils are shipped directly to the stamping plant. Service centers purchase coils from producing mills and supply small users either with these coils directly, after slitting the large coils into smaller widths, or after cutting the coils up into stacks of flat sheets.

Technological improvements frequently come only after years of development, and then sometimes unexpected improvements result. For example, the use of magnesium coke to remove the sulfur from the hot metal after tapping a blast furnace gives a steel that requires less manganese to prevent hot shortness; and as a bonus, the steel has superior formability in the final sheet product. The use of pure oxygen to refine metals in basic oxygen furnaces, vacuum degassing, and continuous casting have been known for many years, but these techniques were not employed until recently.

In the steel manufacturing process the use of continuous cast slabs, around 9 in. (230 mm) thick and up to 72 in. (1800 mm) wide, represents current technology. This approach eliminated the ingots, the need for a stock of molds, the soaking pits, and the slabbing mill. Work is underway to develop a process to cast slabs 0.1 in. (2.5 mm) or thinner, possibly 0.05 in. (0.1 mm). This would eliminate the expensive hot-rolling step of the processing. It would still be necessary to cold-roll the thin-strip casting to develop a structure in the metal to maximize the forming properties.

For most forming applications the properties of a hot-rolled product are generally the same regardless of how much hot-rolling reduction is done. There is a pronounced improvement in the forming capability of low-carbon steel sheet products when they have been cold-reduced (thickness reduced by cold rolling) from 50 to 70% and then properly annealed to resoften and develop a preferred orientation of the internal structure.

Roll texturing. Two recent improvements in the finish and life of rolls used in cold reduction and the skin pass (temper mills) have been electric-discharge texturing and laser texturing of roll surfaces. The finish of the roll is imprinted on the coil being processed. Improvements include the elimination of shot blasting equipment, more uniform and consistent texture, and better performance of the sheet product in stamping plants. Another new development that improves the life of a roll is chromium plating of the surface.

Process control. One of the most frequent causes of problems in automobile stamping plants is nonuniformity of the dimensions of the blank. The nonuniformity may involve a thickness variation, a width variation, or improper setting of the length cut in the initial blanking

operation. Current refinements in rolling mills include computerized processing control to produce coils with constant thickness and width, as well as equal amounts of reduction throughout the coil length.

Structure of metals. Metals in the form of commercial alloys solidify with a crystal structure. Three common structures have been classified as body-centered cubic, face-centered cubic, and hexagonal close-packed. Each metal and alloy has its own distinctive crystal arrangement. Crystals can be randomly oriented or they can have a preferred orientation. In some cases the orientation relationship to the product, for example sheet surface and rolling direction, can yield a material with improved properties. The basic structure depends on the alloy composition. Preferred orientation can be produced by certain processing techniques. For cubic structures in steel and orientation described as cube-on-edge is a most desirable configuration for sheets that must be deep-drawn to shape. The overall crystallographic orientation can be determined by x-ray analysis. This preferred orientation plays a critical role in the formability of metals.

Plastic strain ratio. The effects of crystallographic orientation on formability are measured by resistance to thinning of a test strip pulled beyond the yield point in a tension test. This is the plastic strain ratio (r value). Metal sheets that have average r values of near 1 when tested in various directions relative to the coil-processing direction are isotropic and have average formability. When they are formed, they change shape equally throughout their structure. For an r value approaching 3 the metal resists changes in thickness, which improves its formability. Resistance to thinning means the metal can be deep-drawn into cup-shaped parts without breaking as easily as similar metals with lower r values. The addition of alloy elements, such as aluminum, columbium (niobium), and silicon, affects the r value. It has been found that cold reductions during processing to final gage of around 70% are necessary, followed by proper temperatures and times in controlled annealing atmospheres, to obtain high r values.

Strain hardening. Annealing, together with the chemical composition, determines the strain-hardening characteristic, which is the effect of locally strengthening the metal in the formed area and transferring subsequent deformation to adjacent less cold-worked areas. This property is measured by the strain- (work-) hardening exponent (n value). A material with a high n value (above 0.20 for steel) has improved biaxial stretchability.

Strength. If the final part does not have sufficient strength, it is not usable regardless of how easily it can be formed. The yield strength, tensile strength, and elongation are important properties for materials that are used in structural components. The elongation is a quick measure of the ductility of a metal. For aluminum, an elongation of around 30% is expected; for low-carbon drawing steels, elongations of around 40% are usual; and for some stainless steels, elongations of 60% are common. None of these has been related to the values for the plastic strain ratio, the strain-hardening exponent, and the strain-rate sensitivity of the flow stress. Thus, each property is distinct, and yet all contribute to metal formability.

Strain analysis studies. A better understanding of metal-forming processes can be gained through strain grid analysis. This procedure permits developing a blank shape that requires fewer hold-down beads, consumes less sheet metal in each stamping, and reduces the occurrence of breakage. The Society of Automotive Engineers has developed a procedure for evaluating stampings using a severity curve. Associated with this curve is a test being developed by automobile manufacturers and steel producers using a 4-in.-diameter (100-mm) ball punch to grade metal strips. The test is designed to produce strains near the bottom of the severity curve; this is called the limiting dome height (LDH) test.

For background information *SEE CRYSTAL STRUCTURE; GRAIN BOUNDARIES; METAL FORMING; METAL ROLLING; PLASTIC DEFORMATION OF METAL* in the McGraw-Hill Encyclopedia of Science and Technology.

John R. Newby

Bibliography. American Society for Testing and Materials, *1985 Annual Book of ASTM Standards*, Sec. 3: Metals Test Methods and Analytical Procedures, vol. 03.01: Metals—Mechanical Testing, Elevated and Low Temperature Tests, pp. 119–151, 618–625, 729–738; Society of Automotive Engineers, *1985 SAE Handbook*, vol. 1: Materials, SAE J863: Methods of Determining Plastic Deformation in Sheet Metal Stampings, 1:3.17–19.

Metallic glasses

Since about 1960, amorphous metallic alloys have been produced primarily by rapid-solidification techniques. Since 1983, new methods of synthesis of amorphous metallic alloys have been developed that are based on solid-state reactions. In contrast to the rapid-solidification techniques, the solid-state amorphizing reactions (SSAR) are performed at constant temperatures (below the crystallization temperature of the amorphous alloy) and over long periods of time. These new techniques may make it possible to manufacture amorphous alloys over wider composition regimes (not accessible with the rapid-solidification technique) and in bulk form.

Rapid solidification and SSAR. In nature, metallic alloys are always found in crystalline states where the atoms of the various species are arranged periodically over many interatomic dis-

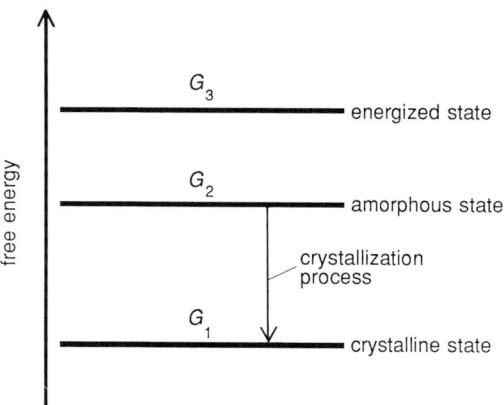

Fig. 1. Free-energy levels at which an alloy may be found in nature and produced artificially.

tances. From the many different crystalline configurations possible, the one that minimizes the alloy's free energy always takes place in nature. **Figure 1** illustrates the free energy of an alloy in three states. In the crystalline state the free energy has the lowest possible value, G_1. In the amorphous state the alloy lacks long-range ordering and thus has a free energy G_2, which is larger than G_1. As long as the amorphous alloy is kept at a temperature lower than its crystallization temperature T_x, the alloy remains in metastable equilibrium in the state G_2. However, if the alloy is heated above T_x, the increased atomic mobility allows the amorphous alloy to crystallize and thus to lower its free energy to G_1.

A large number of amorphous metallic alloys have been artificially synthesized. One possible way to reach the metastable state of free energy G_2 is to start from a state of even higher free energy, G_3, and then to remove the excess energy, $G_3 - G_2$, in such a way as to avoid the occurrence of a reaction into the state G_1. The different methods of synthesis of amorphous alloys differ in how they achieve this goal. In the rapid-solidification methods of synthesis, the initial high-energy state G_3 is the liquid phase of the alloy. In this state the alloy has the same chemical composition and roughly the same atomic structure as the amorphous alloy to be produced. If the excess thermal energy, $G_3 - G_2$, is removed at a very fast rate, there is insufficient time for the atoms in the alloy to rearrange over long distances in order to achieve the crystalline state of energy G_1, so that the alloy is trapped in the metastable amorphous state. The solid-state amorphizing reaction methods differ from the rapid-solidification methods in that the reaction from state G_3 to state G_2 is carried out at a constant temperature and the excess free energy, $G_3 - G_2$, is not thermal. Various versions of the solid-state amorphizing reaction are described below.

Interdiffusion metals from thin films. It was shown in 1983 that a solid-state amorphizing reaction takes place at clean (oxide-free) bound-

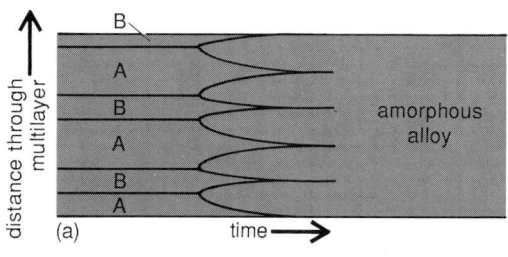

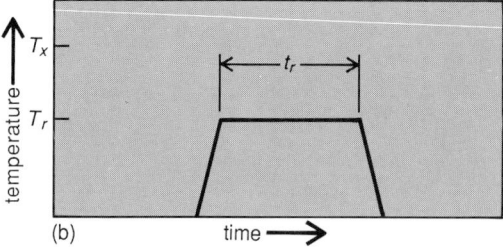

Fig. 2. Reaction of thin films by interdiffusion to form a homogeneous amorphous alloy. (a) Time evolution of a stack of thin films, A and B. (b) Temperature histogram for the annealing of the films. The annealing temperature T_r must be lower than the crystallization temperature T_x. The annealing time is t_r. (After R. B. Schwarz, Formation of amorphous metallic alloys by solid state reactions, Bull. Mater. Res. Soc., 11:55–58, 1986)

aries between pairs of metals, A and B, that have a large chemical affinity. It is characteristic of these systems that in equilibrium they form a large number of crystalline intermetallic compounds of different composition and that some of these compounds melt congruently. One way to achieve the oxide-free boundaries between metals A and B is to condense vapors of A and B alternatively onto a cold substrate inside a high-vacuum chamber. The multilayer arrangement obtained is shown schematically in **Fig. 2a**. The thickness of each layer is of the order of 100 nanometers and the relative thicknesses of the A and B films are those required to give the amorphous alloy the desired composition. In contrast to the rapid-solidification methods, for the solid-state amorphizing reaction, the initial state G_3 in Fig. 1 has a composition and atomic structure quite different from that of the amorphous product to be formed. The excess free energy, $G_3 - G_2$, is chemical. If the thin-film multilayer is annealed at a constant temperature for several hours, an amorphous alloy develops at the A/B interfaces (Fig. 2a). Certainly, the reaction temperature T_r must be below the crystallization temperature T_x of the amorphous alloy to be formed (typically 180°F or 100°C below) but must be sufficiently high to allow for the interdiffusion of A and B in a reasonable time t_r (Fig. 2b). The solid-state amorphizing reaction has been observed at the interfaces of gold/lanthanum, gold/yttrium, cobalt/zirconium, nickel/zirconium, nickel/titanium, nickel/hafnium, iron/zirconium, and cobalt/tin. All these systems are characterized by

negative heats of mixing in the amorphous phase which provide the chemical driving force for the reactions.

Research is being undertaken to understand the nucleation and growth of the amorphous alloy by interdiffusion at the A/B interfaces and to explain the obstacles to the formation of the more stable crystalline intermetallic compounds. The nucleation of the amorphous phase is thought to be assisted by defects (dislocations and grain boundaries) near the A/B boundaries in either or both pure crystalline films. Observations by transmission electron microscopy show that following nucleation the amorphous film grows uniformly and parallel to the A/B interfaces. Therefore the growth of the amorphous phase requires that at least one of the atomic species diffuse across the amorphous alloy film already formed. In all the binary systems listed above where the solid-state amorphizing reaction has been observed, the first element has a smaller atomic volume than the second. Recent diffusion measurements in amorphous nickel zirconium (NiZr) alloys show that nickel (the smaller atom) diffuses much faster than zirconium (the larger atom). There are reasons to believe that this is a general result and that in all binary amorphous alloys the smaller atom has a much larger diffusivity than the larger atom. This disparity in diffusivities may indeed be the underlying reason behind the solid-state amorphizing reactions. For such a reaction to occur, it is necessary that the two species mix, which can be achieved even if only one of the atomic species is mobile. On the other hand, the formation of crystalline intermetallic compounds, which usually have large unit cells of symmetry vastly different from that of both pure metals, should require the relative motion of both atomic species.

Mechanical alloying of powders. A particularly promising method for the manufacture of homogeneous amorphous powders in large quantities is based on the mechanical alloying of powders. Mechanical alloying is a high-energy ball milling technique that has been used extensively in industry to prepare composite metal-base powders with controlled microstructures. In 1983 it was found that the mechanical alloying of a mixture of powders of two pure metals results in an amorphous alloy powder. The binary systems that have been amorphized by this technique include tin-niobium, nickel-niobium, nickel-titanium, iron-zirconium, cobalt-yttrium, cobalt-gadolinium, silver-gadolinium, and nickel-gadolinium. Similar to the systems that have been amorphized by solid-state amorphizing reactions in thin films, these pairs of elements have negative heats of mixing in the amorphous state (that is, a large chemical affinity for each other) and the first element in each pair has a smaller atomic volume than the second. The mechanical alloying process is very simple. A few hardened-steel balls and the powders to be mechanically alloyed are

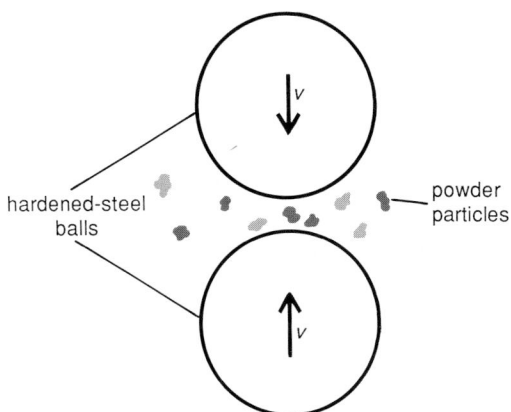

Fig. 3. Mechanical alloying process in which crystalline powders of pure elements are trapped between balls colliding with velocity *v*.

put into a hardened-steel container. The container is shaken repeatedly, causing the balls to collide. The powder particles that are trapped between balls are subjected to plastic deformation at high strain rates (**Fig. 3**). The particles fracture and cold-weld repeatedly. In order to enhance the fracture, the container is externally cooled. Typically, a fully amorphous alloy powder is obtained after 20 h of mechanical alloying.

The first mechanism proposed to explain the formation of an amorphous state by mechanical alloying was based on a rapid-solidification process. It was thought that localized plastic deformation (at the particle surfaces or at plastic shear bands) was sufficient to produce melt pools. These melts would solidify rapidly as heat was removed by conduction into the cooler (less deformed) regions of the particles. Recent calculations have shown that the peak temperatures reached in the particles trapped between colliding balls are far below the melting temperature. It is presently believed that the formation of an amorphous state during mechanical alloying occurs through a mechanism similar to the solid-state amorphizing reaction in thin films, discussed above. After only about 1 h of mechanical alloying, the powder particles attain a finely layered structure formed by alternating films of the two starting elements. This morphology resembles that of the unalloyed thin films in Fig. 2. Chemical interdiffusion at these boundaries is thought to be assisted by the point defects and lattice defects created by plastic deformation.

Other recently developed methods of synthesis of amorphous alloys are based on the swaging of rolls of alternating thin foils of two pure metals or of metal tubes containing a mixture of two metallic powders. As in the thin-film and ball milling techniques, this method works only for metal pairs that have a large negative heat of mixing in the amorphous state, which provides the thermodynamic driving force for the solid-state amor-

phizing reactions at their interfaces.

In another related discovery, it has been found that amorphous alloy powders can also be synthesized by mechanical alloying starting from powders of a crystalline intermetallic compound such as for the nickel-titanium system, the compound $NiTi_2$. This process differs from those previously discussed in that the starting material is at its lowest free-energy state (level G_1 in Fig. 1). The mechanical alloying raises the free energy of the system until the system is trapped in the metastable energy level G_2. This is thought to occur through the accumulation of point and lattice defects and the chemical disordering of the crystalline lattice of $NiTi_2$.

Hydrogen diffusion in alloys. Hydrogen diffuses rapidly in many metals and alloys; thus it can be driven into and out of these systems with ease at relatively low temperatures. In this version of the solid-state amorphizing reaction, the hydrogen is used as a tool to transform a crystalline intermetallic compound into the amorphous state. The starting product is a metastable crystalline intermetallic compound, such as zirconium rhodium (Zr_3Rh), where only zirconium has a large affinity for hydrogen. When hydrogen is allowed to enter this compound at a temperature of 437°F (225°C), it preferentially seeks the zirconium atoms, having little affinity for the rhodium atoms. When sufficient hydrogen is forced into the compound, the hydrogen distorts the lattice of the compound to the point of destroying its crystalline order and creating an amorphous hydride with the composition $Zr_3RhH_{5.5}$. This type of solid-state amorphizing reaction has been studied much less than those that occur in thin films or during the mechanical alloying of powders.

Advantages of SSAR. It has been shown that the products of a given solid-state amorphizing reaction are those predicted by thermodynamic calculations in which the amorphous phase is treated as an undercooled liquid. In binary metallic systems the range of compositions of the single-phase amorphous products obtained with the solid-state amorphizing reaction is, in general, wide and is centered near the equiatomic composition. In contrast, the amorphous alloys prepared by rapid solidification have narrower composition ranges that are centered near deep eutectics in the binary phase diagram. This is because at these compositions the alloys melt at relatively low temperatures, and it is thus easier to rapidly cool the alloy to below the crystallization temperature T_x. For some systems, such as nickel-titanium, amorphous alloys with the composition $Ni_{50}Ti_{50}$ can be formed only by the solid-state amorphizing reaction method.

The discovery that thermodynamically metastable alloys can be produced by isothermal solid-state reactions has opened a new window into the synthesis of novel materials. In order to take full advantage of this potential, methods must be developed for the synthesis and handling of clean submicrometer-size powders of pure elements to allow for a direct solid-state amorphizing reaction from a mixture of powders into bulk amorphous alloy products (the reaction shown in Fig. 2 using powders instead of thin films). Methods are being studied for the consolidation of amorphous powders (and metastable powders in general) that do not destroy their metastability.

For background information SEE ALLOY; AMORPHOUS SOLID; METALLIC GLASSES in the McGraw-Hill Encyclopedia of Science and Technology.

Ricardo B. Schwarz

Bibliography. R. B. Schwarz and W. L. Johnson, Formation of an amorphous alloy by solid-state reaction of the pure polycrystalline metals, *Phys. Rev. Lett.*, 51(5):415–418, 1983; R. B. Schwarz, R. R. Petrich, and C. K. Saw, The synthesis of amorphous Ni-Ti alloy powders by mechanical alloying, *J. Non-Cryst. Solids*, 76:281–302, 1985; X. L. Yeh, K. Samwer, and W. L. Johnson, Formation of an amorphous metallic hydride by reaction of hydrogen with crystalline intermetallic compounds: A new method of synthesizing metallic glasses, *Appl. Phys. Lett.*, 42(3):242–244, 1983.

Metallochlorin

A metallochlorin is a metal-containing chlorin macrocycle, that is, a porphyrin which has been reduced at one pyrrole ring. Chlorins are familiar as the functional unit of the green plant pigment, chlorophyll. Whereas the color of human blood arises from the presence of the iron porphyrin (heme) group, protoporphyrin IX (**Fig. 1**), at the oxygen-binding site of hemoglobin, the brilliant green color of leaves and grass is due to the chlorins chlorophyll and pheophytin. Surprisingly, chlorins are found in humans and other mam-

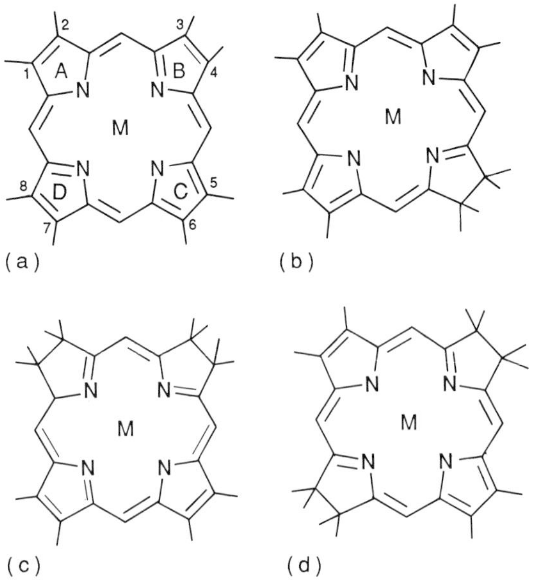

Fig. 2. Tetrapyrrole ring structures. (a) Porphyrin. (b) Chlorin. (c) Isobacteriochlorin. (d) Bacteriochlorin. M = central metal ion.

mals, as well as in bacteria and fungi. Examples include the enzyme myeloperoxidase of the white blood cells, and the green heme proteins recently found in human and bovine spleen.

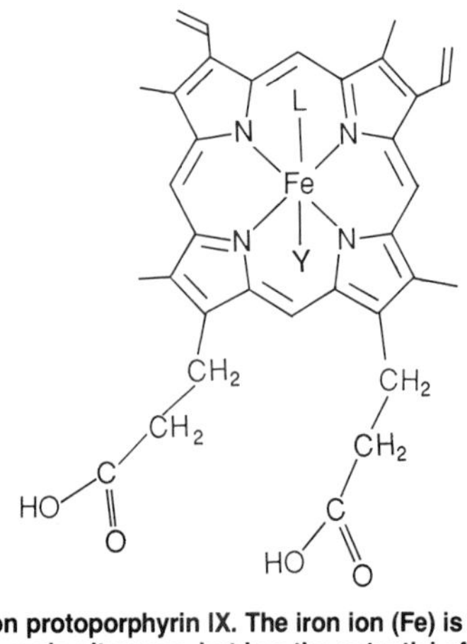

Characteristics. Chlorins differ from porphyrins primarily by the loss of one double bond at one of the pyrrole rings. When only one pyrrole ring has been reduced, the resulting macrocycle is called a chlorin (dihydroporphyrin). Reduction of two pyrrole rings leads to either an isobacteriochlorin (iBC) or a bacteriochlorin (BC), depending on whether the reduced (pyrroline) rings are cis or trans to one another, respectively (**Fig. 2**). For reduced porphyrins a much wider structural diversity occurs than is seen for the porphyrins, including variations in the central metal ion (M), in the state of ring oxidation, and in the type of substituents on the macrocycle (1–8, Fig. 2a), as well as the addition of isocyclic rings. Examples include Mg(II)-chlorophylls (chlorins) and bacteriochlorophylls (BC); Fe(II and III)-*Escherichia coli* heme *d* terminal oxidase, a chlorin; Fe(II and III)-*Pseudomonas aeruginosa* cytochrome cd_1 nitrite reductase (iBC); the Fe(II and III)-sulfite reductases and nitrite reductases (siroheme iBC's); and the even more highly reduced macrocycles of vitamin B_{12} and factor F-430 of methanogenic bacteria.

However, the apparently simple structural modification induced by the loss of one or more pyrrole ring double bonds has a profound effect on the electronic, chemical, and spectroscopic properties of the macrocycle. An example is the characteristic chlorin band at 616 nanometers in the electronic absorption spectrum of Cu(II)-tetraphenylchlorin, as compared with Cu(II)-tetraphenylporphyrin (**Fig. 3**). Some of the differences

Fig. 1. Iron protoporphyrin IX. The iron ion (Fe) is bound to the four pyrrole nitrogens, but has the potential of binding two additional ligands. L is the fifth ligand site, where the iron is bound to an amino acid from the protein. Y is the sixth ligand site, where O_2, CO, or substrates bind.

Comparison of metalloporphyrin and metallochlorin properties

Property	Porphyrin	Chlorin
Color	Red/brown	Green
Biological metal ions	Iron	Iron and magnesium
Planarity	Planar pyrrole rings	Loss of planarity at pyrroline (reduced pyrrole)
Metal-N bonds	All equal	Opposite pairs unequal
Optical activity	Optically inactive	Optically active
Methine hydrogen atoms	Nonexchangeable	Exchangeable
Methine carbons	Electron density equivalent for all	Increased electron density at two methines next to pyrroline ring
Ligand affinity	Lower	Higher
Symmetry	Effective D_{4h} symmetry	$C_2(x)$ symmetry; x bisects pyrroline
Core size	Smaller	Larger
Other		Evidence of faster electron transfer
		More easily oxidized by about 200 mV

between metalloporphyrins and metallochlorins are shown in the **table**.

Substituents. The macrocyclic substituents of chlorins and more highly reduced porphyrins also vary far more widely than is typical for porphyrins. The loss of a pyrrole ring double bond leads to the possibility of up to four substituents on each pyrroline (reduced) ring, in contrast to two pyrrole substituents for analogous porphyrins. The absolute configuration is also relevant, since chirality may be induced if the new substituents are inequivalent. Thus, the elucidation of the structure of a given metallochlorin requires precise location of the pyrroline ring on the macrocycle, and identification of the substituents to the macrocycle and their precise location, as well as assignment of the stereochemistry for the pyrroline substituents. Whereas for most iron-porphyrin proteins the prosthetic group is readily extracted, for the reduced biological porphyrins such methods often result in alteration of the structure that existed in the organism. Partly as the result of this lability, the precise chemical structure is known only for a very few of the biologically occurring chlorins.

Structural identification. Approaches to the study of metallochlorin structures and function are diverse, ranging from spectroscopic analysis of biological and model chlorins (nuclear magnetic resonance, resonance Raman, infrared, electron paramagnetic resonance, magnetic, and natural circular dichroism spectroscopy) to x-ray crystallography, electrochemistry, and investigation of ligand-binding properties. Resonance Raman spectroscopy is of particular value in the study of biological and model metallochlorins,

both because the method is nondeleterious and requires relatively low sample concentrations and volumes, and because diagnostic spectral properties for the chlorin prosthetic group have recently been established. The fact that resonance Raman spectroscopy can distinguish between chlorins which differ only in their pyrroline ring substituents also provides a foundation for precise structural identification of unknown biological chlorins.

Much of the ongoing work depends on the efforts of synthetic chemists, who prepare structurally defined chlorin and porphyrin complexes to model the biological macrocycles. The isolation and purification of chlorin-containing proteins from biological systems is also a key effort; where possible, chlorin prosthetic groups are extracted for structural evaluation. Efforts to reconstitute heme proteins with chlorin model complexes are also in progress.

Sulf-heme proteins. Under certain pathological conditions the iron protoporphyrin IX prosthetic group of the oxygen-binding heme proteins myoglobin and hemoglobin is converted to a sulfur-modified iron chlorin prosthetic group (sulf-form). The presence of this sulf-form of hemoglobin in the blood of humans leads to a type of anemia (sulfhemoglobinemia). Generation of the sulf-chlorin re-

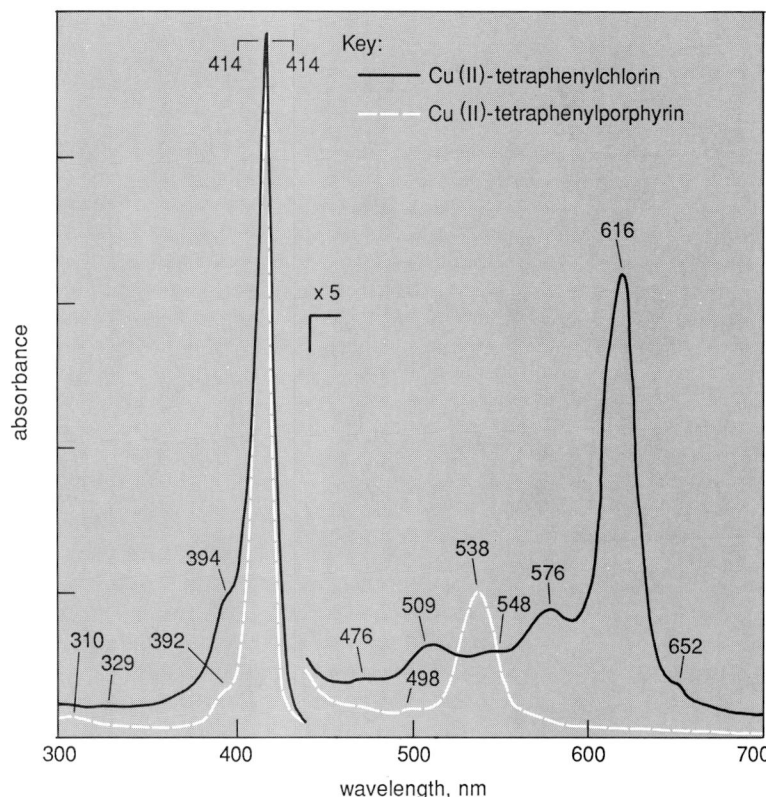

Fig. 3. Electronic absorption spectra of Cu(II)-tetraphenylporphyrin and Cu(II)-tetraphenylchlorin. The symbol with ×5 denotes that the data on the right have been enlarged by a factor of five. (*After L. A. Andersson et al., Resonance Raman spectroscopy of metallochlorins: 2. Properties of meso-substituted systems, J. Amer. Chem. Soc., 108:2908, 1986*)

sults from a reaction between a ferryl [Fe(IV)=O] complex of iron protoporphyrin IX–containing protein and an exogeneous sulfur compound. The resulting chlorin has a sulfur moiety on the pyrroline ring, but the structure of the sulf-chlorin in sulf-myoglobin (the most thoroughly studied sulf-heme protein) is still under investigation. It has recently been suggested that the irreversible inactivation of thyroid peroxidase and lactoperoxidase which occurs upon treatment of thyroid diseases with the sulfur-containing drugs used to treat thyroid disease (goitrogens) is also an example of sulf-heme formation.

The electronic absorption spectral properties of sulf-thyroid peroxidase and sulf-lactoperoxidase differ from those of sulf-myoglobin and sulf-hemoglobin. Since all of the native proteins have iron protoporphyrin IX as a prosthetic group, these observations suggest that the sulf-chlorin prosthetic groups of thyroid peroxidase and lactoperoxidase must differ structurally from those of hemoglobin and myoglobin. The requirements for sulf-heme formation are as yet unclear, since it is not possible to generate a sulf-chlorin from either protoporphyrin IX–containing horseradish peroxidase or protein-free iron protoporphyrin IX. Synthesis of model sulf-chlorins is now in progress, and the sulf-chlorin prosthetic group of sulf-myoglobin has recently been stabilized and shown to be extractable, providing means for further structural analysis. Both nuclear magnetic resonance and resonance Raman spectroscopy have been of value in the investigations of sulf-heme systems.

Other complexes. Whereas the sulf-heme proteins are atypical, other biological chlorin complexes occur naturally and, as such, can be considered to have been selected by nature for a specific functional role. For example, the chlorin heme *d* of the *E. coli* terminal oxidase is produced only under conditions of oxygen limitation. This enzyme has a remarkably high affinity for oxygen. The heme *d* chlorin prosthetic group has been shown to be related structurally to protoporphyrin IX, with vinyl substituents at the 2 and 4 positions and either the C or D ring reduced. However, the identity of the pyrroline ring substituents is still unclear: both diol and lactone structures have been proposed on the basis of mass, nuclear magnetic resonance, infrared, and resonance Raman spectroscopic analysis of the extracted prosthetic group and model chlorin complexes. The present consensus is that the heme *d* chlorin in the organism has two hydroxy substituents on the reduced ring which will readily cyclize to form a lactone upon extraction and handling. Iron chlorin model complexes for heme *d* (bearing the two vinyl substituents and a diol on the pyrroline ring) have recently been synthesized and are under investigation.

Resonance Raman spectroscopy of bovine spleen green heme protein and leukocyte myeloperoxidase strongly supports the presence of iron chlorin prosthetic groups in these systems. Again, the precise identity of the macrocyclic structures and their peripheral substituents are not yet known. However, spectral similarities between the two iron chlorin proteins suggest a high degree of structural similarity. The catalase found in the fungus *Neurospora crassa* was also shown to have an iron chlorin prosthetic group, in contrast to the iron protoporphyrin IX macrocycle found in all other known catalases. The *N. crassa* chlorin is not yet structurally defined.

Future research. In many respects the investigation of biological and model metallochlorins is still in its infancy. While the structural and chemical properties of metalloporphyrins are fairly well defined, this is not the case for metallochlorins at present. Furthermore, the functional advantage of the chlorin macrocycle relative to that of porphyrins is not clearly understood. A significant new direction in research on metallochlorins and other reduced porphyrins is the study of isobacteriochlorins, bacteriochlorins (bacteriochlorophylls), and the even more highly reduced tetrapyrrole macrocycle of the coenzyme F-430, found in methane-producing bacteria. It was recently shown that the green heme d_1 prosthetic group of the *Pseudomonas aeruginosa* nitrite reductase is not a chlorin, as had been assumed for nearly two decades, but a novel dioxo-isobacteriochlorin. New chlorin-containing proteins are being discovered at an increasing rate, and the number of scientists studying the structural, chemical, spectroscopic, and functional properties of metallochlorins is rapidly expanding. SEE SPECTROSCOPY.

For background information SEE CHLOROPHYLL; HEMOGLOBIN; PORPHYRIN; PYRROLE; RAMAN EFFECT; STEREOCHEMISTRY in the McGraw-Hill Encyclopedia of Science and Technology.

Laura A. Andersson

Bibliography. L. A. Andersson et al., Resonance Raman spectroscopy of metallochlorins: Models for green heme protein prosthetic groups, *J. Amer. Chem. Soc.*, 107:182–191, 1985; L. A. Andersson et al., Sulfmyoglobin: Resonance Raman spectroscopic evidence for an iron-chlorin prosthetic group, *J. Biol. Chem.*, 259:15340–15349, 1984; M. J. Chatfield et al., Multiple forms of sulfmyoglobin as detected by ^{1}H nuclear magnetic resonance spectroscopy, *Biochem. Biophys. Res. Commun.*, 135:309–315, 1986; R. Timkovich et al., Proposed structure of heme *d*, a prosthetic group of bacterial terminal oxidases, *J. Amer. Chem. Soc.*, 107:6069–6075, 1985.

Metalloprotein

More than half the elements in the periodic table are metals, and many, both transition and nontransition metals, are important to the structure and function of proteins. Three transition metals, iron, zinc, and copper, are known as the trace elements. These are the most abundant transition metals found in biological systems, and occur at gram levels in mammalian organisms in a ratio of approximately 2:1:0.1. Other transition elements necessary to biological systems, but in much smaller quantities (milligram levels), are known as the ultratrace elements. This article focuses on proteins containing one or more of the trace elements, and models for these metalloproteins.

Iron. In many iron-containing proteins an iron ion coordinated by the four nitrogens in the center of a large planar macrocycle called a porphyrin (in many cases, protoporphyrin IX) is referred to as a heme group. The most studied heme protein is hemoglobin, the vertebrate dioxygen-transporting protein in the blood which contains four heme groups per molecule and carries one O_2 molecule per heme group. Myoglobin, a similar molecule found in muscle tissue, stores oxygen but contains only one heme group and, consequently, one O_2 molecule per molecule. Other proteins containing heme groups are peroxidase and catalase, which are important in peroxide metabolism, and the cytochromes. The cytochromes *a*, *b*, and *c*, unlike most heme proteins, contain iron which is six-coordinate, and they transfer electrons in cell respiration rather than catalyze chemical reactions. Four of the coordinating groups are the nitrogens from the porphyrin, and two groups are from the protein. The protein groups are usually histidines or sulfur-containing amino acids. The reduction potential of the Fe(II)/Fe(III) couple depends on the fifth and sixth ligands on the protein. Cytochrome *c* oxidase is an unusual molecule because it contains two heme groups and two coppers; one copper and one heme seem to be coupled through an imidazole bridge. The protein as a whole can transfer four electrons. How electrons are transferred over quite large distances in and between proteins is a current area of research.

Another cytochrome with a different function is cytochrome P-450. A sulfur of a cysteine occupies the fifth site. In the active form a dioxygen molecule occupies the sixth site, is converted to an oxide during the reaction, and as a result converts the heme group to the Fe(IV)-porphyrin radical. The active oxygen is used to oxygenate a nonpolar substrate to make it more soluble in aqueous solution and more excretable by the organism. A water molecule is the other product of the reaction. Cytochrome P-450 belongs to the class of enzymes called mono-oxygenases where one atom of a dioxygen molecule is added to a substrate and the other is converted to water. *See* TOXICOLOGY.

The iron storage and transport proteins, ferritin, hemosiderin, and transferrin, are nonheme iron proteins. Another important class of nonheme-containing proteins is composed of iron-sulfur proteins. These contain variable numbers of irons and sulfurs and are electron-transport proteins. Depending on the oxidation states and numbers of the irons, the potentials of these proteins can vary from +350 to −600 mV. Other nonheme iron proteins include certain ribonucleotide reductases, which are important in ribose-to-deoxyribose conversion, as well as dioxygenases, which oxidize substrates like catechol.

Zinc. The second most abundant trace element is zinc, found in over 200 different enzymes. Because it exists only in the 2+ oxidation state, it can act as a Lewis acid and participate in the catalytic function of the enzyme without the interference of oxidation or reduction. Catalytic zinc has been found in all classes of enzymes: hydrolases, isomerases, ligases, lyases, oxidoreductases, and transferases. Other than catalysis, zinc may be important for the structure of an enzyme; or it may participate in the regulation of an enzyme. Its function may not be known, in which case it is referred to as noncatalytic zinc. Carbonic anhydrase and carboxypeptidase were the first and most studied zinc-containing enzymes. Both contain catalytic zinc sites.

Other enzymes such as alcohol dehydrogenase and alkaline phosphatase contain more than one kind of zinc. Each contains a catalytic zinc and at least one other type. A group of enzymes important in recombinant deoxyribonucleic acid (DNA) work are the zinc-containing ribonucleic acid (RNA) and DNA polymerases. A major technique for studying zinc in proteins has been metal substitution. Zinc(II) is spectroscopically silent, and it needs to be replaced by other metals for the metal site of the protein to acquire spectroscopic properties. In general, zinc is coordinated to proteins through histidine, carboxyl groups, or sulfur-containing amino acids. The catalytic zinc is coordinated by three of these protein side chains and a water molecule, and the coordination site is found to be distorted from tetrahedral geometry. By contrast structural zinc is generally found to be tetrahedrally coordinated by four groups from the protein. Zinc is transported and stored in the body by metallothionein, macroglobin, transferrin, and albumin. A copper-zinc protein which contains a catalytic copper and a structural zinc is superoxide dismutase.

Copper. The copper in metalloproteins is classified into four categories. The type I copper is found in copper blue proteins, which appear to have an intense charge-transfer band at 600 nanometers from a sulfur of a coordinated cysteine to the copper(II) center and an unusual electron paramagnetic resonance spectrum. These blue proteins, plastocyanin, azurin, stellacyanin, and others, have electron-transfer functions. Type II copper proteins contain copper which behaves spectroscopically like copper(II) coordinated by ordinary, non-sulfur-containing inorganic ligands. This type II copper occurs in copper enzymes with a number of different functions, for example, dopamine beta-hydroxylase, superoxide dismutase, and galactose oxidase. Type III nonblue copper proteins contain pairs of copper(I) which, when oxidized to copper(II), give no electron paramagnetic resonance spectrum. This type of copper is important for oxygen metabolism with hemocyanin, an oxygen transport protein in mollusks and crustaceans, and tyrosinase, a mono-oxygenase belonging to this group. Tyrosinase is responsible for the brown color in bruised fruit and the black noses and tail tips in Siamese cats. The fourth type of copper protein, the copper oxidases, contains all of the other three. Laccase, ascorbate oxidase, and ceruloplasmin belong to this class of copper proteins.

Characterization. Studying trace elements in proteins is a difficult task. Metalloproteins have an enormous range of molecular weights, from the smallest iron sulfur proteins with molecular weights of about 6000 daltons to very large proteins with molecular weights near 200,000 daltons for proteins like cytochrome oxidase. In many of the proteins the metal concentrations are very small, in the part-per-thousand range, and consequently the bioinorganic chemist studies very small amounts of metal ion in a very large protein matrix. One strategy that has been used to study the metal in metalloproteins is to make models (molecular weight less than 2000 daltons) of the metal-containing sites. Once a metalloprotein has been isolated and the physical and chemical properties of the metal-containing site determined, the production of a model or synthetic analog can start. This may involve designing special ligand systems. After the synthetic analog has been made, isolated, and shown to have structural and spectroscopic properties similar to the metalloprotein, its chemical reactivity is compared to the real system. If these match also, then more detailed comparisons and studies with the model can be made to investigate further the extent of the function of the protein matrix.

In model systems the parameters such as ligation, substitution on ligands, solvent polarity, substrate, and hydrogen bonding can be varied in a controlled manner; this allows the experimenter to improve the model, to find the limits of the model, and to gain a better understanding of the reactivity of the model and ultimately the metalloprotein. Another strategy, if the structure of the protein is not known, is to make models which have similar functions, and to infer the structure of the metal site in the protein from the model.

Modeling. Of the many metalloproteins mentioned above, a large number of them, particularly the iron proteins, have been modeled. Hemoglobin, one of the easiest metalloproteins to obtain, and myoglobin have been modeled extensively. Several successful models of the dioxygen (O_2)-binding site in hemoglobin and myoglobin have been made. All contain a heme group, with the iron in some kind of pocket to prevent

autoxidation of the iron(II) when it binds to dioxygen. As long as a second iron cannot get close to the first to form an Fe-O-O-Fe bridge, autoxidation cannot occur but O_2 can bind.

Synthetic analogs with bulky protecting groups on the porphyrin are referred to as picket fence porphyrins, capped porphyrins, cyclophane hemes, and lacunar complexes (**Fig. 1**). All mimic the dioxygen-binding and spectroscopic properties of the natural proteins. By using more specific variations of the models, it was possible to determine what factors allow the protein to discriminate between carbon monoxide group (CO) and dioxygen binding. Since CO could bind strongly enough to the iron site to poison the organism, natural proteins have means for preventing this. The observed difference in binding between CO and O_2 is that CO binds linearly to the iron but the O_2 binds to the iron at about a 130° angle. The distal histidine in hemoglobin may interfere with the linearly bound CO, and this has been tested with capped models.

Picket fence models. Investigations with picket fence models in varied solvents have shown that increasing solvent polarity reduces the rate at which the oxygen comes off the iron and increases the rate of CO loss. Hydrogen bonding from the distal histidine may also slow the O_2 loss, but not the CO loss. Other subtle factors have been investigated as well. Currently being modeled are cooperativity (increased ease of binding of dioxygen molecules after the first one binds to hemoglobin) and the difference in the forms of the iron-binding site; the so-called tensed, or T, state, with the iron out of the plane of the porphyrin and toward the fifth ligand (the proximal histidine) when O_2 is not bound; and the so-called relaxed, or R, state, when O_2 is bound and the iron is in the plane of the porphyrin. By varying the substitution on the imidazole group (the fifth ligand) or attaching a variable-length tail to this ligand and attaching it to the porphyrin, the iron can be prevented from moving into the plane of the porphyrin; this weakens the O_2 binding and mimics the T state. Modeling for these proteins continues to be very successful.

Cytochrome P-450. Another heme-containing protein for which several properties have been successfully modeled is cytochrome P-450. One model consists of an Fe(III)-containing porphyrin, 5,10,15,20-tetramesitylporphyrin (TMP; Fig. 1), attached to a sulfur-containing fifth ligand. In the presence of an oxide ligand, this model could then be oxidized to an unstable green species which showed spectroscopically that it contained an Fe(IV)-porphyrin radical. This intermediate is also postulated in the P-450 reaction. The model also mimics the function of P-450. With iodosylbenzene (OIPh, where Ph = C_6H_5) as the oxide ligand source, the model will transfer an oxygen to an alkene to form an epoxide. If bulky groups are attached to the porphyrin, then the reaction

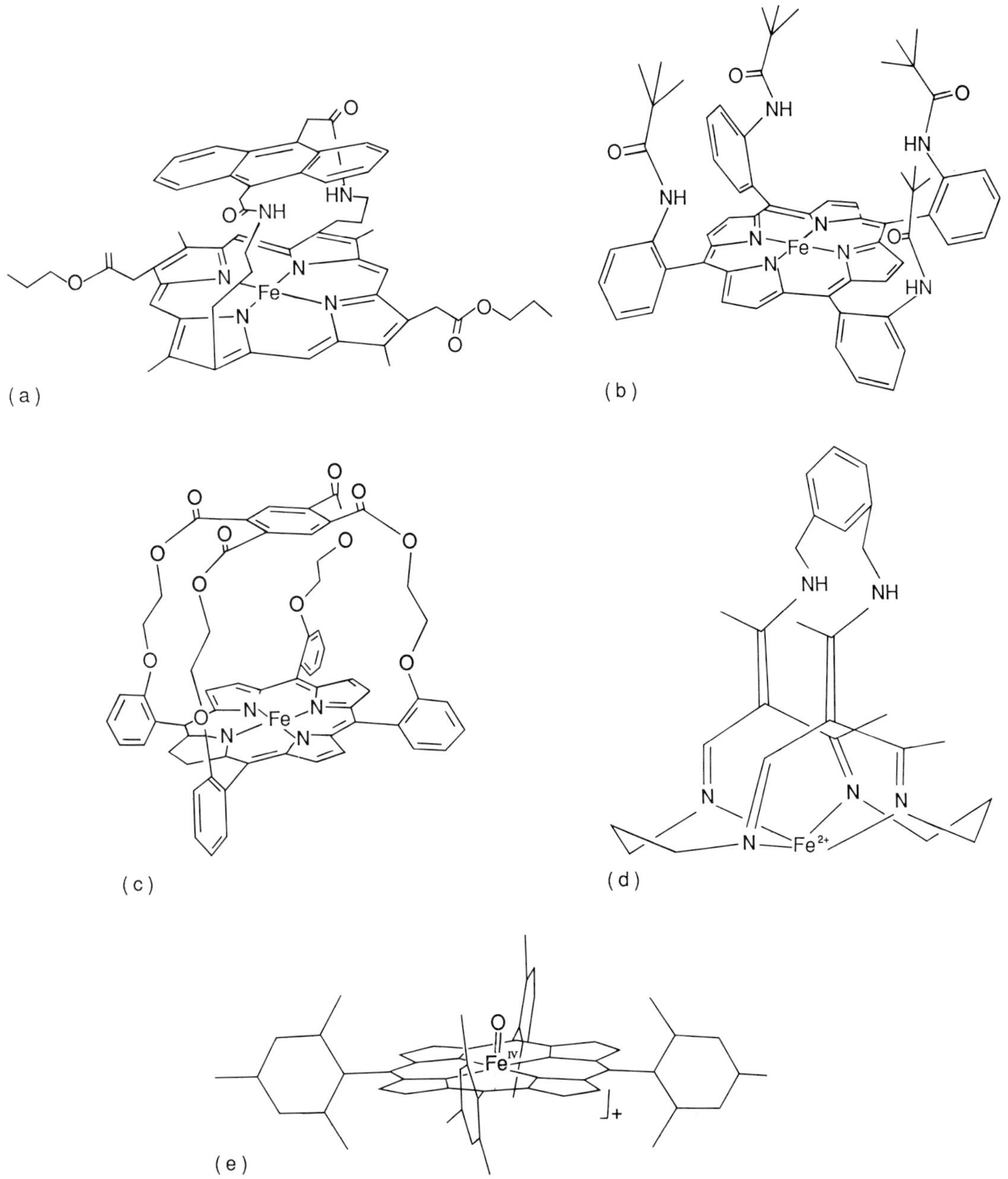

Fig. 1. Heme models. Hemoglobin and myoglobin models: (*a*) cyclopropane heme, (*b*) picket fence porphyrin, (*c*) capped porphyrin, (*d*) lacunar complex (*after K. S. Suslick and T. J. Reinert, The synthetic analogs of O₂-binding heme proteins, J. Chem. Ed, 62(11):974–983, 1985*). (*e*) Model for cytochrome P-450: 5,10,15,20-tetramesitylporphyrin (*after J. T. Groves, Key elements of the chemistry of cytochrome P-450, J. Chem. Ed., 62(11):928–931, 1985*).

with alkenes can be made stereoselective. The model shows the same stereoselectivity as P-450, while models made with other metals give different stereoproducts. This model is undergoing further modifications to obtain more information about the P-450 reaction.

Iron proteins. Extensive modeling has been done for the nonheme-containing iron proteins, the iron sulfur proteins. These proteins contain either one iron bound tetrahedrally by four cysteines; two irons, two inorganic sulfurs, and four coordinated cysteines; three irons, three inorganic sulfurs and six cysteines; or four irons and four inorganic sulfurs arranged cubically and held to

the protein by four cysteines (**Fig. 2**). *See Coordination chemistry*.

Zinc proteins. Although the method which has been used most extensively for studying zinc binding in zinc proteins has been metal substitution, two kinds of modeling have been done. One has been to design a ligand which mimics the metal-binding site in the protein. (For zinc and copper, ligand design is more difficult than in the models that have already been discussed, because not only are both metals held to proteins by amino acid side chains, but the protein imposes a certain geometry on the metal-binding site.) The other type of modeling is functional modeling,

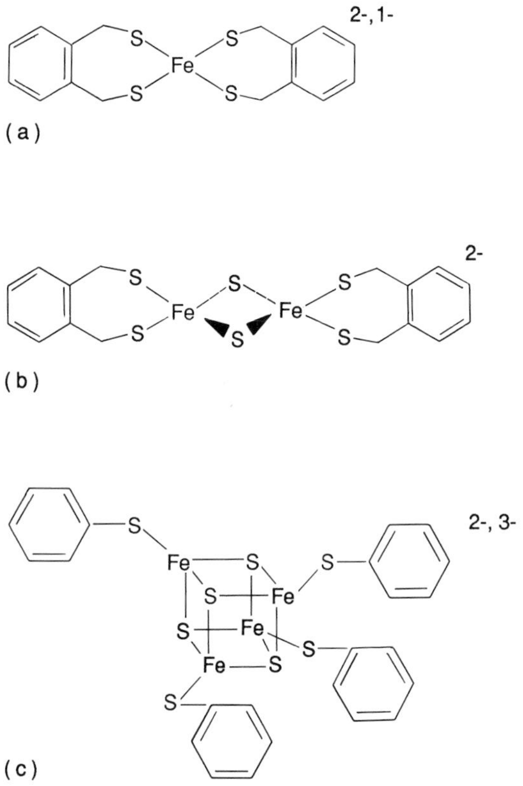

Fig. 2. Iron-sulfur protein models. (*a*) One-iron-site model, Fe(S-orthoxylene)₂. (*b*) Two-iron-site model, Fe₂S₂(S-orthoxylene)₂. (*c*) Four-iron-site model, Fe₄S₄(S-C₆H₅)₄. (*After R. H. Holm, Synthetic approaches to the active sites of iron-sulfur proteins, Acc. Chem. Res., 10:427–434, 1977*).

(II)'s and an imidazole bridge (Fig. 3). The coppers in this complex are pentacoordinate with three sites occupied by the ligand, one site by the bridging imidazole, and one site from one more imidazole for each copper.

For type III copper proteins a number of ligands have been synthesized which will coordinate two copper(I) ions and react with dioxygen, but few will bind dioxygen reversibly. Recently a ligand was made which seems to mimic the function of hemocyanin (Fig. 3), but in many other models the copper is irreversibly oxidized to copper(II). What is probably needed is a model ligand which can allow for variable coordination numbers, depending on the oxidation state of the coppers, and in which the coppers can transfer some electron density to the O_2, but not be totally oxidized to copper(II) oxide or oxidize the ligand. Models have not been made which totally reproduce the spectrum or function of these type III copper proteins.

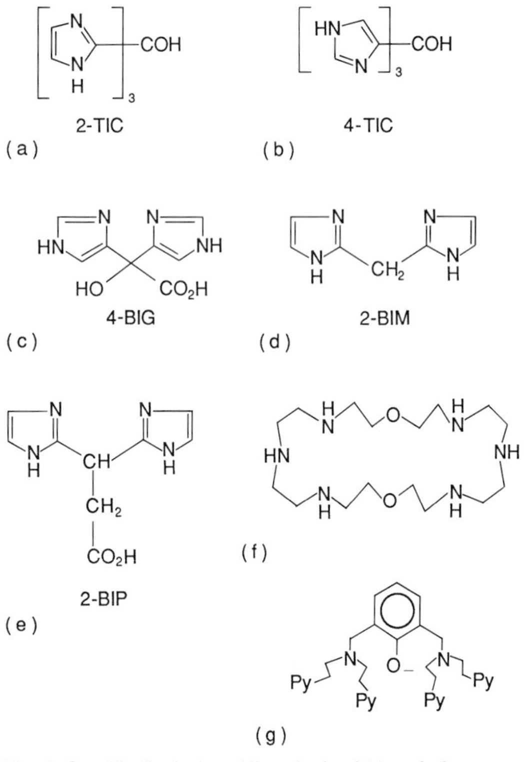

Fig. 3. Specifically designed ligands. (*a–e*) Ligands for binding zinc(II) (*after C. C. Tang et al., Model for metal binding sites in zinc enzymes: Synthesis of tris[4(5)-imidazolyl]carbinol (4-TIC), tris(2-imidazolyl)carbinol (2-TIC) and related ligands, and studies on metal complex binding constants and spectra, J. Amer. Chem. Soc., 100:3918–3922*). (*f*) Polyamino ligand for binding two coppers bridged by an imidazole as a model for superoxide dismutase (*after P. K. Coughlin et al., Synthesis and structure of the imidazoalate bridged dicopper(II) ion incorporated into a circular cryptate macrocycle, J. Amer. Chem. Soc., 101:265–266, 1979*). (*g*) Ligand designed to bind copper and model the properties of hemocyanin. Py = pyridine (*after K. D. Karlin and Y. Gultneh, Bioinorganic chemical modeling of dioxygen-activating copper proteins, J. Chem. Ed., 62(11):983–990, 1985*).

where the Lewis acid catalysis by the zinc is mimicked. To mimic the protein catalytic zinc site of two or three histidines and perhaps a carboxyl group, a number of ligands such as 2- or 4-tris(imidazolyl) carbinol (TIC), or 2- or 4-bis(imidazolyl) methane (BIM), or bis[4(5)-imidazolyl] glycolic acid (BIG) and 3-bis(2,2′-imidazolyl) propionic acid (BIP) [**Fig. 3**], were synthesized and tested for their metal-binding properties with dipositive first-row transition-metal ions. None of the synthetic ligands were as selective for zinc as either carbonic anhydrase, the CO_2-hydrating enzyme, or alkaline phosphatase, which hydrolyzes orthophosphate monoesters at alkaline pH. The complexes that formed with these ligands tended to be six-coordinate instead of four-coordinate.

The functional models for Lewis acid catalysis have tended to use substitution inert cobalt(III) complexes as the Lewis acids and then have extrapolated to the behavior of zinc enzymes. For example, the cobalt complex studies show increased rates (by orders of magnitude) of hydrolysis compared to uncatalyzed hydrolysis rates, and this is also expected to be true for zinc-promoted hydrolysis.

Copper proteins. As with the zinc proteins, ligand design plays an important role in creating models of copper proteins. Superoxide dismutase, which contains an imidazole which bridges the copper and zinc, has been modeled with a cyclic polyamino ligand which can accommodate two copper

For background information SEE COMPLEX COMPOUNDS; COORDINATION CHEMISTRY; CYTOCHROME; ENZYME; PORPHYRIN; PROTEIN; SPECTROSCOPY; TRANSITION ELEMENTS in the McGraw-Hill Encyclopedia of Science and Technology.

Nancy Rowan Gordon

Bibliography. Bioinorganic chemistry issue, *J. Chem. Ed.*, vol. 62, no. 11, 1985; H. B. Dunford et al. (eds.), *The Biological Chemistry of Iron*, 1982; M. N. Hughes, *The Inorganic Chemistry of Biological Processes*, 1981; T. G. Spiro (ed.), *Zinc Enzymes*, 1983.

Metallurgy

Conventional extraction methods for the refractory metals are complex and expensive, and recent developments in the modeling and design of plasma reactors show promise of being economically attractive in the production of these materials. The techniques are simple, and the plants promise to be small (low-capital) and efficient, requiring little labor for high productivity. Operation can be continuous, instantaneous startup or shutdown is possible, and relatively low working capital is required. There should be few environmental problems since the units are totally enclosed, require very little gas, and provide nearly noiseless operation. Plasma technology is playing an increasingly important role in extractive metallurgy, particularly for refractory high-melting-point corrosion-resistant metals requiring highly endothermic reactions for their production.

Recent advances in plasma processing have improved the prospects for large-scale implementation of this technology by the specialty metal, iron and steel, extractive metallurgical, and ceramic industries. Although few designs for electric plasma devices have been developed to production scale, an increasing number of companies are conducting research in this technology.

Melting. The Soviet Union, East Germany, and Japan have pioneered the commercial use of plasma reactor systems for melting refractory materials. A Japanese company has developed top-entry remelting reactors for alloy steels, superalloys, and nonferrous metals. Major advantages have been demonstrated by this technology over the commonly used electric-arc furnace. They include use of a greater than 50% scrap charge, the capability to melt a wide range of alloys, a high level of cleanliness, good slag refining, lower melting costs, and higher yields. The only disadvantage is the relatively high cost of the argon gas at required flow rates.

In East Germany, commercial primary plasma melting reactors use a multiplicity of plasma torches projecting through the furnace sidewalls to attain the necessary energy input not yet available in single torches. At a plant in Freital, 10- and 30-ton (9- and 27-metric-ton) commercial plasma melting units are in continuous operation melting iron-and-nickel–based alloys, stainless steels, and high-alloy tool steels in an argon environment. A unit of similar design is in operation in the Soviet Union at Alma Ata.

A direct-current plasma-arc furnace has been developed to replace conventional arc furnace melting of scrap at Vasteras, Sweden. The new design decreases electrical arc noise and electrode consumption while melting finely divided dusts without normal agglomeration. The only device of this type in commercial operation in the United States is a 5-ton (4.5-metric-ton) titanium scrap agglomerating unit located in Detroit, Michigan, to produce bulk weldable electrodes for vacuum-arc refining.

Combined melting and casting. Both consumable- and nonconsumable-electrode plasma units combining melting and ingot casting are in operation in the Soviet Union and Japan for the melting and remelting of weldable electrodes for the steel industry, and scrap reclamation and consolidation, respectively. Commercial consumable-electrode furnaces for the production of nitrogen-alloyed steels are already in use in the Soviet Union, but the most recent advances have been made with units incorporating continuous ingot withdrawal in cold-crucible nonconsumable-electrode systems.

A continuous ingot withdrawal system is under development in Japan for the purification and casting of titanium alloys made from sponge, scrap, and pure titanium. Called the plasma progressive casting reactor, it uses a single thermal plasma source. A plasma skull casting reactor, using multiple plasma sources, is also under development for production of castings of reactive metals and alloys, for example, intermetallic compounds that have application for hydrogen storage. In such a reactor, the plasma torch is used in a water-cooled container in which a layer of frozen metal (skull) protects the container from attack by the molten alloy.

Extractive metallurgy. Applications of plasma technology to extractive metallurgical operations are multiplying at an accelerated rate and could revolutionize certain conventional operations. In the iron and steel industry, plasma-arc systems of 100 MW are already in use.

Advanced designs of transferred-arc reactors offer unique opportunities in the production of refractory metals such as zirconium, tungsten, niobium, tantalum, chromium, molybdenum, and vanadium. Furnaces of 100-ton (90-metric-ton) capacity using six 3.5-MW plasma torches for power input are being used in the Soviet Union to smelt ferroalloys and produce hot metal to specifications by using a combined charge of steel scrap and direct-reduced iron. With these furnaces, a continuous alloy-steel production operation from ore concentrate to final product is being developed.

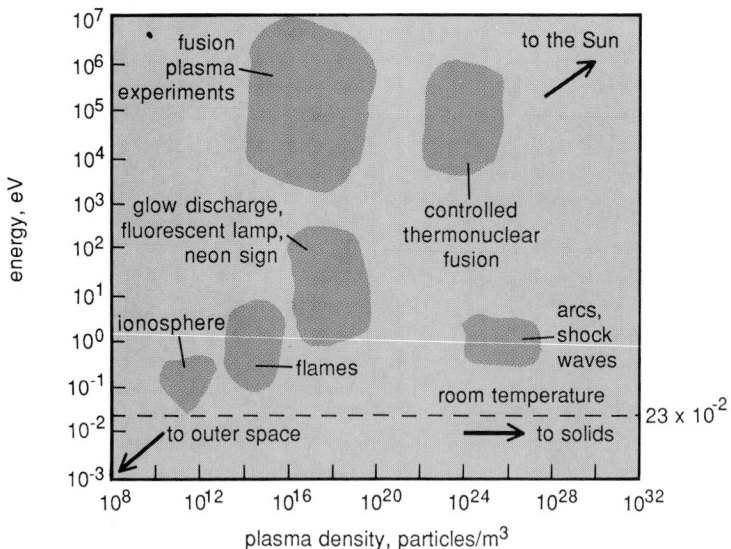

Fig. 1. Energy-density diagram of the plasma state of matter. An energy of 1 eV is equivalent to a temperature of 11,605 K.

Plasma technology. More than 99% of the matter in the known universe exists as plasma. A gas in this form has a sufficiently high energy content to cause a large fraction of the species to become ionized, resulting in a mixture of electrons, ions, and neutral species, which gives rise to high electrical conductivities. In the equipment used for metallurgical purposes, the plasma state is generally created and sustained by heat from an electric current passed through the ionized gaseous medium.

Figure 1 shows the relationship between electron densities and electron temperatures in some artificial and natural plasmas. The electron density scale can be viewed as a continuum ranging from a vacuum on the left to the solid phase on the right. In general, this gives rise to two types of plasmas: equilibrium or thermal plasmas and nonequilibrium or cold plasmas. In glow discharges, fluorescent lamps, and flames, there are too few collisions between the high-temperature electrons and the heavy particles (ions, atoms, molecules) to transmit sufficient energy to equilibrate the gas thermally. High-pressure thermal plasmas, characterized by welding arcs, arc furnaces, plasma torches, and controlled nuclear fusion, have sufficient numbers of collisions between the high-temperature electrons and heavy particles to approach closely thermodynamic equilibrium.

There are three principal methods for generating thermal plasmas (**Fig. 2**). A transferred arc is shown in Fig. 2a, where the work or the material to be heated is one electrode. In a plasma torch or nontransferred arc (Fig. 2b), gas is heated by passing it through or near an arc discharge. Thermal plasmas can be generated by means of radio-frequency discharges (Fig. 2c), but such devices are difficult to scale up, and in commercially available large-scale furnaces the plasma is generated by a high-intensity arc. The transferred arc is used when the work itself can be used as an electrode and gas usage is detrimental to the process. A plasma torch is more often used when gas is not detrimental or is preferred.

Although the multicomponent nature of actual plasma systems and their geometric complexity make mathematical modeling extremely difficult, simplified models have been developed that greatly assist in the understanding of the basic trends and parameters governing the behavior of such systems and in the primarily empirical development of workable furnace designs. Comparisons between predicted and measured isotherms in free-burning high-intensity arcs show reasonable agreement between predictions and measurements.

Commercial equipment. **Figure 3** is a schematic representation of a combined plasma- and induction-heated furnace. The Japanese have pioneered this concept of combining plasma transferred-arc heating with induction heating. This technique of integrating a thermal plasma device with induction melting provides a number of advantages, including a protective inert argon atmosphere over the molten pool and supplemental power to keep the thin layer of slag molten. The 2-ton (1.8-metric-ton) version of the plasma- and induction-heated furnace system incorporates 600 kW of primary induction heating, 400 kW of plasma transferred-arc heating, and an additional 200 kW strategically placed for production stirring, providing a total that exceeds 1 MW. Power consumption is between 1000 and 1300 kWh/ton (900 and 1200 kWh/metric ton) of melted material with about a 3-h melting time. The induction coils operate at a frequency of 150

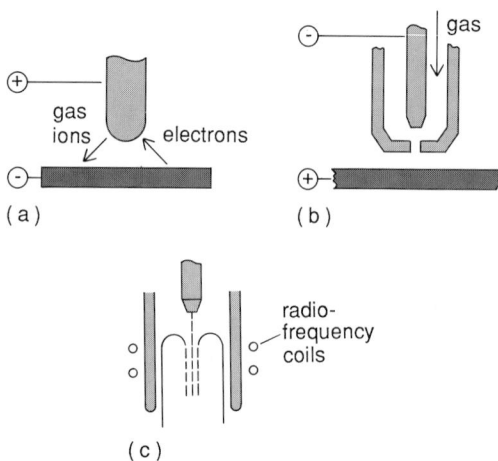

Fig. 2. Three principal methods of generating thermal plasmas. (a) Transferred arc. (b) Plasma torch (nontransferred arc). (c) Radio-frequency discharge.

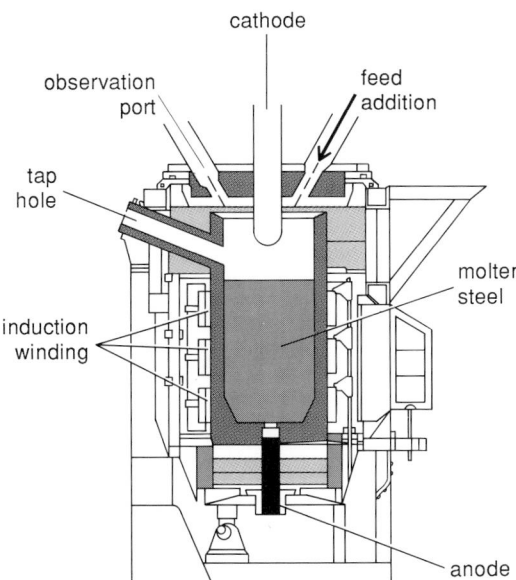

observation port

cathode

feed addition

tap hole

induction winding

molten steel

anode

Fig. 3. Combined plasma- and induction-heated furnace.

Hz and have a nominal capacity of about 200 tons (180 metric tons) per month. This furnace is capable of melting titanium alloys and making resistance alloys such as Kovar, Permalloy, and superalloys, producing a product quality equal to or better than that produced by commonly used vacuum induction furnaces at significantly lower cost and with better conservation of the alloying elements, as shown in the **table**.

The extraction of metals from their naturally occurring compounds involves either chemical reaction with a substance that can remove the compounding agent (oxygen or sulfur) or electrolysis where the compounding agent is separated in solution. In the former case, metals such as iron, lead, and copper may be reduced at moderately high temperatures in moving-bed reactors (blast furnaces). In the electrolytic process, recovery of metals, such as aluminum and magnesium, that form particularly stable oxides

Comparison of induction furnace systems		
Characteristic	Vacuum induction furnace	Plasma- and induction-heated furnace
Volatilization	High	Low
Mechanical property	Good	Good
Workability	Excellent	Good
Yield (melting) percent	98	98.9
Product yield (total)	75%	85%
Melting cost (index)	100	59
Oxygen	5–15 ppm	8–25 ppm
Nitrogen	10–30 ppm	10–50 ppm
Hydrogen	1 ppm	2–5 ppm
Desulfurization	10%	50–85%
Decarburization	90%	70%

can be achieved. The so-called refractory metals, which include chromium, titanium, molybdenum, niobium, vanadium, tungsten, zirconium, and tantalum, because of their free-energy-of-formation characteristics and high melting points, are not easily produced by either of these two processes. These metals are in great demand as alloying agents because of their high strengths and good corrosion resistance, but they are costly because of the elevated processing temperatures required for their recovery. Recently, considerable attention has been given to the possibility of using plasmas as a means of more economical production of these metals.

Only a transferred-arc type of furnace using a nonconsumable electrode and the product feed as the second electrode appears capable of meeting the required residence time and energy input for complete reaction of the ore with its reductant. There is high resistance of plasma zones to the penetration of the particulate ore to be reduced as well as to the short time of flight in the small high-temperature volume of the plasma reactor.

A unique design of a transferred-arc reactor aimed at the solution of these problems has been under way in Canada as a cooperative effort between commercial enterprises and McGill University. A 125-kW version of this new design has been demonstrated, and a 1-MW reactor is under test at Stockton-on-Tees, England; research continues in Canada in the 50–250-kW range. **Figure 4** shows the main features of this reactor, as well as the details of the cathode assembly. In this reactor the powdered feed may vary from a few micrometers to several millimeters and is introduced in a high-velocity cold carrier gas tangentially against the inner wall of a cylindrical water-cooled sleeve. There it is exposed to the intense radiation of the plasma column as it passes between the cathode assembly, with the molten product acting as an anode. The heat is so intense that the feed begins to melt as soon as it hits the inner wall, and it flows down in the form of a molten film. All the radiant heat generated by the plasma is absorbed by this falling film, which moves slowly enough to greatly increase the residence time of the reactants at high temperature and allows the reaction to go nearly to completion before the product forms into droplets 200–250-micrometers in diameter and falls into the molten bath below. As the product falls, it continues to absorb radiant heat; any additional residence time required to complete the reaction is provided in the molten bath before the product is tapped. This longer residence time distinguishes this reactor from the earlier falling-film reactor work carried out in the United States, where the cylinder on which the film was falling was the anode and the molten product bath was heated by convection only.

This design offers some important advantages

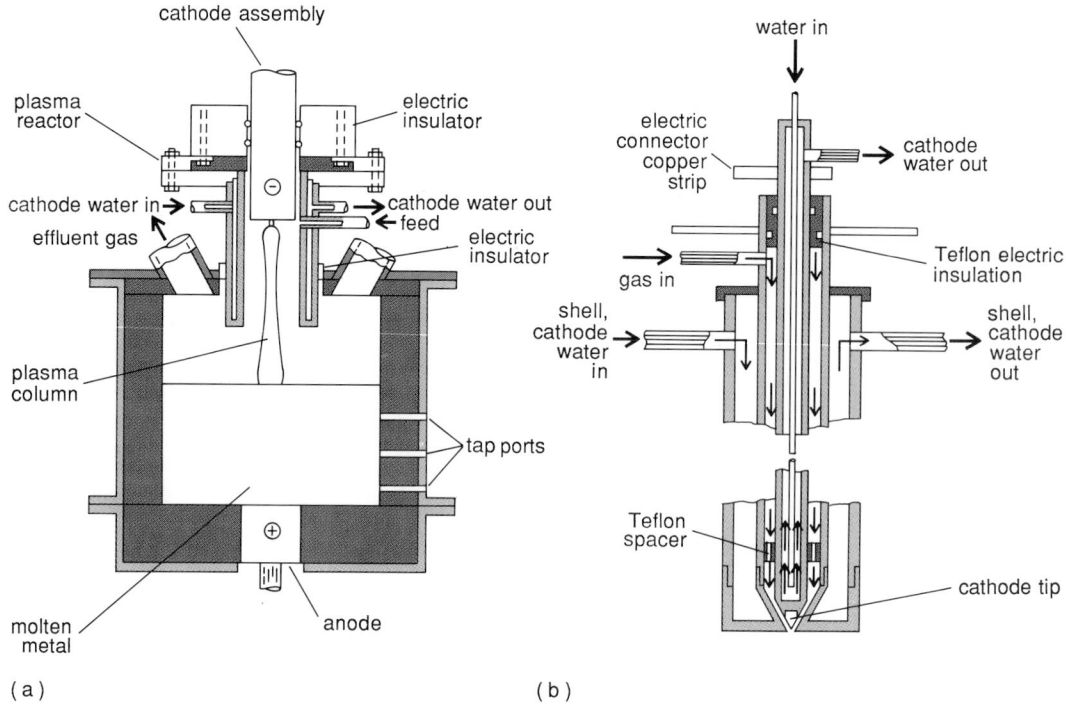

Fig. 4. Transferred-arc reactor. (*a*) Diagram of unit. (*b*) Cathode assembly detail.

such as the complete stabilization of the plasma column by the high tangential velocity of the carrier gas, allowing greater arc lengths and more efficient power introduction through higher voltages and lower currents. The high-efficiency utilization of the electrical energy eliminates the need for water cooling of the reactor, and further heat efficiency is obtained as the molten film continues to absorb radiation as it falls into the molten bath while it shields the walls of the reactor and reduces cooling requirements. The carrier gas absorbs little heat from the plasma by radiation and only a small amount by convection from the molten film on the sleeve, allowing it to help cool the back of the sleeve and the roof of the reactor as it leaves through the exit ports.

Other reactor designs address the penetration of the particulate reactants into the plasma zone and the retention necessary to complete the reactions. One Swedish reactor uses the technique of introducing the particulate reactants through the hollow cathode in a transferred-arc system. A reactor developed in Farrington, England, uses a gimbaled rotating cathode that expands the volume of the high-temperature plasma zone by describing a three-dimensional cone in space as it rotates. A Canadian design uses three-phase (alternating-current) hollow electrodes. The arc is stabilized by feeding argon, hydrogen, or carbon monoxide gas through the hollow electrodes, while the reactants are heated as they fall through a chimney countercurrent to the effluent gases; the reactants then fall through the arc into the molten layer at the bottom of the crucible. All of

these reactors have successfully produced one or more of the refractory metals previously listed.

For background information *SEE ALLOY; METAL-LURGICAL ENGINEERING; METALLURGY; PLASMA PHYSICS* in the McGraw-Hill Encyclopedia of Science and Technology.

Noel Jarrett

Bibliography. F. R. Carr et al., The design and operation of transferred arc plasma systems for pyrometallurgy, *Application Proceedings of the 6th International Symposium on Plasma Chemistry*, Montreal, Quebec, July 1983; Materials Research Society, *Plasma Processing and Synthesis of Materials*, Symposia Proceedings, vol. 30, pp. 1–11, 61–75, 78–79, 1984; National Materials Advisory Board, National Research Council, *Plasma Processing of Materials*, Publ. NMAB-415, pp. 5–16, 53–73, 1985.

Microscope

Recent advances include the development of the scanning tunneling microscope, a new type of electron microscope that shows the configuration of individual atoms; and the fluorescence digital imaging microscope, an optical microscope that combines the sensitivity and selectivity of fluorescence microscopy with digital imaging techniques.

SCANNING TUNNELING MICROSCOPY

Scanning tunneling microscopy, a surface physics technique developed recently by G. Binnig and H. Rohrer and based in part on earlier work by R. Young, promises to be an exciting tool since it shows what the smallest bits of everyday matter, atoms, look like and reveals their configuration. The scanning tunneling microscope exploits the basic principles of quantum physics. A cousin to traditional electron microscopes such as the transmission electron microscope and the scanning electron microscope, the tunneling microscope eliminates the high energies and electron lenses of both those devices by bringing the electron source so close to the sample that electrons can be induced to effectively tunnel through the vacuum barrier that separates the source from the sample. *See Nobel prizes.*

Operation. A fine metal tip, typically tungsten, is brought to within a few tenths of a nanometer of the surface of the material to be studied (**Fig. 1**). This is accomplished with piezoelectric transducers arranged to form an orthogonal coordinate system: one transducer is used to move the tip in and out above the sample, and the other two are used to translate the tip laterally over the sample surface. The conduction electrons in both the tip and the sample are exponentially attenuated as they propagate away from their respective surfaces into the vacuum; this property is crucial in determining the operating characteristics of the scanning tunneling microscope. A small bias voltage V_{bias} ranging from a few millivolts to a few volts is applied across the gap between the tip and sample, and the tip is brought up to the sample until a small current begins to flow, typically of the order of a nanoampere. A feedback loop is then employed to measure the tunneling current, compare it with a fixed reference value, and servo the tip in and out to maintain this current. Since the tunneling current varies by about a factor of 10 for every 0.1-nm change in spacing between the tip and sample, the scanning tunneling microscope can measure heights to within 0.01 nm. It has a lateral resolution of about 0.2 nm, roughly corresponding to the size scale of the atoms composing the tip and sample.

A tunneling image of the sample surface is obtained by applying a raster signal to the x and y lateral piezotransducers, thereby moving the tip in a pattern over the sample surface. The error signal from the feedback circuit V_z, corresponding to heights, is then recorded at each position. These x-y-z positions are suitably displayed, depicting an image which contains a combination of both the geometric and electronic structure of the sample surface. The apparent heights depend both on the physical location of the surface atoms and the electronic states accessible at the chosen bias voltage.

The entire device is a few centimeters in size and is typically housed in a high-vacuum chamber, although progress is being made in both ambient air and liquid operation. The scanning tunneling microscope, because of its very small gap between tip and sample and the exponential

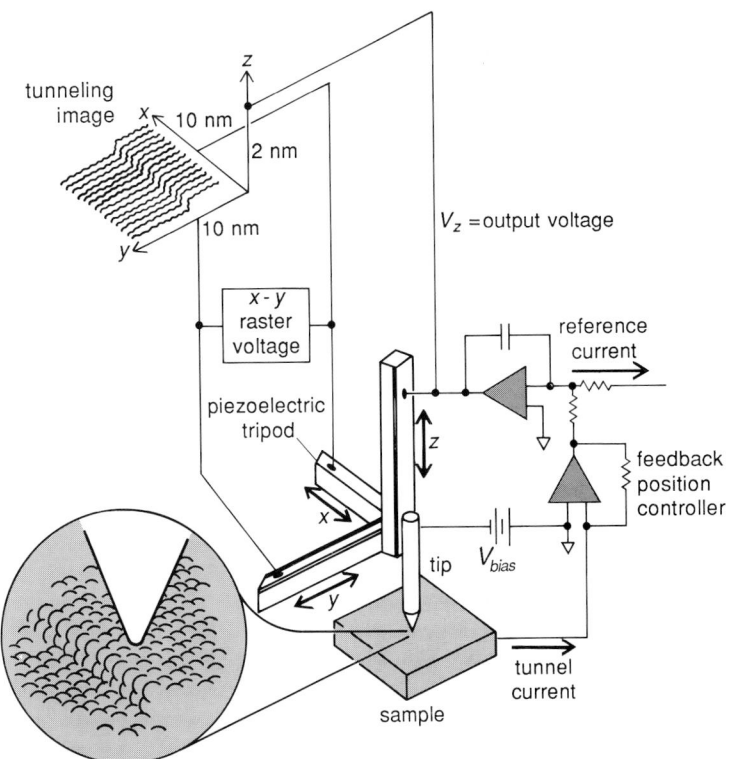

Fig. 1. Schematic diagram of a scanning tunneling microscope, showing probe tip, sample, piezoelectric transducers, feedback control, and raster scanning electronics. (*After J. A. Golovchenko, The tunneling microscope: A new look at the atomic world, Science, 232:48–53, 1986*)

behavior of the tunneling current, is highly susceptible to vibrations. For this reason, it is typically isolated from the outside world by various combinations of spring stages and acoustic filters.

Application to surface studies. To date, the scanning tunneling microscope has been most fruitfully employed in studying the configuration of atoms at crystal surfaces. In general, the atoms at the surface of a crystal are not in the positions that they would occupy within the crystal since they do not experience the same forces as the atoms in the bulk crystal. In many cases, they undergo a process called reconstruction, where the crystallographic periodicity of the surface is

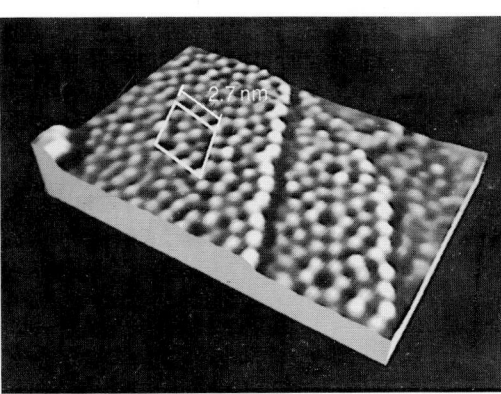

Fig. 2. Tunneling image of atomic steps on a silicon (111) 7 × 7 surface. White depicts local high points, while black denotes depressions. A 7 × 7 unit cell is outlined in white.

changed from that of the bulk. The more covalent the bonding between the crystal atoms, the more they tend to reconstruct. Semiconductors such as silicon and germanium have surface periodicities much different from that of the bulk material. This was discovered in the late 1950s and early 1960s by using the technique of low-energy electron diffraction. The (111) face of silicon in particular displays a spectacular diffraction pattern that has seven times the periodicity of the bulk; elucidating the nature of this surface occupied a number of scientists until very recently, with many models proposed to explain the information obtained from diverse techniques such as photoemission, high-energy electron diffraction, ion scattering, and Auger electron spectroscopy.

Silicon (111) surface. **Figure 2** shows a high-resolution tunneling image of a silicon (111) 7×7 surface. In this picture, local high points are rendered white, while low points are black. The large, hexagonal pattern of bumps is the silicon surface reconstruction, of which one unit cell is shown outlined. Twelve high places are seen in each cell, with the corners delineated by deep depressions. These high spots are atoms on the surface spaced about 0.7 nm apart, and serve to tie up the leftover bonds of the atoms underneath. The 7×7 periodicity is relative to the underlying bulk crystal, and runs from deep depression to deep depression as shown. Two atomic steps, the fundamental defect on a surface, are seen running across the surface. The step risers are locked into the periodicity of the 7×7 reconstruction rather than occurring at arbitrary positions, and the steps themselves are reconstructed in a manner that renders them chemically inert. This observation by the scanning tunneling microscope may explain the difficulty encountered in growing material epitaxially on the silicon $\times 7$ surface.

Germanium (111) surface. **Figure 3** shows a high-resolution tunneling image of a germanium (111) surface. In this case the surface symmetry follows a 2×8 pattern which does not show the hexagonal symmetry displayed by the silicon (111) surface. Rather, the germanium surface is made up of rows of two smaller subunits known

Fig. 4. Tunneling image of a laser-annealed silicon (111) surface. Regions of 9×9, 7×7, 4×2, and $\sqrt{3} \times \sqrt{3}$ reconstructions are indicated.

as 4×2 and 2×2, illustrated in the figure, which are stacked in an ordered fashion to produce the overall 2×8 structure. The surface atoms are here about 0.8 nm apart and again are arranged primarily to satisfy the dangling bonds from the atoms below. Unlike the silicon (111) surface, the germanium (111) surface displays a multiplicity of reconstructions. The very high specificity of the scanning tunneling microscope can be employed to examine each of these as well as the phase boundaries between them. As in the case of the silicon image, a single atomic step is seen traversing the germanium surface. Unlike the silicon steps, which are both ordered and reconstructed, the steps on the germanium (111) surface occur at arbitrary positions, and are not reconstructed.

High specificity. The tremendous power of the scanning tunneling microscope comes from its very high specificity; it does not depend on a large number of repeated structures, as do diffraction methods for determining atomic configurations. This is illustrated in **Fig. 4**, which shows a tunneling image of a silicon (111) surface after laser annealing. In laser annealing, a short pulse of very intense laser light is used to melt the surface of a material, which then subsequently recrystallizes on a time scale of nanoseconds to microseconds. The (111) surface of silicon, which normally displays a 7×7 reconstruction, is known to revert to a 1×1 diffraction pattern after laser annealing. The nature of this laser-stabilized surface was probed by many techniques, but no conclusion had been reached. The silicon surface shown in Fig. 4, annealed at low temperature after laser processing, shows many different reconstructions, some of which are outlined. There are regions of 7×7, 9×9, 4×2,

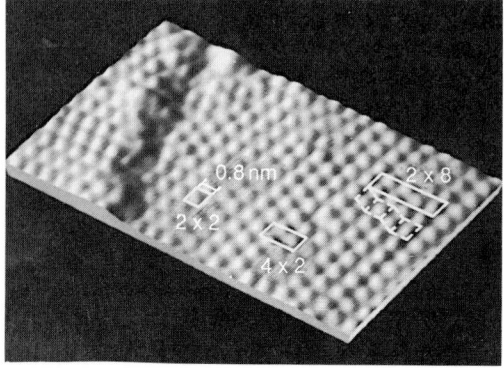

Fig. 3. Tunneling image of a germanium (111) 2×8 surface. A 2×8 unit cell is indicated, as are regions of 4×2 and 2×2 reconstructions. A single atomic step is imaged on the upper left.

and a small area of $\sqrt{3} \times \sqrt{3}$. None of these areas are individually large enough to be probed by diffraction methods, or other techniques which depend on averaging areas large on the scale of the surface features. The scanning tunneling microscope, working in real space and utilizing its atomic lateral resolution, has now allowed scientists to conduct observations on structures only a few atoms in extent.   *Russell S. Becker*

Fluorescence Digital Imaging Microscopy

Fluorescence microscopy differs from other optical light microscopy in that the observed image derives from the emission of light from the object at wavelengths longer than those used to illuminate it. The image may be the result of fluorescence evoked from intrinsic fluorophores (such as chlorophyll in plant cells) or from substances that have been introduced to the object (such as fluorescence lipids which partition in cellular membranes). The great advantage of fluorescence microscopy over other forms of light microscopy lies in its sensitivity and selectivity. Very small discrete objects in the cell, such as microtubules, can be observed even though their dimensions are below the inherent spatial resolution of the light microscope. This is because the fluorescing point object emits light which the objective collects and images against a dark background. A true absorption image of biological samples is often difficult to achieve since very small intensity differences may arise in the short path lengths encountered in the microscope. The signal from a fluorescent molecule, in contrast, is measured above a zero reference level and is thus much easier to detect and quantitate. The selectivity in fluorescence microscopy is directly related to the fact that the molecules to be looked at can be controlled by the appropriate selection of the excitation wavelength (the light with which the object is illuminated) and the emission (observation) wavelength.

As noted above, some biological molecules are inherently fluorescent, although most are not. However, other molecules or macromolecules such as DNA can be tagged or labeled by binding fluorescent compounds covalently or noncovalently to them. Thus, it is possible to look at a few or several cellular components in the absence of interfering signals from many other systems at the same time.

Advances. A new field has emerged from the combination of fluorescence microscopy with digital imaging techniques: fluorescence digital imaging microscopy (FDIM). Digital imaging brings two great advances to microscopy: detection of very low light levels and quantitation of the fluorescence intensities in the images. The purpose of digital imaging is to generate an image of the original object in digital form. Such an image can be manipulated with digital techniques: (1) to restore the true image by eliminating background, nonlinearity, and distortions of the sensor, thus focusing or defocusing problems; (2) to enhance the image by filtering to reduce the contribution from noise, by thresholding and contrast stretching to selectively emphasize certain features; and (3) to analyze the image for the fluorescence intensity (gray-value quantization) and thus determine the absolute amount of a substance in the object, or to analyze the spatial and temporal distribution of the signal in two or three dimensions.

More recently, sophisticated spectroscopic techniques have been coupled to fluorescence

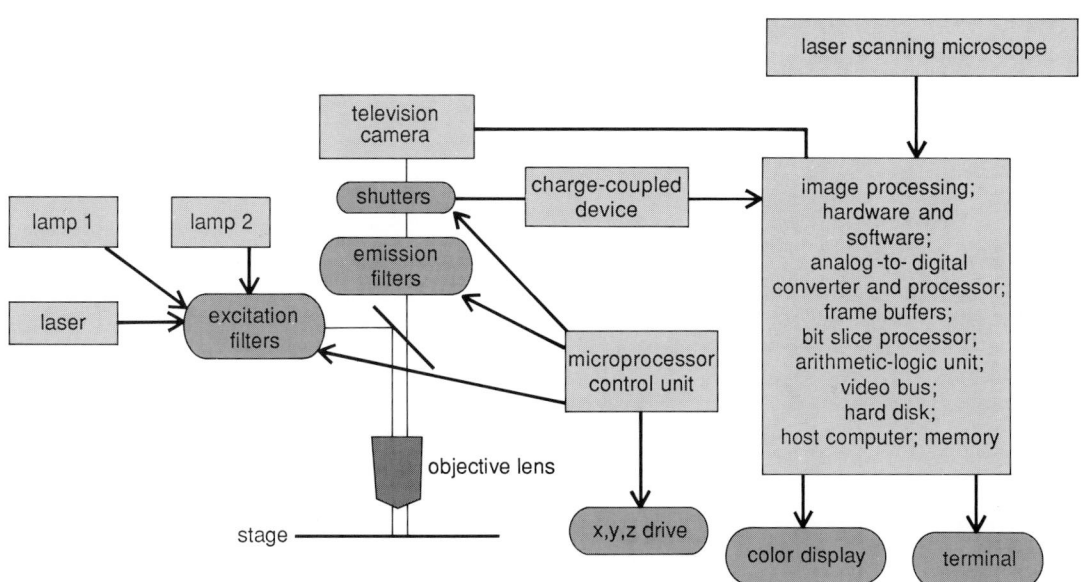

Fig. 5. Block diagram of the components of a fluorescence digital imaging microscopy.

digital imaging microscopy to address certain problems. For example, the distance between macromolecules in the cell membrane can be measured by using fluorescence resonance energy transfer. In this technique the illumination of one molecule, the donor, at a particular wavelength will lead to transfer of its energy to another molecule, an acceptor, if the latter is very close and if its excitation spectrum overlaps that of the donor emission spectrum. Characteristically, the acceptor molecule will fluoresce or emit light at a wavelength higher than that of the original donor molecule.

A number of devices can be used to acquire the image, and these can be divided into two classes: a camera which will image the entire object at one time onto a sensor, or a scanning system in which the exciting light (laser) is scanned across the object and the emission is acquired in a time-correlated fashion by a photomultipler. The former system may use a conventional vidicon television camera or an intensified television camera or a solid-state sensor such as a charge-coupled-device array. The intensified television systems and charge-coupled devices allow the image to be generated with very low light levels. The sensors are coupled via an analog-to-digital converter to a computer image processor which stores the image, provides the software for manipulating the image as discussed above, and converts the digitized signal back to analog form for display. **Figure 5** shows a diagram of the components of a fluorescence digital imaging microscopy system.

Advantages. There are a number of advantages of the fluorescence digital imaging microscopy system over conventional fluorescence microscopy for biological samples: (1) quantitation rather than just observation of molecular components or molecular fluxes in a cell; (2) use of very-low-light-level excitation since the emission may be enhanced by amplification at the sensor; (3) measurement of the accumulation of fluorescent substrates from outside the cell to inside specific organelles by using a subtraction of an image taken at zero time from subsequent images; (4) measurement of ratio images, for demonstrating the distribution of a particular component, for example, the free calcium concentration within a cell (in this case, the information is contained in the ratio of the emission images obtained by excitation at different wavelengths); (5) production of optical sections of an object (by confocal laser scanning or by optical sectioning and image correction) and storage of the images for three-dimensional reconstruction of objects; and (6) quantitation of kinetic processes such as tracking of membrane components during cell motility.

Fluorescence digital imaging microscopy is a tool for the biologist. The objects of interest may vary in size from single DNA molecules, to single cells, to whole insect embryos or mammalian organs. Living cells may be investigated and kinetic processes monitored (for example, Ca^{2+} gradients observed during nerve cell growth or excitation). Fixed and modified preparations may be quantified (for example, the localization and quantification of DNA structures and sequences on chromosomes; **Fig. 6**). Whole arrays of molecules within a structure may be localized and the structure reconstructed in three dimensions (for example, the ordering of α-actinin bodies within a smooth muscle cell to determine the force lines in the contractile apparatus). It is anticipated that fluorescence digital imaging microscopy will be applied increasingly to both cellular and molecular biology for quantitation of simple objects such as electrophoresis gels as well as for the analysis of such complex structures as developing embryos and brain networks. SEE PROTON MICROSCOPY.

For background information SEE CRYSTAL GROWTH; CRYSTAL STRUCTURE; ELECTRON MICROSCOPE; FLUORESCENCE MICROSCOPE; MICROSCOPE; NONRELATIVISTIC QUANTUM THEORY; SURFACE PHYSICS; TUNNELING IN SOLIDS in the McGraw-Hill Encyclopedia of Science and Technology.

D. J. Arndt-Jovin

Bibliography. D. J. Arndt-Jovin et al., Fluorescence digital imaging microscopy (DIM) in cell biology, *Science*, 230:247–256, 1985; R. S. Becker et al., *Phys. Rev. Lett.*, 57:1020, 1986; R. S. Becker, J. A. Golovchenko, and B. S. Swartzentruber, *Phys. Rev. Lett.*, 54:2678, 1985; G. Binnig et al., *Phys. Rev. Lett.*, 50:120, 1983; P.

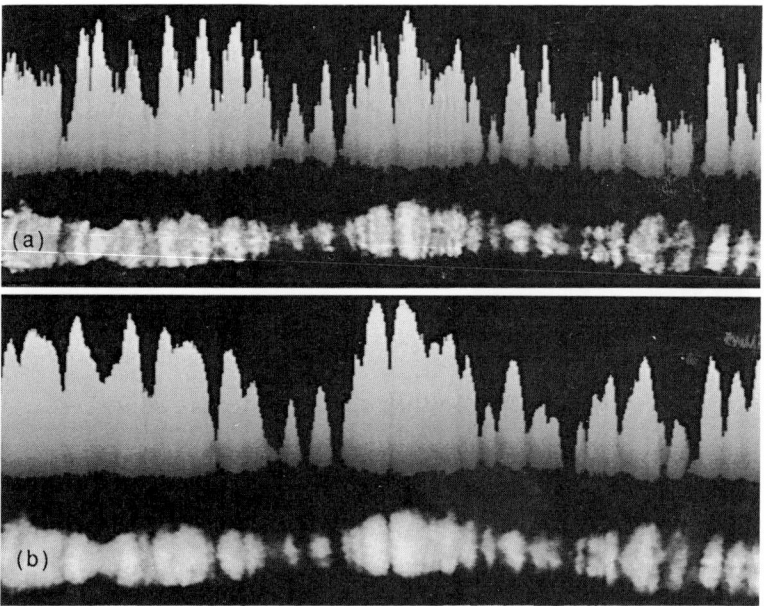

Fig. 6. Images produced by laser-scanning fluorescence digital imaging microscopy. The localization and quantitation of DNA sequences in (a) a left-handed helical conformation (Z-DNA) is compared with (b) the distribution of the total DNA shown for the right end of the insect polytene II chromosome from *Chironomus thummi*. In a is seen the total DNA distribution detected by using a DNA-specific dye, mithramycin; the image is shown below the three-dimensional representation of the fluorescence intensity. In b is seen the image and three-dimensional intensity distribution for the immunofluorescence detected at a different excitation and emission wavelength from the same chromosome stained with an antibody specific for the left-handed helical conformation of DNA.

de Weer and G. Salzburg (eds.), *Optical Methods in Cell Physiology*, 1986; S. Inoué, *Video Microscopy*, 1986; D. L. Taylor et al., *Applications of Fluorescence in the Biomedical Sciences*, 1986; R. Young, J. Ward, and F. Scire, *Phys. Rev. Lett.*, 27:922, 1971.

Microwave measurements

A program of intercomparison of national standards for radio-frequency and microwave measurements has been carried out since 1965 under the auspices of the International Committee of Weights and Measures. By this means, measurements of all the important quantities up to 35 GHz or higher made in different countries can be related.

International harmonization of standards. For purposes of science, trade, and defense, it is important that measurements made in different countries should be harmonized; that is, measurements of the same quantity should agree within acceptable uncertainty wherever they are made. To achieve this objective, the International Bureau of Weights and Measures (BIPM) was set up at Sèvres, near Paris, in 1875. Its work is steered by the International Committee of Weights and Measures (CIPM). The BIPM maintains standards of most of the International System (SI) base units (the kilogram, the meter, and so forth) and of some other quantities. Primary standards maintained by national standards laboratories are accurately related to the BIPM standards either by physical comparison by means of intermediate traveling standards or by relation to agreed values of fundamental physical constants. For electrical measurements it is inconvenient to maintain standards of the base unit, the ampere; instead, primary standards of the volt and the ohm are maintained by the BIPM and by national laboratories.

Intercomparison of microwave standards. In most fields of measurement it is a relatively straightforward matter to provide in any one country chains of instrument calibrations for measurement of derived quantities which are traceable ultimately to the national primary standards. Measurements can then be related, through the BIPM standards, to those made in other countries with similar traceability. In a few fields, of which microwaves is a prime example, providing traceability to national standards of the base units is complex and difficult, and it has been found necessary to arrange direct intercomparisons of national standards of the more important radio-frequency and microwave quantities.

In the period following World War II, rapid developments in microwave techniques and systems took place in a number of countries. Accurate measuring instruments were developed, which led to a need for comparisons of different measurement methods developed in different laboratories. To assist in meeting this need, the International Union of Radio Science (URSI) decided in 1952 to arrange a comparison of microwave power standards. As a result, starting in 1957, several comparisons of power standards at 3 and 10 GHz took place between laboratories in a number of different countries. These were followed by comparisons, also sponsored by the URSI, of noise, attenuation, and permittivity measurements. It became clear that there was a need for an extensive program of such intercomparisons. The URSI had neither the facilities nor the formal status to arrange such a program. The possibility of the BIPM setting up a laboratory to maintain international microwave standards was considered, but the cost would have been prohibitive, and an alternative arrangement was adopted. In 1965 the CIPM Consultative Committee on Electricity (CCE) set up a Working Group on Radio-Frequency Measurements, which took on responsibility for arranging and monitoring a continuing program of direct comparison—through the medium of transfer standards—of national standards of microwave quantities, including those outstanding comparisons originally sponsored by the URSI. The intercomparisons are now formally sponsored by the BIPM, an arrangement which gives certain practical and legal advantages. Detailed arrangements are made in each case by a pilot laboratory appointed by the Working Group. The URSI continues to take an active interest in the scientific aspects of the program and is represented on the Working Group.

Since its first meeting in 1965, the Working Group has met at roughly 3-year intervals. At its second meeting in 1968, it recognized two distinct reasons for undertaking international comparisons. The first is to enable the most advanced laboratories that are developing new measurement techniques to compare them with corresponding but different techniques elsewhere, a primarily scientific purpose. The second reason is to enable other national laboratories to monitor the accuracy of their standards. Such monitoring provides the basis for the international acceptance, for trade and similar purposes, of calibration certificates issued in all countries. Calibration services provide traceability to national standards for instruments covered by such certificates.

About 25 intercomparisons have now been completed, 10 are in progress, and 15 are actively planned. The quantities concerned and their frequency ranges are listed in the **table**. Laboratories in 18 countries—Australia, Canada, China, Czechoslovakia, Denmark, the Federal Republic of Germany, Finland, France, the German Democratic Republic, Hungary, Italy, Japan, the Netherlands, Poland, the Soviet Union, Sweden, the United Kingdom, and the United States—have taken part in at least one comparison. Such participation makes heavy demands on laboratory effort: Complete coverage of the microwave

Electrical quantities compared and their frequency ranges	
Quantity	Frequency range*
Voltage in coaxial line	1 MHz–1 GHz
Power in coaxial line and in waveguide	30 MHz–100 GHz
Impedance, including reflection coefficient and scattering parameters	1 MHz–40 GHz
Attenuation	30 MHz–40 GHz
Thermal noise	4–12 GHz
Field strength or power flux density	100 MHz–10 GHz
Antenna parameters	8–12 GHz
Permittivity	9 GHz
Laser power	10.6-μm and 0.85–1.55-μm wavelengths
Attenuation in optical fibers	1-μm-wavelength region

* Approximate and not necessarily continuous. Includes measurements being carried out or planned in the current round of intercomparisons.

spectrum requires measurements in coaxial-line or waveguide systems in a number of frequency bands and may involve each laboratory in hundreds of measurements. For this reason, no laboratory has taken part in all comparisons, and some have taken part in only one. But the overall result has been to provide a reasonably dense network of reference points, enabling microwave measurements of all kinds to be related.

Quantities compared. Voltage or power, impedance and related quantities, attenuation, and receiver noise level are fundamental quantities for the characterization of microwave systems. The majority of intercomparisons have been concerned with these quantities, but a few special comparisons of other quantities have also been arranged.

Voltage. Radio-frequency voltage can be meaningfully defined only at the lower frequencies. The intercomparisons arranged have been entirely concerned with voltages in coaxial lines at frequencies up to 1 GHz, mostly at the level of 1 V, but with limited measurements at 1 mV and 100 V as well.

Power. The measurement of microwave power is of basic importance. A number of intercomparisons have been concerned with this quantity, at the level of 1–10 mW and at frequencies up to 18 GHz in coaxial lines and up to 100 GHz in waveguides.

Impedance. Impedance is another quantity of fundamental importance for the design of microwave systems. It can be measured in various aspects such as voltage standing-wave ratio, reflection coefficient, and scattering parameters, in addition to the conventional low-frequency two-terminal form. Several of these aspects have been the subjects of intercomparisons at frequencies up to 18 GHz, as well as phase shift at about 10 GHz.

Attenuation. A number of intercomparisons of attenuation have been carried out in coaxial lines

and in waveguides at frequencies up to 35 GHz. Most of these have been limited to values of attenuation of about 60 dB or less. In one case, high attenuations of 100 dB were compared at 30 MHz.

Noise. Thermal noise sets the limit of sensitivity in receivers used, for example, in radar or telecommunications. The accurate measurement of noise is therefore of considerable importance. Noise standards are maintained by laboratories both at normal effective temperatures of a few hundred kelvins and, for purposes such as satellite communications where extreme sensitivity is needed and is achievable, at cryogenic temperatures. Intercomparisons have been made of low-temperature noise in the 4-GHz band; other comparisons are under way at higher temperatures in the 8–12-GHz region.

Electromagnetic field strength. The measurement of the strength of an electromagnetic field is important both for interference measurements and for the assessment of microwave radiation hazards. It is commonly expressed as field strength (volts per meter) at the lower radio frequencies and as power flux density (watts per square meter) in the microwave region. Intercomparisons have been made of field strength at 100 MHz and power flux density at 2.45 GHz; further comparisons to cover 300–1000 MHz and the 10-GHz region are under way.

Antenna parameters. The accurate measurement of antenna gain and related parameters is important in the design of microwave radar and communication systems, but presents practical difficulties. An intercomparison is presently being made in the 8–12-GHz band of measurements of the gain and transverse polarization ratio of a horn antenna for use as a reference standard.

Future program. The Radio-Frequency Working Group met at the BIPM in September 1986 to review the program of international comparisons under way and to discuss future policy. It concluded that there is likely to be a continuing need for a program of intercomparisons for the foreseeable future. As measurement techniques and accuracies improve and as additional countries set up national microwave standard facilities, it may be necessary to repeat some of the earlier intercomparisons. But the main thrust of the program will derive from the need to cover new types of quantities and new frequency bands.

The development of computer-controlled, broadband microwave measuring systems in recent years has revolutionized techniques for routine measurements. These systems require new types of reference standards whose properties can be accurately characterized over wide frequency bands. This has led the Working Group to initiate a comparison of scattering parameter measurements over the band from 2 to 18 GHz. Other broadband comparisons will doubtless follow, particularly as the use of coaxial lines is extended to higher frequencies.

Power in the 100-GHz region is already being compared. The Working Group foresees that further comparisons of power and transmission parameters in the millimeter and submillimeter-wavelength regions will be necessary as the use of these bands increases.

At still higher frequencies, recent technological developments using lasers and fiber-optic communications have made the provision of standards of power, attenuation, and other transmission parameters necessary. Although these developments involve frequencies above the radio and microwave ranges, the quantities to be measured and the techniques used are so similar that it has been appropriate for the Radio-Frequency Working Group to take responsibility for harmonizing national standards. Some intercomparisons of power at 10.6 micrometers, which is the wavelength of the carbon dioxide laser, are already under way, and at its 1986 meeting the Working Group launched comparisons of power and attenuation in optical fibers at several wavelengths of interest in the vicinity of 1 micrometer.

For background information SEE ANTENNA (ELECTROMAGNETISM); ATTENUATION (ELECTRICITY); ELECTRICAL UNITS AND STANDARDS; MICROWAVE; MICROWAVE FREE-FIELD STANDARDS; MICROWAVE IMPEDANCE MEASUREMENT; MICROWAVE MEASUREMENTS; MICROWAVE NOISE STANDARDS; MICROWAVE POWER MEASUREMENT; RADIO-FREQUENCY IMPEDANCE MEASUREMENTS in the McGraw-Hill Encyclopedia of Science and Technology.

A. Earle Bailey

Bibliography. A. E. Bailey, International harmonisation of microwave standards, *Proc. IEE*, part H, 127:70–73, 1980; A. E. Bailey et al., International organization in electromagnetic metrology and international comparison of RF and microwave standards, *Proc. IEEE*, 74:9–14, 1986; Comité Consultatif d'Electricité, *Rapport de la 17ᵉ Session*, 1986; M. C. Selby, International comparison of HF and microwave electromagnetic quantities, *Proc. IEEE*, 55:745–747, 1967.

Mining

The collapse of base-metal prices over the period 1980–1986, followed by the slump in hydrocarbon and fossil fuel prices, led to changes in the technology of mining. It had often been said that little had changed in mining since Agricola wrote *De Re Metallica* in the sixteenth century. Now, however, although the fundamentals of mining are still recognizable, the methodology has become radically different. The major difference lies in the need to reduce the cost of the mined material. In the high-cost United States and also in countries experiencing lower costs, wages can account for more than 60% of the actual mining cost. Costs can be reduced by using two basic approaches: reduction of personnel by increased mechanization, or increase in productivity by increasing output using bulk mining techniques.

Small-scale mining. In small mines the emphasis has been on mechanization, and there have been many new developments. Rubber-tired load-haul-dump (LHD) vehicles are now used to load and haul minerals, either alone or in combination with truck haulage. Sometimes these vehicles are controlled remotely. Fixed and shiftable conveyor belts have been designed for bulk haulage. Self-propelled hydraulic drill carriages (hydraulic drilling machines) are used for both development and production drilling, while self-propelled vehicles are used for all associated work, such as charging, roof bolting, shotcreting, and personnel transport. Surface shaft borers (large surface drilling machines) are available for shaft development, and raise borers and roadheaders are used for underground development. Other recent developments in mechanization include in-mine small-scale crushing and grinding and the remote control of motorized machinery.

The benefits of mechanization in small-scale mining are numerous. The most important is that correct mechanization yields optimum productivity and maximum cost effectiveness; this aspect includes optimum sizing of units. Since the ore body is small and depletion time is short, equipment must be transferable to another site or to alternative use. Other benefits are rapid access to the ore body with benefits to cash flow, centralization of mining activities at a limited number of high-productivity stopes instead of a large number of low-productivity stopes, and reduction in accidents through reduced personnel requirements and reduced exposure. Retraining programs following the introduction of new equipment have been found to improve motivation.

Introduction of mechanization modifies the mining method. For example, in underground mines the major effect of using load-haul-dump vehicles with or without truck haulage is that ramps and spirals connecting levels at 80–100 ft (25–30 m) or less replace raises in the conventional mine layout. The limiting inclination for load-haul-dump equipment is 8–10°.

Traditionally drifts have been in the country rock. Because of the demand to combine development with exploration and to reduce drifting costs, there is now a tendency, where feasible, to drive development drifts in the ore body. This is particularly important in small, low-grade ore bodies. Drifting in the ore body has always been the practice in deposits inclined to the horizontal; it is now being used in deposits inclined to the vertical, with an inclined drift replacing raises. Design of the development layout is determined by the value of the minerals in the ore body as well as its shape, size, and structure. If the ore body is valuable, development may have to be in the country rock to ensure maximum recovery.

In less valuable deposits, pillars may reasonably be left in the ore body.

All of the traditional underground mining methods may be modified for development in the ore body. Sublevel or open stoping with residual pillars, or possibly fill pillars, are most easily adapted. Other methods such as cut-and-fill stoping and sublevel caving can also be modified with relative ease, and there are examples of development in the ore body in sublevel and block caving operations.

In another important development the methods of rock mechanics are being used in underground design. Usually, analytical and numerical methods to isolate potential weaknesses or zones of stress concentration are combined with observation, monitoring, and empirical modification in order to determine the best mining method and optimum requirements for stope and pillar size and support. This process, which has been developed for large, deep mines in the United States and elsewhere, is particularly useful in small mines where rapid development is essential.

Combinations of numerical design methods with probabilistic methods, can also be used in mine planning. An example is a decision support system, modeling the technical and financial environment of a mine in order to assess the effects of various alternatives on the overall economics of the mine system.

Bulk mining. Mechanization and planning are equally important in bulk mining, with the emphasis placed on increasing output. The most discussed methods of bulk mining in recent years have been surface open-pit operations utilizing shovels and trucks of increasing size. These techniques still make up the most important bulk mining operations, although introduction of in-pit crushing and conveying is leading to modification of these methods.

Traditional underground bulk mining operations have utilized room-and-pillar and longwall methods in stratified deposits and block caving, sublevel caving, and blasthole stoping in massive and vein deposits. Recent developments have included the expansion of mechanized longwall mining in coal mining and the introduction of the vertical crater retreat (VCR) method in massive deposits.

Mechanized longwall mining. In this type of mining, a cutting machine removes slices from the face of a rectangular block of coal in the coal seam. The block is usually 660 ft (200 m) wide and is isolated between parallel access entries usually more than 2600 ft (800 m) long. There are two or three entries at each side of the working area, separated by chain pillars and supported by roof bolts or square sets. These are used for access and exploration, for transport of cut coal and supplies, and for ventilation. The longwall coal face is supported by self-advancing hydraulic shields, which can be controlled along with the cutting and bulk transport system to provide virtually remote-controlled mining. The roof is caved behind the shields.

Despite the high investment required, mechanized longwall mining systems are capable of producing up to 33,000 tons (30,000 metric tons) of coal per unit per week. This output is comparable to that of a medium-size mine using traditional extraction methods. The productivity and profitability are high, even when prices are depressed.

Vertical crater retreat. This method, introduced in Canada, shares some characteristics with longwall mining. A block of ore in a massive ore body is isolated by access raises and entries and drilled with blastholes from the top downward (see **illus.**). These holes are blasted upward from the base as the block is excavated by removing horizontal slices. Mining continues in contiguous blocks. Stope sizes vary, but an average excavation may be 50 ft by 100 ft by 300 ft high (15 m by 30 m by 100 m high) from which outputs of 2200 tons (2000 metric tons) per week can be obtained.

The most important part of the vertical crater retreat method is the drilling and blasting operation. A series of crater tests is performed before the blast is designed. Crater testing consists of a series of single-hole shots. The type and amount of explosive is kept constant, but the depth of burial of explosive is varied. Spherical or nearly spherical charges are used. After each blast the volume of crater produced is measured; analysis of the data following a series of experiments yields the optimum depth of burial required for the spherical charge to produce the largest crater. The application of crater testing to the vertical crater retreat method has been made possible by introduction of large-diameter (6-in. or 160-mm) drill holes to underground mining. A drill hole pattern for optimum fragmentation is designed from crater testing results. Holes are drilled from the top sill or from a cut over the stope or block down into the undercut. Charges are placed in these holes at that distance from the undercut equal to the optimum depth of burial found from crater testing. A layer of ore will be blasted downward into the previously mined area.

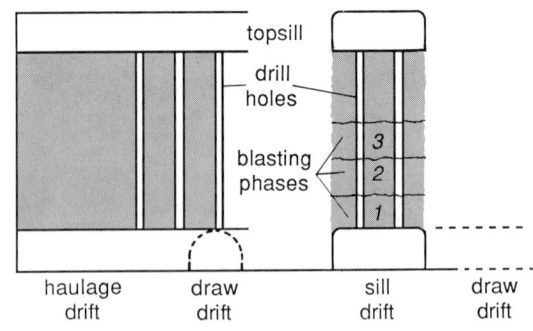

Vertical crater retreat layout.

The vertical crater retreat method has many advantages. Application of large diameter holes reduces the cost of drilling and makes it possible to use low-density and cheaper explosives, for example ANFO, if mining in soft rock. In addition, uphole drilling and slot cutting can be eliminated, and damage to sidewalls and pillars due to blasting is reduced. If vertical crater retreat is performed properly, the mixture of backfill material and ore is minimized. Finally labor and time required by the vertical crater retreat method is significantly less than that required by other conventional mining methods.

For background information SEE MINING; SURFACE MINING; UNDERGROUND MINING in the McGraw-Hill Encyclopedia of Science and Technology.

Ian W. Farmer

Bibliography. W. Hustrulid (ed.), *Underground Mining Methods Handbook*, 1982; L. C. Lang, R. J. Roach, and M. N. Osoko, Vertical crater retreat: An important new mining method, *Can. Min. J.*, pp. 69–76, September 1977; S. S. Peng, *Longwall Mining*, 1984; *2d International Conference on Small Scale Mining: Small Mine Economics and Expansion*, Helsinki, 1983.

Molecular complexation

Biological systems accomplish complicated tasks through highly structured complexation. The design, synthesis, and study of highly structured organic molecular complexes that mimic biological complexes has been called host-guest complexation chemistry by D. J. Cram. Complexes are two or more compounds bound to one another in a definable structural relationship by forces such as hydrogen bonding, ion pairing, metal ion-to-ligand attractions, electron donor-acceptor interactions, van der Waals attraction, and the entropic component of desolvation. Complexing partners are classified as hosts and guests. Upon complexation, solvent molecules solvating the binding sites of free host and guest are displaced into the bulk solution, which represents an entropically favorable process. Complexes are structured by contacts at multiple binding sites between host and guest. In complexes, the binding sites of the host converge. Hosts are the synthetic counterparts of the receptor sites of enzymes, genes, antibodies, neuroreceptors, or ionophores. Guests possess divergently arranged binding sites and are the counterparts of substrates, inhibitors, cofactors, drugs, antigens, or neurotransmitters.

Targets of host-guest chemistry are multiple. The same driving forces of complexation are effective in both synthetic and biological systems. Studies of the principles that govern host-guest complexation increase the fundamental understanding of molecular interactions in both biotic and abiotic complexes. New reagents and methods for the analytical detection or the separation of ions, isotopes, and organic compounds, especially of enantiomers, are developed. Particularly challenging are the possibilities of designing and preparing hosts that mimic enzymes and catalyze organic reactions in highly structured complexes.

Complexation of organic and inorganic ions. Crown ethers (hosts) and their property of forming stable complexes with alkali metal cations (guests) were first reported in 1967. Since then a large number of hosts that ligate alkali metal and ammonium and alkylammonium cations have been prepared. Investigations of the structure-bonding relationships in complexes have provided evidence that stereoelectronic complementarity between the host and the guest is a prerequisite for strong complexation. The principle of stereoelectronic complementarity has proven valid in the entire field of host-guest chemistry. The synthesis and study of the spherands such as structures (I) and (II) [the Li^+ complexes are shown] and the cryptahemispherands, for example, structure (III) [the Cs^+ com-

(I)

(II)

(III)

plex is shown], led to the defining of a second principle similar in general importance for the design of complexes. The spherands and cryptaspherands form extraordinary stable complexes with alkali metal cations and primary ammonium ions. The free energy of complexation ΔG and the association constant K_a are related by Eq. (1), where R is the gas constant and T is the tem-

$$\Delta G = -RT \ln K_a \qquad (1)$$

perature. Examples of free energies of complexation ΔG at 293°C (559°F), standard state 1 M, in chloroform saturated with water are: for the Li^+ complex (I), $-\Delta G = >23.0$ kcal·mol^{-1}, and for the other Li^+ complex (II), $-\Delta G = 18.2$

kcal · mol^{-1}; for the complexes with structure (III), the Na$^+$ form, $-\Delta G = 21.0$ kcal · mol^{-1}; the Cs$^+$ form, $-\Delta G = 21.7$ kcal · mol^{-1}; and the ammonium (NH$_4^+$) form, $-\Delta G = 20.2$ kcal · mol^{-1}.

The decomplexation rates of these complexes are slow on the human time scale at 25°C (77°F). The association constants of these complexes are higher than those of complexes of cryptands and of monocyclic polyethers (crown ethers) that had been previously described. An analysis of the complexation strength as a function of the geometries of free and complexed hosts obtained in a large series of x-ray crystallographic studies led to the conclusion that preorganization is a central determinant of binding power. The principle of preorganization states that the more highly hosts are organized for binding prior to their complexation, the more stable will be their complexes. In a highly organized host the binding sites of the host are largely collected and oriented prior to complexation. The crystal structure comparisons of hosts and their complexes in **Fig. 1** show the relative states of preorganization of different hosts. The structures of the free and complexed spherand (Fig. 1c) are identical. This spherand is fully preorganized and shows the best binding. The cryptand (IV) and especially the 18-crown-6 (V) shown in Fig. 1 must organize their cavities during complex-

ation at the cost of free binding energy; consequently they form weaker complexes.

The spherands and the cryptaspherands show very high specificity in complexation; for example, the association constant K_a [reaction (2)] of

$$[\text{Host}] + [\text{guest}] \overset{K_a}{\rightleftharpoons} [\text{host-guest complex}] \quad (2)$$

the Na$^+$ complex with structure (I) is more than 10^{10} times larger than the association constant of the K$^+$ complex with the same structure.

Approximately two-thirds of all enzymes act as hosts for anionic substrates and cofactors. The importance of anionic molecules for biological processes has led to the development of synthetic hosts for organic and inorganic anions in aqueous solution. J.-M. Lehn observed the strong complexation of organic anions by protonated macrocyclic polyamines in aqueous solution. For example, the fully protonated macrocyclic host (VI)

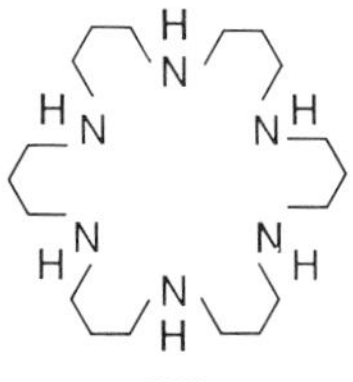

(VI)

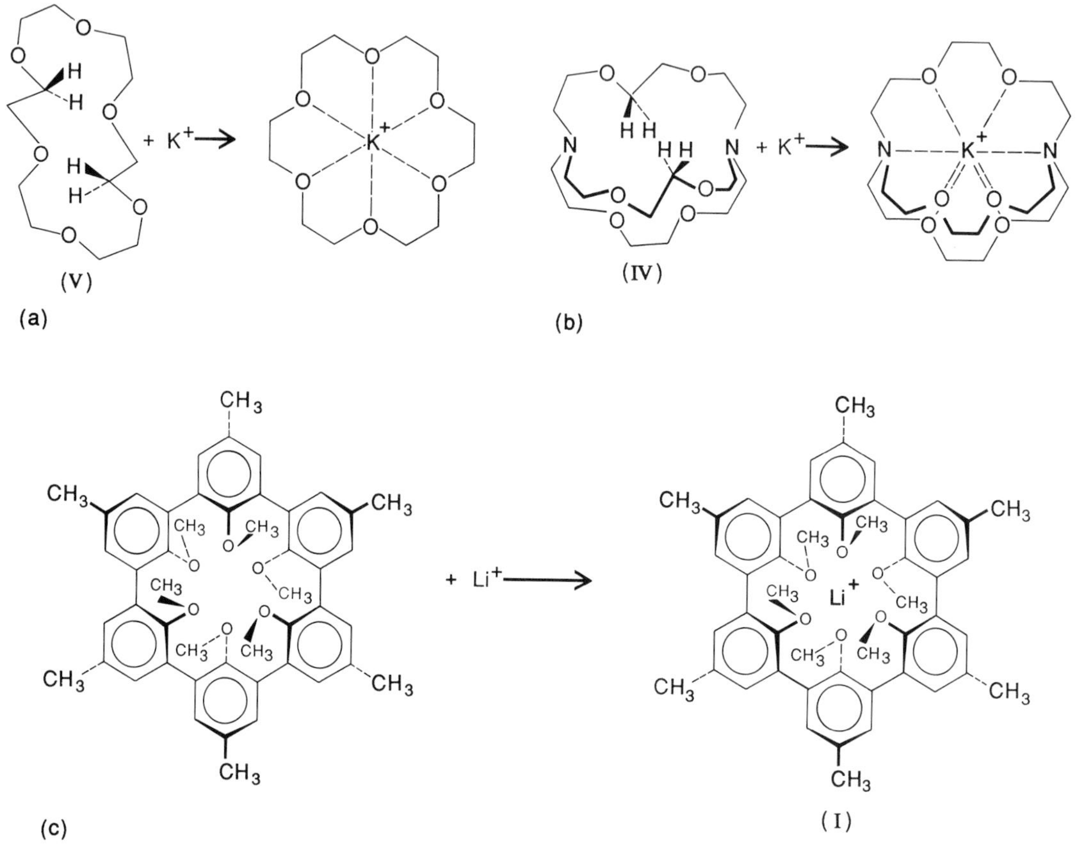

Fig. 1. Crystal structure comparison of hosts and their complexes show the relative states of preorganization of monocyclic polyethers. (a) Crowns. (b) Cryptands. (c) Spherands. (*After D. J. Cram, Preorganization: From solvents to spherands, Angewandte Chem. Int. Ed. Engl., 25:1039–1057, 1986*)

complexes (via ion pairing and hydrogen bonding interactions) the dicarboxylates oxalate, malonate, succinate, and tartrate (association constants of the 1:1 complexes: $K_a \approx 10^2$–10^4 liter · mol^{-1}) as well as the biological substrates citrate^{3-}, the trianion of citric acid ($K_a \approx 5 \times 10^4$ L · mol^{-1}), and the phosphates AMP^{2-}, the dianion adenosinemonophosphate ($K_a \approx 2.5 \times 10^3$ L·mol^{-1}), ADP^{3-}, the trianion adenosinediphosphate ($K_a \approx 3.1 \times 10^6$ L · mol^{-1}), and ATP^{4-}, the tetraanion adenosinetriphosphate ($K_a \approx 7.9 \times 10^8$ L · mol^{-1}). J.-M. Lehn found that protonated macrocyclic polyamines are efficient catalysts for the hydrolysis of ATP to orthophosphates and ADP and subsequently to AMP. At pH = 2.5–8.5, rate enhancements up to a factor of 10^3 are observed. The first step of the catalytic process involves the formation of a 1/1 host-guest complex between the host and ATP.

Recently Lehn described a molecular host which, in a very elegant way, combines both porphyrin rings and azacrown units (which contain nitrogen as well as oxygen donor atoms) as guest selective binding subunits. Simultaneous binding of $^+$H$_3$N-(CH$_2$)$_9$-HN$_3^+$ and of two Zn^{2+} cations by this host yields the very interesting complex (VII) as a model for metalloenzyme

(VII)

complexes. The simultaneous complexation of organic and inorganic guests offers the opportunity to induce or adjust physical and chemical interactions and reactions (catalysis) between the cobound species.

Complexation of neutral organic guests. Complexation of neutral organic guests has for a long time exclusively been observed in aqueous solutions, where solvent participation largely contributes to the free binding energy. Only recently has the formation of stable complexes between neutral hosts and guests been observed in organic solutions.

Aqueous solution. The cyclodextrins are macrocyclic oligosaccharides in which six (α) [structure (VIII)], seven (β), and eight (γ) α-D-glucopyranose units, respectively, are connected by 1,4-glycosidic bonds. Crystal structure and spec-

(VIII)

troscopic analysis has shown that the cyclodextrins in the solid state as well as in solution exist in a roundish conical shape (**Fig. 2**) with a wall height of approximately 0.8 nanometer and an inner cavity diameter varying between 0.5 nm (α) and 0.75 nm (γ). These natural compounds which are formed by the enzymatic catabolism of starch were for decades the only macrocyclic hosts known to form complexes with suitably sized neutral organic guests in aqueous solution and in the solid state. The cyclodextrins most often form 1:1 complexes in which the guests are located inside the conical cavity.

Cyclodextrins are versatile enzyme models, and a large number of reactions are catalyzed by natural and modified cyclodextrins. Like enzymes, cyclodextrins complex their substrates in a fast and reversible equilibrium prior to the catalytic event. Rate accelerations of reactions in the presence of cyclodextrins can be impressive; as a consequence of the chirality of these natural hosts, many reactions proceed under asymmetric induction. For example, one enantiomer of a chiral nitrophenyl ester substrate acylates β-cyclodextrin [structure (IV)] 5,900,000 times faster in aqueous dimethylsulfoxide than it hydrolyzes under the same conditions; the other enantiomer is 62 times slower.

In recent years the successful design and preparation of water-soluble hosts with apolar binding cavities to mimic the hydrophobic binding sites in

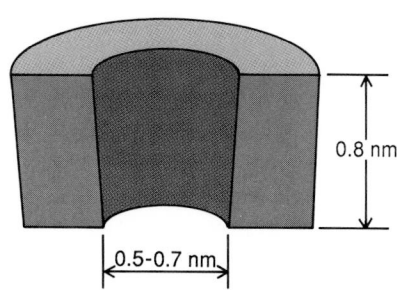

Fig. 2. Shape of cyclodextrins both in solid state and solution. (*After J. Franke and F. Vögtle, Complexation of organic molecules in water solution, Top. Curr. Chem., 132:135–170, 1986*)

biological systems has been reported. Stoichiometric 1:1 complexes form between aromatic guests and the cyclophane-hosts [structures (IX) and (X)] in aqueous solutions. The x-ray crystal

(IX)

(X)

structure of the durene complex (**Fig. 3***a*) which precipitates from acidic aqueous solution demonstrates the cavity inclusion of the guest. According to ^{1}H nuclear magnetic resonance (NMR) analysis in aqueous solution, all complexed aromatic guests are located in a specific plane in the binding cavity of host (X). Figure 3*b* shows a schematic representation of the favored geometry of the pyrene complex which is supported by ^{1}H nuclear magnetic resonance. Species (X) forms complexes with perylene, pyrene, and fluoranthene either by extracting the solid guest with an aqueous solution of host or by extracting the guest from a hexane solution into the aqueous host solution. The amount of these extremely water-insoluble polycyclic aromatic hydrocarbons that can be extracted into aqueous solution is increased dramatically by the complexation. The association constants of the 1/1 complexes formed between (X) and arenes of various size demonstrate the importance of the principle of stereoelectronic complementarity. Perylene has the highest complementarity to the binding cavity and forms the most stable complex ($K_a = 1.6 \times 10^7$ L · mol^{-1}; $T = 293$ K or 68°F), followed by fluoranthene ($K_a = 1.2 \times 10^6$ L · mol^{-1}) and pyrene ($K_a = 1.1 \times 10^6$ L · mol^{-1}). Considerably weaker complexation is observed with the smaller arenes that do not completely fill the

binding cavity, for example, biphenyl ($K_a = 2.2 \times 10^4$ L · mol^{-1}), naphthalene ($K_a = 1.5 \times 10^4$ L · mol^{-1}), and durene ($K_a = 2.0 \times 10^3$ L · mol^{-1}). Hydrophobic and attractive van der Waals interactions are the major driving forces of complexation. With structure (X), the inverted form of the transport of ions from one aqueous solution to another aqueous solution through a lipophilic membrane could be observed for the first time. Host (X) acts as a very efficient molecular carrier and accelerates the transport of arenes such as pyrene and fluoranthene from one lipophilic (hexane) phase to another lipophilic phase through an aqueous phase containing the host.

Organic solvents. The displacement upon complexation of water molecules from a lipophilic binding cavity and from the surface of a lipophilic guest is an energetically favorable process. This desolvation process, known as the hydrophobic effect, represents the major driving force for the complexation between neutral binding partners in aqueous solution. Since similar interactions favoring host-guest binding are weaker or even completely absent in organic solvents, it was believed for a long time that the efficient complexation between neutral, nonhydrogen-bonding partners is not possible in nonaqueous solution. In recent years, however, the synthesis of highly preorganized hosts has led to the observation of strong binding of neutral guests in a variety of organic solvents. The macrobicyclic host (XI),

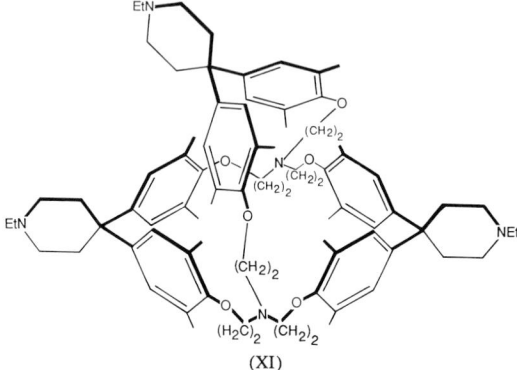

(XI)

where Et represents (C_2H_5), forms stoichiometric complexes with pyrene, fluoranthene, perylene, and other large arenes in organic solvents. ^{1}H nuclear magnetic resonance analysis shows that the guests are located in the plane of the binding cavity passing through the three spiro carbon atoms of the three diphenylmethane units (a spiro carbon atom is common to two rings). With a given guest, host (XI) forms 1:1 complexes of very similar geometry in all organic solvents. However, considerable differences in complexation strength are observed in various solvents which cannot be explained by differences in attractive interactions in complexes of very similar geometry. Instead, the differences in binding are due to solvation-desolvation processes.

Strong complexation of pyrene occurs in alcoholic solvents ($K_a = 4.4 \times 10^4$ L · mol^{-1} in methanol; $T = 303$ K or 86°F), which can be explained by favorable desolvation similar to the hydrophobic effect in aqueous solution. Considerable binding of pyrene is observed in the dipolar aprotic solvents acetone, dimethylsulfoxide, and dimethylformamide ($K_a = 1.2 \times 10^3$ L · mol^{-1} in acetone), which indicates that these solvents do not have a large affinity to the binding site and that their desolvation by the entering guest does not require much energy. Chloroform, benzene, and carbon disulfide seem to interact favorably with the macrobicyclic binding site, and their displacement from the cavity of (XI) by the entering pyrene requires much of the binding energy gained from the attractive interactions in the complex. Consequently, weaker binding is observed ($K_a = 12$ L · mol^{-1} in benzene). These results clearly show how important it is to take into account solvation processes in any computer modeling of host-guest interactions if correct association constants are to be calculated and predicted.

High selectivity in complexation of halogenated compounds in halogenated solvents was observed with spherical hosts such as (XII). Host

(XII)

(XII) complexes chloroform ($K_a = 470$ L · mol^{-1}) in deuterated 1,1,2,2-tetrachloroethane as a solvent, which does not fit into the binding cavity. The complexes of dichloromethane and of bromoform, which are guests with lower geometric complementarity to the binding site of (XII), form complexes that are less stable than the chloroform complex by $\Delta G \approx 1.0$ and 1.4 kcal · mol^{-1}, respectively.

The recent successful development of molecular hosts that complex efficiently in organic solvents is of importance for the design of enzymelike catalysts. Although enzymes are water-soluble and bind their substrates mostly from aqueous environment, the catalytic processes occur in rather nonaqueous conditions at their active sites. The dipolar-aprotic character of the active sites contributes largely to the observed high rates of enzymatic reactions. A complete shielding from water of the reaction and binding sites of synthetic enzymelike catalysts that act in

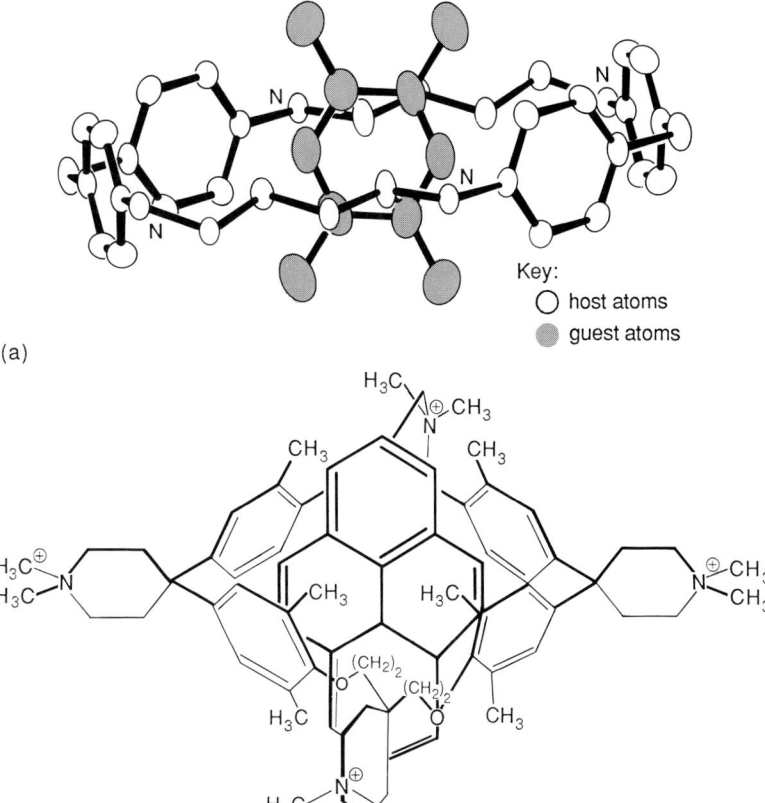

Key:
○ host atoms
● guest atoms

(a)

(b)

Fig. 3. Complexes formed between aromatic guests and cyclophane hosts. (*a*) X-ray crystal structure of a durene complex formed with host structure (IX) (*after K. Odashima et al., Host-guest complex formation between a water soluble polyparacyclophane and a hydrophobic guest molecule, J. Amer. Chem. Soc., 102:2504–2505, 1980*). (*b*) Schematic representation of the highly favored geometry of the pyrene complex formed with host structure (X) in aqueous solution (*after F. Diederich and K. Dick, ¹H NMR investigations of host-guest complexation between a macrocyclic host of the cyclophane type and aromatic guests in aqueous solution, J. Amer. Chem. Soc., 106:8037–8046, 1984*).

aqueous solution can require the expensive construction of sophisticated large host molecules. An easier alternative could be the construction of hosts that bind and catalyze in organic solutions. Finally, the comparison of the thermodynamics and kinetics of host-guest complexation in water and in organic solvents will increase the understanding of the specific properties of water that are responsible for life.

For background information SEE CHELATION; CHEMICAL THERMODYNAMICS; COMPLEX COMPOUNDS; COORDINATION CHEMISTRY; CROWN ETHERS; ENZYME; WATER in the McGraw-Hill Encyclopedia of Science and Technology.

François Diederich

Bibliography. R. Breslow, Artificial enzymes, *Science*, 218:523–537, 1982; J. Canceill, L. Lacombe, and A. Collet, New cryptophane forming unusually stable inclusion complexes with neutral guests in a lipophilic solvent, *J. Amer. Chem. Soc.*, 108:4230–4232, 1986; D. J. Cram, Cavitands: Organic hosts with enforced cavities, *Science*, 219:1177–1183, 1983; D. J. Cram, Preorganization: From solvents to spherands, *Angew.*

Chem. Int. Ed. Engl., 25:1039–1057, 1986; F. Diederich and K. Dick, A new water-soluble macrocyclic host of the cyclophane type: Host-guest complexation with aromatic guests in aqueous solution and acceleration of the transport of arenes through an aqueous phase, *J. Amer. Chem. Soc.*, 106:8024–8037, 1984; F. Diederich, K. Dick, and D. Griebel, Complexation of arenes by macrocyclic hosts in aqueous and organic solutions, *J. Amer. Chem. Soc.*, 108:2273–2286, 1984; J.-M. Lehn, Supramolecular chemistry: Receptors, catalysts, and carriers, *Science*, 227:849–856, 1985.

Molecular modeling

The use of computers for the simulation of chemical entities and processes is known as molecular modeling. The discipline actually includes several diverse areas, with the computer as the common link. This article discusses four of the methods that are particularly useful: computer-assisted organic synthesis, calculation-based modeling, statistics-based modeling, and graphics-based modeling or molecular graphics. Although they are treated separately here, the methods have the potential to be integrated into very powerful systems which could completely change the way in which research is conducted, making it much more cost-effective. In the future, such integrated systems may be able to predict the outcomes of experiments before they are performed and to guide scientists in choosing experiments that will have the greatest likelihood of success.

Computer-assisted organic synthesis. The computer is being used in two general ways to assist in the synthesis and cataloging of new organic compounds. One is to analyze the structure of a target molecule in a stepwise manner and on this basis suggest schemes by which the compound could be synthesized. Such programs often rely on files of individual synthetic transformations. When several of these analytical steps are chained together, a retrosynthetic analysis for the compound is produced. This is a scheme which, if reversed, should result in a viable synthesis of the compound. Since there may be dozens of schemes by which to synthesize a particular compound, this approach is analogous to the computer-based chess programs that analyze a large series of chess moves and pick the most efficient one. The best-known program of this type is LHASA, but several others exist, some of which are more interactive than others.

The second category comprises the so-called passive or database-oriented programs. These programs do not analyze molecules but simply catalog them by structure or reaction type. The most functional of these programs are commercially supported. With these programs, it is possible to search for all occurrences of a particular substructure, or for all known ways to perform a particular synthetic transformation. In this way, such programs serve as a valuable automated resource and window on the chemical literature.

Calculation-based modeling. This area relies heavily on theoretical chemistry and the mathematical expressions that describe the structure of molecules. It seeks to predict the most favorable conformation of a molecule or to calculate the energy of an interaction between two molecules. There are also two types of approaches in this area, based on quantum mechanics and on molecular mechanics.

The most complete expression for the structure of a molecule is based on quantum mechanics. For the complex molecules of current interest, exact solutions to the associated quantum-chemical equations are beyond the capacities of even the largest and fastest of present computers. Because of this limitation, strategies have been devised which make use of approximations to reduce the complexity of the equations and yet maintain an acceptable degree of accuracy. Generally, the quantum-chemical programs fall into two categories, semiempirical and ab initio. The semiempirical programs use greater numbers of approximations to limit the number of interactions that must be considered between electrons in the molecule. For example, they consider only bonding electrons and assume that lower orbital electrons form an invariant core. Since all the electrons of a molecule do in fact interact with each other, each type of bond is assigned empirical parameters (that is, correction factors) which will bring the calculated results for a standard interaction into agreement with observed chemical behavior. This is the origin of the term semiempirical. Because the parameters are based on normal bonding values, an understanding of the limits of the method is necessary if significant results are to be obtained.

The ab initio methods use a minimal number of approximations to calculate the electronic interactions. This requires correspondingly much greater computer resources. For example, the calculation of the minimum energy conformation treatment of the hydrocarbon propane was carried out by using both the semiempirical MNDO program and the ab initio GAUSSIAN82 program. The ab initio program required 456 times more computer time than the semiempirical one, yet the results were very similar. For this case, the semiempirical program was quite adequate, in part because it is well parameterized for this type of molecule. Other applications may not produce such close agreement, and then the cost of the calculations would need to be weighed against the degree of accuracy needed to solve the problem at hand.

The second type of calculation-based modeling, the molecular mechanics or force field approach, is truly empirical and is based on classical mechanics. The assumption is that for any two atoms there is an optimum interatomic distance at which the forces of repulsion and attraction are balanced. Any change in this distance (for example, the stretch or compression of a bond) requires energy, much as a system of two weights connected by a spring does. Over a narrow range of distances, the equation for the energy of bond stretch or compression is closely approximated by Hooke's law. For example, Eq. (1) could predict the potential energy of the bond

$$V = \frac{k(r - r_0)^2}{2} \qquad (1)$$

V for a radius r if the proportionality constant k and the optimum bonding distance r_0 are known. Empirical constants have been determined for many types of atoms and bonds which produce results that closely agree with reality. If the range over which the constants are defined is not exceeded, the method gives excellent results for calculating molecular structure and energies.

To arrive at an accurate prediction of the energy of a molecule, several terms are considered in the calculation, including bond stretching, bond angle bending, torsional interactions to reflect adjacent nonbonded interactions, and van der Waals–type interactions. Electrostatic and dipole-dipole interactions can also be considered. The most recent version of the program, MMP2, also considers delocalized (that is, conjugated) systems. The absence of this capability was a shortcoming of earlier versions.

The computational efficiency of this approach is much better than the quantum-chemical methods. Continuing the above example, MMP2 used only 8% of the computer time that MNDO used. In fact, the molecular mechanics approach is so efficient that it can be carried out on a microcomputer, at least for nonconjugated molecules of moderate size.

Statistics-based modeling. Since the mid-1960s a large effort has been under way to statistically model the interaction of chemicals and organisms. Attempts have been made to relate a compound's action to one or more physicochemical or structural parameters. From such correlations, insights into the requirements for improved biological activity and the mechanism of action of the substance would follow.

This field is commonly known as the study of quantitative structure-activity relationships. In essence, the classical quantitative structure-activity relationship approach is to seek statistical correlations between the potencies of a series of structurally related compounds and one or more quantitative structural parameters such as lipophilicity (log P), polarity (σ), and molecular size (expressed as molar refractivity, MR) by using multilinear regression analysis. Equation (2) represents the form of a typical result. Here

$$\log (1/C) = a(\log P)^2 + b(\log P) \\ + c(\sigma) + d(MR) + e \qquad (2)$$

C is the concentration of the substance that produces a defined biological response, and the coefficients a–e represent the best least-squares fit for the data. Analysis of the sign and magnitude of the coefficients, the calculated error limits, and the goodness-of-fit parameters collectively provides the means by which the validity of the statistical analysis is judged.

More sophisticated statistical methods are also being used to analyze the complex relationships between chemical compounds and biological activity. Some of the nonparametric methods (discriminant analysis, principal component analysis, and factor analysis) have found new applications in quantitative structure-activity relationships, both for analysis of data and selection of derivatives to test in the first place. The choice of appropriate derivatives can maximize the efficiency of the drug design process, which is important in light of the increasing costs of organic synthesis and biological testing.

Graphics-based modeling. The use of computer graphics to display and manipulate chemical structures is known as molecular graphics. Implicit in this process is the ability to represent the molecules in three dimensions, with sufficient depth cueing and accuracy that bond distances and angles may be displayed and reported. Generally, the ability to "dock" or fit together two or more molecules is also included in serious molecular graphics packages. These abilities distinguish molecular modeling packages from those software packages that simply draw or display organic structural formulas.

Current software permits large complex molecules such as deoxyribonucleic acid (DNA) fragments and enzymes to be examined and manipulated. This provides some unique abilities. It is very expensive and unwieldy to construct physical models of such structures, and if such models were built their bulk would prevent many of the most interesting comparisons. For example, it is often instructive to superimpose two or more molecules, so that their structural similarities and differences may be examined. The degree and ease with which this can be accomplished with physical models are quite limited, compared with computer-based models.

The most sophisticated molecular graphics packages are currently run on mainframe computers or dedicated minicomputers with special high-resolution interactive display screens. The best known of these is based on the Evans & Sutherland Picture System. Many commercial vendors now use the Evans & Sutherland technology, and several pharmaceutical companies have also based their own very sophisticated modeling systems on it. Structures are simulated as colored lines on a vector display tube. With the support of a dedicated computer, the images can be rotated, translated, and otherwise manipulated in real time with depth cueing, although space-filled or ball-and-stick models are not as easily handled.

A second type of modeling uses raster technology. This approach is similar to that used in color television and can produce strikingly realistic images. This approach is even more hardware-intensive, and the computer resources necessary to manage images like these in real time are not yet widely available.

Since 1983 at least two molecular modeling packages for use on microcomputers have appeared. While they do not have the high resolution and speed of manipulation of their larger antecedents, they do offer qualitatively the same operations as the larger packages at a small fraction of their cost. These packages have placed the ability to perform highly sophisticated molecular manipulations into the hands of anyone with access to a desktop microcomputer.

For background information SEE ARTIFICIAL INTELLIGENCE; COMPUTER GRAPHICS; MOLECULAR STRUCTURE AND SPECTRA; QUANTUM CHEMISTRY; SIMULATION in the McGraw-Hill Encyclopedia of Science and Technology.

James G. Henkel

Bibliography. U. Burkert and N. L. Allinger, *Molecular Mechanics*, ACS Monogr. 177, 1982; T. Clark, *A Handbook of Computational Chemistry*, 1985; N. J. Hrib, Recent developments in computer assisted organic synthesis, in D. M. Bailey (ed.), *Annual Reports in Medicinal Chemistry*, vol. 21, pp. 303–311, 1986; J. G. Topliss (ed.), *Quantitative Structure-Activity Relationships of Drugs*, 1983.

Monoclonal antibodies

Recombinant deoxyribonucleic acid (DNA) and hybridoma technologies have made possible the large-scale production of monoclonal antibodies and have rekindled enthusiasm for an antibody-mediated approach to cancer therapy. There are a variety of ways in which monoclonal antibodies may be used in cancer treatment.

Passive therapy. There are three general strategies for using monoclonal antibodies alone in cancer patients. They involve cytotoxic, regulatory, or immunization mechanisms.

Cytotoxic mechanism. In this approach the antibody mediates an antitumor cytotoxic effect. Complement-mediated cytotoxicity involves the fixation of complement serum proteins to the nonimmunoreactive portions of immunoglobulin known as the Fc region. Such binding leads to activation of the proteins and results in membrane damage and cell destruction. Most murine monoclonal antibodies are relatively ineffective in fixing human complement in the laboratory.

One interesting exception is that certain monoclonal antibodies of the IgG3 subclass directed against antigens characterized chemically as disialogangliosides (gD2, GD3), which are expressed on melanoma and neuroectodermal cells, are able to effect complement-mediated cytotoxicity with human complement in the laboratory. Because complement fixation occurs on the Fc portion of antibody, this approach requires intact immunoglobulin.

The second cytotoxic mechanism is antibody-dependent cell-mediated cytotoxicity. Again, the Fc portion is required for binding of cytolytic cells via their Fc receptors. Many lymphocytes, monocytes, tissue macrophages, and granulocytes have such receptors and are cytolytic in laboratory cultures. Collectively, these are referred to as effector cells. Antibody may bind to cellular antigen, which is followed by effector cell binding to the Fc segment. Alternatively, the effector cells may bind to the antibody and be carried to the tumor target. In cultures, both methods produce cytotoxicity, and the first approach has produced antitumor responses in animal models. Circulating target cells coated with antibody are removed in the reticuloendothelial system, presumably because of Fc receptors on tissue macrophages in those sites. For human antibodies the ability to bind specifically to macrophages of the reticuloendothelial system is confined to IgG1 and IgG3 antibodies, and is absent from IgM and other IgG subclasses. Murine IgG2A antibodies are superior to other IgG subclasses in antibody-dependent cell-mediated cytotoxicity with human effector cells, although murine IgG3 anti-GD2 and anti-GD3 antibodies are effective as well.

Both antibody-dependent cell-mediated cytotoxicity and complement-mediated cytotoxicity rely on the host immune system to provide either efficient complement proteins or effector cells to produce an antitumor effect. The use of various biological response modifiers may augment the host immune system. To maximize efficiency of antibody–effector cell binding, some investigators have utilized leukophoresis (selective removal of large quantities of white blood cells while returning other cells to the donor) to harvest leukocytes and have incubated these cells with antibodies to enrich the antibody–effector cell population. Various lymphokines (chemical mediators produced by lymphocytes), such as interleukin-2 (IL-2), a chemical mediator crucial for activation and proliferation of T lymphocytes which is also known as T-cell growth factor, may be useful to enhance the cytotoxic activity of certain effector cells. Both antibody-dependent cell-mediated cytotoxicity and complement-mediated cytotoxicity occur only above a threshold of bound monoclonal antibodies. Lymphokines, such as alpha-interferon, may be useful in enhancing target antigen expression, while gamma-interferon increases Fc receptor on effector cells. There are good examples from model animals suggesting that various biologicals may be used in concert to optimize an antitumor effect mediated by monoclonal antibodies.

Regulatory mechanism. Tumor cells have various receptors which are important for growth or proliferative advantages. Antibodies directed against growth factor receptors may down-regulate (cause decreased expression of) the number of receptors available on the cell surface or competitively inhibit ligand binding, thus impairing a cell's ability to differentiate or divide, and ultimately resulting in cell death. Such antigens could also serve as targets for monoclonal antibodies producing cytotoxicity via antibody-dependent cell-mediated cytotoxicity or complement-mediated cytotoxicity, or as targets for immunoconjugates, that is, antibodies attached to other substances.

Probably the most thoroughly studied system for a regulatory approach is the idiotype–anti-idiotype network. Malignant B lymphocytes (B-cells) produce a specific immunoglobulin with a unique hypervariable region with antigenically distinct molecules called determinants which collectively constitute the idiotype of that immunoglobulin. Under normal conditions, a subset of thymic-derived lymphocytes, known as memory T cells, may trigger B cells to produce immunoglobulin via the idiotype receptor when the receptor is expressed on the B-cell surface. It has also been hypothesized that other B cells may produce antibodies which bind to the idiotype (anti-idiotype antibodies) which can inhibit production of the idiotype. The network theory of idiotype–anti-idiotype interaction postulates that a monoclonal antibody directed against a B-cell lymphoma idiotype might suppress that clone back to its baseline state. Such an idiotype represents the most tumor-specific antigen so far defined.

Immunization mechanism. The third application of monoclonal antibodies alone for cancer therapy is also based on the idiotype network theory and consists of passive immunization. The idiotype of an Ig is quite immunogenic, and successful immunization with a mouse antibody directed against a patient's tumor antigen could lead to production of anti-idiotype antibodies directed against the immunizing protein. Subsequently, this anti-idiotype antibody should induce an anti-anti-idiotype antibody which, because of the structural relationship between antigen and antibody, will have the same binding specificity as the original mouse monoclonal antibody, except that it will be a human antibody produced endogenously. This theoretical approach has been demonstrated in animals and may have application in both cancer and autoimmune diseases. *See Vaccination*.

Immunoconjugates. The other major monoclonal antibody–mediated anticancer approach uses immunoconjugates. Shortcomings of the host immune system in cancer patients may limit the efficacy of passive monoclonal antibody therapy. However, it is still possible to take advantage of the tumor specificity of a monoclonal antibody by conjugating a cytotoxic substance to it, that is, chemically attaching it to the antibody. Substances which may be attached include radioisotopes, chemical toxins, cytotoxic drugs, and other biological agents.

Radiolabeled antibodies. Radiolabeled monoclonal antibodies may offer the best therapeutic potential inasmuch as cytotoxic activity can be delivered to a region rather than to only a single cell. This approach might overcome certain problems such as the heterogeneity of antigen expression and diminished blood supply. Most therapeutic work has focused on potential beta emitters for this reason. Alpha particles transmit a greater amount of local energy, while beta emitters produce a lower energy level but over a wider area. Pure gamma emitters are less desirable because they irradiate too large an area. The selection of a specific isotope for therapy must take into account the half-life of the isotope, the radiation emitted, and the ability to conjugate the isotope to an antibody without seriously affecting its immunoreactivity.

Chemical toxins. There are numerous naturally occurring toxins, such as ricin from the castor bean, which exist in nature as two glycopeptide chains (identified as A chain and B chain) linked by a disulfide bond. The ricin A chain blocks protein synthesis on ribosomes leading to cellular toxicity, while the B chain is required for actual binding to cell surfaces and appears to facilitate the penetration of the A chain through cellular membranes. In nature, the cytotoxic effect of these toxins is nonspecific because the B chain binds to any cell-membrane galactose residue. However, there are two means by which monoclonal antibodies could make such toxins more selective for cancer cells. The first involves splitting the disulfide bond to free the A chain, which is then reconjugated to the monoclonal antibody. Alternatively, the entire ricin molecule can be bound to a monoclonal antibody, with sugars or other chemicals used to block the galactose-binding site. Such whole ricin and ricin A-chain immunoconjugates are extremely effective in laboratory cultures.

Cytotoxic drugs. A number of chemotherapeutic agents, including chlorambucil, doxorubicin, daunomycin, vindesine, methotrexate, and cis-platinum, have been successfully conjugated to antibodies, either directly by chemical means or indirectly via various chemical linkers so that drugs are attached to the side chain linker rather than directly to the antibody. The indirect technique has the advantage of attaching more drug molecules without altering the immunoreactivity of the antibody. In addition, certain acid-labile linkers (covalent bonds which are stable at pH

7.4 but labile at acid pH) allow conjugation, which blocks the cytotoxic effect of the drug unless the drug is released from the linker. Such acid-labile linkages are split at the low pH present in lysosomes. In animal tumor models with human xenografts, a variety of drug conjugates appear to be superior to a drug alone, or combinations of drug and antibody, both in terms of antitumor effect and in terms of diminished toxicity.

Clinical trials. Most monoclonal antibodies which have been studied are not effective in assays of complement-mediated cytotoxicity or antibody-dependent cell-mediated cytotoxicity, and they produce limited effects clinically. However, such studies have confirmed the ability of monoclonal antibodies to bind to tumor cells with acceptable toxicity; they have also delineated the kinetics of human antimouse formation, and the kinetics of murine monoclonal antibodies in human serum. Promising antitumor effects were seen with IgG3 monoclonal antibodies directed against GD3 or GD2 in patients with melanoma, osteosarcoma, and neuroblastoma. In a phase I study in which doses of anti-GD3 were escalated in successive patients to determine dose/toxicity relationships, responses were seen in 4 of 21 patients. In a similar trial in which doses were escalated in successive patients, responses were noted in 7 of 19 patients receiving anti-GD2 at another institution. Again, it should be emphasized that, in contrast to most of the early antibody pilot and phase I studies, these monoclonal antibodies were effective with human complement and effector cells in assays of complement-mediated cytotoxicity and antibody-dependent cell-mediated cytotoxicity.

The largest experience with a possible regulatory approach has been with anti-idiotype therapy. Investigators from Stanford University have reported responses in 6 of 11 patients who have received this therapy. The best responses were seen in patients with nodular, poorly differentiated lymphoma, particularly in those patients who had a substantial T-cell infiltration of their B-cell tumor prior to therapy. Recently, investigators at the National Cancer Institute have observed responses in acute T-cell leukemia in patients treated with an antibody to the interleukin-2 receptor.

Most of the work in radioimmunotherapy has been done with animal antisera rather than murine monoclonal antibodies. Encouraging results have been reported for patients with hepatoma and Hodgkin's disease who have received antiferritin antibodies conjugated to iodine. Trials have also recently started with yttrium-labeled antibodies. In lung cancer patients, preliminary studies with a methotrexate immunoconjugate directed against adenocarcinoma appear promising in terms of feasibility and lack of toxicity. Doxorubicin immunoconjugate studies have also been initiated in pa-

tients. A large phase I study has been completed in melanoma with a ricin A-chain immunotoxin. Additional studies are in progress to determine efficacy of this approach. Other studies with ricin A-chain immunotoxins have begun in humans with chronic lymphocytic leukemia, T-cell lymphomas, and colorectal carcinoma.

It is appropriate that human clinical trials have been initiated for the various monoclonal antibody–mediated anticancer approaches. In many instances, it is clear that the products used for therapy need improvement. Human monoclonal antibodies may be more effective in passive approaches because of better activity in antibody-dependent cell-mediated cytoxicity or complement-mediated cytoxicity, and they should produce less of an endogenous immune response. The use of only the two immunoreactive fragments of antibody [F(ab′)₂, that is, the antibody without the Fc fragment] should decrease nonspecific binding of immunoconjugates, thereby decreasing potential nonspecific toxicity. Bifunctional monoclonal antibodies have been produced in which the same antibody molecule has two different determinants. Chimeric monoclonal antibodies have been produced that take advantage of the ability of a mouse monoclonal antibody to bind with a human Fc receptor. Modifications of Ig carbohydrates have been used to decrease immunogenicity. There is intensive interest in utilizing the carbohydrate region of the Fc segment to attach various linkers which can make the delivery of cytotoxic therapy even more specific. Animal and human clinical studies will be needed with all of these products to determine the optimal approach.

For background information SEE ANTIBODY; ANTIGEN; CANCER (MEDICINE); IMMUNOGLOBULIN; IMMUNOLOGY; MONOCLONAL ANTIBODIES in the McGraw-Hill Encyclopedia of Science and Technology.

Robert O. Dillman

Bibliography. R. O. Dillman et al., Continuous infusion of T101 monoclonal antibody in chronic lymphocytic leukemia and cutaneous T cell lymphoma, *J. Biol. Response Mod.*, 5:394–410, 1986; A. N. Houghton et al., Mouse monoclonal IgG3 antibody detecting GD3 ganglioside: A phase I trial in patients with malignant melanoma, *Proc. Nat. Acad. Sci. USA*, 82:1242–1246, 1985; T. C. Meeker et al., A clinical trial of anti-idiotype therapy for B cell malignancy, *Blood*, 65:1349–1363, 1985; S. E. Order et al., Iodine 131 antiferritin, a new treatment modality in hepatoma: A radiation therapy oncology group study, *J. Clin. Oncol.*, 3:1573–1582, 1985.

Mycobacteria

Mycobacterium avium and *M. intracellulare* are very closely related species which are grouped taxonomically as *M. avium-intracellulare* or, preferably, as *M. avium* complex.

These organisms are related to the tubercle bacillus, *M. tuberculosis*, which causes classical human tuberculosis. The mycobacteria are distinguished from other bacteria mainly by their unusual cell wall, which is strikingly rich in lipids and waxes. This cell wall is responsible for several of the special characteristics of the organisms, including their staining properties (they are slow to take up microbial dyes but once they do so, they retain them avidly, even resisting decolorization with acid alcohol; hence, the designation acid-fast bacilli). During the first 60 years of the twentieth century, there were sporadic and uncommon reports of human disease due to *M. avium* complex. However, over the last 25 years there have been increasing numbers of patients with documented disease due to these organisms. Recent advances have helped clarify the transmission, pathogenesis, and treatment of human disease due to *M. avium* complex.

Species, strains, and serotypes. *Mycobacterium avium* was the designation given near the turn of the century in Europe to mycobacteria associated with a tuberculosislike illness in poultry. *Mycobacterium intracellulare* was the name assigned to organisms associated with human pulmonary disease in the United States in the late 1950s. Subsequently, microbiologists found that these bacteria were so similar in conventional laboratory testing that they were lumped as *M. avium-intracellulare*, now *M. avium* complex. Microbiologists have employed an intricate serotyping sytem to distinguish the organisms for many years. Recent investigations examining the homology of deoxyribonucleic acid (DNA) of the various strains have confirmed that there are indeed two species, but that serotyping had falsely classified certain strains as *M. intracellulare* rather than *M. avium*. It is of particular interest that these misclassified organisms include several of the serotypes which are most frequently associated with human disease.

Distribution in nature. Unlike the tubercle bacillus, which has no significant reservoirs in nature other than infected humans, *M. avium* complex organisms are widely distributed in the environment, as well as in numerous animals and fowls; organisms of this complex have been recovered from dust, soil, fresh water, coastal salt water, swine, cattle, primates, and domestic and wild fowl. These organisms have been found throughout Europe, Japan, and Australia, as well as North America; very likely, they would be found commonly in all regions were the surveys to be performed there.

Recently a series of surveys and analyses were made of the *M. avium* complex organisms in the environment in the southeast Atlantic coastal region. The researchers found that, while various *M. avium* complex strains were ascertained to be nearly ubiquitous in the soil and water, the strains sharing the most characteristics with the organisms associated with human disease occurred in highest concentration in sprays generated from brackish coastal water. They also demonstrated that such strains are preferentially aerosolized in an artificial wave-generating system. They have suggested that such airborne microbes might be the vector for cases of human pulmonary infections. While this hypothesis is appealing in explaining the strikingly high prevalence of *M. avium* complex disease and skin test reactivity in the southeast United States, it does not explain the increasing numbers of cases found more widely spread across the country.

Human disease. Many of the patients with lung disease due to *M. avium* complex have underlying pulmonary disorders which appear to increase the risk of infection by these microbes of relatively low virulence. The most common underlying conditions are chronic bronchitis and emphysema related to cigarette smoking. Other associated conditions include diffuse lung injury associated with inhaled dust disorders (pneumoconioses), idiopathic lung scarring or fibrosis, and prior tuberculosis.

In addition to the gradually increasing numbers of cases of lung disease, there has been a dramatic upswing in the number of extrapulmonary infections due to *M. avium* complex in the last decade. This is due mainly to the striking vulnerability of victims of acquired immunodeficiency syndrome (AIDS) to *M. avium* complex organisms. Roughly one-half of AIDS patients who have come to autopsy have evidence of overwhelming, disseminated infection by these mycobacteria. The source of these infections is not clear, but circumstantial evidence points to the alimentary tract as the most likely portal. In addition, patients with impaired immunity due to organ transplantation or cancer chemotherapy are also susceptible.

In addition to these ominous instances of extrapulmonary infection, there is a benign localized infection of the lymph glands of the neck, a form of extrapulmonary disease seen almost exclusively in young children. It is believed to reflect spread from the tonsillar lymphatic tissue which is infected by oral ingestion of the microbes. While it is reported across the United States, it is most common in the Southeast, where *M. avium* complex is most prevalent in the environment.

The combination of increased numbers of pulmonary cases along with disseminated infections in individuals with impaired immunity has raised *M. avium* complex from the level of an uncommon nuisance to a major infectious pathogen. If current trends continue, it is possible that human disease due to *M. avium* complex may become more prevalent in the United States and Canada than classical tuberculosis within the next 20–30 years. However, on a worldwide scale, given the endemic nature of tuberculosis, disease due to the tubercle bacillus will remain vastly greater for the foreseeable future.

Treatment of disease. Chemotherapy is the mainstay of treatment. Like classical tuberculosis, *M. avium* complex disease is very slow to respond and may require many months or years of medication. Standard antituberculosis drugs such as isoniazid, streptomycin, rifampin, and ethambutol are effective against a limited number of strains; however, the patterns of drug resistance of the different strains is extremely variable; very often, second-line antituberculosis drugs such as ethionamide or cycloserine must also be employed. Experimental drugs such as ansamycin LM-427 or clofazimine show considerable promise.

Recent advances in chemotherapy include the first demonstration that laboratory testing for drug susceptibility has value in predicting response to therapy. Such a correlation has been shown for *M. tuberculosis*, but its relevance to *M. avium* complex disease was considered doubtful. Also of interest is the observation that there are synergistic drug interactions against a number of *M. avium* complex strains. While it is too early to state whether this laboratory phenomenon applies to the response to therapy in the organism, there are promising preliminary data to support this notion. Such synergy (suppression of a strain by concentrations of two or more of the drugs which are not active singly) may be responsible for heretofore inexplicable cures. It is important to contrast this rationale for using multiple drugs against *M. avium* complex disease with that for *M. tuberculosis*. Multiple drugs are required in tuberculosis almost entirely to prevent the emergence of drug-resistant mutants; there is no evidence for synergy of drugs against the tubercle bacillus.

Surgery for pulmonary disease due to *M. avium* complex remains controversial. Unfortunately, there have been no controlled clinical trials to evaluate the efficacy of resecting diseased lung tissue. Recently, an expert group reached the consensus that resectional surgery is probably a valuable adjunct to chemotherapy for selected patients with localized pulmonary disease.

Mycobacterium avium complex disease involving the lymph glands of healthy children is almost always a self-limited disease which usually requires neither drugs nor surgery to resolve. However, involvement of the lymph glands in immunocompromised patients, notably AIDS victims, is a marker of disseminated disease with a much less favorable prognosis.

For background information *SEE MYCOBACTERIAL DISEASES; OPPORTUNISTIC INFECTIONS* in the McGraw-Hill Encyclopedia of Science and Technology.

Michael D. Iseman

Bibliography. K. L. Fry, P. S. Meissner, and J. L. Falkinham III, Epidemiology of infection by nontuberculous mycobacteria: VI. Identification and use of epidemiological markers for studies of *Mycobacterium avium, M. intracellulare,* and *M. scrofulaceum, Amer. Rev. Respir. Dis.,* 134:39–43, 1986; L. B. Heifets, Synergistic effect of rifampin, streptomycin, ethionamide, and ethambutol in *Mycobacterium intracellulare, Amer. Rev. Respir. Dis.,* 125:43–48, 1982; M. D. Iseman, Disease due to *Mycobacterium avium-intracellulare, Chest,* 87(2):139S–149S (suppl.), 1985; U. G. Mason et al., *Characteristics of Patients with Pulmonary Mycobacterium avium-intracellulare Infection,* abstract presented at Annual Meeting, American Thoracic Society, Miami, Florida, May 20–23, 1984.

Neutrino

The neutrino holds a unique position in the present picture of fundamental particles and their interactions. It was first postulated by W. Pauli in 1930 to explain the so-called missing energy observed in the radioactive decay of certain nuclei producing electrons (originally known as beta rays). Soon after, E. Fermi developed a quantitative description of this beta-decay process and named the new particle the neutrino, with zero electric charge and a mass less than that of the electron. Since the first observations of interactions of reactor neutrinos in matter by F. Reines and C. Cowan in 1956, experimental studies involving the neutrino have been important in the development of the theory of particles and their interactions.

The question of whether the neutrino has a nonzero mass has been a key issue in many of these studies. Over the past several years, many particle physicists have concentrated on this topic. On the theoretical side, scientists developing new models of particles and their interactions have uncovered hints that neutrinos may be massive. Neutrinos with small masses would also fit into the current cosmological models of the evolution of the universe. These suggestions have inspired experimental physicists to probe for indications of finite neutrino masses in laboratory experiments. While the results so far show no conclusive evidence for massive neutrinos, several experiments have presented preliminary positive indications. A positive result would be of great importance, and many groups are therefore continuing these studies with more sensitive experiments.

Particle physics and neutrino mass. The current understanding of particle physics is contained in the so-called standard model of particles and interactions. In this model, there exist 24 elementary particles grouped into two types, quarks and leptons, and a corresponding number of antiparticles, grouped into antiquarks and antileptons. These particles and antiparticles can interact with each other through three forces, the strong, the weak, and the electromagnetic interactions. The neutrino has the unique property of participating only in the weak interaction. For this interaction the particles are grouped into pairs (or doublets) that can be transformed into each other through weak (or radioactive) decays. The neutrinos come in three types or flavors: ν_e, ν_μ, and ν_τ. These correspond to the associated charge particle or lepton: the electron, muon, and tau lepton.

Particle physicists have been dissatisfied with the standard model for some time, even though the model does explain all of the current experimental data on particle interactions and decays. The dissatisfaction stems from the large number of parameters left unspecified in the model. These include the neutrino, quark, and lepton masses, and the strengths of the three fundamental forces. Grand unification theories have recently been proposed to address the deficiency. In these theories the three interactions become identical at some very high unification energy. New X particles with masses of the order of this unification energy (currently estimated to be 10^{23} eV) mediate the new unified interaction and can lead to new phenomena such as proton decay. This new interaction is consistent with neutrinos having a finite mass; for example, it would predict masses of 10^{-12}, 10^{-7}, and 10^{-3} eV for the electron, muon, and tau neutrino respectively.

More recently, particle theorists have been trying to develop a so-called theory of everything, including gravity. Much of the current work has concentrated on superstring theories. In these theories the world is ten-dimensional when probed at extremely small distances, of the order of 10^{-35} m. All elementary particles correspond to different excitation modes of a single fundamental string, and the three interactions along with the fourth, gravity, are unified into one common interaction. For both the superstring theories and the grand unification theories, neutrino mass measurements are important because they allow glimpses of interactions at the very small distance and high-energy scales of unification. In most of these theories the neutrino does have a finite mass but, as given in the example, the predicted value may be very small. SEE ELEMENTARY PARTICLE.

Cosmology and neutrino mass. The big bang model, the standard model of the development of the universe, has been very successful in explaining many of the current astronomical observations. These observations include the universal expansion rate, the 2.7 K cosmic microwave background radiation, the matter-antimatter asymmetry, and the cosmic abundance of deuterium and helium.

Neutrinos with masses greater than a few electronvolts would play a crucial role in this model. The number of relic neutrinos from the big bang is predicted to be large (approximately 3000/in.^{3} or 200/cm^3) and, if the neutrino has a mass near 10 eV, would dominate the baryonic (neutron and proton) mass of the universe. The measured expansion rate of distant objects in turn limits the maximum possible neutrino mass. If any neutrino type would have a mass greater than 100 eV, the mutual gravitational attraction of these neutrinos throughout the universe would restrict the expansion to less than current observations. On the other hand, massive neutrinos with masses of a few tens of electronvolts might help solve some of the so-called missing mass problems associated with galaxies and galactic clusters. These problems stem from the inconsistency between the mass of a galaxy that is inferred from the amount of emitted light, and the much larger mass necessary to hold the galaxy together gravitationally as it rotates. Massive neutrinos could provide the so-called dark matter necessary to satisfy the gravitational constraints, but the details of how these neutrinos are captured in galaxies are still not understood.

A final constraint imposed by cosmology relates, not to the mass of neutrinos, but to the number of different types. As the universe expands, the temperature decreases and the protons and neutrinos slow down. About 200 s after the big bang the protons and neutrons start to coalesce into deuterium and helium; that is, nucleosynthesis begins. The present abundance of helium and deuterium can be calculated in this model depending on the number of neutrino types, N_ν, since the expansion rate at the time of nucleosynthesis depends on N_ν. The current best observational measurements of the helium-4 (^{4}He) abundance give a ratio by weight to the total baryonic mass of 23 to 26%. This ratio in turn restricts the number of neutrino species to be less than or equal to four and allows at most one new type beyond the three known ones.

Neutrino mass measurements. Experiments to measure neutrino masses are based on studies of the weak decay products of other particles. Since the weakly interacting neutrino is very difficult to detect (neutrinos can easily penetrate the Earth without interactions), these studies measure the momentum and energy of the other decay products and infer the properties of the neutrino by using momentum and energy conservation. For example, in the decay of a neutron n into a proton p with the emission of an electron e^- and an electron antineutrino $\bar{\nu}_e$, $n \rightarrow p + e^- + \bar{\nu}_e$, measurements of the electron and proton momenta would make it possible to deduce the $\bar{\nu}_e$ mass from the known n, p, and e^- masses. Many studies of this type have been performed. All experiments to measure the $\bar{\nu}_e$ mass use the beta (or electron) decay of tritium, ^{3}H$^0 \rightarrow {}^3$He$^+ + e^- + \bar{\nu}_e$, which is sensitive to a finite neutrino mass and easily incorporated into an experiment.

The current experimental situation is somewhat unclear. A group at the Institute of Theoretical and Experimental Physics (ITEP) in Mos-

Present neutrino mass limits	
Neutrino type	Mass limit
Electron antineutrino ($\bar{\nu}_e$)	$m_{\bar{\nu}_e} < 18$ eV
	(17 eV $< m_{\bar{\nu}_e} <$ 40 eV; ITEP)
Muon neutrino (ν_μ)	$m_{\nu_\mu} < 490 \times 10^3$ eV
Tau neutrino (ν_τ)	$m_{\nu_\tau} < 76 \times 10^6$ eV

cow has reported several results, starting in 1980, indicating that the $\bar{\nu}_e$ has a finite mass greater than 17 eV. Recently, a group based in Zurich presented data consistent with a zero $\bar{\nu}_e$ mass, limiting the value to less than 18 eV. Both of these studies have used complex sources, with the tritium forming part of a larger molecule (ITEP) or implanted in another substance (Zurich). Theoretical uncertainties in the molecular and solid-state effects introduce possible uncertainties in the final $\bar{\nu}_e$ mass results. For this reason, a group at Los Alamos National Laboratory is performing a similar experiment using a beam of atomic tritium. At present, their results limit the $\bar{\nu}_e$ mass to less than 27 eV, but should soon be sensitive to masses below 20 eV. In line with common practice in experimental science, the positive ITEP results will not be assumed correct until substantiated or refuted by several other experiments. In late 1986, almost a dozen studies of tritium decay were under way with conclusive results expected in 1 to 2 years.

For the muon neutrino the most precise mass limits come from the studies of the decay of a pion π^+ into a muon μ^+ and a muon neutrino ν_μ, $\pi^+ \rightarrow \mu^+ + \nu_\mu$. Only the muon momentum is measured. Using this technique, a Swiss group at the Schweizerisches Institut für Nuklearforschung (SIN) laboratory limited the ν_μ mass to below 490 $\times 10^3$ eV. Limits on the mass of the tau neutrino ν_τ come from studies of tau-lepton decays. Copious tau leptons are produced at the PEP (Positron Electron Project) colliding beam accelerator at Stanford, California. The High Resolution Spectrometer experiment at PEP, using studies of tau-lepton decays into multiple pions plus a tau neutrino, has limited the ν_τ mass to below 76×10^6 eV. The **table** summarizes the current mass limits on the three known types of neutrino.

Neutrino oscillation measurements. Many of the grand unification theories predict that neutrino types are mixed. For example, the neutrino produced when a pion decays into a muon, $\pi^+ \rightarrow \mu^+ + \nu_\mu$, called a muon neutrino (ν_μ), is really a mixture of several neutrinos with different masses. This mixing comes about if the interaction responsible for giving the neutrino a mass does not conserve neutrino type or lepton number. The experimental consequence of mixing is a phenomenon, referred to as neutrino oscillations, in which one neutrino type changes into

another as the neutrino traverses some distance. How rapidly one neutrino will change into another depends on both the neutrino masses and the amount of mixing. Neutrino oscillation experiments are important because these measurements probe very small neutrino masses even for the muon and tau neutrino types. The additional requirement, though, that a given neutrino type is a mixture of several neutrinos of different masses makes these measurements somewhat less general than the direct decay measurements discussed above.

Experiments look either for a deficiency of neutrinos from a known source (disappearance experiment) or for another neutrino type in a pure beam of another species (appearance experiment). For all of the appearance experiments, detectors have been placed in a muon neutrino beam produced from pion decays. The experiments took place at Fermilab near Chicago, Brookhaven National Laboratory near New York, and CERN (European Center for Nuclear Research) in Geneva, Switzerland. One experiment at CERN has observed a slight excess of electron neutrinos but has not shown that these are inconsistent with the expected beam contamination. All other experiments observe no anomalous production of electron or tau neutrinos and limit the most massive neutrino in the muon neutrino mixture to below 1 eV for mixings greater than 1%.

The disappearance experiments are of two types. For the first, a neutrino detector is placed at several distances (between 30 and 230 ft or 10 and 70 m) from the core of a power reactor. The reactor produces electron antineutrinos that penetrate the shielding and show up in the detector. The experiments compare the number of antineutrinos detected to that expected at each distance; any deficiency would indicate oscillations. Two groups have recently reported measurements. One group, using the Goesgen reactor in Switzerland, has reported results for three distances, 124, 151, and 213 ft (38, 46, and 65 m), that are consistent with the absence of oscillations. These results would limit the electron neutrino mass to below 0.2 eV for mixings larger than 20%. On the other hand, another group using the Bugey reactor in France has observed a small difference in the number of neutrinos detected at 45 and 60 ft (13.5 and 18 m). This difference would correspond to the most massive neutrino in the electron neutrino mixture having a mass of 0.5 eV with a mixing of 30% to another neutrino type, but the uncertainties connected with the measurement must be reduced before the result can be taken as conclusive.

For the second type of disappearance study, two detectors are placed at different distances from a source of muon neutrinos. Measurements have been carried out in three experiments at Fermilab and CERN. In all three a similar number of neutrinos is observed in the two detectors, limiting the most massive neutrino in the muon neutrino mixture to below 1 eV for mixings larger than 10%.

Solar neutrinos. Since 1976 R. Davis and collaborators have been measuring the flux of electron neutrinos from the Sun. The detector, located in the Homestake Gold Mine in South Dakota, uses the chlorine-to-argon transition to indicate a captured solar neutrino. The number of neutrinos detected is consistently about a third of that expected from standard calculations of the dynamics of the Sun. This discrepancy may in fact be due only to inaccuracies in the solar model calculation, but most experts believe that a factor of 3 would be too large an error. Recently, a mechanism was suggested whereby the electron neutrinos could oscillate into muon neutrinos as they emerge from the Sun. The mechanism relies on a quantum-mechanical effect caused by the electron neutrinos interacting with the electrons in the solar plasma. The observed neutrino rate could be explained if the muon neutrino mass were between 0.01 and 0.1 eV, even for mixings as small as 0.1%. The unique property of this mechanism is that the small mixing can be enhanced by the interactions with the electrons in the Sun, giving the large observed reduction. There is much interest in probing this mechanism further by detecting lower-energy solar neutrinos by using gallium as a detection medium. Three experiments are currently being planned in Europe, the United States, and the Soviet Union, with the European experiment expecting first results in 1992. The gallium experiments are much less sensitive to the solar model uncertainties and should provide incontrovertible evidence as to whether or not solar neutrinos oscillate.

For background information *see* COSMOLOGY; ELEMENTARY PARTICLE; FUNDAMENTAL INTERACTIONS; GRAND UNIFICATION THEORIES; LEPTON; NEUTRINO; QUARKS; SOLAR NEUTRINOS in the McGraw-Hill Encyclopedia of Science and Technology.

Michael Shaevitz

Bibliography. T. Ferbel (ed.), *Techniques and Concepts of High Energy Physics III*, 1985; J. Tran Thanh Van (ed.), *Proceedings of the 4th Moriond Workshop on Massive Neutrinos*, 1984.

Nitrogen fixation

The global contribution of symbiotic N_2 fixation has been estimated to be 90 million metric tons of nitrogen per year. In comparison, global nitrogenous fertilizer production was 61 million metric tons (67 million tons) in 1985. Symbiotic N_2 fixation provides a unique and valuable nitrogen input system in legumes, which has not evolved in the other major economic food plant group, the cereals. Although transfer of the ability to fix N_2 symbiotically to nonlegumes has been a widely considered and desired goal, much needs to be learned about the basis of host specificity, and about the plant's role in initiating and maintaining the symbiosis in legumes, before the mentioned goal can be achieved. Current research focuses on optimizing the symbiotic performance in agriculturally significant legume species and their bacterial partners.

Symbiotic nitrogen (N_2) fixation occurs in root nodules of legume species, such as soybean and alfalfa, and results from a species-specific interaction between a host plant and a bacterial strain; most of the bacterial strains belong to the *Bradyrhizobium* and *Rhizobium* genera. When symbiosis is established, carbon products from the host are exchanged for nitrogenous products from the bacteria in the root nodule; both organisms benefit from the association.

Carbon is supplied from photosynthetic carbon dioxide fixation by the plant, and nitrogen is supplied through fixation of atmosphere molecular nitrogen (N_2) into organic nitrogen compounds by the bacteria in the nodule. This relationship provides a productivity advantage for legume plants over nonlegumes (grasses, cereals, and so forth) under conditions of limited soil nitrogen. The dependence of the bacterial counterpart on obtaining carbon from the host plant dictates that any factors which adversely affect plant photosynthesis (photosynthate production) will in turn limit symbiotic N_2 fixation. The limitation of photosynthate availability is accentuated by the process of N_2 fixation, which is energy-intensive, and by the numerous biochemical reactions within the plant which compete for photosynthate. Thus, environmental factors (temperature, light, and so forth) which affect photosynthesis play a key role in rates of symbiotic N_2 fixation. Likewise, since the various parts of the plant (seeds, leaves, stems, roots, and nodules) are also competing for available photosynthate, the temporal pattern of symbiotic N_2 fixation is dependent on the morphological and physiological growth stage of the plant. The amount of atmospheric N_2 fixed by a legume is inversely related to soil nitrogen level, since nitrate (NO_3^-), the primary form of inorganic soil nitrogen, inhibits the development of an effective symbiosis. The events leading to the establishment of an effective symbiosis, and the dependence of N_2 fixation on the environment, soil fertility level, and stage of plant development, have been subjects of recent research.

Root nodule development. Establishment of an effective N_2-fixing symbiosis requires a kind of communication between a specific rhizobial species and particular root cells of its host plant. This recognition results in a complex developmental process whereby rhizobia adhere to and penetrate the root. Although nodulation of most plant species is through the roots, and generally through root hairs, a few species (notably *Sesbania rostrata*) can nodulate through lenticels on the stem.

A model that has been suggested to explain the mechanism of recognition between the bacterial symbiont and the plant host involves the specific binding of compatible rhizobia to host-plant root lectins. Lectins are noncatalytic proteins, prevalent in the plant kingdom, which are capable of binding specific microbial cell-surface carbohydrates. Recent studies have shown that specific flavonoid molecules released from plant roots may also be involved in this recognition process through induction of bacterial gene expression necessary for early phases of infection. Recognition between the bacteria and the root, and subsequent penetration of the bacteria into the root through an infection thread, is accompanied by proliferation and differentiation of cortical root cells into a definable structure termed a

nodule. The shape and size of nodules is dependent upon the host-plant species. Nodules can be elliptical, rod-shaped, or branched and generally range from 0.5 to 3.0 mm (0.02 to 0.12 in.) in diameter, although much larger nodules have been reported.

Following infection, the bacteria are released into many of the host-plant cells. A membrane of plant origin surrounds the bacteria as they continue to divide and pack the interior of the nodule cells. They then differentiate into a terminal, nondividing form termed the bacteroid. It is in this form that conversion of atmospheric N_2 to ammonium (NH_4^+) occurs via the nitrogenase enzyme (see **illus.**). The NH_4^+ is then incorporated into organic nitrogen compounds through reaction with plant-supplied carbon compounds. This process occurs in specialized cells within the nodule which do not contain bacteroids but function to transform and transport products of N_2 fixation.

The primary forms of organic nitrogen compounds resulting from symbiotic N_2 fixation are ureides and amides. The ureides are acyl derivatives of urea, such as allantoin and allantoic acid; they are characteristic products of tropical legumes (such as soybean). The amides are compounds containing the —NH_2 group, such as asparagine; they are characteristic of temperate legumes (such as pea). These nitrogenous compounds are subsequently translocated to other parts of the plant for incorporation into products useful in human and animal nutrition. These are primarily seed proteins in grain legumes or leaf proteins in forage legumes.

Environmental effects. Leguminous species have adapted their symbiotic capabilities to a broad spectrum of temperature conditions, ranging from quite cold tundra conditions where soil temperatures are near the freezing point, through tropical environments where soil temperatures reach 35°C (95°F). Such adaptations involve the ability of specific bacteria and host plants to thrive in the free-living state as well as their ability to establish the symbiosis under extreme conditions. Temperature exerts an additional influence through its effect on the activation energy of enzymatic processes of the host plant as well as on the nitrogenase enzyme itself.

In field environments, it is often difficult to separate the effects of temperature on N_2 fixation from those of light, since both vary diurnally. Although both factors play a role, studies carried out in controlled environments indicate that temperature fluctuations may alter the activity of the nitrogenase enzyme more quickly than any change in light level. Over extended periods (days), light is certainly essential to provide the photosynthate needed to supply the energy requirements for the N_2-fixation process. However, short-term demands of the N_2-fixation process for carbohydrate can be met by mobilization

Diagrammatic representation of conversion of atmospheric nitrogen (N_2) to ammonium (NH_4^+). (*a*) Nodulated legume plant. (*b*) Expanded section of a root nodule. (*c*) Scheme of reactions taking place in the root nodule. Photosynthate from the leaves is translocated to the root nodule via the phloem vascular tissue to support bacteroid respiration, to serve as an energy source for nitrogenase activity, and to serve as a carbon source for synthesis of amino acids and other nitrogenous compounds. In return, the bacteria-produced NH_4^+ is converted to ureides or amides and translocated via the xylem vascular tissue to various parts of the plant to support growth and protein synthesis.

of storage carbohydrate from other plant parts in the absence of current photosynthesis. The importance of photosynthesis to nodulation and N_2 fixation is exemplified by observations that carbon dioxide–enriched environments which stimulate photosynthetic processes will in turn stimulate N_2 fixation.

Soil factors. Among various soil factors which influence nodulation and N_2 fixation, the level of available nitrogen is the most crucial. The available nitrogen is primarily in the form of NO_3^- in agricultural soils. High levels of soil NO_3^- inhibit virtually all phases of nodulation, including successful bacterial infection, subsequent nodule development, and function of the nitrogenase enzyme which reduces N_2 to NH_4^+. The mechanisms by which NO_3^- inhibits N_2 fixation are not known, although two primary hypotheses exist. First, NO_2^-, which is the initial product of NO_3^- reduction, is known to be a potent inhibitor of the nitrogenase enzyme. However, the level of NO_2^- accumulation in the nodule seems to be inadequate to account for the level of inhibition noted when a nodulated plant is exposed to NO_3^-. Second, the metabolism of NO_3^- requires energy and carbon skeletons, and thus it competes with the nodulation process for available photosynthate. This unfavorable carbohydrate balance in plants grown on NO_3^- could then limit the N_2-fixation process.

Soil fertility can affect the N_2-fixation process inasmuch as nutrient deficiencies are detrimental to vigorous crop growth, which is required to maintain carbohydrate availability to support N_2 fixation. The nitrogenase enzyme itself requires molybdenum and iron, and cobalt appears essential for optimal functioning of the nodulation process. Calcium is also essential for nodulation, probably through its requirement for cell wall synthesis. Other nutrient deficiencies are more indirectly involved in N_2 fixation through an effect on the host plant. Soil pH is also important to N_2 fixation. At pH levels below about 5.0, bacterial survival in the soil may be limited and the infection process impaired. Optimum pH for symbiotic N_2 fixation by legumes is generally between 5.5 and 6.5, although certain tropical legumes and associated rhizobia are more tolerant of lower pH (more acidic) than are temperate legumes.

Techniques for measurement. Several techniques have been developed for determining N_2 fixation. The most definitive, but also the most costly and time-consuming, is the use of ^{15}N, the stable isotope of nitrogen. By labeling soil nitrogen with ^{15}N, or less commonly by labeling atmospheric N_2, an accurate determination can be made of the fraction of nitrogen derived from the labeled source and hence a calculation of the amount of N_2 fixed. A second technique is to compare total nitrogen accumulation by lines which are similar (genetic isolines) in all respects except for the genetic capability either to form

nodules or to suppress nodules. By subtracting the total nitrogen accumulated in the nonnodulating line from that in the nodulating line, the amount of N_2 fixation can then be calculated. This technique is limited to species where such isolines exist and requires the assumption that the two lines assimilate the same amount of nitrogen from the soil. A third technique involves the quantification by gas chromatography of the reduction of acetylene to ethylene. This reaction is catalyzed by the nitrogenase enzyme and is quite rapid and inexpensive to monitor, relative to the ^{15}N technique. However, this may be criticized for making only a single measurement while N_2 fixation occurs over the growing season, and for requiring assumptions concerning stoichiometry of acetylene reduction to ethylene, in relation to N_2 reduction to NH_4^+. Although the acetylene reduction technique has serious limitations in determining absolute values for N_2 fixation, it has been used extensively and satisfactorily for comparing treatments.

Temporal sequence. The time of onset of N_2 fixation is somewhat species-dependent, with small-seeded legumes undergoing more rapid infection and functional nodule development than large-seeded legumes. Nodulation in all legumes appears sensitive to the soil nitrogen level, which in large part dictates the time of onset of symbiotic N_2 fixation. Considering soybean as an example, functional nodule development is delayed some 30 days from planting because of residual NO_3^- in the soil, whereas nodule function is detectable after 9–10 days from planting under nitrogen-free control conditions. In field environments, maximum N_2 fixation for grain legumes is often associated with the reproductive growth stage. For forage legumes, such as alfalfa, N_2 fixation increases until the time the vegetative material is harvested. Following harvest, nodule activity declines markedly until shoot regrowth occurs and again begins supplying carbohydrate assimilates to the roots and nodules. The annual rates of N_2 fixation may range upward of 200 kg/hectare (180 lb/acre) for some forage legumes, while more typical values for grain legumes such as soybean are generally on the order of 80 to 100 kg/hectare (70 to 90 lb/acre). SEE SOIL NITROGEN.

For background information SEE LEGUME; NITROGEN CYCLE; NITROGEN FIXATION; RHIZOSPHERE in the McGraw-Hill Encyclopedia of Science and Technology.

J. E. Harper

Bibliography. L. J. Halverson and G. Stacey, Host recognition in the *Rhizobium*-soybean symbiosis, *Plant Physiol.*, 77:621–625, 1985; U. D. Havelka, M. G. Boyle, and R. W. F. Hardy, Biological nitrogen fixation, in F. J. Stevenson (ed.), *Nitrogen in Agricultural Soils*, ASA-CSSA-SSSA, pp. 365–422, 1982; D. G. Jones and D. R. Davies (eds.), *Temperate Legumes: Physiology, Genetics, and Nodulation,*

1983; D. A. Phillips and T. M. Dejong, Dinitrogen fixation in leguminous crop plants, in R. D. Hauck (ed.), *Nitrogen in Crop Production*, ASA-CSSA-SSSA, pp. 121–132, 1984.

Nobel prizes

For 1986 there were 11 recipients of the Nobel prizes, awarded by the Swedish Royal Academy.

Medicine or physiology. The prize for medicine was shared by Rita Levi-Montalcini (United States–Italy) and Stanley Cohen (United States). Levi-Montalcini is a senior scientist at the Institute of Cell Biology, Rome, and Cohen is on the faculty of Vanderbilt University School of Medicine. They were honored for work on nerve growth factor at Washington University, St. Louis, in the 1950s. Since then, Cohen has identified epidermal growth factor, a protein that controls cell development in mammalian skin. The Nobel Committee cited the researchers for their contribution to basic science, with applications in the study of developmental malformation, degenerative changes in senile dementia, delayed wound healing, and tumor diseases.

Physics. The award was shared by Ernst Ruska (West Germany) and the team of Gerd Binnig (West Germany) and Heinrich Rohrer (Switzerland). Ruska, who is an affiliate of the Fritz Haber Institute of the Max Planck Society, West Berlin, was honored for his development of the electron microscope in 1931–1933.

Binnig and Rohrer, both members of the IBM Research Laboratory, Zurich, were honored for the development of the tunneling microscope. The tunneling microscope is based on the fact that some electrons escape from the surface of a sample and tunnel into the "forbidden" space above the surface. The microscope includes a needle having an extremely fine tip and carrying a potential on the order of millivolts. As the needle is brought within a fraction of a nanometer of the sample surface, it encounters tunneling electrons, and nanoampere currents are induced. When the needle is moved across the surface of the sample with a constant tunnel current, it creates an atom-by-atom image of the surface that resembles a topographic map. Resolution can be as good as 0.2 to 0.3 nanometer. Applications for the tunneling microscope are mainly in surface physics. *See* Microscope.

Chemistry. The award was shared by Dudley R. Herschbach (United States), Harvard University; Yuan T. Lee (United States; born in Taiwan), University of California at Berkeley; and John C. Polanyi (Canada), University of Toronto.

Herschbach was cited for work in the 1960s, when he developed a technique for tracing chemical reactions that resembles particle acceleration in physics. In this "crossed molecular-beam technique," molecules of different substances are made to collide in a vacuum chamber. Lee subsequently joined Herschbach and has extended the technique to many kinds of reactions. The result has been great insight into exactly how a chemical reaction takes place.

Polanyi has referred to the "molecular choreography" of chemical reactions. His award was for his studies of chemiluminescence, in which the energy states of excited molecules are revealed by their emission of infrared radiation. This work demonstrates the way molecules are born, and the forces that impel them into existence. Polanyi's work is significant for the development of chemical lasers, where reactions substitute for electrical energy to produce pulses of coherent light.

Economics. James M. Buchanan (United States), a faculty member of George Mason University, in Virginia, received the award for his development of public choice theory. Buchanan's focus, which crosses the line between economics and political science, is on government spending policies. He believes that elected officials act out of self-interest rather than for the public good. Since reelection is usually their goal, politicians vote for spending programs and against raising taxes. The outcome has been an enduring government deficit.

Literature. Wole Soyinka (Nigeria), a playwright and political activist who is visiting professor of theater at Cornell University, is the first African to win the prize for literature. Soyinka writes in both English and Yoruba. His works include *Ake: The Years of Childhood*, a memoir of a Christian-Yoruba family, and *A Shuttle in the Crypt*, a collection of poems written in prison.

Peace. Elie Wiesel received the peace prize for his witness to the Holocaust, his efforts on behalf of religious freedom for the Jews in the Soviet Union, and his concern for victims of political oppression in South Africa and Cambodia. A lecturer and author of 26 books, Wiesel was born in Romania and is now a United States citizen. His best-known books include *The Night*, a chronicle of the concentration camps, and *The Jews of Silence*, which reports his visit to the Soviet Union. His fictional works include *A Beggar in Jerusalem*, based on the Arab–Israeli war of 1967.

Nondestructive testing

The period since the late 1970s has been fruitful for research in nondestructive testing and has resulted in development and evaluation of a number of novel approaches and techniques. The first section of this article surveys recent developments in nondestructive testing. The second section gives a detailed discussion of a computed tomography technique in industry.

The production of high-resolution radiographic equipment and the introduction of computer enhancement of radiographic images have both widened the scope of radiographic methods of nondestructive testing. In ultrasonics, accuracy and reliability in defect sizing have improved for the majority of inspections, and real progress in defect characterization has been achieved. In parallel with these developments, more quantitative thermographic techniques have been developed. These developments are described in this section, and other important advances are outlined.

High-resolution radiography. The resolution of a radiographic image is limited by the physical size of the source of radiation, the ideal being a point source. In addition, the quality of the image places a limit on the amount of useful image magnification that can be achieved. This limitation has been attacked by two complementary approaches. Clearly, one route to improved resolution is to reduce the size of the source. Unfortunately, this is not a trivial step. For radioactive sources the masking required to achieve a significantly smaller source would result in unacceptably low count rates or the need to use unacceptably large amounts of radioactive material in the source. The situation for electronic x-ray sources is better, but the operation of these devices results in the release of energy at the source, and the problems of damage to the anode of the system are increased as the source is reduced in size.

For these reasons, progress in high-resolution equipment is steady rather than spectacular, but equipment available in 1987 is substantially improved over that of a few years previously. For example, the high-resolution Microdyne system can operate over the energy range from 20 to 170 keV and has an effective source size of only 10 to 15 micrometers, compared with typical values of 1000 to 1500 μm in the usual equipment. This makes possible greatly enhanced image resolution. The improvement obtained is exemplified by estimates of the radiographic sensitivity (the minimum detectable fractional loss of material), as measured by using a wire image-quality indicator, of 0.6% compared with 1.6% for a high-quality contact radiograph.

An alternative approach to the improvement of resolution is to use data-processing techniques to reduce the blurring introduced by the constraints of the equipment. A large amount of work in this general area of image resolution enhancement has been carried out in fields other than nondestructive testing, such as processing of the widely publicized data from interplanetary probes. In essence the aim is to construct an algorithm which produces sufficient resolution improvement without introducing artifacts, since the processing techniques are generally unstable if

pushed too far. In radiography it is also important to ensure that the processing techniques operate as well on typical defects as they do on artificial defects or markers.

This technique could equally well be applied to data collected by using conventional or high-resolution x-ray equipment. Thus the two approaches to resolution enhancement could be combined in suitable circumstances to increase the potential resolution still further. The microfocus equipment can also be applied to fluoroscopy.

Accurate and reliable ultrasonic testing. Significant limitations in ultrasonic nondestructive testing have been the relative inaccuracy in sizing defects and the difficulties of ensuring that all defects are likely to be detected. The first limitation arises because the parameter mainly employed to estimate defect height, that is, the extent of the defect through the material, is the amplitude of the reflected pulse. This amplitude may be influenced by many parameters other than the size of the reflector, leading to an unsatisfactory correlation with defect height. The estimation of defect length (and width when required in C-scans, that is, area scans) is more satisfactory, since these dimensions are usually both greater and less significant from a fracture mechanics viewpoint.

A significant step toward the improvement of the precision of estimating defect height has been the introduction of techniques in which defect dimensions are measured by differences in the transit times of ultrasonic pulses. In these techniques the estimate of the defect dimensions is now based on a parameter which is little affected by experimental variations, and precisions of from 0.1 to 1.0 mm are becoming routine. The measurement of the time of arrival of the ultrasonic pulses diffracted from the tips of a defect results in the powerful time-of-flight diffraction technique which can be applied to both defect location and sizing. This technique is most commonly employed in a generalized two-probe version, and the technique has been used successfully in many inspections and trials.

Because time-of-flight diffraction relies on the geometric relationship between the probes, the specimen, and the defect, the basic technique must be modified when the specimen geometry becomes complex, as in the case of the complex node joints of offshore structures. In addition, the particular geometry of the specimen may allow the use of other confirmatory measurements. For example, techniques for reactor pressure-vessel inspections have been developed in which a good deal of complementary data is added to what is basically a single-probe time-of-flight diffraction measurement.

The search for greater precision is expected to continue since even the accuracies quoted above are inadequate in monitoring the early stages of fatigue crack growth, for example. This has already led to developments in higher-frequency and compact pulse transducers so as to improve timing accuracies. To complement this, research is also being carried out on the technique of pulse compression to enhance the effective time resolution of the data at the processing stage. This procedure is inherently similar to the enhancement of radiographic images discussed above.

The synthetic-aperture focusing technique is now widespread in ultrasonic analysis. This technique is used to reduce the blurring of ultrasonic images introduced by the width of the ultrasonic beam. To do this, the effect on the ultrasonic image of the geometry of the inspection system is taken into account, and the image is processed so as to reduce the effective width of the probe beam. Because the effects of beam spread and of probe movement are normally seen only over dimensions comparable to the probe diameter and the defect range, the technique is usually of most use in enhancing C-scans or correcting estimates of defect length.

Ultrasonic defect characterization. A persistent question in nondestructive testing concerns identifying the nature of the defect which has been discovered, and, in particular, establishing whether indeed it is a genuine defect that has been observed. To answer this question, it is essential to attempt to gain information on the nature of the reflector. Ultrasonic inspectors can generally offer useful information on the nature of the reflectors they have observed, but this information seems to be related as much to commonsense interpretation of defect position as to the interpretation of the echo. Thus, an echo located near the weld edge is likely to indicate lack of sidewall fusion and will usually be interpreted as such.

Recent work, concentrating on the reliable identification of cracklike defects, has demonstrated the possibility of a general approach based upon the identification of a few parameters which have been shown experimentally to vary significantly as a function of defect character. In this way it has been possible to select three parameters which, so far, have distinguished six classes of cracklike flaw. Because the identification of suitable parameters has been carried out once and for all, it should be possible to use the technique on a very large number of specimens without recourse to further research. The role of the inspector would be to ensure that the appropriate parameters can be adequately determined from the specimen under test. Although this task is not trivial, especially if the specimen attenuates or scatters ultrasonic radiation, it is definable, and the likelihood of success of the measurement can be determined before inspection begins.

Pulsed video thermography. The inspection of large areas of complex materials such as fiber composites poses special problems. Careful ultrasonic measurements are able to locate many defects, but high-speed application of ultrasonics is difficult. Thermographic techniques can cover large areas, but typical defects are difficult to detect because their interference with the establishment of quasi-static heat patterns is negligible. In this type of inspection problem the new approach of pulsed video thermography shows considerable promise. The basic idea is to apply a source of heat to an area of the specimen, but for a very short time duration. The heat diffuses through the specimen, but its progress is blocked by many defects. For a short time, therefore, the regions close to defects appear hotter or colder than their surroundings. This anomaly may be picked up by a suitably rapid infrared detection system and results in a clear image indicating the presence of a flaw, which may then be analyzed in greater detail by other techniques.

The application described above is not the only one in which pulsed video thermography may be useful, and the technique has been shown to operate in other materials such as metals and ceramics. The limitations to use are provided by the response time of the data collection system, the conductivity of the specimen (and whether this is uniform), the response of the heat source, and the specimen geometry.

Other techniques. Noncontacting generation techniques for ultrasound, such as the electromagnetic acoustic transducer and laser techniques, have undergone considerable development. While these techniques remain less efficient than the use of conventional transducers, they are beginning to find application in many inspection tasks in which ultrasonic couplants are impossible to deploy or undesirable.

An important development is the use of multifrequency techniques in eddy-current testing. These techniques allow the response due to the defect being sought to be more reliably differentiated from spurious responses due to changes in material properties or probe lift-off.

Developments which could have far-reaching consequences have taken place in the use of ultrasonic and magnetic techniques for the measurement of residual stresses. Although not as developed as some of the other techniques discussed, this work holds a real promise of portable ultrasonic or magnetic equipment for the measurement of residual stresses.

An important development has been the greater use of computers and computing techniques in the analysis of nondestructive testing data. From being regarded as esoteric, computer-based equipment has become routine in many field applications.

Other developments. The growth of nondestructive testing is not limited to the developments in technique described above. For example, there is a growing need for more accurate and more automatic use of nondestructive testing procedures in fitness-for-purpose assessment and defect-severity assessment. Although technique development may be an essential part, it is also necessary to link the gathering of nondestructive testing data more closely with the assessment of the potential defects located by the inspection. This type of development is under way but not yet generally available or employed. The increase of the potential precision of nondestructive testing also makes it even more essential that the requirements of the inspection process are taken fully into account at the design stage. These developments would lead to improved design and would enhance final quality.   *M. G. Silk*

COMPUTED TOMOGRAPHY IN INDUSTRY

Computed tomography is well established as a diagnostic technique in medicine. Though not yet well established in industry, computed tomography has broad potential as a technique of nondestructive evaluation. This section briefly reviews the principles of computed tomography, points out the differences between medical and industrial uses of the technology, and discusses current trends and applications.

Principles. The concept upon which computed tomography is based is the reconstruction of an object from its projections. Here the term projection means the integral, or sum, of some property of an object along a specified path. In the case of x-ray computed tomography, the property is x-ray attenuation and the paths are straight lines from an x-ray source to x-ray detectors. In most x-ray computed tomography, all the paths lie in a single plane. The part of the object that lies in this plane, that is, a slice of the object, is reconstructed from the projections.

Various reconstruction algorithms have been developed to convert such projections into a map of the attenuation coefficient of the object. The most frequently used type of algorithm, called filtered back-projection, can be very efficiently programmed to run on an array processor, a type of special-purpose computer.

Medical versus industrial uses. In medical scanners, projections in different directions are usually obtained by rotating an x-ray source and detector bank around the patient. This expensive and cumbersome arrangement is required in order that the patient remain immobile but comfortable. A great deal of effort has been devoted to optimizing the scanning and reconstruction process for minimum x-ray exposure and maximum quality of diagnostic information.

In industrial computed tomography the basic ideas are the same, but the construction of a suitable apparatus is not hindered by the medical constraints. The need to minimize radiation exposure to the object being studied is much less stringent, though in exceptional cases radiation damage may be a concern. The means of achieving different projection directions can also be made simpler in most cases. In particular, it is often practical to rotate the object to be inspected

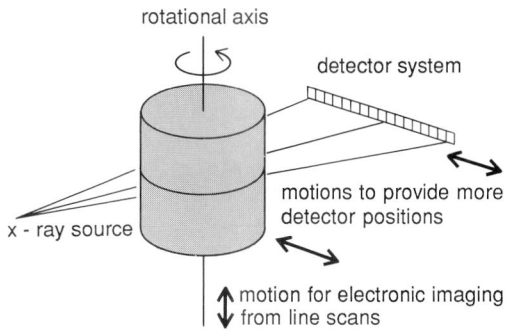

Fig. 1. Typical geometrical arrangement for industrial computed tomography equipment.

rather than the source and detector system.

Geometrical arrangement. Most industrial computed tomography scanners have a geometrical arrangement similar to that of **Fig. 1**. The source energy varies markedly depending on the size and attenuation properties of the object. Moderate-energy x-rays are usually generated by electron bombardment of tungsten targets. High-energy photons (x-rays or gamma rays) may be derived from radioactive isotopes or from an accelerator. The detector system is a crucial part of the system, as it must be fairly efficient and must be appropriate to the energy of the radiation. Several firms now promote proprietary developments in detector technology.

Rotation of the object about a single axis is the fundamental motion of computed tomography scanning. A set of projection integrals is gathered at each rotational position. Each path between the source and a detector provides a projection integral. In some cases, additional projection paths are generated by translating either the detector system or the object.

Translation of the object in the direction of the

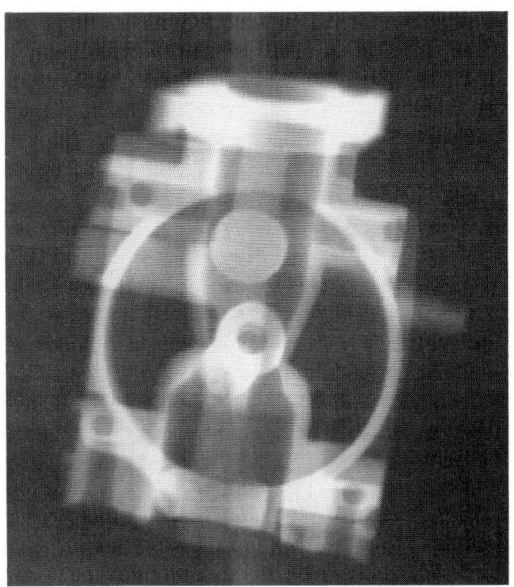

Fig. 2. Projection image of aluminum casting, approximately 2.5 in. (6 cm) wide and 3 in. (7.5 cm) high. Bright regions correspond to x-ray paths of greater attenuation. (*Bio-Imaging Research, Inc.*)

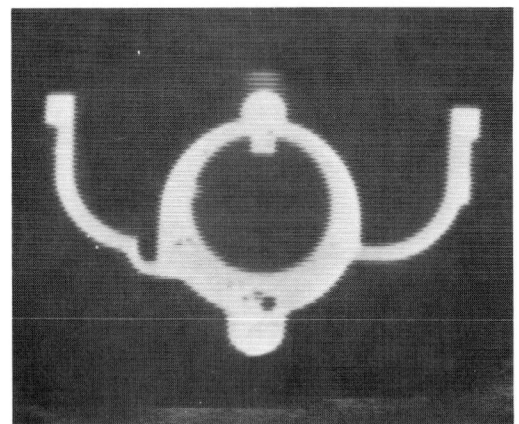

Fig. 3. Tomographic reconstruction of one slice of aluminum casting whose projection image is shown in Fig. 2. Several flaws are visible. (*Bio-Imaging Research, Inc.*)

rotational axis is used to select the slice for reconstruction. In addition, it is possible to synthesize a projection image by stacking the detector outputs taken at a succession of positions. This technique, which has many variations, is sometimes called electronic imaging or digital fluoroscopy.

A projection image of an aluminum casting generated in this way is shown in **Fig. 2**. The overall shape of the object can be visualized here, and further detail could be brought out by image processing. However, much greater detail, including flaws in the casting, can be seen in **Fig. 3**, which shows a single reconstructed slice of the object.

Applications. The task of applying computed tomography in industry is difficult. The barriers fall into two categories, technical and economic, and there is a close relationship between them. While there are many potential applications, no single application that is powerful enough to motivate the production of a standard industrial computed tomography scanner has emerged. The very diversity of applications that should eventually prove a great strength has so far caused the market for the technology to be fragmented. For example, systems based on existing medical scanner designs have not dominated the market. Certain applications for which the x-ray energies available from such scanners would be appropriate, such as the inspection of plastics, composites, and ceramics, require spatial resolution far superior to that easily achieved. For other applications the resolution is adequate, but much greater photon energy is required. Consequently, the current approach is to adapt systems specifically for particular tasks, frequently with elaborate software to extract further information from the reconstructions.

Economic considerations have contributed to the slow emergence of computed tomography in nonmedical areas. For any technology to be successful in industry, it must usually be justified rigorously on a cost/benefit basis. (This is in contrast to application of technology in medicine, where benefit to the patient can outweigh even relatively high cost.) It is natural, therefore, that

the first production applications have involved inspection of high-value parts (for example, turbine blades) or parts in which the consequence of failure is particularly serious (for example, solid-propellant rocket motors).

Many demonstration and developmental applications have been described. These include inspection of wooden power and telephone poles (for insect damage and rotting), raw timber prior to cutting, steel pipe (for flaws), structural ceramic materials (for flaws), and cylindrical concrete piers (for cracks, voids, and aggregate distribution). In addition, a great deal of developmental work is under way.

Rocket motor inspection. Two militarily funded efforts have resulted in systems to inspect aerospace components. The larger of these systems was designed to handle rocket motors 6.5 ft (2 m) in diameter. Structural details and the packing of propellant are of interest. This type of application places a great demand on sensitivity to small density variations. In absolute terms the spatial resolution required is not too different from that of medical scanners, but in relation to the scale of the object it is highly demanding. High-energy sources ranging up to a 16-MeV linear accelerator are used, placing a great burden on the detector system. The geometry of Fig. 1 applies here. Translation of the object is used to generate a large set of ray paths with a modest number of discrete detectors. To achieve both high resolution and high sensitivity to small density changes, it is necessary to use many (of the order of 1000) angles of observation, closely spaced ray paths, and an x-ray detection system with a large dynamic range.

Automotive engine gaging. In this application the objective is precise dimensional information on the inside of a massive metallic structure. Computed tomography reconstruction can provide, for example, cylinder bore dimensions and measurement of eccentricity. The key idea is the direct comparison of the measurement to specifications, made possible by special-purpose software that produces a report based on the reconstruction. A tie-in with the computer-aided design (CAD) process is also possible.

Computed tomography of structural composites. Efforts are being made to relate the results of tomographic studies of composite materials to their mechanical performance. One way is to use reconstruction output as the input to finite element analysis calculations. This requires that a relation be developed between the computed tomography–measured attenuation coefficient and the mass densities of the materials that compose the structure. This connection between structural information derived from computed tomography and mechanical performance predicted by finite element analysis is the subject of much current interest for a wide range of materials and structures.

Turbine blade inspection. Turbine blades present a challenging inspection task. An approach recently described combines tomographic capability with electronic imaging. The latter produces images whose dynamic range is superior to those on film and which can be processed by computer to generate direct measurement of various features, such as wall thicknesses or positions and depths of holes.

Tomographic data acquisition is carried out by rotating the blade at a translational position fixed relative to the detector array. From such data, a cross-sectional slice reconstruction of the blade is obtained. To save time, tomographic analysis is carried out only in regions where analysis based on electronic imaging suggests irregularities.

The strategy of employing electronic imaging, accompanied by some image enhancement, together with computed tomography seems likely to be followed in turbine blade inspection and many other industrial applications as well. A generalization will be the combination of real-time two-dimensional imaging, commonly available by using x-ray image intensifiers, with direct three-dimensional tomographic reconstruction. This technique has been shown to be feasible.

For background information SEE COMPUTERIZED TOMOGRAPHY; IMAGE PROCESSING; MICRORADIOGRAPHY; NONDESTRUCTIVE TESTING in the McGraw-Hill Encyclopedia of Science and Technology.

Lee A. Feldkamp

Bibliography. H. Berger and D. Kupperman, Microradiography to characterise stuctural ceramics, *Mater. Eval.*, 43:201–205, 1985; S. F. Burch and N. K. Bealing, A physical approach to the automated characterisation of buried weld defects in ferritic steel, *NDT Int.*, 19:145–150, 1986; G. T. Herman, *Image Reconstruction from Projections*, 1980; P. Reimers, W. B. Gilboy, and J. Goebbels, Recent developments in the industrial application of computerized tomography with ionizing radiation, *NDT Int.*, 17:197–207, 1984; W. N. Reynolds and G. M. Wells, Video compatible thermography, *Brit. J. NDT*, 26:40–44, 1984; M. G. Silk, The use of diffraction-based time-of-flight measurements to locate and size defects, *Brit. J. NDT*, 26: 208–213, 1984; D. O. Thompson and D. E. Chimenti (eds.), *Review of Progress in Quantitative Nondestructive Evaluation*, vol. 5, 1986.

Nuclear hydrodynamics

A heavy ion is an atom of high mass number A that is stripped of most of its electrons, for example, Nb^{30+}. Such an ion has a positive charge, which can be used to accelerate the nucleus to a speed greater than 50% of the speed of light by using the voltage of an accelerator. A beam of many of these heavy ions (projectiles) interacts with a target in a nuclear collision. In this high-speed interaction the nuclei of the projectile and target can combine to form a hot, dense glob of nuclear matter. High-speed heavy-ion collisions offer the unique opportunity of probing the properties of nuclear matter at extremely high densities and temperatures at which novel, exotic forms of matter may exist.

High pressure, density, and temperature are created in the region where the projectile and the target interact. From studying such interactions, it may be possible to extract the nuclear equation of state, which is the relationship between pressure, density, and temperature for nuclear matter. A knowledge of the nuclear equation of state, in addition to the academic interest, is of prime importance for the understanding of parts of the universe remote in space and time, and the processes therein, such as the big bang, supernova explosions, and the interior of neutron stars.

Nuclear matter in high-energy collisions. Direct investigation of the nuclear equation of state has begun, made possible by the availability of beams of heavy nuclei with energies up to 1 GeV per nucleon at the bevalac accelerator at Lawrence Berkeley Laboratory. These nuclei are moving so fast that the nucleons participating in a collision cannot escape from the interaction volume. They therefore pile up, and the nuclear medium becomes highly excited and strongly compressed so that nuclear shock waves are formed. The reaction of argon and lead nuclei (Ar+Pb) in the nuclear fluid-dynamic model is shown in **Fig. 1**. In 10^{-23} s, the nuclei come into contact and reach a stage of high density and temperature. Nuclear matter is splashed to the side in a Mach shock wave.

The high pressure generated in a collision results in a rapid expansion and fragmentation of the nuclear matter, yielding large numbers of pions, nucleons, and light nuclei. The distributions of these emitted particles can be used to diagnose the characteristics of the original hot, dense matter. Observable compression effects are the collective sideward flow and the dependence of the pion multiplicity on the nuclear compression energy predicted theoretically by nuclear fluid dynamics. Both effects are the result of the high-pressure buildup, which causes a large transverse momentum transfer and a change in the temperature of the system.

Recently, both effects have been observed experimentally. A new multiparticle electronic detector system called the plastic ball was developed, which enables the identification of all the particles emitted in a particular collision event, as well as their respective energies, thus allowing for complete reconstruction of the event. The ball makes possible an event-by-event analysis: an individual projectile nucleus interacts with one target nucleus, and the fragments from only that event are identified. This is repeated for many events to obtain a large amount of data.

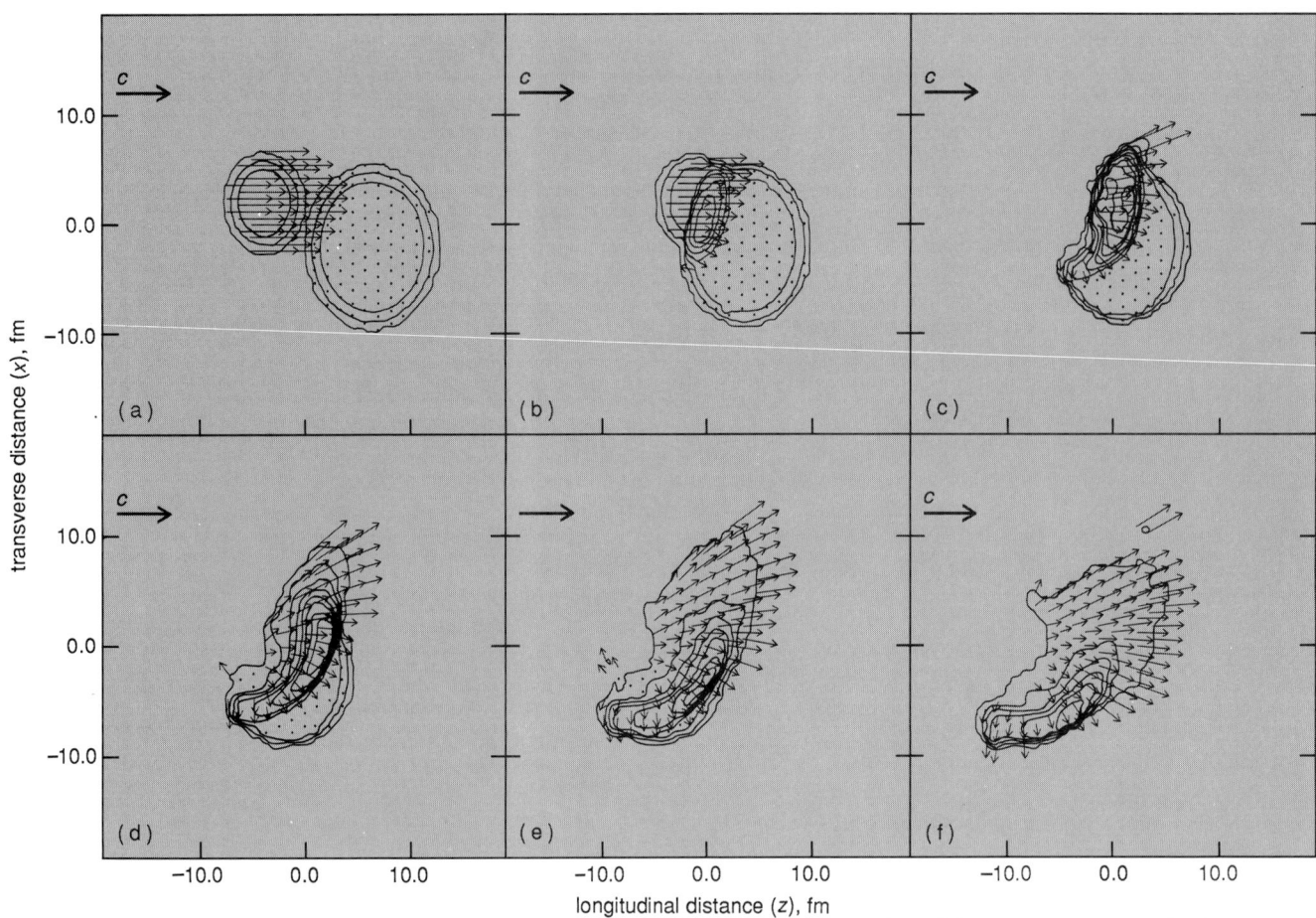

transverse distance (x), fm

longitudinal distance (z), fm

Fig. 1. Computer simulation, using the nuclear fluid-dynamic model, of the collision of an argon nucleus having an energy of 770 MeV per nucleon with a lead nucleus (Ar+Pb) at an impact parameter b of 4 femtometers. Reaction plane, defined by the beam direction and impact parameter, is shown. Velocities are indicated by arrows with a scale given by the arrows labeled c, the speed of light. Collision is shown at a sequence of times t. (a) $t = 0$. (b) $t = 1.69 \times 10^{-23}$ s, $ct = 5.05$ fm. (c) $t = 3.37 \times 10^{-23}$ s, $ct = 10.11$ fm. (d) $t = 5.06 \times 10^{-23}$ s, $ct = 15.16$ fm. (e) $t = 6.74 \times 10^{-23}$ s, $ct = 20.21$ fm. (f) $t = 8.43 \times 10^{-23}$ s, $ct = 25.26$ fm.

High multiplicity and sideward emission. A puzzling difference was found between near-central collisions, in which many particles are ejected, and reactions with larger impact parameters, which have a smaller number of ejectiles. In collisions of two niobium nuclei (Nb+Nb), at energies from 100 to 1000 MeV per nucleon, a clear sideward emission pattern was observed for the near-central collisions, while the less central reactions resulted in angular distributions that peaked in the forward direction.

The niobium nucleus has a radius of about 5 femtometers. Thus, for an impact parameter b of 10 fm, the nuclei just touch; the event is said to be peripheral. For a more central event, with an impact parameter of, say, 2 fm, the two nuclei overlap and interact very strongly. Matter with densities greater than twice the normal density can be formed. The hot, dense nuclear matter fragments into many droplets. Furthermore, in the central event the high pressure in the interaction region (due to the many closely interacting nucleons) thrusts the matter to the side. The impact parameter and the beam direction define a

plane, the reaction plane. The direction of the sideward thrust is in this plane and is perpendicular to the beam direction.

The sideward thrust gives rise to a momentum change, again in the transverse direction. The large transverse momentum obtained by the peripheral particles (those on the outskirts of an event) is called the bounce-off effect, and that obtained by the matter in the hot interaction zone is called collective flow.

Quite analogous results were obtained in a streamer chamber experiment for the reaction of argon and lead nuclei (Ar+Pb). More recently, collisions of two gold nuclei (Au+Au) were studied, and collective flow effects were found to be even more strongly pronounced in this system because of the larger mass number. These experiments showed that nuclei exhibit a collective, fluidlike behavior that serves as the primary mechanism for the creation of dense, hot matter in high-energy nuclear collisions.

Microscopic theory. The data represent a challenge to any microscopic theory. The intranuclear cascade approach neglects the repulsive compression potential and assumes that nuclear collisions proceed via a sequence of independent nucleon-nucleon collisions. In this model the nucleons are thought of as particles moving in straight-line trajectories and colliding like billiard balls. Even in central collisions, the intranuclear cascade model predicts angular distributions peaked in the forward direction (corresponding to zero flow angle), in contrast to the data. The individual collisions of billiard-ball-like particles simply do not produce enough sideward thrust.

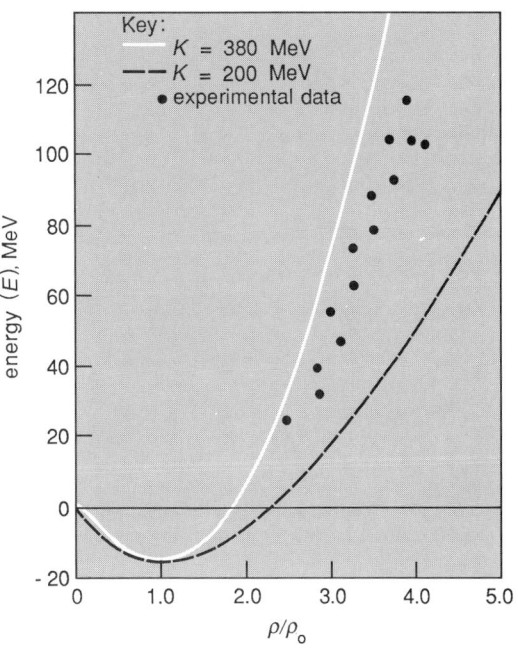

Fig. 3. Nuclear equation of state as used in the Vlasov-Uehling-Uhlenbeck theory, with compressibility of nuclear matter $K = 200$ MeV and $K = 380$ MeV, compared with values extracted from experimental pion yields. Energy of nuclear matter E is plotted as a function of the ratio of density ρ to the normal density ρ_0.

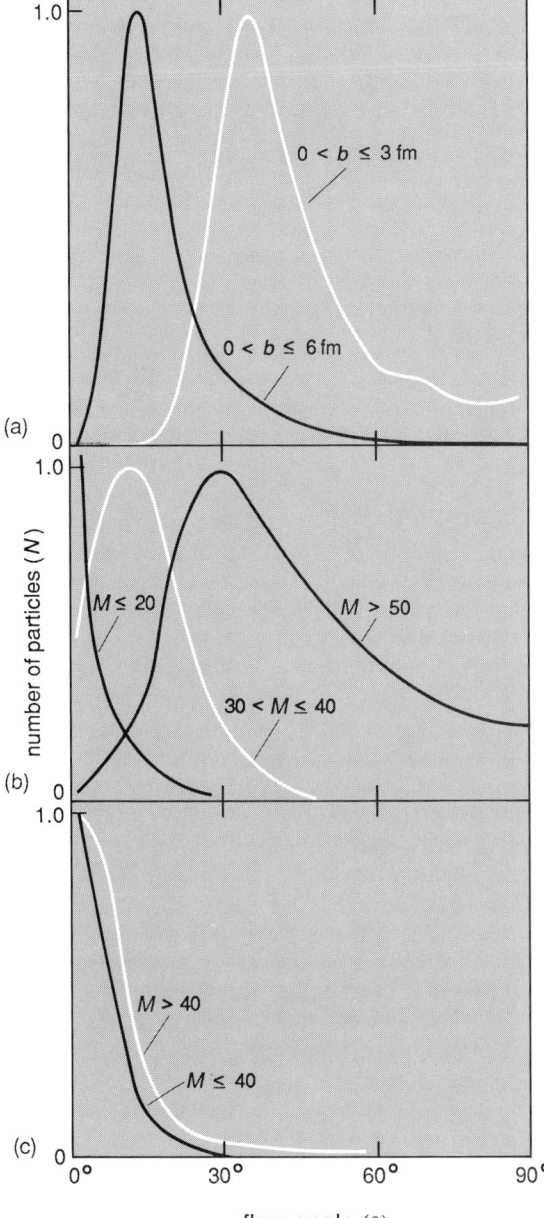

Fig. 2. Kinetic-energy flow-angle distributions for collision of a niobium nucleus having an energy of 400 MeV per nucleon with another niobium nucleus (Nb+Nb). Number of particles is normalized to maximum number in each peak. (a) Predictions of the nuclear fluid-dynamic model for various values of the impact parameter b. (b) Experimental data for various values of the multiplicity of particles emitted M. (c) Predictions of intranuclear cascade model.

Figure 2 shows the flow angle distribution for the intranuclear cascade model, experimental data, and the nuclear fluid-dynamic model. The macroscopic nuclear fluid-dynamic theory explains qualitatively the flow effect.

Recently a microscopic theory of heavy-ion reactions has been developed which incorporates a stiff equation of state, that is, one with high energy values at high density (**Fig. 3**). This microscopic theory is based on the Vlasov-Uehling-Uhlenbeck (VUU) equation, the quantum-mechanical equation of the kinetic theory of gases. The theory shares the successful ideas of the intranuclear cascade model, for example, the interaction of particles in collisions; also included, however, is the nuclear equation of state (here a relation between energy and density) and the Pauli principle. According to the quantum-mechanical Pauli principle, only a certain number of particles can be in a given region of space with a certain momentum. The Vlasov-Uehling-Uhlenbeck model accounts for the large sideward flow angles observed experimentally. As in the macroscopic nuclear fluid-dynamic model, the high values of density imply a high compressional energy. The nuclear matter thus expands like a spring in collective flow. The predictions are in accordance with the data of Fig. 2.

Additional evidence for the stiff equation of state comes from a novel momentum analysis performed for collisions of argon ions with potassium chloride (Ar+KCl) at an energy of 1.8 GeV per nucleon (**Fig. 4**). The momentum p_x perpendicular to the beam axis in the reaction plane is plotted against the rapidity y, which is related to the longitudinal momentum. The experimental data show a maximum transverse momentum p_x

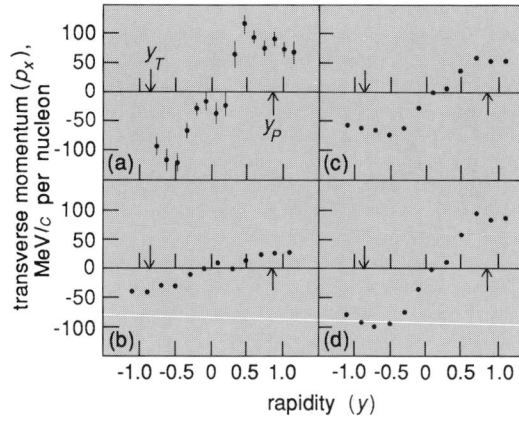

Fig. 4. Transverse momentum spectra for collision of argon nuclei having an energy of 1800 MeV per nucleon with potassium chloride (Ar+KCl). Momentum p_x perpendicular to the reaction plane is plotted against rapidity y, which is related to the longitudinal momentum. The quantity y_T is the negative rapidity corresponding to the velocity of the target, and y_P is the positive rapidity corresponding to the velocity of the projectile. (*a*) Experimental data. (*b*) Prediction of intranuclear cascade model. (*c*) Prediction of Vlasov-Uehling-Uhlenbeck model with soft equation of state ($K = 200$ MeV), and (*d*) stiff equation of state ($K = 380$ MeV).

of about 100 MeV/c per nucleon (Fig. 4*a*), where c is the speed of light. The intranuclear cascade model lacks compressional energy and hence predicts only 25 MeV/c per nucleon (Fig. 4*b*). The Vlasov-Uehling-Uhlenbeck model with a soft equation of state (Fig. 4*c*) gives 50 MeV/c per nucleon. Only the stiff equation of state in the Vlasov-Uehling-Uhlenbeck model (Fig. 4*d*) with a Pauli principle can explain the experimental data.

Further support for a stiff equation of state comes from the pion multiplicities that have been measured event by event in the streamer chamber for near-central collisions of argon ions with energies from 0.4 to 1.8 GeV per nucleon in potassium chloride (Ar+KCl). The pion yields calculated from the cascade model (which neglects the compression potential) drastically overestimate the data. It has been conjectured that the difference between the measured pion yields and the cascade simulations is due to the inherent neglect of the compression energy in the cascade approach and that the cascade approach may therefore be used to extract the nuclear equation of state at high densities. In fact, the nuclear equation of state (Fig. 3) extracted from this discrepancy rises rapidly in the density regime accessible in the experiments. A comparison with the microscopic theory again substantiates the need for a stiff equation of state.

Much progress has been made both experimentally and theoretically in the methods of studying relativistic heavy-ion collisions. The first glimpse of the nuclear equation of state seems to reveal surprisingly strong repulsion at densities about two to four times the ground-state density. Over the next decade, even more difficult problems will be dealt with when collisions in the ultrarelativistic regime are studied with accelerators at the European Center for Nuclear Research (CERN) and Brookhaven National Laboratory.

For background information SEE FLUID DYNAMICS; KINETIC THEORY OF MATTER; NUCLEAR REACTION; NUCLEAR STRUCTURE in the McGraw-Hill Encyclopedia of Science and Technology.

Joseph J. Molitoris; Horst Stöcker

Bibliography. G. Buchwald et al., Kinetic energy flow in Nb plus Nb, *Phys. Rev. Lett.*, 52:1594–1596, 1984; W. Greiner and H. Stöcker, Hot nuclear matter, *Sci. Amer.*, 252(1):76–87, January 1985; H. A. Gustafsson et al., Collective flow observed in relativistic nuclear collisions, *Phys. Rev. Lett.*, 52:1590–1593, 1984; H. Kruse, B. Jacak, and H. Stöcker, Microscopic theory of pion production and sidewards flow in heavy-ion collisions, *Phys. Rev. Lett.*, 54:289–292, 1985; J. J. Molitoris and H. Stöcker, Evidence for a stiff nuclear equation of state, *Phys. Rev.*, C32:346–348, 1985.

Nuclear power

A graphite, tube-type reactor at Chernobyl, near Kiev in the south-central European part of the Soviet Union, sustained a destructive accident on April 26, 1986. The initial indication was the expulsion of flaming fragments that disrupted the upper part of the reactor building structure and parts of the surrounding buildings, and a subsequent fire. Radioactivity continued to be emitted from the ruins of the reactor for 9 days, apparently fueled in part by the burning of the graphite, zirconium, and hydrogen in the reactor. The fire and further release of radiation were eventually smothered by the application of several thousand tons of lead, sand, clay, dolomite, and boron compounds. Severe levels of radioactivity were experienced at distances of 6 to 18 mi (10 to 30 km). Varying wind patterns over the following 2 weeks caused detectable deposition of radioactive particles over most of Europe and over both Asian and European parts of the Soviet Union, and traces were found as far away as Japan and some parts of the United States.

Casualties and effects of accident. At least 31 people died (as of March 1987) as a result of acute burns and radiation exposures that occurred within the reactor complex. Over 200 people from the reactor site were hospitalized, suffering from burns or symptoms of radiation exposure. All but five have since been released. It is estimated that long-term effects will increase the incidence of cancer between 0.4 and 1.6% for the 135,000 people evacuated after April 28 from an area that extends up to 18 mi (30 km) from the plant. Most of this zone—about 200 mi² (500 km²)—may not be usable for general agriculture for several decades, although some localities in

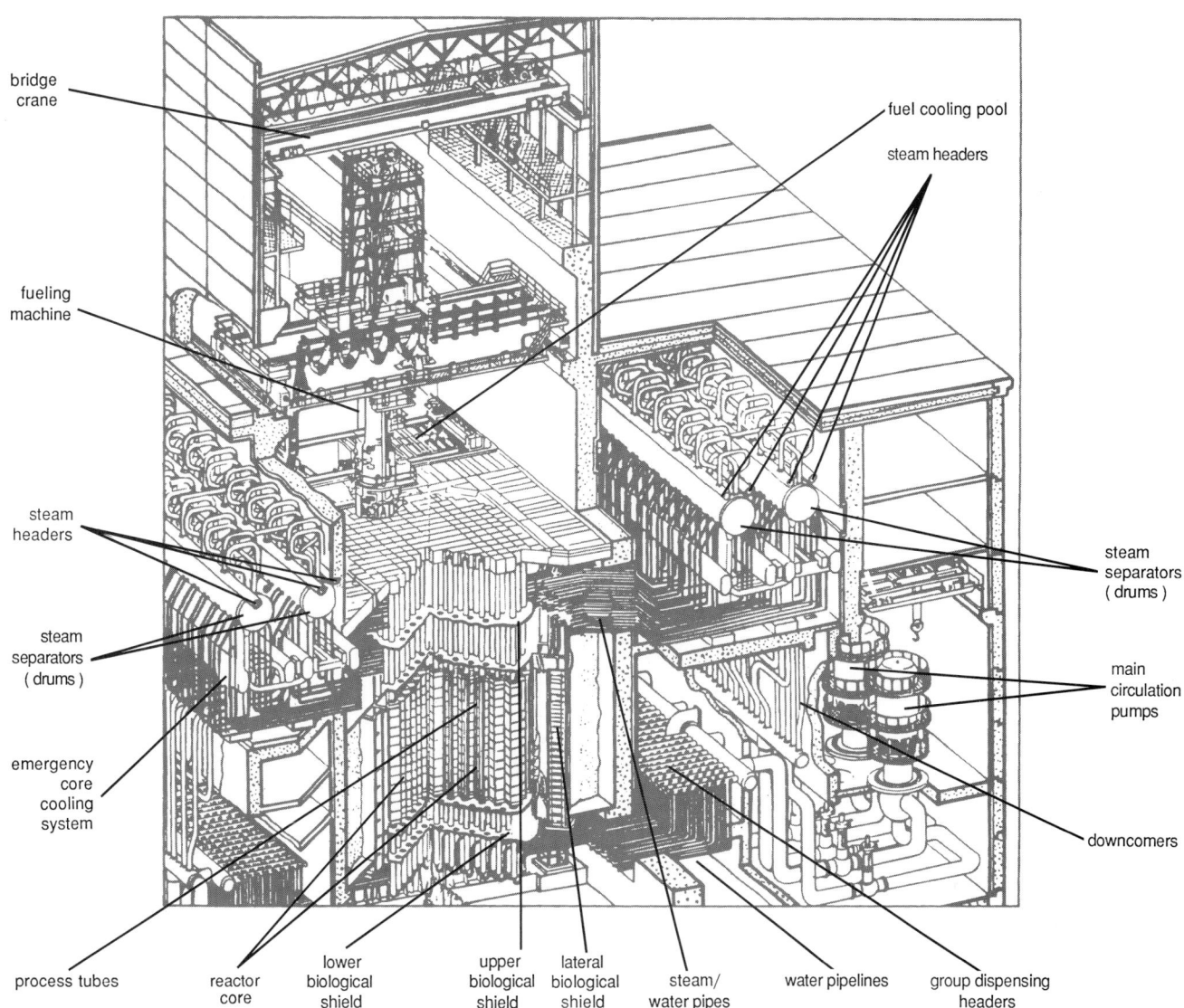

bridge
crane

fueling
machine

steam
headers

steam
separators
(drums)

emergency
core
cooling
system

fuel cooling pool

steam headers

steam
separators
(drums)

main
circulation
pumps

downcomers

process tubes

reactor
core

lower
biological
shield

upper
biological
shield

lateral
biological
shield

steam/
water pipes

water pipelines

group dispensing
headers

Sectional view of Chernobyl-type (RBMK) reactor.

this zone have been recovered. Other long-term increases of cancer—ranging from less than 0.03 up to 0.07% of the normal incidence—have been estimated for the general population of the European portion of the Soviet Union, and up to 0.03% for western Europe.

Vienna conference. A conference on the accident was held on August 25–29, 1986, in Vienna, under the auspices of the International Atomic Energy Agency. An extensive written and oral report detailing the accident and its causes and consequences was presented by Soviet representatives. Fifteen other reactors similar to Chernobyl (providing about 5% of Soviet electricity) have been temporarily shut down or are scheduled to be shut down to make a series of changes intended to eliminate some of the design features that made a severe accident likely. The Vienna meeting also opened additional channels for the timely communication of accidents and perhaps

for safety reviews of Soviet reactors.

Chernobyl-type reactor design. The reactor core (see **illus.**) is made up of 1800 metric tons (2000 short tons) of graphite blocks, arranged in a cylinder, 23 ft (7 m) high and 39 ft (12 m) in diameter. The blocks have vertical holes for 1661 zirconium alloy process tubes, each 4 in. (10 cm) in diameter with $\frac{5}{32}$-in.-thick (14-mm) walls. These are connected to horizontal water-supply manifolds at the bottom end, and to individual water-steam exit pipes.

A cluster of 18 fuel rods fits into each process tube. The fuel rods are 0.53 in. (13.5 mm) in diameter and are made of uranium dioxide at 2% enrichment, in uranium-235, encased in zirconium tubing with a wall thickness of 0.032 in. (0.8 mm).

Water is supplied to two sets of recirculating pumps (six operating pumps and two spares), each pump using a motor of 5700-hp (4.25-MW) output. The water enters the reactor tubes at

518°F (270°C) and exits as a mixture of steam and water at 544°F (284°C) and 990 psi (6.8 megapascals) pressure. The steam from an average channel is 14% by weight of the mixture and 78% by volume. Four steam drums, each 100 ft (30 m) long, feed steam to two separate turbines and generators, each rated at 550 MW. Steam and water from the turbine pass to condensers and flow to feedwater storage and then to feedwater pumps (five pumps of 5600 hp or 4.18 MW each). Water is then reheated to 334°F (168°C), and reinjected to the steam drum to return to the main recirculating pumps. The pumps and most of the main piping are in steel-lined compartments that are intended to direct steam-water leaks to double-deck condensation pools under the reactor compartment.

Control of the reactor is provided by 187 top-entry control rods in six independent groups (plus 24 bottom rods for power shaping). Twenty-four of the top rods are designated as safety rods; they have a normal entry speed of 1.3 ft/s (0.4 m/s).

Reactor accident event. The Chernobyl reactor suffered a violent steam pressure burst, characterized in the Soviet report as a thermal explosion. This apparently was similar to the sudden rupture of a steam boiler in that steam pressures were generated that exceeded the bursting pressure of the process tubes and the reactor vessel.

The Soviet report that was presented in Vienna noted six operator errors that brought into play the basic design features that gave the reactor the capability for a runaway power and pressure increase. Reactor control makes use of the fact that about 0.65% of the neutrons produced by fission of uranium-235 are not produced instantaneously but are delayed for various times between 0.25 and about 56 s. A reactor in normal operation is said to have a multiplication constant K of 1.000. By raising top-entry control rods (out of the reactor), the multiplication constant can be increased. Normally control changes are made with small increments in K. For example, an increase of K to 1.001 causes reactor power to increase at a rate of about 2% per second. If the extra reactivity for K reaches a value that exceeds the delayed neutron fraction, extremely rapid rises in power can occur. For example, with a K of 1.0065, whose deviation from unity is just equal to the delayed neutron fraction, reactor periods between 0.1 and 0.3 s occur. (The reactor period is the time required for the power to increase by a factor of 2.72.)

A critical quantity in the operation of a nuclear reactor is the void coefficient, the increase in K that results from a 1% increase in steam void. The Chernobyl reactor had a void coefficient of +0.02%, making possible a K of 1.020, a value whose difference from unity is several times in excess of prompt critical. The heat-up of fuel associated with a power increase introduces some prompt negative reactivity due to the Doppler effect, but this effect is too small to override a sufficiently large power pulse. If the extra production of steam leads to still further increases in K, uncontrollable or runaway conditions result, and power increases of up to several hundred times normal full power can occur in 1 or 2 s.

The Soviets presented a series of calculations, plotted as curves, to represent the changes in void fraction and the resulting changes in reactivity and reactor power that occurred at the time of the accident. These calculations show two overpower excursions, the first one rising to 110 times full power for about 1 s, and producing fuel temperatures of about 2000°F (1100°C) and rising. (At temperatures above 2000°F or 1100°C, zirconium reacts extremely rapidly with steam, producing hydrogen gas.) This power pulse was stated to have been terminated by the Doppler effect. The brief energy pulse from the first power transient, with a slight delay for heat to flow from the fuel, led to ejection of the remaining water from the tubes in about 0.2 s.

The increased void produced a second increase in reactivity, to about 1.014, according to the Soviet report. This in turn produced a very large power pulse, estimated at 470 times full power, that lasted for about 0.5 s. The Doppler effect was not large enough to override this second power pulse. The heat released raised the pressure of the remaining steam and produced sufficient energy to rupture the process tubes and the reactor vessel.

This rupture released high-pressure steam to the graphite structure and the surroundings. The graphite, in turn, was contained in a light cylindrical vessel that ruptured. The force of the steam acting on the large area of the massive shielding deck (weighing about 1000 metric tons or 1100 short tons) above the reactor was sufficient to lift the deck and throw it to one side. As the deck above the reactor rose, it also tore loose the piping to the fuel channels, releasing additional steam to the building over the reactor. The structure directly above the reactor lid was a flat-walled conventional building, not designed to withstand excess pressure. It was torn apart by the force of the expanding steam, and an overhead crane carrying a 300-metric-ton (330-short-ton) refueling machine collapsed on the reactor. Portions of the graphite core blocks and fuel were forcibly ejected into the upper reactor building, and some were ejected out of the building.

Role of human error. As mentioned above, the Soviet report noted six major operator errors that set the stage for the rapid voiding, reactivity increase, and sudden large energy output. Errors were committed in conducting a test that was intended to determine whether enough interim power to operate pumps could be obtained from the flywheel energy of the spinning turbine as it coasted down after the steam supply was cut off. The errors involved were not simply errors in manipulation, but basic conceptual errors involving faulty planning for the test, and supervisory and management failure to control the operation for a period of 24 h during which the series of increasingly serious violations, involving the blocking of normal safety systems, took place. This situation was explained by the Soviets as due to turning over the running of the experiment to specialists on generator control, the evident absence of reactor-trained supervision, and their apparent lack of knowledge or awareness of the dangers inherent in this reactor design.

Role of reactor design. Unlike all other civilian power reactors, the Chernobyl-type reactors have been operated with a large positive void coefficient. That is, an increase in power leads to an increase in steam production that increases void and leads to a further increase in power. This inherent instability is normally kept in check by automatic regulating rods for the small changes in void content that occur at high-power operation.

The total reactivity represented by the difference between a voided tube and a partly voided tube at full power is normally maintained at slightly less than beta, the amount of reactivity corresponding to the delayed neutron fraction. Also, under normal operating conditions, rods are positioned to be able to insert that amount of negative reactivity in about 1 s. However, at low power, with slight overcooling, channels are nearly full of water, and control rods are almost entirely out of the reactor. Under these conditions the potential reactivity from voiding tubes is several times beta, and the control rods take many seconds to offset this amount of excess reactivity if the steam content increases rapidly.

The major hazard evident in this design is that dangerously large and rapid increases in reactivity are available when the reactor is operating with tubes that are nearly full of water. Any of several changes in power, water flow, or pressure can increase the volume of steam in the reactor. Increase in the volume of steam increases the reactor power output. This increase becomes runaway or autocatalytic (self-accelerating) since more power produces more steam, and more steam increases the power rate further.

Role of containment. The Chernobyl design did not have a complete containment building over the reactor. The pressurized-water reactors with 1000 MW net electrical power built recently in the Soviet Union have large, high-pressure containment buildings fully surrounding the reactor, apparently along the same general containment principles as used in the Western world. In the event of an accident involving release of radioactivity, the presence of a sturdy containment building prevents the spread of large amounts of radioactivity to the environment. For example, the 1979 accident at Three Mile Island near Harrisburg, Pennsylvania, also resulted in disruption and partial melting of the reactor core and a later hydrogen ignition and low-order explosion, following a loss-of-coolant accident. However, the large, high-pressure containment building retained all but traces of the biologically important radioactive substances.

Safety record. The Chernobyl reactor sustained a severely damaging accident after only 90 plant years of experience with this design. The accident was approximately 1000 times more severe in terms of spread of long-lived radiation than the next most severe reactor accident; one that occurred in a small, uncontained graphite reactor at Windscale in England in 1958. The Chernobyl accident was more than 1 million times more severe than the accident at Three Mile Island in terms of the spread and deposition of measurable radioactivity in the environment.

Future of nuclear power. As of June 1986, 52 additional power reactors were under construction or firmly planned in the Soviet Union, 9 of them similar to Chernobyl. The Soviets have announced that pressurized-water reactors with large containment buildings will continue to be built, with the objective of more than a 400% increase over present Soviet nuclear capacity by the mid-1990s.

In the United States, it appears likely that no additional reactors, beyond those already nearing completion, will be built before the year 2000. This is due to a combination of factors: increased public apprehension, especially after the Chernobyl accident; the improved price and availability of oil and natural gas; and the improving technology for clean utilization of coal. Future commitments of nuclear capacity may be considered in the mid-1990s, involving complex trade-offs of the costs and environmental concerns of fossil fuels, and the achieved and perceived safety and costs of future models of nuclear reactors. *See* Electrical utility industry.

Elsewhere in the world, nuclear capacity continues to grow and be committed, and it is expected to exceed three times the total United States capacity in the 1990s. Fourteen countries already have larger percentages of nuclear capacity than the United States does. A total of 26 countries currently operate significant nuclear power programs, and a total of 39 countries have announced plans or have projects under initial stages of construction. Several of these have been put on hold pending a full study of Chernobyl. Sweden, Germany, and Switzerland are considering reduced reliance on nuclear units. However, more than half of the countries that have nuclear programs have very limited indigenous fuel or energy resources (for example, France, Japan, Belgium, and Taiwan) and have announced plans for continued growth of capacity. In 1985, 32 new reactors started up, a 19% increase in capacity outside the United States, and commitments to new projects increased by 7%. In 1986, 21 reactors started operation.

Table 1. Comparison of reactor design safety attributes

Hazard or challenge	Capability provided to cope	
	RBMK reactors	Pressurized-water reactors
Containment of largest line break	*	*
Containment of ruptures of several process tubes or steam generator tubes	†	*
Containment of large escape of radioactive materials from core-damaging accident	†	*
Rapid insertion of maximum available reactivity overriden by inherent negative coefficients	†	*
Provision for containment of hydrogen ignition or explosion after severe core damage	†	*
Prolonged loss of external power or water supply	†	*
Delay or partial failure in inserting control rods on scram signal	†	*
Prolonged loss of feedwater pumps	†	*
Failures of several shutdown systems (by provision for redundant and diverse systems)	†	*
Operator errors or control failures causing large disturbances in power level or power distribution	†	*
Operator confusion and error in complex situations (avoided by training, operator aids for key data, and symptom-based response procedures)	†	*
Confusion in external response to emergencies (avoided by provision for rapid communications from plant to civil authorities)	†	*

* Demonstrable assurance of the capability is provided.
† Capability is not provided or is doubtful.
SOURCE: Compiled primarily from A. A. Abagyan et al., *Report of the Accident at the Chernobyl Atomic Energy Station and Its Consequences*, International Atomic Energy Agency Expert Conference on the Accident at Chernobyl, August 25–29, 1986, Vienna, and V. A. Legasov et al., supplementary data presented at the conference; N. A. Dollezhal' and I. Ya. Yemel'yanov, Channel-type nuclear energy reactor, *Atomizdat*, Moscow, 1980; information from approximately 200 Soviet technical publications available on the RBMK reactor.

Comparison of reactor types. **Table 1** lists the principal safety characteristics of the Chernobyl design, known as the RBMK reactor (large channel-type reactor), and those of the most common reactor type in the Western world, the pressurized-water reactor. Boiling-water reactors parallel pressurized-water reactors for most attributes that are applicable. The entries for pressurized-water reactors refer to reactors that are licensed and operated according to the standards and regulations of the United States. The recent models of Soviet pressurized-water reactors also appear to meet many of these challenges.

The entries in Table 1 are typical of the kinds of faults that can arise from a combination of mechanical failures and human errors. Licensing requirements in the United States, and for most of the Western world, have required a showing by analysis and demonstration by system tests that the reactor design and operation can cope with such multiple faults, or failing that, can contain most of the radioactivity safely. In addition to the power excursion that occurred in the Chernobyl accident, most of the other hazards listed in Table 1 can also threaten the integrity of the reactor vessel and shielding deck and can lead to a high probability of escape of radiation to the environment.

The principle that reactor cores must be designed so that the prompt power coefficient has a negative value has been widely recognized since the early 1950s and is stated in basic textbooks on reactor safety. All of the power reactor designs that have achieved widespread use, apparently with the exception of the RBMK type, have observed this fundamental safety criterion. The close relation of the RBMK design to the design of military production reactors and its evident dual-purpose capability may have caused bypassing of the concerns and disciplines appropriate for its use in a civilian environment.

Comparison of institutional features. **Table 2** lists some of the main institutional differences between the situation at the Chernobyl plant and that prevailing in the United States (and in most countries of the Western world). The 1979 accident at Three Mile Island dramatized the possibility of severe core-damaging accidents. By law and by regulation, such contingencies now must be covered rigorously—by training, management, emergency control systems, containment, and civil emergency response capabilities, as well as by design, operation, and maintenance—to make such accidents as unlikely as possible. After 1979 a massive upgrading of the features perceived to be involved in assurance of safety, with emphasis on the human factors (operator training, supervision, and management) was undertaken in the United States. The practice of having continuing surveillance by on-site inspectors and periodic intensive audits of practices was instituted by the Nuclear Regulatory Commission. In addition, a separate audit and review and tutorial function by the industry-supported Institute for Nuclear Operation was started in 1980.

The need for intensified oversight of safety also has been recognized in the Soviet Union. A safety authority was established in 1983, and general enabling edicts were issued in 1984. A new All-Union Atomic Energy Authority was announced in September 1986. However, prior to the Chernobyl accident, independent safety audits apparently were not common.

Containment buildings, simulator training of operators, and redundant and diverse safety shutdown systems have been instituted for the more recent pressurized-water reactors in the Soviet Union, but are only now being considered or planned for reactors of the RBMK type. This lag suggests the existence of a major cultural,

Table 2. Comparison of reactor institutional safety attributes

Safety attribute	Capability provided	
	RBMK reactors	United States reactors
Systematic risk analysis of all known and conceivable damage sequences using probabilistic risk analysis	†	*
Defense-in-depth for low-probability events involving multiple human errors and mechanical failures, using probabilistic risk analysis	†	*
Periodic system hardware upgrading, by legal requirement, with independent audits and enforcement	†	*
Simulator training for all operators at all plants	†	*
Training for responses to severe accidents	†	*
Continuous retraining of operators and supervisors (one shift always in retraining)	†	*
Periodic independent, adversarial, safety reviews and audits; mandatory correction of deficiencies	†	*
Mandatory timely reporting and analysis of all incidents, plus real-time dedicated communication to local, state, and federal civil authorities, and annual emergency drills	†	*
On-site, full-time, independent safety inspectors	†	*
In-plant safety review groups for continuous feedback of operating experience	†	*
Detailed on-line incident data and timely remedy analysis using on-line information network	†	*
Comprehensive, accessible, on-line data base on reliability and safety of components and systems, with timely reporting	†	*

* Capability is provided for pressurized-water reactors and boiling-water reactors in the United States (and most other countries of the Western world).

† Capability is not evidenced in the available literature, or not provided. Some of these capabilities are now planned in the Soviet Union.

technical, and managerial gap between the RBMK and pressurized-water reactor sectors.

The Soviets have announced their intention to continue operation of the remaining RBMK units with various changes. These involve administrative limits on control rod positions, prohibition of operation at low power, and improvements in training. Longer-term changes are planned that may take several years to implement. These include raising the enrichment of the fuel to 2.4% and later possibly to 3.0%. Such a change should result in near-zero or negative void coefficients for all operating conditions. Other announced plans for changes include the provision of a diverse, fast-acting shut-down system that cannot be defeated, detection of low subcooling and pump cavitation, and the development and use of plant simulators for training. Broad changes in supervision, management, and "technological discipline" are also indicated. The possible addition of a containment building has not been mentioned. The announced changes, when and if fully implemented, may reduce the likelihood of another Chernobyl-type accident somewhat, but they still appear to fall substantially short of the safety standards evident in the recent Soviet pressurized-water reactors, as well as in light-water reactors in the Western world.

For background information *SEE* N*UCLEAR POWER;* N*UCLEAR REACTOR;* R*EACTOR PHYSICS* in the McGraw-Hill Encyclopedia of Science and Technology.

Edwin L. Zebroski

Bibliography. A. A. Abagyan et al., *Report on the Accident at the Chernobyl Atomic Energy Station and Its Consequences*, prepared for the International Atomic Energy Agency Expert Conference on the Accident at Chernobyl, August 25–29, 1986, Vienna, and V. A. Legasov et al., supplementary data presented at the conference (English transl. by IAEA, August 1986); N. A. Dollezhal' and I. Ya. Yemel'yanov, Channel-type nuclear energy reactor, *Atomizdat*, Moscow, 1980 (English transl. by Foreign Broadcast Information Service, 1981); M. Goldman et al., *Assessment of the Dosimetric and Health Consequences of the Chernobyl Reactor Accident*, Office of Health and Environment, U.S. Department of Energy, 1987; E. E. Lewis, *Nuclear Power Reactor Safety*, 1977.

Nuclear structure

Measurement of the nuclear response function is the natural extension of the macroscopic senses of sight and touch to the microscopic realm of atomic nuclei. In the macroscopic world the shape of an object is perceived by the way the object scatters light from a source of illumination into the eye. Its mechanical properties are determined by the way it responds to external stimuli. For example, when a bell is struck, it vibrates at a specific frequency for a long time, whereas a chunk of soft putty shows no response after the stimulus is removed.

In the microscopic world the shapes and properties of atomic nuclei are discerned in an analogous manner. Instead of visible light, the wavelength of which is much too large to resolve nuclear dimensions of the order of 10^{-15} m, or macroscopic probes, which are too large to excite nuclei, beams of extremely high-energy electrons, protons, or other fundamental particles are used to measure nuclear properties. In these experiments the energy and momentum of the scattered particles are measured and compared with the known energy and momentum of the original incident beam. The probability that the probing particle scatters from the nucleus with a given energy transfer ω and momentum transfer $\vec{q}$ can then be determined. This scattering probability, denoted as $\sigma(\vec{q},\omega)$, is called the nuclear response function and reveals a wealth of information concerning the rich and diverse properties of atomic nuclei.

Elastic scattering. When an electron is scattered elastically from a nucleus, no energy is transferred to the nucleus and the nucleus remains in its quantum ground state. This process is analogous to observing the shape of a macroscopic object by the way the object scatters light, and thus allows scientists to determine accurately the shape and radial distribution of matter in a nucleus. The great advantage of the process of electron scattering is the fact that the interaction of electrons with nuclei is both weak and precisely known, so that measurement of $\sigma(\vec{q},0)$ essentially specifies the Fourier transform of the charge distribution, from which the charge distribution itself can be determined to an accuracy of better than 1%.

Such measurements of charge distributions have revealed much about the structure of nuclei. For example, it is known that they behave approximately like liquid drops; having a nearly constant interior density and a sharp surface with the same surface thickness for elements throughout the periodic table. Many nuclei, like oxygen or lead, are spherically symmetric, whereas others, such as erbium or uranium, are highly deformed. Such precision measurements provide striking confirmation of mean field theory, which approximates the complicated many-nucleon interactions and correlations within a nucleus by assuming that each nucleon moves independently in the average field generated by all the other nucleons. The case of lead-208 (^{208}Pb) [**Fig. 1**] shows the level of quantitative agreement between theory and experiment, which now ranges over 11 orders of magnitude.

Low-energy discrete states. Nuclei exhibit a wide variety of discrete excited states. Single-particle excitations correspond to removing a nucleon from a normally occupied state to a normally unoccupied state. Vibrational states correspond to a coherent collective oscillation of the entire density distribution relative to its ground-state configuration. Rotational states, analogous to the rotational bands observed in diatomic molecules, correspond to the rotation of a deformed nucleus with a specific angular momentum. All these states show up as sharp peaks in the response function $\sigma(\vec{q},\omega)$ when the energy transfer ω is precisely equal to the excitation energy of the discrete state. Typical excitation energies are a few MeV. The $\vec{q}$ dependence of the response at this value of ω then specifies the transition charge density for the particular excitation. As in the case of elastic scattering, the same quantitative confirmation of the mean field theory as seen in Fig. 1 is observed for selected single-particle, vibrational and rotational states throughout the periodic table.

Giant resonances. At a slightly higher energy transfer ω, on the order of 10 MeV, the response function displays large broad peaks called giant resonances, each of which reveals different properties of the nuclear medium. To understand these different resonances, it is useful to think of the nucleus as being made up of four distinct fluids: spin-up protons, spin-down protons, spin-up neutrons, and spin-down neutrons. In the isoscalar monopole mode, all four fluids undergo spherically symmetric oscillations together, so that the nucleus alternately compresses and ex-

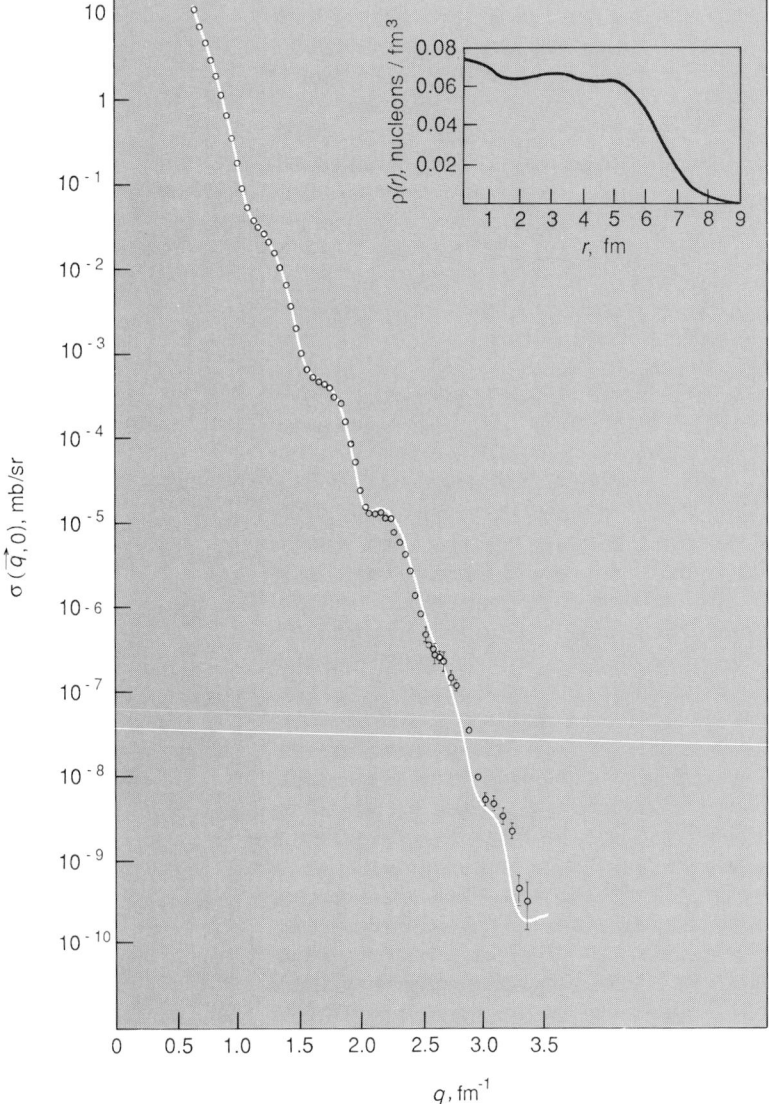

Fig. 1. Response function for elastic scattering of electrons from lead-208 (^{208}Pb) at incident energy of 502 MeV. Circles and error bars denote experimentally observed probability of elastic scattering $\sigma(\vec{q},0)$ at the specified momentum transfer $\vec{q}$, and solid line shows mean field theory prediction corresponding to the density distribution shown in the inset, which plots density ρ(r) against distance r from center of nucleus. The quantitative agreement confirms this theoretical density distribution. (Momentum transfer q is measured in terms of $\vec{q}/\hbar$, where $\hbar$ is Planck's constant divided by 2π; 1 fm = 10^{-15} m; 1 millibarn = 10^{-31} m².)

pands. Since this motion gives the appearance of "breathing," the mode is also called the breathing mode. The energy of this mode depends upon the compression modulus of nuclear matter—that is, the stiffness of the matter with respect to compression—and provides the most direct measure of this important quantity characterizing the equation of state of this matter. In the giant dipole mode, protons and neutrons oscillate in opposite directions, so that the total density is constant, but the ratio of protons to neutrons at any given point in the nucleus varies sinusoidally in time. This mode measures the symmetry energy of nuclear matter—that is, the stiffness of the matter with respect to variation of the ratio of neutrons to protons. Other possibilities include oscillations of up and down nucleons or even combinations of up-down and proton-neutron oscillations. In addition to specifying the energies at which these giant resonances occur, the peaks in the response function convey other important information. The width of the peak for energy transfer ω indicates the lifetime of the excitation, thus distinguishing between the long-lived ringing of a bell mentioned above and the totally damped response of a chunk of putty. The height of the peak indicates just how large or "giant" the resonance is compared with the theoretical maximum strength that such an excitation could have because of general sum rules specified by quantum mechanics. As in the previous cases, the dependence of the response on $\vec{q}$ specifies the detailed spatial distribution of the transition density and poses a strict quantitative test of microscopic theory.

Quasi-free scattering from nucleons. At still higher energy transfer, the nuclear response function evidences scattering from individual nucleons instead of from the entire nucleus. If momentum $\vec{q}$ is transferred to a nucleon at rest, it will recoil with energy ω given by Eq. (1), where

$$\omega = \frac{\vec{q}^2}{2m_n} \tag{1}$$

m_n is the mass of the nucleon. Thus, at any fixed $\vec{q}$ the response function will have a peak at the value of ω given by Eq. (1). In the Fermi gas model the nucleus may be viewed as a collection of nucleons of different momenta, so that no nucleon has a momentum greater than a maximum value called the Fermi momentum k_F and so that all values less than k_F are equally probable. The nuclear response function in this model is the sum of the responses for each momentum less than k_F and gives rise to a peak centered at energy ω given by Eq. (1), with width qk_F/m_n (**Fig. 2**). The experimentally measured response functions for nuclei shown in Fig. 2 are qualitatively similar to the result from the Fermi gas or shell model but differ in quantitative detail. Thus, mean field theory, which was so accurate at low energies, begins to break down in this case. It is currently an important, unresolved question

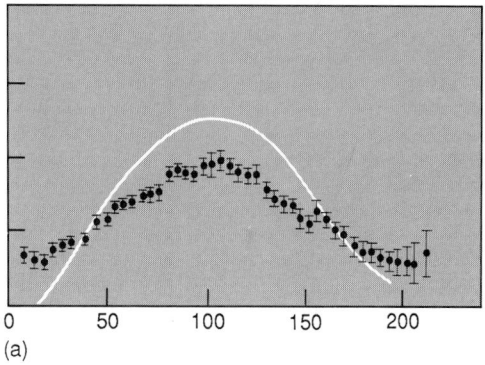

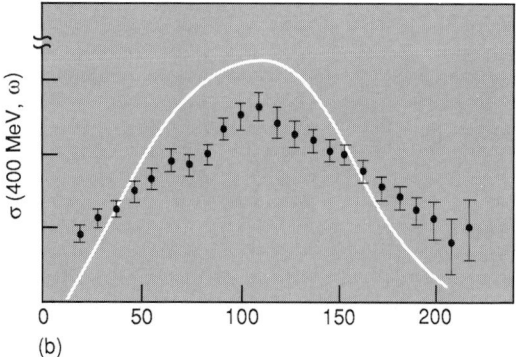

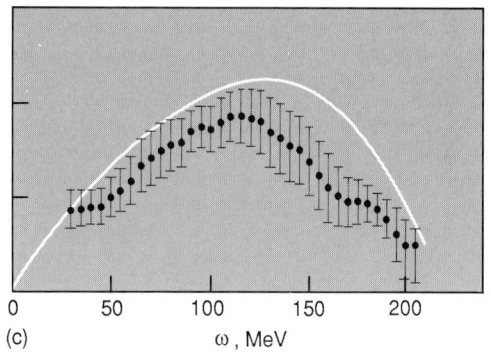

Fig. 2. Response functions (measured in arbitrary units) for (a) calcium-40 (^{40}Ca), (b) iron-56 (^{56}Fe), and (c) uranium-238 (^{238}U) at momentum transfer $\vec{q}$ = 400 MeV/c (where c is speed of light; $\vec{q}/\hbar$ = 2 fm^{-1}). Error bars denote experimental results, and solid lines show typical results using mean field and Fermi gas approximation.

whether this breakdown reflects other, more complicated correlations in the nuclear medium, or signifies a more dramatic alteration of the structure of the nucleon itself when it is immersed in the nuclear medium.

Quasi-free delta production. The nucleon is not a structureless point particle but is itself composed of constituents called quarks. Just as a nucleus can be excited when the energy transfer exceeds the energy of the first excited nuclear state, so too the nucleon may be excited when the energy transfer exceeds 293 MeV, the energy of the first excited nucleon state called the delta (Δ) resonance. Thus, above 293 MeV a new quasi-free process is possible. When momentum $\vec{q}$ and energy as given by Eq. (2) are transferred to a

$$\omega = 293 + \frac{\vec{q}^2}{2m_\Delta} \tag{2}$$

nucleon at rest, where m_Δ is the mass of the delta resonance, the first 293 MeV of energy is used to convert the nucleon to a delta resonance, and the rest of the energy provides the recoil energy of the delta resonance, $q^2/2m_\Delta$. Experimentally, a peak in the nuclear response function is then observed, analogous to the quasi-free peak but shifted up in energy transfer by 293 MeV. In addition to its dependence on the Fermi gas physics seen in Fig. 2, the quasi-free delta peak depends upon differences between the properties of a delta resonance in the nuclear medium and in free space. Hence, as in the case of quasi-free scattering, this process probes questions of nucleon structure as well as nuclear structure.

Scattering from quarks. Finally, when energy and momentum transfer reach the domain of several thousand MeV, the response function probes the physics of quarks within nucleons and nuclei. One of the unique features of quantum chromodynamics, the fundamental theory of quarks and gluons which is believed to describe the structure and properties of nucleons and nuclear forces, is the fact that certain processes become especially simple at high momentum transfer. In particular, instead of depending upon $\vec{q}$ and ω separately, the response function at very high momentum transfer depends only upon the single scaling variable x given by Eq. (3). Physi-

$$x = \frac{\vec{q}^2 - \omega^2}{2m_n\omega} \tag{3}$$

cally, this variable corresponds to the fraction of the total momentum carried by a single quark, so that the response function for a single nucleon specifies the momentum distribution of quarks in the nucleon. A fundamental question in contemporary nuclear physics is how a nucleon is modified when it is in the nuclear medium. In particular, it is important to know whether the quark distribution becomes less localized and thus has more low-momentum components when the nucleon interacts with other nucleons or whether the nucleon is very stiff and compact and negligibly altered by the medium. A partial answer to this question is provided by the ratio of the response function for a nucleus to the response function for a single nucleon, plotted in **Fig. 3**. Since this ratio is very different from unity in many cases, there is clearly a substantial difference between the momentum distribution of quarks in a single nucleon and in a nucleus, and it is a major challenge to understand this difference quantitatively.

Prospects. As the above discussion indicates, at higher energy, study of the nuclear response function addresses fundamental questions concerning the role of quarks and subnuclear degrees of freedom in nuclear structure. A central question is whether a nucleon is in fact altered in the nuclear medium, and if so, whether it is possible to understand both the deviation from Fermi gas behavior in Fig. 2 and the surprising ratio of quark momentum distributions in Fig. 3. A thorough exploration of these questions requires new kinds of experimental information. Measurements of the response function discussed here are inclusive measurements in the sense that they include all possible excited states at a given energy and momentum transfer. To understand these excited states in even more detail, it is necessary to perform exclusive measurements of the probability of exciting specific states while experimentally excluding all others. These exclusive experiments require detecting particles coming from the excited states in coincidence with the original particle scattered from the nucleus. Since the probability for observing two or more particles simultaneously is much lower than the probability of observing a single particle by itself, these coincidence experiments require a whole new generation of accelerators with much more intense beams, and detectors containing much larger angular ranges. Such facilities in the United States are under development at the Continuous Electron Beam Accelerator Facility (CE-BAF) at Newport News, Virginia, and the Massachusetts Institute of Technology (MIT) Bates Linear Accelerator in Boston, Massachusetts.

For background information *SEE* DELTA RESONANCE; GIANT NUCLEAR RESONANCES; NUCLEAR STRUCTURE; QUANTUM CHROMODYNAMICS; QUARKS; SCATTERING EXPERIMENTS (NUCLEI) in the McGraw-Hill Encyclopedia of Science and Technology.

John W. Negele

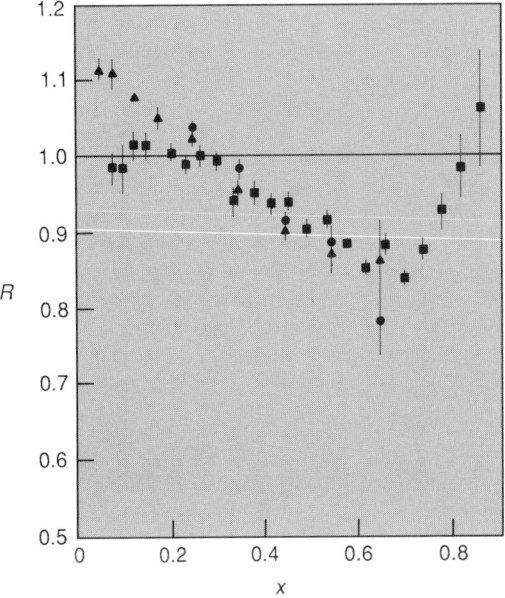

Fig. 3. Ratio R of the response function in the iron nucleus to that of a free nucleon as a function of the dimensionless scaling variable x. Deviation from unity displays difference between quark momentum distribution in a nucleus and in a nucleon. Agreement between data at different energies, denoted by triangles, circles, and squares, demonstrates that the response function is a unique function of the scaling variable at all but the lowest values of x.

Bibliography. C. Blatchley, Quasielastic electron scattering from ^{238}U, *Phys. Rev. C,* 34:1243–1247, 1986; A. Bohr and B. Mottelson, *Nuclear Structure,* 1975; B. Frois et al., Elastic scattering from ^{208}Pb, *Phys. Rev. Lett.,* 38:152–155, 1977; R. L. Jaffe, The EMC effect today, in S. Brodsky and E. Moniz (eds.), *Quarks and Gluons in Particles and Nuclei,* pp. 215–237, 1986; Z. E. Meziani et al., Coulomb sum rule for ^{40}Ca, ^{48}Ca and ^{56}Fe, *Phys. Rev. Lett.,* 52:2130–2133, 1984; J. W. Negele, The mean field theory of nuclear structure and dynamics, *Rev. Mod. Phys.,* 54:913–1015, 1982.

Oceanography

Quantitative information about ocean waves and currents is of increasing importance in a number of activities, notably offshore engineering, ship operations, coastal sea defenses, and pollution control. Such activity has revealed the shortcomings in conventional data sources, ship reports, and point measurements from buoys, and has directed interest to remote-sensing methods.

Remote sensing, involving the measurement of the characteristics of the ocean surface from a distant point, becomes a particularly powerful tool when radar is used, since radio wavelengths, unlike optical and infrared wavelengths, can penetrate clouds, and since radar can measure range and velocity and hence delineate the wavelengths, height, speed, and direction of ocean waves and the direction of surface currents.

The power of remote-sensing techniques can be illustrated by the global mapping of average wave height achieved with the radar altimeter on the *Seasat* satellite. The number of measurements made with this single instrument while it was operating exceeded by a factor of 17 the total number of ship reports made worldwide in the same period; the measurements were also distributed over the whole ocean area, including the Antarctic Ocean for which data from ship reports were very sparse.

Two main classes of radar sensor will be discussed: those using microwaves and limited to line-of-sight ranges, so that an elevated platform such as a satellite or aircraft is needed to achieve large-area surveillance; and those using decametric waves, which make it possible to effectively see around the curve of the Earth from a fixed location on the surface.

Microwave radar sensors. These sensors use radio wavelengths between 300 and 10 mm, corresponding to frequencies between 1 and 30 GHz. The sensors can receive echoes obliquely from the sea surface at distances not greater than the distance to the horizon D, given by Eq. (1a), where both D and the height h are expressed in meters. The corresponding relation between the distance to the horizon D' and the height h', both expressed in feet, is given by Eq. (1b).

$$D = 3570 \sqrt{h} \qquad (1a)$$

$$D' = 6470 \sqrt{h'} \qquad (1b)$$

Radar altimeter. The simplest radar instrument to visualize is the altimeter, which sends at regular intervals a very short pulse from a narrow beam directed vertically downward. If the sea surface is smooth, a single echo returns as if from a mirror at the location of the surface, and has the same shape as the transmitted pulse. The time delay thus gives the height of the satellite above the local surface; this height, if the orbit is very precisely known, yields the precise radius from the center of the Earth to the sea surface. This radius will vary, of course, with the tidal cycle at a particular location, but if the variations are found by regular observations and corrected there are other, more interesting effects. The shape of the ocean surface, the so-called ocean geoid, then portrays the shape of the sea bottom (including seamounts and trenches) through the influence of the gravitational pulls exerted by these massive features on the water overhead. The altimeter also reveals humps and dips as the satellite passes over features such as the Gulf Stream. *Seasat* in 1978 measured such details to an accuracy of 2 in. (50 mm) from an orbit known to an accuracy of 20 in. (500 mm). The first dramatic revelation of a surface feature was the existence of a 60-ft (18-m) drop in mean sea level over a 120-mi (200-km) track due to the deep ocean trench off Puerto Rico.

When the ocean is disturbed by waves, specular echoes are also received obliquely from suitably tilted facets on the undulating surface, within a circular patch illuminated by the antenna beam. These oblique echoes return slightly delayed relative to truly vertical echoes, so that a collection of pulses, overlapping in time, are received. The effect is that the sharp leading edge of the pulse as transmitted becomes smeared into a more gradual rise on reception, and the slope of the rise is a measure of the surface roughness. It has been possible to calibrate this slope in terms of the measure of surface roughness used in oceanography, the significant wave height (equal to four times the root-mean-square wave amplitude). The agreement obtained after careful calibration is within the accuracy of the wave buoys used to measure roughness from the ocean surface.

Scatterometer. The altimeter described above is not able to sense the direction of waves. To achieve this, a radar beam is directed obliquely downward at the ocean surface. The mechanism of scattering of the radio waves back to the radar is now no longer specular, which demands normal incidence, but involves diffraction and resembles the scattering of light from a grating. The radio waves incident on the surface at an angle θ are preferentially scattered back along the direction of incidence by ocean waves traveling directly away from or toward the radar, whose wavelength is such that the scatter from successive wave crests adds in phase for the backscatter direction. The condition for this to occur is given by Eq. (2), where λ is the radio wavelength and

$$L \sin \theta = \lambda/2 \qquad (2)$$

L is the water wavelength. A sinusoidal water wave whose length satisfies this equation diffracts energy only in this backscatter direction; waves of other lengths send energy in other directions. This mechanism is known as Bragg resonance by analogy with the somewhat similar phenomenon in x-ray diffraction. (Breaking waves spread energy more widely by a different mechanism and make a small contribution at all angles.)

The airborne or satellite-borne scatterometer uses this mechanism and senses the amount of energy scattered back to the radar from a large patch of sea at a frequency of about 20 GHz (15 mm wavelength). This energy is a measure of the amplitude of the short wind waves (ripples) raised on the sea surface directly by the action of local surface wind, and proves to be a good measure of surface wind speed. Typically, four radio beams are used, one pair forward to the left and right of the vehicle track, and the other pair rearward to the left and right. Since the ripples travel predominantly in the direction of the wind, comparison of the backscattered energies yields a good value for the surface wind direction; the typical accuracy is ±20° for direction and ±4 knots (±2 m/s) for wind speed.

Wavelength and wave direction sensors. The scatterometer is able to sense the direction of the short wind waves, but the greatest part of the wave energy is contained in the longest waves present on the sea surface. These are waves which, if locally generated, approach the wind velocity; for a sustained local surface wind of 20 knots (10 m/s), waves 160 ft (50 m) in length would be present, very many times the length of the Bragg resonant ripples detected by a microwave radar. To delineate the wave structure by radar, a different technique is used in which a radar with a very narrow beam and a short pulse is employed so that individual long waves can be distinguished. The planned European satellite *ERS 1* is to have a resolution of 100 ft (30 m) both in range and in the transverse (azimuthal) direction.

In this mode of operation the long waves can be identified on a radar image of the sea by the variation in the intensity of the image; more energy is returned, for instance, from the crests of waves where the ripples are larger in amplitude because of exposure of water to the wind and to hydrodynamic strains. These and other mecha-

nisms, by which the ripple properties (amplitude, tilt, and motion) vary over the long-wave structure, are very complex, and a full quantitative explanation has not been given. Other complex processes govern the relationship between these processes and the image intensity. Nevertheless, radar images obtained from space have been used to derive and plot the distribution with direction and wavelength of wave energy at locations on the ocean surface along a satellite track. (This distribution is the so-called directional wave-energy spectrum as measured by a pitch-roll wave buoy.)

Examples of sensors that use these ripple modulation principles are the tower-mounted, short-range Norwegian sensor MIROS (which also uses the Doppler effects due to wave motion and surface current) and the satellite-mounted synthetic aperture radar. The latter achieves its 100 × 100 ft (30 × 30 m) spatial resolution by using, for range discrimination, short pulses, and for azimuthal discrimination, a synthesized antenna aperture, many times the radar antenna dimension, that is formed by combining successive observations made as the satellite moves along its path.

Decametric radar sensors. The limitations of microwave sensors, when used on the Earth's surface, to short line-of-sight paths aroused interest in a different technique, high-frequency or decametric radar, operating in the 3–30-MHz (10–100-m) wavelength band. In its ground-wave version, this radar makes use of the very low attenuation of vertically polarized radio waves transmitted over the sea from the coast.

As in the oblique microwave sensor, Bragg-resonant backscatter from radially traveling ocean waves is employed. However, the resonant length is now in the range from 16 to 160 ft (5 to 50 m), much closer to the length of the energy-carrying long waves, so that interpretation is less indirect than with the centimetric ripples detected by microwave radar.

The advantage of ground-wave, high-frequency radar for oceanographic applications is that it is based at a fixed location, so that surface current patterns or the development of so-called wave directional spectra may be observed continuously over an area of some 90 × 90 mi (150 × 150 km). (Two radars observing the area from different directions are required to eliminate directional ambiguities.) In contrast, a satellite revisits one area of sea, typically, at 3-day intervals; its strength is the global survey feature. Accuracies of ±0.04 knot (±20 mm/s) for surface current and ±15% for significant wave height have been reported.

Larger ranges may be obtained with high-frequency sky-wave radar employing ionospheric reflection, a coverage span from 560 to 1740 mi (900 to 2800 km) being achievable. This mode is particularly effective for plotting surface wind direction over ocean areas, and has been demonstrated in hurricane tracking. Because of ionospheric motion, wave-height measurement is feasible for only a portion of sky-wave radar observations.

For background information *see* Applications satellites; Ocean waves; Radar; Remote sensing in the McGraw-Hill Encyclopedia of Science and Technology.

E. D. R. Shearman

Bibliography. T. D. Allan, (ed.), *Satellite Microwave Remote Sensing*, 1983, E. D. R. Shearman, Radio science and oceanography, *Radio Sci.*, 18:299–320, 1983; Special issue on high-frequency radar for ocean and ice mapping and ship location, *IEEE J. Ocean. Eng.*, OE-11 (2): 145–344, April 1986; Special issue on *Seasat, J. Geophys. Res.*, 87(15):3173–3438, 1982.

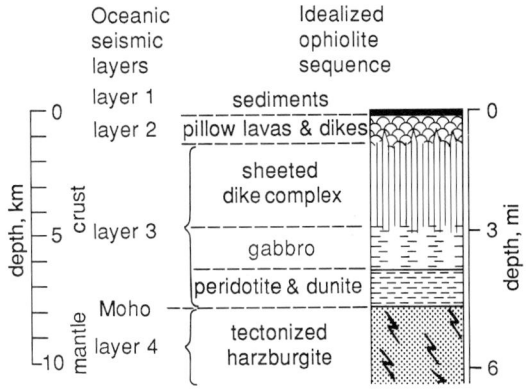

Fig. 1. An idealized ophiolite sequence compared with the seismically defined layering of the oceanic lithosphere in place. *(After I. G. Gass, The Troodos massif: Its role in the unravelling of the ophiolite problem and its significance in the understanding of constructive plate margin processes, in A. Panayiotou, ed., Ophiolites, Proceedings of the International Ophiolite Symposium, Cyprus, 1979, pp. 23–35, Cyprus Geological Survey, 1980)*

Ophiolite

Ophiolites are fragments of oceanic lithosphere that have been uplifted and emplaced on continental margins during periods of mountain building. Although the submarine origin of these bodies has long been accepted, there is still considerable disagreement as to exactly where in the ocean basins they formed. Do ophiolites represent oceanic lithosphere similar to that being generated today at mid-ocean ridges, or did they originate in island-arc regions or back-arc basins? How are these thick, dense slabs of oceanic lithosphere uplifted and transported to the continental margins? The results of several recent investigations suggest that most ophiolites formed above subduction zones and that the processes of formation and emplacement are closely linked.

Lithologic sequence. Complete ophiolites have a specific assemblage of rock types (**Fig. 1**). From the base upward they contain: deformed ultramafic rocks (tectonized harzburgite) thought to represent depleted mantle material; a sequence of layered plutonic rocks ranging from olivine-rich dunite at the base, through olivine- and pyroxene-bearing peridotite, to plagioclase-rich gabbro at the top; homogeneous gabbros which grade upward into quartz-bearing plagiogranites; a sheeted dike complex consisting of 100% dikes; and an upper sequence of pillowed and massive lava flows. Overlying many ophiolites are fine-grained manganese-rich sediments and deep-sea chalks.

The lithologic sequence outlined above corresponds closely to the seismically defined layering of the oceanic lithosphere in place (Fig. 1). The deep-sea sediments correspond to seismic layer 1 of the oceanic crust, the pillow lavas and upper dikes to layer 2, the sheeted dikes and plutonic rocks to layer 3, and the tectonized harzburgite to layer 4 or the mantle.

The presence of sheeted dike complexes, in which dikes intrude dikes without any intervening material, indicates that ophiolites form in extensional environments where rifting is accompanied by upward movement of molten magma to create new crust.

Origin. The inferred similarity of ophiolites to the oceanic lithosphere in place, the presence

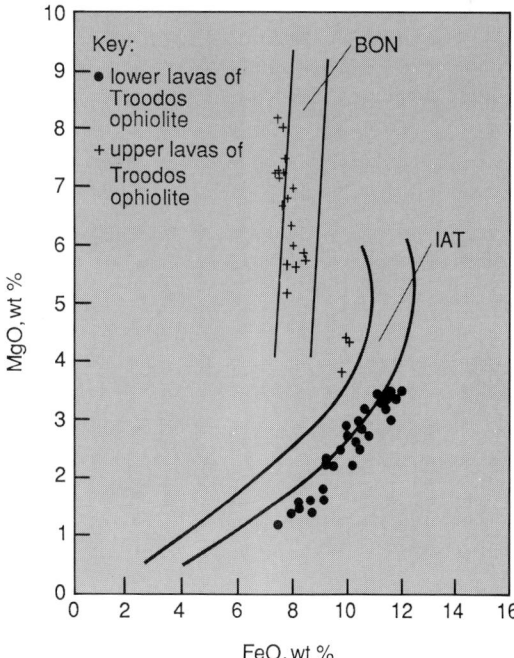

Fig. 2. Magnesium oxide–iron oxide (MgO–FeO; total iron expressed as FeO) variation diagram showing volcanic glass compositions of the Troodos ophiolite in relation to island-arc tholeiites and boninites of the Marianas Islands. BON field indicates Marianas arc boninite suite. IAT field indicates island arc tholeiite suite. *(After P. T. Robinson et al., Volcanic glass compositions of the Troodos ophiolite, Cyprus, Geology, 11:400–404, 1983)*

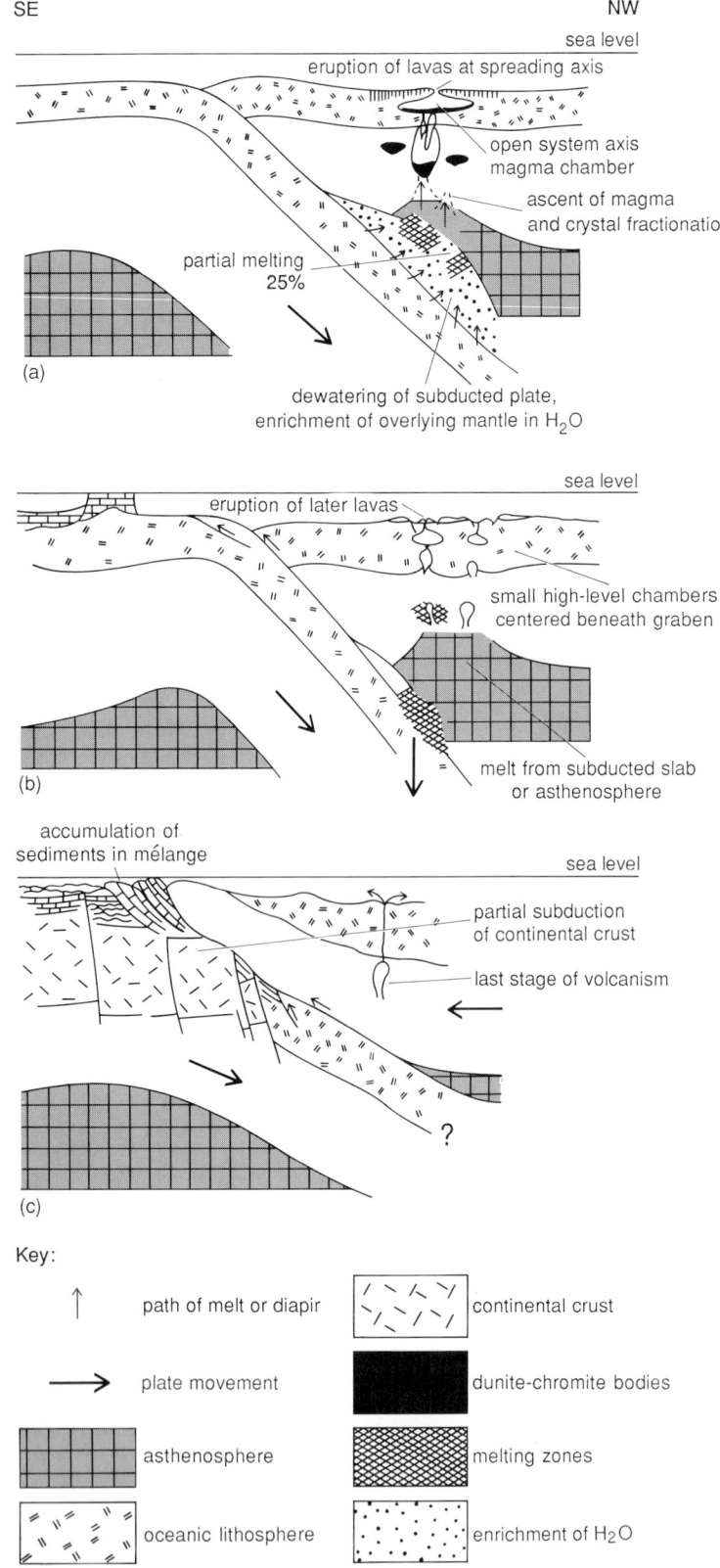

Key:

↑   path of melt or diapir

⟶   plate movement

asthenosphere

oceanic lithosphere

continental crust

dunite-chromite bodies

melting zones

enrichment of H_2O

Fig. 3. Possible stages in the formation of an ophiolite in a subduction zone environment. (*a*) Igneous activity associated with sea-floor spreading forms a typical ophiolite section above the subduction zone. (*b*) Later melting of depleted mantle material produces another volcanic event which follows the spreading event. (*c*) Subduction ceases as continental crustal material is dragged beneath the ophiolite. Continental margin sedimentary and volcanic rocks form an accretionary wedge on the inner trench wall. Later isostatic rebound of the continental crustal material may cause uplift of the ophiolite and emplacement on the continental margin. (*After T. Alabaster, J. A. Pearce, and J. Malpas, The volcanic stratigraphy and petrogenesis of the Oman ophiolite complex, Contrib. Mineral. Petrol., 81:168–183, 1982*)

of overlying deep-sea sediments, and the evidence for formation in a spreading environment led early workers to suggest that ophiolites formed at mid-ocean ridges such as the modern Mid-Atlantic Ridge or East Pacific Rise. This view was widely accepted, although some ophiolites such as Betts Cove in Newfoundland and the Smartville complex of California were recognized as having formed in an island arc.

Detailed investigations of several ophiolites in the last few years have provided new insights into the origin of these bodies and their mechanisms of emplacement. The most useful information has come from studies of the Troodos ophiolite of Cyprus and the Semail ophiolite of Oman. In both of these bodies the volcanic rocks are only moderately altered, and in the Troodos ophiolite fresh volcanic glass is widely preserved. Alteration of the lavas was mostly by interaction with cold sea water, and the irregular nature of the alteration reflects the variable permeability of the lavas. Analysis of the freshest rocks has allowed accurate determination of their original chemical compositions. The volcanic glasses are particularly useful in this regard because they represent quenched samples of the original magmas (**Fig. 2**).

The newly obtained geochemical data indicate that both ophiolites have multistage magmatic histories involving several different parental magmas. Five geochemical and stratigraphic units, ranging in composition from tholeiitic basalt to highly evolved rhyolite, are recognized in the extrusive sequence of the Semail ophiolite. In the Troodos ophiolite, three magmatic series have been defined: an andesite-dacite-rhyodacite assemblage, a depleted basalt–basaltic andesite assemblage, and a highly depleted basaltic assemblage. None of these volcanic rocks are compositionally equivalent to mid-ocean ridge basalts, but compare closely with lavas erupted in modern subduction zone environments such as those of the western Pacific. The ophiolitic lava compositions suggest eruption in a number of tectonic environments ranging from island arc through fore-arc to back-arc. In the Semail ophiolite, eruption of early marginal basin lavas was followed by localized seamount activity. This was followed after a short interval by eruption of basalts in rifts or graben between the major seamounts. The early magmatism ended with collision of the Arabian continental margin and the subduction zone. The final magmatic event involved lavas formed from the subducted lithosphere, that is, mantle material underlying the Arabian continent.

In the Troodos ophiolite, early andesitic and dacitic magmatism formed an embryonic arc which was entirely submarine. These lavas were followed by olivine-bearing basalts and picrites which probably formed in a similar tectonic environment. A series of highly depleted lavas on the south flank of the Troodos ophiolite is compositionally similar to rocks found in the Bonin Islands and may represent a fore-arc environment. Preliminary studies of other ophiolites such as the Hatay ophiolite of Syria and Turkey and the Coast Range ophiolite of California indicate lava compositions similar to those of the Troodos and Semail complexes, suggesting formation in similar tectonic environments.

Although subduction zones are generally considered to be regions of compression, crustal spreading is known to occur in back-arc basins and in some arcs, such as the Marianas Islands, where the arc has split and the two halves rifted apart. Presumably, the intervening, newly created lithosphere contains a sheeted dike complex which records this rifting event. Such complexes in which dike intrudes dike without any intervening material are thought to form only in spreading or rifting environments.

Recent models. Recent models for the origin of ophiolites have attempted to reconcile the new geochemical evidence for formation in subduction zone environments with the geologic evidence suggesting formation in deep ocean basins. One such model suggests that the Andaman Sea region in the eastern Indian Ocean may provide a modern example of ophiolite formation. In this region, spreading occurs along short ridge segments which overlie an active subduction zone. Subduction of the Indian Plate beneath Asia is very oblique, and the island arc is discontinuous. In particular, there are no large stratovolcanoes over the subduction zone in the region of active spreading. Thus, the newly formed oceanic lithosphere is overlain by pelagic sediments, as are the Troodos and Semail ophiolites.

The recognition that many ophiolites originated in a subduction zone environment has led to new interpretations of their mechanisms of emplacement. Ophiolites are emplaced during periods of mountain building and commonly mark suture lines which represent ancient plate boundaries. Thus, they are believed to represent continent-continent collisions resulting from subduction of the intervening oceanic lithosphere. The ophiolite forms above an active subduction zone in which oceanic lithosphere is being consumed (**Fig. 3**). Continental crust carried along on the subducting plate collides with the subduction zone and is dragged down beneath the ophiolite. Because the continental crust has a relatively low density, it cannot be carried far down the subduction zone and thus subduction effectively ceases at this location. Later isostatic uplift of the partially subducted continental slab causes uplift of the overlying ophiolite, which may then be transported onto the continental margin along low-angle faults.

For background information SEE CONTINENTAL MARGIN; CONTINENTS, EVOLUTION OF; LITHOSPHERE; OROGENY in the McGraw-Hill Encyclopedia of Science and Technology.

P. T. Robinson

Bibliography. T. Alabaster, J. A. Pearce, and J. Malpas, The volcanic stratigraphy and petrogenesis of the Oman ophiolite complex, *Contrib. Mineral. Petrol.,* 81:168–183, 1982; E. M. Moores et al., Model for the origin of the Troodos massif, Cyprus and other mideast ophiolites, *Geology,* 12:500–503, 1984; A. Panayiotou (ed.), *Ophiolites,* Proceedings of the International Ophiolite Symposium, Cyprus, 1979, Cyprus Geological Survey, 1980; P. T. Robinson et al., Volcanic glass compositions of the Troodos ophiolite, Cyprus, *Geology,* 11:400–404, 1983.

Optical detectors

The detection of infrared radiation of wavelength greater than 8 micrometers is important for the passive sensing and imaging of objects with temperatures at or below 300 K (80°F). Advances in the technology for growing and fabricating semiconductor materials now permit the fabrication of small semiconductor detectors that are optimized for high-sensitivity detection in the long-wavelength infrared. These detectors can be fabricated as large linear or area arrays which can be hybridized with large-scale integrated silicon signal processors and scanned optically or electrically to produce high-resolution infrared images.

Since the atmosphere has transmission windows in the mid-wavelength (3–5 μm) and long-wavelength (8–14 μm) infrared bands, terrestrial or airborne objects can be imaged in either band. However, because the peak of the spectrum radiated by a 300 K (80°F) blackbody is near 10 μm, thermal imaging of objects near ambient temperature is most effectively realized in the 8–14 μm long-wavelength infrared band, where the power radiated by the object is maximum.

Detection mechanisms. Although there are a number of methods for detecting infrared radiation, the methods of highest sensitivity involve the use of photon detectors. These are detectors which respond incrementally to the absorption of each photon, rather than to the total energy absorbed by the detector. In the long-wavelength infrared, semiconductor photon detectors utilize either impurity absorption or intrinsic absorption. In either case, the absorption of a photon produces mobile electric charge which is detected as an electric current. This current is produced either by the application of an external bias voltage (in which case the device is referred to as

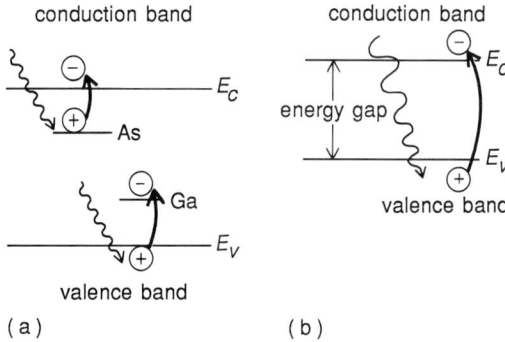

Fig. 1. Band diagrams of infrared detectors. (*a*) Extrinsic detector, which utilizes absorption by impurity states in the band gap, near the edges of the conduction band (E_c) and valence band (E_v). A process is shown in which a photon (indicated by wavy arrow) excites a negatively charged electron from the arsenic (As) impurity state to the conduction band, leaving behind a positive charged donor impurity. Also shown is the excitation of an electron from the valence band to a gallium (Ga) impurity state, which is equivalent to excitation of a hole into the valence band. (*b*) Intrinsic detector, which employs absorption across the band gap.

a photoconductor) or by the presence of the built-in field produced by a *pn* junction in the material (in which case the device is referred to as a photodiode).

Extrinsic detectors. Impurity absorption is the mechanism employed in the extrinsic photoconductor (**Fig. 1***a*). A suitably pure wide-gap semiconductor, usually silicon or germanium, is doped with an elemental impurity which introduces a shallow neutral impurity state into the band gap adjacent to one of the band edges. Photons with energy greater than the difference between this state and the band edge can photoexcite an electron into the conduction band or

Properties of photoconductors used in long-wavelength infrared detectors		
Extrinsic detectors		
Material	Excitation energy, eV	Cutoff wavelength, μm
Si:Tl	0.26	4.8
Si:In	0.16	7.8
Si:Bi	0.069	18.0
Si:Ga	0.065	19.1
Si:As	0.049	25.3
Si:Sb	0.039	31.8
Intrinsic detectors		
Material	Energy gap, eV	Cutoff wavelength, μm
Si	1.08	1.15
Ge	0.66	1.88
PbS	0.41	3.02
InAs	0.388	3.2*
InSb	0.225	5.5*
$Hg_{0.79}Cd_{0.21}Te$	0.103	12.0*

* At an operating temperature of 80 K (−316°F).

a hole into the valence band. This produces an increase in the carrier density, increasing the conductivity of the material by an amount proportional to the product of the photogeneration rate and the lifetime of the excess carriers. A number of dopants have been utilized in silicon (see **table**). The best-developed systems are indium, gallium, bismuth, and arsenic.

Intrinsic detectors. Intrinsic detectors utilize the generation of electron-hole pairs across the semiconductor band gap (Fig. 1*b*). While there are a number of elemental and compound semiconductors with energy gaps corresponding to the short- and mid-wavelength infrared, the narrowest gap found in an elemental or binary-compound semiconductor is that of indium antimonide (InSb), which has an energy gap corresponding to about 5.5 μm at its usual operating temperature (80 K or −316°F). The production of intrinsic materials for low-wavelength infrared detection has thus required the creation of new material systems. Ternary alloys of two binary-compound semiconductors are used to achieve band gaps tailored to the application. The most highly developed system of this sort is $Hg_{1-x}Cd_xTe$, which is an alloy of the wide-gap semiconductor cadmium telluride (CdTe) with the semimetal mercury telluride (HgTe). By adjusting the mole fraction x of cadmium, the energy gap may be tuned from zero ($x = 0.15$) to 1.6 eV ($x = 1.0$). Intrinsic detectors can be operated as photoconductors by the application of a bias, or as photodiodes by the formation of a *pn* junction in the material.

Operating temperature. The signal-to-noise ratio of a long-wavelength infrared detector is determined principally by competition between the photogenerated signal current and the noise caused by fluctuations in currents from competing mechanisms such as photogeneration by the infrared background, thermal generation, tunneling, and surface leakage. To minimize thermally generated dark currents, long-wavelength infrared detectors are operated at very low temperatures. Generally the detector is operated at a temperature sufficiently low that the noise from the infrared background dominates. Such a detector is said to be background-limited. At high backgrounds (for example, in terrestrial imaging over a wide spectral bandwidth), the operating temperature will be relatively high, whereas against low backgrounds (for example, in astronomical imaging from a satellite) it will be quite low. For moderate to low background applications, detectors are usually operated at a temperature T such that the thermal energy kT (where k is Boltzmann's constant) is less than about ⅟₂₅ of the minimum energy for photoexcitation of a carrier in the case of an intrinsic detector, and ⅟₅₀ of this energy in the case of an extrinsic detector. Thus, for example, an intrinsic detector with an energy gap corresponding to response out to 12 μm (that is, a cutoff wavelength of 12 μm) will be operated below 45 K (−379°F) for low-background applications. This requirement puts a high premium on obtaining a detecting material with a cutoff wavelength no longer than that required by the application, as longer cutoff wavelengths require lower operating temperatures. This consideration works against the extrinsics, as no suitable dopant exists which provides a cutoff wavelength in the 8–14-μm band.

Detector performance. Extrinsic photoconductors demonstrate the highest sensitivities at long wavelengths and low backgrounds. However, they have some important limitations. Their performance is determined by material properties, such as compensating impurity concentration, which are hard to measure and control. The short lifetime of excess carriers, caused by the large cross section for capture of photoexcited carriers by ionized impurities, requires lower operating temperatures than those used with intrinsics.

Because the electric field in the detector depends upon the ionized impurity space charge, the detector response varies as this space charge changes due to dielectric relaxation. Since the rate of dielectric relaxation depends upon the mobile charge carrier concentration, and hence on the infrared background, the response of an extrinsic photoconductor depends upon the background and can be very slow at low backgrounds. This makes extrinsic photoconductors difficult to calibrate and unsuitable in systems where wide-bandwidth response is essential.

Blocked impurity-band detectors. Improvements in semiconductor growth by metallo-organic chemical vapor deposition and molecular-beam epitaxy permit the fabrication of a new type of extrinsic detector referred to as a blocked impurity-band detector (**Fig. 2**). In this device the concentration of the dopant impurity is increased to the level at which conductivity occurs as a result of the hopping of carriers beween the impurities. The increase in concentration of absorbing impurities decreases the volume of the detector required to achieve high quantum efficiency and permits rapid relaxation of the ionized impurity space charge. To prevent dark current from flowing in the impurity band, the band is blocked by the epitaxial growth of an undoped intrinsic layer on the heavily doped substrate. Compensating acceptor impurities in the heavily doped layer ionize some of the donors in the impurity band and produce hopping conductivity in that band. The intrinsic blocking layer allows the device to be biased like a metal-insulator-semiconductor photodiode so that the impurity band is depleted of mobile carriers. Impurity-band detectors may be illuminated through a substrate that is transparent to infrared radiation by incorporation of a thin, heavily doped transparent contact beneath the active layer. Impurity-band detectors utilizing arsenic-doped silicon exhibit high sensitivity and excellent frequency response.

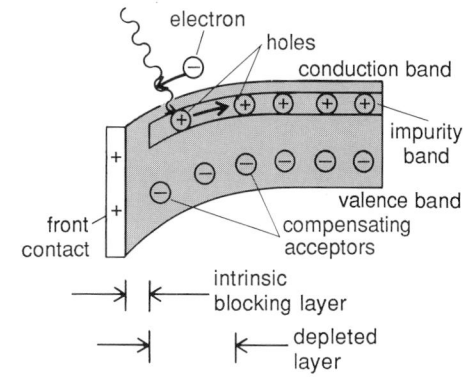

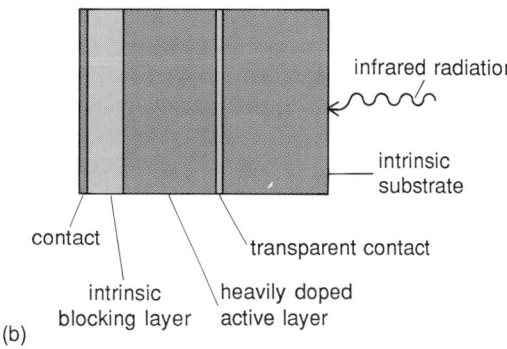

Fig. 2. Blocked impurity band detector, in which the intrinsic blocking layer is grown on a heavily doped extrinsic photoconductor. (*a*) Band diagram. Positive charges on front contact indicate application of a positive bias to that contact. (*b*) Physical structure (reverse-illuminated).

Performance of intrinsic detectors. Most of the development of intrinsics for use in the long-wavelength infrared has involved $Hg_{1-x}Cd_xTe$. The tunability of the band gap allows operation at the maximum operating temperature for a given application. Material is grown in bulk boules from the melt or by liquid-phase epitaxy on an infrared-transparent cadmium telluride substrate. High-performance photoconductors have been fabricated for long-wavelength infrared applications, although they achieve high photoconductive gain by trapping minority carriers and thus have characteristics similar to extrinsic photoconductors. Photodiodes are fabricated by ion implantation of p-type material to create n-type damage centers or by impurity doping of the liquid-phase epitaxy melt. These diodes are satisfactory for use in the medium-wave infrared. However, most long-wave infrared diodes have high reverse leakage currents that are believed to originate in direct band-to-band and impurity-assisted tunneling in the high-field region of the diode.

Heterojunction devices. Much of the current research effort in long-wavelength infrared materials is directed at the use of liquid-phase epitaxy and molecular-beam epitaxy to produce structures which incorporate heterojunctions between layers of $Hg_{1-x}Cd_xTe$ of different composition x. In the simplest of these structures, the tunneling currents inherent in the $Hg_{1-x}Cd_xTe$ long-wavelength infrared photodiode are reduced by the liquid-phase epitaxial fabrication of a hetero-

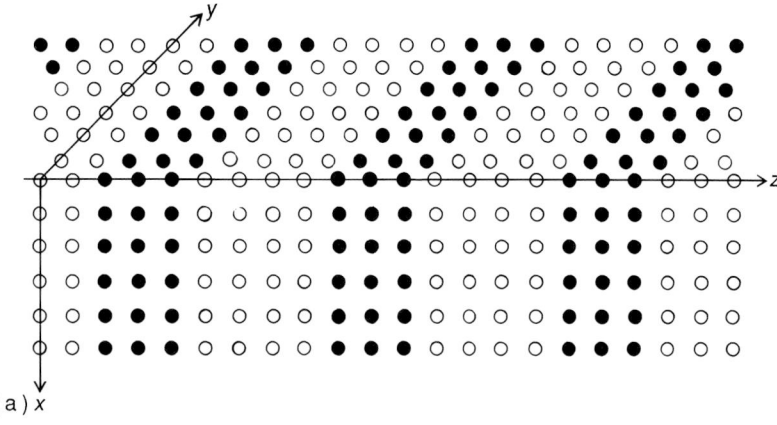

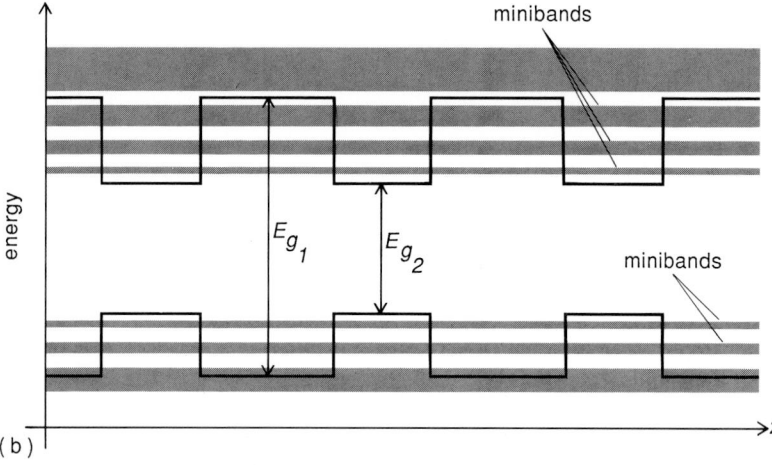

Fig. 3. Compositional superlattice, consisting of alternating layers of a wide-gap (E_{g_1}) and a narrow-gap (E_{g_2}) semiconductor. (a) Physical structure. (b) Band diagram.

TECTORS; PHOTOCONDUCTIVITY; PHOTODIODE; SEMICONDUCTOR; SEMICONDUCTOR HETEROSTRUCTURES in the McGraw-Hill Encyclopedia of Science and Technology.

Michael J. Daugherty

Bibliography. P. R. Bratt and T. N. Casselman, Potential barriers in HgCdTe Heterojunctions, *J. Vac. Sci. Technol.*, A3:238–245, 1985; R. J. Keyes et al., *Optical and Infrared Detectors*, vol. 19 of *Topics in Applied Physics*, 1977; C. R. McCreight et al., *Proceedings of the Second Infrared Technology Workshop*, August 13–14, 1985, NASA Tech. Mem. 88213, 1985; D. L. Smith, T. C. McGill, and J. N. Schulman, Advantages of the HgTe-CdTe superlattice as an infrared detector material, *Appl. Phys. Lett.*, 43:180–182, 1983.

Optical information systems

The use of light to process and store information offers the possibility of constructing information systems with a speed and capacity far exceeding those of present computers. Furthermore, such systems would be naturally suited to parallel processing. The first section of this article discusses the development of the transphasor, an optical switch, analogous to the transistor, in which light controls light. This and related devices would form a complete family of digital circuit elements, the basic components of an optical digital computer.

Optical data storage already offers the greatest storage density of any current technology. The second section of this article discusses the possible use of an extra dimension, optical frequency, to increase this density by a factor of approximately 1000.

Optical digital circuits. Two discoveries during the 1970s are now leading to a new technology in which information can be processed by virtue of devices in which light controls light. The first important step was the observation of the phenomenon of optical bistability, in which a device can show two output states for one value of an input light intensity. The second step was the discovery of a giant optical nonlinearity at wavelengths near the band gap of a semiconductor. In this work, nonlinearities were discovered as much as 10^9 times greater than expected. The combination of these two achievements has resulted in the development of a device in which a power (or holding) light beam can be modulated by the application of a second control beam to yield an amplified output analogous to that of the electronic transistor. The device, consisting of two parallel partially reflecting mirrors sandwiching a nonlinear optical material, works by transference of phase thickness (referenced to the wavelength of light) from one beam to another, and is therefore called a transphasor. The original

structure consisting of a wide-gap layer grown on top of narrow-gap material. If the heterojunction and the *pn* junction are made coincident, tunneling currents can be minimized without sacrificing quantum efficiency. However, the routine production of high-performance heterojunction devices remains elusive. One promising new approach involves the fabrication of a compositional superlattice (**Fig. 3**) consisting of many thin alternating layers of a narrow-gap material (for example, mercury telluride) with a wide-gap material (for example, cadmium telluride) to produce a structure with a tunable energy gap. When many alternating layers of materials of different band gap are grown, the conduction and valence bands split into minibands, the position and width of which can be controlled by the thickness of the layers rather than by their composition. Calculations suggest that such materials may be easier to control than comparable alloy systems and may exhibit significantly reduced tunneling currents. Active development of infrared-sensitive superlattice materials is under way in a number of laboratories.

For background information SEE ARTIFICIALLY LAYERED STRUCTURES; ELECTRICAL NOISE; OPTICAL DE-

experiments demonstrated the property of amplification, necessary for the realization of optical circuits, in which one device can give an optical output strong enough to drive a succeeding device. With simple adjustments to the initial conditions, the same basic nonlinear interferometer can give a complete family of optical circuit elements. These can be arranged to provide optical logic gates such as AND, OR, NAND, NOR, and a bistable read/write optical memory, as well as the amplifying function of the transphasor.

Optical nonlinearity. The optical nonlinearity that is used in the devices is usually the change of refractive index with light intensity. However, nonlinear absorption can also be used. The refractive case is particularly sensitive since it makes use of interfering light beams that require only a small change of refractive index to switch from on to off. Such a nonlinearity can be induced in materials by at least two mechanisms. The first involves the excitation of electrons in the material into certain quantum states so that the refractive index is directly changed through a quantum-mechanical process. In the second mechanism the entire material can be heated with a similar effect. The first process is intrinsically very fast, and short laser pulses can induce changes on a picosecond time scale. The second process can be surprisingly fast if the sample volume is restricted to a few cubic micrometers, and effects as fast as nanoseconds have been observed.

Mechanism. An optical interferometric device, such as a Fabry-Perot etalon, acts through the interference of multiple-reflected beams between two partially reflecting parallel mirrors. For the case of such an interferometer, filled with a medium whose optical thickness is intensity-dependent, the operation can be as follows: the interferometer will show wavelengths of maximum transmission when the optical thickness is such that constructive interference takes place (usually when this thickness is an integral number of half-wavelengths). The device can be illuminated by a laser beam whose wavelength is slightly displaced from the condition for maximum transmission. On increasing the intensity, the optical thickness will change and, for example, the peak of the transmission curve will move toward the laser wavelength. This in turn will increase the transmission and hence output of the device so that the input-output curve becomes nonlinear. However, as resonance is approached, more of the incident light circulates within the interferometer. It is this circulating internal intensity that causes the nonlinear change of refractive index. Thus, as the interferometer sweeps toward resonance with the laser, a positive feedback effect occurs which increases intensity, thus increasing the nonlinear change and so further increasing the internal intensity, and so on. A point is soon reached in which the input-output characteristic shows a negative slope and more light has been put into the device than is necessary to switch it. This gives an optical hysteresis loop and causes the device to operate as a read/write optically bistable memory.

The amount of initial detuning of the laser wavelength from that of maximum transmission of the interferometer determines the shape of the input-output curve. Thus, it is easy to set up a whole series of optical circuit elements varying from a power limiter, through an optical transphasor with gain, to the bistable memory condition.

Compatibility and speed. The attraction of devices that use photons to process information lies in their compatibility with high-speed communication devices. Photons are already used to transmit information in optical fibers, but also may be required in future information processing devices to overcome an intrinsic limitation of electronics: the product of resistance and capacitance limits the time response of a circuit. Thus, beyond a certain point, microelectronic circuits do not become faster as they become smaller, and optics promises to overcome this limitation.

Parallelism. A further advantage of optics lies in its natural parallelism. This property may enable arrays of optical logic devices to process complete images in parallel, with an increase in processing rate equal to the number of channels. Bistable optical elements operating as logic devices have already been constructed with dimensions of a few micrometers, which implies that massively parallel processing (involving 10^4 to 10^8 channels) will be possible in the future. Such devices provide, theoretically, a method of overcoming a bottleneck that ties electronic computing to sequential operation, namely the inability of electronic circuits to access all parts of a memory simultaneously.

Demonstration device. In 1986 a first optical finite-state machine was demonstrated to show that primitive building blocks of an optical digital parallel processing device could be built. (The concept of a finite-state machine was introduced in early discussions of the theory of binary digital computers, but only with this device was it first realized in practice.) Within this machine a new computer architecture called lock-and-clock was utilized to provide an optical delay. Such a delay was necessary because communication between the elements at the speed of light was more rapid than the switch delays. With this procedure it has been possible to show that restoring optical logic can be achieved and extended indefinitely without error. (Restoring logic implies that even with an imperfect input pulse a standard output occurs at the exit of each successive logic gate.)

Although much work will be required before anything resembling an optical computer can be built, this new technology has applications in image processing, broadband switching, and displays, as well as suggesting new directions for the solution of computational problems.

Optical computer architecture. Now that it has been shown that all-optical logic decision making can be achieved, and the results stored and passed on by purely optical communication, a considerable amount of research is in progress to create a new computer architecture which can exploit the advantages of optics. It already seems possible to use a single instruction to control a massively parallel optical data stream; this stream then evolves in an adaptive manner. As a result, differential multiple control of such an image flow of logical information will be provided. This leads naturally to the concept of symbolic substitution which can be adapted with only a few basic rules to undertake all the tasks of digital arithmetic. Individual optical computing subsystems, such as a digital full adder which provides sum and carry, have already been invented.

S. Desmond Smith

Frequency-domain optical storage. In conventional laser-disk optical recording, information is stored by modifying some property of a two-dimensional storage medium with a laser beam which is focused to a spot about 1 micrometer in diameter. This property should be one which can be read optically by the same or a similar laser at lower power, such as the property of reflectivity or Faraday rotation. This leads to storage densities approaching 10^8 bits/cm^2, limited by the fundamental physical phenomenon of diffraction.

Principles. Frequency-domain optical storage is an exploratory technology, not yet in development, which has the potential to go well beyond this limit by adding an extra dimension, optical frequency or color, to the process. A large number (approximately 10^3) of frequency bits are stored at each spatial point. This technique relies on the physical process of persistent spectral holeburning (PHB), discovered in organic materials by two groups of Soviet scientists (A. A. Gorkhovskii and B. Kharlamov and coworkers) in the mid-1970s. Fundamental to this process is the fact that at temperatures approaching absolute zero many of the optical absorption bands (so-called zero-phonon lines) of defect centers

such as ions, molecules, or color centers in solids (optical centers) are inhomogeneously broadened by the unavoidably imperfect nature of all real solids. This broadening is especially pronounced in noncrystalline materials such as glasses and polymers, but is present to a lesser extent even in the best-quality single crystals. This produces a range of environments for the optical centers, and with it a range of different resonant optical frequencies or colors at which they absorb light (**Fig. 1a**). The resulting inhomogeneous linewidth Γ_{inh} varies from several GHz in many crystals to about 10^4 GHz in glasses or polymers. A narrowband laser source can be made to interact with typically 1000 or more distinguishable sets of optical centers, each with its own characteristic frequency. If these centers undergo a light-induced change due to photochemistry or a photophysical rearrangement of the local environment, the absorption of the set resonant with the laser can be bleached, producing a spectral hole (Fig. 1b). This is the mechanism for recording a bit of information in the frequency domain.

Several bleaching processes have been identified and it is likely that more will be found. In glasses, which are characterized by a broad distribution of local environments around an optical center, each configuration is separated by energy barriers. Optical excitation can result in a change in these local configurations which shifts the optical transition frequency. The lifetime of the resulting hole is determined by the height and width of the barriers to motion of the atoms or molecules in the glass, and by the temperature. Estimates give lifetimes of years at liquid helium temperatures. In crystalline materials, as well as in glasses, two other important mechanisms are known. Light-induced proton motion (tautomerism) in hydrogen-bonded organic systems, such as phthalocyanines and porphyrins, again produces frequency-shifted absorption of the selectively excited molecules, and hence holeburning. Photoionization of molecules or ions also clearly leads to bleaching because the ionized species has a different electronic structure and absorption. In this case the hole persists because the photon-released electrons are trapped in deep traps.

The lifetime of the holes, that is, the lifetime of the stored data, can be months or years, but so far no direct measurements of very long lifetimes have been made. The width of a hole is at least twice the homogeneous linewidth Γ_h of the selected group of optical centers. At low temperatures, this is typically tens of megahertz and is the width experienced by all centers because of time-varying thermal and magnetic fields in the medium and because of the radiative lifetime of the optically excited state of the center. An upper limit on the number of frequency bits that can be recorded is the ratio of the inhomogeneous width to the homogeneous width (Γ_{inh}/Γ_h). At low tem-

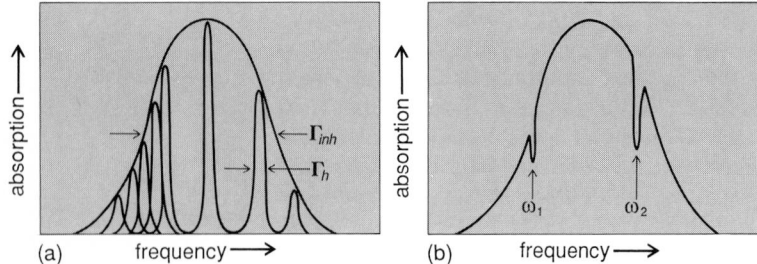

Fig. 1. Persistent spectral holeburning. (*a*) Optical absorption band of the defect centers in a solid, with inhomogeneous linewidth Γ_{inh}, and absorption bands of some of the sets of optical centers that interact with a narrowband laser source, each with homogeneous linewidth Γ_h. (*b*) Effect on the absorption band of bleaching sets of optical centers at frequencies ω_1 and ω_2, producing spectral holes.

peratures, this upper limit varies between 10^3 and 10^5 for most materials exhibiting persistent spectral holeburning. Since Γ_h is strongly dependent on temperature above a few kelvins, this is essentially a cryogenic technology in which the recording medium is immersed in liquid helium.

Reading the holes is achieved by scanning a probe laser of lower intensity through the inhomogeneous profile and monitoring the absorption or a quantity proportional to the absorption. At each frequency where a bit (a binary 1) was written, a decrease in optical absorption is measured, while at other frequencies (where there is a binary 0) there is no change. Thus, by using a tunable laser such as a semiconductor laser and a means of addressing different spatial locations, data can be written in the frequency domain at each spatial spot. For, say, 10^3 frequency bits, this leads to an overall density of $10^3 \times 10^8 = 10^{11}$ bits/cm^2. This phenomenal increase in density is what makes frequency-domain storage so attractive, and has stimulated efforts to produce a practical device.

Device configuration. A possible device configuration is shown in **Fig. 2**. A laser addresses an arm of the system consisting of several (say 16 to 64) chips, each containing 10^6 spatial memory locations. These chips may be addressed in parallel by using

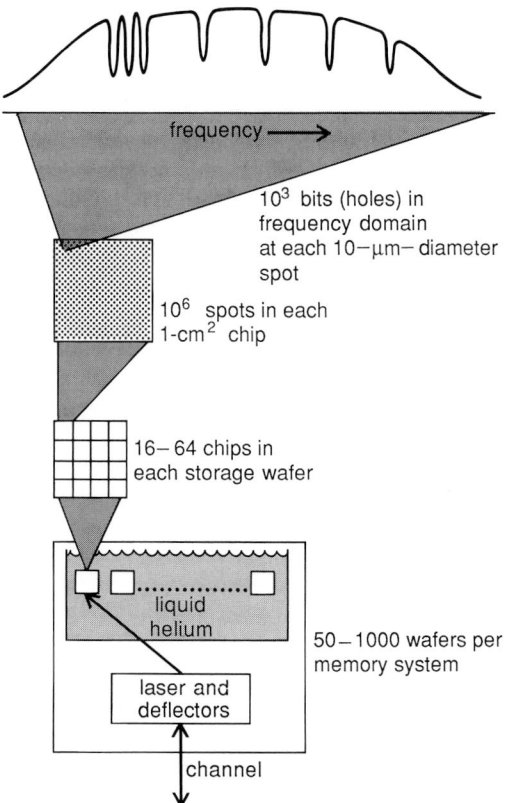

10³ bits (holes) in frequency domain at each 10−μm− diameter spot

10^6 spots in each 1-cm^2 chip

16− 64 chips in each storage wafer

liquid helium

laser and deflectors

channel

50 − 1000 wafers per memory system

Fig. 2. Storage organization of a memory system based on persistent spectral holeburning. Each arm has one wafer with $10^3 \times 10^6 \times$ (16–64) bits = 16–64 gigabits = 2–8 gigabytes. Memory system has (2–8) × (50–100) gigabytes = 100–8000 gigabytes in a volume approximately 8 ft × 8 ft × 4 ft (2.4 m × 2.4 m × 1.2 m).

beam-splitting optics. The points on the chip can be randomly accessed in about 2 ms by a light-beam deflector consisting of a galvanometer-mounted mirror which can be activated to tilt in the *x*-*y* plane. This gives random access that is an order of magnitude faster than can be achieved with a mechanically rotating disk as in conventional optical storage. The limitation that the mirror mounted on an *x*-*y* galvanometer deflector can address only 10^6 separate spots is not a problem here because 10^3 frequency bits can be read or written rapidly at each spot. This is done by ramping the frequency of the laser, for example, by current-tuning a semiconductor laser through the inhomogeneous absorption profile and recording the presence or absence of spectral holes. As noted above, this can be done in parallel for many chips at a time.

Device parameters. A desirable set of parameters for such a device will now be given. The material should exhibit a holewidth (or $2\Gamma_h$) of approximately 100 MHz, an inhomogeneous absorption band approximately 100 GHz or more in width, and a hole lifetime of years. Read and write times per bit of about 30 nanoseconds are required, and a random-access time to 10^9 bits of 2 ms. One arm of the system would be driven by a continuous-wave, red or near-infrared gallium aluminum arsenide (GaAlAs) laser with an output power of tens of milliwatts and would contain approximately 5 gigabytes (40 gigabits) of information. For example, a system with 200 arms would contain a total of 1000 gigabytes. These parameters have not all been demonstrated simultaneously in prototypes. In the above description the frequency domain has been added to two spatial dimensions. Such units could be stacked or, perhaps ultimately, integrated into a third dimension. In another configuration the frequency domain could be added to a three-dimensional holographic storage by recording a series of holograms at slightly different frequencies by using persistent spectral holeburning.

Photon-gated materials. One of the main impediments to the implementation of frequency-domain storage is the lack of a storage material satisfying all of the constraints enumerated above. A further difficulty is that a high reading-beam flux is required in order to get a readout with an acceptably high signal-to-noise ratio (approximately 30 dB) with a 20-MHz bandwidth. If the photobleaching process is a single-photon one, that is, linear in the laser power, then bleaching also occurs during reading, and this limits the number of reads that are possible before it becomes necessary to refresh the medium. Recently, significant progress has been made on this problem by the discovery of photon-gated materials. In these the writing step occurs only in the presence of a second enabling light beam that produces bleaching in conjunction with the narrow-band laser. With the enabling or gating light off, no bleaching occurs. Hence the reading step can be

Materials that exhibit persistent spectral holeburning

Type of material	Representative examples
Organic molecules in crystals	Free-base porphin and phthalocyanine in n-alkanes
	Pentacene in benzoic acid
Organic molecules in glasses and polymers	Tetracene in ethanol glass, dyes such as resorufin and cresyl-violet in ethanol and polymethylmethacrylate
	Carbazole in boric acid glass*
Color centers produced by radiation damage	Centers in alkali halides
	Centers in diamond
Inorganic ions in crystals	Divalent samarium in barium fluorochloride* and calcium fluoride*
	Divalent cobalt in lithium gallium spinel*
Inorganic ions in glasses	Trivalent rare-earth ions such as neodymium and praseodymium in silicate and phosphate glasses

* Exhibits photon-gated holeburning.

performed many times without degradation of the signal. All of the materials known to exhibit this effect (see **table**) use two-step photoionization for the gating process, but other forms of gating using external electric fields, for example, can be envisaged. The table lists the most important groups of materials that have been shown to exhibit persistent spectral holeburning, and representative examples of each.

The use of the frequency domain for information storage is still a long way off. There is tremendous scope for innovation in the solution of outstanding problems, many of which are still unknown. The characteristics of frequency-domain storage invite, for example, new approaches to computer architecture, and the related use of the frequency domain for signal processing is still in its infancy.

For background information SEE COLOR CENTERS; COMPUTER STORAGE TECHNOLOGY; INTEGRATED OPTICS; LINEWIDTH; NONLINEAR OPTICAL DEVICES; NONLINEAR OPTICS; OPTICAL BISTABILITY; OPTICAL COMMUNICATIONS; OPTICAL INFORMATION SYSTEMS; SUPERCOMPUTER in the McGraw-Hill Encyclopedia of Science and Technology.

R. M. Macfarlane

Bibliography. H. M. Gibbs, S. L. McCall, and T. N. C. Venkatesan, Differential gain and bistability using a sodium-filled Fabry-Perot interferometer, *Phys. Rev. Lett.*, 36:1135–1138, 1976; A. A. Gorokhovskii, R. K. Kaarli, and L. A. Rebane, Holeburning in the contour of a pure electronic line in a Shpol'skii system, *JETP Lett.*, 20:216–218, 1974; A. Huang, Architectural considerations involved in the design of an optical digital computer, *Proc. IEEE*, 72:780–786, 1984; B. Kharlamov, R. I. Personov, and L. A. Bykovskaya, Stable gap in absorption spectra of solid solutions of organic molecules by laser irradiation, *Opt. Commun.*, 12:191–193, 1974; W. H. Lee et al., Two color photon-gated spectral holeburning in organic material, *Chem. Phys. Lett.*, 118:611–616, 1985; W. Lenth et al., High-density frequency domain optical recording, *SPIE Proc.*, 695:216–223, 1986; D. A. B. Miller and S. D. Smith, Two beam optical amplification and bistability in InSb, *Opt. Commun.*, 31:101–104, 1979; W. E. Moerner, Molecular electronics for frequency domain optical storage: Persistent spectral holeburning—A review, *J. Mol. Electr.* 1:55–71, 1985; S. D. Smith, The demonstration of restoring optical logic, *Nature*, 325:27–31, 1987; S. D. Smith et al., Room temperature, visible wavelength optical bistability in ZnSe interference filters, *Opt. Commun.*, 51:357–362, 1984; A. Winnacker, R. M. Shelby, and R. M. Macfarlane, Photon-gated holeburning: A new mechanism using two-step photoionization, *Opt. Lett.*, 10:350–352, 1985.

Organic chemistry

During the evolution of organic chemistry, perhaps the most significant advance involved the recognition that the orbital symmetry characteristics of the reactants and products controlled the relative reactivities and the stereochemistry of concerted reactions. In these reactions, there occur simultaneous making and breaking of bonds. This discovery was made during the course of work by R. B. Woodward and colleagues at Harvard University in the late 1950s. Their total synthesis of vitamin B involved the chemical transformations shown in reaction scheme (1), where Me represents the methyl group (CH_3). The thermal ring closure of structure (I) gave exclusively (II), while the thermal ring

(1)

closure of (III) gave only (IV). Structures (II) and (IV) differ in absolute configuration only at C_6 (asterisk). The photochemical ring-opening process, which was originally thought to be the reverse of the thermal ring-closure process, gave instead the isomer having the opposite stereochemistry about the double bond. The difference in these reactions is in the direction of the rotation of the terminal carbon atom of the double bond and its substituents in the ring-closure and -opening reactions. There was an interesting contrast in the direction of rotation in these reactions compared with that in ring-opening reaction (2) involving a

(2)

(V)

cyclobutene (V); here rotation occurred in a sense opposite to that in (II) and (IV). (The terminology used to describe the directions of rotation are conrotatory when both ends of the π-electron system rotate in the same direction, and disrotatory when they rotate in opposite directions.) Thus it was demonstrated that the stereochemical course of a reaction is controlled by the symmetry properties of the molecular orbitals of the reactant and product. Woodward and R. Hoffmann developed the concept of the conservation of orbital symmetry, which states that a reaction can occur readily when there is congruence between the orbital symmetry characteristics of reactants and products; that is, orbital symmetry is conserved in concerted reactions.

Symmetry. As all chemical reactions are controlled by interactions between the molecular orbitals of the reactants, it is necessary to know what the symmetry characteristics of the molecular orbitals are. The orbitals of a molecule must contain the same symmetry elements as the molecule itself. The symmetry elements are (1) the identity element (E) possessed by all objects; (2) axis of rotation (C_n); (3) plane of symmetry (σ); (4) improper or alternating axis of rotation (S_n); and (5) center of inversion (i). These are illustrated in (VI) by using the molecule ethylene (C_2H_4) which possesses three twofold axes of rotation indicated as C_2's which on rotation 180° produces an identical orientation of the molecule in space. Two vertical planes of symmetry indicated as σ's exist perpendicular to the plane of the molecule as well as one horizontal plane in the plane of the molecule. There is also a center of inversion, i, at the center of the carbon-carbon double bond. (Ethylene also possesses an alternating axis of symmetry which is equivalent to one of the other symmetry elements.)

(VIa)

(VIb)

(VIc)

A number of reactions that can be used to illustrate the effect of symmetry on reactivity and stereochemistry involve mostly π-electron systems; in such cases the principles of symmetry apply only to the π molecular orbitals. In ethylene the π bond is formed by the interaction of the p atomic orbitals on the two carbon atoms. For ethylene molecular orbitals the two (always equal to the number of constituent p atomic orbitals in a system) can be constructed according to the method of the linear combination of atomic orbitals (LCAO), Eq. (3), where ψ_i is the

$$\psi_i = \pm c_1\phi_1 \pm c_2\phi_2 \qquad (3)$$

ith molecular orbital, and c_1 and c_2 are coefficients indicating the contributions of the atomic orbitals ϕ_1 and ϕ_2 to the ith ψ. For ethylene the lowest-energy molecular orbital is structure (VII) and is the π-bonding molecular orbital; the second molecular orbital is represented in (VIII),

(VII)

(VIII)

and is the π* molecular orbital. Although this is a relatively trivial system, it is important to note the π molecular orbital is symmetric with respect to the vertical axis of rotation and the vertical plane, while the π* molecular orbital is antisymmetric (the mathematical signs change) with respect to same axis and plane. In 1,3-butadiene (IX), which contains one axis of rotation and one vertical and horizontal plane, it is somewhat more complicated. Four π molecular orbitals are possible. Note that the placement of the nodes (regions where a change in mathematical sign takes place) are such that the molecular orbitals are either completely symmetric or antisymmetric about the various symmetry elements. (In linear systems the number of nodes increases monotonically in the higher-energy molecular orbitals.) In 1,3-butadiene (IX), ψ_1 and ψ_2 contain

(IXa) ψ_2

(IXb) ψ_4

(IXc) ψ_1

(IXd) ψ_3

the four π electrons, while ψ_3 and ψ_4 are vacant. The highest occupied molecular orbital (ψ_2) is termed the HOMO, and the lowest-lying unoccupied molecular orbital (ψ_3) is termed the LUMO.

Conservation of orbital symmetry. The application of conservation of orbital symmetry requires consideration of the symmetry of all of the molecular orbitals involved in a reaction.

Electrocyclic reactions. In reaction (2) of the cyclobutene (V) the σ- and π-bonding and antibonding molecular orbitals must be considered, while in the butadiene the four π molecular orbitals shown in (IX) must be considered. A symmetry element must be chosen that is common to both the cyclobutene and butadiene along the reaction coordinate; for the disrotatory motion, shown in the upper portion of notation (4), the symmetry element is a plane of symmetry. In the cyclobutene the π^* and σ^* molecular orbitals are antisymmetric. The four electrons involved in the ring-opening process occupy the σ and π molecular orbitals. In the butadiene, ψ_1 and ψ_3 are symmetric with respect to the plane of symmetry, and ψ_2 and ψ_4 are antisymmetric. The four electrons in the butadiene occupy ψ_1 and ψ_2. During the ring-opening process the two symmetric, doubly occupied molecular orbitals of the cyclobutene must be transformed into doubly occupied symmetric orbitals of the butadiene. For this to happen, the butadiene will be formed having ψ_1 and ψ_3 doubly occupied—a doubly excited electronic state of the butadiene. The formation of the butadiene in the doubly excited state will require the input of a great deal of energy, far more than is available by thermal means. This process is thus said to be thermally forbidden.

In the conrotatory process, shown in the lower portion of notation (4), the symmetry element present along the reaction coordinate will be a twofold rotation axis. In this case the σ bond of the cyclobutene will be symmetric, but the π bond will be antisymmetric. In the butadiene, ψ_1 will be antisymmetric and ψ_2 will be symmetric. In this process the doubly occupied symmetric and antisymmetric molecular orbitals of the cyclobutene are transformed into those of the butadiene which is in the ground electronic state. In

this process there is no requirement for the input of a large amount of thermal energy for the activation of the process, and the reaction is said to be thermally allowed. (This terminology simply implies that a symmetry-allowed process will occur more easily than a forbidden process. There is no absolute forbiddenness.)

$$
\begin{array}{ll}
\sigma^*A \ \text{——} & \text{—— } A \ \psi_4 \\
\pi^*A & S \ \psi_3 \\
\pi \ S \ \underline{\perp\!\!\perp} & \underline{\perp\!\!\perp} \ A \ \psi_2 \\
\sigma \ S \ \underline{\perp\!\!\perp} & \underline{\perp\!\!\perp} \ S \ \psi_1
\end{array}
\qquad (4)
$$

$$
\begin{array}{ll}
\sigma^*A \ \text{——} & \text{—— } A \ \psi_4 \\
\pi^*S \ \text{——} & \text{—— } S \ \psi_3 \\
\pi \ A \ \underline{\perp\!\!\perp} & \underline{\perp\!\!\perp} \ A \ \psi_2 \\
\sigma \ S \ \underline{\perp\!\!\perp} & \underline{\perp\!\!\perp} \ S \ \psi_1
\end{array}
$$

In more complicated systems the application of such symmetry correlations can be rather complicated. However, a simpler rule can be formulated: Transformations are symmetry-allowed when the σ bonds can be broken in such a way that the resulting p orbitals will have the symmetry of the HOMO of the product (photochemical processes have opposite symmetry requirements).

In the ring-opening of structure (V) the cleavage of the σ bond in a conrotatory manner will result in an antisymmetric reflection in the vertical plane, while a disrotatory motion will result in a symmetric reflection. As the HOMO of 1,3-butadiene is antisymmetric with respect to reflection in the vertical plane, the conrotatory motion is symmetry-allowed, while the disrotatory motion is symmetry-forbidden, as in notation (5). In

$$
\xleftarrow{\text{disrotatory}} \qquad \xrightarrow{\text{conrotatory}} \qquad (5)
$$

symmetric   antisymmetric

the reactions of structures (II) and (IV), disrotatory breaking of the σ bond results in a symmetric reflection which corresponds to the symmetry of the HOMO of a 1,3,5-triene. (The molecular orbitals of a triene can be constructed by using the principles outlined above for the construction of the molecular orbitals of 1,3-butadiene.)

Cycloadditions. Long before the discovery of the principles of conservation of orbital symmetry, it was known that 1,3-dienes and alkenes undergo facile cycloaddition reactions to generate cyclohexenes as in reaction (6). However, the cycload-

$$
(6)
$$

dition of one alkene with another alkene [reaction (7)] was not a facile reaction. This lack of reac-

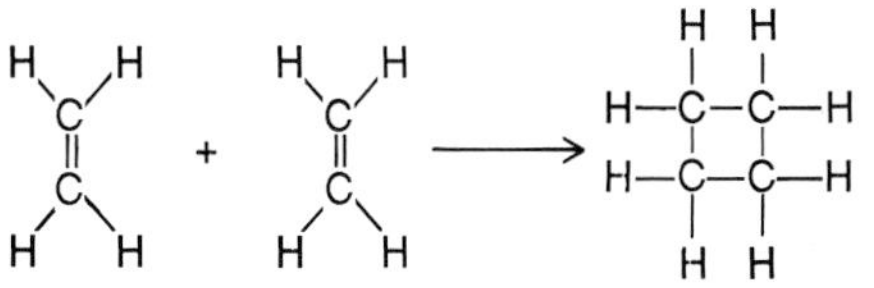

(7)

tivity is not due to the strained nature of the four-membered ring in that the cycloaddition of two alkenes is an exothermic reaction; that is, the lack of reactivity is kinetic in origin and not thermodynamic. The development of the concept of conservation of orbital symmetry provided an immediate explanation of these observations. Cleavage of the two σ bonds in cyclobutane, represented as the linear combination of two σ bonds, will produce two molecules of ethylene, one having the π molecular orbital occupied (the HOMO) and the other having the π* (the LUMO) occupied [reaction (8)]. Occupancy of the

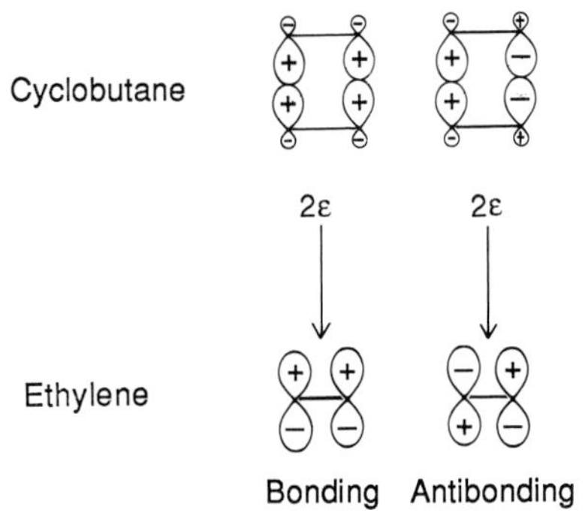

(8)

LUMO makes the cycloreversion and, its reverse, the cycloaddition reaction, thermally forbidden. The cleavage of the two σ bonds in cyclohexene to form 1,3-butadiene and ethylene [reaction (9)] produces the antisymmetric HOMO

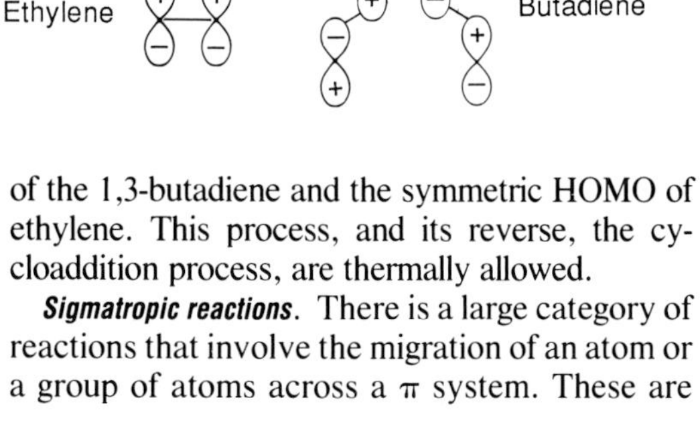

(9)

of the 1,3-butadiene and the symmetric HOMO of ethylene. This process, and its reverse, the cycloaddition process, are thermally allowed.

Sigmatropic reactions. There is a large category of reactions that involve the migration of an atom or a group of atoms across a π system. These are

known as sigmatropic reactions; two examples are reactions (10) and (11). The relative reactivity

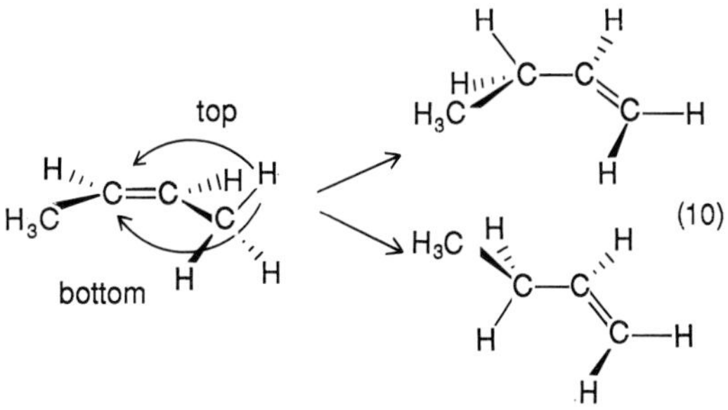

(10)

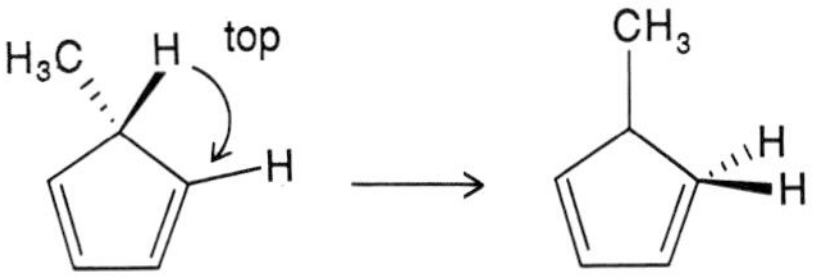

(11)

and the stereochemistry of these reactions are also governed by the principles of orbital symmetry. The application of the conservation of orbital symmetry to these reactions is, however, much more complicated; the results of such considerations can be illustrated by using two very simple reactions. The first reaction (10) illustrates the migration of an hydrogen atom across a double bond. Orbital symmetry considerations show that the migration across the top of the molecule is symmetry-forbidden. Migration to the underside of the molecule is symmetry-allowed; however, this process is geometrically impossible. As a result, simple alkenes do not undergo this type of rearrangement. This is indeed fortunate, for otherwise the chemistry of alkenes would be tremendously complicated and their use rather limited. In the second reaction (11), migration of the hydrogen across the top of the molecule is symmetry-allowed, and to the bottom, symmetry-forbidden. In this case the symmetry-allowed process is also geometrically feasible. This sometimes complicates the chemistry of a diene, particularly when the molecule is constrained in the proper conformation for the reaction such as that illustrated in reaction (11).

Although the sigmatropic rearrangements were illustrated by using the migration of an hydrogen atom, many other groups can participate in such rearrangements. Some of these rearrangements occur with great facility, and organic chemists have used their imagination to devise molecules capable of undergoing very rapid rearrangements. Such an example is a molecule known as bullvalene, in which the degenerate rearrangement [reaction (12)] occurs very fast at room

(12)

temperature; this gives rise to approximately 10^6 structures identical with the original but in which the bonding relationships between the carbon atoms is different.

Relative reactivity and stereochemistry. The role of orbital symmetry in controlling the relative reactivity and stereochemistry in concerted organic reactions has been discussed thus far in terms of rather simple organic molecules, except for structures (I) and (III). In most cases the reactant molecules contain other functions which may exert a significant effect on the relative reactivity of these reactions which has to do with the finer details of the molecular orbital energies and atomic orbital coefficients. Modern theory has provided many insights into the nature of these substituent effects, further increasing the utility of the theory.

The ability to understand the factors which control the relative reactivity and stereochemical features of reactions is of great value in the rational design of procedures for the synthesis of complicated organic molecules. It makes possible the intelligent development of strategies for synthesizing such molecules without having to rely on guesses.

For background information SEE CHEMICAL BONDING; ELECTRON CONFIGURATION; MOLECULAR ORBITAL THEORY; ORGANIC CHEMISTRY; ORGANIC REACTION MECHANISM; STEREOCHEMISTRY; WOODWARD-HOFFMAN RULE in the McGraw-Hill Encyclopedia of Science and Technology.

Daniel J. Pasto

Bibliography. T. L. Gilchrist and R. C. Storr, *Organic Reactions and Orbital Symmetry*, 1979; R. B. Woodward, *Aromaticity*, Chem. Soc. Spec. Publ. (Lond.) 21, pp. 217–249, 1967; R. B. Woodward and R. Hoffman, *Angew. Chem. Int. Ed. Engl.*, 8:781, 1969.

Organic evolution

The genus *Drosophila* has developed most remarkably in the Hawaiian Islands, and its evolution there is one of the world's most striking examples of rapid and adaptive radiation. Approximately one-fourth of the described species of this genus of fruit flies is endemic to the Hawaiian Archipelago. Not only have numerous species developed in a limited area, but they exhibit the greatest diversity of form and behavior known for any group of *Drosophila*. A great many structural peculiarities are also found that apparenty do not occur in other drosophilid faunas. The total fauna is conservatively estimated to consist of at least 800 species, leaving an additional 150 species of *Drosophila* remaining to be described in Hawaii. In spite of their morphological, anatomical, ecological, and ethological diversity, the Hawaiian *Drosophila* species constitute a closely related evolutionary group.

Early studies indicated that Hawaiian *Drosophila* followed an evolutionary pattern in which the species were formed on different islands. Almost all of the species (98%) are restricted to single islands, with the most closely related forms often on adjacent islands. For example, each species of the *crassifemur* complex is endemic to one of the four major islands of the Hawaiian Archipelago. From a phylogenetic study of the giant chromosomes of larvae from the complex, it appears that the four species in the complex contain a linear biogeographic sequence of accumulated inversions which parallel the age of the islands (**Fig. 1**). This, in turn, suggests that the evolutionary sequence begins with the primitive species found on Kauai, the oldest island (formed about 5 million years ago), and progresses to the most derived species on Hawaii, the youngest island (less than 1 million years old). Concomitant with speciation were changes in the amount and distribution of heterochromatin which can best be identified in metaphase chromosomes.

Chromosome evolution. The study of karyotypes of Hawaiian *Drosophila* has been carried out by several workers. Fortunately, powerful tools are now available for studying both the chromosomal and the molecular evolution of this organism in natural populations. For example, in the nuclei of salivary gland cells of larvae, there are giant polytene chromosomes which are exceptionally well suited for examining the structural changes. The *Drosophila* genome is made up of one short and five long elements. The principal type of chromosomal variation among these Hawaiian species is the paracentric inversion, which is relatively easy to detect (**Fig. 2**). Furthermore, most of these inversions have become fixed in the populations of the one or more species in which they are found. The phylogenetic relationships among many species have been established in great detail. Thus far, 326 inversions have been detected in the 165 species of Hawaiian *Drosophila* which have been studied. This means that, on the average, two inversions per species have been fixed during its evolution. SEE CHROMOSOME.

In early studies of Hawaiian *Drosophila* a chromocenterlike configuration was found in polytene nuclei of several species in addition to the true chromocenter. Only specific sites of certain chromosomes join to give this characteristic configuration, which was designated the pseudochromocenter. These sites are apparently composed of intercalary heterochromatin (chromosomal segments of heterochromatin that are localized between the centromere and terminal in position) and represent multiple points where nonhomologous association (unusual pairing between chromosomes which contain dissimilar genes) occurs. Based on several lines of cytogenetic evidence, a model involving the pseudo-

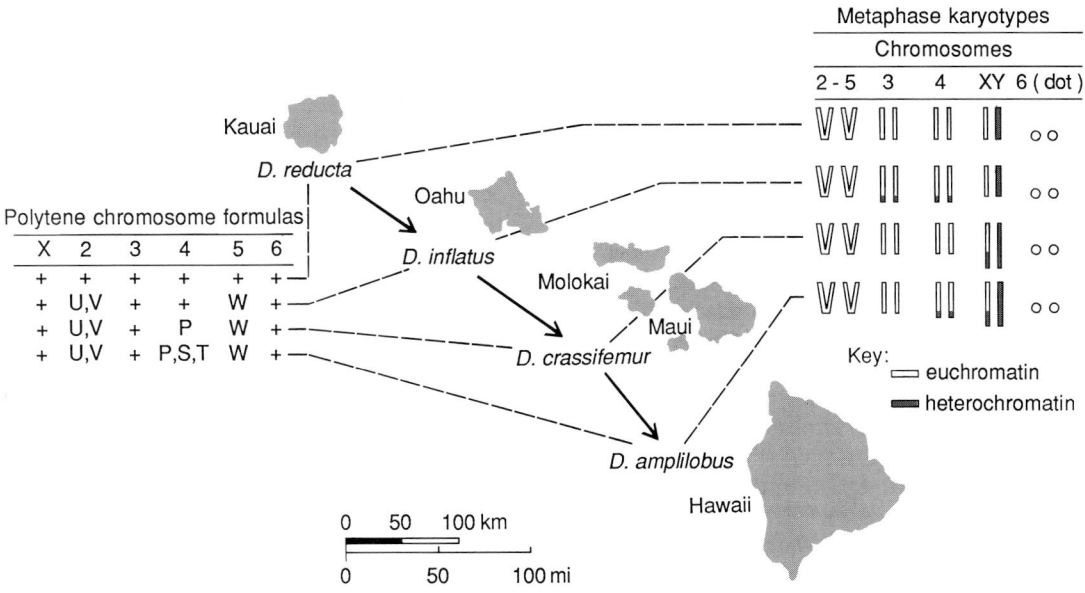

Fig. 1. Phylogenetic relationships of the *Drosophila crassifemur* complex and the presumptive colonization steps among the Hawaiian Islands. In the polytene chromosome formulas, + indicates the standard gene order; the letters P, S, T, U, V, and W indicate the inversions acquired after each colonization.

chromocenter in the production of chromosome rearrangements has been proposed.

The wealth of cytological comparisons possible among the many relatively closely related species of Hawaiian *Drosophila*, and the radical changes in heterochromatin between these species, have been exploited in various ways. It has been suggested that centromeric heterochromatin may play a significant role in the evolution of the karyotype of both plants and animals, being associated with changes in chromosomal morphology through breakage of chromosomes and reunion of their segments. Early studies investigated the phylogenetic changes of heterochromatic sites in polytene chromosomes of Hawaiian *Drosophila*, as well as the total heterochromatin associated with individual metaphase chromosomes. The distribution of heterochromatin also appears to be related to speciation. Several phy-

logenies in the Hawaiian *Drosophila* exhibit an accumulation of heterochromatin on the so-called dot (micro-) chromosome, thereby converting it to a rod. These large accumulations of heterochromatin on microchromosomes have been observed only in species which appear to have been recently derived from, but which show no differences in gene sequence from, their nearest relatives. More detailed analysis of the metaphase and polytene chromosomes of 165 species (445 strains) of Hawaiian *Drosophila* has yielded the following facts and interpretations: (1) There are 14 species with large accumulations of heterochromatin on the microchromosome which convert it into a large acrocentric chromosome. (2) Each of these 14 species is a member of a different set of species. Species within each set have identical polytene chromosome karyotypes. (Such species are said to be homosequential.) (3) Each of these 14 sets of homosequential species is situated at what appears to be a terminal position of a phylogeny and, therefore, may represent the most recently derived species of that lineage. This provides a means of determining the direction of evolution in several instances.

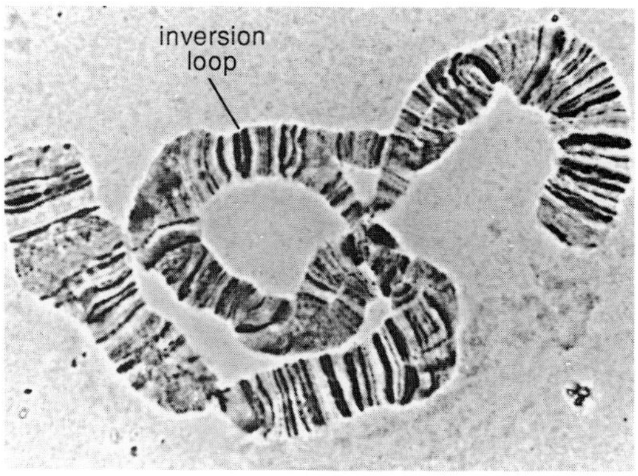

The facts also suggest that localized accumulations of heterochromatin are correlated with the absence of chromosome rearrangements. In other words, concentrations of heterochromatin in localized regions may signal that there is a reduced potential for rearrangements to occur elsewhere in the genome. This suggests that until the intercalary heterochromatin sites form, the accumulation of inversions may be delayed, contributing to the accumulation of homosequential species at the phylogenetic termini. In addition to the effect on structural changes, the location or

Fig. 2. Inversion loop in a polytene chromosome of a fly heterozygous for inversion.

amount of heterochromatin may serve as a factor in differentiating newly formed species.

Chromosome changes and speciation. Many evolutionary geneticists have suggested that evolution at the organismic level (adaptive evolution) depends chiefly on mutations in genes that regulate other genes, and that gene arrangement on the separate chromosomes determines how they function as regulators. Therefore, rearranging the order of genes on a chromosome can be an important mechanism for achieving altered patterns of gene regulation. Regulatory mutations influence the expression of genes that encode proteins without affecting the structure of proteins. Therefore, it has been proposed that speciation of some animal groups may proceed primarily by chromosomal rearrangements, and that shifts in the location of chromosomal segments create a new arrangement of genes which can become established in only a few generations if it provides some selective advantage. In addition, it is possible that changes in gene regulation are associated with the karyotypic changes of heterochromatin discussed above. The chromosome rearrangements not only would alter the regulation of gene expression presumably by position effects (including those of heterochromatin), but also would reduce the reproductive viability of hybrid offspring by interfering with the meiotic pairing process. Thus, the rearranged chromosomes would both create a new species difference and provide the mechanism for maintaining it.

New populations are often started by small numbers of individuals who carry only a fraction of the genetic variability of the parental population and hence may be underrepresentative. If chance operates in the selection of such founder individuals, the new populations will differ from the parental population and from each other. This founder principle (or founder effect) is of potential importance in the origin of species through such founder events. The evidence from chromosome analyses and other available data suggests that much of the explosive speciation of Hawaiian *Drosophila* has been due to the transportation of founder individuals from one island to another. For example, the fact that two species from different islands, with common gene arrangements, are chromosomally polymorphic for certain inversions implies that the two species have an ancestral-descendant relationship or sibling relationship, and thus provides support for the hypothesis of a single founder event.

Recent studies in the Hawaiian Drosophilidae further support, both quantitatively and qualitatively, the hypothesis that chromosomal rearrangements play an important role in the process of speciation. In *Drosophila*, because of the existence of polytene chromosomes, karyotypic changes can be studied in greater detail than is possible with vertebrates. Quantitative studies of speciation rates and chromosomal evolution in

mammals and in *Drosophila* have been carried out which contribute to a better understanding of the mechanism for evolutionary changes. Furthermore, these findings suggest chromosome rearrangement mechanisms involving intercalary heterochromatin and the formation of the pseudochromocenter, whereby rapid genome changes take place. These changes include both alterations of polytene chromosome banding (gene) sequences and metaphase modifications.

The rates of both chromosomal and molecular evolution in Hawaiian *Drosophila* have recently been estimated. It appears that karyotypic evolution and morphological evolution have been more rapid and their effect more profound than evolution which has occurred through substitutions of deoxyribonucleic acid (DNA) bases. It may be concluded, therefore, that chromosome (gene) rearrangements and the resulting effects on gene regulation may account for the most profound differences between species.

For background information *SEE CHROMOSOME; CHROMOSOME ABERRATION; GENE ACTION; MUTATION; ORGANIC EVOLUTION; SPECIATION* in the McGraw-Hill Encyclopedia of Science and Technology.

Jong Sik Yoon

Bibliography. M. Ashburner, J. N. Thompson, Jr., and H. L. Carson (eds.), *The Genetics and Biology of Drosophila*, vol. 3, 1982; L. V. Giddings, K. Y. Kaneshiro, and W. W. Anderson (eds.), *Genetics, Speciation, and the Founder Principles*, 1986; J. S. Yoon et al., Evolution in Hawaiian Drosophilidae: Chromosomal phylogeny of the *Drosophila crassifemur* complex, *Evolution*, 29:249–256, 1975; J. S. Yoon and R. H. Richardson, A mechanism of chromosomal rearrangements: Role of heterochromatin and ectopic joining, *Genetics*, 88:305–316, 1978.

Peritonitis

When as a result of various kidney diseases the kidneys reach their end stage (end-stage renal disease) the patient dies within a short period. However, at present patients with end-stage kidney disease can be maintained alive either with a kidney transplant or with various types of dialysis. Of the latter there are two main types, hemodialysis or continuous ambulatory peritoneal dialysis. Both forms of dialysis have advantages and disadvantages. Patients managed on continuous ambulatory peritoneal dialysis are prone to develop peritonitis.

In hemodialysis, blood circulates outside the body and is cleansed through a filter. In peritoneal dialysis, waste products are removed through the body's natural membrane, the peritoneum. A dialysis solution is infused into the peritoneal cavity through a permanent catheter.

Chronic peritoneal dialysis can be intermittent or continuous. In the intermittent mode, an appropriate volume of dialysate (the liquid which will come in contact with the membrane) is infused into the peritoneal cavity and is kept there for 20–30 min before being drained out. The patient receives 40 h of dialysis per week on the average, usually divided into two 20-h sessions.

In continuous ambulatory peritoneal dialysis the dialysate is infused into the peritoneal cavity and allowed to remain in place for 6–8 h before it is drained and replaced with fresh solution. This type of dialysis has many social and medical advantages over the intermittent treatment; however, these must be balanced against the risk of peritonitis, which remains one of the method's most serious complications. The incidence of peritonitis in patients on intermittent peritoneal dialysis is low. When continuous ambulatory peritoneal dialysis was first introduced, the incidence of peritonitis averaged one episode every 10 to 12 weeks. The introduction of the Toronto Western Hospital technique for continuous ambulatory peritoneal dialysis has lowered the infection rate to one episode every 10 to 12 months. In this modification the solution is contained in a plastic bag rather than in a bottle, the connection tubing between the bag and catheter is used for 4 weeks, and the fluid drains back into the original load bag. Many innovations have been tried to further decrease the incidence of peritonitis, but significant decreases have not been achieved, with the exception of the Y-shaped tubing set (described below).

In surgical peritonitis, that is, peritonitis that develops after abdominal operations, the peritoneal cavity is contaminated with large numbers of enteric bacteria and enteric contents; in contrast, in most episodes of continuous ambulatory dialysis peritonitis, the peritoneal cavity is invaded by small numbers of bacteria, usually skin organisms such as *Staphylococcus epidermidis* or *S. aureus*.

Routes of bacterial invasion. Because organisms from routine cultures of the skin and those isolated from peritonitis show a similar distribution, it appears that most infecting organisms are from the endogenous flora. **Table 1** shows the various routes of infection, the organisms usually responsible, and a speculative frequency of distribution, which is an estimate of the frequency of each route.

Intraluminal infections. Occurring within the catheter lumen, intraluminal infections stem from contamination which occurs during the connection and disconnection procedures. Most of the technological innovations in continuous ambulatory peritoneal dialysis are designed to reduce the rate of such contamination.

Periluminal infections. These occur around the catheter and are usually secondary to infection at the site where the catheter exits the skin. Despite improvements in catheter technology, the junction between catheter and skin is never sealed

Table 1. Routes of infection in patients on peritoneal dialysis

Route	Organisms	Frequency of distribution, %
Intraluminal	Coagulase-negative staphyloccus *Staphyloccus aureus* *Acinetobacter* *Pseudomonas*	30–40
Periluminal	Coagulase-negative staphylococcus *Staphyloccus aureus* Fungi *Pseudomonas* sp. Diphtheroids	20–30
Transmural	Enteric organisms Anaerobic organisms	25–30
Hematogenous	*Streptococcus* sp. *Mycobacterium tuberculosis*	5–10
Ascending (vaginal)	*Candida* Enteric organisms	2–5

and is often susceptible to infection. It has been suggested that a subcutaneous Dacron cuff on the catheter might facilitate prevention of infection. One controlled study showed that the cuff had no effect in reducing the incidence of infections at the skin exit site, whereas uncontrolled retrospective reviews suggested that the cuff might perform this important role. Recently, a group investigating the problem indicated that downward exit of the catheters may be significant in the prevention of infections at the skin site.

Transmural infections. When more than one organism or an anaerobe is isolated from the peritoneal effluent of patients with peritonitis, fecal contamination is indicated. Such contamination usually is the result of diverticulitis, especially in patients with diverticulosis; however, bacteria can also penetrate the bowel wall during tissue damage or during episodes of severe diarrhea. Fecal peritonitis may develop after colonoscopy; prophylactic antibiotics should be administered during such a procedure.

Hematogenous infections. *Streptococcus viridans* peritonitis and tuberculous peritonitis are the only two types that seem to follow bacteremia.

Ascending infections. It is rare but possible for peritonitis to develop following contamination from the vagina and through the fallopian tubes. However, despite the theoretically free communication between the vagina and peritoneal cavity, women do not have a higher incidence of peritonitis than men.

Peritoneal defenses. Recently, it has been demonstrated that small numbers of bacteria frequently invade the peritoneal cavity during continuous ambulatory peritoneal dialysis but do not lead to peritonitis because they are removed by endogenous peritoneal defenses, either humoral or cellular.

The main humoral defenses in the peritoneal fluid are immunoglobulins and complement. It has been suggested that patients with frequent episodes of *S. epidermidis* peritonitis have low levels of opsonins in the peritoneal cavity and that these patients may be helped by intraperitoneal gamma globulins.

Normally, bacteria are cleared from the peritoneum by macrophages, cells normally found in the peritoneal cavity. During peritonitis a large number of polymorphonuclear cells enter the cavity and begin to remove bacteria. Their function may be neutralized temporarily by an initial acid pH and high osmolality of the dialysis solution. However, once these values reach normal, usually within 1 or 2 h, the function of polymorphonuclear cells is restored.

Lymphocytes may appear in the peritoneal cavity during episodes of tuberculous peritonitis. Occasionally, some patients show large numbers of eosinophils (a type of white blood cells). The mechanism of this benign and self-limiting peritonitis is unknown.

Diagnosis. In diagnosing peritonitis in a patient on continuous ambulatory peritoneal dialysis, the physician seeks to establish the presence of at least two of the following conditions: abdominal pain (mainly rebound tenderness), cloudy fluid due to an increased number of polymorphonuclear cells (over 100 per cubic millimeter), and the presence of organisms identified either by a Gram stain or culture. Occasionally, the patient may have nausea and diarrhea.

Although visual inspection of cloudy fluid reveals an increase in the number of white blood cells, the number of cells should be counted in the laboratory so that the course of the infection can be assessed daily. The initial differential may demonstrate the presence of neutrophils (large granular leukocytes), eosinophils (eosinophilic peritonitis), or lymphocytes (in tuberculous peritonitis).

Gram stain may show organisms in 30–40% of these episodes. Similarly, because of the small numbers of organisms present in the fluid, only a small percentage of routine cultures are positive. Identification of the organisms requires large volumes of the peritoneal effluent after centrifugation, filtration, or a radiometric detection technique. **Table 2** shows the organisms most frequently found in peritonitis episodes among patients on continuous ambulatory peritoneal dialysis.

The term relapsing is applied to peritonitis which recurs with the same organism 2–4 weeks after the end of antibiotic treatment. It is caused by incomplete treatment, a contaminated catheter, or an abscess around the catheter. Sterile or aseptic peritonitis usually means that the culture techniques were ineffective. Cloudy effluent not due to peritonitis may be seen during episodes of diarrhea, in the presence of endotoxin, during

Table 2. Distribution of organisms responsible for peritonitis in patients undergoing continuous ambulatory peritoneal dialysis

Organisms	Frequency, %
Staphylococcus epidermidis	30–40
Staphylococcus aureus	10–20
Streptococcus sp.	10–15
Gram-negative organisms	15–20
Fungi	2–10
Culture-negative	0–30

fibrin formation, after intercourse, during pancreatitis/cholecystitis, during ovulation, or during menstruation. In the last case the fluid is usually bloody.

Treatment. It is imperative to begin treatment as soon as the clinical symptoms become apparent. Since the results of culture and sensitivity are not available for at least 24–48 h, the initial treatment usually consists of an antibiotic combination, which is effective against the organisms most frequently observed in peritonitis. The antibiotic of choice frequently is cephalothin, which kills *S. epidermidis*, most strains of *S. aureus*, and some gram-negative organisms. However, if it is judged necessary to cover all the gram-negative organisms, an aminoglycoside should be added, preferably tobramycin because of its lower ototoxicity. If the patient is allergic to cephalosporins, an excellent alternative is Vancomycin, which some medical centers use as the drug of first choice.

After the organism has been identified and antibiotic sensitivity determined, appropriate adjustments should be made. With an initial loading dose in the first exchange, adequate blood levels can be achieved within 3 h, making intravenous administration of antibiotics unnecessary. A maintenance dose is added to subsequent exchanges. A 2000-unit dose of heparin is added to each bag because of the tendency for fibrin-clot formation.

An initial lavage with three rapid exchanges reduces the pain. Subsequent exchanges can be prolonged for 3 to 6 h. This scheme may have an advantage over the continuous lavage, because it does not deplete the defenses of the peritoneal cavity and the cells are not exposed continuously to the acidic and hypertonic dialysis solutions.

The effectiveness of antibiotic treatment can be monitored by a white blood cell count of the effluent and by the presence or absence of bacteria in the cultures. In uncomplicated cases, the culture becomes sterile within 2 to 3 days. In these circumstances, treatment is continued for 7 days after the first negative culture.

If the peritonitis does not respond to appropriate antibiotics for 4 to 5 days, it may become necessary to consider catheter removal. The catheter should be removed in those patients with fungal, *Pseudomonas*, fecal, and tuberculous peritonitis, and also in the presence of severe exit-site infection. In all these cases, treatment is ineffective without catheter removal. A new catheter can be installed after 5 to 20 days, depending on the severity of the infection.

In patients with *S. aureus* peritonitis, it may be necessary to consider administration of cloxacillin intraperitoneally and Rifampicin by mouth or intravenously. Surgery may be required when the problem is fecal peritonitis.

Complications. Severe episodes of peritonitis, such as those caused by *S. aureus*, *Pseudomonas*, or fecal organisms may lead to the formation of adhesions and of abscesses, making continuation of continuous ambulatory peritoneal dialysis impossible. Occasionally, such patients develop sclerosing peritonitis, a condition characterized by marked thickening of the peritoneal membrane, which eventually encapsulates the bowel loops in the form of a cocoon (encapsulating peritonitis). This is usually a fatal complication. However, sclerosing peritonitis has been associated mainly with the use of dialysates containing acetate or with the use of the antiseptic chlorhexidin.

The mortality rate of individuals with peritonitis has been reported as high as 3 to 4%. However, many deaths during peritonitis are not attributable directly to the infection but may hve been triggered by the hospitalization and the added stress.

Prevention. A major contribution toward the prevention of peritonitis has been made by the development of devices which reduce contamination during connection of the catheter. It has been reported that the Y-shaped tubing set developed by U. Buoncristiani in Perugia, Italy, a component which is filled with a sodium hypochlorite solution between exchanges, has reduced peritonitis to one episode for every 35 to 40 patient months. However, antibiotic prophylaxis has not been effective in decreasing the incidence of peritonitis.

Patient selection. Most peritonitis episodes (50–60%) develop in a small percentage of patients (15–20%), and it is difficult to identify these high-risk patients in advance. Through analysis of patient characteristics and peritonitis episodes, several workers have reported identification of some predictive characteristics in certain patients such as noncompliance and apparent depression. Blacks, especially those older than 60, may be subject to a greater number of episodes. Elderly patients with extensive diverticulosis may be susceptible to the development of fecal peritonitis.

Future prospects. It seems that the reduction in the incidence of peritonitis has reached a plateau of one episode every 12 to 16 patient months. Further reduction may come from the develop-

ment of new catheters which will prevent skin exit-site infections and catheter colonization with bacteria, improved connection devices, better understanding of the patient's endogenous defenses, better selection criteria, improved diagnostic and therapeutic methods in the management of peritonitis, and some sort of chemical or drug prophylaxis.

For background information SEE PERITONEUM; PERITONITIS in the McGraw-Hill Encyclopedia of Science and Technology.

D. G. Oreopoulos

Bibliography. S. I. Vas, Microbiological aspects of chronic ambulatory peritoneal dialysis, *Kidney Int.*, 23:83–92, 1983; S. I. Vas, D. E. Low, and D. G. Oreopoulos, Peritonitis, in K. D. Nolph (ed.), *Peritoneal Dialysis*, pp. 344–365, 1981.

Pharmacognosy

Derivatives from plants are a major component of many drugs sold in the United States. Approximately 25% of all prescriptions in 1973 contained natural products such as emetine, digitoxin, quinine, atropine, vincristine, and reserpine. Many of these products have been developed as a result of the scientific study of remedies traditionally employed in various cultures. Despite these successes, research on drugs derived from folk medicine waned in the 1970s. Recently the interdisciplinary approaches of a new subfield, ethnopharmacology, and attempts to demonstrate the effectiveness of folk remedies in terms acceptable to Western medicine have invigorated this area of research. Ethnopharmacology may provide a variety of therapeutic agents in the future.

There are between 250,000 and 500,000 higher plants on the Earth. It would be a very expensive and wasteful process to screen plants randomly for therapeutic activity. However, plants which are used in folk medicine have already been screened. There is an increased probability that such plants will not be harmful if used properly and that some therapeutic activity will be present. An analysis completed at the National Cancer Institute's study of plants and other natural substances collected randomly and studied for antitumor activity is shown in the **table**; if antitumor screening had been guided by the knowledge of medicinal folklore and poisonous plants, the yield of active species would have been greatly increased.

One of the biggest success stories in cancer therapy resulted from the National Cancer Institute's study of periwinkle (*Catharanthus roseus*), a folk antidiabetic remedy. Vincristine and vinblastine obtained from this plant are two of the most useful clinical antitumor agents in current use, and account for the dramatic drop in mor-

Antitumor activity of various substances	
Types	% active*
Plants collected at random	10.4
Plants used against cancer	19.9
Anthelmintics	29.3
Fish poisons	38.6
Plants poisonous to humans	50.0
Arrow poisons	52.2

* Extracts show a significant inhibitory effect in experimental tumor systems.

tality from Hodgkin's disease. Other recent successes have been the introduction of tubocurarine, used as an adjunct in chest surgery, and reserpine, used in modern medicine for reducing hypertension. Curarine is derived from the South American arrow poison curare, and reserpine is derived from *Rauwolfia serpentina*, an Indian hypertensive.

Acceptance of folk medicine. There has been a problem in gaining acceptance by Western medicine of folk medicine as a source. In order for remedies to be accepted readily, they must be appropriate in terms of currently accepted theories concerning disease mechanisms and drug action. Even efficacious treatments for diseases have been rejected or ignored in the past because they did not fit contemporary theories. For example, in the late 1800s several reports showed that high doses of aspirin were very effective in relieving the pain, swelling, and stiffness of rheumatoid arthritis. However, at that time medicine had just accepted a theory that rheumatoid arthritis was an infectious disease. It did not make sense for aspirin, a fever and pain medicine, to have any real effect on a chronic infectious process. Thus aspirin was totally ignored as a treatment for arthritis until 1950. Today, high aspirin doses are the treatment of choice for this ailment.

A second factor which may impede acceptance exists when the mechanism or biochemical pathway of the folk remedy has not yet been discovered and thus is not available as a possible rationale. For example, the use of acupuncture in China for anesthesia and pain relief was rejected as invalid in the West until the discovery of endorphins, locally generated pain-killing proteins in the brain, and a demonstration that acupuncture can mobilize these substances.

Use of folk remedies has been found to be more sophisticated than previously thought. An example is the use of a hallucinogen called ayahuasca by South American Indians. One component of the drink is a plant which contains tryptamine derivatives. These are hallucinogenic but are not effective orally because a liver enzyme, monoamine oxidase, inactivates them. The Indians, however, prepare ayahuasca as a mixture containing a second hallucinogen, harmine, which is an inhibitor of monoamine oxidase and thus allows tryptamines to be effective. The preparation of ayahuasca is an example of synergism; that is, the effect of the mixture is greater than that of either of the separate components.

New approaches. Ethnopharmacology has developed a more rigorous methodology. An accurate botanical identification of the plant, which was often omitted in the past, is now included, and an anthropological investigation of precisely how and when the native people use the remedy is undertaken. There are variations in the potency of plants depending on the time and season of collection, as well as variations in the mode of preparation, dosage, and time of administration. These variations can dramatically change the effect of plant medicines. Additionally, synergistic effects may be detected, which might not be found if a single plant were tested in the laboratory. Other aspects of the methodology of ethnopharmacology include tests of physiological activity in animals, isolation and characterization of the chemical components responsible for the activity, and efforts to elucidate physiological pathways for the observed activity that are congruent with accepted theories of disease etiology.

The discovery of potent new substances in the human body has made it possible to provide biochemical explanations for the actions of folk medicines. Examples are the discovery of prostaglandins and of endorphins. Prostaglandins, which are present in many body tissues, mediate a large number of biological effects. Endorphins are the endogenous anesthetics. These remedies could not be rationalized in a scientifically acceptable manner before these discoveries were made.

Examples. Examples of folk medicine which show potential or have been adapted in modern medicine include feverfew, qinghaosu, jararaca venom, naturally occurring polysaccharides, and hernandulcin.

Feverfew. *Tanacetum parthenium*, commonly called feverfew, has been used in Europe since monastic times as a remedy for fever, arthritis migraine, and psoriasis. Until very recently, this herb was considered to be ineffective by physicians. However, experiments now show that extracts of feverfew prevent the synthesis of prostaglandins. This would explain why feverfew is effective in the treatment of fever and migraine, acting as a "medieval aspirin," which also functions through this pathway. It has been further shown that extracts of feverfew inhibit the release of damaging materials from white cells. This discovery supports the claims made for feverfew as being efficacious in the treatment of arthritis and psoriasis.

Qinghaosu. Malaria is still one of the great scourges of the world. Millions of cases occur yearly, and many, particularly in children, result in fatalities. The first widely accepted cure was the natural product cinchona bark, brought to

Europe by Jesuit missionaries in the sixteenth century. Quinine, the alkaloid isolated from this bark, became the main treatment for malaria until the 1930s, when synthetic antimalarials were developed. Unfortunately, a number of strains of the malaria parasite have developed resistance to these synthetic drugs. The herb *qinghaosu* (*Artemisia annua*) has been used in Chinese folk medicine as a treatment for fever and malaria for many centuries. In 1971 Chinese chemists isolated artemisinin from this plant and have successfully treated thousands of malarial patients, including those with strains of malaria which were resistant to chloroquine. Water-soluble derivatives of this compound show little toxicity and promise to be a totally new class of antimalarials. At present they are the subject of much research in the United States.

Jararaca venom. The hormone bradykinin has been identified as the most potent of the pain producers. It causes the release of histamine and attracts white blood cells to the site of an injury. One of the new approaches to the development of a new generation of analgesics is to find substances which oppose the action of bradykinin. This area of study, in the forefront of medical research, has close ties to Brazilian folk medicine. The jararaca (*Bothrops jararaca*) is a very poisonous Brazilian snake. The venom releases bradykinin from precursor molecules in the blood, and it also contains a small protein which prevents the normal destruction of this hormone. This protein, peptidyl dipeptase, is found normally in humans and assists in raising the blood pressure. Inhibitors of peptidyl dipeptase would therefore be expected to lower blood pressure. The orally active peptidyl dipeptase inhibitor, Captopril, has become a widely prescribed antihypertensive drug. The folk medicine antidote for the bite of the jararaca is the root of *Mandevilla velutina*. Tests on extracts of this plant on standard assays show that it inhibits the activity of bradykinin. Further research on the identity and structure of the active compound may lead to a new general type of analgesic.

Naturally occurring polysaccharides. The Aztecs were involved in constant warfare and developed effective wound treatments. A recent investigation showed that their remedy, the concentrated sap of the century plant (*Agave* sp.), was effective in destroying bacteria that usually infest wounds. The sap consists of concentrated polysaccharides (sugars) and kills the bacteria by dehydration due to osmotic pressure, which tries to equalize the concentration of sugars on both sides of the bacterial cell wall. A high concentration of saccharides in honey and the osmotic pressure they cause is also the reason for the effectiveness of honey as a wound treatment; it was used for this purpose in ancient Greece. This characteristic is also the basis for the recent use of granulated sugar in Argentina to pack infected

surgical incisions, and also in Paris to prevent infections after open heart surgery.

Hernandulcin. Aztec plants are also involved in another interesting problem. Artificial sweeteners, such as saccharin, have been found to be carcinogenic. There is a large potential market for an effective, inexpensive, and safe sugar substitute. An investigation of a sixteenth-century compilation of Aztec plants discovered that a plant called very sweet herb (*Lippia dulcis*) by the Aztecs contained a relatively simple molecule, hernandulcin, which was 1000 times sweeter than sucrose (ordinary table sugar). By comparison, saccharin is 550 times sweeter and aspartame is 200 times sweeter than sucrose. Hernandulcin has passed one of the common screening tests for carcinogenicity. This sweetener has great potential as the sugar substitute of the future.

For background information SEE *PHARMACOGNOSY* in the McGraw-Hill Encyclopedia of Science and Technology.

Bernard Ortiz de Montellano

Bibliography. J. R. Davidson and B. R. Ortiz de Montellano, The antibacterial properties of an Aztec wound remedy, *J. Ethnopharma.*, 8:149–161, 1983; S. Heptinstall et al., Extracts of feverfew inhibit granule secretion in blood platelets and polymorphonuclear leucocytes, *Lancet*, 1:1071–1074, 1985; D. L. Klayman, Qinghaosu (artemisinin): An antimalarial drug from China, *Science*, 228:1049–1055, 1985; R. W. Spjut and R. F. Perdue, Plant folklore: A tool for predicting sources of antitumor activity, *Cancer Treat. Rep.*, 60:979–985, 1976.

Photoreception

Sensory photoreceptors are widely distributed in nature, and range from single eyespots of unicellular microorganisms to the highly differentiated neural photoreceptor, the eye of vertebrates. In order to understand how biological systems transform light energy, extensive interdisciplinary studies are being conducted on rhodopsin, the visual pigment, and related pigments of animal or bacterial origin, exemplified by bacteriorhodopsin.

Photoreceptor pigments are transmembrane proteins (molecular weights of 28,000–40,000 daltons) to which is bound retinal, the chromophore derived from carotene. The structures of many photoreceptor molecules have been elucidated in recent years either by protein-sequencing methods or by genomic means, but no three-dimensional structure has as yet been fully elucidated. Moreover, understanding of crucial problems in photoreception is still very obscure on a molecular structural basis. These problems include changes in protein and chromophore structures initiated by absorption of light, subsequent transformations of structures, and the induced events leading to transduction, phototaxis, and translocation of protons or chloride ions across membranes. The two most thoroughly investigated pigments involved in sensory processes are rhodopsin, the pigment responsible for vision, and bacteriorhodopsin, the purple pigment that changes light energy into chemical energy. The studies practically encompass the entire range of science, including biology, chemistry, biochemistry, physics, biophysics, theoretical calculations, and psychophysics. Organic chemical investigations of photoreceptor molecules which are summarized below offer unique opportunities to explore these most fundamental biological processes.

Characteristically, the native pigment is bleached, or stripped of its natural chromophore by irradiation with light, and the resulting apoprotein, or opsin, is then recombined with an artificial chromophore to give a synthetic pigment analog. The advantage of this approach is that the structure of each retinal analog is tailored to yield a pigment analog which answers specific questions.

Rhodopsin. Most rhodopsin studies are carried out with the bovine rod pigment, bovine rhodopsin, because of its availability, but pigments from other sources are also used. However, in view of the homologies with human cone pigments (responsible for color vision) and other pigments, it is clear that they are all structurally similar.

In bovine rhodopsin, 11-*cis*-retinal (I; R = CHO) is bound to the terminal amino group of lysine-296 of the opsin by a protonated Schiff-base linkage as in structure (I; R = CH=

(I)   (II)

$\overset{+}{N}$H-opsin). Although bovine rhodopsin (I; R = CH=$\overset{+}{N}$H-opsin) absorbs at 500 nanometers the protonated Schiff base (I; R = CH=$\overset{+}{N}$H-*n*-butyl) formed from (I; R = CHO) and *n*-butylamine absorbs only at 440 nm, the difference between the two maxima clearly being due to some effect operating within the protein-binding cavity; in other naturally occurring pigments the absorption maxima vary from 440 to 640 nm. Study of a series of visual pigments regenerated from dihydroretinals (III; dotted

(III)

lines are saturated bonds) made it possible to

check the validity of theoretical models proposed earlier, and led to the external point charge model based on experimental data and theoretical calculations [structure (IV)]. According to this mod-

(IV)

el, negative charges reside near the protonated nitrogen and in the vicinity of carbon atoms 12 and 14, and the absorbing wavelengths of visual pigments are regulated in a subtle manner by such charged groups from amino acids. Recent findings with human pigments show that charged groups do exist near the chromophore.

Visual transduction is the process by which rods and cones convert light energy into a neural signal which is transmitted to the brain via the optic nerve. The process is initiated by the absorption of light by rhodopsin, which is converted in a few picoseconds into the primary photoproduct, bathorhodopsin (λ_{max}, 548 nm). According to resonance Raman spectroscopy and Fourier-transform-difference infrared spectroscopy, the chromophore in bathorhodopsin is the protonated Schiff base of a "distorted" *trans*-retinal, but details are unknown. Microcalorimetry shows that bathorhodopsin is 35 kilocalories higher than rhodopsin, and it is this high-energy distortion of the chromophore and protein which eventually leads to the expulsion of the chromophore as *trans*-retinal (II; R = CHO) and to a cascade of enzymatic reactions culminating in the hydrolysis of cyclic guanylic acid (GMP). This process is called bleaching because the initial orange-red color of the pigment is discolored into a pale yellow. Various spectroscopic data of the bleaching intermediates of rhodopsin have been measured, and it is now understood that the species directly involved in transduction is an intermediate called metarhodopsin II; however, the chemical basis of the transduction process is still very obscure.

Rhodopsin analogs have also proven to be useful in clarifying certain aspects of the primary event. For example, 11,12-dihydroretinal (V; R = CHO), which lacks the crucial double

(V)

bond responsible for bleaching, gave a pigment absorbing at 315 nm. As expected, this pigment was stable to light and could not be bleached, because the double bond between carbon-11 and carbon-12 is absent; that is, the pigment gave an inactive rhodopsin. Also, intraperitoneal injection of the tritium-labeled (T) retinal analog (V; R = CTO) into vitamin A–deficient rats resulted in incorporation of the retinal analog in the retina and loss of vision.

Another analog which conclusively showed that photochemical cis-to-trans isomerization of the retinal moiety is a prerequisite for visual transduction is structure (VI; R = CHO), in

(VI)

which the isomerization is blocked by a seven-membered ring. As demonstrated by the tritiated analog (VI; R = CTO), application of this analog to rats and bullfrogs also led to decreased vision resulting from its incorporation into the retina.

Chlamydomonas. *Chlamydomonas* is an aquatic organism of 10-micrometer diameter with a single eyespot and two flagella which guide the swimming cell toward or away from light. This unicellular eukaryote is estimated to have separated from the vertebrate evolutionary line at least a billion years ago. While *Chlamydomonas* is widely used for the study of basal bodies, circadian rhythms, flagellar architecture and function, and so forth, the nature of its photoreceptor was unknown. Retinal analog incorporation studies have shown that the photoreceptor of this unicellular eukaryote is homologous to bovine rhodopsin. Blind mutants of *Chlamydomonas*, which contained no eye-spot chromophore because of blockage of carotenoid biosynthesis, were incubated with various retinal analogs, including the dihydro series (III) and (I). This led to a restoration of phototaxis, with the action spectrum maxima similar to the absorption maxima of corresponding bovine rhodopsin analogs. These findings, together with recent results obtained by deoxyribonucleic acid (DNA) hybridization techniques, reveal that photosensory pigments in a wide variety of species are homologous and arise from a common precursor. *Chlamydomonas* promises to be an important addition to the various systems used in photoreceptor studies because of the availability of numerous mutants, and most importantly because of the economy in time, sample amount, and ease of experiments.

Bacteriorhodopsin and sensory rhodopsin. Whereas rhodopsin serves as a light detector for the sensory system, *Halobacterium halobium*, an ancient bacterium estimated to be 1.3 billion years old, uses the pigment bacteriorhodopsin

(bR) for photosynthesis. This prokaryotic cell lives in salt water, and under anaerobic conditions begins to develop patches of purple membrane; it contains as its sole protein source bR, which absorbs solar energy and converts it into chemical energy. The pigment bR, composed of 248 amino acids of known sequence, is another transmembrane protein consisting of seven α-helical rods spanning the lipid bilayer, but as with other rhodopsins the detailed tertiary structure yet has to be elucidated. Because bR can be readily cultured without contamination and is the simplest of the transmembrane proteins engaged in the important function of transmembrane proton pumping, it has been widely investigated since its discovery in the late 1960s.

The purple membrane has an orderly structure comprising three bR molecules, each of which contains one retinal molecule as its chromophore. In its so-called dark-adapted resting state, bR statistically consists of a 1:1 mixture of all-*trans*-retinal and 13-*cis*-retinal attached to lysine-216 via a protonated Schiff-base linkage as in structures (VII; R = CH=$\overset{+}{\text{N}}$H-opsin) and (VIII;

(VII)   (VIII)

R = CH=$\overset{+}{\text{N}}$H-opsin) [the ring structures are the same as in (I) and (II), but are depicted differently to reflect recent results]. Upon irradiation, the dark-adapted species (γ_{max}, 560 nm) converts to the light-adapted species (γ_{max}, 570 nm) containing structure (VIII; R = CH=$\overset{+}{\text{N}}$H-opsin) as its chromophore. The absorption maximum of bR is even more redshifted than in bovine rhodopsin and appears purple. As rhodopsin, the incorporation of various retinal analogs, including *trans*-dihydroretinals (IX; dotted lines = saturated

(IX)

bonds), yielded the corresponding bR analogs; their absorption maxima, coupled with further model studies, led to the external point charge model shown in structure (X), where an external

(X)

negative-positive ion pair resides close to the ring, in addition to the counter anion.

The light-adapted bR is only a transient species, and irradition immediately changes it to the primary photoproduct, K intermediate, which according to vibrational spectral data has a distorted 13-*cis*-retinal protonated Schiff-base chromophore. As in the case of rhodopsin, charge separation between the positively charged nitrogen of the Schiff base and its negative counter anion leads to energy storage, which induces a series of nonphotochemical transformations through intermediates L → M → O to reconvert the pigment to the original light-adapted bR. The net result of this photoinduced cycle is the vectorial transport of two protons from the cytoplasm to the external medium. Analogs of bR prepared from retinals (XI) and (XII) which pos-

(XI)

(XII)

sess, respectively, trans-locked and 13-cis-locked structures due to the five-membered ring moiety in the side chain were completely inactive in terms of their proton-translocating activity. This provides direct chemical proof that a 13-trans to 13-cis isomerization is necessary for proton pumping. Likewise, the series of dihydroretinals (IX) or more drastically modified retinals such as structures (XIII), (XIV), and (XV) were incorpo-

(XIII)

(XIV)

(XV)

rated into bR, and their physical and biochemical properties were compared with those of reconstituted native bR. Such studies will clarify the mode of chromophore-protein interaction within the binding site as well as the mechanism of proton pumping.

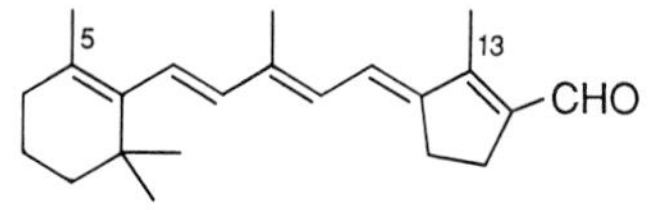

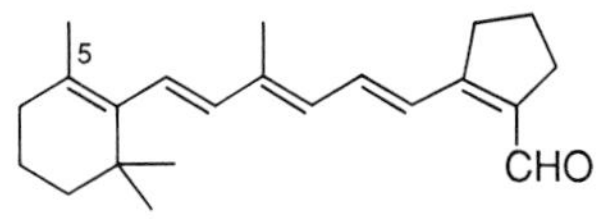

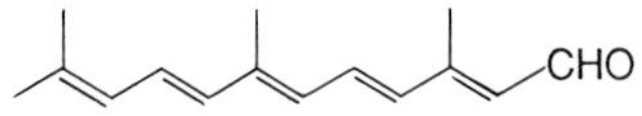

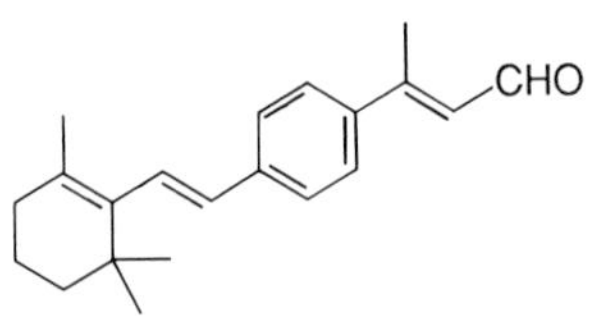

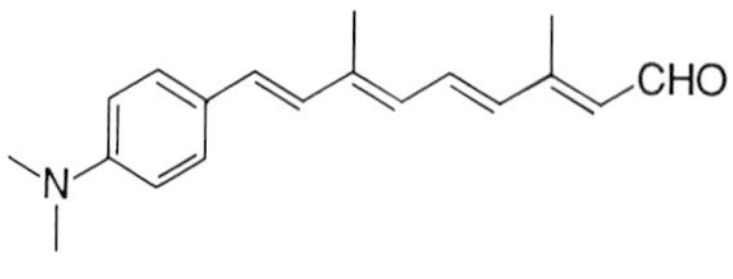

It is possible to incubate *H. halobium* in a medium containing isotopically labeled amino acids and thus to exchange all amino acids with the labeled species. Careful comparisons of Fourier-transform-difference infrared spectra between the various labeled and nonlabeled intermediates are clarifying the state of protonation of key amino acids such as aspartic acid, glutamic acid, and tyrosine in the various intermediates. It has also become possible to change specific amino acids by site-specific mutations and then to measure the proton-translocating ability. Such powerful techniques will play crucial roles in clarifying the mechanism of proton pumping, a central problem in biosciences.

Similar investigations are being pursued in the phototactic sensory rhodopsin and the chloride-pumping halorhodopsin, two additional pigments recently discovered in *H. halobium*. Although this ancient bacterium may have no direct bearing on humans, ongoing active and multidisciplinary studies are reminiscent of the studies with *Escherichia coli* and *Drosophila melanogaster*. See Color vision.

For background information See Amino acids; Color vision; Photoreception in the McGraw-Hill Encyclopedia of Science and Technology.

Fadila Derguini; Koji Nakanishi

Bibliography. F. Derguini et al., Studies with retinal pigments: Modified point charge model for bacteriorhodopsin and difference Fourier transform infrared studies, *Pure Appl. Chem.*, 58:719–724, 1986; F. Derguini and K. Nakanishi, Synthetic rhodopsin analogs, *Photobiophys. Photobiol.*, 3:259–282, 1987; K. W. Foster et al., A rhodopsin is the functional photoreceptor for phototaxis in the unicellular eukaryote *Chlamydomonas*, *Nature*, 311:756–759, 1984; H. Shichi, *Biochemistry of Vision*, 1983.

Photorespiration

Plant productivity can be greatly altered by environmental conditions during growth. The use of fertilizers to increase mineral nutrient availability, intensive plant cultivation management (such as pruning), and chemical pest control have all contributed substantially to increasing plant yield.

While in the past most of the increases in yield came from improved cultural practices and improved genotypes from breeding techniques, it is believed that little increase in plant productivity will be made by further use of these techniques. Therefore in recent years research efforts have been directed toward understanding the biochemical components that affect plant yield as well as external cultural effects. With the advent of biotechnology and genetic engineering, understanding and identifying the biochemical mechanisms of yield have become important factors for increasing plant productivity and producing superior plant genotypes. Two important processes of plant growth and determinants of yield are photosynthesis and photorespiration.

Net photosynthesis. Photosynthesis is the biochemical mechanism by which plants fix carbon dioxide from the atmosphere and produce sugars that are utilized for plant growth. Photosynthesis has been the focus of research on plant productivity, because the sugars made by photosynthesis support plant growth as well as fruit development (reproduction). Much of the research on photosynthesis relating to productivity has focused on quantitating and increasing photosynthetic rates.

Photosynthetic rate (as measured by carbon dioxide assimilation) is affected by many biochemical and environmental processes and is commonly reported as a net rate. Photosynthesis (or photosynthetic carbon reduction), photorespiration (or photosynthetic carbon oxidation), dark respiration, stomatal resistance to carbon dioxide (CO_2) diffusion, and transpiration interact in controlling the net carbon gain or loss in leaves (see **illus.**). In C_3 plants (plants that produce 3-phosphoglyceric acid as the first photosynthetic product), photorespiration is the largest component affecting net photosynthesis and can decrease by approximately 33–50% the net carbon fixed in normal atmospheres at room temperature (25°C or 77°F). Photorespiration resides in the "capacity" of the primary photosynthetic enzyme, ribulose 1,5-bisphosphate carboxylase, to act as an oxygenase or to use oxygen (O_2) when the concentration of O_2 increases (to atmospheric levels) or the CO_2 concentration de-

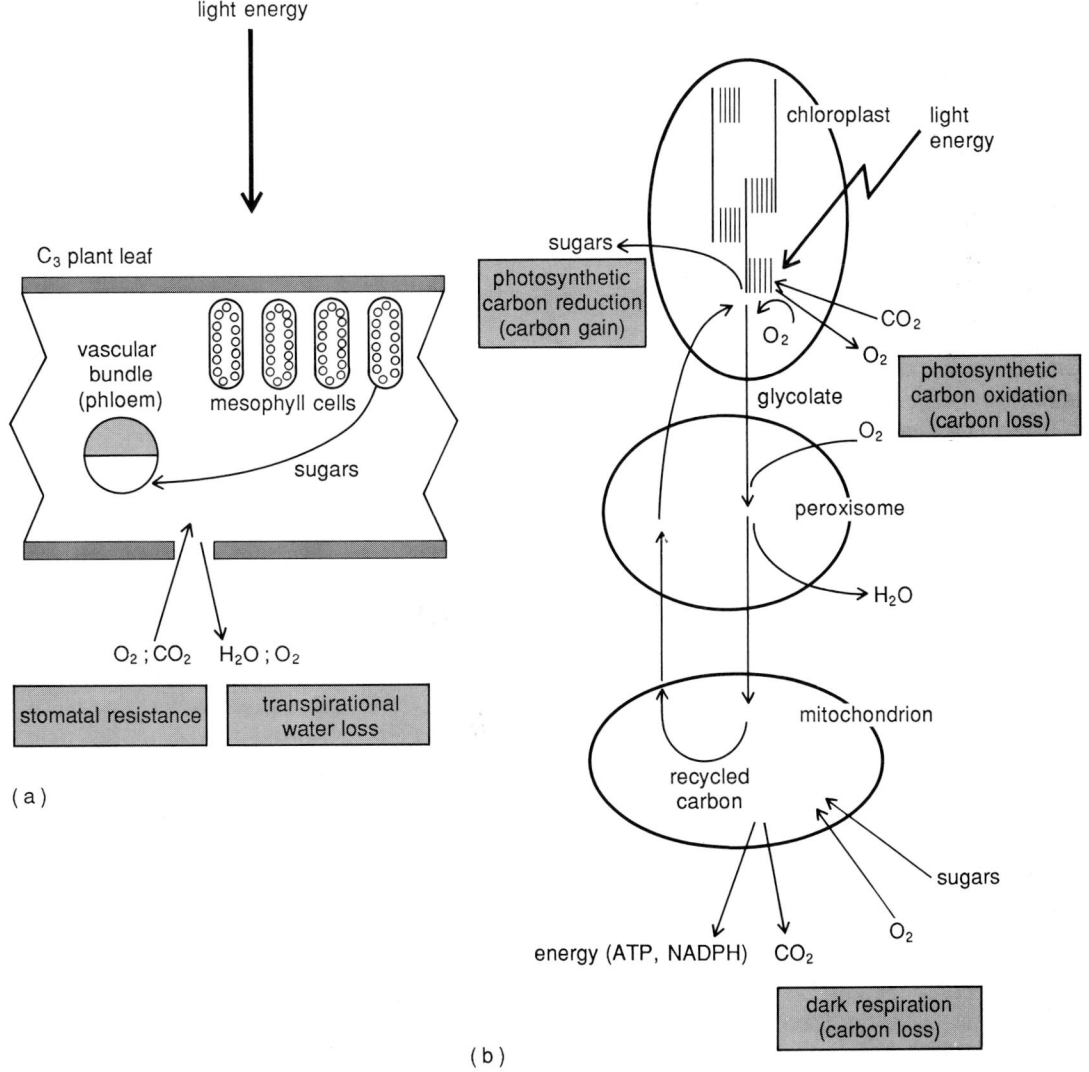

Interactions of photosynthesis (photosynthetic carbon reduction), photorespiration (photosynthetic carbon oxidation), dark respiration, stomatal resistance, and transpiration in (a) C_3 plant leaf and (b) mesophyll cell within the leaf.

creases (such as during water-stressed conditions when the stomata close).

As an oxygenase, ribulose 1,5-bisphosphate carboxylase utilizes O_2 and carbon to form phosphoglycolate. Phosphoglycolate is oxidized to glyoxalate and is subsequently converted to 1-phosphoglyceric acid in a complex series of reactions involving the mitochondria, peroxisomes, and chloroplasts (see illus.). 1-Phosphoglyceric acid can be utilized directly in the photosynthetic carbon reduction reactions. Thus, one of the functions of the photosynthetic carbon oxidation cycle seems to be to recycle up to 75% of the carbon lost due to photorespiratory processes.

In C_4 plant species (which produce 4-carbon organic acids such as malate and aspartate as the initial products of photosynthesis), photorespiration constitutes less than 10% of net photosynthesis. The greater efficiency of C_4 plants is partially due to the existence of a second carbon fixation enzyme, phosphoenolpyruvate carboxylase, that has a greater affinity than ribulose 1,5-bisphosphate carboxylase for CO_2; thus carbon fixation at lower CO_2 concentrations is possible. In these plants, CO_2 is fixed as malate or aspartate in one location in the leaf and diffuses to another location, where the organic acid is decarboxylated to release CO_2 near the site of the photosynthetic carbon reduction reactions. This process has the effect of increasing the concentration of CO_2 and thus reduces the net carbon loss due to the photosynthetic carbon oxidation reactions.

Plants exhibiting low photorespiration (C_4 plants) outcompete C_3 plants in certain environments (such as high light intensity and low water availability). This indicates that the lack of photorespiration may confer an ecological advantage for certain plants.

Dark respiration also decreases photosynthetic rates. Dark respiration, which occurs in the light as well as in the dark, typically decreases net photosynthetic rates by 10% in most plants and therefore has about one-third or one-fourth the impact of photorespiration on net photosynthesis in C_3 plants.

Stomatal resistance affects photosynthetic rates indirectly by regulating the passage of carbon dioxide from the atmosphere into the leaf. High stomatal resistance indicates closure of the stomata and, in turn, decreased internal CO_2 concentrations and photosynthetic rates. While low stomatal resistance is desirable for rapid diffusion of CO_2 from the atmosphere into the leaf, the concurrent diffusion of water (H_2O) out of the leaf via transpiration provides a regulating balance between CO_2 influx and H_2O efflux from the leaf.

Photosynthesis and fruiting. Many investigators have shown the existence of some control between photosynthesis and fruit growth in many plants, with photosynthetic promotion or reduction according to the presence or absence of fruit. Experiments on sweet pepper, pea, soybean, pecan, and eggplant showed significantly higher net photosynthesis in fruiting plants over nonfruiting plants. Strawberry has also recently been reported to have enhanced photosynthetic rates of 60–80% over that of deblossomed plants. Also, many investigators have examined the effect of photorespiration, dark respiration, and stomatal resistance on plant productivity. Experiments on soybeans and lucerne, utilizing partial shading or defoliation to alter the source (leaf) to sink (pod) ratio, showed that increased net photosynthesis in the remaining or nonshaded leaves was not due to a decrease in photorespiration rates. While there have been reports of decreased photorespiration in fruiting *Citrus* trees, experiments on apple trees (that exhibited increased photosynthetic rates during fruiting) did not show decreased photorespiration. Although most studies have weakened the possibility of a role for photorespiration in enhancing photosynthesis under fruiting conditions, the regulating processes that control photosynthetic enhancements have yet to be elucidated.

Leaf dark respiration rates have generally not been shown to be affected by fruiting in many plants, including apple, *Citrus*, and strawberry. Because of the relative magnitude of the effect of photorespiration on net photosynthesis compared with dark respiration, it is anticipated that any change in dark respiration rates would have minimal impact on net photosynthetic rates.

Stomatal resistance, like dark respiration, has been shown not to be directly affected by the presence of fruit. There are reported instances of decreased stomatal resistance during fruiting; however, these results have not been consistent even within the same plant species.

While photosynthesis determines the amount of sugars available for growth, the sink (fruit) is generally believed to have a role in the regulation of photosynthesis that is equivalent to or greater than the role of the source (leaf). Fruits, such as beans, are hypothesized to supply hormonal control over leaf photosynthesis by modulating the concentrations of such plant hormones as indoleacetic acid, kinetin, and gibberellic acid. Just as the presence of fruit has an impact on photosynthesis, so does the stage of fruit development. Many of the early studies on the effect of fruiting on photosynthesis were not comprehensive; often these studies consisted of a few measurements conducted once during the growing season. Subsequent studies on photosynthesis in fruiting versus nonfruiting plants throughout the growing season have indicated that enhancements of photosynthesis may occur only at specific stages of fruit development, particularly during flowering and also during rapid fruit growth. The impact of the sink on photosynthesis is related to the sink size (fruit size) as well as the activity (metabolism) of the sink. In addition, fruiting has been reported to delay leaf senescence in some plants, hence extending the photosynthetic capacity of the leaves.

Interacting with the dynamics of fruit growth is the process of expansion and development of new leaves, which also occurs during early fruit growth, especially in fruit trees where flowering and early leaf development coincide. Leaves commonly achieve their highest photosynthetic rate at full expansion, thus complicating the interaction of photosynthesis and fruit effects with leaf development. Another process that modulates the correlation between fruiting and photosynthetic rates is the storage of photosynthate in roots and stems in many perennials.

Role of photorespiration. Many studies on photorespiration and its effects on photosynthesis and plant yield have been conducted with the view that a reduction of photorespiration would correlate with an increase in net photosynthetic rates. While the role of photorespiration is still unclear, it has been hypothesized to serve the function of preventing chloroplasts from photoinhibition during conditions of high irradiance combined with stomatal closure (such as in water-stress conditions).

Photoinhibition is damage to the light-harvesting reaction centers of the photosynthetic appartus caused by excess light energy trapped by the chloroplast. The condition of low internal CO_2 levels in the leaf would induce photoinhibition if photorespiration did not provide a sink for the photosynthetic electron transport pathway. Photorespiration may thus serve an energy-dissipating function for the plant. This hypothesis is equivocal, but it does suggest a physiological purpose for photorespiration. While breeding for low photorespiration rates has thus far proved to be futile, boosting net photosynthetic rates by increasing atmospheric CO_2 concentrations has been shown to increase growth and yield in some C_3 crops. This result indicates that growth may be photosynthetically limited in some environments and that increasing net photosynthetic rates may increase assimilate (carbohydrate) levels and possibly plant productivity.

For background information SEE PHOTORESPIRATION; PHOTOSYNTHESIS; PLANT METABOLISM in the McGraw-Hill Encyclopedia of Science and Technology.

Jo Ann A. Fujii

Bibliography. M. R. Badger, Photosynthetic oxygen exchange, *Annu. Rev. Plant Physiol.*, 36:27–53, 1985; R. M. Gifford and L. T. Evans, Photosynthesis, carbon partitioning, and yield, *Annu. Rev. Plant Physiol.*, 32:485–509, 1981; H. J. Kende and J. A. D. Zeevart (eds.), *Partitioning of Assimilates*, American Society of Plant Physiologists, May 7–9, 1979; W. L. Ogren, Photorespiration: Pathways, regulation, and modification, *Annu. Rev. Plant Physiol.*, 35:415–442, 1984.

Plant metabolism

Major advances in understanding plant metabolism have been made recently. Research has provided insights into the regulation of sugar metabolism and the physiological significance of protein degradation in plants.

Plant sugar metabolism. This process has been studied for many years. It was believed that the mechanisms by which plants utilize sugar had been completely characterized. However, two discoveries have led to new investigations of plant sugar metabolism. The first was the discovery of the presence of an enzyme, inorganic (PP_i-PPK), which utilizes inorganic pyrophosphate (PP_i) as a phosphate donor and energy source. Previously, adenosinetriphosphate-dependent phosphofructokinase (ATP-PFK) was thought to be the only enzyme responsible for the conversion of fructose-6-phosphate (Fru-6-P) to fructose 1,6-bisphosphate (Fru 1,6-P_2).

The second discovery involved the role in plant metabolism of fructose 2,6-bisphosphate (Fru 2,6-P_2). This sugar bisphosphate was initially found to regulate the activity of ATP-PFK from animal tissues and also the enzyme fructose 1,6-bisphosphatase. Fru 2,6-P_2 has since been found to regulate PP_i-PFK activity in plants as well as other enzymes involved in the various pathways of sugar metabolism.

The process most affected by these discoveries are glycolysis and gluconeogenesis, the sequences of reactions that convert glucose to pyruvate and pyruvate to glucose, respectively.

Discovery of PPi-PPK. PP_i-PPK was discoverd in the leaves of pineapple plants and has since been found in all higher plants, seedless vascular plants, bryophytes, and algae that have been investigated. It has also been found in some bacteria and protozoa but not in animal tissues.

There exist several major differences between PP_i-PPK and ATP-PFK found in plant tissues, although their catalysis functions are similar, as shown in reactions (1) and (2). One difference

$$\text{Fru-6-P} + PP_i \rightleftharpoons \text{Fru 1,6-}P_2 + P_i \qquad (1)$$

$$\text{Fru-6-P} + \text{ATP} \longrightarrow \text{Fru 1,6-}P_2 + \text{ADP} \qquad (2)$$

concerns localization within the plant cell. PP_i-PFK is found only in the cytoplasm of the cell, whereas ATP-PFK is found in both the chloroplast and cytoplasm. Another difference involves the enzyme's function within the cell. PP_i-PFK catalyzes the interconversion of Fru-6-P to Fru 1,6-P_2, while ATP-PFK functions only in the conversion of Fru-6-P to Fru 1,6-P_2. This indicates that PP_i-PFK can function in both glycolysis and gluconeogenesis, while ATP-PFK has a role only in glycolysis. A third difference concerns the sensitivity of the two enzymes to Fru 2,6-P_2. While ATP-PFK from animals is stimulated by Fru 2,6-P_2, ATP-PFK from plant tissue sources is unaffected by this sugar bisphosphate and PP$_i$-PFK is stimulated.

Fru 2,6-P$_2$ in plants. At approximately the same time that PP$_i$-PFK was discovered in plants, Fru 2,6-P$_2$ was obtained from animal tissues as a low-molecular-weight fraction isolated from ATP-PFK by gel filtration. This sugar bisphosphate was found to activate ATP-PFK from animals as well as to inhibit the action of Fru 1,6-P$_2$ase.

Fru 2,6-P$_2$ was first isolated from plants by using mung bean seedlings. Upon anion exchange chromatography of a crude extract, two peaks of activity were found in the bisphosphate region. One peak was associated with Fru 1,6-P$_2$, while the other was identified as Fru 2,6-P$_2$ by its ability to stimulate PP$_i$-PFK. Since its discovery, Fru 2,6-P$_2$ has been found in all plant tissues examined, although the concentration of this bisphosphate varies widely.

The location of Fru 2,6-P$_2$ in plant cells has been found to coincide with that of PP$_i$-PFK; thus Fru 2,6-P$_2$ is found predominantly in the cytoplasm and rarely in the chloroplast.

Synthesis and degradation of Fru 2,6-P$_2$. If Fru 2,6-P$_2$ is to act as an effective regulator of plant sugar metabolism, there must be a way to alter the amount of bisphosphate present in the cell at any given time. It has been found that plants contain a fructose 6-P$_2$,2-kinase that synthesizes Fru 2,6-P$_2$ from Fru-6-P, and a Fru 2,6-P$_2$,2-phosphatase that degrades Fru 2,6-P$_2$, as shown in reactions (3). While these activities were thought to be the

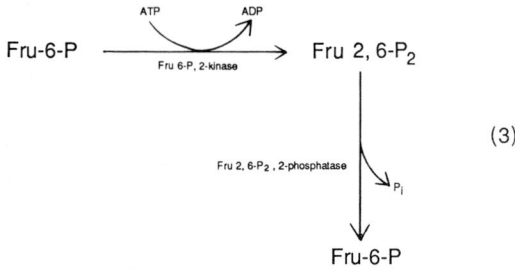

$$(3)$$

result of one enzyme with two active sites in animals, little is known about their structure or regulation in plants.

The synthesis and degradation of Fru 2,6-P$_2$ in animals is affected by glucagon and insulin, which are not operative in plants. Inorganic phosphate (P$_i$) and triose phosphates appear to be the only known substances that affect the activity of Fru-6-P$_2$,2-kinase and Fru 2,6-P$_2$,2-phosphatase in plants.

Factors affecting Fru 2,6-P$_2$ levels. At present, investigations are under way in a number of laboratories to ascertain what environmental signals and signal receptors are responsible for the modulation of Fru 2,6-P$_2$ levels in plant tissues. Recent studies have shown that, upon flooding, there is an increase in Fru 2,6-P$_2$ to a half-maximal level in a matter of minutes, reaching maximum levels in 5 min. This is followed by a gradual decrease to preflood levels over a period

of several hours. At first it was thought that the cause was ethylene production by the submerged plant, but later studies showed that the Fru 2,6-P$_2$ level increased much sooner than the start of ethylene production. Other environmental factors, such as light, day length, and temperature, are being investigated to see if they could have an effect on Fru 2,6-P$_2$ levels.

Modulation of PP$_i$ by Fru 2,6-P$_2$. PP$_i$-PFK has been found to exist in both a small and a large form. The small form consists of an α and a β subunit, while the large form is made up of 2 α and 2 β subunits. These two forms, large and small, are modulated by Fru 2,6-P$_2$.

The enzyme is found in the large form in the presence of Fru 2,6-P$_2$, while it dissociates to the smaller form in the absence of Fru 2,6-P$_2$. When the enzyme is in its large form, Fru 2,6-P$_2$ has little affect on its activity, while the small form is greatly affected by Fru 2,6-P$_2$. Also, the large form is more active in glycolysis, and the small form is active in gluconeogenesis, based on the ratio of the glycolytic-to-gluconeogeneic activity of the two forms of the enzyme.

This discovery was made during a study of PP$_i$-PFK in developing pea seedlings. When pea seeds were first imbibed, the PP$_i$-PFK was sensitive to Fru 2,6-P$_2$. After 3 to 5 days the enzyme was no longer sensitive to the bisphosphate; but several days later the enzyme was once again sensitive. Gel filtration has shown that the enzyme is in the small form initially, associates into the large form when Fru 2,6-P$_2$ insensitivity is exhibited, and once again assumes the small form when Fru 2,6-P$_2$ sensitivity recurs.

Other factors influencing enzyme form are PP$_i$ content, acid environment, and presence of denaturing compounds.

PP$_i$ in plant tissues. PP$_i$ is produced during the formation of polymers such as starch, glycogen, sucrose, cellulose, protein, deoxyribonucleic acid, and ribonucleic acid, and is hydrolyzed to P$_i$. The levels of PP$_i$ in plants have been measured enzymatically by using PP$_i$-PFK, the levels ranging from 5 to 39 nanomoles per gram of fresh tissue weight. This calculates to a concentration of 100 to 200 micromolar (μM) considering the cytoplasm to be 10% of the cell volume. This concentration is more than enough to serve as a substrate for PP$_i$-PFK activity.

Since there are enzymes known as pyrophosphatases to break down the pyrophosphate present in plant cells, it was thought that there may be a competition between them and PP$_i$-PFK for the available PP$_i$ in the cell. However, at pH 5.8 (the pH of the cytoplasm), pyrophosphatase activity is quite low. Thus competition between PP$_i$-PFK and pyrophosphatase is very unlikely, leaving more than enough PP$_i$ available to serve as a substrate for PP$_i$-PFK.

Other enzymes affected by Fru 2,6-P$_2$. While the main emphasis of research thus far has been on how Fru 2,6-P$_2$ affects PP$_i$-PFK, it has been demonstrated that other enzymes are also affected by this sugar bisphosphate, as shown in reaction scheme (4), where (+) indicates Fru

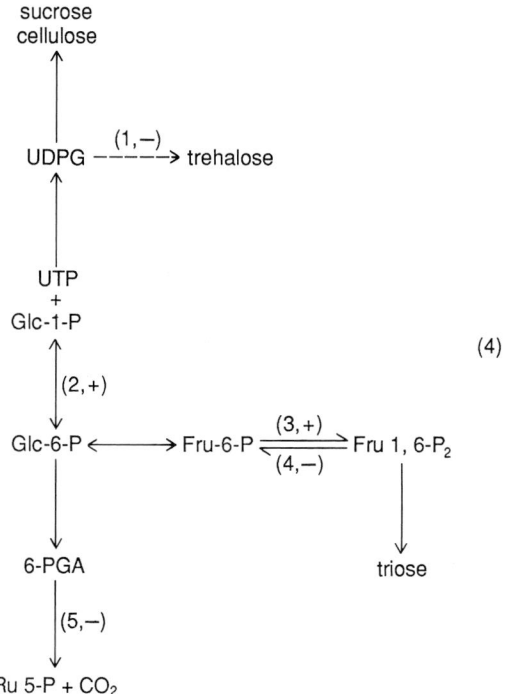

(1) = trehalose phosphorylase
(2) = phosphoglucomutase
(3) = PP$_i$-PFK
(4) = Fru 1,6-P$_2$ase
(5) = 6-phosphogluconate dehydrogenase

2,6-P$_2$ activation and (−) indicates inhibition. Studies have shown that 6-phosphogluconate dehydrogenase, trehalose phosphorylase, and Fru 1,6-P$_2$ase are inhibited by Fru 2,6-P$_2$. Phosphoglucomutase and UDP-glucose phosphorylase, however, have been found to be stimulated. The levels of Fru 2,6-P$_2$ necessary to affect the enzymes listed above vary by three orders of magnitude, suggesting that as levels of Fru 2,6-P$_2$ in tissues change, the different enzymes are affected. The regulation of sugar metabolism by Fru 2,6-P$_2$ is quite complex and, as yet, not totally understood.

Cynthia M. Galloway

Protein degradation. Although the concept of protein turnover was introduced years ago, only recently has protein degradation been accepted as an integral component in the regulation of plant metabolism. As a consequence, the specific mechanisms used by plants to degrade proteins have been poorly studied, and remain obscure. Recently a major advance in understanding how plants turn over protein has come from evidence that plants, like animals, degrade many intracellular proteins by using the highly conserved protein, ubiquitin.

Physiological significance. At first glance, the breakdown of functional proteins appears to be a wasteful event, but after a closer look its enormous physiological value becomes evident. For all types of cells, protein degradation is necessary for maintaining homeostasis and for controlling growth, morphogenesis, and adaptation to new environmental conditions. It provides this control by modulating (in conjunction with protein synthesis) the levels of important enzymes. Indeed, a survey of the list of enzymes that catalyze rate-limiting reactions for vital metabolic pathways demonstrates that almost all have very short half-lives, indicative of rapid turnover rates. Degradation is also important for the so-called general housekeeping of the cell. It accomplishes this by removing either functional proteins no longer needed at a particular developmental stage or abnormal proteins produced by denaturation or biosynthetic errors. If not removed, these substances would eventually clutter the internal milieu as a cell develops or ages. During periods of growth or starvation, turnover also provides a source of amino acids and energy by breaking down proteins that are expendable.

For plants in particular, protein catabolism is involved in several important aspects of growth and metabolism. During both seed germination and leaf senescence, the mobilization of amino acids to areas of growth and storage is accomplished by the highly coordinated catabolism of protein. In germinating seeds, for example, degradation involves synthesis of specific proteases that break down a special class of proteins stored in the protein body for this purpose. In both cases, degradation can be modulated by environmental conditions and several plant hormones. The most striking example is the ability of cytokinins to delay protein loss in senescing leaves.

Turnover of specific proteins is associated with the availability of certain minerals (for example, rapid degradation of nitrate reductase when plants are fed ammonia) and with morphogenic changes that occur during de-etiolation (greening; for example, rapid degradation of the morphogenic photoreceptor, phytochrome, upon transfer of dark-grown plants into the light). In the chloroplast, protein turnover is essential for maintenance of the photosynthetic machinery by removing proteins damaged by photooxidation. Also, proteolysis is involved in controlling the accumulation of subunits for chloroplast holoenzymes (enzyme-cofactor complex) that are encoded in the nucleus. For example, the concentration of the nuclear-encoded small subunit of ribulose bisphosphate carboxylase/oxygenase in the chloroplast is regulated in part by degradation of the unassembled protein when present in excess of the large subunit. This regulation was originally thought to reside in the coordinate regulation of gene transcription for both the chloroplast- and nuclear-encoded subunits.

Studies with bacterial, animal, and plant systems indicate that the mechanisms used to degrade proteins are highly complex, involve at least several pathways, and are remarkably selective. In fact, degradation is so selective that proteins in the same cell compartment can have turnover rates ranging from 10 min to 100 h. How the cell recognizes which proteins to degrade, and how fast, is unclear; however, specific structural domains on the proteins themselves are at least partially involved. Two observations which appear to be universal are that the complete disassembly of individual proteins is very rapid and that all degradation in plants requires energy, usually in the form of ATP. The energy requirement had not been predicted, since peptide bond hydrolysis is an energy-liberating reaction.

Ubiquitin. In the last few years a major advance in understanding how animal cells degrade protein was made with the identification of a specific pathway involving a small (76 amino acids), heat-stable protein called ubiquitin. The protein was first isolated accidentally by G. Goldstein and coworkers. Because it was detectable by radioimmunoassay in all the eukaryotes they examined, including several plant species, they called this protein ubiquitin. Amino acid sequence analyses subsequently revealed that ubiquitin is one of the most conserved proteins yet identified, being invariant in organisms as diverse as mammals (including humans), amphibians, insects, fishes, and birds. This widespread occurrence and unparalleled sequence conservation led to the conclusion that this protein must be important in cell physiology, even before its function was revealed.

The actual function of ubiquitin was determined in the course of research investigating how reticulocytes degrade protein. They found that the energy-dependent degradation system in this cell type could be chromatographically separated into two components; one of these was unusual because it was small and heat-stable but sensitive to proteases. Further characterizations revealed that this proteinaceous factor (originally designated APF-1) was actually ubiquitin and that in the presence of ATP it would form complexes with a variety of proteins prior to their catabolism. It is now known that the function of ubiquitin is to bind covalently to proteins destined for breakdown, forming an intermediate committed for degradation. Ligation occurs by an unusual peptide linkage between free alpha and epsilon amino groups on the target protein and the terminal glycine carboxyl group of ubiquitin. The series of reactions involved in conjugation is catalyzed by three enzymes, and requires an energy-rich thiol-ester intermediate of ubiquitin generated by ATP hydrolysis in which the carboxy terminus of ubiquitin is linked to specific sulfhydryl groups on the conjugating enzymes. Reaction scheme (5) shows a pathway for ubiquitin (UBQ) dependent proteolysis that has been proposed by A. Hershko and A. Ciechanover, where E1, E2, and E3 are the enzymes required

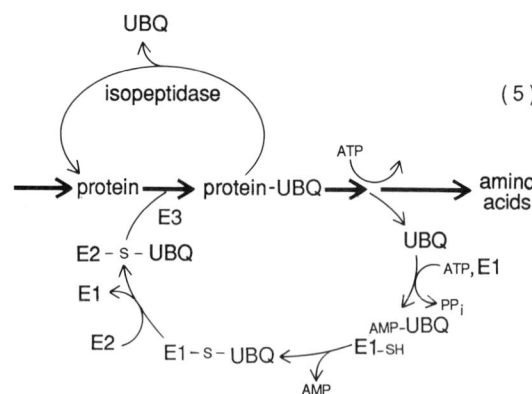

(5)

for conjugation. Once a protein is tagged with one or more molecules of ubiquitin, it is then broken down by a large multienzyme complex specific for ubiquitin conjugates. Only the target protein is degraded, and free, functional ubiquitin is released. As a result of this cycle, ubiquitin serves as a reusable recognition signal for selective proteolysis. While the reactions involved in conjugate formation and degradation are at least partially resolved, the key reactions that specify which proteins should serve as targets for ubiquitin conjugation, and hence govern the selectivity of turnover, are unknown.

The importance of the ubiquitin-dependent proteolytic pathway has become apparent recently. Studies with animal cells containing a temperature-sensitive defect in ubiquitin conjugation indicate that more than 90% of the short-lived proteins (half-lives less than 5 h) are degraded via ubiquitinated intermediates. Since most regulatory proteins have short half-lives, it follows that this pathway would be of importance in cell regulation. This pathway appears to be the major route for degrading abnormal proteins, indicating that it also fills the necessary role as the cell's "housekeeper."

Until recently, little was known about how plants degrade intracellular protein. Although a large number of plant proteases have been identified, the specific roles that these enzymes fill or how the enzymes are controlled remain a mystery. Based on studies suggesting that plants have a protein which cross-reacts with antibodies against animal ubiquitin, it appeared reasonable that plants have ubiquitin. Based on the strong amino acid sequence conservation and immunological crossreactivity of the protein, antibodies generated against human ubiquitin were used to probe for a similar protein in crude plant extracts subjected to polyacrylamide gel electrophoresis. With such a procedure, a protein was detected by using immunological methods in oat (*Avena sativa*) extracts that had a molecular mass identical to that of animal ubiquitin. Subsequent biochemical and immunological characterizations of the purified protein demonstrated that it was ubiquitin. In fact, amino acid sequence analysis revealed that the oat protein differs by only 3 residues out of 76 from the sequence determined for animal ubiquitin. The molecule was present in every tissue type examined, including green leaves, etiolated shoots, and dry seeds. In addition to ubiquitin, plants have the enzymes necessary for the formation and degradation of ubiquitin-protein conjugates. Similar to the pathway in animals, conjugation of ubiquitin to other proteins requires ATP and proceeds through a thiol-ester intermediate.

The function of ubiquitin in plants is not yet resolved. Nevertheless, based on the homologies observed thus far, it seems likely that ubiquitin conjugation serves the same functions in plants as it does in animals. When crude plant extracts were subjected to polyacrylamide gel electrophoresis and then probed with antibodies against ubiquitin, a substantial percentage of ubiquitin was found conjugated to a variety of plant proteins in the plant. Whether these conjugates represent intermediates awaiting degradation has not been confirmed, but studies with a specific protein, phytochrome, would indicate that they do. Phytochrome is a photoreversible chromoprotein that exists in two forms, a red light–absorbing form, Pr, and a far-red light–absorbing form, Pfr. The molecule is very stable as Pr (half-life greater than 100 h), but upon photoconversion from Pr to Pfr it is rapidly degraded (half-life 1 to 2 h). As recently observed, after this photoconversion Pfr is rapidly conjugated with ubiquitin, and these ubiquitinated forms are rapidly catabolized. Clearly, the connection between degradation of Pfr and ubiquitination suggests that phytochrome is degraded via the ubiquitin-dependent proteolytic pathway.

The discovery of ubiquitin and ubiquitin conjugation represents a major advance in understanding how plants degrade intracellular protein. As more information on this pathway is collected, it is likely that ubiquitin will attain a prominent status as an important regulator of plant metabolism. SEE UBIQUITIN.

For background information SEE CARBOHYDRATE METABOLISM; ELECTROPHORESIS; ENZYME; PLANT METABOLISM in the McGraw-Hill Encyclopedia of Science and Technology.

Richard D. Vierstra

Bibliography. C. C. Black et al., Fructose 2,6-bisphosphate roles in regulating plant sugar metabolism, in D. Randall, D. G. Blevins, and R. L. Larsen (eds.), *Current Topics in Plant Biochemistry and Physiology*, vol. 4, pp. 156–165, 1985; N. W. Carnal and C. C. Black, Pyrophosphate dependent 6-phosphofructokinase: A new glycolytic enzyme in pineapple leaves, *Biochem. Biophys. Res. Comm.*, 86:20–26, 1979; A. Hershko and A. Ciechanover, Mechanisms of intracellular protein breakdown, *Annu. Rev. Biochem.*, 51:335–364, 1982; D. Finley and A. Varshavsky, The ubiquitin system: Functions and mechanisms, *Trends Biochem. Sci.*, 117:343–347, 1985; J. Shanklin, M. Jabben, and R. D. Vierstra, Red light-induced formation of ubiquitin-phytochrome conjugates: Identification of possible intermediates of phytochrome degradation, *Proc. Nat. Acad. Sci. USA*, 84:359–363, 1987; D. A. Smyth and C. C. Black, Discovery of a new pathway of glycolysis in plants, *What's New Plant Physiol.*, 15:13–16, 1984; R. D. Vierstra, S. M. Haas, and A. L. Haas, Purification and initial characterization of ubiquitin from the higher plant *Avena sativa*, *J. Biol. Chem.*, 260:12015–12021, 1985; M.-X. Wu, D. A. Smyth, and C. C. Black, Regulation of pea seed pyrophosphate dependent phosphofructokinase: Evidence for the interconversion of two molecular forms as a glycolytic regulatory mechanism, *Proc. Nat. Acad. Sci. USA*, 81:5051–5055, 1984.

Plant mineral nutrition

Recent developments in analytical instrumentation and experimental methods have created the opportunity for plant nutritionists to make significant advances in their understanding of: (1) the functions of mineral nutrients in higher plants; (2) the processes controlling mineral nutrient uptake, translocation, deposition, and remobilization during plant development and maturation; (3) the processes controlling root growth within the soil; and (4) the identification of those factors which regulate or control the movement of mineral elements to the root surfaces, the interaction of these elements at the root surfaces, and their transport across root-cell membranes. Additionally, new elements may be added to the list of essential micronutrients for plant growth, and their metabolic functions may be determined through the application of several recently introduced experimental techniques.

A brief summary of several of these modern technologies is presented here, along with some of their potential applications in plant mineral nutrition research. The techniques discussed include nuclear magnetic resonance (NMR) spectroscopy, electron spin resonance (ESR) spectroscopy, and inductively coupled argon-plasma mass spectrometry.

Nuclear magnetic resonance spectrometry. This nondestructive technique is one of the most productive spectroscopic analytical methods available to chemists. NMR spectroscopy is possible because some atomic nuclei exhibit a property known as spin. All isotopes that have a nucleus with a magnetic dipole moment (or nonzero spin angular momentum) can be studied by using NMR methods. Usually, odd-atomic-number isotopes possess this property.

The availability of sensitive NMR spectrometers with high-field superconducting magnets and computerized routines for radio-frequency magnetic pulsing, Fourier transform, and data collection has made it possible for plant scientists to apply NMR spectrometry techniques to a number of difficult research problems. These include—all in the living plant—the quantitative measurement of mobile metabolites, the study of cellular compartmentation of metabolites, and the measurement of the rates of adenosinetriphosphate (ATP) synthesis and degradation in cells; as well as, in plant tissues and cells, the investigation of the metabolism of atoms and chemical bonds.

Recently, NMR spectroscopy has been used to study various aspects of plant metabolism and physiology. However, NMR sensitivity (signal-to-noise ratio) and resolution are still inadequate to allow the study of micronutrients and trace elements at concentrations that normally occur in plant tissues. The sensitivity of modern NMR spectrometers for different nuclei capable of producing NMR spectra in plant tissues varies, depending on a number of complex factors. An awareness of the sensitivity of NMR spectrometers for a specific type of nucleus can be obtained from NMR measurements on the stable phosphorus isotope, ^{31}P. ^{31}P-NMR spectra are readily acquired at plant tissue phosphorus concentrations of between 0.01 and 0.1 millimolar. Fortification of a plant's growth environment with highly enriched sources of stable isotopes can be used to improve NMR serviceability in plant nutrition research. Stable isotope enrichment can overcome the insensitivity problem created by low naturally occurring abundances of some nutritionally important stable isotopes. The **table** lists magnetic dipole nuclei of some stable isotopes that have or may have applications in plant mineral nutrition research employing NMR techniques.

NMR spectroscopy should become an increasingly important part of research directed at developing an understanding of various mineral nutrient functions in plant metabolism, ion transport processes in plant-cell membranes, and the role of certain mineral nutrients, for example, calcium, zinc, and boron, in the maintenance of cellular membrane integrity.

Nuclear magnetic resonance imaging. Computed tomography (CT) is a nondestructive, computer-assisted method used to reconstruct, from projection data, the cross-sectional distribution of some physical property of an experimental object. The method has found wide application in medicine and industry, for example, x-ray computed tomography scanning. Recently, NMR computed tomography (that is, NMR imaging) has become a very important clinical diagnostic tool in medicine.

Imaging is another emerging NMR technique that may find many applications in plant mineral nutrition. For example, NMR imaging has been used to measure the spatial distribution of radish (*Raphanus sativus*) and broad bean (*Vicia faba*) root systems and water in soil without removing the plants from the soil or disturbing the soil environment. The NMR signal from hydrogen can be related to water potential (that is, the difference between the chemical potential of water in the soil or in plant tissue and that of pure

Some useful* or potentially useful† stable isotopes for NMR spectrometric studies of soil-plant systems

Isotope	Spin	% Natural abundance	Sensitivity relative to hydrogen-1
^{1}H*	1/2	99.98	100.0
^{2}D†	1	0.016	1.5×103^{-4}
^{11}B†	3/2	80.42	13.3
^{13}C†	1/2	1.1	1.6×10^{-2}
^{14}N*	1	99.6	1.0×10^{-1}
^{15}N†	1/2	0.36	3.7×10^{-4}
^{19}F*	1/2	100.0	83.0
^{23}Na*	3/2	100.0	9.3
^{25}Mg†	5/2	10.13	2.7×10^{-2}
^{27}Al†	5/2	100.0	20.6
^{29}Si†	1/2	4.70	3.7×10^{-2}
^{31}Pa*	1/2	100.0	6.6
^{33}S†	3/2	0.76	1.7×10^{-3}
^{35}Cl*	3/2	75.4	3.5×10^{-1}
^{39}K*	3/2	9.1	4.6×10^{-2}
^{43}Ca†	7/2	0.145	9.3×10^{-4}
^{51}V†	7/2	99.76	38.1
^{55}Mn†	5/2	100.0	17.5
^{63}Cu†	3/2	69.09	6.4
^{65}Cu†	3/2	30.91	3.5
^{67}Zn†	5/2	4.11	1.2×10^{-2}
^{77}Se†	1/2	7.58	5.3×10^{-2}
^{113}Cd†	1/2	12.26	1.3×10^{-1}

* Isotopes have been recommended for biological studies.

† Isotopes may be useful if their enrichment is substantially increased in the experimental system or NMR instrumentation sensitivities are increased in the future.

Source: Data after D. G. Gadian, *Nuclear Magnetic Resonance and Its Applications to Living Systems*, Clarendon Press, 1982; and R. K. Harris, *NMR and the Periodic Table*, Academic Press, 1978.

free water). Therefore, in addition to water content of plants, it may be possible to quantify water potential spatially in plants via this technique. Theoretically, in addition to imaging the protons in water, it should also be possible to obtain images from phosphorus-containing metabolites, from sodium ions, and from any other NMR-active isotope in a soil-plant system, if the isotope is adequately enriched and the instrument has high enough sensitivity.

Unfortunately, the use of NMR imaging to solve plant nutrition problems is currently restricted to those plant scientists that have access to suitable instrumentation. Presently, this frequently requires the willingness of medical institutions (where many of the current instruments are housed) to cooperate with plant scientists who are interested in pursuing this type of research. The demand by the medical community for access to this technology is great, and so there is generally little time available to plant scientists to carry out their studies on these instruments. Thus, progress in plant nutrition research requiring NMR imaging instrumentation likely will be slow.

Electron spin resonance. Unpaired electrons in an atom or molecule placed in a magnetic field absorb electromagnetic radiation at resonance with the externally applied magnetic field. This is the basis for electron spin resonance spectroscopy. The method is applicable to many paramagnetic transition-metal ions which contain unpaired electrons in their d or f atomic orbital (for example, the Fe^{3+} cation). Unfortunately, most molecules in cell membranes are diamagnetic rather than paramagnetic and cannot produce ESR spectra. To circumvent this problem, paramagnetic probes (spin labels) can be incorporated into membranes at various depths within the membrane bilayer, depending on the chemical form of the probe used. The use of free-radical spin labels [usually derivatives of fatty acids, phospholipids, or sterols containing a nitroxide radical ($-N{\overset{\bullet}{-}}O$) in ESR investigations can yield information on protein structure, cytoplasmic water viscosity, and membrane structure and function. Spin labels have also been used to measure bioenergetic parameters in various cell systems (for example, cell volumes, pH, electrical gradients across membranes, one-electron oxidation-reduction gradients within cellular compartments, and electrical surface and boundary potentials at membrane surfaces). Membrane fluidity measurements can also be made by utilizing the ESR spectra generated from numerous types of spin labels incorporated into cell membranes.

ESR techniques have not yet been used extensively in plant mineral nutrition studies. However, such techniques could yield important information on the effects of mineral nutrient deficiencies and heavy-metal toxicities on meta-bolic processes and especially on membrane structure and function in living plant cells. Recently, ESR imaging of free radicals and nitroxide spin labels has been achieved in organs of living plants. This may open even more doors for the use of ESR techniques in plant mineral nutrition research.

Inductively coupled argon-plasma mass spectrometry. This technique merges the high ionization efficiency of inductively coupled argon plasma with the spectral simplicity and very high sensitivity of a mass spectrometer. Multielement analyses of most of the elements in the periodic table are possible by this technique at detection limits ranging from 0.1 to 10 parts per billion depending on the element. The method is usable over a wide concentration range of at least six orders of magnitude and is rapid (that is, 30 elements per minute can be analyzed when their concentrations in the sample exceed 10 times the detection limit). The technique can also be used to determine isotope ratios and isotopic abundances in samples with a precision of 0.5%.

The recent availability of inductively coupled argon-plasma mass spectrometers should revolutionize the use of stable isotopes in plant mineral nutrition research. Such instrumentation will make possible the routine use of stable isotopes for tracing the movement of mineral nutrients and of trace elements through the soil-plant-animal or -human food chain. The high sensitivity of this technique will allow for the analysis of elements in plant tissues at trace concentrations which were very difficult or impossible to measure with other spectrophotometric techniques (such as atomic absorption and emission spectrophotometry). Thus, plant nutritionists will be able to test for the presence of trace elements in plant tissues and nutrient media at levels which could not practically be measured in the past. These types of measurements could lead to the discovery of additional essential trace elements for higher plants.

For background information SEE COMPUTERIZED TOMOGRAPHY; ELECTRON PARAMAGNETIC RESONANCE (EPR) SPECTROSCOPY; FOOD WEB; MASS SPECTROMETRY; NUCLEAR MAGNETIC RESONANCE (NMR); PLANT MINERAL NUTRITION; PLANT-WATER RELATIONS; PLASMA CHROMATOGRAPHY; ROOT; SPIN LABEL in the McGraw-Hill Encyclopedia of Science and Technology.

Ross M. Welch

Bibliography. J. S. Cohen (ed.), *Magnetic Resonance in Biology*, vol. 2, 1983; D. G. Gadian, *Nuclear Magnetic Resonance and Its Application to Living Systems*, 1982; K. G. Heumann, Trace determination and isotopic analysis of the elements in life sciences by mass spectrometry, *Biomed. Mass Spectrom.*, 12:477–488, 1985; J. K. M. Roberts, Study of plant metabolism *in vivo* using NMR spectroscopy, *Annu. Rev. Plant Physiol.*, 35:375–386, 1984.

Plant movements

Geotropism, also known as gravitropism, is the movement of a plant with regard to the direction of gravity. Generally shoots are negatively geotropic, growing upward, while roots are positively geotropic, growing downward. Recently plant physiologists have intensified their efforts to decipher the mechanisms regulating the three phases of geotropism: perception of gravity, transduction of the stimulus to the site of response, and the growth response orienting the plant axis.

Whereas past research had focused on the roles of hormones in geotropism, calcium has now emerged as the central factor in regulation. While the ubiquity or variation of the nature of root and shoot response mechanisms is yet to be ascertained, it is useful to study the geotropism of roots.

Statolith theory. How plant cells detect the gravitational force and discern the relative vector is a mystery. The statolith model, common for many animal systems, involves heavy bodies which move and settle with respect to the gravitational vector, their displacement initiating an asymmetric response within the sensing cell, tissue, or organ. For plant roots responsive to gravity, the site of gravity perception appears to be the root cap, a mucilage-producing protective covering of cells over the meristem, the zone in which new cells of both the root and the cap are generated (**Fig. 1***a*). The root cap cells normally contain starch-filled plastids, or amyloplasts, which are pulled by gravity to the lower side of cells located in the center, or columella, of the root cap (Fig. 1*a*, *b*). Similar orientation of amyloplasts occurs in shoot tissues responsive to gravity. Many plant physiologists believe that these amyloplasts act as statoliths for the detection of gravity by exerting pressure on the nearby lower cell membrane. This hypothesis is controversial, however, because when starch is carefully removed from columella cells, rendering the amyloplasts "weightless," roots and shoots remain geotropic.

Recent research using a starchless single-gene mutant of *Arabidopsis* has demonstrated definitively that amyloplast movement is not required for normal sensing of gravity. Very small (less than 0.8 in. or 2 cm) *Arabidopsis* seedlings (roots and shoots) of the wild and mutant types show similar geotropic responses, provided that the plants are grown under continuous light to maintain the health of the mutants. To show that the starchless amyloplasts do not move in the mutant plants, the distribution of amyloplasts in columella cells was determined by electron microscopy of serial sections of root caps taken from roots oriented vertically for 1.5 h with the apex either up or down prior to preparation for electron microscopy (Fig. 1*b–d*). The thin, lightweight, starchless amyloplasts of the mutant retain a random distribution regardless of root orientation and thus cannot act as statoliths (Fig. 1*d*). Yet the geotropic response of mutants (indistinguishable from the response of the wild type) is unaffected both in the onset and in the magnitude of curvature. Therefore, the settling of starch-filled amyloplasts in the wild type cannot be necessary for the detection of gravity in this

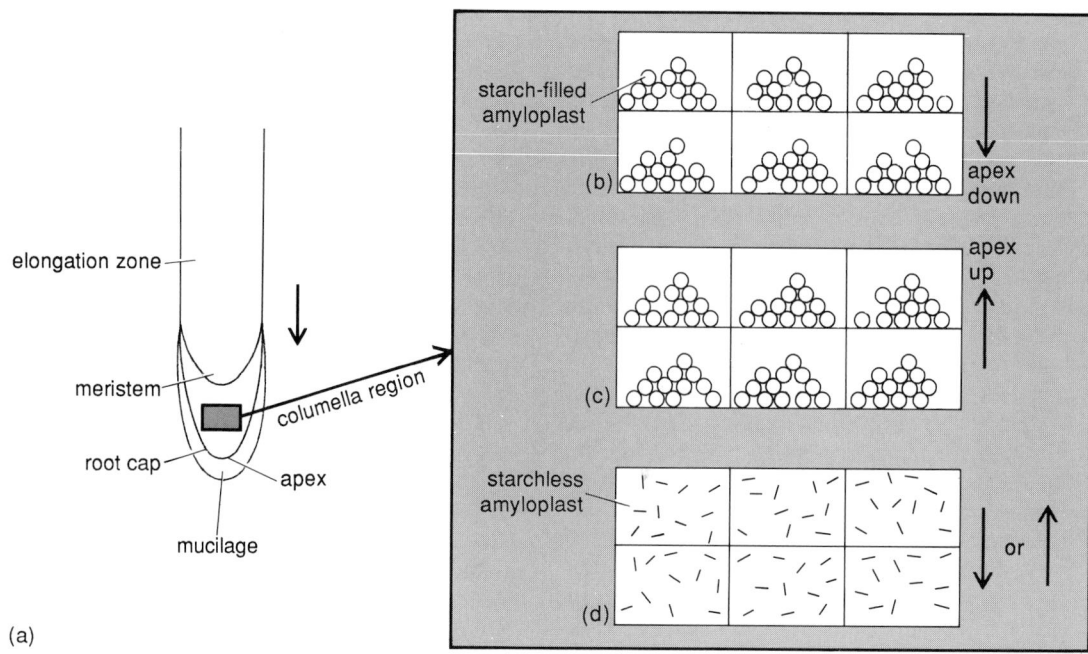

Fig. 1. Amyloplast movement in *Arabidopsis* and starchless mutant following root inversion. (*a*) Apical portion of seedling root tip. (*b*) Wild type, root tip down. (*c*) Wild type, root tip up. (*d*) Starchless mutant, root tip up or down.

plant. That amyloplast settling is not a requirement for sensing gravity in other seedling roots and shoots is supported by the fact that *Arabidopsis*, although a small plant, has geotropic kinetics comparable to those of crop plants such as corn, wheat, pea, and tomato. Furthermore, when amyloplasts of wheat were carefully depleted of starch, geotropism was unaffected. To date, no other organelle of columella cells has been characterized to serve as a possible statolith. Additional doubt is cast by the fact that settling of amyloplasts is slower than the minimal time required to detect and remember a change in orientation to gravity, a mere 0.5 s. Some of the best documentation shows 12 s for the first amyloplast to reach the lower side following horizontal placement of the root, and 4 min for the majority.

Stretch-activated channels. If the statolith model is not the mechanism for sensing gravity, another explanation is required. Calculations show that the energy in the hydrostatic pressure across a cell resulting from a 90° or even 180° change in cell orientation is so small that it would be impossible to distinguish this energy from the energy of normally vibrating molecules (thermal noise) making up the cell itself or from the high pressure already exerted on the cell membrane by turgor. However, if the energy were focused on a few responsive points in the cell, the energy at those points could be well above thermal noise. A testable theory has been proposed based on a model to account for the sensitivity of some animal cells to energy inputs at the level of thermal noise.

In a recent study, cultured chick muscle cells were found to contain a protein which traverses the cell membrane, a channel protein that is responsive to membrane stretch. That is, when the tension in the cell membrane is increased slightly, there is a conformational change in the channel protein which makes the membrane more likely to open, allowing ions to pass into or out of the cell. These stretch-activated channels can be responsive to stretch energy below the thermal noise of the cell. It is believed that these channels must respond to energy gathered over a large area of membrane in order to surmount thermal noise. This could be achieved if the membrane channels were linked by a rigid rodlike protein which would be strained when the membrane was stretched. The tension in the linker protein would result in pull on and distortion of the channel proteins, enhancing their frequency of opening. Ion flow through these channel proteins could in turn initiate a sequence of biochemical regulatory events. The function of these stretch-activated ion channels is unknown, but they could be involved in osmoregulation, for example.

Recently plant cells have been shown to have similar stretch-activated channels. Such channels could be sensors for a number of physical stimuli

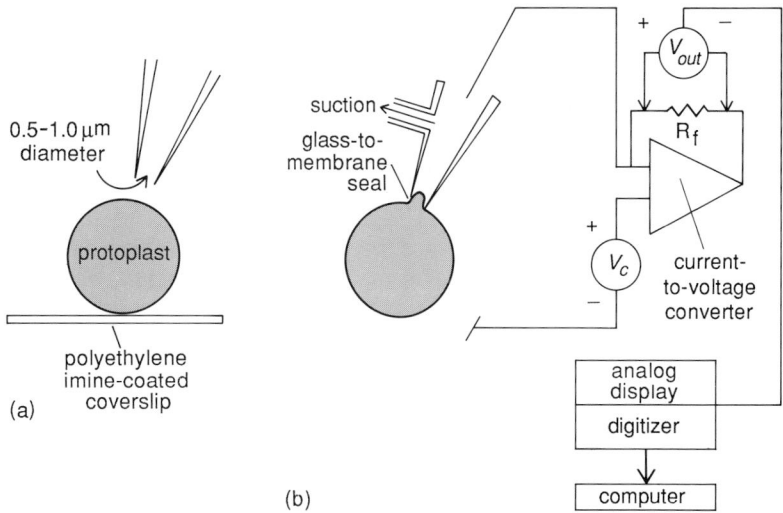

(a)

(b)

Fig. 2. Formation of the membrane seal in the patch clamp method. (*a*) Approach of the patch pipette to the protoplast adhered to the coverslip. (*b*) A patch of membrane is suctioned to the glass pipette, the seal forms, and electrical recording of ion flow across the patch can begin. R_f = current measuring resistance.

to which plants respond: touch in thigmotropism; friction as the root grows through the soil, following a path of least resistance; flexure of stems and branches resulting in a strengthened architecture or, in cases where a stem or branch is bent or leaning, producing a distinctive change in anatomical character of the wood, called tension wood in angiosperms; turgor pressure in osmoregulation, cold resulting in cell changes to increase hardiness; pressures from neighboring cells, initiating cell division or differentiation; and gravity in geotropism.

Stretch-activated channels can be studied with special methods developed to study channels in animal cells in close detail. A high-resolution technique called the patch clamp method, perfected in 1982, allows the measurement of ion current through single ion channels in membranes. A small area of cell membrane is sealed to a salt-filled glass pipette 0.5–1 micrometer in diameter by suction applied to the pipette (**Fig. 2**). The nature of the seal is unknown, but the association between the glass and the membrane has a very high resistance, in the gigohm range. With no leakage at the edge of the pipette, only ions flowing through channels present in the isolated membrane patch will be be measured. The voltage across the membrane can be selected and held constant (clamped; V_c). **Figure 3** shows a patch clamp record from an intact tobacco-stem cell protoplast. At 0 mV clamp voltage, there are no channels opening (Fig. 3, top trace), but when the membrane is stretched by applying suction to the pipette, discrete jumps of current are observed showing the activity of single ion channels opening (Fig. 3, middle trace). As suction is increased, so are the number of channels opening (Fig. 3, lower trace). Primarily, these chan

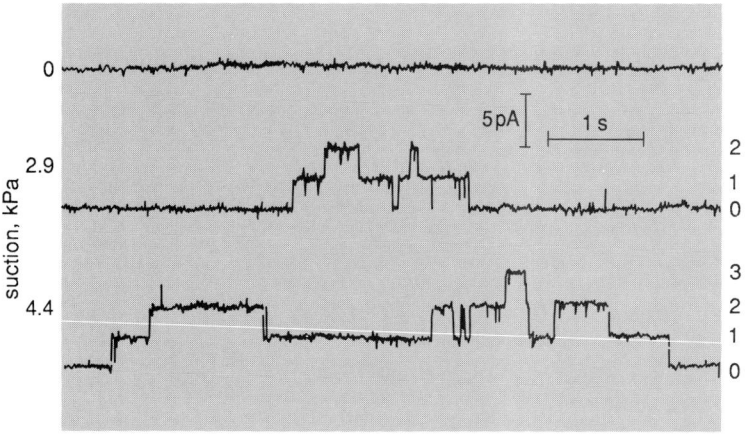

Fig. 3. Patch clamp recording demonstrating stretch-activated single ion channels in the cell membrane of a tobacco protoplast. 1 kPa = 0.145 psi.

nels pass anions such as chloride.

It has been proposed that similar channels will be found in root cells, where they function to detect gravity and friction as the root grows through the soil. Gravitational energy might be focused on these channels via both a cytoskeletal protein linker system and through rodlike cell-wall proteins, such as a glycoprotein rich in hydroxyproline, which may be linked to the stretch-activated channel proteins. When the cells are fully turgid, the cell wall would experience shear forces as the root is reoriented in the plane of gravity; this shear force could be focused on the membrane channels via rodlike proteins traversing the wall. This would provide a highly sensitive system for gravity detection while not requiring statoliths. This appealing idea is presently under investigation.

Transduction. Once a change in orientation to gravity is detected, a cascade of biochemical events must take place that produces asymmetric growth and curvature. For geotropic roots, this process entails the formation of a message or messages in the root cap that are transferred to the site of growth response, the elongation zone several millimeters behind the cap, and the translation into the appropriate growth response (Fig. 1a). Calcium is thought to be important in all of these events. The calcium content of the cell cytoplasm is very low (1 micromole), while that of the vacuole, endoplasmic reticulum, mitochondria, cell wall, and root cap mucilage is comparatively high. In some systems, small changes in the calcium level of the cytoplasm activate calcium-requiring enzymes (that is, protein kinase C and calmodulin), calcium pumps, and so forth, which in turn change metabolic pathways. If calcium is immobilized or calcium gradients are prevented in the root cap, geotropism is inhibited. Calcium and calcium transport asymmetries have been found in root caps and in the elongation zone of roots responding to gravity.

Determining the sequence of biochemical events remains open to the imagination. A number of likely scenarios are being considered. The sensing of gravity by stretch-activated channels could allow the entry of calcium into the cytoplasm from the cell wall. The increased calcium could stimulate an asymmetric pumping of protons, which is evident from pH-sensitive dye experiments and changes in membrane potential of upper and lower columella cells 8 s after gravity stimulation. Several plant physiologists believe that this directional change in the electric current across the root cap would serve to drive free calcium in the outer mucilage of the cap and within the cell walls downward across the root cap as well as backward toward the root meristem and possibly even to the elongating cells of the root. The calcium concentration would thus increase in the cell walls on the lower side of the root and root cap and may also increase in the cytoplasm of these cells. The presence of more calcium in the elongation zone might alter cell sensitivity to the growth hormone auxin and thereby reduce the growth rate on the lower side resulting in downward curvature of the root. In addition, a small increase of calcium in the cytoplasm could mediate changes in metabolic pathways via the activation of calmodulin found in the root cap. Activated calmodulin is a requirement for a number of enzymes in animal cells and some similar interactions have been found in plants. The precise nature of the steps that would follow calmodulin activation in the root cap is unknown at this time.

The role of the meristem, the intervening tissue between the root cap and the elongation zone, in transduction is little understood or studied. However, some recent work has shown that red light, a requirement for switching on geotropic sensitivity of some roots, boosts calmodulin and ATPase messenger-RNA activity and stimulates the release of bound hormone, abscisic acid, as well. Shortly following exposure to light, an abscisic acidlike substance accumulates in the meristem from the root cap. These light-induced events are likely to be important in preparing these tissues for gravity perception and transduction.

The picture is becoming intricate, a far cry from the simple statolith model and the Cholodny-Went theory for auxin regulation of curvature. Recent molecular technologies, such as the patch clamp method, molecular genetics, and the vibrating probe, however, will undoubtedly be providing breakthroughs in this area over the next several years.

For background information *SEE AUXIN; CALMODULIN; PLANT MOVEMENTS; PLANT PHYSIOLOGY; ROOT* in the McGraw-Hill Encyclopedia of Science and Technology.

Kathryn L. Edwards

Bibliography. K. L. Edwards and B. G. Pickard, Detection and transduction of physical stimuli in plants, in H. Greppin, B. Millet, and E. Wagner (eds.), *The Cell Surface in Signal Transduction, NATO Advanced Research Workshop, Besanson, France, March 1986*, 1986; R. Moore and M. L. Evans, How roots perceive and respond to gravity, *Amer. J. Bot.*, 73:574–587, 1986; B. G. Pickard, Roles of hormones, protons and calcium in geotropism, in R. P. Pharis and D. M. Reid (eds.), *Encyclopedia of Plant Physiology*, NS vol. 11: *Hormonal Regulation of Development, III: Role of Environmental Factors*, 1985; F. Sachs, Biophysics of mechanoreception, in *Membrane Biochemistry: Proceedings of a Symposium of the Membrane Subgroup of the Biophysical Society, 1985*, 1986.

Plant pathology

The severity of soil-borne diseases of agronomic and horticultural crops is directly influenced by the surrounding environment. Grain legumes (peas, *Pisum sativum*; beans, *Phaseolus vulgaris*) are particularly sensitive to an adverse soil environment. One environmental parameter is soil compaction. Compaction results from wheel traffic, tillage, and planting operations or occurs naturally in some soils. It is not easily recognized or measured.

The interrelationships of soil compaction and root disease severity caused by soil-borne, fungal, and plant pathogens involve the following factors: root function and physiology; soil water, temperature, aeration, and fertility; and the ecology and biology of root pathogens.

Root function and physiology. Plant roots perform important functions, including anchoring the plant in soil; absorbing and translocating water and nutrients; synthesizing and transporting growth regulators and other organic compounds; and acting as storage organs for carbohydrates. When roots function as absorbing organs, the important characteristics are those that affect the absorbing surface area, such as root length and density, number of root hairs, and mycorrhizal associations. Under normal growing conditions the root environment is seldom optimum for extensive, effective growth. Several of the root functions and the environmental factors can interact, and an unfavorable condition caused by any one factor can affect the response of a plant to several other factors. Thus, root growth is impeded by excess or deficient water, low-oxygen stress, nonoptimal soil temperatures, compaction, imbalanced nutrient supply, and pest and pathogen problems.

Soil aeration. It has been found that in a rhizosphere containing less than 10% oxygen, there will be reduced root growth. Large pores between aggregates contribute more to soil-air exchange than do small pores within aggregates. Soil compaction reduces the number of large pores as well as the soil's total porosity. When large pores become blocked with water, problems with air exchange are even more acute. As the roots are deprived of oxygen, they are forced into anaerobic respiration. Insufficient energy is produced for normal growth and resistance to infection, and severe injury or death may result. In addition, at least 17 hydrocarbons have been reported to be synthesized and exuded by roots under anaerobic or near-anaerobic conditions. These compounds include ethylene, methane, fatty acids, alcohol, ketones, and aldehydes; their production is a severe drain on the photosynthate assimilates required for normal plant nutrition. In addition to possible phytotoxicity, these compounds stimulate microbial activity in the rhizosphere. Reduced oxygen levels and soil compaction occurring together can seriously reduce root elongation rates; the oxygen requirements for root penetration of a compact soil are higher than for less compact soils.

Plant pathogen–root interactions. The interactions between host root and pathogen are not easily defined because of the interrelated effects of the soil environment. The pea pathogen *Fusarium solani* f. sp. *pisi* grows rapidly on the surface of pea roots placed in an environment containing insufficient oxygen. Even though *Fusarium* is classified as an obligate aerobe, it is able to grow profusely on the host root surface because of the increased amounts of carbon compounds exuded by the host root during anaerobic metabolism. Further, the nonbean pathogen *F. solani* f. sp. *pisi* injures bean roots and permanently reduces water absorption and plant growth when the root is exposed to temporary low-oxygen stress. The injury is less, however, than that caused by the bean pathogen *F. solani* f. sp. *phaseoli*. The effect of poor soil aeration is so dramatic that even bean lines bred for resistance to root rot caused by *F. solani* f. sp. *phaseoli* are susceptible when exposed to short periods of poor aeration. Even when soil moisture is not excessive or nonoptimal, soil temperatures can seriously affect root growth and disease severity in compact soil.

Compaction and alleviation. Previous research has shown that fracturing compact layers created by tillage implements should be accomplished immediately prior to planting. In general the yield response is greater for a cultivar susceptible to root rot than for one that is rot-resistant. The plant response to fracturing tillage pans (subsoiling), however, is expressed only when a pathogen or moisture stress is present. Equipment traffic is also an important cause of soil compaction. Increases of 26% in dry bulk density have been observed on farms using high-powered tractors with tire contact pressures ranging from 5.5 to over 7 psi (38 to over 50 kilopascals). Positive linear correlations have been reported when increased soil bulk densities at the 6–8-in. (15–20-cm) depth were compared to root rot severity of green peas.

Compaction predisposes roots to increased disease severity and also influences the survival and distribution of soil inoculum. A field survey completed in southeastern Washington and northeastern Oregon found a characteristic hard, compacted layer (plow pan) at about 8 in. (20 cm) depth in both wheat and pea fields. It was determined that inocula of *Pythium ultimum* and *F. solani* f. sp. *pisi* had soil depth distributions determined primarily by environmental optima for each pathogen and water distribution related to compaction. *Fusarium* was present in the plow layer and as deep as the pea roots penetrated. In contrast, *Pythium* was found almost exclusively in the cultivated soil level. The greatly reduced movement of water through the compacted layer created by tillage caused excess water in the cultivated layer during the winter months, with a corresponding drier sublayer. This condition is favorable for larger *Pythium* populations above the tillage pan and for *Fusarium* inoculum extending deep into the underlying subsoil. This same distribution was produced by traffic both during secondary tillage and when moldboard tillage was compared to sweep-type primary tillage. Root disease of peas under the above conditions was readily explainable by the depth distribution of *Pythium* and *Fusarium* inoculum.

Increased soil compaction greatly reduces root growth rates, directly affects branching patterns, and increases the probability of host-pathogen contacts. Soil-borne root pathogens exist typically as dormant structures. The first requirement for infection is germination, which is induced by root and seed exudates. The distance from the root at which exudates will influence microbial activity in soil (rhizosphere) is about 0.08 in. (2 mm). The mobility of dormant fungal spores is considered negligible when compared to the extensive growth of plant root systems. Exudation from healthy roots is greatest near the root tip and the zone of elongation. Roots grow in well-aerated, friable soil at a rate of about 0.016 in. (0.4 mm) per hour. With the greatest rhizosphere influence only 0.04 in. (1 mm), a pathogen propagule will detect the approaching root tip only a few hours before its arrival. Consequently, fungal propagules requiring 6 to 10 h for germination and growth toward a nutrient source will miss the root tip and contact the region of maturation where exudation is minimal and the chance for infection is reduced. Increased available soil water results in faster diffusion of exudates from roots and seeds into bulk soil. This will permit stimulation of propagules to germinate from greater distances if the water flux to the root does not significantly offset the diffusion gradient for

the exudate. To ensure contact with a root tip advancing at a rate of 0.016 in. (0.4 mm) per hour, a dormant fungal propagule must germinate within 1 to 3 h after receiving the exudate signal. Such germination times have been recorded for the pythiaceous fungi, that is, *Pythium* spp., which are root tip pruners.

Fusarium chlamydospores cannot germinate as quickly, and this helps to explain why *Fusarium* tends to have a high proportion of epiphytic colonies which grow on the root surface. This slow germination time also helps to explain why *F. solani* f. sp. *pisi* tends to attack the cotyledonary attachment area of peas (peas are hypogeal) and the epicotyl and hypocotyl, which are stationary in soil. With adverse growing conditions, root growth will be impeded and *Fusarium* will invade the entire root system. Infection by epiphytic colonies of *F. solani* f. sp. *pisi* on the root surface due to root stress (water and temperature stress, fruit or seed maturation) and epidermal breakdown has a role in damage of peas by *Fusarium* late in the growing season.

The distribution of soil-borne inoculum is generally related to the root proliferation pattern. In a structured or layered soil the primary and some of the secondary roots may be confined to the same cracks or fissures year after year. Consequently, this concentration of inoculum will assure more successful infections per unit of inoculum density than is found in frequently mixed soil. Mixing and tillage reduce clustering and, therefore, the numbers of inoculum to which the root is exposed. The activity of soil-borne pathogens and interactions with competitive microorganisms and host roots can be controlled to some extent by tillage, which affects the physical environment; the position of the crop residue occupied by the pathogen, that is, whether the residue is buried or on the surface; and the position of the pathogen itself. Most root pathogens complete all of their life cycles in soil. Consequently, how and when the soil is tilled will have a significant impact on the survival and distribution of these organisms.

For background information *SEE AGRICULTURAL SOIL AND CROP PRACTICES; PLANT PATHOLOGY; RHIZOSPHERE; ROOT; SOIL MICROBIOLOGY* in the McGraw-Hill Encyclopedia of Science and Technology.

J. M. Kraft

Bibliography. R. R. Allmaras, J. M. Kraft, and A. J. M. Smucker, Soil compaction and crop residue management effects on root disease of annual food legumes, in R. J. Summerfield (ed.), *World Crops: Cool Season Feed Legumes*, 1987; J. M. Kraft and R. R. Allmaras, Pea root pathogen populations in relation to soil structure, compaction, and water content, in C. A. Parker et al. (eds.), *Ecology and Management of Soil-borne Plant Pathogens*, American Phytopathology Society, pp. 203–205, 1985; D. E. Miller, Root systems in relation to stress tolerance,

HortScience, 21(4):963–970, 1986; A. J. M. Smucker, Carbon utilization and losses by plant root systems, in S. A. Barber and D. R. Bouldin (eds.), *Roots, Nutrient and Water Influx, and Plant Growth*, ASA Spec. Publ. 149, 1984.

Plant physiology

Weeds cause billions of dollars of economic loss throughout the world every year. This loss is manifested in weed control costs, reduced crop yield and quality, and environmental damage by herbicides and weeds. Understanding the physiology of weeds and how herbicides affect it is increasingly important in developing more economical, efficient, and environmentally safe strategies for the control of these pests.

The physiological differences between crop and weed species are often much smaller than the physiological differences between crops and other types of pests (for example, insects or nematodes). For this reason, fundamental knowledge of plant physiology is important in the design of weed control strategies. In the near future the rapidly increasing knowledge of comparative weed and crop physiology and the development of new biotechnological capabilities promise to provide scientists with powerful new tools in their efforts to control weeds.

Weed seed dormancy. One way in which weeds differ from crops is that they must tolerate year-round environmental conditions. This toleration is frequently accomplished by the production of seeds that remain dormant but viable in the field through the harsh conditions of a winter or a dry season. Because weed seeds are generally small, they are often buried during cultivation at soil depths at which germination would be lethal for them. Seeds of many weed species remain dormant in the soil, often for many years, until returned by cultivation to a more favorable environment near the soil surface. Because of the longevity of weed seeds in the dormant state in the field, it usually takes many years to completely rid a field of weeds, even with complete weed control. New information on the dormancy and breaking of dormancy in weed seeds may provide farmers with the means to eliminate weed problems for years in advance by killing the bank of dormant weed seeds in the soil.

A common mechanism of dormancy is production of a seed coat that is impermeable to water. Without absorption of water, normal metabolism and growth do not occur. Near the soil surface, in environments more subject to freeze-thaw or wet-dry cycles, these seed coats are more apt to break, allowing water into the seed and releasing dormancy. In some species the breaking under these conditions usually occurs in a certain site composed of thin-walled cells in the seed coat (**Fig. 1**). Recent studies have shown that forma-

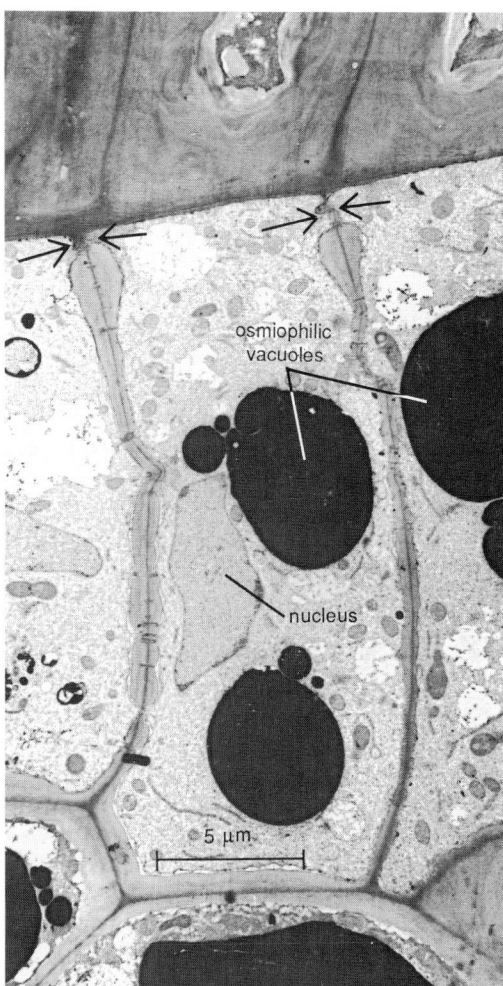

Fig. 1. Transmission electron micrograph of subpalisade region of prickly sida (*Sida spinosa*) seed. Arrows point to thin regions in the cell wall where future breakage occurs prior to uptake of water. (*Courtesy of R. N. Paul*)

tion of water-impermeable seed coats in a number of weed species is accomplished by polymerization of phenolic compounds by the enzyme peroxidase to form the hard polymer lignin. This information on the dormancy mechanism of these species suggests several sites of vulnerability, including synthesis of phenolic compounds and peroxidase activity during seed development. For example, a chemical inhibitor of peroxidase should prevent lignin formation and, thus, formation of a water-impermeable seed coat.

Discovery of an effective method for releasing weed seed dormancy at a time (for example, winter) or place (for example, deep in the soil) that would be unsuitable for survival of the weed seedling has been an objective of considerable research. Light is one key environmental factor that triggers germination when the seed is returned to the soil surface. Phytochrome, the photoreversible pigment responsible for photomorphogenesis, is the detector of light in many weed seeds. Scientists now believe that one of the early steps in the induction of germination by phytochrome is alteration of cell membrane properties by the pigment. Other chemicals that affect membranes, such as ethylene and certain anesthetics, can also induce germination, reducing or eliminating the requirement for light to break dormancy. It is not yet clear whether these chemicals can substitute for phytochrome or lower the requirement for the active form of the pigment (the far-red-light-absorbing form). A more complete understanding of phytochrome interactions with membranes may lead to a safe and effective chemical soil treatment to eradicate the seed bank by induction of germination of weed seeds before the crop is planted.

Herbicide mechanisms of action. In the search for new herbicides the procedure of randomly screening chemicals for herbicidal properties has reached the point of diminishing returns. One strategy for the rational design of herbicides is to first determine the molecular site of action of a phytotoxic (toxic to plants) compound and to then tailor more carefully the molecule to fit that site. Knowledge of the site of action is also a prerequisite for genetic engineering to produce herbicide-resistant crops.

For instance, the herbicide glyphosate [N-(phosphonomethyl)glycine] was shown to inhibit the enzyme 5-enolpyruvylshikimate-3-phosphate synthase (EPSP synthase), an enzyme required for the synthesis of the essential aromatic amino acids, phenylalanine, tyrosine, and tryptophan. No higher plant species are known to be resistant to this highly effective and environmentally safe herbicide. This knowledge of the mode of action of glyphosate has been used in two different approaches to produce herbicide-resistant crop species.

In one approach, the transformation methods of biotechnology were used to move the gene for a glyphosate-resistant EPSP synthase from a bacterium to tobacco plants (**Fig. 2**). The plants produced by this transformation are highly tolerant to levels of glyphosate that are lethal to untransformed plants. In a second approach, glyphosate-resistant petunia plants have been produced by greatly increasing the number of copies of the gene coding for the glyphosate-susceptible form of EPSP synthase. Thus, the petunia plants overproduce the EPSP synthase enzyme, and much more of the herbicide is required to shut down the biosynthetic pathway for aromatic amino acids. Another strategy for the production of herbicide-resistant crop plants is to transform them with bacterial genes which code for enzymes that break the herbicide down to nonphytotoxic derivatives. Similar utilization of new knowledge of the molecular site of action of other herbicides is being made to produce herbicide-resistant crops. *See* Forestry.

Since the mid-1970s, more has been learned of the molecular site of action of herbicides than in all previous years. Not only is this information useful in further herbicide design and in agricultural biotechnology, but it has also provided considerable new knowledge of the fundamental physiology and biochemistry of plants.

Herbicide resistance. Long-term use of a herbicide in the same location can lead to the appearance of weed biotypes (genetic variants within a species) that are resistant to the herbicide, much in the same manner that microorganisms become resistant to antibiotics. In weeds, however, development of resistance usually takes years because weeds generally have a life cycle of at least 1 year and the soil bank contains a long-lived population that may never have been exposed to the herbicide. Still, biotypes of weeds that are resistant to several types of herbicides have arisen throughout the world. Characterization of the physiological and biochemical causes of herbicide resistance is leading to a better understanding of how herbicides work and to the production of herbicide-resistant crop plants by genetic manipulation.

For instance, the appearance of weeds resistant to s-triazine herbicides has led to detailed knowledge of how this large class of herbicides kills plants, and to the commercial production of crop species that are resistant to this group of herbicides and other herbicides that act at the same site of action. Much of this knowledge was obtained through the use of a radiolabeled form of the s-triazine herbicide atrazine that irreversibly binds to the molecule closest to it during exposure to ultraviolet radiation (^{14}C-azido-atrazine). Atrazine was known to inhibit photosynthetic electron transport. Thus, the membranes where photosynthetic electron transport is known to occur (the thylakoid membranes of the chloroplast) of atrazine-resistant and -susceptible weed biotypes were incubated in this radiolabeled azido-atrazine. After incubation, they were exposed to ultraviolet radiation. The

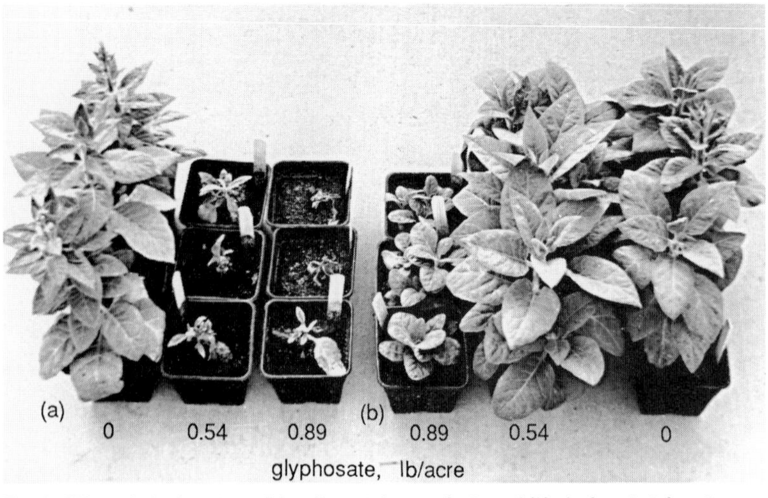

Fig. 2. Effect of glyphosate on (*a*) ordinary tobacco plants and (*b*) glyphosate-tolerant tobacco plants transformed with the EPSP synthase from glyphosate-resistant *Salmonella typhimurium*. The plants were photographed 40 days after being sprayed with 0, 0.54, or 0.89 lb/acre (0, 0.6, or 1.0 kg/ha) of glyphosate. (*L. Comai, CalGene*)

(a)   0   0.54   0.89   (b)   0.89   0.54   0
glyphosate, lb/acre

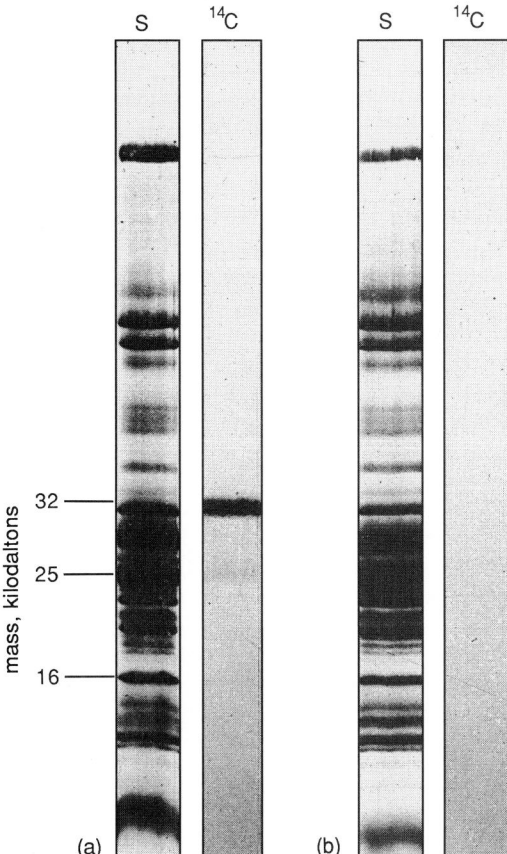

(a)   (b)

Fig. 3. Thylakoid proteins of (*a*) atrazine-susceptible and (*b*) atrazine-resistant biotypes of pigweed (*Amaranthus hybridus*) separated by polyacrylamide gel electrophoresis. The lane on the left (labeled S) for each biotype has been stained for protein, and the lane on the right (labeled ^{14}C) is a fluorograph of proteins labeled with ^{14}C-azido-atrazine. (*K. Pfister, Ciba-Geigy*)

proteins were then removed from the membranes with a detergent and separated by polyacrylamide gel electrophoresis. A protein with a molecular mass of 32 kilodaltons was found to be labeled with the ^{14}C-azido-atrazine in the susceptible biotype; however, no protein was labeled in the atrazine-resistant biotype (**Fig. 3**). These data, along with other information, led to the discovery of the crucial function of the 32-kilodalton protein in photosynthetic electron transport. The chloroplast gene that codes for the form of this protein that does not bind atrazine has subsequently been introduced into crop plants. Now varieties of certain crops that are resistant to *s*-triazine herbicides are available to farmers.

One problem in transforming crop species with genes that confer herbicide resistance is that such transformed plants are rarely as productive as the herbicide-sensitive, untransformed plant. Just as in nature, genetic alterations only rarely result in a more fit form or, in this case, a more productive form of the species. Nevertheless, plant scientists are overcoming this problem and others to produce agriculturally

useful, herbicide-resistant crops through advances in biotechnology.

Natural compounds as herbicides. Natural products have provided some of the most important weapons against disease and insects. A recent development in the search for new herbicides is the active consideration of naturally occurring phytotoxic compounds for use against weeds. Natural compounds have several advantages over synthetic phytotoxins. These include potentially fewer environmental hazards. For instance, lower amounts of these compounds are likely to accumulate in the environment because lower amounts probably will be used due to higher efficacy and greater specificity. Also, these compounds are more likely to rapidly degrade to harmless substances than are synthetic compounds. Another advantage is that natural compounds are a reservoir of thousands of phytotoxic chemicals and chemical classes that would require enormous effort from synthetic chemists to produce. This vast, untapped source of herbicide templates is beginning to be recognized by the herbicide industry.

For many years, scientists have known that many plants engage in a form of chemical warfare called allelopathy. Many of the phytotoxic compounds (allelochemicals) with which plants battle their competitors are now being identified and evaluated as herbicides. Several plant-produced phytotoxins have been patented as herbicides, and only a small fraction of the hundreds of thousands of compounds produced by plants have been evaluated.

One of the most promising herbicides produced by plants is δ-aminolevulinic acid (ALA). This colorless compound is harmless to animals and is synthesized by all plants as a precursor to chlorophyll. When applied to higher plants at high doses, however, it causes a massive buildup of green chlorophyll precursors. This results in an inability of the plant to contend with the light energy absorbed by these pigments, resulting in the production of toxic light-generated radicals. Hence, ALA has been termed the laser herbicide. Other compounds produced by plants which more directly cause light-generated toxic radicals have also been patented as herbicides. The phytotoxins that microorganisms produce are even more promising. Toxins produced by plant-disease-causing microorganisms are of particular interest because of their high potency and often because of their desirable selectivity. For instance, tentoxin, produced by the fungus *Alternaria alternata*, is toxic at very low levels to virtually every weed species that can cause losses of soybeans and corn, but is harmless to these crops. Japanese scientists have patented many microbial compounds as herbicides, and several are being developed commercially. These include herbicides derived from anisomycin and bialophos, both products of *Streptomyces* species.

The known molecular mechanisms of action of microbial phytotoxins are generally very different from those of synthetic herbicides. This information is valuable to scientists in their search for sites of vulnerability in plants which do not exist in animals or in certain other plants. This same information, applied to the chemical structure of the phytotoxin, allows scientists to modify the structure to alter selectivity, potency, and other characteristics in predictable ways. For instance, bialophos kills plants by inhibiting the enzyme glutamine synthetase, but simpler and more easily synthesized compounds based on the chemical structure of bialophos are even more potent as herbicides. See Allelopathy.

For background information see Allelopathy; Dormancy; Genetic engineering; Herbicide; Plant physiology in the McGraw-Hill Encyclopedia of Science and Technology.

Stephen O. Duke

Bibliography. L. Comai et al., Expression in plants of a mutant *aro* A gene from *Salmonella typhimurium* confers tolerance to glyphosate, *Nature*, 317:741–744, 1985; S. O. Duke, Naturally occurring chemical compounds as herbicides, *Rev. Weed Sci.*, 2:15–44, 1986; S. O. Duke (ed.), *Weed Physiology*, vols. 1 and 2, 1985; *Pesticide Resistance*, National Academy Press, 1986; K. Pfister et al., Photoaffinity labeling of an herbicide receptor protein in chloroplast membranes, *Proc. Nat. Acad. Sci. USA*, 78:981–985, 1982.

Plant-water relations

Plants have evolved to survive in a wide range of habitats. Hydrophytic plants grow in wet habitats, mesophytes in habitats with intermediate water levels, and xerophytes in dry habitats. For a plant to grow and reproduce, positive net photosynthesis (net carbon fixation) must occur. Short periods of net carbon loss during or immediately after exposure to dry conditions, or to any stress, do not harm the plant; but positive net photosynthesis must be regained after the stress is relieved. Xerophytic higher plants, which evolved in dry habitats, have numerous morphological adaptations to protect their cells from exposure to desiccating conditions and to prevent damage to the photosynthetic apparatus. Mesophytic higher plants vary in their ability to survive dry conditions, but their cells cannot survive exposure to desiccating conditions. Xerophytic lower plants, however, have physiological and metabolic adaptations which allow the cells to survive such exposure.

In the context of this article the term desiccation-resistant is used to denote any plant which can survive exposure to desiccating conditions, regardless of the survival mechanism. Plants which cannot survive such conditions are known as desiccation-sensitive. Dehydration-resistant plants have mechanisms which prevent or impede water loss from the plant. Plants which survive exposure of their cells to desiccating conditions but are not dehydration-resistant at the whole-plant level are known as desiccation-tolerant.

Adaptations in higher plants. The leaves of higher plants are generally composed of several cell layers with internal water-saturated air spaces through which carbon dioxide (CO_2) diffuses, outer epidermal layers with stomates controlling gas exchange, and a waxy cuticle covering the remaining leaf surface and reducing evaporation. Under dry conditions, stomatal closure occurs, slowing water loss and decreasing CO_2 uptake. If the CO_2 concentration inside the leaf decreases, the photosynthetic rate may also decrease. Several mechanisms have evolved to increase the ability of the plants to function when water is in limited supply. In some plants a modified carbon fixation pathway with higher water-use efficiency has evolved. These plants, which include corn and other grains, are called C_4 plants because the initial product of the pathway contains four carbon atoms (the usual number of carbon atoms in the initial product is three). In addition to the modified biochemical pathway, C_4 plants exhibit differences in leaf anatomy and increased differentiation of functions in the leaf cells. In the past few years, plants which seem to be intermediate between the C_3 and C_4 in respect to anatomy and biochemical pathways have been discovered and are being studied intensively. A third pathway called CAM (for crassulacean acid metabolism) occurs primarily in desert plants. It is similar to the C_4 pathway, but the plants conserve water by opening their stomates only at night.

In addition to physiological and biochemical adaptations to dry habitats, xerophytic plants have elaborate morphological adaptations for dehydration resistance. These adaptations serve to decrease water loss or increase water uptake. Water loss can be retarded significantly by lowering the ratio of surface area to volume. In cacti, for example, the leaf surface area has been greatly reduced by the evolution of extremely narrow leaves (spines). In this case the stem, which has become enlarged to store water, carries out most of the photosynthetic activity. A thick, waxy cuticle covering the stem surface also slows evaporation. The stomates in xerophytes are sunken below the surface; this arrangement increases the boundary-layer resistance to gas exchange and is another mechanism for lessening water loss. Some desiccation-resistant higher plants have deep root systems which can tap the water table; despite dry conditions, these plants have access to water. Other xerophytes can obtain water from dry or saline soil.

Most of the adaptations in higher plants prevent the exposure of the plant cells to desiccation, making these plants dehydration-resistant. Only a few higher plants are truly desiccation-tolerant; for the most part, plants with this ability are lower plants.

Despite the adaptations in higher plants that serve to prevent exposure of the cells to dryness, desiccation damage, as shown by low photosynthetic rates, frequently occurs. In higher plants, two processes may contribute to the decrease in photosynthesis during drying: stomatal closure may limit the photosynthetic rate by lowering the internal CO_2 concentration, and metabolic functions in the chloroplast may decrease. If carbon fixation decreases, the light energy absorbed by the chloroplast for photosynthesis cannot be efficiently utilized and may cause damage.

This damage may lead to photo-oxidation of the photosynthetic pigments, which can lead to further membrane damage. Loss of the pigments does not necessarily limit photosynthesis; in certain cases the pigments seem to be in excess of the amount needed for the plant to have net positive photosynthesis under its normal growth conditions. Chlorophyll breakdown can be a symptom of desiccation damage, however. Certain desiccation-resistant higher plants seem to lose their chlorophyll before any photo-oxidation has occurred. This may be a mechanism to limit absorption of light energy and limit light-induced damage.

Desiccation resistance in mosses. Mosses, which occupy a wide range of habitats from wet to extremely dry, are excellent lower plants in which to study desiccation resistance, because they are anatomically much simpler than higher plants. Mosses lack an elaborate root system and are unable to obtain water from their substrate. Most moss leaves are composed of a single layer of cells without stomates and with only a thin, protective, waxy cuticle. Although some mosses have hollow dead cells which could retain water, most mosses, with the exception of the genus *Sphagnum*, are unable to store water.

On the basis of anatomy the mechanisms responsible for desiccation resistance in mosses must differ from those operating in higher plants. Individual plants of both resistant and sensitive mosses lose loosely bound water during drying at the same rate and to the same extent. In the field, however, some mosses grow in tight vertical clumps or horizontal mats. This clustering increases boundary-layer resistance and shading, while capillary action between the stems of individual plants can increase water uptake and storage. Although these characteristics contribute to dehydration resistance in the field, studies show that after desiccation and subsequent rehydration, shoots of desiccation-resistant mosses separated from their clumps have high survival rates. Desiccation resistance in mosses, therefore, is based not only on interactions between shoots but also on physiological and metabolic adaptations.

In certain mosses the cells may store sugars or other metabolites which could increase binding of water within the cells. With other mosses, water that is bound to the cell wall could be accessible to the cells even under very dry conditions. It seems, however, that in many cases the moss cells actually become dehydrated when the plants are exposed to dry conditions, suggesting that these plants are desiccation-tolerant.

Desiccation may damage the cell in many different ways. Membranes which maintain cell compartmentalization and function under normal conditions may become disorganized during desiccation or rehydration, leading to cell damage or death. Studies on the desiccation-tolerant moss *Hedwigia ciliata* and the desiccation-sensitive moss *Mnium cuspidatum* show that tolerant mosses have higher cell survival after desiccation and rehydration than sensitive ones, indicating that tolerant mosses are damaged less easily or repaired more efficiently than sensitive ones.

Measurements of photosynthetic rate during drying show that photosynthesis decreases or stops in both desiccation-tolerant and -sensitive plants at a certain level of dryness. After 24 h of rehydration, however, desiccation-tolerant plants regain positive net photosynthesis, often up to the level of undesiccated plants, while desiccation-sensitive plants do not. Respiratory rates increase slightly in both tolerant and sensitive plants after this treatment, but not enough to account for the change in photosynthetic rate. Experiments with *Hedwigia* and *Mnium* indicate that loss of chlorophyll after desiccation and rehydration is not necessarily an indicator of desiccation sensitivity. Just as in some higher plants, photosynthetic rates in *Hedwigia* remain unaffected by even a significant loss of chlorophyll. If the plants are illuminated during desiccation, however, *Mnium* loses chlorophyll in a manner related to the light intensity, while chlorophyll loss in *Hedwigia* is independent of light intensity. This suggests that absorption of excess light energy by desiccation-sensitive species during drying may cause damage which does not occur to the same extent in desiccation-tolerant species.

Acclimation. Both higher and lower plants are able to acclimate to changes in the moisture level of their environment. Exposure to mild drought, for example, often leads to increased desiccation resistance when the plants are then exposed to more severe conditions. In some higher plants the leaves can become succulent, which decreases evaporation by decreasing the ratio of leaf surface to volume. The desiccation resistance of a certain plant often varies seasonally, with low resistance in the spring when water is available and higher resistance in the summer when water may be scarce. Different varieties or races of species may differ in their desiccation resistance at any one time and in the range of conditions to which they can acclimate. Acclima tion may affect the optimal water content for photosynthesis or may decrease the water content at which permanent damage occurs. Acclimation in higher plants can affect both the water level at which stomatal closure occurs and the optimal water content at which the chloroplasts photosynthesize effectively.

For background information SEE LEAF; PHOTOSYNTHESIS; PHYSIOLOGICAL ECOLOGY (PLANT), PLANT-WATER RELATIONS in the McGraw-Hill Encyclopedia of Science and Technology.

V. J. Streusand

Bibliography. J. D. Bewley and J. E. Krochko, Desiccation-tolerance, in O. L. Lange et al. (eds.), *Encyclopedia of Plant Physiology*, 1982; M. A. Matthews and J. S. Boyer, Acclimation of photosynthesis to low leaf water potentials, *Plant Physiol.*, 74:161–166, 1984; M. C. F. Proctor, Diffusion resistance in bryophytes, in E. D. Ford and J. Grace (eds.), *Plants and Their Atmospheric Environment*, Symposium of the British Ecological Society, pp. 219–229, 1980; F. B. Salisbury and C. W. Ross, *Plant Physiology*, 3d ed., 1985; V. J. Streusand, J. A. Weber, and H. Ikuma, Desiccation-tolerance in mosses, II. Differences in the responses of *Hedwigia ciliata* and *Mnium cuspidatum* to desiccation and rehydration, *Can. J. Bot.*, 64:2393–2398, 1986.

Plants, saline environments of

Saline soils and waters are major detriments to crop production in many areas. Low concentrations of salts suppress plant growth; higher concentrations cause death. Yet in the dry regions of the world, brackish waters that contain up to 10,000 parts per million of total dissolved solids are being increasingly used in place of fresh water for irrigation. For example, Israel uses 95% of its known fresh-water resources, and any additional agricultural development must make use of alternative water resources such as the moderately saline waters beneath the Negev desert. For this reason, crop plants that are tolerant to the salts in these brackish waters are being increasingly sought.

Mechanisms of salinity tolerance. The important characteristics which contribute to salinity tolerance are the regulation of the transport and distribution of ions within tissues and organs of the plant, and regulation of the ion concentration of the cytoplasm within individual cells. Crop plants enhanced for these characteristics have improved tolerance to salts.

Whole plants. At the whole-plant level, salinity adversely affects growth of the root and shoot, and this in turn increases the shoot/root ratio by reducing root growth relatively more. The fruits produced are often small and distorted, as exemplified by the tomatoes shown in the **illustration**,

Tomato fruits produced hydroponically in a greenhouse. The control fruit on the left was grown in a complete half-strength Hoagland solution, while the two small fruits were grown in Hoagland solution supplemented with 200-millimolar sodium chloride (NaCl).

where the fruits of the salt-stressed plants are notable not only for their small size but also for the presence of blossom-end rot caused by loss of calcium from the tissue. These defects ruin the market value of a crop.

It is apparent that tolerance to salt in higher plants involves the integration of the many activities of its cells, tissues, and the whole plant. Salinity tolerance is not simply a cellular phenomenon. This interrelationship of the plant parts vastly complicates the development of saline-tolerant crops by introducing multiple factors which must be manipulated in order to achieve a significant change in whole-plant tolerance.

Tissues. The cells which compose a tissue are affected by their interrelationships, that is, the contacts, arrangements, and special activities (a leaf, for example, is composed of many specialized cells). Thus, the distribution of solutes and water within a tissue is affected by a variety of anatomical and physiological factors that influence the solute concentrations within the individual cells.

Some of the anatomical factors in roots of grasses have recently been described and illustrate the complexity that exists in a plant struggling to survive under saline conditions. A root is composed essentially of three regions: the epidermis, cortex, and stele. The epidermis includes the root hairs which absorb the soil solution. The cortex is separated from the stele by a thin endodermis. The cortex contains much of the metabolic machinery of the root, while the stele contains the conductive tissue for transport of solutes, sugars, and water between the root and shoot.

Selective transport through the root is determined by barriers to the movement of ions. The most important of these are the suberized cell walls of the endodermis which force solutes moving from the cortex to the stele to pass through the plasmalemma and into the symplasm. In the stele, active-transport processes are involved in loading solutes into the xylem stream—potentially another site for selectivity—and transfer cells may further regulate active transport out of the xylem. In some of the most salt-sensitive crops (for example, beans, squash, and tomatoes), salt may be retransported from the xylem and exported from the roots.

Cells. Under saline conditions when tissue concentrations exceed 200 millimoles (corresponding to 0.9 megapascal or 9 atm of osmotic pressure), the large quantities of salt [mostly sodium chloride (NaCl)] that are absorbed into leaves are accumulated mainly in the vacuoles of individual cells which compose the tissue. While the concentration of inorganic ions in the cytoplasm normally is held by the plant in the range from 100 to 200 millimoles, at hyperosmotic conditions (osmotic pressures greater than 0.9 MPa or 9 atm), the maintenance of osmotic equilibrium across the tonoplast requires the accumulation in the cytoplasm of nontoxic organic solutes (compatible solutes), the chemical nature of which varies with the crop. Most of these solutes are derivatives of polyols or nitrogen dipoles (see **table**).

Little is known about the mechanisms for

Taxonomic distribution of possible compatible solutes*		
Solute	Structure	Distribution
D-Sorbitol	CH₂OH / H–C–OH / HO–C–H / H–C–OH / H–C–OH / CH₂OH	Plantaginaceae Rosaceae
D-Mannitol	CH₂OH / HO–C–H / HO–C–H / H–C–OH / H–C–OH / CH₂OH	Combretaceae Myrsinaceae Rubiaceae
D-Pinitol		Leguminoseae Rhizophoraceae Caryophyllaceae
L-Quebrachitol		Euphorbiaceae
Glycinebetaine	(CH₃)₃N⁺CH₂COO⁻	Chenopodiaceae Amaranthaceae Asteraceae Solanaceae Gramineae Avicenniaceae
B-Alaninebetaine	(CH₃)₃N⁺CH₂CH₂COO⁻	Plumbaginaceae
Proline		Juncaginaceae Asteraceae Gramineae
Prolinebetaine (stachydine)		Labiateae Capparidaceae Leguminoseae
3-Dimethylsulfonio-propionate	(CH₃)₂S⁺CH₂CH₂COO⁻	Asteraceae Gramineae

*After J. Gorham, R.G. Wyn Jones, and E. McDonnell, Some mechanisms of salt tolerance in crop plants, in D. Pasternak and A. San Pietro (eds.), *Biosalinity in Action: Bioproduction with Saline Water*, pp.15 - 40, 1985.

solute compartmentation within a cell, whereby salt is sequestered in a vacuole so that it does not poison the cell; nor is it understood how the cell controls the synthesis of the compatible solutes in response to salinity stress, despite improved knowledge of the biosynthetic pathways involved. The plant hormone abscisic acid may have a role in regulating solute compartmentation because it regulates ion fluxes in plants.

Breeding for salt tolerance. Successful breeding generally requires the existence of variability, and a means for stable transfer of a character from one individual to another. However, breeding plants for tolerance to salinity, while simple in principle, actually is difficult because the criteria used to identify successful individual plants are physiological rather than morphological. A saline-tolerant plant cannot be distinguished by its appearance. Furthermore, variation of salinity within the field where the plants are grown, and interactions between environmental factors, especially heat and drought, contribute to large errors in the selection process. Since most crop plants were developed to provide high yields of grain or fiber, they are poorly tolerant of salinity and lack sufficient variability among individuals for development of salt tolerance. Efforts in Israel to breed salt-tolerant tomatoes have resulted in the development of genetic lines useful for the commercial production of tomato paste, but breeding successes have been few. SEE TOMATO.

Wild relatives of crop plants offer the hope that crosses between species in different genera (called wide crosses) will provide the required tolerance. Crop improvement through crossing with related wild species is a methodology which has already been successful for developing disease-resistant crops; for example, a gene for resistance to leaf rust has been introduced into wheat from a wild grass.

This approach of using wide crosses to develop salt-tolerant plants has aleady had some success, even though the very complexity of salt tolerance implies that more than one gene must be transferred before tolerance can be achieved. For example, several species of the wild grasses in the genus *Elytrigia* are adapted to naturally saline environments and are considerably more salt-tolerant than the most salt-tolerant cultivars of wheat, *Triticum aestivum*. It has been found necessary to transfer five complete chromosomes and a telosome (chromosome fragment) from *Elytrigia pontica* (syn. *Agropyron elongatum*) in order to achieve salt tolerance in wheat. Progeny with fewer *Elytrigia* chromosomes (and the genes which they contain) are poorly tolerant to salt. Thus salinity tolerance is a characteristic which is expressed only when many genes act together, a fact which vastly complicates breeding strategy.

Biotechnology. The rapid development of recombinant deoxyribonucleic acid (DNA) methodology opens up new possibilities for improving the success rate of breeding plants tolerant to salt stress, and many studies now support the concept that the expression of multiple genes is induced or enhanced in response to an imposed salinity stress. Studies have also shown that the expression of other genes apparently ceases. For example, in a study of the protein pattern of cultured tobacco (*Nicotiana tabacum*) cells that were adapted to a medium containing 170-millimolar sodium chloride, it was found that the synthesis of at least eight polypeptides is enhanced while that of four is reduced. Synthesis of the polypeptides begins with a period of osmotic adjustment and growth.

While nothing is yet known about the roles of these proteins, or of controls over the expression of the genes responsible for their synthesis, the fact that the genes are expressed suggests that emerging gene-transfer techniques might soon be used for the development of salt-tolerant crop plants. Now that it is possible to transfer groups of genes and to achieve identifiable salinity tolerance, as shown in the studies with wheat discussed above, the new techniques of molecular genetics should soon provide an improved understanding of the genetics of salt tolerance, and in turn improved success in the development of salt-tolerant crops.

For background information *SEE* B*REEDING* (*PLANT*); C*ELL MEMBRANES*; G*ENE ACTION*; G*ENETIC ENGINEERING*; P*LANTS, SALINE ENVIRONMENTS OF*; R*OOT* in the McGraw-Hill Encyclopedia of Science and Technology.

Richard C. Staples

Bibliography. D. Pasternak and A. San Pietro (eds.), *Biosalinity in Action: Bioproduction with Saline Water*, 1985; J. Dvorak, K. Ross, and S. Mendlinger, Transfer of salt tolerance from *Elytrigia pontica* (Podp.) Holub to wheat by the addition of an incomplete *Elytrigia* genome, *Crop Sci.*, 25:306–309, 1985; N. K. Singh et al., Proteins associated with adaptation of cultured tobacco cells to NaCl, *Plant Physiol.*, 79:126–137, 1985.

Potato, sweet

The sweet potato (*Ipomoea batatas*) is widely accepted as native to tropical northwestern South America, but may have been transported to Pacific Oceania in pre-Columbian times. The species is often described as a cultigen, that is, a plant not found except as cultivated or one recently escaped from cultivation. Among other members of the genus, only *I. aquatica* (water convolvulus, Kang Kong) is commonly cultivated for food. Several species are often cultivated as ornamental plants, commonly called morning glories (*I. tricolor*, *I. nil*).

Worldwide, sweet potatoes are ranked the seventh most important food crop in terms of annual production (**Table 1**). Of the total produc-

Table 1. Average (1982–1984) annual worldwide production of the 10 leading food crops*

Crop	10^6 short tons	10^6 metric tons
Wheat	529	480
Rice	472	428
Maize	459	416
Potatoes	315	286
Barley	177	161
Cassava	140	127
Sweet potatoes	120	109
Soybeans	96	87
Sorghum	75	68
Grapes	73	66

* *FAO Production Yearbook*, vol. 38, 1984.

tion, more than 90% is in Asia, with the People's Republic of China alone accounting for more than 80% of the annual world production. Slightly more than 100,000 acres (40,000 hectares) produce about 1.4×10^9 lb (6×10^8 kg) annually in the United States. Commercial production areas are mainly in the southern tier of states from California to Florida and along the eastern seaboard as far north as New Jersey. North Carolina and Louisiana together account for a majority of the production.

In commercial channels in the United States, sweet potatoes are often called yams, although they are not botanically related to true yams (*Dioscorea* spp.).

Production practices. Technically, the sweet potato is a perennial plant but is usually cultivated as a short-season (90–150-day) annual. In commercial production the growth is terminated by harvest due to impending change of seasons (cold or rainy) or by achievement of optimum storage root size. The plantings are established from sprouts obtained from stored roots in temperate areas or from vine cuttings in warmer regions. Because the production of storage roots does not require flowering or pollination, and the crop does not mature, adverse conditions have the effect of delaying or reducing harvests, and crop failures are rare. A variety of soil types may be used; however, light sandy soils are preferred for ease of harvest. After harvest, the roots may be marketed at once (as is commonly practiced in the tropics) or "cured" for a week at 85°F (29°C) and then stored for up to a year at 55°F (13°C).

Attributes. A number of factors contribute to the worldwide importance of sweet potatoes. Perhaps foremost is the energy yield, since the sweet potato has the highest value among the 10 leading food crops (**Table 2**). Even so, sweet potatoes have received relatively little research attention to improve yield through breeding or cultural practices in the developing countries.

Sweet potato roots are also high in nutritional quality. The data presented in **Table 3** are expressed as nutrient density (that is, the percentage ratio of nutrient supplied per unit energy

Table 2. Yields and energy production of 10 leading food crops

Crop	Yield*		Energy content[†]			Energy yield		
	lb/acre	kg/ha	Cal/lb	Cal/kg	10^6 J/kg	Cal/ft^2	Cal/m^2	10^6 J/m^2
Wheat	1907	2141	1514	3330	13.95	66.3	713	2.985
Rice	2764	3103	1550	3410	14.25	98.4	1058	4.425
Maize	2962	3325	1586	3490	14.60	107.9	1160	4.855
Potatoes	12,918	14,503	323	710	2.95	95.7	1030	4.310
Barley	1892	2124	1486	3270	13.70	64.6	695	2.905
Cassava	8095	9088	450	990	4.15	83.6	900	3.765
Sweet potatoes	12,765	14,332	455	1000	4.20	133.2	1433	5.995
Soy beans	1520	1707	1782	3920	16.40	62.2	669	2.800
Grapes	6311	7086	177	390	1.65	25.7	276	1.155
Sorghum	1231	1382	1555	3420	14.30	43.9	473	1.980

* *FAO Production Yearbook*, vol. 38, 1984.
[†] Food composition table for use in East Asia, FAO, Rome, 1972.

supplied to the nutrient required per unit energy required). Thus if the entire energy requirement of the person cited in the table were met by sweet potato roots, 73% of the protein requirement and 4160% of the vitamin A requirement would be met. The nutritional quality of the protein is high as measured by amino acid analyses. For most people in developing countries, additional consumption of sweet potatoes may serve to improve the diet.

Utilization. The sweet potato is a versatile crop that is used not only as a food but also as an ingredient in other foods, as an animal feed, and as a feedstock for the production of starch and ethanol. In the future, it is expected that the germ-plasm resources will be increasingly utilized for the improvement of the crop.

Food. Sweet potatoes may be substituted in recipes for carrots, squash, or pumpkins and may improve culinary and nutritional properties. They may be preserved by canning, freezing, or dehydration. Among the various products, sweet potato flour produced from dried sweet potatoes is perhaps the most versatile. It can be substituted for 10 to 25% of wheat flour (depending on the product) with no loss of culinary quality and

Table 3. Nutritive value of sweet potato roots*

Nutrient	Value[†]
Protein	73
Vitamin A	4160
Vitamin C	825
Niacin	81
Riboflavin	91
Thiamin	167
Phosphorus	139
Iron	161
Calcium	95

* After *Composition of Foods*, Handb. 8, USDA, 1975; Recommended Daily Allowances, NAS/NRC, 1980.
[†] Expressed as the percentage ratio of nutrient supplied per unit energy supplied to nutrient required per unit energy required. Based on a male, age 23–50, weight 154 lb (70 kg), height 5 ft, 6¾ in. (172 cm).

sometimes with a considerable increase in nutritional quality.

Tender vine tips and leaves are consumed in many parts of the world as a pot herb. They provide a good source of pro-vitamin A, calcium, iron, vitamin B_2, and vitamin C.

Feed. The high energy production and nutritional quality of the sweet potato also contribute to its importance as an animal feed, especially in the tropics, where many temperate-zone feed grains are not well adapted. A number of studies have indicated that dried roots and dried or fresh vines are very acceptable feeds for dairy cows, beef cattle, and sheep. Dried roots are less suitable for swine and poultry production and should not exceed 25% of the energy provided, because of the presence of low levels of trypsin inhibitors and poorly digestible starch. Cooked roots are much better feeds for monogastric animals.

Industrial products. Because the dry matter and energy content of sweet potato roots are largely due to starch, the crop has been used as a feedstock for the production of starch and ethanol. The procedures for producing starch are similar to those used for other root crops such as cassava or canna. The starch is of fairly good quality with an amylose:amylopectin ratio between 20:80 and 30:70, and can be used for food or for industrial products such as glues and sizing. The high energy yields seem to suggest a good potential for the use of sweet potatoes for the production of ethanol. However, the energy balance is not as favorable for sweet potatoes as for some other crops. This indicates that greater efficiency and higher yields are important research priorities.

Subsistence agriculture. Because of the nature of the crop and the fact that sweet potatoes will generally produce a crop even under adverse conditions, sweet potatoes have often been used as an "insurance" crop. In traditional Chinese and other Asian cultures, sweet potatoes are considered a survival food—an important staple food in times of war or adverse weather. In the

United States, land devoted to sweet potato production reached an all-time high of 1.06×10^6 acres (4.3×10^5 ha) in 1932, which was more than 10 times the current acreage. The association of sweet potatoes with hard times has led in many countries to a preference not to eat sweet potatoes as economic conditions improved. Thus, in future years, the most accepted utilization technologies in developing countries may be in the areas of food ingredients, feeds and feeding systems, and improved efficiencies for industrial production.

Germ-plasm resources. Plant and human factors contribute greatly to a large germ-plasm base. Sweet potato plants flower sparingly and usually require cross pollination. Inheritance patterns are complex, and progeny rarely resemble parents closely. Either chance seedlings or mutations, if superior to parents, can easily be preserved by vegetative reproduction. As a result, a wide gene pool is presumed to exist. The germ plasm has not been adequately evaluated, but the work that has been done suggests considerable variability for most traits studied. The lack of importance of sweet potatoes in many temperate countries and the lack of status in tropical developing countries have resulted in low priorities for the collection and preservation of germ-plasm resources. This situation has recently been addressed by the International Board for Plant Genetic Resources (supported through the Consultative Group for International Agricultural Research and housed by the Food and Agricultural Organization of the United Nations), which has formulated plans for the systematic collection and preservation of sweet potato germ-plasm resources. It is expected that the previous neglect of sweet potato germ-plasm resources will be remedied by the implementation of these plans. Presently, over 6000 accessions of sweet potato and 500 of related species have been collected and are being maintained.

The utilization of germ plasm has also presented problems. Often the most important germ plasm must be vegetatively propagated. Since viruses are also likely to be carried in vegetative materials, many countries have strict quarantine requirements. Infected materials generally have been destroyed. Recent advances in virus detection and elimination and in biotechnology are expected to alleviate problems and will greatly enhance the timely exchange of vegetative materials.

Once this germ plasm can be easily exchanged across national borders, it may be used in a number of ways. Some clones, developed or discovered in one region, may be found to be useful in another and can be increased and distributed to growers. This may occur particularly for cultivars to be used for industrial feedstocks or animal feeds where appearance and culinary properties are less important.

Researchers will be able to evaluate a large cross section of sweet potato germ plasm to discover useful traits such as disease resistance or tolerance to an adverse condition such as saline soils. Selected plants would serve as parents for improved cultivars through the use of traditional breeding techniques.

Genes from related species may be incorporated into the sweet potato gene pool by modern techniques such as somatic hybridization, followed by crossing with existing sweet potato clones. This approach seems particularly promising because sweet potatoes are hexaploid (six sets of 15 chromosomes) and many relatives are either diploid (two sets of 15) or tetraploid (four sets of 15). Thus, a large number of combinations of diploid and tetraploid species could be somatically hybridized to produce hexaploid lines, some of which might be sexually compatible with sweet potatoes. A better understanding of the phylogeny and evolution of the species would be gained in addition to the exploitation of genes presently outside the available gene pool. It appears, therefore, that in the near future the genetic resources will be collected, preserved, exchanged, and utilized for the improvement of the crop.

For background information SEE AGRICULTURAL SCIENCE (PLANT); ANIMAL FEEDS; ETHYL ALCOHOL; FOOD; POTATO, SWEET; STARCH in the McGraw-Hill Encyclopedia of Science and Technology.

John C. Bouwkamp

Bibliography. J. C. Bouwkamp (ed.), *Sweet Potato Products: A Natural Resource for the Tropics*, 1985; J. C. Bouwkamp, Sweet potatoes: Potentials and problems, *World Crops*, 36(2):59–62, 1984; J. B. Edmond and G. R. Ammerman, *Sweet Potatoes: Production, Processing, Marketing*, 1976; R. L. Villareal and T. D. Griggs (eds.), *Sweet Potato: Proceedings of the 1st International Symposium*, AVRDC, Shanhua, Tainan, Taiwan, 1982; D. E. Yen, *The Sweet Potato and Oceania*, Bernice P. Bishop Mus. Bull. 236, Honolulu, 1974.

Pregnancy

The fundamental immunologic paradox of mammalian pregnancy has been its apparent success. In an outbred population typified by humans, the fetus inherits genes from its father for major and minor histocompatibility antigens that are foreign to the mother. The implantation of the fetus in the uterus establishes a host-graft (mother-fetus) relationship. Grafts of other types of tissue bearing so-called strong transplantation alloantigens [that is, major histocompatibility complex (MHC) antigens such as the HLA antigen in humans] are rejected even when placed in the uterus. In contrast, approximately 50% of the fertilized eggs that implant in the human uterus gestate successfully to term, and studies in experimental animals have shown that the fetal allograft survives even when the female has been deliberately immunized against the MHC antigens of her mate. Indeed, there is now evidence to suggest that maternal immunity to certain paternal antigens may favor successful pregnancy and prevent spontaneous abortion.

MECHANISMS ENSURING IMMUNOLOGIC SUCCESS OF THE FETUS

Four hypotheses have been proposed to explain the success of the fetus as a transplant: the fetus might not be antigenic; the mother's immune response might be suppressed; the uterus might function as a privileged site; and the placenta and fetal membranes that interpose themselves between maternal tissues and the fetus might act as a protective barrier.

Fetal nonantigenicity. The concept of fetal nonantigenicity is rendered untenable by the fact that antibodies to HLA antigens are obtainable from a proportion of pregnant and parous women. The **illustration** summarizes current information about antigen expression on tissues of the fetoplacental unit. Information is included from studies on the laboratory mouse for several reasons. The mouse has a hemochorial type of placenta, as do humans; there is great similarity between the murine and human immune systems; and there is homology of antigen expression by the fetoplacental unit of both species.

It is important to distinguish between the cells of the fetus itself and the cells of the fetal trophoblast which forms the fetomaternal interface. The cells of the fetus express paternal MHC antigens, non-MHC transplantation antigens, and organ-specific and blood group antigens to which the mother's immune system can respond. Those trophoblast cells which contact maternal blood in the placenta, however, do not express paternal MHC. Nevertheless, these trophoblast cells do bear other types of antigens against which the mother is thought to respond immunologically, and it has recently been shown that antitrophoblast antibodies appear in maternal blood within a few weeks of implantation. The illustration shows that a second type of trophoblast cell (cytotrophoblast in humans and spongiotrophoblast in mice) does express a certain type of paternal MHC called class I MHC. Why therefore is the conceptus not rejected?

Several experimental observations concerning the immunology of trophoblast tissue are relevant to understanding the potential meaning of trophoblast-borne antigens. Naturally, most of these data have been obtainable only by study of pregnancy in animals.

At the time of implantation, MHC expression on trophoblast tissue is shut off and reappears only with formation of the placenta. Deliberate immunization with trophoblast tissue does not elicit transplantation immunity, and the anti-HLA antibody response seen in pregnant women is believed to result primarily from the

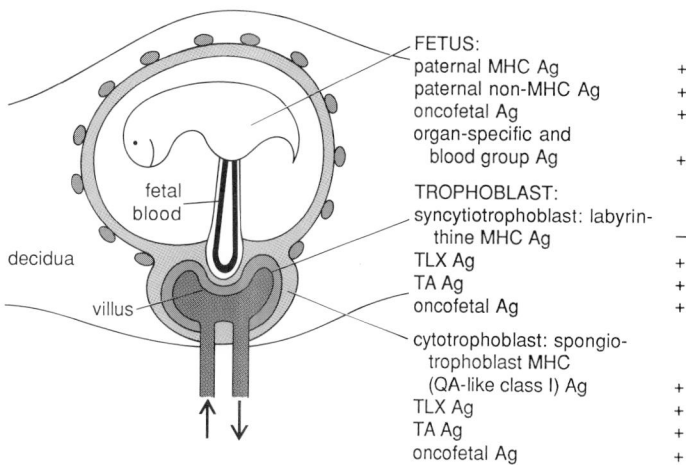

FETUS:	
paternal MHC Ag	+
paternal non-MHC Ag	+
oncofetal Ag	+
organ-specific and blood group Ag	+
TROPHOBLAST:	
syncytiotrophoblast: labyrinthine MHC Ag	−
TLX Ag	+
TA Ag	+
oncofetal Ag	+
cytotrophoblast: spongiotrophoblast MHC (QA-like class I) Ag	+
TLX Ag	+
TA Ag	+
oncofetal Ag	+

Antigen expression by cells of the implanted mammalian fetoplacental unit. Data obtained from studies of laboratory mice and from humans are shown. Ag = antigen. TLX = trophoblast and lymphocyte cross-reactive; antigens shared by both trophoblasts and lymphocytes. TA = trophoblast associated. QA = gene locus near MHC that makes a similar molecule. + = present. − = not present. (After D.A. Clark et al., Immunoregulatory molecules of trophoblast and decidual suppressor cell origin at the materno-fetal interface, Amer. J. Reprod. Immunol. Microbiol., 10:101, 1986)

entry of cells from the fetus into the maternal circulation across the placenta; fetal cells are more immunogenic than are class I MHC-positive trophoblast cells since the latter remain in the uterus. Limiting exposure of the mother to cells of the fetus (by limiting the number of fetal cells that cross the trophoblast into the mother) minimizes the maternal response. Only occasionally does immunization by pregnancy lead to sensitization of maternal cytotoxic effector T cells that participate in rejection of grafts bearing alloantigens (allografts). In most instances the type of immunity elicited by successful allopregnancy (pregnancy where the male is allogeneic with respect to the female) is of a nonharmful variety.

Antibody responses against blood group antigens where the maternal antibodies cross freely into the fetal circulation and cause hydrops foetalis (a condition usually resulting in fetal death) represent one documented harmful maternal antifetal immune reaction. Anti-HLA antibodies (in humans) and anti-H-2 antibodies arising from natural allopregnancy in mice do not lead to obvious harm. Therefore it seems that while the cells of the fetoplacental unit bear antigens and while immune responses occur to some of these antigens, neither fetal cells nor trophoblast cells stimulate the type of immune response that leads to rejection. However, this does not explain why the fetoplacental unit is resistant to deliberate immunization of females against paternal MHC antigens.

Some of the cells of the fetus and trophoblast are susceptible to immune damage. Fetal mouse-fibroblasts become susceptible to lysis by cytotoxic T lymphocytes approximately 7 days after implantation, and entry of a sufficient number of maternal lymphocytes into the fetus has been reported to result in graft-vs.-host disease and immunodeficiency in surviving neonates. Trophoblast might be resistant to rejection, but its outgrowths are representative only of MHC antigen–negative tissues. Murine spongiotrophoblast (MHC antigen–positive) may be susceptible to damage by alloantibody in the organism and by antibody (assisted by xenogeneic rabbit complement) and cytotoxic T lymphocytes in laboratory preparations. Autologous complement is insufficiently potent to cause serious damage. While sensitized maternal T cells such as cytotoxic T lymphocytes may damage fetal tissues and certain subpopulations of trophoblast cells, there is clearly some mechanism operative in the uterus during pregnancy to prevent such damage.

Maternal immune response. There is some nonspecific decrease in systemic maternal immune responses during pregnancy. However, the degree of systemic suppression is markedly less than that required to prevent rejection of surgical transplants. Pregnant females do not become highly susceptible to infection as do immunosuppressed transplant patients. The slight increase in susceptibility to mycobacterial and viral infections may be related to altered macrophage function. There is no evidence that generation of cytotoxic T cells that could cross-react with paternal MHC antigens on the fetus is specifically impaired during first allopregnancy. Suppressor T cells specific for paternal MHC antigens do develop as a result of repeated allopregnancy, but have not proven capable of preventing allograft rejection even with allografts placed in the uterus of experimental animals. A viral infection could generate cross-reactive antifetal cytotoxic T cells, but neither fetal nor trophoblast tissues usually stimulate such a reaction.

Uterus as a privileged site. The concept that the uterus serves as a privileged site could almost be dismissed on the basis that human pregnancy can occur at ectopic sites outside the uterus. However, many such pregnancies begin within the uterus where they are associated with maternal decidua, and the success of experimental ectopic pregnancy where preimplantation murine blastocysts have been transplanted to ectopic sites is only 1%.

Certain types of immunity appear capable of damaging the conceptus even in a uterine environment. Antisperm antibodies have been associated with postimplantation abortion, perhaps because some sperm antigens are expressed in the membrane of fertilized eggs. Antiembryo antigen immunity also seems capable of producing spontaneous abortion. Therefore, maternal immunity to certain types of antigens can be deleterious even to pregnancy occurring at its physiologic site. Since sperm and embryo tissue bear antigens capable of eliciting these reactions, the absence of these harmful types of immune responses to non-MHC antigens during natural

pregnancy cannot be explained solely by poor immunogenicity of the embryo.

There is now good evidence that the uterus affords a degree of privilege to allografts, but this protection requires decidualization. The endometrial cells of the uterus of hormonally primed females transform into decidua in response to implantation or local trauma. Grafts placed on decidua survive longer than expected because of failure to stimulate transplantation immunity. Nevertheless, immunological privilege associated with prevention of immunization by the graft cannot explain failure of rejection of the fetal allograft hosts preimmunized to paternal MHC. Two distinct mechanisms protecting the fetal allograft may be conceived. The first prevents the generation of harmful types of immune responses to non-MHC antigens. The second prevents expression of transplantation immunity against paternal MHC that appears with development of the placenta. Such a mechanism would appear to be ineffective against immunity to non-MHC either because the effector mechanism in this case is different from the mechanism of rejection of MHC-bearing grafts, or because it develops too late in pregnancy to prevent damage directed toward the non-MHC antigens expressed early during development of the embryo.

An initial step in rejection of MHC-incompatible grafts is invasion of host effector lymphocytes into the graft. The ability of maternal lymphocytes to enter the fetal circulation is markedly impaired in comparison to erythrocyte traffic. However, maternal lymphocytes invade both aborting human embryos and xenogeneic embryos in equids. In the latter situation, resulting from xenopregnancy (pregnancy where the male is of a different species from the female), there is evidence for an immune attack on the trophoblast, discouraging its growth in the uterus. This may be due to xenoantigens (species antigens) which, unlike paternal MHC, are readily demonstrable as targets for rejection in trophoblast transplant studies. During xenopregnancy in mice, maternal lymphocytes invade through the trophoblast without causing trophoblast lysis, and subsequently maternal cytotoxic T lymphocytes can be isolated from the dying embryo.

Protective barrier. Data from a number of studies suggest that the barrier function of trophoblast results from active local inhibition of maternal effector-cell activation that prevents migration of cytotoxic cells into the fetus.

Important support for the hypothesis of local suppression was provided by the observation that alloimmunized rats rejected skin allografts placed on the decidua, but not those placed at the trophoblast-decidual junction. This inhibition could be explained by production of suppressor molecules by trophoblast that diffuse outward into the decidua and inhibit sensitized effector cells such as cytotoxic T lymphocytes, or could be due to a similar type of suppressor activity arising in decidua. Studies in allopregnant mice have demonstrated a population of small lymphocytic non-T-suppressor cells in decidua of successfully pregnant animals. These cells release a soluble inhibitor blocking interleukin 2 (IL-2), a molecule required to activate cytotoxic T lymphocytes in the graft rejection process. Antibody against the receptor on cytotoxic T lymphocytes for IL-2 markedly prolongs survival of grafts of adult tissue, and the decidual suppressor-cell factor can block expression of transplantation immunity in MHC-sensitized mice. These suppressor cells differ from those in the normally primed endometrium and are dependent for recruitment on signals from trophoblast. These trophoblast-dependent suppressor cells are deficient at the implantation sites in mice that will abort their xenogeneic or allogeneic embryos.

Local suppressor-cell recruitment by human or murine trophoblast thus appears to require the uterine environment, where it leads to a chemical barrier of suppressor activity that inhibits maternal immunity to paternal MHC through blocking of the response to IL-2.

ANTIGENIC SIMILARITY OR DIFFERENCE AND PREGNANCY OUTCOME

The information in the previous section indicates that xenogeneic pregnancy and some allogeneic pregnancies may fail because of maternal immune rejection of the fetoplacental unit. However, many types of xenopregnancy in equids and in laboratory mice show moderate success without evidence of rejection. These successes may be attributable to sufficient local suppressor-cell activation to protect the conceptus. At the opposite end of the genetic spectrum, it is known that mating may occur between individuals of the same species which share MHC antigens. Inbreeding classically leads to reduced fecundity and loss of reproductive vigor so that only a few lines of "syngeneic" animal strains have been developed. Similarly, it has been noted that some human couples suffering from recurrent spontaneous abortion showed an unexpected degree of MHC sharing, perhaps most notably in class II MHC antigens. Since class II antigens potently stimulate immune responses, the observation of HLA sharing led to the idea that an active immune response by the female against male antigens might be necessary for optimum success of pregnancy. However, class II antigens are not expressed on trophoblast, and the occurrence of HLA sharing is far from universal in couples with recurrent abortion. Two models have been put forward to explain these findings.

Model I states that couples with recurrent abortion constitute a heterogeneous population, and those with HLA sharing differ from those without sharing. Those with HLA sharing are primary aborters who fail to have any live births with their mate because of lack of a pregnancy-enhancing immune response. Secondary aborters show no HLA sharing, may have had a live-born fetus, and represent examples of abortion due to a harmful type of immune response by the mother, though a number of mechanisms may lead to abortion, and human recurrent spontaneous abortion is not a single type of disease.

Model II states that the association of HLA sharing and abortion is explained not by HLA alleles but by the linkage of HLA to other genes. Two types of linkage relevant to spontaneous abortion have been proposed. Model II(a) states that flanking MHC is a collection of alleleic loci belonging to the T/t locus. Mating between MHC-matched males and females enhances the likelihood of homozygosity at a recessive lethal T/t locus. In Model II(b) an antigen shared by trophoblast cells and lymphocytes (TLX antigen) may be linked to MHC. Women whose husbands have similar MHC may be unable to develop immune responses to TLX which enhance the success of pregnancy.

The argument for recessive lethals in MHC sharing has considerable experimental and epidemiologic support, but the finding that many HLA-sharing couples can have successful non-aborting pregnancies after immunotherapy using lymphocytes and platelets from the husband or a third-party male has challenged the recessive-lethal hypothesis. Further, such treatment appears effective in couples whether or not there is MHC antigen sharing. If there were sharing of the same TLX antigen (trophoblast-lymphocyte cross-reactive antigen) identity between the male and female, no form of immunization except that which induces autoimmunity would be expected to elicit an anti-TLX response.

In the CBA/J mouse strain mated to the DBA/2 mouse strain, 30–65% of the embryos abort at mid-pregnancy, and these abortions are associated with development of maternal cytotoxic cells in the fetus. Deficiency of non-T-suppressor cells in the decidua at the implantation site has been correlated with subsequent abortion. The abortion rate can be reduced and litter size increased by preimmunizing the female CBA mice against paternal MHC antigens. However, the DBA/2 strain does not effectively immunize, and splenocytes (spleen cells) from Balb/c strain mice (which have the same MHC as DBA/2) must be used. It is clear that Balb/c elicits a more vigorous anti-H-2 antibody response and that these anti-MHC antibodies mediate protection against abortion. However, further work with recombinant strains has shown that there is no correlation between the anti-H-2 antibody response and the efficacy in preventing abortion. Therefore, immunity elicited by non-MHC antigens may be very important in determining pregnancy outcome.

The immunologic and genetic mechanisms underlying these phenomena remain unclear, however, since in aborting human couples, as well as in mice, stimulation of several different types of antibody response has been associated with successful prevention of abortion. Studies in the mice have shown that vaccination that protects against abortion increases local suppressor-cell activity in the uterus, whereas vaccination that increases the abortion rate decreases suppressor-cell activity. By contrast, no suppressor cell deficiency in the decidua of mice with lethal T gene alleles appears to have preceded fetal death nor, conversely, have there been reports of an increased frequency of babies with major congenital anomalies born to women treated with immunotherapy for spontaneous abortion. Nevertheless, it has been suggested that immune surveillance eliminates abnormal infants in the uterus in the case of neural tube defects and for minor anomalies in babies gestating in aged mice.

Thus, some investigators using vaccination for unexplained abortion have reported cases of intrauterine growth retardation and suspected neonatal immunodeficiency. It has been suggested that these infants represent maternal lymphocyte colonization with graft-vs.-host disease. However, couples with a history of unexplained recurrent abortion and at least one live-born child manifest a higher than expected rate of intrauterine growth retardation. Therefore, such effects may not be related to deliberate immunization against alloantigens as occurs with immunotherapy.

Effect of Maternal Immune System on the Placenta

Spontaneously occurring examples of maternal lymphocytic infiltration and placental damage include idiopathic villitis (inflammation at the fetal villi that form the placenta), and the villitis accompanying herpes gestation, in which antibodies occur that cross-react with placental and cutaneous basement membrane. A direct stimulatory effect of alloantibody on trophoblast was shown in adoptive transfer of serum to unimmunized pregnant females, and this effect was amplified in the organism by xenogeneic complement. Increased trophoblast mass was also seen when mice were successfully vaccinated for prevention of spontaneous abortion. However, a distinction must be made between stimulation of trophoblast growth that is a repair response to trophoblast damage, and proliferation that represents nontoxic growth stimulation with enhanced placental function.

How might beneficial stimulation of trophoblast be achieved with immunization against abortion? Direct boosting of local suppressor-cell activity reduces immune damage to trophoblast. Supernatants prepared from murine decidua also exert a stimulatory effect on trophoblast in laboratory preparations, as do certain T-cell lymphokines, soluble mediators secreted by T cells. It is possible that alloantibody or immune complexes may bind to decidual suppressor cells, perhaps via their receptors that bind the Fc end of antibody molecules, and increase the release of factors that stimulate trophoblast growth; these growth factors might also suppress immunity. Precedent for such a hypothesis is provided by the finding that epithelial growth factor that stimulates epithelial-cell proliferation is immunosuppressive when applied to lymphocytes. However, decidual suppressor cells are dependent upon soluble factors from trophoblast cells, and it is possible that successful vaccination increases local suppression by a direct immunostimulating action on trophoblast growth or function.

The major unsolved problems are definition of various suppressor and growth-promoting factors in the placental bed, what distinguishes helpful types of maternal immunity from harmful varieties, and how immunity acts to alter trophoblast and suppressor-cell function. Available data have implied that antibodies can be beneficial or harmful, but sensitized T cells are mainly harmful. T-cell lymphokines may also stimulate trophoblast growth, and there may be T-cell responses that are beneficial to pregnancy. Identification of these different types of effects is a prerequisite for treatment of a variety of disorders of reproduction where maternal immunity may play a role.

For background information *see* Histocompatibility; Immunology; Pregnancy in the McGraw-Hill Encyclopedia of Science and Technology.

D. A. Clark

Bibliography. J. Bulmer and P. M. Johnson, Antigen expression by trophoblast populations in the human placenta and their possible immunological relevance, *Placenta*, 6:127–140, 1986; D. A. Clark, Maternofetal relations: A minireview, *Immunol. Lett.*, 9:239–247, 1985; D. A. Clark and B. A. Croy (eds.), *Reproductive Immunology 1986*, Proceedings of the 3d International Congress of Reproductive Immunology, 1986; F. Zuckermann, Ph.D. thesis, University of Texas at Dallas, 1986.

Printing

A number of advances have been made in the field of graphic arts. This article discusses electron-beam engraving of gravure cylinders, new color standards for web offset and gravure publication, gray component replacement, and the North American color survey.

Electron-Beam Engraving

The gravure printing process is used for printing large quantities of catalogs and weekly mag-

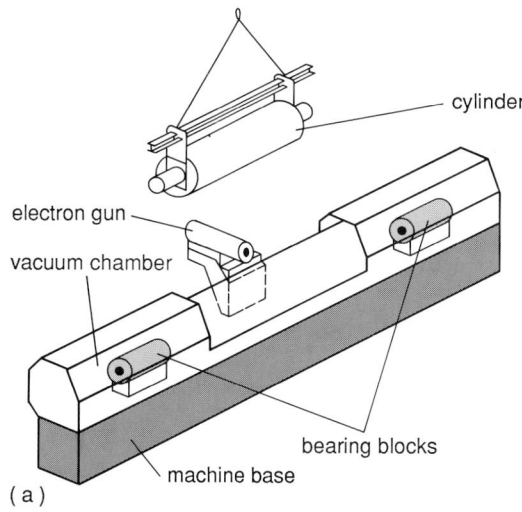

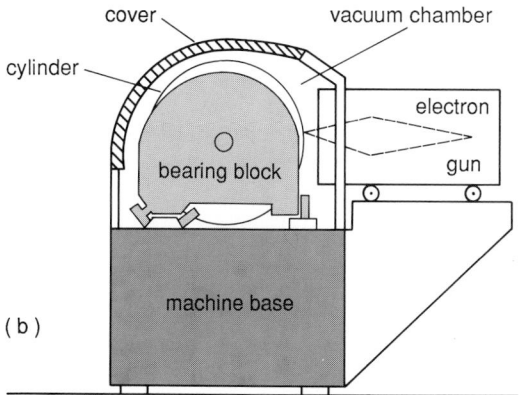

Fig. 1. Electron-beam engraving machine. (*a*)
Three-dimensional view. (*b*) **Cross section.**

azines with high-quality color reproduction. The
printing forms used today are large copper-plated
cylinders with up to 48 pages per cylinder. The
images of text, line art, and pictures are set into
the cylinder surface in the form of tiny gravure
cells—up to 30,000 cells per square inch (1800
cells per square centimeter). The number of cells
per unit area and their volume determine the
amount of ink transferred to paper during the
printing process.

The first technology applied to generate these
gravure cells was chemical etching. This process
utilized a carbon tissue layer whose permeability
for the etching liquid had been influenced by a
kind of photographic process dependent on the
text and image content of the pages. However,
only a few companies, mainly in the packaging
and decorative printing market, are still using this
etching method.

In the 1960s, another engraving method began
to replace chemical etching. The gravure cells are
engraved by an electromechanically driven diamond stylus which is capable of cutting up to
4000 gravure cells per second. By using several of
these engraving heads in parallel, one per circum-

ferential engraving ribbon, a cylinder is engraved
in about 1 h. This process has been widely
accepted in the gravure industry and is being
used internationally. It guarantees high quality
and stability. Since the engraving tool can be
controlled by electrical signals, the interfacing to
computerized text-and-image processing systems
in the prepress areas is feasible.

A new technique of engraving gravure cylinders has been developed that utilizes the power
of an electron beam. As engraving speeds of up
to 150,000 cells per second have been achieved
by using only one electron-beam gun, this new
technology has the potential of higher speed and
productivity for the gravure industry.

Design of engraving unit. The electron-beam engraver consists of a heavy, cast-iron machine
base with a vacuum chamber in which there are
two bearing blocks that chuck and move the
rotating printing cylinder in the cross-feed direction (**Fig. 1**).

The electron-beam gun has a fixed position on
the machine bed. The remaining machine components are the electronic cabinets, the high-
voltage supply, and the vacuum pumps. The
mechanical design of the electron-beam engraving machine allows a fully computer-controlled
engraving process with automatic cylinder loading, a precondition for a short setup time and high
overall productivity. The prototype model of the
electron-beam engraving machine design is 24 ft
(7.3 m) long. It allows a maximum engraving
length of up to 8.5 ft (2.6 m) and cylinder circum-
ferences of up to 4.9 ft (1.5 m).

Electron beam: a thermal tool. The heart of the
electron-beam engraving machine is the electron-
beam gun with its two main components, the
beam generator and the electrical and magnetic
beam optical system (**Fig. 2**).

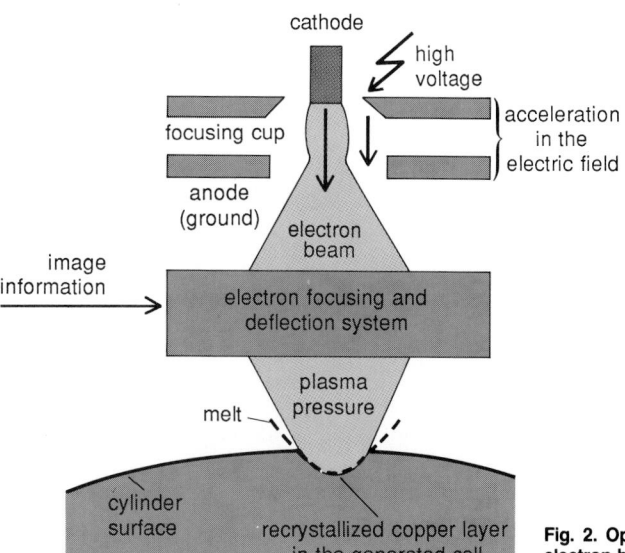

**Fig. 2. Operation of the
electron-beam engraving
tool.**

The beam generator contains a heated tungsten cathode which produces electrons by thermionic emission. These electrons are accelerated in an electric field, attaining a speed of about 3×10^8 ft (10^8 m) per second. This corresponds to one-third of the speed of light, 186,000 mi or 300,000 km per second. The beam accelerator and modulator employs a design similar to that used for television cathode-ray tubes, but the power transported by the electron beam in this process is much higher, in the range of 2 kW.

The electron beam works like a thermal tool. In the high-voltage electric field the electrons gain speed and kinetic energy. The kinetic energy is converted into heat when the focused beam hits the copper surface of the gravure cylinder. The high-energy electrons penetrate deeply, approximately 0.0002 in. (5 micrometers), into the copper and are decelerated in the electric field of the copper atoms. They deliver their entire kinetic energy to the copper and produce a superheated copper melt.

A typical gravure cell, with a diameter of 0.0044 in. (110 μm) and a depth of 0.001 in. (25 μm), has a volume of 6.1×10^{-9} in.³ (1×10^5 μm³). Because of this small volume, the material contained in the cell is melted within a few microseconds. A high plasma pressure is generated by the electron beam itself, caused by copper and air ions and produced by primary beam electrons and by electrons reflected from the copper surface. This pressure forces the liquid copper out of the cell at the circumferential edge of the melt zone. The pulses employed are rather short; only about 6 microseconds are required to produce the largest cell volume (one pulse = one cell). This pulse time is sufficiently short to prevent much heat from migrating out of the cell area into its surroundings. Otherwise the copper material in the direct neighborhood of the cell would become weak, and the plasma pressure would affect the formation of the walls around the cell. These nonuniform walls would make the gravure cylinder unusable for high-quality printing.

After the electron-beam pulse has been switched off, a small quantity of melted copper remains in the cell after the shot. The surface tension of this copper melt causes the extremely thin melt zone to recrystallize to a smooth surface. During recrystallization a small thin burr is left around the edge of the cell. This burr is cut away during the engraving process by using a diamond shaver positioned at the side of the cylinder opposite the electron gun.

The shape of the cell resembles a spherical cap and therefore ensures good ink transfer characteristics (**Fig. 3**). The cell volume is controlled by the electron focusing and reflection system. The electron pulse time and the dynamic zooming and focusing by means of magnetic coils allow independent control of cell diameter and depth. This gives the flexibility required to achieve a tonal value scale from highlight to shadow, best suited for the gravure printing process.

The high circumferential speed of the cylinder during engraving, in the range of 70 to 100 ft (20 to 30 m) per second, causes the molten droplets of copper that have been forced out of the cells to fly away from the rotating cylinder, mostly in a tangential direction. The copper droplets cool very slightly during their flight through the vacuum and then adhere to a special shield.

Most of the cell volume leaves the cell in molten form. Evaporation into gaseous copper would need much greater energy. This is the reason for the high efficiency of the electron beam as an engraving tool for copper as compared to a laser-beam tool. The photons of a laser beam do not really penetrate into the material; most of them are deflected from the copper surface, which acts as a mirror. The remaining photons evaporate the copper with a low efficiency in terms of energy usage.

Productivity and quality. Besides the substantially higher productivity in gravure cylinder production, the electron-beam engraving process includes several significant quality improvements. The almost ideal shape of the cells guarantee a good and stable ink transfer to paper in

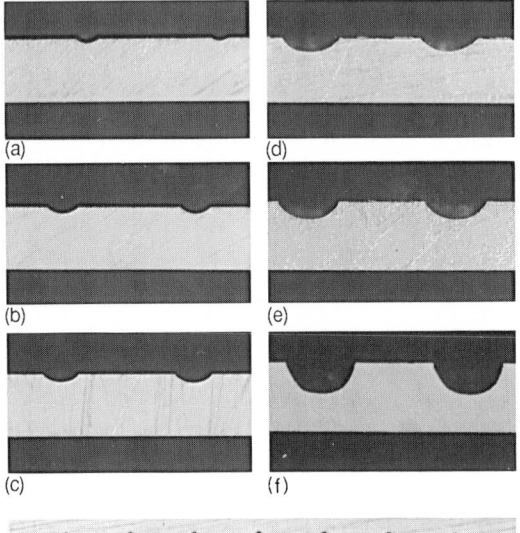

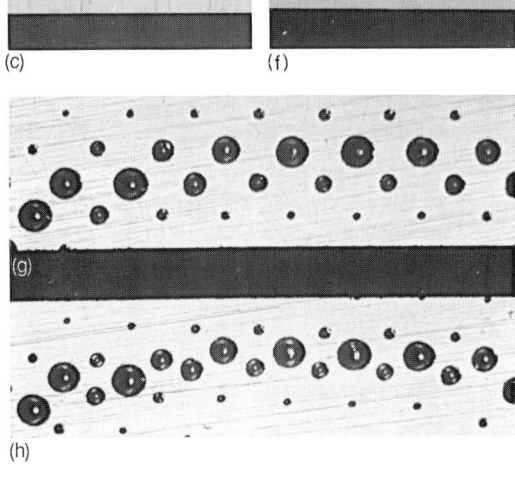

Fig. 3. Electron-beam gravure cells. (*a–f*) The shapes of the cells, showing the different diameters and depths for the reproduction of highlight to shadow tones. (*g–h*) A comparison of cell arrangements without and with the EBG cell shift technique for contour enhancement.

the printing process. The possibility of deflecting an electron beam by means of electric and magnetic fields allows engraving of any desired screen angle by shifting the cells a predefined amount in reference to the desired screen angle. Special combinations of screen angles and rulings within the four colors avoid problems of moiré and color drift during printing. A special contour-enhancement technique using the electron-beam engraver improves the reproduction quality of fine text and line work. The small and medium-size cells near the contours are shifted toward the image boundary. This increases image sharpness and readability remarkably (Fig. 3). The shift vectors are calculated in real time during engraving, depending on the image content to be engraved.

Data input. The technology of electron-beam engraving allows an extremely rapid engraving speed, up to 150,000 cells per second. This is five times as fast as the existing electromechanical engraving machines, with eight engraving heads working simultaneously in parallel.

Production electron-beam engraving machines will operate in computerized, highly automated environments. The pages to be engraved onto the printing cylinders will be prepared in prepress digital image processing systems.

The amount of digital data in a standard catalog color page is in the range of some 40 megabytes. To achieve the economical advantages of the electron-beam engraving process and to make use of its high productivity, the data of all the pages are managed in a suitable data base management system of several gigabytes. The different pages are called out of this data base when needed, sorted in accordance to the cylinder layout, and delivered to the output device, the electron-beam engraving machine.

During the process of engraving using this image data, all necessary parameters for positioning the cells and controlling their specific shape and volume are calculated in real time. It was anticipated that a publishing house would start a pilot operation with an electron-beam engraving machine during 1987–1988. All tests and proofs up to now confirm that this new technology can be used to develop a commercial machine that will be capable of providing a high degree of productivity and quality. This new technology will be especially important for magazine and catalog printing.   *Wolfgang Boppel*

Color Standards for Web Offset and Gravure

Since 1975, under the SWOP (Specifications for Web Offset Publications) program, carefully selected printed swatches of the process inks have been supplied by the International Prepress Association [formerly Platemakers Educational and Research Institute (PERI)] as guides for control of solid ink amounts in ink-paper offset proofing. In 1985 the gravure industry, reacting to the growth of halftone gravure, accepted the same proofing ink hues and ink references used in the SWOP program for offset proofs of halftone material for gravure cylinder-making. This brings closer the day when both offset and gravure publications can use the same halftone input material, the so-called universal film.

Background. In any form of rotary printing with pages printing in sequence around the printing cylinder, input standards must be imposed to assure that images can be reproduced correctly when the pages are given equal inking by the press. This situation was first addressed on an industry-wide basis in the mid-1930s, when heat-set web letterpress became the predominant production method for magazine printing. Advertising engravings were produced by many different photoengravers. It became obvious that all of the plates must be proofed with the same inks on the same paper and in the same amounts or the printed result would not match the customer-approved proof.

Individual publications created their own standards, furnishing proofing inks and paper to engravers, and specifying that these were to be used in preparing material to be printed in their publications. Standard ink amounts were illustrated by printed color bars which each publication furnished to the engraver. The system was destined to fail, since advertisers ordered the same plates to be used in several magazines, even though the individual publications had different input standards.

By 1971, through the work of committees of advertisers, publishers, printers, and photoengravers, a single set of specifications for advertising engravings for web letterpress printing was developed. These called for a standard proofing paper, proofing ink from a single source, and a standard color bar design and printed reference color bars, showing the acceptable ink amounts for proofing, produced by PERI.

Shortly after this system was in place, web offset became the major process used in magazine reproduction, and it again became necessary to create specifications for furnished advertising material. Based on letterpress experience, proofing paper, proofer's color bar design, proofing ink hues and tinctorial strength, and solid printed ink densities were specified. The printing ink specifications were incorporated into printed ink references, again prepared and selected by PERI. The system, later to be known as SWOP, was introduced in 1975. Leading trade and technical associations endorsed the program and formed a review committee charged with the responsibility of revising and updating the specifications as necessary.

Ink references for offset lithography. The term color references implies standardization of hue; the references of the SWOP program display both hue and density standard values. Originally, ink references were printed on a four-color sheet-fed offset press, and were selected to show density values within ±.03 density units of the agreed standard. The original specifications stated that limits of acceptability were ±.05 from the values displayed on whatever ink reference the proof-press operator had at hand.

This system had two serious drawbacks. The first involved the total allowable variance, combining the ±.03 tolerance in producing the references and the ±.05 deviation allowed the proofer created a ±.08 limit. If a proofer's reference was at one extreme of the tolerance limits and that used by the inspector at the other, confusion ensued. The other drawback arose from the variations in sensitivity exhibited by different densitometers in measuring differences between ink-on-paper references. The ±.05 tolerance limits as measured on one instrument could measure ±.03 or ±.08 on another instrument.

Hi-Lo references. In 1984 the SWOP Review Committee accepted as standard a new Hi-Lo ink reference, developed and produced by PERI. These references (**Fig. 4**) consist of ink-on-paper printed swatches selected to display density values of ±.05 density units from standard, as read in the PERI laboratory. The instruments used in selecting the references are regularly checked against retained samples of printed color references and against ceramic tile color standards.

These ink references have been standard for the SWOP program since 1984. In 1986, with a revision of the printed specifications, the tolerance limits for solid ink densities were broadened to ±.07. This change was based on agreement among members of the review committee that controlling dot gain in printing and maintaining correct "gray balance" between the printed colors were of first-order importance in quality control, while solid ink density might be considered almost a second-order variable by comparison.

Halftone gravure. Since the late 1950s the gravure industry followed standards for input material developed by the Gravure Technical Association (GTA). These were based on the conventional gravure process, using continuous-tone positives with a specified density range (.35–1.65) and ink hues and amounts which had been customary in the gravure trade. GTA published a set of color standards, including prints of gravure process colors, printed from etched cylinders made from film with specified density values. Thus, a particular printed color value, if all cylinder-making operations were under control, could be related to a particular transmission density of a continuous-tone film.

With the growth in popularity of electromechanical cylinder engraving systems, the use of chemically etched cylinders rapidly decreased. In Europe and more recently in the United States, "halftone gravure" systems became popular, and by 1985 became the commonly accepted cylinder-making system for publication printing. In this system a screened image, resembling that used in lithographic reproduction, is scanned by an optical-electronic device, and an engraving stylus is operated to create the corresponding gravure printing cells. In effect, it became possible to use essentially the same halftone images for gravure and lithographic printing surface preparation.

However, the ink hues long used for gravure publications (designated by GTA as Group V) were far from theoretically ideal, and differed greatly from lithographic inks. Unless the hues were changed, there could be no universal (lithography-gravure) input material. A GTA committee, consulting with ink manufacturers, discovered that it would be economically feasible for gravure publication inks to be formulated to match lithographic inks in hue; in mid-1985, new Group VI Gravure Standards were adopted. These call for proofing of halftone gravure input material on offset presses using the standard paper, inks, and ink amounts specified under the SWOP program for lithographic input material. SWOP Hi-Lo ink references were designated the standard for halftone gravure proofing for publications.

Newspaper supplements printed by gravure are still produced according to different ink hue and amount specifications; it is hoped that this situation will eventually meet a rational solution.

Hue standards: lithography. The original proofing ink standards were based on inks prepared by a single manufacturer, which furnished wet-ink samples to other manufacturers wishing to produce standard inks. In time it was realized that wet-ink samples needed to be referenced to a printed result, and in 1985 the National Association of Printing Ink Manufacturers worked with the Graphic Arts Technical Foundation to set up an ink certification program. This involves comparison of candidate inks to wet reference samples, and use of spectrophotometric analysis of prints of the candidate and standard inks to

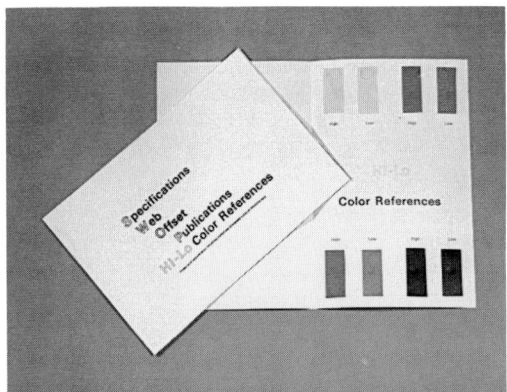

Fig. 4. The Hi-Lo ink references, standard for SWOP program since 1984.

determine whether the candidate ink corresponds to the accepted standard. As a consequence of the correlation of the SWOP and GTA standardization programs that are now in place, there exist a standard proofing paper for lithography and halftone gravure publication; standard ink-paper references for both prepared by the International Prepress Association; and a reference service involving the Graphic Arts Technical Foundation and the National Association of Printing Ink Manufacturers to assure conformity to standards by inks used for lithographic and gravure publication proofing.   *Paul F. Borth*

GRAY COMPONENT REPLACEMENT

This is a new process for improving the quality, increasing the consistency, and reducing the cost of color reproduction. In gray component replacement the graying component of colors is replaced by black. Wherever three colors are overprinted, the two predominant ones determine what color it will be (its hue—red, green, blue, and so forth); the lesser third color determines its purity, grayness, saturation, or chroma. When combined with equivalent amounts of the other two colors, the third color produces gray, which has no color, or is achromatic. This amount of gray, which is called the gray component of the color, can be removed from the color and replaced with black ink. The final color is the same as the original color, but it consists of less of the two process colors, none of the third color, and a little more black. Gray component replacement is illustrated in **Fig. 5**. This is called achromatic color reduction in Germany, where the modern techniques for its use were developed, and is called gray component replacement almost everywhere else.

Replacing the gray component of colors with black results in many obvious and some not-so-obvious benefits and advantages in color reproduction. Sharper printing is achieved due to all detail being in the black. There is greater consistency in the printed color, and brighter colors can be printed on poorer grades of paper. There is stable gray balance, less sensitivity to color ink fluctuation, reduced metameric (color) variations under different light sources, and reduced ink consumption. Drying problems are diminished, and less energy is required for drying ink. Other advantages include shorter makeready time, higher printing speed, the use of lighter-weight papers, reduced spray powder to prevent smearing in sheet-fed printing, better ink receptivity, fewer trapping or ink transfer problems, and less dot gain and higher print contrast. All of the advantages derived from gray component replacement benefit the printing operations of the color reproduction process. There are no advantages in the preparatory operations. In fact, some problems can arise in subjects for which complete substitution of black is used, since local

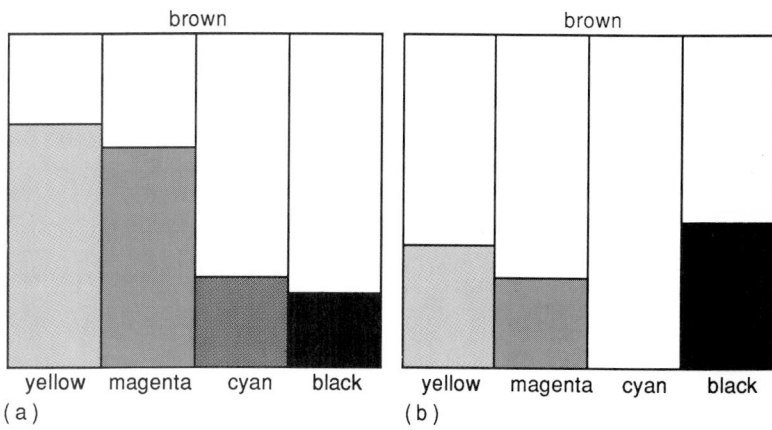

Fig. 5. Principle of gray component replacement. (*a*) The color brown is composed of about equal amounts of yellow and magenta, a smaller amount of cyan, and a small amount of black. (*b*) The same brown color reproduced by about equal amounts of yellow and black, a smaller amount of magenta, and no cyan.

corrections which involve hue changes cannot be accomplished. The advantages in the printing operation, however, are so numerous and significant that the concept has spread rapidly throughout the printing industry.

Development. The principle of gray component replacement is based on a study involving undercolor removal reported in 1940. It was suggested that black be substituted for the graying component in colors consisting of the three subtractive primaries—yellow, magenta, and cyan. For example, a brown color composed of about equal amounts of yellow and magenta and a smaller amount of cyan could be reproduced by removing the cyan, using equal amounts of yellow and magenta, and substituting an equivalent amount of black or gray for the neutral component removed. This is illustrated in Fig. 5.

The first electronic color scanner, introduced in 1950, used this principle of color synthesis. With the inks and computational techniques available at the time, color hues could not be controlled and the darker colors lost saturation. The original concept of undercolor removal was modified and restricted to the substitution of black for part of the neutral component of the darker colors. It was at this time that 240% undercolor removal was adopted for letterpress printing; later 280% undercolor removal was adopted for web offset and sheet-fed lithography.

Gray component replacement remained essentially dormant until 1976, when a theory of achromatic color reduction was published which included color charts using achromatic synthesis or gray component replacement.

In 1980 a modern concept of gray component replacement was proposed using a modification of the Neugebauer equations (relationships that define color mixtures of halftone images), a computer, and look-up tables of printed color charts to convert colors from the red, green, and blue

(RGB) values read on densitometers and scanners to gray component replacement values. Other uses soon followed, and gray component replacement came into general use in 1984.

Actually, gray component replacement is a very complex process—so complex, in fact, that it is practically impossible to accomplish photographically. One of the main obstacles is the impurity of the pigments in the printing inks. Another obstacle is additivity failure—when the strength or density of two or three overprinting inks is less than the sum of the densities of the individual inks printed singly on the paper. A third important factor is the effect of the paper, press, and halftone screen on graying colors.

A perfect cyan ink should absorb all the red light and transmit or reflect all the blue and green light. Most cyan inks transmit or reflect more blue light than green light and some red light. A perfect magenta should absorb all the green light and transmit or reflect all the red and blue light. The best magenta inks transmit or reflect more red light than blue and some green light. Yellow, which is the most efficient of the process inks, transmits or reflects almost all of the green and red light and also a small amount of the blue light which it should absorb.

Therefore, when two colors such as cyan and magenta are overprinted, most of the red and green in the light is absorbed; but instead of all the blue light being transmitted or reflected, as much as 50% of it is absorbed, making the overprint appear as dark as if it contained as much as 50% black. The overprint of 100% yellow and cyan can result in a murky green with as much as 40% grayness; and the overprint of yellow and magenta produces a red with as much as 20% grayness or black. The addition of a third color to the two-color combination imparts additional absorption of the color that should be transmitted or reflected. In the reproduction of color tones or tints, the absorption of light in the paper between the dots causes graying of the color in the halftone tints, a phenomenon known as optical dot gain. In addition, during printing, the physical dot gain on the press can vary the relative amounts of colors by changing their dot sizes.

It is practically impossible to compensate for all these factors when calculating the amount of black needed for gray component replacement in photography and on conventional scanners with analog computers. Now that digital computers have become available on the newer scanners, it has become practicable to use gray component replacement efficiently. While all the scanner manufacturers are trying to simplify color reproduction by substituting black for the graying components in overprinting colors, each has approached it differently and has obtained slightly different results.

Limitations. Full or 100% gray component replacement provides the most benefits in consistency, cost reduction, increased productivity, and quality of color reproduction, but it also has some limitations and disadvantages. Full gray component replacement produces images in which all the detail is carried in the black image with no more than two colors and black in any part of the image. Changes in ink feed on the press affect mainly the value (lightness and darkness) of the colors in the image; the effect on their hues is much smaller. Color correction artists and scanner operators have used this principle for a long time in the reproduction of skin tones, especially those of dark-skinned people. Skin tones consist mainly of tints of yellow and magenta, and the third color controlling saturation can be cyan or black, or both. When cyan is used, any variation in dot gain on the press can cause shifts in the hues of skin tones from reddish to greenish. These objectionable hue shifts in printing have been eliminated by substituting black for the cyan.

Another example of the limitations is in the reproduction of pearls. Bluish pearls are less expensive than pinkish pearls. Printing from normal separations could shift the color of the pearls from pink to blue. Gray component replacement separations with no cyan could eliminate this problem, as well as offer fewer trapping or ink transfer problems, shorter drying times, and ink savings on the press.

There are, however, disadvantages to gray component replacement. In some cases, local editorial changes may be impossible. In the example of the pearls, if the pearls in the transparency are bluish, with 100% gray component replacement there would be no way to make the pearls look pink, as there is no magenta in the pearls in the separation. There are other problems with gray component replacement. Unless a very strong black ink is used, contrast may suffer because it may not extend the range enough in the gray scale, and it will have less gloss than a color made with three or four inks. Also, misregister with 100% gray component replacement produces white lines or borders around objects in the image. For these reasons, some amount of the three colors is run under the black in shadow areas, which is designated as less than 100% gray component replacement and is also known as undercolor addition. One hundred percent gray component replacement is generally satisfactory on poor papers like newsprint; on magazine and coated papers 50% gray component replacement is used almost exclusively.   *Michael H. Bruno*

NORTH AMERICAN PRINT SURVEY

A 2-year survey of North American printing completed in 1984 has established performance levels for industry segments. These data are used by individual printers to judge their own performance and by industry organizations to help develop specifications for high-quality printing.

Measurements in printing. The rapid change in printing technology since about 1970 has presented both the opportunity and the need to understand the printing process better through careful measurement of printing properties. Printers with multi-million-dollar presses cannot afford long makeready periods during which acceptable color is established on press. Using standardized printing conditions can cut makeready time and yield better quality more quickly; however, this procedure requires that input films and plates be made according to specifications.

Part of the need for specifications arises from the lack of vertical integration in the printing industry. The industry is made up of many segments: color trade shops create color separations and halftone films, and they often print proofs of these films on specialized presses; commercial printers may make their own halftone separations or receive them from a few trade shops and then print on either sheet-fed or web-fed presses; and publication printers receive film from many sources and then print on high-speed web presses. Each segment has its own needs and quality goals.

As discussed in the second section of this article, several industry organizations have been examining the range of printing performance and the need for voluntary standards or specifications. The Graphic Communications Association (GCA), a division of Printing Industries of America, has concentrated on the publication segment. In a study spanning a number of years, member companies have donated press time, paper, and other supplies to define the operating window for high-speed web presses. This information has served as part of the basis of discussion for the SWOP committee. The SWOP committee's goal is to formulate guidelines by which films can be made and proofed by various trade shops and then printed with few problems by various printers. Similar efforts are under way internationally.

An extensive survey of North American lithographic printing was conducted bu DuPont. More than 700 press sheets were measured over 2 years as part of a printing analysis service offered to customers. Of the several printing properties that were analyzed, two will be discussed here: solid ink density and dot gain. For comparison purposes, the industry was divided into four major segments: sheet-fed presses used for proofing or production, and web-fed presses used for commercial printing or publication printing. The results for the web publication segment were largely obtained in cooperation with GCA in their print properties tests.

The survey used a special test form which contained both test strips and pictorial elements. Printers were asked to print normally, which in most cases meant adjusting the press until the picture quality was good. Measurements were made with a densitometer equipped with narrow-

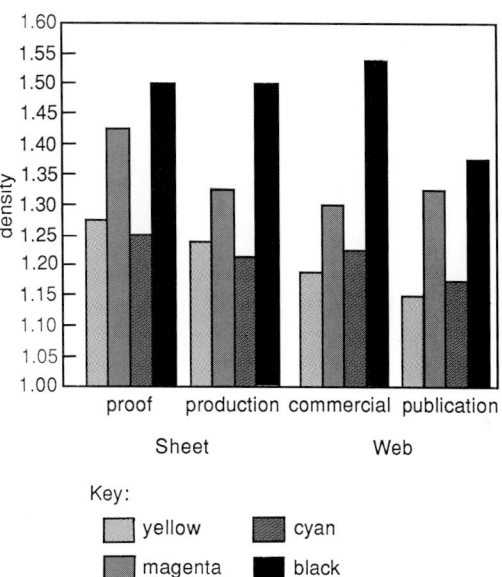

Fig. 6. Solid ink densities for each industry segment. Sheet-fed proofing runs are at higher densities than high-speed web publication printing.

band filters to give the highest sensitivity to density variations in all four printing colors.

Solid ink density. The maximum depth of color that can be achieved in printing is referred to as the solid ink density. Many factors affect solid ink density, including the amount of ink fed to the plate through the inking system, the pigment concentration or strength of the ink, and the type of paper being used. High solid ink density gives good color saturation in solid areas, such as in a red fire engine, whereas low solid ink density leads to a washed-out appearance. **Figure 6** summarizes solid ink density performance in each industry segment.

As expected, densities are higher in sheet-fed printing, especially in proofing, and are lower in web printing. Sheet-fed printing is used for short-run, high-quality work such as art books, annual reports, and limited circulation magazines. However, the standard deviation in the densities between printers is 0.10, so there is a fair amount of overlap between the segments.

Dot gain. Dot gain refers to the change in the size of the halftone dots from the film original to the final printed sheet. The physical transfer of ink from plate to blanket to paper in offset printing always causes the dot to be larger on the paper than on the printing plate; in addition, optical effects of the ink lying on the surface of the paper make it appear larger than it actually is. The net effect is that printed halftone patterns look heavier than predicted by the original dot size in the film.

Whereas printers have traditionally measured solid ink density, dot gain has come under examination only relatively recently. Dot gain can be controlled on press by adjusting ink rheology (for example, ink film thickness, strength, viscosity, or tack); changing pressure between plate, blan-

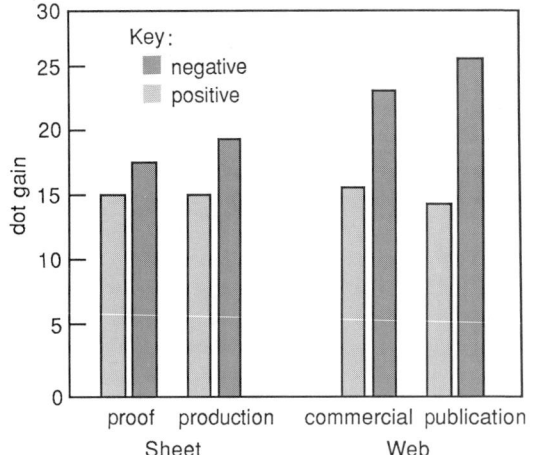

Fig. 7. Dot gain at the 50% dot (midtone dot gain). Two types of printing are shown: (1) using plates made from negatives and (2) using plates made from positives. The results are averaged over all four printing ink colors, but show the difference between positive and negative plates.

working plates and positive-working plates.

The data says that a 50% (midtone) dot in sheet-fed negative proofing, for example, appears on paper to be a 68.6% dot. The range in dot gain is wide: some proofing was less than 10% gain (and thus all optical dot gain), while some web publication gain was in excess of 40% (indicating that the midtones were printing almost as a solid). Theoretical considerations indicate that printing from positive plates should be sharper (have less dot gain) than from negative plates.

Dot gain itself is not undesirable. It is a fact of offset lithographic printing and must be measured and controlled. The differences between industry segment and type of printing plate should be factored into the color separation process. The tone reproduction curve used to make separations can compensate for the dot gain in printing, if that dot gain is known and is consistent.

Applications. These data have been used to help individual printers judge their performance relative to others in their market segment. Increasing interest in printing properties has led to the appearance of several printing analysis products in which a computer with densitometer or spectrophotometer interface accumulates printing data, provides analysis of individual sheets, and establishes a data base for tracking performance trends. **Figure 8** shows one of the new devices for measuring printing quality; it is a self-sufficient electronic work station providing computerized print analysis to printers and trade shops so that they can diagnose quality-control problems in the plant.

ket, and impression cylinders; and choosing different paper. When dot gain is not the same for each ink, and varies differently between colors, strong shifts in color result. Excessive dot gain will cause the printed image to appear muddy and to lose shadow detail.

The amount of dot gain depends on the type of dot being printed. It increases with increasing screen ruling (dot gain for a ruling of 150 dots/in. or 60 dots/cm is twice that for a ruling of 75 dots/in. or 30 dots/cm) and for a given screen ruling is largest for the midtone region (40–60% dots) and less in the highlight and shadow areas. The results reported in **Fig. 7** are for a 50% dot on a 150-dot/in. (60-dot/cm) target. The dot gain results given in Fig. 7 are averaged across all four colors and show the distinction between using negative-

Another use has been to support the discussions of SWOP and GCA in establishing specifications for intersegment exchange of films and proofs. These specifications are designed to prevent the situation in which an advertising agency may buy a good clean proof (dot gain of 13%) with lots of "punch" (solid ink densities above 1.4) and then be disappointed when the magazine run looks muddy (dot gain 28%) and washed out (solid ink densities around 1.1–1.2).

Increased use of measurement and adherence to specifications are leading to higher quality and profitability in the printing industry.

For background information SEE INK; PRINTING in the McGraw-Hill Encyclopedia of Science and Technology.

Alan R. Muirhead

Bibliography. International Prepress Association, *Recommended Specifications for Web Offset Publications (SWOP)*, 1986; H. Kueppers, *Color Atlas*, 1978; H. Kueppers, *Logic of Color*, 1976; A. R. Muirhead, M. Burgstein, and R. H. Fahr, North American Print Survey, *TAGA* Technical Association of Graphic Arts *Proc.*, pp. 585–601, 1985 (also available from E. I. DuPont de Nemours and Co., Photosystems Department, Wilmington, Delaware, Publ. E-77435); J. Yule, Four color processes and the black printer, *J. Opt. Soc. Amer.*, 30:322, 1940.

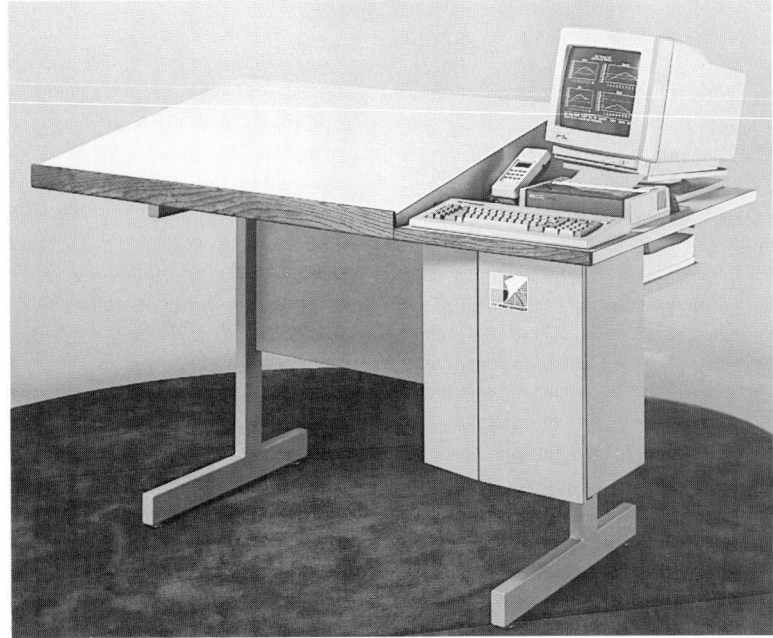

Fig. 8. DuPont Print Manager. (*E. I. DuPont de Nemours and Co.*)

Programming languages

Recent advances in the area of programming languages include (1) the development of fourth-generation languages, higher-level languages which automatically perform many of the functions that must be specified by detailed instructions in traditional languages, and can obtain results with an order-of-magnitude fewer instructions; and (2) programs that perform symbolic mathematical computation, the manipulation or combination of symbols according to mathematic rules, such as those of algebra or calculus.

FOURTH-GENERATION LANGUAGES

Fourth-generation languages are higher-level programming languages that automate many of the basic functions which had to be spelled out in third-generation languages such as COBOL or PL/1. They can obtain results with an order-of-magnitude less coding than the traditional languages because of their richer content of commands. Fourth-generation languages provide the programmer (who frequently is an end user) with the ability to execute complex functions with only a few commands. They provide speed, flexibility, and ease of use, and they also move the systems development process back toward the user.

Motivation. Fourth-generation languages were created so that nonprogrammers could obtain results from computers, and the programming process could be greatly speeded up. There was a need to allow applications to be created, modified, and enhanced much more rapidly with fewer hours of professional expertise. Fourth-generation languages are easier to use; they place a powerful tool in the hands of end users, who require little training to master them. They substantially reduce the complexity of developing online and data-base inquiry systems. The required level of programming expertise is reduced, making all software developers more productive, whether system professionals or end users. In many cases, users have reported a 10-fold increase in productivity over conventional programming using COBOL or PL/1.

Fourth-generation languages were also created to permit skilled professionals to create more complex systems, enabling them to conceptualize the required actions and control structures.

Using the languages. Some fourth-generation languages, such as those that function as compiler languages or interpreters, require that the "programs" or language parameters be entered, validated, and then executed. Other fourth-generation languages, for example, data-base inquiry systems, are interactive; that is, commands are entered into the system individually and executed instantly. Most query languages provide the ability to store commands and command strings for future use.

Procedural and nonprocedural languages. Fourth-generation languages can be categorized as procedural or nonprocedural. Some fourth-generation languages combine procedural and nonprocedural facilities. A procedural language specifies how something is accomplished. A nonprocedural language specifies what is accomplished but not in detail how it is accomplished. The traditional languages such as COBOL, FORTRAN, and PL/1 are procedural. Programmers must give precise detailed instructions for how each action is to be accomplished. Many fourth-generation languages, such as an application generator, where users fill in forms to tell the machine what to do, are nonprocedural. The user merely specifies what is to be done, and is not concerned with the detailed procedure for how it is done.

The fourth-generation languages that contain both procedural and nonprocedural statements differ widely in their syntax and capability. (No standards have yet been accepted.) Most fourth-generation languages are linked to a data-base system and allow users to create their own personal data bases. Some people create and employ relational data bases because these provide more powerful and flexible user commands than do the traditional data bases. Fourth-generation languages are user friendly, so that people can become competent at obtaining useful results after only a short training course.

Intelligent capabilities. The fourth-generation language compiler or interpreter makes intelligent assumptions about the user's perceived needs. A relatively large amount of computer power is used for compiling or interpreting so that the user does not have to specify details of procedures. For example, the interpreter may automatically select a useful format for a report, put page numbers on it, select chart types for graphics displays, put labels on axes or on columnar headings, and query the user in an understandable manner when the interpreter needs more information. The user need not describe in code the format of a report or display screen. The interpreter selects a reasonable format, which the user can adjust if desired. Results can be obtained with brief nonprocedural statements such as:

LIST BY CUSTOMER AVERAGE (INVOICE TOTAL).

This is a complete program. It leaves the software to decide how the list should be formatted, when to skip pages, when to number pages, how to sort items into CUSTOMER sequence, and how to compute an average. If these so-called intelligent default assumptions are not correct, the user can adjust the results easily and quickly.

Program generators. These advances in programming languages have made possible the information center concept, whereby the user can develop applications quickly and easily. The application development tools that support automatic program generation are sometimes known as program generators. They are typically designed to provide logical view support through a central dictionary so that programmers need not know how to navigate through a data base in order to access information contained within it; the programmer simply needs to know the data required for a particular application. These packages are online and interactive, a characteristic that, together with logical view support, is vitally necessary for the application development tools to be fully effective. While these tools may yield somewhat inefficient code and require considerable resources, they generally provide impressive savings in personnel costs and are much less expensive than procedural languages.

Generators may be used for most commercial data-processing applications, sometimes with the help of specially programmed subroutines. Database application generators may provide software for data entry, the creation of output, accuracy and validity checks, and restart and recovery after failures, as well as a simple means of creating the application logic. The best generators allow maintenance changes to be made quickly and easily. The result is that the applications may be adjusted constantly to meet the user's needs.

The increase in productivity, or in application development speed, with application generators is spectacular. They clearly play a vitally important role in the future of data processing and will greatly influence the systems analysis process.

Prototyping. Fourth-generation languages improve the productivity of systems professionals with much of the development of the problem solution going to the end user. Frequently, fourth-generation languages are referred to as prototyping tools, and the term prototyping is interchanged with software development using fourth-generation languages. Prototyping presents a different way of approaching systems development, and it addresses some of the problems associated with traditional approaches.

Prototyping allows users to become actively involved in the system design process; this involvement encourages them to make changes and helps ensure that the system is designed correctly. The low initial cost of developing prototypes invites experimentation with new applications of information systems. The user and the analyst typically work together to create a prototype of what the user has asked for, and together they refine it in a step-by-step fashion.

Prototypes were not used prior to the 1980s in any great number because it took almost as long to program the prototype as it did to program the live system. With the advent of fourth-generation languages, the use of prototypes became cost-justifiable and rapid. Using a good fourth-generation language, the analyst can create a prototype faster than by writing the specifications for it. An analyst working with an end user can create and demonstrate dialogues for database queries, report generation, and manipulation of screen information.

Characteristics. In summary, a fourth-generation language should have the following characteristics: (1) It is user-friendly. (2) A nonprofessional programmer can obtain results with it. (3) It employs a data-base management system directly. (4) Its procedural code requires an order-of-magnitude fewer instructions than COBOL or PL/1. (5) Nonprocedural code is used where possible. (6) The language makes intelligent default decisions about what the user wants. (7) It enforces or encourages structured code. (8) A user can easily understand and maintain another person's code. (9) The language is designed for interactive and online operation. (10) Users who are not data-processing professionals can master it in a very short period of training, 2 days or less. (11) It is designed for easy debugging of application software.

DMS. These properties of fourth-generation languages will be illustrated by discussing a particular language: IBM's DMS (Development Management System). It is a tool for generating interactive applications using DL/1 data bases. Often, complete applications are generated with it; sometimes it is used in conjunction with routines programmed in COBOL, PL/1, Assembler Language, or RPGII (Report Program Generator II). It forms a valuable tool for the systems analysis process when it is employed efficiently. Often, the systems analyst creates a prototype which needs to be tuned by a DMS expert or to have programmed routines written to supplement it. In either case, the systems analyst works with the end users and refines the results repeatedly until the users are satisfied with them and can use them efficiently.

DMS uses a so-called fill-in-the-blanks technique that makes it possible for online applications to be developed with little or no programming. Preprogrammed facilities are selected for functions such as data entry, inquiry, update, dialogue processing, and message switching.

The user describes data files, display screens called panels, and the application processing, either interactively through a terminal or offline through the use of a series of forms. With the interactive facility, DMS prompts the user through each step of the application definition. Specifications are validated to reduce errors. A HELP facility is provided which gives additional information and explains how data are to be entered.

Functions provided in DMS, in addition to the ability to perform calculations and editing, include those for data-base operation (search/browse, inquiry, update, amend/insert, deletion), data routing, program-to-program control, user processing, data validation, calculation (arithmetic and simple logic), application generation through forms, restart/recovery operations, audit control, and security and integrity functions. Data fields may be edited for validity by using such techniques as table lookup and comparison of one value with another. The user can select editing where it is needed or alter the sequence of processing as a result of any errors that are encountered.

The system can build an end user dialogue employing multiple screen panels which are logically linked. The user can capture data from a set of logically linked panels and construct from this logical file records of a data-base path involving up to four different files or data bases. The updating or creation of the data records occurs after the last panel of the set is processed.

Users in distributed locations may generate and test applications at a central point. Restart and recovery functions are provided to allow the user to save data and instructions previously entered if a system failure occurs. Audit procedures are available in which a log is kept, in some detail, of all terminal errors, allowing the auditors to check for inconsistencies.

Security is provided at various levels, including password protection at the terminal level. Integrity controls are provided to allow the user to identify and verify the contents of records. Search requests can produce listings of keys or records.

In general, systems produced with DMS are easier to design, avoid most of the need for programming, are easier to test, are self-documenting, and are much easier to maintain than conventionally produced systems. The analyst can specify the screens that will be used in an interactive dialogue, how the user may respond in the dialogue, and how the responses are processed.   *Bartow Hodge*

SYMBOLIC MATHEMATICAL COMPUTATION

Symbolic mathematical computation refers to the manipulation of symbols, representing variables, functions, and other mathematical objects, and combinations of those symbols, representing formulas, equations, and expressions, according to mathematical rules (for example, the rules of algebra or calculus). It is distinct from numerical computation, in which operations can be carried out only on numbers. Symbolic mathematics is used in all fields of science and engineering to solve problems exactly when possible, to generate approximate solutions, and to perform or check tedious manipulations too large (perhaps thousands of pages long) to compute with certainty by hand. Often, a single small formula may result whose meaning is much easier to understand than a large table of numerical values of some approximation to it.

There exist several general-purpose computer programs for symbolic mathematics that are running and undergoing active development on a variety of computers. The examples in this article were generated by using SMP. Programs begun earlier include MACSYMA, REDUCE, SCRATCHPAD, and muMATH, while a newer program is MAPLE. A number of special-purpose programs have also been written.

Examples. The **illustration** is a transcript of an SMP session. SMP asks for the first input expression with the prompt **#I[1][::** . Whatever the user types will be the first element in a list of input expressions that can be referred to later, and whatever SMP calculates from that input will be the first element of a list of output expressions named **#O**. In this case, a trigonometric expression, $\mathbf{Sin[1/(x + 1)]}$, is differentiated with respect to $\mathbf{x}$, producing the result $-\mathbf{Cos[1/(1 + x)]/(1 + x)^2}$. The function **D** performs differentiation of its first argument with respect to the second, the arguments being delimited by square brackets and separated by commas. The function treated here is highly oscillatory near $\mathbf{x = -1}$; this behavior is shown by the graph produced by SMP in response to input line 2, where the user has passed an expression, the variable, and the lower and upper limits of its range to the **Graph** function. An attempt to numerically differentiate the function at $\mathbf{x = -.98}$ by constructing the slope between two nearby points, say $\mathbf{x = -.985}$

```
# I [1] : :  D [Sin[1/(x+1)],x]

                   1
           -Cos [-----]
                  1+x
# O [1] :  ---------------
                     2
             (1 + x)

# I [2] : :  Graph [Sin [ 1/ (x+1) ],x, - .99, -.97]

# O [2] :
```

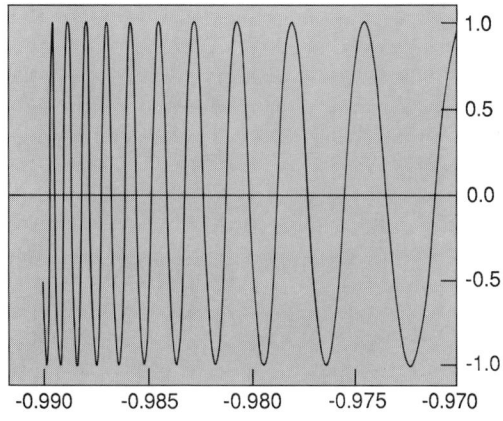

```
# I [3] : :  N [S [#0 [1], x -> -.98], 12]

# O [3] : :  -2412.41507123

# I [4] : :  Sol [{0=x y - c y - d x+ c d,0=x y - 3  y+7  x-21},{x,y}]

# O`[4] : :  {{x -> 3,y -> d},{y -> -7,x ->   c}}

# I [5] : :  flb [1] : flb [2] : 1 ;   flb [$n_= Natp [$n]] : flb [$n-1]+flb [$n-2] ;

# I [6] : :  Tan [a] : 2-z ;   flb [3+2z+2Tan [a]]

# O [6] :  13
```

Transcript of a session with the SMP computer program for symbolic mathematics.

and $x = -.975$, gives grossly inaccurate results. Evaluating the symbolic formula for the derivative at some value of x can be easily done to arbitrary accuracy as shown in line 3 of the session, in which $-.98$ is substituted (using the function **S**) for **x** in the result of the first calculation, and the result is numerically evaluated (using the function **N**) to 12 places.

At the fourth input prompt, the user types Sol [{0 = x   y − c   y − d   x + c   d, 0 = x y − 3y + 7x − 21}, {x,y}] to direct SMP to solve a set of simultaneous nonlinear equations for **x** and **y**, producing two families of solutions, not obtainable numerically. (Since symbols and variables can have multiletter names, the terms of a product must be separated by a space or asterisk.)

Input line 5, **fib[1]** : **fib[2]** : **1;fib[$n__ = Natp [$n]] : fib[$n − 1] + fib[$n − 2]**, defines the terms of the Fibonacci series: the first two terms are unity, any other term is the sum of the two previous terms. The last part of the definition restricts it to arguments that are natural numbers ("**__=**" can be read as "such that"). This requirement ensures that any use of this recursive (self-invoking) definition will always terminate by reaching one of the forms **fib[2]** or **fib[1]**. Given only this definition, SMP would return exactly the user's input if asked for **fib[5.7]**, but if SMP is now told that the tangent of **a** has a particular value, **2 − z**, and **fib [3 + 2z + 2Tan[a]]** is then requested, the result is 13, the value of **fib[7]**. The quantity **fib[7]** is defined in terms of **fib[6]** and **fib[5]**, and so on, down to **fib[1]** and **fib[2]**.

This result, and all others, is obtained by a general process called simplification, which consists of replacing parts of the input expression by their values. The values can be obtained by internal algorithms (for example, the sum of the terms after replacing **Tan[a]** above), by values assigned to symbols or expressions by the user, or by general or specific transformation rules defined by the user (or found in libraries of routines written by the program's developers and other users). Usually, the arguments to a function such as **fib** are simplified before the function definition is used; that is why the definition of **fib** for natural number arguments can be used. None of the examples can be performed by numerical programs.

Most general-purpose symbolic mathematics programs contain facilities to evaluate expressions to specified numerical precision. Many also have graphical capabilities, and the capability, after performing desired transformations of symbolic expressions, of translating them into programs in a numerical programming language such as FORTRAN.

Nature of systems. The ideal of designers of symbolic mathematics systems is to provide a program that can carry out any mathematical operation, including one unknown to the designer or not yet invented, preferably in a notation similar to that used by a mathematical practitioner. All existing systems imperfectly realize that goal, falling short in different ways and degrees.

Operational systems. Existing designs of symbolic mathematics systems fill a spectrum between two distinct design philosophies. The first is mechanistic, or operational. The language provided by the symbolic mathematics program allows the user to define a transformation rule which takes combinations of symbols satisfying a given set of conditions as input and then produces the desired new combination of symbols as output. If the language is rich enough to describe any possible transformation of symbols by some sequence or combination of its primitives, if the program can recognize any given expression or combination of symbols that satisfies the necessary conditions, and if the language allows facilities that control where the rule should be used, any mathematical operation can be carried out. SMP is a language designed mainly in accordance with this approach, but all the languages contain access to some pure symbol manipulation primitives, perhaps by direct use of the lower-level language used to implement the program. SMP's pattern recognizer is perhaps the most deeply embedded of those available.

SCRATCHPAD program. The SCRATCHPAD program goes farthest toward the other extreme of design philosophy, using deep mathematical knowledge more heavily than unconstrained symbol manipulation. An operation in SCRATCHPAD is not fully specified until the algebraic domains from which its operands are drawn are specified. In different parts of a computation, a single expression may be viewed as an element of several different domains. (For example, the number 3 may be treated as an integer, a real or complex number, polynomial in x of degree 0, and so forth.)

The allowed operations of each domain are fully specified, and the program can perform legal conversions between domains and proper manipulations of expressions that belong to composite domains of great complexity, using knowledge of the hierarchical relations between domains. New domains can be added by the user, or built up from already included ones, to implement new areas of mathematics. These new domains inherit much machinery from their components.

For example, if two matrices are multiplied, SCRATCHPAD automatically checks that the elements of the matrices are from consistent domains (which can be coerced into a common domain), and forms the elements of the product matrix by using appropriate multiplication and addition rules for the elements. For example, to multiply a matrix of complex numbers by a matrix of polynomials in the variable z with integer coefficients, SCRATCHPAD will determine that the elements of both matrices are from consistent domains; in particular, they can be regarded to be members of the domain of, and be coerced automatically into SCRATCHPAD's internal representation for, polynomials in z with complex coefficients. The multiplication and addition operations for that common domain are used to carry out the matrix multiplication. Thus, any operations defined properly are guaranteed to produce mathematically correct results for any inputs upon which they are allowed to work.

Comparison of systems. The user of SMP employs ingenuity to specify a mechanism for producing the answer, while a SCRATCHPAD user often employs ingenuity to find a set of known operations and domains in which the answer is automatically derived. It is easy to find problems in which one approach or the other leads most conveniently to the desired result. For this reason the implementation of each system is not as radical as its design approach; each adopts features of the other for the convenience of users. In particular, SCRATCHPAD contains facilities for using rewrite rules such as SMP's, and SMP contains some ability to implement algebraic domains and hierarchies. A result can usually be more directly justified rigorously by reviewing the transcript of a SCRATCHPAD session than an SMP session.

A user approximating **p**, a polynomial of many terms in **x**, for x near zero, may decide to ignore (as small and inconsequential) terms with high powers of x. Before further manipulations occurred, a person doing the problem with pencil and paper would probably cross out all such terms. Similarly, a user of SMP can type **S[p,x^ ($n__=$n>5)->0]**, meaning "Everywhere x is raised to some power greater than 5 in the expression p, substitute zero." A user of SCRATCHPAD would probably more likely tell the system to regard p as a power series in x truncated at fifth order. (SMP can also produce the result in this way.)

In a slightly more complicated problem, each term of the polynomial p may be multiplied by an unknown function $f(x)$. The user may be intending to eventually integrate this expression over all real values of x, and may know enough about the physical laws governing the behavior of whatever physical system the expression models to know that integrals of terms with odd powers of x multiplying $f(x)$ vanish, and that integrals of terms with large powers of x (greater than 5) are small and can be ignored. An SMP user could type **S[p,x^ ($n__=($n>5| Oddp[$n]))->0]** to substitute 0 for powers of x that are either odd or greater than fifth power before the integral is carried out. The SCRATCHPAD user would probably take a similarly nonrigorous approach to avoid a more cumbersome sequence of commands needed to achieve the same result by setting up an appropriate domain for the problem.

Use of advanced algorithms. In principle, all standard mathematical operations could be built out of symbol manipulation primitives. In practice, however, all of these systems provide the standard operations of algebra and calculus, including highly sophisticated recent algorithms for polynomial factorization, exact solution of algebraic equations, and indefinite integration, much more powerful and efficient than the methods known by most users. The developers of many of the programs heavily emphasize these components of the systems and refer to their products as computer algebra systems, or algebra engines.

Research. Research continues in more powerful and complete algebraic algorithms. Development of graphical user interfaces to symbolic mathematics programs is beginning to make them easier to use, at least on high-resolution bitmap terminals. Study of the important engineering problem of improving the performance of automatically generated programs using symbolic transformations of the mathematical expressions those programs represent is in progress.

For background information SEE DATA-BASE MANAGEMENT SYSTEMS; PROGRAMMING LANGUAGES; SYSTEMS ENGINEERING in the McGraw-Hill Encyclopedia of Science and Technology.

Jeffrey M. Greif

Bibliography. B. Caviness (ed.), *Eurocal '85: European Conference on Computer Algebra*, vols. 1 and 2, 1985; B. Char (ed.), *SYMSAC '86: Proceedings of ACM Symposium on Symbolic and Algebraic Computation*, Association for Computing Machinery, 1986; C. Cole and S. Wolfram, SMP: A symbolic manipulation program, in P. S. Wang (ed.), *Proceedings of SYMSAC '81*, Association for Computing Machinery, pp. 20–22, 1981; B. Hodge, Implementing 4GL yields healthy DP environment, *Data Management*, 24(3):22–23, 1986; IBM, *Development Management System/Customer Information Control System/Virtual Storage: General Information Manual*, IBM Man. GH20-2195-2, 1980; R. D. Jenks, A primer: 11 keys to new Scratchpad, in J. Fitch (ed.), *Proceedings of EUROSAM '84*, pp. 123–147, 1984; D. Stoutemeyer and D. Y. Y. Yun, Symbolic mathematical computation, in *Encyclopedia of Computer Science and Technology*, vol. 15, pp. 235–310, 1980.

Protective coloration

Many conspicuously colored insects and some other invertebrates exhibit similarities in their color pattern involving mimicry. A spectrum of unpalatability underlies mimetic resemblances. At one extreme, species which are toxic or distasteful to predators derive mutual benefit from Müllerian mimicry. At the other extreme, one species which is palatable derives protection from predators because it is a Batesian mimic of another species which is toxic and exhibits the same color pattern. Studies of ladybugs provide support for such a palatability spectrum within a single group. Furthermore, particular species may be palatable to some predators but distasteful to others. An invariable pattern of warning coloration is probably characteristic of ladybug species which have a highly developed chemical defense. These insects form the basis of a series of so-called Müllerian mimicry rings comprising species with similar color patterns. Other ladybugs which are probably less well protected chemically are apparently polymorphic between different rings, with some forms deriving a Batesian form of protection from one ring and others deriving a similar protection from another ring. Such species are effectively intermediate in nature between the classical definitions of Müllerian and Batesian mimics. Many other forms of selection influence local variability in the frequency of the forms of such polymorphic species. In the two-spot ladybug, *Adalia bipunctata*, thermal effects are locally critical with high frequencies of melanic forms favored in regions of low levels of sunshine.

Mimicry. One area of current research on animal coloration is the investigation of the ecological and evolutionary processes which are the basis of striking similarities in the color patterns of different species. Such similarities are characteristic of the phenomenon of mimicry. Two basic forms of mimetic resemblance between individuals of species of insects and some other invertebrates have long been recognized.

Müllerian mimicry occurs between two or more species (called a mimicry ring), each of which has the same or a closely similar warning, or aposematic, coloration. Their color patterns act as conspicuous signals to visually hunting predators, especially birds and lizards, of some form of toxicity, noxiousness, or distastefulness. Many laboratory tests have confirmed that when so-called naive predators, which have been raised without any experience of a particular color pattern, are presented with such unpalatable prey they rapidly learn to avoid all individuals with the same pattern. Müllerian mimics exhibiting the same warning coloration each derive some mutual benefit in terms of lower mortality because the efficiency of avoidance learning by predators increases with the rate of encounter of that particular color pattern.

In contrast, Batesian mimicry involves one (or more) unpalatable species that exhibit warning coloration and another species that has the same color pattern, although it is perfectly palatable to predators. The latter species derives some protection because predators sometimes mistake it for the unprofitable prey. The former species acts as a model for the latter, which is the mimic. Some Batesian mimics, such as the African butterfly *Papilio dardanus*, are polymorphic, with forms mimicking different unpalatable models.

Color patterns in ladybugs. The adults of many species of coccinellid beetles or ladybugs have brightly colored head capsules and wing cases involving classic aposematism, with two or three very distinct colors in simple, contrasting, and conspicuous patterns. The most typical pattern is a red or orange background of carotenoid pigments with black spots or marks of melanin. The numerous species of ladybugs found in Europe tend to show one or other of a small number of different color patterns, the most abundant of which are red with black spots, black with red marks, yellow with black spots, and brown with white or cream spots. These assemblages represent putative mimicry rings, each based on a number of more or less invariable or monomorphic species.

Interestingly, the degree of resemblance between members of a particular ring is often not nearly as close as that characteristic of Müllerian mimicry rings in *Heliconius* butterflies of the South American rainforest. Thus, the size and number of spots usually differ between the ladybugs. It is not known whether such imperfect resemblance is sufficient to gain full mutual protection from bird predators. Such predators are predicted to exert stabilizing selection resulting in close convergence of the color patterns of Müllerian mimics. The patterns of variation found in ladybugs are in many ways more comparable to those of mimicry rings in North American bumblebees.

Chemical ecology. Studies of the chemical ecology of ladybugs, feeding experiments with captive or wild birds, and investigations of the contents of bird stomachs provide strong overall evidence that these beetles have a high or moderate degree of unpalatability and related chemical protection from most or many bird predators. However, certain birds do feed more or less freely on some ladybugs, at least during periods of food scarcity.

The general picture is one of a diversity in predator response, both within and between species, to ladybugs with warning colorations. The beetles have a characteristic odor and bitter taste. The most important defensive chemicals so far identified in ladybugs are a group of closely related alkaloids, many of which have been synthesized in the laboratory. Other compounds associated with toxic or repellent qualities include histamines, cardiac glycosides, pyrazines, and quinolenes. Some of these are probably synthesized completely by the beetles, while others are sequestered from plants via their aphid prey. Feeding experiments suggest that ladybug species differ in their chemical unpalatability to birds. Variability in the development and effectiveness of chemical defense in these beetles is currently being investigated. There is, however, substantial support for a chemical basis for the existence of mimicry rings.

Polymorphism in ladybugs. A number of abundant species of ladybugs are highly polymorphic, exhibiting a variety of color forms. In Europe the two-spot ladybug, *Adalia bipunctata*, and the ten-spot ladybug, *A. decempunctata*, have forms which are red with black spots and others which are black with red spots (**Fig. 1**). Such color polymorphisms in ladybugs are controlled by a series of alleles at a single gene locus or perhaps at a group of tightly linked loci, or so-called supergene. In the context of Müllerian mimicry, such diversity is unexpected. The predominantly red forms are similar in appearance to the mimicry ring of species which are red with black spots, especially the very common *Coccinella septempunctata*. In contrast, the melanic black forms are closely similar to a more distantly related group of ladybugs in the subfamily Chilocorinae, which are black with red marks.

An explanation is required for the global stability or maintenance of the polymorphism, that is, for the failure of each polymorphic species to converge on the best-protected pattern, the one which is most abundant or most distasteful. There are two possibilities. First, some predators could generally avoid eating ladybugs while others may prey more freely on them. The former will tend to make more mistakes, resulting in beetle deaths, when encountering the rarer color patterns, while the latter will tend to concentrate their feeding on the commoner patterns. Theoretically, such a system could act in a balancing or frequency-dependent way to maintain the variation. Second, and probably more realistically for ladybugs, several polymorphic species might be less well protected by chemical defenses, and each set of their color forms would then benefit from a form of Batesian mimicry of more protected models in the Müllerian mimicry rings. The associated prediction of a relationship between color pattern diversity and distastefulness to predators is currently being tested in detail. A similar complex of species probably exists in the soldier beetles, or Cantharidae.

Geographic variation and thermal melanism. If mimicry is the paramount selective form influencing color polymorphism, a predicted consequence of spatial variation in comimics is that sympatric species of ladybugs exhibiting similar sets of forms will show parallels in their geographical variation. Early survey work supported the existence of parallel long-range changes in the distribution of predominant color patterns, with centers of light and dark pigmentation tending to occur in arid and humid regions, respectively. The prediction is, however, not fulfilled on a more local scale for the two species of *Adalia* in Europe since *A. decempunctata* shows a general uniformity in the frequency of the forms while *A. bipunctata* exhibits marked geographical differentiation.

Natural selection has been considered to influ-

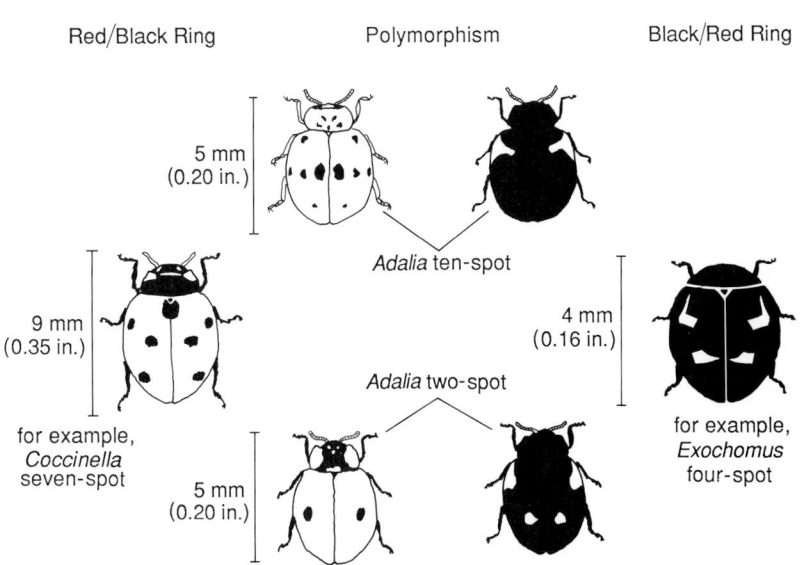

Fig. 1. The most abundant melanic and nonmelanic forms of two *Adalia* ladybugs in western Europe, together with examples of monomorphic species of Müllerian mimicry rings. Red areas of ladybugs are depicted in white. (*After P. M. Brakefield, Polymorphic Müllerian mimicry and interactions with thermal melanism in ladybirds and a soldier beetle: A hypothesis, Biol. J. Linn. Soc., 26:243–267, 1985*)

ence polymorphism in *A. bipunctata* in many different ways (**Fig. 2**). Several lines of evidence show that melanics exhibit a more efficient absorption of solar radiation and an increased activity in the spring. This is referred to as thermal melanism. The differences in thermal properties between the red and black forms have been confirmed in the laboratory and are similar to

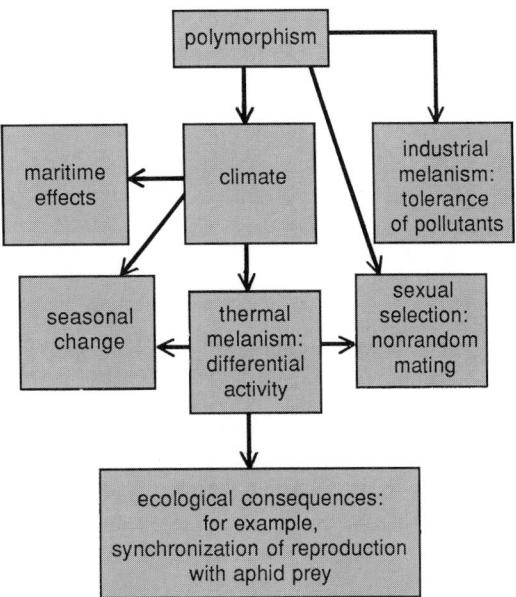

Fig. 2. Ways in which natural selection has been considered as an influence on polymorphism in the two-spot ladybug, other than through mimicry. Arrows indicate relationships. (*After P. M. Brakefield, Polymorphic Müllerian mimicry and interactions with thermal melanism in ladybirds and a soldier beetle: A hypothesis, Biol. J. Linn. Soc., 26:243–267, 1985*)

those measured in some other polymorphic species and to results from comparisons between species. High frequencies of the melanic forms of *A. bipunctata* are associated with regions of low sunshine, and declines in melanic frequency in urban areas have followed the introduction of smoke-control legislation in the United Kingdom. Melanic beetles in natural populations in the Netherlands mate more frequently than non-melanics and tend to reproduce earlier in the spring following hibernation. At this stage of the life cycle, most adults are on low-growing shrubs exposed to the effects of direct sunshine. The effects of thermal melanism on reproduction can account for some local patterns of geographical variation. The life cycle of *A. decempunctata* does not include a spring period on shrubs, but is restricted to trees on which such thermal effects are likely to be much less significant.

Thermal effects and mimicry. Thus, a particular aspect of life history may result in some species responding to certain environmental factors and, so to speak, escaping the tendency of mimetic selection to promote parallel spatial change. Escape is made more likely if the form concerned resembles the pattern of another mimicry ring, as in *A. bipunctata*, rather than being unique, since the species will continue to derive some mimetic protection. Interactions or compromises between thermal effects and mimetic selection have also been postulated for other insects. For example, a situation comparable to *Adalia* ladybugs has been described in North American *Nicrophorus* burying beetles. Some races of species of *Heliconius* butterflies at higher altitudes in regions of the Andes bordering the Amazon basin have substantially larger black areas on the wings. Similarly, one of the North American bumblebees is polymorphic, with a darker race in a region of cooler climate where appropriate co-mimics are also found. Seasonal changes in the frequency of paler and darker forms of some *Eristalis* hoverflies also probably reflect a balance between selection for a Batesian mimetic resemblance to bumblebee models and seasonal changes in requirements for thermoregulation. The discovery of such complexity in the webs of natural selection influencing certain striking color polymorphisms in insects parallels the situations which are coming to light for other examples of genetic variation in ecologically important traits.

For background information SEE CHEMICAL ECOLOGY; COLEOPTERA; PROTECTIVE COLORATION in the McGraw-Hill Encyclopedia of Science and Technology.

Paul M. Brakefield

Bibliography. R. S. Anderson and S. B. Peck, Geographic patterns of colour variation in North American *Nicrophorus* burying beetles, *J. Nat. Hist.*, 20:283–297, 1986; P. M. Brakefield, Polymorphic Müllerian mimicry and interactions with thermal melanism in ladybirds and a soldier beetle: A hypothesis, *Biol. J. Linn. Soc.*, 26:243–267, 1985; R. C. Plowright and R. E. Owen, The evolutionary significance of bumble bee color patterns: A mimetic interpretation, *Evolution*, 34:622–637, 1980; J. R. G. Turner, Mimicry: The palatability spectrum and its consequences, in R. I. Vane-Wright and P. R. Ackery (eds.), *The Biology of Butterflies*, pp. 141–161, 1984.

Protein

Protein molecules are among the most interesting and complicated of all the known biological macromolecules. They not only contribute to an organism's structure but are known to catalyze most of the complex chemical reactions in living systems. Numerous recent studies have broadened the understanding of the structure, dynamics, and thermodynamics of proteins. All three of these characteristics contribute to the understanding of the biological function of proteins.

The first structural pictures of proteins were provided in the 1960s by high-resolution x-ray diffraction experiments. The general clarity of the individual atoms and groups of atoms given by those studies was striking. In fact, proteins were long considered to be rather rigid, inflexible molecules with highly localized atoms. However, a cursory examination of many proteins, such as those of the globin family, indicates that there would be no way for the reactant molecules to enter or leave the vicinity of the active site if the structure were truly rigid.

It is now widely appreciated that proteins are rather flexible and undergo a wide range of motions. The dynamical motions are a manifestation of thermal energy or the heat in the surroundings. The fluctuations of each atom reflect not only the motions of its neighbors but the magnitude of its interactions with its neighbors. In fact, the simulated thermal motions of a protein, obtained from solving Newton's equations of motion on a large digital computer for the protein and its aqueous surroundings, can be used to sample the interactions relevant to calculating many thermodynamic quantities.

Methods of study. Considerable progress has been made in the experimental field of x-ray crystallography, especially in the area of large molecular assemblies. A good part of the current understanding of dynamics and thermodynamics comes from theoretical studies. By far the most powerful theoretical technique in common use is the computer simulation of molecular dynamics alluded to above. An evaluation of the forces on each atom due to its neighbors and solution of the equations of motion can produce a very detailed view of the atomic motions and some of the collective motions in proteins. The time scales for simulations attainable by current technology are short (about 10^{-9} s) compared with events in the macroscopic world but are sufficient to allow a great deal of detail at the atomic level. A number of experimental methods, such as nuclear magnetic resonance and picosecond pulsed laser experiments, are also being developed to probe short time fluctuations.

Structure. The most impressive recent advances in structure have come from x-ray crystallographic studies of large assemblies of proteins and other macromolecules. Most notable in this category are the studies on a virus for the common cold and the polio virus. These viruses belong to the picornavirus family, of which hepatitis A and foot-and-mouth disease are also members. Each virus is over 30 nanometers in diameter. The outer coat is composed of four capsid protein subunits arranged in roughly triangular form (see **illus**.), with 60 of the triangles completing the icosahedral surface of the virus. On the inside of this hollow protein ball is a single strand of ribonucleic acid (RNA) with more than 7000 nucleotides linked to a single small interior protein. Platinum atoms were used to provide reference points in the electron density maps. This is a common technique in protein crystallography and allows a reference point to aid in the interpretation of the nonplatinum-soaked crystal structure. A great portion of the proteins are in the so-called beta-sheet configuration. That configuration consists of long linear strands of amino acids which run either parallel or antiparallel to each other and are linked by hydrogen bonds. These beta sheets give the hollow capsid ball a remarkable stability at certain salt concentrations and values of pH.

Now that the structure of viral protein coats is known, it is possible to work on the details of specific drug interactions with the proteins. This may lead to a long and productive era for the design of antiviral compounds.

Dynamics. In the mid-1970s, experimental dynamics studies were concerned primarily with time scales on the order of tens of nanoseconds and longer, while theoretical studies were available only for times shorter than tens of picoseconds. The gap between these time intervals has gradually decreased, and there is now a considerable overlap in the times accessible by theory and experiment.

Some of the advances in pulse laser technology have been central to the development of experiments which probe events on the picosecond (and subpicosecond) time scale. In general, the laser experiments are of the so-called pump-probe type. In the pump stage an initial laser pulse is used to prepare a certain state of the protein system. Then succeeding laser pulses at later times are used to probe (measure) the dynamic time evolution of the system.

The dynamics of ligands, such as oxygen (O_2) and carbon monoxide (CO), binding to (or releas-

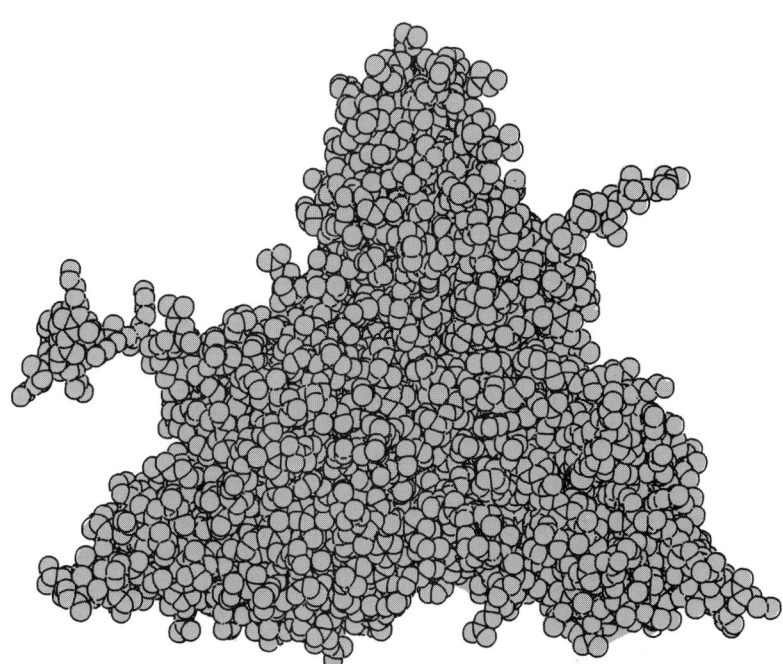

A space-filling picture of the four capsid proteins in the viral coat of the common cold virus. Only nonhydrogenic atoms are shown. Each atom is roughly 0.3 nm in diameter. This roughly triangular face is 1/60 of the entire surface of the virus.

ing from) hemoglobin (Hb) and myoglobin have been studied by these laser techniques. Starting with the bound complex, say HbCO, photodissociation is induced by an 80-femtosecond pulse. The kinetics of the system is then followed by probing the system with 100-fs pulses at a series of later times to produce a set of time-slice pictures of the spectrum of the system.

The information on the dynamics of the central iron atom to which the ligand molecule is bound indicates that the iron atom moves out of the plane of the prophyrin group embedded in the protein matrix in about 350 fs. This particular motion of one atom is only about 0.05 nm, but it triggers a set of events which moves other atoms by more than 0.1 nm and changes the conformation and the relative position of the four subunit proteins which constitute each molecule of hemoglobin.

This one event, which happens on the subpicosecond time scale, is the trigger for motions of considerably longer time duration. The conformational changes which hemoglobin undergoes are important to its ability to bind oxygen cooperatively. That is, after hemoglobin binds one oxygen molecule, the next one is easier to bind and so on, up to the maximum of four oxygen molecules (one for each subunit). The critical first step in this process has now been well characterized; much remains to be done, however, on the succeeding complex events of long duration.

Thermodynamics. A number of theoretical methods have recently been developed that can provide information on the energies (or free energies) of ligand binding and the changes in binding for both different ligands and mutant proteins. These techniques are based on statistical mechanical theories for connecting the microscopic (atomic) averages and fluctuations to the macroscopic thermodynamics for the system. The needed atomic averages and their fluctuations may be calculated from a computer simulation of the motions.

Occasionally, the consideration of systems which are unphysical intermediates has been found to be useful. For example, in the study of the binding of a new drug candidate to a protein, rather than attempt to actually simulate the encounter and capture involved in the binding process, which may be an infrequent event, it is possible to simulate a thermodynamically equivalent unphysical event which is less costly in terms of computer time. The essential idea is that measured thermodynamic differences depend only on the initial and final states of the system—for example, ligated and unligated. Thus, if the initial state is unligated, the thermodynamics can be computed by following a path of growing or gradually creating the ligand in the proper position in the protein to reach the final state.

This sort of computer alchemy has proven very effective in reproducing observed binding thermodynamics, and is rapidly being applied to the design of new drugs and pharmaceuticals. By starting with the structure for a known ligand-protein complex, either the ligand can be metamorphosized into a new drug which binds more readily to the protein, or the amino acids of the protein can be changed into different amino acids to produce a better substrate protein.

This approach has been applied to the binding of benzamidine to trypsin. Benzamidine is a known inhibitor of the enzyme trypsin. In the first such study the relative free-energy change between benzamidine and *p*-fluorobenzamidine, a modified inhibitor, was calculated. The free-energy difference obtained, about 1 kcal/mol at 300 K, was very nearly that obtained from experiment. Besides the obvious value of the calculation as a predictive tool, these calculations have led to an interesting mechanistic insight into the molecular details of binding. It was found that the major contribution to the free-energy change between binding benzamidine and the *p*-fluoro analog was due to the work done in desolvating the ligands (inhibitors), and not due to specific enzyme-inhibitor interactions. In a related series of thermodynamic calculations the change in free energy upon binding benzamidine to trypsin versus a mutant of trypsin was calculated. Specifically, one of the amino acids of trypsin near the binding site was changed from a glycine to an alanine. Again the free-energy change was found to be in accord with experiment. This proved the validity of the method for describing mutant protein-substrate interactions.

These approaches offer a far brighter future for the specific design of proteins and the molecules that bind to them than the old, rigid, "hand-in-glove" models of ligand binding.

For background information SEE ANIMAL VIRUS; HEMOGLOBIN; OPTICAL PULSES; PROTEIN; SIMULATION; STATISTICAL MECHANICS; X-RAY CRYSTALLOGRAPHY in the McGraw-Hill Encyclopedia of Science and Technology.

Bernard Montgomery Pettitt

Bibliography. J. L. Martin et al., Spectral evidence for subpicosecond iron displacement after ligand detachment, *Europ. Mol. Biol. Org. J.*, 2:1815–1819, 1983; B. M. Pettitt, C. L. Brooks, and M. Karplus, Proteins: A theoretical perspective of structure, dynamics and thermodynamics, *Adv. Chem. Phys.*, vol. 81, 1987; M. G. Rossmann et al., Structure of a human common cold virus, *Nature,* 317:145–153, 1985; C. F. Wong and J. A. McCammon, Dynamics in the design of enzymes and inhibitors, *J. Amer. Chem. Soc.*, 108:3830–3831, 1986.

Proton microscopy

Any form of radiation that can be focused can be used in principle for microscopic examination of a specimen. The information obtained will depend on the way in which the radiation interacts with the specimen. A focused beam of protons or charged particles of similar mass (alpha particles, deuterons, and so forth) can provide a particularly sensitive and versatile microprobe for determining the elemental content of microregions or the spatial distribution of elements in thin specimens or surfaces. The sensitivity of the several interaction modes is associated with the mass of the beam particles, which is comparable with that of the specimen atoms.

The scanning proton microprobe is similar in operation to a scanning electron microscope. A beam of protons or other ions from a charged-particle accelerator is focused to a small spot. This spot is scanned over the surface of a specimen, and a battery of detectors, associated electronics, and an on-line computer records the nature and energy of all the radiation events generated in the specimen by the incoming beam. The computer builds up distribution maps of the elements present and elemental spectra of any regions of interest. Elemental sensitivity can be 1 part per million (ppm), depth resolution about 1 nanometer, and spatial resolution 1 micrometer, but these values will improve with advances in ion optics.

The major application of proton microscopy is in elemental microanalysis, where its sensitivity, quantitative accuracy, versatility, and wide application supplement the spatial information provided by other microscopical techniques.

Operation. A beam of particles from a charged-particle accelerator is first focused onto a slit or diaphragm which selects a central portion of the beam, about 20 μm in diameter. A probe-forming lens several meters downstream then forms a demagnified image of this object diaphragm. Most proton microprobes operate with a particle energy of about 3 MeV. Protons or heavier particles of this energy are not easily deflected by magnetic or electric fields. Strong-focusing magnetic quadrupole lenses (multiplets of two, three, or four lenses) are normally used for the probe-forming lens, the focal length typically being about 0.2 m (8 in.). Immediately before or after the lens, two sets of magnetic field coils or electrostatic field plates are used to steer the beam. The beam tubing is under high vacuum to avoid scattering of the beam particles by gas molecules, and several detectors surround the specimen in the vacuum chamber.

The beam can be maintained in a fixed-spot mode, but generally it is scanned across the specimen. When one of the detectors receives radiation from the specimen, the position of the beam spot (x deflection and y deflection), the identity of the detector, and the energy of the radiation are all recorded as a single event by the computer, generally on magnetic tape. The computer sorts all recorded events into a three-dimensional block of data for each detector (two spatial dimensions and one energy dimension). From a block of sorted data the computer can retrieve information on any element within the scanned area of the sample and generate the results that would have been obtained from spot analysis, line scans in any direction, or two-dimensional maps of any element. An example of an elemental distribution map is shown in **Fig. 1**, where the computer has extracted information on the potassium distribution within a red blood cell that is 7 μm in diameter and has plotted the intensity of the potassium signals as three-dimensional contours. The computer extracts such information for all elements simultaneously.

Signals. The data shown in Fig. 1 were obtained with an x-ray detector. When a charged particle, such as an electron or proton, passes close to an electron in an inner atomic shell, the electrostatic repulsion or attraction can remove the shell electron from the atom (inner-shell ionization). An electron from a neighboring shell then fills the vacancy, and the difference in binding energy is emitted in the form of an x-ray (or an Auger electron). The energy of the x-ray is characteristic of the atomic element and provides a clear identification of the target atom. Inner-shell ionization is most likely to occur when the beam particle has a velocity close to the orbital velocity of the shell electron. For this reason, protons or heavier beam particles must have energies of several MeV. The beam particle also emits radiation as it is deflected by the electrostatic fields of atomic nuclei. This radiation (known as bremsstrahlung) provides a continuous background of x-ray energies and limits the sensitivity of elemental detection. Protons and heavier particles are less easily deflected than electrons; for this reason the sensitivity of proton-induced x-ray emission analysis is about 1000 times greater than that of electron-induced x-ray emission analysis. The limiting sensitivity of approximately 1 ppm is set by radiation from the secondary electrons knocked out by the beam particles.

The secondary electrons can also be collected by a positively charged detector, and since they are produced in copious numbers, they provide a good signal with which to image the specimen. Such images resemble those from a scanning electron microscope and are used for identification of surface features of the specimen and positioning of the beam. For thin specimens the instrument can be operated as a scanning transmission ion microscope in which the focused beam passes through the specimen into a detector and the energy loss or scattering yield is a measure of local specimen density. This technique may be used with bright-field or dark-field imaging and is very efficient. It is analogous to scanning transmission electron microscopy.

Information on depth distributions of trace elements can be obtained by measurement of the energy lost by a beam particle that has been backscattered from the nucleus of a target atom. Since protons (or alpha particles) have masses comparable to those of specimen atomic nuclei, they exchange measurable fractions of their energy with those nuclei in elastic collisions, the fraction exchanged depending on the ratio of the colliding masses and the angle of scatter. Therefore, an energy spectrum of beam particles backscattered from a surface into a detector displays a

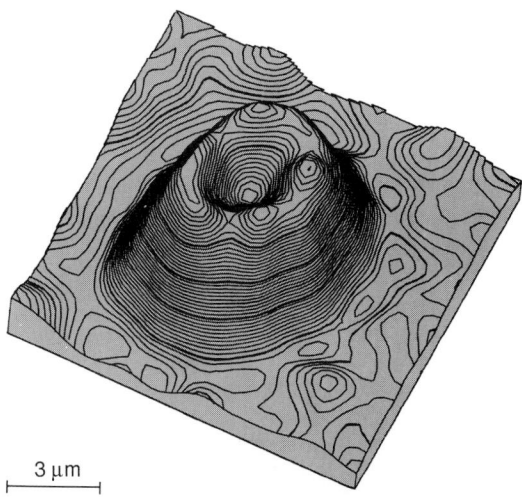

3 μm

Fig. 1. Proton-induced x-ray micrograph of potassium distribution in red blood cell. (*Courtesy of P. M. O'Brien*)

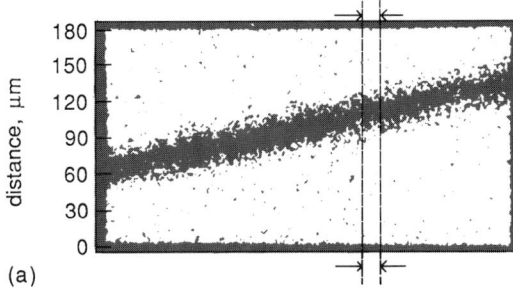

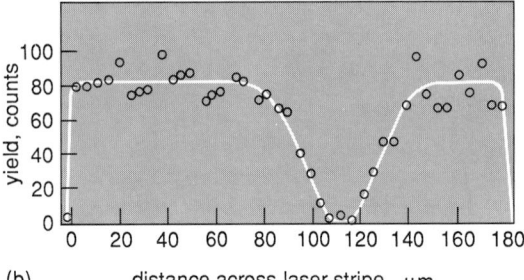

Fig. 2. Channeling contrast microscopy. (a) Micrograph of silicon crystal surface, showing narrow stripe annealed by passage of laser beam. (b) Yield of alpha particles backscattered from silicon atoms as a function of distance across the stripe, for the strip between the arrows in part a, showing decrease in annealed region. (From J. C. McCallum et al., Channeling contrast microscopy: Application to semiconductor structures, Appl. Phys. Lett., 42:827–829, 1983)

number of peaks denoting different elements in the specimen surface. If a target atom lies below the surface, the scattered particle loses energy on passing in and out through the overlying material of the specimen. Hence a narrow elemental peak in the energy spectrum spreads to lower energies, denoting the distribution of that element in depth beneath the surface of the sample (the so-called depth profile).

It is also possible to align the direction of the microbeam with an axis of symmetry in a crystalline sample. Then beam particles channel down between the rows of atoms, with a much reduced probability of escape from the crystal when they eventually collide with an atom. **Figure 2a** is the image of a silicon crystal surface that had been implanted uniformly with antimony ions to a dose of 10^{15} cm^{-2}. The image was formed with alpha particles that had been backscattered by silicon atoms from a microbeam of 2 MeV energy. The distribution of silicon atoms was of course uniform, but a black stripe appears across the image, indicating a region of very low backscattering yield. The implantation process had destroyed the crystallinity of the semiconductor surface, leaving it amorphous. Later a laser beam had been used to locally heat and thereby anneal a narrow stripe on the surface. It is this stripe which is detected in the image. The microbeam was aligned with a silicon crystal axis and therefore channeled deeply into the silicon in

this region of good crystallinity. The degree of annealing or crystal regrowth is shown in Fig. 2b, where the computer used the same data to plot the backscatter yield as a function of distance across the stripe for the strip between the arrows in Fig. 2a.

Nuclear reactions can also be induced by energetic beams of protons and heavier particles. These are of particular use in the identification of light elements. For example, carbon, nitrogen, and oxygen in metals are identified by the characteristic energies of protons emitted by each element when bombarded by deuterons of given energy.

Limitations and advantages. The spatial resolution obtained with any microscope is limited by the wavelength of the radiation and by the aberrations of the lens system that is used to focus the radiation. The wavelength of a particle beam is inversely proportional to the particle momentum. Protons of energy 3 MeV have a wavelength of 2×10^{-14} m. This is less than typical nuclear dimensions, as would be expected from the fact that protons of comparable energies are found within a nucleus. Electron microscopy can now approach atomic resolution, but proton microscopy, despite the small wavelength and hence small diffraction limit of resolution, is limited at present to a resolution of a few tenths of a micrometer, comparable with that of optical microscopy. The chromatic and spherical lens aberrations which provide the present practical limit are associated with the relatively low brightness of proton sources, the energy instabilities of proton accelerators, and the difficulty of constructing a lens of short focal length for high-energy protons. These are all areas of current activity in ion optical research. *See* MICROSCOPE.

For background information *see* CHANNELING IN SOLIDS; ELECTRON MICROSCOPE; PROTON-INDUCED X-RAY EMISSION (PIXE); SURFACE PHYSICS in the McGraw-Hill Encyclopedia of Science and Technology.

George J. F. Legge

Bibliography. J. A. Cookson, A. T. G. Ferguson, and F. D. Pilling, Proton microbeams, their production and use, *J. Radioanalyt. Chem.*, 12:39–52, 1972; G. J. F. Legge et al., Microbeam imaging at micron and submicron resolution, *Nucl. Instrum. Meth. Phys. Res.*, B15:669–674, 1986; G. J. F. Legge, Microprobe analysis, in J. R. Bird and J. S. Williams (eds.), *Ion Beams for Materials Analysis*, 1987.

Pyroxene

Ureyite, NaCrSi$_2$O$_6$, is a mineral in the pyroxene family and is the chromium analog of jadeite, NaAlSi$_2$O$_6$. Chromium has a strong crystal-field light absorption that makes ureyite emerald green in color and very conspicuous in even minor amounts of solid solution (the solid equivalent of liquid solutions) with other pyroxenes of monoclinic symmetry. In spite of this noteworthy coloring capacity, ureyite has been virtually undescribed, even as a minor pyroxene component, in terrestrial rocks until recently, although emerald-green jadeite jades have been known since the end of the nineteenth century.

Ureyite was first described (as the mineral kosmochlor) as a rare accessory mineral from the Mexican iron meteorite Toluca by H. Laspeyres in 1897, and then again, with some redefinition and a new name (after Harold Urey), when it was found in the iron meteorite Coahuilla in 1965. The International Mineralogical Association has approved the use of the mineral name kosmochlor as the official name.

Rare accounts of terrestrial ureyitic pyroxene are to be found in the literature up to 1981, but no definitive descriptions or petrogenetic models were offered except for some critical data published in 1930. Experimental efforts have been more productive in determining the solid miscibility of ureyite in diopside, $CaMgSi_2O_6$, and jadeite. Only a few percent diopside can dissolve in ureyite and only 10 to 20% ureyite can dissolve in diopside, depending on temperature and pressure. For jadeite, a mineral stable only at relatively high pressure, there is complete solid solubility at 18 kbar (1.8 GPa) and 800°C (1470°F), with the solubility decreasing with decreasing pressure to a minimum of about 10% jadeite in ureyite. This information is important when trying to understand the conditions of formation of ureyitic pyroxene.

Among the lapidary "jadeite" jades are rare dark-emerald-green rocks that variously are called Maw-sit-sit and Tawmawite, referring to Burmese mine occurrences. Recently, there have been reports of small portions of these Burmese rocks with up to 86 mol % $NaCrSi_2O_6$. Similar occurrences, with minor variations of the mineral assemblage, have now been found in jadeite-rich rocks from Italy and Guatemala and in Mesoamerican jade artifacts.

Burmese occurrence. The jades of Burma are primarily jadeite rock (jadeitite) that occurs as inclusions in serpentinite (a rock composed predominantly of serpentine, a hydration product of olivine and pyroxene rich in magnesium) along the Sagang Fault in northern Burma (often referred to as Mogaung, Burma). Light to brilliant emerald-green jade is due to 1 to about 10 mol % ureyite in solid solution with jadeite. Ureyite as a mineral species (hence greater than 50 mol %) occurs in dark-green rocks with black specks or blotches. These rocks consist mainly of pyroxene with specks of chromite (a spinel mineral principally of $FeCr_2O_4$) and variable amounts of sodic Mg-Al amphibole as lighter-colored veins or zones. The ureyite (reaching 86 mol %) occurs as fine, radiating coronal aggregates up to 0.5 mm (0.02 in.) thick surrounding and invading corroded

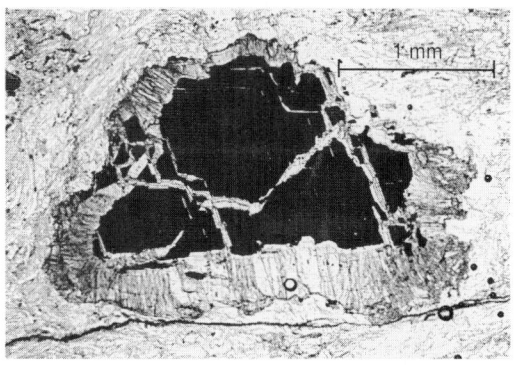

Fig. 1. Micrograph made with plane-polarized light of a fractured, corroded chromite (black) with surrounding ureyite (dark gray) in a matrix of less ureyitic omphacite textures in ureyitic jadeitites from the jade mines, Burma.

chromites or as concentric bands enclosing chromite (**Fig. 1**). Ureyitic pyroxene appears to replace chromite grains, particularly where veins and shear zones (common in the jades) intersect the chromites. The highest Cr/(Cr+Al) ratio of these pyroxene grains correlates with that of the adjacent chromite. Most pyroxene grains in the rocks have much lower ureyite content (from 2 to 30 mol %), and they range in composition from a nearly binary jadeite and ureyite solid solution to a quaternary mixture with usually less than 20 mol % diopside and lesser acmite ($NaFe^{3+}Si_2O_6$; **Fig. 2**). One sample has been found to be substantially calcic, being primarily ureyitic omphacite (a discrete pyroxene intermediate to jadeite and diopside in composition) with lesser jadeite-ureyite.

Italian occurrence. One small dark-green rock from Mocchie, Susa, Italy (presumably from a serpentinite body in the Lonzo massif in the western Alps), contains, in addition to ureyite-rich pyroxene (up to 87 mol % $NaCrSi_2O_6$), chromite, glaucophane, chlorite, and sphalerite. The high-ureyite pyroxene occurs as lenses (up to 3 mm or 0.1 in. wide) of micrometer-size grains that appear to be a replacement of chromite. The ureyitic lenses occur in rock which contains less ureyitic jadeite and glaucophane amphibole. Pyroxene compositions are low in calcium but occasionally have up to 26 mol % acmite.

Guatemalan occurrence. Jadeites occur associated with serpentinite in Guatemala north of the Motagua Fault, primarily in the districts of El Progreso and Zacapa. In contrast to Burmese jadeitites, the Guatemalan jadeitites have produced no emerald-green jade in historical times. However, two specimens and a number of fragments from the site of a Maya workshop have been found that have up to 37 mol % ureyite in the pyroxene (Fig. 2), which is consistent with their pale to vivid emerald-green color in a hand specimen. The ureyitic pyrox-

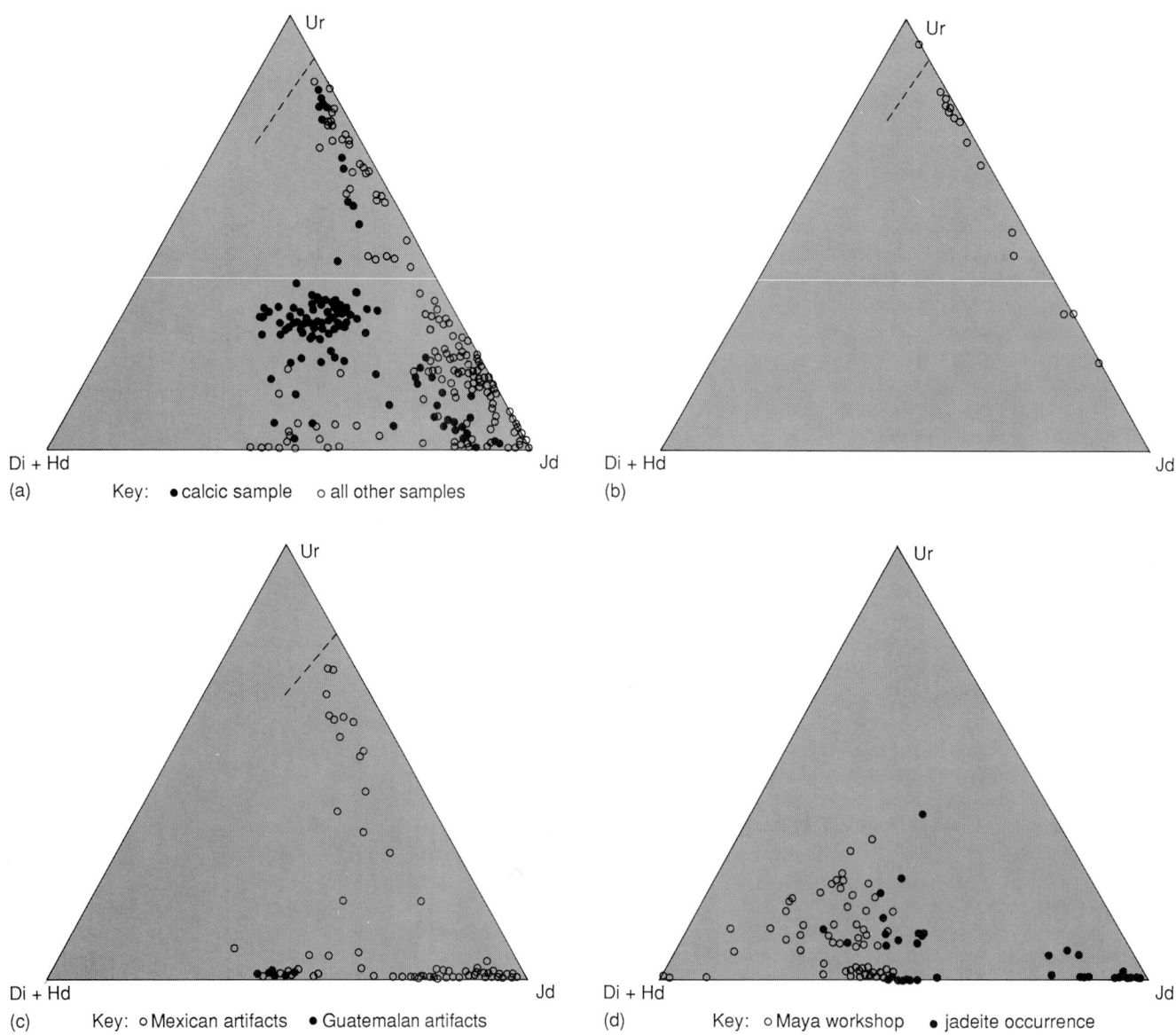

(a)   Key: ● calcic sample   ○ all other samples

(b)

(c)   Key: ○ Mexican artifacts   ● Guatemalan artifacts

(d)   Key: ○ Maya workshop   ● jadeite occurrence

Fig. 2. Molar-component ternary plots of ureyitic pyroxene compositions from different sources. Corners represent 100% of the compound in the pyroxene solid solution. Broken lines represent the Cr/Al ratio for the coexisting chromite. (*a*) All Burmese samples. (*b*) Italian sample. (*c*) Mexican and Guatemalan artifacts. (*d*) Guatemalan samples. Di + Hd = diopside + hedenburgite; Ur = ureyite; Jd = jadeite (projected from acmite).

enes follow two distinct trends: one group is highly jadeitic with only a small amount of ureyite, and the other group is omphacitic with appreciable ureyite. The primarily jadeite rock is typical of Guatemalan jadeitite in that it contains variable-size jadeite grains (up to 6 mm or 0.2 in. long) and paragonite as a minor (about 10%) primary phase. The ureyitic regions are small and extremely pale green; they appear to be correlated with slightly more diopsidic and acmitic zones of pyroxene growth. The omphacitic samples consist of intergrown lathy pyroxene with minor broken grains of chromite and some intergranular albite and chlorite. Ureyitic compositions are particularly associated with chromite but without any visible reaction zone.

Mexican and Central American artifacts. Many Mesoamerican jade artifacts have obvious emerald-green color. Artifacts that have been studied are predominantly jadeitic to omphacitic pyroxene (about 90%), with emerald-green spots showing 2 to 42 mol % ureyite (Fig. 2), and minor albite or white mica or both. An Olmec bead which was sacrificed for study by petrographic means consists of about 80% jadeitic pyroxene, with accessory muscovite (about 15%), chromite, and ureyite. Emerald-green spots (about 0.2 mm or 0.008 in. across) with up to 70 mol % ureyite enclose smaller corroded chromites and exhibit narrow zones of reaction texture. It appears likely that these artifacts were made from Guatemalan jade based on the similarity of composition and mineral assemblage.

Paragenesis. In igneous (and metamorphic) rocks, sodium and aluminum are highly correlated and crystal-chemically "tied" to one another so that they tend to crystallize together, usually as albite feldspar but also in other Na-Al silicates, relatively late in igneous events. Chromium, on the other hand, while crystal-chemically similar to aluminum, has a greater preference for oxide minerals and is usually fractionated early as chromite from silicate liquids. Consequently, Na-Cr silicates are uncommon because of petrogenetic separation of sodium and chromium rather than because of crystal-chemical incompatibility.

In meteorites, ureyite is presumed to have crystallized from a very sodic silicate melt in equilibrium with metallic iron. Experimental work on pressure dependence of jadeite-ureyite solid solution is consistent with crystallization of ureyite at very low pressures in the metallic meteorites. Even though ureyite coexists with albite in meteorites, it contains no jadeite component.

In terrestrial rocks the textural association, that is, corroded chromites surrounded by fine-grained ureyite in most ureyite-bearing samples, suggests a reaction in which chromite is consumed and ureyite is produced. All described jadeitites are hosted in serpentinite adjacent to major faults. Chromite rock also commonly occurs in serpentinites, and in Burma and Guatemala is found adjacent to jadeitites. Presumably, chromite-bearing ureyitic rocks incorporated their chromite from surrounding serpentinite. However, the reaction to form ureyite requires a source of sodium. It is extremely unlikely that sodium will be decoupled from aluminum to make ureyite from jadeite or albite, so an external source is required. Allowing for mobility of Na^+, Fe^{2+}, and Mg^{2+}, in the presence of water (that is, in an open system) and albite, it is possible to write the following simple reaction involving ureyite for the Burmese assemblage (considering only the Mg^{2+} end members):

$$[MgCr_2O_4 + 4NaAlSi_3O + 2Na^+ \rightleftharpoons$$

Cr-spinel   Albite

$$Mg^{2+} + 4NaAlSi_2O_6 + 2NaCrSi_2O_6$$

Jadeite   Ureyite

It is also possible to produce ureyite more directly if silica, SiO_2, is available in solution. The main point is that, given the observed minerals and assuming the presence of albite (or other silica source), an influx of Na^+ and escape of Mg^{2+} is required to explain the chromite-ureyite reaction. A hydrous fluid would facilitate such a reaction and is consistent with the presence of serpentinite, as long as the pressure is high enough to avoid the breakdown of jadeite. By this argument, the chromite-ureyite reaction and the presence of sodic Mg-Al amphibole in Bur-

mese samples imply a sodic-fluid-dominated reaction and crystallization process. Metasomatism is the term usually applied to such a process, particularly where monomineralic and bimineralic rocks such as jadeitite are produced. In addition, the restriction of jadeitites to serpentinites suggests a connection between serpentinization and jadeitization. An important implication of ureyite paragenesis is the necessity of a sodic fluid in jadeitites.

Pressure, temperature, and pyroxene crystallization. The conditions under which jadeitites form are not well constrained. Estimates range from 2 kbar (0.2 GPa) and 250°C (480°F) for New Idria (California) jadeitites to pressures greater than 10 kbar (1 GPa) and temperatures near 500°C (930°F) for jadeitites in the Soviet Union. It is known, however, that ureyite can be crystallized from a silicate melt at 1 bar (100 kPa) pressure. Hence the presence of ureyite in jadeitites probably places no additional restrictions on required pressure-temperature conditions for jadeitite petrogenesis. The compositions shown in Fig. 2 are generally consistent with the experimental work, in that there is a big gap in points between ureyite and diopside and a continuum between jadeite and ureyite. The middle of the triangle has not been worked on experimentally, but the data show the limits for solid solubility at the pressure and temperature of jadeitite formation.

Considering that jadeitites are uniformly related to tectonized serpentinites, it is reasonable to expect some emerald-green pyroxene (ureyitic) in all jadeitite occurrences. This could be of some economic significance because emerald-green jade can be extremely valuable. Other occurrences of ureyitic pyroxene described in the literature, such as in metamorphosed gabbros and in exotic rocks brought up from the Earth's mantle by kimberlites (the volcanic rocks that yield diamonds), are likely to be similarly affected by an outside source of Na^+ and metasomatic fluids.

For background information SEE JADEITE; METASOMATISM; PYROXENE; SOLID SOLUTION in the McGraw-Hill Encyclopedia of Science and Technology.

George E. Harlow

Bibliography. W. A. Deer, R. A. Howie, and J. Zussman, *Rock Forming Minerals*, vol. 2A: *Single-Chain Silicates*, pp. 520–527, 1978; C. Frondel and C. Klein, Ureyite, $NaCrSi_2O_6$: A new meteoritic pyroxene, *Science*, 149:742, 1965; G. E. Harlow and E. P. Olds, Observations on terrestrial ureyite and ureyitic pyroxene, *Amer. Mineralog.*, 72:126–136, 1987; H. Laspeyres, Mittheilungen aus dem mineralogischen Museum der Universitat Bonn, VII, Theil, *Z. Krystallog. Mineralog.*, 27:586–600, 1897; C. M. D. Yang, Terrestrial source of ureyite, *Amer. Mineralog.*, 69:1180–1183, 1984.

Quantum mechanics

New data on the electrical properties of very small, normal-metal structures at low temperatures have produced several unexpected results that have changed scientists' concept of electrical resistance in solid-state systems. Recent theoretical and experimental work indicates that in a loop of wire the electrical resistance oscillates periodically as the magnetic flux threading the loop is increased. The period of oscillation is h/e, the normal-metal flux quantum, where h is Planck's constant and e is the charge of the electron. In single wires the electrical resistance fluctuates randomly as a function of magnetic flux. Both observations are due to an Aharonov-Bohm effect and indicate that flux quantization effects exist in nonsuperconducting systems. They have opened up a new field of condensed-matter physics because they are evidence that the electrical properties are dominated by quantum-mechanical behavior of the electrons, and the usual rules of classical physics are no longer operative.

These experiments were possible because the technology for fabricating small samples has advanced significantly in the last few years. Devices can now be made with all dimensions less than 10^{-6} m, and the smallest feature that can be routinely fabricated is less than 3×10^{-8} m. An example of a very small metallic ring which was fabricated from a polycrystalline gold film is shown in **Fig. 1**. This sample is about 100 atoms wide and 100 atoms thick, and contains only about 10^8 atoms. It is expected that samples of these dimensions and smaller will allow a wide variety of new quantum-mechanical phenomena to be studied.

Mesoscopic physics. Classical statistical physics deals with the properties of a large collection of particles, and averages are taken over all possible states of a system in order to obtain the

bulk properties of a sample. The electrical resistance of a wire is an example of an average property that can be successfully computed by using these ensemble-averaging concepts. The basic idea is to assume that the electrons in the sample move without scattering for some distance ℓ (the mean free path), scatter off impurities or boundaries (specularly, elastically, or inelastically), and repeat this process many times. The electrical resistivity can be accurately computed by averaging over all possible scattering sites and taking the limit as the volume V of the system approaches infinity with the density of electrons N/V a constant, where N is the number of electrons. The quantum-mechanical properties of the individual electrons are known to be important over microscopic distances of the order of 10^{-9} m or a few lattice constants. On this microscopic scale, both the wavelike nature of the electron and the details of the individual scattering centers must be understood in order to compute the resistance. Generally it is not necessary to understand the microscopic details because, once the size of the sample becomes much larger than this characteristic scale, the bulk classical properties are achieved.

It has recently been discovered that at temperatures below a few kelvins the quantum properties of electrons in normal metals and semiconductors are important over a much larger length scale. Theoretical and experimental results clearly show that once the phase-coherence length L_Φ (the distance the electron travels before losing phase memory in its wave function) is greater than the size of the system, the measured transport properties depend crucially upon the individual details of the scattering, and the phases of all the electrons in the sample can become highly correlated. In addition, dissipation or electrical resistance arises not from elastic scattering of electrons from impurity sites but from contact to thermal reservoirs. The transport properties must be computed by using quantum mechanics, and classical ensemble-averaging approaches must be avoided.

This intermediate size regime between microscopic and macroscopic is now frequently referred to as the mesoscopic range. It is a length scale that is defined as the region where a large number of particles interact in a quantum-mechanically correlated fashion. There are a growing number of examples where at low temperatures this effective length scale L_Φ can be 100–10,000 times larger than the characteristic microscopic scale and can involve more than 10^{13} particles.

Aharonov-Bohm effect. In 1959 Y. Aharonov and D. Bohm suggested an experiment where a beam of electrons traveling in a vacuum would be split so that half would travel clockwise and the other half counterclockwise around a region containing a magnetic flux. If the beams were then recombined, the intensity of the resulting beam

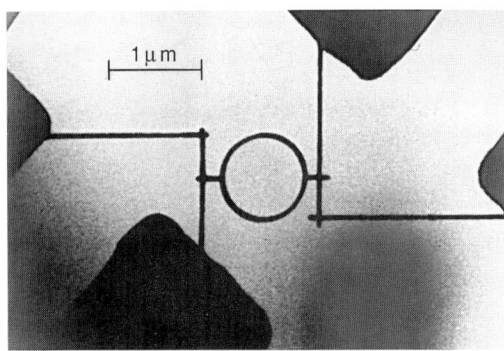

Fig. 1. Photograph of very small ring fabricated from a gold film 3.8×10^{-8} m thick. The width of the lines forming the ring are 4×10^{-8} m, and the average diameter is 8.2×10^{-7} m. The electrical resistance is measured by applying a constant current through two of the leads that connect the sample and measuring the voltage difference between the two other leads. (*From R. A. Webb et al., Observation of h/e Aharonov-Bohm oscillations in normal-metal rings, Phys. Rev. Lett., 54:2696–2699, 1985*)

would oscillate periodically as the enclosed flux was changed. The period of this oscillation would be $\Phi_0 = h/e$ and would represent patterns of constructive and destructive interference occurring between the two beams. This effect would be present in the absence of a magnetic field or flux in the path of the electrons. The magnetic field $\vec{B}$ can be expressed as a curl of a vector potential $\vec{A}$, as in Eq. (1). The vector field $\vec{A}$ exists in all

$$\vec{B} = \Delta \times \vec{A} \tag{1}$$

space, while the magnetic field need not (an infinite solenoid is an example). The quantum-mechanical wave function for the electrons can be written as Eq. (2), where δ is a phase factor

$$\Psi = Ce^{i\delta} \tag{2}$$

and C is an amplitude. In general the phase of the wave function can be written as Eq. (3), where

$$\delta = \delta_0 + \frac{2\pi e}{h} \int \vec{A} \cdot \vec{dl} = \delta_0 + \frac{2\pi e}{h} \Phi \tag{3}$$

δ_0 is an arbitrary phase factor and the integral of the scalar product is taken over the electron path. The quantity Φ is the magnetic flux enclosed by the electron path (B times an effective area S). With increasing flux through the center of the loop, the phase of the electrons in the upper branch will increase, and the phase of the electrons in the lower branch will decrease. The resulting intensity of the recombined beam will be given by Eq. (4), where I_0 is the intensity when

$$I = I_0 \cos (2\pi\Phi e/h) \tag{4}$$

$\delta = 0$. Within a year of this prediction the Aharonov-Bohm effect was observed in electron-beam experiments, and it has been verified in numerous subsequent experiments. Before these experimental observations, it was thought that the vector potential was a convenient mathematical representation by which a magnetic field could be described but classically could not have any measurable consequences. The Aharonov-Bohm effect clearly demonstrates that if the quantum-mechanical properties of the system are maintained, changes in the vector potential can have physically realizable effects.

Effect in normal metals. It was believed that both the effects of finite temperature and the scattering inherent in any large condensed-matter system would randomize the phases of the wave function and prevent the observation of the Aharonov-Bohm effect in bulk samples. If the thermal energy $k_B T$, where k_B is Boltzmann's constant and T is the thermodynamic temperature, is larger than the separation between electron energy levels, the phases of individual electron wave functions become uncorrelated because of thermal smearing. This separation is inversely proportional to the volume of the sys-

tem. For the very small ring displayed in Fig. 1, the temperature at which the thermal energy is equal to the electron separation energy is 0.0001 K, or 3×10^6 times colder than room temperature. Fortunately, the nearest energy levels are correlated, and the averaging over different uncorrelated levels occurs only at much higher temperatures (0.03 K for this sample).

The first observation of the Aharonov-Bohm effect in very resistive normal metals was made in long (1 mm), 1-micrometer-diameter magnesium cylinders. Surprising features about the measured resistance were that it oscillated periodically with a period of $h/2e$ rather than h/e and that very small magnetic fields destroyed the oscillations. Both observations actually confirmed earlier theoretical predictions and can be understood in terms of averaging over uncorrelated regions of the cylinder (the length of the cylinder was much longer than L_Φ). The h/e oscillations average to zero for a long sample, while the second harmonic, $h/2e$, does not. Shortly after these observations, samples in which L_Φ was larger than the size of the system were fabricated. **Figure 2** displays the electrical resistance of the gold ring shown in Fig. 1 as a function of magnetic field over a small field range at a temperature of 0.01 K. The electrical resistance oscillates with a characteristic period of $B = 0.0076$ tesla. From measurements of the average area of the ring, the period is equal to the normal-metal flux quantum $\Phi_0 = h/e = BS$. These data prove that the h/e Aharonov-Bohm effect exists in resistive samples when L_Φ is larger than all dimensions of the sample. With the resistance of the ring measured as 29.5 ohms, the

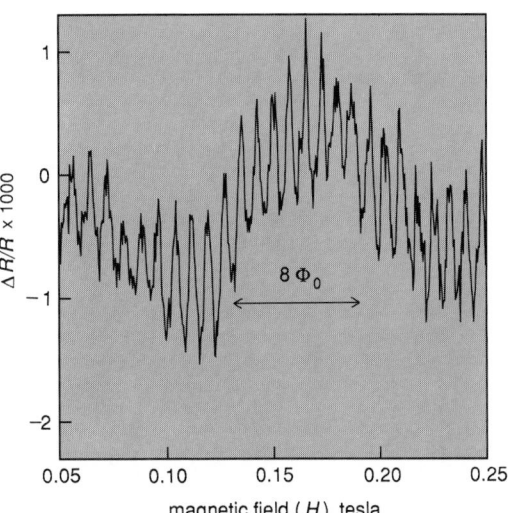

Fig. 2. Resistance oscillations observed in the ring of Fig. 1 at a temperature of 0.01 K over a very small magnetic field range. Vertical scale measures deviation ΔR of resistance R from its mean value of 29.5 ohms. The period of the h/e Aharonov-Bohm oscillations in a magnetic field is 0.076 T (76 gauss). The arrows show the field scale for eight flux quanta ($8\Phi_0$) in the average area of the ring.

mean free path is calculated to be 100 times smaller than the sample size. This means that each electron experiences about 10^4 collisions while circling the ring, thus clearly demonstrating that elastic scattering does not destroy the phase memory of the electron wave functions in a condensed-matter system.

It has been shown that an Aharonov-Bohm effect in a condensed-matter system implies that flux quantization also exists. Many of the properties of the electrons (such as the energy and phase of the wave function) will be periodic in h/e. While these effects are well-known properties of superconductors, the surprising result here is that they occur in a sample that exhibits finite electrical resistance. One unexpected prediction is that this gold ring should exhibit persistent currents. That is, in this ring, in the absence of any applied current or voltage, there should be a circulating current that does not decay over long time scales. This means that there is no energy dissipation. Although to date these persistent currents have not been observed, they must exist if the present understanding of quantum mechanics is correct.

Universal conductance fluctuations. As seen from Fig. 2, the h/e resistance oscillations are superimposed upon a randomly fluctuating background magnetoresistance. Both periodic h/e and aperiodic fluctuations persist unattenuated to 16 T in this sample. The aperiodic background fluctuations can be understood by realizing that these experiments are not performed in the same way as the electron-beam experiments. The magnetic field in the ring experiments is uniform in space. In addition to a field inside the loop that gives rise to the periodic h/e oscillations, there is a magnetic field present over the path of the electrons. This is the origin of the aperiodic background signal. Theoretical calculations have shown that

this random signal can be understood by considering electrons traveling on the inner and outer perimeters and by realizing that upon encircling the ring they acquire a different phase shift because their paths enclose a different amount of flux. In a real sample the motion of the electrons is highly diffusive and there are many different paths possible. Thus, the magnetic field introduces phase-dependent contributions to the conductivity which add incoherently and give rise to a random fluctuation. The aperiodic background fluctuations are really due to an Aharonov-Bohm effect and exist in any small wire.

Perhaps the most surprising result of recent theoretical and experimental work is that the root-mean-square value ΔG of the conductance fluctuations is universal, independent of the material from which the sample is fabricated, and independent of the electrical resistance of the sample. This value is given by Eq. (5), where A is

$$\Delta G = \frac{\Delta R}{R^2} = A\,\frac{e^2}{h} \qquad (5)$$

a constant of order 1 that depends on the geometry of the sample, R is the electrical resistance of the device, and ΔR is the root-mean-square value of its fluctuations. This result should apply to both the periodic and aperiodic signals, and is valid only when L_Φ is comparable to the sample size.

An example of the universal conductance fluctuations is shown in **Fig. 3** for the gold ring of Fig. 1 over a magnetic field range of 8 T. The h/e periodic oscillations are shown in the three insets over selected 0.1-T ranges. The root-mean-square value of the random conductance fluctuations for this sample is $0.7e^2/h$ and $0.2e^2/h$ for the h/e oscillations. Although the background fluctuations look like noise, they are very reproducible. Exactly the same trace would be obtained if the measurement were to be repeated after several days. If another sample of exactly the same dimensions were to be fabricated, the root-mean-square value of the fluctuations would be the same, but the details of the individual fluctuations would be entirely different.

For background information SEE ELECTRICAL CONDUCTIVITY OF METALS; POTENTIALS; QUANTUM MECHANICS; STATISTICAL MECHANICS; SUPERCONDUCTIVITY in the McGraw-Hill Encyclopedia of Science and Technology.

Richard A. Webb

Bibliography. B. L. Altshuler et al., Observation of the Aharonov-Bohm effect in hollow metal cylinders, *JETP Lett.*, 35:588–591, 1982; M. Büttiker et al., Generalized many channel conductance formula with application to small rings, *Phys. Rev.*, B31:6207–6215, 1985; P. A. Lee and A. D. Stone, Universal conductance fluctuations in metals, *Phys. Rev. Lett.*, 55:1622–1625, 1985; S. Washburn and R. A. Webb, Aharonov 13-ohm effect in normal metal: Quantum coherence and transport, *Adv. Phys.*, 35:375–422, 1986.

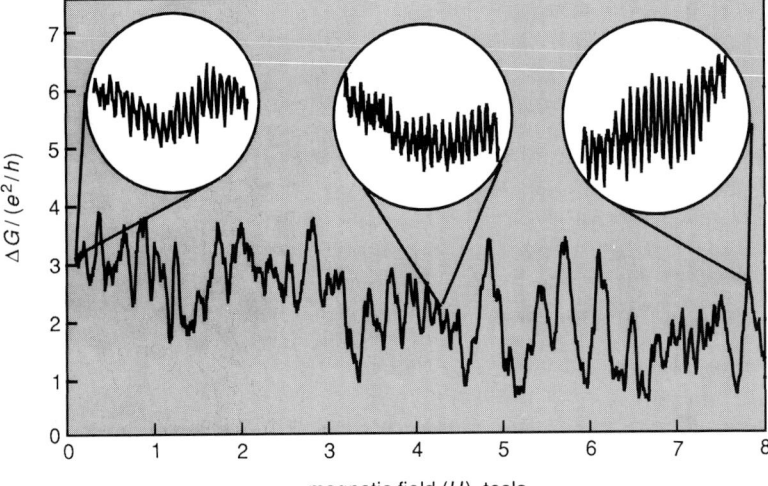

Fig. 3. Aperiodic universal conductance fluctuations measured in the gold ring of Fig. 1 at a temperature of 0.01 K over an 8-T magnetic field range. The vertical coordinate of the figure is the conductance variation of the sample in units of e^2/h. The three insets are selected 0.1-T-long segments which display the h/e oscillations, demonstrating the persistence of the Aharonov-Bohm effect to very large fields.

Radio paging systems

Paging systems transmit radio signals to portable, pocket-sized one-way receivers. These systems allow travelers to receive a variety of messages from radio pagers, commonly called beepers. Companies can contact employees, and household members can contact each other, saving the time and money necessary to telephone their offices or home for messages.

These systems utilize national networks of telephone lines, microwave transmissions, and satellite links to send data to major cities. Advances in microminiaturization of computer chips are resulting in smaller pagers with expanded message capabilities.

Local paging. Local radio paging has been offered by private companies (called radio common carriers) and telephone companies (called wireline common carriers) since about 1950. The basic radio paging system includes a terminal, an antenna, and pagers. The terminal is a sophisticated, microprocessor-controlled switch that can receive paging messages of callers from the public switched telephone network.

The terminal contains such basic components as an interface device connected to telephone lines, a central processing unit, a memory base (for storing pager identification numbers, for example), and an encoder for translating the telephone data into a form that can be sent to the antenna tower. The terminal sends the pages to one or more antenna towers, which broadcast the signals over the air—by using specific frequency bands assigned by the Federal Communications Commission (FCC)—to the paging receivers. The pagers decode the signal and provide the messages in audio or data formats to the user.

Pagers. There are four major types of pagers: tone-only, tone-and-voice, numeric display, and alphanumeric display.

A tone-only pager broadcasts a beep tone to alert the paging user to call a specific telephone number, such as the user's office or home. Some models feature up to four different types of beeps that correspond to different telephone numbers.

A tone-and-voice pager broadcasts a 7- to 20-s voice message over the pager's speaker. Callers speak the message into a telephone, from which it is sent through telephone lines to the paging terminal.

Numeric display and alphanumeric display devices are the most advanced pagers. The pager contains a liquid-crystal display which shows a group of numbers or letters (**Fig. 1**). The most common type of message sent to a numeric pager is a 7- or 10-digit telephone number, which the paging subscriber may then call. Numeric messages can also be sent to an alphanumeric pager.

As the most sophisticated type of pagers, alphanumeric devices can display any type of text and numeric message, such as "YOUR CONTRACT HAS

Fig. 1. Numeric display scanning pager. (*DiversiCom*)

BEEN APPROVED" or "CALL 212-555-1234. COMPANY'S EQUIPMENT MUST BE REPAIRED TODAY." Alphanumeric pagers have not been as commercially successful as the other types of units. Although they offer enhanced capabilities, the devices are larger and heavier and more expensive to purchase or lease.

Also, callers usually cannot transmit a message to an alphanumeric pager from a telephone. A computer keyboard and modem is required. To send a message, a caller must either call the paging company's operator who will type in the message, or directly type the message from a keyboard and send it to the terminal via a modem and communications program.

Tone-only and numeric display pagers are currently available for nationwide systems (Fig. 1). By the end of 1987, it is expected that alphanumeric pagers also will be offered. In addition, nationwide systems offer a voice mail option. In essence, a voice mail terminal is a large, sophisticated telephone answering machine. A caller leaves a message for the paging subscriber on the voice mail terminal, and the terminal automatically "pages" the subscriber, indicating that a message has been stored.

Nationwide paging. Nationwide paging systems contain the same components as local and regional systems. However, they also utilize more advanced communications software and computer switching techniques that permit transmission of the paging signal to dozens of cities. Currently, there are two national systems in the United States and they operate in two different frequency bands: 900-MHz land mobile channels and frequency-modulation (FM) broadcast radio channels.

900-MHz system. The FCC used a lottery to choose three licensees from 16 applicants for the 900-MHz band. Only one nationwide 900-MHz applicant is now operating. The second licensee has received an extension of time in which to establish a system. The third licensee has been turned down by the FCC for an extension of time, and another licensee is expected to be selected.

System using FM broadcasting channels. The second nationwide system in operation uses frequency bands allocated to FM radio stations. These stations broadcast their main-channel programs of music, news, sports, and so forth over only a portion of the 200-kHz bandwidth generally assigned to them. Monaural stations generally use the first 20 kHz for their main programs, while stereo stations use a maximum of the first 53 kHz. It is possible to insert additional transmissions, called subchannels or subcarriers, in the unused portions of the frequency bands. Generally, a monaural FM station can use frequencies from 20 to 99 kHz above the lower end of its assigned frequency band, while a stereo station can use frequencies from 53 to 99 kHz. Within this range, stations may broadcast virtually any type of audio, text, or graphics service, including radio paging. About 40–60 companies offer local paging over these FM subcarriers. One company offers nationwide service, using a so-called scanning numeric display pager (Fig. 1).

The nationwide operator pays radio stations a fee for leasing the rights to use a specific subcarrier, for example, the subcarrier centered at 57 kHz above the lower end of the assigned frequency band. In some cities, the operator may lease the same subcarrier frequency, such as 57 kHz, from several FM stations to provide better paging coverage. The pagers scan a specific subcarrier in cities where the nationwide system is operating. In cities where the operator has leased several subcarriers, the pager searches for the strongest transmission and locks onto that signal.

System configuration. Because national systems have been developed only recently, there is no typical configuration. A sample configuration is illustrated in **Fig. 2**. Pages are routed either directly to a central paging terminal or to a local paging terminal which then transmits the page to a central terminal. The page is uplinked to a communications satellite, which sends it to multiple satellite receiving facilities. Each satellite downlink station transmits the page to decoding equipment, which sends it to a tower or towers for broadcast to the pager.

Operation. Generally, to initiate a numeric display page, a caller would use a tone telephone to dial a local or toll-free number. By following a series of tone prompts or voice-synthesized commands, the caller would enter the paging subscriber's identification number. The paging terminal would determine whether the identification number is valid and would also record the transaction for billing purposes.

The caller would then enter a 10- or 12-digit telephone number to appear on the pager's display. Finally, the caller would push the tele-

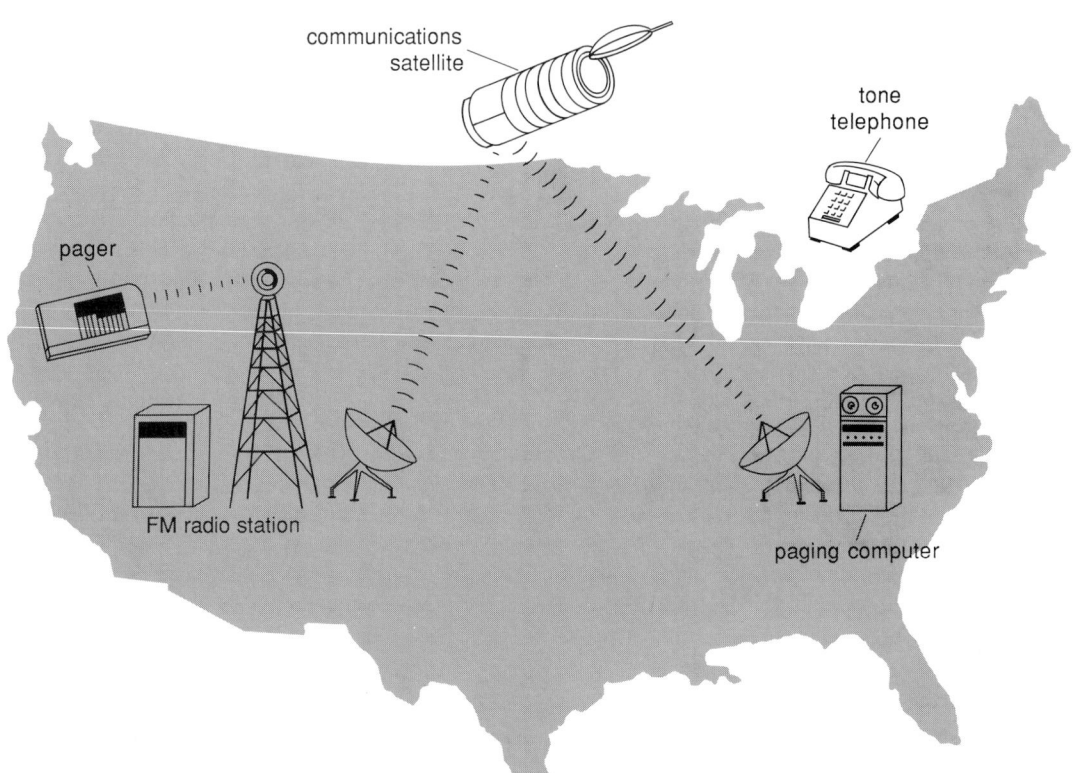

Fig. 2. Configuration of nationwide paging system that uses FM radio channels. A message and access code are entered on a tone telephone by dialing a toll-free number. The message is sent to a paging computer, routed by a communications satellite to FM radio stations nationwide, picked up by these stations, and transmitted along with their normal FM signals. A pager automatically locks onto the proper frequency and receives the message. (*DiversiCom*)

phone's pound (#) key and hang up. Within about 15 s, the number would be broadcast to the pager. Numeric pagers used in nationwide systems typically can store four to six messages in memory.

Regional service. Although nationwide systems can send pages over the entire system in every city where they operate, some systems also can offer regional or multiregional service. For example, a user who travels primarily in the northeastern United States could instruct thesystem to send pages only in the region covering Boston to Washington, D.C. This not only saves the user money but also reduces unnecessary pager traffic over the entire network.

User expenses. The use of nationwide paging is relatively inexpensive compared with that of many other forms of portable communications, such as cellular radiotelephones. National paging subscribers must buy or lease the paging device and pay for air time. Some companies include an unlimited number of messages for a set fee, while others charge a per-message fee. Customers who purchase their own pagers do not have to pay equipment rental fees. In addition, some companies charge a one-time setup fee.

150- and 450-MHz systems. In the United States, two other nationwide systems may go into operation in 1987. The systems are based on two new scanning pagers that function differently from the scanning receivers used by FM radio subcarrier operations. Instead of searching for only one frequency in the FM radio band, the two pagers under development utilize bands already assigned to radio common carriers and telephone paging companies offering local and regional— but not nationwide—service. One unit scans all paging channels in the 150-MHz band. The other unit scans paging channels in the 150- and 450-MHz bands.

The advantage of this method is that a significant part of the nationwide network already exists. Any paging company that currently broadcasts signals at 150 or 450 MHz could join one of the two nationwide paging networks. Nationwide operators on the 900-MHz band, however, must construct new systems designed only for that bandwidth.

Systems outside United States. Although local radio paging is available in about a dozen countries outside the United States, only a few countries offer national coverage. Great Britain provides sophisticated local and nationwide service. Sweden offers national numeric paging service over FM subcarrier bands.

Usage. Some 4.5–5.5 million pagers are now in service, of which fewer than 10,000 operate over nationwide networks. Studies estimate that 3–10% of current paging subscribers will use nationwide service. In addition, it is expected that many persons who never used pagers because of limited range will subscribe to these wider-area systems. By 1990, a total of 8–10 million pagers could be in operation in the United States, of which several hundred thousand could be operating over national systems.

For background information SEE COMMUNICATIONS SATELLITE; FREQUENCY-MODULATION RADIO; RADIO PAGING SYSTEMS in the McGraw-Hill Encyclopedia of Science and Technology.

Alan A. Reiter

Bibliography. L. Anderson, Scanning VHF, *Telocator*, 10(8):22–35, August 1986; D. Baker, 900 MHz: Down to the wire, *Pers. Commun. Technol.*, 4(7):28–30, July/August 1986; Frost & Sullivan, Inc., *The Radio Paging Market*, A1355, 1984.

Radioactivity

Two kinds of cluster radioactivities, in which a nucleus heavier than an alpha particle is emitted, were discovered in 1984 and 1985: spontaneous emission of carbon-14 from radium isotopes with mass numbers 222, 223, 224, and 226; and neon-24 emission from thorium-230, protactinium-231, and uranium isotopes 232 and 233.

Nuclear decay processes. About 300 different nuclear species (or nuclides), each characterized by an atomic number Z and a mass number A, have been found in nature; their natural occurrence is due to the fact that they are stable or their half-lives are sufficiently long compared with the age of the universe. More than 1700 other nuclides have been artificially produced, mainly through nuclear reactions, and it is expected that about 5000 to 6000 new nuclides will be identified in the future.

A common feature of the 1700 artificially produced nuclides is their transmutation or decay. The phenomenon, which takes place spontaneously in nature, was discovered in 1896 by H. Becquerel in uranium compounds, and was named radioactivity soon thereafter. After some years of research, it was established that there are three modes of nuclear decay: alpha (helium-4 nuclei), beta (electrons or positrons), and gamma (electromagnetic waves). The first information about atomic nuclei was derived from the study of radioactivity; particularly useful was alpha decay, not only as an object of study but also as a source capable of initiating nuclear reactions, before particle accelerators were developed.

Classical theoretical physics was not able to explain alpha decay. Only in 1928, after the development of the quantum mechanics, was it shown that alpha decay is a tunneling process through the nuclear potential barrier. Induced fission, discovered in 1939, was explained in the same way, as was spontaneous fission from nuclear ground states, predicted and confirmed in 1940. Theory (1966) also followed experiment (1962) in the case of fission isomers. But the proton radioactivity of an isomeric state (1970) and of the ground state (1981) were discovered long after their predictions (in 1960 and 1951, respectively).

Extremely asymmetric fission. Both in alpha decay and in the spontaneous fission process, the initial (parent) nucleus A,Z is split into two fragments, A_1,Z_1 and A_2,Z_2, conserving the hadron numbers: $A = A_1 + A_2$ and $Z = Z_1 + Z_2$. In the former process, one of the fragments (the emitted cluster $A_2 Z_2$) is very small compared with the other (the daughter nucleus or the heavy fragment $A_1 Z_1$). In other words, the mass asymmetry of the two fragments, $\eta = (A_1 - A_2)/A$, is very large, of the order of 234/238, in alpha decay, and is much smaller, of the order of 42/238, for the most probable split in fission.

The traditional theory of alpha decay was developed in the framework of the R-matrix treatment of nuclear reactions, supplemented by the single-particle shell model. In contrast, for many years (1939–1966) the phenomenological liquid-drop model, considering essentially only the collective properties of the nucleons, dominated fission theory, in spite of its failure to explain the small but well-established asymmetry in the distribution of the fragment masses, particularly for nuclei in the actinide region. Only by simultaneously taking into account the collective and the individual character of the nucleon motion, in the framework of a hybrid macroscopic-microscopic theory, was it possible to explain the fission asymmetry as due to shell effects.

Much theoretical effort has been devoted to unifying the treatment of various processes, extending the fission theory on the asymmetry axis η beyond the region usually studied, which corresponds to masses of the light fragment A_2 greater than about 70, and going down to alpha decay, for which $A_2 = 4$. Four theoretical approaches have been used to predict a very large number of new kinds of radioactivities by spontaneous emission of clusters heavier than an alpha particle but lighter than fission fragments. Both the mechanism of a superasymmetric fission process and that of a preformed cluster emission have been considered. In these studies, ^{14}C, ^{24}Ne, magnesium-28, calcium-48, and other kinds of radioactivities appeared quite naturally in the range of mass values $4 < A_2 < 70$, which until then had been unjustifiably ignored.

Figure 1 shows a mass fragment distribution for the fission of nobelium-252 at three excitation energies, $E^* = 0$, 10, and 25 MeV, calculated by using fragmentation theory and the asymmetric two-center shell model. Besides the well-known low asymmetry leading to a heavy fragment not far from the doubly magic tin-132 nucleus ($^{132}_{50}$Sn$_{82}$), a new peak appears at a considerably larger asymmetry, due to spontaneous emission of ^{48}Ca which corresponds to a daughter, lead-204, not far from the doubly magic lead-208 ($^{208}_{82}$Pb$_{126}$) nucleus.

Energies, half-lives, and branching ratios. The important quantities which can be experimentally determined with presently available techniques in order to confirm or refute the theory are the kinetic energies of the emitted clusters E_k, and the half-lives T of the parent nuclides with respect to the new radioactivities. It is preferable to measure a branching ratio relative to alpha decay, $b = T_\alpha/T$, because T is usually much greater than T_α, and the half-life for alpha decay T_α is well known.

Each of the approximately 2000 known nuclides is presumptively suspected to be a good candidate to emit any of the approximately 200 nuclides that are possible emitted clusters. Consequently, a systematic study of the new radioactivities involves a large amount of calculations, on account of the necessity to check all the possible combinations and to select only the most probable ones. For this purpose, an analytical superasymmetric fission model has been developed that can solve the problem with a reasonable computer running time.

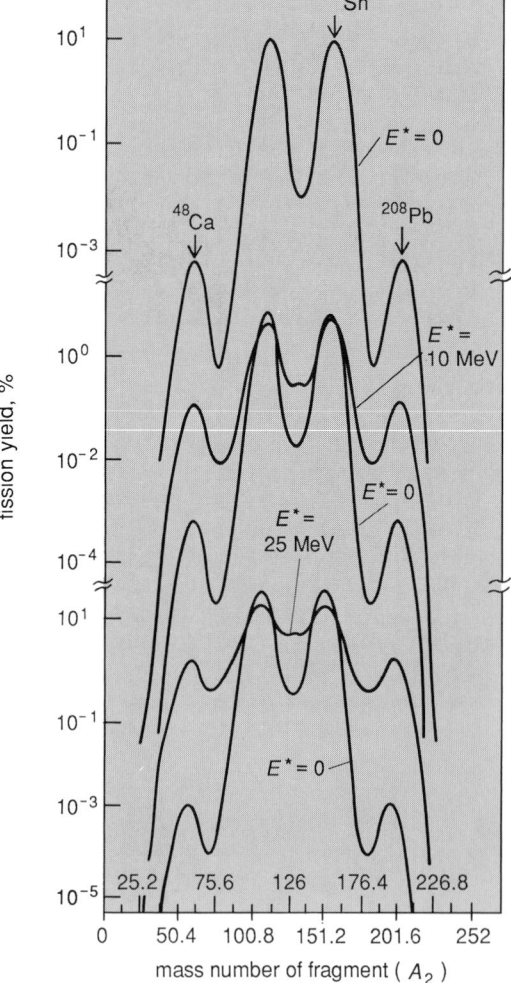

Fig. 1. Calculated mass fragment distributions for fission of nobelium-252 at excitation energies E^* of 0, 10, and 25 MeV. For purposes of comparison, the curve for spontaneous fission from the nuclear ground state ($E^* = 0$) has been superimposed on the curves for $E^* = 10$ and 25 MeV. *(After A. Săndulescu, D. N. Poenaru, and W. Greiner, New type of decay of heavy nuclei intermediate between fission and α-decay, Sov. J. Part. Nucl., 11:528–541, 1980, translated by American Institute of Physics)*

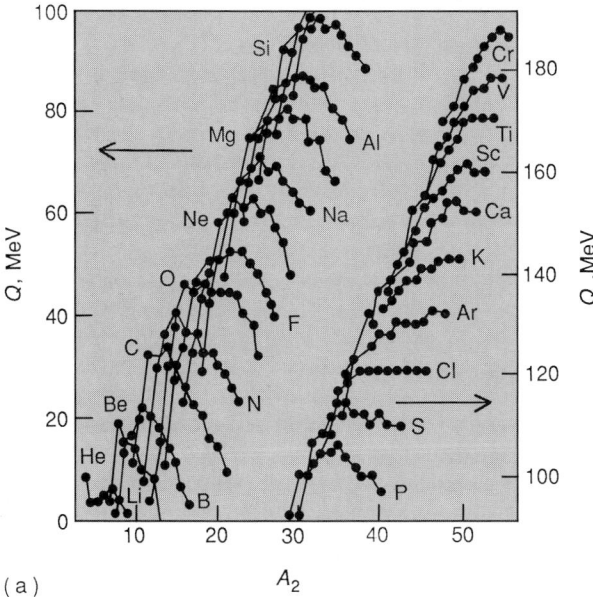

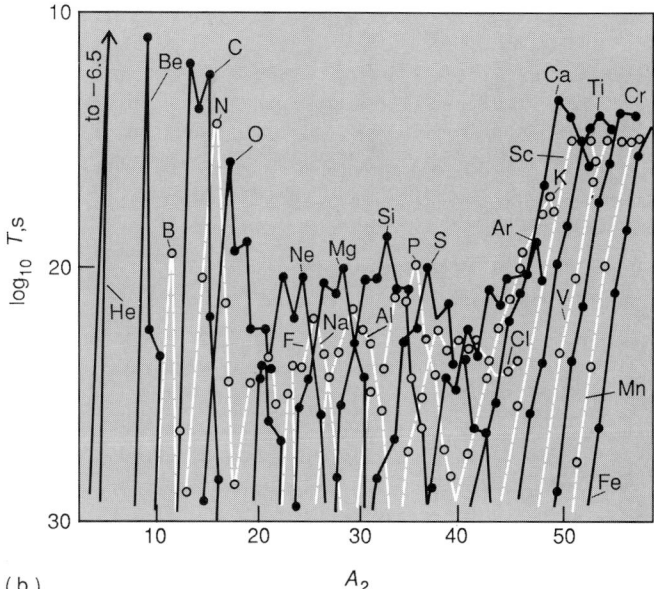

(a) A_2 (b) A_2

Fig. 2. Extremely asymmetric fission processes of various parent nuclei in which the daughter nucleus or heavy fragment is the doubly magic lead-208. The symbols that label the curves indicate the element of the emitted cluster or light fragment, and the horizontal scale indicates the mass number A_2 of the emitted cluster. (*a*) Released energy Q. The released energies for emitted clusters of lighter elements (He through Si) are measured on the left-hand scale, while those for heavier elements (P through Cr) are measured on the right-hand scale. (*b*) Decimal logarithms of lifetimes T, in seconds. To aid in distinguishing elements, even-numbered elements are shown with solid lines and solid circles, while odd-numbered elements are shown with broken lines and open circles. (*After D. N. Poenaru et al., Atomic nuclei decay modes by spontaneous emission of heavy ions, Phys. Rev., C32:572–581, 1985*)

Results of these calculations are exhibited in **Fig. 2***a* for the released energy in the process, Q, and in Fig. 2*b* for the half-life. Usually the most favorable cases (lower values of the half-life T) correspond to those combinations leading to a doubly magic daughter, $^{208}_{82}\text{Pb}_{126}$, or another nucleus with proton number and neutron number close to the magic ones, $Z_1 = 82$ and $N_1 = 126$. Figure 2*a* and *b* refer to this privileged combination. More than 140 new decay modes, including the already discovered ^{14}C and ^{24}Ne radioactivities, can be identified on these two figures by the chemical symbol of each element and the mass numbers of the emitted cluster A_2. The heavier the emitted cluster, the larger is the released energy Q and the corresponding kinetic energy $E_k = QA_1/A$.

A reversed scale for the decimal logarithm of the half-lives was adopted in Fig. 2*b* in order to make the peaks in the figure correspond to processes with the highest emission rate (which is inversely proportional to the lifetime). A large variety of cluster radioactivities, which are waiting for experimental confirmation, are apparent in the figure.

Experimental observation. The first of eight observed cluster radioactivities was discovered with a standard silicon telescope ΔE-E detector. This type of detector allows identification of the atomic number and measurement of the total energy of the particles passing through it. It consists of two silicon surface-barrier detectors

of different thickness that measure the ionization rates of correlated and noncorrelated events. Actinium-227, in secular equilibrium with ^{223}Ra, was used as a source, and the charge, energy, and emission rates of the decay products were measured. With a more refined experimental setup, using a superconducting magnetic spectrometer which makes it possible to reject the enormous background of alpha particles, another group confirmed the discovery and reported ^{14}C radioactivity of ^{226}Ra. The ISOLDE on-line mass separator made various sources available. From data taken with polycarbonate track-recording films which are sensitive to heavier clusters but not to alpha particles, ^{14}C radioactivity of ^{222}Ra and ^{224}Ra was measured. This technique of using track-recording films to suppress the alpha-particlebackground has proved to be very promising, and was subsequently applied by different groups in the discovery of ^{24}Ne radioactivities.

Fission with compact shapes. There is experimental evidence showing that two kinds of mechanisms with different features contribute to spontaneous fission phenomena. One of the mechanisms is the well-known process in which nuclear shapes during the fission are very deformed and the total kinetic energy of the fragments is smaller than the released energy Q, because the fragments are internally excited. They deexcite shortly after scission by emission of gamma rays and neutrons.

The other mechanism, studied only during

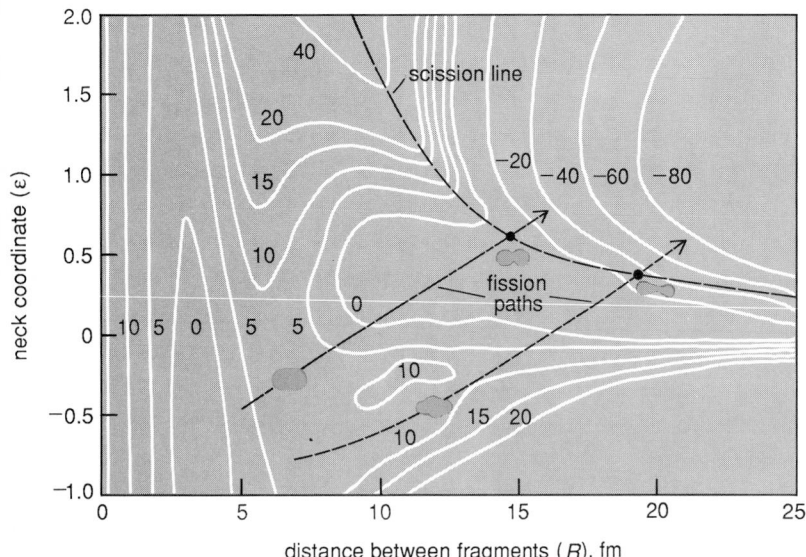

Fig. 3. Potential surface of fermium-258, showing two fission paths (bimodal fission). The neck coordinate measures the squeeze-in of the nucleus as it fissions. The numbers labeling the equipotential contours give the potential energy in MeV. Shaded figures along the path show the nuclear shapes. *(After K. Depta et al., Bimodal fission in ^{258}Fm, Mod. Phys. Lett., A1:377–381, 1986)*

recent years in some actinides (uranium, plutonium, fermium, mendelevium, nobelium, and element 104), is characterized by compact shapes; high total kinetic energy, almost exhausting the Q-value; and consequently no excitation of the fission fragments (cold fission). This process can be considered the inverse of fusion.

Potential energy surfaces calculated by using the shell correction method with a two-center shell model clearly show the possibility of this bimodal fission, and also account for the observed difference in total kinetic energy between the two fission modes. Thus, **Fig. 3** shows two fission paths, the upper path leading to two spherical fragments of high kinetic energy at the scission line, the lower path leading to a very elongated neck between fragments of low kinetic energy. In the framework of the analytical model, this bimodal process is interpreted as a cluster radioactivity with a heavy fragment corresponding to the most probable emission having neutron and proton numbers in the neighborhood of magic ones, $N_1 \approx 82$ and $Z_1 \approx 50$. Such bimodal fission has been experimentally observed. Thus, processes that appear to be very different, such as alpha decay, heavy-cluster radioactivity, and spontaneous fission, can be described theoretically in a unified way.

For background information SEE MAGIC NUMBERS; NUCLEAR FISSION; NUCLEAR STRUCTURE; RADIOACTIVITY in the McGraw-Hill Encyclopedia of Science and Technology.

Walter Greiner

Bibliography. K. Depta et al., Bimodal fission in ^{258}Fm, *Mod. Phys. Lett.*, A1:377–381, 1986; E. K. Hulet et al., Bimodal symmetric fission observed in the heaviest elements, *Phys. Rev. Lett.*, 56:313–316, 1986; P. B. Price et al., Discovery of radioactive decay of ^{222}Ra and ^{224}Ra by ^{14}C emission, *Phys. Rev. Lett.*, 54:297–299, 1985; H. J. Rose and G. A. Jones, A new kind of natural radioactivity, *Nature*, 307:245–247, 1984; A. Săndulescu, D. N. Poenaru, and W. Greiner, New type of decay of heavy nuclei intermediate between fission and α-decay, *Sov. J. Part. Nucl.*, 11:528–541, 1980.

Reproductive system disorders

Human papillomavirus (HPV) has been known as the causative agent of warts of the human skin and condylomata of the genitalia for over half a century. Nevertheless, the causal relationship between condylomata and female genital cancer has not been studied until recently because this virus could not be propagated in a laboratory. Before the carcinogenesis of human papillomavirus drew the attention of investigators, type II herpes simplex virus (HSV-II) was suspected as the causative agent of uterine cervical cancer. Extensive investigation, however, failed to establish this strain of the herpes virus as the cause of female genital cancer.

The process of female genital cancer, notably squamous cell (epidermoid) carcinoma of the uterine cervix, starts with a premalignant condition, or dysplasia, and progresses to cancer

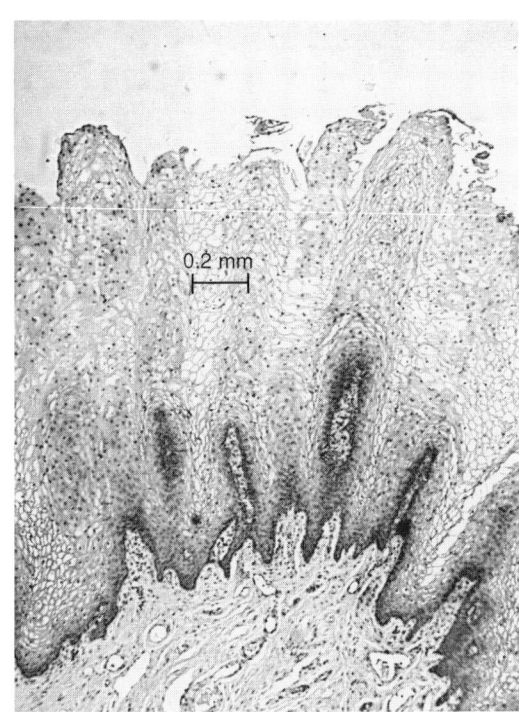

Fig. 1. Condyloma of the uterine cervix, a benign lesion produced by human papillomavirus.

through a stage known as carcinoma in situ. These two stages of premalignancy, dysplasia and carcinoma in situ, are referred to as cervical intraepithelial neoplasia, based on an assumption that the cells have already transformed into cancer cells but have not started invading into the adjacent tissues.

In 1976, after a review of the cases of condyloma acuminatum by means of morphologically identifiable koilocytotic cells in Papanicolaou smears, it was suggested that cervical intraepithelial neoplasia might be preceded by condyloma acuminatum of the uterine cervix. Change of normal squamous cells to koilocytotic cells in Papanicolaou smears is characterized by a clear zone around the nucleus and irregular shrinkage of the nucleus, and it is now considered a morphological change in the cells due to human papillomavirus infection. Subsequent histological studies of the uterine cervix showed that 40–80% of cervical intraepithelial neoplasia were flanked by condylomata acuminata (**Fig. 1**), thus suggesting a common cause for both lesions. In a method known as Southern blotting, radiolabeled deoxyribonucleic acid (DNA) attaches to DNA of a similar base sequence. This development added a highly sensitive method for detecting human papillomavirus DNA in the tissues (**Fig. 2**). In fact, it has been demonstrated by a number of research groups that the tissues of cervical intraepithelial neoplasia and invasive cancer contain human papillomavirus DNA (**Fig. 3**). In 1983 a German group of investigators identified and cloned human papillomavirus type 16 and then in 1984 type 18 from the tissues of genital cancers. Human papillomavirus DNA appears to be present in 20–70% of invasive cervical cancer.

Human papillomavirus as carcinogen. The relationship between the premalignant and malignant lesions of the vagina and the vulva is analogous to those of the uterine cervix. Vulvar cancer is preceded by the premalignant lesions Bowenoid atypia (Bowenoid papullosis or Bowenoid dy-

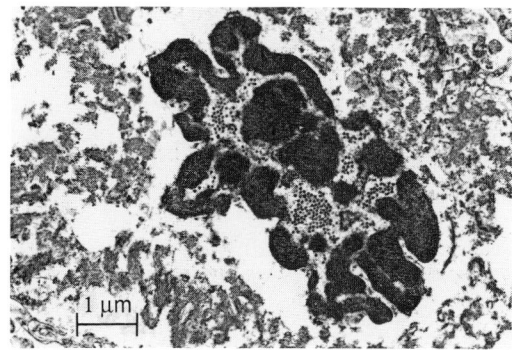

Fig. 3. Mature virions of human papillomavirus are packed in a cell nucleus of cervical intraepithelial neoplasia.

splasia) and Bowen's disease, now categorically termed vulvar intraepithelial neoplasia. Dysplasia and carcinoma in situ of the vagina are categorized as vaginal intraepithelial neoplasia in a similar fashion. A high proportion of premalignant and malignant diseases of the vagina and vulva also contain human papillomavirus.

Genital human papillomavirus infection is believed to be transmitted by sexual contact. More than 40 types of human papillomavirus are currently known. Types 6, 16, 18, 31, and 33 infect genital organs. Human papillomavirus types that are present in invasive cancer of the external female genitalia (the cervix, vagina, and vulva) are usually types 16 and 18, and occasionally 31 and 33. In contrast, the majority of benign lesions, that is, condylomata, contain type 6 human papillomavirus. Type 6 and type 16 human papillomaviruses are equally frequent in premalignant lesions. From this observation, some investigators postulated that the benign or premalignant lesions containing type 16, 18, 31, or 33 are of high risk, whereas the lesions containing type 6 human papillomavirus are of low risk. What complicates the matter, however, is that 10–20% of benign and premalignant lesions show mixed infections of human papillomavirus of more than one type, usually type 6 and type 16. Some researchers have observed that type 6 infection evolved to type 16 infection over a period of time in a clinical followup. In other words, the lesion which started as a low-risk lesion produced by type 6 human papillomavirus might have evolved to a high-risk lesion by the second infection by human papillomaviruses of other types.

There are a number of questions that must be resolved before human papillomavirus is identified as the definite cause of cancer of the external female genitalia. In the majority of premalignant lesions (cervical intraepithelial neoplasia, vaginal intraepithelial neoplasia, and vulvar intraepithelial neoplasia), human papillomavirus is not integrated into the host cell DNA, which is considered essential for activation of oncogenes. Integration of human papillomavirus into the host cell DNA is not a consistent finding in invasive cancer of the external

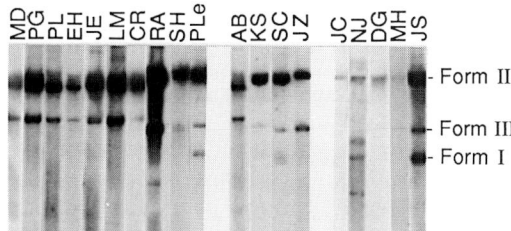

MD PG PL EH JE LM CR RA SH PLe   AB KS SC JZ   JC NJ DG MH JS

- Form II
- Form III
- Form I

Fig. 2. Southern blot hybridization of intact (uncut) human papillomavirus DNA isolated from various degrees of premalignant lesions of external female genitalia. Radiolabeled DNA was used to attach to a similar DNA band of electrophoresis. Dense bands indicate human papillomavirus DNA separated on an agarose gel. Initials at each column of gel electrophoresis represent individual patients. (*From T. Okagaki et al., Identification of human papillomavirus DNA in cervical and vaginal intraepithelial neoplasia with molecularly cloned virus-specific DNA probes, Int. J. Gynecol. Pathol., 2:153–159, 1983*)

female genitalia, either. For this reason, some investigators have postulated a synergistic relationship between human papillomavirus and another agent. The human papillomavirus may act as a promoter instead of a carcinogen, and an additional carcinogen may be necessary for the initiation of cancer. Although *ras* and *myc* oncogenes are reportedly activated in some cases of cervical cancer, the precise mode of their activation in relation to the splicing of the host genes by human papillomavirus is not known.

Genital neoplasm-papilloma syndrome. Genital condyloma is usually treated with podophylin (a caustic agent) application, liquid-nitrogen application, laser surgery, or surgical excision. The premalignant conditions are treated by cryosurgery, laser surgery, or surgical excision. Invasive cancer is treated by surgery and radiotherapy. In the majority of patients, these standard therapies are effective. However, recurrence of another human papillomavirus–related lesion (any of the benign, premalignant, or malignant lesions) in 20% of patients is a significant problem.

It has been shown that nearly 20% of genital carcinoma is associated with premalignant or malignant human papillomavirus–containing lesions of the vagina and the vulva simultaneously (synchronously) or over a span of time (metachronously). Similarly, nearly 20% of vulvar cancer is accompanied by premalignant or malignant human papillomavirus–containing lesions of the cervix and the vagina. Further, benign condylomata acuminata are frequently associated with neoplasms of the external female genitalia. These associations were previously explained by a postulate known as the field theory. Field theory assumes that there is an unknown common etiological factor exerting an influence on the uterine cervix, vagina, and vulva, and it precipitates development of multiple premalignant and malignant conditions in this area. Human papillomavirus is now believed to be this long-sought factor in field theory.

A higher preponderance of cervical intraepithelial neoplasia is observed in organ-transplant (immunosuppressed) patients. The immunosuppressed patients are also prone to have extensive condylomata. In a recent study, it was demonstrated that patients with multicentric human papillomavirus infection (multiple benign, premalignant, or malignant lesions of the cervix, vagina, and vulva) without known previous immunosuppressive therapy actually had reduced immunocompetence as shown by a low T4/T8 lymphocyte ratio and a reduced mitogenic response of the T lymphocytes. That is, individuals who had an underlying idiopathic immunoincompetence developed genital neoplasm-papilloma syndrome represented by multicentric benign, premalignant, and malignant human papillomavirus–containing lesions of the external genitalia. Such patients account for 20% of those with

cancer of the uterine cervix, vagina, and vulva. This syndrome involves all ages of the adult female population and occurs synchronously or metachronously. It may be somehow related to sexual promiscuity, multiple pregnancy, or cigarette smoking—known associative factors of genital cancer. But the precise cause of the subclinical immunoincompetence is not yet known.

The multicentricity of human papillomavirus–produced lesions in the genital neoplasm-papilloma syndrome, both synchronous and metachronous, poses a therapeutic challenge. After topical or surgical therapy of one lesion, the second benign neoplastic lesion may develop in other parts of the genitalia. Multiple surgeries for benign or neoplastic lesions of the external genitalia may be considerably disfiguring. An effective therapy for this syndrome probably requires a systemic approach. American, Scandinavian, and Chinese investigators reported a partial response of cervical intraepithelial neoplasia to topical interferon therapy. Topical interferon therapy was also shown to be effective in treating vulvar condylomata. However, clinical use of systemic interferon therapy with patients with genital neoplasm-papilloma syndrome has not been initiated because of the possible side effects of interferon. Another possibility is immunization of a high-risk population against human papillomavirus. Molecularly engineered antibody against bovine papillomavirus has been shown to be effective in prevention and therapy of papillomatosis of cattle. Until an effective systemic therapy against human papillomavirus infection is developed, the topical interferon therapies seem to be the only choices.

For background information SEE ONOCOLOGY; REPRODUCTIVE SYSTEM DISORDERS; TUMOR VIRUSES in the McGraw-Hill Encyclopedia of Science and Technology.

Takashi Okagaki

Bibliography. C. P. Crum et al., Human papilloma type 16 and early cervical neoplasia, *New Engl. J. Med.*, 310:880–883, 1984; M. Dürst et al., A papillomavirus DNA from a cervical carcinoma and its prevalence in cancer biopsy samples from different geographic regions, *Proc. Nat. Acad. Sci. USA*, 80:3812–3815, 1983; T. Okagaki, Female genital tumors associated with human papilloma virus infection, and the concept of genital neoplasm-papilloma syndrome (GEN-PS), *Pathology Annual, Part II*, 19:33–62, 1984.

Respiratory system disorders

The adult respiratory distress syndrome (ARDS) is a shock lesion of the lung that was first defined in the mid-1960s, when it was observed that some patients with pulmonary failure who died after open-heart surgery had microemboli of platelet aggregates in their lungs. Within a few years, platelet emboli were also found in severely injured patients who developed lung dysfunction after they were resuscitated. Later still, a similar mechanism for pulmonary dysfunction was suggested for patients who developed pulmonary failure after sepsis. (In the case of sepsis, the emboli were thought to be primarily white blood cells, rather than platelets.) Measurements of cardiac filling pressures (a measurement of the amount of fluid in the circulatory system) demonstrated that the pulmonary failure was not caused by fluid administration, and experimental work demonstrated that it was not caused by low plasma oncotic pressure (a measurement of the osmotic pressure of plasma proteins). Rather, the pulmonary dysfunction and pulmonary edema were caused by increased capillary permeability with extravasation of intravascular contents into the interstitium and alveoli of the lung; the increased permeability, in turn, was caused by microemboli of platelets and white blood cells to the pulmonary vasculature.

Clinical features. Patients who have been subjected to shock, major surgery, trauma, or sepsis may show signs of adult respiratory distress syndrome within 24–48 h. Early in the process the patient may develop tachypnea (an abnormally rapid rate of breathing, defined as a respiratory rate greater than 25 breaths per minute) and become short of breath. Analysis of arterial blood usually reveals hypoxemia (oxygen deficiency). Edema forms in the lungs, first in the interstitium and then in the alveoli and small airways. The lungs become stiff. If the patient cannot maintain adequate oxygenation, mechanical ventilation through an endotracheal tube is required.

Even with fairly severe adult respiratory distress syndrome, the lungs will sound clear on auscultation, and there will be few secretions. The chest radiograph will initially be normal, but gradually it will develop fluffy infiltrates and consolidation, indicating collapse and accumulation of mucus in the lungs. As the process continues, the lungs become even less capable of providing oxygen for the blood, and it becomes necessary to use higher and higher concentrations of oxygen in the ventilator.

Pathogenesis. The underlying physiologic change in the adult respiratory distress syndrome, regardless of the many different causes, is disruption of the pulmonary capillary endothelium. This increased permeability can develop after major surgery, after a severe injury, or with sepsis. The adult respiratory distress syndrome is likely to be especially severe if the patient has been in severe shock or has suffered extensive soft tissue injury. The endothelial disruption is not limited to the lungs, and generalized body edema and fluid retention is frequently associated with this syndrome.

The adult respiratory distress syndrome usually develops only after resuscitation. Restoration of normal circulation washes debris out of the injured or infected tissues, thus embolizing microaggregates of platelets and white blood cells, along with other products, into the pulmonary vasculature. As these particles and substances reach the lungs, they activate tissue factors, which are short-lived, locally acting compounds generally made by endothelial and inflammatory cells. These include arachidonic acid metabolites such as prostaglandins, thromboxanes, and leukotrienes and also histamine, serotonin, and complement.

Initially, the pulmonary lymphatic system can remove the protein-rich interstitial fluid from the lungs, but eventually this system is overwhelmed, and interstitial edema increases. The pulmonary alveoli fill with fluid, and because these unventilated areas of the lung are still perfused with blood, hypoxemia occurs.

Prevention and treatment. Prevention of the adult respiratory distress syndrome requires prompt initial resuscitation, prompt control of hemorrhage, debridement of devitalized tissue, fixation of major fractures, correction of bleeding problems, and treatment of sepsis. If these measures are successful, and if pulmonary dysfunction is still absent at 1 week, it is unlikely that the patient will develop the problem.

Treatment for patients who are developing or have developed the adult respiratory distress syndrome includes intensive-care-unit monitoring, supplemental oxygen and mechanical ventilation, and Swan-Ganz catheterization.

Intensive care. Patients who are developing the adult respiratory distress syndrome need to be monitored in an intensive-care unit. Aggressive pulmonary toilet, such as coughing, deep breathing, and tracheal suctioning, should be instituted to prevent collapse of the lung (atelectasis) and pneumonia. Close observation will also allow elective endotracheal intubation if necessary, and institution of mechanical ventilation.

Mechanical ventilation. Positive-pressure mechanical ventilation through endotracheal tubes is usually required in established cases of the adult respiratory distress syndrome. This ventilation is provided by devices, known as mechanical ventilators, that deliver a certain volume of air into the lungs at a predetermined frequency. The volume, frequency, and percent of oxygen can be changed as needed to maintain adequate oxygenation and ventilation.

Oxygen at higher than normal values (a given volume of ambient air is 21% oxygen) is frequently required. Occasionally, in severe cases of adult respiratory distress syndrome, concentrations approaching 100% oxygen are needed, but usually patients can be managed with lower values.

Positive end-expiratory pressure can be used to prevent airway pressures from decreasing to atmospheric levels. At the end of expiration, in patients who are intubated, pressure in the alveoli drops to zero, and the alveoli can collapse. The increased pressure at end expiration, produced in the circuitry of the mechanical ventilator, helps prevent alveoli from collapsing and can reopen collapsed alveoli. This increases the functioning part of the lungs and stabilizes the airways, thus decreasing arteriovenous shunting and improving oxygenation.

Several complications result from using positive end-expiratory pressure and positive-pressure mechanical ventilation, including depression of cardiac output and development of pneumothoraces. Positive end-expiratory pressure and positive-pressure ventilation increase intrathoracic pressure and decrease venous return, thus decreasing cardiac filling and the amount of blood pumped by the heart. If positive end-expiratory pressure is used at high values, then cardiac output should be determined, by using a Swan-Ganz catheter, at various end-expiratory pressures to maintain the best cardiac function. High values of positive end expiratory pressure can also rupture alveoli and collapse a lung, causing a pneumothorax. Reexpansion of the lung in this circumstance requires insertion of a tube into the chest so as to withdraw the air introduced from the ruptured alveoli.

Swan-Ganz catheterization. Patients with severe adult respiratory distress syndrome usually require a Swan-Ganz catheter. This long, thin tube is placed percutaneously in a large neck or chest vein and traverses the right heart and lodges in a distal pulmonary artery. The catheter measures central venous pressure, pulmonary artery wedge pressure, cardiac output, systemic vascular resistance, pulmonary vascular resistance, and mixed venous oxygen. These values are important for determining if resuscitation is adequate and for maintaining proper intravascular fluid volume. The catheter is especially valuable in patients on mechanical ventilators with positive end-expiratory pressure.

For background information SEE CARDIOVASCULAR SYSTEM; RESPIRATORY SYSTEM DISORDERS in the McGraw-Hill Encyclopedia of Science and Technology.

Jeffrey Moore; James W. Holcroft

Bibliography. F. W. Blaisdell and F. R. Lewis, *Respiratory Distress Syndrome of Shock and Trauma*, 1977.

Retrovirus

The existence of the type of viruses now called retroviruses was discovered 75 years ago in a chicken sarcoma and in leukemic chickens. Since then, this type of virus has been strongly associated with malignancies of the blood-forming cells such as leukemia and lymphoma and implicated in other disorders of the hematopoietic (blood-forming) system, such as immune deficiency states and anemia.

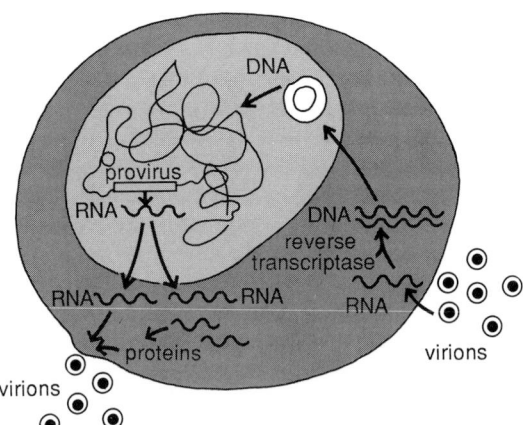

Fig. 1. Retrovirus life cycle. Virions bind and penetrate the cell membrane and enter the cytoplasm, where the viral RNA is copied into linear double-stranded DNA. The viral DNA enters the nucleus, circularizes, and is integrated into the host cell genome. The provirus is transcribed into viral RNA, which is translated into viral proteins. (Both processes use cellular enzymes and factors.) Viral RNA and proteins are then assembled into virions. (*After P. H. Levine et al., Issues in the seroepidemiology of human retroviruses, in R. C. Gallo et al. (eds.), Viruses and Human Cancer, Alan R. Liss, Inc., 1987*)

Reverse transcriptase. The modern era of retrovirology began in the early 1970s with the discovery of the enzyme reverse transcriptase. This enzyme, which reverses the normal flow of genetic information, deoxyribonucleic acid → ribonucleic acid → protein, is exclusively associated with the retroviruses. The discovery of the enzyme permitted the establishment of a sensitive assay for these viruses and resulted in the understanding of the retroviral life cycle (**Fig. 1**). This life cycle includes a stage of integrated viral DNA, called the provirus, in which the DNA copy of the viral RNA is linked directly to the host's DNA, where it can remain latent for long periods, but subsequently becomes replicated with the host genome upon cell division.

Specific and sensitive assays for reverse transcriptase were developed to test the possibility that some human leukemias and related diseases, like the animal counterparts, might be caused by retroviruses. Although highly suggestive evidence for the presence of reverse transcriptase was obtained for some leukemias, in the absence of demonstrable replicating virus these results were not conclusive and could not be extended at that time.

Human T-lymphotropic virus type I (HTLV-I). The failure to detect human retroviruses was thought to have been related to low levels of viral replication. It seemed possible that if a large amount of a pure population of a particular cell type (for example, leukemic cells) could be grown, then the virus might be detected. Moreover, leukemia in cattle had been shown to have been caused by a retrovirus which remained latent until the leukemic cells were cultured. This observation sug-

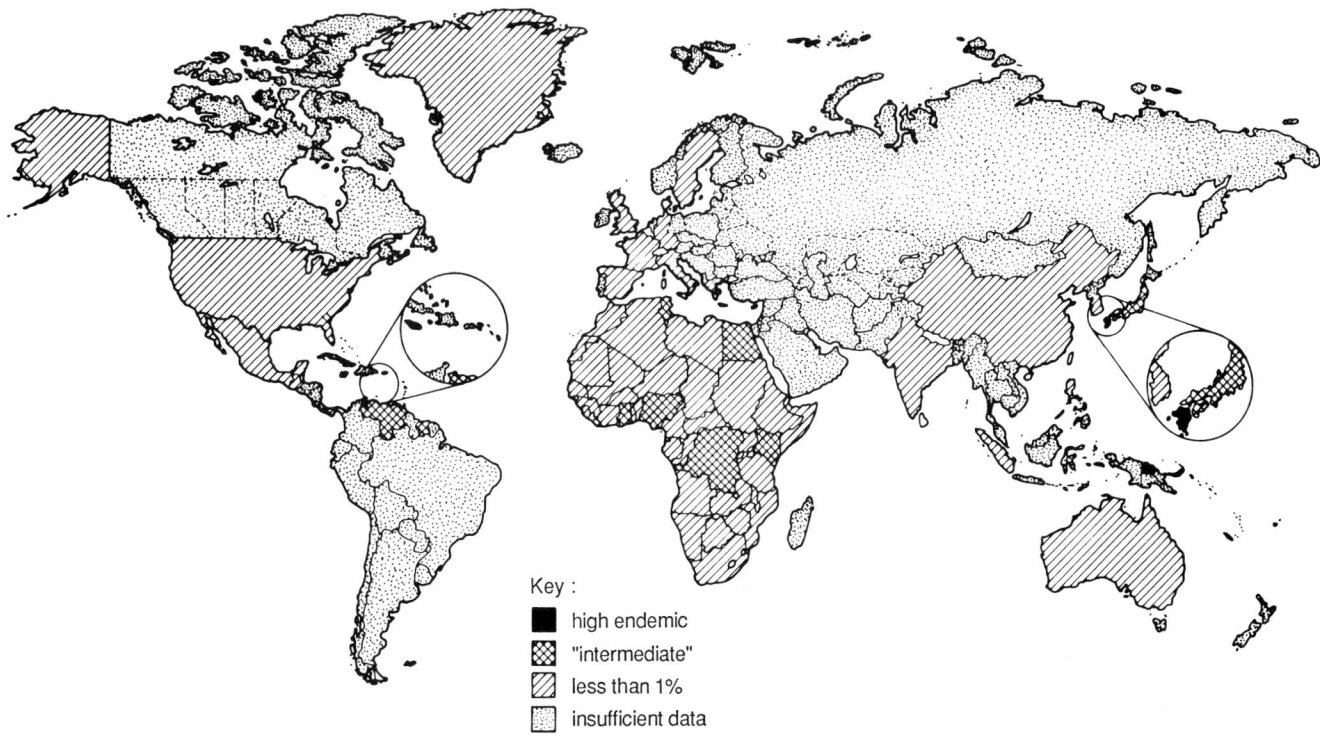

Key :
- ■ high endemic
- ▨ "intermediate"
- ▨ less than 1%
- ▨ insufficient data

Fig. 2. Global pattern of HTLV-I infection; highly endemic areas are southwestern Japan and the Caribbean basin. (*After P. H. Levine et al., Issues in the seroepidemiology of human retroviruses, in R. C. Gallo et al. (eds.), Viruses and Human Cancer, Alan R. Liss, Inc., 1987*)

gested that actual growth of the relevant cells could be important. Therefore, studies were undertaken in which factors allowing the growth of hematopoietic cells were sought. In 1976 such a factor for T lymphocytes was identified. Called T-cell growth factor, or later interleukin-2 (IL-2), this factor allowed the growth of malignant T cells; and along with sensitive and specific reverse transcriptase assays, contributed to the isolation of the first human retrovirus, now called human T-lymphotropic virus type I (HTLV-I), where lymphotropic means having a preference for growth in T lymphocytes.

Studies in the United States and Japan linked HTLV-I to a form of leukemia called adult T-cell leukemia/lymphoma (ATLL). The disease and the virus show a close correlation by seroepidemiology (the study of the distribution of viral serum antibodies) and are endemic in clustered regions throughout the world (**Fig. 2**). Infection of T cells by HTLV-I in the test tube transforms and immortalizes the target cells (that is, allows their unrestricted proliferation), mimicking in many respects the process of leukemic change in the living organism. Many more people are infected by HTLV-I than develop leukemia: the lifetime risk for infected persons is between 1 in 100 and 1 in 1000. Thus, it is evident that other factors (such as virus dosage at first infection and the host immune response) come into play during leukemogenesis by HTLV-I. Further outcomes

of infection by HTLV-I are becoming recognized, however. One is an indirect relationship with B-cell leukemia (indirect because, unlike the T-cell leukemias, the virus is not present in the malignant B cells). HTLV-I had also been linked to a multiple sclerosis–like disease in regions of the Caribbean, South America, and Japan. Immunosuppression is also associated with HTLV-I, including opportunistic infections like those seen in the acquired immunodeficiency syndrome (AIDS).

HTLV-II. The second human retrovirus, HTLV type II (HTLV-II), was discovered by using probes (radiolabeled proteins and nucleic acids) from HTLV-I. HTLV-II was isolated from a relatively benign type of leukemia known as hairy cell leukemia. It has been found to be related to and have biological properties similar to those of HTLV-I, but its disease association and extent of distribution are not yet clear. This problem is due partly to the fact that many assays do not discriminate between HTLV-I and HTLV-II. It has been shown, however, that HTLV-II, like HTLV-I, is currently spreading in drug addict populations, presumably owing to needle sharing. Both are also transmitted by sexual contact, by blood, and from mother to fetus during pregnancy. It is likely that HTLV-I originated in Africa since there are highly related viruses (called STLV-I) in various primates there. It is also likely that spread of the virus was initiated by Portuguese traders to Japan and by the slave trade to the Caribbean. Since worldwide travel and distribution of blood products have risen substantially, these and similar viruses now have the opportunity to spread globally.

HTLV-III. The third human retrovirus, HTLV type III (HTLV-III), also called lymphadenopathy virus (LAV) and human immunodeficiency virus (HIV), is the causative agent for AIDS. In 1981, cases of opportunistic infection and Kaposi's sarcoma were reported in homosexuals in New York and Los Angeles and were soon associated with a generalized immunodeficiency disease marked by a depletion of a subset of T cells characterized by an antibody called OKT4. These T4+ T cells functionally are called helper T cells and regulate many aspects of the immune response.

The disease was apparently transmitted by sexual contact and by blood. This, along with the involvement of T4+ T cells, the same target cells as those for HTLV-I and HTLV-II, suggested that a retrovirus could be involved. Such a virus was first isolated by Luc Montagnier at the Pasteur Institute in France, but the inability to grow the virus (which he called LAV) precluded detailed studies or a definitive characterization. A research group in the United States was able to identify a T-cell line which, while permissive for virus replication, was (unlike other T cells) not killed by this virus, named HTLV-III. The virus (which was later shown to be identical with LAV) could be cultured from most patients with AIDS or members of risk groups.

The availability of large amounts of virus permitted its complete characterization and the development of serological tests for evidence of viral infection (that is, antibodies) which are currently used by blood banks and as a diagnostic tool. There is a virtual 100% correlation of HTLV-III infection with AIDS and related conditions. Virus could be demonstrated in the brain, explaining the origin of the dementia frequently seen as a part of the clinical picture of AIDS. Infection of T4+ T cells in culture results in their death, and thus HTLV-III, like HTLV-I, mimics in the test tube the disease it causes in the organism. It was also possible to show that HTLV-III had much greater genetic heterogeneity than HTLV-I, particularly in the gene for the envelope proteins.

The recent finding of neutralizing antibodies to HTLV-III raises the hope of developing an effective vaccine. Most likely the safest type of vaccine would consist of peptides constructed to resemble portions of the viral envelope proteins. The peptides would be produced by recombinant DNA techniques, for example, by essentially harmless infectious agents which could be introduced into, and allowed to reproduce in animal cells or in immunized people. This would avoid the risk of infection attendant upon using killed or attenuated HTLV-III as a vaccine. There are several problems to consider. These include the considerable envelope gene heterogeneity of different HTLV-III isolates, the possible necessity for presenting viral proteins in their native state and in the same processed form as is present in infected people, and the possiblity that the human immune system may not be intrinsically able to elaborate an effective anti-HTLV-III response. Nonetheless, the at-risk population is sufficiently large to justify the strongest possible efforts at vaccine development.

A vaccine will not help the 1–2 million people in the United States currently estimated to be infected with HTLV-III. For these people, drugs with antiviral activity offer the greatest promise. Recently, encouraging results were reported by several research groups involving inhibition of viral infection in the test tube and treatment of AIDS patients clinically with deoxyribonucleotide derivatives substituted at the 3′ OH group. One such drug, azidothymidine, is now being made available to AIDS patients. Further work on similar drugs is of great possible usefulness. *See Virus chemoprophylaxis.*

Recent studies have identified a virus in wild African green monkeys, called simian T-lymphotropic virus type III (STLV-III), which is related to HTLV-III but readily distinguishable. Closely related to STLV-III are new human retroviral isolates called HTLV-IV and LAV-2. The significance of these latter viruses to the current AIDS pandemic and their exact relationship to each other are currently under study.

Other forms. The identification of the first four human retroviruses, two of which have been linked to diseases, has taken place within a 5-year period. How many more retroviruses remain to be identified in humans? A recent report suggests the involvement of T-lymphotropic retrovirus in Kawasaki syndrome, a disease of children involving severe vasculitis which is endemic to certain areas of Japan and the United States and appears to be spreading. It is also likely that as the ability is gained to grow other types of specialized cells routinely and in quantity, such as myeloid cell types, further retroviruses will be found. The availability of these human retroviruses and the ability to identify others, along with the employment of existing powerful techniques in genetic manipulation, present the possibility of dealing effectively with a variety of human diseases. *See Kawasaki syndrome.*

For background information *see Acquired immune deficiency syndrome (AIDS); Retrovirus; Reverse transcriptase; Virus* in the McGraw-Hill Encyclopedia of Science and Technology.

Robert C. Gallo; Marvin S. Reitz, Jr.

Bibliography. S. Broder and R. C. Gallo, A pathogenic retrovirus (HTLV-III) linked to AIDS, *N. Eng. J. Med.*, 311:1292–1297, 1984; R. C. Gallo, The AIDS virus, *Sci. Amer.*, 256(1):46–56, January 1987; R. C. Gallo, The first human retrovirus, *Sci. Amer.*, 255(6):88–98, December 1986; F. Wong-Staal and R. C. Gallo, Human T-lymphotropic retroviruses, *Nature*, 317:395–403, 1985.

Rift valley

One of the tenets of the plate tectonic theorem is that large pieces of the Earth's outer skin (lithosphere) occasionally break apart along sutures that are termed rifts. Juvenile rifting is a stage of continental breakup that occurs prior to the onset of actual spreading (that is, prior to generation of new oceanic lithosphere). New rifts apparently can develop in a myriad of different plate settings (**Fig. 1**), but the surface scars produced by juvenile rifting are almost always spectacular. Nowhere is this more true than in East Africa, where rifting has produced great holes in the ground that are filled with water, sediments, or volcanic rocks. These holes are flanked by some of the most rugged mountains on the continent. Together, the rift valleys and shoulders straddle 600-mi-wide (1000-km) bulges in the lithosphere.

Even if the morphology of rifts were not so spectacular, the scientific importance of these features still would be great. One reason is that rifts are the embryos of future oceans. For example, the Atlantic Ocean probably passed through a juvenile structural stage comparable to that of the present-day Red Sea, from an original continental tear that was much like the East African Rift. (It is quite likely that a string of great rift lakes, such as Lakes Tanganyika and Malawi in Africa, existed along the American east coast about 180 million years ago.) The process of creating oceans from rifts is essentially one of accreting new lithosphere at spreading ridges (Fig. 1); hence a discussion of the significance of rifts is tantamount to a discussion of the renewal process for the upper 60 mi (100 km) of the Earth. Another point to consider is that the constitution of the hydrosphere, the origin of some mineral deposits, and the Earth's thermal budget can be strongly influenced by events or conditions at rifted plate edges. It is even possible to extrapolate the relevance of rifts to archeology, because a strong geographic correlation exists between continental rifts and the discoveries of early hominid remains.

Figure 1 illustrates the diversity of rift types, ranging from classic continental rift valleys, to mid-ocean ridges, to back-arc spreading systems. So much research progress has been made on so many fronts with respect to these rift types that it is impractical to discuss more than a few advances. This article will focus on some recent breakthroughs in regard to the mechanics of continental rifting.

Seismic profiles. Among the many types of data that have contributed information concerning continental rifts, multichannel seismic profiles have played one of the key modern roles. This technique is a way of imaging the Earth's interior in a fashion that is analogous to how a

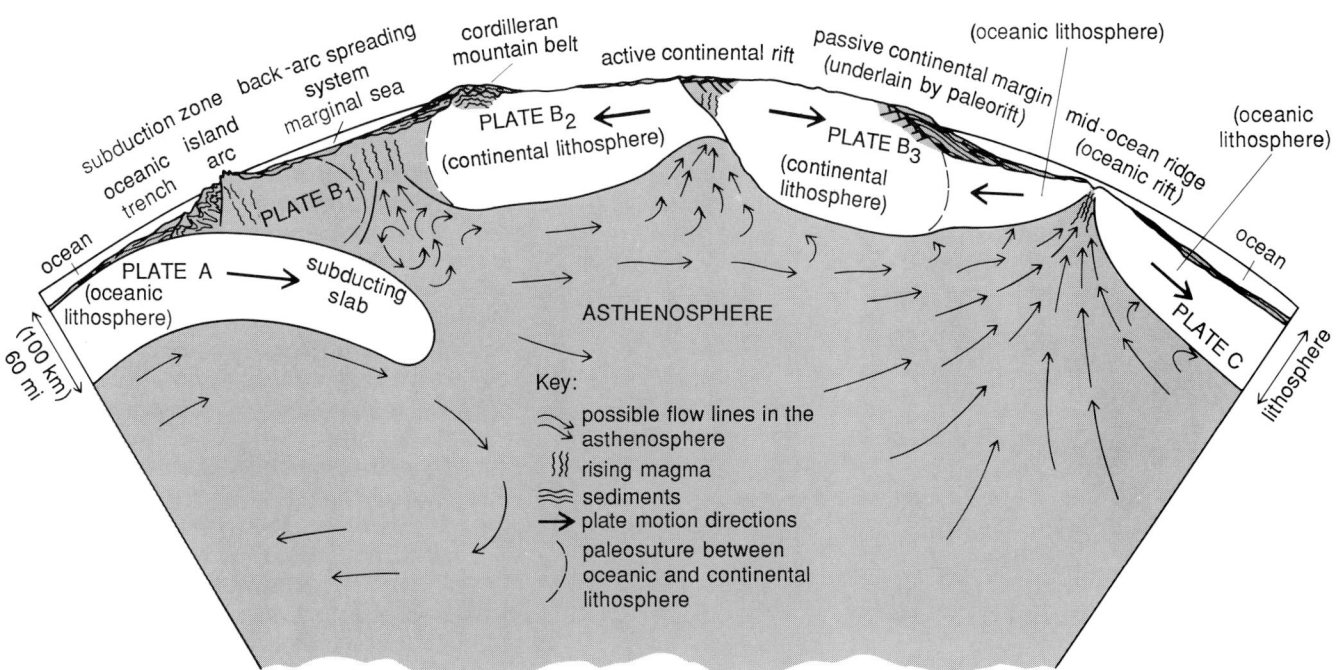

Fig. 1. Cross section showing some of the geologic structures associated with plate tectonics and rifting. Plates A and C are composed entirely of lithosphere generated at mid-ocean ridges (oceanic rifts). Plate B is in the process of being subdivided into three smaller plates (B₁, B₂, and B₃) by back-arc spreading and continental rifting. Note that the passive continental margin is a remnant of a continental rift that successfully evolved into the oceanic rift shown on the right.

lens focuses light beams. Seismic rays generated by some synthetic source (for example, explosives, airguns, thumpers, or giant vibrator trucks) impinge on a common point in the Earth from many different paths. Typically, 24 to 96 different paths are used.

If the common point lies along an acoustic discontinuity, which is the usual case at the boundary between different rock types, the rays are reflected to the surface and detected by geophones or hydrophones (land and marine listening devices, respectively). The incoming reflections are then subjected to various types of filtering and gain modifications, turned into digital signals, and finally recorded on magnetic tape in some multiplexed format. These tapes are sent back to a main-frame computer facility for data processing, which ordinarily involves such signal-processing routines as demultiplexing, deconvolution, stacking, and migration. Most of these tasks pertain to correctly imaging where the rays are coming from and adding the different paths together in ways that enhance the geological information. Ideally, the end result is a cross section through the rift that is the sonic equivalent of what a human eye would see if a deep trench had been dug across the rift.

The sonic picture, called a seismic profile, gives an especially clear view of the sedimentary layers that are deposited as the rift subsides. This basin fill can reach 2.4 mi (4 km) or more in some continental rifts. Seismic profiles also are beginning to reveal what happens in the crystalline rocks below the sedimentary parts of rifts, sometimes as far down as 12 mi (20 km) or more. Taken together, these different seismic scales and perspectives are providing a graphic picture of rifting.

New discoveries. Seismic profiling shows that virtually all continental rifts are composed of discrete structural units called half-graben. In cross section these are asymmetric troughs that are bounded by master fault systems on their deep sides. In map view the border faults tend to form arcuate or orthorhombic systems, giving the ideal half-graben a crescentic shape (**Fig. 2**).

One of the more useful new discoveries emerging from studies of the East African Rift is that rift zones are created by linking of half-graben units in various modes. One common linking mode occurs where adjacent half-graben switch polarities (that is, their direction of asymmetry) along strike, resulting in a sinusoidal alternation of the border fault systems from one side of the rift zone to the other (Fig. 2). Such a geometry can be recognized now in most continental rifts, including the Gulf of Suez, Rhinegraben, Oslo Rift, Rio Grande Rift, and of course in the African rifts. Adjacent half-graben with opposite polarities also canoverlap each other, creating an apparent full-graben within the region of overlap. Some type of internal flexure is usually associated with such geometries (Fig. 2). This is a common morphology in the Tanganyika Rift, and it probably characterizes a relatively advanced stage of continental rift deformation. For several reasons, seismic profiles from certain rifts show that adjacent half-graben do not always alternate polarities. Instead, they can face in the same direction, like a vertical column of parentheses when viewed on a map. The Triassic rift zones along the eastern seaboard of the United States are an example.

Seismic profiles show that no two rifts are exactly alike, even those which are genetically and geographically related. However, a growing body of evidence suggests that two broad categories of continental rifts exist. One category (type 1) is characterized by relatively uniformly narrow zones of rifting (typically 24 mi or 40 km); relatively planar and high-angle faults, at least at shallow depths; and considerable vertical uplift of the rift flanks and surrounding region some time during the rifting process. Examples include the East African rift zones, Rhinegraben, and the Rio Grande, Oslo, Baikal, and Suez rifts. These

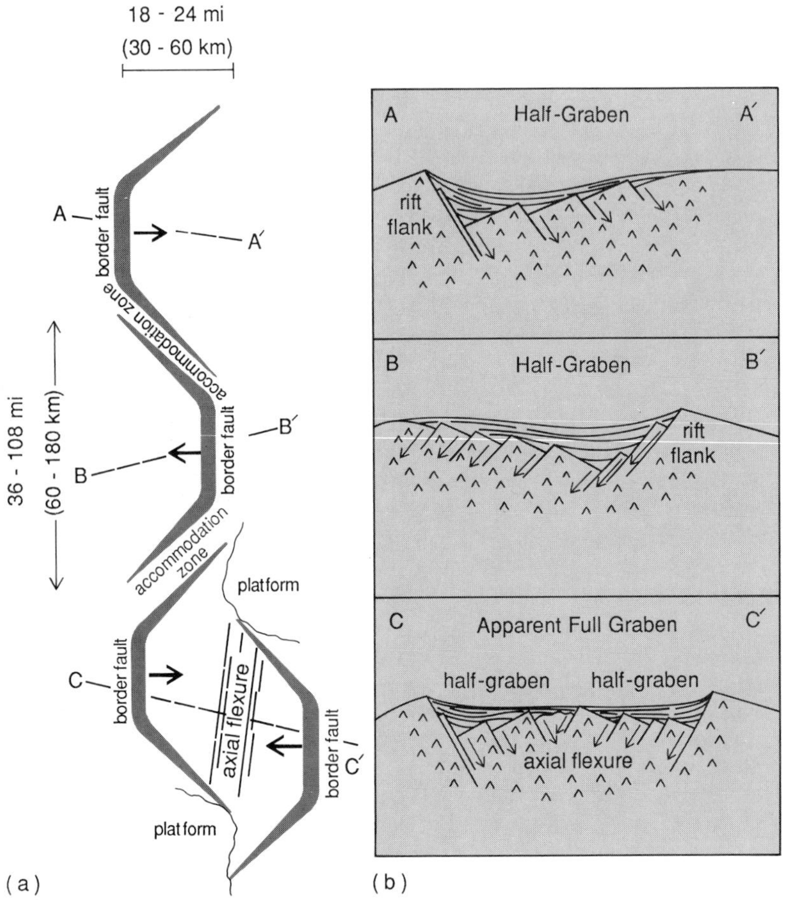

Fig. 2. Sketch showing several ways in which half-graben units can link to make rift zones. (a) Map view. (b) Cross sections. A–A' and B–B' show simple half-graben that alternate polarity along strike. C–C' shows a typical morphology across an overlapping, facing arrangement of half-graben.

zones typically show half-graben that alternate polarities along strike. The other category of continental rift (type 2) is characterized by wide zones of rift deformation (for example, hundreds of miles); relatively low-angle faults that often flatten with depth; and little or no vertical uplift, at least regionally. This category includes the Triassic basins along the United States east coast, the Basin and Range Province of the American southwest, and the South China Sea. Both types of rift zone can evolve into oceans, as observed in the Red Sea and Gulf of California, which began as the first category of rift, and the Galicia Bank area of the eastern Atlantic, which was a type 2 rift. Both types also can fail (that is, fossilize) at any point in their evolutionary progressions, even after commencement of successful sea-floor spreading.

An ongoing debate concerns the mechanisms of rift genesis and maintenance. Most workers agree that rifts are produced by predominantly horizontal tensile forces, but considerable disagreement exists about what causes these stresses. Some researchers believe that the tension is derived from the upward buoyant forces of a rising plume of partially molten material (termed the active mechanism), whereas other workers favor a process of lithospheric stretching associated with plates being pulled apart (the passive mechanism). It is possible that the active and passive rifting mechanisms correlate to the respective type 1 and 2 rift categories described above. However, it should be pointed out that both categories of rift could be produced by a passive, stretching-type mechanism. For example, the difference between the Tanganyika Rift (type 1) and the Triassic basins of the eastern seaboard (type 2) can be entirely explained by rheological differences in the materials being stressed (**Fig. 3**). In the former case the underlying rifted terrain is mainly cold, brittle, Precambrian crystalline rocks with few systematic lines of preferred fracture. The Triassic basins, on the other hand, are wedges created by reactivation of low-angle thrust sheets associated with formation of the Appalachians. Because the sheets usually dip eastward, the resulting half-graben tend to show like polarities. Also, because these preexisting sheets act much like an inclined deck of playing cards, they can distribute stress across the entire surface area of stacked sheets. The stress on the base of the deck can be quite far removed from the loci of surface expression (Fig. 3).

The next decade will see a continuation of seismic profiling in rifts, in large measure because much of the petroleum yet to be found resides in environments that are or were related to continental rifting. Along with improved understanding of the kinematics of rifting, this work will probably resolve the controversy over the dynamics of ripping continents apart.

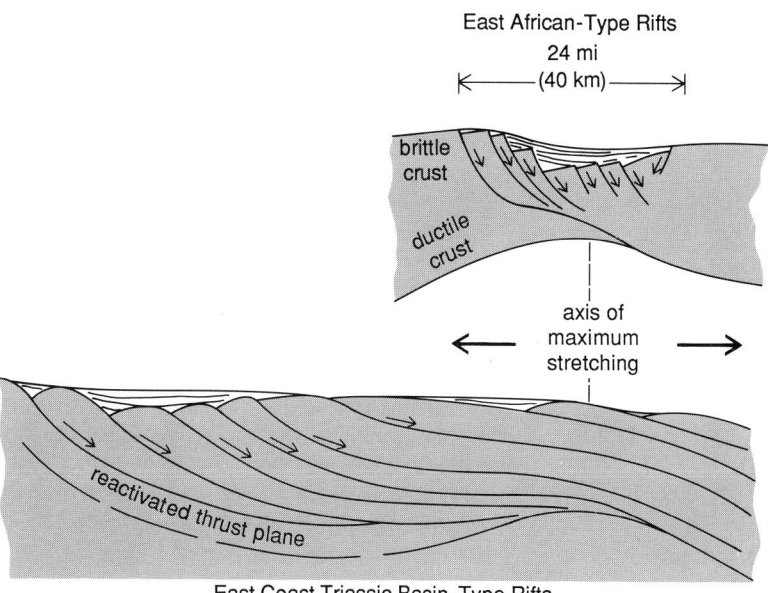

Fig. 3. Hypothesized contrast between two different types of continental rift. The controlling factor probably relates to the existence of low-angle thrust sheets formed during a plate collision (that is, episode of mountain building), not too long before rifting occurred. Where present, such sheets can transfer rifting stresses over relatively long ranges via reactivation of motions along the original fault planes. If these low-angle fault planes are absent or have been given enough time to heal, rifting probably produces relatively narrow half-graben that tend to alternate polarity along strike.

For background information SEE FAULT AND FAULT STRUCTURES; GEOPHYSICAL EXPLORATION; GRABEN; MARINE GEOLOGY; PLATE TECTONICS; RIFT VALLEY in the McGraw-Hill Encyclopedia of Science and Technology.

Bruce R. Rosendahl

Bibliography. A. W. Bally, Musings over sedimentary basin evolution, *Phil. Trans. Roy. Soc. Lond.*, 305:325–338, 1982; B. R. Rosendahl et al., Structural expressions of rifting: Lessons from Lake Tanganyika, Africa, in *Sedimentation in the African Rifts*, Spec. Publ. GSSP 23, Quart. J. Geol. Soc. Lond., 1986; B. Wernicke and B. C. Burchfiel, Modes of extensional tectonics, *J. Struct. Geol.*, 4:105–115, 1982.

Rock mechanics

The science of the behavior of rocks under various natural conditions is known as rock mechanics. Ground control is the engineering discipline in which the principles of rock mechanics are used to solve mining engineering problems. In this context, ground control deals with the design or evaluation of structures that use the existing rock strata. In performing these functions, two sets of parameters must always be determined: the applied stresses in various parts of the mine structures, and the properties and behavioral characteristics of the rock strata of which the mine structures are made. In order to design or evaluate the mine structures successfully, both sets of parameters must be known precisely in advance.

In underground coal mines, blocks of coal are left unmined to support the overburden rock strata; those blocks are called pillars. Because of improper design or mining methods, they are sometimes crushed, and consequently the openings are destroyed. Rib roll is the peeling off the sides of the opening. It occurs frequently when the mining height exceeds 8–10 ft (2.4–3 m). Floor heave occurs when soft floor is squeezed horizontally and buckled at the center of the opening. In severe cases, the buckled floor will close up the opening. Roof fall is the unintentional fall of any rock from the roof, and generally leaves a cavity in the roof. Roof fall has been the main killer ever since the beginning of coal mining. Cutter roof is a large-scale roof fall involving the whole roof across the top of the opening. In longwall mining, the first time that the roof caves in the gob is known as the first caving. Thereafter the roof caves periodically; this is the periodic caving. Floor digging occurs in a soft floor that is so weak that the support penetrates instead of walking on it. If the roof is hard and strong, it does not cave easily and, as a result, remains overhanged in the gob. The overhanged roof, once broken, will cause an earthquake of various degrees.

Nearly all theories developed for ground-control applications assume that rock strata are homogeneous, are linearly elastic, and are distributed uniformly everywhere. In practice this is seldom the case. Geological changes take place frequently. Many problems of entry stability are due to the presence of geological anomalies. Therefore a thorough understanding of the geological conditions is a prerequisite for sound ground-control design.

Competency analysis. Competency is a measure of the desirability for rock strata to be used as roof and floor of the opening. Competent strata are stable, whereas incompetent strata tend to cause ground-control problems. The competency of the rock strata follows these principles: (1) homogeneous lithology is superior to heterogeneous ones, (2) a massive stratum is superior to laminated strata, and (3) quartzitic and calcareous compositions are superior to argillaceous ones.

Geological anomalies that contribute to entry failures are sandstone channels, slickensides, clay veins, rider coal seams, faults, joints, and cleats. Some surface features such as streams and large topographic relief also affect underground opening stability.

The measured in-place vertical stress indicates that it is equal to the weight of the overburden rock strata. However, the measured horizontal stresses are considerably larger than those which are commonly assumed to be due purely to Poisson's effect; that is, horizontal stress is approximately one-third of the vertical stress. In fact it is at least twice the vertical stress. In some places it is up to six times.

The finite element method has been used extensively for coal structural analyses. The most serious problem arising from finite element modeling involves material properties. Nearly all physical properties (that is, strengths and Young's modulus) that have been measured with laboratory-prepared specimens have produced results that are incompatible with field data. Thus so-called "back-in" analysis is used in most cases. This consists of accepting the field-measured data as the desired results and obtaining the proper "physical properties" by trial and error. The results show that a reduction factor of $\frac{1}{4}$ to $\frac{1}{6}$ is needed for physical properties that are determined in the laboratory. In this respect a better model of material behavior is needed.

Pillar design. Coal pillars are made up of that portion of the coal left in place to support the overburden and to maintain the integrity of openings. As underground coal mining is practiced in the United States, pillar design is the most important factor, frequently the only one, in ground-control design. A logical pillar design must take into account the expected load history, pillar strength, and the interaction between roof, pillar, and floor.

The expected load history includes pillar loading in the development openings and mining-induced abutment pressures during pillar-retreat mining. Pillar loading during development is generally determined by using the tributary area concept, while abutment pressures during retreating equal the sum of the pillar loading determined by the tributary area concept and the abutment pressure from one or more sides.

Pillar strength depends on size and geometry. In general, it decreases with size and increases with the ratio of width to height of the pillar.

Roof bolts. Roof bolting is used today in nearly all of the underground coal mines in the United States. It has become the primary support system in underground coal mines, and over 100 million roof bolts are used each year. There are two types of roof bolts. One type is point-anchored and the other is full-length-anchored. The point-anchored bolt is a tensioned bolt anchored at one end with either an expansion shell or a resin. At the other end a bearing plate is inserted between the bolt head and the roofline at the mouth of the borehole. In the full-length-anchored bolt the annulus between the bolt and the hole wall is grouted with resin throughout the full length of the hole. Roof bolts are usually installed in 4 × 4 ft (1.2 × 1.2 m) patterns. This system can be used to suspend the weak lower roof strata to the upper competent roof. The applied tension in the bolts or the resin annulus prevents slippage of the strata and thus maintains the stability of the openings.

Coal bursts. These are sudden, violent bursts of coal from a pillar or pillars, usually accompanied by an air blast and great clouds of pulverized coal dust in suspension. Release of gas may also occur. Coal bursts occur when the following conditions exist: a strong and brittle coal, an overburden height greter than 500 ft (150 m), a strong overlying roof stratum, a strong floor, and a mining sequence that causes high stress concentration in the coal.

There are several methods employed for predicting the time and place of coal bursts. The most practical, yet effective one is known as drilling yield testing. In this method a hole 2 to 3 in. (5 to 7 cm) in diameter is drilled in the coal seam ahead of the working place. As drilling moves from the rib to the interior of the coal blocks, the quantity of cuttings ejected and any dynamic effects (for example, audible knocking, or grating and jamming of the drill rod in the hole) are observed. Highly stressed zones are represented by increased dynamic effects, and thus zones of high and low stress can be determined. The closer a highly stressed zone is to the ribside, the greater is the danger of burst. When the highly stressed zone is located at a distance less than three times the mining height from the ribside, it is likely to burst out.

The common method for prevention is to drill a hole into the highly stressed zone. A light explosive is inserted into the hole and detonated. The explosion fractures the surrounding coal and reduces the stress to a very low level. This procedure eliminates the danger of a coal burst.

Stress and strain determination. Since ground control deals with the stability of mine structures by analyzing the imposed stress (load configurations) and the induced strain (deformation) in each structural element, knowledge of the magnitude and distribution of stress or strain in each element is important. Stress and strain can be obtained by analytical modeling or field measurement. Field measurements involve two types of instrumentation. One type measures load or stress, while the other type measures deformation or strain. Regardless of the measurement being made, the instruments must be inexpensive, rugged, and simple, yet capable of good precision. They must also be highly resistant to dust, dirt, water, drift, shock, and vibration.

There are two types of load-measuring devices. Flat jacks are thin-walled, hydraulic-oil-filled metal containers. A pressure gage is generally attached to each flat jack for pressure reading. The load cell consists of a metal annular ring with at least four electrical-resistance strain gages mounted on the inner surface of the ring. The load-sensing direction is parallel to the axis of the annular ring.

The borehole deformation gage is most commonly used to determine the in-place stress. In this method a small hole (1½ in. or 3.8 cm in diameter) is drilled to a depth approximately 1 ft (30 cm) past the point where the measurement is intended, and the gage is inserted. Then a larger, 6-in.-diameter (15-cm) hole is drilled concentrically with the smaller hole to release the core

between the two boreholes. As the core is separated from the surrounding rock, its pressure is also released. During this time the changes in diameters in three different directions in the smaller hole are measured. These diametral changes are used to determine the in-place stresses perpendicular to the borehole.

For background information SEE COAL MINING; ELASTICITY; FINITE ELEMENT METHOD; MINING; ROCK MECHANICS; UNDERGROUND MINING in the McGraw-Hill Encyclopedia of Science and Technology.

Syd S. Peng

Bibliography. S. S. Peng, *Coal Mine Ground Control*, 1986.

Satellite communications

The increased use of geostationary satellites to relay communications signals between different points has brought about several concomitant developments. As the number of satellites in geostationary orbit increases, various techniques are being developed to deal with interference between signals intended for relay by different satellites. At the same time a variety of Earth stations are being built to provide access to satellites. Prominent among these stations are those located at teleports, which are planned business development areas that feature direct and economic access to a large number of domestic and international satellites for users in the surrounding area, with the aid of a regional distribution network. SEE COMMUNICATIONS SATELLITE.

GEOSTATIONARY SATELLITE ORBIT INTERFERENCE

The geostationary orbit, located approximately 22,300 mi (35,900 km) above the Equator, is the ideal location for satellites to relay radio communications signals between distant points on or near the Earth's surface. Currently, over 100 geostationary satellites are in orbit with more than 150 additional satellites planned for operation. For these satellites to operate successfully, some means is required for discriminating between the desired signals and the interfering signals. The primary methods of achieving this have involved the use of different frequency bands, different polarizations, and discrimination in the Earth-station antenna for satellite separation. Additional discrimination of signals can also be achieved through discrimination in the satellite antenna and modulation techniques, such as pseudorandom coding.

Nature of interference. Because of the limited number of frequency channels suitable for satellite network operations, each satellite in geostationary orbit must share frequencies with a large number of other satellites. This sharing involves limiting the amount of interference that enters any individual receiver to a level that does

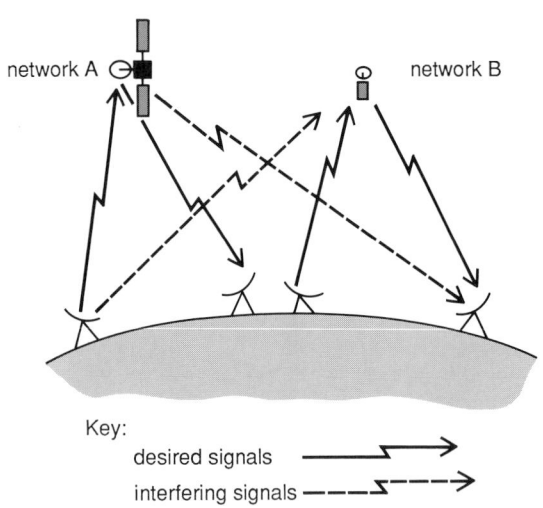

network A

network B

Key:

desired signals ──────▶

interfering signals ╌╌╌▶

Fig. 1. Potential interference paths between two satellite networks.

not seriously degrade its functions. Interference from one satellite network can enter another in two basic ways (**Fig. 1**). The emissions from a transmitting Earth station of network A can enter the satellite receiver of network B, at which point they are retransmitted to the network B Earth station. The satellite emissions (downlink) of network A can also enter the Earth-station receiver of network B directly. The total interference effect received at the Earth station is the sum of the interference generated on the uplink and the downlink.

If the ratio of the desired signal power level to interfering signal power level (S/I) is too small, the performance of the receiving system is degraded. This degradation takes the form of additional noise in analog transmissions such as telephone and television, or an increased number of errors in digital transmissions. In cases of severe interference, it may not be possible to receive the desired signal at all. The required level of S/I will vary, depending upon the nature of the interfering and desired signals.

The frequency bands that are currently available for communications satellites include 6/4 GHz (6 GHz uplink/4 GHz downlink), 8/7 GHz, 14/12 GHz, and 30/20 GHz. A satellite operating in one band would not interfere with a satellite network operating in another because the satellites can filter out signals that are not within the same frequency band as their desired signals. Within each band a given satellite network will involve virtually all of the available frequencies to make maximum communications use of the extremely expensive satellite. For this reason, frequency separation between satellite networks operating in the same band generally cannot be achieved, and other methods of controlling interference must be used.

Earth-station antenna discrimination. For most satellite communications networks operating at the same frequency, the primary method of sep-

arating the signals of one network from those of another is through discrimination in the Earth-station antennas. An antenna that is pointed at a given satellite will have a certain level of antenna gain toward that satellite. For a typical 30-ft-diameter (10-m) paraboloidal antenna receiving at 4 GHz, this main-beam gain would be about 50.5 dBi (decibels relative to an isotropic antenna). The gain of this same antenna toward a satellite separated by 2° of longitude from the intended satellite would be about 21.5 dBi, producing an interfering power level in the receiver that is 29 dB below the power level of the desired satellite signal. An interfering signal this far below the desired signal would not normally constitute harmful interference to the receiver.

The trend in Earth-station antennas is toward using the smallest possible diameter to complete the communications link to reduce Earth-station costs. Systems operating with 14.8-ft-diameter (4.5-m) and 16.4-ft-diameter (5-m) antennas are common. Even receive-only Earth stations using antennas 2 ft (0.6 m) in diameter are being implemented. Because of the smaller antennas, these Earth stations have a lower main-beam gain and provide less antenna discrimination toward adjacent satellites. A 15-ft-diameter (4.5-m) antenna would have a gain of about 44.0 dBi and would provide only about 22.5 dB of discrimination between the desired signal and the interfering signal. This protection ratio is barely sufficient to protect some signal types and would not provide protection to others.

Polarization discrimination. Radio waves, like light waves, can be polarized. The commonly used polarizations for transmissions to or from a satellite are horizontal linear, vertical linear, right-hand circular, and left-hand circular. Satellite and Earth-station antennas are able to separate signals of one polarization from signals of the opposite polarization, thereby providing polarization discrimination. Satellites are designed to take advantage of this discrimination by receiving and transmitting two sets of signals on each frequency, one set using one polarization, say horizontal, and the other set using the opposite polarization, vertical in this case. This permits a satellite to reuse the frequency band in which it operates, relaying the equivalent of 24 television channels in a radio-frequency band that would normally support only 12.

In addition, satellites use a slight frequency separation between the two sets of cross-polarized signals. The signal paths, or channels, through the satellite are offset in frequency slightly so that the center of a channel of one polarization falls on the boundary between the channels of the opposite polarization. A common frequency-polarization plan, shown in **Fig. 2** for network A, would have vertically polarized channels, 36 MHz wide, centered at frequencies of 3720 and 3760 MHz. A horizontally polarized

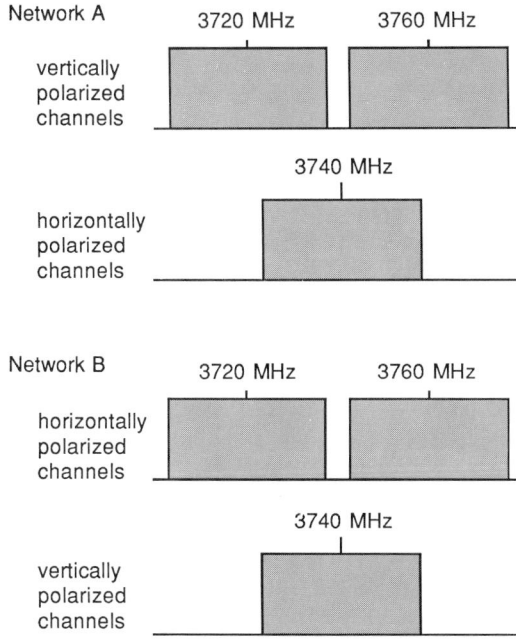

Network A

3720 MHz   3760 MHz

vertically polarized channels

3740 MHz

horizontally polarized channels

Network B

3720 MHz   3760 MHz

horizontally polarized channels

3740 MHz

vertically polarized channels

Fig. 2. Example of satellite frequency-polarization plans.

channel would then be inserted at 3740 MHz. This places the center portion of the signal, which has the highest concentration of power, between the signals of the opposite polarization, where the receiving system would be relatively insensitive to interference. A second satellite network located 2° of longitude away from network A could use this same frequency-polarization plan, but with the polarizations reversed as shown in Fig. 2. The use of opposite polarizations for cofrequency channels on adjacent satellites produces additional discrimination between potentially interfering signals. For satellites using opposite linear polarizations (horizontal and vertical), this discrimination is about 10 dB when the satellites are separated by 2° of longitude. For satellites using opposite circular polarizations (right-hand and left-hand), this discrimination would be about 6 dB.

For satellite systems operating in the 14/12-GHz frequency band and higher, the combination of Earth-station antenna discrimination and polarization discrimination is normally sufficient to protect adjacent satellites from mutual interference at a 2° spacing. This is because, as frequency increases, the main-beam gain of an antenna of a given size increases, allowing smaller, high-gain antennas to be built at higher frequencies. In the 6/4-GHz band, however, these two factors are not sufficient to protect satellite networks from adjacent satellite interference under all circumstances because some types of satellite signals are more sensitive to interference than others. It appears possible, however, to operate co-frequency satellites at a spacing of only 2° longitude if, in addition to the Earth-station antenna and polarization discrimination, there is coordination between adjacent satellites as to the types of communications signals, or traffic, carried on each frequency. If the polarizations used cannot be controlled so as to provide discrimination, or if traffic coordination between satellites is not possible, then the discrimination between satellite networks is normally supplied only by the Earth-station antennas. Satellite spacings of 3 or 4° are then required in the 6/4-GHz frequency band to reduce interference to acceptable levels.

Satellite antenna techniques. In cases where adjacent satellites are serving different parts of the Earth's surface, additional discrimination can be provided by the satellite antenna. This concept is demonstrated by the directional antennas used on the *INTELSAT V* satellites, which use separate high-gain antennas to cover different parts of the Earth, such as North America and Europe. Adjacent satellites using the same type of high-gain antennas can serve separate areas of the Earth's surface from virtually the same location on the geostationary orbit without interfering with each other.

Pseudorandom noise code techniques. Extremely small Earth-station antennas provide transmissions with relatively low data rates between Earth stations and satellites. Paraboloidal antennas with a 2-ft (0.6-m) diameter are being used for Earth stations receiving digital information. Hand-held transmitter-receivers being planned for certain mobile communications applications will use omnidirectional antennas only 0.4 in (1 cm) long. These small units will be able to discriminate between signals from different satellite networks by using pseudorandom noise code modulation techniques. In these techniques the digital signal that is to be transmitted is encoded by using a very long series of binary digits known as a pseudorandom noise code, which has a much higher data rate than the original data and occupies a much larger bandwidth. This encoded signal is then transmitted to the receiving system, where it is decoded by using an identical copy of the original pseudorandom noise code. At the same time the interference effects of any received signals not encoded with an exact copy of the pseudorandom noise code are greatly reduced. This can provide sufficient discrimination for the desired signals to be received, even in the presence of strong interfering signals from a different satellite network. The pseudorandom noise coding technique is especially useful in those applications where digital information is being transmitted at low data rates through extremely small antennas that cannot provide any antenna discrimination between satellites. SEE SPREAD SPECTRUM COMMUNICATION.   *Alan Rinker*

TELEPORTS

A teleport is an economic regional information infrastructure that integrates enhanced office facilities, communications access to satellites and terrestrial long-distance networks, and a regional distribution system for flexible and real-time communications service requirements.

The information explosion that began in the 1970s has accelerated significantly. Of the United States work force, 46% is employed in the information industries. In the New York metropolitan area, 55% of all business is generated from such industries.

This information economy is driven by an explosion in communications technology. By the late 1990s, there will be a substantial increase in the number and types of communications satellites, a major increase in fiber-optic network distribution, and the development of increasingly sophisticated intelligent premise switches and private branch exchanges.

The basis for teleports is the integration of communications technology with information processing and the need for enhanced real estate. Teleports address the needs of corporations within a region for access to terrestrial and satellite communications facilities.

Teleports provide satellite and terrestrial services that form the infrastructure to meet the telecommunications needs of businesses in the region. The concept of a "port" with all its implications—exchange of goods, commercial transactions, and so on—is expanded to include communications access as the critical ingredient for continued viability.

A teleport provides three essential telecommunications services: multisatellite access, a regional distribution network, and enhanced office-park tenant services. From a telecommunications point of view, a teleport is a network that clusters multiple satellite-access facilities (**Fig. 3**) away from the congested and interfering air microwaves of central cities. It also supplements the present Earth-station access facilities, which are insufficient to meet the growing needs of satellite communications. Finally, a teleport provides the tenants of an office park with telecommunications-enriched facilities and ease of access to all types of communications and computational facilities.

Satellite access. Teleports are a necessary addition to the communications infrastructure because of the increase anticipated in the number of satellites and transponders aboard them. With the ever-increasing number of transponders available for communication, access to these transponders throughout a region becomes a critical issue. The need is not just for access to a limited number of transponders, but for access to all satellites and transponders. A crucial factor in selecting a teleport Earth-station site is its ability to access the full United States domestic satellite arc as well as the Atlantic international arc.

Proximity. Proximity provides significant terrestrial communications cost savings for businesses and residents in a region. For example, in the New York City region there are approximately four common carriers that utilize C-band satellites for their communications offering. Their domestic satellite Earth-station facilities are located more than 50 mi (80 km) from the city, versus the teleport's 11 mi (18 km). The additional distance results in an estimated cost increase of 300% to the user or common carrier.

The impact of proximity and satellite access on international satellite services is even more dramatic. At present, all international satellite access from the New York region is from Earth stations located at Etam, West Virginia, or Andover, Maine, both more than 300 mi (500 km) from New York City.

Ease of access. A teleport user will have access to many domestic and international common-carrier satellites, reducing the need for many terrestrial facilities to access multiple common-carrier Earth stations. This will provide significant savings to the user, be it a private corporation or a common carrier.

Satellite services. A teleport offers unique satellite Earth-station services. These services are flexibly designed to meet the needs of business. They range from protected environments for Earth stations to voice, data, facsimile, computation, and video communications services.

To meet the various needs of the region and industry, teleports will accommodate shared or dedicated Earth stations. Options are available to permit a private corporation or common carriers

Fig. 3. Teleport Communications's three Earth stations at the Teleport in New York City. There are currently seven Earth stations in operation, with three more under construction; the 11-acre site can accommodate 27. The wall surrounding the site protects the Earth stations from microwave interference.

to install and operate their own Earth-station facilities. Customers can also share Earth stations available from the teleport operators.

Regional distribution network. The transmission and receipt of satellite communications does not involve only the satellite Earth station. Of great importance to a teleport is the high-capacity regional distribution network. This system provides satellite users end-to-end service at various transmission speeds. Examples of regional distribution networks are a fiber-optic network of Teleport Communications, which operates a teleport in the New York City area, and a microwave network in the San Francisco Bay Area. These systems enable satellite services to be distributed from the teleport to the business districts within the region.

Advantages of fiber optics. Fiber-optic installation costs are greater than those for microwave installation, but the many superior attributes of optical fiber far outweigh the initial costs. The microwave congestion issue is resolved, and the ever-increasing improvements in optical-fiber technology promise enormous increases in capacity. Improvements in the quality of the optical-fiber and the electronic terminal capacity have increased the transmission capacity of a single pair of hair-thin fibers to 560 megabits/s, which is the equivalent of more than 8000 simultaneous voice conversations, by a conservative estimate. Further improvements are expected to provide for 1.2-gigabits/s capacity in the near future.

Facilities and services. Teleport Communications's 150-mi (250-km) regional network spans the New York and New Jersey region with interconnection to the teleport itself. The network provides for full redundancy of electronics and standby optical-fiber pairs to ensure continued service to its users. Routes with full physical diversity are employed from Manhattan to the teleport site. An optical-fiber ring network in Manhattan is designed to ensure that the system does not fail to function during potential cable outages between the strategically located fiber-optic network operation centers. The initial network interconnects the teleport with Manhattan, Brooklyn, and Queens in New York and Newark, Jersey City, North Brunswick, and Princeton in New Jersey (**Fig. 4**).

To meet the demand of users, optical-fiber links are constructed from the fiber-optical network operation centers to user locations. This ensures user service from customer premises to the satellite Earth station at the teleport and to other locations within the region.

Teleport optical-fiber services are offered at basic DS-1 (1.544-megabits/s) and DS-3 (45-megabits/s) transmission speeds. The services are offered to users anywhere in the region to the teleport for domestic and international Earth-station access, from user premise to user premise, and from user premise to an interstate

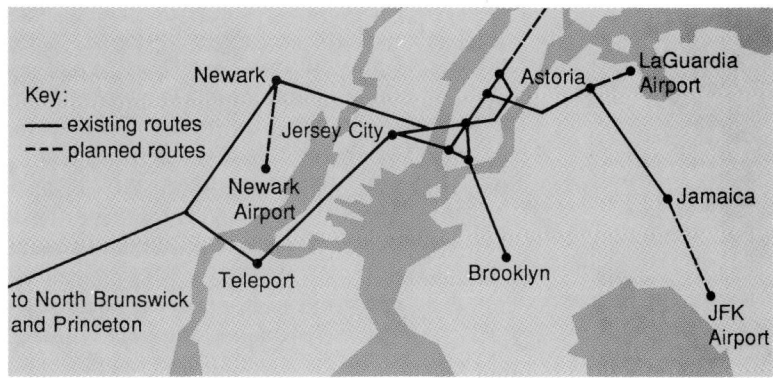

Fig. 4. Fiber-optic regional distribution network of Teleport Communications.

or international common carrier's point of presence in the region. The flexibility of the fiber-optic network provides the options that common carriers and businesses require to control their communications and the diversity essential to ensure reliable communication services. The network thus provides the region a telecommunications infrastructure to meet business needs. It also relieves the bottleneck imposed by local facilities on the flow of communications from interstate and international common-carrier operating offices to the end users.

Advantages of direct access. Businesses require high capacity and direct access to the many interstate and international common-carrier services they have to choose from. Direct access from the user's location to the common carrier permits cost savings and flexibility, not only to the corporate user or developer utilizing communications, but also to the common carrier. Direct access allows common carriers to streamline network designs, whether they are using the local telephone company or an alternate supplier of local facilities.

Direct access also provides a type of client control for the interstate or international common carrier by virtue of the large-capacity service commitment. Direct access, along with plug-in electronics technology, reduces installation intervals and provides the communication capacities required for various services ranging from low-speed data and voice to video conferencing. Direct access is also more economical because of streamlined installation and network design.

Enhanced office facilities. A significant response to industry needs is the telecommunications-enhanced suburban office park. Shared multitenant telecommunications services are offered to tenants of the teleport, allowing them to utilize the latest technologies to supplement their office facilities.

The shared multitenant system concept benefits all users by the sharing of capital costs and by achieving high-value performance. Users save office space and the costs of preparing this space for electric, heating, ventilation, and air-conditioning use by a communications system.

Of the many park services offered, three basic systems are voice management, data management, and building services management. Each system is designed for ease of use by individual tenants and is partitioned to simulate a separate control system for each user. The shared systems for tenants at the teleport optimize economies of scale and complement the needs of businesses for enhanced real estate.

Voice management system. A voice management system will provide basic voice-communication switching services. Intelligent terminals, provided by the teleport or the user, can be connected to the system. Standard enhanced features such as conference calling and detailed accounting for recording of user calls will be provided. Of significance to the voice management system will be the economic call-routing program. This program will pool all multitenant calling requirements and provide the most economic medium and routing for call completion over a variety of interstate and international carriers. The combining of the multitenant calling needs will clearly save tenants communications costs over subscribing individually for communications from interstate and international carriers.

Data management system. Data management systems at the teleport will permit a tenant to again save communications costs through utilization of modem pooling and the sharing of computational facilities and access lines to sophisticated data networks for switched 56-kilobits/s and 1.544-megabits/s service. With the utilization of adjunct processors, terminal emulation will be performed, which allows dissimilar types of communications terminals and word processors to interact with other devices on the park premises or even in other cities.

Building management system. The building management system will utilize a protected processor to provide such services as a central message center and electronic mailbox, an electronic directory, and security alarm services. Security services will employ various devices, such as employee identification readers and sensor technology for monitoring of security entrances and doors.

Prospects. As the teleport establishes a telecommunications infrastructure for the region, new regional and international telecommunications needs and pressures begin to materialize. Airports, seaports, trade centers, and business districts will require more of a central infrastructure for manipulation and transmission of information within the region and to the world. Teleports can respond to developing a range of enhanced telecommunications and information services that utilize the regional telecommunications network and the facilities of the teleport office park. In addition, they can provide access to satellite communications, microwave communications, coaxial cables, fiber-optic networks, and undersea international cable systems, thereby becoming the gateway for communications to and from the region.

A teleport can be responsible for both national and international transmission standards to ensure worldwide service. In addition, teleports can provide transaction protocol standards for communication with other teleports that can improve the efficiency of international communications. The characteristics of the various national and international protocols used for telecommunications and information processing differ; some protocols are better suited to satellite transmission, while others are better suited to terrestrial or undersea cable transmission. Teleports will be called upon to provide the manipulation or translation of these protocols that may be required to interface with other teleports and Earth stations around the world. This conversion service can expedite and provide ease of national and international communications.

For background information SEE ANTENNA (ELECTROMAGNETISM); COMMUNICATIONS SATELLITE; OPTICAL COMMUNICATIONS; POLARIZED LIGHT; PRIVATE BRANCH EXCHANGE; SPREAD SPECTRUM COMMUNICATION in the McGraw-Hill Encyclopedia of Science and Technology.

Robert Annunziata

Bibliography. R. Annunziata, Satellite communications: What is a teleport?, *Telecommun. Prod. Technol. (TPT)*, 4(11):24–28, November 1986; R. C. Dixon, *Spread Spectrum Systems*, 1976; D. M. Jansky and M. C. Jeruchim, *Communications Satellites in the Geostationary Orbit*, 1983; A. D. Lipman, A. D. Sugarman, and R. F. Cushman (eds.), *Teleports and the Intelligent City*, 1986; J. Martin, *Communications Satellite Systems*, 1978.

Schizophrenia

Schizophrenia remains the most puzzling of psychiatric disorders. Symptoms include bizarre delusions, hallucinations, incoherent thought, inappropriate affect, and deterioration of social functioning. Although symptoms often can be lessened dramatically with drug treatment, there is no cure. Data from sophisticated genetic and biochemical studies that have been completed over a period of more than 20 years provide compelling evidence that schizophrenia is a brain disease; however, no specific brain dysfunction has been identified as a likely cause. Increasingly, researchers are assuming that schizophrenia may have multiple causes and that distinct subtypes with different causes exist. To date, no distinct subtypes have been identified unequivocally, and this remains a major research goal. New findings with brain imaging technology make the frontal lobes of the brain a target of detailed study.

Difficulties of brain research. The brain has been a difficult organ to study for a number of reasons: (1) the skull limits direct access; (2) x-rays of brain tissue do not show detail well; (3) different structures and tissue within the brain have very different functions; and (4) brain events often occur in micro amounts, in milliseconds, and in the smallest of synaptic spaces. Therefore, researchers have depended on indirect measures of brain activity from blood, urine, spinal fluid, and scalp electroencephalography. Recent advances in imaging techniques, however, allow more direct access. As a result, a radical change is taking place in brain research in schizophrenia because the brain can now be visualized, quantified, and mathematically dissected noninvasively, with little discomfort or risk to human patients.

Brain imaging techniques. Four different brain imaging techniques have advanced schizophrenia research: cerebral blood flow (CBF), positron emission tomography (PET), electroencephalography (EEG) topography, and magnetic resonance imaging (MRI).

Cerebral blood flow. Blood flow in the brain is directly related to activity of neurons and therefore can be used as an indirect measure of brain functioning. Several ways to measure blood flow in the brain have been used. In the basic technique the subject breathes xenon-133 gas. When it is inhaled, this inert radioactive gas enters the bloodstream quickly and has no physiological effect itself, acting solely as a marker or label. External detectors on the scalp monitor the arrival and disappearance of the radioactive gas, and blood flow at each detector can be calculated. Simultaneous monitoring at different scalp locations allows the regional distribution of flow to be determined. To the extent that neuron activity changes with cognitive activity, blood flow is an indirect measure for localizing cognitive functioning. Areas as small as 0.4 in. (10 mm) can be seen.

Early blood flow studies (1974) showed that schizophrenics had decreased blood flow in the frontal lobes, a pattern called hypofrontality, both during a resting state and during the performance of cognitive tasks. Other investigators have found similar so-called hypofrontality in schizophrenia with blood flow techniques. A specific cognitive test, the Wisconsin card sort, is reported to be a strong inducer of hypofrontality in schizophrenics. This test is a relatively simple matching task that requires logical thinking, and it is regarded as a task that is specific to frontal-lobe function. Normal individuals doing the Wisconsin card sort show marked increases in frontal-lobe blood flow; schizophrenics do not.

Positron emission tomography. A more direct measure of brain metabolism can be made with the technique of positron emission tomography. Like blood flow, glucose metabolism is directly related to neuronal activity. Positron emission tomogra-phy uses a special kind of glucose (or other brain-active substance) labeled with a positron-emitting isotope (fluorine-18 fluorodeoxyglucose). This glucose analog is special in that it becomes fixed after one metabolic step so that it stays wherever it is taken up in the brain. If one region of the brain is working harder than another, more glucose will be taken up in the hardworking part and more positrons will be emitted from that area. About 35 min are required for this glucose to be taken up and fixed in the brain. Since it is metabolically fixed, the person need not be in the scanner during the uptake period. An important consideration is that the subject do a well-controlled task during the uptake. It is the brain's functioning to this task that will show up as the positron emission tomography image, expressed in units of glucose metabolism. After the task, as the subject lies in the scanner, positrons decay into gamma rays which are detected by a ring of monitors around the head. The resulting pattern of gamma-ray detection can be turned into slice images of the brain with mathematical reconstruction. These slices show brain function during the 35-min uptake period and therefore highlight the areas and structures most active in the specific task. Areas as small as 0.2 to 0.3 in. (5 to 7 mm) can be seen with today's scanners.

Since the positron emission tomography scan shows brain activity during the uptake period, an appropriate psychological task is necessary to challenge those brain areas believed to be important for the problem under study. In schizophrenia a difficult test of attention, the Continuous Performance Test (CPT), has been used in some positron emission tomography research. The test requires the subject to view numbers flashing rapidly on a screen and to press a button every time a zero appears. This requires sustained attention for successful performance.

Several studies using the basic positron emission tomography technique have reported less

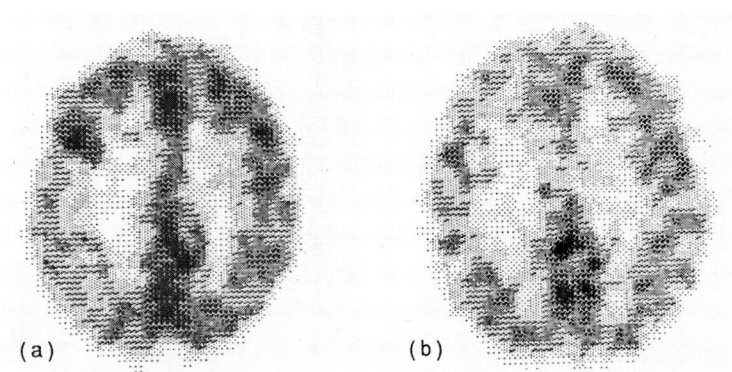

(a)   (b)

Fig. 1. Two positron emission tomography scans are shown at the supraventricular level. (*a*) A normal control shows glucose activity during the Continuous Performance Test (CPT). (*b*) Glucose activity during the Continuous Performance Test is shown for a schizophrenic patient off medication. The most activity is shown in black; the least in white. Note that the schizophrenic has decreased glucose use, especially in the frontal lobes.

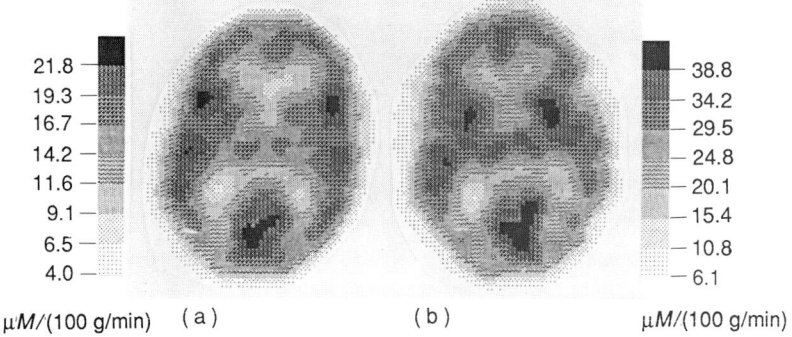

Fig. 2. **These two positron emission tomography scans show the midventricular slice in the same 18-year-old male schizophrenic on two occasions.** (*a*) **Scan after 19 drug-free days.** (*b*) **Scan after 8 months of neuroleptic treatment. In the on-medication scan, note the glucose activity increase in the central brain areas. These areas contain masses of cells called the basal ganglia. The neurotransmitter dopamine is an important chemical transmitter in these areas. Electrical shock stimuli to the right arm were used during each glucose uptake period.**

activity in the frontal lobes of schizophrenics compared with controls (**Fig. 1**). Schizophrenics also show decreased activity in the basal ganglia, areas important in the regulation of the neurotransmitter dopamine. Hypofrontality and decreased basal ganglia activity, however, do not occur in all schizophrenics. Not enough cases have been studied to confirm whether this kind of hypofrontality or decreased activity in basal ganglia can be used to divide schizophrenics into subgroups. There is some evidence that activity in the basal ganglia increases in schizophrenics after treatment with antipsychotic drugs (**Fig. 2**).

Other studies of schizophrenia have used dopamine analogs instead of glucose in positron emission tomography research, since the antipsychotic drugs used effectively in schizophrenia act on the dopamine system. No definitive results are yet available, but this research may help clarify the role of the frontal lobes; at least one type of dopamine receptor is concentrated there and another type is concentrated in the basal ganglia.

Electroencephalography. The brain's activity continually generates electrical signals. These minute electric currents, known as brain waves, change with cognitive activity and different states of consciousness; they can be recorded and interpreted by a technique known as electroencephalography. Electroencephalography is also very sensitive to drug effects in the brain. The measurement is made by electrodes pasted at different locations on the head. Some investigators use simultaneous EEG recordings from 8 to 32 locations. A map or topographic distribution of EEG parameters over the brain cortex can be inferred by interpolating values among the exact electrode locations. Unlike positron emission tomography, EEG is an indirect method, but it has the advantages of being inexpensive and totally noninvasive so it can be repeated in the same subject many times. Electroencephalographic measurements are made millisecond to millisecond, so the time resolution for watching brain events unfold is very short compared to the 35-min averaging required by positron emission tomography with the glucose analog. Several EEG studies indicate that decreased electrical activity is found in frontal-lobe areas in schizophrenics.

Evoked potentials are measured with a special EEG technique. A brief stimulus like a light flash is presented many times. Recording of the electroencephalography is carried out continuously while the light is flashed; however, with computer control, only the segment of EEG from the time the light goes on to the time it goes off is saved. These segments, one for each flash, are averaged. Since the brain is doing many things in addition to responding to the light, extraneous electrical activity is averaged to zero when the EEG from many flashes is summed. The resulting waveform is inferred to be the brain's specific response to the light flash. As with the standard EEG, parameters of the evoked potential (that is, amplitude and latency) are sensitive to cognitive states and drug effects. Serial topography suggests that antipsychotic medication may activate the frontal lobes in schizophrenics (**Fig. 3**).

Magnetic resonance imaging. Early studies with the x-ray computer axial tomography (CAT) scan showed that many schizophrenics had enlarged ventricle spaces in the brain, suggesting shrinkage. Magnetic resonance imaging shows brain structure much more clearly than the x-ray CAT scan. Magnetic resonance imaging uses powerful magnets to align atoms in tissue according to the magnetic field. Then a short pulse of radio waves is introduced which temporarily realigns the atoms. As a result, energy is released, detected, and, in a matter of minutes, reconstructed mathematically as an image. Since the hydrogen atom is particularly sensitive to this technique, tissue with high hydrogen density (for example, water), like that in the brain, shows up most clearly. Magnetic resonance imaging studies confirm that many schizophrenics have enlarged ventricles. The especially sensitive spatial resolution of magnetic resonance imaging (less than 0.04 in. or 1

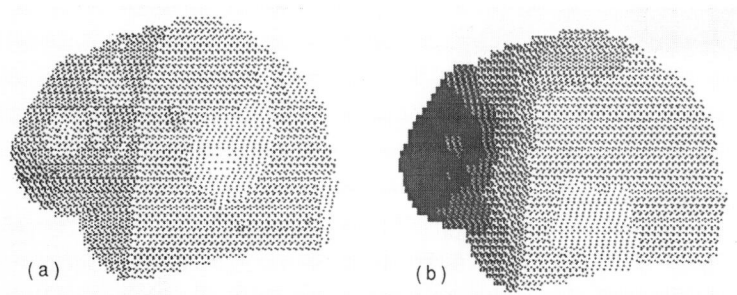

Fig. 3. **Maps of the cortical distribution of evoked potential amplitudes 200 milliseconds after bright light flashes.** (*a*) **Map showing the left hemisphere of a schizophrenic patient off medication.** (*b*) **Map showing the same patient after 2 weeks of antipsychotic medication. The highest amplitudes are shown in black; the lowest in white. Note the increases after medication in the frontal lobes.**

mm) has been used to survey brain structures in schizophrenia. New studies report that some schizophrenics have relatively small frontal lobes, providing possible anatomical evidence for the functional hypofrontality noted with other techniques.

Conclusion. Much remains to be learned about the role of the frontal lobes in understanding the diagnosis and the cause of schizophrenia. Even when frontal-lobe abnormalities exist, it is still unknown whether they are the cause of schizophrenia or the result of another brain dysfunction. The new brain imaging technologies have confirmed some older ideas and have generated new ones. Imaging techniques, even when used in conjunction with neurochemical challenges, may be only as good as the cognitive probes (such as the Continuous Performance Test or the Wisconsin card sort) used along with them. Unraveling the schizophrenia puzzle will require continued cross-discipline collaboration.

For background information *SEE BIOPOTENTIALS AND IONIC CURRENTS; COMPUTERIZED TOMOGRAPHY; MEDICAL IMAGING; NEUROBIOLOGY; SCHIZOPHRENIA* in the McGraw-Hill Encyclopedia of Science and Technology.

Richard J. Haier

Bibliography. M. S. Buchsbaum et al., Cerebral glucography with positron tomography: Use in normal subjects and in patients with schizophrenia, *Arch. Gen. Psychiat.*, 39:251–259, 1982; D. H. Ingvar and G. Franzen, Abnormalities of cerebral blood flow distribution in patients with chronic schizophrenia, *Acta Psychiat. Scand.*, 50:425–462, 1974.

Sea-level fluctuations

The many processes that lead to variations in sea level have been under study for a long time. Recent research has yielded significant insights into methods for determining the difference between ancient and present sea levels, and the dependence of changes in sea level on the various processes that control the volume of the ocean basins.

MEASUREMENT OF PAST SEA LEVELS

Sea level is the relation between the elevation of the sea surface and the elevation of the Earth's crustal surface. However, parts of the crust, as well as the sea surface, have experienced variations in elevation. Therefore, the choice of the specific crustal surface to use as a reference frame against which to measure past sea levels becomes a critical consideration. The main difficulty arises in distinguishing eustatic sea-level changes from epirogeny (vertical movements) of the crust. Three possible frames of reference are the basis for the measurement techniques of seismic stratigraphy, continental hypsometry, and shoreline deposits on continental cratons. Although other methods exist

for determining sea-level changes, such as paleontology (planktonic/benthonic ratios) and delta or shelf sediment package thicknesses, these methods are difficult to calibrate or are of only local significance.

Seismic stratigraphy. This technique is based primarily on a continental margin reference frame, and uses low-frequency acoustic waves reflected off sediment layers possessing contrasting acoustic impedance (speed of sound). These surfaces are taken to reflect time horizons, and they are interpreted to represent the boundaries of depositional stratal packages which are used to map subsurface seismic stratigraphic sequences. When such data are correlated worldwide, the character of transgressive and regressive sequences deposited in nearshore (continental-margin) environments reflects eustatic sea-level changes. Long-term eustatic sea-level changes can be quantified only after other long-term effects such as passive margin subsidence are taken into account. The departure from an established set of calculated subsidence curves is taken to be a result of eustatic sea-level change.

It is important in this analysis to use accurate subsidence curves, as the eustatic sea-level change–induced departures from the curves are small in amplitude relative to the total thermal subsidence normally experienced by a passive margin (a continental margin which was formed by rifting during continental breakup). The subsidence of any point on a passive margin depends on time, position (landward or seaward) on the margin, sedimentation rate and flexural strength of the lithosphere, and rate of sea-level change. Only when all of these factors are known (except sea level) can the subsidence history of a passive margin be calculated. The remaining difference between the calculated curve and the seismic stratigraphic interpretation of onlap and offlap is then taken to reflect eustatic sea-level changes. A recent analysis using seismic stratigraphy shows a mid-Cretaceous (Cenomanian) sea level about 1000 ft (300 m) above present (**Fig. 1**).

Hypsometry. Another type of crustal area that could be used as a reference frame against which to measure sea-level changes is the average coastal areas through time, with application of a hypsometric method. A hypsometric curve is related to the cross-sectional profile of a continent (**Fig. 2**) but is actually a plot of elevation vs. cumulative area (Fig. 2*c*). For example, a continent consisting mostly of lowlands, with a narrow belt of mountains, will have an initially shallow hypsometric curve, steepening at the end. The curve for a continent with large areas of highland plateaus, with few lowland areas, will be steep initially and level off subsequently. In the hypsometric method, the present-day hypsometric curves and percentages of continental areas flooded in the past are used to determine eustatic sea-level changes and differential continental-

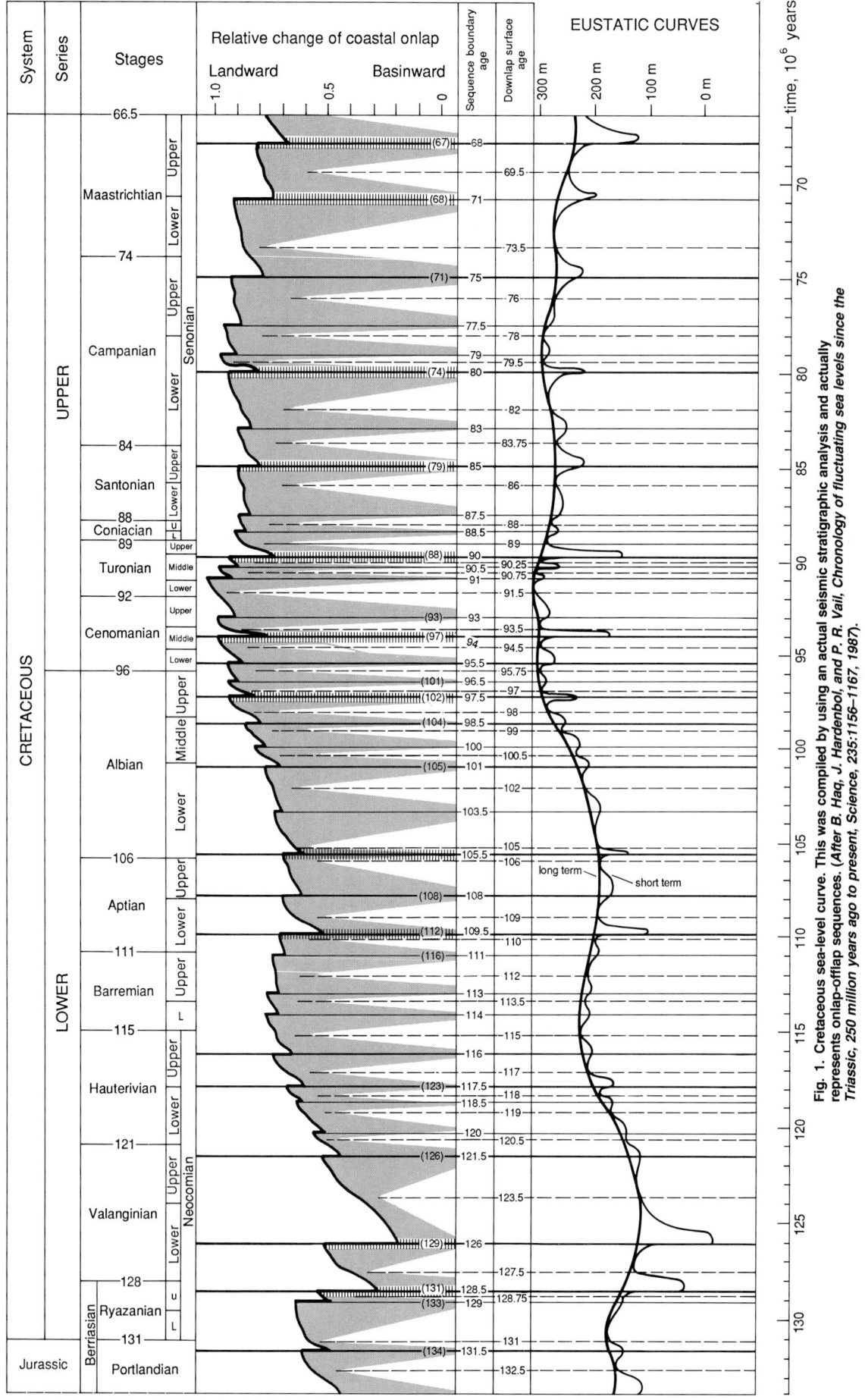

Fig. 1. Cretaceous sea-level curve. This was compiled by using an actual seismic stratigraphic analysis and actually represents onlap-offlap sequences. (*After B. Haq, J. Hardenbol, and P. R. Vail, Chronology of fluctuating sea levels since the Triassic, 250 million years ago to present, Science, 235:1156–1167, 1987*).

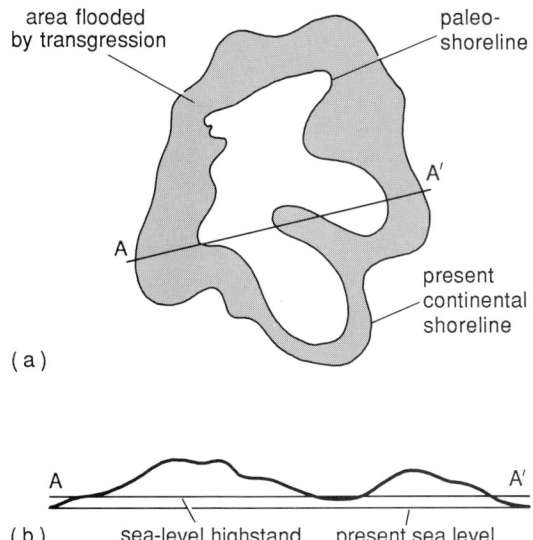

(a)

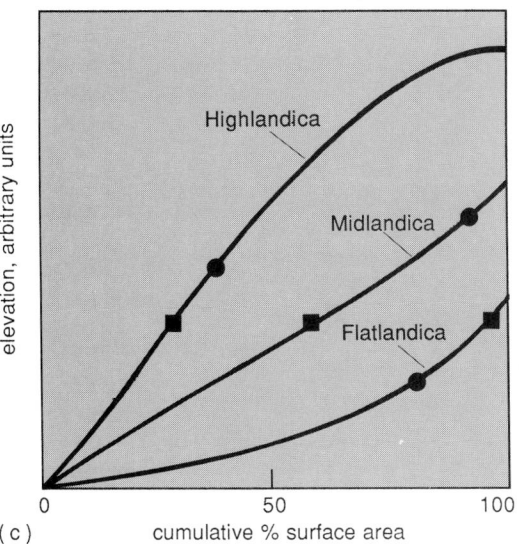

(b)

(c)

Key:

■ Percentages of continental areas flooded at some time in the past if they were to be caused by eustatic sea level change.

● Percentages of continental areas flooded at some time in the past if they were to be caused by differential epeirogenic motions.

Fig. 2. Hypsometric curves. (a) Areal view of hypothetical continent showing area flooded some time in the past during high sea level. (b) Cross section of continent in a along A-A' indicating the amount of sea-level rise. (c) The hypsometric curves for three hypothetical continents: Flatlandica has large areas of lowlands with few mountains; Midlandica has more mountains and less lowlands; and Highlandica has extensive high plateau areas and few lowlands. The squares and circles represent two different scenarios. First, one can consider flooding of the continents as indicated by the squares. More of Flatlandica was flooded than Highlandica. Since the flooded areas shown on the curves represent the same elevation, the implication is a high eustatic sea level, and no differential continental epeirogenic motions. Then, one can consider that, instead of the squares, geologic data showed that the flooding of the continental areas was as illustrated by the circles. Since the analysis uses present-day hypsometric curves, it must assume that the continental hypsometric curves have remained unchanged through time.

epeirogenic motions. This is done by plotting a point which represents the percentage of the continental area which was flooded at a particular time in the past (as determined from paleogeographic maps such as that shown in **Fig. 3**) on a continent's present-day hypsometric curve. The magnitude of the sea-level rise required to flood that percentage of continental area can then be directly read off the vertical axis of Fig. 2c. When this is done for each continent, agreements (or disagreements) as to the amplitudes of the sea-level changes result. Those continents that require the same amount of sea-level rise to flood the continental areas appropriate for a certain time in the past are taken as a reference frame against which to measure the sea-level changes. The rest of the continents are then considered to have experienced some epeirogenic motions of their own.

Estimates of the sea-level difference between the present and various times in the past have been made by considering flooding percentages and the hypsometry of each continent. Although great variation exists in the indicated sea-level difference between continents, with Africa and South America indicating a sea level 100 million years ago of about 1080 and 330 ft (330 and 100 m) above the present level, respectively, the eustatic sea level was taken to be the average of that indicated by all continents, which would have been 560 ft (170 m) above present. This is probably an underestimate, because the passive margins on many continents have subsided to varying degrees since then, causing greater flooding to offset the effect of the drop in eustatic sea level. A correction for this was estimated by allowing a steeper hypsometric curve for the mid-Cretaceous continents, thus increasing the amount of sea-level change indicated between then and now to 920 ft (280 m).

Stable continental cratons. A third approach to the problem of determining paleo-sea level is to examine the stratigraphic record of transgressions which have inundated the most epeirogenically stable areas of the continental interiors, the stable cratons (Fig. 3). In this approach a locality on a stable craton far from tectonic effects such as thrusting, rifting, or passive margin subsidence is defined as a "fixed" basis for a frame of reference against which to measure sea-level changes (as well as other continental epeirogenic motions). The present elevation of shoreline deposits of a certain age at that locality will then indicate the amount of sea-level change since the time of deposition.

A good locality for this purpose is in Minnesota, which was at the eastern limit of the extent of the Western Interior Seaway (Fig. 3) at the end of the Cenomanian Stage of the mid-Cretaceous (94–102 million years BP). At this locality, distinctive shoreline and nearshore sediments were deposited in shallow inlets in the valleys of the preexisting Cenomanian topography. The Ceno-

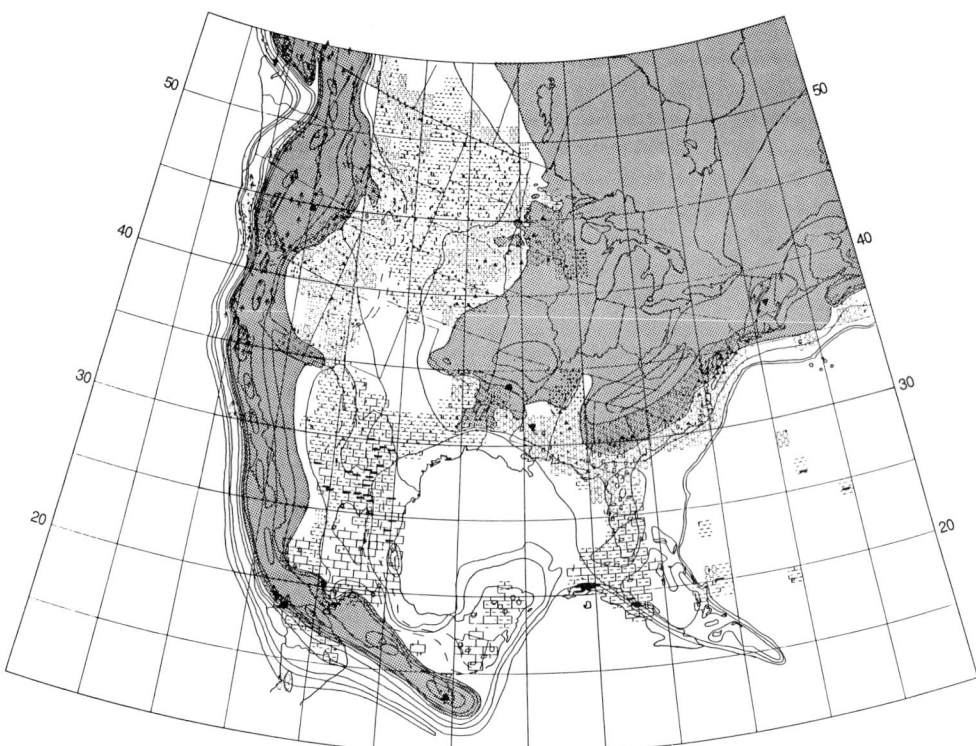

Fig. 3. Late Cenomanian paleogeography of North America. Land areas are shaded. Topographic contours are at 200, 1000, and 2000 m; bathymetry contours are at −50, −200, −4000, and −6000 m (1 m = 3.3 ft). Note the broad epicontinental Western Interior Seaway, and the offsetting of present state boundaries due to the absence, during the Cenomanian, of the Rocky Mountains and Great Basin. The various patterns indicate the types of sediment, and so forth, being deposited during the Cenomanian. (*After A. M. Zeigler et al., Paleogeographic interpretation: With an example from the mid-Cretaceous, Annu. Rev. Earth Planet. Sci., 13:385–425, 1985*)

manian deposits now lie at an elevation of 1300 ft (390 m) above present sea level. The locality is far from the eastern (East Coast) and southern (Gulf Coast) regions which have experienced passive margin subsidence. It is also far from the thrusting tectonic activity which characterized the Late Cretaceous and early Tertiary Laramide Orogeny to the west. In Minnesota the Cenomanian shoreline sediments (as well as underlying strata) are essentially flat-lying, and have very little vertical relief—less than 300 ft (90 m) over more than 60 mi (100 km) lateral extent—with what little there is probably being the result of differential compaction and paleotopography at the time of deposition. The flat-lying nature of the Cenomanian sediments provides further confidence in the assertion that the area is appropriate as a sea-level reference frame.

Although far from tectonic effects, the Minnesota locality has not remained totally undisturbed. The accuracy and reliability of this frame of reference can be improved if some additional geologic factors are taken into account. One consideration is the effect of Pleistocene glacial loading. The locality lies on what was a forebulge (an uplift just south of the glacier) caused by the response of the elastic lithosphere to the weight of the glacial ice mass to the north. This bulge has not yet completely disappeared, and has approx-

imately 230 ft (70 m) farther to subside (as indicated by regionally averaged free air gravity anomalies) before reaching the isostatic equilibrium that would have existed in the Cretaceous. An additional correction can be made to account for the erosion of the paleotopographically higher ground around the areas of deposition. At least 250 ft (75 m) of erosion is required in the Precambrian highlands in order to have removed the thick weathered zone that is known to have existed in the Precambrian rocks. The removal of these rocks since the Cenomanian would have resulted in isostatic uplift of about 160 ft (50 m) due to unloading. Taking the above geologic factors into account, the sea level at the end of the Cenomanian was 1300−230−160=910 ft (390−70−50=270 m) above present sea level.

Despite the need for geologically known corrections, the Minnesota locality was chosen as the basis for a sea-level reference frame because of its great distance from tectonic activity, the effects of which are much more difficult to quantify. Since this locality defines a reference frame, it requires no agreement from other areas. It is interesting to note, however, that in Africa, on two separate cratonic areas—one in Egypt and the other in Mali, Algeria, and Mauritania—flat-lying Cenomanian shoreline deposits occur at an altitude between 820 and 980 ft (250 and 300 m).

Comparisons. Each of the above methods of determining paleo-sea level has advantages and disadvantages, varying with each paleogeographic situation. Uncertainties in subsidence history calculations may introduce errors in the quantification of long-term (100-million-year) sea-level changes as calculated from seismic stratigraphy. Although the difficulty of long-term calibration is a drawback in using seismic stratigraphy for quantifying paleo-sea levels, a great advantage is that it makes available a practically complete record of sea-level changes through time, with excellent short-term resolution when all of the contributing factors are taken into account.

The hypsometric method has been used with some success, but it is not without difficulties. Included in the areas to be used as a reference frame (flooded areas) are subsiding or otherwise active continental margins. These regions would artificially alter the record of paleo-sea level because no account is made for their activity, as it is in the seismic stratigraphic method. Another source of error comes from the use of present-day hypsometric curves with the percentages of present-day continents that were flooded in the past. More appropriate would be to use paleohypsometries (which are unknown to the required accuracy) as hypsometry changes with time in response to rifting, passive margin subsidence, collisional orogeny, hot-spot activity, and plateau uplift, as well as erosion.

An advantage of the cratonic definition method is that for times of high sea level (when data are available due to the transgression of epicontinental seas over the stable cratons) it is probably more accurate than the other methods because it does not need to account for a variety of poorly known variables or to average out conflicting signals. Further, because individual localities are considered, specific epeirogenic motions of local continental areas can be resolved, most reliably distinguishing eustatic sea-level changes from epeirogenic motions. However, because data are available only for times of high sea level when epeiric seas transgress across stable continental interiors, the utility of this method is severely limited. Ideally, all approaches should be used in concert, each for its strengths, so that the greatest amount of information can be brought to bear on the determination of sea level.   *Dork Sahagian*

PLATE TECTONICS

Long-term sea-level variations are commonly attributed to changes in the rate at which sea floor is generated. Recent theoretical work emphasizes that sea level will also be affected by changes in the pattern of sea-floor consumption. One study suggests that sea-level variations are a natural result of the breakup of supercontinents and subsequent formation of new ocean basins.

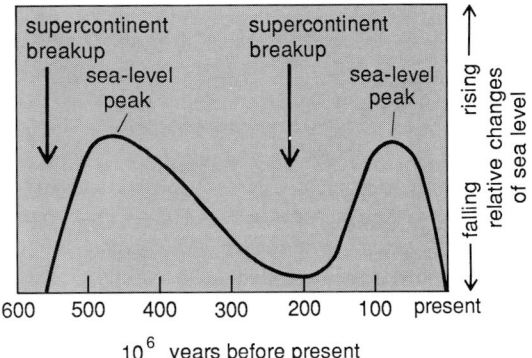

Fig. 4. Schematic diagram showing relative sea-level variations during Phanerozoic time from studies of marine sediments on various continents. Sea-level highs may be causally related to supercontinent breakup.

Sea-level variation. The record of relative sea-level variation is well documented for most of Phanerozoic time. As shown in **Fig. 4**, the variation is approximately sinusoidal with a period of 400 million years. Sea level has peaked twice, once during Ordovician time (470 million years ago) and, more recently, during Cretaceous time (80 million years ago). In both cases the breakup of a supercontinent preceded the sea-level high by approximately 100 million years.

Various techniques have been used to study the sea-level fluctuations. One type involves studies of the distribution and thickness of marine sediments on continental margins; these suggest a sea-level decline of approximately 350 ft (107 m) since the Cretaceous.

Independent estimates, made by another technique in which changes in sea-floor generation rates are considered, predict sea-level falls of about 750 ft (230 m). Problems exist with both techniques. The first assumes that hypsometric curves (which involve measurements of elevation with respect to sea level) for the continents do not vary through time. It is difficult to justify this assumption. The accuracy of the second technique depends on knowing the area-age distribution of the sea floor over the past 80 million years. As much of the sea floor from this period has already been consumed, much of the area-age distribution must be estimated. A recent analysis of the inaccuracies and unknowns in this technique shows that sea level may have decreased by as much as 1200 ft (365 m) or as little as 150 ft (45 m) during the past 80 million years. Given the uncertainties inherent in each technique, the quantitative agreement between the various estimates is surprisingly good.

Ocean basin volume. Sea-level variations result from fluctuations in the volume of sea water as well as changes in the volumes of the world's ocean basins. Fluctuations in water volume are probably linked to variations in the volume of continental ice sheets, which occur over geologically short periods of time, and are unlikely to affect sea level over longer periods. The volume of the ocean basins is affected primarily by changes in the proportion of young to old sea floor. This is because old sea floor, on average, lies at a greater depth than young sea floor. The explanation of the subsidence of the sea floor as it ages and moves away from the spreading ridge has been one of the great successes of plate tectonics. As the lithosphere, underlying the sea floor, cools and thickens, it becomes more dense and, because of isostasy, it subsides. The sea floor subsides in proportion to the square root of its age for the first 70 million years and exponentially thereafter, with a time constant of about 60 million years. If the amount of young sea floor increases, as a result of higher spreading rates, the average depth of the sea floor would decrease and sea level would rise.

The total volume V of the ocean basins can be calculated by using Eq. (1), where t is time, τ is

$$V(t) = \int d(\tau) \frac{dA}{d\tau} d\tau \qquad (1)$$

age, $d(\tau)$ is the depth-age relation discussed above, and $dA/d\tau$ is the area-age distribution of the sea floor, a function which specifies how much sea floor of a given age exists. Sea-level variations Δh, measured from $t = 0$, are directly related to changes in volume according to Eq. (2),

$$\Delta h(t) = -\frac{\rho_m - \rho_s}{A_0 \, \rho_m} [V(t) - V(0)] \qquad (2)$$

where A_0 is the total ocean area, ρ_m is the density of the mantle, and ρ_s is the density of the sea water. Densities in Eq. (2) account for the isostatic deflection of the sea floor as sea level varies.

Sea-floor variations. Equations (1) and (2) show that sea-level variations are directly related to changes in the area-age distribution of the sea floor. The area-age distribution is affected by changes in both sea-floor generation rates and the pattern of sea-floor consumption. **Figure 5** shows idealized distributions for the global ocean and the Atlantic Ocean; they indicate how much sea floor of a given age is present. The shape of the area-age distributions is determined by a balance between the processes of sea-floor generation and consumption. Variations in the global area-age distribution cause sea level to fluctuate.

Two independently conducted studies (1974 and 1982) have demonstrated that the area-age distribution of the ocean floor is roughly triangular (Fig. 5a), with more young than old sea floor present. A triangular distribution requires equal consumption rates for young and old sea floor. A 1982 study found that present-day consumption rates for sea floor of various ages vary considerably about the mean required by the area-age distribution; the consumption rates may vary

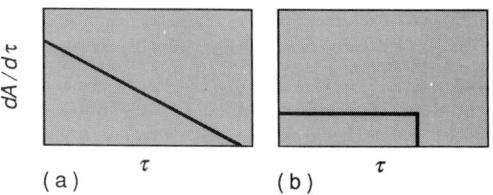

Fig. 5. Idealized area-age distributions for (a) the global ocean and (b) the Atlantic Ocean. τ = age. $dA/d\tau$ = area/unit age.

about the mean through time, affecting the area-age distribution enough to cause sea-level fluctuations on the order of 250 ft (75 m). The balance of the sea-level fall since the Cretaceous would be due to decreasing plate generation rates and the reduction in sea-water volume resulting from the waxing of continental ice sheets.

The 1974 study reported that the Atlantic sea floor should show a rectangular area-age distribution (Fig. 5b) because no subduction occurs, and any sea floor generated in the past is still present. Also considered was how the area-age distribution might be affected by the breakup of a supercontinent and the formation of the Atlantic Ocean. It was assumed that prior to breakup there was a single Pacific-type ocean bordered by subduction zones. This ocean would have a triangular area-age distribution because consumption was occurring in it. As the Atlantic opens, the area of the older basin diminishes in order to conserve total ocean area. As the Atlantic ages, the amount of young sea floor first increases and then decreases. This result suggests that sea level should first rise and then fall after the breakup of a supercontinent.

In more recent research, completed in 1985, a third group made quantitative calculations of the sea-level fluctuations expected from this simple scenario. They found that sea level would rise by approximately 130 to 200 ft (40 to 60 m), reaching a peak about 55 million years after breakup, and then fall by 200 to 330 ft (60 to 100 m) to the present. **Figure 6** shows sea-level variations cal-

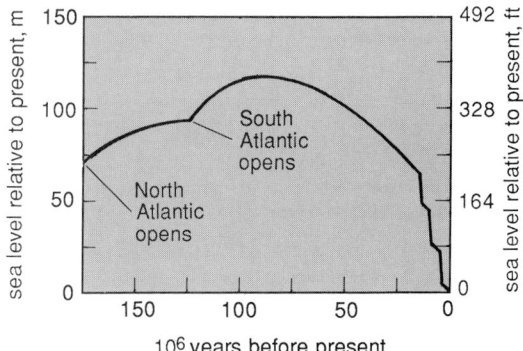

Fig. 6. Sea-level variations calculated for a two-stage opening of the Atlantic Ocean following breakup of the supercontinent Pangea.

culated for a two-stage opening of the Atlantic Ocean following the breakup of the supercontinent Pangea, with the North Atlantic preceding the South Atlantic. The North Atlantic begins to spread at 175 million years ago, whereas opening of the South Atlantic is delayed until 125 million years ago. Sea level reaches a peak of 384 ft (117 m) above present at 90 million years ago. Included in this curve are three 52-ft (16-m) drops at 15, 10, and 5 million years ago representing the effects of post-Miocene increases in continental ice sheets. A correction of 157 ft (48 m) for the increase in volume of the continental ice sheets has been included. This sea-level history is in excellent agreement with that obtained by a study in 1979 that used the sedimentary record. Thus, sea-level fluctuations may indeed be a natural consequence of supercontinent breakup.

Other issues. Although the theoretical and observational sea-level histories are in good agreement, important differences still are to be resolved. Based upon reconstructions of the area-age distribution over the past 120 million years, the Cretaceous sea-level high has been attributed to a strong increase in sea-floor generation rates at that time. The present area-age distribution shows no surfeit of Cretaceous-aged sea floor. If these reconstructions are realistic, they imply that Cretaceous sea floor has been preferentially consumed. Why there would be preferential consumption of Cretaceous sea floor is not clear. In contrast, another recent model predicts a modest decrease in sea-floor generation rates since the opening of the South Atlantic 125 million years ago. It may be necessary to turn to independent measures of sea-floor generation rates, such as variations in the bulk chemistry of sea water, to resolve these differences.

For background information SEE BASIN; CONTINENTAL MARGIN; PLATE TECTONICS; SEA-LEVEL FLUCTUATIONS; SEISMIC STRATIGRAPHY in the McGraw-Hill Encyclopedia of Science and Technology.

Charles L. Angevine

Bibliography. W. H. Berger and E. L. Winterer, Plate stratigraphy and the fluctuating carbonate line, *Spec. Publ. Int. Ass. Sedimentol.*, 1:11–48, 1974; G. Bond, Speculations on real sea-level changes and vertical motions of continents at selected times in the Cretaceous and Tertiary Periods, *Geology*, 6:247–250, 1978; C. Harrison et al., Continental hypsography, *Tectonics*, 2:357–377, 1983; P. L. Heller and C. L. Angevine, Sea-level cycles during the growth of Atlantic-type oceans, *Earth Planet. Sci. Lett.*, 75:417–426, 1985; M. A. Kominz, Oceanic ridge volumes and sea-level changes: An error analysis, in J. S. Schlee (ed.), *Interregional Unconformities and Hydrocarbon Accumulation*, Amer. Ass. Petrol. Geol. Mem. 36, pp. 109–127, 1984; B. Parsons, Causes and consequences of the relation between area and age of the ocean floor, *J. Geophys. Res.*, 87:289–302, 1982; D. Sahagian, Epeirogeny and eustatic sea level changes as inferred from Cretaceous shoreline deposits: Applications to the central and western U.S., *J. Geophys.Res.*, vol. 92, 1987; A. B. Watts and M. S. Steckler, Subsidence and eustasy at the continental margin of eastern North America, *Amer. Geophys. Union, Maurice Ewing Ser.* 3:218–234, 1979; A. M. Ziegler et al., Paleogeographic interpretation: With an example from the mid-Cretaceous, *Annu. Rev. Earth Planet. Sci.*, 13:385–425, 1985.

Secretory structures (plant)

The external surfaces of leaves, flowers, and young stems of many kinds of plants are covered with specialized secretory structures known as glandular trichomes or glandular hairs. Depending upon the plant or plant part, they secrete a variety of substances, including lipophilic materials, such as essential oils, resins, waxes, and flavonoids; hydrophilic substances, such as carbohydrates and proteins; and mineral salts. Although the function of a particular type of glandular trichome is often elusive, trichomes in general (both glandular and nonglandular) are considered necessary for the plant to interact and cope with its environment successfully. Trichomes of many plant species secrete essential oils and resins—substances familiar to most people. The fragrance of leaves and flowers is often from essential oils secreted by glandular hairs. Resins secreted by trichomes make leaves and stems of many plants sticky. These natural products are composed primarily of terpenoids (or terpenes, as they are often called). In recent years, significant progress has been made toward understanding how terpenoids are biosynthesized and elucidating the structure and function of the glandular trichomes producing them.

Structure of glandular trichomes. Most glandular trichomes that secrete terpenoids are composed of two distinctive parts, the head and stalk (**Fig. 1**). The head is secretory or glandular in function and may consist of one or more cells derived from the epidermis; the stalk may be short or long, comprising one or more cells cytologically distinct from head cells and generally not secretory. In some trichomes, stalk cells are derived exclusively from the epidermis; in others, subepidermal tissue, including vascular tissue, may contribute to the development of the stalk. The number and arrangement of head and stalk cells vary extensively, as exemplified by the large array of shapes and sizes of trichomes characteristic of a particular plant or plant part. Frequently, more than one type of glandular hair is found on the same plant organ.

As secretion commences in terpenoid-secreting trichomes, the cuticle surrounding the head detaches from the cell walls, forming a

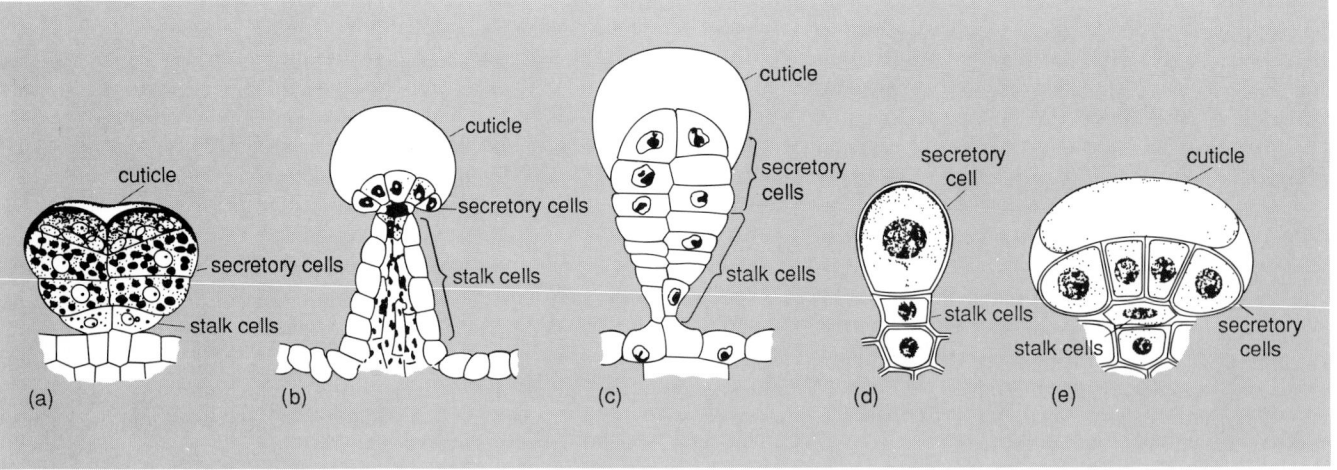

Fig. 1. Terpenoid-secreting glandular trichomes. (*a*) Gland type found on leaves or flowers of *Chrysanthemum*, *Artemisia*, and *Tanacetum*. (*b*) Capitate-stalked gland from bracts of *Cannabis sativa* (the stalk is composed of epidermal and subepidermal cells). (*c*) Glandular hair from young stems of *Arcticum lappa*. (*d*) Young capitate glandular hair from leaves of *Mentha piperita* (cuticle not distended). (*e*) Mature peltate glandular hair from *M. piperita*. (*Part a after G. Haberlandt, Physiological Plant Anatomy, Macmillan, London, 1914; parts b–e after A. Fahn, Secretory Tissues in Plants, Academic Press, 1979*)

cavity where oils or resins accumulate. As secretion continues, the cuticle may balloon out into a sac. In many types of glands the cuticular sac is delicate and ruptures on its own; in other gland types the sac is easily ruptured by physical contact with insects or animals. In either case, the consequence is the release of secretory products. The examples of glandular trichomes shown in Fig. 1 are but a few in which oils or resins accumulate in a subcuticular space.

Some glandular trichomes have a different way of accumulating secretory products. For example, in potato plants (*Solanum tuberosum*) the secreted oil simply accumulates in an intercellular cavity situated in the center of four secretory head cells.

Ultrastructure of terpenoid-secreting glandular trichomes. Several cytological features are common among trichome cells actively secreting essential oils and resins. These include a large nucleus (in some cases polyploid), dense ribosome-rich cytoplasm, relatively numerous mitochondria, and usually an extensive network of tubular smooth endoplasmic reticulum. In a number of species, plastids are prominent and usually ensheathed by a proliferation of the cisternae or tubules of the smooth endoplasmic reticulum. In addition to plastids, chloroplasts (plastids containing chlorophyll) are present in the secretory cells of closely related members of the family Compositae—for example, *Artemisia* (sagebrush) and *Chrysanthemum*. In *Chrysanthemum*, where critical ultrastructural observations have been made, a close association between the smooth endoplasmic reticulum and chloroplast is evident.

An unusual feature of plastids and chloroplasts of secretory cells is that they contain membrane structures or crystalline inclusions not found in typical leaf or stem chloroplasts. Also, starch grains, which are usually abundant in normal chloroplasts, are conspicuously absent.

Stalk cells of glandular trichomes do not have ultrastructural features characteristic of secretory cells; they generally have larger vacuoles, smaller nuclei, fewer mitochondria, and moderately developed cytoplasmic membrane systems. However, in some glandular trichomes, stalk cells have unique intra- and extracellular features, indicating that these cells probably have specialized functions.

Histochemistry. Various reagents, dyes, and stains, including sulfuric acid, osmium tetroxide, sudan dyes, and Nile blue—all of which produce characteristically colored reaction products with lipids—have revealed the lipophilic nature of substances which accumulate in extracellular cavities of terpenoid-secreting glandular trichomes (**Fig. 2**).

The use of histochemical procedures has shown that some stalk-cell walls are impregnated with either suberin or cutin, which are both water-impermeable substances. These so-called barrier cells presumably prevent a backflow of secretory product to underlying tissues and slow down the outflow and evaporation of H_2O from the stalk.

In a number of gland types, marked chemical differences in the cell walls of adjacent secretory cells and differences between head and stalk cells have been demonstrated by staining them with toluidine blue. Recently, it was discovered that chloroplasts in glandular head cells of trichomes on sagebrush leaves do not stain with toluidine blue; in contrast, mesophyll chloroplasts stain intensely. This indicates that there must be remarkable chemical differences between the two

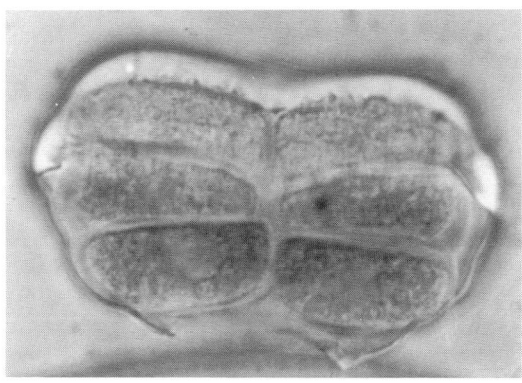

(a)

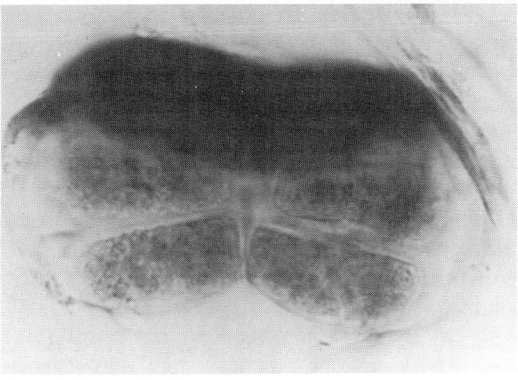

(b)

Fig. 2. Glandular trichome from sagebrush leaves (a) before and (b) after treatment with osmium tetroxide. Note the presence of darkly stained osmiophilic material, most likely terpenes, under the distended cuticle. Only head cells are shown; stalk cells become detached during preparation.

several subclasses based on the number of isoprene units: monoterpenes contain two isoprene units; sesquiterpenes contain three units; and diterpenes, triterpenes, and tetraterpenes contain four, six, and eight units, respectively. Terpenoids with larger numbers of isoprene units are called polyterpenes or polyisoprenes. Natural rubber is an example of a polyterpene and contains hundreds of isoprene units arranged in linear order. While tetraterpenes and polyterpenes are not major constituents of glandular cells, several commonly found in plant cells are of major biological significance; these include β-carotene (a tetraterpene involved in light harvesting and a precursor of vitamin A) and ubiquinone and plastoquinone (polyterpenes which are electron carriers for biological oxidation).

Terpenes occur in acyclic and cyclic forms; some are mixtures of both. The isoprene units are linked, almost universally, in head-to-tail fashion. The structure of isoprene and several examples of terpenes secreted by glandular trichomes are shown in **Fig. 3**. Of the monoterpenes, menthol and camphor are perhaps the two most familiar; they are major components of mint oil and camphor oil, respectively. Cumambrin-B is a sesquiterpene lactone and a constituent of the oil produced by glandulartrichomes on sagebrush leaves and flowers. This terpene is toxic to certain seedlings and may prevent surrounding plants from invading a sagebrush community.

types of chloroplasts. Unfortunately, the nature of these differences is not known, because the nature of the reactions and of the colored reaction products producing the various staining patterns is not well understood.

Although histochemical techniques generally do not offer definitive qualitative or quantitative answers about cell chemistry, they often provide scientists with the only means of identifying and appreciating significant chemical differences between various cell types. These chemical differences reflect the high degree of specialization that can occur in cells within the same tissue, and the glandular trichome is a prime example.

Types of terpenoids. With literally thousands having been identified in plants, terpenoids represent one of the largest and most diverse classes of natural products. As a whole, they are extremely widespread, occurring among all members of the plant kingdom. Terpenoids specifically sequestered in glandular structures, however, occur primarily among vascular plants.

The basic unit of all terpenoids is isoprene, a 5-carbon hydrocarbon. Terpenes are divided into

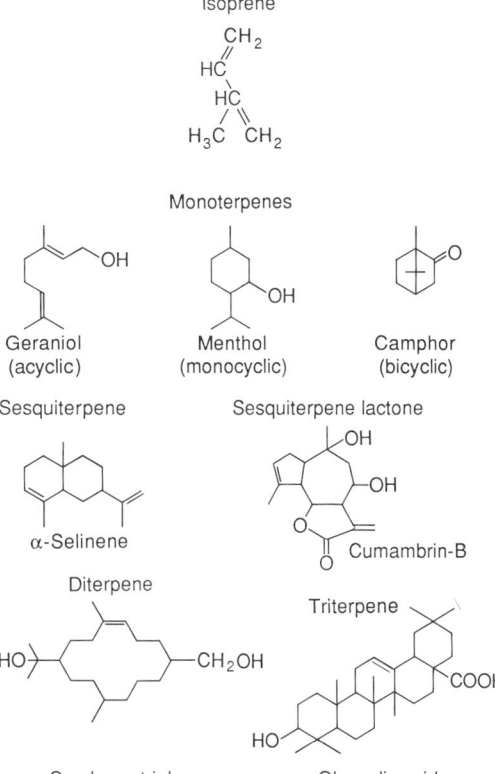

Fig. 3. Isoprene, the building block, and examples of terpenes secreted by glandular trichomes.

Oleanolic acid is a component of the surface resin secreted by glandular trichomes on leaves of *Newcastelia viscida*, a shrub growing in arid regions of Western Australia.

Biosynthesis of terpenoids. The key precursor of terpenoids, as for all lipids, is acetate in the form of acetyl-coenzyme A. Enzymatic condensation of three molecules of acetyl-coenzyme A, with the concomitant loss of coenzyme-A, produces mevalonic acid (**Fig. 4**). Mevalonic acid is then phosphorylated twice by adenosinetriphosphate (ATP) to form the pyrophosphate ester, mevalonic acid pyrophosphate. Decarboxylation of this particular form of mevalonic acid then produces isopentenyl pyrophosphate (IPP), which isomerizes to dimethyallyl pyrophosphate (DMPP). These two compounds are the actual isoprene building blocks of terpenes. As shown in Fig. 4, head-to-tail condensation of one, two, and three molecules of IPP with only one molecule of DMPP results in the formation of geranyl pyrophosphate (C_{10}), farnesyl pyrophosphate (C_{15}), and geranylgeranyl pyrophosphate (C_{20}). Larger isoprenyl pyrophosphates are built by further addition of IPP. These pyrophosphorylated compounds are the parental structures from which monoterpenes, sesquiterpenes, diterpenes, and so forth are constructed by trans-

formations involving the attachment of functional groups, such as hydroxyl, carbonyl, carboxylic acid, and lactone. In addition, cyclization reactions are required to form cyclic terpenes.

Some of the enzymes participating in the earlier steps of terpenoid biosynthesis (that is, from acetyl-coenzyme A to the isoprenyl pyrophosphates) have been purified and characterized extensively. However, in recent years researchers in this field have focused on understanding the cyclization process of monoterpenes.

The formation of cyclic monoterpenes occurs initially via the activity of enzymes called monoterpene cyclases. They catalyze the conversion of the acyclic precursor, geranyl pyrophosphate, to cyclic products which are then further modified by other enzymes to form particular types of monoterpenes. Several cyclases have been partially purified, including fenchol cyclase from fennel (*Foeniculum vulgare*), α-pinene cyclase from sage (*Salvia officinalis*), and the enzyme (also from sage) that cyclizes geranyl pyrophosphate to bornyl pyrophosphate, a precursor of camphor. The general properties (molecular weight, pH optimum, cofactor requirement, and so forth) of these enzymes have been characterized, but perhaps the most interesting recent discovery is that α-pinene cyclase can efficiently catalyze

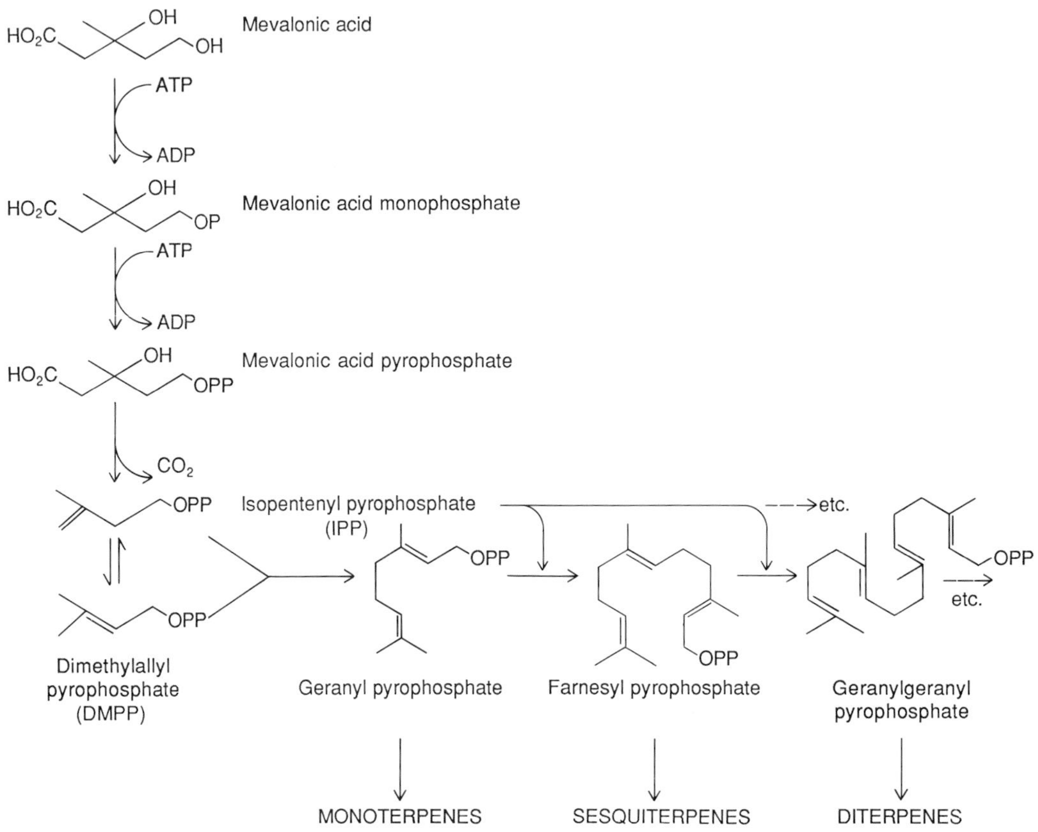

Fig. 4. Pathway of terpene biosynthesis starting from mevalonic acid. ATP = adenosinetriphosphate; ADP = adenosinediphosphate.

cyclization reactions in a medium predominantly containing hexane—an unexpected phenomenon, since hexane and other organic solvents occur in low concentrations or are not present at all in the cells, and high concentrations generally are not compatible with enzymes and metabolic processes.

Other recent findings of interest concern the effects of artificial growth regulators, such as Phosfon and Cycocel, on essential oil yield of certain plants. Foliar application of these growth regulators increases the amount of essential oil produced by sage and peppermint (*Mentha piperita*) leaves. Oil composition, that is, the proportions of the different kinds of terpenes that make up the oil, varies depending upon the concentration of the growth regulator. Phosphon D and Cycocel appear to affect the level or activity of terpene cyclases, thus changing the amount and composition of essential oils. Glandular trichomes are the main structures producing these oils in sage and peppermint.

Site of synthesis. The secretory cells of trichomes that accumulate oils and resins not only secrete terpenoids but also appear to biosynthesize them. That is, the terpenoids are not synthesized elsewhere in the plant and then transported to these cells. The likely sites of terpene synthesis are in the smooth endoplasmic reticulum and plastids. Thus, the abundance of smooth endoplasmic reticulum, the distinctiveness of plastids and chloroplasts (that is, the proliferation of their internal membranes, abundance of osmiophilic bodies, crystalline inclusions, and lack of starch), and the close association of smooth endoplasmic reticulum with plastids are special ultrastructural features providing evidence that secretory cells are capable of terpene biosynthesis. However, this evidence is still circumstantial because the isolation of large quantities of glandular trichomes needed to purify these organelles has not been achieved to date.

Somewhat more supportive evidence has been obtained by demonstrating that isolated glandular trichomes and strips of epidermis with trichomes are capable of incorporating exogenous precursors, such as [^{14}C] acetate, into terpenoid products. Unfortunately, because of the small number of trichomes tested and the apparent compartmentation of terpenes within secretory cells, only low levels of precursor incorporation have been achieved. Efforts to obtain large quantities of isolated glandular trichomes are currently under way in several laboratories. Some promising results were obtained recently by using an isotonic-density-gradient medium to separate secretory cells from other cells based on their differences in buoyant density.

Function of glandular trichomes. Considering that as much as 20% of the dry weight of some plants is composed of trichomal terpenes—5 to 10% being very common—and considering the widespread occurrence of glandular trichomes producing terpenes, their significance to the survival of these plants is undeniable. Trichomal terpenes are toxic to, and avoided by, a variety of insect larvae and adults, as well as larger animals. Terpenoids also provide a defense against fungi and bacteria. Resins on the surface of leaves prevent water loss from plants growing in arid habitats. Thus, the placement of trichomes on the surface of organs provides a plant with a tactical advantage in protecting itself against herbivory and environmental extremes.

Animals and pathogens are not the only organisms affected by terpenes. Plants, too, are affected. Cineole and camphor from leaves of *Salvia leucophylla* and *Artemisia californica* are toxic to seedlings of neighboring plant species. These plants and many others which secrete glandular terpenes tend to dominate their habitat by releasing the toxic terpenes to the environment. This particular phenomenon is part of a general phenomenon known as allelopathy, which refers to any biochemical interaction between plants.

Recently, two sesquiterpene lactones isolated from sunflower leaves were shown to specifically inhibit auxin-induced elongation growth of parts of oat and sunflower seedlings. It was hypothesized that these terpenes bind to an auxin-bound receptor protein, consequently preventing sustained auxin-induced growth.

During the last decade there has been a renewed interest in investigating natural plant products because of their potential use as insecticides and herbicides; also it is possible that plants may provide a renewable, alternative source of hydrocarbon fuels. The function of many terpenoid-secreting trichomes is still not clearly understood, and glandular trichomes probably will be the focus for new and continued investigations.

For background information SEE ALLELOPATHY; ESSENTIAL OILS; PLANT PHYSIOLOGY; SECRETORY STRUCTURES (PLANT); TERPENE in the McGraw-Hill Encyclopedia of Science and Technology.

J. Henry Slone

Bibliography. R. Croteau, Biosynthesis of terpenoids in glandular trichomes, *Biology and Chemistry of Plant Trichomes*, pp. 133–185, 1984; R. Croteau, Biosynthesis and catabolism of monoterpenes, in W. D. Nes, G. Fuller, and L.-S. Tsai (eds.), *Isopentenoids in Plants: Biochemistry and Function*, pp. 31–64, 1984; R. G. Kelsey, G. W. Reynolds, and E. Rodriguez, The chemistry of biologically active constituents secreted and stored in plant glandular trichomes, in E. Rodriguez, P. L. Healey, and I. Mehta (eds.), *Biology and Chemistry of Plant Trichomes*, pp. 187–241, 1984; J. H. Slone and R. G. Kelsey, Isolation and purification of glandular secretory cells from *Artemisia tridentata* (ssp. *vaseyana*) by Percoll density gradient centrifugation, *Amer. J. Bot.*, 72:1445–1451, 1985.

Semiconductor

In a normal piece of semiconductor, such as a crystal of silicon, electrons and waves of vibration (referred to as phonons) can move in any direction. Their physical properties are then described by equations in three dimensions, usually the cartesian coordinates x, y, and z. All the physical properties of the semiconductor, such as its conductivity, magnetic susceptibility, and specific heat, depend on this three-dimensional description. If the electrons could be confined so that they were free to move in only a plane (two dimensions or 2D) or in only a line (one dimension or 1D), the semiconductor properties would be changed, sometimes dramatically. It is, indeed, now possible to put together different semiconductors, essentially atom by atom, in such a way that electrons (and to a lesser extent phonons) are confined to narrow layers, lines, or even dots. Assemblies of semiconductors which confine the electrons in this way are sometimes known as low-dimensional structures, and exhibit unusual, useful, and occasionally extraordinary properties.

Two-dimensional electron gas. A simple way to confine electrons to a plane is to use a single heterostructure consisting of two different semiconductors with a very flat interface between them. A frequently used structure has gallium arsenide (GaAs) on one side of such an interface and gallium aluminum arsenide (GaAlAs) with a low concentration of silicon atoms as impurities on the other. Most of the silicon atoms ionize, each giving up one electron which moves away, leaving behind a positively charged (donor) ion. The electrons move into the gallium arsenide layer because their energy is lower there, but stay as close as possible to the gallium aluminum arsenide because of the Coulomb attraction of the ions. Thus the electrons are confined by these opposing forces to a thin planar region at the edge of the gallium arsenide. In the direction perpendicular to the interface, they are bound into quantum-mechanical states reminiscent of the electron levels of the hydrogen atom. They remain free to move along the interface, and do so, scattering off each other and forming a so-called two-dimensional electron gas (2DEG). The two-dimensional electron gas behaves very differently from electrons in any normal semiconductor crystal. The best example of this difference is the quantum Hall effect.

A simpler but more directly useful property of the reduced dimensionality of the heterostructure follows from the separation of the electrons and their parent ions onto different sides of the interface. Contrary to the situation in a crystal, where electrons and ions share the same space, the electrons in the two-dimensional electron gas can move unimpeded by the ions; thus their mobility, or rate of movement along the layer, is greater

than it can be in a crystal. This property can be used to design a fast electronic device, the high-electron-mobility transistor (HEMT), in which the two-dimensional electron gas carries a current between source and drain contacts, and the current is altered by the electric field of a gate electrode placed above the doped layer. Such devices, made from gallium arsenide and gallium aluminum arsenide, are much faster than silicon devices of a similar size, but are difficult to manufacture in integrated circuit form because of the unhelpful chemistry of gallium arsenide. Even so, circuits containing over 25,000 high-electron-mobility transistors have been fabricated. *See Transistor.*

Layered structures. In a layered structure the potential energy of those electrons that are free to move, called the conduction electrons, varies so little in a direction parallel to the layers that along them the potential may be treated as flat. The motion of the electrons may then be decoupled into two parts, parallel and perpendicular to the layers. **Figure 1** illustrates the potential experienced by the conduction electrons in a quantum well consisting of a thin layer of gallium arsenide (typically between 1 and 10 nanometers thick) sandwiched between two cladding layers of gallium aluminum arsenide. The motion of the electrons in the direction perpendicular to the layers (z in Fig. 1) is then described by the quantum mechanics involving that single direction only. This is exactly equivalent to the simplest form of quantum mechanics, which is discussed in introductory texts and courses on the subject. What was merely an academic exercise for 50 years now describes real systems. If the potential well is strong enough to hold a bound state, an electron in that state can move parallel to the layer but has a fixed wave function in the z direction.

Quantum wells can be used to produce a laser that emits light when an electron jumps from the bound state to fill a vacancy in the valence band. The frequency of the light emitted corresponds to the energy change of the electron. Varying the well width alters the bound-state energy, and thus the frequency of the laser can be chosen

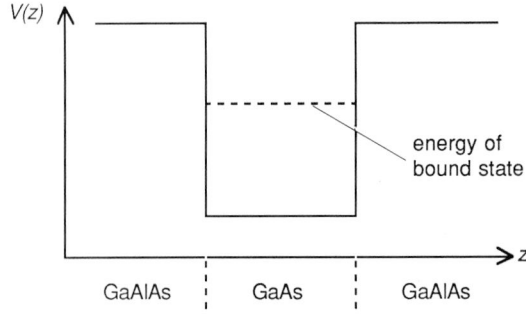

Fig. 1. Potential energy $V(z)$ of a quantum well structure as a function of the depth z perpendicular to the layers. Bound-state energy is indicated.

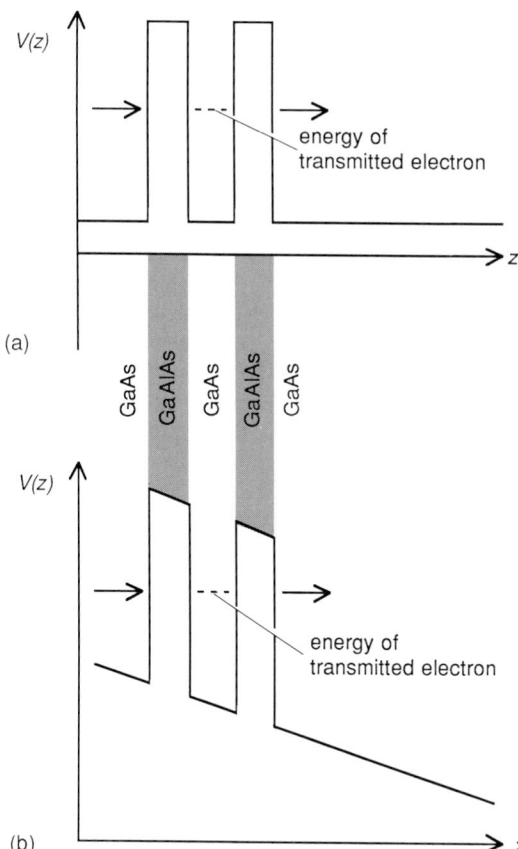

(a)

(b)

Fig. 2. Potential energy of a resonant tunneling structure (a) with no applied field and (b) with an applied field. Energies of transmitted electrons are indicated.

When an electron tunnels through to the right, it has a kinetic energy which can be large compared to the usual small energy of the conduction electrons. It initially travels rapidly in a straight line, in so-called ballistic motion, but eventually gives up energy to the lattice vibrations of the structure. In bulk solids this energy loss is very rapid, but the layered structure can be designed so that the electron retains its energy for much longer. This property offers the possibility of another very fast device since the electrons move rapidly, with speeds up to 3×10^5 ft/s (10^5 m/s), and pass through the layer, a distance of the order of 10–100 nm, rather than along it, which would involve traveling a distance of 1–10 micrometers in a typical device. In practice, ballistic devices of this kind have proved difficult to construct.

Quantum wires. To produce one-dimensional structures, the surface of a layer is modified by the use of etching and lithography to remove all but a thin strip. Through the use of electron-beam lithography, strips thinner than 10 nm can be produced. These so-called quantum wires show most unusual properties, starting with the simplest experiment: measurement of current as a function of voltage. Unlike the straight line given by the graph of current against voltage for a large resistor, the quantum wire gives a jagged current-voltage curve with an immense amount of minute detail on it. This detail is entirely reproducible with one wire, but varies from wire to wire. The explanation is that in the wire there are no quantum-mechanical electron states extending from one end to the other and available to carry current. Instead, all the states are of finite extent and the current is carried by electrons which hop from site to site. As the applied voltage changes, so do all the states, causing changes in the hopping current and leading to the observed complicated curve.

If the wire is drawn in the configuration of **Fig. 3**, a further phenomenon comes into play. In quantum mechanics the electron wave travels around both sides of the loop and the two components interfere with each other when they combine at the end of the loop. Maxima and minima occur in the current-voltage curve when the interference is constructive or destructive, respectively, again giving a curve with considerable structure. Further, if this experiment is carried out at a fixed voltage, the current is not

within a limited range by using wells of appropriate width. Wells of different semiconductor materials give different ranges of laser frequencies.

Figure 2*a* shows the conduction electron potential for a more complicated layered structure which yields the property known as resonant tunneling. The well between the two barriers would have a bound state for the electron if the barriers were infinitely high, but in quantum mechanics, electrons can pass through finite barriers by the tunneling process. The probability that an electron, incident from the left, will pass through the barriers is very small except when the electron has an energy equal to the energy of a bound state in a well with infinite barriers. The process by which the electron passes through from left to right is called resonant tunneling. If an electric field is applied across the layers, the potential is changed (Fig. 2*b*); this changes the bound-state energy and thus changes the energy of the transmitted electrons. In this way the electron energy can be selected by using the applied potential, a technique which allows electron energy distributions in solids to be studied.

Another interesting phenomenon in low-dimensional structures can also be illustrated by using the resonance tunneling device of Fig. 2*b*.

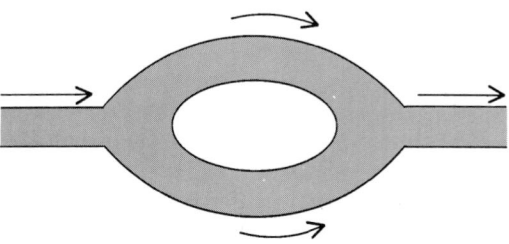

Fig. 3. Quantum wire loop for the observation of electron wave coherence. Arrows indicate electron motion.

constant but is observed to jump irregularly between two levels. This extraordinary behavior arises because the electron state nearest the Fermi energy (the highest energy of filled electron states) is sometimes filled and sometimes empty. As it fills or empties, the potential acting on the electron wave changes, altering the interference condition. Thus in this experiment an individual electron site can be observed as it fills and empties. SEE QUANTUM MECHANICS.

Optical properties. It is not only the electronic properties which change with the dimensionality. The emission and absorption of light depend on the electron wave functions involved and therefore change not only in energy but also in character. Thus, quantum well lasers have fewer higher harmonics in their output, which is of great benefit for fiber-optic transmission. There are other, more complex phenomena involving light. In some layered structures the dielectric constant can be changed by intense illumination, thus altering the refraction of a second, signal beam and allowing the possibility of an optical switch. It is not yet known exactly how such properties could be exploited to produce an optical computer, or indeed whether such an optical computer would be superior in any way to an electronic one, but low-dimensional structures offer many options. Very efficient coupling between optical and electronic systems also may be possible through the use of diffraction gratings drawn lithographically onto the surface of a layered structure. This could allow integration of optical and electronic components within computers and communications equipment, avoiding the wiring difficulties which arise as circuits become more complex. SEE OPTICAL INFORMATION SYSTEMS.

The many variables involved and the many possible structures, together with the constant improvement in production techniques, ensure that there will be further discoveries of unusual and useful properties in the science of low-dimensional structures.

For background information SEE ARTIFICIALLY LAYERED STRUCTURES; LASER; NONRELATIVISTIC QUANTUM THEORY; OPTICAL BISTABILITY; OPTICAL INFORMATION SYSTEMS; QUANTUM MECHANICS; SEMICONDUCTOR; SEMICONDUCTOR HETEROSTRUCTURES; TRANSISTOR in the McGraw-Hill Encyclopedia of Science and Technology.

J. L. Beeby

Bibliography. T. Ando (ed.), *Electronic Properties of Two-Dimensional Systems*, 1986; L. L. Chang and B. C. Giessen (eds.), *Synthetic Modulated Structures*, 1985; M. J. Kelly and C. Weisbuch (eds.), *The Physics and Fabrication of Microstructures and Microdevices*, 1986; K. S. Ralls et al., Discrete resistance switching in submicrometer silicon inversion layers: Individual interface traps and low-frequency (1/f ?) noise, *Phys. Rev. Lett.*, 52:228–231, 1984.

Ship powering and steering

Historically, water jets were used for marine propulsion before the screw propeller, which now propels most ships and small watercraft, both commercial and naval, throughout the world. The earliest propulsive device to use mechanical power seems to have been of the water-jet type, involving a prime mover and a pump which received water through an inlet duct and discharged it through a nozzle at increased velocity to produce propulsive thrust. Patents for this device were granted in 1661. The water jet was used by James Rumsey in 1782 to propel an 80-ft (24-m) passenger boat.

The first use of the screw propeller is generally assumed to have been in 1804 for a steam-driven boat; however, a proposal to use a screw propeller appears to have been made in 1680. Following the first Atlantic crossing of a screw-propeller-driven steamer in 1845, the propeller gained great popularity.

Description of water-jet systems. The water jet and screw propeller are similar devices, differing only in design detail. Both produce propulsive thrust (reactive force) by the acceleration of water in accordance with Newton's law which states that the force acting on an object equals its rate of change of momentum. In a typical screw propeller propulsion system (**Fig. 1***a*), the propeller is generally located in the open water beneath the aft end of the hull where the inflow velocity is relatively uniform and a rudder for steering can be located in the high-velocity slipstream. The propeller shaft, support strut, skeg, and rudder all produce drag, and this resistance must be overcome by the thrust. This appendage drag may be as high as 30% of the total ship resistance and offsets the propeller's efficiency advantage over the water jet.

The water-jet propulsion system is usually located within the hull to minimize appendage resistance. The water is ducted to the pump through various inlet configurations (Fig. 1*b–d*) for different types of ships. The flush inlet results in low drag and is ideal for planing craft and high-speed displacement hulls. The ram inlet is used for hydrofoils and surface-effect ships where the hull and pump are above the water surface. The semiscoop inlet can have advantages on lower-speed displacement hulls and some planing craft when the inlet velocity ratio (inlet exit velocity/inlet entrance velocity) is near unity and no acceleration or diffusion occurs.

The velocity of the jet leaving the nozzle must be of the order of 1.5 to 2.0 times the velocity of the craft for good propulsive efficiency. The discharge velocity from a nozzle varies with the square root of the pressure. Thus, the type of pump, flow rate, and pressure rise selected are dependent on the design speed of the craft. The required input power is proportional to the flow rate times the pump pressure rise. Therefore, a low-speed craft is characterized by a high-flow-rate, high-velocity jet. The pressure measured in feet or meters of water at the jet nozzle may vary from 15 ft (4.5 m) on a 10-knot (18.5-km/h) craft to 1700 ft (520 m) on a 100-knot (185-km/h) craft. Corresponding jet velocities leaving the water-jet nozzle are 30 and 330 ft/s (9 and 100 m/s) respectively.

The water-jet impeller which rotates within the pump case may be of the axial flow design, similar to a screw propeller, for low-speed craft requiring low jet velocity. The water flows through such a pump in an axial direction parallel to the shaft. Pressure is generated by vane curvature which induces tangential velocity in the cylindrical pump case. Stationary vanes direct the flow back to the axial direction, resulting in a pressure increase and increased axial velocity.

For higher-speed craft in the 20–40-knot (37–74-km/h) range, a mixed-flow pump (**Fig. 2**) or a multistage axial flow pump is commonly used. The multistage axial pump may consist of two or more similar stages mounted in a row on a common shaft, each imparting an additional pressure rise to the water. Water jets whose stages differ in design are also built; the stages are mounted on separate shafts and operate at different speeds of rotation.

Mixed-flow water-jet pumps are characterized by impeller water passages which increase in radius from entrance to exit. The water has velocity components in both the axial and radial directions. Tangential velocity and pressure are increased by vane curvature, as in the axial pump, plus the centrifugal action from the increase in radius. The higher presssure risk makes the mixed-flow pump ideal for pleasure craft and larger boats up to speeds of 40 knots (74 km/h). For higher-speed crafts, multistage mixed-flow or radial-discharge centrifugal pumps are usually employed.

A typical water-jet installation, using a flush inlet and a mixed-flow pump suitable for a 40-knot (74-km/h) craft, is shown in Fig. 2. The jet discharges through a steering deflector, which can be rotated up to 30° to either side to vector the thrust. Reverse thrust is produced by closing the reverse gate, which directs the jet forward beneath the transom.

Comparison of water jet with propeller. For a given input power, the diameter of an optimum water jet will be of the order of 60% of that of a propeller for the same application. The number of input revolutions per minute (rpm) will be approximately 1.5 times that of the propeller. This higher rotation rate is a definite advantage, since it reduces the size and weight of shafting, gearing, and support structure.

Rapid switching to full-power reverse thrust, without reducing engine rotation rate, is possible with the water jet. The load on the engine is

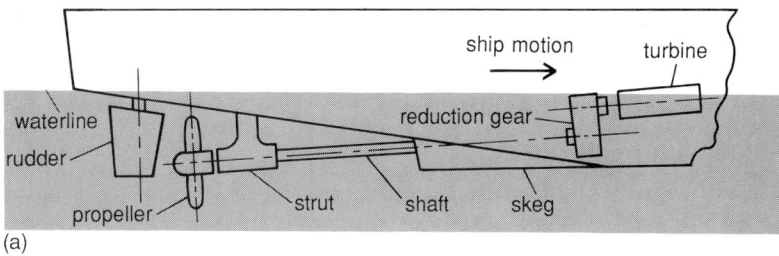
(a)

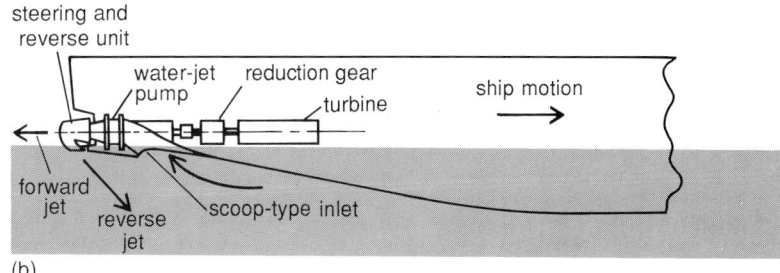
(b)

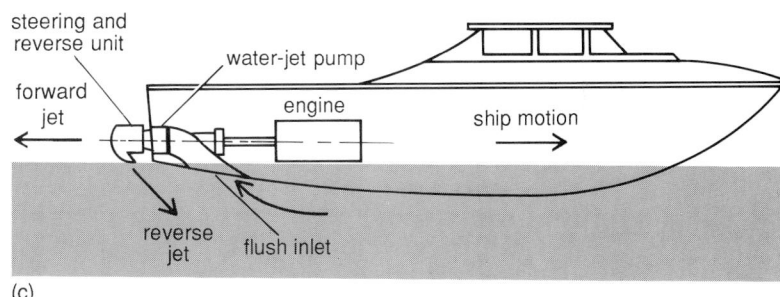
(c)

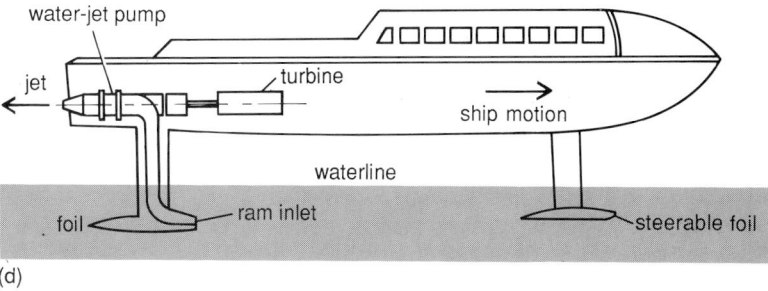
(d)

Fig. 1. Comparison of water-jet and propeller systems. (*a*) Propeller-driven displacement ship. (*b*) Water-jet-propelled displacement ship. (*c*) Water-jet-propelled planing craft. (*d*) Water-jet-propelled hydrofoil.

essentially independent of ship speed, position of the reverse gate (forward or reverse), and steering angle. The steering side force of a water jet is also essentially independent of ship speed. A side thrust equal to 50% of the unvectored thrust is readily attainable with a 30° rotation of a tapered steering deflector to either side of the forward direction, even at zero speed.

With a conventional propeller-rudder system, steering is highly dependent on ship speed, and in general, shaft speed must be reduced or reversed for reverse thrust. Allowable engine power in reverse is also reduced with a propeller, particularly if the ship is still moving forward.

When compared to a conventional screw pro-

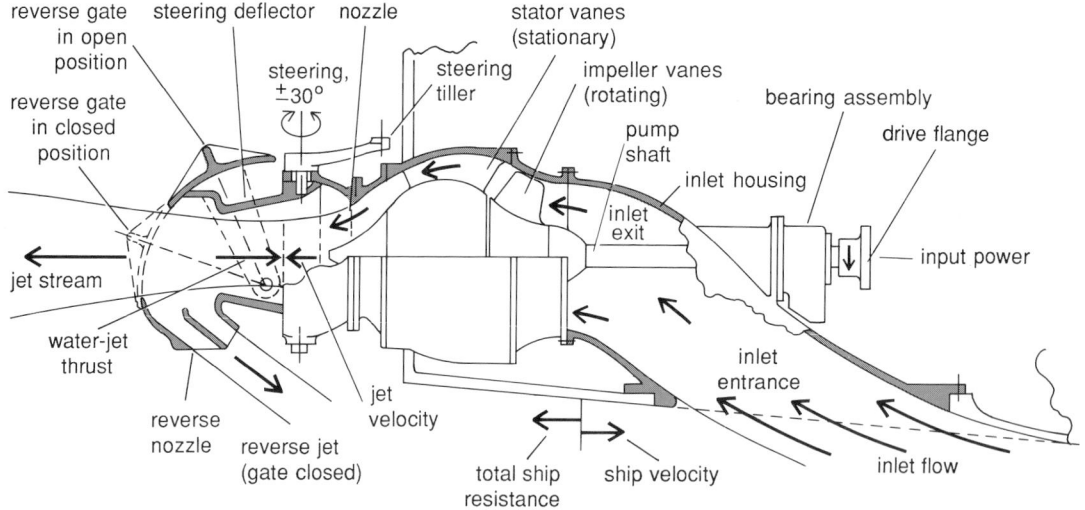

Fig. 2. Typical water-jet installation, using a flush inlet and a mixed-flow pump.

peller, the water jet may be characterized by its reduced size and weight; reduced drag (resistance which must be overcome by the propulsive thrust); higher input shaft speed (that eliminates or reduces reduction gearing); reduced hazard (due to the use of enclosed rotating parts); reduced draft and vulnerability to damage in shallow water; higher reliability and reduced maintenance; improved maneuvering and reverse capability; reduced underwater noise (which is important in naval ships); elimination of the need for reverse-gear or reversible-pitch propeller blades; and reduced propulsion machinery cost.

It is widely believed that water jets are less fuel-efficient than conventional propellers, but this view is not entirely justified. The input power P_d required to propel a ship is given by the equation below, where R is the total ship resis-

$$P_d = \frac{RV_s}{\text{P.E.}}$$

tance, V_s is the ship velocity, and P.E. is the propulsive efficiency. While the inherent efficiency of the water jet is less than that of the propeller for most applications, the total resistance of the craft is also less for the water-jet system, and the power calculated from the equation is essentially the same for both systems. There have been major advances in water-jet propulsion since about 1960, and these have led to studies, test programs, and actual water-jet applications in which the input power and fuel consumption of the water jet are equal to or less than those of a propeller system. Because of the many other advantages of the water jet, it can still be a valid choice of propulsion, even in installations where the power required exceeds that of a propeller system.

Typical performance characteristic curves of a water-jet-propelled displacement ship are shown

in **Fig. 3**. The propulsive efficiency value shown, slightly greater than 0.6, is typical of what would be obtained with a large water-jet-propelled displacement hull such as a naval destroyer.

Recent water-jet development. The water jet gained some popularity, particularly on small pleasure craft, starting about 1960. Safety to swimmers, high maneuverability, and ability to operate in shallow water were attractive features.

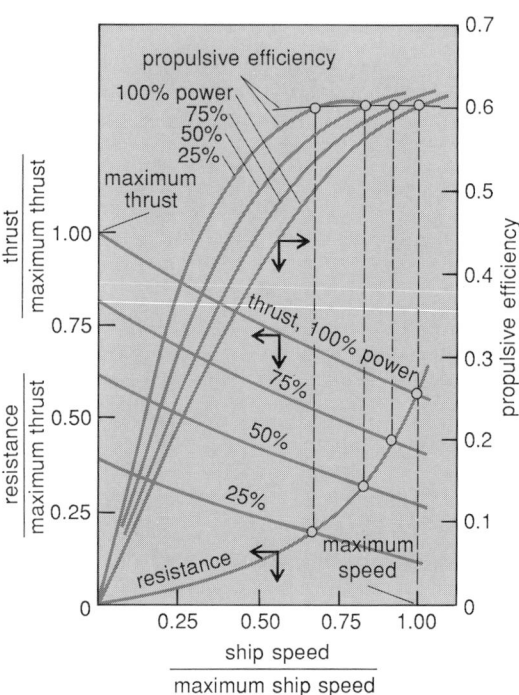

Fig. 3. Typical performance characteristics of water-jet-propelled displacement ship. The top line labeled "propulsive efficiency" indicates the maximum propulsive efficiency that can be attained at a given ship speed by adjusting the input power so that the thrust just balances the resistance.

Because fuel prices were low, the relatively poor efficiency of the units at that time was not overly important, and little effort was made to adopt improved technology. Since the oil price rise of the mid-1970s, the water jet has lost some of its popularity on pleasure craft because of its high fuel consumption.

Water jets were also widely used on patrol boats operating in shallow water during the Vietnam conflict. Many of these craft are still in use throughout Southeast Asia.

The water jet can be used with varying degrees of performance to propel any ship. It is particularly attractive for high-speed ships for which improved propulsive efficiency and reduced propulsive system drag result in power requirements comparable to those of propeller systems, which require a more complex and less reliable mechanical drive system.

The U.S. Navy developed an interest in water jets for propulsion of advanced surface-effect and hydrofoil ships around 1965. Prior to this time, water-jet development was restricted mainly to the traditional marine industry. Hydrofoil and surface-effect ships are supported above the water surface, and hence must be light in weight if a useful payload is to be carried. The hydrofoil ship uses underwater hydrofoils for support. The surface-effect ship is supported by an air bubble pumped beneath the hull. The technology for developing these new types of lightweight, high-speed ships was not available in the marine industry, but was in use in the aerospace industry for aircraft and rockets. Several aerospace companies were early participants in the development of the Navy's new-type high-speed ships. These companies contributed much new technology to the marine field and were instrumental in greatly improving water-jet propulsion systems.

The SES-100A surface-effect ship, delivered to the Navy in 1972, is 88 ft (28 m) in length, weighs 100 tons (91 metric tons), and achieves a speed of over 85 knots (157 km/h), propelled by twin 6000-shaft-horsepower (4.5-MW) water jets. These two-stage water jets are unique in that they utilize two impellers operating at different shaft speeds to obtain the required performance at minimum weight. Each 6000-shp (4.5-MW), 22-in.-diameter (56-cm) water jet weighs 1600 lb (725 kg), including reduction gear and steering and reverse devices. Use of conventional water-jet technology would have resulted in an unacceptably large and heavy four-stage unit of approximately twice the diameter and five times the weight of the one actually used. The new type of water jet met all performance requirements and proved highly reliable during extensive ship testing. The SES-100A water jet experienced no cavitation erosion damage, a common major problem with earlier water jets and propellers used on high-speed ships.

Based largely on the success of the SES-100A

marine jet, water jets were selected for propulsion of the patrol hydrofoil missile (PHM) Pegasus-class hydrofoil. Six of these ships were delivered to the Navy between 1974 and 1980, and are currently operational. These ships, which have a speed in excess of 55 knots (102 km/h) while foil-borne, are propelled by a single 20,000-shp (14.9-MW) water jet. These 48-in.-diameter (1.2-m) water jets are the most powerful in existence, and are of the two-stage, two-speed type similar in design to the earlier SES-100A units. While hull-borne, the ship is propelled by twin 900-shp (670-kW) water jets, similar to the one shown in Fig. 2.

In addition to the military applications of water jets, a large number of commercial hydrofoil ferries, known as jetfoils, which are propelled by twin 3500-shp (2.6-MW) water jets, have been produced.

In Europe, most high-power water-jet propulsion systems are produced in Sweden. The largest unit to date is a 13,400-shp (10-MW) water jet developed to propel a 230-ton (209-metric-ton) yacht at 45 knots (83 km/h).

Water-jet technology developed for these advanced high-speed naval applications is readily adaptable for use on other naval ships, many commercial ships, and craft for water sports. Aided by current lower fuel prices and advanced technology, a new popularity for water-jet propulsion is starting to develop, and much wider use of water jets for both military and commercial applications is anticipated.

For background information SEE FERRY; HYDRO-FOIL CRAFT; NAVAL SHIP; PROPELLER (MARINE CRAFT); SHIP POWERING AND STEERING in the McGraw-Hill Encyclopedia of Science and Technology.

Herman C. Schlappi

Bibliography. D. J. Berg, W. S. Jones, and H. W. Marron, Why waterjets, *Nav. Eng. J.*, pp. 779–783, 1967; J. P. Comstock (ed.), *Principles of Naval Architecture*, Society of Naval Architects and Marine Engineers, 1967; H. C. Schlappi, An innovative energy saving propulsion system for naval ships, *Nav. Eng. J.*, pp. 200–213, 1982; R. Svensson, Experience with waterjet propulsion in the power range up to 10,000 kW, Society of Naval Architects and Marine Engineers Power Boat Symposium, Miami, Florida, February 1985.

Sickle cell disease

The prenatal diagnosis of human genetic diseases by deoxyribonucleic acid (DNA) analysis has made significant progress in the last several years. Unlike biochemical and serological tests which detect the absence of a normal protein or the presence of an abnormal one, DNA-based diagnostics directly examine the genes that encode those proteins. This type of analysis is particularly valuable in situations where the expression of the gene does not normally occur in the fetus or where the gene is expressed in locations that are difficult to sample. Because DNA analysis looks only at the genes themselves and does not rely upon their expression, the DNA can be extracted from any readily available source of fetal cells. Such cells can be safely obtained from amniotic fluid by amniocentesis or from chorionic villi by chorionic villus biopsy. However, these samples generally yield only a few micrograms of DNA, and the development of reliable techniques capable of routinely utilizing such small amounts of nucleic acid has been one of the most difficult and challenging aspects of DNA-based prenatal diagnostics. Two recently introduced methods, one for the selective amplification of a specific gene fragment and the other or the direct detection of specific DNA sequences, have facilitated the evolution of these tests and have been successfully applied to the prenatal diagnosis of sickle cell anemia.

Genetic cause. Sickle cell anemia is a debilitating blood disease that afflicts approximately 1 in 625 black Americans, usually as a severe, life-threatening anemia. The disease is caused by a mutation in the human β-globin gene, one of two genes whose protein products make up the predominant form of normal adult hemoglobin. The actual mutation within the DNA sequence of the gene is located in the second position of the sixth codon, where a thymidine base (T) replaces an adenine base (A). This results in the substitution of a valine amino acid for a glutamic acid in the β-globin protein (**Fig. 1**). The sickle cell defect is an autosomal-recessive mutation; an individual must be homozygous for the sickle cell allele to be affected by the disease. Heterozygotes possessing one copy of the normal β-globin gene (β^A) and one copy of the sickle cell allele (β^S) generally present no clinical symptoms. The disease follows a simple mendelian mode of inheritance. Only those couples in which each mate carries at least one copy of the β^S gene are at risk for a child

afflicted with sickle cell anemia. In the United States, there are about 3000 of these at-risk pregnancies each year.

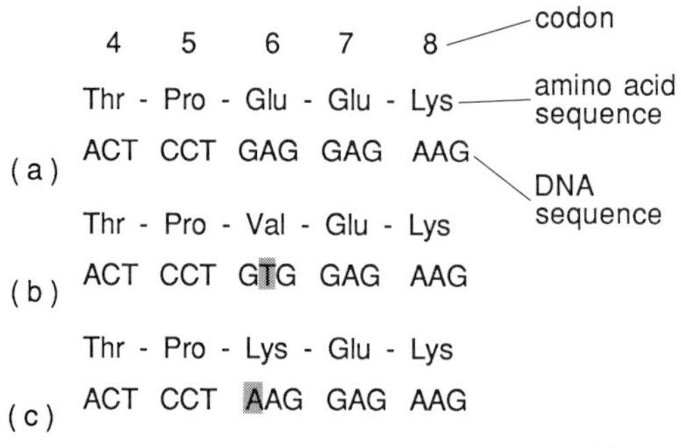

Some heterozygous combinations in the β^S allele with other abnormal β-globin genes can also cause serious disease. The defective gene associated with hemoglobin C disease is an example of this second allele. The genetic mutation responsible for hemoglobin C disease has been traced to the sixth codon of the β-globin gene as well. In this case, an adenine replaces the guanine (G) normally present at the first position of the codon, resulting in the substitution of a lysine in the protein product (Fig. 1). Like β^S, the hemoglobin C disease allele (β^C) is autosomal-recessive. But in contrast to the situation with sickle cell anemia, individuals homozygous for β^C are only mildly anemic. When both β^S and β^C alleles are present in the same individual, however, severe anemia typically results. For this reason, it is important to be able to test for the presence of this disease allele when such compound heterozygosity is possible.

Prenatal diagnosis. The recently developed procedure for determining the β-globin genotype of a DNA sample for prenatal diagnosis involves two steps. First, a small segment of the β-globin gene containing the sixth codon is amplified. Second, the amplified region is hybridized to small synthetic DNA molecules which detect only those sequences to which they are perfectly matched.

DNA amplification. The technique of DNA amplification is known as the polymerase chain reaction. It uses two synthetic DNA fragments, called primers, and a DNA-replicating enzyme, DNA polymerase, to increase the amount of a particular piece of genomic DNA up to a million-fold. The process for the amplification of the β-globin gene is diagrammed in **Fig. 2**. The genomic DNA is heated to separate the two complementary strands, commonly referred to as (+) and (−), and the two primers are allowed to bind to the DNA. These primers, PC03 and PC04, are designed to anneal to specific areas of the DNA on opposite strands—PC03 to the (−) strand and PC04 to the (+) strand. In this manner the primers define the segment of DNA between them as the region to be amplified. DNA polymerase is then added to initiate the synthesis of complementary strands at the bound primers. The result of this reaction is the creation of two double-stranded copies of the region flanked by the primers. Moreover, if these four strands are separated again, four more primer molecules will be able to reanneal to those strands [for example, PC03 to the original and the newly synthesized (−) strands]. After extension of the primers by polymerase, there will be four double-stranded copies of the DNA between the primers. Each time the reaction is repeated, the number of copies of the target DNA is essentially doubled.

Fig. 1. DNA sequences and corresponding amino acid sequences of (*a*) normal [β^A], (*b*) sickle cell [β^S], and (*c*) hemoglobin C [β^C] β-globin genes from the fourth to eighth codons. The single-base DNA mutations of the β^S and β^C alleles are indicated by boxes.

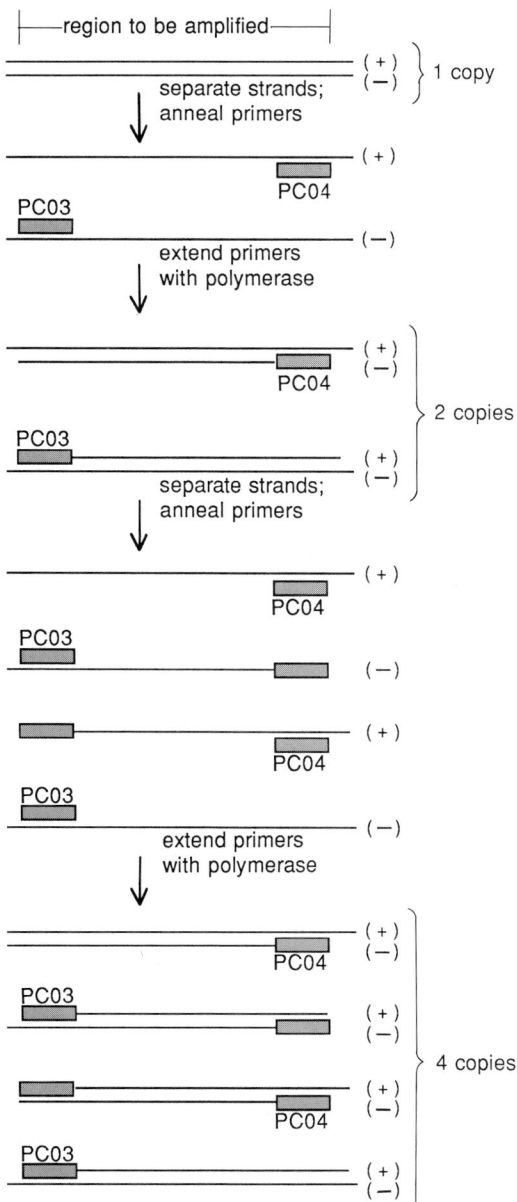

region to be amplified

PC03 PC04

separate strands; anneal primers

extend primers with polymerase

Fig. 2. Polymerase chain reaction amplification of a specific segment of the β-globin gene. The genomic DNA is strand-separated, annealed to the polymerase chain reaction primers PC03 and PC04, and extended with polymerase. The result is a doubling of the number of copies of double-stranded DNA in the region bounded by the primers. Two cycles of amplification are shown.

After 25 such repetitions, or cycles, a millionfold amplification of the β-globin sequence can be achieved. This high degree of amplification substantially simplifies the next step of analysis in which the DNA is tested for the presence of the β^S and β^C alleles.

Detection of single-base mutations. Synthetic DNA fragments can be used to detect single-base mutations in genomic DNA. The basis for the method is that under carefully controlled conditions a short, radioactively labeled DNA probe

(usually 19 bases in length) will anneal only to those sequences to which it is perfectly matched. A single-base mismatch is sufficiently destabilizing to prevent the hybridization of the probe to the genomic DNA. By preparing two probes, one exactly complementary to the normal gene and the other to its mutant counterpart, it is possible to determine which alleles are present in the DNA sample. This approach has been successfully applied to the prenatal diagnosis of several human genetic diseases, including sickle cell anemia, β-thalassemia, and α-1 antitrypsin deficiency. These applications, however, did not use polymerase chain reaction amplification to enrich for the particular gene of interest. Consequently, special measures had to be taken to maximize sensitivity and ensure detection of the very weak radioactive signals that indicated the presence of the different alleles.

Results. By combining both methods, a fast and sensitive diagnostic test was obtained. **Figure 3** presents an example of test results obtained with this procedure. Six DNA samples containing all possible diploid combinations of the β^A, β^S, and β^C alleles were subjected to polymerase chain reaction amplification and spotted onto three nylon filter membranes, which bind the DNA and immobilize it. Three radioactively labeled synthetic DNA probes, identical in sequence to each of the β-globin alleles, were then incubated with one of the filters. After exposure to x-ray film to localize the radioactive probes, the resulting autoradiogram shows that the probes have bound only to those DNA samples that have at least one

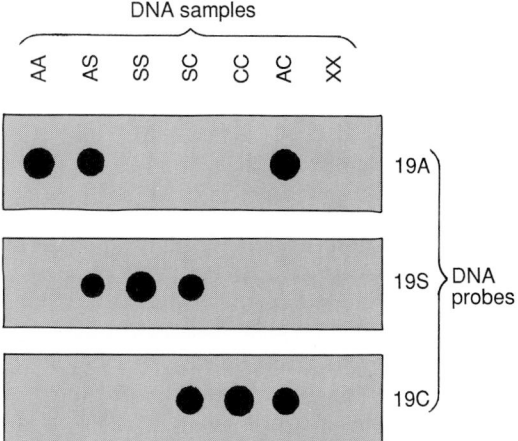

DNA samples

AA AS SS SC CC AC XX

19A
19S DNA probes
19C

Fig. 3. Analysis of the β-globin genotype of DNA samples using sequence-specific DNA probes: β^A-specific (19A), β^S-specific (19S), and β^C-specific (19C). The genotypes of the samples were: AA, $\beta^A\beta^A$; AS, $\beta^A\beta^S$; SS, $\beta^S\beta^S$; SC, $\beta^S\beta^C$; CC, $\beta^C\beta^C$; and AC, $\beta^A\beta^C$. The sample XX is a negative control which contains DNA that should not bind any of the probes. (*After R. K. Saiki et al., Genetic analysis of enzymatically amplified β-globin and HLA-DQα genomic DNA with allele-specific oligonucleotide probes, Nature, 324:163–166, 1986*)

copy of the allele to which the probes are perfectly matched. The β-globin genotype of each sample is easily determined by examination of the pattern of hybridization. For example, sample AS binds the β^A-specific probe, 19A, as well as the β^S-specific probe, 19S, confirming the presence of both alleles in that patient. Similarly, sample SS binds only the 19S probe, indicating that the DNA was extracted from an individual homozygous for the β^S mutation and afflicted with sickle cell anemia. The other disease genotype, the β^S-β^C compound heterozygote, is present in sample SC, which anneals to the 19S probe and to the β^C-specific 19C probe.

It is significant that this approach to prenatal DNA analysis is a general one, capable of detecting the mutations that lead to other genetic diseases. With the eventual introduction of automated devices and nonradioactive probes, the procedure should find applications in a wide variety of important genetic analyses.

For background information SEE DEOXYRIBONU- *CLEIC ACID (DNA); GENETIC CODE; HUMAN GENETICS; SICKLE CELL DISEASE* in the McGraw-Hill Encyclopedia of Science and Technology.

Randall K. Saiki

Bibliography. B. J. Conner et al., Detection of sickle cell β^S-globin by hybridization with synthetic oligonucleotides, *Proc. Nat. Acad. Sci. USA*, 80:278–282, 1983; R.K.Saiki et al., Genetic analysis of enzymatically amplified β-globin and HLA-DQα genomic DNA with allele-specific oligonucleotide probes, *Nature*,' 324: 163–166, 1986.

Silicate phase equilibria

Rocks occur both in and on the Earth and other planetary bodies; they are made up of individual minerals that are uniquely characterized by chemistry and structure. The chemical composition as well as the type of constituent minerals of a specific rock are governed by the pressures and temperatures of the formation environment and the chemical composition of the rock itself. The chemistry and structure of the rock-forming materials can be determined by laboratory analyses. This information forms the basis of silicate phase equilibrium experiments that are aimed at determining the pressure, temperature, and composition history of these materials.

The physical and chemical properties of the substances that make up the present solar system reflect the conditions that existed in the early solar nebula some 4.5–5.0 billion years ago. After the nucleosynthesis that led to the formation of the elements and their individual isotopes, these species combined and precipitated from the original hydrogen-rich gas cloud to form the initial minerals and mineral assemblages. Records of many of these initial events are completely or partially preserved in chondritic meteorites.

Chemical and petrographic evidence from chondrites indicates that the formation of rock-forming minerals in the solar nebula resulted from processes that included repeated evaporation and condensation of melt droplets and crystals. Chondritic meteorites have two particularly important features: the chondrules, from which the term chondrite is derived, and the Ca,Al-rich inclusions, known as CAIs. The chondrules show evidence of having once existed as a melt. The Ca,Al-rich inclusions exhibit compelling chemical and textural evidence of a history where mineral phases condensed directly from a gas or where evaporation of already-condensed minerals resulted in a refractory residue which may be found as the inclusions.

Laboratory data. The bulk compositions of individual grains (condensate) and agglomerate (multiple grains) that eventually accreted to form terrestrial planets depend critically on whether vapor-crystal equilibria or vapor-liquid-crystal equilibria took place. For a given composition, there is a minimum pressure above which a crystalline phase cannot condense directly from a gas phase or evaporate directly to a gas without an intervening liquid field (**Fig. 1**). The pressure and temperature that define the transition from crystal-liquid and liquid-gas equilibria to crystal-gas equilibria is termed the triple point (Fig. 1). The chemical compositions of the rock-forming materials in the solar system formed by these processes will differ substantially depending on whether the minerals were formed by crystallization directly from a gas, or liquid was first formed by condensation from the gas and then

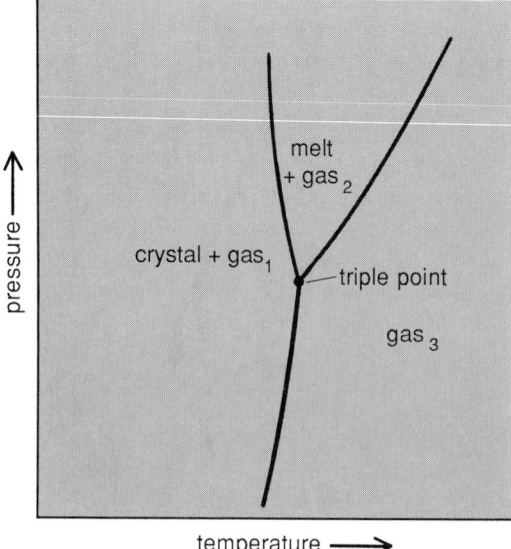

Fig. 1. Schematic pressure-temperature diagram to illustrate the principal phase relations around the triple point of a compound. The terms gas$_1$, gas$_2$, and gas$_3$ represent gases of different compositions.

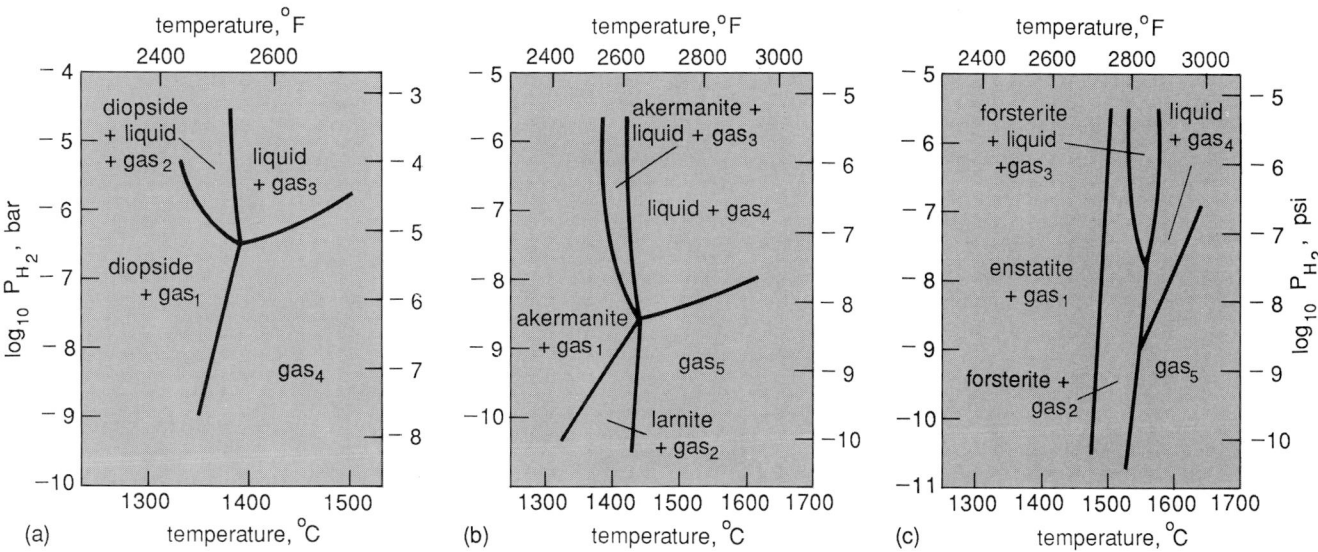

Fig. 2. Relationships of mineral phases, liquid, and gas for three systems. (*a*) $CaMgSi_2O_6$ (diopside)–H_2. (*b*) $Ca_2MgSi_2O_7$ (akermanite)–H_2. (*c*) Mg_2SiO_4–SiO_2–H_2. The terms gas$_1$, gas$_2$ and so forth represent gases with different silicate proportions in hydrogen-rich gas.

crystallization from the liquid occurred.

Laboratory experiments on silicate materials relevant to these processes require duplication of controlled pressures ranging from 10^{-11} to 10^0 bar (10^{-6} to 10^5 Pa) and temperatures between 1000 and 2000°C (1800 and 3600°F). Chemical and petrographic information from chondritic meteorites is the principal guide in selecting the minerals and mineral assemblages for experimental investigation. Among the silicate minerals, the most important include olivine, pyroxenes, and melilite-group minerals.

Experimental results are now available for some of these mineral phases (**Fig. 2**). The most simple relationships have been observed for the pyroxene diopside ($CaMgSi_2O_6$), where diopside crystals, liquid, or gas is the only stable phase (Fig. 2*a*). Diopside exists together with hydrogen-rich gas to temperatures slightly below 1400°C (2550°F) in the pressure range from 10^{-10} to 10^0 bar (10^{-5} to 10^5 Pa). At pressures below 4×10^{-7} bar (4×10^{-2} Pa) crystalline diopside sublimates directly to a gas, whereas at higher pressures the crystal melts before the temperature at which the melt evaporates to a gas phase is reached. The temperature and pressure at which diopside crystals, liquid, and gas coexist correspond to the simple triple point (Fig. 1).

Evidence from chondritic meteorites suggests that melilite-group minerals were among the first to crystallize in chondrites, because they are found, for example, among the phases in Ca,Al-rich inclusions. These melilites can be described chemically as solid solutions that involve two end-member melilite minerals, akermanite ($Ca_2MgSi_2O_7$) and gehlenite ($Ca_2Al_2SiO_7$). The phase diagram of one of these end members, akermanite (Fig. 2*b*), shows some differences

from that of diopside (Fig. 2*a*). In contrast to diopside, which at pressures below the triple point sublimates directly to gas, for akermanite there is an intervening phase field at temperatures below that of complete vaporization where another mineral, larnite (Ca_2SiO_4), coexists with a gas enriched in silica and magnesium compared with the chemical composition of akermanite itself. Therefore, akermanite cannot crystallize directly from a hydrogen-rich gas with calcium, magnesium, and silicon in the appropriate proportions. However, akermanite remains stable up to higher temperatures than diopside does. The data also reveal that a melt of akermanite composition remains stable down to pressures two orders of magnitude lower than a melt of diopside composition does.

The pyroxene-group mineral enstatite ($MgSiO_3$) is ubiquitous in chondritic meteorites

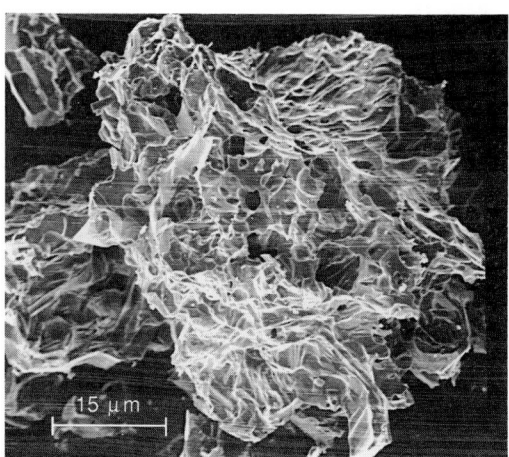

Fig. 3. Electron scanning micrograph of forsterite crystals formed by fractional evaporation of enstatite to form forsterite and silica-rich vapor. (*Courtesy of I. Kushiro and B. O. Mysen*)

and is often closely associated with forsterite (Mg_2SiO_4) in chondrules. Therefore, experiments in the chemical system Mg_2SiO_4–SiO_2 are necessary to characterize the pressure and temperature conditions of chondrule formation. The experimental results (Fig. 2c) show that enstatite breaks down to forsterite (**Fig. 3**) and a silica-rich vapor at temperatures above 1450–1500°C (2640°–2700°F) in the pressure range 10^{-11} to 10^{-3} bar (10^{-6} to 10^2 Pa). Pure silica (SiO_2) and enstatite condenses out of the vapor at temperatures below about 1200°C (2200°F) in the same pressure range. Forsterite and gas melts to liquid and gas at pressures greater than about 10^{-9} bar (10^{-4} Pa) near 1550°C (2822°F). Complete transformation of forsterite and silica-rich gas to a gas of enstatite composition occurs at temperatures near 1550°C (2822°F) at pressures below the triple point.

Interpretations relevant to solar nebula. The temperatures, pressures, and their gradients in time and space in the early solar nebula are uncertain and controversial. It is frequently concluded that at the time in the early history of the solar nebula when temperatures exceeded 1000°C (1800°F) the pressures may have ranged from 10^{-3} to nearly 10^{-6} bar (10^2 to nearly 10^{-1} Pa) with higher pressures accompanying higher temperatures. From the experimental data (Fig. 2) it is reasonable to conclude that condensation of the nebular gas cloud commonly resulted in liquid droplets. Thus all the experimental data are consistent, with chondrules probably being formed originally as a liquid from which minerals were subsequently produced. The overall chemical fractionation of elements in the early nebula probably was governed extensively by vapor-to-liquid-to-crystal equilibria. Thus as the condensation of liquid droplets was increasingly important, the distance to the Sun decreased.

The experimental data also indicate that the temperatures and pressures necessary for direct condensation of silicate minerals from a gas cloud are such that processes of this nature probably were not very important for the refractory phases found in Ca,Al-rich inclusions in chondritic meteorites. These materials could, however, be formed by fractional evaporation of preexisting materials. In a fractional evaporation process the refractory minerals would remain behind (see Fig. 3) while the more volatile components (for example, alkali metals, silica, and some alkaline earths) boiled off. Evidence from chondrites suggests that such processes may have occurred repeatedly in the early stages of formation of the solar system.

For background information SEE METEORITE; SILICATE PHASE EQUILIBRIA; SOLID SOLUTION; TRIPLE POINT in the McGraw-Hill Encyclopedia of Science and Technology.

Bjorn Mysen

Bibliography. A. G. W. Cameron, Physics of the primitive solar accretion disk, *The Moon and the Planets*, 18:5–40, 1978; L. Grossman, Refractory inclusions in the Allende meteorite, *Annu. Rev. Earth Planet. Sci.*, 8:559–609, 1980; B. O. Mysen, D. Virgo, and I. Kushiro, Experimental studies of condensation processes at low pressures and high temperatures. I. Phase equilibria in the system $CaMgSi_2O_6$–H_2 in the temperature range 1200–1500°C and the pressure range (P_{H2}) 10^{-6} to 10^{-9} bar, *Earth Planet. Sci. Lett.*, 75:139–146, 1985; A. E. Ringwood, *Origin of the Earth and the Moon*, 1979.

Silurian

Soft-bodied organisms, such as annelid worms and jellyfish, are presently abundant and diverse in oceans, and probably always were. The fossilization potential of these organisms is quite small, however, and the fossil record is biased heavily toward organisms that possess mineralized skeletal elements such as shell and bone. Only in rare cases are soft-bodied and lightly mineralized organisms preserved as fossils. Even so, preservation of these organisms is very important because it provides the only indication of the true diversity of ancient life. Such preservation also documents the evolutionary history of soft-bodied groups and reveals the nature of soft tissues in organisms otherwise known only from fossilized hard parts. In each case of exceptional preservation, it is essential to determine whether the soft-bodied organisms were typical and widespread but rarely fossilized, or whether they represent a unique biota that lived in an unusual environment.

A new biota containing soft-bodied and lightly sclerotized organisms has been uncovered in Lower Silurian (430 million years ago) marine rocks in the Brandon Bridge beds at Waukesha, Wisconsin (see **illus.**). This is the most extensive soft-bodied Silurian biota known in the world, and it helps to fill a 100-million-year gap between other known major cases of exceptional preservation in the Middle Cambrian Burgess Shale of British Columbia and the Lower Devonian Hunsrück Slate of Germany. The Waukesha biota contains the earliest known examples of several marine invertebrate groups and a few new organisms, and it reveals that the diversity of Silurian marine invertebrates is greater than realized previously.

Arthropods. The Waukesha biota is dominated by arthropods, and it may be the only biota in the geologic record known to contain representatives of all four major arthropod groups (trilobitomorphs, crustaceans, chelicerates, and uniramians).

Trilobites. The most abundant and diverse of the Waukesha arthropods, trilobites, are represented by 13 genera, of which an undescribed dalmanitid is the most numerous. Many of the

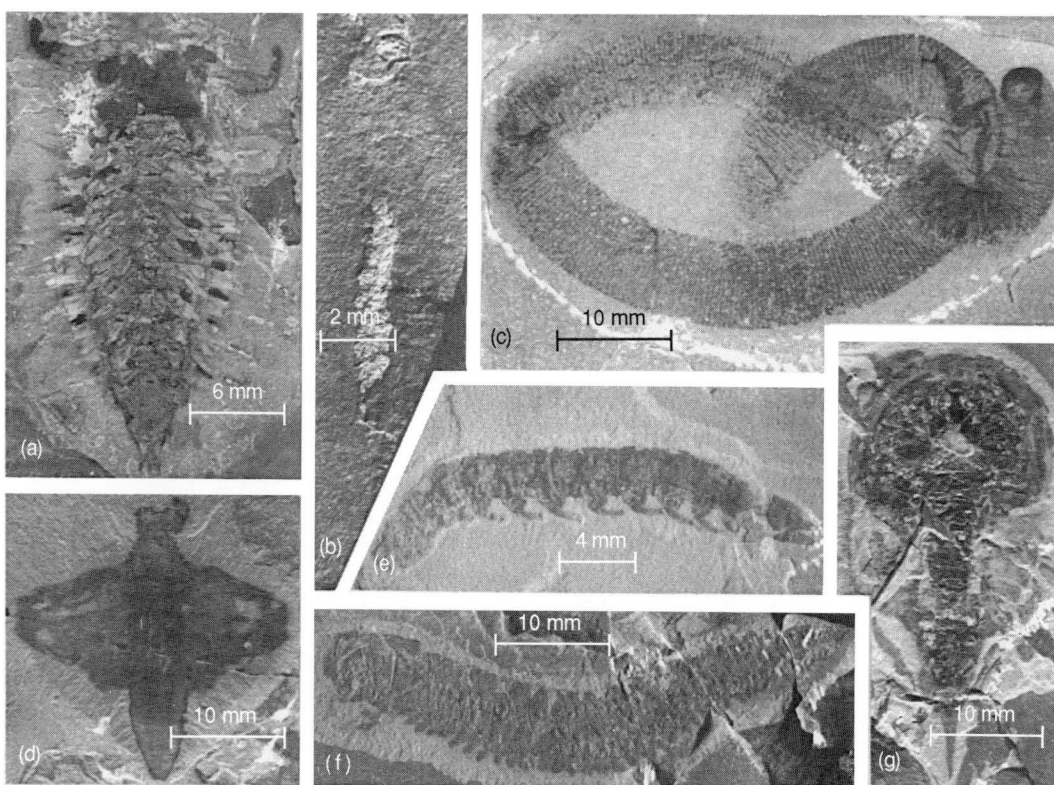

Representative soft-bodied and lightly mineralized animals from the Lower Silurian Waukesha biota. (*a*) ?Branchiopod crustacean, ventral view. (*b*) Conodont animal, dorsal view, showing apparatus anteriorly and faint transverse segmentation of body. (*c*) ?Leech. (*d*) Arthropod with trangular "valves," dorsal view. (*e*) Myrapodlike arthropod, left lateral view. (*f*) Wormlike arthropod, left lateral view. (*g*) Xiphosure, dorsal view. Question marks indicate tentative identification. (*From D. G. Mikulic, D. E. G. Briggs, and J. Kluessendorf, A Silurian soft-bodied biota, Science, 228:715–717, 1985*)

trilobite specimens are preserved with articulated dorsal exoskeletons, but all lack soft parts, appendages, and hypostomata. These factors suggest that the trilobite specimens are molted exoskeletons that have been transported and concentrated at the site.

Crustaceans. These are represented by ostracodes ("seed shrimp"), phyllocarids (primitive malacostracans), and other organisms that are assigned tentatively to this group. The assigned organisms include a large bivalved animal belonging to the order Concavicarida. This order previously was known to range from the Early Devonian to the Jurassic; therefore, the Waukesha discovery extends its range about 40 million years. These arthropods possess a lightly mineralized bivalved carapace that covers the body and most of the appendages. Large protruding eyes and massive spiny grasping appendages below the head suggest that this organism was an efficient predator.

One of the most characteristic of the Waukesha arthropods displays similarities to both branchiopod and remipede crustaceans. The earliest verifiable branchiopod in the fossil record is from Lower Devonian rocks in Scotland. Remipedes, a recently discovered extant group, are unknown as fossils. Large compound eyes and a pair of large, possibly grasping, head appendages suggest that this Waukesha arthropod also was a predator.

Chelicerates. At Waukesha, chelicerates are represented by the earliest well-preserved xiphosure, a distant relative of the modern horseshoe crab or king crab. This is only the second known Paleozoic xiphosure with preserved appendages and the only one to bear evidence of chelae. Eurypterids, the characteristic chelicerates in younger Silurian rocks elsewhere, have not been found at Waukesha.

Uniramians. A group which includes millipedes, centipedes, and insects, uniramians may be represented by a myriapodlike arthropod in the Waukesha biota. Although all modern myriapods live in terrestrial environments, they are thought to have been derived from aquatic ancestors in the Middle Paleozoic; therefore, the Waukesha arthropod may have evolutionary significance as the earliest marine myriapod.

Other forms. Several of the Waukesha arthropods cannot yet be assigned to any of the four major arthropod groups. The most common of these problematic forms has an elongate wormlike segmented body with a head shield or carapace at one end, in front of which is a large compound eye. Unfortunately, the morphologic characteristics displayed by this organism are considered primitive for many arthropod groups and thus are not useful for determining taxonomic affinities.

The most enigmatic of the Waukesha animals appears to be an arthropod. It possesses a short segmented body which is covered partially by a pair of triangular or semicircular structures (or bivalved carapace) that extend laterally beyond the trunk, giving the animal a spread-wing appearance. Most of the Waukesha arthropods seldom exceed 2 in. (5 cm) in length; however, some disarticulated valves of this organism have been found that alone are as much as 8 in. (20 cm) wide.

Worms. Annelid worms are common members of the Waukesha biota. Papillate annelids are most numerous and best preserved. None has been found with jaws, and the guts appear to be filled with sediment. This suggests that these annelids may have lived within the soft sea-floor sediment, where they swallowed mud and extracted nutrients as it passed through their guts. No soft parts of polychaete worms have been found at Waukesha yet, but this group is represented by their microscopic jaw elements (scolecodonts). The rarest of the Waukesha worms is a large annulate form which bears a circular structure at one of its tapering ends. This circular structure resembles the sucker disk of a leech. If this Waukesha worm is a leech, it would extend the range of the class Hirudinea back nearly 300 million years, from the Upper Jurassic. This worm may have preyed on small marine invertebrates, as do modern nonparasitic leeches.

Conodont animal. A single specimen of a conodont animal (*Panderodus*) has been uncovered at Waukesha. Although poorly preserved, this specimen is very important because it is only the second known occurrence in the fossil record of preservation of a conodont animal's soft body. The only other specimens, which are similar in appearance to the Waukesha find, are from younger Lower Carboniferous rocks of Scotland. The Waukesha conodont animal possesses an elongate segmented body, which is incompletely preserved, and an apparatus of 11 conodont elements (microscopic phosphatic dentate structures). Conodonts are known primarily from these fossilized elements, which are common in Cambrian to Triassic marine rocks throughout the world. Because preservation of their soft parts is so rare and the function of the elements is poorly understood, the taxonomic affinities and lifestyle of the conodont animal are uncertain. Most commonly, conodonts have been assigned to the chordates (the group that includes sea squirts, lancelets, and vertebrates) or to the chaetognaths (arrowworms); however, at present they are assigned to the separate phylum Conodonta.

Other organisms. Graptolites and conulariids are locally common at Waukesha, but with the exception of the arthropods and worms most other organisms are quite rare. Algae, possible hydroids, monaxonid sponges, nautiloid cephalopods, and brachiopods are among these rarer forms. In addition to conodont elements, the microfauna contains common agglutinated foraminiferans.

Preservation. Although preservation of the soft-bodied organisms is most noticeable, another unusual feature of the Waukesha biota is the near absence of hard-shelled nonmobile organisms that lived on the sea floor (sessile benthos). These organisms, such as corals, brachiopods, mollusks, and echinoderms, normally dominated the typical Silurian marine fauna. At Waukesha, dendroid graptolites and conulariids are the only common sessile benthos.

The organisms with calcareous shells, such as trilobites, have undergone decalcification at Waukesha and, consequently, are poorly preserved. The soft-bodied and lightly mineralized organisms either are preserved as a thin organic film or have been infilled with diagenetic fluorapatite, which may yield exquisite preservation.

Setting. The Waukesha biota is very restricted in distribution, both areally and stratigraphically. It is found at only one localized site in laminated carbonate rocks. The rhythmic bedding of these laminae suggests cyclicity of deposition, which may be due to tidal or seasonal variations. These laminated rocks occur at the base of the Brandon Bridge beds along the foot of a 24-ft-high (8-m) scarp, which may have been a submarine limestone cliff when these sediments were being deposited. Distribution and preservation of the biota appear to have been controlled by the environment adjacent to this scarp.

Preservation of undisturbed laminae suggests that quiet conditions prevailed on the sea floor during and after deposition. The presence of laminae also indicates the lack of disruption of the sedimentary layers (bioturbation) by organisms moving on or through the sediment. Absence of bioturbation by mobile organisms and the absence of sessile benthos suggest unfavorable environmental conditions at the sediment-water interface, perhaps a lack of oxygen.

Anoxic conditions also could explain the preservation of soft-bodied organisms because the lack of oxygen would have excluded aerobic bacteria and scavenging organisms which would have destroyed any organic remains. The Waukesha biota probably was transported a short distance to the site of deposition. Gentle current transport is implied because almost all of the organisms are lightweight, free-living forms, most of which are in an articulated, unabraded condition.

Although study of this biota is still in its early stages, the biologic and sedimentologic features suggest that the soft-bodied and lightly sclerotized organisms preserved at Waukesha were

more common and widespread than is implied by
the scarcity of their fossils, with the explanation
being their poor fossilization potential. Unique
environmental conditions at Waukesha were responsible for their preservation. The Waukesha
organisms are not thought to represent a unique
biota that was preserved under normal depositional conditions within its habitat.

For background information *see* ANNELIDA; AR-
THROPODA; CONODONT; SILURIAN in the McGraw-
Hill Encyclopedia of Science and Technology.

Joanne Kluessendorf

Bibliography. D. G. Mikulic, D. E. G. Briggs,
and J. Kluessendorf, A new exceptionally preserved biota from the Lower Silurian of Wisconsin, U.S.A., *Philos. Trans. Roy. Soc. London*,
B311:75–85, 1985; D. G. Mikulic, D. E. G.
Briggs, and J. Kluessendorf, A Silurian soft-
bodied biota, *Science*, 228:715–717, 1985.

Soil

A great deal of attention is given to soil compaction in modern soil research. This article
discusses the sources and mechanisms of soil
compaction, and research involving compaction alleviation.

SOURCES AND MECHANISMS

Soil compaction may result from natural causes, animal or human traffic, or high-intensity rain
or irrigation, but the problem confronting many
farmers worldwide is deep soil compaction resulting from mechanization. A vast array of mechanical equipment is paraded across farms, and
with few exceptions it is supported by tires.
When farming was done with animals, the soil
was compacted by hooves, usually only to depths
within the plow layer. Mechanization brought
steel wheels and tracks, then rubber tires; as
machinery grew larger, so did the tires. Traffic-
induced soil compaction has become a major
problem. Research has shown that (1) tire and
track footprints that are long and narrow in the
direction of travel give better traction and less
soil compaction than a contact surface with the
same area but a different shape; (2) deep compaction (subsoil), which is very expensive to
remove, is caused largely by the total load carried
on the axle, and is little influenced by the use of
dual tires or large flotation tires; and (3) most
compaction in tilled soils takes place on the first
pass of a tire.

Extensive research shows that tires need firm,
dry soil for effective and efficient usage, while
crops generally need moist, friable soil for good
growth. Thus the nearly opposite needs of crops
and tires must be satisfied in the same field. To
this end, a concept called controlled traffic has
been introduced which zones fields into rootbeds
and roadbeds. Permanent traffic lanes (roadbeds)

result in firm, dry surfaces which improve traction, flotation, and timeliness of operations. Un-
trafficked crop zones (rootbeds), once deep-
tilled, tend to stay uncompacted with high
infiltration rates, and store water for crops rather
than produce runoff and erosion.

History of compaction problem. Soil compaction
by tires of heavy machinery became a problem in
the Southeast in the 1950s. Complaints spread
through the Mississippi Delta and the Southern
Plains into California by the early 1960s. In the
1970s and 1980s the Corn Belt and Lake States
became victims of soil compaction.

The initial concern was that soil compaction
in the crop zone was reducing yields. In the
1950s many farmers and researchers tried to
relieve the problem by subsoiling (deep tillage).
Some years on some soils, subsoiling worked
well. In other years it was worthless in the
same fields. Many farmers gave up in frustra-
tion, as did some researchers.

Much of the early subsoiling research was
done by people with a poor understanding of
plant physiology. The weather continuously interacts with human attempts to determine cause-
and-effect relationships in soil compaction. A
plant responds as a living system, and its development is influenced by its ability to obtain
sufficient and timely physiological requirements
from its environment. It is quite possible to
produce good plant yields in soils that are in poor
physical condition. If compaction severely limits
the flow of water and air and thwarts extensive
root growth, a plant may still do well if its
physiological needs are met. Nature might meet
these needs by providing rain daily during critical
growth periods 1 year in 5, even though average
conditions call for rain at 7-day intervals.

Problems such as poor stands and irregular
growth of plants, wet and cold soil, and ineffective tile drainage are often recognized as symptoms of soil compaction. High-bulk density
causes mechanical impedance of root growth and
excessive energy use for tillage; however, the
continuity of pore space needed by plants for
exchange of air and water is probably a more
important indicator of soil compaction.

Compaction management. Uncompacted, moist
soils suitable for good crop growth are generally
too soft to support modern farm machinery, and
the excessive sinkage of the wheels reduces fuel
efficiency. Conversely, firm dry soils suitable for
efficient machinery operations are not conducive
to good crop growth. Road-building contractors
till soil intensely and then compact it, sometimes
with pneumatic tires. Many crop production systems follow the same sequence. A crop production system employing permanent traffic lanes
has great potential for improving both the crop
zone and the traffic lanes.

Soil compacts easily when its structure is
weakened, as by tillage, or when it becomes wet.

Fig. 1. Spanner used for controlled-traffic research.

Tillage destroys the structure, and traffic following tillage quickly recompacts the soil. A cycle of tillage-traffic-tillage-traffic has developed. The deeper the tillage is done to remove compaction, the deeper the next wheel compacts the weakened soil. In general, this cycle requires slightly deeper tillage each year to remove subsoil compaction.

Deep tillage used annually to correct subsoil compaction is very expensive and treats only the symptoms, not the basic problem. Management, not elimination, of compaction is needed: intensive compaction for traffic lanes, very little compaction for root zones, and intermediate-level compaction for seedbeds. Control of traffic-induced soil compaction canbe done with conventional tractors, but some researchers use spanners such as the one shown in **Fig. 1** for definitive experiments.

Soil-machine relationships. To fully appreciate the controlled-traffic concept for soil compaction management, there are certain soil-tire, or soil-machine, relationships that need to be understood. If these relationships are used as guides,

there are many ways to approach and implement the concept.

One basic soil-machine relationship is the effect of the shape of the traction device footprint on compaction. An experiment was conducted with a pneumatic track and a pneumatic tire with equal footprint areas; however, the track was twice as long and half as wide as the tire (**Fig. 2**). These devices were operated at equal loads and inflation pressures in the same soil conditions. The shape of the footprint was the only difference. The track was far superior in both drawbar pull (pull ratio) and tractive efficiency. The peak vertical pressures, measured 8 in. (20 cm) beneath the center line, were almost twice as high under the tire as they were under the track. Soil bulk densities measured at 0–2-in. (0–5-cm) depths beneath each device were significantly greater under the tire. This research confirmed that the long, narrow traction device footprint desired for controlled-traffic systems to conserve cropping areas also gives better traction and higher tractive efficiency while reducing both the volume and intensity of soil compaction.

Another basic soil-machine relationship investigated was the effect of total load on the axle as distinguished from average soil pressure beneath the tire. Two tractor tires of different size and load capacity were operated in a homogeneous soil condition. When soil pressure at the surface was equalized under the two tires, pressure readings at 7-, 12-, and 20-in. (18-, 30-, and 50-cm) depths were consistently higher under the large tire. While the pressure in the upper layers of soil is determined by the specific pressure at the surface, which depends upon inflation pressure and soil deformation, this research indicates that the pressure in deeper soil layers is influenced by the total amount of the load.

A third basic soil-machine relationship is the multipass effect of tires on unstructured soil. A pneumatic tire was operated for four passes in the same rut on several freshly tilled soils. The drawbar pull (pull ratio) and the tractive efficiency were significantly lower on the first pass than on subsequent passes, indicating a great loss of power. Examination of soil properties revealed that 75% of the total change in bulk density and 90% of the soil sinkage occurred in the first pass. For practical field systems, this work demonstrates the adverse effects of random wheel traffic on freshly tilled soils. The first pass is the critical one, so repeated passes should be run in the same rut.

System compatibility. In addition to increased yields from the crop zone and improved mobility from permanent traffic lanes, there is another aspect of the controlled-traffic concept worthy of attention: the enhanced compatibility with automated and nonrandom operations. It is likely that fields will be mapped in the near future for several functions, including yields, nutritional

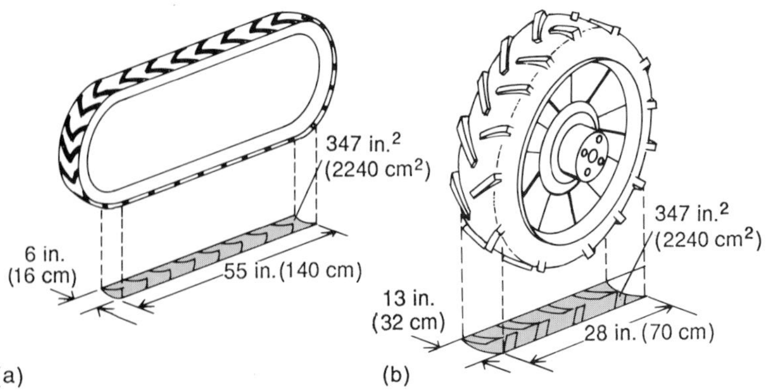

Fig. 2. Shapes of traction device footprints. (a) Pneumatic track, which has long, narrow shapes that improve drawbar pull and tractive efficiency while reducing soil compaction. (b Pneumatic tire, which has short, wide footprints that give more flotation.

needs, weeds and pests, and soil moisture. Such mapping requires more exact knowlege of the location of any specific point in the field than can be obtained with random operations. Also, trickle irrigation lines have already been buried in fields by cotton farmers as well as nursery operators. Burial of any such objects or erection of permanent stakes or wires for crop support dictates the dropping of all random operations for the time that such structures are in place. Great savings of human labor can be realized by placement of such objects for longer periods of time.

In the beginning, researchers desired only to eliminate traffic-induced compaction from the crop zone; with the introduction of basic traction and mobility research data, management, not elimination, of compaction became the objective. Improved mobility on compacted traffic lanes is expected to significantly enhance timeliness of operations, and the change from random field culture to fixed patterns lays the foundation for the future industrialization of agriculture.   *James H. Taylor*

COMPACTION ALLEVIATION

Soil compaction in cropland resulting from activities of humans and, to some extent, animals can be minimized by restricting field activities to periods when the soil is least compactable, and by using equipment and practices that result in relatively little compaction. Soils are least compactable when dry; hence, performing operations when soils are as dry as practical minimizes compaction. Other techniques for minimizing compaction include: (1) reducing the number of field trips by combining operations and by using reduced tillage farming systems; (2) using lighter equipment, as with conservation tillage systems; (3) reducing pressures on the surface by using more or wider tires and inflating them to pressures as low as practical; (4) using lighter tractors that transmit power to implements by mechanical, hydraulic, or electrical means, rather than heavier tractors that are needed for pulling some high-draft implements; and (5) limiting loads on such equipment as pesticide and fertilizer applicators, harvesters, and produce-hauling equipment.

While compaction can be minimized by use of such techniques, major compaction may be unavoidable under some conditions. In such cases, it can be managed by restricting field traffic to specified paths, which may result in severe compaction in the traffic zone but minimal compaction between the traffic zones.

When compaction is severe enough to affect plant growth and yield adversely, alleviation of the compaction generally is desirable. Methods of alleviating compaction depend on such factors as depth and degree of compaction, soil type, equipment available, and crop to be grown.

Mechanical means. Depth of compaction depends on the nature of compactive forces and soil conditions at the time the forces are applied. The type of operation most suitable to alleviate compaction depends on soil type and condition. On any soil the operation should adequately loosen the compacted zone. Operations that alleviate compaction in the tillage zone are termed primary tillage. A moldboard plow is the most effective implement for soil loosening in the 0–12-in. (0–30-cm) depth range. Other primary tillage implements for relatively shallow soil loosening are chisel plows, cultivators, disks, harrows, and sweep plows. These implements are effective on coarse- and fine-textured soils. In-row subsoiling or ripping to open a slit in compacted subsoil layers of coarse-textured soils has received much attention in the Southeast. It is performed with machines that loosen the compacted layers below the depth of primary tillage. Roots of crops planted over the slit readily penetrate the loosened soil; thus, plants can extract water and nutrients from deeper soil. Yields of nonirrigated crops were increased from 20 to 57% in several studies in the Southeast because of in-row subsoiling. Increases generally were much lower under irrigated conditions.

In contrast to producing narrow slits through compacted subsoils with in-row subsoiling in coarse-textured soils, loosening a greater soil volume often is the goal on medium- and fine-textured soils. Moldboard and chisel plows are effective for depths to about 12 in. (30 cm). For deeper loosening, subsoilers or large moldboard or disk plows have been used.

Subsoiling has been practiced for many years. Subsoiling should be performed when the soils are dry to achieve the most effective loosening, but subsoiling at water contents suitable for good plowing has resulted in soil shattering similar to that achieved at much lower soil-water contents. Based on reported results, subsoiling depth should be about 1.5 times the expected depth of primary tillage in order for the subsoiler tine to develop enough upward force to cause shear failure in the trench above the tine.

The importance of depth of operation also was demonstrated by research which showed that there is a critical working depth for all rigid subsoiler tines. At greater depths, tines do not achieve sufficient lifting force to cause shear failure in the trench, and the subsoiler forms only a narrow slot. Improved loosening is achieved by using shallow chisels to loosen the surface soil layers ahead of the subsoiler or by using a closer tine spacing. These results suggest that several passes at progressively greater depths may be required to obtain deep loosening with subsoilers.

A primary tillage implement that has received considerable attention in recent years is the Paraplow, which is a slant-legged implement that loosens the soil through a lifting action. In one study, water infiltration was higher, soil bulk density tended to be lower, but corn yields were lower with the Paraplow treatment than with moldboard plowing. Almost as much residue was retained on the surface with the Paraplow treatment as with a no-tillage treatment; therefore, such plowing offers good protection against erosion. In other studies, yields were identical with moldboard and Paraplow treatments, but were slightly higher than with a no-tillage treatment in both cases. The Paraplow treatment, as compared with no-tillage, resulted in an improved soil physical condition as indicated by higher water-infiltration rates, lower penetrometer resistances, and greater root development. Despite the improved soil conditions in the above studies, the Paraplow treatment failed to increase yields above those with moldboard plowing and only slightly above those with no-tillage. This suggests that soils on which the studies were conducted did not have a mechanical impedance that restricted plant growth and development. On more restrictive soils a yield advantage over no-tillage may have occurred with the Paraplow treatment.

Natural means. Besides mechanical means, compaction may be alleviated by natural means, such as freezing and thawing, shrinking and swelling, root action of certain crops, and earthworm and other soil fauna activity. Freezing and thawing play a major role in alleviating compaction, but they alone may not fully alleviate it, are effective only in limited climatic zones, and are highly variable from year to year.

On soils containing high amounts of montmorillonitic and illitic clays, shrinking and swelling associated with soil drying and wetting may play a major role in alleviating compaction. As these soils dry, cracks develop that disrupt the compacted zone. Subsequent compaction may be reduced because of roots that grow through the compacted zone via the cracks.

Some plants have greater potential than others for penetrating dense soil layers. These include legumes such as alfalfa, sweet clover, and guar, which have resulted in better yields of subsequent crops. The reasons for better yields are not clear because factors such as increased root penetration, nitrogen supply, aeration, and water infiltration may be involved. However, nonleguminous bahia grass roots effectively penetrate the dense layer of sandy soils that impede cotton roots. Cotton planted where bahia grass was plowed under with a moldboard 3 years before still yielded more than where the soil was chiseled to a 14-in. (35-cm) depth.

On soils containing sufficient organic matter, earthworm activity may be adequate to overcome the adverse effects of compaction. Earthworm activity generally is greater in no-tilled than in conventionally tilled soil. Undoubtedly, other soil fauna could aid in alleviating compaction if conditions were otherwise favorable for their activity.

For background information SEE SOIL in the McGraw-Hill Encyclopedia of Science and Technology.

Paul W. Unger

Bibliography. C. B. Elkins, R. L. Haaland, and C. S. Hoveland, Grass roots as a tool for penetrating soil hardpans and increasing crop yields. *Proceedings of the 34th Southern Pasture and Forage Crops Improvement Conference*, Auburn, Alabama, pp. 21–26, 1977; D. C. Erbach et al., *Soil Conditions and Corn Growth Response to Paraplowing*, Amer. Soc. Agr. Eng. Pap. 84–1013, 1984; W. Soehne, Distribution of pressure in the soil and soil deformation under tractor tires, *Grundlagen der Landetechnik, Heft*, 4:49–63, 1953; G. Spoor and R. J. Godwin, An experimental investigation into the deep loosening of soil by rigid tines, *J. Agr. Eng. Res.*, 23:243–258, 1978; J. H. Taylor, Benefits of permanent traffic lanes in a controlled traffic crop production system, *Soil Tillage Res.*, 3:385–395, 1983; J. H. Taylor and W. R. Gill, Soil compaction: State-of-the-art report, *J. Terramech.*, 21:195–213, 1985; W. B. Voorhees, Relative effectiveness of tillage and natural forces in alleviating wheel-induced soil compaction, *Soil Sci. Soc. Amer. J.*, 47:129–133, 1983.

Soil microbiology

The intentional application of low-input technology in farming defines the origin of agriculture. Low-input farming implies subsistence-level farming practices. A subsistence farming operation can be characterized as involving a low-acreage farm (usually infertile), minimal working capital, and labor-intensive (low-technology) methods.

Originally, all agriculture was low-technology in nature. As populations increased and concentrated in cities, the need for mass production of food evolved. Science and industry responded by providing the high-technology methods required to elevate farming gradually from a family-support operation to a profitable business. This has been a natural and necessary evolution in farming practice throughout the world. However, this evolution has proceeded at dramatically different rates in different countries, depending on such factors as soil, water, climate, natural resources, and political and cultural considerations. In the developed countries, it can be generalized that the many are fed by the few. However, even in many of these countries, as in developing countries, subsistence farming is still predominant and high-technology methods are inappropriate or unavailable. Countless numbers of subsistence farmers are finding that their traditional practice of agriculture is no longer a satisfying or successful means of family support. This results in abandonment of small farms and large-scale migration of unskilled labor to urban centers which are not prepared to assimilate them. The resulting socioeconomic effects are ultimately devastating. The pressures of civilization combine to present subsistence farmers with problems which can no longer be solved by individuals using traditional low-technology farming methods. These methods may be more appropriately characterized as sustainable, low-input "management practices" such as mulching, composting, cover crops, green manuring, and conservation tillage.

Management practices. These involve the management of soil organic-matter content and the presence and activity of various indigenous soil microorganisms. The observed beneficial (or harmful) effect of soil microorganisms on crop production is often related to provision and availability of organic matter (plant residues) as a soil amendment. Organic matter favors moisture retention, provides and retains mineral nutrients, and provides a food source for a great variety of microorganisms. In that sense, subsistence farmers have traditionally used low-input (low-technology) practices based on established concepts, principles, and knowledge of soil microbes and their activities in the soil. By these means, the farmers unwittingly manipulate the microbes indirectly so that crop production is benefited.

Two situations provide examples in which an existing crop-production problem might feasibly be resolved by manipulation of populations of soil microbes. In one instance, potentially beneficial microbes (microbial processes) are either absent from the soil or present at an insignificant level. In the other instance, microbes are present which are detrimental and need to be reduced or eliminated. In either case, a microbial process or population is manipulated by management of a soil component directly related to survival, growth, and activity of the microbes. If potentially beneficial microbes are absent, which is a rare situation, or are present at an insignificant level, then the system may be manipulated by addition of the microbe to the system (inoculation) in the former case or by the addition of a food source (organic matter) in the latter case.

Soil factors. In seeking to solve such problems by manipulating, that is, increasing or decreasing, microbial populations, it is important to bear in mind that the kinds and numbers of microbes present in the soil are ultimately determined by soil factors. The microbes do not select their environment; rather, the environment selects the microbes which will establish and function. The environment will also determine the population size of the microbes and the expression of various physiological activities. It follows that the microbes may often be most successfully controlled by manipulating their environment in specific ways. This approach may or may not be feasible in practice.

If an undesired microbe is present at a level detrimental to crop production, it can be inferred that the environment is unduly favorable. A possible approach to this type of problem is to change the soil environment without changing soil properties in a way that would result in the crop production system being adversely affected. For example, the incidence of a plant pathogen (severity of disease) may be significantly reduced by altering a soil property, such as soil acidity, in order to create an environment that is inimical to pathogen (disease) development, yet is still conducive to crop growth.

Changing other soil environmental factors, such as oxygen availability, can be effective in altering soil microbe populations. A brief period of flooding prior to planting may markedly reduce numbers of undesirable microbes which require oxygen for growth.

Sanitation. A facet of microbial technology which is often applicable to sustained low-input farming is sanitation. This refers to management practices aimed specifically at minimizing the populations (effects) of plant pathogenic microorganisms. In cases where such organisms survive in diseased crop residues from one growing season to the next, it is advantageous to burn the residues at harvest. This destroys a large part of the pathogenic organisms and thus reduces infection in the succeeding crop. Mineral nutrients are retained but vital organic matter is lost.

Addition of organic matter to soil has another beneficial effect related to the activity of soil microorganisms. This may be described as biological buffering. A soil which is biologically well buffered supports a large and highly diverse community of microorganisms, the components of which are kept in balance by common, naturally occurring mechanisms of antagonism (for example, antibioses, lysis, or competition). Addition of organic matter may also provide a preferred food source for some potential or opportunistic plant pathogens, thus reducing crop infection. This, in essence, is a method of manipulating a microbial population by changing the biological environment. This practice has the advantage that organic matter, a vital and often limiting component of the soil, is returned to the soil rather than removed from it.

Plant pathogenic populations may also be manipulated by such practices as crop rotation or fallowing. In such cases the numbers of detrimental microorganisms may be reduced by natural (biological) means by forcing them to survive in a hostile soil environment over a prolonged period of time in the absence of a host plant (crop) required for reproduction.

Soil inoculation. The preceding discussion has considered ways in which populations of soil microorganisms may be manipulated indirectly by altering the soil environment in various ways (for example, food source, pH, or moisture). These unsophisticated methods use locally available, inexpensive resources (materials and labor) so that soil-microbe processes are intentionally influenced in ways which benefit plant growth (crop production). The extent to which these low-technology farming practices can be followed by subsistence farmers is highly variable and may be severely limited by specialized local conditions of plant growth.

There are also direct methods of manipulating soil-microbe populations of the soil–plant environment, such as by inoculation with a desired organism. This approach represents an alteration in the biological environment of the plant and may be employed when a desired (beneficial) microbe is absent from the soil or is present in only very low numbers.

The efficacy of soil inoculation, in concept, has gained widespread acceptance. This can be attributed largely to the historic success of the inoculation of legumes with symbiotic, nitrogen-fixing bacteria (*Rhizobium*). When this inoculation is practiced on nitrogen-deficient soils, otherwise suitable for growth of the legume, the resulting plant–bacteria association customarily fixes significant amounts of atmospheric nitrogen, which is then utilized for growth by the legume host. This technology is widely used in large-scale farming (for example, soybean, or peanuts) in developed countries, but it is largely unknown, unavailable, or ineffectively used as a low-input technology by subsistence farmers in developing countries.

Inoculation may also be used as an approach to reducing a detrimental microbial population. In this way, biological control of the pathogen is achieved by activities of an introduced (inoculated) specific microbe which is antagonistic to the pathogen. This approach is rapidly gaining credibility, but a disadvantage is that it fails in the field a majority of the time. In practice, it requires extensive knowledge of the basic biology of an interaction between microbes, complex methodology, and strictly defined management practices, which are not characteristically available in subsistence farming systems. In most cases, failures in attempts to apply biological control result from insufficient knowledge of the system or inability to control the environment to the extent required for the system to work.

Blue-green bacteria. Current studies of the potential benefit of inoculating with blue-green bacteria to enhance nitrogen fixation are yielding promising results under rather specific aquatic conditions. A similar situation exists in attempts to enhance nonsymbiotic nitrogen fixation in the root zone of grasses (cereals) by inoculation with root-associated, free-living soil bacteria. Observed yield increases may be the result of enhanced root growth and nutrient uptake resulting from the activity of plant growth–promoting substances produced by the bacteria. However, knowledge of the functioning and management of the systems is incomplete, and practical application of this system is not yet possible as a simple, inexpensive standard procedure. On the other hand, enhancement of nitrogen fixation by inoculation of certain woody nonlegumes (*Alnus*) with the nitrogen-fixing actinomycete *Frankia* is now a commercially and agronomically feasible practice.

Fungus–root association. Certain fungi infect plant roots and form an intimate fungus–root (mycorrhizal) association which functions symbiotically. The plant provides the fungus with food, and the fungus provides needed essential mineral nutrients (mainly phosphorus) to the plant. Ectomycorrhizal fungi are easily grown on a large scale and are used successfully as inoculants to enhance survival and growth of commercially grown seedlings of woody plants (pine). Endomycorrhizal fungi infect virtually all agronomic crops, and their beneficial role in plant nutrition has been firmly established in limited field trials. However, potential large-scale application of inoculation awaits development of means for producing inoculant in large quantity.

Other research. Additional innovative research involving the concept of inoculation includes (1) reducing frost injury in crops by spraying with ice-nucleating *Pseudomonas* bacteria, (2) increasing soil phosphorus availability to plants by inoculating with phosphate-solubilizing bacteria, (3) increasing availability of essential iron to plants by inoculation with microbes producing siderophores (iron-binding organic molecules), (4) reducing plant disease by inoculation with microbes obtained from so-called disease-suppressive soils, and (5) removing environmental pollutants by inoculation with microbes having specific biodegradation capabilities.

It can be seen from the given examples that soil microbes associate with plants in various ways, at different levels of intimacy (contact), and with different results, such as symbiosis or disease. Characteristics which can enhance plant growth, usually under special conditions, include nitrogen fixation and production of plant growth–promoting substances. Microorganisms having these capabilities are readily isolated from soil and are of great importance as potential inoculants. However, the practice of inoculation frequently fails, even under seemingly optimum conditions.

Failure of a microbe to survive, establish, and function according to plan when inoculated into a new soil environment can often be attributed to the inherently inhospitable nature of the receiving soil environment. The inoculated microbe will soon die unless the soil environment is altered to provide a suitable setting. As stated earlier, the environment selects the microorganism.

Plant genetics. This discipline provides a highly promising alternative to inoculation as a means of enhancing beneficial plant growth effects that result from activities of soil microbes. This involves the development, perhaps by genetic engineering, of plant varieties having genetic information which determines plant characteristics conducive to establishment of native soil microbes with ability to enhance plant growth by some mechanism. *See* Ecological interactions; Fertilizer.

For background information *see* Mycorrhizae; Soil microbiology in the McGraw-Hill Encyclopedia of Science and Technology.

David H. Hubbell

Bibliography. J. M. Lynch, *Soil Biotechnology*, 1983; W. R. Oschwald (ed.), *Crop Residue Management Systems*, Amer. Soc. Agron. Spec. Publ. 31, 1978; B. A. Stewart (ed.), *Soil Conditioners*, Soil Sci. Soc. Amer. Spec. Publ. 7, 1975; N. R. Usherwood (ed.), *Transferring Technology for Small-Scale Farming*, Amer. Soc. Agron. Spec. Publ. 41, 1981.

Soil nitrogen

The process by which nitrogen (N) is lost from the soil surface as ammonia (NH_3) gas is termed ammonia volatilization. Ammonia volatilization may occur when ammoniacal or ammonium-forming fertilizer is applied to the soil surface or when crop residues of high nitrogen content are returned to the soil surface without incorporation. These are common management practices with sod crops and in no-tillage crop production.

Until about 10 years ago, ammonia volatilization had been studied primarily under controlled laboratory conditions. More recently, however, NH_3 volatilization research has been conducted under natural field conditions. This has required development of methodology to measure NH_3 volatilization under field conditions. As a result, a much improved understanding of NH_3 volatilization and the factors which influence it in a dynamic field environment has been achieved.

Methodology and magnitude of NH_3 losses. The four techniques for field measurement of ammonia volatilization are forced-draft techniques; micrometeorological techniques; recovery of ^{15}N, a stable, naturally occurring nitrogen isotope; and plant response. The forced-draft and micrometeorological techniques involve direct measurement of volatilized NH_3, while the ^{15}N recovery and plant response methods are indirect methods.

Under controlled laboratory conditions, it is possible to have an optimum environment for maximum NH_3 losses. However, in a dynamic field environment the conditions necessary for maximum losses seldom, if ever, exist. For this reason, reported losses under field conditions are often less than those for the same or similar soil type under laboratory conditions. The magnitude of losses under laboratory conditions commonly ranges from 30 to 70% of the applied nitrogen, but can be as great as 90% for sandy soils with very low buffering capacity. Losses under field conditions generally are in the range of 10 to 50% of the applied nitrogen, but they can be as great as 60 to 70%. Since fertilizer nitrogen application rates are usually in the range of 100 to 300 kg/hectare (90 to 270 lb/acre), NH_3 volatilization represents a substantial nitrogen loss from the soil-plant system.

Soil factors influencing NH_3 volatilization. Factors in the soil that influence the volatilization of NH_3 include soil pH, soil hydrogen ion (H^+) buffering capacity and cation-exchange capacity, cation and anion effects, and urease activity.

Soil pH. When ammonium salts or ammonium-forming fertilizers are added to soil, ammonium ions (NH_4^+) and ammonia are in equilibrium in the soil solution according to reaction (1). The

$$NH_4^+ \rightleftharpoons NH_3 + H^+ \qquad (1)$$

activity of H^+ ions, or the pH, largely determines the ratio of NH_3 to NH_4^+. At pH 9.3, the ratio is about 1:1. At pH values greater than 7.0, significant NH_3 volatilization can occur, but as long as pH remains less than 7.0, NH_3 losses are generally insignificant.

Soil pH is subject to rapid change following fertilizer applications, resulting from microbial activity and from other chemical factors. A soil property more important than initial soil pH in determining NH_3 loss potential is the capacity of the soil to resist changes in pH, or its H^+ buffering capacity.

Soil H^+ buffering capacity and cation-exchange capacity. A soil's ability to resist an increase in pH is most closely related to its titratable acidity. Titratable acidity in soil generally results from exchangeable acidity, organic matter, and iron and aluminum oxides and hydroxides. Buffering capacity is well related to cation-exchange capacity.

The influence of soil buffering capacity on NH_3 volatilization is illustrated in **Fig. 1**. Such data indicate that the amount of buffering between the initial soil pH and a pH of about 7.5 would be more directly related to the expected NH_3 loss than to the initial soil pH.

Soil cation-exchange capacity is also important because it allows NH_4^+ ions to be removed from soil solution, thereby reducing the total amount of NH_3 subject to volatilization. Soils with high pH but also a high cation-exchange capacity actually may lose only a small amount of NH_3.

Cation and anion effects. The addition of calcium cation (Ca^{2+}) to urea-containing fertilizers has been shown to reduce NH_3 losses. The mech-

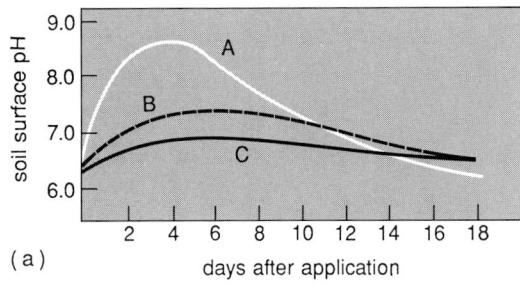

(a)

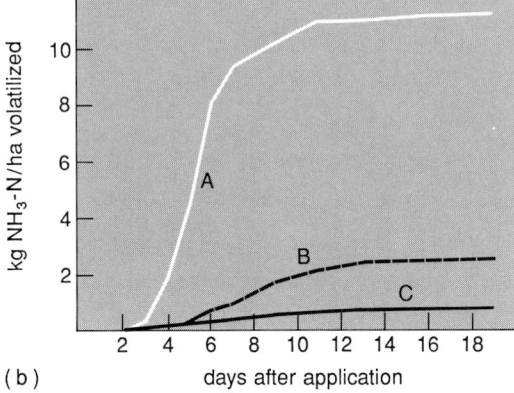

(b)

Fig. 1. Buffering capacity of three different soils. (*a*) Soil surface pH. (*b*) Cumulative NH₃-N loss. 1 kg/ha = 0.9 lb/acre. A = low buffering capacity; B = medium buffering capacity; C = high buffering capacity. (*After R. B. Ferguson et al., Ammonia volatilization from surface-applied urea: Effect of hydrogen ion buffering capacity, Soil Sci. Soc. Amer. J., 48:578–582, 1984*)

anisms by which this reduction occurs include one or more of the following: (1) precipitation of calcium carbonate ($CaCO_3$) in neutral and alkaline soils which stabilizes the NH_4^+ either as a soluble salt such as ammonium nitrate (NH_4NO_3) or ammonium chloride (NH_4Cl) or as an adsorbed cation on the soil solid phase; (2) reduction in pH in both calcareous and acid soils due to the salt effect, as soil pH is lower in salt solutions than in water and goes down as salt concentration increases because of the displacement of H^+ and Al^{3+} ions from exchange sites; and (3) reduction in urea hydrolysis rate. The effect of Ca^{2+} has been measured in both calcareous and noncalcareous soils. Also, the addition of other cations, such as potassium (K^+) and magnesium (Mg^{2+}), has resulted in reduced NH_3 losses, primarily by increasing the soil solution Ca^{2+} concentration via cation exchange.

Anion effects are particularly important in calcareous soils. The generalized reaction of an ammonium salt with $CaCO_3$ can be represented as reversible reaction (2), where X represents the

$$NH_4X + CaCO_3 + H_2O \rightleftharpoons$$
$$NH_4HCO_3 + CaX + OH^- \quad (2)$$

anion of the NH_4^+ salt. The reaction product, ammonium bicarbonate (NH_4HCO_3), is unstable

and decomposes, producing NH_3 and carbon dioxide (CO_2) gases. Ammonium compounds whose anions tend to form insoluble products with calcium generally result in greater NH_3 losses than those which form soluble products with calcium. This is because the insoluble products precipitate, driving the reaction to the right and increasing the amount of NH_4HCO_3 formed. Examples of anions which form sparingly soluble products include phosphate (HPO_4^{2-}) and sulfate (SO_4^{2-}), while examples of those which form soluble products include nitrate (NO_3^-) and chloride (Cl^-).

Urease activity. Urease is the enzyme which converts urea to NH_3 and CO_2. It is produced by both higher plants and microorganisms and occurs in soil both in living organisms and as free enzyme external to living organisms. Abundant urease and rapid urea hydrolysis results in an increased potential for NH_3 loss, since this generally results in increased pH and greatly increased concentrations of NH_4^+ ions. The following factors have been found to influence urease activity: (1) Soil organic matter content: as organic carbon increases, urease activity also increases. (2) Soil pH: the optimum soil pH for urease activity is between 7 and 9, with reduced activity at very acid (< 5.0) and very alkaline (> 9.0) pH values. (3) Soil temperature: generally, urease activity increases as temperature increases over the range of 50 to 149°F (10 to 65°C) and decreases above 149°F (65°C). (4) Soil water content: urease activity generally increases with increasing moisture content up to field capacity; at soil water potentials less than −15 atm (−15 × 10⁵ pascal), urease activity is considerably reduced; a negative effect at moisture contents above field capacity has also been reported. (5) Urea concentration: an increase in urease hydrolysis rates usually accompanies high rates of urea application.

Although urease activity varies considerably from soil to soil and under various climatic regimes, urease concentration seldom limits NH_3 volatilization under field conditions. Generally, environmental factors are more important than urease concentration in determining total NH_3 losses under field conditions.

Environmental factors influencing NH₃ losses. Environmental influences on the loss of NH_3 include temperature, soil water content, and air exchange.

Temperature. Under controlled conditions, NH_3 losses increase with increasing temperature over the range of 41 to 113°F (5 to 45°C). This is due to temperature effects on chemical and biochemical reactions, especially on urease activity (**Fig. 2**). However, the influence of temperature on NH_3 loss under field conditions is quite complex and is also closely related to soil water content.

Soil water content. The influence of soil water content has been most difficult for scientists to

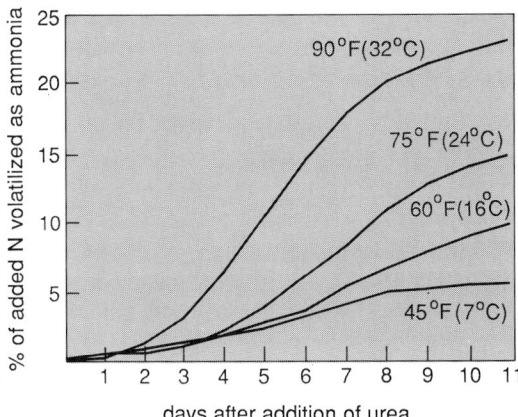

Fig. 2. Influence of temperature on cumulative NH_3 loss from addition of urea to Dickson silt loam. (*After J. W. Ernst and H. F. Massey, The effects of several factors on volatilization of ammonia formed from urea in the soil, Soil Sci. Soc. Amer. Proc., 24:87–90, 1960*)

quantify in a consistent manner under dynamic field conditions, especially where the soil water content fluctuates widely on a diurnal basis because of the combined effects of dew formation and evaporation. However, from research conducted under field conditions, several relationships have emerged. (1) For maximum NH_3 loss rates, the soil water content must be at or near field capacity at the time of fertilizer application. If the soil is dry, dissolution of dry fertilizer materials is very slow, and thus the biological and chemical reactions necessary for NH_3 volatilization are very slow or do not occur at all (see **table**). (2) On a diurnal basis, maximum NH_3 loss rates occur when the soil surface is drying. As the soil dries, the concentration of NH_4^+ and NH_3 in the soil solution increases, resulting in NH_3 being lost to the atmosphere to maintain equilibrium. (3) If the soil surface dries but is not rewetted by dew or light rainfall, NH_3 loss is reduced because of insufficient moisture for the necessary biological and chemical reactions. The exact moisture content or water potential at which this occurs is not well defined. (4) If rainfall is sufficient to

move the nitrogen source into the soil profile, NH_3 volatilization ceases. The exact amount of rainfall necessary to ensure no NH_3 loss depends on soil type, but for most soil types is 1 to 2 cm (0.4 to 0.8 in.).

Considering all of these relationships, the influence of soil water content can be summarized as follows: On a diurnal basis, high NH_3 loss rates are generally associated with periods of rapid soil drying as long as the soil is not too dry initially, but high total NH_3 losses are associated with an initially moist surface soil followed by several days with no or small rainfall events.

Air exchange. The importance of air exchange has been overlooked by many researchers, thereby compromising experimental results. In laboratory systems, it was found that relative NH_3 losses were maximum when the air exchange rate was at least 10 chamber volumes per minute (the air above the soil was completely replaced ten times per minute).

Under field conditions, NH_3 loss has been shown to increase linearly with wind speed and the partial pressure of NH_3 in the floodwater of a rice field (**Fig. 3**).

Management factors influencing NH_3 losses. As discussed above, nitrogen source influences losses of NH_3. Generally, NH_3 losses from both NH_4^+ salts and urea increase with increasing nitrogen rate. This may be linear or exponential so that the relative loss (as percent of the applied nitrogen) may decrease, remain constant, or increase with increasing application rates.

Application method. Maximum NH_3 losses can occur when nitrogen fertilizers are broadcast on the soil surface. However, incorporation of nitro-

Effect of initial soil water content and rainfall regime on estimated NH_3 losses (based on ^{15}N recovery)

Initial soil water content, %	Estimated NH_3 loss, % of applied N
31.3	28.8
37.0	26.2
24.3	21.0
21.2	21.3
Rainfall regime	
High	12.1
Medium	22.3
Low	38.7

* After R. J. B. Bouwmeester, P. L. G. Vlek, and J. M. Stumpe, Effect of environmental factors on ammonia volatilization from a urea-fertilized soil, *Soil Sci. Soc. Amer. J.*, 49:376–381, 1985.

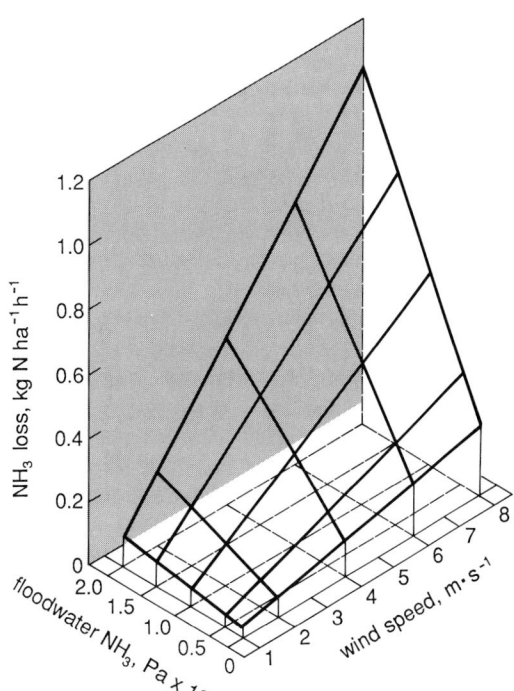

Fig. 3. Effect of wind speed and partial pressure of NH_3 in floodwater on ammonia flux. 1 kg/ha = 0.89 lb/acre; 1 Pa = 10^{-5} atm; 1 m · s^{-1} = 3.28 ft · s^{-1}. (*After I. R. P. Fillery, J. R. Simpson, and S. K. DeDatta, influence of field environment and fertilizer management on ammonia loss from flooded rice, Soil Sci. Soc. Amer. J., 48:914–920, 1984*)

gen fertilizers results in much reduced or no NH_3 loss. The depth to which incorporation is necessary depends on the soil type, but for most soils is from 5 to 10 cm (2 to 4 in.).

Fertilizer form is also important in determining NH_3 losses. When broadcast, liquid sources tend to lose more NH_3 than dry forms. Surface band application has sometimes resulted in less NH_3 loss from urea-based fertilizers than broadcast applications. The reason for this reduction is that by concentrating urea in a band it is exposed to less urease and thereby subject to slower hydrolysis. However, incorporation is necessary to ensure no NH_3 loss.

Crop residues. Ammonia volatilization is generally increased by crop residues present on the soil surface, which is often the case in sod-based or conservation crop production systems. Crop residues increase NH_3 losses in several ways: they form a physical barrier between the nitrogen source and the soil; they have a high urease activity; and they prevent the soil surface from completely drying. For these reasons, NH_3 losses are commonly less for a bare surface soil than for a soil surface with crop residues.

Fertilizer modification. Recently, several chemical methods have been explored to attempt to reduce the potential for NH_3 volatilization. These include acidification, salt addition, or urease inhibition. Acidification can be achieved by combining urea with sulfuric or phosphoric acid to make fertilizer products of low pH. These fertilizers reduce NH_3 losses by maintaining a low pH in the vicinity of the fertilizer granule. Calcium salts and magnesium salts, individually or combined, have been added to urea to form insoluble carbonates and thereby stabilizing NH_4^+ salts. Urease inhibition can be achieved by using one of several chemicals that inhibit the hydrolysis of urea via urease. Examples include phenyl mercuric acetate and phenylphosphorodiamidate. However, none of these methods have consistently proven to be successful under field conditions. See FERTILIZER; NITROGEN FIXATION.

For background information SEE AGRICULTURAL SOIL AND CROP PRACTICES; FERTILIZER; NITROGEN CYCLE; NITROGEN FIXATION; SOIL in the McGraw-Hill Encyclopedia of Science and Technology.

W. L. Hargrove

Bibliography. L. B. Fenn and L. R. Hossner, Ammonia volatilization from ammonium or ammonium-forming nitrogen fertilizers, *Adv. Soil Sci.*, 1:123–169, 1985; L. B. Fenn and D.E. Kissel, Ammonia volatilization from surface applications of ammonium compounds on calcareous soils, I. General theory, *Soil Sci. Soc. Amer. Proc.*, 37:855–859, 1973; W. L. Hargrove and D. E. Kissel, Ammonia volatilization from surface applications of urea in the field and laboratory, *Soil Sci. Soc. Amer. J.*, 43:359–363, 1979; D. E. Kissel, H. L. Brewer, and G. F. Arkin, Design and test of a field sampler for ammonia volatilization, *Soil Sci. Soc. Amer. J.*, 41:1133–1138, 1977.

Solar cell

The modern solar cell using high-purity semiconductor material was demonstrated in 1954, when the first device made of single-crystal silicon was able to convert 4% of the incident sunlight into electricity. By the early 1960s cell conversion efficiency had increased to over 10%, and by 1975 silicon cells with nearly 15% efficiency had been demonstrated. Since that time, advances in conversion efficiency have continued while the cost of solar cells has dropped. As progress continues, solar cells will probably become a significant source of electricity for terrestrial power applications. Solar cells have already become the preferred power source of many consumer products such as calculators and watches, and are being used increasingly for power applications such as refrigeration and water pumping in remote areas that are not readily accessible to the conventional electric power distribution system.

Wide acceptance of photovoltaics as a competitive power source depends on both increasing the conversion efficiency and reducing the cost of present solar cells. During 1975–1985, costs of photovoltaic devices dropped dramatically, partly because of the tremendous growth of the electronics and semiconductor industries. At the same time, solar-cell conversion efficiency continued to make steady progress. Recently, important research progress has been achieved in high-efficiency devices (**Fig. 1**). Laboratory solar cells fabricated from highly refined single-crystal silicon semiconductor material can now convert into electricity up to 22% of the sunlight that is incident on their active area. In concentrated sunlight (such as light that is focused on a cell by

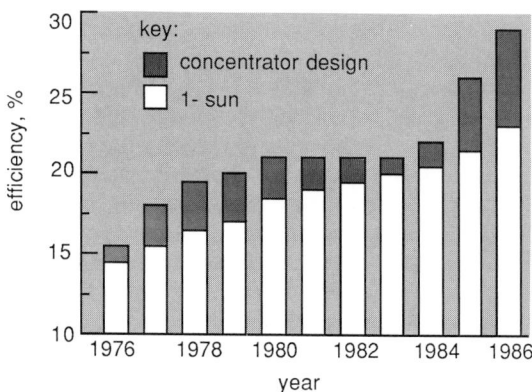

Fig. 1. Silicon solar-cell solar-to-electric conversion efficiency plotted for the years 1976–1986 for both 1-sun (without a concentrator) and concentrated-sunlight designs. Efficiency is given for peak conditions, that is, noon on a clear sunny day at sea level.

a lens) these silicon cells can convert almost 28% of the light to electricity. Commercially available solar cells are typically 15% efficient when operated in light having normal solar intensity (by definition, light with an intensity of 1 sun) and 20% efficient when used with concentrators.

Several alternative solar-cell approaches are being pursued to further reduce the cost of energy generated. One alternative is thin-film photovoltaic devices made of single-crystal, polycrystalline, or amorphous materials which can be fabricated by low-cost processes but suffer from low conversion efficiency. Another alternative requires using devices that are relatively expensive but have very high efficiency (near 30% or even greater). In this approach, it is sometimes advantageous to use inexpensive lenses to focus sunlight onto these cells. In either of the photovoltaic approaches, the use of multiple layers of different semiconductor materials to boost possible efficiencies to 20% for thin films and 40% for high-efficiency options appears to be promising and is a topic of current research.

Band gaps of semiconductors. The amount of energy required to generate electron-hole pairs in a semiconductor, known as the energy gap or band gap, is dependent on the electronic properties of the material. Many semiconductor materials can be considered for solar applications. The range includes germanium with a band gap of 0.7 eV, silicon with 1.1 eV, gallium arsenide with 1.4 eV, and cadmium sulfide with 2.4 eV. The value of the band gap of the semiconductor is important because photons with less than the material band-gap energy are not capable of generating electron-hole pairs, and photons with greater than the band-gap energy are only partially utilized because any excess energy beyond the material band gap is lost to the crystal as heat. Since the solar spectrum photon distribution is known, it is possible to determine the optimum semiconductor band gap for solar applications. For terrestrial use the optimum single band gap is near that of gallium arsenide at 1.4 eV. However, even in this optimum case over 55% of the energy in the solar spectrum cannot be used. Single-band-gap devices suffer from this fundamental limitation.

Multijunction cells. Theoretical analysis indicates that substantial increases in solar-cell conversion efficiency are possible through the use of multi-band-gap structures. Such structures are realized by using solar cells made up of two or more materials having band-gap energies carefully selected to match the solar spectrum so that each cell can be designed to be optimally efficient over a limited spectral range. One configuration receiving attention is the cascade approach, where solar cells of decreasing band gaps are stacked either monolithically or mechanically. The high-band-gap cell at the top of the stack absorbs all the photons with energies equal to or

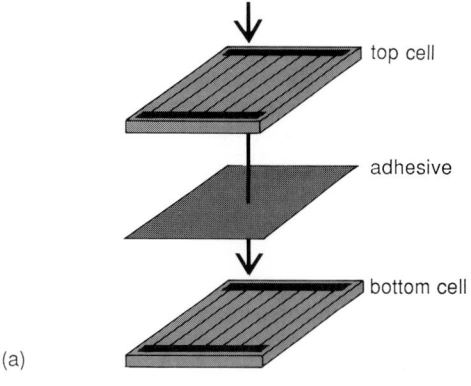

(a)

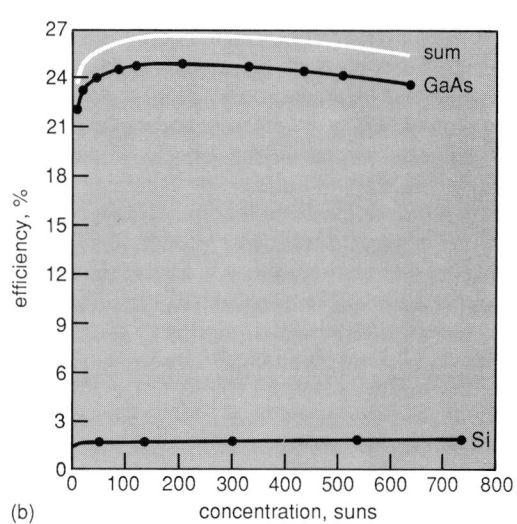

(b)

Fig. 2. Mechanically stacked multiple-junction cascade cell consisting of a gallium arsenide top cell and a silicon bottom cell. (*a*) Structure of a cell. (*b*) Efficiencies of each component cell and their sum, the efficiency of the combined cell, as a function of light concentration

greater than its band gap while it transmits the remaining photons to the cell or cells below. The second cell and any subsequent cells absorb photons with energies equal to or greater than their band gaps and transmit the remaining lower-energy photons to the cells below. In principle a large number of cells can be stacked in cascade with a theoretical efficiency approaching 60%. However, in practice only two-cell stacks (with a theoretical efficiency of over 40%) have been investigated at any length. A mechanically stacked device has been produced whose efficiency was measured at nearly 27% at 300 suns intensity (that is, in light whose intensity is 300 times that of normal sunlight; **Fig. 2**). The device consisted of a gallium arsenide top cell and a silicon bottom cell coupled by an optically clear adhesive. Recent advances with technically advanced silicon cells have given back to silicon the efficiency lead for concentrated sunlight; however, continued research on multiple-junction cells is expected to produce devices with greater than 30% efficiency in concentrated sunlight in the near future.

High-efficiency silicon cells. Past research with silicon cells for concentrated sunlight applications typically used a high concentration of impurity atoms in the base and a planar *pn* junction.

This approach, which has produced cells having 20% efficiency, appears to be limited by a trade-off of current versus voltage. In any solar cell, the conversion power output is determined by the product of the device short-circuit current, the open-circuit voltage, and the fill factor, which is a measure of the resitance losses and junction quality. A fill factor of 0.8 is considered excellent. Limits on the cell voltage and current are determined by the semiconductor band gap, the electronic properties of the material, and the available solar spectrum input. In conventional cells, heavy impurity doping decreases the resistance to charge carrier flow and, up to a point, increases the voltage. However, heavy impurity doping shortens the carrier lifetime, which directly impacts the cell current. Thus a tradeoff of current versus voltage occurs.

In recent advanced designs an effect known as conductivity enhancement has been exploited. In this case, silicon with a very small concentration of impurity doping is used, and such silicon normally has high resistance to charge carrier flow. However, it has been shown that at high illumination levels the great number of photogenerated carriers present overcome this resistance to flow. The carrier lifetime is very long in this material because there are fewer impurities with which charge carriers can interact.

The challenge in this approach has been to contain the carriers after they are generated until they can be collected by the junction. Areas of high recombination such as at the surface and under the metal contact terminals must be carefully treated with layers of insulating materials such as silicon dioxide. It has been found that ultraclean processing and novel methods of device design, such as reducing the interaction area of the metal contacts and the semiconductor, can greatly ameliorate the recombination problem. Reduction of this interaction area is employed in the cell shown in **Fig. 3**, where the junction is an array of thousands of tiny point diffusions that are interconnected with interdigitated metal contacts on the back of the cell. Sunlight enters the cell from the bottom of Fig. 3a through a surface that is purposely textured to help trap light in the cell. Such laboratory silicon cells, 0.20 by 0.12 in. (5 by 3 mm) in size, have been measured at nearly 28% conversion efficiency at a light intensity corresponding to 100 suns (Fig. 3b). It is projected that these structures will eventually reach 30% efficiency. In addition, although this approach was not expected to work well at normal illumination levels, it has produced the highest 1-sun efficiency using silicon: over 22%.

For background information SEE SOLAR CELL in the McGraw-Hill Encyclopedia of Science and Technology.

D. E. Arvizu; D. G. Schueler

Bibliography. J. C. C. Fan et al., Optimal design of high efficiency tandem cells, *Proceedings of the 16th IEEE Photovoltaics Specialist Conference*, pp. 692–701, 1982; M. A. Green, *Solar Cells: Operating Principles, Technology, and System Application*, 1982; R. J. Schwartz, Review of silicon solar cells for high concentrations, *Solar Cells*, 6(1):17–38, 1982; R. M. Swanson, Point contact solar cells: Theory and modelling, *Proceedings of the 18th IEEE Photovoltaics Specialist Conference*, pp. 604–610, 1985.

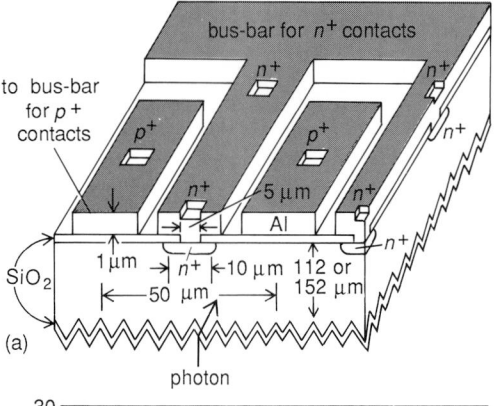

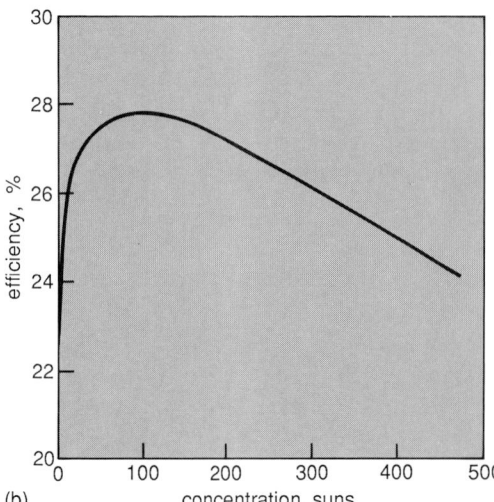

Fig. 3. High-efficiency advanced-design silicon cell. (*a*) Structure of a cell. (*b*) Efficiency as a function of light concentration.

Sonochemistry

Sonochemistry is the study of the chemical changes which occur in the presence of sound or ultrasound. The chemical and biological effects of ultrasound were first reported by A. L. Loomis more than 50 years ago. Within 15 years, industrial applications of ultrasound included soldering, dispersion, emulsification, disinfection, refining, cleaning, extraction, flotation of minerals, and degassing of liquids. Industrial applications have been extended to include welding, pasteurization, and production of gas-liquid sols. However, until recently the use of ultrasound within the chemical community has been sporadic. With the recent advent of inexpensive and reliable sources of ultrasound, there has been a resurgence of interest in its chemical applications.

Mechanisms. Ultrasound spans the frequencies of roughly 20 kHz to 10 MHz (human hearing has an upper limit of less than 18 kHz). Since the velocity of sound in liquids is approximately 5000 ft/s (1500 m/s), ultrasound has acoustic wavelengths of roughly 7.5 to 0.015 cm. Clearly, no direct coupling of the acoustic field with chemical species on a molecular level can account for sonochemistry. Instead, the chemical effects of ultrasound derive from several different physical mechanisms, depending on the nature of the system.

The most important of these mechanisms is cavitation, the formation of gas bubbles in a liquid which occurs when the pressure within the liquid drops significantly below the vapor pressure of the liquid. Cavitation can occur from a variety of causes: turbulent flow, laser heating, electrical discharge, boiling, radiolysis, or acoustic irradiation. When sound passes through a liquid, it consists of expansion (negative-pressure) waves and compression (positive-pressure) waves. This can cause the formation, growth, and rapid recompression of vapor bubbles in the liquid (**Fig. 1**). The implosive bubble collapse generates localized heating and associated high-energy chemistry. The dynamics of cavity growth and collapse are strikingly dependent on local environment, and cavitation in a homogeneous liquid should be considered separately from cavitation near an interface. The symbol $\overset{)))}{\rightarrow}$ will be used in this article to indicate ultrasonic irradiation (sonication or insonation).

In homogeneous media the generally accepted

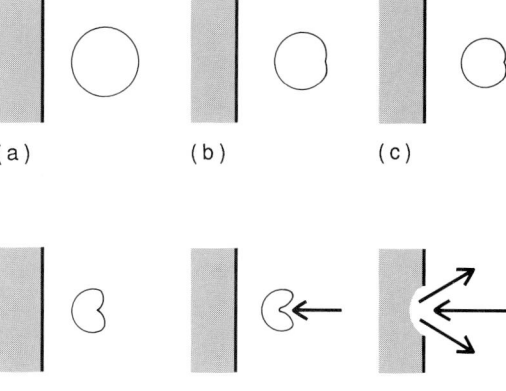

Fig. 2. Sequence *a–f* of a single bubble collapsing during cavitation near a surface.

sonochemical mechanism involves intense localized heating which is produced by the implosive collapse of a bubble during cavitation. A recent measurement of the temperature generated during this implosive collapse established that the effective temperature in the gas-phase reaction zone is approximately 5200 K with pressures of hundreds of atmospheres.

When a liquid–solid interface is subjected to ultrasound, cavitation still occurs, but with major changes in the nature of the bubble collapse. No longer do cavities implode spherically. Instead, a markedly asymmetric collapse occurs, which generates a jet of liquid directed at the surface (**Fig. 2**). The jet velocities are greater than 330 ft/s (100 m/s). The origin of this jet formation is essentially a shaped-charge effect. The impingement of this jet can create a localized erosion (and even melting), responsible for surface pitting and ultrasonic cleaning. Enhanced chemical reactivity of solid surfaces is associated with these processes. The cavitational erosion generates unpassivated, highly reactive surfaces; causes short-lived high temperatures and pressures at the surface; produces surface defects and deformations; forms fines and increases the surface area of friable solid supports; and ejects material in unknown form into solution. Finally, the local turbulent flow associated with acoustic streaming improves mass transport between the liquid phase and the surface, thus increasing observed reaction rates. In general, all of these effects are likely to occur simultaneously.

Homogeneous systems. A variety of novel reactivity patterns are beginning to emerge which are distinct from either normal thermal or photochemical activation. Most of the reactions which have been reported are stoichiometric in terms of a consumed reagent, but a few examples of true sonocatalysis have also appeared.

The sonochemistry of water is the earliest and most thoroughly studied. One group of researchers has recently provided definitive evidence for the formation of hydrogen atoms and hydroxyl

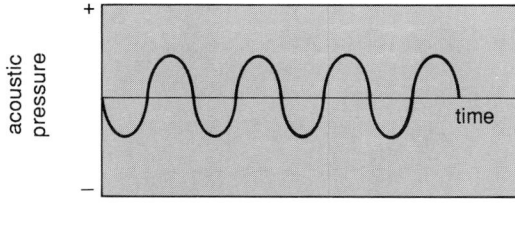

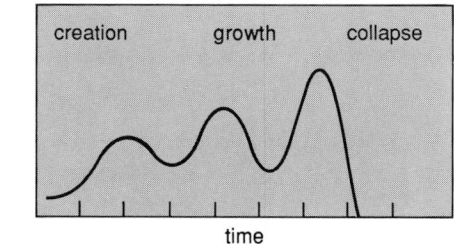

Fig. 1. Idealized representation of bubble growth and collapse during acoustic cavitation. (*After K. S. Suslick, Organometallic sonochemistry, in F. G. A. Stone and R. West, eds., Advances in Organometallic Chemistry, vol. 25, pp. 73–119, 1986*)

radicals (OH·) during the ultrasonic irradiation of water. In separate research, secondary redox reactions due to these highly reactive species have been exhaustively examined.

The sonochemistry of nonaqueous solutions and of discrete organometallic complexes was first reported in 1981. The transition-metal carbonyls were chosen for these initial studies because their thermal and photochemical reactivities have been well characterized. Unusual reactivities were observed, including controlled multiple-ligand substitution and clusterification; examples are shown in reactions (1) and (2).

$$Fe(CO)_5 \xrightarrow{)))} Fe_3(CO)_{12} \qquad (1)$$

$$Fe(CO)_5 + L \xrightarrow{)))} Fe(CO)_{5-n}L_n$$
$$(n = 1, 2, 3; L = \text{Lewis base}) \qquad (2)$$

Initiation of homogeneous catalysis. Having demonstrated that ultrasound can induce ligand dissociation from organometallic complexes, the initiation of homogeneous catalysis by ultrasound becomes practical. The potential advantages of such sonocatalysis include (1) the use of low ambient temperatures to preservethermally sensitive substrates and to enhance selectivity; (2) the ability to generate high-energy species unobtainable from photolysis or simple pyrolysis; (3) the mimicry, on a microscopic scale, of bomb reaction conditions; and (4) the potential ease of scale-up. The transient, coordinatively unsaturated species produced from the sonication of metal carbonyls are likely candidates as homogeneous catalysts, since similar species produced photochemically are among the most active catalysts known. A variety of metal carbonyls upon sonication will catalyze the isomerization of 1-alkenes (terminal olefins) to the internal alkenes. Initial turnover rates are as high as 100 per hour, and represent rate enhancements of approximately 10^5 over thermal controls. Results are consistent with the generally accepted mechanism for alkene isomerization in analogous thermal and photochemical systems. This involves the formation of a hydrido-π-allyl intermediate and alkene rearrangement via hydride migration to form the thermodynamically more stable 2-alkene complex, as shown in general scheme (3).

$$M(CO)_n \xrightarrow{)))} M(CO)_m + (n-m)CO$$

$$M(CO)_m + \text{1-alkene} \rightarrow M(CO)_x(\text{1-alkene}) + (m-x)CO$$

$$M(CO)_x(\text{1-alkene}) \rightarrow M(CO)_x(H)(\pi-\text{allyl})$$

$$M(CO)_x(H)(\pi-\text{allyl}) \rightarrow M(CO)_x(\text{2-alkene})$$

$$M(CO)_x(\text{2-alkene}) + \text{1-alkene} \rightarrow$$
$$M(CO)_x(\text{1-alkene}) + \text{2-alkene} \quad (3)$$

Heterogeneous systems. One of the major industrial applications of ultrasound is emulsification. The first reported and most studied liquid–liquid heterogeneous systems have involved ultrasonically dispersed mercury. The use of such emulsions for chemical purposes has been delineated by extensive investigations. The effect of the ultrasound in this system appears to be due to the large surface area of mercury generated in the emulsion.

The effects of ultrasound on liquid-solid heterogeneous organometallic reactions has been a matter of intense recent investigation. The first use of ultrasound to prepare organometallic complexes of the main-group metals (for example, lithium, magnesium, and aluminum) from organic halides, however, originates in work reported in the 1950s. Much of the recent interest was initiated by a 1980 report of the use of an ultrasonic cleaner to accelerate lithiation reactions. Various research groups have dealt with extremely reactive metals, such as lithium (Li), magnesium (Mg), or zinc (Zn), as stoichiometric reagents for a variety of common transformations. Examples are shown in reactions(4)–(7), where R represents an organic functional group. Ultrasonic cleaning of the reactive metal surface to remove passivating impuritites (for example, water, hydroxide, metal halide, or organolithium) are likely to be important in the origin of these effects.

$$RBr + R'R''CO \xrightarrow[\text{Li, }))]{} RR'R''COH \qquad (4)$$

$$RBr + (CH_3)_2NCHO \xrightarrow{)))} RCHO \qquad (5)$$

$$R_2SiCl_2 \xrightarrow{)))} \text{cyclo-}(R_2Si)_3 \qquad (6)$$

$$RR'CO + BrCH_2CO_2R'' \xrightarrow{Zn,))}$$
$$RR'C(OH)CH_2CO_2R'' \quad (7)$$

The activation of less reactive metals continues to attract major efforts in heterogeneous catalysis, metal-vapor chemistry, and synthetic organometallic efforts. Given the extreme conditions generated by acoustic cavitation at surfaces, analogies to autoclave conditions or to metal-vapor reactors may be appropriate. It has been found that the use of ultrasonic irradiation facilitates the reduction of a variety of transition-metal salts to an active form that will react with carbon monoxide to give good yields of the carbonyl anions for vanadium, niobium, tantalum, chromium, molybdenum, tungsten, manganese, iron, and nickel, even at 10°C (50°F) and 1 atm (10^5 pascals) carbon monoxide. The reduction of transition-metal halides with lithium has been recently extended for Ullman coupling (benzyl halide to bibenzyl) by copper or nickel. Another recent application of ultrasound to the activation of transition metals has been reported in which an extremely reactive magnesium species was used to reduce metal salts in the presence of various ligands to form their metal complexes.

The effects of ultrasound on heterogeneous systems are quite general, and ultrasonic rate enhancements for many nonmetallic insoluble reagents also occur. A wide range of organic reactions has been studied by various groups, including reductions with lithium aluminum hydride ($LiAlH_4$) or sodium hydride (NaH), oxidations with potassium permanganate ($KMnO_4$), and thioamidation with phosphorus pentasulfide

(P_4S_{10}). Ultrasound has been used to enhance the rates of mass transport near electrode surfaces, and thus to enhance rates of electrolysis. This has had some useful synthetic applications for the production of both organic and inorganic chalcogenides.

Applications to heterogeneous catalysis. Ultrasonic irradiation can alter the reactivity observed during the heterogeneous catalysis of a variety of reactions. Sonication has shown such behavior by altering the formation of heterogeneous catalysts, by perturbing the properties of previously formed catalysts, or by affecting the reactivity during catalysis. However, ultrasonic rate enhancements of heterogeneous catalysis have usually been relatively modest (less than tenfold). The effect derived from irradiating operating catalysts is often due simply to improved mass transport. In addition, increased dispersion during the formation of catalysts under ultrasound will enhance reactivity, as will the fracture of friable solids (for example, noble metals on carbon or silica). It has recently been discovered that hydrogenation of alkenes by nickel powder is dramatically enhanced (more than 10^5-fold) by ultrasonic irradiation. The effect of ultrasonic irradiation is to remove the passivating oxide coating to which nickel is quite prone. It appears likely that the changes in surface morphology, in the degree of aggregation, and in the surface's elemental composition originate from interparticle collisions caused by the turbulent flow and shock waves created by high-intensity ultrasound.

For background information SEE CATALYSIS; CAVITATION; HETEROGENEOUS CATALYSIS; HOMOGENEOUS CATALYSIS; ULTRASONICS in the McGraw-Hill Encyclopedia of Science and Technology.

Kenneth S. Suslick

Bibliography. P. Boudjouk, Synthesis with ultrasonic waves, *J. Chem. Ed.*, 63:427–429, 1986; K. S. Suslick, Organometallic sonochemistry, *Adv. Organometal. Chem.*, 25:73–119, 1986; K. S. Suslick, Synthetic applications of ultrasound, *Mod. Synth. Meth.*, 4:1–60, 1986; K. S. Suslick (ed.), *Ultrasound: Its Chemical, Physical and Biological Effects*, 1987.

Space biology

The space shuttle has enhanced opportunities to study biological systems in space, but only a few biological experiments have been flown by the United States. Four experiments involving higher plants have been flown.

Higher plants were first flown as part of the second Space Shuttle Transportation System (STS) mission in 1981. The flight package was a bioengineering test designed to determine which soil moisture content was satisfactory for germination and growth of young sunflower (*Helianthus annuus*) seedlings in microgravity. The *STS 2* mission was cut short, and the test package was reflown on *STS 3* in 1982. The *STS 3* flight also contained a plant experiment designed to test experimental hardware and to establish the effect of microgravity on lignification. The two experimental packages flown on *STS 3* were precursors to subsequent experiments flown as part of *Spacelab 1* and *Spacelab 2*. The follow-up experiment to the bioengineering test package was the HEFLEX experiment flown on *Spacelab 1* in November 1983. This experiment was designed to determine whether plants continued to follow a growth pattern known as circumnutation when they were grown away from Earth in the microgravity environment.

The follow-up of the *STS 3* experiment was flown on *Spacelab 2* in July 1985. The principal objective of this experiment was to establish the effect of microgravity on plant lignification and to examine other parameters related to plant growth and development.

HEFLEX bioengineering test on STS 2 and 3. The flight hardware used for the HEFLEX test was designed as part of the flight hardware for *Spacelab 1*. The hardware, known as the plant carry-on container, was located in a locker in the middeck of the shuttle. The plant carry-on container consisted of a support platform fitted with openings in which tubelike modules could be inserted. The modules (85 in total) were filled with a soil mixture and different amounts of water. The water content of the soil mixture was varied from 55 to 77% for the *STS 2* test and from 60 to 81% for the *STS 3* test. Sunflower seeds were planted in the modules approximately 15 h prior to the scheduled launch, and the plant container was carried onto the shuttle as a late-access package. After the flights, stem heights were measured and plotted against soil moisture content. Results showed that optimum moisture content for germinating and growing sunflowers was about 70%.

HEFLEX experiment on Spacelab 1. A young seedling on Earth does not grow upward in a truly straight line. Instead it grows in an elliptical path around the growth vector; this pattern is known as circumnutation. The frequency and amplitude of the elliptical path is dependent upon species and age. It was generally believed that gravity was the principal force causing circumnutation. The objective of the HEFLEX experiment was to determine whether circumnutational oscillations common to plants grown on Earth occurred in plants grown away from Earth. The experiment utilized the plant carry-on container located in the middeck and additional flight hardware located in *Spacelab 1*. The *Spacelab 1* hardware consisted of light-tight plant modules that contained infrared-transparent windows. These windows allowed plants to be photographed periodically. Photographic equipment was positioned above the plants, viewing the position of the growing point. The video images taken during flight provided investigators with a record of stem position at different time intervals.

For the *Spacelab 1* experiment, sunflower seeds were planted in modules of the plant carry-on container 1, 2, and 3 days before scheduled launch. An additional 24 seeds were planted in space. The *Spacelab* hardware was stowed on the shuttle payload bay several days prior to flight, but the plant carry-on container with the germinating sunflower seeds was carried to the launch pad for late-access loading. The flight crew transported seedlings from middeck to *Spacelab* for analysis. The video recordings in flight were made on 4-day-old plants. Results of experiments showed that circumnutation still existed in space. Amplitude of elliptical motion, however, was about 50% of that observed on Earth. The plants exhibited a mean oscillation period of about 110 min. The conclusion was that circumnutation in growing plants is not dependent on gravity.

Lignification experiment on STS 3. The plant lignification experiment was originally selected for flight on *Spacelab 2*, but it also was included in the earlier *STS 3* flight. The experimental objectives were to: (1) test the function and effectiveness of the flight hardware to be used in the *Spacelab 2* experiment, (2) measure the effect of microgravity on lignification in high plant species, and (3) search for overall changes in plant growth.

The flight hardware for the lignification experiment was designed around the concept of a small growth chamber that would support plant seedling growth and development. Two units of hardware, designated plant growth units, were designed and built for this purpose. Each plant growth unit, measuring 20 in. × 14 in. × 11 in. (51 cm × 36 cm × 27 cm), replaced a locker on the forward bulkhead of the orbiter middeck. The plant growth units contain a cavity for growing plants that is fitted with support and regulatory components. The units are equipped with three 15-W plant growth lamps, a timer toprovide day-night cycling, temperature sensors, electronically controlled fans, a data acquisition system, and internal batteries. The plants are grown in plant growth chambers which fit into the cavity of the plant growth unit. The plant growth chambers are made of a metal alloy base and a Lexan top that fit together between a rubber gasket to form an airtight chamber. The chamber bases are fitted with a temperature probe and two gas-sampling ports. In the *STS 3* experiment, one plant growth unit fitted with six chambers was flown, and the other unit was used as a real-time 1-*g* control.

Results from the *STS 3* experiment showed that the plant growth unit functioned well throughout flight and could support the growth of young seedlings. The oat, mung bean, and pine seedlings exhibited some reduction in lignin content. Only in mung beans, however, was the difference in lignin content statistically significant. Roots were also seen growing upward, and stems were shorter in flight seedlings than in control seedlings.

Lignification experiment on Spacelab 2. The results of the *STS 3* experiment were not only encouraging but also helpful in designing a more extensive experiment for flight on *Spacelab 2*. Objectives of the experiment were to establish the effect of microgravity on lignin production and related parameters in higher plant species, and to study overall plant growth and development in microgravity.

Preflight activities for the *Spacelab 2* experiment included germinating pine seedlings to be 4 and 10 days old at launch, and planting mung bean and oat seeds on the evening prior to launch. Two plant growth units containing a total of 12 chambers were loaded on the space shuttle *Challenger* about 12 h prior to the scheduled launch. During flight the hardware was checked periodically, and the plants were photographed. Once the *Challenger* had landed, the plant growth units were returned to the principal investigator team at the landing site. The plant growth chambers were removed, and the plants were examined, photographed, measured, sectioned, and packaged for return to the principal investigator laboratory for analysis. Within a week after the shuttle landing, pine seed germination was initiated to begin a control experiment. Flight hardware and pertinent flight data (time periods and temperature profiles) were used in conducting the control experiment.

The results showed that all three plant species grown aboard the space shuttle contained less lignin than control seedlings grown at 1 *g*. Reduction in lignin in flight seedlings ranged from 7 to 15% in slow-growing pine seedlings to about 25% in faster-growing mung beans and oats seedlings. The activities of three enzymes involved in the lignin biosynthetic pathway (phenylalanine ammonia lyase, μ-glucosidase, and peroxidase) also were significantly lower in flight than in control seedlings. Cellulose, the other major structural polymer in higher plant cells, also was reduced in flight seedlings compared to control seedlings, but not as much as lignin.

Certain physiological and growth parameters also were affected by microgravity. Microgravity did not affect germination success, but germination rates were slower, especially for mung beans. Flight seedlings exhibited a reduction in both stem length and mass and leaf area and mass. Total chlorophyll was slightly reduced in space-grown seedlings but was 5–10% more on a leaf-area basis. Thirty to forty percent of oat and mung bean roots were growing upward in flight seedlings but none in control seedlings grown at 1 *g*. One laboratory also reported that some of the oat seedlings exhibited considerable aberrations in cell division.

For background information *see Space biology; Space shuttle* in the McGraw-Hill Encyclopedia of Science and Technology.

Joe R. Cowles

Bibliography. A. H. Brown and D. K. Chapman, Circumnutation observed without a significant gravitational force in spaceflight, *Science*, 225:230–232, 1984; A. H. Brown and D. K. Chapman, A test to verify the biocompatibility of a method for plant culture in a microgravity environment, *Ann. Bot.*, 54(Suppl. 3):19–31, 1984; J. R. Cowles et al., Growth and lignification in seedling exposed to eight days of microgravity, *Ann. Bot.*, 54(Suppl. 3):33–48, 1984; A. D. Krikorian and S. A. O'Connor, Karyological observations, *Ann. Bot.*, 54(Suppl. 3):49–63, 1984.

Space flight

Future historians of space flight may well regard 1986 as a watershed year. But unlike 1957 (which saw the launch of *Sputnik 1*) or 1969 (the first lunar landing), this year was marked not only by success but also by a series of failures and by tragedy.

The most spectacular event, and perhaps the one with most far-reaching implications, was the explosion aboard the space shuttle *Challenger* on January 28, 1986, with a total loss of the shuttle craft and seven crew members. The accident prompted a reevaluation of the United States shuttle program, leading to a policy change that may severely restrict, if not eliminate, the role of NASA in the commercial launch field. It remains to be seen whether this event will stimulate the growth of a private launch industry in the United States or will result in forfeiting a major share of the satellite launch market to overseas competitors.

Other failures and untoward events included: the destruction of a Delta expendable launch vehicle on May 3, 1986, after it headed off course, immediately following launch from Florida's Kennedy Space Center; the loss of two Titan 34D vehicles carrying military reconnaissance satellites, one on April 18, 1986, on the launch pad at Vandenberg Air Force Base, Cali-

Table 1. Some significant space launches, November 1985–November 1986

Payload	Date	Payload Country	Purpose or outcome
Atlantis STS-61B	Nov. 27, 1985	United States	Shuttle; launched satellites for Mexico and Australia, experiments in assembling structures with extravehicular activities
Columbia STS-61C	Jan. 12, 1986	United States	Shuttle: Materials Science Laboratory; infrared imaging experiment; blood storage experiment; Shuttle Student Involvement Program
Challenger STS-51L	Jan. 28, 1986	United States	Shuttle; destroyed 60 s after launch, 8 mi (13 km) altitude, speed just under 2000 mi/h (3200 km/h); five male and two female astronauts killed in the shuttle explosion
Mir	Feb. 19, 1986	Soviet Union	Initial step in establishing a space station; module features three sets of docking ports, allowing expansion of the space station or possible use by transient spacecraft; based on the *Mir* concept, Soviet Union could have a six-module space station assembled in orbit and weighing more than 100 tons (90,000 kg) by 1991
Cosmos 1738	Apr. 4, 1986	Soviet Union	Communications satellite providing continuous telephone, telegraph, and television links within the Soviet Union and extending support for the *Mir* space laboratory
Cosmos 1740	Apr. 15, 1986	Soviet Union	Military photoreconnaissance; recovered after 13 days in orbit
Progress 26	Apr. 23, 1986	Soviet Union	Uncrewed cargo ferry carried supplies and equipment for the *Mir* space laboratory
Polar Bear	Nov. 13, 1986	United States	Polar Beacon Experiments and Auroral Research satellite; carried three experiments to photograph aurora and sample electrical particles and magnetic fields over the Earth's poles

Table 2. Payloads launched, November 1985–November 1986

Country or organization	Number
United States	41
Soviet Union	279
Australia	1
Brazil	1
European Space Agency	2
France	1
Japan	6
Mexico	1
People's Republic of China	2
Sweden	1

fornia, and another on August 18, 1986; the failure of a NASA-launched Nike Orion sounding rocket with scientific instruments aboard on April 25, 1986, following 55 consecutive successful flights; the cancellation of an Air Force contract for delivery of the Centaur, an upper-stage rocket scheduled to be deployed by the space shuttle to boost payloads into high orbit; and the failure of an Ariane launch on May 30, 1986, which requires extensive redesign of the vehicle's third-stage ignition system.

A more hopeful development was the release by NASA of plans for a new space station to provide a permanent base for United States astronauts by 1994. February 19, 1986, saw the launch of *Mir*, a Soviet space station featuring marked design improvements over seven previous Salyut stations.

Significant space launches in the period from November 1985 to November 1986 are listed in **Table 1**. The total numbers of payloads launched by various countries during this period are given in **Table 2**, and the numbers of payloads in orbit in **Table 3**.

Table 3. Payloads in orbit*

Country or organization	Number	Country or organization	Number
United States	532	International	
Soviet Union	958	Telecommunication	
Australia	3	Satellite	
Brazil	1	Organization	35
Canada	14	Italy	1
European		Japan	31
Space		Mexico	2
Agency	15	NATO	6
France	14	People's Republic of	
France/West		China	4
Germany	2	Saudi Arabia	2
West Germany	5	Spain	1
India	7	Sweden	1
Indonesia	3	TOTAL	1637

* In November 1986; includes many multiple satellite launches utilizing a single launch vehicle.

The seven-member crew lost in the *Challenger* explosion was made up of five astronauts, Francis R. (Dick) Scobee, Michael Smith, Ronald E. McNair, Ellison S. Onizuka, and Judith A. Resnik; and two payload specialists, Gregory Jarvis, representing the Hughes Company, and Christa McAuliffe. McAuliffe, a 38-year-old schoolteacher from Concord, New Hampshire, had been selected by NASA for the Teacher in Space Project to serve as an inspiration to youth and also to demonstrate the safety and presumably routine nature of shuttle flights. Overconfidence in the safety of space flight, bred by the remarkable successes of NASA over many years, thus served to heighten the public shock over the tragedy.

A special presidentially appointed commission, headed by former Secretary of State William P. Rogers, was convened to determine the cause of the accident. The commission's report cited failure of the O rings joining segments of the solid-propellant booster stage of the launch vehicle as the proximate cause, but went on to deliver a severe indictment of NASA management practices as an important contributing factor.

The accident, and the Rogers Commission Report, have led to a major delay in the United States space program. The disaster resulted in the cancellation of a shuttle mission which would have included observation of Comet Halley from space. The accident also delayed several scientific missions previously planned for 1986, including launch of two space vehicles to Jupiter, and placement in orbit of the Hubble space telescope. This loss to science is, of course, in addition to losses sustained by NASA owing to its inability to fulfill pending contracts and obligations for some 90 shuttle payloads. *SEE HALLEY'S COMET* .

With support from President Reagan, Congress has authorized a replacement for the *Challenger*. NASA conducted a test drill with five astronauts aboard the *Atlantis* shuttle on November 18, 1986. This was the first shuttle exercise with astronauts aboard since the *Challenger* failure. This concluded a 7-week series of simulated launches of the *Atlantis* to test the shuttle itself and its launch support systems. The final countdown was conducted without shuttle engines and took place on Pad 39-B, from which the doomed *Challenger* lifted off.

Role of private industry. On August 15, 1986, the President issued a directive restricting future NASA shuttle flights to military and scientific payloads. Administration officials believe that the President's decision to get NASA out of the business of launching commercial satellites will help convince United States industry to move into the commercial launch field and will encourage industry to take on foreign competition. Skeptics argue that even if the United States government removes itself from the commercial launch market, private industry may not jump in quickly, as it still must compete with firms that are owned or subsidized by other governments.

Nevertheless, many within the space community strongly endorse the President's action barring commercial payloads for the shuttle. They argue that the earlier NASA decision, in the 1970s, to launch satellites from the shuttle instead of using expendable launch vehicles was a mistake. They point out that, at present, proven expendable launch vehicles are available to accomplish satellite launch missions more economically than the shuttle, and such vehicles present a practical alternative, well within the financial and technical capability of private industry.

Department of Transportation officials estimate that seven to ten United States firms are now actively considering entry into the commercial launch market. Among these is Martin Marietta, whose Denver Aerospace division manufactures the Titan rocket. Martin Marietta is already moving ahead with production of the Titan 3, with an eye toward the commercial market. This vehicle is capable of carrying payloads weighing as much as 9000 lb (4000 kg) into geostationary transfer orbit, and payloads up to 32,000 lb (14,500 kg) into low Earth orbit. A tentative agreement was announced in September 1986 between Martin Marietta and the Federal Express Corporation calling for launch of a relay satellite from Cape Canaveral, Florida, in 1989.

The Titan 4, which is being delivered by Martin Marietta under an Air Force contract, also has possibilities as a commercial launch vehicle. The obsolete Titan 2, which is now being phased out of the Air Force ballistic missile inventory, is being considered for conversion to a space vehicle, capable of placing 7700-lb (3500-kg) payloads into low Earth orbit.

McDonnell Douglas Corporation's Delta rockets (**Fig. 1**) are also contenders for service as commercial launch vehicles; unlike Martin Marietta, however, McDonnell Douglas may choose not to do its own marketing. Transpace Carriers of Lanham, Maryland, is seeking to become a systems contractor to launch Delta payloads of up to 4000 lb (1800 kg) into low Earth orbit.

General Dynamics announced its intention to restart its Atlas Centaur production line, and has indicated that it could produce rockets for commercial launches by 1989. In the past the Air Force and NASA have used the Atlas Centaur to carry payloads as large as 11,300 lb (5100 kg) into low Earth orbit.

In addition to these aerospace giants, nearly two dozen smaller firms are attempting to establish a place in the future space market. Perhaps the best known, Houston-based Space Services,

Fig. 1. Lift-off of Delta expendable launch vehicle from the Kennedy Space Center, Florida, on September 5, 1986.

Inc., hopes to carry 2000-lb (900-kg) payloads into low Earth orbit by the end of 1987.

Shuttle flights. The *Challenger* tragedy overshadowed two notable successes for the NASA shuttle which occurred in late 1985 and early 1986. *Atlantis STS-61B*, the twenty-third shuttle flight, launched from the Kennedy Space Center on November 27, 1985, successfully deployed two Hughes 367 communications satellites. These include the *Morelos-B*, providing television, telephone, and wire services for Mexico; and the *Aussat 2*, which provides domestic communications for Australia's population of 15 million. The *Atlantis* mission also featured two experiments designed to study extravehicular construction methods in space. One experiment, termed EASE (Experimental Assembly of Structures in Extravehicular Activity), involved construction of an inverted pyramid composed of a few large beams and nodes. The other, termed ACCESS (Assembly Concept for Construction of Erectable Space Structures), is in effect a high-rise tower composed of many small struts and nodes. In the EASE experiment, crew members were free to move about during construction, but in ACCESS the crew members worked from fixed positions.

The payload of *Columbia STS-61C*, launched on January 12, 1986, included the Materials Science Laboratory 2, featuring experiments with a

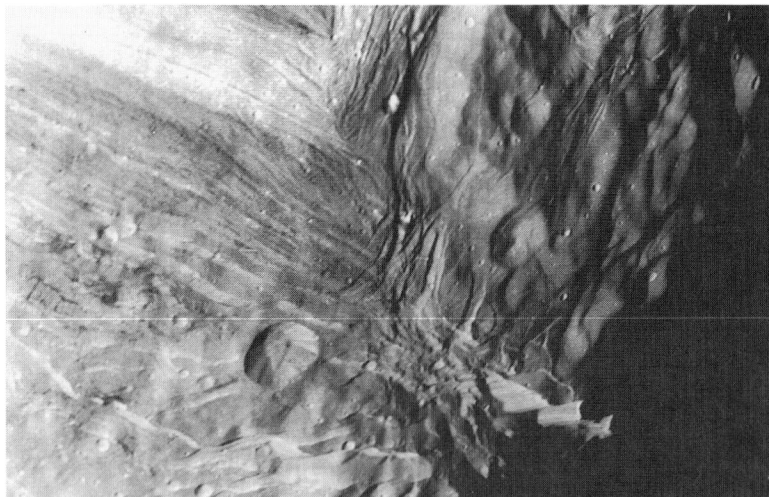

Fig. 2. *Voyager 2* image of Miranda, from distance of 22,000 mi (36,000 km). Image is about 150 mi (250 km) across. Rugged, higher-elevation terrain with numerous craters, to right, shares this area with younger, lower, striated terrain that shows evidence of geologic activity.

variety of materials that were carefully observed as they were melted and solidified in zero gravity; the Infrared Imaging Experiment, which acquired radiometric pictures of selected terrestrial and celestial targets; and the Initial Blood Storage Experiment, which was designed to gain a better understanding of factors limiting the storage of human blood, including comparison of blood that had experienced weightlessness with Earth-bound blood samples. *Columbia* also carried three experiments designed by high school students, under NASA's Shuttle Student Involvement Program.

Voyager 2 flyby of Uranus. Since launch on August 20, 1977, NASA's *Voyager 2* has reached Jupiter in July 1979 and Saturn in August 1981. The next encounter, at Neptune, is scheduled for August 1989. *Voyager 2* will eventually leave the solar system and enter interstellar space.

Voyager 2 encountered Uranus, seventh planet from the Sun, in January 1986. At the point of closest approach, on January 24, the spacecraft flew 50,000 mi (81,000 km) above the planet's cloudtops and 66,500 mi (107,000 km) from its center. It returned thousands of images and voluminous amounts of other scientific data on Uranus's satellites and rings, its atmosphere and interior, and the magnetic environment surrounding the planet.

Voyager 2 images of the five large previously known satellites revealed complex surfaces indicative of varying geologic pasts (**Fig. 2**); ancient, heavily cratered surfaces are apparent on some of the satellites, while other satellitesshow strong evidence of internal geologic activity. They appear to be about 50% water ice, 20% carbon- and nitrogen-based materials, and 30% rock. The cameras also detected 10 previously unseen satellites. The largest of these, temporarily designated 1985Ul, is about 100 mi (170 km) in diameter, which is larger than most asteroids.

All nine previously known rings of Uranus were photographed (**Fig. 3**) and measured, as were other new rings and ringlets in the Uranian system. Radio measurements indicated that the outermost ring, epsilon, is composed mostly of boulder-size chunks of ice several feet across. However, a very tenuous distribution of fine dust also seems to be spread throughout the ring system. Two new distinct rings have been positively identified. The first, 1986U1R, was detected between the outermost of the previously known rings—epsilon and delta—at a distance of 31,000 mi (50,000 km) from Uranus's center. It is a narrow ring like the others. The second, designated 1986U2R, is a broad region of material perhaps 1900 mi (3000 km) across and 24,000 mi (38,000 km) from the planet's center.

As expected, the dominant constituents of Uranus's atmosphere were determined to be hydrogen and helium. But the abundance of helium—about 15%—was considerably less than had been suggested by some Earth-based studies. Methane, acetylene, and other hydrocarbons exist in much smaller quantities. Voyager images showed that the atmosphere is arranged into clouds running east-west, an orientation similar

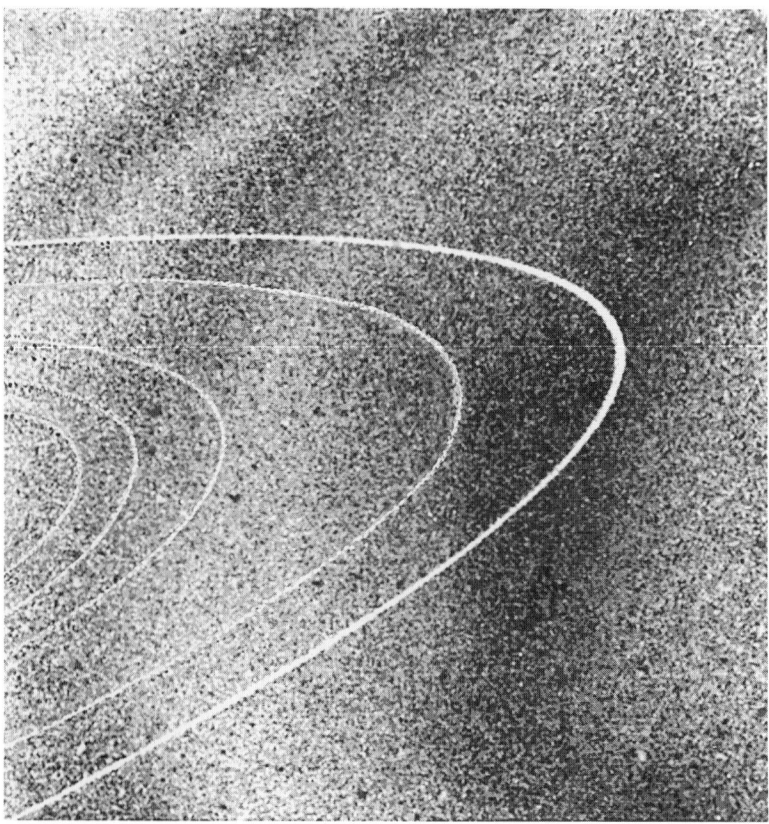

Fig. 3. *Voyager 2* image of outer Uranian rings, taken 11 min before spacecraft passed through ring plane. Bright outer ring is designated epsilon, and inside it lie (from right to left) newly discovered 1986UR1, delta, gamma, and eta.

to the clouds on Jupiter and Saturn. Winds on Uranus blow at speeds of 90 to 360 mi/h (40 to 160 m/s) in the same direction as the planet rotates.

Radio emissions detected several days before *Voyager 2*'s closest approach to Uranus provided the first conclusive evidence that the planet possesses a magnetosphere—an intense magnetic field whose axis is tilted at a 60° angle to the rotational axis. The intensity of the magnetic field at Uranus's surface is roughly comparable to that of Earth. The field is probably generated by an electrically conductive, superpressurized ocean of water and ammonia presumed to lie between the core and the atmosphere.

The Uranus encounter officially ended on February 25, 1986. Eleven days earlier, project engineers took a major step toward the encounter at Neptune by commanding *Voyager 2* to fire its thrusters for a course-correction maneuver lasting more than 2½ h.

Completion of the maneuver allows mission planners the widest range of choices for the spacecraft's precise flight path during its flyby of Neptune in August 1989. One possibility, known as the polar crown mission, would take *Voyager 2* over the north pole of Neptune at a flyby distance of only 800 mi (1300 km). A second plan calls for a more distant flight past the planet. The final choice will depend on further Earth-based studies and the completion of the next course correction sometime in 1987.

Polar Bear satellite. On November 13, 1986, the United States Air Force successfully launched a Polar Beacon Experiments and Auroral Research satellite, designated by the acronym *Polar Bear*, from Vandenberg Air Force Base, California. The *Polar Bear*, which was placed in a 625-mi-high (1000-km) polar orbit, will study space communications interference caused by solar flares and increased auroral activity. This transit-type spacecraft, built by Johns Hopkins Physics Laboratory, was displayed for 8 years in the Smithsonian's National Air and Space Museum before it was refurbished into the *Polar Bear* configuration at a savings in development costs.

Strategic Defense Initiative. A space experiment, conducted on September 5, 1986, culminated in the deliberate destruction of an orbiting Delta rocket stage (Fig. 1). The experiment provided critical data to researchers for design of weapons of small kinetic energy capable of defending against ballistic missiles in the Strategic Defense Initiative. The orbiting Delta rocket stage carried advanced infrared and other sensors to detect the plumes of attacking missiles, and the first laser radar ever flown in space.

Space station development. Progress continued toward developing a United States space station, proposed by President Reagan in 1984, and intended to be permanently crewed by 1994. On July 31, 1986, the NASA administrator called for a 90-day internal review of the space station program, with recommendations for any needed changes and improvements to the original plan.

As a result of the review, NASA released details of a new design for the space station, some of which addressed important technical issues that were criticized in the earlier plan. Among these were concerns relating to the amount of spacewalking that was associated with the orbiting station in the original design, and the lack of a so-called lifeboat to rescue astronauts in the event of an emergency. The new plan leaves open the possibility of rescue craft, and decreases the need for spacewalking by locating certain control devices inside the space station.

The redesigned space station is smaller than first planned, and it may be uncrewed at times during the early operational phases. It will consist of an array of giant metal trusses about 330 ft (100 m) long and 165 ft (50 m) wide, and will feature two vertical keels 360 ft (110 m) long, connected by 1460-ft (445-m) horizontal booms. There will be five locations on the structure for attaching payloads, and a facility for servicing free-flying spacecraft. The operational altitude will be 285 mi (460 km).

SOVIET SPACE ACTIVITY

The Soviet space program in 1986 was characterized by a heavy schedule of satellite launches, including many for military purposes.

Mir space station. On February 19, 1986, the Soviet Union launched its eighth space station, called *Mir*, constituting a milestone in space exploration. The *Mir* represents a considerable advance over earlier Salyut designs, particularly from the standpoint of allowing for expandability, which is vital for space station growth and the ability to evolve to accommodate changing missions.

The *Mir* features a port which may be attached to a new docking module. The docking module features three sets of docking ports, with each set aligned along a different axis. This module is the key to the *Mir*'s new capability for station expansion. The docking module is permanently attached to the *Mir*, and five ports are open for docking other space station modules or transient spacecraft. With the single module now in orbit attached to *Mir*, the Soviets may be able to assemble a modular space station as much as six times the size of previous stations, with six ports for docking crewed or uncrewed spacecraft. With two of these modules the station complex could nearly double in size, and with four almost double again. If launch rates continue at their present pace, the Soviets could, by 1991, have a six-module space station assembled in orbit, weighing more than 100 tons (90,000 kg).

Launch services. In June 1986 the Soviet Union formally announced its interest in providing commercial launch services. Also in June, officials of Glavkosmos, the Soviet space technology agency, announced an agreement with India to launch a satellite sometime in 1987, designed for study and research relating to natural resources.

Proton launch vehicle. A principal Soviet launch vehicle continues to be the Proton, whose variants have made approximately 120 successful launches over the past 20 years, including the launch of the *Mir* space station. The largest known Proton vehicle has a capacity considerably higher than the Ariane 4, now being developed by the European Space Agency. The Soviets are said to be developing a powerful new launcher with five times the payload of present United States launchers and seven times the payload capacity of the Proton now in use.

EUROPEAN SPACE ACTIVITY

Arianespace Inc., the commercial arm of the European Space Agency, has been the immediate beneficiary of NASA's near-removal from the commercial space market. The company has signed contracts with many customers who had

Fig. 4. One of a series of Ariane launches from the Kourou, French Guiana, launch facility of the European Space Agency.

intended to use the NASA shuttle, including several United States firms. Arianespace is now booked to capacity for launches through 1990.

The Ariane rocket fleet (**Fig. 4**) used by Arianespace was grounded during much of 1986 as a result of accidents, including the failure of its last launch. The completion of modifications to the Ariane 2 third-stage ignition system and the resumption of launches were expected in 1987.

The company is proceeding with plans for expansion of its services. In January 1986 it unveiled a second launch pad at Kourou, French Guiana, and it is presently developing larger vehicles, capable of matching the shuttle's lift capacity. These include the Ariane 4, which could carry payloads of up to 5555 lb (2520 kg) into geostationary orbit, and the Ariane 5, which is planned for delivery in the mid-1990s and designed to lift 33,000-lb (15,000-kg) payloads into low Earth orbit.

The Space Media Network, a Swedish firm with offices in Stockholm, Paris, and New York, began offering satellite photographs for sale to the news media. Photos that were taken from a civilian satellite, *SPOT*, launched by Arianespace on May 19, 1986, revealed details of construction on a Soviet space shuttle base at Tyuratam, in central Asia. The *SPOT* heralds a new era in remote sensing, with commercial applications in the areas of mineral exploration, agriculture, forestry, water resources, and cartography.

Planning and design work continued on two European reusable, crewed space shuttles, the British *HOTOL*, and the French *Hermes*, a delta-winged spaceplane to be launched atop the expendable Ariane 5.

ASIAN SPACE ACTIVITY

Japan and the People's Republic of China are both making an effort to capture a share of the future business in space services. The Japanese space program attracted attention in August 1986 with the successful launch of the new-generation H-1 rocket, which relies heavily on the technology of the Delta rocket developed by McDonnell Douglas. The H-2, now under development, will feature all-Japanese technology and is designed to lift 2-ton (1800-kg) payloads into orbit, in direct competition with the Ariane. The H-2 is planned for test launch in 1992.

The Chinese have deployed a total of 19 satellites since 1970, including one in October 1986. Most of the launches have involved medium-size payloads—up to 4000 lb (1800 kg)—on Long March 1 and 2 rockets. The Long March 3, said to be capable of carrying 5000-lb (2250-kg) payloads, has been launched three times, with one failure.

China has reached agreements with Swedish and Brazilian firms to provide launch services, and is reported to be holding discussions with at least a dozen others, including potential customers in the United States. Western Union became China's first declared United States customer by signing a letter of intent to launch a communications satellite, *Westar 6-S*, originally scheduled for a June 1986 shuttle flight. Telesat Canada, which has reserved two 1990 Ariane slots, is also exploring the possibility of Chinese launches.

The Indian Space Research Organization announced that it intends to establish a National Rural Telecommunications Project in northeastern India, augmenting the existing telecommunications network. The region would be served with portable terminals that would communicate directly with satellites. The 4500 portable terminals used 10 years ago in connection with India's Satellite Instructional Television Experiment are still available and could be used in the rural telecommunications project.

INTERNATIONAL COOPERATION

In an event of major potential significance, United States and Soviet negotiators have agreed on a new space pact which calls for cooperation in life-science, astronomical, astrophysical, and solar-terrestrial research. The pact, which is more specific than the 1972 agreement which led to the Apollo/Soyuz Project, includes the prospect of cooperative Mars landing missions. The pact may become effective when signed by President Reagan and General Secretary Gorbachev in 1987.

For background information SEE COMMUNICATIONS SATELLITE; MAGNETOSPHERE; REMOTE SENSING; SPACE BIOLOGY; SPACE FLIGHT; SPACE PROBE; SPACE PROCESSING; SPACE SHUTTLE in the McGraw-Hill Encyclopedia of Science and Technology.

Larry G. Hastings; Robert J. Griffin, Jr.

Bibliography. British Interplanetary Society, *Spaceflight*, vol. 28, no. 7, July/August 1986; NASA, *Satellite Situation Report*, vol. 26, no. 3, September 30, 1986; NASA Goddard Space Flight Center, *Spacewarn Bulletin*, September 30, 1986; National Space Society, *Space World*, vol. W-9-273, September 1986.

Spectroscopy

Magnetic circular dichroism spectroscopy is used extensively to characterize the structure and magnetic properties of the metal-binding centers (active sites) of metalloproteins. Because the magnetic circular dichroism properties of cobalt complexes are very sensitive to the number of coordinated ligands and to the overall geometry, magnetic circular dichroism spectroscopy has frequently been used to assign the cobalt structure in cobalt-substituted metalloproteins. Another useful application of magnetic circular dichroism spectroscopy relies on the so-called fingerprinting capability of the technique for the characterization of structurally unknown heme iron sites through spectral comparisons with structurally defined heme iron protein derivatives and model metal complexes. The most sophisticated application of magnetic circular dichroism spectroscopy involves the use of ultralow temperatures and very high magnetic fields to saturate the intensity of certain magnetic circular dichroism signals as a way of investigating the magnetic properties of the metal in the protein.

Faraday effect. The phenomenon of magnetic circular dichroism was discovered in 1845 by Michael Faraday, and is also known as the Faraday effect. Light-absorbing molecules (chromophores) exhibit magnetic circular dichroism spectra because they absorb left and right circularly polarized light unequally in the presence of a magnetic field. Spectral features known as molecular circular dichroism "A" and "C" terms result from the magnetic field–induced lifting of excited-state or ground-state degeneracies, respectively (**Fig. 1**). Magnetic circular dichroism "B" terms occur whether or not degeneracies exist and are therefore observable for all chromophores. As shown in Fig. 1 on the right, the derivative-shaped magnetic circular dichroism A term is made up of a peak and a trough centered at the energy ν_0 of the original absorption transition. The intensities of magnetic circular dichroism B terms, like those of A terms, are temperature-independent. However, B terms can be distinguished from A terms by their shape, which consists of only either a peak or a trough centered at the energy of the absorption transition. C terms (Fig. 1, right) are shaped like B terms, but are temperature-dependent as a result of population differences in the two states that are resolved by the magnetic field. Thus, it is possible to separate a magnetic circular dichroism spectrum into its A, B, and C components based on an analysis of the shapes and temperature dependence of the spectral features.

In trying to understand magnetic circular dichroism spectroscopy, it is important to emphasize that although magnetic circular dichroism and the related technique of natural circular dichroism measure the same physical parameter, namely the difference in light absorption by left and right circularly polarized light, the origins of the two techniques are distinct. Circular dichroism spectra are observed only for asymmetric molecules or molecules in an asymmetric environment, while the magnetic circular dichroism effect results from the specific influence of the magnetic field on the electronic structure of any chromophore. In fact, magnetic circular dichroism and natural circular dichroism signals are additive; for this reason, it is necessary to subtract the natural circular dichroism signal from the observed magnetic circular dichroism signal in studying asymmetric chromophores to obtain the inherent intensity of the magnetic circular dichroism. Because magnetic circular dichroism

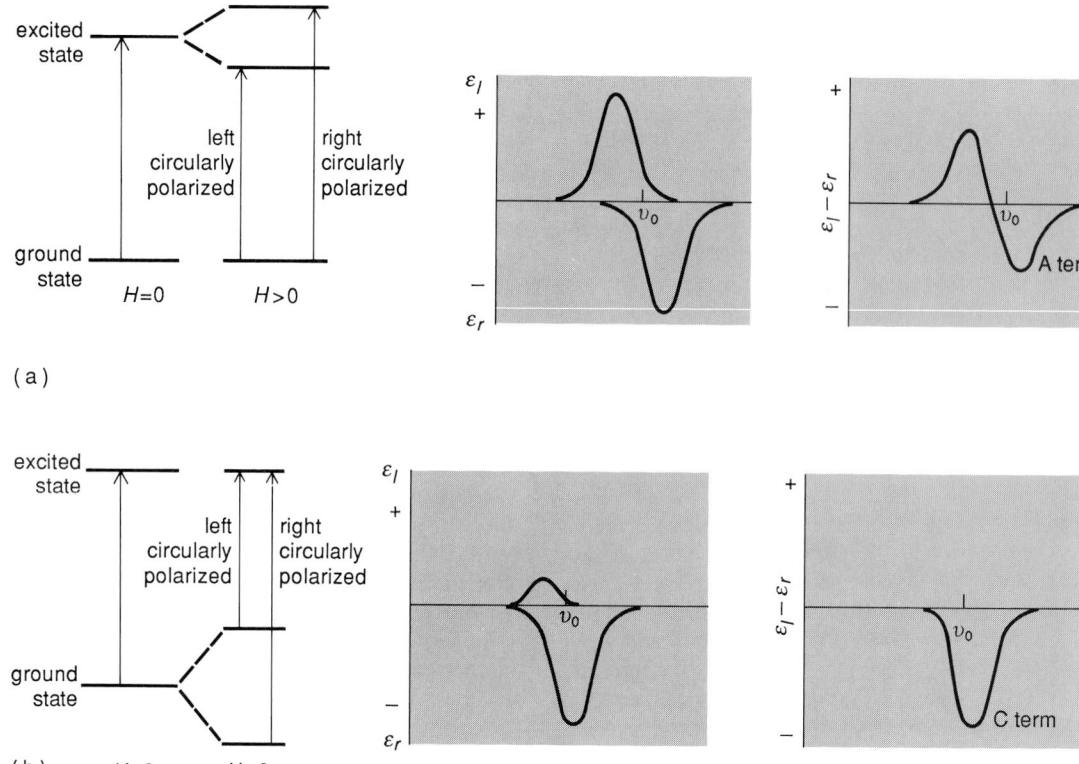

Fig. 1. Magnetic circular dichroism spectra. (*a*) The transitions and expected spectrum for a magnetic circular dichroism A term. (left to right) The transitions; the spectra for left (ϵ_l) and right (ϵ_r) circularly polarized light; the magnetic circular dichroism spectrum ($\epsilon_l - \epsilon_r$). (*b*) The transitions and expected spectrum for a magnetic circular dichroism C term. (left to right) The transitions; the spectra for left and right circularly polarized light; the magnetic circular dichroism spectrum. (*After R. S. Drago, Physical Methods in Chemistry, W. B. Saunders, 1977*)

signals result from the effect of the magnetic field on the electronic structure of a chromophore, they are more sensitive to changes in the physical structure of the chromophore than are natural circular dichroism spectra.

Magnetic circular dichroism spectroscopy can be used to assign particular electronic transitions, to establish the symmetries of ground and excited states, to measure interactions between molecules, and to probe the magnetic properties of molecules. Because the spectral features of magnetic circular dichroism can be either positive or negative in sign, considerably more fine structure is seen than in corresponding absorption spectra. This increased fingerprinting capacity, coupled with the sensitivity of magnetic circular dichroism to the physical structure of a given chromophore and its relative insensitivity to environmental factors, makes this type of spectroscopy an extremely useful empirical tool for structure determination.

Thus there are several advantages to the use of magnetic circular dichroism spectroscopy in studying metal sites in proteins. First, measurements can be made on all oxidation states of the metal under investigation. Second, magnetic circular dichroism spectra need not be measured at cryogenic temperatures, as is often necessary

with other methods, although magnetic circular dichroism studies at low temperatures are quite feasible and provide additional and very useful information. Third, there is no need for isotopic substitution. Finally, the concentrations necessary for magnetic circular dichroism studies are the same as those required for optical absorption spectroscopy, typically 10–1000 micromolars. Three recent examples will serve to illustrate the use of magnetic circular dichroism spectroscopy to examine metal sites in protein systems.

Cobalt-substituted metalloproteins. Although cobalt is not a metal frequently encountered in proteins, it can be substituted for essentially all of the more commonly found metals, including copper, iron, and zinc. The use of cobalt substitution in the study of zinc proteins has been particularly productive, since the chromophoric properties of zinc complexes are very limited. In addition, the cobalt-substituted enzymes are usually active catalytically, suggesting that the interchange of cobalt for zinc has not disturbed the native structure. It has been demonstrated that the magnetic circular dichroism spectra of four-coordinate tetrahedral, five-coordinate octahedral, and six-coordinate octahedral cobalt complexes can be distinguished. Thus, the magnetic circular dichroism spectrum of a structurally undefined cobalt-substituted metalloprotein can be used to infer the native structure of the unsubstituted protein. For cobalt-substituted zinc proteins the tetrahedral or near-tetrahedral geometry is the most frequently observed. Examples include carboxypeptidase, D-lactate dehydrogenase, superoxide dismutase, and thermolysin. Five coordination or sites of low symmetry have been inferred in several cases. Although octahedral coordination is comparatively rare in zinc metalloproteins, it has been shown to occur in alkaline phosphatase, pyruvate kinase, glyoxalase I, and most recently in aminopeptidase and phospholipase C. Aminopeptidase is unusual in having both tetrahedral and octahedral zinc sites; phospholipase C is unique in having two octahedral zinc sites.

Heme iron proteins. Magnetic circular dichroism spectroscopy has proved to be a powerful method for use in determining the active-site structures of heme iron proteins. By using the fingerprinting capabilities of the technique as discussed above, it is often possible to establish the structure of a new heme iron system by comparison of its magnetic circular dichroism spectrum with spectra of structurally defined heme iron chromophores. As a result of the sensitivity of the magnetic circular dichroism spectrum to the physical structure of the metal complex, direct comparison between two different heme iron proteins or between a particular heme iron protein and a heme iron model complex can be made without having to worry about the effect of solvent or environment. For the approach to be successful, however, each study must be as extensive as possible, with numerous comparisons considered in order to avoid false conclusions.

An example of this experimental approach is found in the magnetic circular dichroism spectra of ligand complexes of ferric cytochrome P-450 and chloroperoxidase with a water-soluble phosphine [bis(hydroxymethyl)methyl-

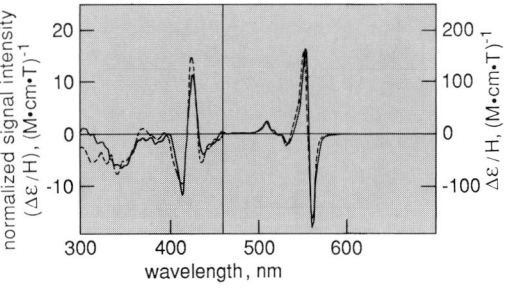

Fig. 3. Magnetic circular dichroism spectra of ferrous cytochrome b_5 (solid line) and bis-imidazole ferrous protoporphyrin IX dimethyl ester (broken line) in CH_2Cl_2. Note the different scales: 300–460 nm region, left scale; 460–700 nm region, right scale. (*After E. W. Svastits and J. H. Dawson, Models for ferrous cytochrome b_5: Sign inversions in the magnetic circular dichroism spectra of bis-imidazole ferrous porphyrin systems, Inorg. Chim. Acta, 123:83–86, 1986*)

phosphine; **Fig. 2**]. The exceptionally close similarity between the two magnetic circular dichroism spectra has been used to argue that the coordination structures of the two enzymes are identical, with a thiolate sulfur donor as the endogenous protein-derived heme iron ligand for chloroperoxidase as had been previously established for P-450. As part of a study of imidazole ligation to ferrous heme iron systems, the magnetic circular dichroism spectra of ferrous cytochrome b_5, a bis-histidine (imidazole) ligated heme iron protein, was compared to a bis-imidazole heme iron model complex (**Fig. 3**). As can be seen, when the coordination structures are the same, the magnetic circular dichroism spectra can be virtually superimposable.

Magnetization studies. For magnetic circular dichroism A and B terms, the intensity of the signal increases linearly with increasing magnetic field. In the case of C terms (Fig. 1*b*), however, signal intensity will increase as a function of magnetic field strength until the difference in energies of the previously degenerate ground states is large enough that only the lower energy state is populated. At that point, further increases in magnetic field will not lead to any further increases in signal. A similar effect occurs upon lowering the temperature at constant magnetic field strength—that is, as the temperature is lowered, the C-term signal intensity reaches a maximum. This phenomenon, known as saturation or magnetization, is always observed in metal complexes with unpaired electrons, that is, in paramagnetic systems.

It has been found that the rates of magnetization of paramagnetic systems as revealed in plots of the signal intensity vs. magnetic field strength of the magnetic circular dichroism signals at a given temperature (saturation or magnetization curves) differ considerably, depending on the magnetic properties of the ground state. Several significant advantages result from the study of

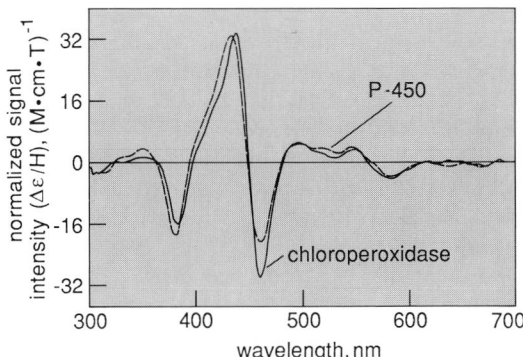

Fig. 2. Magnetic circular dichroism spectra of the ferric chloroperoxidase and P-450-CAM complexes with bis(hydroxymethyl)methylphosphine. (*After M. Sono, J. H. Dawson, and L. P. Hager, Phosphine binding as a structural probe of the chloroperoxidase active site: Spectroscopic evidence for endogenous thiolate ligation to the heme iron, Inorg. Chem., 24:4339–4343, 1985*)

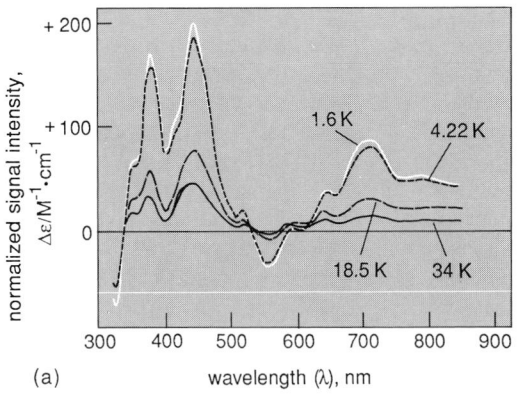

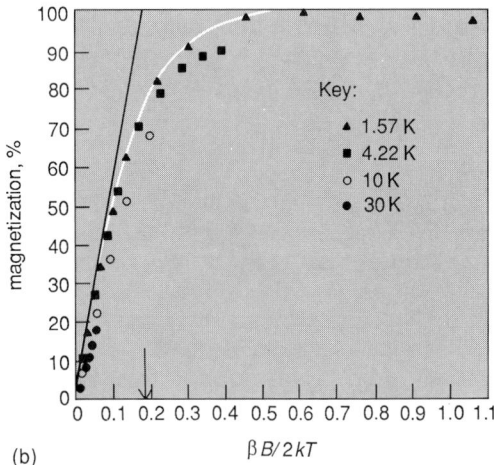

Fig. 4. Dithionite-reduced aconitase. (*a*) Ultralow-temperature magnetic circular dichroism spectra of dithionite-reduced aconitase. (*b*) Magnetic circular dichroism magnetization plot for dithionite-reduced aconitase. Conditions of measurement: wavelength, 710 nm; magnetic fields, 0–4.91 teslas; *k*, Boltzmann constant. (*After A. J. Thomson et al., Characterization of the Fe-S cluster in aconitase using low temperature magnetic circular dichroism spectroscopy, J. Biol. Chem., 259:2274–2282, 1984*)

study of temperature-dependent C terms.

Recent work on the important citric acid cycle enzyme aconitase illustrates the use of magnetic circular dichroism spectroscopy at ultralow temperatures to examine metalloproteins. Although aconitase was known to contain an iron-sulfur center of some type, considerable controversy had existed over the number of iron atoms, the structure of the iron site, and the apparent need to activate aconitase by treatment with additional iron. Extensive work on a large number of iron-sulfur proteins had shown that all known types of iron-sulfur centers could be distinguished by using magnetic circular dichroism spectroscopy, provided that at least two different oxidation states of a particular iron-sulfur protein could be examined. The aconitase study, therefore, focused on both the oxidized and reduced states of the enzyme.

The effect of temperature on the magnetic circular dichroism spectrum of dithionite-reduced aconitase is shown in **Fig. 4** along with the magnetization curve derived from analysis of the data. Comparison of these data on reduced aconitase and on the oxidized enzyme to data on proteins known to contain a three-iron center clearly showed that aconitase, as isolated, contains such a center. This conclusion is based not only on the shapes of the magnetic circular dichroism spectra but also on the magnetization curves. For example, with dithionite-reduced aconitase the rapid rise in magnetization seen in Fig. 4b is interpreted as evidence for a ground state having four unpaired electrons as in other reduced three-iron proteins. Although significant enzymatic activity was found in the three-iron form of aconitase, maximal catalysis was observed with enzyme that had been converted to a four-iron form by treatment with additional iron.
See Metalloprotein; Toxicology.

For background information *see* Bioinorganic chemistry; Faraday effect; Spectroscopy in the McGraw-Hill Encyclopedia of Science and Technology.

John H. Dawson

Bibliography. B. Holmquist et al., Spectral and kinetic studies of metal-substituted *Aeromonas* aminopeptidase: Nonidentical, interacting metal-binding sites, *Biochemistry*, 24:5350–5356, 1985; B. Holmquist et al., A spectral study of cobalt(II)-substituted *Bacillus cereus* phospholipase C, *Biochemistry*, 25:4219–4223, 1986; M. Sono, J. H. Dawson, and L. P. Hager, Phosphine binding as a structural probe of the chloroperoxidase active site: Spectroscopic evidence for endogenous thiolate ligation to the heme iron, *Inorg. Chem.*, 24:4339–4343, 1985; E. W. Svastits and J. H. Dawson, Models for ferrous cytochrome b_5: Sign inversions in the magnetic circular dichroism spectra of bis-imidazole ferrous porphyrin systems, *Inorg. Chim. Acta*, 123:83–86, 1986; A. J. Thompson et al., Characterization of the Fe-S cluster in aconitase using

temperature-dependent C terms. First, it is possible to obtain information about the magnetic properties of specific sites in metalloproteins rather than being forced to measure magnetic properties of the bulk solution. Second, because of the dramatic increase in magnetic circular dichroism intensity for C terms at low temperature, it is often possible to identify electronic transitions that are not evident in the ambient-temperature ultraviolet-visible absorption spectrum. For example, the optical absorption bands due to molybdenum(V) have recently been located for the first time in xanthine oxidase. Third, in contrast to electron paramagnetic resonance spectroscopy, which can generally be applied only to half-integer spin systems, the magnetic circular dichroism magnetization method can be used to study all paramagnetic spin systems. Fourth, it is possible to obtain evidence for low-lying thermally accessible excited states from the

low temperature magnetic circular dichroism spectroscopy, *J. Biol. Chem.*, 259:2274–2282, 1984; A. J. Thompson et al., Detection by low-temperature magnetic circular dichroism spectroscopy of optical absorption bands due to molybdenum(V) in the form of xanthine oxidase giving the desulfo inhibited EPR signal, *Biochem. J.*, 233:107–110, 1986.

Spread spectrum communication

The use of commercial satellite communication networks employing spread spectrum transmission is rapidly increasing. These networks transmit financial, weather, and other information in real time to large numbers of users and can also carry intermittent transmissions from users at a large number of geographically dispersed locations to a central facility.

The receiving stations in a spread spectrum satellite network employ antennas only 2 ft (0.6 m) in diameter, although these antennas must receive signals relayed from satellites in geostationary orbit at an altitude of 22,300 mi (35,800 km). These personal satellite earth stations can now be purchased at commercial outlets. A dramatic example of this technology is its use by professional financial investors. For an ongoing monthly fee, a user equipped with a personal satellite earth station, a personal computer, and appropriate software can receive information about current prices on stock exchanges and other financial markets as the trading takes place. Information can be provided about transactions that occurred less than 2 s previously.

Spread spectrum technique. Spread spectrum modulation has been used for many years by the military to achieve reliable, secure communication despite enemy attempts to intercept or interfere with the signals. It has also been used by radio astronomers to investigate distant planets by transmitting signals to the planet and listening to the extremely weak return echoes. For example, spread spectrum signaling techniques have been used to map the surface of Venus and to study the properties of the rings of Saturn.

Spread spectrum takes advantage of one of the conclusions that can be derived from the fundamental theory of information. In theory, a message can be transmitted reliably by a weak signal subject to substantial interference, provided there is enough redundancy in the message to compensate for the weakness of the signal and the strength of the interference. In conventional communication signaling, each bit (binary digit) of information is communicated by a single symbol (for example, a pulse of energy that is on or off). If that symbol is interfered with or is too weak to be detected, the message is lost.

In spread spectrum communication, each bit of information is represented by a long coded sequence of symbols in a prearranged pattern called a pseudorandom noise (PN) sequence. The weaker the signal or the greater the interference, the longer the pseudorandom noise code sequence must be. One pseudorandom noise sequence may be used to represent a 1 bit, and a different pseudorandom noise sequence may be used to represent a 0 bit. The receiver must know which pair of codes is being used by the transmitter in order to perform a pattern recognition analysis to determine whether the incoming sequence of symbols more nearly resembles a 1-bit pattern or a 0-bit pattern. If the patterns are distinct enough and long enough, the receiver can always distinguish between them.

Application to satellite networks. Without spread spectrum transmission, the small personal earth stations could not function. The amount of satellite signal energy that an earth station collects is proportional to the area of the collection surface, so that a 2-ft-diameter (0.6-m) antenna collects only one-quarter of the energy of a 4-ft (1.2-m) antenna. Furthermore, the smaller the antenna, the larger the angle of view, just as a small telescope has wider field of view than a large telescope. Since communication satellites are spaced 2° apart in the geostationary orbital arc over the Equator, the 2-ft (0.6-m) antenna also receives signal energy from the two satellites on either side of the one at which it is pointed (see **illus.**). Spread spectrum techniques permit the reception of the weak desired signal despite interference from the adjacent satellites. Spread spectrum also protects against interference from ground-based microwave signals which share the same frequency.

This reception of digital data on personal earth stations (sometimes called micro-earth stations) differs from proposed but not yet implemented plans for use of satellites to broadcast television

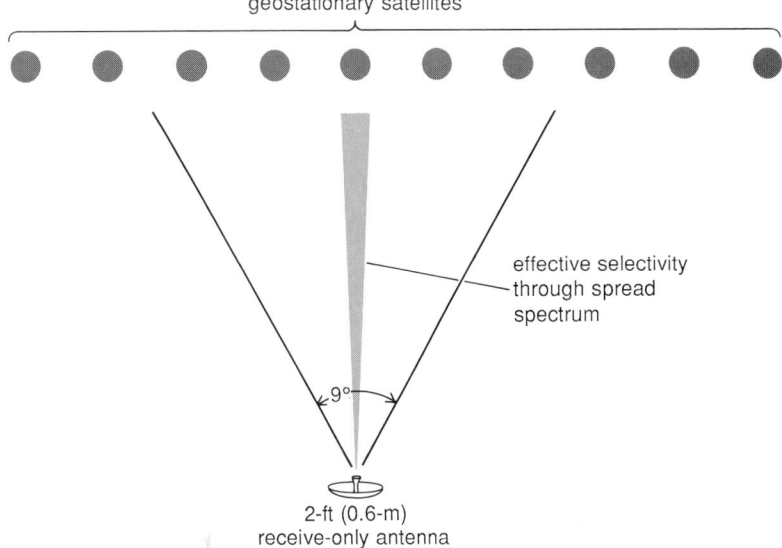

geostationary satellites

effective selectivity through spread spectrum

9°

2-ft (0.6-m) receive-only antenna

Reception by antenna from geostationary satellites in a satellite communication network using spread spectrum transmission. Receive-only antenna, operating in the C band (4–6 GHz), has normal receiving beamwidth of 9°, but is given effective selectivity through spread spectrum. Spacing between adjacent satellites is 2°.

directly to small antennas. All the plans for direct broadcast satellites (DBS) require extremely powerful satellites using signal frequencies different from the conventional satellite communication frequencies. In order to avoid adjacent satellite interference, such satellites were planned to be 20° apart in orbit, compared with conventional satellites spaced 2° apart, so that only one-tenth as many satellites could be used. By early 1986 there were no direct broadcast satellite earth stations in commercial use. In contrast, more than 30,000 personal earth stations were being used for reception of financial, sports, weather, general news, and specialized business information services. Most newspapers and radio stations in the United States receive their news stories from the wire services on the tiny spread spectrum earth stations.

Code division multiple access (CDMA). This technique is another use of spread spectrum in commercial satellite communication networks that permits intermittent transmissions from a large number of geographically dispersed earth stations to be efficiently multiplexed together in the same frequency channel. One business application currently using the technique is transmission directly by satellite from a bank's automatic teller machine to the bank's central computer facility. A small personal earth station with an antenna about 4 ft (1.2 m) in diameter is located at the automatic teller machine. When the user inserts a bank card, keys in a code, or depresses buttons to indicate how much cash is to be withdrawn from the account, those few bits of information are transmitted directly to the satellite and relayed from a large central receiving earth station to the bank computer facility. Return communications from the central computer to authorize the dispersing of cash to the user make use of the same spread technique as the receive-only earth stations discussed previously.

The number of bits transmitted per second from an automatic teller machine is very small, even when the machine is used to capacity. Nevertheless, in order to provide fast response time when information bits are ready for transmission, it is necessary to have a transmission channel ready. However, reserving a dedicated satellite channel for the few bits of information from each automatic teller machine would be an expensive waste of satellite capacity.

In code division multiple access, a different pair of spread spectrum codes is assigned to each micro-earth station sharing a common satellite channel (one code for a 1 bit and a different code for a 0 bit). The codes are typically more than 2000 symbols long for each bit of information. The different codes are selected mathematically to be independent of each other. The result is that, to a receiver at the central facility listening for the particular pair of codes assigned to a particular micro-earth station, all the other codes transmitting on the same channel at the same time sound like random noise.

Even a busy automatic teller machine may require the transmission of fewer than 12 bits per second at its peak busy period. When those bits are transmitted at a rate of 1200 per second, each automatic teller machine is using the channel capacity less than 1% of the time. The code division multiple-access network is designed to permit more than 100 micro-earth stations to share the satellite channel at the same time. Therefore several thousand earth stations can be assigned the same transmitting channel, with each allowed to transmit whenever a user depresses the keys of the automatic teller machine to which it is attached. Other satellite multiplexing techniques are possible, but none provides both fast response time and efficient channel utilization for earth stations that transmit data only a small fraction of the time.

An additional advantage of code division multiple access when used with conventional satellite frequencies is that very little transmitter power is required for each micro-earth station. (Typically a 1-W power amplifier is used.) When this power is spread over the wider frequency bandwidth needed to transmit the long code sequences, the power level per unit of bandwidth is extremely small. This makes it practical to locate micro-earth stations in urban areas without causing interference to nearby radio receivers sharing the same frequency for other uses.

For background information SEE COMMUNICATIONS SATELLITE; DIRECT BROADCASTING SATELLITE SYSTEMS; SPREAD SPECTRUM COMMUNICATION in the McGraw-Hill Encyclopedia of Science and Technology.

Edwin B. Parker

Bibliography. M. L. Olson, D. Johnson, and P. L. Arst, Interactive networking with a satellite, *Data Commun.*, 15(4):195–202, April 1986; E. B. Parker, Satellite micro earth stations: A small investment with big returns, *Data Commun.*, 12(1):97–102, January 1983.

Surface physics

The wave-particle duality of quantum mechanics assigns particle properties (quanta) to waves as well as wavelike properties to particles such as electrons. Plasmons are the quanta associated with waves propagating in matter through the collective motion of large numbers of electrons, while surface plasmons are associated with similar surface waves. In recent years, surface plasmons have been shown to produce many striking optical effects and are being investigated for potential applications in holography, solar cells, miniature light sources, new types of microscopy, ultrasensitive chemical detection instruments, submicrometer physics, and other areas.

Observation of plasmons. For several decades, quantum theory was almost exclusively applied to single-particle behavior. With the development of t he theory of many-body behavior in the 1950s, the plasmon concept evolved in an attempt to explain the experimental data described below. The name plasmon reflects the concept of a metal as a high-density plasma with enormous numbers of mobile electrons. The data of **Fig. 1** show energy losses to plasmons by fast electrons transmitted through a thin metal film. Excitation of one plasmon results in the energy loss peak at 14.2 eV, while excitation of larger numbers of plasmons results in peaks at approximately integral multiples of the one-plasmon loss peak. Data of this type helped establish the plasmon concept in the 1950s.

Similar energy-loss peaks may be obtained for excitation of surface plasmons if a grazing incidence beam is used. In this case the energy loss peaks are observed in the spectrum of reflected electrons. Also, modern scanning-transmission electron microscopes obtain high resolution by utilizing a narrow electron beam, which has proved useful in exciting surface plasmons on particulates on the 10-nanometer scale of sizes. The atomic resolution obtainable in the new field of scanning-tunneling microscopy is used to advantage in stimulating surface plasmons on even smaller particulates. Since the energy of the surface plasmon is characteristic of the type of material, as well as the particulate's shape, chemical identification is made of the smallest amounts of material known by measuring the surface plasmon energy with scanning-tunneling microscopy. *See* MICROSCOPE.

Physical basis of plasmons. If the time-varying electric field of light or of a moving ion or electron acts upon a thin foil, particulate, or other bounded sample, the field polarizes the sample with a time-varying electric charge density. The induced charges on a surface act upon one another electrically in a collective manner to attempt restoration of equilibrium. If the applied field is removed, the rush back toward neutral equilibrium gives the charges more kinetic energy, which causes them to overshoot the equilibrium point. The process subsequently reverses, and so oscillations are established. Surface plasmons are the quanta associated with these collective oscillations. Under the continued action of a time-varying applied field, the surface plasmons appear as resonances in the energy removed or scattered from the source.

The electrodynamic response of a material to a time-varying field depends upon frequency ω. This is codified in a frequency-dependent dielectric function $\epsilon(\omega)$ which denotes the factor by which an applied electric field is reduced inside the material. The surface plasmon frequency is often obtained by solving for the frequency at which $\epsilon(\omega)$ has a characteristic value. This is further elucidated in the discussion below.

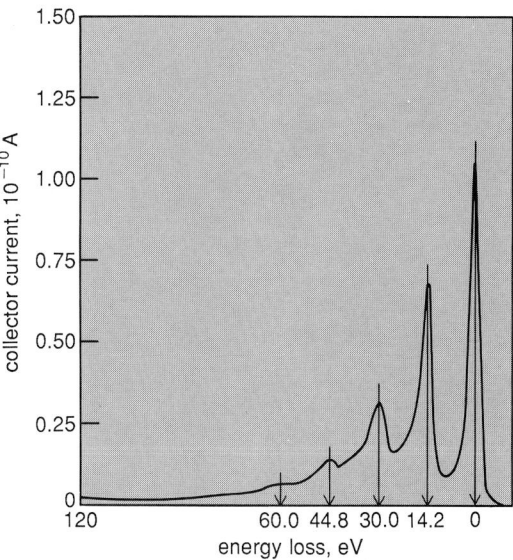

Fig. 1. Current of electrons which have lost energy as a function of the size of the energy loss undergone in traversing a thin foil of aluminum.

If n_0 is the number of dynamical electrons per volume, e is the electron charge, and m is the electron mass, then the volume plasmon energy is $\hbar\omega_p$, where $\hbar$ is Planck's constant divided by 2π, and ω_p is given in gaussian units (commonly used by physicists in this field) by the equation below. In SI units, the right-hand side of this

$$\omega_p^2 = \frac{4\pi n_0 e^2}{m}$$

equation is divided by $4\pi\epsilon_0$, where ϵ_0 is the vacuum permittivity. For a plane-bounded sample, the surface plasmon energy is $\hbar\omega_p \sim \sqrt{2}$, which for most metals is comparable to that of an ultraviolet photon. For a sphere much smaller than the wavelength of light, the surface plasmon energy is $\hbar\omega_p \sim \sqrt{3}$ in the case of a so-called simple metal (a metal in which the electrons are basically free to move throughout the volume).

In practice, for most metals many atomically bound electrons participate in dynamical processes, and the surface plasmon energy is not so simply related to the volume plasmon energy. It turns out that ω_p satisfies $\epsilon(\omega_p) = 0$ in general. Also, the surface plasmon energy on a plane is generally obtained by solving for the frequency at which $\epsilon(\omega) = -1$, while for a sphere the frequency is given by the solution to the equation $\epsilon(\omega) = -2$. These equations are essential when dealing with nonsimple metals.

Surface plasmon optical effects. The surface plasmon energy is always less than the volume plasmon energy and may lie in the range of visible light. This leads to very striking optical effects. Indeed, the connection between electrodynamic effects and optical effects can be quantified in familiar optical parameters such as the index of

Fig. 2. Reflection of four different incident beams from a quartz slide coated with microparticulates of silver. Surface plasmons excited on the microparticulates give color (shown as shades of gray in this black-and-white reproduction) to the reflected light. The most violet-colored reflected beam (the darkest in this figure) is the one in which the minimum amount of reflection comes from the quartz itself, the incident beam being at Brewster's angle in this case.

refraction, a quantity which for nonmagnetic materials is the square of $\epsilon(\omega)$. A particularly simple optical effect of surface plasmons is the light red color of gold sols ("drinkable gold"), which have been known since the studies by alchemists during the Middle Ages. This color is now known to be due to excitation of surface plasmons on the small gold particles by the ambient light. The portion of incident white light absorbed has a frequency given by solving the equation $\epsilon(\omega) = -2$, where $\epsilon(\omega)$ is the dielectric function of gold relative to water. The remaining light is colored due to the absence of the absorbed frequency.

Figure 2 shows some color effects (rendered in shades of gray in this black-and-white reproduction of a color photograph) due to surface plasmon excitation on small silver particles placed on a quartz slide. Four different beams of white light are shown incident on the sample, each beam making a given angle of incidence. For all but one beam the reflectivity of the quartz partially obscures the light reflected from the silver particles. However, one of the beams is incident at Brewster's angle for quartz, the angle at which a minimum amount of light is reflected by the quartz. For this beam the reflected light is a deep-violet color (a darker shade in the figure),

the color seen for white light minus the frequency corresponding to the energy of the silver surface plasmon on the particles.

Applications involving particulates. Samples of this type are very useful in surface-enhanced Raman scattering, which is a technique for chemically identifying ultralow levels of molecules on surfaces. Surface-enhanced Raman scattering can detect less than 1% of a molecular monolayer. It relies upon the high electric fields produced by surface plasmons on microstructures.

Microparticulates of the type used in the example of Fig. 2 have also been deposited on metal-oxide-metal tunnel junctions to produce a new type of light source. In the junctions a small applied voltage permits the quantum-mechanical phenomenon of electron tunneling to occur, thus establishing a current which would be prohibited if classical physics were valid. The tunneling excites surface plasmons on the particulates, and the surface plasmons decay by emitting visible photons. The process is inefficient at present, as is the inverse process (which allows new types of solar cells to be made), but research is under way to improve the efficiency.

Attenuated total reflection apparatus. Unfortunately, surface plasmons cannot be directly stimulated by light on a smooth, flat surface because the laws of conservation of energy and momentum cannot be simultaneously satisfied in this process. However, by utilizing a prism with a flat metal film on the base, the situation is remedied. In this case the velocity of light at a given frequency is reduced by the index of refraction of the prism. As the light passes into the prism at various angles normal to one of the sides, it makes various angles of incidence on the base. At the critical angle, the angle at which the light is usually totally internally reflected, the presence of the foil allows energy transfer. Since the momentum transfer is only the vector component tangential to the surface, the conservation laws can be satisfied. This device is called an attenuated total reflection apparatus. By using a thin film of photoresist on the base, holography can be done with such an apparatus, the main advantage being the high-energy density attained in the thin layer. Holography requires long exposures in the usual experimental setup using photographic film, but the attenuated total reflection apparatus allows much briefer exposures, a definite practical advantage.

Other potential applications. In the future, many new applications will be found for surface plasmons. On general grounds, this can be inferred from the fact that electrons are responsible for most electrical and optical effects, and from the importance of surfaces in nature. Some recent research areas with potential applications (besides those already mentioned) include photon absorption microscopy, integrated optics, optical computers, and nonlinear optics. New types of lasers in which nonlinear processes involving surface plasmons are used to advantage may be produced eventually. At present, considerable basic research remains to be done before technological products are obtained, but many as yet unforeseen applications may be developed.

For background information SEE FREE-ELECTRON THEORY OF METALS; HOLOGRAPHY; REFLECTION OF ELECTROMAGNETIC RADIATION; SURFACE PHYSICS in the McGraw-Hill Encyclopedia of Science and Technology.

Thomas L. Ferrell

Bibliography. D. Bohm and D. Pines, A collective description of electron interactions, *Phys. Rev.*, 108:1175–1204, 1951; T. L. Ferrell, T. A. Callcott, and R. J. Warmack, Plasmons and surfaces, *Amer. Sci.*, 73:344–353, 1985; R. H. Ritchie, Plasma losses by fast electrons in thin films, *Phys. Rev.*, 106:874–881, 1957.

Time

For centuries, time and time transfer have been of vital importance for navigation and for land survey. With the advent of radio or wireless, many countries established shortwave stations to broadcast time signals, and these signals, with an accuracy of a fraction of a second, became available throughout the world. Since the 1920s the combination of accurate time signals and astronomical tables has produced a very useful positioning system.

The development of atomic clocks, particularly cesium clocks, has made possible fully electronic navigation systems. Timing accuracies of 1 microsecond are readily achieved, and with the velocity of radio waves of 1.86×10^5 mi/s (3×10^8 m/s), 1 μs represents an uncertainty in position of only 914 ft (300 m).

Loran C. Loran (long-range navigation) C is a navigation system that covers much of the Earth. It now consists of 13 chains, each having a master station and a number of slave stations. The master station transmits a series of pulses on a frequency of 100 kHz, starting exactly on the second. The slave stations transmit similar pulses, but each slave is delayed by a known number of microseconds.

A ship at sea has a receiver which measures the difference in the arrival time of the pulses from the master and a slave station, which places the ship somewhere on a hyperbolic curve. A similar measurement between the master and a second slave station gives a second hyperbolic curve, and the intersection of the two gives the position.

Loran C also performs a timekeeping function. Each standards laboratory measures the time of arrival of loran C pulses against each of its cesium clocks. These measurements are reported to the Bureau International de l'Heure (BIH) in Paris. The BIH has records of the time delay from the loran C stations to the various laboratories, measured by portable cesium clocks, and the timing of all the loran C chains. This information, and the measurements of some 125 clocks in 25 laboratories, are entered into a computer and used to calculate an average that becomes International Atomic Time (TAI). The time scale used by all countries is Coordinated Universal Time (UTC), which has the same rate as TAI, but which is stepped by adding a leap second where necessary, to keep UTC within 0.8 s of the astronomical time UT1, which is determined by the rotation of the Earth. The BIH reports back to each laboratory how much the laboratory's time differs from UTC. Thus, each country's time is coordinated with the international time UTC, and will normally be only a few microseconds in error.

Global Positioning System. The Global Positioning System (GPS) is a satellite navigation system of the U.S. Department of Defense. When it becomes fully operational, it will have 18 satellites in six orbits with 12-h periods, inclined to the Equator by 55°. While for military purposes the signals will be coded for the highest accuracy of position determination, there is at present a lower-grade signal called the C/A code in the L-band at 1575 MHz that is available to the public. The C/A code has accuracies specified as 305 ft (100 m) in horizontal position, 475 ft (156 m) in height, and 250 nanoseconds in time.

These errors can be reduced by averaging over many satellite passes but, for comparison of time between two laboratories, the so-called common-view technique developed by the National Bureau of Standards is more accurate. When two sites record a satellite's signals at the same time and compare results, any clock error in the satellite is canceled, and errors in satellite position cancel or are much reduced depending on

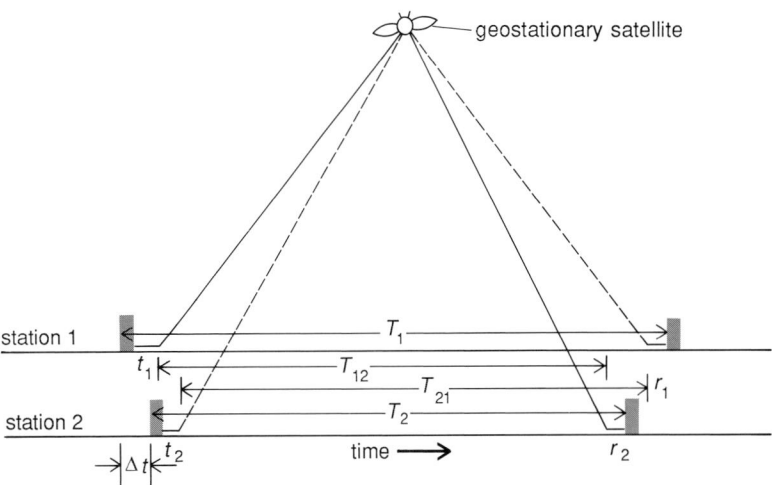

Fig. 1. Satellite paths and time intervals involved in deriving the equation for two-way time transfer using a geostationary satellite.

the geometry of the satellite and the two sites. In addition, variations of uncertainties in the propagation delays in the ionosphere and troposphere can be significantly reduced by the common-view operation.

In 1986 more than half of the 125 cesium clocks reporting to the BIH used the Global Positioning System. The uncertainties in the time transfer are of the order of 10 ns. This is a marked improvement over using the loran C system, where for some sites seasonal variations of over 1 μs in propagation delay have been observed.

Two-way satellite time transfer. Numerous experiments have shown that two-way time transfer using a geostationary satellite has the capability of subnanosecond precision and nanosecond accuracy. In this transfer, each station starts a time interval counter with the seconds pulse as the pulse is transmitted to the satellite, and stops the counter with the seconds pulse received from the partner (**Fig. 1**). The measured time intervals T_1 and T_2 are given by Eqs. (1) and (2), where Δt is

$$T_1 = \Delta t + t_2 + T_{21} + r_1 \qquad (1)$$

$$T_2 = -\Delta t + t_1 + T_{12} + r_2 \qquad (2)$$

the time difference between the two station clocks, T_{12} and T_{21} are the propagation times over the paths from station 1 to station 2 and from station 2 to station 1 respectively, and t_i and r_i are the transmitter and receiver delays in the two stations. If the propagation times over the two paths are identical, Eqs. (1) and (2) can be combined to yield Eq. (3) for the time difference between the station clocks.

If the two stations are identical, the transmitter and receiver delays will cancel, but this is un-

$$\Delta t = \frac{(T_1 - T_2)}{2} + \frac{(t_1 - r_1)}{2} - \frac{(t_2 - r_2)}{2}$$

$$= \frac{(T_1 - T_2)}{2} + \frac{(t_1 - t_2)}{2} - \frac{(r_1 - r_2)}{2} \qquad (3)$$

likely to happen. The simplest way to calibrate the stations is to transport a portable satellite Earth station to each site, a procedure that would be worthwhile for an operational network.

As noted above, Eq. (3) assumes that the propagation times over the two paths are identical. Small errors can arise from different ionospheric and tropospheric delays, but these are much reduced for the 14/12 GHz band now being used.

Figure 2 gives the results of two-way time transfer between the standards laboratory of the National Research Council (NRC) in Ottawa, Ontario, Canada, and that of the Physikalisch-Technische Bundesanstalt (PTB) in Braunschweig, Germany, using the *Symphonie* satellite. The time difference between the clocks at the two laboratories (which are the basis of the national time scales of the respective countries) is plotted as a function of the Mean Julian Day (MJD), a continuous day count used by astronomers. The slope of the curve formed by connecting the data points gives the rate at which the time difference between the clocks varies with time, a dimensionless quantity that is also equal to the fraction by which the time intervals measured by the two clocks differ. Straight lines have been placed on the graph to indicate slopes corresponding to rates of -5×10^{-14} and -2×10^{-14}. The change in rate of 5×10^{-14} at Mean Julian Day 44,500 was the result of a correction applied at the laboratory in Ottawa after an evaluation of its primary cesium standard. During this period the agreement between the time intervals measured by the primary cesium standards of NRC and PTB was well within the combined uncertainty of 7×10^{-14} claimed for the standards in the two laboratories.

Efforts are now being directed toward the establishment of operational networks for time transfer using commercial communications satellites. Standards laboratories have purchased pseudorandom-noise code modems which make it possible to operate with small antennas, 10 ft (3 m) in diameter, and transmitter powers of less than 1 W. The satellite Earth stations can therefore be installed at the laboratory sites, with direct connections to the national time standards. The first network, which should be operational in 1987, will be between the United States Naval Observatory in Washington, D.C., the National Bureau of Standards in Boulder, Colorado, and the National Research Council in Ottawa. The establishment of such a network with nanosec-

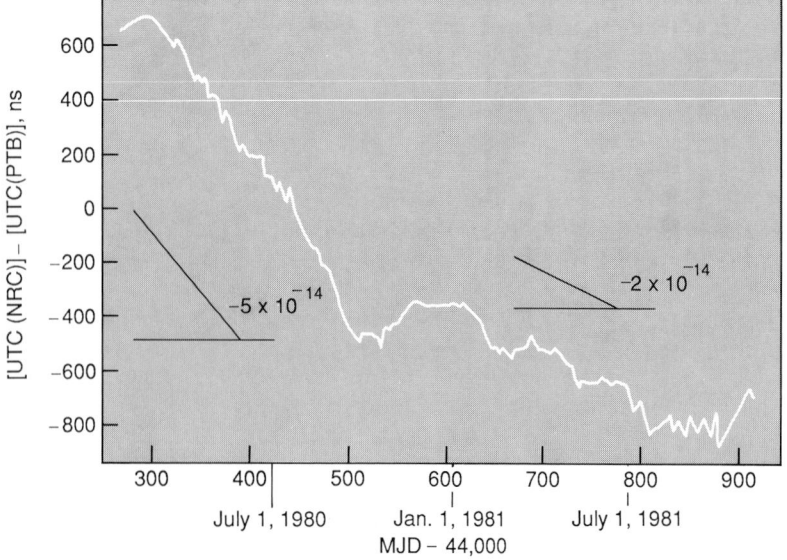

Fig. 2. Difference in the national time scale of Canada, determined by the National Research Council (NRC), and that of West Germany, determined by the Physikalisch-Technische Bundesanstalt (PTB), as measured by using the *Symphonie* satellite. The Mean Julian Day (MJD) is a continuous day count used by astronomers.

ond synchronization will be invaluable in evaluating other systems, such as the satellites of the Global Positioning System, and in the development of new systems of precise time and frequency distribution. SEE SPREAD SPECTRUM COMMUNICATION.

Applications of accurate time signals. Any advance in precision or accuracy in the fields of time, time transfer, and frequency standards has immediate application for commercial exploitation and for public safety. There is a need for a one-way, nonmilitary time distribution system with microsecond accuracy. One important group that would use such a service is the electric power companies. With precise time available at many points along power transmission lines, more accurate measurement of impedance and the phase of voltage and current could be made, leading to increased efficiency of power distribution. Since every percentage point gained represents millions of dollars saved, there is little doubt that some system now under development will become operational.

Accurate time and frequency signals are of increasing importance in communications. Most networks use commercial cesium standards which enable communication to take place between networks without one becoming the slave of other. The frequency matching is particularly important for high-speed data transmission, since otherwise large buffer memories are needed to store the overflow data.

Accurate time signals are necessary for the development of time-sharing systems. Since the communication networks are needed for time transfer, and since the networks need accurate time signals, the mutual self-interest of those involved in establishing and maintaining time standards and those responsible for operating the networks should lead to even better cooperation than in the past.

There are continuing efforts to develop more accurate systems for navigation. A continuing goal is to develop systems with a time accuracy of 1 ns, equivalent to a position accuracy of 1 ft (30 cm). Thus, the symbiotic relation between the fields of timekeeping and navigation will certainly continue.

For background information SEE ATOMIC CLOCK; ATOMIC TIME; COMMUNICATIONS SATELLITE; LORAN; SATELLITE NAVIGATION SYSTEMS; TIME in the McGraw-Hill Encyclopedia of Science and Technology.

C. C. Costain

Bibliography. International Radio Consultative Committee (CCIR), *Standard Frequencies and Time Signals*, Rep. 518, vol. 7, 1982; 1984 Conference on Precision Electromagnetic Measurements, *IEEE Transac. Instrum. Meas.*, vol. IM-34, no. 2, June 1985; Special issue on radio measurement methods and standards, *Proc. IEEE*, vol. 74, no. 1, January 1986.

Tomato

Tomato germ plasm is a collective term that includes the living genetic reserves of this vegetable. The most common type of tomato germ plasm is represented by commercial varieties, such as Marglobe, Rutgers, and Bonny Best. These varieties, or cultivars, represent the culmination of breeding programs designed to accumulate many genes for good quality into a single plant. This plant will then pass these genes on to its offspring. These offspring are essentially identical to the parent because of the uniformity (homozygosity) of many of the specific gene loci (specific sites on chromosomes) occurring in that parent.

Storage of germ plasm. The most common storage form for tomato germ plasm is seed. Since all the genetic information of the parents is contained in the embryo within the seed, the seed is the most efficient storage mechanism for germ-plasm preservation. Under proper conditions (a temperature of 41°F or 5°C and a relative humidity of 35%) seeds stored for over 30 years will produce healthy tomato plants with little, if any, changes in form or quality.

Other storage forms for tomato germ plasm include live plants, meristems, and tissue cultures. These forms require more sophisticated equipment and techniques. Because of their cost, they are used mainly for special purposes.

Genetic diversity. Tomato breeders have produced an enormous number of different cultivars since the 1930s. These varieties were made possible by the great genetic diversity available in the tomato (**Fig. 1**).

The tomato belongs to the potato family and has the taxonomic name *Lycopersicon esculentum*. Tremendous genetic diversity has been found and exploited within this species and its small fruited botanical variety, *L. esculentum* var. *cerasiforme*. Not until the 1930s and 1940s, however, was the tomato germ-plasm pool expanded by the production of interspecific crosses

Fig. 1. Various types of tomato, indicating its genetic diversity, from small-fruited wild types to large-fruited commercial types.

between *L. esculentum* and its wild relatives. These crosses, especially with *L. pimpinellifolium* (red currant tomato) and *L. peruvianum*, greatly broadened the genetic diversity available to tomato breeders. Most of the modern tomato varieties now include some genes from these wild species, usually genes for disease resistance.

Germ plasm in breeding programs. In the early 1900s, tomato breeders utilized only *L. esculentum* germ plasm. Enough variability existed in this species, however, to allow the development of many important varieties, such as Pritchard, Marglobe, and Marvel.

In the 1930s, a new era in the use of tomato germ plasm was inaugurated when disease resistance from the wild species *L. pimpinellifolium* was transferred into *L. esculentum*. From that time on, breeders could look for desirable new genes in the entire genus *Lycopersicon*, instead of just in *L. esculentum*.

Disease resistance. After the importance of the *Fusarium* wilt resistance gene I from *L. pimpinellifolium* was realized, almost all succeeding varieties carried this wild germ plasm. This germ plasm was collected in Peru in 1929, and in 1940 was reported to have completely dominant resistance to *Fusarium* wilt. Another cross between these two species led to the spotted wilt–resistant varieties Pearl Harbor and Oahu, developed in Hawaii. Other disease resistance attributed to various accessions of the red currant tomato includes bacterial wilt, leaf mold, bacterial speck, bacterial canker, collar rot, *Stemphyllium*, and *Septoria* leaf blight.

Other wild species have also been rich sources of valuable germ plasm. Resistance to the omnipresent tobacco mosaic virus was located in *L. peruvianum* (TM-2^a gene), as well as resistance (Mi gene) to root-knot nematode. Crosses between *L. esculentum* and *L. peruvianum* are difficult to make, usually requiring some form of embryo rescue, a technique developed in California in the 1940s. In this procedure the wild species is used as the male parent. Approximately 1 month after pollination, the embryos are excised from normally developing fruits, and placed on an artificial medium. Continued growth occurs in only a few (less than 10%) of these rescued embryos, but none survive if left in the fruit. After root and shoot development, the small plants are potted in greenhouse soil for continued growth. However, only a small portion of the great genetic diversity available in *L. peruvianum* has been successfully transferred to commercial tomato varieties.

Curly top virus resistance was derived from *L. peruvianum*. *L. peruvianum* var. *humifusum*, and *L. hirsutum* (**Fig. 2**). Inheritance of this resistance is complex, involving many genes that are closely linked (located nearby on the chromosome) with undesirable genes. After more than 40 years of constant selection and crossing, enough of these linkages were broken so that high curly top virus resistance occurred in lines having commercial quality. This program was initiated in Utah and reached fruition when the four curly top virus–resistant varieties, namely Rowpac, Columbia, Saladmaster, and Roza, were released in 1976.

Enhanced flesh color. In addition to disease resistance, high vitamin C content and enhanced flesh color have been derived from *L. peruvianum*. The discovery of the latter characteristic was a surprise since the trait is invisible in the green fruits of the wild species. Only when genes coding for this characteristic were incorporated into the red-fruited species did the trait become visible.

Insect resistance. Insect resistance has also been located in various parts of the tomato germ-plasm gene pool. One of the first reports was of the resistance of *L. hirsutum* and *L. hirsutum* f. *glabratum* to red spider mites (members of a group closely related to insects) in 1941. By 1968, *L. hirsutum* had also been reported to be resistant to white flies and leaf miners, and *L. hirsutum* f. *glabratum* resistant to tobacco flea beetles and fruit worm.

Tolerance to environmental stresses. Recent studies have shown remarkable tolerance to various environmental stresses in the wild species. For example, an accession of *L. cheesmanii*, collected in 1971, was growing only 16 ft (5 m) from the tide line on Isabela Island in the Galápagos. Subsequent studies in California demonstrated that this accession could tolerate up to 100% sea water in hydroponic culture, and could even grow slowly under those conditions. No other tomato germ plasm has endured such high osmotic concentrations. SEE PLANTS, SALINE ENVIRONMENTS OF.

Another remarkable species is *L. pennellii*. It is found in desert environments in Peru and Chile

Fig. 2. Comparison of commercial types of tomato killed by curly top virus with breeding lines that have acquired resistance to this virus from genes from wild species.

killed commercial types

resistant breeding lines

and possesses unusual drought tolerance. Cuttings left on greenhouse benches for 24 h will often revive when subsequently placed in a rooting bed. Other tomato species would not survive such water stress.

Cold tolerance is found in the species *Solanum lycopersicoides*. Even though botanically classified as a member of the potato genus, it is crossable with tomato. Additional tolerance to cold temperatures is found in collections of *L. hirsutum* from high-altitude sites (up to 10,000 ft or 3000 m). Growth at temperatures of 50°F (10°C) is encountered in some of this germ plasm.

Identification of most useful germ plasm. Some of the more recently recognized tomato relatives (*L. chmielewskii*, *L. parviflorum*, and *L. pennellii*, for example) have not yet been adequately evaluated. Due to difficulty in crossing with tomato, *L. peruvianum* germ plasm has not been extensively utilized. In 1975 a summary of the disease, insect, mite, nematode, and chemical resistance reported on the tomato germ-plasm collection maintained by the U.S. Department of Agriculture at Ames, Iowa, showed that over 200 individual accessions in the collection had been reported to have at least some resistance to two or more pests. The Ames collection numbered over 3000 at that time. The breakdown of accessions with resistance to two or more pests was: 72 accessions of *L. esculentum*, 15 of *L. hirsutum*, 52 of *L. peruvianum*, 38 of *L. pimpinellifolium*, and 57 of *L. esculentum* × *L. pimpinellifolium* hybrids. A few accessions showed resistance to more than ten pests each. Only two species, *L. hirsutum* (3 accessions) and *L. peruvianum* (9 accessions) showed this broad range of pest resistance. Leading the list was an accession of *L. hirsutum* with resistance to eight diseases, two nematodes, and five insects. Closely following was an accession of *L. peruvianum* with resistance to eleven diseases, two nematodes, and one insect. It is likely that additional research will expand this impressive list of pest-resistant wild germ plasm.

Genetic engineering potential. From the above discussion it appears that whatever genes may be needed for development of new tomato varieties can be located somewhere in the genus *Lycopersicon*. However, some gene transfers are accomplished only with great difficulty and much perseverance. New tools provided by genetic engineers may prove very useful in accomplishing some of these difficult transfers.

Work with gel electrophoresis of alloenzymes has provided a tool to monitor quickly, and very precisely, the inheritance of associated genes. The extremely wide range of well-documented genotypes in tomato and the extensive chromosome mapping of these genes have provided a resource to genetic engineers unequaled in any other crop species. Application of genetic engineering techniques to tomato germ plasm will undoubtedly lead to additional advances in the development of this important vegetable crop.

For background information *see* Breeding (plant); Genetic engineering; Plant pathology; Tomato in the McGraw-Hill Encyclopedia of Science and Technology.

Raymond L. Clark

Bibliography. H. J. Otto et al., *Plant Germplasm Preservation and Utilization in U. S. Agriculture*, Counc. Agr. Sci. Technol. Rep. 106, 1985; C. E. Peterson, Plant introductions in the improvement of vegetable cultivars, *HortScience*, 10:575–579, 1975; C. M. Rick, Conservation of tomato species germplasm, *Calif. Agr.*, 31:32–33, 1977; C. M. Rick, Tomato mutants: Freaks, anomalies, and breeders' resources, *HortScience*, 21:918, 1986.

Toxicology

Many substances in the environment that are taken in by an organism can have harmful effects on the organism. This article first discusses various aspects of biological detoxification and then treats drug toxicity in the mammalian fetus and newborn.

Biological Detoxification

Detoxification can be defined as the removal of the toxic properties of a chemical substance. This process is seemingly ubiquitous in living organisms, since it occurs in bacteria, plants, and insects as well as in animals and humans. Existing detoxification mechanisms do not provide absolute protection. Indeed, if they did, the present armamentarium of antibiotics, insecticides, herbicides, rodenticides, and so forth would be useless; and chemical carcinogens, environmental pollutants, and drug toxicity would be of little concern. However, the detoxification mechanisms available have to be considered overwhelmingly successful in an era when living creatures are each year subjected to innumerable new xenobiotics, chemical substances not naturally found in living organisms.

Animals and humans. Vertebrates possess the most diverse and the most thoroughly studied detoxification mechanisms. They will be discussed first because they illustrate many of the mechanisms used by lower organisms as well. The detoxification reactions are often classified as phase I and phase II reactions. Phase I reactions are nonsynthetic and usually convert the chemical compound into a more water-soluble (hydrophilic) metabolite by means of oxidation, reduction, or hydrolysis. Phase II reactions, which are primarily synthetic, conjugation reactions, also often result in an increase in watersolubility. The increased water solubility of many products of detoxification facilitates their equilibration with body water and thereby renal excretion. Other routes of excretion can include elimination in the bile, by the intestine, and sometimes by the lungs. The major detoxification systems participating in these reactions are the hepatic microsomal enzymes and the soluble (cytoplasmic) enzymes.

Microsomal enzyme systems. Microsomes are particles 50–150 nanometers in diameter which are formed when cells are ground or homogenized. They represent fragments of the endoplasmic reticulum, a complex membrane system found in the cytoplasm. These membranes are the sites of numerous important detoxification reactions, notably those catalyzed by the cytochrome P-450 family of enzymes and by the glucuronyltransferases. This latter enzyme conjugates (by a phase II reaction) glucuronic acid with a variety of alcohols and carboxylic acids. For example, by using uridinediphosphate glucuronic acid as one of the substrates, phenol is converted to phenylglucuronide, and bilirubin is converted to a variety of bilirubin glucuronides by glycosylation of the propionic acid side chains. This latter reaction is important for the detoxification of the approximately 250 mg of bilirubin generated per day in adults, the accumulation of which would lead to jaundice and other more serious problems such as kernicterus (brain damage from bilirubin accumulation).

The cytochrome P-450 enzymes were so named because they are pigmented hemoproteins with an absorption maximum near 450 nm when combined with carbon monoxide. The heme portion of this cytochrome is ferriprotoporphyrin IX, the same found in hemoglobin. Cytochrome P-450 uses the electrons donated to it from nicotinamide adenine dinucleotide (NADH) or nicotinamide adenine dinucleotide phosphate (NADPH) [via a flavoprotein, cytochrome P-450 reductase] and molecular oxygen to bring about the oxidation of numerous compounds The cytochromes P-450 exist as a number of isoenzymes, all with molecules having a mass of approximately 50,000 daltons, that vary between species, strains, and individuals. These isoenzymes differ in their substrate specificities, physical properties, and inducibility. Recent genetic studies have shown that the cytochrome P-450 apoproteins have several regions of amino acid and deoxyribonucleic acid (DNA) homology, allowing their classification into several families within a multigene superfamily. The genetics of these enzymes is being studied in great detail; they may be one factor in determining why, for example, some smokers develop lung cancer and other smokers do not. These genetic families may also provide an explanation for the different induction patterns seen with the enzymes. The most commonly studied inducer chemicals are phenobarbital and polycyclic aromatic hydrocarbons (such as benzo[a]pyrene and 3-methylcholanthrene), and they can increase the quantity of some P-450

enzymes by 20-fold or more. *See Metalloprotein.*

The cytochromes P-450 are also referred to as mixed-function oxidases or xenobiotic monooxygenases. They catalyze a variety of reactions, including the hydroxylation of aliphatic and aromatic compounds and the generation of epoxides as shown in the **illustration**. Bromobenzene can be converted to the 2,3-oxide or the 3,4-oxide. These epoxides are highly reactive electrophiles and can exert toxic (including carcinogenic) effects by reacting with essential macromolecules such as DNA and protein. They can rearrange nonenzymatically to the respective 2- and 4-bromophenols or participate in the other pathways shown. The P-450 system can also catalyze dealkylation, that is, removal of alkyl groups coupled to nitrogen, oxygen, or sulfur molecules. These reactions probably proceed by a hydroxylated intermediate. An example of *N*-dealkylation is shown in reaction (1). Codeine

| Phenacetin | *p*-Hydroxyacetanilide |

(methylmorphine) is converted to morphine by *O*-dealkylation. The P-450 enzymes also catalyze *N*- and *S*-oxidations [reaction (2)], desulfuration

$$(CH_3)_3N \longrightarrow (CH_3)_3N{=}O \tag{2}$$

Trimethylamine   Trimethylamine *N*-oxide

(for example, parathion → paraoxon), and deamination (for example, amphetamine → phenylacetone + NH_3).

The versatility of the P-450 system is fully matched by the capabilities of the soluble enzymes of detoxification. These enzymes are referred to as soluble because they are found mainly in the soluble phase of disrupted cells and tissues. However, they may have been subtly organized within the cell, and not simply diffused freely in the cytoplasm. Although there are numerous soluble enzymes of detoxification, only a few of the major ones will be discussed here.

Glutathione S-transferases. The glutathione *S*-transferases are major soluble enzymes of detoxification. These enzymes are found in humans and other animals, plants, and insects, and in some bacteria. Their widespread distribution and, in many cases, high concentration (for example, up to 10% of the soluble protein of rat liver) argue for an important role.

All of the glutathione transferases have molecules with masses of 40,000 to 50,000 daltons and consist of two subunits of approximately equal size. Combination of different subunits can lead to the formation of a family of enzymes, each with differing catalytic properties. Over 10 forms of these enzymes have been found in humans and in rats. At least some of the rat and human transferases exhibit regions of homologous amino acid sequences, suggesting a common ancestral gene.

The glutathione transferases conjugate the tripeptide glutathione (γ-glutamylcysteinyl-glycine) with numerous reactive electrophiles. The liver is unusually rich in glutathione (GSH); rat liver, for example, contains about 5 millimoles of glutathione per kilogram.

The ionized sulfhydryl group of cysteine is the reactive portion of glutathione and the enzymes catalyze the attack of this thiolate anion (GS^-) on an electrophilic carbon, nitrogen, oxygen, or sulfur atom. The glutathione conjugate is often converted subsequently to a mercapturic acid (an N-acetylcysteine derivative) by other enzymes which remove the glycine and glutamate residues and acetylate the cysteine.

The illustration depicts some important principles. The potentially toxic bromobenzene epoxides can react with glutathione to form a harmless conjugate. In some animals, up to 70% of a subtoxic dose of bromobenzene is excreted as the mercapturic acid. The degree of toxicity of bromobenzene increases greatly if the level of tissue glutathione is reduced below a threshold value either by treatment with other chemicals which deplete glutathione or block its synthesis, or by treatment with large doses of bromobenzene, which itself depletes the available glutathione. For this reason, potentially lethal overdoses of the popular analgesic acetaminophen, which is metabolized to a thiol-reactive intermediate, are sometimes treated with agents that replenish hepatic thiol levels.

The transferases occur as many different isoenzymic forms in the same animal. These enzymic forms exhibit some differences in catalytic capabilities, but in general the transferases possess very broad and overlapping specificities for the electrophilic substrate. This lack of a tightly defined specificity allows them to catalyze the reaction of glutathione with an extremely large range of second substrates.

Another example of the transferase-catalyzed conjugation of glutathione with an electrophilic carbon is shown in reaction (3). Chlorodinitro-

acid, a diuretic, is also a substrate for many of these enzymes. In this case the GS^- attacks an electrophilic alkene carbon.

The nitrate ester vasodilators (and explosives) such as nitroglycerin and erythrityl tetranitrate are metabolized by the transferase-catalyzed attack of GS^- on an electrophilic nitrogen atom. This reaction is thought to proceed through a reactive intermediate [reactions (4)].

$$RONO_2 + GSH \xrightarrow{\text{enzymatic}} ROH + [GSNO_2]$$

$$[GSNO_2] + GSH \xrightarrow{\text{non-enzymatic}} GSSG + HNO_2$$

$$\overline{RONO_2 + 2GSH \longrightarrow ROH + GSSG + HNO_2} \quad (4)$$

Nitrate ester	Glutathione		Alcohol	Glutathione disulfide	Nitrous acid

In a similar two-step reaction the attack of GS^- on an electrophilic oxygen can result in conversion of an organic hydroperoxide to an alcohol. The reaction of glutathione with the sulfur atom of organic thiocyanates results in the formation of hydrogen cyanide (HCN) and an asymmetric disulfide. Some organic thiocyanate insecticides may be

benzene is a good substrate for nearly all of the known glutathione transferases; hence it is widely used to study these enzymes. Ethacrynic

(3)

1-Chloro-2,4-dinitrobenzene   S-(2,4-Dinitrophenyl)glutathione

Bromobenzene illustrates the multiple pathways of reactive electrophiles.

toxic to animals because of this transferase-catalyzed release of cyanide. In this case the enzyme is not functioning as a detoxification mechanism.

In addition to the catalytic capabilities of the glutathione transferases, there is evidence that they may play a role as binding proteins for a variety of nonsubstrate molecules. For example, an abundant liver protein was purified and named ligandin because of its ability to bind bilirubin, an azodye carcinogen, and a cortisol metabolite. Only later was this protein shown to be a glutathione transferase. It also appears that in some cases (for example, binding of an azodye carcinogen and paracetamol metabolites) the transferases themselves may function as sacrificial targets for reactive electrophiles, thereby sparing other more critical intracellular macromolecules.

Additional evidence of the role of these enzymes in detoxification comes from the finding of elevated levels of mercapturic acids in the urine of workers exposed to chemicals. While the major role of the transferases appears to be detoxification, there is evidence that these enzymes also participate in the metabolism of some normal endogenous steroids, leukotrienes, and lipid peroxides.

Epoxide hydrolase and others. Epoxide hydrolase, an enzyme which catalyzes the conversion of epoxides to dihydrodiols (glycols), is found in both soluble and microsomal forms. The reaction shown in the illustration is the conversion of bromobenzene epoxide to a diol. The microsomal location of some of this enzyme may facilitate rapid detoxification of reactive epoxide products formed nearby by the microsomal P-450 enzymes.

Glutathione peroxidase converts organic hydroperoxides and hydrogen peroxide to alcohols and water, respectively. This enzyme and the glutathione transferases are thought to play an essential role in protection against lipid peroxidation. Another important enzyme of detoxification is arylsulfotransferase, which converts a variety of phenolic compounds to more readily excretable arylsulfates. Numerous other interesting and functionally important detoxification systems are known.

Plants. Atrazine is one of the most widely used herbicides for the control of annual weeds in corn and sorghum. It functions so well because these crops appear to be resistant to atrazine (2-chloro-4-ethylamino-6-isopropyl-amino-*s*-triazine) because of their ability to metabolize this herbicide to a glutathione conjugate. Plants contain both glutathione and glutathione *S*-transferases. Interestingly, the gene encoding for one of the corn glutathione transferases has been sequenced, but no homology with animal transferases has been detected despite several similarities in the enzymes, for example, molecular weight, subunit structure, and substrates.

The toxicity of another herbicide, glyphosphate [*N*-(phosphonylmethyl)glycine], has been experimentally reduced by a novel method. Glyphosphate is a potent inhibitor of 5-enolpyruvylshikimate-3-phosphate synthase, an enzyme essential for production of certain aromatic amino acids in plants. Resistance to glyphosphate has been achieved by greatly increasing the quantity of this enzyme by using genetic engineering methods. In this way, even in the presence of some glyphosphate, adequate synthase activity remains for normal plant growth. This apparent tolerance conferred by gene duplication can also occur naturally and is functionally analogous to the excess capacity found in many animal organs. For example, many events (for example, drug and alcohol use) result in some toxic effects on the host, but these are often not noticeable because the remaining undamaged tissue allows normal body functions to continue.

Bacteria. Bacteria are often resistant to the effects of antibiotics because of their ability to detoxify these drugs. Penicillin, for example, is toxic to many gram-positive bacteria because of its ability to interfere with cell wall synthesis. Some bacteria, however, can produce β-lactamase, an enzyme which hydrolyzes the β-lactam ring of penicillin to produce a less active product.

Chloramphenicol

+

Acetyl coenzyme A   (5)

3-Acetoxy chloramphenicol

+

CoA—SH

Coenzyme A

As another example, bacteria which produce chloramphenicol acetytransferase are resistant to chloramphenicol. This enzyme uses acetyl coenzyme A to acetylate the C-3 hydroxyl group of the antibiotic, thus rendering it inactive [reaction (5)].

William Habig

Drug Toxicity in Fetus and Newborn

The newborn mammal is anatomically, functionally, and metabolically different from the adult. As a consequence, the capacity to handle drugs is unique in the newborn; in general, more severe adverse effects are observed. Neonatal adverse effects can arise either from direct drug administration or from exposure via the mother during pregnancy. The modern toxicologist is becoming aware that drug toxicity in the newborn must be regarded as a separate field of study.

Drug adverse effects. Drugs are chemical agents that under normal circumstances are consumed for therapeutic reasons. They enter the mammalian system by ingestion or injection and then undergo various processes, including absorption, distribution, metabolism, and finally excretion. In this ideal situation the drug produces a desired effect and is then eliminated. However, drugs can cause adverse or toxic effects even if applied to combat a disease. In the case of cancer chemotherapy, drug adverse effects can sometimes worsen the condition, as normal tissue is subject to severe damage.

The ability of mammals to cope with drugs and eliminate these substances by excretion via the kidneys is partially dependent on the rates at which drugs are metabolized to inactive compounds. The major sites of metabolism by mammalian enzymes are the liver and gastrointestinal tract, with lesser activity present in the lungs, kidneys, and skin. Essentially, drugs are biotransformed to intermediates, which are then conjugated or combined with substrates present in mammalian cells. The conjugated compounds are excreted, usually the final step in the drug detoxification process. The drug-metabolizing enzyme system has been studied extensively in the adult and is characteristically different from that of the newborn.

With the realization that age plays a critical role in the responsiveness of organisms to drugs, attention has recently shifted to the study of the effects of drugs on the fetus and newborn. In the last few years the field of teratology has expanded, with a resultant increase in knowledge of the development of the fetus. As a natural progression, a new field is emerging that follows fetal development into the newborn state and studies the consequences of drug treatment in newborns.

In comparison with the effects in adults, drugs in newborns are metabolized more slowly and the activities of many drug-metabolizing enzymes are considerably lower. With the onset of parturition, the newborn is thrust into a new environment where it must assume the function of drug metabolism. Consequently, the drug-metabolizing enzyme systems develop rapidly, attaining adult levels within a few days or weeks. In terms of these differences in drug-metabolizing enzyme systems, the responsiveness of neonates to drugs is unique and must be considered prior to therapy.

The period of drug administration in the neonate plays an important role in adverse reactions. Drug toxicity in the newborn can be manifested by the exposure of the pregnant mother to the chemical; an example is the tranquilizer thalidomide, which resulted in deformed infants. Toxicity can also be produced directly in the newborn; the direct administration of chloramphenicol to newborns resulted in cardiovascular collapse, coma, and cyanosis. Conversely, newborns can display a decreased susceptibility to adverse drug reactions in comparison to adults; exposure to oxygen produces less pulmonary damage in neonates than in adults. The structural and functional characteristics of newborns govern the unique responses to drugs.

Maternal drug exposure. The agents categorized as drugs of abuse include alcohol and the opiate narcotics such as heroin, morphine, and codeine. In general, these drugs are not used for therapeutic reasons, and infants suffer the toxic consequences as a result of maternal exposure. Infants born to heroin-dependent mothers have an increased incidence of low birth weight and a low Apgar score, a numerical expression of the condition of the newborn involving assessment of heart rate, respiration, color, and reflexes. Other common postnatal problems include jaundice, infection, pneumonia, hyaline membrane disease, and congenital malformations. Opiates produce a phenomenon of dependence. Hence removal of the drug, at birth, can result in neonatal abstinence syndrome (NAS). In the case of infants born to heroin-addicted mothers, neonatal abstinence syndrome occurs within minutes to hours after birth and can persist for several months. The features of neonatal abstinence syndrome include hyperirritability, respiratory distress, tremors, high-pitched cries, increased muscle tone, ineffectual and uncoordinated sucking and swallowing reflexes, and dehydration. While the mechanism of opiate-induced toxicity at the cellular level has not been resolved in the infant, it has been demonstrated that the metabolizing enzymes required to inactivate morphine are virtually absent in the liver of the newborn for the first 2 weeks of life.

Excessive intake of alcohol during pregnancy gives rises to fetal alcohol syndrome (FAS). Infants with fetal alcohol syndrome exhibit growth retardation, irritability, poor motor coordination, hyperactivity in childhood, and retarded mental development, as well as abnormal facial characteristics. At present, it is believed that alcohol causes fetal hypoxia; insufficient amounts of oxygen reach the fetus, leading to behavioral abnormalities. As in the case of opiates, the activity of the enzyme necessary to inactivate alcohol is very low at birth, and this could account for enhanced toxicity in the newborn.

Barbiturates are used therapeutically as

Table 1. Effect of maternal drug ingestion on the neonate*

Drug (class)	Terato-genesis	Abnormal postnatal growth	Impaired sexual reproductive capacity	Mental retardation	Metabolic disturbances	Fetal death	Hemato-logical changes
Phenytoin (anticonvulsant)	+	+	·	+	·	·	+
Diazepam (antidepressant)	+	+	·	+	·	·	·
Dicumarol (anticoagulant)	+	·	·	+	·	+	+
Chlorpropamide (oral hypoglycemic)	·	·	·	·	+	+	·
Cytarabine (antineoplastic)	+	·	+	·	·	·	+
Propranolol (antihypertensive)	·	·	·	·	+	·	·
Corticosteroids (anti-inflammatory)	+	+	·	·	+	·	·
Salicylates (anti-inflammatory)	·	·	·	·	+	+	+

* + denotes a positive reaction.

sedative-hypnotics and as anticonvulsants in epilepsy. However, the abuse of these compounds is associated with toxicity in newborns. Infants born to mothers who have ingested excessive amounts of barbiturates display a depression of central nervous system function. Since barbiturates produce dependence, removal of newborns from the maternal drug source results in barbiturate withdrawal syndrome. The symptoms associated with barbiturate withdrawal syndrome include irritability, tremulousness, high-pitched cry, hyperphagia, and hyperacusia. The incidence and severity of drug toxicity is far more prominent in newborns than in adults. Since the ability of the newborn to metabolize drugs is lower than that of the adult, compounds tend to accumulate and cause prolonged and more severe toxicity.

The preceding examples are extensively studied drugs which, when taken during pregnancy for nontherapeutic reasons, cause adverse reactions in the newborn. However, virtually all drugs are chemicals foreign to the mammal and, even when administered therapeutically during pregnancy, can exert a toxic reaction in the neonate. The adverse reactions of some drugs commonly used during pregnancy are listed in **Table 1**.

Newborn exposure. Sometimes drugs must be administered directly to newborns for the prevention of certain diseases. Gentamicin is an antibiotic commonly used to combat gram-negative infections. It has been established that administration of gentamicin produces nephrotoxicity in mammals and that accumulation of phospholipid in the kidneys is a necessary step in the pathogenesis of this toxic effect. Surprisingly, the newborn is less susceptible (or more resistant) than the adult to gentamicin-induced nephrotoxicity. Since inhibition of lysosomal phospholipases is responsible for phospholipid accumulation, it is conceivable that gentamicin does not block these enzymes in the newborn. It is also possible that phospholipases are not yet developed in the newborn and hence cannot be inhibited by gentamicin. This phenomenon of

decreased susceptibility to drug-induced toxicity has also been seen in newborns given chlorphentermine, an appetite suppressant.

Infant respiratory distress syndrome is a condition associated with premature birth where there is insufficient surfactant to stabilize the alveoli and allow inflation. Consequently, there is a marked dyspnea with cyanosis followed by death. Oxygen replacement is utilized in the therapy of infant respiratory distress syndrome. However, the use of hyperoxia in infant respiratory distress syndrome has been found to result in lung damage and blindness. In contrast to premature infants, normal full-term babies, if exposed to oxygen, develop a resistance to oxygen toxicity. As in the case of gentamicin, newborns are less susceptible to oxygen-induced lung damage than adults. Mammals possess lung enzymes capable of counteracting oxygen toxicity which appear to function more efficiently in normal newborn infants.

In general, drug toxicity is more severe in newborns in comparison to adults (**Table 2**). The use of theophylline is undesirable in the treatment of apnea, because it may cause spastic diplegia, hearing loss, and mental retardation in the newborn. In the treatment of infantile status epilepticus, diazepam must be used cautiously; this drug can alter the level of growth hormone, resulting in abnormal development of the brain

Table 2. Comparative susceptibility to drug toxicity of newborns and adults

Newborns less susceptible	Newborns more susceptible
Gentamicin	Chloramphenicol
Chlorphentermine	Theophylline
Oxygen	Diazepam
	Salicylates
	Phenytoin
	Phenobarbital
	Acetaminophen
	Digoxin
	Chlorcyclizine
	Corticosteroids

and behavioral deficits. In infants the use of analgesic, antipyretic agents such as salicylates is limited, as the half-life is two to six times longer in the young. This leads to more severe metabolic disturbances, central nervous system intoxication, and finally death. In neonates with cardiac failure caused by a patent ductus arteriosus, acetaminophen has proved effective in duct closure. However, the primary limitation is renal toxicity and liver injury due to conversion of the drug to active, toxic metabolite. The use of phenytoin as an anticonvulsant in the treatment of epilepsy must be monitored cautiously, as this drug is cleared more slowly from the neonate. This prolonged half-life of phenytoin predisposes the neonate more than the adult to adverse drug toxicity characterized by central nervous system disturbances.

For background information SEE *BACTERIAL PHYSIOLOGY AND METABOLISM; BIOLOGICAL OXIDATION; CYTOCHROME; ENZYME, LIVER; METABOLISM; PHARMACOLOGY; PLANT METABOLISM; TOXICOLOGY* in the McGraw-Hill Encyclopedia of Science and Technology.

Sam Kacew

Bibliography. Z. Benet and L. B. Sheiner, Pharmacokinetics, the dynamics of drug absorption, distribution and elimination, in A. G. Gilman et al. (eds.), *The Pharmocological Basis of Therapeutics*, 7th ed., pp. 3–34, 1985; W. H. Habig, Glutathione S-transferases: Versatile enzymes of detoxication, in O. F. Nygaard and M. G. Simic (eds.), *Radioprotectors and Anticarcinogens*, pp. 169–190, 1983; R. M. Hill and L. Stern, Drugs in pregnancy: Effects on the fetus and newborn, *Drugs*, 17:182–197, 1979; S. Kacew and S. Lock (eds.), *Toxicologic and Pharmacologic Principles in Pediatrics*, 1987; S. Kacew and M. J. Reasor, *Drug Toxicity in the Newborn Symposium, Federation Proceedings*, 44:2301–2334, 1985; S. Kacew and M. J. Reasor (eds.), *Toxicology and the Newborn*, 1984; D. M. Shah et al., Engineering herbicide tolerance in transgenic plants, *Science*, 233:478–481, 1986.

Transducer

Chemical and biological transducers differ from physical transducers in several respects. One distinguishing feature is the enormous diversity of parameters in chemical or biological systems. The parameters in question are essentially the concentrations of possible chemical and biochemical substances, but also include ionic strength, enzyme activity, ligand affinity, and so forth. This diversity in turn implies that there are a considerable number of interfering substances. Obtaining selective sensitivity is therefore a key problem in developing chemical and biological transducers.

Most applications of these transducers involve reliable operation in aqueous media such as blood, urine, foods, drinks, and wastewater. An encapsulation of the transducer that is chemically resistant but allows access to the active part of the device is therefore crucial to operational reliability.

Not many chemical and biological transducers are as yet fully developed, and the very few that are in wide use are far from ideal in terms of reliability and versatility. There is considerable interest in this area, and especially in transducers for the sensing of biochemical substances, often called biosensors. Biosensors are expected to have a significant impact, particularly on the medical field.

Electrochemical transducers. Electrochemical transducers are essentially of two types: potentiometric and amperometric electrodes.

Potentiometric transducers. Potentiometric electrodes, such as ion-selective electrodes, produce a voltage which is logarithmically dependent on the concentration of a selected ionic substance. The pH electrode, which measures the activity of hydrogen ions, is the classic example of an ion-selective electrode. Electrodes that are selective to a wide range of other ions have also been commercially available for a number of years. The limitations of such transducers are their relatively large size in some applications, their voltage drift, and their fragility. Most of the devices contain an inner electrolytic solution separated from the test solution by a thin, ion-selective membrane based on glass or plasticized polyvinyl chloride, and are therefore not robust enough to be used in, for example, many industrial applications. Attempts are being made to develop small solid transducers whose readings will be sufficiently stable and repeatable so that they will not need regular calibrations.

Amperometric transducers. Amperometric transducers are based chiefly on chemically inert materials such as platinum, gold, or carbon. If an electrode composed of one of these substances is kept at a selected electrical potential with respect to the test solution, it will oxidize or reduce chemical substances in the solution. An electrical current is thus produced between this so-called working electrode and a counterelectrode, whose magnitude is related to the concentration of the substance in question. The voltage dependence of the electrochemical reaction gives a certain degree of selectivity. However, other means of discriminating against interfering substances are also often essential to the performance of an amperometric transducer, such as the introduction of semipermeable membranes and so-called redox mediator substances. Considerable progress has been made in the area of modified electrodes, where mediators, promoters, and electrochemically active materials are attracted to the electrode surfaces.

Enzyme-sensitized transducers. There is a wide

range of biochemical substances that are not electrochemically oxidizable or reducible and are therefore not directly detectable by electrochemical electrodes. The combination or integration of enzymes with electrochemical transducers has turned out to be a viable approach to biochemical transducers for these substances. Enzymes such as oxidases and dehydrogenases can produce or consume electroactive species by catalyzing reactions involving the biochemical substances for which transducers are required. The application of such an enzyme to an electrochemical transducer results in a new transducer which is selectively sensitive to a specific substance. Such transducers have been made and demonstrated for numerous substances, including glucose, cholesterol, urea, and creatinine. The terms enzyme electrode or biosensor are often applied to these transducers, but biosensor has increasingly become a general term for a transducer that detects or monitors any biologically relevant parameter.

Solid-electrolyte gas transducers. A possibly less versatile but still technically very important type of electrochemical transducer is the solid-electrolyte gas transducer. By far the most developed of these is the zirconium oxide–based oxygen transducer. A suitably doped zirconium oxide material is heated to a high temperature, typically to several hundred degrees Celsius, thus becoming permeable to oxygen ions. Exposure to the test gas on one side and a reference gas, for example, fresh air, on the other side results in a diffusion voltage across the zirconium oxide material. This voltage is a logarithmic function of the difference in oxygen partial pressure on the two sides. Various transducers based on this principle have been developed, and are being used for oxygen monitoring, especially in combustion control systems. There is considerable interest in this type of transducer in the automobile industry. Other solid electrolytes, such as polymer-based ones, are also being developed for new gas transducers sensitive to other gases such as hydrogen.

Chemically sensitive FETs. Field-effect transistors (FETs) are widely used in modern electronic circuits, and the manufacturing technology for these devices is well established. The possibilities for making new chemical and biological transducers by modifying field-effect transistors were recognized during the 1970s. The basic concept involves the replacement of the ordinary gate electrode with a chemically selective membrane. Both ion-sensitive field-effect transistors (ISFETs) and gas-sensitive field-effect transistors have been demonstrated.

The advantages of ion-selective field-effect transistors over conventional ion-selective electrodes are their very small size, low impedance output, suitability for direct integration with electronic circuits, and low-cost mass-production possibilities. All these properties are believed to increase the applicability of chemical transducers significantly, especially in the medical area. One example of such an application is the *in vivo* monitoring of body electrolytes during surgery. Catheter-type transducers and multitransducer flow-through cells are currently being tested.

However, the use of ion-selective field-effect transistors has been delayed because of problems arising in the mating of semiconductor technology to electrochemistry. The application of ion-selective membranes to transistors and the reliable encapsulation of the devices are still relatively expensive procedures, but mass-production methods compatible with conventional semiconductor technology are being developed.

Gas-sensitive field-effect transistors have been developed for gases such as hydrogen, ammonia, ethanol, and carbon monoxide. Hydrogen-sensitive transistors using palladium as a highly selective gate electrode material have been successfully used in certain applications in recent years.

Experiments have been made with enzyme-sensitized transistors for monitoring nonionic biochemical substances such as urea, glucose, and penicillin. There are high expectations for these devices, but they are still in a very early stage of development. The sensitization of transistors with antibodies for immunochemical testing has been suggested, but strong doubts exist about the feasibility of such transistors, known as immuno-FETs.

Optochemical transducers. Optical techniques such as photometry and fluorometry have been widely used for chemical analysis for many years. Miniaturization of these and the development of new techniques are expected to give rise to a range of versatile chemical transducers in the near future. The main advantage of using optical techniques for chemical sensing are the chemical selectivity of light and the immunity to electrical and electrochemical interference, for example, during invasive monitoring.

A parameter such as pH has been monitored by entrapping a pH indicator dye at the tip of an optical fiber and by analyzing the light that has been guided through the fiber, reflected at the tip, and guided back again. Such a pH transducer is extremely small and is potentially very inexpensive. Dissolved oxygen in blood has been measured by means of a fluorescent dye at the end of a fiber. The effect of oxygen on the fluorescence is monitored by an optoelectronic instrument at the other end of the fiber. Compared with electrochemical transducers for the same substance, the fiber-optic transducer has the advantage of not needing any reference electrodes or counterelectrodes.

Various waveguide phenomena are being studied for eventual use in optochemical transducers. There is particular interest in optical phenomena associated with molecular binding to specific antibodies attached to an optical surface. Direct detection of molecules picked up by antibodies on a surface has been achieved by monitoring the modulation of surface plasmon resonance phenomena in a thin metal film on the same surface. A surface plasmon resonance gives rise to a significant reduction of the light reflectance at the surface, and the chemically induced modulation is readily monitored with a photodiode. SEE SURFACE PHYSICS.

Other methods involve the use of fluorescently labeled molecules that are competing with the analyte molecules. Transducers in which the fluorescent molecules radiate into a waveguide when bound to a selective surface are under development.

Piezoelectric transducers. Piezoelectric devices can be made to oscillate at very high frequencies. Adsorption of even small amounts of material onto such a device gives rise to a detectable decrease in oscillation frequency. Gas transducers are made by depositing selectively gas-absorbing layers onto piezoelectric devices. Quartz crystals of the same type as those used in clocks make very reliable and reasonably sensitive gas transducers. Even higher sensitivities are provided by surface acoustic-wave (SAW) devices, in which high-frequency waves are generated at the surface of a piezoelectric material. While ordinary quartz crystal transducers respond only to changes in mass, the surface acoustic-wave devices respond also to changes in viscosity and electrical conductivity of the absorbing layer.

Attempts are being made to operate piezoelectric sensors in aqueous solutions. The main problem with this approach is energy loss into the water phase, but this can be partially overcome in certain arrangements.

Conductivity-based transducers. The conductivity of certain semiconducting materials can be very sensitive to exposure to various reactive gases. The most commonly used gas transducers are based on a sinter of a semiconducting metal oxide such as tin oxide. Oxygen diffuses into the material and adsorbs chemically onto the surface of the particles. The chemical adsorption of oxygen localizes electrons at the surface, thereby increasing the contact resistance between adjacent particles. This effect is simply monitored as an increase in device resistance. Even a very small concentration of a combustible gas consumes the adsorbed oxygen and is therefore observed as a resistance decrease. These very simple devices are widely used in gas alarms for explosive gases, such as hydrogen, methane, and butane. High sensitivity and reversibility are obtained by heating the material to a few hundred degrees Celsius. The selectivity of these devices is generally not very high, but considerable research is being done on the addition of catalytic materials to selectively increase the sensitivity. Thick-film technology offers convenient methods for the fabrication of these transducers.

Conducting polymers such as polypyrrole and organometallic compounds such as phthalocyanides offer new possibilities for the development of gas transducers. Research is being carried out on the mechanisms behind the observed effects of compounds such as ammonia on the conductivity of these materials. Improved understanding of the physics and the chemistry of conducting organic materials is expected to lead to simple and cheap gas transducers in the future. Attempts to operate conductivity-type transducers in aqueous media have so far not been very successful because of chemical degradation problems and because of the conductivity of the aqueous medium itself.

For background information SEE ENZYME; FLUOROMETRIC ANALYSIS; ION-SELECTIVE MEMBRANES AND ELECTRODES; OPTICAL FIBERS; PIEZOELECTRICITY; SURFACE ACOUSTIC-WAVE DEVICES; TRANSDUCER; TRANSISTOR in the McGraw-Hill Encyclopedia of Science and Technology.

Claes Nylander

Bibliography. H. Freiser (ed.), *Ion-Selective Electrodes in Analytical Chemistry*, vol. 1, 1978, vol. 2, 1980; C. Nylander, Chemical and biological sensors, *J. Phys. E: Sci. Instrum.*, 18:736–750, 1985; H. Wohltjen, Chemical microsensors and microinstrumentation, *Anal. Chem.*, 56:87A–103A, 1984; J. Zemel and P. Bergveld, *Chemically Sensitive Electronic Devices*, 1981.

Transistor

The need for fast electron devices is constantly increasing. These devices are used to improve the performance of microwave amplifiers in communication equipment as well as to enhance switching speed in large-scale computers. A high-speed field-effect transistor (FET) called the high-electron-mobility transistor (HEMT) has been developed to meet such requirements. It switches on or off in about 10 picoseconds with the maximum oscillation frequency exceeding 100 GHz. Since 1980 considerable research has been done on the fabrication and characteristics of the device as well as on its microwave and digital applications. The device is also called by various other names: two-dimensional electron gas FET (TEGFET), selectively doped heterojunction transistor (SDHT), and modulation-doped FET (MODFET), all descriptive of various aspects of the operation or structure of the device.

Structure and operating principles. The device is made of gallium arsenide (GaAs) and gallium aluminum arsenide (GaAlAs) with a Schottky metal contact on the gallium aluminum arsenide layer and two ohmic contacts penetrating into the gallium arsenide layer, serving as the gate, the source, and the drain, respectively (**Fig. 1**). Schottky and ohmic contacts exhibit rectifying

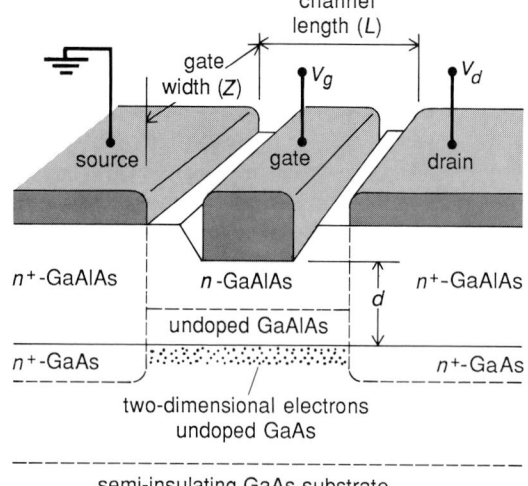

Fig. 1. Cross section of a high-electron-mobility transistor (HEMT). The source, gate, and drain are metal electrodes; the source and gate are usually gold-germanium alloys and the drain is typically aluminum.

and nonrectifying characteristics, respectively. The device's fast response to the signal is attributed to the fact that two-dimensional electrons (discussed below) at the interface between gallium arsenide and gallium aluminum arsenide

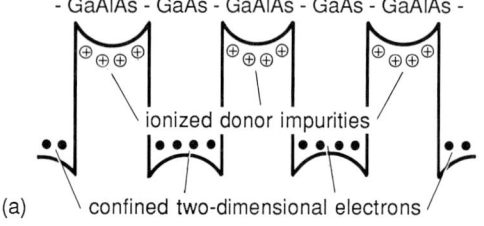

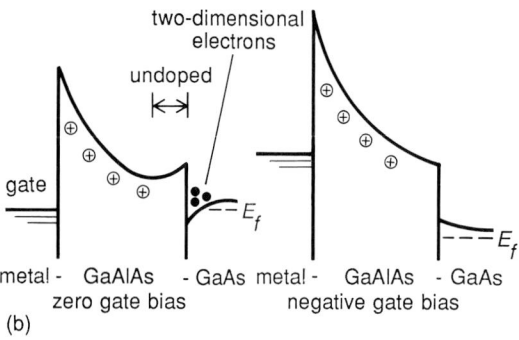

Fig. 2. Energy diagrams of modulation doping.
(a) Superlattice.
(b) Depletion-mode HEMT (D-HEMT), without and with negative gate bias voltage.
E_f = Fermi level.
(c) Enhancement-mode HEMT (E-HEMT), without and with positive gate bias voltage.

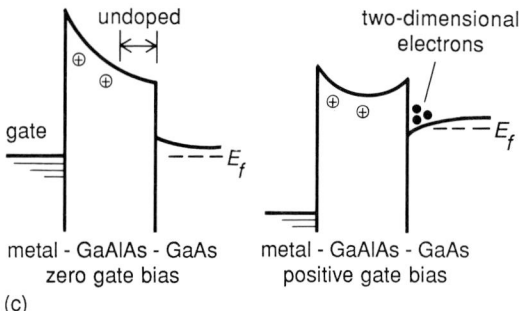

have high mobilities and move very rapidly from source to drain.

The high-electron-mobility transistor evolved from work on gallium arsenide–gallium aluminum arsenide superlattices, periodic structures prepared by deposition of alternating ultrathin layers of gallium arsenide and gallium aluminum arsenide (**Fig. 2**a). Usually free carriers—electrons and holes created in a semiconductor by impurity doping—suffer from impurity scattering. In superlattices, however, it is possible to spatially separate free carriers and their parent impurity atoms by doping impurities in the potential barriers of gallium aluminum arsenide. Electrons from donor impurities transfer to the potential well of gallium arsenide, forming a layer of so-called two-dimensional free electrons (Fig. 2a) that are free to move in the two dimensions parallel to the layers but are confined to bound quantum states in the well with respect to motion perpendicular to the layers. Such two-dimensional electrons suffer less from impurity scattering because of their spatial separation from the impurities, and thus have high mobilities. This concept of modulation doping was proposed together with the superlattice concept in 1969, and was first successfully implemented in 1978, resulting in the observation of high electron mobilities in gallium arsenide. Modulation doping also can be used to achieve high mobility for two-dimensional electrons confined in the potential well formed at the interface between gallium arsenide and gallium aluminum arsenide where, again, electrons transfer from donor impurities in gallium aluminum arsenide to the potential well of gallium arsenide. *See Semiconductor.*

The high-electron-mobility transistor employs this modulation-doping technique to obtain high electron mobilities: impurities are doped only in the gallium aluminum arsenide. Figure 2 shows the energy diagrams for two types of high-electron-mobility transistors: depletion-mode (D-HEMT; Fig. 2b) and enhancement-mode (E-HEMT; Fig. 2c) without and with negative and positive bias voltages for the gate Schottky barrier, respectively. A narrow portion of gallium aluminum arsenide close to the gallium arsenide is kept undoped. This further separates impurities and electrons, resulting in more enhancement on electron mobilites.

The high-electron-mobility transistor structure (Fig. 1) is fabricated on semi-insulating gallium arsenide substrates by the advanced thin-film growth techniques of molecular beam epitaxy or metal organic (or organometallic) chemical vapor deposition. Such techniques enable growth of high-quality heterostructures having designed potential profiles and impurity distributions with dimensional control close to interatomic spacing (approximately 0.2 nanometer), and with virtually defect-free interfaces in a lattice-matched pair such as gallium arsenide–gallium aluminum arsenide. This great precision in epitaxy is required in order to fabricate both D-HEMTs and E-HEMTs on the same wafer and achieve precise control of their threshold voltages.

The principle of operation of the high-electron-mobility transistor is similar to that of the silicon metal-oxide-semiconductor FET (MOSFET). In D-HEMTs, the thickness of the doped gallium aluminum arsenide under the gate is great enough for two-dimensional electrons to be present when there is no bias voltage (the device is normally on), and a negative gate bias depletes them (Fig. 2b); in E-HEMTs, that thickness is small enough for two-dimensional electrons to be absent when there is no bias voltage (the device is normally off), and a positive gate bias induces them (Fig. 2c). In practice, a small amount of recessing of the gate electrode (Fig. 1) is done by etching; the amount of the recess determines the type of device, either depletion-mode or enhancement-mode.

Performance characteristics. In the linear region, at low drain voltages, the transconductance g_m is given by Eq. (1), where Z and L are the gate

$$g_m = \frac{Z}{L}\mu C V_d \qquad (1)$$

width and the channel length, respectively (Fig. 1); μ is the electron mobility; the capacitance C is given by ϵ/d (where ϵ is the permittivity and d is the thickness of gallium aluminum arsenide under the gate); and V_d is the drain voltage. The high-electron-mobility transistor exhibits a high transconductance because of high electron mobilities as well as a large value of the relative permittivity, 1.3 for gallium arsenide as compared with 3.9 for silicon dioxide (SiO_2). At high bias voltages where the electron velocity saturates, the transconductance is given by Eq. (2), where v_s is the

$$g_m = ZC v_s \qquad (2)$$

saturation electron velocity that is characteristic of gallium arsenide, which is considerably faster than that of silicon. Typical values for L and d are 1.2 micrometers and 35–40 nm, respectively.

For high-speed operations, there are two factors that limit the frequency response of a field-effect transistor, the RC time constant and the transit time. The RC time constant is a measure of how rapidly the charge associated with the capacitance C of the device can be acquired and lost. The high electron mobility helps to reduce the resistance R and thus reduces the RC time constant. The transit time τ is the time required for electrons to travel from source to drain. In the linear region it is given by Eq. (3), and when the

$$\tau = \frac{L^2}{\mu} V_d \qquad (3)$$

velocity is saturated it is given by Eq. (4). The

$$\tau = L/v_s \qquad (4)$$

high mobility, high saturation velocity, and short channel length all contribute to improvements in the frequency response. The electron mobilities in the high-electron-mobility transistor increase considerably at low temperatures. Thus, the device performance at −321°F (77 K; liquid-nitrogen temperature) is substantially better than that at 80°F (300 K; room temperature).

Applications. D-HEMTs for microwave low-noise amplifiers have been developed and are commercially available, for instance, in satellite communication systems. A noise figure of 1.8 dB and associated gain of 8 dB at a frequency of 20 GHz have been reported.

E-HEMTs have been used for switching, because of their low power dissipation, in high-speed integrated digital circuits, where D-HEMTs with gate shorted to source serve as two-terminal loads. Three basic circuits, a NOR gate, a NAND gate, and a cross-coupled flip-flop memory cell, are illustrated in **Fig. 3**, where so-called direct-coupled FET logic circuit config-

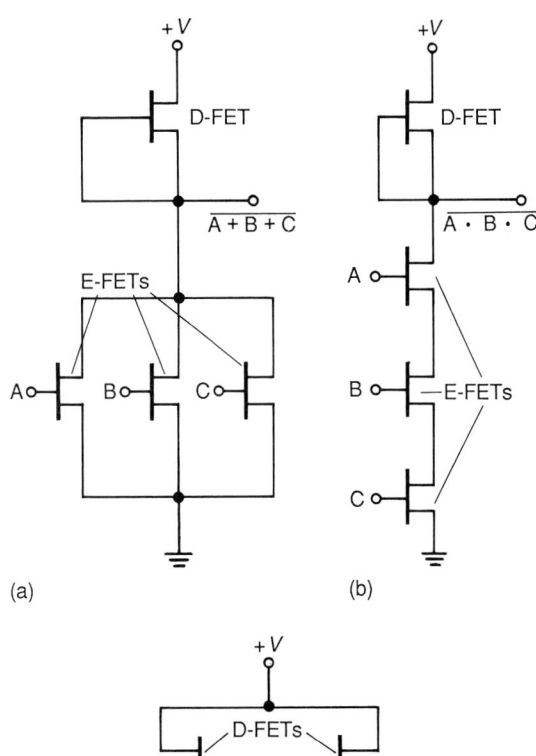

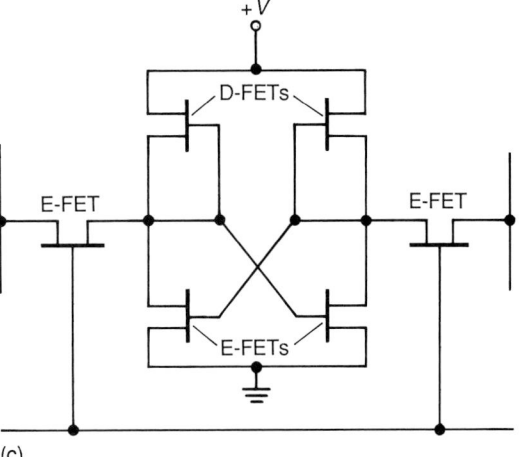

Fig. 3. Basic circuits employing high-electron-mobility transistors. (a) NOR gate with inputs A, B, and C, and output A + B + C. (b) NAND gate, with inputs A, B, and C, and output A · B · C. (c) Cross-coupled flip-flop memory cell.

urations are implemented. By using the flip-flop cell, a 4000-bit static random access memory (SRAM) has been reported with minimum address access time of 2.0 nanoseconds at $-321°F$ (77 K) and 4.4 ns at 80°F (300 K). By using the NOR gate, a 1500-gate array has been reported with multiplication time of 3.1 ns at $-321°F$ (77 K) and 4.9 ns at 80°F (300 K) in an 8×8 parallel multiplier. These figures, particularly those at $-321°F$ (77 K), probably represent the highest speed recorded so far for any switching device.

Recently, high hole mobilities with gallium arsenide and gallium aluminum arsenide heterostructures have been achieved by a modulation-doping technique similar to that described above, except that doping is carried out with acceptors such as beryllium (Be) instead of donors such as silicon (Si). An attempt was then made to fabricate a high-hole-mobility transistor, which would complement the high-electron-mobility transistor.

Obviously, the production of high electron or hole mobilities by the modulation-doping technique is not limited to gallium arsenide and gallium aluminum arsenide heterostructures and, indeed, has been obtained with heterostructures such as gallium indium arsenide–indium phosphide (GaInAs-InP) and gallium indium phosphide–gallium arsenide (GaInP-GaAs). It is certainly possible to fabricate high-electron-mobility transistors with those heterostructures, some of which may outperform the conventional high-electron-mobility transistor for some purposes.

In summary, the high-electron-mobility transistor possesses many of the attributes required for high-speed operation. With more advanced material and fabrication technologies, better performance and larger integration for supercomputers can be expected in the near future.

For background information SEE ARTIFICIALLY LAYERED STRUCTURES; INTEGRATED CIRCUITS; LOGIC CIRCUITS; MICROWAVE SOLID-STATE DEVICES; SEMICONDUCTOR HETEROSTRUCTURES; SEMICONDUCTOR MEMORIES; TRANSISTOR in the McGraw-Hill Encyclopedia of Science and Technology.

Leo Esaki

Bibliography. R. Dingle et al., Electron mobilities in modulation-doped semiconductor heterojunction superlattices, *Appl. Phys. Lett.*, 31:665–667, 1978; L. Esaki and R. Tsu, Superlattice and negative conductivity in semiconductors, *IBM J. Res. Dev.*, 14:61–65, 1970; H. Morkoc and P. M. Solomon, The HEMT: A superfast transistor, *IEEE Spectrum*, 21(2):28–35, 1984; P. M. Solomon and H. Morkoc, Modulation-doped GaAs/AlGaAs heterojunction field-effect transistors (MODFETs), ultra high-speed device for supercomputers, *IEEE Trans. Electr. Devices*, ED-31:1015–1027, 1984.

Transplantation biology

Human bone marrow contains the precursor cells for all developing blood and bone marrow as well as the immune system. Transplanted marrow from normal twin (syngeneic) or tissue type–matched (allogeneic) donors can correct abnormalities of the immune system in children with immunodeficiency; restore blood cell production in patients with aplastic anemia or those subjected to accidental radiation exposure; and allow both hematologic and immunologic reconstitution in patients who have received high-intensity chemotherapy and radiation treatment for various malignant diseases. Though this therapy is complex and is delivered only at great risk, it can offer curative treatment for a significant fraction of patients with certain otherwise-lethal diseases. The major obstacle to more widespread application of such treatment is the problem of donor availability. Matched sibling donors are available for only 30 to 40% of patients who could benefit from marrow transplantation. Recent advances, such as autologous marrow transplantation (after laboratory purging of potential malignant cell contamination of the bone marrow), partially matched transplantation (for those without a suitably matched sibling donor), and, most recently, the use of closely matched, unrelated bone marrow donors, provide techniques which can extend the applicability of this new therapy.

Bone marrow transplantation. The success of bone marrow transplantation since its origin in the 1950s has depended on new knowledge and its technical application in three areas. The first is the delineation of the inherited genetic structure which regulates human tissue typing—the HLA major histocompatibility complex. The second is the development of intensive chemotherapy and radiation therapy techniques that are potent enough to eliminate residual malignant tissue and immunosuppressive enough to allow engraftment of matched but not genotypically identical bone marrow. The third comprises advances in supportive care techniques which can protect patients from the major hazards of infection and hemorrhage in the period between initiation of chemoradiotherapy and hematologic and immunologic reconstitution accompanying successful growth of the donor bone marrow. Progress in each of these areas has led to wide use of matched donor allogeneic bone marrow transplantation for patients with leukemia, including acute lymphocytic leukemia, acute nonlymphocytic leukemia, and chronic myelogenous leukemia.

Marrow transplantation is usually reserved only for patients whose illnesses apparently cannot be cured by conventional chemotherapeutic techniques, which are limited by their toxicities. The major dose-limiting toxicities of nontransplant chemotherapies are reflected in bone marrow suppression, lowered blood counts, and sus ceptibility to infections. Transplanted marrow is infused after the patient's chemotherapy and radiation have been completed, thus circumventing such concerns. Therefore marrow transplantation allows application of more intense, more effective antileukemic treatment, usually by using both antileukemic drugs and total body irradiation. Additionally, immunologically potent cells developing from the donor marrow may recognize and destroy tumor cells which persist after chemotherapy—the so-called graft-vs.-leukemia effect. Overall, by using various drug combinations and radiation dose schedules in preparing for allogeneic bone marrow transplantation, many medical research centers have reported potential cure rates of 40 to 50% for patients with otherwise lethal leukemias. Improvements in supportive care designed to reduce the hazards of transplantation will surely increase success rates of transplantation.

Techniques. The techniques of marrow harvest and infusion have been well established. Under general anesthesia, 12–60 in.^{3} (200–1000 ml) of bone marrow is aspirated from the donor's pelvic bones and suspended in anticoagulated tissue culture media. The marrow is infused intravenously into the recipient promptly after collection. There are few major risks to the donor, and these are primarily associated with general anesthesia.

Recipients are prepared to receive the graft by suppression of their immune systems (to prevent rejection) as well as concurrent ablation of cancerous cells for patients with malignancies. Immunosuppression most often includes high doses of cyclophosphamide; those patients with malignancies usually are given total body irradiation as well. Patients receiving marrow grafts for nonmalignant disease (for example, severe aplastic anemia, immunodeficiencies, or inborn errors of metabolism) may be prepared for transplantation with varying doses of cyclophosphamide either alone, in combination with other immunosuppressive agents, or with limited radiation. These agents all induce profound immunosuppression in the recipient. The purpose is to overcome the first major hazard of marrow transplantation—poor marrow function or frank rejection of the transplant. With suitably intense preparation and with marrow from a fully matched sibling, rejection is encountered only rarely.

Complications. The immunosuppression is long-lasting. Two to four weeks are required for successful marrow growth before production of white blood cells for initial infection defense develops. Several months are required for antibody production to return, and 6 to 12 months are required for full immunologic reconstitution and restoration of defense against infection. Marrow graft recipients are, therefore, extraordinarily vulnerable to serious or fatal infection—the most common complication following transplantation.

Various protective isolation techniques to prevent person-to-person, bloodborne, or airborne infection have been developed; in addition, there are prophylactic and aggressive therapeutic antibiotic regimens for minimizing the mortality associated with infection in the first three posttransplant months. Careful monitoring of the immune system's development after engraftment allows identification and recognition of the specific immune deficits present at varying times during the posttransplant period, and the application of suitable infection prophylaxis as the patient recovers immunologic competence. Despite these measures, marrow recipients remain vulnerable to numerous bacterial, viral, fungal, and parasitic infections throughout their treatment course.

A unique complication associated with allogeneic (nonidentical twin) marrow transplantation is graft-vs.-host disease. In this syndrome, immunocompetent cells within the donor bone marrow recognize the immunologic disparity between donor and recipient tissues and induce an immune attack on recipient tissue, usually skin, gastrointestinal tract, or liver. Graft-vs.-host disease is a major barrier to successful transplantation and occurs in up to 50% of all allogeneic transplants. This syndrome occurs more frequently and with greater severity in transplantation involving only partial matching between donor and recipient tissue types. Graft-vs.-host disease is often accompanied by delayed immunologic maturation of the transplanted marrow and additional complicating infections as well.

Graft-vs.-host disease can be prevented by two methods. One involves careful donor selection and attention to the best tissue match available. The other consists of pharmacologic suppression of the aggressor donor marrow lymphocytes in order to minimize their attack on recipient tissue and to facilitate the integration of the transplanted immune and blood-forming system into the environment of the new host's body. New techniques aimed at depletion of the cells that cause graft-vs.-host disease before transfusion of the donor bone marrow have been promising in reducing the incidence and the severity of graft-vs.-host disease in transplantation involving matched donors and recipients. The donor T lymphocytes that cause the disease are depleted from marrow by identifying them, usually with monoclonal antibodies, and either destroying them with cytotoxic reagents or mechanically removing the cells from the marrow before infusion into the recipient. Refinements of these techniques may also prevent or control the otherwise severe graft-vs.-host disease accompanying mismatched marrow transplantation.

Expansion of the therapy. Since matched sibling donors are unavailable for 60 to 70% of the population, new techniques for marrow grafting are being developed by using alternate marrow sources. In the past, graft rejection has often followed transplantation involving partially matched donors and recipients. However, by using the techniques of T-lymphocyte depletion, there have been reports of successful transplantation with use of partially matched related donors without either marrow rejection or overwhelming graft-vs.-host disease. Numerous centers are now actively studying this problem.

Also, studies have begun using unrelated, but coincidently closely HLA tissue type–matched, volunteer donors identified through computer banks of potential donors. At several institutions, researchers, through clinical trials, are attempting to determine if unrelated donors can be efficiently identified and if transplantation using matched unrelated donors is as effective as that with matched, genetically related donors.

Finally, techniques for depleting the patient's own bone marrow of residual malignant cells may allow the harvesting and later reinfusion of the recipient's own autologous bone marrow. This technique of autologous bone marrow transplantation is applicable to a broad range of malignant diseases, especially those which spare the bone marrow. Autologous transplantation also avoids both graft-vs.-host disease and its accompanying immunodeficiency and risks of infection. This greater safety margin allows autologous bone marrow transplantation to be used for older patients (45 to 60 years of age), thought unsuitable for the riskier allogeneic transplant.

Marrow purging of contaminating malignant cells makes use of chemotherapy drugs (more toxic to malignant than to normal marrow cells) as well as monoclonal antibody techniques to find and then destroy residual tumor cells within the harvested marrow. Success in autologous bone marrow transplantation is contingent on eradication of tumor cells from the marrow, but, like allogenic transplantation, also demands effective therapy for the whole patient. A potential disadvantage of autotransplantation is the absence of the putative graft-vs.-leukemia effect, in which the new immune system can immunologically recognize and potentially destroy any residual malignant tissue remaining after chemoradiotherapy. The clinical importance of this potential immunologic antimalignancy effect is uncertain.

For background information *see Histocompatibility; Immunology; Immunosuppression; Monoclonal antibodies; Transplantation biology* in the McGraw-Hill Encyclopedia of Science and Technology.

Daniel Weisdorf

Bibliography. P. B. McGlave, N. K. C. Ramsay, and J. H. Kersey, Allogeneic and autologous bone marrow transplantation, in A. V. Hoffbrand (ed.), *Recent Advances in Haematology 4*, pp. 171–198, 1985; R. J. O'Reilly, Allogeneic bone marrow transplantation: Current status and future directions, *Blood*, 62:941–964, 1983; C. D. Petz and K. G. Blume, *Clinical Bone Marrow Transplantation*, 1983.

Transposons

The results of both population surveys and theoretical analyses suggest that the view of transposons as intragenomic parasites provides a satisfactory explanation for their maintenance within host populations. This view does not preclude the occasional utilization by selection of favorable mutations induced by the insertion of transposons, although there is no direct evidence for this at present. In bacteria the transfer into plasmids of useful genes picked up by transposons occurs, but its significance for evolution under natural conditions is unclear. The ultimate evolutionary origin of transposons remains obscure.

Structure and taxonomic distribution. A significant fraction of the genome of both prokaryotes and eukaryotes is composed of families of deoxyribonucleic acid (DNA) sequences which are present in multiple copies, dispersed throughout the genome. Several lines of molecular and genetic evidence indicate that such "middle repetitive" sequences are capable of self-replication and movement to new positions within the host's genome, and they are now generally referred to as transposable elements or transposons. Members of a given family share a high degree of sequence homology, as well as general structural features such as the presence or absence of inverted or direct terminal repeats (**Fig. 1**). Transposable elements normally code for one or more proteins involved in their replication and transposition. In some cases, notably the maize

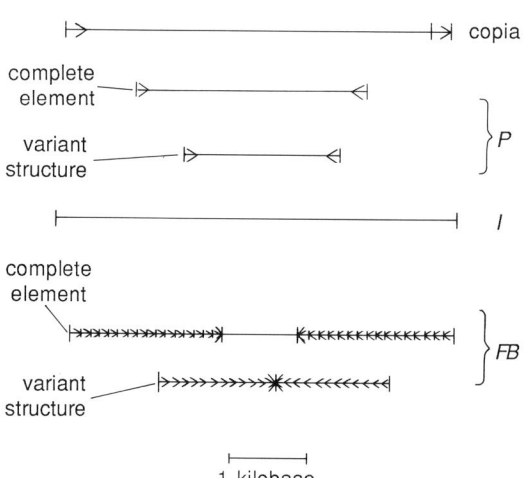

Fig. 1. General structures of the copia, *P*, *I*, and *FB* families of *Drosophila* transposable elements. The arrows represent the orientations of homologous repeated sequences within the transposons. Variant structures of the *P* and *FB* elements are shown, as well as the complete elements.

element *Ac-Ds* and the *Drosophila P* and *FB* elements, defective elements carrying large deletions of the coding sequences are found in the same host genome as complete elements, and depend on the latter for transposition.

Relatively little is known at present about the taxonomic distribution of a given element family with respect to the host species in which it is found, but in *Drosophila* it is known that the same family may be found in different species of the genus. The frequency with which significant homology can be detected between elements from different host species tends to decline with taxonomic affinity, suggesting that they are usually transmitted through the germ line of the host, rather than by infection, since transspecific infection events would tend to obscure similarities due to descent. There are exceptions to this behavior, however; the *P* element is present in *D. melanogaster* and in members of distantly related species but not in its nearest relatives, suggesting the occurrence of an interspecies transmission event. Vertebrate retroviruses are capable of both infectious and genomic transmission; elements in yeast and some *Drosophila* elements are very similar to retroviruses in their structure, and probably replicate via ribonucleic acid (RNA) intermediates which are then reverse-transcribed into DNA. Other elements, such as bacterial *IS* sequences, various maize elements, and *Drosophila P* elements, replicate directly.

Evolutionary significance. The questions of the significance of transposable elements for the organisms that host them, and the nature of the forces responsible for maintaining them in the host population, have been the subject of much debate. Three main hypotheses have been proposed and are discussed below. The first is that they have a role in coding for products needed for host functions such as the regulation of gene activity, as originally suggested by Barbara Mc-Clintock for the maize elements. The second is that the process of transposition causes mutations and chromosome rearrangements that can be utilized by natural selection, and that elements are retained in populations in order to accelerate adaptive evolution. The third is the selfish DNA hypothesis, which proposes that elements are maintained purely as a result of their own powers of replication within the host genome, and that, if anything, their presence lowers the fitness of the host. In other words, they behave as intragenomic parasites.

Provision of host functions. The evidence for the first hypothesis is weak. It comes partly from the observation that a significant fraction of messenger RNA (mRNA) in the host cells is transcribed from transposable elements, and partly from the fact that transposition in flies and maize occurs at specific times and places in development. Furthermore, strains of the bacterium *Escherichia coli* carrying a single copy of the *IS5* or *IS50* sequences have a competitive advantage over non-*IS* strains in laboratory culture, under conditions where transposition is not occurring. The first two facts are consistent with the other interpretations of the role of transposons. The relevance of the third observation to natural conditions is not clear, since many natural isolates of *E. coli* lack *IS* sequences.

Induction of favorable mutations. The evidence for this hypothesis comes primarily from the fact that the activities of transposons induce mutations and chromosome rearrangements, which, in rare instances, could be selectively advantageous, especially in changed environments. An experiment with *E. coli* has demonstrated that a strain carrying the *Tn10* transposon can have a competitive advantage over a strain without this transposon in a novel environment, due to the induction of favorable mutations by the movement of *Tn10*. In addition, many bacterial transposons were first isolated from plasmids carrying drug- or metal-resistance genes introduced by the transposon. This association suggests that the frequency of the transposons in the population has increased as a result of their having accidentally picked up the resistance genes and caused their insertion into the plasmids, which then transmit the genes from host to host.

The weakness of this hypothesis is that the vast majority of new mutations with noticeable phenotypic effects are selectively disadvantageous, even in a new environment where some mutations may be favorable. In the experiment in *Tn10*, the strain had to be present initially at a moderately high frequency in order to have an advantage, because advantageous mutations have a good chance of appearing and being established by selection only in a large population; if the *Tn10* strain is initially rare, all the new mutations that appear will be deleterious and hence will reduce its fitness relative to the normal strain. Similarly, strains of *Drosophila* with active *P* elements suffer greatly reduced fitnesses, even though their elevated mutation rate sometimes permits a higher rate of response to artificial selection. It is therefore difficult to see how the net abundance of transposons within a population could be increased from a low initial level purely as a result of the induction of favorable mutations. Another possibility that could be imagined is that some populations of a species acquire large numbers of copies of a transposon family by random sampling effects due to finite population size (genetic drift), even in the face of weak counterselection due to the induction of deleterious mutations by transpositions. If the presence of transposons were to cause more rapid evolution of the high-copy number populations in a changing environment, it is possible that more rapid extinction of populations with low copy numbers would cause the species to evolve higher copy numbers by a process of group selection.

Evidence against the favorable mutation hypothesis is provided by the fact that this hypothesis predicts that elements should be raised, at least occasionally, to high frequencies at individual chromosomal sites by selection on the favorable effects of their insertion at the sites in question or (in the case of the group selection model) by genetic drift. This is not observed in those cases where good population data are available. For example, a study of three families of *D. melanogaster* elements from a natural population showed that element frequencies at individual sites on the X chromosome are always low. This seems to be generally true in *Drosophila*, from the results of surveys of variation at the DNA level for several limited regions of the genome. In *E. coli*, there is a good deal of variation between individual bacterial isolates in the number and location of *IS* sequences. Little evidence is at present available for other organisms, apart from a study in yeast which gave very similar results. An apparent exception is provided by the mammalian repeated sequences such as *Alu*, which appear to have reached very high frequencies at sites where they are present in the genome. There is, however, some question as to the status of these sequences as transposable elements.

Intragenomic parasitism. The population data provide strong support for the view that transposons are essentially intragenomic parasites, maintained in the population by virtue of the fact that their rate of intragenomic replication exceeds that of loss by excision or of elimination by selection (**Fig. 2**). If the probability of replication per element per host generation u exceeds the probability of excision v, the mean number of elements per individual of a sexually reproducing host population will increase by approximately $\bar{n}(u - v)$ per host generation. This raises the question as to what maintains the balance be-

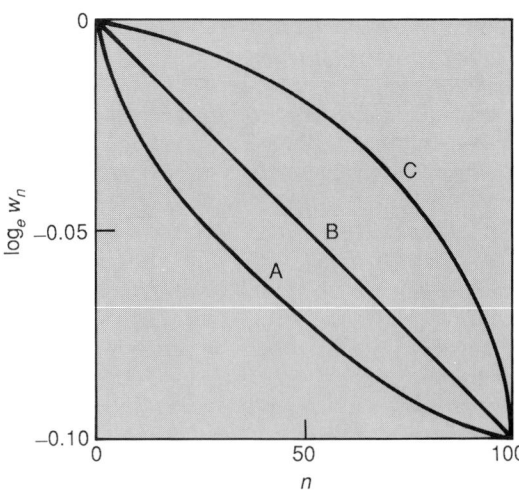

Fig. 3. Some possible relationships between the natural logarithm of host fitness (log$_e$ w_n) and the number of copies per host individual of a family of transposons (*n*). Only curve C will permit a stable equilibrium of mean copy number, in the absence of self-regulated transposition.

tween the rate of increase by transposition and the rate of loss, resulting in the low frequencies typically observed at individual chromosome sites and the consequent wide variation between individuals in the positions at which transposons belonging to a given family are located. Mathematical models of large sexually reproducing populations have shown that two forces are capable of producing stable equilibria (genetic exchange between individuals is necessary for the spread of elements within a population by transpositional replication, since otherwise copy numbers can increase only within a single clone). An approximately Poisson variation around this mean is expected, which agrees well with the *Drosophila* data.

The first force is the regulation of transposition rate in response to an increase in the number of homologous elements in the same cell. This has been clearly documented in bacteria, *Drosophila*, and maize. Specific molecular mechanisms for regulation are known to exist. If regulation causes the net rate of transposition per element to fall off as mean copy number $\bar{n}$ increases, then a stable equilibrium can be generated, even in the absence of selection, at a value of $\bar{n}$ at which $u(\bar{n})$ drops to approximately v, where $u(n)$ represents the functional dependence of u on n.

The second force is selection; if the logarithm of fitness falls off more steeply than linearly with an increase in the number of copies carried by an individual (**Fig. 3**), then selection can stabilize copy numbers at an intermediate level. The mean copy number per individual at equilibrium satisfies the approximate equation below, where $w_{\bar{n}}$ is

$$\frac{d(\log_e w_{\bar{n}})}{d\bar{n}} = v - u$$

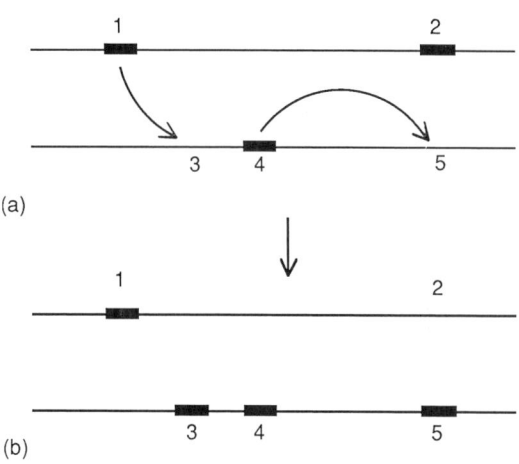

Fig. 2. Two host chromosomes (*a*) before and (*b*) after transposition and loss events. The elements at sites 1 and 4 produce replicates, which become inserted into sites 3 and 5. The element at site 2 is excised. There is thus a net increase of one in the number of elements in the genome.

the population's mean fitness. Such a relationship between fitness and copy number could be generated if insertions of transposons caused mutations with minor deleterious effects on fitness, and is consistent with data from *Drosophila* on the relationship between fitness and number of mutations carried by an individual. Another possible mode of selection is the generation of deleterious chromosome rearrangements by exchanges between homologous elements located at different places in the same cell; such rearrangements have been detected in yeast and *Drosophila*.

It is far from clear at present which of these forces is more important in determining copy numbers. The fact that insertions of transposons in natural populations of *Drosophila* are essentially confined to noncoding regions of the genome suggests that selection eliminates individuals with insertions that disrupt gene activity. Since transposition rates are generally very low (10^{-4} or less per element per host generation), only a weak pressure of selection is needed to balance transposition. The two mechanisms are not, of course, mutually exclusive, and theoretical models show that transpositional regulation can evolve in response to the induction of deleterious mutations by the movement of elements. Such a selection pressure may also mold the specificity of time and place of transposition referred to above.

For background information *SEE GENE; GENE ACTION; POPULATION GENETICS; TRANSPOSONS* in the McGraw-Hill Encyclopedia of Science and Technology.

Brian Charlesworth

Bibliography. L. Chao et al., Transposable elements as mutator genes in evolution, *Nature*, 301:633–635, 1983; B. Charlesworth, The population genetics of transposable elements, in T. Ohta and K. Aoki (eds.), *Population Genetics and Molecular Evolution*, pp. 213–232, 1985; D. E. Dykhuizen et al., Joint distribution of insertion elements *IS4* and *IS5* in natural isolates of *Escherichia coli*, *Genetics*, 111:219–231, 1985; E. A. Montgomery and C. H. Langley, Transposable elements in Mendelian populations, II. Distribution of copia-like elements in natural populations, *Genetics*, 104:473–483, 1983.

Ubiquitin

Ubiquitin is a very small protein with a molecular weight of approximately 8000 daltons. The initial studies in which ubiquitin was identified by G. Goldstein and colleagues showed that it was largely conserved from single-cell to multicellular organisms. Indeed, the ubiquitins from barley, yeast, and mammalian species differ only in two or three amino acids. Ubiquitin was aptly named, because it really does occur ubiquitously and exhibits a high order of conservation.

Although a role for free ubiquitin has not yet been defined specifically, the ubiquitin moiety has received increasing attention as a marker of biochemical events. In the cell nucleus, ubiquitin is conjugated with specific histones. In the cytoplasm, its conjugates relate to protein degradation. On the cell surface, ubiquitin conjugation was found for lymphocyte homing receptors. The increasing number of proteins to which ubiquitin binds suggests that the cell utilizes this peptide as a recognition signal or flag.

Ubiquitin contains only 76 amino acids; other than its amino terminal methionine, there is little that is special about its overall amino acid sequence. Of greater interest are the two glycine residues on its C-terminal end which serve as activation and linkage sites. The molecule appears naturally either with no glycine, one glycine, or two glycine residues. However, it is not activated unless both glycines are present; they are required for formation of the unusual isopeptide linkage.

Although the overall sequence offers no special features, the three-dimensional packing of ubiquitin is unique. Ubiquitin is extremely resistant to digestion by trypsin, despite the fact that the ubiquitin molecule contains 11 potential cleavage sites for this enzyme. Apparently, the packing or coiling of the structure is so tight that trypsin is unable to penetrate the structure. That is, ubiquitin is packaged (configured) so that access by proteolytic agents is sterically hindered.

Free ubiquitin exists in cells. Although it may serve a role in trout testis as a nonhistone chromatin protein, its function is still unclear. Studies have shown that in trout chromatin 50% of the isolated ubiquitin had glycine as the C-terminus, and the other 50% of the isolated product had a free arginine C-terminus. Thus far, ubiquitin has been found in the free form only in trout chromatin.

Ubiquitination. Ubiquitination is the addition of the ubiquitin moiety to a protein. Ubiquitin conjugation has been demonstrated for histones and surface proteins.

Histone conjugates. The ubiquitin-histone complex (Ub-2A or protein A-24) was first described for a histone complex referred to as protein A24 (see **illus.**). This was one of many new proteins discovered when it became possible to separate nuclear proteins on two-dimensional gels. Although protein conjugates had not been described earlier, protein Ub-2A (A24) clearly contained the entire sequence of histone 2A and the entire sequence of ubiquitin. With precise chemical analysis of the linking structures between the two proteins, it was shown that the junction between histone 2A and ubiquitin was on the amino acid lysine (the E-amine) at position 119 of histone 2A. No other conjugated site was identified. The linkage site on the ubiquitin chain was the C-terminal glycine residue mentioned above.

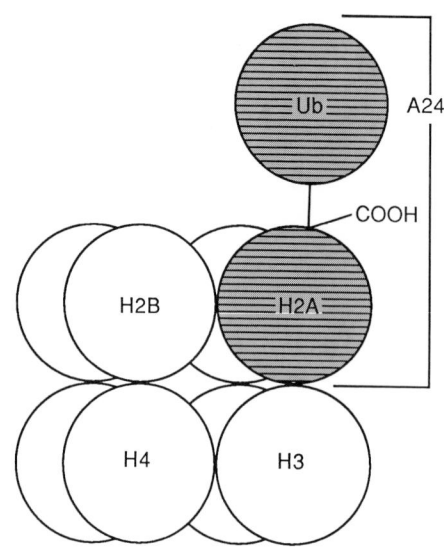

Diagram of a branched-protein A24 molecule (shaded spheres) fitted into an octameric nucleosome core. In this nucleosome the histone 2A molecule in the core and the ubiquitin (Ub) is outside. H2A, H2B, H3, and H4 are specific histone subunits of the octameric nucleosome.

Interestingly, this linkage between the ubiquitin and its conjugated protein was the first characterized and is general for all types of ubiquitin complexes.

Once this Y-shaped complex was identified, it was quickly shown that a similar conjugate exists for another histone, 2B. Quantitatively 10% of histone 2A is conjugated with ubiquitin, but only 1% of histone 2B is conjugated. The reasons for this difference and its physiological consequences are still not apparent.

Proteolysis. Ubiquitination has been found to promote proteolysis of reticulocyte proteins. Reticulocytes are intermediate cells on the path to formation of mature red blood cells; following extrusion of the nucleus, there is rapid destruction of excess proteins which are not required for the mature red cell. By using elements of the reticulocyte system, multiple ubiquitin conjugates were formed with protein substrates, including lysozymes, globin, and lactalbumin.

Homing receptors. The term homing is applied to the migration of lymphocytes from the blood to particular lymphoid sites. The homing receptors are the specific structures which the lymphocytes recognize and adhere to in lymphoid organs. The lymph-node entry sites are postcapillary venules which contain a high-walled endothelium; these venules are known as HEV, and their interactions with recirculating lymphocytes are studied in the laboratory. Other blood vessels do not bind these lymphocytes. The B lymphocytes specifically bind to HEV in Peyer's patches (aggregated lymphatic nodules found in the lining of intestine walls), and the T lymphocytes specifically bind to HEV in peripheral nodes.

Recently, studies on these homing receptors have been simplified by the development of mo-noclonal antibodies, which react specifically with epitopes on portions of these structures. One research group used immunoselection for genes expressing the antigens that react with the homing receptors. Two genes were identified, one for ubiquitin and the other for a unique polypeptide. It was concluded that the lymphocyte homing receptor is a ubiquitinated cell-surface glycoprotein. Moreover, a variety of other cell-surface proteins were found to be ubiquitinated, and the number of possible surface-protein-binding sites is greatly increased by ubiquitination.

Recent reports have suggested that a lymphocyte growth factor is a protein that contains a single ubiquitin conjugate. The role of this growth factor, an autocrine (self-regulatory) hormone, is under study.

Mechanism of conjugation. Exactly how and when the ubiquitin moiety is added to the surface proteins is not clear, but the systems involved are likely to be similar to those involved in other protein conjugation. Several steps have been demonstrated to be intermediates in the formation of ubiquitin conjugates. These include adenylation of the glycine C-terminus, formation of a thiol (—SH) ester on the glycine C-terminus, transesterification from one enzyme to the other, and conjugation to the substrate or target. These reactions are catalyzed by special enzymes designated E1, E2, and E3. Depending on the particular substrates, either a single ubiquitin or several ubiquitin molecules are linked. In the case of histones and the surface proteins, only one ubiquitin moiety is conjugated to the protein. However, for the products to be proteolyzed, several ubiquitin moieties may be conjugated to one protein. Ubiquitin-dependent proteolysis can be blocked by proteins with blocked amino groups, which indicates that they may compete for binding to ubiquitin-protein complexes for recognition elements.

The exact number of enzymes available for ubiquitin conjugation has not yet been completely defined. Several different fractions with differing activity to specific proteins have been identified. Accordingly, it is probable that the systems are quite complex, and sorting of these activities will require a good deal of further experimentation. Another element which was reported to be in the ubiquitin pathway is a transfer or acceptor RNA for histidine (tRNA^{His}) which may be a selective requirement for a special ubiquitin pathway.

Cleavage enzymes. Cleavage of ubiquitin from conjugated sites requires other enzymes known as deconjugases, isopeptidases, or hydrolases. Apparently, such mechanisms have been developed to conserve ubiquitin for further conjugation reactions. Studies on the mechanism of this cleavage have suggested that the enzyme-binding site has properties of an aldehyde which can be blocked with reagents.

Genes. Studies have now been made on ubiquitin genes in plants and animal cells. Initial results showed that there are multiple copies of these genes and that they are juxtaposed consecutively. Although ubiquitin is turned over relatively slowly, initial studies on its rate of turnover in the histone 2A complex showed that it had a higher turnover rate than the histones themselves. The exact relationship of its synthesis to the rates of other cellular reactions has not been demonstrated, nor has the role of the control factors for genes that are involved in its synthesis. There may be specific elements that differ in these genes, but information on this point is not currently available.

An interesting feature of the ubiquitin genes in yeast, humans, and plants is that multiple ubiquitin-coding repeats are present in a head-to-tail arrangement. Accordingly, as the genes are read, a long proprotein (protein precursor) is produced in which the glycine-methionine peptide bonds are joined directly between the last and the first residues. In some species, as many as 12 of these units are joined in tandem. One question is whether this tandem arrangement is associated with compartmentalization of ubiquitin or its precursors. Such compartmentalization could be important to the specific roles of ubiquitin in the nucleus and the cytoplasm.

Many questions arise from work on ubiquitin. Why should there be a polyubiquitin precursor? How many ubiquitin genes are there, and do they all code for the same molecule? What is the relationship of these synthetic products to the final products in the cell? Ubiquitin chemistry is clearly an area of growing interest as information is gained concerning gene controls, ubiquitin structure, localization, and function. SEE PLANT METABOLISM.

For background information SEE AMINO ACIDS; CELL NUCLEUS; GEL PERMEATION CHROMATOGRAPHY; GENE; MONOCLONAL ANTIBODIES in the McGraw-Hill Encyclopedia of Science and Technology.

Harris Busch

Bibliography. H. Busch and I. L. Goldknopf, Ubiquitin-protein conjugates, *Mol. Cell. Biochem.*, 40:173–187, 1981; M. Gallatin et al., Lymphocyte homing receptors, *Cell*, 44:673–680, 1986; A. Hershko and A. Ciechanover, Mechanisms of intracellular protein breakdown, *Annul. Rev. Biochem.*, 51:335–364, 1982; P. L. Lund et al., Nucleotide sequence analysis of a cDNA encoding human ubiquitin reveals that ubiquitin is synthesized as a precursor, *J. Biol. Chem.*, 260:7609–7613, 1985; C. A. Pickart and I. A. Rose, Mechanism of ubiquitin carboxyl-terminal hydrolase, *J. Biol. Chem.*, 261:10210–10217, 1986; D. Schlessinger and G. Goldstein, Molecular conservation of 74 amino acid sequence of ubiquitin between cattle and man, *Nature*, 255:423–424, 1975; D. C. Watson, B. Levy, and G. H. Dixon, Free ubiquitin is a non-histone protein of trout testis chromatin, *Nature*, 276:196–198, 1978.

Underground mining

Recent subsidence research in the United States has been focused primarily on meeting the regulatory requirements of the Surface Mining Control and Reclamation Act (Public Law 95-87, August 3, 1977). As in the past, the bulk of the research and the resulting technology is directed at predicting the location and magnitude of surface displacements and strains resulting from the collapse and closure of the openings of underground coal mines. There has been emphasis on improvement of the procedures for predicting the extent of surface subsidence, evaluation of the effects of subsidence on groundwater regimes, and research programs seeking methods of minimizing the detrimental effects of subsidence on human-made structures. The temporal aspects of mine subsidence, particularly with respect to the amount of time required for subsidence features to reach the surface over abandoned mines, have also experienced a modest increase of research interest.

Predicting subsidence effects. This requires consideration of surface displacements and strains, time of subsidence development, and effects of subsidence on groundwater.

Surface displacements and strains. In longwall mining, the total extraction of large blocks (panels) of coal is accomplished through use of special mining procedures. Sustained efforts in instrumentation and monitoring of longwall mine panels have resulted in the development of a considerable library of American experience in relation to controlled subsidence. Substantial agreement appears to have been reached within the research community that British experience, which has been widely accepted in the past, is not directly applicable in the United States. Predictions of degree of surface displacements and strains over mines in the United States, based on British experience, have been shown to be consistently greater than actually encountered. Subsidence troughs in the United States appear to be generally narrower and shallower than over comparable mined panels in the United Kingdom. As a result, empirical prediction of the size and shape of anticipated subsidence now tends to be based upon data derived from the United States and upon figures from the immediate coal province rather than from United Kingdom experience.

An increasingly important alternate approach to the prediction of subsidence trough geometry is the use of empirical influence functions to describe ground movements over longwall mines or second-mined room-and-pillar mine panels. In second mining, the pillars left for roof support in an underground coal mine are removed as the mining operations move back out of the previously mined area. The room-and-pillar mine panels are those large blocks of coal which are left for roof support as the coal is mined.

Various mathematical equations can be used to describe the amount of influence that an elementary unit of extraction (an extraction element) has on the vertical component of movement (the subsidence of a point on the surface). These relationships are known as subsidence basin influence functions. The most commonly used influence function investigated has been the Knothe function (or one of its variants), although good results have been obtained with others.

An example of this research approach is a study reported in 1985 on the application of a hyperbolic-tangent profile function and the Budryk-Knothe influence function to subsidence prediction. Both functions were adapted to fit subsidence data collected over a number of longwall and room-and-pillar panels. The hyperbolic-tangent method appeared to be most suitable for quick estimation of subsidence displacements along lines across the mine panel, whereas the influence function provided the capability of calculating displacements at almost any location across the panel. Another example of this direction of research was the development of a subsidence prediction method based on the Weibull distribution and on the elastic theory. In this case, the subsidence basin influence function was applied to eight sets of subsidence data from mines in the northern Appalachian Coal Field with results that agreed very closely with observed values. In another subsidence prediction method, SPASID (Subsidence Prediction and System Identification), the influence function has been applied to subsidence data from three longwall panels in the Illinois Coal Basin; again, the results agree closely with observed values. The general conclusions are that, with appropriate empirical adjustments for regional geologic conditions, each of the above three computer programs, which are based on influence functions, has been demonstrated to be capable of developing subsidence displacement curves which compared favorably with actual displacements. However, the results of this line of research have not yet provided a universally applicable procedure for subsidence prediction, and it is still difficult to incorporate the effects of regional geology into the analysis.

Time of subsidence development. Recently, there has been renewed interest in the development of methods for assessing the risk of delayed subsidence over abandoned room-and-pillar coal mines. A study completed in 1986 involved the time sequence of subsidence events and geologic conditions along the Front Range of Colorado. It was concluded that 94% of the subsidence features observed along the Front Range had occurred within 20 to 40 years after mining and that chimney (pit) subsidence events were confined to those areas where the ratio of overburden thickness to mine height was less than 10. The ratio of caving height to mined height corresponds to a

bulking factor of 1.15 and compares favorably with previous calculations. The bulking factor is the relationship between the original volume of material and the volume after the material has been disturbed.

The specific conditions existing at a mine and the specific procedures and techniques used during the mining process are known as site geologic and mine parameters. These were used in a study completed in 1986, in which an analysis of these parameters related to 80 subsidence events over abandoned room-and-pillar mines throughout the United States indicated that both the types of subsidence and the delay between mining and surface distress were affected by the relative thickness of competent material in the lower portion of the mine overburden. This material, usually limestone or sandstone, consists of rock strata that have sufficient strength to span a mine opening without failure. Increased competence of the overburden led to greater probability of sag subsidence rather than pit subsidence. The longest delays associated with sag subsidence were encountered at those sites where there was less than 20% competent material (limestone and sandstone) in the lower part of the mine overburden. However, the prediction of the time required for subsidence features to develop at the surface still is limited by the data base of subsidence events suitable for analysis, and the significance of regional geologic conditions remains difficult to include in the prediction equations.

Effects of subsidence on groundwater. The effects of mine subsidence on local and regional groundwater resources have been recognized as significant environmental problems, leading to considerable research. Monitoring of piezometric levels above a retreat pillar mining panel in West Virginia has shown that fluctuations in piezometric levels in shallower formations were only slightly disturbed during both development and retreat. The deeper piezometric surfaces, however, were significantly disturbed during the retreat mining. All fluctuations in this study ceased after mining was completed and the piezometric surfaces were stabilized at levels near or below those prior to mining. Additional studies of groundwater levels over three longwall panels in the Northern Appalachians have shown water-level fluctuations in excess of 50 ft (15 m) in borings over the panels; smaller fluctuations were observed outside the panel rib lines. In five drill holes monitored, all water levels failed to return to their premining elevations. Studies of the effects of subsidence on the groundwater regime have yielded inconclusive results. Further work is required to develop methodology that will successfully predict the response of the groundwater regime to subsidence activity.

Subsidence damage prevention. Primarily this requires mine backfilling but also may require structural modifications.

Mine backfilling. The primary preventive technology for dealing with threatened subsidence of valuable real estate is the backfilling of mine voids. In hydraulic backfilling, solid material is placed in a mined-out void space through a process that includes the use of water as a transporting mechanism. In pneumatic backfilling, the solid material is placed in a mined-out void space through a process that includes the use of air as a transporting mechanism.

The conventional approach has been hydraulic backfilling. However, pneumatic backfilling of a mine underlying a high school in Burgettstown, Pennsylvania, was considered necessary, since hydraulic backfilling might have intensified acid drainage problems at the site. Pneumatic techniques have been employed to place 14,000 yd^3 (11,000 m^3) of granular fill in the voids; the consequences of the backfilling process have yet to be evaluated.

Structural modifications. The response of structure foundations to subsidence-induced ground movements is another area of research for the prevention of subsidence damage. Recently, the use of posttensioned rock anchors to stabilize a residence located in the tension zone over a collapsing room-and-pillar section of a coal mine in Fairmont, West Virginia, was studied. The anchors were tensioned to create a zone of compression in the bedrock beneath the house, which is intended to prevent further damage to the structure. In addition, recent research has suggested the incorporation of expansion joints in typical residential structures and proper alignment of those joints with respect to anticipated subsidence displacements. If this approach is effective, it could develop into a design procedure that could mitigate, if not eliminate, major structural damage in areas prone to subsidence.

For background information SEE COAL MINING; MINING; UNDERGROUND MINING in the McGraw-Hill Encyclopedia of Science and Technology.

John D. Rockaway; Alphonse C. Van Besien

Bibliography. K. A. Heasley and L. W. Saperstein, Practical subsidence prediction for the operating coal mine, *Abstracts: 2d Workshop on Surface Subsidence Due to Underground Mining*, West Virginia University, pp. 24–30, 1986; M. Karmis, P. Schilizzi, and P. Jarosz, The development of ground subsidence above underground coal mines in the Appalachian Coal Field and its prediction using empirical techniques, *Proceedings of the 2d Conference on Ground Control Problems in the Illinois Coal Basin*, Southern Illinois University—Carbondale, pp. 127–137, 1985; G. M. Matheson and A. D. Eckert-Clift, Characteristics of chimney subsidence sinkhole development from abandoned underground coal mines along the Colorado Front Range, *Abstracts: 2d Workshop on Surface Subsidence Due to Underground Mining*, West Virginia University, pp. 64–71, 1986; A. C. Van Besien and J. D. Rockaway, Subsidence events over room and pillar coal mines, *Application of Rock Characterization Techniques in Mine Design*, A.I.M.E., Society of Mining Engineers, pp. 250–254, 1986.

Vaccination

The immune system recognizes certain chemical structures on bacteria or viruses as targets to fight infections. In response to the encounter with pathogens, it produces antibodies, which are specialized proteins circulating in the bloodstream. Antibodies interact with the antigenic targets on infectious agents and can initialize complicated effector mechanisms which destroy or neutralize the pathogens. Immunologically speaking, the antigens on the vaccine and the pathogen must be similar enough so that immune responses against the vaccine must also be effective against the pathogen.

The working principle in vaccination is induction of protective immunity against disease without the host's going through the disease process itself. Classically this approach depends on the use of nonpathogenic materials which are derived from the disease-causing agent or organism. Successful vaccination therefore requires a sufficient similarity in immunogenic properties between the vaccine and the disease-causing entity.

Idiotypes. Idiotypes are antigens on antibodies. They consist of several individual antigenic determinants called idiotopes. They are self-antigens or autoantigens that play an important role in the regulation of the immune response. By recognizing idiotypic determinants on other antibodies and cellular antigen receptors, the immune system is able to balance responses to individual challenges and to maintain a steady state of readiness in the absence of challenge. In the early 1970s, N. Jerne and J. Lindenmann formulated a network theory for the immune system based on the mutual interactions of idiotypes and anti-idiotypes. In this concept a large repertoire of complementary three-dimensional shapes configured by highly variable antibody structures and antigen receptors on B and T lymphocytes is constantly engaged in interactions which balance stimulatory and suppressive signals. While the exact mechanisms of these network interactions are still under investigation, the network concept is already being exploited to manipulate the immune response in specific ways or to induce specific immunities without using immunizations with environmental antigens or pathogens. The rationale for these experiments is the fact that anti-idiotypic antibodies can mimic the structure of antigens and therefore can be used as substitutes for antigens in vaccines.

Immune network. In structural terms, idiotypes are created by the highly variable regions of the antibody surfaces. Because the structural diversity of the hypervariable antibody region is extremely large, it must be expected that some of these different structures resemble structures of bacterial or viral antigens. From statistical analysis the necessity arises that many of these idiotypic structures resemble the three-dimensional shapes of external antigens. Therefore, it should be possible to use these internal antigenic structures expressed by idiotypes as surrogate antigens. The method of making so-called idiotype antigens or vaccines mimics a naturally occurring sequence of immune responses in which the antibody against the original antigen induces a second anti-antibody, which in turn can induce a third anti-anti-antibody. First an antibody is made against the pathogen in question. The first antibody (Ab1) is then isolated and used to induce a second antibody (Ab2, anti-idiotypic antibody). Some of these Ab2 antibodies express structures which mimic the three-dimensional shape of the original antigen. According to the network hypothesis, this second antibody, which expresses the internal image of the original antigen in the immunization cascade, should be able to induce a third antibody response (Ab3). If the Ab2 is a faithful surrogate antigen, the induced Ab3 response will be very similar to the Ab1 response induced by the original antigen. Thus, with this network cascade experiment it is possible to induce a protective immunity without using the pathogen or material derived from the pathogen.

Mimicry. The structural basis of the molecular mimicry by idiotypes is poorly understood. Detailed crystallographic studies of antibody molecules expressing an internal antigen image are not available. Furthermore, it is still difficult to assign a particular idiotypic structure to a given molecular portion of the antibody molecule, though it is believed to reside in the antigen-binding region whose composition varies with each antibody, and is therefore termed hypervariable. A promising approach to the understanding of internal antigens, besides x-ray crystallographic analysis of idiotype–anti-idiotype complexes, is the use of synthetic peptides, which are constructed to resemble particular hypervariable amino acid sequence regions of initial antibodies. These peptides are then used to evoke the production of corresponding antibodies, directed against the idiotypes of the initial antibodies. Subsequent ability of the two sets of antibodies to interact with each other confirms that the sequence region which provided the blueprint for the peptide synthesis has also been responsible for the expression of the given idiotype specificity (or the expression of the internal antigen). However, this method is limited to idiotypes whose three-dimensional configuration is adequately represented by the continuous amino acid sequence and spontaneous folding assumed by the synthetic peptide. Clearly, it cannot be used directly for idiotypes that are created by a juxtaposition of amino acids located in different portions of the primary chain, which becomes folded so as to bring about a particular spatial orientation.

Vaccine production. As noted above, idiotype vaccines are made by a sequential immunization protocol which starts with the induction of an antibody against a given pathogen or toxin. It is important that this first antibody, Ab1, should be capable of conferring protection. In most instances, this Ab1 must be able to neutralize the pathogen or inactivate the toxin. It is also desirable that the Ab1 be a monoclonal antibody made by the hybridoma fusion technique. In most cases, the mouse is the host for this Ab1. A purified Ab1 is then injected into a second mouse or rabbit to make an anti-idiotypic antibody, Ab2. In this instance, it is also best to make monoclonal anti-idiotypic antibodies. Immunization with Ab1 induces the production of Ab2 antibodies, which are also called anti-idiotypic antibodies. The next step is to select among the Ab2 antibodies those which can be inhibited in their binding to the Ab1 by the corresponding antigen. Because the antigen inhibits the binding of Ab1 to Ab2, the Ab2 may mimic the structure of antigen. These antigen-inhibitable Ab2 antibodies may therefore be the internal antigens mentioned above. The antigen-inhibitable anti-idiotypes are candidates for internal image anti-idiotypes. The proof for this property comes from using this Ab2 as antigen to induce a specific antibody which should react with the original antigen or pathogen. Another important criterion for true internal antigen images is their binding to specific antibodies from different species. For optimal immunization with anti-idiotypes, it has been helpful to increase the antigenicity by coupling carrier molecules to the Ab2 anti-idiotypic antibody.

Applications. Anti-idiotype vaccines have been made and tested in several animal models. The first report of a protective anti-idiotype was for *Trypanosoma*, which induces sleeping sickness. The anti-idiotype was made in mice and induced protection against infection. However, this anti-idiotype worked only in certain strains of mice

Summary of animal idiotype vaccines		
Disease agent	Anti-idiotype	Protection
Trypanosoma	Polyclonal	Partial
Escherichia coli	Monoclonal	Complete
Streptococcus	Monoclonal	Complete
Sendai virus	Monoclonal	Partial
Hepatitis virus	Polyclonal	Complete
Poliovirus	Monoclonal	Partial
Reovirus	Monoclonal	Complete

and thus did not appear to be an exact internal image anti-idiotypic antibody. Other laboratories subsequently made rabbit and mouse monoclonal anti-idiotypes which could induce partial or complete protection against bacterial, viral, and parasitic diseases. A list of the experimental systems is given in the **table**.

Advantages. It is now commonly accepted that the idiotype approach to vaccine production represents a viable alternative approach for vaccines. However, the idiotype approach will be practical and economical only in special cases where conventional vaccines do not work well, cannot be made, or are uneconomical. In these instances, idiotype vaccines may be advantageous.

If a pathogen cannot be safely attenuated or the relevant antigenic material is difficult to purify in large quantities, it will be easier and more practical to produce monoclonal anti-idiotypes which function as vaccines. An example is a tumor vaccine in which immunization with tumor material is not without risk to the patient or where purification of the tumor antigen is impractical.

Idiotype vaccines might also be superior in cases where conventional vaccines work poorly. For example, available vaccines against the bacteria *Influenza hemophilus* or *Pneumococcus* do not work well in infants because the immune system is immature. In idiotype vaccines the relevant antigenic structure is presented to the immune system as a protein antigen, in contrast to carbohydrate antigens of the bacteria, and the immune response to proteins matures earlier than that to carbohydrate antigens. A similar rationale for idiotype vaccines applies to tumor-associated antigens, to which the cancer patient has become more or less tolerant. By using an antitumor idiotype vaccine in immunotherapy, the cancer-associated antigen is presented in a different molecular form or carrier environment, and therefore might break through the existing state of tumor tolerance. Examples of breaking tolerance by changing the molecular environment in the antigen are known from work in animals.

Conclusion. The concept of inducing specific immunity without using the original antigen or antigen-derived material has been demonstrated by using anti-idiotypic antibodies. Such antibodies mimic relevant structures on the original antigen. They are produced by a sequential immunization protocol of making antibodies against the antigen first, followed by a series of immunizations in which antibodies against the pathogen are made first and used to make anti-antibodies, that is, anti-idiotypic antibodies. Such idiotype vaccines have potential application in situations where no protective immunity can be achieved with conventional vaccines. The structural basis of this molecular mimicry by antibodies is poorly understood so far, and more information is needed to develop a rationale and a general approach in the development of idiotype vaccines. See Monoclonal antibodies.

For background information *see* Acquired immunological tolerance; Antibody; Antigen; Immunity; Monoclonal antibodies; Vaccination in the McGraw-Hill Encyclopedia of Science and Technology.

Heinz Köhler

Bibliography. N. K. Jerne, The immune system, *Sci. Amer.*, 229:52–60, 1973; R. K. Kennedy, G. R. Dreesman, and H. Kohler, Vaccines utilizing internal image anti-idiotypic antibodies that mimic antigens of infectious organisms, *Biotechniques*, 3:404–408, 1985; M. K. McNamara, R. E. Ward, and H. Kohler, Monoclonal idiotype vaccine against *Streptococcus pneumoniae* infection, *Science*, 226:1325–1326, 1984; A. Nisonoff and E. Lamoyi, Implications of the presence of an internal image of the antigen in anti-idiotypic antibodies: Possible application to vaccine production, *Clin. Immunol. Immunopathol.*, 21:397–406, 1981.

Virus chemoprophylaxis

The relatively brief history of antiviral chemotherapeutic drugs can be summed up as a search for agents with progressively higher therapeutic indices. Such drugs are more and more selectively effective on viruses or virus-infected cells and increasingly sparing of noninfected host cells. Although historically the basis for the development of antiviral drugs has been primarily serendipitous, it is rapidly becoming possible to direct research more rationally. Most current work involves structural modification of established antiviral agents (lead compounds) with the hope of achieving a higher therapeutic index. The finding of drugs more potent and selective than the ones currently available for the prevention and treatment of viral infection will be possible as understanding of virus-host interactions increases. The recent advances in cellular and viral biochemistry and in viral genetics will undoubtedly hasten this process.

Research objectives. Much remains to be learned about the structural requirements, chemical nature, and composition of such compounds and their targeted enzymes. However, as knowledge of the biochemistry of virus replication and host cell–virus interaction increases, targets unique to the virus are emerging. For example, there are already several antiviral agents that have high therapeutic indices for herpes simplex virus (HSV). These compounds include several acylic nucleosides such as 9-(2-hydroxyethoxymethyl)guanine (acyclovir, structure I) and 9-(1,3-dihydroxy-2-propoxymethyl)guanine (DHPG), the 5-substituted nucleosides such as 5-ethyl-2'-deoxyuridine (EdU), several 2'-fluoroarabinopyrimidine nucleosides such as 1-(2-

Licensed and investigational new drugs for the treatment of human viral diseases						
Herpes simplex virus	Varicella zoster virus	Cytomegalovirus	Hepatitis B	Human immunodeficiency virus	Influenza A	Respiratory syncytial virus
Idoxuridine*	FMAU	DHPG	FMAU	3'-Azido-3'-deoxythymidine (AzT)*	Amantadine*	Ribavirin*
Trifluoridine*	FEAU	Foscarnet (PFA)		2',3'-dideoxycytidine	Rimantadine	
Vidarabine*	BVaraU			Ansamycin		
Acyclovir*				Foscarnet (PFA)		
6-Deoxyacyclovir				AL 721†		
FMAU				Antimoniotungstate (HPA-23)		
FEAU				Ribavirin		
Ribavirin						
EdU						

* Drugs licensed by the Food and Drug Administration.
† A lipid mixture of neutral glycerides, phosphatidylcholine, and phosphatidylethanolamine in a 7:2:1 ratio.

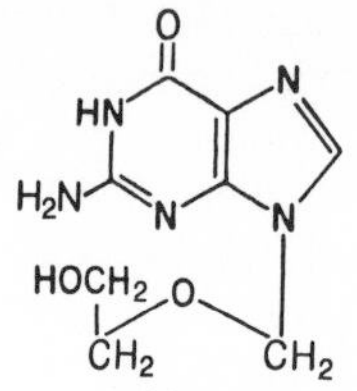

(I)

deoxy-2-fluoro-β-D-arabinofuranosyl)-5-methyluracil (FMAU) and its 5-ethyl analog (FEAU), the carbocyclic nucleosides such as cyclaradine, and recently a whole new series of 2-acetylpyridine thiosemicarbazones. These compounds take advantage of the known potentially exploitable biochemical targets unique to this virus for either drug activation (such as HSV thymidine kinase) or inhibition (such as HSV-DNA polymerase or HSV-ribonucleoside diphosphate reductase).

With the advent of 9-β-D-arabinofuranosyladenine (vidarabine, II) and acyclovir for the

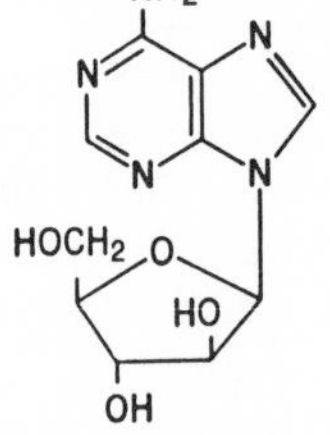

(II)

treatment of herpetic infections, great strides have been made in the development of antiviral drugs. However, much remains to be done to improve drug delivery of these agents in order to increase their bioavailability to the target organ, and to find drugs with different mechanisms of action in order to combat drug-resistant variants. Most important, there is a need to uncover more potent and selective antivirals or a combined modality that can decrease or prevent the significant morbidity still resulting from severe infections despite current antiviral therapy.

Other antiviral drugs that have been licensed and are in current use prophylactically and therapeutically for the treatment of viral infections

include amantadine (III), idoxuridine (IV), trifluridine (V), and ribavirin (VI). The **table** shows

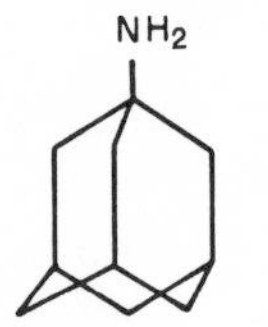

(III)

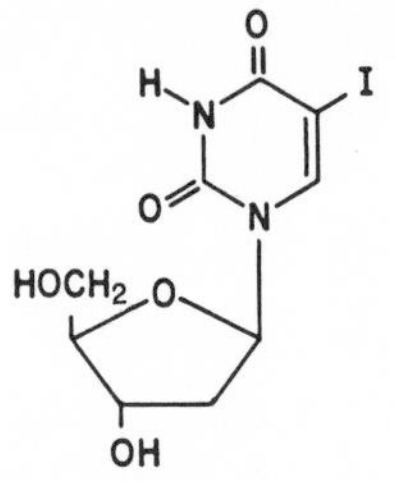

(IV)

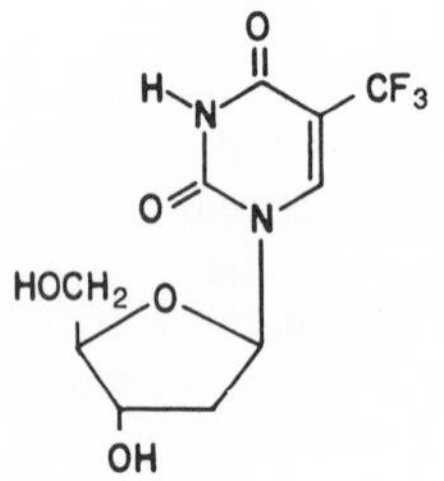

(V)

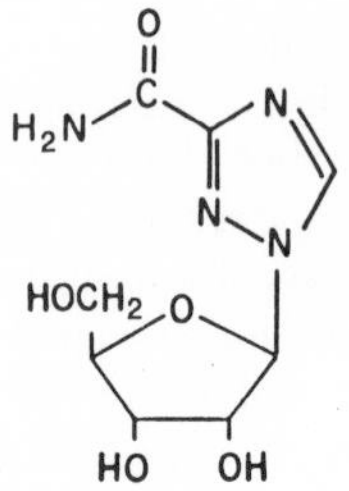

(VI)

newer agents that are undergoing extensive evaluation in animal models and in humans.

Influenza A. Clinical trials have documented that amantadine hydrochloride and its analog rimantadine hydrochloride are effective drugs for the prophylaxis and treatment of influenza A virus infection. However, widespread use of amantadine has been limited in part by concern about its side effects. Rimantadine has been considered to be better tolerated in humans than amantadine, although conflicting results have been reported from studies of their relative toxicities.

Amantadine is not virucidal and has no effect on the adsorption of virus to host cells, but it does inhibit virus penetration or uncoating, and thus the viral genome is not expressed.

Herpes simplex virus (HSV). There are currently several drugs that are licensed for the treatment of herpetic conditions. These include 5-iodo-2'-deoxyuridine (idoxuridine), 5-trifluoromethyl-2'-deoxyuridine (trifluridine) and vidarabine for ocular infections, vidarabine for herpes encephalitis and neonatal herpes, and acyclovir for prophylaxis and treatment of mucocutaneous infections, including genital herpes.

Ocular infections. As described above, there are three drugs approved by the Food and Drug Admimistration for the treatment of HSV-keratitis, and these appear to be more effective in conjunction with corneal debridement. For more severe eye infections such as stromal keratitis, trifluridine is used in combination with a steroidal anti-inflammatory drug. More potent compounds such as 5-*E*-(2-bromovinyl)-2′-deoxy-uridine (BVDU) and FMAU are currently being developed in the United States and Europe for this and other less severe infections. Acyclovir is still not licensed in the United States for the treatment of ocular herpetic infections.

Brain infections. Controlled studies in neonates or older children and adults with herpes simplex virus encephalitis have indicated that intravenous vidarabine decreases mortality and morbidity significantly, particularly if treatment is started early. Results of recent controlled studies comparing vidarabine to acyclovir for herpes simplex virus encephalitis, out-side the neonatal age group, indicate that intravenous acyclovir is preferable to vidarabine. However, until results of similar comparativedrug studies in the neonate are analyzed, acyclovir is reserved for neonates in whom vidarabine may be too toxic.

Recent studies in several animal models for herpes simplex virus encephalitis have indicated that FMAU and DHPG are superior to either vidarabine or acyclovir. Similarly, combined treatment with these drugs in these models was more effective than single drug therapy.

There is still some uncertainty about the primary site of inhibition by vidarabine. Its triphosphate, ara-ATP, is a competitive inhibitor of HSV-induced DNA polymerase with respect to deoxyadenosine triphosphate (dATP). The affinity of ara-ATP for viral DNA polymerase is greater than for the host cell DNA polymerase; however, this specificity depends on the source of the viral enzyme. Recent studies also indicate that ara-ATP is incorporated in internucleotide linkage of not only viral DNA but also cellular DNA. Vidarabine also inhibits the enzyme *S*-adenosylhomocysteine hydrolase, the consequence being inhibition of essential methylation reactions, such as that of messenger RNA. The role of arabinosyl hypoxanthine (ara-H), the main metabolite of vidarabine, in the antiviral activity of this drug is not clear.

The selective antiviral activity of acyclovir depends on the virus-induced thymidine kinase for the initial phosphorylation. The 5′-triphosphate form of the drug (ACVTP) is a competitive inhibitor of the utilization of deoxyguanosine triphosphate (dGTP) by HSV-induced DNA polymerase and is also incorporated into viral DNA as a chain terminator. The viral DNA polymerase becomes irreversibly inactivated during the process, and therefore ACVTP is considered to produce "suicide" inhibition of the virus; this is a reaction in which a compound that resembles the normal substrate for an enzyme but interacts with the enzyme, forms a covalent (strong) bond, thus inactivating the enzyme. Once the ACVTP-enzyme complex is formed, the process is irreversible, dGTP (the normal substrate) can no longer compete, and the enzyme becomes inactive. This inactivation does not occur with the cellular α-DNA polymerase.

Genital herpes infections. Topical, oral, and intravenous forms of acyclovir have been shown to be effective in the treatment of primary genital herpes (confirmed by the absence of preexisting antibodies). These formulations are licensed for the treatment of initial genital herpes (the first clinically manifest genital infection occurring in an individual with prior exposure to nongenital herpes). Because oral acyclovir can be administered without the need of hospitalization, its use for the treatment of various herpetic conditions has been increasing. In patients with a mean of one genital herpes recurrence per month before therapy, oral acyclovir suppressed recurrences; but recurrence rates returned to their pretreatment frequencies once medication was discontinued.

A prodrug is an analog of a drug that is metabolized in cells or in the body and is converted biochemically to the known drug. A prodrug (precursor) of acyclovir, 6-deoxyacyclovir (BW515), produces higher plasma levels after oral treatment; therefore it is being considered for prophylaxis and treatment of herpes simplex virus infections. This drug is activated by the enzyme xanthine oxidase which converts BW515 to acyclovir.

Respiratory syncytial virus. Infections caused by this virus in children are often fatal or result in considerable morbidity. Ribavirin in an aerosol formulation was recently licensed for the treatment of respiratory syncytial virus infections in infants and children. It was shown to be effective in adult volunteers and also in infants and children with bronchiolitis and pneumonia.

Ribavirin, which structurally resembles guanosine, is converted to its active form by phosphorylation. In the cell, ribavirin is metabolized to the 5′-monophosphate; this product inhibits inosinate dehydrogenase, which ordinarily converts inosinate to xanthylate. In other words, ribavirin interferes with the formation of guanosine monophosphate, upon which both DNA and RNA synthesis depend. The 5′-triphosphate derivative of ribavirin (RTP) is a potent inhibitor of influenza virus RNA polymerase, and this may be responsible for the antiviral activity against this virus. With certain RNA and DNA viruses, RTP inhibits the viral-specific messenger RNA capping enzymes, guanylyl transferase and N^7-methyl transferase, so that either no viral protein is transcribed or, if viral protein is synthesized, it is nonfunctional.

Acquired immune deficiency syndrome (AIDS). There are several antiviral agents that appear to be selective in cell culture for the human immunodeficiency virus (HIV), which causes AIDS, and several classes of compounds are at different stages of evaluation in humans (see table). Of the compounds that have so far been reported to inhibit selectively the HIV reverse transcriptase and replication, 3'-azido-3'-deoxythymidine (AzT) and several 2',3'-dideoxy nucleosides appear to be the most potent. AzT, developed for the treatment of AIDS and related diseases, is a compound originally synthesized in 1964 and reported in early 1970 to have activity against Friend virus. More recently, AzT has been shown to have potent activity against other retroviruses, including HIV and murine or feline leukemia virus in cell culture. This antiviral activity is prevented by thymidine, suggesting that it can behave as a thymidine analog. The 5'-triphosphate form of the drug is a competitive inhibitor of the HIV reverse transcriptase, but it is a poor inhibitor of the human αDNA polymerase. The 5'-triphosphate of AzT binds tighter to the enzyme than does thymidine triphosphate (dTTP). The compound accumulates primarily as the 5'-monophosphate in cells infected with HIV.

When administered to humans, AzT presents three problems: (1) its half-life is relatively short (about 1 h); (2) it is extensively metabolized and excreted in the urine as 3'-azido-3'-deoxy-5'-glucuronylthymidine (the enzyme responsible for this bioconversion is probably UDP-glucuronyl transferase); (3) the drug markedly reduces the white blood cell count. AzT was recently shown to be very toxic to human bone marrow precursor cells in culture. Although AzT is now in clinical trials and looks promising, it is still too early to evaluate its clinical role in the treatment of patients with AIDS or pre-AIDS, or what the long-term toxicities may be.

It has recently been reported that 2',3'-dideoxy derivatives of inosine, adenosine, thymidine, and cytidine are potent inhibitors of HIV replication. The cytidine analog is about 50 times more potent than the dideoxy derivatives. Clinical trials with 2',3'-dideoxycytidine for the treatment of AIDS are planned. *See* Retrovirus.

Varicella zoster virus infections. With the exception of acyclovir and vidarabine, which lead to only moderate clinical improvement, there is currently no treatment for this virus, which causes chicken pox and herpes zoster in humans. In African Green monkeys infected with simian varicella zoster, the drugs FMAU, FEAU, and 5-*E*-(2-bromovinyl)-arabinosyluracil (BVaraU) showed excellent activity when administered at low doses.

Cytomegalovirus infections. Infection with the virus or its reactivation is a major cause of mortality in immunocompromised hosts such as children with malignant tumors, organ transplant patients, or AIDS patients. Trials with acyclovir, vidarabine, or interferons have shown variable and transient reduction in viral production or clinical symptoms. DHPG and phosphonoformate (foscarnet, PFA) are currently the two most prominent candidates that appear to be effective for the treatment of various infections caused by this virus. However, because these drugs have serious side effects, their use may be limited to individuals with life-threatening conditions.

DHPG is an analog of acyclovir and requires phosphorylation before it can interfere with the cytomegalovirus-DNA polymerase. The drug inhibits cellular DNA polymerase at levels about 10 times higher than that required to inhibit the viral-induced enzyme. PFA is thymidine kinase–independent and interacts directly with the pyrophosphate binding site of the DNA polymerase.

Hepatitis B infections. Although an effective vaccine is available for these infections, there is currently no drug that can be used therapeutically. FMAU has recently shown good results in the treatment of woodchuck hepatitis virus and therefore may prove effective against human hepatitis virus. Recently, in an uncontrolled study FMAU was shown to be very effective in humans infected with Hepatitis B virus; hopefully, an ongoing controlled study with this drug will confirm this finding.

Other infections. Ribavirin has been in clinical trials for a number of viral infections which include genital herpes, herpes zoster, influenza A, as well as exotic diseases such as Lassa fever. It may be of value for influenza A and B infections, adenovirus pneumonia, exotic infections caused by human Korean hemorrhagic fever, or sandfly fever virus.

Recently a number of analogs of viral polypeptides which specifically inhibit viral replication in cell culture have been prepared. A number of these oligopeptides were found to inhibit several RNA and DNA viruses, including measles and influenza A virus. The synthetic oligopeptides resemble the F protein of influenza viruses involved in virus penetration, cell fusion, and hemolysis. Such studies not only may result in the discovery of compounds that may be useful to treat human viral diseases but also may provide information on the mechanism of virus–cell membrane fusion.

Latency and antiviral drugs. With the exception of acyclovir, none of the compounds that have been tested can prevent virus reactivation. Research into antiviral chemotherapy aimed at controlling recurrences by interfering with mechanisms of reactivation of the virus is continuing. Animal studies have indicated that a certain combination of drugs can reduce the rate of isolation of herpes simplex virus from the trigeminal ganglion. Other investigators have used a procedure called iontophoresis, whereby ionized drugs may be transported by an electric field.

Thus an antiviral drug can be made to penetrate a defined area of involvement. More recent work involves the use of retrograde axoplasmic transport (transport of drugs back into the neurons) for delivery of drugs selectively to neurons innervating the skin and mucosa involved in recurrent herpes simplex virus infections.

For background information SEE VIRUS; VIRUS CHEMOPROPHYLAXIS in the McGraw-Hill Encyclopedia of Science and Technology.

Raymond F. Schinazi; William H. Prusoff

Bibliography. Y. Iwasaka et al., Eradication of herpes simplex virus persistence in rat trigeminal ganglia by retrograde axoplasmic transport, *J. Virol.*, 59:242–248, 1986; E. H. Kass and R. Platt (eds.), *Current Therapy in Infectious Diseases*, pp. 126–132, 1984; W. H. Prusoff et al., *Antiviral Res. Suppl.*, 1:1–10, 1985; R. K. Robins, Synthetic antiviral agents, *Chem. Eng. News*, pp. 28–40, January 27, 1986; R. F. Schinazi and A. J. Nahmias, Herpes simplex virus infections: Current therapy, *Current Therapy in Infectious Diseases*, 1983–1984; R. F. Schinazi and W. H. Prusoff, Antiviral agents, *Ped. Clin. N. Amer.*, 30:77–92, 1983.

Virus satellite

Some plant viruses with single-stranded ribonucleic acid (RNA) genomes have other associated small RNAs called satellites or virus satellites. The latter RNAs are dependent for their replication on the principal viruses, which are called helper viruses. Virus satellites usually have little, if any, nucleotide sequence homologies with genomes of their helper viruses or plant hosts. Some satellites, usually referred to as satellite viruses, encode their own coat protein to form capsids which are morphologically and antigenically distinct from those of their helper viruses. Other satellites, usually referred to as satellite RNAs, have no coding capacity for coat protein and rely on their helper viruses for encapsidation by coat protein. Virus satellites can have profound effects on disease symptoms induced by their helper viruses on plants. The term virus satellite includes both satellite viruses and satellite RNAs.

Satellite viruses. The most extensively studied satellite virus is satellite tobacco necrosis virus (STNV). A number of strains of both tobacco necrosis virus (TNV) and STNV have been described, all of which are transmitted by zoospores of the chytrid fungus *Olpidium brassicae* in the soil. However, the transmission of STNV is dependent on the correct combination of the satellite and helper virus strain, fungus race, and host plant species. In the presence of STNV, the lesions formed by TNV on the inoculated leaves of host plants are significantly smaller. The RNA of STNV consists of 1239 nucleotides whose sequence is known. It has less than 2% homology

with the RNA of TNV, and hence the RNAs cannot have a common sequence longer than 30 nucleotides. Both RNAs terminate with ppAp-GpUp . . . at their 5′-terminus and a hydroxyl group at their 3′-terminus. The RNA of STNV codes for a single polypeptide of 195 amino acids, 60 molecules of which assembled around the RNA to form an icosahedral particle about 17 nanometers in diameter with $T = 1$ symmetry. The molecular weight of each particle is 1.64×10^6. The icosahedral helper TNV particles have $T = 3$ symmetry and a diameter of about 30 nm.

Particles similar to those of STNV have been shown to be associated with panicum mosaic and maize white-line mosaic viruses. The satellite nature of the small particles associated with panicum mosaic virus has been demonstrated, but no conclusive data about that of maize white-line mosaic virus are yet available.

Tomato black-ring virus satellite RNA. There are a number of satellite RNAs (sat-RNAs) of molecular weight about 0.5×10^6 associated with viruses belonging to the Nepovirus group. The RNAs have a single open-reading frame which is expressed but whose function is not known. The sat-RNAs rely on their helper viruses for encapsidation. The most extensively studied of these is the sat-RNA of tomato black-ring virus (TBRV).

Like other satellites, the sat-RNA of TBRV is helper-specific, with different strains of the virus having their own sat-RNAs. The complete sequence of 1375 nucleotides of a TBRV sat-RNA isolate has been determined and shown to act as messenger RNA for a protein of 424 amino acids. Like the two RNAs of its helper virus, the sat-RNA has a genome-linked protein (VPg) at its 5′-terminus and a poly A sequence at its 3′-terminus. The sat-RNA of TBRV does not appear to affect the replication of its helper virus or to modify the symptoms it induces. However, it has been shown to reduce the number of local lesions on the inoculated leaves of *Chenopodium amaranticolor*.

Tobacco ringspot virus satellite RNA. Small single-stranded RNAs of molecular weight about 1×10^5 have been shown to be associated with several Nepoviruses. By far the most extensively studied of these is the sat-RNA of tobacco ringspot virus (TRSV). Variants of this sat-RNA have been detected in a number of TRSV isolates, and some specificity between the sat-RNAs and their helper viruses have been observed.

The sat-RNA of TRSV is a linear molecule of 359 nucleotides with a hydroxyl group at the 5′-terminus and a cytosine-2′:3′-cyclic phosphodiester at the 3′-terminus. This is unlike the ends of the helper virus RNAs which have a VPg and a poly A tail at their 5′- and 3′-termini, respectively. The sat-RNA of TRSV lacks *in vitro* messenger activity, and there is no evidence that it is translated *in vivo*. Satellite RNA can markedly diminish the amount of TRSV RNA in infected plants, and as much as 90% of the RNA encapsidated by the helper virus can be sat-RNA. The severity of the disease symptoms induced by the virus can also be reduced by sat-RNA.

Cucumber mosaic virus satellite RNA. Satellite RNA has been detected in numerous isolates of cucumber mosaic virus (CMV) belonging to the Cucumovirus group, and this satellite has sometimes been referred to as cucumber mosaic virus–associated RNA5 (CARNA5). At least 10 CMV sat-RNA isolates have been sequenced and shown to consist of linear single-stranded RNA molecules varying between 334 and 386 nucleotides with a high degree of sequence homology. Like their helper virus isolates, the sat-RNAs are capped with M^7Gppp at their 5'-terminus and have a hydroxyl group at the 3'-end. The sat-RNAs of CMV are highly base-paired and their secondary structure appears to be transfer RNA-like. Although some isolates of CMV sat-RNA have been translated *in vitro*, there is no evidence of a functional protein being expressed *in vivo*.

The amount of sat-RNA encapsidated by CMV varies considerably and appears to be dependent on the helper virus strain as well as the host plant. Whereas many sat-RNAs attenuate symptoms induced by CMV, some actually exacerbate disease. One sat-RNA isolate, which was found to attenuate symptoms on most plant species tested, induces a necrotic reaction on tomato which often kills young seedlings. Tomato aspermy virus (TAV), another Cucumovirus, can also act as helper virus to CMV sat-RNA, although much less efficiently.

A sat-RNA of peanut stunt virus (PSV), also belonging to the Cucumovirus group, has been reported. The sat-RNA consists of 393 nucleotides. It is similar in some respects to the CMV sat-RNAs; however, it also shows some differences and failed to be helped by CMV.

Viroidlike satellite RNAs. Four putative Sobemoviruses have been shown to encapsidate small sat-RNAs which are reminiscent of viroids. They are usually referred to as viroidlike RNAs or virusoids. Like a viroid, the sat-RNA of each virus has two single-stranded RNA components, a covalently closed circle (RNA2) and another molecule which is of the same size (324–388 nucleotides) and sequence but which is linear (RNA3). Also like viroids, the sat-RNAs have a high degree of base pairing and lack messenger activity *in vitro*. However, whereas viroids replicate autonomously and appear to replicate in the nucleus, the sat-RNAs depend for replication on their helper viruses and appear to be synthesized in the cytoplasm.

Velvet tobacco mottle virus (VTMoV) and solanum nodiflorum mottle virus (SNMV) and their viroidlike sat-RNAs are closely related and infect principally solanaceous plants. Lucerne transient streak virus (LTSV) and subterranean clover mottle virus (SCMoV) infect almost exclu-

sively leguminous plants and are remotely related to each other but not to either VTMoV or SNMV. In some instances the sat-RNAs do not appear to have any significant effects on disease severity induced by their helper viruses, but in others they exacerbate symptoms.

Satellite RNAs of Tombusviruses. Satellite RNAs have been shown to be associated with tomato bushy stunt virus (TBSV) and several other closely related viruses belonging to the Tombusvirus group, but have not been characterized in detail. They are about 700 nucleotides long and have extensive nucleotide sequence homology among themselves. They attenuate disease symptoms induced by their helper viruses. There is evidence that when sat-RNA-free TBSV is passaged through *Nicotiana benthamiana* the virus acquires the satellite.

Another sat-RNA of TBSV of about 450 nucleotides has been reported. It also attenuates disease symptoms but differs from the other sat-RNAs in that it has some sequence homology with the RNA genome of its helper virus. A similar-size sat-RNA has been found associated with turnip crinkle virus (TCV) which has particles similar to those of TBSV but which has some significant differences and remains as yet unclassified. This sat-RNA exacerbates disease symptoms in turnip plants with whose genome it appears to have some base sequence homology.

Satellite RNA of tobacco necrosis. A unique sat-RNA has been reported to be associated with TNV and its satellite virus, STNV. The sat-RNA is about 600 nucleotides long and has no significant nucleotide sequence homology with either TNV or STNV. It is dependent on TNV for its replication and on STNV for its encapsidation. Thus it appears to be a molecular parasite of two viruses involved in a satellite-helper relationship.

Conclusions. All satellites of plant viruses are dependent for their replication on conventional viruses (helper viruses). Some are encapsidated by proteins for which they themselves code (satellite viruses), whereas others also rely on their helper viruses for encapsidation (sat-RNAs). Although all the satellites mentioned here are either satellite viruses or sat-RNAs helped by viruses with small polyhedral particles, a sat-RNA of a virus with rod-shaped particles has recently been reported. The small RNA associated with tobacco mosaic virus (TMV) appears to be a true sat-RNA in that it is dependent on TMV for both replication and encapsidation. However, the details for its properties have not been reported. It is interesting that no satellite-helper relationships similar to those described here have been reported for viruses of organisms other than plants.

Satellites can be looked upon as molecular parasites capable of modifying disease symptoms induced by their helper viruses in plants. The satellites can therefore be important to agriculture. The significance of base sequence homologies of some sat-RNAs with the genomes of their host plants and others with the genomes of their helper viruses is at present obscure. However, it points to the possibility that they may have different origins. Some may have evolved from parts of plant genomes and others from the single-stranded RNAs of conventional viruses. It has been suggested that some sat-RNAs as well as viroids may have evolved from introns of plant genomes. It is also plausible that the satellite viruses and some sat-RNAs, especially those with messenger activity, may have originated from conventional viruses by loss of some of their coding capacity.

For background information SEE DEPENDOVIRUS; VIRUS in the McGraw-Hill Encyclopedia of Science and Technology.

R. I. B. Francki

Bibliography. R. I. B. Francki, Plant virus satellites, *Annu. Rev. Microbiol.*, 39:151–174, 1985; J. M. Kaper, Plant disease regulation by virus-dependent satellite-like replicating RNAs, in E. Kurstak and R. G. Marusyk (eds.), *Control of Virus Diseases,* pp. 317–343, 1984; R. E. F. Matthews, *Plant Virology*, 2d ed., 1981; A. F. Murant and M. A. Mayo, Satellites of plant viruses, *Annu. Rev. Phytopathol.*, 20:49–70, 1982.

Volcanology

Basaltic volcanism is a fundamental manifestation of the separation of the great tectonic plates along the mid-ocean ridges and the means by which almost two-thirds of the Earth's crust has formed. The scientific advance of the theory of plate tectonics has led to a picture of a dynamic Earth, constantly in motion. The globe has a radius of 3960 mi (6370 km) and is covered by the lithosphere, a thin rigid skin composed largely of a rock called peridotite. The lithosphere is some 60 mi (100 km) thick and constitutes the Earth's eight major tectonic plates. Beneath these plates lies a hotter ductile peridotitic layer, the asthenosphere, on which the lithosphere floats. The plates are continuously in motion, with new lithosphere forming along accretionary plate boundaries marked by the ocean ridges, and old lithosphere destroyed at destructive plate margins, termed subduction zones, where the plates collide.

Ocean ridge basalts. The separation of the lithosphere at the ocean ridges causes the underlying asthenosphere to rise up continuously between the plates. As it rises, the asthenosphere spontaneously undergoes partial melting to form basaltic magma. This happens as the melting point of rock decreases rapidly with decreasing pressure; material which is solid at depth, if hot enough, will become partially molten as it rises. The asthenosphere rising between the spreading plates forms new lithosphere as it cools. However, the lighter and less viscous basaltic melt rises out of the asthenosphere and erupts over the new lithosphere, forming its upper layer, the crust. This crust is quite variable in thickness; while it can thin locally to zero, it is typically some 4 mi (6 km) thick. Ocean ridges wend their way in a curvilinear fashion along the sea floors. They contain a series of short rift segments some 12 to 30 mi (20 to 50 km) long, where the ocean crust is continuously splitting apart and new crust is formed by volcanic action. Periodically fracture zones offset individual rift segments along zones of strike-slip motion. In other cases, the boundary between segments may be simply a depression or other irregularity along the axis of the ridge. This segmentation has been found to be a statistically regular feature apparently related to the dynamics of spreading. On a human time scale, spreading is slow, with new crust forming at about the same rate as the growth of a human thumbnail. On a geologic time scale, this spreading is rapid, generally adding up to tens of miles or more of new crust every million years. The rate varies greatly from ridge to ridge, with about 0.6 in. (1.6 cm) per year of new crust forming along the Southwest Indian Ridge as compared to nearly 8 in. (20 cm) per year locally along the East Pacific Rise. SEE RIFT VALLEY.

The basalt composing the crust is the world's most abundant lava. Compared to the underlying peridotitic residue in the mantle, the basalt is rich in the elements which are preferentially partitioned into the liquid during melting such as alumina, calcium, sodium, titanium, and various trace elements such as the rare-earth elements. Mantle peridotite, named after its major constituent (olivine or the gemstone peridotite), on the other hand, is rich in elements which tend to enter the solid phase during melting, such as magnesium, chrome, and nickel.

Understanding the evolution of ocean ridge basalts is the key to understanding the fundamental processes of plate tectonics, and may provide insight into the composition of the deeper layers of the Earth from which it arises. Recent studies of these lavas have concentrated on the mechanisms by which they form and segregate from the asthenosphere, and on their spatial chemical variability along ocean ridges. The studies have provided advances in understanding basalt genesis along ocean ridges, geodynamic processes operating beneath the ridges, and the chemical variability of the underlying source region.

Local variability of ridge basalts. As a consequence of the particular geometry of ocean ridges, great interest exists in the variability of ocean ridge basalts along the axis of individual ridge segments. The lithosphere must neck beneath an individual ridge segment, and the asthenosphere must rise to nearly the base of the crust. Thus, where the ridge axis is offset, upwelling asthenosphere is juxtaposed against old cold lithosphere formed previously at an adjacent ridge segment.

As a consequence, the rising asthenosphere should cool more rapidly near fracture zones and undergo less melting. This would affect the basalt composition, since the formation of a melt from a rock is chemically complex. Initially, at low degrees of melting, the melt composition is very different from that of the rock from which it forms. Only at high degrees of melting, as the rock is totally consumed, does the melt approach the composition of the original rock. Thus from low to high degrees of melting there is a continuous change in the chemistry of the melt formed. Basalts sampled recently near some fracture zones are richer in titanium dioxide (TiO_2) and other incompatible elements and contain higher iron(II) oxide (FeO) content at a given magnesium oxide (MgO) content, characteristics attributed to lower degrees of melting of the underlying mantle source.

Certain chemical features of ocean ridge basalt chemistry, including some characteristic trace-element and isotopic ratios, cannot be substantially modified through melt formation or crystallization processes. These variations must reflect differences in the initial composition of the mantle source region. Earlier workers have shown that these ratios and isotopes vary systematically along ocean ridges with proximity to the shallowest portions of the world ocean ridges where the latter cross over or near hot spots. Hot spots are regions marked by extraordinarily effusive volcanism and thickened ocean crust. In such regions the isotopic and trace-element composition of the basalts indicates that the lavas come from a mantle different from the typical source of ocean ridge basalt. The typical source mantle for ocean ridge basalt is termed depleted mantle, because it has chemical characteristics which indicate that it has undergone a previous episode of melting in Earth history. This episode depleted such mantle in the incompatible or magmophile elements, particularly certain radioisotopes. The magmas erupting in proximity to hot spots do not show this effect, and are believed to come from more primordial or primitive mantle material which did not undergo the earlier melting event.

Recent more detailed examination of rocks dredged from individual ridge segments has revealed that a very similar total variation in trace-element chemistry can be found at a single ridge segment. For example, one research group found specimens spanning the entire range from hot-spot basalt chemistry to depleted mid-ocean ridge basalt chemistry in a single dredge haul from a ridge segment on the Southwest Indian Ridge. This only modifies the former conclusions, because the average composition of the basalts from individual ridge segments are still found to vary systematically with proximity to hot spots. This local variability is taken to show that both primitive and depleted mantle sources underlie ridge segments everywhere. Small amounts of hot-spot-type basalt can be erupted with basalt from a depleted mantle source because of contamination of that source with melts from a primitive source. Many workers now speculate that the shallow mantle is the depleted mantle source, but it is veined by small amounts of material that has crystallized from melts which leaked upward from the deeper mantle.

Until recently the complementary parent-daughter nature of ridge basalt and the underlying mantle has been inferred largely on the basis of plate tectonic theory. This has changed dramatically because of the recent publication of the first systematic global studies of the compositions of peridotitic rocks dredged form the world's ocean ridge system. Residual peridotites—that is, samples of the underlying mantle residue of melting and magma genesis—have been dredged from many locations along the world's ocean ridge system. The ocean ridges possess a relatively thin crust; they are tectonically very active and are thus strongly disrupted by faulting and fissuring. This process has resulted in samples of mantle peridotite being emplaced to the surface along faults directly to the sea floor. The material is generally highly altered because its high-pressure mineralogy is not particularly stable under sea-floor or near-sea-floor conditions. Nonetheless, it contains enough of relict primary mineral grains to demonstrate its prealteration chemistry. A number of geologists have now found that the primary mineralogy of such dredged peridotites varies systematically along the ocean ridges, from areas where normal ridge basalts are generated to hot-spot regions such as the Azores or Iceland regions of the Mid-Atlantic Ridge. As the hot spots are approached, the peridotites contain less of the pyroxene minerals diopside and enstatite (from which most of the basaltic components are drawn during melting) and more of the mineral olivine, which is systematically enriched in the residues of melting. At the same time, the composition of the minerals changes systematically as well, containing less and less of the incompatible magmophile elements near hot spots: elements which are removed preferentially from the peridotitic mantle during melting. These chemical variations have also been demonstrated to be complementary to variations in basalts. Thus, for the first time it was demonstrated that the mantle has undergone higher degrees of melting in hot-spot regions, and that ocean ridge basalts are, as previously assumed, generated from the mantle peridotites emplaced as upwelling asthenosphere to the base of the ocean crust. This higher degree of melting of the shallow oceanic mantle could be a consequence of hotter mantle upwelling at ocean ridges near hot spots, or of contamination and fluxing of the shallow mantle by melt drawn from an underlying more primitive mantle, which could cause excessive melting of the shallower mantle.

Formation and segregation of oceanic basalt lavas. No subject in the study of basalts has been more mysterious and unconstrained than how melt forms, segregates, and passes out of the mantle. Because the melt is generated from within the mantle rock during ascent, it must form originally as disseminated films on grain boundaries. Because its density and viscosity are lower than that of the enclosing rock, its buoyancy should force it to rise to the surface. This subject has been recently illuminated by the work of a number of geophysicists who have concentrated on the physics of melt migration. Their work provides significant insights into, and new constraints on, how melting works. A major new perception is that the amount of melt that can be held in the mantle at any one time is small, a matter of a few percent or less. The degree of melting which is thought to produce ocean ridge basalt is large (10–30%); once the mantle starts to melt, the basalt will continue to be produced as the mantle rises to the surface. These findings have led to new models for the generation of abyssal basalts, which concentrate on the realization that melts pass upward through the melting mantle rapidly in comparison to the rising mantle itself. Thus, any one batch of melt reaching the surface will possess contributions, not from a single small source region, but from the entire ascending column. It has been proposed that individual batches of melt erupting to the surface will possess a compatible element composition in equilibrium with the residual mantle at the top of the ascending column of melting asthenosphere. The same batch of melt, however, will possess an incompatible element composition (comprising those elements which are largely removed from the solid at the earliest stages of melting), reflecting the initial composition of the mantle at the bottom of the ascending column of asthenosphere.

New insight into the process by which dispersed melt can collect and segregate from the mantle has also been gained from the discovery that large areas of the ocean ridges around some fracture zones are almost devoid of crustal rocks. There are regions where little or no basalt is dredged; almost nothing but highly depleted altered residual mantle peridotite is exposed. Because this peridotite is nearly completely drained of basaltic components, containing practically none of such essential components of basaltic lava as sodium or titanium, it is clear that the peridotites underwent high degrees of melting and that the melt has been removed. The only possibility is that the melt which formed the peridotite during its ascent migrated within the ascending asthenosphere beneath the fracture zone laterally toward the adjacent ridge segments. The concept that basaltic melts might migrate laterally over long distances, rather than float up directly to the surface, may seem strange initially; it is a common phenomenon, however,

in the study of fluid dynamics. A layer of partially molten mantle ascending beneath an ocean ridge approximates a layer of light, less viscous liquid floating up through a denser liquid in a tank. In this case, it is observed commonly that the layer will become unstable, draining into protrusions at regularly spaced instability points. Because the liquid in a partially molten layer of asthenosphere will migrate faster than the solid residue by porous flow, it will rapidly increase in volume as such instability points are reached. When it exceeds about 30% of the rock content, simple crystal settling alone will cause it to separate and rise to the surface through the overlying mantle. Thus ocean ridge volcanism may closely resemble island arc volcanism, with a series of regularly spaced magmatic centers (volcanoes) overlying the instability points in a linear zone of ascending partially molten mantle. The principal difference is that, unlike the situation in island arcs, ocean ridges are under continuous extension; thus major volcanic edifices such as the chains of ocean island volcanoes are never built.

For background information SEE ASTHENO-SPHERE; BASALT; FLUID DYNAMICS; LITHOSPHERE; OCEANIC ISLANDS; PERIDOTITE; VOLCANOLOGY in the McGraw-Hill Encyclopedia of Science and Technology.

Henry J. B. Dick

Bibliography. H. J. B. Dick, R. L. Fisher, and W. B. Bryan, Mineralogic variability of the uppermost mantle along mid-ocean ridges, *Earth Planet. Sci. Lett.*, 69:88–106, 1984; C. H. Langmuir and J. F. Bender, The geochemistry of oceanic basalts in the vicinity of transform faults: Observations and implications, *Earth Planet. Sci. Lett.*, 69:107–127, 1984; A. P. leRoex et al., Petrology and geochemistry of basalts from the American-Antarctic Ridge, Southern Ocean: Implications for the westward influence of the Bouvet mantle plume, *Contrib. Mineral. Petrol.*, 90:367–380, 1985; J.-G. Schilling et al., Hotspot-migrating ridge interaction in the South Atlantic, *Nature*, 313:187–191, 1984; J. A. Whitehead, H. J. B. Dick, and H. Schouten, A mechanism for magmatic accretion under spreading centres, *Nature*, 312:146–148, 1984.

Weather forecasting and prediction

The complexity and variability of atmospheric motion and its manifestation in all kinds of weather phenomena present a continuous challenge. The weather affects human affairs and influences most aspects of society, both directly and indirectly. Prediction of the weather requires a fundamental understanding of the forces which govern the atmosphere, and of the many complex feedback processes involving the oceans and land surfaces. Observations are required for initiation of any such prediction. Great progress has been made recently in this field, and the forecasting domain has been extended to encompass the whole globe.

Complexity. Weather prediction is a well-defined deterministic problem. Starting from a given initial state, any future state can be obtained by integrating the classical Navier-Stokes equation forward in time. Therefore a weather forecast can, in principle, be calculated in the same way as the motion of the planets or the trajectory of a missile.

However, a more thorough analysis of forecasting problems reveals an enormous complexity. Atmospheric processes span a large range of scales. The largest scale encompasses the whole atmosphere, while the smallest components, such as cloud droplets, have sizes of a few micrometers. The sizes of the cloud droplets and their distribution are of importance for the calculation of radiative processes, and are crucial components in the mechanism for release of precipitation. This range of atmospheric motion covers a range of 13 orders of magnitude (10^{-6} to 10^{7} m). There are thunderstorms with a horizontal extension of a few miles, traveling frontal depressions in middle latitudes stretching over 600–1200 mi (1000–2000 km), and, finally, the largest circulation systems of global scale forced by the large mountain ranges and by the distribution of oceans and continents. Every weather system has its characteristic time scale, which generally increases with the spatial scale. This time scale is about an hour for a thunderstorm and up to a few months for the largest circulation system such as the monsoon.

There is continuous interaction between all these scales of motion. In the tropical belt the trade winds provide heat and moisture to feed the convective system in the equatorial zone. The latent heat released by all the many convective systems in this zone in turn maintains the trade winds. In a similar way the microscale processes in the form of heat and moisture fluxes over land and sea are the driving forces for the large-scale circulation systems of the Earth.

This nonlinear interaction between different scales of motion is the fundamental reason for the difficulty in predicting atmospheric flow. Inevitable errors in observing the smallest scales of motion sooner or later contaminate the larger scales and finally destroy the accuracy of any prediction. Weather prediction can therefore be seen as an unstable problem in that small initial differences can have large final effects. Although the prediction problem as such is deterministic, it is for practical reasons nondeterministic, since the initial stage can never be perfectly known. Theoretically, weather prediction has much in common with more general nondeterministic problems involving economic and social systems. Atmospheric prediction models may also serve as useful prototypes for better understanding of a more general class of problems where the dynamical laws are not yet well understood.

Atmospheric models. Because of the complexity of atmospheric processes, progress in predicting the weather has gone hand in hand with development of computers. The ratio in speed between the supercomputers of today and the first computers used for numerical prediction around 1950 is more than a million, or the same ratio as between the speed of light and the speed of sound. Meteorology is one of the major users of supercomputers, and is generally considered to be second only to the defense industry in its use of the fastest computers available. Computer models of the atmosphere are based on the hydrodynamical equations as used in fluid mechanics, with certain extra approximations associated with the shallow depth of the atmosphere; 75% of the mass of the atmosphere and most of what is known as weather is contained in the lowest 10 km (6 mi). The models also include equations for the gas law and the transfer of heat and moisture.

Limitations in speed and memory of the first computers required that the early models be very simplified. They (1) included only approximations to the equations of motion (the quasi-geostrophic approximation); (2) were essentially adiabatic, that is, there were no exchanges of heat with land or ocean; (3) had a coarse vertical and horizontal resolution, for example, with only one or three levels in the vertical and with an interval of 5° × 5° in the horizontal; and (4) were integrated (solved) only over a limited domain, for example, part of the Atlantic and western Europe.

Gradually, as more powerful computers were developed, these constraints have been relaxed. The model developed at the European Centre for Medium Range Weather Forecasts (ECMWF) is

Fig. 1. Vertical levels of the ECMWF 19-level global model. 1 m = 3.3 ft; 1 hectopascal = 9.87×10^{-4} atm.

considered to be the most advanced model available and represents the state of the art in numerical weather prediction. The ECMWF model is global; values of basic variables such as wind, temperature, and moisture are specified at 19 levels in the vertical with spacing in the lower atmosphere of about 500 m (1600 ft) [**Fig. 1**]. The horizontal structure is specified by a series of wavelike functions (Fourier components and Legendre polynomials) with 106 waves in each direction. At any given time, about 5 million numbers represent the state of the atmosphere. The basic equations are then used to obtain estimates for the rate of change of each of the variables; the values of those variables can be predicted for a time typically 15 min later. By repeating this process step by step, the state of the atmosphere at some future time can be forecast. Each day, ECMWF produces forecasts up to 10 days ahead, requiring about 1000 successive steps. In practice, the calculations are quite complicated, because many physical processes have to be considered. The basic driving force of the atmosphere, solar radiation and the loss of heat from the atmosphere to space, is strongly influenced by factors such as cloud cover, the surface of the Earth, and the water vapor of the atmosphere. The way the atmosphere is heated (or cooled) must be known in great detail, since the driving force of the atmosphere is not the heating itself but the difference of the heating. Similarly, the frictional process at the surface which provides the ultimate drag on the motion of the atmosphere must be known in great detail. Consequently, the surface orography and roughness must be described accurately. The way different physical processes are included in the model must be carefully judged, since they easily put excessive demands even on the most powerful supercomputer. In the operational ECMWF model, they increase the number of computations per time step to about 6×10^9, so that a 10-day forecast requires about 7×10^{12} operations and takes a little more than 2½ h on a supercomputer.

Observations. To initiate a forecast model, the state of the atmosphere must be known with sufficient accuracy. Wind, temperature, and moisture must be observed for the whole globe and through the whole depth of the atmosphere, but has not as yet been achieved because of tremendous economic and technical problems. One basic component of the global observing system is a set of upper-air soundings established at the end of World War II. They are made twice daily from about 700 stations in the Northern Hemisphere and 100 stations in the Southern Hemisphere. For economic and practical reasons, these stations tend to concentrate in midlatitude land areas where, in the Northern Hemisphere, they are spaced about 500 km (300 mi) apart. Soundings are made by releasing balloon-borne sensors at midnight and at noon Greenwich

time. The sondes rise through the atmosphere, telemetering values of temperature, pressure, and humidity back to the ground. Wind velocities are obtained by tracking the sondes. Although the accuracy of these measurements is high, the sparsity over oceans and remote land areas seriously limits the accuracy with which the state of the atmosphere can be determined.

To obtain more uniform spatial coverage, satellite-based sensors have been developed to provide temperature soundings. These sensors measure infrared radiances at a set of sharply defined wavelengths and are sensitive to the temperature in different layers of the atmosphere. Their vertical resolution and accuracy are still less than available from balloon-borne sensors, but this is compensated by a superior horizontal resolution of about 50 km (30 mi) and an almost global coverage.

Wind information is also obtained from satellites through the tracking of cloud elements. This is accomplished from satellites in geostationary orbits, and constitutes the most important data source for tropical and subtropical latitudes. There remain some uncertainties about the reliability of clouds as tracers of air motion and the accuracy with which satellites can determine the vertical location of clouds.

Data assimilation. The present global observing system is still incomplete, and substantial efforts have been made to determine the state of the atmosphere in areas where there are considerable "holes" in the observing network. Furthermore, the observations also represent small scales of motion which cannot be described by the model. Such subgrid-scale noise must be filtered before the observations are inserted in the model. ECMWF, as well as other large forecasting centers, has developed a statistical-dynamical method to analyze global observations. In this technique the observations are combined with the prediction model into a four-dimensional data assimilation system. Observed data are thereby checked against predicted data to determine their acceptability. The predicted values and the observations are then combined to give the most probable state of the atmosphere at the required time. In terms of the computations required, the process of combining the observed and predicted data is almost as demanding and sophisticated as the forecasting process itself.

A particular problem is related to the very general character of the forecasting equations, which can represent a much wider range of flows than those of interest to meterologists. An observation, if not carefully inserted into the model, can generate spurious gravity waves instead of the meterological features which the model is set out to represent. This problem is handled by a special filtering procedure, whereby the normal modes associated with nonmeterological features are excluded. These new methods for data analysis and initialization have led to a significant improvement in the specification of the initial state of the forecast, and have played an important role in the improvement in global weather prediction which has been achieved over the last years.

Parameterization problems. Observational and computational restrictions make it necessary to confine the computer model to a description of phenomena larger than a certain scale. Although this scale has gradually become smaller following computer developments and improvement in the global observing system, the minimum scale is still significantly larger than that of many weather systems. These so-called subgrid-scale processes must be related to the macroscale (more than 100 km or 60 mi) currently resolved by the global models. This step is known as parameterization, since the subgrid-scale processes are described in terms of the parameters of the large-scale flow. Of particular importance is the description of transfers of heat and moisture from the ground, and the associated heating of the atmosphere through turbulent transfer processes and the release of latent heat by convective systems.

Of equal importance is description of the role of the Earth's orography in blocking and deflecting the flow and producing the intense dissipation of momentum in areas of steep and rugged terrain. An improper description of the thermal and orographic forcing in numerical models gradually changes the predicted flow pattern and leads to systematic errors. If the model forcing is too weak (the most common deficiency), a typical effect is to produce a westerly circulation which is stronger than that observed in middle latitudes. Over Europe the erroneous result is that the Atlantic cyclones penetrate too deeply over the continent instead of being deflected toward the northeast over the Norwegian Sea. The cause of this error has been traced to an improper description of major mountain ranges such as the Rockies and the Alps, which deflect and steer weather systems around the Earth. In the models, because of the coarse resolution the mountains are smoothed and reduced in height. At ECMWF this has been alleviated by assuming that mountain passes and valleys are mostly filled by stagnant air. The average height of the model mountains is thus increased, and the blocking effect is enhanced. This technique is known as envelope orography.

Of great importance is the incorporation of the drag caused by orographic obstacles. This drag affects the flow of the atmosphere in great depth through vertical propagation of gravity waves. Through the process of rapid vertical propagation, these waves remove momentum from the upper troposphere and lower stratosphere. The incorporation of envelope orography and gravity-wave drag in global models has reduced systematic model errors and improved the prediction of important phenomena such as blocking.

Global forecasts. Global centers, like ECMWF, produce daily forecasts a week or more in advance. Generally such forecasts are interpreted by trained meteorologists before they are given to the public. Quality is assessed both by objective, statistical scores and by systematic evaluation of their practical usefulness. Judged by any of these measures, the forecast from ECMWF for day 3 is more accurate than the forecast for day 1 which was available at the beginning of the 1970s, and the forecast for day 6 is more accurate than the forecasts for day 3 ten to fifteen years ago. **Figure 2** compares the correlation between the observed and predicted anomaly (deviation from the normal climate) among 12 January forecasts carried out at the Geophysical Fluid Dynamics Laboratory (GFOL) in Princeton, New Jersey, during 1968–1972 and the forecasts from ECMWF for the winters of 1979/1980 and 1985/1986. The scores refer to the circulation at a height of about 3 mi (5 km), and to the Northern Hemisphere. Using 60% as a measure of predictive skill, the forecasts at ECMWF for the first winter extended the skill from 3½ to 5½ days, and to almost 7 days for the last winter. The skill varies between summer and winter and is somewhat lower during the summer.

The significant improvement in predictive skill has led to a large increase in practical use of the forecasts. As confidence in them has gradually increased, new applications and new users are being found constantly. The selling of forecasts or forecast services to special customers is becoming a substantial source of revenue for several meteorological services. However, the increased usefulness of the forecasts is not due only to improved skill in predicting the large-scale atmospheric flow. Of equal importance is the prediction of parameters such as wind and temperature in the boundary layer as well as cloudiness and precipitation. As an example, **Fig. 3** shows an ECMWF forecast of meteorological parameters that can be used as guidance in the

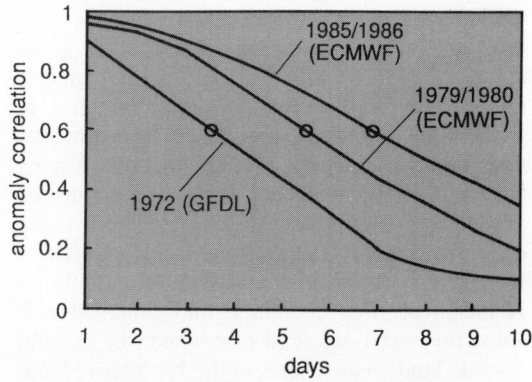

Fig. 2. Anomaly correlation for Northern Hemisphere winters, showing the improvement in forecast quality over the period 1972–1986. The circle on each curve represents the 60% value used to measure predictive skill.

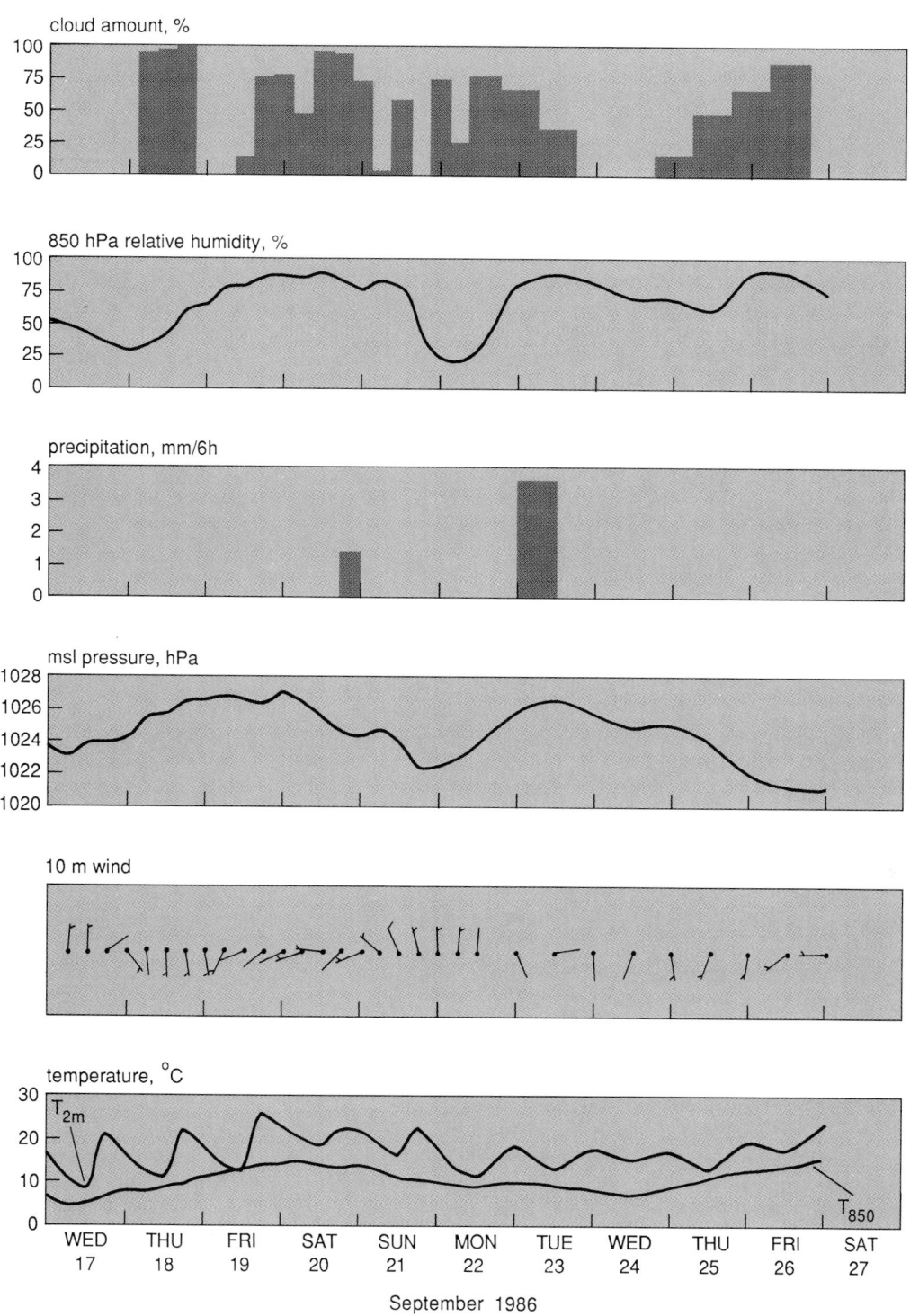

Fig. 3. Direct model output of a 10-day weather forecast of meteorological parameters for Washington, D.C., beginning September 17, 1986. Relative humidity given at height where air pressure is 850 hPa (850 mbar). MSL = mean sea level. 10 m wind = wind at a height of 10 m (33 ft) above the model surface with speed indicated in knots (1 knot = 0.5 m/s). T_{2m} = 2 m (6.4 ft) above the model surface. T_{850} = height at which air pressure is 850 hPa (850 mbar). °F = (°C × 1.8) + 32.

provision of forecast information of direct interest to the end users. Such forecasts are available daily for every part of the globe via the ECMWF data base. However, it must be kept in mind that this type of weather prediction should be considered only as the weather of the macroscale.

Interpretation of these forecasts into local weather, influenced by small-scale features not resolved by the model, must be carried out.

Recent studies indicate that the limit of useful predictive skill can be extended to about 2 weeks if model errors alone can be eliminated. Further extension is possible if more observations can be obtained with higher accuracy. A reduction of the initial error by half is expected to add another 2 days of useful predictive skill. Atmospheric predictability is not a constant factor, and it varies from time to time and from area to area. There are certain episodes during which the atmospheric circulation is trapped into very stable regimes like atmospheric blocking. During such regimes, error growth is limited and is generally confined to smaller scales. Other regimes are highly unstable, and even the largest scales of motion are influenced by errors emanating from smaller scales. The problem of atmospheric predictability is an area of exciting research in which interesting work is taking place. While the prediction of the instantaneous weather is limited to 2 weeks or so, forecasts for mean conditions, such as weekly or monthly averages, will be made in the future for longer periods, particularly at lower latitudes. This problem is being attacked on a broad scale with close international cooperation.

For background information SEE ATMOSPHERIC GENERAL CIRCULATION; FLUID FLOW; NAVIER-STOKES EQUATIONS; SATELLITE METEOROLOGY; SUPERCOMPUTER; WEATHER FORECASTING AND PREDICTION in the McGraw-Hill Encyclopedia of Science and Technology.

L. Bengtsson

Wind shear

Wind shear is any change of vertical or horizontal wind divided by the distance over which the change is measured. Mathematically, wind shear is a tensor whose nine components correspond to the three orthogonal directions in which variations can occur combined with the three directions from which the wind can be blowing.

Wind shear is important to aviation because it has caused serious accidents. Aircraft are sustained by their motion through the air, and wind variations consequently affect flight.

Causes. Wind shear of great significance to aviation is associated with horizontal temperature contrasts in the temperature latitudes and with thunderstorms. Always present in Earth's atmosphere, wind shear constitutes a flight hazard only occasionally.

Nonthunderstorm shear. Outside of thunderstorms, wind shear is prominent in temperate latitudes as a result of horizontal contrasts of temperature. This nonthunderstorm shear, manifested as variation of horizontal wind with height and often culminating in jet streams of great altitude, is most intense in frontal zones, where temperature variations are concentrated. On December 17, 1973, sixteen persons were injured in an accident at Boston's Logan Airport. At the time, there was a warm front nearby. The wind at the ground was from the northwest, but wind only 500 ft (150 m) above the surface was from the south-southwest, and visibility was hampered by rain and fog. In this case, the unusually strong directional change of wind and the increase of head wind encountered by the descending jetliner during its approach to the airport from the southeast were causal factors in its landing short of the runway.

Aircraft accidents have resulted from encounters with strong wave motions and turbulence in clear air, often associated with fast airflow over mountain ranges or extraordinary vertical variations of the horizontal wind near jetstreams.

Thunderstorm shear. Most wind shear that is a serious hazard to aircraft occurs in and near thunderstorms. As shown in the **table**, there have been fatal accidents attributed to various phenomena that accompany thunderstorms, including hail, turbulence, lightning, the shear associated with outflowing winds, and excessive precipitation, which restricts visibility and may degrade engine performance when large amounts of water are ingested by the engines. Prior to the 1970s, most storm-related accidents began at considerable altitude with the aircraft encountering a storm which lay along the air route.

Effects on aircraft. In steady level flight, drag created by motion of an aircraft through the air is balanced by thrust created by engines. Likewise, flow of air over the wings creates lift, balanced by gravity. When a zone of wind shear is encountered, these balances are altered. For example, a sudden reduction of head wind decreases the speed of air over the wings, reducing the lift so that the aircraft pitches down and descends. The lower airspeed also lessens the drag; thrust produced by the engines then exceeds drag, and with an additional boost from gravity, aircraft speed increases again until balances between thrust and drag and between lift and gravity are restored.

Where wind changes are confined to small distances, up to a few times the size of the aircraft, such a sequence is hardly begun before the wind on the aircraft changes again. This is a situation of turbulence (so-called rough air), a series of bumps and jolts which in extreme cases has destroyed aircraft. Where substantial values of wind shear prevail over greater distances, aircraft altitude can be changed significantly; serious consequences can ensue during takeoffs or landings, because the pilot may not compensate with changes of engine thrust and other controls in time to avoid collision with the ground.

Although it has not always been possible to measure precisely the role of each different factor operative in combination, shear manifested as turbulence and vertical drafts has always been important in aircraft accidents involving thunderstorms along air routes. Much was learned from these costly tragedies. Aircraft structures were improved, systems to quench explosions in fuel

Fatal United States air-carrier accidents with thunderstorms and low-altitude wind shear as cause or major factor*

Date	Aircraft type	Location	Aircraft position	Weather condition or cause	Fatalities
1 September 1940	DC-3 (2-engine prop)	Virginia	En route	Lightning	25
29 May 1947	DC-4 (4-engine prop)	New York	Takeoff	Thunderstorm outflow	43
29 August 1948	Martin 202 (2-engine prop)	Minnesota	En route	Thunderstorm combined effects	37
25 June 1950	DC-4 (4-engine prop)	Lake Michigan	En route	Thunderstorm combined effects	58
28 April 1951	DC-3 (2-engine prop)	Indiana	En route	Thunderstorm outflow	11
15 February 1953	DC-6 (4-engine prop)	Gulf of Mexico	En route	Thunderstorm combined effects	46
12 May 1959	Viscount (4-engine jetprop)	Maryland	En route	Lightning and turbulence	31
27 June 1959	Constellation (4-engine prop)	Italy	En route	Lightning	68
19 January 1960	Viscount (4-engine jetprop)	Virginia	En route	Thunderstorm combined effects	48
12 February 1963	B-720B (4-engine jet)	Florida	En route	Thunderstorm drafts	43
2 July 1963	Martin 404 (4-engine prop)	New York	Takeoff	Thunderstorm shifting wind and rain	7
8 December 1963	B-707 (4-engine jet)	Maryland	En route	Lightning	81
7 August 1966	BAC-111 (2-engine jet)	Nebraska	En route	Thunderstorm turbulence	42
3 May 1968	Electra (4-engine jetprop)	Texas	En route	Thunderstorm turbulence	85
23 July 1973	F-27 (2-engine jetprop)	Missouri	Landing	Thunderstorm outflow, lightning, and rain	38
30 January 1974	B-707 (4-engine jet)	Samoa	Landing	Thunderstorm outflow and rain	96
24 June 1975	B-727 (3-engine jet)	New York	Landing	Thunderstorm outflow and rain	112
4 April 1977	DC-9 (2-engine jet)	Georgia	En route	Thunderstorm, heavy rain, and hail	70
12 June 1980†	SA-226 (2-engine jetprop)	Nebraska	En route	Thunderstorm turbulence and rain	13
9 July 1982	B-727 (3-engine jet)	Louisiana	Takeoff	Thunderstorm outflow and rain	153
2 August 1985	L-1011 (3-engine jet)	Texas	Landing	Thunderstorm outflow and rain	135

* After J. T. Lee and W. B. Beckwith, Thunderstorms and aviation, in E. Kessler (ed.), *The Thunderstorm in Human Affairs*, pp. 125–145, 1983; National Research Council (1983); Reports of the National Transportation Safety Board; and New York Times Index.
† Commuter airline, not classified as an air carrier.

tanks were developed, and use of radar on commercial aircraft was emphasized as a tool for detecting and avoiding storm centers often otherwise hidden from view.

During the 1960s, aircraft powered by jet engines came into wide commercial use. These aircraft fly faster than the aircraft powered by reciprocating engines with propellers, their engines do not respond as rapidly to pilot inputs, and the wings on jet aircraft are typically not as large in relation to aircraft weight as on the slower craft. These qualities reduce the time available for the pilot to respond to any dangerous situation, and they produce greater sensitivity of the aircraft to wind changes over longer distances. During the 1970s and 1980s, accidents caused by thunderstorms en route diminished, largely through emphasis on avoidance. But accidents caused by thunderstorm wind shear during takeoffs and landings became more prominent than before, as shown in the table.

Detailed analyses of the related weather situations have shown that aircraft traversed regions of thunderstorm outflow with dimensions of just a few miles. Often, the first untoward event was a head-wind increase, as shown in the **illustration**. Rapid increase of head wind causes a plane to pitch up and may deceive a pilot into reducing engine power. Then there may be a head-wind decrease; the consequent reduced lift, reduced power, and downdraft all contribute to bringing the plane to the ground in the wrong place. Also, it has been shown that loss of visibility in heavy rain and cloud or fog is a significant contributor to accident potential during landings and takeoffs.

Safety measures. A number of actions begun during the 1970s and 1980s are aimed at reducing wind shear accidents near air terminals. A system of anemometers on airport grounds, the Low Level Wind-Shear Alert System (LLWAS), presents a display in the control tower of winds at various points around the airport. When the wind difference between an outlying anemometer and the center-field anemometer attains a critical value, visual and aural alarms are activated, and the information is transmitted to inbound and outbound pilots. This system had been installed at 85 airports in the United States by mid-1986,

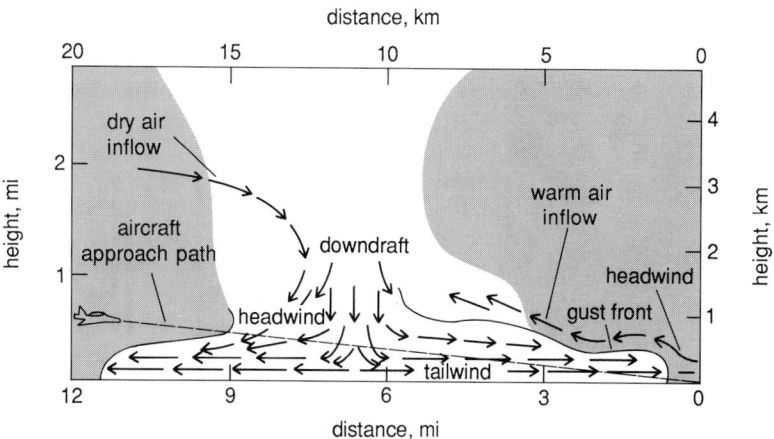

Path of winds associated with a gust front, thunderstorm downdraft, and outflow (downburst). Note change from headwind to tailwind along the path of an aircraft's approach to landing at the right edge. Possibly disastrous effects on aircraft flight depend on the intensity and scale of wind shift phenomena and details of pilot input and aircraft response. (*After J. T. Lee and W. B. Beckwith, Thunderstorms and aviation, in E. Kessler, ed., The Thunderstorm in Human Affairs, pp. 125–145, 1983*).

and 25 additional installations were scheduled. The system as deployed has substantial value, but its effectiveness is not guaranteed because it does not detect wind shifts until they occur within airport boundaries and at the anemometers. Experiments involving increased numbers of anemometers were under way in 1986.

More attention is being given to avoiding thunderstorm cells which lie on approach and departure courses; radars on the ground as well as on the aircraft are used, and meteorologists stationed at air-route traffic control centers send critical weather data to terminal controllers. Although vitally important and valuable, this system is not wholly reliable because information may not be passed along soon enough to be helpful to pilots when storms develop very rapidly, as was the case preceding the crash at Dallas on August 2, 1985. In addition, only the ground radars can assess the intensity of storms by measuring their precipitation reliably. The airborne radars are helpful only for immediate storm avoidance; in order to achieve desired resolution with the small antennas dictated by space limitations in aircraft, they must operate at short wavelengths that cannot penetrate far into heavy precipitation.

Detection systems are being designed for installation on aircraft to alert flight crews immediately upon a wind shear encounter and to indicate appropriate compensatory responses. During 1986 the Federal Aviation Administration certified the first such system for operational use.

Finally, special training courses help pilots learn how to respond more effectively to dangerous shear situations encountered inadvertently or in spite of best efforts at avoidance. The training courses utilize flight simulators, technologically advanced devices which embody computers and realistic displays to mimic flight under a wide variety of conditions. Inputs to a flight simulator include aircraft performance parameters, ambient wind conditions, instrument settings, and pilot inputs, all variable from moment to moment. By varying the numerous simulated conditions and using real pilots to test responses, a full range of strategies can be evaluated in safety. In developing meterological inputs, findings of recent thunderstorm research with Doppler radar have had a critical role. Although thunderstorm downdraft and outflow phenomena have been broadly known since the late nineteenth century and were considerably elaborated in the 1940s, much more detail has become available since 1970. Major field projects have been located at Chicago, Illinois; Denver, Colorado; Huntsville, Alabama; and Norman, Oklahoma. Another focus of these field programs is development of Doppler weather radar and methods for processing Doppler data for use in the aviation weather system of the 1990s and beyond.

When an aircraft accident occurs, the circumstances are carefully examined, and steps are taken to reduce the possibility of recurrence. Notwithstanding certain limitations in the system for navigating aircraft near busy terminals when severe storms are present, accident statistics show that safety has advanced significantly. Indeed, between 1960 and the mid-1980s the average annual number of accidents diminished, while passenger miles flown increased by a factor of 6. During the 1980s there has been less than one fatality per billion passenger miles flown.

For background information SEE FRONT; GEOSTROPHIC WIND; RADAR METEOROLOGY; THUNDERSTORM; WIND in the McGraw-Hill Encyclopedia of Science and Technology.

Edwin Kessler

Bibliography. R. J. Doviak and D. Zrnic, *Doppler Radar and Weather Observations*, 1984; J. T. Lee and W. B. Beckwith, Thunderstorms and aviation, in E. Kessler (ed.), *The Thunderstorm in Human Affairs*, pp. 125–145, 1983; National Research Council, *Low-Altitude Wind Shear and Its Hazard to Aviation*, 1983; National Transportation Safety Board, *Aircraft Accident Rep. NTSB/AAR-86/05*: Delta Airlines at Dallas/Fort Worth International Airport, August 2, 1985 (available from National Technical Information Service, Springfield, Virginia), 1986.

X-ray diffraction

An x-ray incident on an electron is scattered both by the electron's charge and by its magnetic moment. Charge scattering is the dominant mechanism and is the basis for structural investigations of condensed matter by x-ray diffraction. With the availability of intense x-ray beams from advanced synchrotron sources, however, magnetic scattering of x-rays has recently been shown to be a powerful new probe of the magnetic structure and phase transformations of condensed matter. Similar to the well-established technique of magnetic neutron scattering, magnetic x-ray scattering using synchrotron radiation has many complementary features and already is opening new areas of materials investigation. Some recent examples include high-resolution studies of the long-period magnetic modulation in rare-earth metals, studies of interfacial and two-dimensional magnetism in multilayer thin films, and spin-dependent resonance studies of the magnetic cross section in ferromagnetic nickel. Particular features of the x-ray scattering cross section, such as its polarization dependence and unique resonance and interference properties, are interesting in themselves and, in addition, they promise the means for developing new synchrotron techniques. One of the most exciting and challenging prospects for further study is two-dimensional magnetism, such as occurs at the clean surface of a magnetic material, in an adsorbate layer, or at an interface.

Development of magnetic x-ray scattering. Since about 1950, magnetic neutron diffraction has been an extraordinarily fruitful tool for studying the magnetic properties of condensed matter. Because the neutron magnetic moment interacts with the orbital and spin moments of a magnetic atom, a neutron beam incident on a magnetic material is scattered by these moments. The resulting diffraction pattern gives detailed information about the size and orientation of the atomic moments in the solid and characterizes their long-range order.

After the development of the Klein-Nishina quantum theory of photon scattering in 1929, it was widely appreciated that x-ray photons also interact with and are scattered by the magnetic moment of an electron. Because of the weakness of this interaction (and, later, the enormous success of magnetic neutron diffraction), magnetic x-ray scattering was not at first considered a viable technique for the investigation of magnetism in solids. In 1970 a theoretical investigation of the cross section for x-rays scattered from electrons bound to atoms in a solid led to the suggestion that magnetic diffraction could be observed by using modern x-ray tubes. The first experiments were performed on the antiferromagnet nickel oxide (NiO) in 1972. Later, a series of experiments on both ferromagnets and antiferromagnets was carried out, and numerous properties of the cross section were quantitatively demonstrated by using both conventional and synchrotron x-ray sources.

In the 1980s, magnetic x-ray scattering studies have evolved beyond investigation of the cross section itself and to the broader investigation of magnetism in solids. These developments have been aided, in part, by further elucidation of the theory. The crucial element for experimental progress has been the application of high-brilliance synchrotron sources (developed since about 1970). In addition to the high x-ray flux (10^{12}–10^{13} photons/s can be focused on a few square millimeters of sample area), synchrotron radiation is naturally suited to magnetic x-ray scattering studies because of its high momentum-transfer resolution (corresponding to a resolution in reciprocal space of 10^{-5} nm^{-1}), high sensitivity to modulations of the electronic charge density accompanying magnetic ordering, high degree of linear polarization in the median plane of the storage ring, circular polarization out of the median plane, and wavelength tunability.

X-ray scattering cross section. The classical approximation for the interaction of an x-ray with an electron is illustrated in **Fig. 1**. In charge scattering (or Thomson scattering; Fig. 1a) the electric field of the incident x-ray exerts a force on the electronic charge; the electron's subsequent acceleration leads to electric dipole rera-

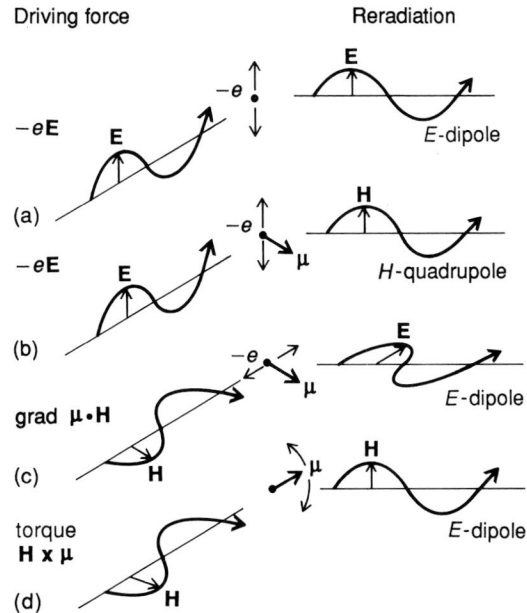

Fig. 1. Four mechanisms of x-ray scattering from an unbound electron in the classical approximation. (a) Charge or Thomson scattering, in which electric field E of incident x-ray exerts a force on electronic charge $-e$, and the electron's resulting acceleration leads to electric dipole reradiation. (b) Magnetic scattering in which acceleration of electron's magnetic dipole moment μ accompanying motion in part a leads to magnetic quadrupole reradiation. (c) Magnetic scattering in which magnetic field H of incident x-ray exerts a force on μ, and the electron's resulting acceleration leads to electric dipole reradiation. (d) Magnetic scattering in which incident magnetic field exerts a torque on μ, and the electron's resulting precession leads to magnetic quadrupole reradiation. (After F. de Bergevin and M. Brunel, Diffraction of x-rays from magnetic materials, Acta Cryst., A37:314–331, 1981)

diation of the scattered field. Although considerably weaker, the electric and magnetic fields of the incident x-ray also interact with the electronic magnetic moment, which gives rise to magnetic scattering (Fig. 1b–d). For a single unbound electron the ratio of magnetic dipole reradiation to electric dipole reradiation is $(\hbar\omega/mc^2)^2$, or approximately 1×10^{-4} for a typical x-ray of wavelength 0.1 nanometer. Here $\hbar\omega$ is the photon energy (where ω is the angular frequency of radiation and $\hbar$ is Planck's constant divided by 2π), m is the rest mass of an electron, and c is the speed of light. This factor is the primary reason that the magnetic cross section is so weak.

The x-ray scattering cross section for electrons bound to atoms in a solid, in the limit of high photon energy, is given to order $(\hbar\omega/mc^2)$ by the equation below. In this expression the initial

$$\frac{d^2\sigma}{dE'\,d\Omega'}\bigg|_{\vec{k},\,\vec{\varepsilon}\to\vec{k}',\,\vec{\varepsilon}'} = \left(\frac{e^2}{mc^2}\right)^2 \bigg| \rho(\vec{Q})A$$

$$-\frac{i\hbar\omega}{mc^2}\,[\tfrac{1}{2}\vec{L}(\vec{Q})\cdot\vec{B} + \vec{S}(\vec{Q})\cdot\vec{C}]\bigg|^2$$

photon momentum and polarization are, respectively, $\vec{k}$ and $\vec{\varepsilon}$; the final momentum and polarization are $\vec{k}'$ and $\vec{\varepsilon}'$. The left-hand side of the equation is the differential cross section per unit interval of energy E' and per unit solid angle Ω' for the scattered photons, evaluated at $\vec{k}$, $\vec{\varepsilon}$, $\vec{k}'$, and $\vec{\varepsilon}''$. The quantity $\hbar\vec{Q}$, which is equal to $\hbar\vec{k} - \hbar\vec{k}'$, is the photon momentum transfer, and e is the magnitude of the electronic charge. The modulus of the first term on the right-hand side of the equation is the usual expression for charge scattering (giving rise to Bragg reflections) and depends on the Fourier transform of the charge density ρ. The modulus of the second term, which is reduced from the first by the factor $(\hbar\omega/mc^2)^2$, describes the moment-dependent scattering and depends on the Fourier transform of the orbital and spin densities ($\vec{L}$ and $\vec{S}$, respectively). This term is the pure magnetic scattering—the term in the cross section analogous to the magnetic neutron cross section. As an example, for 0.1-nm x-rays incident on the rare-earth metal holmium the ratio of the pure magnetic scattering to the charge scattering is of the order of 1×10^{-6}. (Solid-state effects reduce this ratio even further from that for a free electron.) For this reason, pure magnetic scattering is most easily observed in antiferromagnets, where the charge and magnetic peaks occur at different locations in reciprocal space. The matrices A, $\vec{B}$, and $\vec{C}$ in the equation are distinct and summarize the polarization dependence of the scattering.

Because of the differing polarization dependences, the charge scattering, orbital-angular-momentum-dependent scattering, and spin-dependent scattering may be experimentally distinguished. In a neutron scattering experiment, it is not possible to separately measure the orbital and spin contributions to the cross section directly.

A unique property of the cross section is the existence of an interference term, involving the products of the charge and magnetic scattering in the above equation. The interference term is reduced from the charge scattering by only a single factor $(\hbar\omega/mc^2)$, which is of the order of 0.02 for a 0.1-nm x-ray and, in the simplest cases, is directly proportional to the imaginary part of the charge form factor f''. By tuning the incident x-ray energy near an absorption edge, f'' may be made very large. This use of the interference and resonance properties of the cross section has been most productive, so far, in the study of ferromagnets as discussed below. In the more general case the interference term may, in addition, have spin-dependent resonant properties, as have recently been observed in nickel. The use of circular polarization will make possible direct studies of the interference scattering.

Magnetic x-ray scattering studies. Examples of the use of magnetic x-ray scattering techniques for determining magnetic structures are provided by recent studies of the rare-earth metal holmium and of gadolinium-yttrium (Gd-Y) superlattices.

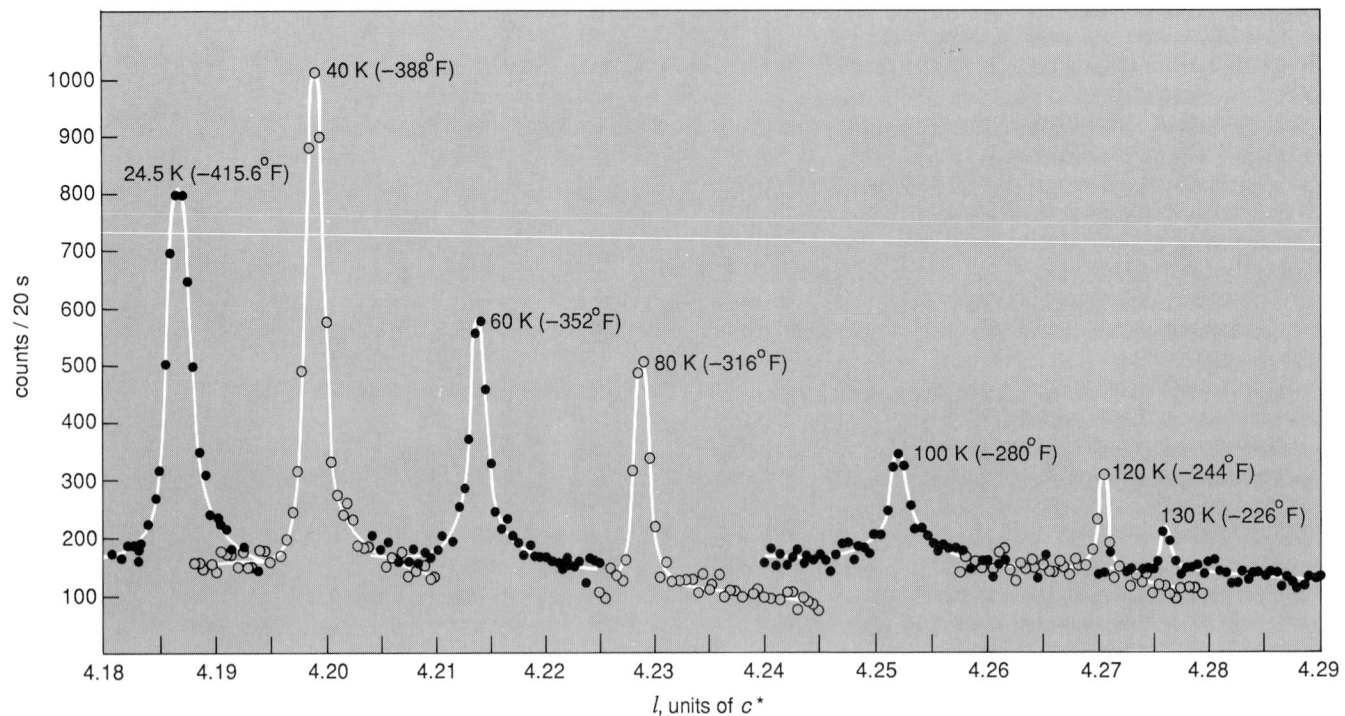

Fig. 2. Temperature dependence of the magnetic satellites surrounding the holmium (004) Bragg reflection, obtained by x-ray scattering. These scans were taken at the Stanford Synchrotron Radiation Laboratory by using an eight-pole wiggler beam line. (*After D. Gibbs et al., Magnetic x-ray scattering studies of holmium using synchrotron radiation, Phys. Rev. Lett., 55:234–237, 1985*)

Holmium. Holmium has a hexagonal close-packed crystal structure and a large spontaneous magnetic moment (approximately 10 μ_B, where μ_B is the Bohr magneton) per atom. Below its magnetic ordering temperature of 131 K ($-224°$F), holmium is a spiral antiferromagnet in which the moments are ferromagnetically aligned within the basal planes (the plane perpendicular to the c axis) but rotate from plane to plane with a wavelength varying between about 7 layers near 130 K ($-226°$F) and about 12 layers near 20 K ($-424°$F). Within the hexagonal planes, there are six equivalent easy directions along which the moments tend to align. The diffraction pattern for this structure consists of magnetic satellites split symmetrically about each of the main Bragg reflections. The detailed magnetic structure and period at a given temperature are determined by the competition among exchange, lattice, and magnetoelastic forces.

Figure 2 shows several scans obtained by x-ray scattering from one of the satellites of the (004) reflection at a sequence of temperatures. At each temperature the number of scattered photons is measured as a function of the momentum transfer. The magnetic wave vector τ_m is the distance in reciprocal space from the (004) Bragg peak to the magnetic satellite $[(00\ell) = (004 + \tau_m)]$. The quantity τ_m is inversely proportional to the number of atomic layers in the magnetic period. The quantity ℓ is the momentum transfer in units of $c^* = 2\pi/c$, where c is the basal-plane layer spacing. As the temperature is raised from 24.5 K ($-415.6°$F), the satellite position shifts to higher momentum transfer ℓ and the intensity decreases, falling continuously to zero near 130 K ($-226°$F). Typical signal rates were approximately 25 counts/s on a background of approximately 10 counts/s. The best momentum-transfer resolution was obtained for satellites of the (002) reflection with a full width at half maximum (FWHM) in reciprocal space of approximately 0.01 nm^{-1}. For comparison, magnetic neutron scattering experiments performed on the same sample gave signal rates of approximately 50 counts/s on backgrounds of 0.1 counts/s but with a resolution FWHM of about 0.05 nm^{-1}.

By virtue of the high momentum-transfer resolution available from synchrotron radiation, new lock-in behavior was observed in the temperature dependence of the magnetic period. Specifically, for certain ranges of the temperature the wave vector τ_m describing the magnetic modulation was observed to be commensurate with the lattice with the rational values 5/27, 2/11, and 1/6 (in units of c^*). Associated with the lock-in behavior at $\tau_m = 5/27$ and $\tau_m = 2/11$ were additional diffraction peaks of intensity comparable with the magnetic peaks observed near the values 2/9 and 2/11, respectively (**Fig. 3**). Because the polarization dependence for magnetic scattering is different from that for charge scat-

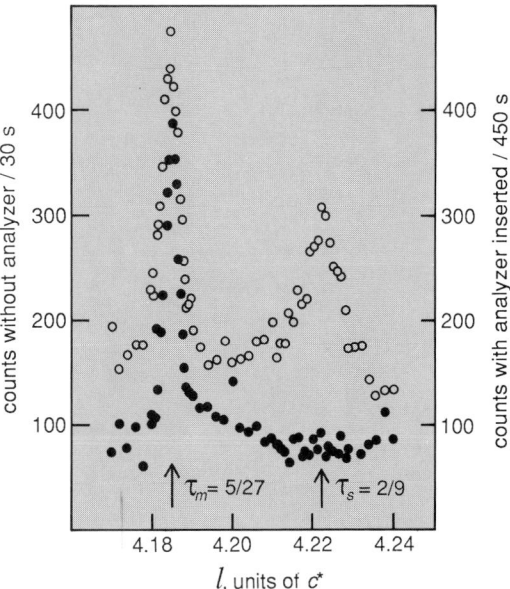

Fig. 3. Scans of holmium magnetic satellite (004 + τ_m) at τ_m = 5/27, and additional charge satellite (004 + τ_s) at τ_s = 2/9, showing effect of using polarization analyzer inserted to suppress charge scattering and accept magnetic scattering. Open circles (scale at left) show scattering without analyzer; solid circles (scale at right) show scattering with analyzer inserted. These data show that the peak at 2/9 is charge scattering, arising from lattice modulations, and the peak at 5/27 is pure magnetic scattering. (*After D. Gibbs et al., Magnetic x-ray studies of holmium using synchrotron radiation, Phys. Rev. Lett., 55:234–237, 1985*)

tering, it was possible to show that the additional peaks arise from charge scattering (Fig. 3). This additional scattering suggests that there is a regular distortion of the lattice by the spin-density wave.

A unified view of these effects has emerged from a new model of the magnetic structure of rare-earth metals based on the concept of spin discommensurations or spin slips. Briefly, spins are arranged in pairs associated with the six easy directions to form a uniform spiral of doublets. A single spin slip in the magnetic spiral is created by associating only one spin, instead of two, with one of the easy directions (**Fig. 4**). In holmium, the lattice distorts at slip positions as a result of magnetoelastic coupling, which gives rise to the additional charge satellites observed in the diffraction pattern. Considered in this way, spin slips are analogous to domain walls or solitons.

Rare-earth multilayers. Recent experiments on gadolinium-yttrium superlattices illustrate the potential of magnetic x-ray scattering techniques for studies of two-dimensional magnetism. Gadolinium and yttrium have closely similar chemical and structural properties, but gadolinium, as it naturally occurs in bulk, is ferromagnetic while yttrium is nonmagnetic. By the use of modern molecular-beam-epitaxy methods, it is possible to grow single crystals of high perfection consisting of regularly spaced, alternating layers of

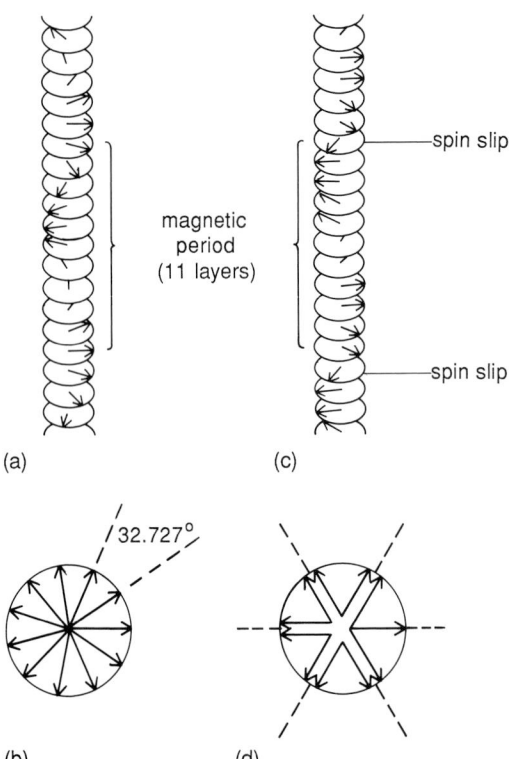

Fig. 4. Idealized τ_m = 2/11 magnetic structures. (a) Front view of the 2/11 simple spiral structure. The turn angle of the moments from plane to plane is 32.727°. (b) Projection of one period of the 2/11 simple spiral structure into the basal plane. (c) Front view of the 2/11 spin slip structure. In this structure there are five doublets for every spin slip. The presence of spin slips gives rise to lattice modulations, which lead to additional peaks in the diffraction pattern. (d) Projection of one period of the 2/11 spin slip structure into the basal plane. The broken lines indicate the six easy directions.

measured upon flipping of the spins in an applied magnetic field. By tuning the incident energy near the L_{III} absorption edge, this flipping ratio could be optimized. In all, flipping ratios from 20 different reflections were measured at 150 K ($-190°F$) and compared with three different models of the spin structure. It was found that the model of the magnetic structure which gave the best fit to the data had a full gadolinium moment at the center of the gadolinium layer with a smooth reduction in the moment as the interface is approached. This dependence follows the modulation of the interlayer strain and is consistent with magnetization measurements made on the same sample.

For background information SEE ARTIFICIALLY LAYERED STRUCTURES; ELECTRON SPIN; HELIMAGNETISM; MAGNETISM; NEUTRON DIFFRACTION; SYNCHROTRON RADIATION; X-RAY DIFFRACTION in the McGraw-Hill Encyclopedia of Science and Technology.

Doon Gibbs

Bibliography. M. Blume, Magnetic scattering of x-rays, *J. Appl. Phys.*, 57:3615–3618, 1985; D. Gibbs et al., Magnetic x-ray scattering studies of holmium using synchrotron radiation, *Phys. Rev. Lett.*, 55:234–237, 1985; K. Namikawa et al., X-ray resonance magnetic scattering, *J. Phys. Soc. Japan*, 54:4099–4102, 1985; P. M. Platzmann and N. Tzoar, Magnetic scattering of x-rays from electrons in molecules and solids, *Phys. Rev.*, B2:3556–3559, 1970; C. Vettier et al., Magnetic x-ray scattering study of interfacial magnetism in Gd-Y superlattice, *Phys. Rev. Lett.*, 56:757–760, 1986.

gadolinium and yttrium. The magnetic properties of such superlattices, since they depend, for example, on chemical composition, on layer thickness and spacing, and on possible interfacial effects, are of fundamental physical interest and, to the extent that they are controllable, may be of great technological importance. As the magnetic layer spacing approaches one layer of atoms, the magnetic properties may be expected to become progressively two-dimensional. In the experiments described here the superlattice consisted of 21 atomic layers of gadolinium and 21 of yttrium repeated 40 times. The questions of primary interest were, first, the magnetic structure of the gadolinium layers, both deep within each layer and at the interface, and second, the role of the interlayer strain modulation.

Because bulk gadolinium is ferromagnetic below room temperature, the strong structural and weak magnetic peaks are coincident. For a ferromagnet the largest moment-dependent term is the interference term involving the product of charge and magnetic scattering. To isolate the interference term, the change in Bragg intensity was

Xiphosurida

Biological rhythms play important roles in the activities of most living creatures. Circadian rhythms are those that complete a cycle every 24 h in natural lighting but persist in continuous darkness and therefore do not simply reflect the direct effects of daylight and darkness. Circadian rhythms in structure and function are a prominent feature of the eye of the horseshoe crab, *Limulus polyphemus*. Recent experimental studies have led to an understanding of how these rhythms are produced and controlled in *Limulus* and how they may benefit the animal's visual performance. These studies show that two-way communication between the eye and brain is essential in how the horseshoe crab's eye works, and may be important in the sensory systems of other animals as well.

Vision and behavior. Adult horseshoe crabs have three visual organs, the lateral, median, and ventral eyes. Only the paired lateral compound eyes (**Fig. 1**a) are organized for the pattern vision necessary to see objects. Until recently the role of pattern vision in *Limulus* behavior was uncertain. Studies of visually guided behavior in natu-

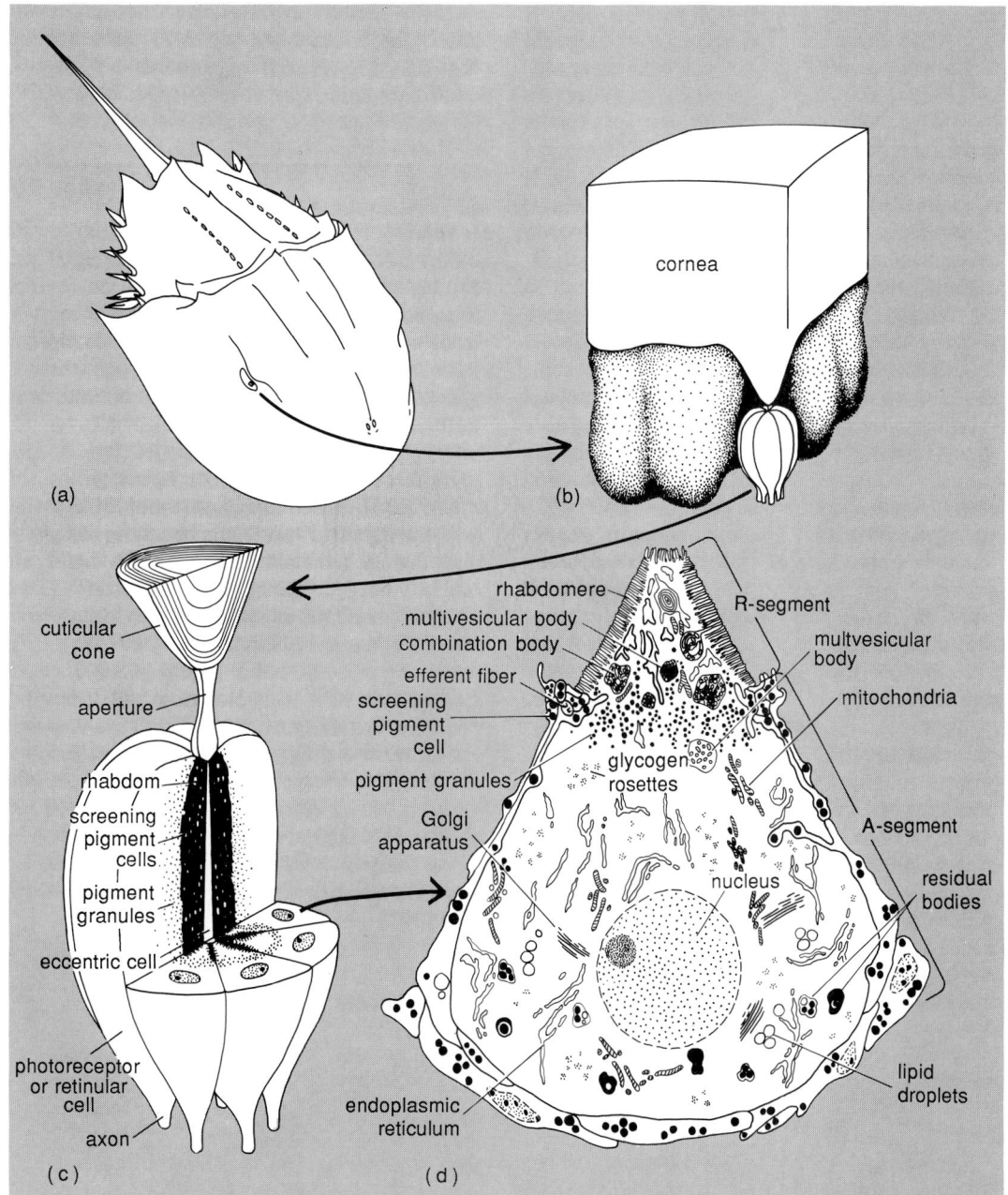

Fig. 1. Anatomy of lateral eye of the horseshoe crab, *Limulus polyphemus*, during the day. (*a*) Position of eye in the adult animal. (*b*) Small block of cornea and retina, showing structure of an ommatidium if the ensheathing pigment cells are stripped away. (*c*) Cutaway view of a single ommatidium. (*d*) Cross section of one photoreceptor labeled to show the rhabdomeral segment (R-segment) and the arhabdomeral segment (A-segment). *(After S. C. Chamberlain and R. B. Barlow, Jr., Transient membrane shedding in Limulus photoreceptors: Control mechanisms under natural lighting, J. Neurosci., 4:2792–2810, 1984)*

rally mating populations of horseshoe crabs along the beaches of Cape Cod in eastern Massachusetts have now established that male horseshoe crabs use vision to locate female horseshoe crabs during the mating season. Since most mating occurs at night in somewhat turbid water, the animal requires eyes that are very sensitive to light but need be of only moderate resolution. In fact, the sensitivity of the eye is increased at night by circadian rhythms in retinal structure and function that improve ability to detect light at the expense of visual resolution.

Eye structure. An adult compound eye consists of an array of about 800 separate lenses, each with its own cluster of photoreceptors called an ommatidium (Fig. 1*b*). About 12 photoreceptors are radially arranged like the segments of a lemon around the central dendrite of an offset, or eccentric, cell (Fig. 1*c*). The external tip of the grouped photoreceptors is separated from the internal tip of the lens or cuticular cone by a transparent aperture surrounded by opaque pigment cells.

The outside of each ommatidium is ensheathed with opaque pigment cells as well.

The central portion of the ommatidium (Fig. 1c) bears an array of specialized light-sensitive membrane called the rhabdom. The rays of the rhabdom surround the central process of the eccentric cell, and screening pigment cells form columns at the end of each ray. During the day, pigment granules in the photoreceptors form a circular band around the peripheral ends of the rhabdom.

Each photoreceptor (Fig. 1d) has a rhabdomeral segment (R-segment), which is covered by the light-sensitive rhabdomere, and an arhabdomeral segment (A-segment), whose surface is covered by pigment cells. The cytoplasm of the R-segment contains endoplasmic reticulum, mitochondria, and structures such as multivesicular bodies and combination bodies that are involved in the degradation of shed light-sensitive membrane, as discussed below. The A-segment cytoplasm contains the nucleus and other organelles common to all neurons, as well as multivesicular bodies. During the day, pigment granules are found at the junction of the R- and A-segments.

Two-way eye-brain connections. The internal tip of the ommatidium gives rise to a bundle of axons, one from each photoreceptor and one from the eccentric cell. These afferent axons pass through a plexus of complex interconnections behind the retina and join the optic nerve to terminate in the brain. The optic nerve also contains a group of efferent axons which originate in the brain and terminate in the retina in efferent fibers that are found among the clusters of pigment cells at the junctions between the R- and A-segments of photoreceptors (Fig. 1d). Innervating all types of cells in each ommatidium, these axons establish and maintain the circadian rhythms in the eye.

Visual processing in eye. Light enters the ommatidium through the cuticular cone, which internally refracts the light rays and concentrates them toward the center of the aperture. The shape of the aperture largely controls the amount of light which enters the photoreceptors and the directions from which it is gathered. Light absorbed by the visual pigment in the rhabdom triggers electrical events in the photoreceptors which, in turn, are electrically coupled to the eccentric cell. The summed analog signals from all of the photoreceptors in an ommatidium are digitized into nerve impulses at the spike initiation zone of the eccentric cell. Collaterals of eccentric-cell axons interact in the lateral plexus to form spatial Mach bands that accentuate edges in the visual image. Thus, the ensemble firing pattern of the 800 eccentric-cell axons represents an encoded version of the animal's visual environment that has already been processed to enhance the edges of objects. The visual processing of the retina is modified by the activity of the efferent fibers from the brain.

Circadian efferent inputs to eye. Recording the neural firing patterns of the efferent fibers which innervate the eye via the optic nerve reveals that efferent activity has a circadian rhythm. The efferent fibers become active in early evening, fire vigorously during the night, cease activity at dawn, and remain silent until the next evening. In natural lighting, the rhythm of efferent activity remains synchronized with the cycle of daylight and darkness. In continuous darkness, the rhythm continues, but slowly drifts out of phase with the original lighting cycle because the period of the unentrained rhythm is slightly shorter than 24 h in some individual horseshoe crabs and slightly longer than 24 h in others.

The circadian efferent activity causes the release of neurotransmitters in the retina, and octopamine is almost certainly one of the substances released. Octopamine is present in the optic nerve and retina. Newly synthesized octopamine is localized to efferent fibers and is released by incubation in high-potassium media. Injection of octopamine into the retina mimics efferent activity. Known agonists and antagonists of octopamine affect efferent neurotransmission. A second efferent neurotransmitter may be a neuropeptide. Work to isolate and characterize additional efferent neurotransmitters is under way.

Retinal effects of efferent activity. The onset of efferent activity each evening immediately initiates a coordinated set of physiological and anatomical changes in the retina and primes metabolic activities which are expressed as additional anatomical changes the following morning. The immediate changes include a reduction in photoreceptor noise, an increase in photoreceptor gain, a decrease in lateral inhibition, widening of the aperture, dispersal of screening pigment, and outward movement of light-sensitive membrane. The delayed changes include light-triggered shedding of light-sensitive membrane, light-driven narrowing of the aperture, and light-driven inward movement of light-sensitive membrane. These changes and their functional significance are summarized in the **table** and are discussed below.

Photoreceptor noise. In constant darkness, random fluctuations in the membrane potential of the photoreceptors produce trains of nerve impulses in the optic nerve fibers. This spontaneous impulse activity represents noise which could obscure the neural responses to light. The onset of efferent activity causes a rapid suppression of random potential fluctuations without reducing the sensitivity to light. This effectively increases the signal-to-noise ratio of the visual input to the brain and thereby improves the animal's ability to detect very dim

Circadian rhythms in the horseshoe crab retina

Measured retinal property	Rhythm	Effect on visual capability
Photoreceptor noise	Low at night, high during day	Improved detection of light at night
Photoreceptor gain	High at night, low during day	Increased sensitivity at night
Lateral inhibition	Low at night, high during day	Maintenance of constant contrast sensitivity day to night
Photon catch	High at night, low during day	Increased sensitivity at night
Acceptance angle	Wide at night, narrow during day	Decreased resolution at night
Photosensitive membrane shedding	Primed at night, triggered during day	Maintenance of photosensitivity
Photomechanical movement	Primed at night, triggered during day	Protection from bright light during day

lights. The resulting circadian rhythm in detectability gives the animal improved visual capabilities at night.

Photoreceptor gain. The capture of light quanta by the light-sensitive rhabdom causes fluctuations in the photoreceptor membrane potential which, in turn, trigger nerve impulse activity in the optic nerve. The size of the membrane voltage change depends upon the photoreceptor gain. Efferent activity increases the gain of the photoreceptors so that the electrical response to a light stimulus of constant intensity will be larger at night than during the day.

Lateral inhibition. Neighboring ommatidia mutually inhibit one another. This lateral inhibition accentuates the response to light-dark boundaries in the visual world and thereby enhances the animal's ability to detect objects. The magnitude of the edge enhancement, or Mach band, is proportional to overall response levels. At night, efferent activity reduces the strength of lateral inhibition to compensate for the increase in retinal responsiveness and thereby maintains the size of the Mach bands and a constant level of visual edge contrast.

Photon catch. With the nightly onset of efferent activity, a major structural reorganization of the retina begins (**Fig. 2***a*). The aperture between the cuticular cone and the rhabdom widens and shortens, the rhabdom array of light-sensitive membrane is compressed outward toward the aperture, and the screening pigment is dispersed away from the rhabdom. Together, these changes increase the amount of light captured by increasing the amount of light that reaches the rhabdom at night. The combined increases in photon catch and photoreceptor gain produce a circadian rhythm in visual sensitivity.

Acceptance angle. The structural changes described above also increase the size of the portion of the visual field detected by each ommatidium. This increase in ommatidial acceptance angle largely results from the widening of the aperture and serves to reduce the resolution of the eye. Thus, the efferent-driven anatomical changes both increase the eye's sensitivity and decrease its resolution.

Turnover of light-sensitive membrane. Cyclic break-

down and rebuilding of light-sensitive membrane is a characteristic common to all photoreceptors. In the photoreceptors of this eye the most observable manifestation of the turnover of the rhabdom is a massive daily burst of membrane shedding triggered by early morning light. At dawn, much of the light-sensitive membrane is stripped out of the rhabdom into multivesicular bodies which are gradually degraded. The rhabdom is rapidly rebuilt and remains well organized throughout the rest of the day and night. The nightly period of efferent input must precede the morning burst of shedding. The metabolic processes of the photoreceptor involved in the breakdown and synthesis of the rhabdom are thus in some manner primed by efferent input, even though the efferent fibers need not be active when the shedding occurs.

Light-driven anatomical changes. In constant darkness the nighttime structural changes described

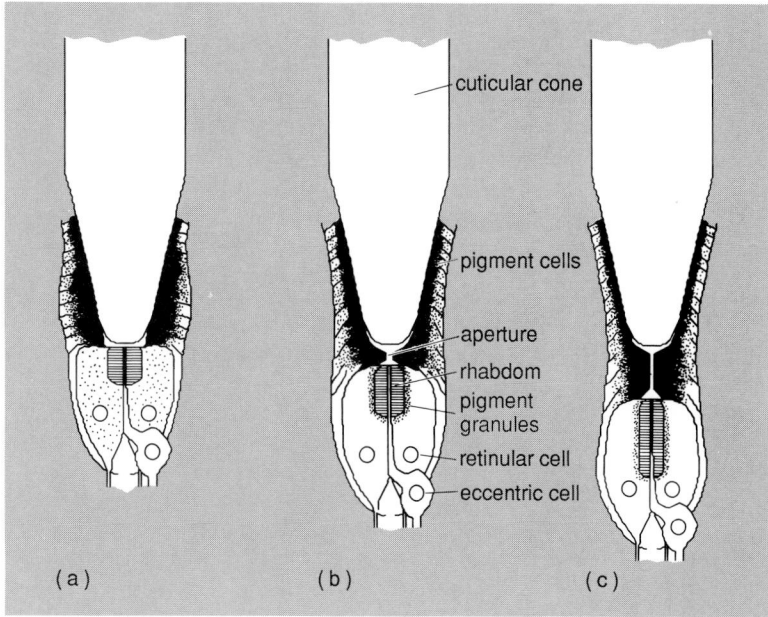

Fig. 2. Longitudinal cross sections, approximately to scale, showing the structure of *Limulus* ommatidia at the extremes of circadian and diurnal rhythms in retinal structure. (*a*) Nighttime darkness. (*b*) Daytime darkness. (*c*) Daytime daylight. (*After S. C. Chamberlain and R. B. Barlow, Jr., Control of structural rhythms in the lateral eye of Limulus: Interactions of natural lighting and circadian efferent activity, J. Neurosci., vol. 7, no. 7, 1987*)

above reverse during the day when the efferents are silent (Fig. 2b). The aperture narrows and lengthens, the rhabdom array of light-sensitive membrane is expanded inward away from the aperture, and the screening pigment is aggregated next to the rhabdom. In natural lighting, these structural changes are accentuated by daylight (Fig. 2c). The additional modifications of the aperture and rhabdom caused by light, however, require prior nighttime efferent input. Just as membrane shedding is somehow primed by efferent activity, so too is the ability of the retinal structure to mechanically adapt to light. Together, circadian efferent input and cyclic lighting produce a coordinated set of major changes in retinal structure in the morning and evening. Efferent input alone produces the same changes but with smaller magnitude. Cyclic lighting alone has no effect on the structure of the aperture or rhabdom.

For background information SEE EYE (INVERTE-BRATE); PHOTOPERIODISM; XIPHOSURIDA in the McGraw-Hill Encyclopedia of Science and Technology.

Steven C. Chamberlain

Bibliography. R. B. Barlow, Jr., S. C. Chamberlain, and L. Kass, Circadian rhythms in retinal function, in S. R. Hilfer and J. B. Sheffield (eds.), *Molecular and Cellular Basis of Visual Acuity*, pp. 31–53, 1984; J. E. Evans, S. C. Chamberlain, and B.-A. Battelle, Autoradiographic localization of newly synthesized octopamine to retinal efferents in the *Limulus* visual system, *J. Comp. Neurol.*, 219:369–383, 1983; W. H. Fahrenbach, The morphology of the horseshoe crab (*Limulus polyphemus*) visual system, VII. Innervation of photoreceptor neurons by neurosecretory efferents, *Cell Tissue Res.*, 216:655–659, 1981; M. K. Powers and R. B. Barlow, Jr., Behavioral correlates of circadian rhythms in the *Limulus* visual system, *Biol. Bull.*, 169:578–591, 1985.

Contributors

Contributors

The affiliation of each Yearbook contributor is given, followed by the title of his or her article. An article title with the notation "in part" indicates that the author independently prepared a section of an article; "coauthored" indicates that two or more authors jointly prepared an article or section.

A

Afsar, Prof. Mohammed Nurul. *Department of Electrical Engineering, The City College, City University of New York.* DIELECTRIC MATERIALS.

Albritton, Dr. Daniel. *Director, Aeronomy Laboratory, Environmental Research Laboratories, National Oceanic and Atmospheric Administration, U.S. Department of Commerce, Boulder, Colorado.* AERONOMY.

Andersson, Dr. Laura A. *Department of Chemical, Biological and Environmental Sciences, Oregon Graduate Center, Beaverton.* METALLOCHLORIN.

Andrews, Dr. Peter. *Bayer AG, Institut für Chemotherapie, Wuppertal, West Germany.* MEDICAL PARASITOLOGY—coauthored.

Angevine, Dr. Charles L. *Assistant Professor, Department of Geology and Geophysics, University of Wyoming.* SEA-LEVEL FLUCTUATIONS—in part.

Annunziata, Robert. *President and Chief Operating Officer, Teleport Communications, New York, New York.* SATELLITE COMMUNICATIONS—in part.

Arndt-Jovin, Dr. Donna J. *Max-Planck-Institut für Biophysikalische Chemie, Abteilung Molekulare Biologie, Göttingen-Nikolausberg, West Germany.* MICROSCOPE—in part.

Arvizu, Dr. Daniel E. *Solar Energy Department, Sandia National Laboratories, Albuquerque, New Mexico.* SOLAR CELL—coauthored.

B

Bailey, A.E. *Formerly, Superintendent of Electrical Science, National Physical Laboratory, London, England.* MICROWAVE MEASUREMENTS.

Bard, Prof. Allen J. *Hackerman/Welch Chair in Chemistry, Department of Chemistry, College of Natural Sciences, University of Texas, Austin.* ELECTROCHEMISTRY.

Barker, Dr. Robert H., Jr. *Program of Tropical Medicine and International Health, Harvard School of Public Health and Harvard Medical School.* MEDICAL PARASITOLOGY—in part.

Barnett, Dr. Richard W. *Research Scientist, Department of Molecular Biology, Allelix, Inc., Mississauga, Ontario, Canada.* GENETIC ENGINEERING.

Barth, Dr. Phillip. *NovaSensor, Fremont, California.* DESIGN ENGINEERING—coauthored.

Bassemir, Dr. Robert W. *Chief Scientist, Corporate Research and Development, Graphic Arts Laboratory, Sun Chemical Corporation, Carlstadt, New Jersey.* INK.

Becker, Dr. Russell. *AT&T Bell Laboratories, Murray Hill, New Jersey.* MICROSCOPE—in part.

Beeby, Prof. J. L. *Programme Coordinator, Low Dimensional Structures Programme, Science and Engineering Research Council, Physics Department, Leicester, England.* SEMICONDUCTOR.

Bellanti, Dr. Joseph A. *Director, International Center for Interdisciplinary Studies of Immunology, Georgetown University.* IMMUNOLOGY—coauthored.

Bengtsson, Dr. Lennart. *Director, European Centre for Medium Range Forecasts, Reading, Berkshire, England.* WEATHER FORECASTING AND PREDICTION.

Bierer, Dr. Barbara. *Division of Pediatric Oncology, Dana-Farber Cancer Institute, Boston, Massachusetts.* CELLULAR IMMUNOLOGY—coauthored.

Boppel, Wolfgang. *Dr.-Ing. Rudolf Hell GMBH, Informationstechnik Elektronik für Satz und Reproduktion, Kiel, West Germany.* PRINTING—in part.

Borth, Paul F. *Technical Director, International Prepress Association, South Holland, Illinois.* PRINTING—in part.

Bouwkamp, Dr. John C. *Associate Professor, Department of Horticulture, Division of Agricultural and Life Sciences, College of Agriculture, University of Maryland.* POTATO, SWEET.

Boynton, Dr. Robert M. *Department of Psychology, University of California, San Diego.* COLOR VISION—in part.

Brakefield, Dr. Paul M. *Department of Zoology, University College, Cardiff, Wales.* PROTECTIVE COLORATION.

Brown, Prof. Brian H. *Sheffield University and Health Authority, Department of Medical Physics and Clinical Engineering, Royal Hallamshire Hospital, Sheffield, England.* MEDICAL IMAGING.

Bruno, Michael H. *Graphic Arts Consultant, Publisher and Editor, "What's New(s) in Graphic Communications," Nashua, New Hampshire.* PRINTING—in part.

Buikstra, Prof. Jane E. *Department of Anthropology, University of Chicago.* ANTHROPOLOGY.

Burakoff, Dr. Steven J. *Chief, Division of Pediatric Oncology, Dana-Farber Cancer Institute, Boston, Massachusetts.* CELLULAR IMMUNOLOGY—coauthored.

Busch, Dr. Harris. *Distinguished Professor and Chairman, Department of Pharmacology, Baylor College of Medicine, Texas Medical Center, Houston.* UBIQUITIN.

C

Carlson, Dr. Richard. *Department of Terrestrial Magnetism, Carnegie Institution of Washington.* LITHOSPHERE—in part.

Castellino, Dr. Francis J. *Dean, Department of Chemistry, College of Science, University of Notre Dame.* BLOOD.

Chamberlain, Dr. Steven C. *Associate Professor of Neuroscience, Institute for Sensory Research, Syracuse University.* XIPHOSURIDA.

Charlesworth, Dr. Brian. *Department of Biology, University of Chicago.* TRANSPOSONS.

Chu, Dr. Nathan M. *Postdoctoral Associate, Department of Biology, Osborn Memorial Laboratories, Yale University.* DEVELOPMENTAL GENETICS.

Clark, Dr. Arthur E. *Associate Director and Program Manager, Bioglass Research Center, J. Hillis Miller Health Center, University of Florida.* BIOMEDICAL ENGINEERING—coauthored.

Clark, Dr. David A. *Professor and MRC Scientist, Mc-Master University, Department of Medicine, Hamilton, Ontario, Canada.* PREGNANCY.

Clark, Dr. Raymond L. *U.S. Department of Agriculture, Regional Plant Introduction Station, Iowa State University.* TOMATO.

Clark, Dr. Robert P. *Storage Batteries, Sandia National Laboratories, Albuquerque, New Mexico.* BATTERY—coauthored.

Clauson, Dr. Barbara L. *Research Assistant, Museum of Natural History, University of Kansas.* MAMMALIA—coauthored.

Cooke, Dr. Howard J. *Department of Zoology, Medical Research Council, Mammalian Genome Unit, University of Edinburgh, Scotland.* CROSSING-OVER (GENETICS).

Cormack, Dr. Gordon V. *Associate Professor, Department of Computer Science, University of Waterloo, Ontario, Canada.* DATA COMPRESSION.

Costain, Dr. C. C. *Electrical and Time Standards Section, Division of Physics, National Research Council of Canada, Ottawa, Ontario.* TIME.

Cowles, Prof. Joe R. *Department of Biology, University of Houston.* SPACE BIOLOGY.

Cox, Dr. Daniel L. *Department of Physics, University of California, San Diego.* KONDO EFFECT.

D

Daugherty, Dr. Michael J. *Director, Electronics Research Laboratory, Laboratory Operations, Aerospace Corporation, Los Angeles, California.* OPTICAL DETECTORS.

Dawson, Dr. John H. *Department of Chemistry, University of South Carolina.* SPECTROSCOPY.

de Bold, Dr. Adolfo J. *Professor of Pathology and Physiology, and Director of Research, University of Ottawa Heart Institute, Ottawa Civic Hospital, Ontario, Canada.* HORMONE.

Deikman, Dr. Jill. *Division of Molecular Plant Biology, University of California, Berkeley.* AMYLASE.

Denison, Dr. Rodger E. *Research Department, Dallas Research Laboratory, Mobil Research and Development Corporation, Dallas, Texas.* MARINE SEDIMENTS—in part.

Derguini, Dr. Fadila. *Department of Chemistry, Columbia University.* PHOTORECEPTION—coauthored.

De Valois, Dr. Russell. *Department of Psychology, University of California, Berkeley.* COLOR VISION—in part.

Dick, Dr. Henry J. B. *Woods Hole Oceanographic Institution, Woods Hole, Massachusetts.* VOLCANOLOGY.

Diederich, Dr. François N. *Department of Chemistry, University of California, Los Angeles.* MOLECULAR COMPLEXATION.

Dillman, Dr. Robert O. *Director, Experimental Clinical Oncology, Scripps Clinic and Research Foundation, La Jolla, California.* MONOCLONAL ANTIBODIES.

Duke, Dr. Stephen O. *Plant Physiologist and Research Leader, Herbicide Interactions in Plants and Soil, U.S. Department of Agriculture, Jamie Whitten Delta States Research Center, Southern Weed Science Laboratory, Stoneville, Mississippi.* PLANT PHYSIOLOGY.

Duren, Prof. Peter L. *Department of Mathematics, University of Michigan.* COMPLEX NUMBERS AND COMPLEX VARIABLES.

E

Edwards, Dr. Kathryn L. *Department of Biology, Washington University, St. Louis, Missouri.* PLANT MOVEMENTS.

Elgin, Prof. Sarah C. R. *Department of Biology, Washington University, St. Louis, Missouri.* CHROMOSOME.

Elias, Dr. Robert W. *Health Scientist, U.S. Environmental Protection Agency, Environmental Criteria and Assessment Office, Research Triangle Park, North Carolina.* FOOD WEB—in part.

Esaki, Dr. Leo. *International Business Machines Corporation, Thomas J. Watson Research Center, Yorktown Heights, New York.* TRANSISTOR.

F

Farkas, Dr. Daniel. *Adjunct Professor, Department of Food Science, University of Delaware, and Vice President, Process Research and Development, Campbell Soup Company, Camden, New Jersey.* FOOD MANUFACTURING—in part.

Farmer, Prof. Ian. *Department of Mining and Geological Engineering, College of Mines, University of Arizona.* MINING.

Feldkamp, Dr. Lee A. *Physics Department, Scientific Research Laboratories, Ford Motor Company, Dearborn, Michigan.* NONDESTRUCTIVE TESTING—in part.

Fendler, Prof. Janos H. *Distinguished Professor of Chemistry and Director of Center for Research in Membranes and Colloid Science, Department of Chemistry, Syracuse University.* MEMBRANE MIMETIC CHEMISTRY.

Ferrell, Dr. Thomas L. *Oak Ridge National Laboratory, Oak Ridge, Tennessee.* SURFACE PHYSICS.

Ferris, Dr. James P. *Department of Chemistry, Rensselaer Polytechnic Institute.* LIFE, ORIGIN OF.

Fisher, Dr. Robert J. *Associate Professor, Reaction Engineering/Separation Technology, Department of Food Science, University of Delaware.* FOOD ENGINEERING—in part.

Francki, Dr. R. I. B. *Head of Department of Plant Pathology, Waite Agricultural Research Institute, University of Adelaide, Glen Osmond, Australia.* VIRUS SATELLITE.

Fujii, Dr. Jo Ann. *Research Associate, Plant Genetics, Inc., Davis, California.* PHOTORESPIRATION.

Fujinami, Dr. Robert S. *Department of Pathology, School of Medicine, University of California, San Diego.* AUTOIMMUNITY.

G

Gallo, Dr. R. C. *Department of Health and Human Services, Public Health Service, National Institutes of Health, Bethesda, Maryland.* RETROVIRUS—coauthored.

Galloway, Dr. Cynthia M. *Department of Botany and Plant Sciences, Citrus Research Center and Agricultural Experiment Station, College of Natural and Agricultural Sciences, University of California, Riverside.* PLANT METABOLISM—in part.

Gibbs, Dr. Doon. *Department of Physics, Brookhaven National Laboratory, Associated Unversities, Inc., Upton, New York.* X-RAY DIFFRACTION.

Giddings, Dr. George G. *Director, Food Irradiation Services, Isomedix, Inc., Whippany, New Jersey.* FOOD MANUFACTURING—in part.

Gill, Dr. Peter. *Central Research Establishment, Home Office Forensic Science Service, Aldermaston, Reading, Berkshire, England.* FORENSIC CHEMISTRY.

Gordon, Dr. Nancy Rowan. *Department of Chemistry, College of Arts and Sciences, The American University, Washington, D.C.* METALLOPROTEIN.

Graham, James A. *Business Manager, Industrial Adhesives Division, Lord Corporation, Erie, Pennsylvania.* ADHESIVE BONDING.

Greif, Dr. Jeffrey. *Inference Corporation, Los Angeles, California.* PROGRAMMING LANGUAGES—in part.

Greiner, Prof. Dr. Walter. *Institut für Theoretische Physik der Universität Frankfurt am Main, West Germany.* RADIOACTIVITY.

Griffin, Dr. Robert J., Jr. *Science Writer, Washington, D.C.* SPACE FLIGHT—coauthored.

H

Haas, Paul R. *Vice President, Corporate Planning, Kearney & Trecker Corporation, Milwaukee, Wisconsin.* COMPUTER-INTEGRATED MANUFACTURING—in part.

Haberkorn, Dr. A. *Bayer AG, Institut für Chemotherapie, Wuppertal, West Germany.* MEDICAL PARASITOLOGY—coauthored.

Habig, Dr. William. *Food and Drug Administration, Office of Biological Research and Review, Bethesda, Maryland.* TOXICOLOGY—in part.

Haier, Dr. Richard J. *Associate Professor, Department of Psychiatry and Human Behavior, California College of Medicine, University of California, Irvine.* SCHIZOPHRENIA.

Hajash, Dr. Andrew. *Associate Professor, Department of Geology, Texas A&M University.* GEOCHEMISTRY.

Hammond, Dr. Lawrence L. *Director, Agro-Economic Division, International Fertilizer Development Center, Muscle Shoals, Alabama.* FERTILIZER—in part.

Hargrove, Dr. W. L. *Associate Professor, University of Georgia College of Agriculture, Experiment Stations—Georgia Station, Department of Agronomy.* SOIL NITROGEN.

Harlow, Dr. George E. *Associate Curator, American Museum of Natural History, Department of Mineral Sciences, New York, New York.* PYROXENE.

Harper, Dr. James E. *Plant Physiologist, U.S. Department of Agriculture, University of Illinois, Urbana.* NITROGEN FIXATION.

Hastings, Dr. Larry G. *U.S. Department of Energy, Washington, D.C.* SPACE FLIGHT—coauthored.

Hayes, William C. *Editor, "Electrical World," McGraw-Hill Publications Company, New York, New York.* ELECTRICAL UTILITY INDUSTRY.

Hench, Prof. Larry L. *Advanced Materials Research Center, College of Engineering, University of Florida.* BIOMEDICAL ENGINEERING—coauthored.

Henkel, Prof. James. *Medicinal Chemistry Department, School of Pharmacy, University of Connecticut.* MOLECULAR MODELING.

Hodge, Prof. Bartow. *Professor, Information Systems, School of Business, Virginia Commonwealth University.* PROGRAMMING LANGUAGES—in part.

Holcroft, Dr. James W. *Professor, Department of Surgery, School of Medicine, University of California, Davis.* RESPIRATORY SYSTEM DISORDERS—coauthored.

Hooke, Dr. Anne Morris. *Assistant Professor, Departments of Pediatrics and Microbiology, and Scientific Director, International Center for Interdisciplinary Studies of Immunology, Georgetown University.* IMMUNOLOGY—coauthored.

Howell, Dr. David G. *U.S. Department of the Interior, Geological Survey, Menlo Park, California.* ACCRETION TECTONICS.

Hubbell, Prof. David H. *Department of Soil Science, Institute of Food and Agricultural Sciences, University of Florida.* SOIL MICROBIOLOGY.

I

Iseman, Dr. Michael D. *Chief, Mycobacterial Disease Service, National Jewish Center for Immunology and Respiratory Medicine, and Associate Professor of Medicine,* *Division of Pulmonary Sciences, University of Colorado School of Medicine.* MYCOBACTERIA.

Iveson, Robert. *Electric Power Research Institute, Palo Alto, California.* ELECTRIC POWER SYSTEMS.

J

Jarrett, Dr. Noel. *Alcoa Technical Center, Aluminum Company of America, Alcoa Center, Pennsylvania.* METALLURGY.

Jayawant, Prof. B. V. *Dean of the School of Engineering and Applied Sciences, and Professor of Electrical and Systems Engineering, University of Sussex, School of Engineering and Applied Sciences, Falmer, Brighton, Sussex, England.* MAGNETIC SUSPENSION AND LEVITATION.

Jegla, Dr. Dorothy E. *Assistant Professor, Department of Biology, Kenyon College, Gambier, Ohio.* FATE MAPS (EMBRYOLOGY).

Johnson, Robert A. *Filter Products, Collins Transmission Systems Division, Rockwell International Corporation, Costa Mesa, California.* ELECTRIC FILTER.

Junkiewicz, Dr. Edmund. *INMARSAT Consultancy Program Manager, Comsat General Corporation, Comsat Space Communications Division, El Segundo, California.* COMMUNICATIONS SATELLITE—in part.

K

Kacew, Prof. Sam. *Pharmacology, Faculty of Health Sciences, School of Medicine, University of Ottawa, Ontario, Canada.* TOXICOLOGY—in part.

Kessler, Prof. Edwin. *Departments of Geography and Meteorology, University of Oklahoma.* WIND SHEAR.

Keymer, Dr. Anne E. *Department of Zoology, University of Oxford, England.* ECOLOGICAL INTERACTIONS—in part.

Kibble, Dr. T. W. B. *Head of Department and Professor of Theoretical Physics, Blackett Laboratory, Imperial College of Science and Technology, London, England.* COSMOLOGY.

Kluessendorf, Dr. Joanne. *Department of Geology, University of Illinois, Urbana-Champaign.* SILURIAN.

Kohler, Dr. Heinz. *Director, Department of Molecular Immunology, Roswell Park Memorial Institute, New York State Department of Health, Buffalo.* VACCINATION.

Koning, Dr. Ross E. *Assistant Professor, Department of Biological Sciences, Nelson Biological Laboratory, Rutgers University.* FLOWER.

Kraft, Dr. John M. *U.S. Department of Agriculture, Agricultural Research Service, Irrigated Agriculture Research and Extension Center, Prosser, Washington.* PLANT PATHOLOGY—in part.

Krugman, Dr. Stanley L. *Director of Timber Management Research, U.S. Department of Agriculture, Forest Service, Washington, D.C.* FORESTRY.

L

Lawton, Robert S. *Manager, ACTS Program, RCA Aerospace and Defense, Astro-Electronics Division, Princeton, New Jersey.* COMMUNICATIONS SATELLITE—in part.

Legge, Dr. George J. F. *Director, Micro Analytical Research Centre, School of Physics, University of Melbourne, Victoria, Australia.* PROTON MICROSCOPY.

Leinonen, Dr. Maija. *National Public Health Institute, Helsinki, Finland.* BRANHAMELLA—coauthored.

Leith, Prof. Emmett N. *Schlumberger Professor of Electrical Engineering, College of Engineering, University of Michigan.* HOLOGRAPHY.

Lenard, Prof. John. *Department of Physiology and Biophysics, College of Medicine and Dentistry, Rutgers University.* ANIMAL VIRUS.

Lines, Dr. Larry R. *Amoco Production Company, Tulsa, Oklahoma.* GEOPHYSICAL EXPLORATION.

Lohr, Dr. Jacob A. *Department of Pediatrics, Children's Medical Center, University of Virginia.* KAWASAKI SYNDROME.

Lovelace, Prof. Eugene. *Department of Psychology, Alfred University, Alfred, New York.* MEMORY.

M

Macfarlane, Dr. Roger M. *International Business Machines Almaden Research Center, San Jose, California.* OPTICAL INFORMATION SYSTEMS—in part.

McLelland, Prof. James M. *Charles A. Dana Professor of Geology, Department of Geology, Colgate University.* ANORTHOSITE.

Mäkelä, Dr. P. Helena. *National Public Health Institute, Helsinki, Finland.* BRANHAMELLA—coauthored.

Maran, Dr. Stephen P. *NASA Goddard Space Flight Center, Greenbelt, Maryland.* HALLEY'S COMET.

Marks, Dr. Jay S. *Associate Professor, Department of Food Science, Purdue University.* FOOD ENGINEERING—in part.

Martin, Prof. Dean F. *Professor of Chemistry and Director, Chemical and Environmental Management Services Center, University of South Florida.* ALLELOPATHY.

Millar, Dr. Michelle M. *Research Associate Professor, Department of Chemistry, State University of New York, Stony Brook.* COORDINATION CHEMISTRY.

Molitoris, Dr. Joseph J. *Institut für Theoretische Physik der Universität Frankfurt am Main, West Germany.* NUCLEAR HYDRODYNAMICS.

Moore, Dr. Jeffrey. *Department of Surgery, School of Medicine, University of California, Davis.* RESPIRATORY SYSTEM DISORDERS—coauthored.

Morgan, Dr. Ronnie G. *Associate Professor of Food Engineering, Department of Agricultural Engineering, Department of Food Science and Human Nutrition, Michigan State University.* FOOD ENGINEERING—in part.

Morse, Prof. John W. *Department of Oceanography, College of Geosciences, Texas A&M University.* MARINE SEDIMENTS—in part.

Muirhead, Alan R. *E. I. Du Pont de Nemours & Company, Inc., Photosystems and Electronic Products Department, Wilmington, Delaware.* PRINTING—in part.

Munn, Prof. R. W. *Professor of Chemical Physics, Department of Chemistry, University of Manchester Institute of Science and Technology, Manchester, England.* ELECTRONICS.

Murphy, Kevin D. *Storage Batteries, Sandia National Laboratories, Albuquerque, New Mexico.* BATTERY—coauthored.

Mysen, Dr. Bjorn O. *Senior Scientist, Carnegie Institution of Washington, Geophysical Laboratory.* SILICATE PHASE EQUILIBRIA.

N

Nakanishi, Dr. Koji. *Department of Chemistry, Columbia University.* PHOTORECEPTION—coauthored.

Negele, Dr. John W. *Department of Physics, Massachusetts Institute of Technology.* NUCLEAR STRUCTURE.

Newby, Dr. John R. *Formerly, Armco Inc., Research and Technology Division; Consultant on Metal Forming, Middletown, Ohio.* METAL FORMING.

Nierzwicki-Bauer, Dr. Sandra A. *Assistant Professor, Rensselaer Polytechnic Institute.* ECOLOGICAL INTERACTIONS—in part.

Nylander, Dr. Claes. *Laboratory of Applied Physics, Linköping Institute of Technology, Linköping, Sweden.* TRANSDUCER.

O

Okagaki, Dr. Takashi. *Stone Professor, Departments of Laboratory Medicine and Pathology and of Obstetrics and Gynecology, Medical School, University of Minnesota.* REPRODUCTIVE SYSTEM DISORDERS.

Orbach, Prof. Raymond. *Department of Physics, University of California, Los Angeles.* FRACTALS.

Oreopoulos, Dr. Dimitrios G. *Professor of Medicine, University of Toronto, and Director, Peritoneal Dialysis Unit, Toronto Western Hospital, Ontario, Canada.* PERITONITIS.

Ortiz de Montellano, Dr. Bernard. *Department of Anthropology, Wayne State University.* PHARMACOGNOSY.

P

Pamp, Dr. Douglas E. *Ruminant Nutritionist, Hubbard Milling Company, Mankato, Minnesota.* FOOD WEB—in part.

Parker, Dr. Edwin B. *Vice President, Network Development, Equatorial Communications Company, Mountain View, California.* SPREAD SPECTRUM COMMUNICATION.

Pasto, Prof. Daniel J. *Department of Chemistry, College of Science, University of Notre Dame.* ORGANIC CHEMISTRY.

Peng, Prof. Syd S. *Professor and Chairman, Department of Mining Engineering, College of Mineral and Energy Resources, West Virginia University.* ROCK MECHANICS.

Petersen, Dr. Kurt E. *Executive Vice President, Technology, NovaSensor, Fremont, California.* DESIGN ENGINEERING—coauthored.

Pettitt, Dr. Bernard Montgomery. *Assistant Professor, Department of Chemistry, University of Houston.* PROTEIN.

Power, Dr. James F. *Research Leader, U.S. Department of Agriculture, Agricultural Research Service, Soil and Water Conservation Research Unit, University of Nebraska.* AGRICULTURAL SOIL AND CROP PRACTICES.

Prusoff, Dr. William H. *Professor of Pharmacology, Yale University School of Medicine.* VIRUS CHEMOPROPHYLAXIS—coauthored.

Purser, Dr. Kenneth. *Chief Scientist, General Ionex Corporation, Newburyport, Massachusetts.* ION IMPLANTATION.

Q

Quigg, Prof. Chris. *Fermi National Accelerator Laboratory, Batavia, Illinois.* ELEMENTARY PARTICLE.

Qureshi, Dr. Shahid. *Senior Director, Research, Transmission Products Division, Codex Corporation, Mansfield, Massachusetts.* ELECTRICAL COMMUNICATIONS.

R

Reiter, Alan A. *Subcarrier Communications, Washington, D.C.* RADIO PAGING SYSTEMS.

Reitz, Dr. Marvin S., Jr. *Department of Health and Human Services, Public Health Service, National Institutes of Health, Bethesda, Maryland.* RETROVIRUS—coauthored.

Rice, Dr. Prudence M. *Associate Professor of Anthropology, and Associate Curator in Archaeology, Florida State Museum, University of Florida.* ARCHEOLOGY.

Rinker, Alan. *Systematics General Corporation, Sterling, Virginia.* SATELLITE COMMUNICATIONS—in part.

Robinson, Dr. Paul T. *Centre for Marine Geology, Dalhousie University, Halifax, Nova Scotia, Canada.* OPHIOLITE.

Rockaway, Prof. John D. *Department of Geological Engineering, University of Missouri, Rolla.* UNDERGROUND MINING—coauthored.

Rosendahl, Dr. Bruce R. *Department of Geology, Duke University.* RIFT VALLEY.

Ross, Dr. Colin. *Topexpress Limited, Scientific and Computer Consultants, Cambridge, England.* ACOUSTIC NOISE.

S

Saenz, Dr. Richard. *Associate Professor, Department of Physics, California State Polytechnic University.* GRAVITATIONAL RADIATION—coauthored.

Sagi, Dr. Amir. *Hebrew University of Jerusalem, and Aquaculture Production Technology (Israel) Limited.* DECAPODA (CRUSTACEA).

Sahagian, Dr. Dork. *Department of Geophysical Sciences, University of Chicago.* SEA-LEVEL FLUCTUATIONS—in part.

Saiki, Dr. Randall K. *Associate Scientist, Department of Human Genetics, Cetus Corporation, Emeryville, California.* SICKLE CELL DISEASE.

Saunders, Dr. Mary Jane. *Assistant Professor, Department of Biology, University of South Florida.* CYTOKININS.

Schinazi, Dr. Raymond F. *Research Scientist, Veterans Administration Medical Center, Atlanta, and Assistant Professor, Emory University.* VIRUS CHEMOPROPHYLAXIS—coauthored.

Schlappi, Dr. Herman C. *Ihgalls Shipbuilding, Pascagoula, Mississippi.* SHIP POWERING AND STEERING.

Schueler, Dr. Donald G. *Manager, Solar Energy Department, Sandia National Laboratories, Albuquerque, New Mexico.* SOLAR CELL—coauthored.

Schwarz, Dr. Ricardo B. *Center for Materials Science, Los Alamos National Laboratory, Los Alamos, New Mexico.* METALLIC GLASSES.

Scott, Dr. John W. *Department Director, Chemical Development, Hoffmann-La Roche, Inc., Nutley, New Jersey.* ASYMMETRIC SYNTHESIS.

Shaevitz, Prof. Michael H. *Associate Professor, Department of Physics, Nevis Laboratories, Columbia University, Irvington, New York.* NEUTRINO.

Shapiro, Prof. Stuart L. *Professor of Astronomy and Physics, Cornell University.* GRAVITATIONAL RADIATION—coauthored.

Shearman, Prof. E. D. R. *Department of Electronic and Electrical Engineering, University of Birmingham, England.* OCEANOGRAPHY.

Shevach, Dr. Ethan M. *Senior Investigator, Laboratory of Immunology, National Institute of Allergy and Infectious Diseases, Bethesda, Maryland.* CYCLOSPORIN.

Silk, Dr. M. G. *National Nondestructive Testing Centre, Harwell Laboratory, Oxfordshire, England.* NONDESTRUCTIVE TESTING—in part.

Sleigh, Prof. M. A. *Department of Biology, University of Southampton, England.* FEEDING MECHANISMS (INVERTEBRATE).

Slone, Dr. J. Henry. *Research Associate, Department of Biology, Washington University, St. Louis, Missouri.* SECRETORY STRUCTURES (PLANT).

Smith, Prof. S. Desmond. *Department of Physics, Heriot-Watt University, Riccarton, Edinburgh, Scotland.* OPTICAL INFORMATION SYSTEMS—in part.

Staples, Dr. Richard C. *Plant Biochemist, Boyce Thompson Institute for Plant Research, Cornell University.* PLANTS, SALINE ENVIRONMENTS OF.

Steffy, David A. *Hughes Aircraft Company, Space and Communications Group, Los Angeles, California.* COMMUNICATIONS SATELLITE—in part.

Stöcker, Prof. Dr. Horst. *Institut für Theoretische Physik der Universität Frankfurt am Main, West Germany.* NUCLEAR HYDRODYNAMICS—coauthored.

Streusand, Dr. Virginia J. *Department of Agronomy, College of Agriculture, University of Illinois, Urbana-Champaign.* PLANT-WATER RELATIONS.

Suslick, Dr. Kenneth S. *Associate Professor, School of Chemical Sciences, Noyes Laboratory, University of Illinois, Urbana.* SONOCHEMISTRY.

Swisher, Dr. M. E. *Zamorano, Escuela Agricola Panamericana, Tegucigalpa, Honduras.* FERTILIZER—in part.

T

Tešanović, Prof. Zlatko B. *Department of Physics, Lyman Laboratory of Physics, Harvard University.* LIQUID HELIUM.

Timm, Dr. Robert M. *Curator, Mammals, Museum of Natural History, University of Kansas.* MAMMALIA—coauthored.

Todd, Dr. Michael J. *College of Engineering, School of Operations Research and Industrial Engineering, Cornell University.* LINEAR PROGRAMMING.

V

Van Besien, Alphonse C. *Associate Professor, Department of Geology and Geological Engineering, University of Mississippi.* UNDERGROUND MINING.

Vierstra, Dr. Richard D. *Department of Horticulture, University of Wisconsin, Madison.* PLANT METABOLISM—in part.

von Boletzky, Dr. S. *Laboratoire Arago, Banyuls-sur-Mer, France.* CEPHALOPODA.

W

Wachtman, Dr. John B., Jr. *Director, Center for Ceramics Research, College of Engineering, Rutgers University.* CERAMICS.

Weaver, Dr. Barry L. *School of Geology and Geophysics, University of Oklahoma.* LITHOSPHERE—in part.

Webb, Dr. Richard A. *International Business Machines Corporation, Thomas J. Watson Research Center, Yorktown Heights, New York.* QUANTUM MECHANICS.

Weisdorf, Dr. Daniel J. *Assistant Professor of Medicine, and Associate Director, Adult Bone Marrow Transplant Program, University of Minnesota Hospital.* TRANSPLANTATION BIOLOGY.

Weise, David. *Marketing Engineer, Magnetic Bearings, Inc., Radford, Virginia.* MAGNETIC BEARING.

Welch, Dr. Ross M. *Plant Physiologist, U.S. Department of Agriculture, Agricultural Research Service, Plant, Soil and Nutrition Laboratory, Ithaca, New York.* PLANT MINERAL NUTRITION.

Whitley, Dr. Richard J. *Professor of Pediatrics and Microbiology, Department of Pediatrics, School of Medicine, University of Alabama, Birmingham.* HERPES.

Wilkins, Dr. Patricia C. *Department of Chemistry, New Mexico State University.* HEMERYTHRIN—coauthored.

Wilkins, Dr. Ralph G. *Department of Chemistry, New Mexico State University.* HEMERYTHRIN—coauthored.

Wisdom, Prof. Jack. *Department of Earth, Atmospheric and Planetary Sciences, Massachusetts Institute of Technology.* CELESTIAL MECHANICS.

Y–Z

Yoon, Prof. Jong Sik. *Department of Biological Sciences, Bowling Green State University.* ORGANIC EVOLUTION.

Zebroski, Dr. Edwin L. *Electric Power Research Institute, Palo Alto, California.* NUCLEAR POWER.

Index

Index

Asterisks indicate page references to article titles.